Daniel Greenberger
Klaus Hentschel
Friedel Weinert
Editors

Compendium of Quantum Physics

Concepts, Experiments, History and Philosophy

Springer

Editors

Daniel Greenberger
Department of Physics
The City College of New York
138th St. & Convent Ave.
New York NY 10031
USA
greenbgr@sci.ccny.cuny.edu
dansnzy@nyc.rr.com

Klaus Hentschel
University of Stuttgart
Section for the History of Science
& Technology
Keplerstr. 17
D-70174 Stuttgart
Germany
klaus.hentschel@po.hi.uni-stuttgart.de

Friedel Weinert
Department of Social Sciences
and Humanities
University of Bradford
Bradford BD7 1DP
UK
f.weinert@bradford.ac.uk

ISBN 978-3-540-70622-9 e-ISBN 978-3-540-70626-7
DOI 10.1007/978-3-540-70626-7
Springer Dordrecht Heidelberg London New York

Library of Congress Control Number: 2008942038

Cover design: deblik, Berlin

Printed on acid-free paper

Springer is part of Springer Science+Business Media (www.springer.com)

Preface

Since its inception in the early part of the twentieth century, quantum physics has fascinated the academic world, its students, and even the general public. In fact, it is – or has become – a highly interdisciplinary field. On a topic such as "the physics of the atom" the disciplines of physics, philosophy, and history of science interconnect in a remarkable way, and to an extent that is revealed in this volume for the first time. This compendium brings together some 90 researchers, who have authored approximately 185 articles on all aspects of quantum theory. The project is truly international and interdisciplinary because it is a compilation of contributions by historians of science, philosophers, and physicists, all interested in particular aspects of quantum physics. A glance at the biographies at the end of the volume reveals author affiliations in no fewer than twenty countries: Australia, Austria, Belgium, Canada, Denmark, Finland, France, Germany, Greece, Italy, Israel, the Netherlands, New Zealand, Norway, Poland, Portugal, Spain, Switzerland, the United Kingdom and the United States. Indeed, the authors are not only international, they are also internationally renowned – with three Physics Nobel Prize laureates among them.

The basic idea and motivation behind the compendium is indicated in its subtitle, namely, to describe in concise and accessible form the essential concepts and experiments as well as the history and philosophy of quantum physics. The length of the contributions varies according to the topic, and all texts are written by recognized experts in the respective fields. The need for such a compendium was originally perceived by one of the editors (FW), who later discovered that many physicists shared this view. Due to the interdisciplinary nature of this endeavor, it would have been impossible to realize it without the expertise and active participation of a professional physicist (DG) and a historian of science (KH). We should not forget, however, that it was brought to life by the numerous contributions of the many authors from around the world, who generously offered their time and expertise to write their respective articles. The contributions appear in alphabetical order by title, and include many cross-references, as well as selected references to the literature. The volume includes a short English–French–German lexicon of common terms in quantum physics. This will be especially helpful to anyone interested in exploring historical documents on quantum physics, the theory of which was developed side-by-side in these three cultures and languages.

The editors would like to thank Brigitte Falkenburg and Peter Mittelstaedt for their initial work on the project. Angela Lahee (at Springer publishers) deserves our gratitude for her unwavering support and patience during the four years it has taken to turn the idea for this compendium into reality.

January 2009

Dan Greenberger
Klaus Hentschel
Friedel Weinert

Contents

Alphabetical Compendium

Aharonov–Bohm Effect 1
Aharonov–Casher Effect 3
Algebraic Quantum Mechanics 6
Angular Momentum 10
Anyons 10
Aspect Experiment 14
Asymptotic Freedom 18
Atomic Model 18
Atomic Models, J.J. Thomson's "Plum Pudding" Model 18
Atomic Models, Nagaoka's Saturnian Model 22
Bell's Theorem 24
Berry's Phase 31
Black Body 36
Black-Body Radiation 39
Bohm Interpretation of Quantum Mechanics 43
Bohmian Mechanics 47
Bohm's Approach to the EPR Paradox 55
Bohr's Atomic Model 58

Bohr–Kramers–Slater Theory ... 62
Born Rule and its Interpretation ... 64
Bose–Einstein Condensation ... 71
Bose–Einstein Statistics ... 74
Bremsstrahlung ... 78
Brownian Motion ... 81
Bub–Clifton Theorem ... 84
Casimir Effect ... 87
Cathode Rays ... 89
Causal Inference and EPR ... 93
Clauser-Horne-Shimony-Holt (CHSH) – Theorem ... 96
Cluster States ... 96
Coherent States ... 106
Color Charge Degree of Freedom in Particle Physics ... 109
Complementarity Principle ... 111
Complex-Conjugate Number ... 114
Compton Experiment (or Compton Effect) ... 115
Consistent Histories ... 117
Copenhagen Interpretation ... 122
Correlations in Quantum Mechanics ... 122
Correspondence Principle ... 125
Counterfactuals in Quantum Mechanics ... 132
Covariance ... 136
CPT Theorem ... 138
Creation and Annihilation Operators ... 139
Creation and Detection of Entanglement ... 145
Davisson–Germer Experiment ... 150

De Broglie Wavelength ($\lambda = h/p$) ... 152
Decay ... 154
Decoherence ... 155
Degeneracy ... 159
Delayed-Choice Experiments ... 161
Density Matrix ... 166
Density Operator ... 166
Diffeomorphism Invariance ... 170
Dirac Equation ... 171
Dirac Notation ... 172
Double-Slit Experiment (or Two-Slit Experiment) ... 174
Effect ... 179
Ehrenfest Theorems ... 180
Eigenstates, Eigenvalues ... 182
Einstein Locality ... 182
Electron Interferometry ... 188
Electrons ... 195
Ensembles in Quantum Mechanics ... 199
Entanglement ... 201
Entanglement Purification and Distillation ... 202
Entropy of Entanglement ... 205
EPR-Problem (Einstein-Podolsky-Rosen Problem) ... 209
Errors and Paradoxes in Quantum Mechanics ... 211
Exclusion Principle (or Pauli Exclusion Principle) ... 220
Experimental Observation of Decoherence ... 223
Fermi–Dirac Statistics ... 230
Feynman Diagrams ... 235

Fine-Structure Constant 239

Franck–Hertz Experiment 241

Functional Integration; Path Integrals 243

Gauge Symmetry 248

Generalizations of Quantum Statistics 255

GHZ (Greenberger–Horne–Zeilinger) Theorem and GHZ States 258

Gleason's Theorem 263

Grover's Algorithm 266

GRW Theory (Ghirardi, Rimini, Weber Model of Quantum Mechanics) 266

Hamiltonian Operator 271

Hardy Paradox 275

Heisenberg Microscope 279

Heisenberg Picture 280

Heisenberg Uncertainty Relation (Indeterminacy Relations) 281

Hermitian Operator 284

Hidden Variables 284

Hidden-Variables Models of Quantum Mechanics (Noncontextual and Contextual) 287

Hilbert Space 291

Holism in Quantum Mechanics 295

Identity of Quanta 299

Identity Operator 304

Ignorance Interpretation of Quantum Mechanics 305

Indeterminacy Relations 306

Indeterminism and Determinism in Quantum Mechanics 307

Indistinguishability 311

Interaction-Free Measurements (Elitzur–Vaidman, EV IFM) 317

Interpretations of Quantum Mechanics 322

Invariance 322

Ithaca Interpretation of Quantum Mechanics 325

jj-Coupling 327

Kaluza–Klein Theory 328

Kochen–Specker Theorem 331

Landé's g-factor and g-formula 336

Large-Angle Scattering 337

Light Quantum 339

Locality 347

Loopholes in Experiments 348

Lüders Rule 356

Mach–Zehnder Interferometer 359

Magnetic Resonance 359

Many Worlds Interpretation of Quantum Mechanics 363

Matrix Mechanics 368

Matter Waves 371

Measurement Problem 373

Measurement Theory 374

Mesoscopic Quantum Phenomena 379

Metaphysics of Quantum Mechanics 384

Mixed State 389

Mixing and Oscillations of Particles 390

Modal Interpretations of Quantum Mechanics 394

Neutron Interferometry 402

No-Cloning Theorem 404

Nonlocality 405

Nuclear Fission 411

Nuclear Models 414

Objectification 417

Objective Quantum Probabilities 420

Observable 425

One- and Two-Photon Interference 428

Operational Quantum Mechanics, Quantum Axiomatics and Quantum Structures 434

Operator 440

Orthodox Interpretation of Quantum Mechanics 444

Orthonormal Basis 447

Parity 450

Particle Physics 455

Particle Tracks 460

Parton Model 465

Paschen–Back Effect 468

Pauli Exclusion Principle 470

Pauli Spin Matrices 470

Photoelectric Effect 472

Photon 476

Pilot Waves 476

Planck's Constant *h* 478

POVM (Positive Operator Value Measure) 480

Probabilistic Interpretation of Quantum Mechanics 485

Probability in Quantum Mechanics 492

Projection 497

Projection Postulate ... 499

Propensities in Quantum Mechanics ... 502

Protective Measurements ... 505

Pure States ... 508

Quantization (First, Second) ... 509

Quantization (Systematic) ... 510

Quantum Chaos ... 514

Quantum Chemistry ... 518

Quantum Chromodynamics (QCD) ... 524

Quantum Communication ... 527

Quantum Computation ... 533

Quantum Electrodynamics (QED) ... 539

Quantum Entropy ... 543

Quantum Eraser ... 546

Quantum Field Theory ... 549

Quantum Gravity (General) and Applications ... 565

Quantum Hall Effect ... 572

Quantum Interrogation ... 591

Quantum Jump Experiments ... 595

Quantum Jumps ... 599

Quantum Logic ... 601

Quantum Mechanics ... 605

Quantum Numbers ... 605

Quantum State Diffusion Theory (QSD) ... 608

Quantum State Reconstruction ... 609

Quantum Statistics ... 611

Quantum Theory, 1914–1922 ... 613

Quantum Theory, Crisis Period 1923–Early 1925 613

Quantum Theory, Early Period (1900–1913) 617

Quantum Zeno Effect 622

Quarks 626

Quasi-Classical Limit 626

Radioactive Decay Law (Rutherford–Soddy) 630

Relativistic Quantum Mechanics 632

Renormalization 637

Rigged Hilbert Spaces in Quantum Physics 640

Rigged Hilbert Spaces for the Dirac Formalism of Quantum Mechanics 651

Rigged Hilbert Spaces and Time Asymmetric Quantum Theory 660

Russell–Saunders Coupling 671

Rutherford Atom 671

Scattering Experiments 676

Schrödinger Equation 681

Schrödinger's Cat 685

Schrödinger Picture 689

Selection Rules 690

Self-Adjoint Operator 692

Semi-classical Models 697

Shor's Algorithm 702

Solitons 702

Sommerfeld School 716

Specific Heats 719

Spectral Decomposition 721

Spectroscopy 721

Spin 726

Spin Echo 731

Spin Statistics Theorem 733

Squeezed States 736

Standard Model 738

Stark Effect 738

States in Quantum Mechanics 742

States, Pure and Mixed, and Their Representations 744

State Operator 746

Statistical Operator 746

Stern–Gerlach Experiment 746

Superconductivity 750

Superfluidity 758

Superluminal Communication in Quantum Mechanics 766

Superposition Principle (Coherent and Incoherent Superposition) 769

Superselection Rules 771

Symmetry 779

Time in Quantum Theory 786

Trace 793

Transactional Interpretation of Quantum Mechanics 795

Tunneling 799

Two-State Vector Formalism 802

Uncertainty Principle, Indetermincay Relations 807

Unitary Operator 807

Vector Model 810

Wave Function 812

Wave Function Collapse ... 813
Wave Mechanics ... 822
Wave Packet ... 828
Wave-Particle Duality: Some History ... 830
Wave-Particle Duality: A Modern View ... 835
Weak Value and Weak Measurements ... 840
Werner States ... 843
Which-Way or Welcher-Weg-Experiments ... 845
Wigner Distribution ... 851
Wigner's Friend ... 854
X-Rays ... 859
Zeeman Effect ... 862
Zero-Point Energy ... 864

English/German/French Lexicon of Terms ... 867
Selected Resources for Historical Studies ... 869
The Contributors ... 871

A

Aharonov–Bohm Effect

Holger Lyre

The Aharonov–Bohm effect (for short: AB effect) is, quite generally, a non-local effect in which a physical object travels along a closed loop through a gauge field-free region and thereby undergoes a physical change. As such, the AB effect can be described as a holonomy. Its paradigmatic realization became widely known after Aharonov and Bohm's 1959 paper – with forerunners by Weiss [1] and Ehrenberg and Siday [2]. Aharonov and Bohm [3] consider the following scenario: A split electron beam passes around a solenoid in which a magnetic field is confined. The region outside the solenoid is field-free, but nevertheless a shift in the interference pattern on a screen behind the solenoid can be observed upon alteration of the magnetic field. The schematic experimental setting can be grasped from the following figure:

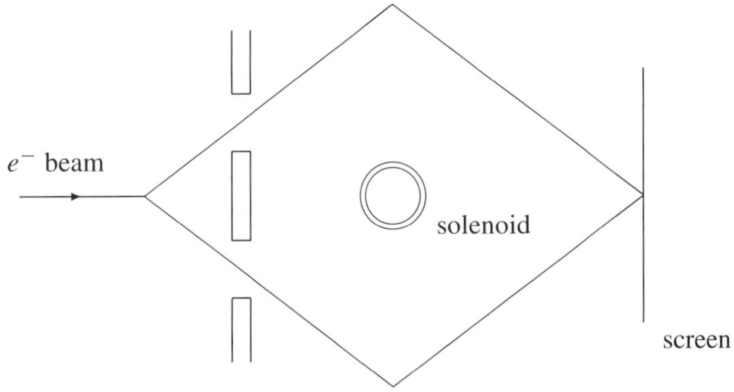

The phase shift can be calculated from the loop integral over the potential, which – due to Stokes' theorem – relates to the magnetic flux

$$\Delta\chi = q \oint_C \mathbf{A}\, \mathrm{d}\mathbf{r} = q \int_S \mathbf{B}\, \mathrm{d}\mathbf{s} = q\, \Phi_{\mathrm{mag}}. \tag{1}$$

Convincing arguments can be given that the effect is no artifact of some improper shielding of the fields involved. On the one hand, the magnetic field can perfectly be confined by the usage of toroidal magnets [15], the unavoidable penetration of the quantum ▶ wave function into the solenoid, on the other hand, is not known to be correlated to any scaling of the effect with the quality of the solenoid's shielding.

While the above experimental setting is called the magnetic AB effect, it is also possible to consider the electric pendant where the phase of the wave function depends upon varying the electric potential for two paths of a particle travelling through regions free of an electric field. Moreover, Aharonov and Casher [4] described a dual to the AB effect, called the ► Aharonov–Casher effect, where a phase shift in the interference of the magnetic moment in an electric field is considered.

The discovery of the AB effect has caused a flood of publications both about the theoretical nature of the effect as well as about the various experimental realizations. Much of the relevant material is covered in Peshkin and Tonomura [14]. The theoretical debate can basically be centered around the questions, whether and in which sense the AB effect is of (1) quantum, (2) topological, and (3) non-local nature.

1. Contrary to a widely held view in the literature, the point can be made that the AB effect is not of a genuine quantum nature, since there exist classical gravitational AB effects as well ([5]; [6]; [7]). A simple case is the geometry of a cone where the curvature is flat everywhere except at the apex (which may be smoothed). Parallel transport on a loop enclosing the apex leads to a holonomy. Also, the second clock effect in Weylian spacetime can be construed as an AB analogue, as Brown and Pooley [8] have pointed out. In Weylian spacetime, a clock travelling on a loop through a field free region enclosing a non-vanishing electromagnetic field undergoes a shift. It has been shown that the AB effect can be generalized to any SU(N) gauge theory ([9]; [10]).

2. The AB effect does not depend on the particular path as long as the region of the non-vanishing gauge field strength is enclosed. It is therefore no instance of the ► Berry phase, which is a path-dependent geometrical quantum phase. It does depend on the topology of the configuration space of the considered physical object (in case of the electric AB effect this space is homeomorphic to a circle). Nevertheless, the AB effect can still be distinguished from topological effects within gauge theories such as monopoles or instantons, where the topological nature can be described as non-trivial mappings from the gauge group into the configuration space (this incidentally also applies to the magnetic AB effect, but generally not to SU(N) or gravitational AB effects).

3. It is obvious that the AB effect is in some sense non-local. A closer inspection depends directly on the question about the genuine entities involved, and this question has been in the focus of the philosophy of physics literature. In the magnetic AB effect, the electron wave function does not directly interact with the confined magnetic field, but since the vector gauge potential outside the solenoid is non-zero, it is a common view to consider the AB effect as a proof for the reality of the gauge potential. This, however, renders real entities gauge-dependent. Healey [11] therefore argues for the holonomy itself as the genuine gauge theoretic entity. In both the potential and the holonomy interpretation the AB effect is non-local in the sense that it is non-separable, since properties of the whole – the holonomy – do not supervene on properties of the parts. As a third possibility even an interpretation solely in terms of field strengths can be given at the expense of violating the principle of local action. The case can be made that this is an instance of ontological underdetermination, where only the gauge group structure is invariant (and, hence, a case in favour of structural realism [12]).

Remarkably, van Kampen [13] has argued that the AB effect is in fact instantaneous, but that this cannot be directly observed since the instantaneous action of the magnetic effect is accordingly cancelled by the electric AB effect. ▶ Also Berry's Phase.

Primary Literature

1. P. Weiss: On the Hamilton-Jacobi Theory and Quantization of a Dynamical Continuum. Proc. Roy. Soc. Lond. A 169, 102–19 (1938)
2. W. Ehrenberg, R.E. Siday: The Refractive Index in Electron Optics and the Principles of Dynamics. Proc. Phys. Soc. Lond. B 62, 8–21 (1949)
3. Y. Aharonov, D. Bohm: Significance of Electromagnetic Potentials in the Quantum Theory. Phys. Rev. **115**(3), 485–491(1959)
4. Y. Aharonov, A. Casher: Topological Quantum Effects for Neutral Particles. Phys. Rev. Lett. **53**, 319–321(1984)
5. J.-S. Dowker: A Gravitational Aharonov–Bohm Effect. Nuovo. Cim. **52**B(1), 129–135 (1967)
6. J. Anandan: Interference, Gravity and Gauge Fields. Nuovo. Cim **53**A(2), 221–249 (1979)
7. J. Stachel: Globally Stationary but Locally Static Space-times: A Gravitational Analog of the Aharonov–Bohm Effect. Phys. Rev. D **26**(6), 1281–1290 (1982)
8. H.R. Brown, O. Pooley: The origin of the spacetime metric: Bell's 'Lorentzian pedagogy' and its significance in general relativity. In C. Callender and N. Huggett, editors. *Physics meets Philosophy at the Planck Scale*. (Cambridge University Press, Cambridge 2001)
9. T.T. Wu, C.N. Yang: Concept of Nonintegrable Phase Factors and Global Formulation of Gauge Fields. Phys. Rev. D **12**(12), 3845–3857 (1975)
10. C.N. Yang: Integral Formalism for Gauge Fields. Phys. Rev. Lett. **33**(7), 445–447(1974)
11. R. Healey: On the Reality of Gauge Potentials. Phil. Sci. **68**(4), 432–455 (2001)
12. H. Lyre: Holism and Structuralism in U(1) Gauge Theory. Stud. Hist. Phil. Mod. Phys. **35(4)**, 643–670 (2004)
13. N.G. van Kampen: Can the Aharonov-Bohm Effect Transmit Signals Faster than Light? Phys. Lett. A **106**(1), 5–6 (1984)

Secondary Literature

14. M.A. Peshkin, A. Tonomura: *The Aharonov-Bohm Effect*. (Lecture Notes in Physics 340. Springer, Berlin 2001)
15. A. Tonomura: *The Quantum World Unveiled by Electron Waves*. (World Scientific, Singapore 1998)

Aharonov–Casher Effect

Daniel Rohrlich

In 1984, 25 years after the prediction of the ▶ Aharonov–Bohm (AB) effect, Aharonov and Casher [1] predicted a "dual" effect. In both effects, a particle is excluded from a tubular region of space, but otherwise no force acts on it. Yet it acquires a measurable quantum phase that depends on what is inside the tube of space from which it is excluded. In the AB effect, the particle is charged and the tube contains a magnetic flux. In the Aharonov–Casher (AC) effect, the particle is neutral, but has a magnetic moment, and the tube contains a line of charge. Experiments in neutron [2], vortex [3], atom [4], and electron [5] interferometry bear out the prediction of Aharonov and Casher. Here we briefly explain the logic of the AC effect and how it is dual to the AB effect.

We begin with a two-dimensional version of the AB effect. Figure 1 shows an electron moving in a plane, and also a "fluxon", i.e. a small region of magnetic flux (pointing out of the plane) from which the electron is excluded. In Fig. 1 the fluxon is in a quantum ► superposition of two positions, and the electron diffracts around one of the positions but not the other. Initially, the fluxon and electron are in a product state $|\Psi_{\text{in}}\rangle$:

$$|\Psi_{\text{in}}\rangle = \frac{1}{2}(|f_1\rangle + |f_2\rangle) \otimes (|e_1\rangle + |e_2\rangle),$$

where $|f_1\rangle$ and $|f_2\rangle$ represent the two fluxon wave packets and $|e_1\rangle$ and $|e_2\rangle$ represent the two electron wave packets. After the electron passes the fluxon, their state $|\Psi_{\text{fin}}\rangle$ is not a product state; the relative phase between $|e_1\rangle$ and $|e_2\rangle$ depends on the fluxon position:

$$|\Psi_{\text{fin}}\rangle = \frac{1}{2}|f_1\rangle \otimes (|e_1\rangle + |e_2\rangle) + \frac{1}{2}|f_2\rangle \otimes (|e_1\rangle + \mathrm{e}^{\mathrm{i}\phi_{\text{AB}}}|e_2\rangle).$$

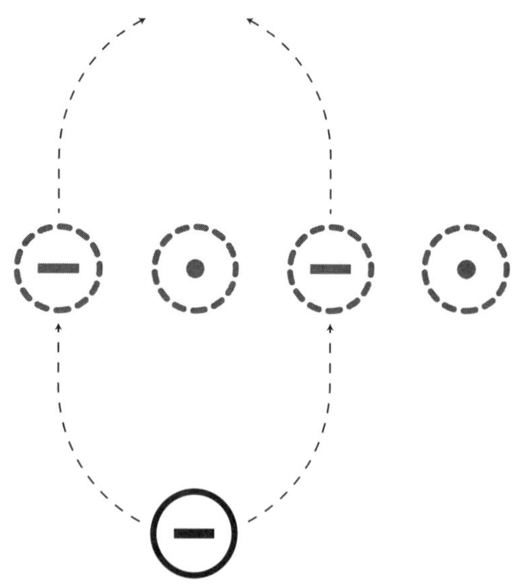

Fig. 1 An electron and a fluxon, each in a superposition of two wave packets; the electron wave packets enclose only one of the fluxon wave packets

Here ϕ_{AB} is the Aharonov–Bohm phase, and $|f_2\rangle$ represents the fluxon positioned between the two electron ▶ wave packets. Now if we always measure the position of the fluxon and the relative phase of the electron, we discover the Aharonov–Bohm effect: the electron acquires the relative phase ϕ_{AB} if and only if the fluxon lies between the two electron paths. But we can rewrite $|\Psi_{fin}\rangle$ as follows:

$$|\Psi_{fin}\rangle = \frac{1}{2}(|f_1\rangle + |f_2\rangle) \otimes |e_1\rangle + \frac{1}{2}(|f_1\rangle + e^{i\phi_{AB}}|f_2\rangle) \otimes |e_2\rangle .$$

This rewriting implies that if we always measure the relative phase of the *fluxon* and the position of the *electron*, we discover an effect that is analogous to the Aharonov–Bohm effect: the *fluxon* acquires the relative phase ϕ_{AB} if and only if the *electron* passes between the two fluxon wave packets. Indeed, the effects are equivalent: we can choose a reference frame in which the fluxon passes by the stationary electron. Then we find the same relative phase whether the electron paths enclose the fluxon or the fluxon paths enclose the electron.

In two dimensions, the two effects are equivalent, but there are two inequivalent ways to go from two to three dimensions while preserving the topology (of paths of one particle that enclose the other): either the electron remains a particle and the fluxon becomes a tube of flux, or the fluxon remains a particle (a neutral particle with a magnetic moment) and the electron becomes a tube of charge. These two inequivalent ways correspond to the AB and AC effects, respectively. They are not equivalent but *dual*, i.e. equivalent up to interchange of electric charge and magnetic flux.

In the AB effect, the electron does not cross through a magnetic field; in the AC effect, the neutral particle does cross through an electric field. However, there is no force on either particle. The proof [6] is surprisingly subtle and holds only if the line of charge is straight and parallel to the magnetic moment of the neutral particle [8]. Hence only for such a line of charge are the AB and AC effects dual.

Duality has another derivation. To derive their effect, Aharonov and Casher [1] first obtained the nonrelativistic Lagrangian for a neutral particle of magnetic moment $\boldsymbol{\mu}$ interacting with a particle of charge e. In Gaussian units, it is

$$L = \frac{1}{2}mv^2 + \frac{1}{2}MV^2 + \frac{e}{c}\boldsymbol{A}(\boldsymbol{r}-\boldsymbol{R}) \cdot (\boldsymbol{v}-\boldsymbol{V}),$$

where $M, \boldsymbol{R}, \boldsymbol{V}$ and $m, \boldsymbol{r}, \boldsymbol{v}$ are the mass, position and velocity of the neutral and charged particle, respectively, and the vector potential $\boldsymbol{A}(\boldsymbol{r}-\boldsymbol{R})$ is

$$\boldsymbol{A}(\boldsymbol{r}-\boldsymbol{R}) = \frac{\boldsymbol{\mu} \times (\boldsymbol{r}-\boldsymbol{R})}{|\boldsymbol{r}-\boldsymbol{R}|^3}.$$

Note L is invariant under respective interchange of $\boldsymbol{r}, \boldsymbol{v}$ and $\boldsymbol{R}, \boldsymbol{V}$. Thus L is the same whether an electron interacts with a line of magnetic moments (AB effect) or a magnetic moment interacts with a line of electrons (AC effect). However, if we begin with the AC effect and replace the magnetic moment with an electron, and all the electrons with the original magnetic moment, we end up with magnetic moments that all point in the same direction, i.e. with a straight line of magnetic flux. Hence the original line of electrons must have been straight. We see intuitively that the effects are dual only for a straight line of charge.[1]

Primary Literature

1. Y. Aharonov, A. Casher: Topological quantum effects for neutral particles. Phys. Rev. Lett. 53, 319–21 (1984).
2. A. Cimmino, G. I. Opat, A. G. Klein, H. Kaiser, S. A. Werner, M. Arif, R. Clothier: Observation of the topological Aharonov–Casher phase shift by neutron interferometry. Phys. Rev. Lett. 63, 380–83 (1989).
3. W. J. Elion, J. J. Wachters, L. L. Sohn, J. D. Mooij: Observation of the Aharonov–Casher effect for vortices in Josephson-junction arrays. Phys. Rev. Lett. 71, 2311–314 (1993).
4. K. Sangster, E. A. Hinds, S. M. Barnett, E. Riis: Measurement of the Aharonov–Casher phase in an atomic system. Phys. Rev. Lett. 71, 3641–3644 (1993); S. Yanagimachi, M. Kajiro, M. Machiya, A. Morinaga: Direct measurement of the Aharonov–Casher phase and tensor Stark polarizability using a calcium atomic polarization interferometer. Phys. Rev. A65, 042104 (2002).
5. M. König et al.: Direct observation of the Aharonov–Casher Phase. Phys. Rev. Lett. 96, 076804 (2006).
6. Y. Aharonov, P. Pearle, L. Vaidman: Comment on Proposed Aharonov–Casher effect: Another example of an Aharonov–Bohm effect arising from a classical lag. Phys. Rev. A37, 4052–055 (1988).

Secondary Literature

7. For a review, see L. Vaidman: Torque and force on a magnetic dipole. Am. J. Phys. 58, 978–83 (1990).

[1] I thank Prof. Aharonov for a conversation on this point.

Algebraic Quantum Mechanics

N.P. Landsman

Algebraic quantum mechanics is an abstraction and generalization of the ► Hilbert space formulation of quantum mechanics due to von Neumann [5]. In fact, von Neumann himself played a major role in developing the algebraic approach. Firstly, his joint paper [3] with Jordan and Wigner was one of the first attempts to go beyond Hilbert space (though it is now mainly of historical value). Secondly, he founded the mathematical theory of operator algebras in a magnificent series of papers [4, 6].

Although his own attempts to apply this theory to quantum mechanics were unsuccessful [18], the operator algebras that he introduced (which are now aptly called von Neumann algebras) still play a central role in the algebraic approach to quantum theory. Another class of operator algebras, now called C^*-algebras, introduced by Gelfand and Naimark [1], is of similar importance in algebraic quantum mechanics and quantum field theory. Authoritative references for the theory of C^*-algebras and von Neumann algebras are [14] and [21]. Major contributions to algebraic quantum theory were also made by Segal [7, 8] and Haag and his collaborators [2, 13].

The need to go beyond Hilbert space initially arose in attempts at a mathematically rigorous theory of systems with an infinite number of degrees of freedom, both in quantum statistical mechanics [9, 12, 13, 19, 20, 22] and in quantum field theory [2, 13, 20]. These remain active fields of study. More recently, the algebraic approach has also been applied to ▶ quantum chemistry [17], to the quantization and ▶ quasi-classical limit of finite-dimensional systems [15, 16], and to the philosophy of physics [10, 11, 16].

Besides its mathematical rigour, an important advantage of the algebraic approach is that it enables one to incorporate ▶ Superselection Rules. Indeed, it was a fundamental insight of Haag that the superselection sectors of a quantum system correspond to (unitarily) inequivalent representations of its algebra of ▶ observables (see below). As shown in the references just cited, in quantum field theory such representations (and hence the corresponding superselection sectors) are typically labeled by charges, whereas in quantum statistical mechanics they describe different thermodynamic phases of the system. In chemistry, the chirality of certain molecules can be understood as a superselection rule. The algebraic approach also leads to a transparent description of situations where ▶ locality and/or ▶ entanglement play a role [11, 13].

The notion of a C^*-algebra is basic in algebraic quantum theory. This is a complex algebra A that is complete in a norm $\| \cdot \|$ satisfying $\|ab\| \leqslant \|a\| \, \|b\|$ for all $a, b \in A$, and has an involution $a \mapsto a^*$ such that $\|a^*a\| = \|a\|^2$. A quantum system is then supposed to be modeled by a C^*-algebra whose self-adjoint elements (i.e. $a^* = a$) form the observables of the system. Of course, further structure than the C^*-algebraic one alone is needed to describe the system completely, such as a time-evolution or (in the case of quantum field theory) a description of the localization of each observable [13].

A basic example of a C^*-algebra is the algebra M_n of all complex $n \times n$ matrices, which describes an n-level system. Also, one may take $A = B(H)$, the algebra of all bounded operators on an infinite-dimensional Hilbert space H, equipped with the usual operator norm and adjoint. By the Gelfand–Naimark theorem [1], any C^*-algebra is isomorphic to a norm-closed self-adjoint subalgebra of $B(H)$, for some Hilbert space H. Another key example is $A = C_0(X)$, the space of all continuous complex-valued functions on a (locally compact Hausdorff) space X that vanish at infinity (in the sense that for every $\varepsilon > 0$ there is a compact subset $K \subset X$ such that $|f(x)| < \varepsilon$ for all $x \notin K$), equipped with the supremum norm $\|f\|_\infty := \sup_{x \in X} |f(x)|$, and involution given by (pointwise) complex conjugation. By the Gelfand–Naimark lemma [1], any commutative C^*-algebra is isomorphic to

$C_0(X)$ for some locally compact Hausdorff space X. The algebra of observables of a classical system can often be modeled as a commutative C^*-algebra.

A von Neumann algebra M is a special kind of C^*-algebra, namely one that is concretely given on some Hilbert space, i.e. $M \subset B(H)$, and is equal to its own bicommutant: $(M')' = M$ (where M' consists of all bounded operators on H that commute with every element of M). For example, $B(H)$ is always a von Neumann algebra. Whereas C^*-algebras are usually considered in their norm-topology, a von Neumann algebra in addition carries a second interesting topology, called the σ-weak topology, in which its is complete as well. In this topology, one has convergence $a_n \to a$ if $\operatorname{Tr} \hat{\rho}(a_n - a) \to 0$ for each density matrix $\hat{\rho}$ on H. Unlike a general C^*-algebra (which may not have any nontrivial projections at all), a von Neumann algebra is generated by its projections (i.e. its elements p satisfying $p^2 = p^* = p$). It is often said, quite rightly, that C^*-algebras describe "non-commutative topology" whereas von Neumann algebra form the domain of "non-commutative measure theory".

In the algebraic framework the notion of a state is defined in a different way from what one is used to in quantum mechanics. An (algebraic) state on a C^*-algebra A is a linear functional $\rho: A \to \mathbb{C}$ that is positive in that $\rho(a^*a) \geqslant 0$ for all $a \in A$ and normalized in that $\rho(1) = 1$, where 1 is the unit element of A (provided A has a unit; if not, an equivalent requirement given positivity is $\|\rho\| = 1$). If A is a von Neumann algebra, the same definition applies, but one has the finer notion of a normal state, which by definition is continuous in the σ-weak topology (a state is automatically continuous in the norm topology). If $A = B(H)$, then a fundamental theorem of von Neumann [5] states that each normal state ρ on A is given by a ► density matrix $\hat{\rho}$ on H, so that $\rho(a) = \operatorname{Tr} \hat{\rho} a$ for each $a \in A$. (If H is infinite-dimensional, then $B(H)$ also possesses states that are not normal. For example, if $H = L^2(\mathbb{R})$ the Dirac eigenstates $|x\rangle$ of the position operator are well known not to exist as vectors in H, but it turns out that they do define non-normal states on $B(H)$.) On this basis, algebraic states are interpreted in the same way as states in the usual formalism, in that the number $\rho(a)$ is taken to be the expectation value of the observable a in the state ρ (this is essentially the ► Born rule).

The notions of pure and mixed states can be defined in a general way now. Namely, a state $\rho : A \to \mathbb{C}$ is said to be pure when a decomposition $\rho = \lambda\omega + (1 - \lambda)\sigma$ for some $\lambda \in (0, 1)$ and two states ω and σ is possible only if $\omega = \sigma = \rho$. Otherwise, ρ is called mixed, in which case it evidently does have a nontrivial decomposition. It then turns out that a normal pure state on $B(H)$ is necessarily of the form $\psi(a) = (\Psi, a\Psi)$ for some unit vector $\Psi \in H$; of course, the state ρ defined by a density matrix $\hat{\rho}$ that is not a one-dimensional projection is mixed. Thus one recovers the usual notion of pure and mixed states from the algebraic formalism.

In the algebraic approach, however, states play a role that has no counterpart in the usual formalism of quantum mechanics. Namely, each state ρ on a C^*-algebra A defines a representation π_ρ of A on a Hilbert space H_ρ by means of the so-called GNS-construction (after Gelfand, Naimark and Segal [1, 7]). First, assume that ρ is faithful in that $\rho(a^*a) > 0$ for all nonzero $a \in A$. It follows that $(a, b) :=$

$\rho(a^*b)$ defines a positive definite sesquilinear form on A; the completion of A in the corresponding norm is a Hilbert space denoted by H_ρ. By construction, it contains A as a dense subspace. For each $a \in A$, define an operator $\pi_\rho(a)$ on A by $\pi_\rho(a)b := ab$, where $b \in A$. It easily follows that $\pi_\rho(a)$ is bounded, so that it may be extended by continuity to all of H_ρ. One then checks that $\pi_\rho : A \to B(H_\rho)$ is linear and satisfies $\pi_\rho(a_1a_2) = \pi_\rho(a_1)\pi_\rho(a_2)$ and $\pi_\rho(a^*) = \pi_\rho(a)^*$. This means that π_ρ is a representation of A on H_ρ. If ρ is not faithful, the same construction applies with one additional step: since the sesquilinear form is merely positive semidefinite, one has to take the quotient of A by the kernel N_ρ of the form (i.e. the collection of all $c \in A$ for which $\rho(c^*c) = 0$), and construct the Hilbert space H_ρ as the completion of A/N_ρ.

As in group theory, one has a notion of unitary (in)equivalence of representations of C^*-algebras. As already mentioned, this provides a mathematical explanation for the phenomenon of superselection rules, an insight that remains one of the most important achievements of algebraic quantum theory to date. See also ▶ operational quantum mechanics; relativistic quantum mechanics.

Primary Literature

1. I.M. Gelfand & M.A. Naimark: On the imbedding of normed rings into the ring of operators in Hilbert space. Mat. Sbornik 12, 197–213 (1943)
2. R. Haag & D. Kastler: An algebraic approach to quantum field theory. J. Math. Phys. **7**, 848–861 (1964)
3. P. Jordan, J. von Neumann & E. Wigner: On an algebraic generalization of the quantum mechanical formalism. Ann. Math. **35**, 29–64 (1934)
4. F.J. Murray & J. von Neumann: On rings of operators I, II, IV. Ann. Math. **37**, 116–229 (1936), Trans. Amer. Math. Soc. **41**, 208–248 (1937), Ann. Math. **44**, 716–808 (1943)
5. J. von Neumann: *Mathematische Grundlagen der Quantenmechanik* (Springer, Berlin 1932). English translation: *Mathematical Foundations of Quantum Mechanics* (Princeton University Press, Berlin 1955)
6. J. von Neumann: On rings of operators III, V. Ann. Math. **41**, 94–161 (1940), ibid. **50**, 401–485 (1949)
7. I.E. Segal: Irreducible representations of operator algebras. Bull. Amer. Math. Soc. **61**, 69–105 (1947)
8. I.E. Segal: Postulates for general quantum mechanics. Ann. Math. **48**, 930–948 (1947)

Secondary Literature

9. O. Bratteli & D.W. Robinson: *Operator Algebras and Quantum Statistical Mechanics. Vol 1: C* and W*-algebras, Symmetry Groups, Decomposition of States; Vol. 2: Equilibrium States, Models in Quantum Statistical Mechanics*, Second Edition (Springer, Heidelberg 1996, 2003)
10. J. Butterfield & J. Earman (ed.): *Handbook of the Philosophy of Science Vol. 2: Philosophy of Physics* (North-Holland, Elsevier, Amsterdam 2007)
11. R. Clifton, J. Butterfield & H. Halvorson: *Quantum Entanglements - Selected Papers* (Oxford University Press, Oxford 2004)
12. G.G. Emch: *Mathematical and Conceptual Foundations of 20th-Century Physics* (North-Holland, Amsterdam 1984)

13. R. Haag: *Local Quantum Physics: Fields, Particles, Algebras* (Springer, Heidelberg 1992)
14. R.V. Kadison & J.R. Ringrose: *Fundamentals of the Theory of Operator Algebras. Vol. 1: Elementary Theory; Vol. 2: Advanced Theory* (Academic, New York 1983, 1986)
15. N.P. Landsman: *Mathematical Topics Between Classical and Quantum Mechanics* (Springer, New York 1998)
16. N.P. Landsman: Between classical and quantum, in *Handbook of the Philosophy of Science Vol. 2: Philosophy of Physics*, ed. by J. Butterfield and J. Earman, pp. 417–554 (North-Holland, Elsevier, Amsterdam 2007)
17. H. Primas: *Chemistry, Quantum Mechanics and Reductionism*, Second Edition (Springer, Berlin 1983)
18. M. Redei: Why John von Neumann did not like the Hilbert space formalism of quantum mechanics (and what he liked instead). Stud. Hist. Phil. Mod. Phys. **27**, 493–510 (1996).
19. G.L. Sewell: *Quantum Mechanics and its Emergent Macrophysics* (Princeton University Press, Princeton 2002)
20. F. Strocchi: *Elements of Quantum mechanics of Infinite Systems* (World Scientific, Singapore 1985)
21. M. Takesaki: *Theory of Operator Algebras. Vols. I-III.* (Springer, New York 2003)
22. W. Thirring: *Quantum Mathematical Physics: Atoms, Molecules and Large Systems*, Second Edition (Springer, New York 2002)

Angular Momentum

► See Spin; Stern–Gerlach experiment; Vector model.

Anyons

Jon Magne Leinaas

Quantum mechanics gives a unique characterization of elementary particles as being either *bosons* or *fermions*. This property, referred to as the ► quantum statistics of the particles, follows from a simple symmetry argument, where the ► wave functions of a system of identical particles are restricted to be either symmetric (bosons) or antisymmetric (fermions) under permutation of particle coordinates. For two spinless particles, this symmetry is expressed through a sign factor which is associated with the switching of positions

$$\psi(\boldsymbol{r}_1, \boldsymbol{r}_2) = \pm\psi(\boldsymbol{r}_2, \boldsymbol{r}_1) \, , \tag{1}$$

with $+$ for bosons and $-$ for fermions. From the symmetry constraint, when applied to a many-particle system, the statistical distributions of particles over single particle states can be derived, and the completely different collective behaviour of systems like ► electrons (fermions) and photons (bosons) (► light quantum) can be understood.

The restriction to two possible kinds of quantum statistics, represented by the sign factor in (1), seems almost obvious. On one hand the permutation of particle coordinates has no physical significance when the particles are identical, which means that the wave function can change at most by a complex phase factor $e^{i\theta}$. On the other hand a double permutation seems to make no change at all, which further restricts the phase factor to a sign ± 1. This is the standard argument used in textbooks like [14].

However, there is a loophole to this argument, as pointed out by J.M. Leinaas and J. Myrheim in 1976 [1]. If the dimension of space is reduced from three to two the constraint on the phase factor is lifted and a continuum of possibilities appears that interpolates between the boson and fermion cases. In [1] these unconventional types of quantum statistics were found by analysis of the wave functions defined on the many-particle configuration space. Other approaches by G.A. Goldin, R. Menikoff, and D.H. Sharp [2] and by F. Wilczek [3] lead to similar results, and Wilczek introduced the name *anyon* for these new types of particles. As a precursor to this discussion M.G.G. Laidlaw and C.M. DeWitt had already shown that a path integral description applied to systems of identical particles reproduces standard results, but only in a space of dimensions higher than two [4].

The difference between continuous interchange of positions in two and three dimensions can readily be demonstrated, as illustrated in Fig. 1a. In two dimensions a two-particle interchange path comes with an orientation, and as a consequence a right-handed path and its inverse, a left-handed path, may be associated with different (inverse) phase factors. In three and higher dimensions there is no intrinsic difference between orientations of a path, since a right-handed path can be continuously changed to a left-handed one by a rotation in the extra dimension. Therefore, in dimensions higher than two the exchange phase factor has to be equal to its inverse, and is consequently restricted to ± 1. This explains why anyons are possible in two but not in three dimensions. Since the statistics angle θ in the exchange factor $\mathrm{e}^{\mathrm{i}\theta}$ is a free parameter, there is a different type of anyon for each value of θ. For

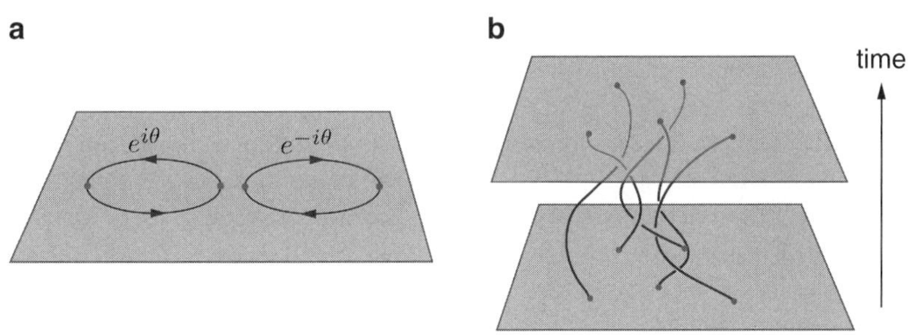

Fig. 1 Switching positions in two dimensions. (**a**) The difference between right-handed and left-handed interchange may give rise to quantum phase factors $\mathrm{e}^{\pm\mathrm{i}\theta}$ that are different from ± 1. (**b**) When many particles switch positions the collection of continuous particle paths can be viewed as forming a braid and the associated phase factor can be viewed as a representing an element of the *braid group*

systems with more than two particles the different paths define more complicated patterns (Fig. 1b), which are generally known as braids, and in this view of quantum statistics the corresponding *braid group* is therefore more fundamental than the permutation group. The generalized types of quantum statistics characterized by the parameter θ is often referred to as *fractional statistics* or *braiding statistics*.

Since anyons can only exist in two dimensions, elementary particles in the world of three space dimensions are still restricted to be either fermions or bosons. But in condensed matter physics the creation of quasi-twodimensional systems is possible, and in such systems anyons may emerge. They are excitations of the quantum system with sharply defined particle properties, generally known as quasiparticles. The presence of anyons in such systems is not only a theoretical possibility, as was realized after the discovery of *the fractional* ► *quantum Hall effect* in 1982. This effect is due to the formation of a two-dimensional, incompressible electron fluid in a strong magnetic field, and the anyon character of the quasiparticles in this system was demonstrated quite convincingly in theoretical studies [5, 6]. Although theoretical developments have given further support to this idea, a direct experimental evidence has been lacking. However, experiments performed by V.J. Goldman and his group in 2005, with studies on interference effects in tunnelling currents, have given clear indications for the presence of excitations with fractional statistics [7].

The discovery of the fractional quantum Hall effect and the subsequent development of ideas of *anyon superconductivity* [15] gave a boost in interest for anyons, which later on has been followed up by ideas of anyons in other types of systems with exotic quantum properties. One of these ideas applies to rotating atomic ► *Bose-Einstein condensation*, where theoretical studies have lead to predictions that at sufficiently high angular velocities a transition of the condensate to a bosonic analogy of a quantum Hall state will occur, and in this new quantum state anyon excitations should exist [8].

Topology is an important element in the description of anyons, since the focus is on continuous paths rather than simply on permutations of particle coordinates [1]. This focus on topology and on braids places the theory of anyons into a wider context of modern physics. Thus, anyons form a natural part of an approach to the physics of exotic condensed matter systems known as *topologically ordered systems*, where the two-dimensional electron gas of the quantum Hall system is a special realization [9]. The braid formulation also opens for generalizations in the form of non-abelian anyons. In this extension of the anyon theory, the phase factor associated with the interchange of two anyon positions is replaced by *non-abelian* unitary operations (or matrices). This is an extension of the simple *identical particle* picture of anyons, since new degrees of freedom are introduced which in a sense are shared by the participants in the braid. In the rich physics of the quantum Hall effect there are indications that such *nonabelions* may indeed exist [10], and theoretical ideas of exploiting such objects in the form of *topological* ► *quantum computation* [11] have gained much interest.

The topological aspects are important for the description of anyons, but at the same time they create problems for the study of many-anyon systems. Even if no additional interaction is present such systems can be studied in detail only when the particle number is small. There are also limitations to the application of standard many-particle methods. For these reasons the physics of many-anyon systems is only partly understood. One approach to the many-anyon problem is to trade the non-trivial braiding symmetry for a compensating *statistics interaction* [1], which is a two-body interaction that is sensitive to the braiding of particles, but is independent of distance. The same type of statistics transformation has also been used in field theory descriptions of the fractional quantum Hall effect, where the fundamental electron field is changed by a *statistics transmutation* into an effective bosonic field of the system [12].

Even if anyons, as usually defined, are particles restricted to two dimensions, there are related many-particle effects in one dimension. The interchange of particle positions cannot be viewed in the same way, since particles in one dimension cannot switch place in a continuous way without actually passing through each other. Nevertheless there are special kinds of interactions that can be interpreted as representing unconventional types of quantum statistics also in one dimension [13]. The name *anyon* is often applied also to these kinds of particles.

For further reading see [15] and [16].

Primary Literature

1. J.M. Leinaas, J. Myrheim: On the theory of identical particles. Il Nuovo Cimento B 37, 1 (1977).
2. G.A. Goldin, R. Menikoff, D.H. Sharp: Representations of a local current algebra in nonsimply connected space and the Aharonov-Bohm effect. J. Math. Phys. 22, 1664 (1981).
3. F. Wilczek: Magnetic flux, angular momentum and statistics. Phys. Rev. Lett. 48, 1144 (1982).
4. M.G.G. Laidlaw, C.M. DeWitt: Feynman integrals for systems of indistinguishable particles. Phys. Rev. D. 3, 1375 (1971).
5. B.I. Halperin: Statistics of quasiparticles and the hierarchy of fractional quantized Hall states. Phys. Rev. Lett. 52, 1583 (1984).
6. D. Arovas, J.R. Schrieffer, F. Wilzcek: Fractional statistics and the quantum Hall effect. Phys. Rev. Lett. 53, 722 (1984).
7. F.E. Camino, W. Zhou, V.J. Goldman: Realization of a Laughlin quasiparticle interferometer: Observation of fractional statistics. Phys. Rev. B 72, 075342 (2005).
8. N.K. Wilkin, J.M.F. Gunn: Condensation of composite bosons in a rotating BEC. Phys. Rev. Lett. 84, 6 (2000).
9. X.G. Wen: Topological orders in rigid states. Int. J. Mod. Phys. B 4, 239 (1990).
10. G. Moore, N. Read: Nonabelions in the fractional quantum Hall effect. Nucl. Phys. B. 360, 362 (1991).
11. A.Yu. Kitaev: Fault-tolerant quantum computation by anyons. Ann. Phys. (N.Y.), 303, 2 (2003).
12. S.C. Zhang, T.H. Hansson, S. Kivelson: Effective-field-theory model for the fractional quantum Hall effect. Phys. Rev. Lett. 62, 82 (1989).
13. J.M. Leinaas, J. Myrheim: Intermediate statistics for vortices in superfluid films. Phys. Rev. B 37, 9286 (1988).

Secondary Literature

14. L.D. Landau, L.M. Lifshitz: *Quantum Mechanics: Non-Relativistic Theory* (Elsevier Science, Amsterdam, Third Edition 1977).
15. F. Wilczek: *Fractional Statistics and Anyon Superconductivity* (World Scientific, Singapore, 1990).
16. A. Khare: *Fractional Statistics and Quantum Theory* (World Scientific, Singapore, Second Edition 2005).

Aspect Experiment

A.J. Leggett

In 1965, John S. Bell proved a celebrated theorem [1] which essentially states that no theory belonging to the class of "objective local theories" (OLT's) can reproduce the experimental predictions of quantum mechanics for a situation in which two correlated particles are detected at mutually distant stations (▶ Bell's Theorem). A few years later Clauser et al. [2] extended the theorem so as to make possible an experiment which would in principle unambiguously discriminate between the predictions of the class of OLT's and those of quantum mechanics, and the first experiment of this type was carried out by Freedman and Clauser [3] in 1972. This experiment, and (with one exception) others performed in the next few years confirmed the predictions of quantum mechanics. However, they did not definitively rule out the class of OLT's, because of a number of "loopholes" (▶ Loopholes in Experiments). Of these various loopholes, probably the most worrying was the "locality loophole": a crucial ingredient in the definition of an OLT is the postulate that the outcome of a measurement at (e.g.) station 2 cannot depend on the nature of the measurement at the distant station 1 (i.e., on the experimenter's choice of which of two or more mutually incompatible measurements to perform). If the space-time interval between the "event" of the choice of measurement at station 1 and that of the outcome of the measurement at station 2 were spacelike, then violation of the postulate under the conditions of the experiment would imply, at least prima facie, a violation of the principles of special relativity, so that most physicists would have a great deal of confidence in the postulate. Unfortunately, in the experiments mentioned, the choice of which variable to measure was made in setting up the apparatus (polarizers, etc.) in a particular configuration, a process which obviously precedes the actual measurements by a time of the order of hours; since the spatial separation between the stations was only of the order of a few meters, it is clear that the events of choice at 1 and measurement at 2 fail to meet the condition of spacelike separation by many orders of magnitude, and the possibility is left open that information concerning the setting (choice) at station 1 has been transmitted (subluminally) to station 2 and affected the outcome of the measurement there. While such a hypothesis certainly seems bizarre within the framework of currently accepted physics, the question of the viability or not of the class of OLT's is so fundamental an issue that one cannot afford to neglect it completely.

In this situation it becomes highly desirable, as emphasized by Bell in his original paper, to perform an experiment in which the choice of what to measure at station 1 is made "at the last moment", so that there is no time for information about this choice to be transmitted (subluminally or luminally) to station 2 before the outcome of the measurement there is realized. Of course, whether or not this condition is fulfilled in any given experiment depends crucially on exactly at what stage the "realization" of a specific outcome is taken to occur, and this question immediately gets us into the fundamental problem of measurement in quantum mechanics (► Measurement Theory); however, most discussions of the incompatibility of OLT's and quantum theory in the literature have been content to assume that the realization occurs no later than the first irreversible processes taking place in the macroscopic measuring device.(For example, in a typical photomultiplier it is assumed to take place when the photon hits the cathode and ejects the first electron, since in practice any processes taking place thereafter are irreversible). Although this assumption is certainly questionable, for the sake of definiteness it will be made until further notice.

The first experiment to attempt to evade the locality loophole was that of Aspect et al. [4] in 1982, and subsequent experiments which continue this approach are often referred to as "Aspect-type". In some sense these experiments are a sub-class of the more general category of "delayed-choice" experiments ► Delayed-Choice Experiment), but they have a special significance in their role of attempting to exclude the class of OLT's. In the original experiment [4], the distance between the detection stations is about 12 m, corresponding to a transit time for light of 40 nsec. At each station, the "switch" which decides which of the two alternative measurements to make is an acousto-optical device; in each case two electro-acoustical transducers, driven in phase, create ultrasonic standing waves in a slab of water through which the relevant photon must pass, with a period of about 25 MHz (the frequency is different for the two stations). The periodic density variation in the wave acts as a diffraction grating: If a given photon ► wave packet (length in time ~5 nsec) arrives at (say) station 1 when the wave has a node (i.e., the density and hence dielectric constant of the water is uniform) it is transmitted rectilinearly through the slab and enters a polarizer set in direction a; if on the other hand it arrives at an antinode (periodic density variation) it undergoes Bragg diffraction and is directed into a polarizer set at a'. (See Fig. 1). Photons (► light quantum) incident at intermediate phases of the wave are deflected into neither polarizer and thus missed in the counting. The period of switching between the alternative choices (a quarter period of the transducers) is about 10 nsec., short compared to the transit time of light between the stations. To the extent, then, that one can regard the switching as a "random" process, the locality loophole is blocked. The data obtained in ref. [4] violate the OLT predictions by 5 standard deviations.

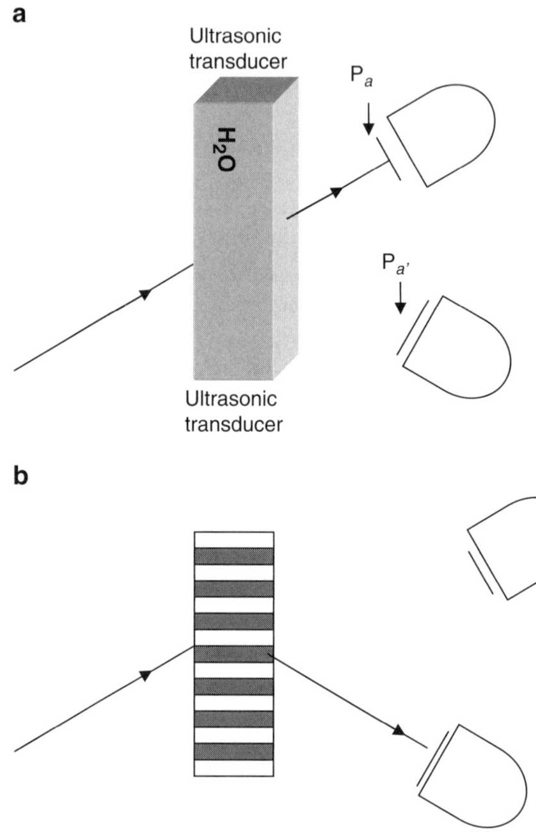

Fig. 1 Schema of switching devices in Aspect experiment. P_a ($P_{a'}$) are polarisers with transmission axis ***a*** (**a**). When a photon arrives at time on ultrasonic cycle when density of H_2O is constant, it is directed into P_a; (**b**) if it arrives at a maximum of the standing wave, into $P_{a'}$

Is the switching in fact a truly random process? On the one hand, since the transducer pairs are driven by different generators at different frequencies, there is no correlation between the choices made at the two stations, and as we have seen no time for information about the choice itself to be transmitted between them. On the other hand, since the driving at each station separately is periodic, a sufficiently determined advocate of OLT's might argue that station 2 has the information to predict what the setting at station 1 will be at a given time in the future and to make arrangements accordingly (and of course vice versa). Thus, while the experiment of ref. [4] is clearly a major advance on the original Freedman-Clauser one, not everyone was convinced that it had definitively blocked the locality loophole.

Of the various Aspect-type experiments performed subsequently to 1982, probably the most notable is that of Weihs et al. [5]. This experiment used a much longer baseline, around 400 m, and the choice of measurement was made by a quantum random number generator (QRNG), with a total switching time of less than 100 nsec.

A further feature of this experiment, unique up to now among the whole class of "Bell's theorem" experiments, is that instead of being channelled to a central coincidence counter the detection outcomes are recorded in situ and compared, with the help of accurate timing, only hours or days later (so that, coming back to the question of the time of "realization", its postponement until the time of comparison, which is not totally implausible in other experiments, would in this case seem distinctly unnatural). The duration of the registration process was such that it is completed well within the signal transit time. The data obtained are consistent with the predictions of quantum mechanics and violate those of the class of OLT's by 30 standard deviations.

One further experiment which has some significance in the present context is that of Tittel et al. [6]. Although there was no in-situ recording, this is otherwise similar in spirit to that of ref. [5], with an even longer base-line (10 km); the difference is that the role of the QRNG which controls the choice of measurement is played by the measured photon itself (it impinges on a beam splitter where the output beams correspond to different choices). Once more good agreement with the predictions of quantum mechanics is obtained.

In the light of these experiments, any attempt to continue to exploit the locality loophole to defend a theory of the OLT class would have either to deny that the QRNG's used work in a genuinely random way, or postpone the realization process for at least 1.3 microsec after the photon enters the photomultiplier (the signal transit time in the experiment of Weihs et al.). A truly definitive blocking of this loophole would presumably require that the detection be directly conducted by two human observers with a spatial separation such that the signal transit time exceeds human reaction times, a few hundred milliseconds (i.e., a separation of several tens of thousand kilometers). Given the extraordinary progress made in quantum communication in recent years, this goal may not be indefinitely far in the future. In the meantime, a small step in this direction might be taken by repeating the experiment of Weihs et al. with inspection of the outcomes by independent human observers before they are correlated, something which was not done in ref. [5].[1]

Primary Literature

1. J.S. Bell: On the Einstein-Podolsky-Rosen paradox. Physics **1**, 195 (1964)
2. J.F. Clauser, M.A. Horne, A. Shimony and R.A. Holt: Proposed experiment to test local hidden-variable theories. Phys. Rev. Lett. **23**, 880 (1969)
3. S.J. Freedman, J.F. Clauser: Experimental test of local hidden-variable theories. Phys. Rev. Lett. **28**, 938 (1972)
4. A. Aspect, J. Dalibard, G. Roger: Experimental test of Bell's inequalities using time-varying analysers. Phys. Rev. Lett. **49**, 1804 (1982)
5. G. Weihs, T.Jennewein, C.Simon, H.Weinfurter and A.Zeilinger: Violation of Bell's inequality under strict Einstein locality conditions. Phys. Rev. Lett. **81**, 5039 (1998)

---

[1] This work was supported by the National Science Foundation under grant no. NSF-EIA-01-21568.

6. W. Tittel, J. Brendel, H. Zbinden and N. Gisin: Violation of Bell inequalities by photons more than 10 km apart. Phys. Rev. Lett. **81**, 3563 (1998)

Secondary Literature

7. J.F. Clauser, A. Shimony: Bell's theorem: experimental tests and implications. Reps. Prog. Phys. **41**, 1881 (1978)
8. F. Selleri (ed.): *Quantum Mechanics versus Local Realism* (Plenum, New York)

Asymptotic Freedom

See ► Color Charge Degree of Freedom in Particles Physics; QCD; QFT.

Atomic Model

See also: ► Bohr's Atomic Model; Rutherford Atom.

Atomic Models, J.J. Thomson's "Plum Pudding" Model

Klaus Hentschel

In 1897, Joseph John Thomson (1856–1940) had announced the discovery of a corpuscle. Others soon called it ► electron, despite Thomson's stubborn preference for his original term, borrowed from Robert Boyle (1627–91) to denote any particle-like structure. Very soon afterwards, Thomson began to think about how to explain the periodicity of properties of the chemical elements in terms of these negatively charged corpuscles as atomic constituents. Chemical properties would thus have to depend on the number and constellations of these corpuscles inside the atom. They would have to have stable positions in it, bound by electrostatic and possibly kinetic forces. Because under normal conditions chemical atoms are electrically neutral, the total electric charge of all these negatively charged electrons had to be compensated for by an equal amount of positive charge. For Thomson it was natural to

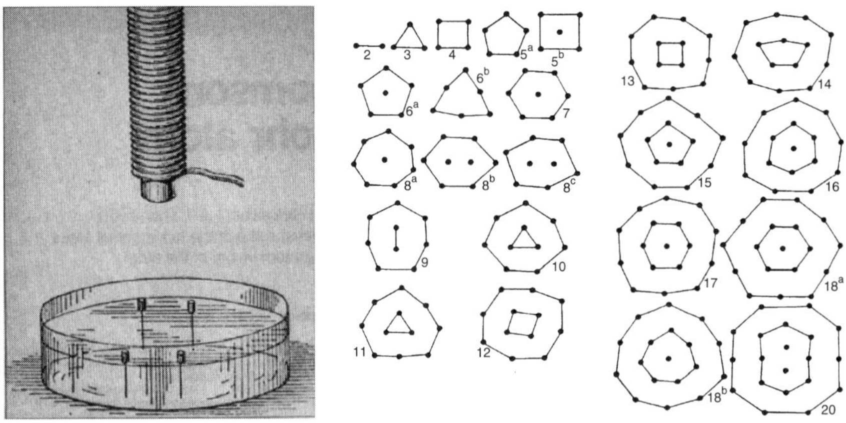

Fig. 1 *Left*: From [1, p. 248]; *right*: from [2, pp. 100–101]

assume that this positive charge was continuously distributed throughout the atom, whose radius was estimated at the time to be around 10^{-12} m. The very small negatively charged electrons (contemporary estimates indicated an order of magnitude of 10^{-15} m) were distributed in the atom like raisins inside a cake or like plums in a pudding, whence the popular nickname for Thomson's atomic model as the "plum pudding model".

In order to get a better idea of the stable configurations of these corpuscles inside the atom, Thomson drew an analogy to experiments by Alfred Marshall Mayer (1836–1897) who had pierced small magnetic needles into corks and watched them float in water below a strong magnet (see Fig. 1, left). In 1878/79 Mayer had observed that the magnetized floating needles quasi-automatically positioned themselves in characteristic configurations depending on their number. With more than six magnetic needles present, a seventh and eighth would inevitably position itself inside the outer ring of six (see the third row of Fig. 1 middle). As the number of floating magnets increased, more and more rings would form. Thomson hoped that a similar ring-structure composed of corpuscules could be found inside chemical atoms, and suspected that each of these rings would be analogous to the chemical periods in the period table of the elements. Specific configurations of the innermost ring would determine the chemical properties of the chemical element at hand. Two chemical elements with differing numbers of outer rings of corpuscles but similar innermost configurations would thus share similar chemical properties, like elements situated beneath each other in a column of the periodic table. To stabilize these configurations, Thomson also assumed that the concentric rings would all rotate around their common center.

Around 1904 Thomson believed each chemical atom would contain a very large number of ► electrons, something in the order of magnitude of 1,000 or more. With such high numbers he hoped to explain the puzzle of the exceedingly many spectral lines in each atom's spectrum and the fact that the masses of atoms proved to be several thousand times the mass of an electron. Radioactive decay (► radioactive decay law) very often correlated with the emission of negatively charged β-rays, turned out to be nothing but highly accelerated electrons, which Thomson thus interpreted as a mechanical instability of these electron configurations. A slight disturbance of the carefully balanced equilibrium position would result in electrostatic repulsion taking over and the expulsion of individual electrons or whole groups of electrons from the atom, where they would be experimentally observable as β-rays. Thomson also tried to explore the atomic structure by using corpuscles/electrons as projectiles in β-ray scattering experiments onto thin foils. The scattering angles observed by him and his students were predominantly very small, with a Gaussian distribution peaking sharply around zero-degree refraction and a width proportional to the thickness of the target layer. This experimental finding was interpreted as evidence for small-angle scattering, with successive layers of matter in thicker foils inducing an increasing, but still relatively small probability of multiple scattering, with occasional larger scattering angles resulting.

When Ernest Rutherford (1871–1937) started to work on ► scattering experiments, he varied Thomson's set-up by also using the positively charged and much heavier α-rays as projectiles. As will be discussed in detail in the entries on ► large-angle scattering and the ► Rutherford atom model, Rutherford's experiments showed that ► large-angle scattering was far more frequent than would be expected on the basis of J.J. Thomson's plum pudding ► atomic models. Rutherford decided to modify J.J. Thomson's atomic model: instead of assuming a continuous smeared-out positive charge, Rutherford postulated a concentrated atomic nucleus model with positive charge surrounded by a diffuse sphere of negative electricity (cf. Fig. 2). Quantitative analysis of his α-ray scattering experiments showed this atomic nucleus model was consistent with his data if the positive charge of the core was of the order of $A/2 \cdot e$, with A being the atomic number of the chemical element and e equal to the charge of J.J. Thomson's corpuscles, the elementary charge quantum. Thus Rutherford's estimate (which proved to be correct) drastically reduced the number of electrons inside atoms compared to J.J. Thomson's.

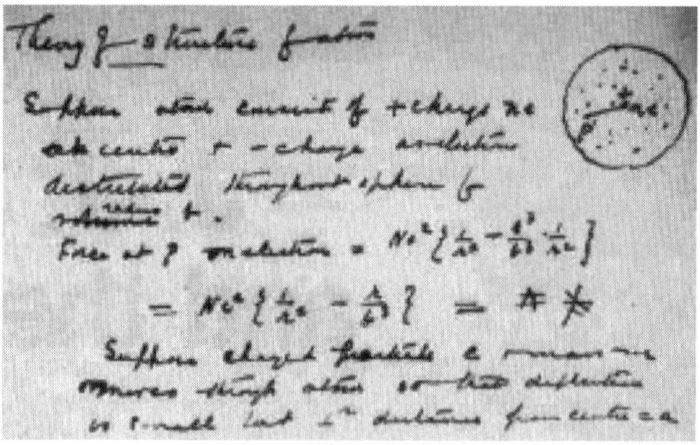

Fig. 2 Rutherford's first calculations on the passage of α-particles through atoms: "Theory of structure of atoms/Suppose atoms consist of + charge ne at centre & – charge as electrons distributed throughout sphere of radius ρ." From the Rutherford papers, Cambridge University Library, reproduced from [7, p. 24]

When the young Niels Bohr (1885–1962) finished writing his Ph.D. thesis at the University of Copenhagen, he obtained a fellowship for postgraduate study abroad. He chose to go to Cambridge, hoping to get to work more closely with J.J. Thomson, who was director of the Cavendish laboratory since 1884. The two personalities did not match, however, and Bohr soon decided to move on to Manchester where Ernest Rutherford introduced him to the intricacies of scattering experiments with α-rays and discussed his brand new nuclear core model of the atom. In the atomic model Bohr introduced in 1913, later refined by Arnold Sommerfeld (1868–1951) and others (► Bohr's atomic model; ► Sommerfeld school), Bohr masterfully merged ideas by J.J. Thomson, Rutherford and Nagaoka (► Atomic models). He also superimposed quantum conditions introduced by Max Planck (1858–1947) in 1900 and first employed in atomic models from 1910 on by Arthur Erich Haas (1884–1941) and John William Nicholson (1881–1955) [cf., e.g. [10], and [8]. While Bohr and Rutherford soon looked back on the older atomic models by J.J. Thomson and others as "a museum of scientific curiosities", J.J. Thomson for his part rejected Bohr's advances as "meretricious superficialities obtained without, or at the price of, an understanding of the mechanism of atoms" [7, p. 23]. Today we know that J.J. Thomson's hope to arrive at an intuitive, quasi-mechanical understanding of the atom was in vain – but at the time no one could be sure.

Primary Literature

1. Alfred M. Mayer: Floating magnets. American Journal of Science **116,** 248–9 (1878), also in Nature **17,** 487–488
2. Alfred M. Mayer: On the morphological laws of the configurations formed by magnets floating vertically and subjected to the attraction of a superposed magnet. Philosophical Magazine (5th series) **17,** 98–108 (1879)
3. Joseph John Thomson: On the structure of the atom. Philosophical Magazine (6th series) **7,** 237–265 (1904)
4. Joseph John Thomson: The structure of the atom (an evening lecture at the Royal Institution of London on March 10, 1905), published in the Proceedings of the Royal Institution of London (1905, 1–15)
5. Joseph John Thomson: *The Corpuscular Theory of Matter* (Constable & Company, London 1907)

Secondary Literature

6. Edward Arthur Davis & Elisabeth Falconer: *J.J. Thomson and the Discovery of the Electron* (Taylor & Francis, London 1997, esp. chap. 6, with a reprint of Thomson 1905 on pp. 215–229)
7. John Heilbron: J.J. Thomson and the Bohr atom. Physics Today **30,** 23–30 (April 1977)
8. Armin Hermann (ed.): *Der erste Quantenansatz für das Atom: Arthur Erich Haas* (Battenberg, Stuttgart 1965) (Dokumente der Naturwissenschaft, Abt. Physik, issue 10)
9. Helge Kragh: The First Subatomic Explanations of the Periodic System. *Foundations of Chemistry* **3**/2, 129–143 (2001)
10. Russell McCormmach: The atomic theory of John William Nicholson. *Archive for History of Exact Sciences* **3**/2, 160–184 (1975)

Atomic Models, Nagaoka's Saturnian Model

Klaus Hentschel

In late 1903, Hantaro Nagaoka (1865–1950) developed the earliest published quasi-planetary model of the atom. This graduate of the University of Tokyo from 1887 spent his postdoctoral period in Vienna, Berlin and Munich before obtaining a professorship in Tokyo to become Japan's foremost modern physicist. Nagaoka assumed that the atom is a large, massive, positively charged sphere, encircled by very many (in order of magnitude: hundreds) light-weight, negatively charged ► electrons, bound by electrostatic forces analogous to Saturn's ring, which is stabilized and attracted to the heavy planet by gravitation and consists of a myriad of small fragments. Thus, Nagaoka's model is also called a saturnian model. (Fig. 1) Even though its basic assumption foreshadowed later models of the atom, such as William Nicholson's (1753–1815) and Niels Bohr's (1885–1962), it differed from ► Bohr's atomic model in crucial points. Unlike Bohr one decade later, Nagaoka thought that

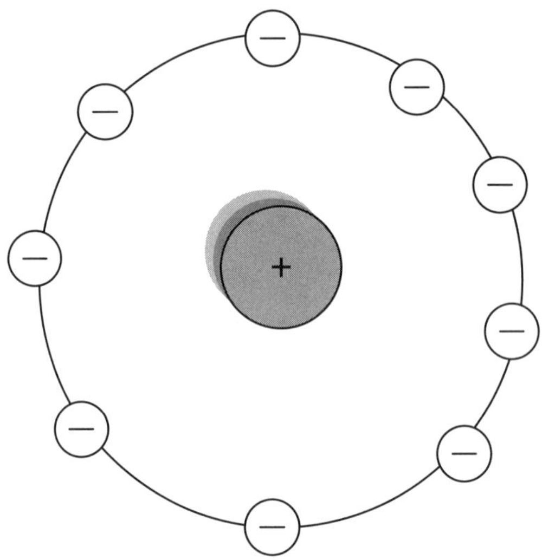

Fig. 1 Nagaoka's 'Saturnian' model: very many electrons move in one ring around a positively charged central body. In Nagoka's own words (1903/04, pp. 445f.): "The system, which I am going to discuss, consists of a large number of particles of equal mass arranged in a circle at equal angular intervals and repelling each other with forces inversely proportional to the square of distance; at the centre of the circle, place a particle of large mass attracting the other particles according to the same law of force. If these repelling particles be revolving with nearly the same velocity about the attracting centre, the system will generally remain stable, for small disturbances provided the attracting force be sufficiently great .... The present case will evidently be *approximately* realized if we replace these satellites by negative electrons and the attracting centre by a positively charged particle"

the observed atomic spectra should be directly correlated with the electron's orbit frequency. Radioactivity was interpreted as an occasional breakdown of saturnian rings, with electrons then being ejected from the atoms as β-rays. Consequently, Nagaoka and others tried to correlate spectral series, bands and other data observed in ► spectroscopy and early research on radioactivity with predictions derived from his model – in vain. Another problem of Nagaoka's and Nicholson's planetary models was a lack of stability of the electron orbits to oscillations orthogonal to the plane of rotation, as J.J. Thomson pointed out, which ultimately led to Nagaoka himself abandoning the Saturnian model in 1908.

Primary Literature

1. H. Nagaoka: Kinetics of a system of particles illustrating the line and the band spectrum and the phenomena of radioactivity. Philosophical Magazine (Ser. 6) **7**, 445–455 (1904) (Read before the Tokyo Physico-Mathematical Society, December 5th, 1903, also available online unter http://dbhs.wvusd.k12.ca.us/webdocs/Chem-History/Nagaoka-1904.html [Sept 3, 2008])
2. E. Yagi: H. Nagaoka's theory of the structure of matter (1904–1905). Actes du XIIIe Congrès International d'Histoire des Sciences 1971, **6**, 91–94 (1974)
3. K. Koizumi: The emergence of Japan's first physicists: 1868–1900. Historical Studies in the Physical Sciences **6**, 3–108 (1975)
4. R. McCormmach: The atomic theory of J.W. Nicholson. Archive for History of Exact Sciences **3**, 160–184 (1966)

B

Bell's Theorem

A.J. Leggett

Bell's theorem, first proved by John Stewart Bell (1928–1990) [1] in 1964, is probably the most celebrated result in the whole of twentieth-century physics. Briefly stated, it demonstrates that a whole class of theories about the physical world ("objective local theories", see below) defined by the conjunction of three apparently plausible general principles, must yield experimental predictions which under certain conditions are inconsistent with the predictions of quantum mechanics. Over the last 35 years a series of experiments motivated by the theorem have shown that under the relevant conditions the experimental properties of the world are consistent with the predictions of quantum mechanics and thus, subject to certain caveats, inconsistent with those of the alternative class of theories, so that the latter must apparently be rejected.

Let's first define an idealized experimental arrangement which is useful for the discussion of the theorem (see Fig. 1). A source emits pairs of particles (let us say for definiteness photons (► light quantum) as is usually the case in the real-life experiments). The photons travel to two different experimental "stations" S_1 and S_2 which are distant not only from the source but from one another, so that the space-time points at which they are detected at the stations are spacelike separated in the sense of special relativity (i.e. there is no time for a light wave, or anything slower, to pass between them). At (say) station 1 the relevant photon (1) encounters a randomly activated switch which directs it into one of two "measurement devices". Each measurement device gives a binary output ("yes" or "no"), but to two different "questions". To put a little flesh on this rather abstract formulation, let us imagine (as is usually the case in practice) that the "measurement" is of photon polarization; then one measurement device (call it M_a) would consist of a polarizer set to transmit photons polarized along direction a in the plane orthogonal to its propagation direction and reflect photons with the orthogonal polarization, together with counters [$C_a^{(+)}$ and $C_a^{(-)}$] to detect both the transmitted and reflected photons. The second measurement device, $M_{a'}$, is similar except that the polarizer now has a transmission axis a' which is different from a. A similar setup is constructed at station 2, with the alternative polarizer axes now b and b'. It is important that the "events" not only of the arrival of the photons at S_1 and S_2 but of the activation of the two switches, i.e. of the "choice" of which of the two alternative measurements to make at each station, be spacelike separated.

It is further assumed that we are able to identify precisely which photon 2 has been emitted in conjunction with a given photon 1 (e.g. by turning down the source intensity to a sufficiently low value). The output of each of the counters is a

D. Greenberger et al. (eds.), *Compendium of Quantum Physics: Concepts, Experiments, History and Philosophy*,

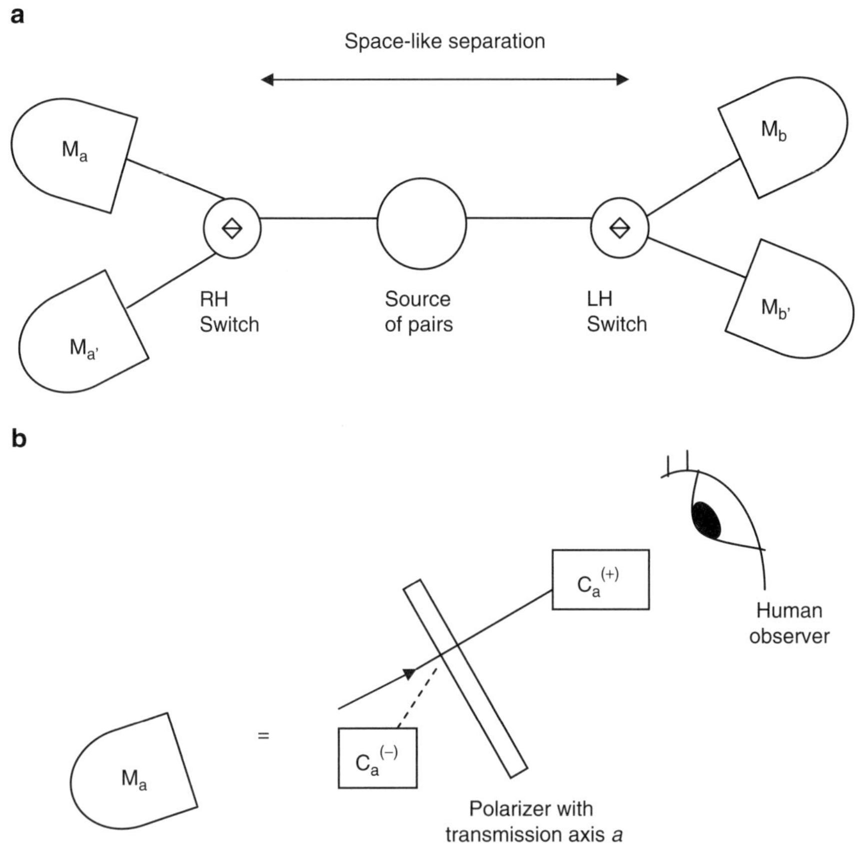

Fig. 1 Schematic setup of experimental arrangement. (**a**) The source and the two measurement stations. (**b**) Details of the measurement apparatus M_a. The apparatuses $M_{a'}$, M_b, $M_{b'}$ are similarly constructed

macroscopic event, e.g. an audible click; for complete idealization we may assume that at each station the click is noted by a conscious human observer (who can later report what he/she heard) and that the spacetime separation between the event of random switching at station 1 and that of conscious observation at station 2, and between the conscious observations at 1 and 2 themselves, is itself spacelike. Needless to say, real-life experiments do not fulfil all of the above requirements, particularly the last, but I will assume them for the sake of a clean discussion.

It is useful to develop a vocabulary to describe the data obtained in such an experimental setup. Consider a given pair of photons 1 and 2 which we are sure have been emitted in conjunction. Let us suppose that on this particular occasion the switch at station 1 has directed photon 1 into counter M_a. Then, if the design is ideal, one of two things will happen: either counter $C_a^{(+)}$ will click while counter $C_a^{(-)}$ remains silent, or vice versa. Let us define a dichotomic variable A so that the measured value of A is *by definition* $+1$ in the former case and -1 in the latter.

Similarly, if we suppose that the switch at station 2 has directed photon 2 into measurement apparatus M_b, we can define a quantity B so that the measured value of B is by definition $+1$ if it is counter $M_b^{(+)}$ which clicks, and -1 if it is $M_b^{(-)}$. Now let us consider a different pair of photons, for which (say) photon 2 is still switched into M_b but photon 1 is now switched into $M_{a'}$. We can define B as previously, but instead of A we must now define a quantity A', which has the measured value A' if $M_{a'}^{(+)}$ clicks, etc. Note that for this second pair, the "measured value" of A is *not defined* (as was not that of A' for the first pair). A quantity B' is introduced in the obviously analogous way. Let us now define the correlation of A and B, $\langle AB \rangle$, by the formula

$$\langle AB \rangle = \frac{\{N_{++}(ab) + N_{--}(ab) - N_{+-}(ab) - N_{-+}(ab)\}}{\{N_{++}(ab) + N_{--}(ab) + N_{+-}(ab) + N_{-+}(ab)\}} \qquad (1)$$

where $N_{++}(ab)$ means the number of occasions on which photon 1 was switched into counter M_a and photon 2 into M_b, and A and B were both measured to be $+1$, etc.; note that the denominator is simply the number of times that 1 was switched into M_a and 2 into M_b, irrespective of the outcome of the measurements. Correlations $\langle A'B \rangle$, $\langle AB' \rangle$, $\langle A'B' \rangle$ are defined analogously. With these definitions it is clear that we can measure $\langle AB \rangle$ on *one subensemble* of the total ensemble of photon pairs, namely that consisting of those pairs for which photon 1 was switched into M_a and photon 2 into M_b. Similarly, we can measure the correlation $\langle AB' \rangle$ on a *different subensemble* (1 switched into M_a, 2 into $M_{b'}$), and so on.

Let us next define the class of "objective local theories" (OLT's) whose predictions are to be compared with those of quantum mechanics and with experiment in situations approximating the idealized one described above. While the details of the definition as presented in the literature tend to vary from one author to another and with Bell's original one, one can summarize by saying that the class of OLT's is defined by the conjunction of three independent general hypotheses about the physical world, which for brevity I will refer to as (1) ► locality (2) induction and (3) realism. (As will be discussed below, some treatments in the literature do not explicitly include (2)). I now discuss these three postulates in turn.

1. *Locality* (sometimes called ► "Einstein locality") is the postulate, central to the special theory of relativity, that events which are spacelike separated cannot causally influence one another. In the experimental arrangement described above, this means that (for example) the outcome of a measurement at station 2 cannot depend on the setting of the switch at station 1.
2. *Induction* means basically our normal assumption about the "arrow of time", i.e. that physical ► ensembles in quantum mechanics (the collections of systems which possess reproducible statistical properties) existing at a time $t > 0$ are defined only by their past experience (e.g. by the initial conditions at time 0 and forces acting between 0 and t), and not by anything which is going to happen at a time later than t. In the relevant experiments this means that the statistical properties of the subensemble consisting of those pairs of photons on which (say)

A and B are measured should be identical to those of the ensemble of photons as a whole (in intuitive language, the photons cannot "know" in advance which polarization components are to be measured on them).

In many papers on Bell's theorem in the literature, postulate (2) is not included explicitly, probably because of a belief that it is subsumed under (1). This is a rather delicate issue: within the context of special relativity without any additional constraints the belief is obviously correct, in the sense that if one considers three events X,Y,Z such that X and Y are spacelike separated but both are in the past light cone of Z, then violation of (2) would allow Z to influence Y, and we assume that X influences Z in the usual way then X can influence Y, in violation of (1).
However, there is no obvious reason why a general OLT should not incorporate, for example, the postulate that such "causal triangles" are forbidden to occur, so that it is useful to incorporate postulate (2) explicitly in the definition of an objective local theory.

3. *Realism* is probably the conceptually trickiest ingredient in the definition of the class of OLT's. In the simplest form (essentially that used by Bell in his original paper) it is the statement that each individual particle (in the described experiment, each individual photon of a given pair) possesses definite properties; for example, each photon 1 carries with it information which determines *both* how it will respond if directed by the switch into M_a, *and* how it will respond if directed into $M_{a'}$. Let's call this assumption the hypothesis of microscopic realism, and denote it (3a). Note that while in his original paper Bell, whose original motivation was the issue of the consistency of "hidden-variable" theories (► Hidden Variables) with quantum mechanics, assumed that the response is deterministic as in most theories of that type, this is not essential; one can perfectly well consider models where there is intrinsic randomness in the outcome of the relevant measurement, provided only that the statistics of the latter is completely determined by information carried by photon 1 alone.

A possible alternative formulation of postulate (3) (call it (3b)) eschews any statement about the properties of microscopic objects (photons) in favor of statements about (actual and possible) directly observed events at the macroscopic level (clicks). Consider for example a case in which photon 1 is actually switched into $M_{a'}$; then, of course, this particular photon cannot be measured by M_a, and consequently the value of the quantity A is not defined. Now imagine, contrary to fact, that this particular photon had been switched into M_a. It is, of course, a (rather trivial) "fact" about the world that under these (counterfactual) conditions either counter $C_a^{(+)}$ would have clicked, giving $A = +1$, or counter $C_a^{(-)}$ would have clicked ($A = -1$). In other words we can presumably agree, referring to the given counterfactual conditions, that

(P_1): It is a fact that either A would have been $+1$, or A would have been -1.

Now consider the subtly different assertion:

(P_2): Either it is a fact that A would have been $+1$, or it is a fact that A would have been -1.

The assertion of (P_2) is called the hypothesis of *macroscopic counterfactual definiteness* (hereafter abbreviated MCFD ▶ Counterfactuals in QM)). In contrast to assertion 1, which makes as it were no particular metaphysical statement, assertion (P_2) claims that the outcome of an unperformed experiment is a fixed property of the world. It should be emphasized that the above formulation of the defining postulates of the class of theories for which Bell's theorem holds is only one of many possible such formulations. The equivalence or not of these alternative formulations, and the advantages and disadvantages of each, has been the subject of an extensive literature.

With these preliminaries we are now in a position to state and prove Bell's theorem. In the literature, the formulation tends to depend on whether the context is a discussion of the conflict of the predictions of the class of "objective local theories" with those of quantum mechanics, or rather of that with the experimental data; in the latter case, an extension of Bell's original theorem (the "CHSH theorem") proved by Clauser et al. [2] a few years after his paper tends to be more directly applicable than the original version. Here I shall present the CHSH theorem, and treat the original theorem proved by Bell as a special case of it.

The CHSH theorem states that, for any choice whatever of the settings a, b, a', b', any theory of the OLT class must predict the inequality

$$K\left(a,b,a',b'\right) \equiv \langle AB\rangle + \left\langle AB'\right\rangle + \left\langle A'B\right\rangle - \left\langle A'B\right\rangle \leqslant 2 \tag{2}$$

(and some related inequalities; in the interests of clarity I state only the first, which is the one most often used in experimental tests). Bell's original inequality is the special case of (2) which is obtained under the additional assumption that for $A = -B'$ (which in the polarization case means that b' is orthogonal to a') the quantity $\langle A'B'\rangle = +1$, as predicted by quantum mechanics for certain states (see below). Relabelling the various quantities so as to make closer contact with Bell's original notation, we find in this case the inequality

$$\langle AB\rangle - \langle CB\rangle <= 1 + \langle AC\rangle \tag{3}$$

which is known as Bell's inequality (or more precisely one of Bell's inequalities; again for clarity I give only one version). The inequalities (2) and (3) do not at first sight seem particularly dramatic, but the crucial point is that for certain states and settings *they are violated by the predictions of quantum mechanics*. For example, if we consider the pair of photons emitted in a so-called $0^+(J = 0,+ \to J = 1, - \to J = 0,+)$ atomic transition like that used in the experiments on C_a, we find that quantum mechanics unambiguously predicts, under ideal conditions, the result

$$\langle AB\rangle = \cos\left(2\theta_{ab}\right) \tag{4}$$

where θ_{ab} is the angle between the settings a and b. Setting $a'=0$, $b=3\pi/8$, $a=\pi/4$ and $b'=3\pi/8$, we find that the quantum mechanical prediction for this choice of settings is

$$K = 2^{3/2}$$

which violates the CHSH inequality by a factor of $2^{1/2}$. Similarly, for a 0^- transition, for which quantum mechanics predicts $\langle AB\rangle = \sin(2\theta_{ab})$, (hence $\langle AB\rangle = +1$ for a and b orthogonal as assumed by Bell, who actually treated explicitly the spin singlet state of two spin-½ particles, which is isomorphic to the 0− photon pair) the inequality (3) is violated by the quantum prediction over a range of settings (this is most intuitively obvious when (e.g.) a and c are both close to zero and b to $\pi/4$, since the LHS of (3) is then fairly obviously linear in θ_{ab} while the RHS is quadratic).

The proof of the CHSH theorem and hence of Bell's theorem as a special case, while conceptually subtle, requires only the most elementary algebra. For definiteness I will take the third postulate defining an OLT as the assumption of MCFD; it is straightforward to adapt the argument so as to substitute the assumption (3a) of microscopic realism. Then a possible derivation of the inequality (3) (one of many!) goes as follows:

1. By assumption (3b), the quantity A exists for each photon 1 and possesses a definite value, independently of whether or not that photon was directed into M_a. Similarly for A', B, B'.
2. By postulate (1), the value of A for any particular photon 1 cannot depend on the choice of what to measure at the distant station 2, nor on the outcome of that measurement. Similarly for A', B, B'.
3. Hence each of the quantities A, A', B and B' exists and takes a value $+1$ or -1 which is, in the case of A, independent of whether it is B or B' which is measured at the distant station, and vice versa. In other words, the value of A which occurs in the product AB is identical to that occurring in AB', etc.
4. It is then a matter of elementary algebra to show that *for any given pair* the quantities AB, etc. must satisfy the inequality

$$AB + AB' + A'B - A'B' \leqslant 2 \tag{5}$$

   (Any reader who doubts the truth of this statement is invited simply to exhaust the 16 possibilities!).
5. It then immediately follows that *when taken on the whole ensemble* of pairs (irrespective of which quantities were actually measured on them) the expectation values $\langle AB\rangle_{\text{all}}$ etc. satisfy an inequality of the same structure as (5).
6. By postulate (2), the statistical properties of each subensemble are identical to those of the complete ensemble. Hence, for example, the average of (AB) over the whole ensemble may be legitimately identified with the *measured* quantity $\langle AB\rangle$, which is of course strictly the average for the ab-ensemble only. Making this identification, we see that the measured correlations satisfy the CHSH inequality (3), QED.

Over the last 35 years, starting with the work of Freedman and Clauser [3] in 1972, a large number – probably hundreds – of experiments based on Bell's theorem have been performed. With a handful of exceptions, these experiments have all obtained results which are consistent with the predictions of quantum mechanics, and prima facie inconsistent with those of the whole class of objective local theories, in some cases by hundreds of standard deviations. However, no existing experiment has conformed entirely to the idealized setup described above, and this gives rise to various so-called "loopholes" in the refutation of OLT's. Generally speaking, these loopholes arise because of doubts about whether the OLT postulates are adequately satisfied by a given real-life experimental setup (for example, whether the relevant "events" of realization are sufficiently separated that one can legitimately invoke the locality assumption) ► Loopholes in Experiments.

Apart from the question of whether or not the conditions to invoke the OLT postulates have actually been satisfied in existing experiments, the implications of Bell's theorem are so disturbing that the theorem itself has been repeatedly challenged; that is, it has been argued that even if it turns out that even when (if?) all the loopholes have been plugged the experimental data still conform to the quantum mechanical predictions, this will not mean that we have to abandon the class of OLT's. In the present author's opinion, all these challenges to Bell's theorem as such have been uniformly unsuccessful: at best they reduce to the claim that one or other of the defining assumptions of an OLT is less overwhelmingly plausible than generally believed, while leaving the theorem itself intact.

If we assume that the loopholes will progressively be blocked and the data continue to conform to the quantum-mechanical predictions, so that we must conclude that the class of OLT's is ruled out, which of the three defining assumptions should we abandon? To abandon postulate (1) would be in prima facie conflict with the basic postulates of the special theory of relativity, and is therefore something that most practising physicists (as distinct from most popular writers on the subject!) would be extremely loath to do. Of course, we cannot rule out the possibility, which has been advocated by some prominent physicists, that (for example) an ultimate theory of ► quantum gravity will reveal special relativity to be only an approximate description of reality, so that postulate (1) might fail, but at present no such theory seems to be developed in a sufficiently concrete way to give us this escape-hatch. To challenge postulate (2) would be to abandon our conventional notions concerning the "arrow of time"; again, it cannot be excluded that future theoretical developments might force us to do just that, but the prospect is certainly not appealing; most of us would not currently know how to do physics without this deeply ingrained assumption. The weakest link would appear to be postulate (3), and that is probably what most practising physicists would choose to sacrifice; that is, they would claim that neither the assumption (3a) of microscopic realism nor that (3b) of MCFD is actually true of the real world. In the words of the late Asher Peres [4], "unperformed experiments have no results"!

While this conclusion is in some sense in the spirit of the Copenhagen interpretation of quantum mechanics, it is still a very surprising and, if one really takes it seriously, alarming fact about the physical world.[1] See also ► Aspect experiment and Section on Bell inequalities in ► Wave function collapse.

---

[1] This work was supported by the National Science Foundation through grant no.NSF-EIA-01-21568.

Primary Literature

1. J.S. Bell: On the Einstein-Podolsky-Rosen paradox. Physics **1**, 195 (1964)
2. J.F. Clauser, M.A. Horne, A. Shimony and R.A. Holt: Proposed experiment to test local hidden-variable theories. Phys. Rev. Lett. **23**, 880 (1969)
3. S.J. Freedman and J.F. Clauser: Experimental test of local hidden-variable theories. Phys. Rev. Lett. **28**, 934 (1972)
4. A. Peres: Unperformed experiments have no results. Am. J. Phys. **46**, 745 (1978)

Secondary Literature

5. N. Herbert: *Quantum Reality: beyond the new physics* (Anchor Press, Garden City, NY 1985)
6. B. d'Espagnat: *Conceptual foundations of quantum mechanics* (Benjamin, Menlo Park, CA 1971)
7. F. Selleri: *Quantum mechanics versus local realism*: the Einstein-Podolsky-Rosen Paradox (Plenum, New York 1988)
8. J. F. Clauser and A.Shimony: Bell's theorem: experimental tests and implications. Reps. Prog. Phys. **41**, 1881 (1978)

Berry's Phase

Daniel Rohrlich

Berry's phase [1] is a quantum phase effect arising in systems that undergo a slow, cyclic evolution. It is a remarkable correction to the quantum adiabatic theorem and to the closely related Born–Oppenheimer approximation [2]. Berry's elegant and general analysis has found application to such diverse fields as atomic, condensed matter, nuclear and elementary ► particle physics, and optics. In this brief review, we first derive Berry's phase in the context of the quantum adiabatic theorem and then in the context of the Born–Oppenheimer approximation. We mention generalizations of Berry's phase and analyze its relation to the ► Aharonov–Bohm effect.

Consider a Hamiltonian $H_f(\boldsymbol{R})$ that depends on parameters $R_1, R_2, \ldots, R_N$, components of a vector $\boldsymbol{R}$. Let us assume that $H_f(\boldsymbol{R})$ has at least one discrete and nondegenerate eigenvalue $E_i(\boldsymbol{R})$ with $|\Psi_i(\boldsymbol{R})\rangle$ its eigenstate; $E_i(\boldsymbol{R})$ and $|\Psi_i(\boldsymbol{R})\rangle$ inherit their dependence on $\boldsymbol{R}$ from $H(\boldsymbol{R})$. If the vector $\boldsymbol{R}$ changes in time, then $|\Psi_i(\boldsymbol{R})\rangle$ is not an exact solution to the time-dependent ► Schrödinger equation. But if $\boldsymbol{R}$ changes slowly enough, the system does not ► quantum jump to another eigenstate. Instead, it adjusts itself to the changing Hamiltonian. A heavy weight hanging on a string illustrates such adiabaticity. Pull the string quickly – it snaps and the weight falls. Pull the string slowly – the weight comes up with it.

"Slowly enough" has the following formal sense. Let $\boldsymbol{R}[t/T]$ evolve over a time interval $0 \leqslant t \leqslant T$; the larger T, the slower the evolution. If at time $t = 0$ the system is in the state $|\Psi_i(\boldsymbol{R}[0])\rangle$, then at time $t = T$ the state is $\mathrm{e}^{\mathrm{i}\phi_i(T)}|\Psi_i(\boldsymbol{R}[1])\rangle$ with probability approaching 1 as T approaches infinity, according to the quantum adiabatic theorem [10]. We obtain the phase $\phi_i(t)$ by substituting $\mathrm{e}^{\mathrm{i}\phi_i(t)}|\Psi_i(\boldsymbol{R})\rangle$ into the time-dependent ► Schrödinger equation,

$$i\hbar\frac{\mathrm{d}}{\mathrm{d}t}\mathrm{e}^{\mathrm{i}\phi_i(t)}|\Psi_i(\boldsymbol{R})\rangle = H_f(\boldsymbol{R}[t/T])\mathrm{e}^{\mathrm{i}\phi_i(t)}|\Psi_i(\boldsymbol{R})\rangle,$$

and projecting both sides of the equation onto $\mathrm{e}^{\mathrm{i}\phi_i(t)}|\Psi_i(\boldsymbol{R})\rangle$:

$$\frac{\mathrm{d}}{\mathrm{d}t}\phi_i(t) = \mathrm{i}\langle\Psi_i(\boldsymbol{R})|\nabla_{\boldsymbol{R}}|\Psi_i(\boldsymbol{R})\rangle\cdot\frac{\mathrm{d}\boldsymbol{R}}{\mathrm{d}t} - \frac{1}{\hbar}E_i(\boldsymbol{R}).$$

Thus

$$\begin{aligned}\phi_i(t) - \phi_i(0) &= \int_0^t \mathrm{d}t'\left[\mathrm{i}\langle\Psi_i(\boldsymbol{R})|\nabla_{\boldsymbol{R}}|\Psi_i(\boldsymbol{R})\rangle\cdot\frac{\mathrm{d}\boldsymbol{R}}{\mathrm{d}t'} - \frac{1}{\hbar}E_i(\boldsymbol{R})\right]\\ &= \int_{\boldsymbol{R}[0]}^{\boldsymbol{R}[t]}\langle\Psi_i(\boldsymbol{R})|\mathrm{i}\nabla_{\boldsymbol{R}}|\Psi_i(\boldsymbol{R})\rangle\cdot\mathrm{d}\boldsymbol{R} - \frac{1}{\hbar}\int_0^t \mathrm{d}t'\ E_i(\boldsymbol{R}).\end{aligned}$$

The integrand $\boldsymbol{A}_B \equiv \langle\Psi_i(\boldsymbol{R})|\mathrm{i}\nabla_{\boldsymbol{R}}|\Psi_i(\boldsymbol{R})\rangle$ is *Berry's connection* for the state $|\Psi_i(\boldsymbol{R})\rangle$. The integral $-\int_0^t E_i dt'/\hbar$ is called the *dynamical* phase.

The overall phase of a quantum state is not observable. But a quantum system may be in a ► superposition of states; the *relative* phase of these states is observable. Consider two paths $\boldsymbol{R}[t/T]$ and $\boldsymbol{R}'[t/T]$ with the same endpoints $\boldsymbol{R}[0] = \boldsymbol{R}'[0]$ and $\boldsymbol{R}[1] = \boldsymbol{R}'[1]$, and suppose that the system evolves in a superposition of states $|\Psi_i(\boldsymbol{R}[t/T])\rangle$ and $|\Psi_i(\boldsymbol{R}'[t/T])\rangle$. At time $t = T$ the relative phase of this superposition contains two parts. One part is the relative dynamical phase. The other part is Berry's phase, the difference between $\boldsymbol{A}_\mathrm{B}$ integrated along $\boldsymbol{R}$ and $\boldsymbol{A}_\mathrm{B}$ integrated along $\boldsymbol{R}'$, i.e. it is the circular integral of $\boldsymbol{A}_\mathrm{B}$ along the *closed* path comprising $\boldsymbol{R}$ and $\boldsymbol{R}'$ with opposite senses. This phase is well defined, because it is gauge invariance (► gauge symmetry): If we multiply $|\Psi_i(\boldsymbol{R})\rangle$ by a phase factor $\mathrm{e}^{\mathrm{i}\Lambda(\boldsymbol{R})}$, it remains the same instantaneous eigenstate of $H_f(\boldsymbol{R})$, but $\boldsymbol{A}_\mathrm{B}$ changes by $-\nabla_{\boldsymbol{R}}\Lambda(\boldsymbol{R})$. Since the change in $\boldsymbol{A}_\mathrm{B}$ is a gradient, the integral of $\boldsymbol{A}_\mathrm{B}$ around a closed loop is unchanged, hence well defined.

As an example of Berry's phase, consider the spin-1/2 Hamiltonian $H_f = \mu\boldsymbol{R}\cdot\boldsymbol{\sigma}$, where σ_x, σ_y and σ_z are the ► Pauli spin matrices. The eigenstate corresponding to the positive eigenvalue $E_+ = \mu R$ is

$$\begin{pmatrix}\cos\dfrac{\theta}{2}\\ \mathrm{e}^{\mathrm{i}\phi}\sin\dfrac{\theta}{2}\end{pmatrix},$$

B

where $R_z = R\cos\theta$ and $R_x + \mathrm{i}R_y = R\mathrm{e}^{\mathrm{i}\phi}\sin\theta$. The Berry connection, expressed as a function of θ and ϕ, is $(\boldsymbol{A}_\mathrm{B})_\theta = 0$, $(\boldsymbol{A}_\mathrm{B})_\phi = (\cos\theta - 1)/2$ and matches the vector potential of a Dirac monopole of strength 1/2 located at the origin $\boldsymbol{R} = \boldsymbol{0}$. The integral of $\boldsymbol{A}_\mathrm{B}$ along any loop in $\boldsymbol{R}$ equals $-1/2$ times the solid angle subtended by the loop at the origin (as an application of Stokes's theorem shows). This example is generic because wherever two nondegenerate energy levels cross at a point in a space of parameters, the Hamiltonian near the point reduces to an effective two-level Hamiltonian proportional to $\boldsymbol{R}\cdot\boldsymbol{\sigma}$, with the degeneracy at $\boldsymbol{R} = \boldsymbol{0}$. Hence an effective magnetic monopole can arise wherever two discrete, nondegenerate levels become degenerate.

The spin-1/2 example also illustrates how Berry's phase can be topological. A loop in $\boldsymbol{R}$ defines *two* solid angles, just as a loop on the surface of a sphere cuts the surface into two parts. Why, then, is Berry's phase not ambiguous? The answer is that the difference between the two solid angles is equal to $\pm 4\pi$. (The two solid angles have opposite signs because their orientations, or the directions of integration of $\boldsymbol{A}_\mathrm{B}$, are opposite.) But a $\pm 4\pi$ difference of solid angle corresponds to a $\mp 2\pi$ difference in phase, which is unobservable. Here Berry's phase obeys a constraint arising from the topology of a sphere.

In the Born–Oppenheimer approximation, the $R_1, R_2, \ldots$ are quantum observables and may not even commute. They evolve according to their own "slow" Hamiltonian H_s, and the overall Hamiltonian is the sum $H = H_f + H_\mathrm{s}$. The eigenvalues of H_f must be discrete, and the adiabatic limit applies if H_s is an arbitrarily weak perturbation on H_f. The weaker the perturbation, the smaller the probability of transitions (▶ quantum jumps) among the eigenstates of H_f. The unperturbed ▶ Hilbert space for H divides into subspaces, one for each eigenvalue E_i of H_f. In the adiabatic limit, the "fast" variables remain in an eigenstate $|\Psi_i(\boldsymbol{R})\rangle$ of H_f, with i fixed, while dynamical and Berry phases of $|\Psi_i(\boldsymbol{R})\rangle$ show up in H as induced scalar and vector potentials.

Born and Oppenheimer multiplied $|\Psi_i(\boldsymbol{R})\rangle$ by a function $\Phi(\boldsymbol{R}, t)$ and obtained an effective Hamiltonian for $\Phi(\boldsymbol{R}, t)$. Here we obtain and simplify their effective Hamiltonian algebraically. Let $\boldsymbol{\Pi}_i$ denote the operator of ▶ projection onto the subspace corresponding to E_i. The subspaces are disjoint and form a complete set: $\sum_i \boldsymbol{\Pi}_i = 1$. In the adiabatic limit, we can replace H_s by $\sum_i \boldsymbol{\Pi}_i H_\mathrm{s} \boldsymbol{\Pi}_i$ to obtain the effective Hamiltonian of Born and Oppenheimer:

$$H_\mathrm{eff} = H_f + \sum_i \boldsymbol{\Pi}_i H_\mathrm{s} \boldsymbol{\Pi}_i .$$

In H_eff there are induced potentials. If

$$H_\mathrm{s} = P^2/2M + V(\boldsymbol{R}),$$

where $P_i = -\mathrm{i}\hbar\partial/\partial R_i$, the sum $\sum_i \boldsymbol{\Pi}_i H_\mathrm{s} \boldsymbol{\Pi}_i$ in H_eff contains products of the form

$$\boldsymbol{\Pi}_i P^2 \boldsymbol{\Pi}_i = \sum_j \boldsymbol{\Pi}_i \mathbf{P} \boldsymbol{\Pi}_j \mathbf{P} \boldsymbol{\Pi}_i .$$

We simplify them by decomposing $\mathbf{P}$ into two parts, $\mathbf{P} = (\mathbf{P} - \boldsymbol{A}) + \boldsymbol{A}$. The first part acts only *within* subspaces; that is, $[\mathbf{P} - \boldsymbol{A}, \boldsymbol{\Pi}_i] = 0$ for all i. Only the second part, $\boldsymbol{A}$, causes transitions among the subspaces. Like a vector potential, $\boldsymbol{A}$ is somewhat arbitrary: we can add to $\boldsymbol{A}$ any term that commutes with the $\boldsymbol{\Pi}_i$. Let us remove this arbitrariness by requiring $\boldsymbol{\Pi}_i \boldsymbol{A} \boldsymbol{\Pi}_i = 0$ for each i. The effective Hamiltonian for the $\boldsymbol{R}$ is then [3]

$$H_{\text{eff}} = H_f + \frac{1}{2M}(\mathbf{P} - \boldsymbol{A})^2 + \frac{1}{2M}\sum_i \boldsymbol{\Pi}_i \boldsymbol{A}^2 \boldsymbol{\Pi}_i + V(\boldsymbol{R}).$$

The sum in i is an induced scalar potential, while $\boldsymbol{A}$ is an induced vector potential: $\boldsymbol{A}$ is Berry's connection $\boldsymbol{A}_{\mathrm{B}}$ in an off-diagonal gauge. For example, let $H_f = \mu \boldsymbol{R} \cdot \boldsymbol{\sigma}$ as in the spin-1/2 example above. The operators of projection corresponding to $E_{\pm} = \pm \mu R$ are

$$\boldsymbol{\Pi}_{\pm} = \frac{1}{2}(1 \pm \boldsymbol{R} \cdot \boldsymbol{\sigma}/R),$$

and the vector potential

$$\boldsymbol{A} = \frac{\hbar \boldsymbol{R} \times \sigma}{2R^2}$$

solves the two conditions $[\mathbf{P} - \boldsymbol{A}, \boldsymbol{\Pi}_{\pm}] = 0$ and $\boldsymbol{\Pi}_{\pm} \boldsymbol{A} \boldsymbol{\Pi}_{\pm} = 0$; $\boldsymbol{A}$ is off-diagonal. The field corresponding to $\boldsymbol{A}$,

$$B_i = \frac{1}{2}\epsilon_{ijk} F_{jk} = \frac{1}{2}\epsilon_{ijk}(\partial_j A_k - \partial_k A_j - \mathrm{i}[A_j, A_k]) = -\frac{\hbar R_i}{2R^4}(\boldsymbol{R} \cdot \boldsymbol{\sigma}),$$

is a monopole field $\mathbf{B} = \mp \hbar \boldsymbol{R}/2R^3$ since the eigenvalues of $\boldsymbol{R} \cdot \boldsymbol{\sigma}/R$ are ± 1.

So far we have taken the eigenvalues of H_f to be discrete and nondegenerate. If H_f has a discrete and *degenerate* eigenvalue, Berry's phase may be non-abelian [4]. The eigenstates belonging to this eigenvalue do not (in the adiabatic approximation) jump to eigenstates belonging to other eigenvalues, but they may mix among themselves. The mixing amounts to multiplication by a non-abelian phase, i.e. a unitary matrix.

Another generalization of Berry's phase is the Aharonov–Anandan phase [5]. Suppose a system evolves according to Schrödinger's equation, but the change in the Hamiltonian is neither adiabatic nor cyclic. Aharonov and Anandan showed that the system can still exhibit a Berry phase; all that is needed is cyclic evolution of the *state* of the system. Cyclic evolution of a state defines a closed path in the Hilbert space of the state. Whether or not this evolution is adiabatic, it leaves the system with a dynamical phase, which depends on the Hamiltonian of the system, and a geometrical phase – Berry's phase – which depends only on the closed path of the state in its Hilbert space. Thus Berry's phase need not be adiabatic (although it is still a correction to the adiabatic theorem).

We have considered evolution consistent with Schrödinger's equation. But as Pancharatnam showed [6], geometric phases can emerge from nonunitary evolution. For example, let an ► ensemble be divided into two subensembles, one of which is subjected to a sequence of filtering measurements (projections). If the subsubensemble that survives this filtering has returned to its initial state, it has a well defined phase (relative to the unfiltered subensemble) which equals a relative dynamical phase plus the Berry phase for this evolution.

Berry's phase has a classical analogue: *Hannay's angle* [7] is a phase effect in a classical periodic system that depends on adiabatically changing parameters. A canonical pair of variables for such a system is an "action" variable I, which is an adiabatic constant of the motion, and a conjugate "angle" variable ϕ. Hannay's angle is an extra shift in ϕ acquired by the system during a cyclic evolution in the space of parameters. When the Hannay angle of a system depends on its action I, the corresponding quantum system acquires a Berry phase during the same cyclic evolution [8].

Although the Aharonov–Bohm effect has no classical analogue, we may treat it as an example of Berry's phase. More generally, however, the Aharonov–Bohm and Berry phases can *combine* in a topological phase [9]. For example, imagine a "semifluxon", something like a straight, heavy, infinite solenoid enclosing flux $hc/2e$ – exactly half a flux quantum – that moves perpendicular to itself. It interacts with an electron ► wave function that has support in two disjoint regions. If the semifluxon moves in a slow circuit, we can ask what phase the electron acquires from this adiabatic cyclic evolution. Figure 1 shows one of the two regions where the electron wave function has support, and two possible circuits for the semifluxon. If the semifluxon evolves along C_1, the electron acquires no relative Berry phase and also the Aharonov–Bohm phase vanishes. If the semifluxon evolves along C_2, the relative Berry phase is π and it is entirely the Aharonov–Bohm phase. If the semifluxon does neither but plows through the electron wave function, we might expect the Berry phase to lie between 0 and π. However, it can be shown (using time-reversal symmetry) that the Berry phase can only be 0 or π. Since the path of the semifluxon is arbitrary, at some point $\mathcal{P}$ the Berry phase must jump from 0 to π, i.e. the electron wave function must become degenerate when the semifluxon is situated at $\mathcal{P}$. Here the Berry phase and the Aharonov–Bohm phase combine in a single topological phase that depends only on the winding number of the semifluxon path around the point $\mathcal{P}$.

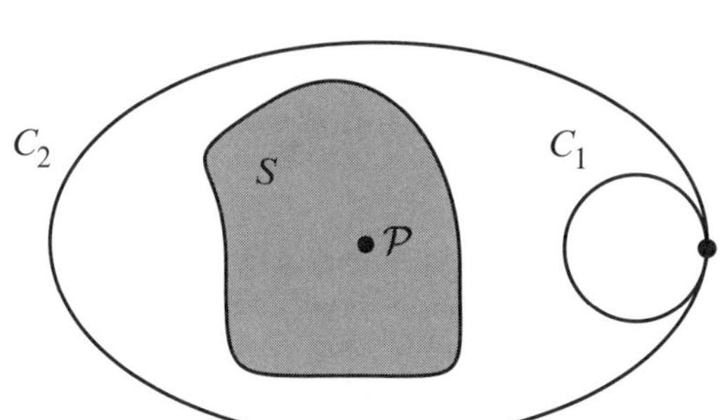

Fig. 1 An electron cloud with support in a region $\mathcal{S}$ (and in disjoint region not shown) and two possible paths, C_1 and C_2, of a semifluxon. At the point $\mathcal{P}$, the semifluxon induces a degeneracy in the energy of the electron

Primary Literature

1. M. V. Berry: Quantal phase factors accompanying adiabatic changes. Proc. R. Soc. Lond. A392, 45–57 (1984).
2. M. Born, J. R. Oppenheimer: Zur Quantentheorie der Molekeln. Ann. Phys. 84, 457–84 (1927); see also C. A. Mead, D. G. Truhlar: On the determination of Born–Oppenheimer nuclear motion wave functions including complications due to conical intersections and identical nuclei. J. Chem. Phys. 70, 2284–96 (1979).
3. Y. Aharonov, E. Ben-Reuven, S. Popescu, D. Rohrlich: Perturbative induction of vector potentials. Phys. Rev. Lett. 65, 3065–67 (1990) and 65, 863 (1992); Born–Oppenheimer revisited. Nucl. Phys. B350, 818–30 (1991).
4. F. Wilczek, A. Zee: Appearance of gauge structure in simple dynamical systems. Phys. Rev. Lett. 52, 2111–14 (1984); A. Zee: Non-Abelian gauge structure in nuclear quadrupole resonance. Phys. Rev. A38, 1–6 (1988).
5. Y. Aharonov, J. Anandan: Phase change during a cyclic quantum evolution. Phys. Rev. Lett. 58, 1593–96 (1987).
6. S. Pancharatnam: Generalized theory of interference, and its applications. Part I. Coherent pencils. Proc. Ind. Acad. Sci. A44, 247–62 (1956).
7. J. H. Hannay: Angle variable holonomy in adiabatic excursion of an integrable Hamiltonian. J. Phys. A: Math. Gen. 18, 221–30 (1985).
8. M. V. Berry: Classical adiabatic angles and quantal adiabatic phase. J. Phys. A: Math. Gen. 18, 15–27 (1985).
9. Y. Aharonov, S. Coleman, A. Goldhaber, S. Nussinov, S. Popescu, B. Reznik, D. Rohrlich, L. Vaidman: Aharonov-Bohm and Berry phases for a quantum cloud of charge. Phys. Rev. Lett. 73, 918–921 (1994).

Secondary Literature

10. See A. Messiah: *Quantum Mechanics*, Vol. II, trans. J. Potter (North-Holland, Amsterdam, 1963), Chap. XVII, Sects. 10–12.

Black Body

Dieter Hoffmann

A black body was first defined by Gustav R. Kirchhoff (1824–87) in 1859 as an object that absorbs all radiation falling upon it. Such a conception of an ideal black body was crucial for understanding heat radiation and its laws. Since a completely black body does not exist in nature, it had to be constructed. Kirchhoff had already suggested that a black body was technically feasible in his famous paper formulating his radiation law: "If a volume is enclosed by bodies of the same temperature and rays cannot penetrate those bodies, then each bundle of rays inside this volume has the same quality and intensity it would have had if it had come from a completely black body of the same temperature and is therefore independent of the constitution and the shape of these bodies and is determined by the temperature alone."

Although Kirchhoff as well as Ludwig Boltzmann (1844–1906) had already experimented with the design of a black body using a heated cavity, most of the first experimentalists trying to verify the radiation laws did not take up Kirchhoff's idea. Instead they made do with metal sheets with specially prepared surfaces or metals – through oxidizing, a layering of lamp black, roughening, etc. – to achieve a maximum of blackness. For instance the Danish physicist Christian Christiansen (1843–1917) had carried out such experiments around 1880. He tested the optical behavior of such powders as soot. He also made the observation, that conical tubes radiate with an emissivity of about 1, which means that they act as "small black spots". All these arrangements had shown that it was possible to make a black body effective for a limited range of wavelengths and temperatures, but a totally black body remained a distant hope.

The turning point for the design of a truly black body was reached in 1895 when Wilhelm Wien (1864–1928) and Otto Lummer (1866–1925) – at that time both fellows of the Physikalisch-Technische Reichsanstalt in Berlin (Imperial Institute of Physics, PTR) – recognized that one "had to disregard artificially blackened metal sheets." Instead "one had to consider the radiation of a black body as the state of thermodynamical equilibrium. . . To use this conception as the basis for a practical method for producing radiation arbitrarily close to that of a black body, one needs to heat a cavity to a uniform temperature and allow the radiation to escape through the opening."

With Wien's and Lummer's description, in principle, of a design for a black cavity radiator, Lummer (together with Ernst Pringsheim (1859–1917) in particular) was able to build a functioning device in 1897/98. First they experimented with small cylindrically and spherically shaped cavities of iron and copper, and later they designed hollow spheres of porcelain or metal, the inner surfaces of which were covered with soot (for lower temperatures) or with uranium oxide (for higher temperatures). To produce a definite and stable temperature, the cavities were immersed in a fluid bath – for instance, liquid air, boiling water, hot saltpeter or other liquids of well-defined temperature. In this way Lummer and Pringsheim materialized a completely black body for the temperature range between -188 and 700 °C, and also for temperatures up to 1200 °C, when they placed the cavity into a gas-heated chamotte oven.

With these apparatus they carried out experiments confirming the Stefan-Boltzmann law and Wien's displacement law. But for further verifications of the radiation laws it was necessary to design a black body for much higher temperatures. Furthermore the cavity temperature of the black body had to be more homogeneous and more manageable. An "electrically glowing completely black body" was finally designed by Lummer and Ferdinand Kurlbaum (1857–1927) in 1898, also at the PTR. It consisted of a platinium sheet, 0.01 mm thick and about 40 cm long. It was rolled into a cylinder 4 cm in diameter, one end of which was squeezed and closed. Both ends had rings for the electrical supply of heat. With a current of about 100 A, one could attain temperatures of about 1500 °C. A porcelain tube with a radiating cavity was inserted inside. A thermocouple was also integrated into this tube to measure the temperature of the cavity. Several diaphragms were also included in the arrangement, which served to shelter the cavity from outer disturbances – for instance, incoming air, etc. The inner surface of the tube was blackened with a mixture of chromium, nickel and cobalt oxide. For insulation purposes, the whole arrangement was surrounded by a second tube of a fire-proof material; the insulation could be improved by extra covering tubes or asbestos sheets.

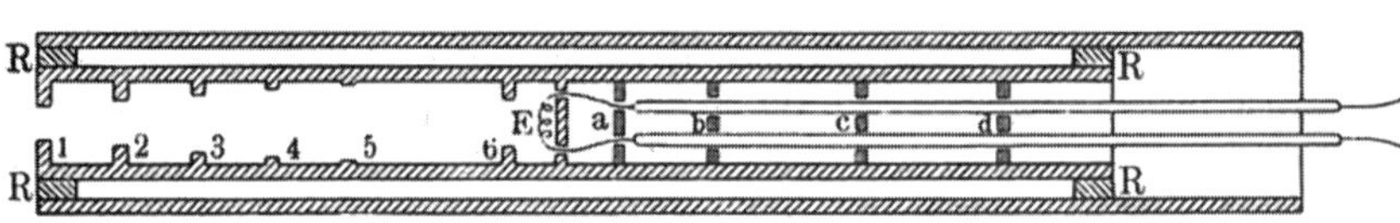

This new black body marked a major step forward in radiation research in general. In particular, the experiments led to Planck's radiation law and the basis for the quantum hypothesis. ► Blackbody radiation the design of a black body for still higher temperatures (already in 1903 Lummer and Pringsheim developed an improved black body on the same principle (but using specific materials and gas atmospheres) for temperatures of about 2100 °C) opened the way to establishing a new definition for temperature on the basis of the Stefan-Boltzmann law.

With the designs by Lummer, Kurlbaum and Pringsheim (1898/1903) the black body attained its more or less final shape and has been used for radiation research in the following decades, remaining occasionally in use to this day.

Primary Literature

1. W. Wien, O. Lummer: Methode zur Prüfung des Strahlungsgesetzes absolut schwarzer Körper. Annalen der Physik, 3rd ser., 56 (1895) 451–456; reprint in: idem: Das Wiensche Verschiebungsgesetz. Die Verwirklichung des schwarzen Körpers. Ostwalds Klassiker vol. 228, Frankfurt/Main 1997.
2. O. Lummer, E. Pringsheim: Die Strahlung eines "schwarzen Körpers" zwischen 100° und 1,300 °C. Annalen der Physik, 3rd ser., 63 (1897) 395–410.
3. O. Lummer, E. Pringsheim: Der electrisch geglühte "absolut schwarze" Körper und seine Temperaturmessung. Annalen der Physik, 4th ser., 17 (1898) 106–111.
4. O. Lummer, E. Pringsheim: Die strahlungstheoretische Temperaturskala und ihre Verwirklichung bis 2,300 °C. Annalen der Physik, 4th ser., 1 (1903) 3–13.

Secondary Literature

5. D. Hoffmann: On the experimental context of Planck's foundation of quantum theory. Centaurus 43 (2001) 240–259.
6. H. Kangro: Early History of Planck's Radiation Law. Taylor and Francis, London 1976.

Black-Body Radiation

Clayton Gearhart

Hot objects give off light and heat in the form of electromagnetic radiation whose character changes with temperature. Black-body radiation is such electromagnetic radiation in equilibrium with its material surroundings. By the late 1800s, it was a lively research topic for both theoretical and experimental physicists. Samuel Pierpont Langley (1834–1906) in the United States, and a group of experimental physicists in Germany centered around the *Physikalisch-Technische Reichsanstalt* (PTR) in Charlottenburg, had developed sophisticated techniques for studying this radiation. Part of their motivation was practical – establishing better absolute temperature scales, and measuring light intensities, at high temperatures (► Black Body).

In December 1900 and January 1901, the German physicist Max Planck (1858–1947) published three short papers in which he derived a new equation to describe black-body radiation—one that ever since has given excellent agreement with observation. This derivation was the culmination of research Planck had begun in the mid-1890s. In a series of lengthy papers, Planck had combined thermodynamics, in which he was an acknowledged authority, with the new electromagnetic theory of James Clerk Maxwell (1831–1879). He considered the electromagnetic field in equilibrium with what he called "resonators" – electric dipoles oscillating in simple harmonic motion – which represented the material cavity containing the field. By late 1899, he had found a new and more rigorous derivation of Wien's law, an equation describing black-body radiation discovered in 1896 by his friend and colleague Wilhelm Wien (1864–1928), and seemingly in good agreement with experiment.

By mid-1900, however, physicists at the PTR had found systematic deviations between Wien's law and their latest experiments. Planck went back to work, and by the end of the year, had produced his new radiation law, which takes the familiar form

$$u_\nu = \frac{8\pi\nu^2}{c^3}\frac{h\nu}{e^{h\nu/kT}-1},$$

where c is the speed of light, and u_ν is the energy density of the electromagnetic field as a function of the frequency ν and the absolute temperature T. This equation also contains two new fundamental constants of nature, h and k – today we call them ▶ Planck's constant and Boltzmann's constant – to which Planck attached the greatest importance. They played a central role in his system of natural units for length, mass, time, and temperature, which as he said in 1899, "necessarily retain their significance for all times and for all cultures, even alien and non-human ones."

However, Planck's derivation was decidedly mysterious. It relied on a 1877 paper by the Austrian physicist Ludwig Boltzmann (1844–1906), relating entropy and probability, now famous but little known in 1900. Today it is summarized in the equation $S = k \log W$, inscribed on Boltzmann's tombstone in Vienna. Boltzmann had begun with a physically unrealistic picture, in which he divided the energy of a gas into finite "energy elements" (as Planck later called them), which he distributed among the molecules of an ideal gas. This step allowed him to use combinatorials to calculate the probabilities of microscopic states and relate them to the entropy of a gas. Planck applied a similar scheme to his resonators, though he persisted in his absolute interpretation of entropy and the second law of thermodynamics, in sharp contrast to Maxwell's and Boltzmann's probabilistic viewpoint.

In 1877, Boltzmann had replaced his artificial scheme with the more realistic one of partitioning molecules among arbitrarily small cells in phase space, thereby recovering the standard description of an ideal gas. Planck, by contrast, could make his derivation work only by retaining these finite "energy elements" and assigning them the specific size $h\nu$. In 1900, he said nothing about the striking differences between the two derivations, though he certainly understood what Boltzmann had done.

Today we call these energy elements "quanta," and over the last century, physicists have developed the strange new theory called quantum mechanics to describe nature at the atomic level. But in 1900, all this was yet to come. The "energy elements," whatever they might be, had no obvious interpretation in the physics of the day. Planck in 1900 said virtually nothing about how to interpret them physically. Both his contemporaries and later historians found it difficult to grasp his meaning.

Over the next decade, scientists slowly came to terms with these new ideas (► Quantum theory, early period). If Planck's energy elements do become arbitrarily small, for example, Planck's law goes over to the Rayleigh-Jeans law, $u_\nu = \left(8\pi\nu^2/c^3\right) kT$, in which the radiation density increases without limit at short wavelengths—an effect Paul Ehrenfest (1880–1933) later dubbed the "ultraviolet catastrophe." Physicists developed an increasingly sophisticated understanding of this theme and its relation to equipartition in the first decade of quantum theory.

Planck contributed to these efforts in his 1906 book, *Lectures on the Theory of Heat Radiation*, in which he presented h as the "elementary quantum of action," since its units were those of action, the product of energy and time. He also showed that h is the size of a finite "elementary domain" in phase space, a step that made his combinatorial assignments of probability more plausible. Hendrik Antoon Lorentz (1853–1928), Paul Ehrenfest, Henri Poincaré (1854–1912) and others also explored the foundations of black-body radiation, and showed that it necessarily involved a sharp and inescapable break with earlier physical theory.

For many years, Planck pointed out the need for a physical interpretation of his theory, but was reluctant to advance one himself. Only in 1909 did he state publicly that the energies of his resonators were restricted to integer multiples of $h\nu$. But in that same year, Lorentz showed that under some circumstances, it would take an implausibly long time to absorb one quantum of radiation from a Maxwellian electromagnetic field. Neither Lorentz, Planck, nor most other physicists were prepared to accept the alternative of "light quanta" that Albert Einstein (1879–1955) had proposed in 1905 (► Light quanta; ► Quantum theory, early period).

In 1911, therefore, Planck proposed what became known as his "second quantum theory," in which resonators absorbed energy continuously, but emitted energy in quanta only when they reached the boundaries of finite cells in phase space, where their energies became integral multiples of $h\nu$. This theory also led Planck to his new radiation law. But in this version, resonators possessed a ► "zero-point" energy, the smallest *average* energy that a resonator could take on. Thus, for the first time, physicists contemplated systems whose energy did not go to zero at the absolute zero of temperature. This zero-point energy soon took on a life of its own, appearing in the early 1920s in the context of both Planck's first and second theories, and after 1925, finally finding a secure home in modern quantum mechanics.

Albert Einstein took perhaps the most radical view of black-body theory, beginning with his famous paper of 1905, in which he suggested that light consists of "a finite number of energy quanta that are localized in points of space, move without dividing, and can be absorbed or created only as a whole." (► light quanta; ► Quantum theory, early period) In succeeding years, black-body radiation and its connection to light quanta remained at the center of Einstein's thoughts. In 1909, for example, it was at the heart of his analysis of fluctuations – random variations in energy and momentum – in which he argued that light sometimes behaved like a wave and sometimes like a particle, and that the dual wave and particle nature of light was inescapable – he spoke of "a kind of fusing of the wave and emission theories of light."

In 1916, he found a new derivation of Planck's radiation law, his famous and influential "*A* and *B* coefficients" argument that involved assumptions on the "stimulated emission" of light and set down the underlying principles of the laser, not invented until decades later. And in 1924, he understood immediately the significance of a paper sent to him by the then-unknown Indian physicist Satyendra Nath Bose (1894–1974), who had found yet another derivation of Planck's radiation law – one that implicitly suggested that Einstein's light quanta were not independent particles. Einstein translated Bose's paper into German and arranged for its publication. He also saw its implications for the seemingly unrelated topic of quantum ideal gases, and published the papers describing what is now known as Bose-Einstein condensation, experimentally confirmed only recently (▶ Quantum statistics, ▶ Bose-Einstein-statistics).

In short, although black-body theory was not the whole of early quantum theory, it remained a continuing source of inspiration and new discoveries. Please see also the Reference ▶ Specific heats.

Primary Literature

1. M. Planck (annotated by Hans Kangro; transl. D. ter Haar and Stephen G. Brush): *Planck's Original Papers in Quantum Physics* (Taylor & Francis, London, 1972; German and English Edition)
2. M. Planck (ed., with an introduction, by Alan A. Needell): *The Theory of Heat Radiation* (American Institute of Physics, College Park, MD, 1988) This work contains both the 1906 edition, never translated into English, and the very different 1913 edition, based on Planck's second theory.

Secondary Literature

3. O. Darrigol: *From c-Numbers to q-Numbers* (University of California Press, Berkeley, 1992); Statistics and combinatorics in early quantum theory. Historical Studies in the Physical and Biological Sciences **19**, 17–80 (1988); The Historians' Disagreement over the Meaning of Planck's Quantum. *Centaurus* **43**, 219–239 (2001)
4. C. A. Gearhart: Planck, the Quantum, and the Historians. Physics in Perspective **4**, 170–215 (2002)
5. M. J. Klein: *Paul Ehrenfest: Vol. 1, The Making of a Theoretical Physicist* (North Holland, Amsterdam, 1970)
6. T. S. Kuhn: *Black-Body Theory and the Quantum Discontinuity*, 1894–1912 (Oxford University Press, New York, 1978)
7. J. J. Prentis: Poincaré's Proof of the Quantum Discontinuity of Nature. American Journal of Physics **63**, 339–350 (1995)
8. R. Staley: On the Co-Creation of Classical and Modern Physics. Isis **96**, 530–558 (2005)

Bohm Interpretation of Quantum Mechanics

B.J. Hiley

The Bohm interpretation aims at providing an interpretation based on the description of the evolution of an actual individual process evolving in space-time. In the case of particles, it accounts for their individual behaviour in terms of their simultaneous positions and momenta, even though these are assumed to be unknown. It is often argued that this view must be untenable owing to the ▶ Heisenberg uncertainty relations. However the uncertainty principle only rules out the possibility of *measuring experimentally* the simultaneous position and momentum. From this principle two conclusions are possible. Either the particle does not have a simultaneous position and momentum to measure, or that it does have a simultaneous position and momentum but it is simply not possible to measure them simultaneously and therefore must remain unknown. There is no direct experimental way to decide which of these two positions is actually correct. The conventional approach adopts the former, the Bohm interpretation adopts the latter. In this latter approach it may be helpful to regard the $(\boldsymbol{x}, \boldsymbol{p})$ as "beables".

Having chosen the latter position, the question is whether it is possible to use the formalism based on the ▶ wave function $\psi(\boldsymbol{r}, t)$ and the ▶ Schrödinger equation to provide a mathematical description of a particle following a trajectory and still reproduce all the statistical predictions of the standard approach. Bohm [1] showed that this was possible contrary to the views of Bohr [2] who argued that such a "picture" was not possible.

The mathematical procedure for a particle that obeys the Schrödinger equation is straight forward. Simply write the wave function in polar form $\psi = R\mathrm{e}^{\mathrm{i}S/\hbar}$ and substitute into the Schrödinger equation. By separating into the real and imaginary parts, we find two equations. The first is

$$\frac{\partial S}{\partial t} + \frac{(\nabla S)^2}{2m} - \frac{\hbar^2}{2m}\frac{\nabla^2 R}{R} + V = 0 \tag{1}$$

The second equation is

$$\frac{\partial R^2}{\partial t} + \nabla . \left(R^2 \frac{\nabla S}{m} \right) = 0 \tag{2}$$

Equation (1) differs by only one term from the classical Hamilton-Jacobi equation

$$\frac{\partial S_{\mathrm{c}}}{\partial t} + \frac{(\nabla S_{\mathrm{c}})^2}{2m} + V = 0 \tag{3}$$

This equation defines a set of trajectories which are identical to those calculated from Newton's law of motion

$$m\frac{\mathrm{d}\boldsymbol{v}}{\mathrm{d}t} = -\nabla(V) \tag{4}$$

Comparing (1) and (3), we see the phase of the wave function has been replaced by the classical action S_c and an extra term

$$Q = -\frac{\hbar^2}{2m}\frac{\nabla^2 R}{R} \tag{5}$$

appears in the quantum case. In the classical Hamilton-Jacobi theory we have two canonical relations, $\boldsymbol{p} = \nabla S_c$ and $E = -\partial S_c/\partial t$. What Bohm did was to assume that these two relations with S_c replaced by S held in the quantum case. This means that the quantum Hamilton-Jacobi equation (1) can be used to provide a set of trajectories that differ from the classical trajectories owing to the presence of the extra term Q. It can be shown that these trajectories can be also be calculated from

$$m\frac{\mathrm{d}\boldsymbol{v}}{\mathrm{d}t} = -\nabla(V+Q) - \nabla(Q) \tag{6}$$

The appearance of Q in this equation suggested that Q be called the *quantum potential*. In some ways (6) is somewhat misleading as it suggests that this "potential" is playing a role similar to that of a classical potential and this has tended to suggest that this interpretation is simply a return to classical physics. Nothing could be further from the truth. The quantum potential is nothing like a classical potential. There is no external source for this potential and should be regarded as a new form of *internal* energy. This becomes more apparent when we realise that (1) is simply an expression of the conservation of energy,

Total energy
= kinetic energy + quantum potential energy + classical potential energy (7)

Although we have the possibility of calculating trajectories for Schrödinger particles, we cannot produce experimentally a particle with a *known* value of $(\boldsymbol{r}, \boldsymbol{p})$ simply because of the restrictions imposed by the uncertainty principle. All we can do is to generate a distribution of initial $\boldsymbol{r}$s and $\boldsymbol{p}$s consistent with the probability being given by the initial wave function $\psi_i(\boldsymbol{r}, t)$. Equation (2) then guarantees that the final probability distribution agrees with the standard quantum predictions provided we assume the probability is still given by $P = R^2$. Equation (2) is then simply an expression for the conservation of probability.

The Bohm interpretation has been applied to many of the usual quantum experiments such as the ► double-slit experiment, the ► Schrödinger cat paradox, the ► delayed-choice experiment, teleportation (► quantum communication) and many other such experiments. The interpretation provides an intuitive picture of what could underlie quantum phenomena without the paradoxes of the standard theory. ► Errors and paradoxes in quantum mechanics for example, each Schrödinger particle goes through one and only one slit, the quantum potential adjusting the trajectories to account for the slit configurations. The Schrödinger cat is either alive or dead and never in a linear supposition of these two contradictory states. There is no measurement problem in this approach. More details of this method can be found in Bohm and Hiley [3] and in Holland [4]. See also ► Bohmian mechanics; Measurement theory; Metaphysics in Quantum Mechanics; Modal Interpretation; Objectification; Projection Postulate.

While this is all straight forward for the Schrödinger particle, we have to generalise the approach to the electromagnetic field where photons (► light quantum) have to be accounted for and a generalisation to apply to Dirac particles is also necessary.

In the case of photons, it is the electromagnetic field, or more accurately, the vector potential field $\psi_\mu(\boldsymbol{r}, t)$ that must be used since it is not possible to attribute a simultaneous $(\boldsymbol{r}, \boldsymbol{p})$ to a photon. The beables in this case are not $(\boldsymbol{r}, \boldsymbol{p})$ but the fields and their conjugate momentum $\psi_\mu(x^\mu)$ and $\pi_\mu(x^\mu)$. We then have a "super-wave function" which is a functional of the field. More details can be found in Bohm, Hiley and Kaloyerou [5], and in Kaloyerou [6].

We can illustrate the mathematical structure of the field approach by using a scalar field $\phi(x^\mu)$. The super-wave function is the functional $\Psi(\ldots\phi(x^\mu)\ldots)$, which is assumed to satisfy the super-Schrödinger equation

$$i\frac{\partial\Psi}{\partial t} = H\Psi \tag{8}$$

where the Hamiltonian is given by

$$H = \frac{1}{2}\int_{\text{All space}}\left[-\frac{\delta^2}{(\delta\phi(x,t))^2} + (\nabla\phi(x,t))^2\right] \tag{9}$$

We then write $\Psi = R[\ldots\phi(x,t)\ldots]\exp\{\mathrm{i}S[\ldots\phi(x,t)\ldots]\}$ and obtain

$$\frac{\partial S}{\partial t} + \frac{1}{2}\int\left[\left(\frac{\delta S}{\delta\phi}\right)^2 + (\nabla\phi)^2\right]\mathrm{d}V + Q = 0 \tag{10}$$

Here the super-quantum potential is

$$Q = -\frac{1}{2}\int\left\{\frac{\left[\delta^2/(\delta\phi(x^\mu))^2\, R\,(\ldots\phi(x^\mu)\ldots)\right]}{R\,(\ldots\phi(x^\mu)\ldots)}\right\}\mathrm{d}V \tag{11}$$

We also obtain a conservation of probability equation

$$\frac{\partial P}{\partial t} + \int\frac{\delta}{\delta\phi}\left[P\frac{\delta S}{\delta\phi}\right]\mathrm{d}V = 0 \tag{12}$$

From (10) using the Hamiltonian (9) the field equation becomes

$$\frac{\partial^2\phi}{\partial t^2} = \nabla^2\phi - \frac{\delta Q}{\delta\phi}. \tag{13}$$

Thus we see that although more involved, the field theory displays a similar general structure to the Schrödinger particle theory only now it is the fields that represent the beables. They have well-defined and continuously changing values. Equation (10) replaces the quantum Hamilton-Jacobi equation (1), while (12) replaces the conservation of probability equation (2). The field equation (13) shows the role played by the super-quantum potential and replaces (6).

The physical picture that emerges from these equations is that the field (the vector potential field, for example) is organised by the super-quantum potential as is clear from the appearance of the last term in (13). This term is generally a non-linear and non-local function of the field ϕ. In the classical limit this term is negligible.

Finally we need to understand how the concept of a photon, a field quanta, emerges from this picture. To do this we must consider the field in interaction with an atom. If the field is in an excited state, the interaction will produce a very complex wave functional of the field together with the atom. During this process the super-quantum potential will change dramatically, producing bifurcation points. These points will correspond to the absorption of quanta by the atom from the field. Suppose the field energy is only sufficient to excite the atom into its first excited state. The super-quantum potential, being non-linear and non-local, sweeps out the energy from the field leaving the atom in its first excited state and the field in its ground state. Since the field takes energy from excited atoms, the energy in the field must be quantised.

In this picture the photon is not localised and does not follow a trajectory. Rather it is the field that evolves in a well defined way and we can regard it as evolving along a "trajectory" defined by a point in the configuration space of the total set of field variables. These ideas have been successfully applied to the photoelectric effect, the Pfleegor-Mandle experiment which involves low intensity interference effects between two independent lasers and to correlated Einstein-Rosen-Podolsky photons (see Bohm and Hiley [3] for more details.)

The interpretation has also been applied to the ► Dirac equation although this equation has presented more difficulties and no successful attempt to construct a quantum potential has been made. The condition $\boldsymbol{p} = \nabla S$ is replaced by the expression for the Dirac current $j^{\mu} = \bar{\Psi}\gamma^{\mu}\Psi$. This has been applied to the two-slit interference experiment where trajectories for electrons have been actually calculated [7]. Application to fermion fields has also presented problems [8].

This approach has produced intuitive pictures lying behind quantum phenomena, but it is not without its own difficulties. The nature of the quantum potential is still unclear in spite of the various attempts that have been made to provide an explanation. Also the quantum potential contains the non-local features which are apparent in the EPR type experiments. Some claim that this is the only interpretation that accounts for this ► nonlocality yet it still sits uncomfortably with special relativity. On the other hand it might be pointing to a deeper a-local structure underlying the quantum phenomena [9].

See also Ignorance interpretation, Ithaca Interpretation, Many Worlds Interpretation, Modal Interpretation, Orthodox Interpretation, Transactional Interpretation.

Literature

1. Bohm, D.: Phys. Rev. 85 (1952) 166–93.
2. Bohr, N.: *Atomic Physics and Human Knowledge*. Science Editions. New York, 1961.
3. Bohm, D., Hiley. B. J.: *The Undivided Universe: an Ontological Interpretation of Quantum Theory*. Routledge, London 1993.
4. Holland, P. R.: *The Quantum Theory of Motion*. Cambridge University Press, Cambridge, 1993.
5. Bohm, D., Hiley, B. J., Kaloyerou, P.: Phys. Rep. 144 (1987) 349–75.
6. Kaloyerou, P. N.: Phys. Rep. **244** (1994) 287–358.
7. Gull, S. F., Lasenby A. N., Doran, C. J. L.: Found. Phys. 23 (1993) 1329–56.
8. Holland, P. R.: Phys. Rep. 224 (1993) 95–150.
9. Hiley, B. J.: Proc. Int. Conf. *Quantum Theory*. Proc. Int. Conf. Quantum Theory: Reconsideration of Foundations. 2, 267–86, ed. Khrennikov, A., Växjö University Press, Växjö, Sweden (2003).

B

Bohmian Mechanics

Detlef Dürr, Sheldon Goldstein, Roderich Tumulka, and Nino Zanghì

Bohmian mechanics is a theory about point particles moving along trajectories. It has the property that in a world governed by Bohmian mechanics, observers see the same statistics for experimental results as predicted by quantum mechanics. Bohmian mechanics thus provides an *explanation* of quantum mechanics. Moreover, the Bohmian trajectories are defined in a non-conspiratorial way by a few simple laws.

Overview. Bohmian mechanics is a version of quantum mechanics for nonrelativistic particles in which the word "particle" is to be understood literally: In Bohmian mechanics quantum particles have positions, always, and follow trajectories. These trajectories differ, however, from the classical Newtonian trajectories. Indeed, the law of motion, see (1) below, involves a ► wave function. As a consequence, the role of the wave function in Bohmian mechanics is to *tell the matter how to move*.

Bohmian mechanics constitutes a *quantum theory without observers*, i.e., a theory that is formulated not in terms of what observers see but in terms of objective events, regardless of whether or not they are observed. Bohmian mechanics provides a consistent resolution of ► errors and paradoxes in quantum mechanics, in particular of the so-called *measurement problem*. In particular, the ► *wave function collapse* (see ► *Projection Postulate*) can be derived from Bohmian mechanics. (On the measurement problem see also ► Measurement theory; Metaphysics in Quantum Mechanics; Modal Interpretation; Objectification; Projection Postulate Measurement theory; Objectification; Projection Postulate).

Bohmian mechanics is sometimes called a ► *hidden variables theory* because it involves variables besides the wave function. However, there is a danger of confusion here because the term "hidden variables theory" is often used to convey the idea that every "quantum measurement" of an "observable" reveals a pre-existing value of that observable, which is not the case in Bohmian mechanics.

Bohmian mechanics is *deterministic*. But the motivation behind Bohmian mechanics is not to obtain a deterministic theory, but rather to obtain a coherent account of the nature of physical reality. In this regard, we note that some variants of Bohmian mechanics, developed by its proponents, are stochastic rather than deterministic, for example Bell's proposal for lattice quantum field theory [4].

Historically, the "Bohmian" law of motion, see eq. (1) below, was first proposed by de Broglie [6]. However, Bohm [5] was the first to recognize that this theory explains all of the phenomena of (non-relativistic) quantum mechanics.

Defining Equations. Bohmian mechanics is a non-relativistic theory governing the behavior of a system of N point particles moving in physical space $\mathbb{R}^3$ along trajectories. Let $\boldsymbol{Q}_i(t) \in \mathbb{R}^3$ denote the position of the i-th particle of the system at time t, and $Q(t) = \big(\boldsymbol{Q}_1(t), \ldots, \boldsymbol{Q}_N(t)\big) \in \mathbb{R}^{3N}$ its configuration.

The trajectories are governed by *Bohm's law of motion* [2,5]

$$\frac{d\boldsymbol{Q}_i}{dt} = \frac{\hbar}{m_i}\mathrm{Im}\frac{\Psi_t^*\nabla_i\Psi_t}{\Psi_t^*\Psi_t}\big(Q(t)\big), \tag{1}$$

where m_i is the mass of particle i, Im denotes the imaginary part, $\Psi_t : \mathbb{R}^{3N} \to \mathbb{C}^k$ (i.e., a function of the configuration with k complex components) is the wave function at time t, $\Phi^*\Psi$ is the scalar product in $\mathbb{C}^k$, and ∇_i is the gradient relative to the 3 coordinates of particle i. (In case $k = 1$, i.e., for complex-valued wave functions, a factor Ψ_t^* cancels on the right hand side of (1).)

The wave function evolves according to the *Schrödinger equation*

$$\mathrm{i}\hbar\frac{\partial\Psi_t}{\partial t} = -\sum_{i=1}^{N}\frac{\hbar^2}{2m_i}\nabla_i^2\Psi_t + V\Psi_t, \tag{2}$$

where $V : \mathbb{R}^{3N} \to \mathbb{R}$ is the potential function. (The potential, while often assumed to be real-valued, may take values in the space of self-adjoint complex $k \times k$ matrices instead of $\mathbb{R}$.) The wave function is postulated to belong to the ► Hilbert space $\mathscr{H} = L^2(\mathbb{R}^{3N}, \mathbb{C}^k)$ of square-integrable functions (and to be sufficiently smooth).

Deterministic Evolution. Since the Schrödinger equation does not involve the particle positions $\boldsymbol{Q}_i(t)$, it can be solved first and determines the wave function Ψ_t for every time t once an initial wave function Ψ_{t_0} is specified for any time t_0 that we choose to regard as the initial time. Next note that the right hand side of (1) consists of the 3 components corresponding to particle i out of the $3N$ components of a vector field v^{Ψ_t} on configuration space $\mathbb{R}^{3N}$. As a consequence, equation (1) for all $i = 1, \ldots, N$ can be summarized by

$$\frac{\mathrm{d}Q}{\mathrm{d}t} = v^{\Psi_t}\big(Q(t)\big). \tag{3}$$

Regarding Ψ_t as known, this is a (time-dependent) ordinary differential equation (ODE) of first order, and as such determines the entire history $t \mapsto Q(t)$ once an initial configuration $Q(t_0)$ is specified. That is why Bohmian mechanics is *deterministic*: once $Q(t_0)$ and Ψ_{t_0} are specified, the entire history is fixed by the equations (1) and (2). This fact also implies that the pair $(Q(t_0), \Psi_{t_0})$ can be regarded as the *state* of the Bohmian particle system at time t_0. Since the choice of t_0 is arbitrary, the state at any time t is the pair $(Q(t), \Psi_t)$, and the phase space of Bohmian mechanics is $\mathbb{R}^{3N} \times \mathscr{H}$.

System or Universe. The equations of Bohmian mechanics could be applied to a familiar system (e.g., an atom) or to the universe as a whole. Of course, one cannot expect that the equations hold for every system, for example for systems that interact with their environments. So let us begin with the system for which the equations are primarily intended: the universe. In this setting, N is the number of particles in the universe, and Ψ_t is the wave function of the universe. To consider such a wave function is unusual; after all, the quantum formalism never refers to a wave function of the universe; the quantum formalism, providing the probabilities for the results of observations performed on a system by an external observer, involves the wave function of that system and not of the entire universe. In the context of Bohmian mechanics, however, the wave function of the universe is not at all a meaningless concept, as it influences the motion of the particles according to (1).

When (1) and (2) hold for the universe, it follows that equations of the same type (but with smaller N) hold for certain subsystems. (We shall assume here for simplicity that $k = 1$, i.e., that we are dealing with spinless particles.) Consider a subsystem of the universe with configuration X (the x-system), so that the configuration Q of the universe is of the form $Q = (X, Y)$ with Y the configuration of the *environment* of the x-system. Then a natural notion of the wave function of the x-system is provided by its *conditional wave function*

$$\psi(x) = \Psi(x, Y), \tag{4}$$

where $\Psi(q) = \Psi(x, y)$ is the wave function of the universe. It is easy to see that the x-system obeys (3) (with $Q = X$ and $\Psi = \psi$).

Moreover, if the x-system is suitably decoupled from its environment, (2) will hold as well. For example, this is the case when there is no interaction between the x-system and its environment, and the wave function of the universe is of the form

$$\Psi(x, y) = \psi(x)\,\varphi(y) + \Phi(x, y) \tag{5}$$

with φ and Φ having macroscopically disjoint y-supports (so that they will never again overlap appreciably), and with Y lying in the support of φ. Such a situation often arises after a "quantum measurement."

Equivariance. If the initial configuration $Q(t_0)$ is chosen at random with probability density $|\Psi_{t_0}|^2$ then the configuration $Q(t)$ at any other time t is random with probability density $|\Psi_t|^2$. (Whenever speaking of probabilities, we assume that Ψ has been normalized, by multiplication by a suitable constant, so that $\langle\Psi|\Psi\rangle = \int |\Psi(q)|^2 dq = 1$.) This fact, known as *equivariance*, follows from the continuity equation

$$\frac{\partial\rho}{\partial t} = -\nabla \cdot (\rho\, v) \tag{6}$$

for $\rho = |\Psi|^2$ and with the Bohmian velocity vector field $v = v^\Psi$ as in (3). The continuity equation (6) is in turn a consequence of the Schrödinger equation; it is usually written (in standard quantum mechanics) in terms of the quantum probability current $J = \rho\, v$.

Identical Particles. Bohmian mechanics can be formulated for identical particles, despite a fact that could be felt to contradict their ▶ indistinguishability, namely that the particle trajectories in $\mathbb{R}^3$ determine "who is who" at different times, i.e., select a one-to-one association between the N points at any time t_1 and the N points at another time t_2. Taking the notion of a particle seriously, as one should in Bohmian mechanics, one recognizes that the configuration space for N identical particles is best regarded as the manifold of all sets of N points in physical space $\mathbb{R}^3$. This manifold has non-trivial topological properties, as its fundamental (homotopy) group is isomorphic to the group of permutations of N objects. On such manifolds there arise several versions of Bohmian mechanics corresponding to the different 1-dimensional representations of the fundamental group; for the permutation group, there are two such representations, corresponding to bosons (with symmetric wave functions on the covering space $\mathbb{R}^{3N}$) and fermions (with anti-symmetric wave functions). Thus, Bohmian mechanics lends support to the modern view that the symmetrization postulate emerges as a topological effect, due to the non-trivial topology of configuration space.

Quantum Equilibrium Hypothesis. This is the assertion that whenever a system has wave function ψ then its configuration is (or can be taken to be) random with probability distribution $|\psi|^2$. Equivariance implies that this hypothesis is consistent with the time evolution of isolated systems, and it is not hard to show that it is also consistent with the time evolution if the system is not isolated, provided we take ψ to mean the conditional wave function. An important consequence of the quantum equilibrium hypothesis is the *empirical equivalence between Bohmian mechanics and quantum mechanics*: For every conceivable experiment, whenever quantum mechanics makes an unambiguous prediction, Bohmian mechanics makes exactly the same prediction. Thus, the two cannot be tested against each other.

Typicality. The quantum equilibrium hypothesis follows from typicality: As shown in [7] using the law of large numbers, results of experiments are as predicted by the quantum equilibrium hypothesis for *typical* initial configurations $Q(t_0)$ of the universe relative to the $|\Psi_{t_0}|^2$ distribution, i.e., for the overwhelming majority, counted using the $|\Psi_{t_0}|^2$ distribution, of the initial configurations.

Operators. Given that it makes the same predictions as quantum mechanics, what is the status in Bohmian mechanics of the non-commuting ► operators of the quantum formalism (the self-adjoint "observables"), with which the predictions of quantum mechanics seem exclusively concerned? The answer is that operators do in fact arise naturally in Bohmian mechanics, but with a different meaning than the one attributed to them in orthodox quantum mechanics (which regards them as more or less the same thing as their classical counterparts: as "► observables" that can be "measured"). Instead, operators in Bohmian mechanics are *mathematical tools encoding statistics*. Let us explain.

The statistics of the random outcome Z of an experiment in a world governed by Bohmian mechanics on a system with wave function ψ can be shown [8] always to be of the form (in ► Dirac notation)

$$\mathrm{Prob}(Z = \alpha) = \langle \psi | E(\alpha) | \psi \rangle, \tag{7}$$

where $E(\alpha)$ is a suitable positive operator. (Together, the $E(\alpha)$ form a positive-operator-valued measure, or ► POVM.) In relevant cases, $E(\alpha)$ is a family of projection operators (► projection) which are mutually orthogonal (a projection-valued measure, or PVM), and thus correspond to the one ► self-adjoint operator

$$A = \sum_{\alpha} \alpha\, E(\alpha), \tag{8}$$

which, by the spectral theorem, contains precisely the same information as the PVM $E(\alpha)$. Thus, operators encode the functional dependence of the outcome statistics on the system's wave function ψ. With this understanding, which is opposite to thinking of operators as representing *quantities* whose values can be "measured," it is no longer surprising that one cannot associate *actual values* with all "observables" in a consistent way. With this understanding, *contextuality* is not surprising either, since it no longer means that the same quantity can choose different values depending on what happens to another system, but rather that, unspectacularly, different experiments can have the same statistics.

► *Wave Function Collapse.* Here is an analysis, for Bohmian mechanics, of an "ideal measurement" of a quantum observable, given by a self-adjoint operator A on the Hilbert space of the relevant system. For simplicity we assume that A has pure point spectrum with non-degenerate eigenvalues α, corresponding to (8) for $E(\alpha) = |\psi_\alpha\rangle\langle\psi_\alpha|$ with normalized eigenstates $\psi_\alpha(x) = |A = \alpha\rangle$. The experiment is implemented by having the system interact with an apparatus in a suitable way. To avoid unimportant complications, we shall assume that the relevant "universe" for the problem at hand consists entirely of the system, with configuration X, and the apparatus, with configuration Y. The measurement begins, say, at time 0, with the initial ("ready") state of the apparatus given by a wave function $\varphi_0(y)$, and ends at time t. The interaction is such that when the state of the system is initially ψ_α it produces a normalized apparatus state $\varphi_\alpha(y)$, that registers that the value found for A is α without having affected the state of the system,

$$\psi_\alpha(x)\varphi_0(y) \xrightarrow{t} \psi_\alpha(x)\varphi_\alpha(y). \tag{9}$$

Here $\xrightarrow{t}$ indicates the unitary evolution induced by the interaction. If the measurement is to provide useful information, the apparatus states must be noticeably different, corresponding, say, to a pointer on the apparatus pointing in different directions. We thus assume that the φ_α have disjoint supports in the configuration space for the apparatus,

$$\operatorname{supp}(\varphi_\alpha) \cap \operatorname{supp}(\varphi_\beta) = \emptyset, \ \alpha \neq \beta. \tag{10}$$

Now suppose that the system is initially, not in an eigenstate of A, but in a general state, given by a ► superposition

$$\psi(x) = \sum_\alpha c_\alpha \psi_\alpha(x). \tag{11}$$

We then have, by (9) and the linearity of the unitary evolution, that

$$\Psi_0(x, y) = \psi(x)\varphi_0(y) \xrightarrow{t} \Psi_t(x, y) = \sum_\alpha c_\alpha \psi_\alpha(x)\varphi_\alpha(y), \tag{12}$$

so that the final wave function Ψ_t of system and apparatus is itself a superposition. The fact that the pointer ends up pointing in a definite direction, even a random one, is not discernible in this final wave function. Insofar as orthodox quantum theory is concerned, we have arrived at the measurement problem.

However, insofar as Bohmian mechanics is concerned, we have no such problem, because in Bohmian mechanics particles always have positions and pointers, which are made of particles, always point—in a direction determined by the final configuration Y_t of the apparatus. Moreover, in Bohmian mechanics we find that the state of the system is transformed in exactly the manner prescribed by textbook quantum theory, as the final wave function of the system, i.e., its conditional wave function at time t, see (4), is

$$\psi_t(x) = \Psi_t(x, Y_t) = \sum_\alpha c_\alpha \psi_\alpha(x)\varphi_\alpha(Y_t) = c_\beta \psi_\beta(x)\varphi_\beta(Y_t) = \mathcal{N}\, \psi_\beta(x) \tag{13}$$

when $Y_t \in \operatorname{supp}(\varphi_\beta)$, i.e., when the value β is registered. (Here $\mathcal{N}$ is a constant that depends upon Y but not on x. According to (13) the wave function of the system at time t, when normalized, is ψ_β.) The probability for this event is, by the quantum equilibrium hypothesis,

$$\int \mathrm{d}x \int_{\operatorname{supp}(\varphi_\beta)} \mathrm{d}y\, |\Psi_t(x, y)|^2 = |c_\beta|^2. \tag{14}$$

B

The upshot of the analysis is this: It is a consequence of Bohmian mechanics that in the course of an ideal measurement of A the (normalized) wave function of the system is transformed from ψ (11) to ψ_β with probability $|c_\beta|^2 = \left|\langle\psi_\beta|\psi\rangle\right|^2$. That is how the ► *projection postulate* arises from Bohmian mechanics. (The fact that the contributions with $\alpha \neq \beta$ will never again overlap with what evolves from $\psi_\beta(x)\varphi_\beta(y)$, and thus will not influence the future motion of the particles, is the reason why they can be ignored from time t onwards, or "collapsed away," without consequences for the trajectories of the particles.)

The Double Slit Experiment. In Bohmian mechanics, ► "wave–particle duality" can be taken literally: there is a wave (ψ) and there are particles. Accordingly, in a ► double-slit experiment the wave passes through both slits, whereas the particle passes only through one slit. Since the motion of the particle depends on the wave, it matters whether or not the other slit is open. The possible trajectories, when both slits are open, are depicted in Fig. 1; by virtue of the quantum equilibrium hypothesis, the actual trajectory will be random with the appropriate $|\psi|^2$ distribution. Thus, the place of the particle's arrival at a screen on the right will have a probability distribution featuring interference fringes. As John Bell commented [10, p. 191]: "This idea seems to me so natural and simple [...] that it is a great mystery to me that it was so generally ignored."

Spin. One may easily get the impression that ► spin cannot be explained in a realist way, given its "non-classical two-valuedness." But actually it can be incorporated into Bohmian mechanics very easily, and Bell discovered how [2]: Do not assume that there is an "actual value" associated with the spin observable $\hat{\sigma}_z$ in the z (or any other) direction! Instead, take the equation of motion (1) seriously, with $\mathbb{C}^k$ the spin space, i.e., $k = (2s+1)^N$ for N spin-s particles. (In particular, it is useful here

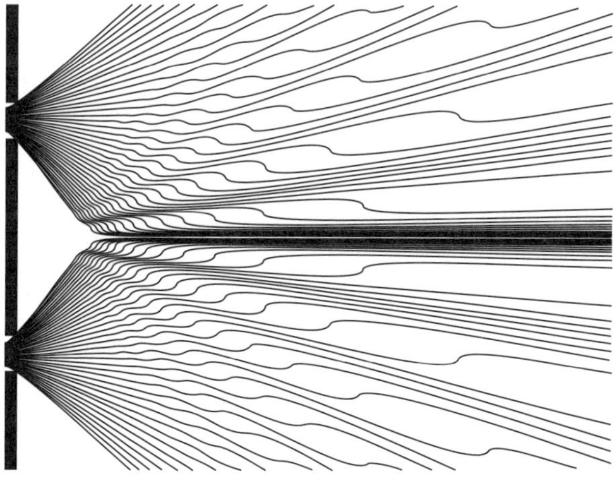

Fig. 1 Possible Bohmian trajectories in the double-slit experiment (from C. Philippidis, C. Dewdney and B.J. Hiley, Il Nuovo Cimento **52**, 15 (1979))

to regard the wave function ψ_t for, say, a single spin-$\frac{1}{2}$ particle not as a function $\psi_t : \mathbb{R}^3 \times \{-1, 1\} \to \mathbb{C}$ of a continuous (position) variable and a discrete (spin) variable, but rather as a spinor-valued function of position, $\psi_t : \mathbb{R}^3 \to \mathbb{C}^2$.)

As a consequence of (1), the motion of a particle with spin is influenced by both the "spin-up" and the "spin-down" component of the wave function. While the particle has an actual position (and a wave function) but no additional actual spin degrees of freedom, these are sufficient to completely account for all quantum phenomena associated with spin.

► *Quantum Field Theory and Relativity.* Bohmian mechanics does not account for phenomena such as particle creation and annihilation characteristic of quantum field theory. This is not an objection to Bohmian mechanics but merely a recognition that quantum field theory explains a great deal more than does nonrelativistic quantum mechanics, whether in orthodox or Bohmian form. There are extensions of Bohmian mechanics to general quantum field theories based on a particle ontology, as well as other approaches. Moreover, like nonrelativistic quantum theory, Bohmian mechanics is incompatible with special relativity, a central principle of physics: it is not Lorentz invariant. Nor can Bohmian mechanics easily be modified to become Lorentz invariant. For an overview of recent proposals aimed at finding a Lorentz invariant extension of Bohmian mechanics, see [13].

Nonlocality. In Bohmian mechanics the motion of a particle may depend on the positions of distant particles, at spacelike separation. This is an instance of ► *nonlocality*. It is worth noting that this dependence is of a kind that does not allow ► superluminal communication. Orthodox quantum mechanics features nonlocality as well, associated with the instantaneous collapse of the wave function for all particles, even distant ones. In 1964, John Bell asked whether nonlocality could be avoided by any version of quantum mechanics, and his celebrated (but often misunderstood) argument [3, 10], involving ► *Bell's theorem*, proves that the answer is no. His argument shows that certain correlations predicted by quantum mechanics (and Bohmian mechanics) and confirmed in experiment [1] cannot be explained in a local way, i.e., without allowing influences travelling faster than light. Thus, nonlocality is a feature of our world.

Primary Literature

1. A. Aspect, J. Dalibard, G. Roger: Experimental Test of Bell's Inequalities using Time-Varying Analyzers. Physical Review Letters **49**, 1804–1807 (1982)
2. J. S. Bell: On the Problem of Hidden Variables in Quantum Mechanics. Reviews of Modern Physics **38**, 447–452 (1966); reprinted as chapter 1 of [10].
3. J. S. Bell: On the Einstein-Podolsky-Rosen Paradox. Physics, **1**, 195–200 (1964); reprinted as chapter 2 of [10].
4. J. S. Bell: Quantum Field Theory Without Observers. Physics Reports **137**, 49–54 (1986); reprinted under the title "Beables for quantum field theory" as chapter 19 of [10].
5. D. Bohm: A Suggested Interpretation of the Quantum Theory in Terms of "Hidden" Variables, I and II. Physical Review **85**, 166–193 (1952)

6. L. de Broglie: In Electrons et Photons: Rapports et Discussions du Cinquième Conseil de Physique tenu à Bruxelles du 24 au 29 Octobre 1927 sous les Auspices de l'Institut International de Physique Solvay, (Gauthier-Villars, Paris 1928); English translation in [9].
7. D. Dürr, S. Goldstein, N. Zanghì: Quantum Equilibrium and the Origin of Absolute Uncertainty. Journal of Statistical Physics **67**, 843–907 (1992)
8. D. Dürr, S. Goldstein, N. Zanghì: Quantum Equilibrium and the Role of Operators as Observables in Quantum Theory. Journal of Statistical Physics **116**, 959–1055 (2004)

B

Secondary Literature

9. G. Bacciagaluppi, A. Valentini: Quantum Theory at the Crossroads, (Cambridge University Press 2009)
10. J. S. Bell: Speakable and unspeakable in quantum mechanics (Cambridge University Press, Cambridge 1987)
11. D. Bohm, B.J. Hiley: The Undivided Universe (Routledge, London 1993)
12. S. Goldstein: Bohmian Mechanics. In E. N. Zalta (ed.), Stanford Encyclopedia of Philosophy, published online by Stanford University (2001) at http://plato.stanford.edu/ entries/qm-bohm/
13. R. Tumulka: The 'Unromantic Pictures' of Quantum Theory. Journal of Physics A: Mathematical and Theoretical **40**, 3245–3273 (2007)

Bohm's Approach to the EPR Paradox

B.J. Hiley

In 1935 Einstein et al. [1] challenged the ► orthodox approach to the quantum formalism by asking whether the formalism was complete or not. The specific point that led them to this conclusion was based on a puzzle that arose when two particles were in an entangled state (► entanglement). These states are characterised by the fact that the ► wave function of the individual particles are not well defined, being ambiguous until the state of one of them was measured. The difficulty arose when the two particles were separated by a large distance and were not interacting with each other through any known classical potential. If a measurement was made on one of the particles, the state of the other became *immediately* well defined, even though it was removed far from the apparatus measuring the state of the first particle. How does this come about?

Einstein et al. chose the position and momentum variables to illustrate the problem, but because the eigenfunctions for these operators were delta functions, $\delta(\boldsymbol{r} - \boldsymbol{r}_0)$, and their Fourier components, the exponentials $\mathrm{e}^{\mathrm{i}\boldsymbol{p}.\boldsymbol{r}}$, it was difficult to see exactly what was happening in these entangled states. Bohm [2] simplified the problem by considering two spin-half particles in an entangled state given by

$$\sqrt{2}\Psi = \psi_{+z}(\boldsymbol{r}_1)\psi_{-z}(\boldsymbol{r}_2) - \psi_{-z}(\boldsymbol{r}_1)\psi_{+z}(\boldsymbol{r}_2)$$

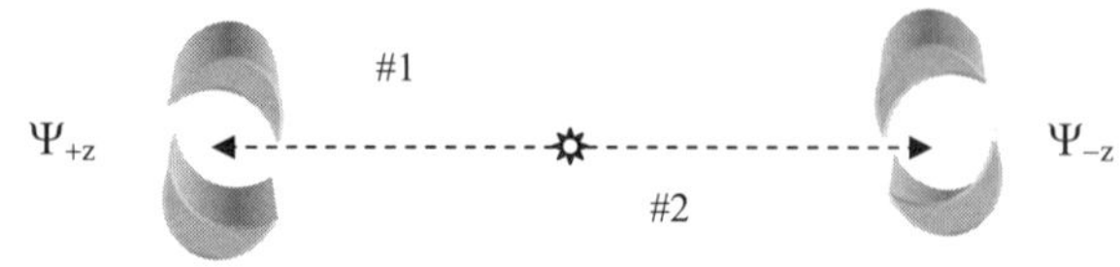

Fig. 1 Two spin-1/2 particles in an entangled state, on which the x-component of spin is measured

Here $\boldsymbol{r}_1$ and $\boldsymbol{r}_2$ refer to the respective positions of the two particles and the suffixes denote the spin states along the z-axis. We can immediately see that the ► spin of each particle is not well defined but ambiguous. When a measurement of the spin in the z-direction is made on particle #1, its state immediately becomes well defined giving either ψ_{+z} or ψ_{-z}. No matter how far away particle #2 is, we immediately know its state. It is either ψ_{-z} or ψ_{+z} respectively (Fig. 1) .

At first sight this appears just like the situation we would have if we had two balls, one red and one blue contained in two separate envelopes. We can then shuffle the envelopes so that we do not know which envelope contains which ball before separating the envelopes. Clearly if we open one envelope we will immediately know which colour ball is in the other envelope. No mystery here then. But the quantum situation is different because the *same* wave function can also be expressed as

$$\sqrt{2}\Psi = \psi_{+x}(\boldsymbol{r}_1)\psi_{-x}(\boldsymbol{r}_2) - \psi_{-x}(\boldsymbol{r}_1)\psi_{+x}(\boldsymbol{r}_2)$$

where the spin components are now in the x-direction. If we had measured the x-component of spin of particle #1 we would have found either ψ_{+x} or ψ_{-x} implying particle #2 was either in the definite state ψ_{-x} or ψ_{+x} respectively. But in quantum mechanics a particle cannot be in the two complementary states, $\psi_{\pm z}$ and $\psi_{\pm x}$, at the same time. How then does particle #2 "know" what direction is being measured when it is far away from particle #1 and there is no known force between the two particles? In other words how does the distant measurement produce the right state for particle #2?

There are two possibilities. Either there are additional "elements of reality" or ► hidden variables that determine the final state of particle #2 independently of what is being measured at particle #1, but not necessarily independently of what is found there. Or there is a "spooky action at a distance" connecting the two particles, a notion that Einstein found so abhorrent that he refused even to consider such a possibility.

When Bohm [3, 4] analysed two-particle entangled states in his interpretation (► Bohm interpretation) he found that the two entangled particles were coupled by the quantum potential. Thus if the entangled state
$\Psi(\boldsymbol{r}_1, \boldsymbol{r}_2, t) = R(\boldsymbol{r}_1, \boldsymbol{r}_2, t)\exp \mathrm{i}S(\boldsymbol{r}_1, \boldsymbol{r}_2, t)$ is substituted into the Schrödinger equation, we find the real part gives

$$\frac{\partial S(\boldsymbol{r}_1, \boldsymbol{r}_2, t)}{\partial t} + \frac{(\nabla_1 S(\boldsymbol{r}_1, \boldsymbol{r}_2, t))^2}{2m} + \frac{(\nabla_2 S(\boldsymbol{r}_1, \boldsymbol{r}_2, t))^2}{2m} + Q(\boldsymbol{r}_1, \boldsymbol{r}_2, t) = 0$$

Here $Q(\boldsymbol{r}_1, \boldsymbol{r}_2, t)$ is the non-local quantum potential, which is non-zero no matter how far apart the two particles are. Thus the Bohm model accounted for the results by providing a non-local, "spooky action at a distance". In the classical limit $Q = 0$, so there are no non-local features in classical mechanics.

Bohm et al. [5] proposed a model for spin in which all the components were defined simultaneously and which reproduced all the results of the conventional model. Here they showed that the separated particles were connected by a quantum potential which produced a non-local torque. Dewdney et al. [6] examined the model in more detail and produced numerical results vividly illustrating the time evolution of the entangled state when one particle had its spin measured. It clearly demonstrates the non-local effect of the quantum torque.

Bell [7] noticed this ► nonlocality in the Bohm model and asked whether all theories that attributed properties to individual particles had this unwelcome feature. Before his first paper appeared in print, he [8] was able to prove under quite general considerations that all theories based on local properties (local hidden variables) must satisfy the Bell inequalities ► Bell theorem, which can be written in the form

$$|P(\hat{a}, \hat{b}) - P(\hat{a}, \hat{b}')| + |P(\hat{a}', \hat{b}') + P(\hat{a}', \hat{b})| \leqslant 2$$

This inequality is violated by certain quantum mechanical entangled states. Furthermore for those quantum states that produce such a violation experiment shows that the inequality is also violated and that predictions of the quantum formalism is, in fact, correct [9].

Thus we are faced with what appears to be a dilemma. On the one hand special relativity tells us that signals cannot travel faster than the speed of light, yet the quantum formalism shows that distant particles in entangled states appear to be connected *instantaneously* with each other while they remain in the entangled states. However Eberhard [10] has shown that it is not possible to use these non-local connections to send signals because they are fragile in the sense that once a measurement is made on one particle, the ► entanglement is destroyed and the particles behave independently from then on. Thus there seems to be a peaceful coexistence between relativity and quantum theory [11].

A good review of the experimental situation regarding the Bell inequality and other similar inequalities see Clauser and Shimony [12]. See also ► Causal Inference and the EPR problem; EPR problem.

Primary Literature

1. A. Einstein, B. Podolsky, N. Rosen: Can Quantum-Mechanical Description of Physical Reality be Considered Complete. Phys. Rev. **47**, 777–780 (1935)
2. D. Bohm: *Quantum Theory* (Prentice-Hall, Englewood Cliffs, NJ 1951, pp. 614–623)
3. D. Bohm: A Suggested Interpretation of the Quantum Theory in Terms of Hidden Variables, I and II. Phys. Rev. **85**, 166–179; 180–193 (1952)
4. D. Bohm, B. J. Hiley: On the Intuitive Understanding of Nonlocality as Implied by Quantum Theory. Found. Phys. **5**, 93–109 (1975)

5. D. Bohm, R. Schiller, J. Tiomno: A Causal Interpretation of the Pauli Equation (A). Nuovo Cim. **1**(supl.), 48–66 (1955)
6. C. Dewdney, P. R. Holland, A. Kyprianidis: A Causal Account of Non-local Einstein-Podolsky-Rosen Spin Correlations. J. Phys. A Math. Gen. **20**, 4717–4732 (1987)
7. J. S. Bell: On the Problem of Hidden Variables in Quantum Theory. Rev. Mod. Phys. **38**, 447–452 (1966)
8. J. S. Bell: On the Einstein–Podolsky–Rosen Paradox. Physics **1**, 195–200 (1964)
9. A. Aspect, J. Dalibard, G. Roger: Experimental Realization of Einstein–Podolsky–Rosen–Bohm Gedankenexperiment: A New Violation of Bell's Inequalities. Phys. Rev. Lett. **49**, 91–94 (1982)
10. P. H. Eberhard: Bell's Theorem and the Different Concepts of Locality. Nuovo Cim. **B 46**, 392–419 (1978)
11. A. Shimony: Events and Processes in the Quantum World, in *Quantum Concepts in Space and Time*, ed. R. Penrose, C. Isham (Clarendon Press, Oxford 1986, pp. 182–203)

Secondary Literature

12. J. F. Clauser, A. Shimony: Bell's Theorem: Experimental Tests and Implications. Rep. Prog. Phys. **41**, 1881–1927 (1978)

Bohr's Atomic Model

Arne Schirrmacher

The model of Niels Bohr (1885–1962) for the atom is since long just the one and only conception for atoms of the vast majority of educated people. The picture of ▶ electrons revolving round a nucleus on select avenues has become the icon of the atomic age. In stark contrast to this omnipresence, historically, the Bohr atom may be identified as the best available theory for the atom only for a period of roughly ten years between 1914 and 1924. For this reason any consideration of Bohr's atom has to take into account both the historical context of its creation and the long and diverse processes of reception within science, education and public that gave rise to much misinterpretation of Bohr's intentions, his actual work and its physical or realistic interpretation.

For the question of the genesis of the Bohr model one has to go back to the beginning of the twentieth century, when it became widely recognized that both atoms contain electrons and at the same time were almost fully penetrable by electron bombardment. Between 1901 and 1905 various physicists and science popularizers draw the analogy between atoms and planetary systems (e.g. Jean Perrin (1870–1942), Wilhelm Meyer (1853–1910), or Hantaro Nagaoka (1856–1950) ▶ atomic models) and some of them immediately realized the difference: Since electric forces were both attractive and repulsive it was hard to understand how stable configurations could result at all. As a consequence in the years before world war I concern with detailed atomic models was not widespread. For this reason also the ► Rutherford atom was largely ignored until it could be reinterpreted as a predecessor of the Bohr atom. The favorite heuristic models for the atom in the years around 1910 also for Bohr was Thomson's that came in various imprecise and at times conflicting variations but was nonetheless able to serve in this way the purpose in helping to conceptualize stability, light emission and the existence of a periodic system of elements.

When Bohr in 1911/1912 went to Cambridge and Manchester to work with Thomson (1856–1940) and Ernest Rutherford (1871–1937), resp., he was mostly interested in extending his doctoral thesis on the electron theory of metals (for which Thomson had been a pioneer). The problem of bound electrons made Bohr looking for special assumptions about their arrangements and motions that could be treated in a Thomsonian manner. The switch to Rutherford then was neither motivated by a discontent with Thomson nor by a particular interest in the Rutherford atom, but by Rutherford's work in radioactivity. Rather by accident in commenting on a theory of α-particle absorption in metals by the Rutherford collaborator Charles G. Darwin (1887–1963) Bohr arrived at discussing atomic structure for the first time, as in this work the problems of bound electrons in metals and atomic structure met. At this stage Bohr conceived of an atomic model that "would not be an indication of the nature of a possibility (like J. J. Thomson's theory) but perhaps a little piece of reality" (letter to Harald Bohr 19th July 1912).

The first version of Bohr's atom in his "Manchester memorandum" than combined Thomsonian modeling with a conviction drawn from his earlier work on electron theory in metals, i.e. that within matter ordinary mechanics and electrodynamics is not sufficient but has to be complemented by some quantum condition (like in the theory of specific heats). In the case of the atom it was the mechanical instability of the models that Bohr wanted to fix by a quantum condition. While he arrived at far-reaching results (explanation of periodic table, though by a wrong calculation) and implemented a quantum condition to relate the kinetic energy of the electrons to the frequency of rotation, $E_{\text{kin}} = K \cdot \nu$ this first version of the Bohr atom would not take off (Fig. 1).

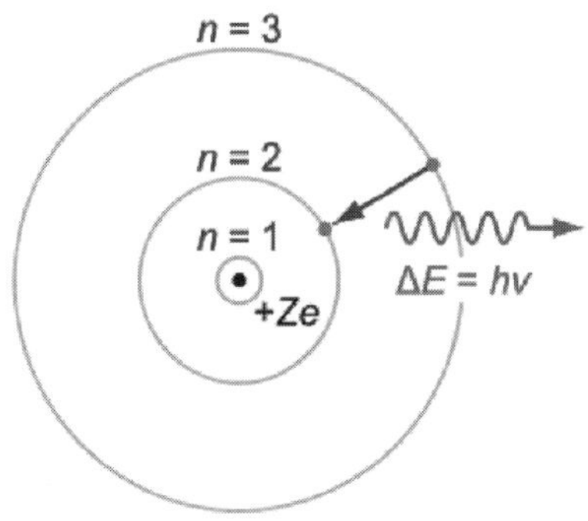

Fig. 1 Bohr model of atom, with quantized energy levels, and electron jumps, accompanied by photon emissions. Source: Wikimedia Commons

Only after Bohr stumbled upon a publications of J. W. Nicholson (1881–1955) late in 1912, who had constructed an comparably immature atomic model also with a quantum condition in order to explain the spectral lines of the solar corona, Bohr realized that ► spectroscopy was the missing link for establishing a sound atomic model. Disregarding spectra was not a particular failure of Bohr, since their complexity and the futile search for explanation rejected most atom builders. Nicholson's work motivated Bohr to combine his initial model with Planck's (1858–1947) quantized oscillator thus postulating series of states with quantized energy. The prize he had to pay was to obscure the nature of the atomic vibrations, or positively turned, this amounted to the most important step towards a quantized atomic model in which the frequency of revolution are disconnected with the frequencies of radiation that simply equate from the energy difference of two atomic states expressed in terms of ► Planck's constant: $E_n E_m = h\nu_{nm}$. With this separation of optical and mechanical frequencies, obviously, the "little piece of reality" the model might claim had become even smaller. However, the good accord with the Balmer series $n\nu_{nm} = Z^2 R[(1/m^2) - (1/n^2)]$ provided irresistible persuasiveness in favor of this new atomic model which amounted to a perfect compromise of general (mechanical) intelligibility and modern (fascinating) quantum properties.

It must have been this attractive combination that made Arnold Sommerfeld (1868–1951) adopting and extending Bohr's model, while Rutherford immediately scolded Bohr for the lack of a mechanism for the electrons to change from one state to another and Thomson just kept on lecturing his atomic theory unchanged. Bohr himself was quite aware of the makeshift character of his theory and appeared pessimistic to many colleagues. This may indicate that besides the spectroscopic success additional factors were necessary for the general recognition of Bohr's achievement, factors that for some reason where most favorable in (war-time) Germany.

While in Göttingen Peter Debye (1884–1966) extended the model to the hydrogen molecule and met experimental results on dispersion convincingly, it was Sommerfeld who took up Bohr's model most forcefully and guided a young generation of German physicists into the refinement of Bohr's theory. Though already mentioned by Bohr only the Munich group worked out the generalization of electron orbits to elliptic ones into a systematic theory and hence introduced a second quantum number for labeling the possible states of the atoms. In combination with relativistic corrections and consideration of the co-movement of the nucleus ► Sommerfeld School mastered the fine-structure of spectral lines to great experimental unison. Further ► quantum numbers and ► selection rules for describing possible transitions between states transformed ► atomic physics to a "number mysticism" while heavy use of pictures for representing complex systems of electron orbits at the same time provided an engineering type of approach to it. Sommerfeld's promotion of the refined Bohr model between 1917 and 1925 would include non-specialized university lectures, articles in popular science journals, wood and brass models for the Deutsches Museum as well as radio programs.

With the older scientists largely skeptical, the Bohr atom won recognition among wider scientific and lay circles by popularization. Although as early as 1916 problems of the theory to account for anomalous dispersion appeared the momentum the pictorial representation of the new understanding of matter developed could not anymore be rescinded. Further progress in atomic theory only developed when Bohr's central postulate of the separation of optical and mechanical frequencies was put aside and Hendrik Kramers (1894–1952) at Bohr's institute associated with each stationary state of Bohr's atom a harmonic oscillator with frequencies equal to those emitted and absorbed. Similarly did Heisenberg (1901–1976) find his way to a quantum mechanical reinterpretation of mechanical relations only after abandoning graphic models and turning to dispersion theory with virtual oscillators.

The Bohr atom has served many scientists, educators and philosophers as exemplar. Notions like "Rutherford–Bohr atom" ► Bohr's atomic model, Rutherford atom are commonplace, logical and rational reconstructions of the (conceived) research have been undertaken and even analyses of Bohr's (idealized) research programs are at hand [8, 10]. All these however, have always to be judged against the rich historical sources that rather provide a complex and coincidental picture of the historical path.

Primary Literature

1. Niels Bohr: *Collected Works*, 12 vols. (Amsterdam 1972–2007; contains in Volumes 2 and 4 in particular the 1912 "Rutherford Memorandum" and the 1913 Bohr "trilogy" On the constitution of atoms and molecules I–III, from Philosophical Magazine, as well as the letters)
2. Arnold Sommerfeld: *Atombau und Spektrallinien* (Braunschweig 1919)
3. Hendrik A. Kramers, Helge Holst: *The atom and the Bohr theory of its structure* (Copenhagen 1923)

Secondary Literature

4. John L. Heilbron, Thomas S. Kuhn: The genesis of the Bohr atom. Historical Studies in the Physical Sciences **1**, 211–290 (1969)
5. Kragh: Niels Bohr's second atomic theory. Historical Studies in the Physical Sciences **10**, 123–186 (1979)
6. John L. Heilbron: Rutherford–Bohr atom. American Journal of Physics **49**, 223–231 (1981)
7. Abraham Pais: *Niels Bohr's Times*, in Physics, philosophy and polity (Oxford 1990)
8. Gerhard Zoubek, Bernhard Lauth: Zur Rekonstruktion des Bohrschen Forschungsprogramms I & II. Erkenntnis **37**(2), 223–247, 249–273 (1992)
9. Eckert: *Die Atomphysiker: Eine Geschichte der Theoretischen Physik am Beispiel der Sommerfeldschule* (Braunschweig 1993)
10. Hinne Hettema: Bohr's theory of the atom 1913–1923: A case study in the progress of scientific research programmes. Studies in History and Philosophy of Modern Physics **26**, 307–23 (1995)
11. Helge Kragh: *Quantum Generations: A History of Physics in the Twentieth Century* (Princeton 1999)

Bohr–Kramers–Slater Theory

Helge Kragh

The Bohr–Kramers–Slater theory (or BKS theory) was proposed in 1924 as an attempt to explain problems in physical optics and to provide a unified picture of the continuous electromagnetic field and the discontinuous quantum transitions in atoms. Although the theory was short-lived it proved most important in the subsequent development of quantum theory, not least because it replaced causal spatio-temporal description of the transitions between stationary states with statistical considerations. Moreover, it followed that energy and momentum was only conserved statistically, not for individual atomic processes.

In early 1924 atomic physics was in a state of crisis (► quantum theory, crisis period), one of the critical problems being the interaction between matter and radiation. In a paper published in *Nature* in February 1924, John Clark Slater (1900–1976) suggested the radical idea that when an atom was in a stationary state, it would "communicate with other atoms… by means of a virtual field of radiation originating from oscillators having the frequencies of possible quantum transitions and the function of which is to provide for statistical conservation of energy and momentum by determining the probabilities for quantum transitions." Note that the field was thought to be emitted by atoms in their stationary states and not, as in Bohr's original theory, during the ► quantum jumps from one state to another.

The idea to conceive the atom as a collection of "virtual harmonic oscillators" had implicitly been suggested by Rudolf Ladenburg (1882–1952) in a paper on dispersion theory from 1924, but it was only with Slater's paper and the subsequent BKS paper that explicit use was made of the idea. Slater provided a picture of emission as well as absorption of radiation inspired by and in qualitative agreement with Einstein's probabilistic radiation theory of 1916–17. He considered his picture to be a reconciliation of the continuous wave theory of the electromagnetic field with the discreteness of light quanta (photons ► light quantum), of whose existence he had been convinced by Arthur Compton (1892–1962) ► Compton experiment.

Slater was at the time a visiting physicist at Niels Bohr's (1885–1962) institute in Copenhagen, and he discussed at length his theory with Bohr and his assistant Hendrik Kramers (1894–1952) who found it interesting but also suggested modifications. Neither Bohr nor Kramers shared Slater's belief in the light quantum. Rather than adopting a theory which harmonized the electromagnetic field with light quanta (Slater's view), they wanted to connect the continuous field responsible for the propagation of light with the discontinuous quantum transitions in the atom. Moreover, the idea of a statistical connection, as proposed by Slater in his *Nature* paper, appealed greatly to Bohr and Kramers who believed that it implied that a causal description of quantum transitions had to be abandoned. If so, they concluded, the conservation laws of energy and momentum could not be strictly valid for individual processes, but should be understood as statistical laws. This idea seems to have been due to Bohr and Kramers rather than Slater. In a general sense it was not new to Bohr, who for some time had been prepared to abandon the validity of energy conservation in the quantum domain.

The result of the discussions in Copenhagen – and the pressure put on Slater to go along with the statistical, non-conservation ideas of Bohr and Kramers – was a joint paper published simultaneously in *Philosophical Magazine* and *Zeitschrift für Physik*. Although jointly authored, the paper reflected Bohr's ideas more than Slater's, and in fact Slater disagreed with much of it. The BKS paper kept to Slater's idea of a virtual radiation field associated with the stationary state of an atom and also incorporated the probabilistic interpretation of transition processes. "The occurrence of a certain transition in a given atom will depend on the initial stationary state of this atom itself and on the states of the atoms with which it is in communication through the virtual radiation field, but not on the occurrence of transition processes in the latter atoms."

Slater had originally conceived the virtual radiation field as a kind of wave-field guiding the light quanta, but in the BKS paper there was no trace of this idea (which was also part of Louis de Broglie's theory (1892–1981)). It remained unclear what the enigmatic virtual oscillators were, except that they were not directly observable. The most radical feature of the BKS theory was the description of atomic processes at the expense of sacrificing the laws of detailed conservation of energy and momentum.

The BKS theory was almost purely qualitative and appealed conceptually to an intuitive understanding of virtual fields and virtual oscillators, but if it was to be taken seriously it had to make testable predictions. Bohr and Kramers (and, nominally, Slater) applied the theory to the ► Compton effect and concluded that the direction of a recoil electron after scattering an X-ray photon would not be uniquely determined, as required by the conservations laws, but display a wide statistical distribution. Even before this prediction could be tested, the theory aroused much attention, if little enthusiasm. Erwin Schrödinger (1887–1961) supported the BKS theory and Bohr's interpretation, but most other physicists either rejected it or expressed reservation. Among those who were opposed to it were Arnold Sommerfeld (1868–1951), Albert Einstein (1879–1955), Compton and Wolfgang Pauli (1900–1958), and it is uncertain if even Kramers supported it.

At any rate, the theory did not last for more than a year. As early as June 1924, Walther Bothe (1891–1957) and Hans Geiger (1882–1945) in Berlin proposed an experiment to test the theory by measuring simultaneously the scattered ► X-rays and the recoil electrons. This was one of the first experiments using electronic coincidence devices, and it was not until April 1925 that they had ready their final result, which was "incompatible with Bohr's interpretation of the Compton effect." Also Compton and Alfred W. Simon, who used a cloud chamber to determine the direction of recoil electrons, concluded in favour of energy and momentum conservation and that experiments had therefore disproved the BKS theory. Karl Popper (1902–1994) later described the experiments of 1925 as a kind of *experimentum crucis*. While this was good news to Slater, it was not to Bohr, who for a year had defended the theory and taken it very seriously. Nonetheless, he accepted the experimental verdict and wrote to Fowler that "there is nothing else to do than to give our revolutionary efforts as honourable a funeral as possible."

In spite of its short lifetime, the BKS theory was singularly important. For one thing, its radically new approach paved the way for a greater understanding that methods and concepts of classical physics could not be carried over in a future quantum mechanics. For another thing, the theory provided the point of departure of Kramers' theory of dispersion of 1924 and its further development into the Kramers–Heisenberg dispersion theory of 1925, the final step before Heisenberg's formulation of quantum or ► matrix mechanics.

Primary Literature

1. J. C. Slater: Radiation and Atoms. Nature **113**, 307–308 (1924)
2. N. Bohr, H. A. Kramers, J. C. Slater: The Quantum Theory of Radiation. Philosophical Magazine **47**, 785–802 (1924)
3. W. Bothe, H. Geiger: Über das Wesen des Comptoneffekts; ein experimenteller Beitrag zur Theorie der Strahlung. Zeitschrift für Physik **32**, 639–663 (1925)

Secondary Literature

4. J. Hendry: Bohr-Kramers-Slater: A Virtual Theory of Virtual Oscillators and its Role in the History of Quantum Mechanics. Centaurus **25**, 189–221 (1981)
5. K. Stolzenburg ed.: *Niels Bohr. Collected Works*. Volume 5 (North-Holland, Amsterdam 1984, 3–218)
6. O. Darrigol: *From c-Numbers to q-Numbers*: The Classical Analogy in the History of Quantum Theory (University of California Press, Berkeley 1992, 214–224)

Born Rule and its Interpretation

N.P. Landsman

The *Born rule* provides a link between the mathematical formalism of quantum theory and experiment, and as such is almost single-handedly responsible for practically all predictions of quantum physics. In the history of science, on a par with the ► Heisenberg uncertainty relations, the ► Born rule is often seen as a turning point where ► indeterminism entered fundamental physics. For these two reasons, its importance for the practice and philosophy of science cannot be overestimated.

The Born rule was first stated by Max Born (1882–1970) in the context of scattering theory [1], following a slightly earlier paper in which he famously omitted the absolute value squared signs (though he corrected this is a footnote added in proof). The application to the position operator (cf. (5) below) is due to Pauli, who mentioned it to Heisenberg and Jordan, the latter publishing Pauli's suggestion with acknowledgment [6] even before Pauli himself spent a footnote on it [8]. The general formulation (6) below is due to von Neumann (see §III.1 of [7]), following earlier contributions by Dirac [2] and Jordan [5, 6].

Both Born and Heisenberg acknowledge the profound influence of Einstein on the probabilistic formulation of quantum mechanics. However, Born and Heisenberg as well as Bohr, Dirac, Jordan, Pauli and von Neumann differed with Einstein about the (allegedly) fundamental nature of the Born probabilities and hence on the issue of determinism. Indeed, whereas Born and the others just listed after him believed the outcome of any individual quantum measurement to be unpredictable in principle, Einstein felt this unpredictability was just caused by the incompleteness of quantum mechanics (as he saw it). See, for example, the invaluable source [3]. Mehra & Rechenberg [20] provide a very detailed reconstruction of the historical origin of the Born rule within the context of quantum mechanics, whereas von Plato [22] embeds a briefer historical treatment of it into the more general setting of the emergence of modern probability theory and probabilistic thinking.

Let a be a quantum-mechanical ▶ observable, mathematically represented by a ▶ self-adjoint operator on a ▶ Hilbert space H with inner product denoted by $(\, , \,)$. For the simplest formulation of the Born rule, assume that a has non-degenerate discrete spectrum: this means that a has an ▶ orthonormal basis of eigenvectors (e_i) with corresponding eigenvalues λ_i, i.e. $ae_i = \lambda_i e_i$. A fundamental assumption underlying the Born rule is that a ▶ measurement of the observable a will produce one of its eigenvalues λ_i as a result. In what follows, $\Psi \in H$ is a unit vector and hence a (pure) state in the usual sense. Then the Born rule states:

> If the system is in a state Ψ, then the probability $P(a = \lambda_i \mid \Psi)$ that the eigenvalue λ_i of a is found when a is measured is

$$P(a = \lambda_i \mid \Psi) = |(e_i, \Psi)|^2. \tag{1}$$

In other words, if $\Psi = \sum_i c_i e_i$ (with $\sum_i |c_i|^2 = 1$), then $P(a = \lambda_i \mid \Psi) = |c_i|^2$.

The general formulation of the Born rule (which is necessary, for example, to discuss ▶ observables with continuous spectrum such as the position operator x on $H = L^2(\mathbb{R})$ for a particle moving in one dimension) relies on the spectral theorem for self-adjoint operators on Hilbert space (see, e.g., [21]). According to this theorem, a self-adjoint operator a defines a so-called spectral measure (alternatively called a projection-valued measure or PVM) $B \mapsto p^{(a)}(B)$ on $\mathbb{R}$. Here B is a (Borel) subset of $\mathbb{R}$ and $p^{(a)}(B)$ is a projection on H. (Recall that a projection on a Hilbert space H is a bounded operator $p : H \to H$ satisfying $p^2 = p^* = p$; such operators correspond bijectively to their images pH, which are closed subspaces of H.) The spectral measure $p^{(a)}$ turns out to be concentrated on the spectrum $\sigma(a) \subset \mathbb{R}$ of a in the sense that if $B \cap \sigma(a) = \emptyset$, then $p^{(a)}(B) = 0$ (hence $p^{(a)}$ is often defined on $\sigma(a)$ instead of $\mathbb{R}$). The map $B \mapsto p^{(a)}(B)$ satisfies properties such as

$p^{(a)}(A \cup B) = p^{(a)}(A) + p^{(a)}(B)$ when $A \cap B = \emptyset$ (and a similar property for a countable family of disjoint sets) and $p^{(a)}(\mathbb{R}) = 1$ (i.e. the unit operator on H). Consequently, a self-adjoint operator a and a unit vector $\Psi \in H$ jointly define a probability measure $P_\Psi^{(a)}$ on $\mathbb{R}$ by

$$P_\Psi^{(a)}(B) := (\Psi, p^{(a)}(B)\Psi) = \|p^{(a)}(B)\Psi\|^2, \tag{2}$$

where $\|\cdot\|$ is the norm derived from the inner product on H. The properties of $p^{(a)}$ just mentioned then guarantee that $P_\Psi^{(a)}$ indeed has the properties of a probability measure, such as $P_\Psi^{(a)}(A \cup B) = P_\Psi^{(a)}(A) + P_\Psi^{(a)}(B)$ when $A \cap B = \emptyset$ (and a similar property for a countable family of disjoint sets) and $P_\Psi^{(a)}(\mathbb{R}) = 1$. Again, the probability measure $P_\Psi^{(a)}$ is concentrated on $\sigma(a)$.

For example, if a has discrete spectrum, then $\sigma(a) = \{\lambda_1, \lambda_2, \ldots\}$ and $p^{(a)}(B)$ projects onto the space spanned by all eigenvectors whose eigenvalues lie in B. In particular, if $\Psi = \sum_i c_i e_i$ as above, then $P_\Psi^{(a)}(\{\lambda_i\}) = |c_i|^2$. In the case of the position operator x as above, $\sigma(x) = \mathbb{R}$ and $p^{(x)}(B)$ equals the characteristic function χ_B, seen as a multiplication operator on $L^2(\mathbb{R})$. The image of $p^{(x)}(B)$ consists of functions vanishing (almost everywhere) outside B, and the measure $P_\Psi^{(x)}$ is given by

$$P_\Psi^{(x)}(B) = \int_{\mathbb{R}} dx\, \chi_B(x)|\Psi(x)|^2 = \int_B dx\, |\Psi(x)|^2. \tag{3}$$

The general statement of the Born rule, then, is as follows:

> If the system is in a state $\Psi \in H$, then the probability $P(a \in B \mid \Psi)$ that a result in $B \subset \mathbb{R}$ is found when a is measured equals

$$P(a \in B \mid \Psi) = P_\Psi^{(a)}(B). \tag{4}$$

For discrete non-degenerate spectrum this reduces to (1). For the position operator in one dimension, (4) yields

$$P(x \in B \mid \Psi) = \int_B dx\, |\Psi(x)|^2 \tag{5}$$

for the probability that the particle is found in the region B.

Note that it follows from the general Born rule (4) that with probability one a measurement of a will lead to a result contained in its spectrum, since $P_\Psi^{(a)}(B) = 0$ whenever $B \cap \sigma(a) = \emptyset$. Curiously, however, the probability $P(a = \lambda \mid \Psi)$ of finding any specific number λ in the continuous spectrum of a is zero! As a case in point, the probability $P(x = x_0 \mid \Psi)$ of finding the particle at any given point x_0 vanishes. Of course, this phenomenon also occurs in classical probability theory (e.g., the probability of any given infinite sequence of results of a coin flip is zero).

The rule (4) is easily extended to n commuting self-adjoint operators $a_1, \ldots, a_n$ [7]:

The probability that the observables $a_1, \ldots, a_n$ simultaneously take some value in a subset $B_1 \times \cdots \times B_n \subset \mathbb{R}^n$ upon measurement in a state Ψ is

$$P_\Psi(a_1 \in B_1, \ldots, a_n \in B_n) = \|p^{(a_1)}(B_1) \cdots p^{(a_n)}(B_n)\Psi\|^2. \quad (6)$$

This version of the Born rule is needed, for example, in order to generalize (5) to three dimensions. Indeed, the ensuing formula is practically the same, this time with $B \subset \mathbb{R}^3$ and x replaced by (x, y, z).

The statement that the expectation value of an observable a in a state Ψ equals $(\Psi, a\Psi)$ is equivalent to the Born rule. To see this, we identify projections with yes-no questions [7], identifying the answer 'yes' with eigenvalue 1 and 'no' with eigenvalue 0. The expectation value $(\Psi, p\Psi) = \|p\Psi\|^2$ of a projection then simply becomes the probability of the answer 'yes'. Taking $p = p^{(a)}(B)$ then reproduces (4), since the probability of 'yes' to the question $p^{(a)}(B)$ is nothing but $P(a \in B \mid \Psi)$. In this fashion, the Born rule may be generalized from pure states to mixed ones (i.e. ► density matrices in the standard formalism we are considering here), by stipulating that the expectation value of a in a state ρ (i.e. a positive trace-class operator with ► trace one) is $\mathrm{Tr}(\rho a)$. For a further generalization in this direction see ► Algebraic quantum mechanics.

Finally, another formulation of the Born rule is as follows:

The transition probability $P(\Psi, \Phi)$ from a state Ψ to a state Φ, or, in other words, the probability of a 'quantum jump' from Ψ to Φ, is

$$P(\Psi, \Phi) = |(\Psi, \Phi)|^2. \quad (7)$$

This related to the first formulation above, in that in standard measurement theory one assumes a ► 'wave function collapse' in the sense that Ψ changes to e_i after a measurement of a yielding λ_i. The transition probability $P(\Psi, e_i)$ is then precisely equal to $P(a = \lambda_i \mid \Psi)$ as stated above.

The *Born interpretation* of quantum mechanics is usually taken to be the statement that the empirical content of the theory (and particularly of the quantum state) is given by the Born rule. However, this is not really an interpretation at all until it is specified what the notions of measurement and probability mean. The pragmatic attitude taken by most physicists is that measurements are what experimentalists perform in the laboratory and that probability is given the frequency interpretation [15, 17] (which is neutral with respect to the issue whether the probabilities are fundamental or due to ignorance). Given that firstly the notion of a quantum measurement is quite subtle and hard to define, and that secondly the frequency interpretation is held in rather low regard in the philosophy of probability [17, 18], it is amazing how successful this attitude has been! Going beyond pragmatism requires a mature interpretation of quantum mechanics, however. Each such interpretation hinges on some interpretation of probability and will contain its own perspective on the Born rule. See Ignorance interpretation, Ithaca Interpretation, Many Worlds Interpretation, Modal Interpretation, Orthodox Interpretation, Transactional Interpretation.

The nature of the Born rule comes out particularly well in the Copenhagen interpretation, ► Consistent Histories; Metaphysics in Quantum Mechanics; Nonlocality; Orthodox Interpretation; Schrödinger's Cat; Transactional Interpretation, especially if this approach is combined with ► Algebraic quantum mechanics. In the algebraic approach, a quantum system is modeled by a non-commutative C^*-algebra of observables. The simplest illustration of this is the algebra M_n of all complex $n \times n$ matrices. This contains the commutative C^*-algebra D_n of all diagonal matrices as a subalgebra. A unit vector $\Psi \in \mathbb{C}^n$ determines a pure state ψ on M_n in the algebraic sense by $\psi(a) = (\Psi, a\Psi)$. The latter may be restricted to a state $\psi_{|D_n}$ on D_n, which turns out to be mixed: if $\Psi = \sum_{i=1}^n c_i e_i$ and $d_\lambda = \text{diag}(\lambda_1, \ldots, \lambda_n)$ is the diagonal matrix with entries $(\lambda_1, \ldots, \lambda_n)$, then

$$\psi_{|D_n}(d_\lambda) = \sum_{i=1}^{n} |c_i|^2 \lambda_i \tag{8}$$

yields the expectation value of d_λ in the state ψ. In particular, if $p_i \in D_n$ is the projection $p_i = \text{diag}(0, \ldots, 1, \ldots, 0)$ having 1 on the i'th diagonal entry and zeros elsewhere, then $\psi_{|D_n}(p_i) = |c_i|^2$ yields the Born probability of obtaining λ_i upon measuring D_λ.

Similarly, one may regard a ► wave function $\Psi \in L^2(\mathbb{R})$ as an algebraic state ψ on the C^*-algebra $B(L^2(\mathbb{R}))$ of all bounded operators on the Hilbert space $L^2(\mathbb{R})$. This C^*-algebra contains the commutative subalgebra $C_0(\mathbb{R})$ given by all multiplication operators on $L^2(\mathbb{R})$ defined by continuous functions of $x \in \mathbb{R}$ that vanish at infinity (roughly speaking, this is the C^*-algebra generated by the position operator). The restriction $\psi_{|C_0(\mathbb{R})}$ of ψ to $C_0(\mathbb{R})$ is given by

$$\psi_{|C_0(\mathbb{R})}(f) = \int_{\mathbb{R}} dx\, |\Psi(x)|^2 f(x). \tag{9}$$

The probability measure $P_{\psi_{|C_0(\mathbb{R})}}$ on $\mathbb{R}$ associated to the functional $\psi_{|C_0(\mathbb{R})}$ by the Riesz representation theorem [21] is just $P_{\psi_{|C_0(\mathbb{R})}} = P_\Psi^{(x)}$, cf. (3). Hence the restricted state $\psi_{|C_0(\mathbb{R})}$precisely yields the Born–Pauli probability (5).

Finally, to recover (4) (assuming for simplicity that the operator $a : H \to H$ is bounded), one considers the commutative C^*-algebra $C^*(a)$ of $B(H)$ generated by a and the unit operator. It can be shown [21] that $C^*(a) \cong C(\sigma(a))$. Hence a unit vector $\Psi \in H$ defines a state ψ on $B(H)$, whose restriction $\psi_{|C^*(a)}$ to $C^*(a)$ yields a probability measure $P_{\psi_{|C^*(a)}}$ on the spectrum $\sigma(a)$ of a. It easily follows that

$$P_{\psi_{|C^*(a)}} = P_\Psi^{(a)}, \tag{10}$$

which reproduces (2).

The physical relevance of these constructions derives from Bohr's doctrine of classical concepts, which is an essential ingredient of the Copenhagen interpretation [24]. In particular, if it is to serve its function, a measurement apparatus has to be described as if it were classical. This implies that if it is used as a measuring device, the apparatus (which a priori is quantum mechanical) has to be described by a commutative subalgebra D of its full non-commutative algebra A of quantum-mechanical observables. Upon the identifications explained above, the Born probability measure then comes out to be just the restriction of the total state on A to the 'classical' subalgebra D thereof that Bohr calls for.

This account does not provide a derivation of the Born rule from first principles, but it does clarify its mathematical and physical origin. In particular, in the Copenhagen interpretation probabilities arise because we look at the quantum world through classical glasses:

> "One may call these uncertainties [i.e. the Born probabilities] objective, in that they are simply a consequence of the fact that we describe the experiment in terms of classical physics; they do not depend in detail on the observer. One may call them subjective, in that they reflect our incomplete knowledge of the world." (Heisenberg [4], pp. 53–54)

In other words, one cannot say that the Born probabilities are either subjective (i.e. Bayesian, or due to ignorance) or objective (i.e. fundamentally ingrained in nature and independent of the observer). Instead, the situation is more subtle and has no counterpart in classical physics or probability theory: the choice of a particular classical description is subjective, but once it has been made the ensuing probabilities are objective and the particular outcome of an experiment compatible with the chosen classical context is unpredictable. Or so Bohr and Heisenberg say...

In most interpretations of quantum mechanics, some version of the Born rule is simply *postulated*. This is the case, for example, in the ► Consistent histories interpretation, the ► Modal interpretation and the ► Orthodox interpretation. Attempts to *derive* the Born rule from more basic postulates of quantum theory go back to Finkelstein [16] and Hartle [19], whose work was corrected and extended in [14]. These authors study infinite sequences of measurements and prove that the ensuing relative frequencies automatically satisfy the Born rule. It is controversial, however, to what extent this argument really derives the Born rule or is eventually circular [11, 12]. In the version of the ► Many worlds interpretation developed by Deutsch [13] and his followers [23, 26], the authors claim to derive the Born rule using arguments from decision theory, but once again the charge of circularity has been raised [9, 10]. See also [27, 25] for a similar debate in the context of ► decoherence. The conclusion seems to be that no generally accepted derivation of the Born rule has been given to date, but this does not imply that such a derivation is impossible in principle.

Primary Literature

1. M. Born: Quantenmechanik der Stoßvorgänge. Z. Phys. **38**, 803–827 (1926).
2. P. Dirac: The physical interpretation of the quantum dynamics. Proc. R. Soc. Lond. **A113**, 621–641 (1926).

3. A. Einstein & M. Born: *Briefwechsel 1916–1955* (Langen Müller, München 2005).
4. W. Heisenberg: *Physics and Philosophy: The Revolution in Modern Science*. (Allen & Unwin, London 1958).
5. P. Jordan: Über quantenmechanische Darstellung von Quantensprügen. Z. Phys. **40**, 661–666 (1927).
6. P. Jordan: Über eine neue Begründung der Quantenmechanik. Z. Phys. **40**, 809–838 (1927).
7. J. von Neumann: *Mathematische Grundlagen der Quantenmechanik* (Springer, Berlin 1932). English translation: *Mathematical Foundations of Quantum Mechanics* (Princeton University Press, Berlin 1955).
8. W. Pauli: Über Gasentartung und Paramagnetismus. Z. Phys. **41**, 81–102 (1927).

Secondary Literature

9. D.J. Baker: Measurement outcomes and probability in Everettian quantum mechanics. Stud. Hist. Philos. Mod. Phys. **38**, 153–169 (2007).
10. H. Barnum, C. M. Caves, J. Finkelstein, C. A. Fuchs, R. Schack: Quantum probability from decision theory? Proc. Roy. Soc. Lond. **A456**, 1175–1182 (2000).
11. A. Cassinello & J.L. Sánchez-Gómez: On the probabilistic postulate of quantum mechanics. Found. Phys. **26**, 1357–1374 (1996).
12. C. Caves & R. Schack: Properties of the frequency operator do not imply the quantum probability postuate. Ann. Phys. (N.Y.) **315**, 123–146 (2005).
13. D. Deutsch: Quantum theory of probability and decisions. Proc. R. Soc. Lond. **A455**, 3129–3137 (1999).
14. E. Farhi, J. Goldstone & S. Gutmann: How probability arises in quantum mechanics. Ann. Phys. (N.Y.) **192**, 368–382 (1989).
15. T.L. Fine: *Theories of Probability* (Academic, New York, 1978).
16. D. Finkelstein: The logic of quantum physics. Trans. N. Y. Acad. Sci. **25**, 621–637 (1965).
17. D. Gillies: *Philosophical Theories of Probability* (Cambridge University Press, Cambridge 2000).
18. A. Hájek: Interpretations of probability. In *The Stanford Encyclopedia of Philosophy*, ed. by Edward N. Zalta, `http://www.science.uva.nl/seop/entries/probability-interpret/`.
19. J.B. Hartle: Quantum mechanics of individual systems. Am. J. Phys. **36**, 704–712 (1968).
20. J. Mehra & H. Rechenberg: *The Historical Development of Quantum Theory. Vol. 6: The Completion of Quantum Mechanics 1926–1941. Part 1: The Probabilistic Interpretation and the Empirical and Mathematical Foundation of Quantum Mechanics, 1926–1936* (Springer, New York 2000).
21. G.K. Pedersen: *Analysis Now* (Springer, New York 1989).
22. J. von Plato: *Creating Modern Probability* (Cambridge University Press, Cambridge, 1994).
23. S. Saunders: Derivation of the Born rule from operational assumptions. Proc. R. Soc. Lond. **A460**, 1771–1788 (2004).
24. E. Scheibe: *The Logical Analysis of Quantum Mechanics* (Pergamon Press, Oxford 1973).
25. M. Schlosshauer & A. Fine: On Zureks derivation of the Born rule. Found. Phys. **35**, 197–213 (2005).
26. D. Wallace: Everettian Rationality: defending Deutsch's approach to probability in the Everett interpretation. Stud. Hist. Philos. Mod. Phys. **34** (2003), 415–438.
27. W.H. Zurek: Probabilities from entanglement, Born's rule $p_k = |\Psi_k|^2$ from envariance, Phys. Rev. **A71**, 052105 (2005).

Bose–Einstein Condensation

A.J. Leggett

Bose–Einstein condensation (BEC) is a phenomenon that occurs in a macroscopic system of bosons (particles obeying ▶ Bose–Einstein statistics) at low temperatures: a nonzero fraction of all the particles in the system (thus a macroscopic number of particles) occupy a *single* one-particle state. This would, of course, happen for a system of distinguishable, noninteracting particles at zero temperature, but in this case the phenomenon disappears as soon as the temperature becomes comparable to the energy splitting between the single-particle groundstate and the first excited state – a quantity which tends to zero with the size of the system. By contrast, in BEC the macroscopic occupation occurs at all temperatures below a transition temperature, usually denoted T_c, which while a function of intensive parameters such as density and interaction strength is constant in the thermodynamic limit.

The fundamental reason for the occurrence of BEC lies in the requirement, which follows from considerations of quantum field theory, that the ▶ wave function of a system of identical bosons should be symmetric under the exchange of any two particles. This has the consequence that states that differ only by such an exchange must be counted as identical, i.e. counted only once. Thus, for example, while for a system of N distinguishable objects, which must be partitioned between two boxes, the number of ways of putting M of them into one box is given by the familiar binomial formula $N!/(M!N-M!)$, for bosons there is exactly one way for each M. The effect is to remove the "entropic" factor, which for distinguishable objects militates against putting a large fraction of them in a single one-particle state.

For noninteracting bosons in thermal equilibrium at temperature T a calculation of the average number of particles $\langle n_i \rangle$ occupying the various single-particle states i is straightforward and was carried out by Albert Einstein (1879–1955) [1] in 1925 on the basis of the statistics derived by Satyendra Nath Bose (1894–1974) [2] a year earlier:

$$\langle n_i \rangle = \{[\exp(\epsilon_i - \mu)/k_B T] - 1\}^{-1} \tag{1}$$

where μ is the chemical potential,which must be fixed by the condition

$$\sum_i \langle n_i \rangle = N \tag{2}$$

where N is the total number of particles present. In order to make sense of (1), it is clear that the chemical potential must be negative (we set the lowest single-particle energy to zero by convention); since the LHS of (2) is an increasing function of μ, it follows that if in it we take the value of $\langle n_i \rangle$ for $\mu = 0$, the equality must be replaced by an inequality. Thus, if we were to replace the sum by an integral and introduce the single-particle density of states $\rho(\epsilon)$ in the standard way, we would find the condition

$$\int_0^\infty \frac{\rho(\epsilon)\mathrm{d}\epsilon}{\exp(\epsilon/k_{\mathrm{B}}T) - 1} \geqslant N \tag{3}$$

However, if $\rho(\epsilon)$ tends to zero with ϵ, as happens for a gas in three-dimensional free space, this condition cannot be fulfilled below a certain "critical temperature" T_{c}, which for 3D free space is given by

$$T_{\mathrm{c}} = 3 \cdot 31 n^{2/3}\hbar^2/m \tag{4}$$

where $n = N/V$ is the density.

What then happens for temperatures $T < T_{\mathrm{c}}$? According to Einstein, while for the states with $\epsilon_i > 0$ the sum can still be legitimately replaced by an integral, the zero-energy state (the single-particle groundstate) must be taken out and handled separately. In fact, the difference – call it N_0 – between the right and left sides of (3), which is proportional to N and for $T < T_{\mathrm{c}}$ is positive, is the number of particles which occupy the groundstate. Thus a single state, in this case the single-particle groundstate, is occupied by a *macroscopic* number of particles – the phenomenon of BEC. Note that for free particles in d dimensions, BEC does not occur for $d \leqslant 2$, since in this case the LHS of (3) is divergent and the equation is trivially satisfied at any nonzero value of T. For a free gas in 3D the condensate fraction is given by the formula

$$N_0(T)/N = 1 - (T/T_{\mathrm{c}})^{3/2} \tag{5}$$

and so tends to 1 as T tends to 0.

Since in real life many-particle systems are rarely noninteracting and in addition may not be in thermal equilibrium, it is desirable to have a more general definition of BEC. Such a definition was formulated by Oliver Penrose (*1929) and Lars Onsager (1903–1976): If we choose any complete ► orthonormal basis (in general time-dependent) of single-particle wave functions $\chi_i(\boldsymbol{r} : t)$, then we can define in this basis the single-particle density matrix $\rho_{ij}(t) \equiv \langle a^\dagger{}_i a_j\rangle(t)$. Since the matrix $\hat{\rho}(t)$ is Hermitian, general theorems guarantee that for any given time t we will be able to find a basis which diagonalizes it, i. e. such that

$$\rho_{ij}(t) = \delta_{ij}\langle n_i\rangle(t) \tag{6}$$

If one and only one[1] of the eigenvalues $\langle n_i\rangle$ (call the relevant value of i 0 by convention) is of order N while all the rest are all of order 1, then we say that the system possesses the property of Bose–Einstein condensation (BEC); the quantity $\langle n_0\rangle$ (often written N_0) is called the "condensate number" (so that N_0/N is the "condensate fraction"), and the associated eigenfunction of $\hat{\rho}(t)$, $\chi_0(\boldsymbol{r})$, is called the "condensate wave function." Note that in the general case both N_0 and $\chi_0(\boldsymbol{r})$ may be functions of time.

---

[1] It is possible, though for various reasons uncommon, for more than one eigenvalue to be of order N. In this case the system is said to possess "fragmented BEC."

There are strong arguments that the occurrence of BEC should lead to the phenomenon of superfluidity (▶ Superfluidity), so that when the latter phenomenon was detected, in 1938, in He-II (the phase of liquid ^{4}He below the so-called lambda-temperature, about 2.17 K), it was almost immediately suggested by Fritz London that BEC is occurring in this phase. This conjecture is now almost universally believed to be correct, and although the strong and mostly repulsive interatomic interactions in liquid helium prevent the direct observation of the onset of BEC which is possible in the alkali gases (see below), it has proved possible (with certain caveats, see e.g. ref. [3]) to observe a nonzero condensate fraction $N_0(T)/N$ by high-energy neutron scattering and other experiments; it increases from zero at the lambda-temperature to about 8% at $T = 0$. (By contrast, the *superfluid* fraction is 100% at $T = 0$). The strong "depletion" of the condensate fraction relative to its value for the free gas is believed to be due to the strong interactions occurring in this high-density system.

A second system in which BEC has been achieved is the bosonic atomic alkali gases[2]. Since (neutral) alkali atoms by definition have an odd number of electrons, odd-A alkali isotopes such as ^{87}Rb, ^{23}Na or ^{7}Li are composed of an even number of fermions and thus behave, as wholes, as bosons; at the densities currently realized the transition temperature T_c to the BEC phase is predicted to be of the order of a microkelvin, a temperature now relatively easily reached by laser cooling and rf evaporation techniques. These gases are normally held in trapping potentials (generated by magnetic fields or lasers) that are harmonic in form, and in such a geometry the effect of the onset of BEC is spectacular: Above T_c the density distribution in the trap is approximately Gaussian, with a large value of the halfwidth. If the atoms were noninteracting, then below T_c a nonzero fraction would occupy the single-particle groundstate of the harmonic potential, which has a very much narrower width. In real life this effect is reduced owing to the repulsive interatomic interactions, but one still sees a sharp "spike" in the density appear below T_c, see e.g. ref. [4]; this is probably the most convincing evidence that BEC is indeed occurring in these systems as theory confidently predicts.

In contrast to liquid helium, the atomic alkali gases are very dilute, and thus the effects of the interatomic interactions are generally rather weak and can be handled by perturbation theory. Thus it has been possible to achieve a very good quantitative understanding of the effects of BEC in these systems.[3]

Primary Literature

1. A. Einstein: On the quantum theory of the perfect gas. Abh. d. Preuss. Akad. Wiss. 1, 3 and 3, 18 (1925).
2. S. N. Bose: Heat-equilibrium in radiation field in presence of matter. Z. Phys. 27, 384 (1924).

[2] BEC has also been realized recently in a few non-alkali bosonic gases.

[3] This material is based upon work supported by the National Science Foundation under Award No. NSF-DMR-03-50842.

3. W. M. Snow, P. E. Sokol: Density and temperature dependence of the momentum distribution in liquid He 4. J. Low Temp. Phys. 101, 881 (1995).
4. M. R. Andrews, M-O. Mewes, N. J. Van Druten, D. S. Durfee, D. M. Kurn, W. Ketterle: Direct, nondestructive observation of a Bose condensate. Science 273, 84 (1996).

Secondary Literature

5. L. P. Pitaevskii, S. Stringari: *Bose–Einstein Condensation*. Oxford University Press, Oxford 2003.
6. A. J. Leggett: *Quantum Liquids: Bose Condensation and Cooper Pairing in Condensed Matter Systems*. Oxford University Press, Oxford 2006.
7. C. J. Pethick, H. Smith: *Bose Einstein Condensation in Dilute Gases*. Cambridge University Press, Cambridge 2003.

Bose–Einstein Statistics

Arianna Borrelli

Bose–Einstein statistics is a procedure for counting the possible states of quantum systems composed of identical particles with integer ▶ spin. It takes its name from Satyendra Nath Bose (1894–1974), the Indian physicist who first proposed it for ▶ light quanta (1924), and Albert Einstein (1879–1955), who extended it to gas molecules (1924, 1925).

Both in classical and in quantum mechanics, the behaviour of systems composed of a large number of particles can be investigated with the help of statistical considerations. If all particles obey the same dynamics, and if their interactions can be neglected in a first approximation, one can determine all possible energy states of a single particle, and then make statistical assumptions on the distribution of the particles among single-particle states, thus computing the average behaviour of the whole system. The usual statistical assumption is that all possible states of the many-particle system (i.e. all configurations) are equally probable. As became clear around the middle of the 1920's, the description of quantum systems of many particles has to be different from that of classical ones, a fact usually described by referring to the ▶ indistinguishability of quantum particles as opposed to the distinguishability of classical ones. Two kinds of ▶ quantum statistics have been found to play a role in quantum mechanics: the statistics of Bose–Einstein and that of ▶ Fermi-Dirac.

Let us consider the classical case first, i.e. a system of N identical, noninteracting particles which are assumed to be distinguishable. The configuration of the system is determined by indicating which particles are in which states, for example particle a in state 1 and particle b in state 2:

particle a	particle b
state 1	state 2

.

Since a and b are distinguishable, this configuration is different from the configuration:

particle a	particle b
state 2	state 1

,

with particle a in state 2 and particle b in state 1.

In quantum statistics, the configurations of the whole system are not described by specifying *which* particles are in which states, but only by saying *how many* particles are in each state. For example:

one particle	one particle
state 1	state 2

,

for a configuration with one particle in state 1 and one in state 2. In the classical case, this description corresponds to two distinct configurations, but in the quantum case there is by definition only one configuration which can be described in this way. This method of counting configurations can be seen as expressing the particles' indistinguishability, although in fact it is the notion of "particle" itself that becomes problematic in quantum statistical systems. Any number of particles following the Bose–Einstein distribution (bosons) can occupy the same state at the same time, while for particles satisfying Fermi-Dirac statistics (fermions) each state can be occupied by at most one particle at a time.

The key difference between Bose–Einstein statistics and the classical way of counting is that a large number of configurations which in the classical case are considered different, in Bose–Einstein statistics count as one. More precisely, when N particles occupy N different single-particle states, all of their $N!$ permutations count as only one configuration. On the other hand, for particles which are in the same state, there is no difference with respect to the classical way of counting: the classical configuration

particle a	particle b
state 1	state 1

with both particles in state 1, counts only once, just like the Bose–Einstein configuration

two particles	no particles
state 1	state 2

.

If, as usually done, it is assumed that all configurations of the many-particle system are equally probable, it follows that, for Bose–Einstein particles, the statistical weight of configurations in which many particles are in the same state is enhanced with respect to the classical case. In other words, it is more probable to find two or more bosons in the same single-particle state than it is the case for classical particles. Because of this, bosons cannot be considered statistically independent from each other even when they are not interacting.

In the limit of high temperatures, i.e. for high average energies, an increasing number of energy states becomes accessible to the particles, and the number of configurations with two or more of them in the same state eventually becomes negligible. The overall effect of Bose–Einstein statistics is then simply a reduction of the statistical weight of any configuration by a factor $N!$ with respect to the classical case. In the low-temperature limit, instead, the number of configurations with two or more particles occupying the same state is not negligible, and those configurations are privileged: at low temperature, a boson has a greater probability than a classical particle of occupying the ground state. Under specific conditions, the formalism predicts the phenomenon of ► Bose–Einstein condensation.

Bose–Einstein statistics had its origin in Max Planck's (1858–1947) formula for the energy density u_ν of ► black-body radiation (1900) of frequency ν at thermal equilibrium at temperature T. To justify his formula, Planck considered the energy density u_ν as associated to N_ν oscillators of average energy $U(\nu, T)$, with

$$u_\nu = \frac{8\pi\nu^2}{c^3} U(\nu, T).$$

This relation was derived from classical electrodynamics. He then assumed that the radiant energy was distributed among the N_ν oscillators in form of P energy elements of value $h\nu$. The configurations of the system were described by giving only the total number of energy elements in each oscillator, without considering the possibility of permuting the energy elements: this method of counting corresponded to what would later be called Bose–Einstein statistics. However, Planck did not regard the energy elements as particles, but only as a computational device whose physical significance remained to be determined.

In the following years, Planck's formula and its possible relationship to Albert Einstein's hypothesis of a ► light quantum (1905) were discussed by a number of authors, whose views have been discussed by Silvio Bergia [10]. In 1911, the Polish physicist Władysław Natanson (1864–1937) noted that Planck's counting method implied the indistinguishability of the energy elements and the distinguishability of the oscillators [1]. The correctness of this assumption, Natanson remarked, was supported only by the agreement of Planck's formula with experiments. In 1914, Paul Ehrenfest (1880–1933) and Heike Kamerlingh–Onnes (1853–1926) underscored that Planck's energy elements were not statistically independent from each other and therefore, in their opinion, could not be regarded as real, independent particles [2].

In 1923, Einstein's light quantum hypothesis was vindicated by the ► Compton experiment. In 1924, Bose, at the time working at Dacca University, showed how Planck's formula could be derived without recourse to classical electrodynamics, but instead assuming the existence of massless light quanta whose position and momentum were quantized by dividing phase-space into cells of volume $(h)^3$ [3, 4]. As in the case of Planck's energy elements and oscillators, Bose's light quanta were distributed among the phase-space cells by specifying only the number of quanta in a cell, without considering permutations. A factor 2 took into account the two possible states of polarisation of light so that, in the end, Planck's radiation formula was recovered. In conclusion, Bose derived Planck's formula by assuming that light quanta existed and satisfied a new kind of statistics. He developed his theory in two papers written in English which he sent to Einstein, whom he did not know, asking for help for the publication in a German journal. Einstein, recognizing the importance of Bose's contribution, translated the papers into German, had them published (1924) and wrote two papers of his own (1924, 1925) in which he extended Bose's statistics to an ideal gas of molecules, making explicit a number of implicit features of the theory [5, 6]. However, it remained open to discussion whether the new statistics would be applicable to particles different from light quanta.

In 1926, after the formulaton of Erwin Schrödinger's (1887–1961) ► wave mechanics, Bose–Einstein statistics was linked to the behaviour of many-particle ► wave functions. This result was obtained by Werner Heisenberg (1901–1976) and, somewhat later but independently, by Paul Dirac (1902–1984). Consider a wave function $\psi(x_1, x_2, \ldots, x_i, \ldots)$ which is a solution of ► Schrödinger's equation for a system of N particles satisfying the same dynamics, with x_i representing the set of coordinates of the i-th particle. A generic ψ will not remain unchanged under a permutation of the indexes i, but, because of the ► identity of the particles, the permuted function shall be a solution of the equation of motion as well. If, following the model of Bose–Einstein statistics, one imposes on the wave function the additional requirement that a permutation of the particles should not change the configuration of the system, it follows that the only physically acceptable ψ's are those which, under a permutation of the indexes i, either remain unchanged (symmetrical wave functions) or change sign (antisymmetrical wave functions). The indeterminacy of the sign derives from the fact that only $|\ \psi\ |^2$ is physically significant.

As both Heisenberg and Dirac noted, the choice of symmetrical wave functions implied the same shift in statistical weights as the one brought about by Bose–Einstein statistics. Choosing antisymmetrical wave functions instead resulted in a system obeying Pauli's ► exclusion principle and satisfying Fermi–Dirac statistics. After initial discussions as to whether particles of matter would obey Bose–Einstein or Fermi-Dirac statistics, it eventually became clear that both alternatives are realised in nature, depending on the spin of the particles: particles with zero or integer spin satisfy Bose–Einstein statistics, while particles of half-integer spin obey Fermi-Dirac statistics (► spin-statistics theorem).

Primary Literature

1. W. Natanson: Über die statistische Theorie der Strahlung. Physikalische Zeitschrift **12**, 659–666 (1911)
2. P. Ehrenfest, H. Kamerlingh-Onnes: Vereenvoudige afleiding van de formule uit combinatieleer, welke Planck aan zijne theorie der strahling ten grondlaag heefd gelegt. Verslag van de gewone Vergadering der wis- en natuurkundige Afdeeling, Koniklijke Akademie van Wetenschappen te Amsterdam **23**, 789–792 (1914)
3. S. N. Bose: Plancks Gesetz und Lichtquantenhypothese. Zeitschrift für Physik **26**, 178–181 (1924a)
4. S. N. Bose: Wärmegleichgewicht und Strahlungsfeld bei Anwesenheit von Materie. Zeitschrift für Physik **27**, 384–393 (1924b)

5. A. Einstein: Quantentheorie des einatomigen idealen Gases. Sitzungsberichte der Preussischen Akademie der Wissenschaften (Berlin). Physikalisch-mathematische Klasse, 261–267 (1924)
6. A. Einstein: Quantentheorie des einatomigen idealen Gases. Zweite Abhandlung. Sitzungsberichte der Preussischen Akademie der Wissenschaften (Berlin). Physikalisch-mathematische Klasse, 3–14 (1925)
7. K. Huang: *Statistical mechanics* (Wiley, New York 1963, 183–205, 253–273, 439–454)

Secondary Literature

8. J. Mehra, H. Rechenberg: *The Historical Development of Quantum Theory 1,2* (Springer, New York 1982, 554–578)
9. F. Hund: *Geschichte der Quantentheorie* (Bibliographisches Institut, Zürich 1984, 152, 163–174)
10. S. Bergia: Who discovered Bose–Einstein statistics? in: M.G. Doncel et al. (eds.), *Symmetries in Physics (1600–1980)*(Servei de Publications Universitat Autònoma de Barcelona, Barcelona 1987, 221–245)
11. O. Darrigol: Statistics and combinatorics in early quantum theory II: Early symptoms of indistinguishability and holism. Historical Studies in the Physical Sciences **21**, 237–298 (1991)
12. J. Stachel: Einstein and Bose. in: J. Stachel, *Einstein from 'B' to 'Z'* (Birkhäuser, Boston 2002, 519–538)

Bremsstrahlung

Bruce R. Wheaton

All charged particles emit radiation when accelerated. Indeed on the Maxwell view, that radiation (which takes energy from the particle) is the "wake" left by that acceleration in an æther of crossed electric and magnetic fields, and the concept underlies Hertz' corroboration of Maxwell in 1888. So when ► cathode-rays were about 1900 identified by most physicists as streams of ► electrons, their impact on the anti-cathode in Röntgen's vacuum tube should produce an irregular sequence of dislocated electromagnetic impulses due to the electrons' deceleration. This is "braking," hence the term Arnold Sommerfeld (1868–1951) coined in 1909 of *Bremsstrahlung*.

Wilhelm Conrad Röntgen (1845–1923) had thought in 1895 he had found the elusive longitudinal e-m wave in his discovery of ► x-rays. But Sommerfeld in 1899 found two species in the new radiation: at the low-energy end periodic waves like ultra-violet light, at the high end a broad spectrum to be expected from discontinuous impulses dissected by Fourier frequency expansion. This distinction was reinforced by Charles Barkla (1877–1944) in 1907: superimposed on the spectrally-spread out x-radiation from electron impacts (Bremsstrahlung) was a series of sharp strong peaks characteristic of the anti-cathode metal (*fluorescent x-rays*) that Barkla

B

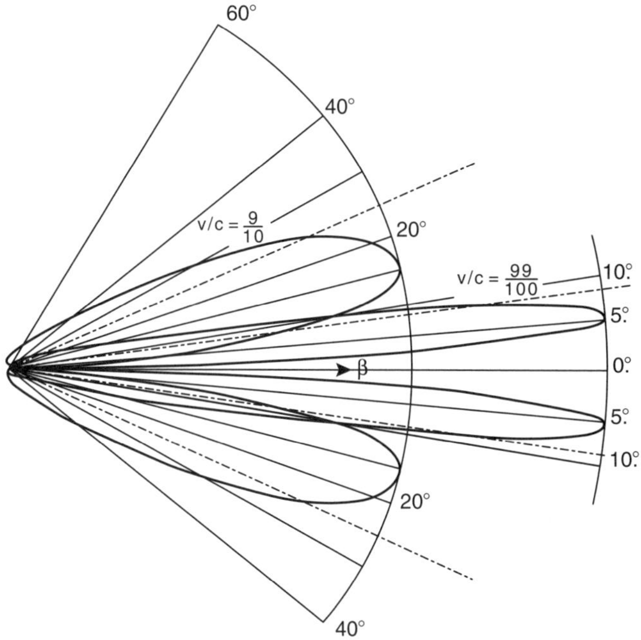

Fig. 1 Sommerfeld's calculated distribution of γ-ray intensity as a function of azimuthal angle. [München *Sb, 41* (1911), 11.] (The two cases for v/c are not to the same scale. Were they, the case for .99c would extend down the hall to your right, a thousand times the other)

showed were polarized. Sommerfeld returned to the issue in 1911 with a non-relativistic analysis of γ-rays (Bremsstrahlung from exiting β-electrons) to show their energy is emitted markedly in the forward direction like "directed radiation," or "needle radiation," see Fig. 1. Niels Bohr had to contend with Bremsstrahlung as fundamental evidence for his atom in 1913, although Joseph Larmor (1857–1942) [1] and J. J. Thomson (1856–1940) [3] had defused the notion of the ► Bohr atom necessarily destroying itself by radiation from orbiting electrons.

With the integration of quantum mechanics in the mid-1920s, and with emerging recognition of the distinction between atomic and nuclear phenomena, came a new understanding of the essential nature of Bremsstrahlung in investigating the nucleus. In particular Dirac's ► relativistic quantum mechanics (1928) predicted *positive* electrons; so the passage of high-energy ($>800\,\mathrm{MeV}/Z$) electrons through matter (of atomic mass Z) can emit photons (► light quantum) of sufficient energy to decay into an e^-e^+ pair, leading to more Bremsstrahlung from the products, resulting in a succession of pairs decreasing in energy, as had been seen in cosmic ray showers using Wilson's (1911) cloud chamber.

When you accelerate charged particles in a cyclotron (1932+) they also radiate and lose energy. This is a particular problem for electrons in a synchrotron, since they have large charge and little mass ($E_r \propto a^2/M^2$), requiring regions in the machine where they can regain energy lost at each turn in order to keep the beam together. This puts constraints on, *i.e.*, storage-rings. In extremely high energy (15 GeV) collisions of e^+e^-, Bremsstrahlung takes the form of hadron jets able to traverse 30 m of air, the least energetic of which can be explained by a quark (▶ QCD) emitting a field particle (gluon); or in the case of neutron scattering by emission of neutrinos.

Perhaps the most pregnant analyses of Bremsstrahlung also came with the accelerator. An accelerated beam of electrons or deuterons that passes through a dense medium might do so with a velocity exceeding the velocity of light in that medium. Its Bremsstrahlung then consists of shock waves, similar to the sonic boom from an airplane traveling above Mach 1. These are constructed periodic wave-phenomena that interact with matter as do particles and were discussed by Cherenkov [10] in 1934. They echo the speculations of Huygens from the seventeenth century about light and of early (1900) views of ▶ x-rays. Here may indeed lie more detailed understanding of ▶ wave-particle duality.

Primary Literature

1. Larmor, J., On the theory of the magnetic influence on spectra; and on the radiation from moving ions, Phil Mag, *44* (1897), 503–12.
2. Sommerfeld, A., Theoretisches über die Beugung der Röntgenstrahlen, Phys Zeit, *1* (1899), 105–11; 2 (1900), 55–60.
3. Thomson, J. J., The magnetic properties of systems of corpuscules describing circular orbits, Phil Mag, *6* (1903), 673–93.
4. Sommerfeld, A., Über die Verteilung der Intensität bei der Emission von Röntgenstrahlen, Phys Zeit, *10* (1909), 969–76.
5. Bohr, N., On the decrease in velocity of swiftly moving electrified particles on passing through matter, Phil Mag, *30* (1915), 581–612.
6. Sugiura, Y., On the transition probability between two states with positive energy in a central field, Phys Rev, *34* (1929), 858–75.
7. Oppenheimer, J. R. & M. S. Plesset, On the production of the positive electron, Phys Rev, *44* (1932), 53–5.
8. Heisenberg, W., Theoretische Überlegungen zur Höhenstrahlung, Ann Phys, *13* (1932), 430–52.
9. Heitler, W. Über die bei sehr schnellen Stöße emitterte Strahlung, ZfP, *84* (1933), 145–67.
10. Cherenkov, P., Vidimoye svyechyeniye chistukh zhidkostyei pod dyeeistviyem γ-radiatsiy, Akad. Nauk. C. C. C. P., Comptes rendus (Doklady), *2* (1934), 451–7, which includes a German translation (without figures.)
11. Frank, I. & I. Tamm, Coherent visible radiation of fast electrons passing through matter, Akad. Nauk. C. C. C. P., Comptes rendus (Doklady), *14* (1937), 109–14.

Secondary Literature

12. Dirac, P. A. M., Principles of quantum mechanics (Oxford: Clarendon, 1930)
13. Sommerfeld, A., Über die Beugung und Bremsung der Elektronen, Ann Phys, *403* (1931), 257–330, virtually a book in itself.
14. Sauter, F., Zur unrelativistischen Theorie des kontinuerlichen Röntgenspektrums, Ann Phys, *18* (1933), 486–96; Über die Bremsstrahlung schneller Elektronen, *20* (1934), 404–12.

15. Bethe, H. A. & W. Heitler, On the stopping of fast particles and on the creation of positive electrons, RSL, Proc., A, *146* (1934), 83–112.
Bethe, H. A. The influence of screening on the creation and stopping of electrons, Camb Phil Soc Proc *30* (1934), 524–39.
16. Heitler, W., The quantum theory of radiation (Oxford: Oxford University Press, 1936, § 17, 161–77).
17. Bethe, H. A. & E. E. Salpeter, Quantum mechanics of one and two systems, in Handbuch der Physik, *35* (*Atoms I*) (Berlin: Springer, 1957, 409–30).
18. Heisenberg, W., Kosmische Strahlung Vorträge... (Berlin: Springer, 1953)
19. Blumenthal, G. & R. Gould, Bremsstrahlung, synchrotron radiation, and Compton scattering of high-energy electrons traversing dilute gases, Rev Mod Phys, *42* (1970), 237–70.
20. Wheaton, B. R., The tiger and the shark: Empirical roots of wave-particle dualism (Cambridge: Cambridge University Press, 1983 parts I & II, 71–195).
21. Schweber, S., QED and the men who made it... (Princeton: Princeton University Press, 1994)
22. Wheaton, B., Inventory of sources for history of twentieth-century physics (ISHTCP): the research database, Physis, *36* (1999), 225–31.

B

Brownian Motion

Charlotte Bigg

Brownian motion is the irregular and perpetual agitation of small particles suspended in a liquid or gas. In 1828 the Scottish botanist Robert Brown (1773–1858) published the first extensive study of the phenomenon. Brown showed notably that this motion equally affects organic and inorganic particles, suggesting a physical rather than a biological explanation [1]. Developments in thermodynamics and the kinetic theory in the second half of the nineteenth century led several scientists to consider Brownian motion as a visible consequence of thermal molecular agitation; but it was not until the early twentieth century that a convincing quantitative description and theoretical explanation of the motion was worked out.

In particular A. Einstein (1879–1955), M. von Smoluchowski (1872–1917) and J. Perrin (1870–1942) demonstrated that the Brownian motion of particles suspended in a liquid is caused by their incessant collisions with the molecules making up the liquid, and they developed new, statistical methods of measuring this motion. Instead for instance of measuring the instantaneous velocity of individual particles, as scientists had previously, finding values widely diverging from those predicted by the kinetic theory, Einstein proposed in 1905 to measure their mean displacement. He found that the mean displacement of a particle on the X axis during a period of time t is proportional to the square root of t:

$$\lambda_x = \sqrt{t} \times \sqrt{\frac{RT}{N}\frac{1}{3\pi k P}}.$$

R is the gas constant, T the absolute temperature, N the number of molecules in a mole (Avogadro's number), k the viscosity of the fluid, and P the radius of the particle. The mean displacement for a given period of time can be thus be calculated when R, N, T, k, and P are known; conversely N or P can be obtained when mean displacement and other factors are known [2].

In a series of experiments on colloidal suspensions that involved careful measurement of the diameter, density and displacement of particles, Perrin supplied evidence in support of this approach (see Fig. 1), and he demonstrated the broad agreement of experimental determinations of Avogadro's number made by himself and others on the basis of a wide range of phenomena [3,4].

Beyond the elucidation of the origin of Brownian motion, the significance of these investigations is twofold. First, they helped clarify two major scientific and epistemological issues of late nineteenth century physical science, about the atomic hypothesis and the relationship between mechanics and thermodynamics. In the introduction to his 1905 paper on Brownian motion, Einstein stated

"In this paper it will be shown that according to the molecular-kinetic theory of heat, bodies of microscopically-visible size suspended in a liquid will perform movements of such magnitudes that they can be easily observed in a microscope, on account of the molecular theory of heat. [...]

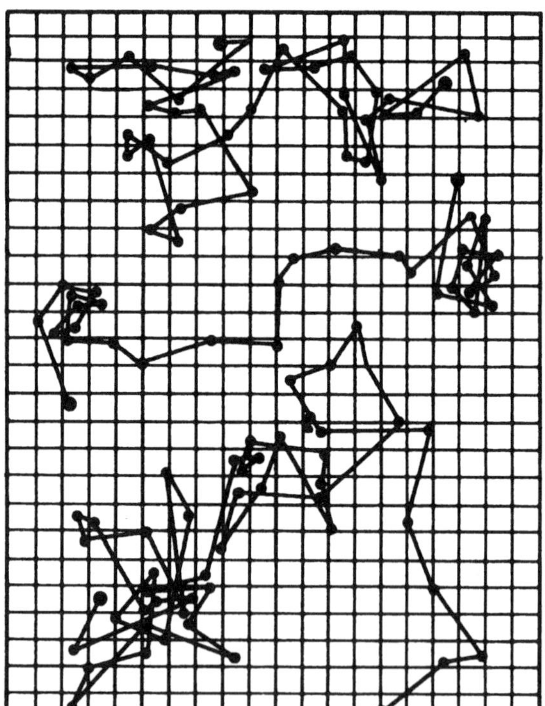

Fig. 1 Measuring the displacement of invidual particles: "three drawings obtained by tracing lines to link the consecutive positions of the same grain of rubber at intervals of 30s" [3,81]

If the movement discussed here can actually be observed (together with the laws relating to it that one would expect to find), then classical thermodynamics can no longer be looked upon as applicable with precision to bodies even of dimensions distinguishable in a microscope; an exact determination of actual atomic dimensions is then possible. On the other hand, had the prediction of this movement proved to be incorrect, a weighty argument would be provided against the molecular-kinetic conception of heat" [2].

Einstein and others' investigations of Brownian motion provided conclusive evidence in favour of the kinetic theory of heat and the existence of atoms, as well as of the statistical nature of the second law of thermodynamics. Perrin was awarded the Nobel Prize in Physics in 1926 for having "put a definite end to the long struggle regarding the real existence of molecules." Secondly, this work announced and prepared the emergence of new fields of investigation in twentieth century physical science: statistical thermodynamics, the study of fluctuation phenomena, and the general theory of stochastic processes, of which Brownian motion continues to constitute the archetypal example.

In the history and philosophy of science, the history of research on Brownian motion is frequently cited as a perfect example of "the failure of experiment and observation, unguided (until 1905) by theory, to unearth the simple laws governing a phenomenon." [6]

Primary Literature

1. R. Brown: A brief account of microscopical observations made in the months of June, July, and August 1827, on the particles contained in the pollen of plants; and on the general existence of active molecules in organic and inorganic bodies. Edinburgh New Philosophical Journal **5**, 358–371 (1828)
2. A. Einstein: Über die von der molekularkinetischen Theorie der Wärme geforderte Bewegung von in ruhenden Flüssigkeiten suspendierten Teilchen. Annalen der Physik **17**, 549–560 *(1905)*. This and subsequent articles by A. Einstein on Brownian motion were reprinted in: A. Einstein: *Untersuchungen über die Theorie der "Brownschen Bewegung" (1905)*. Ed. R. Fürth (Akademische Verlagsgesellschaft, Leipzig 1922) (=Oswalds Klassiker der exacten Wissenschaften, 199) [Engl. *Investigations on the Theory of the Brownian Movement*. Trans. A.D. Cooper (Methuen, London 1926)]
3. J. Perrin: Mouvement brownien et réalité moléculaire, Annales de Chimie et de Physique **18**, 1–114 (1909) [Trans. F. Soddy repr. in *The question of the Atom*. Ed. M.J. Nye (Tomash Publishers, Los Angeles, San Francisco 1986, 505–601]
4. J. Perrin: *Les Atomes* (Alcan Paris, 1913) [engl. (4th edition) *Atoms*. trans. D.LL. Hammick (Constable & Company, London 1916)]

Secondary Literature

5. J. Stachel ed.: *The collected papers of Albert Einstein*, vol. 2: *The Swiss Years: Writings 1900–1909* (Princeton University Press, Princeton 1989, esp. 206–222)
6. S. G. Brush: A History of Random Processes. I. Brownian Movement from Brown to Perrin. Archive for the History of the Exact Sciences **5**, 1–36 (1968–9)

7. M. J. Nye: *Molecular Reality* (Elsevier Publishing Company, Amsterdam 1972)
8. G. Pohl: The theory of Brownian motion – one hundred years old. In: *The Global and the Local*: The History of Science and the Cultural Integration of Europe, ed. by M. Kokowski (Cracow 2007, 419–424). Online as: http://www.2iceshs.cyfronet.pl/proceedings.html

Bub–Clifton Theorem

Jeffrey Bub

The two fundamental 'no go' theorems for hidden variable reconstructions of the ► quantum statistics, the ► Kochen-Specker theorem [4] and ► Bell's theorem [1], can be formulated as results about the impossibility of associating a classical probability space $(X, \mathfrak{F}, P_\rho)$ with a quantum system in the state ρ, when certain constraints are placed on the probability measure P_ρ. The Bub–Clifton theorem [2, 3], by contrast, is a 'go' theorem: a positive result about the possibility of associating a classical probability space with a quantum system in a given state.

If P_ρ is required to satisfy the conditions:

(a) $P_\rho(a, b, \ldots | A, B, \ldots)$ is a classical probability measure defined for all eigenvalues $a, b, \ldots$ of the ► observables $A, B, \ldots$ in some set of observables $\mathcal{E}$.
(b) If $A, A', \ldots \in \mathcal{E}$ commute, then $P_\rho(a, a', \ldots | A, A', \ldots)$ coincides with the quantum mechanical probability assigned by ρ.

then the existence of P_ρ is equivalent to the requirement that the set of numbers:

$$\{P_\rho(a, a', \ldots | A, A', \ldots); A, A' \in \mathcal{E} \text{ commute}\}$$

should satisfy a finite family of inequalities (Boole's 'conditions of possible experience'), so the non-existence of P_ρ entails a violation of at least one inequality (see Pitowsky [6, 7]). If P_ρ exists, then it is a weighted average of pure states (characteristic functions onto 1-element subsets of X or 2-valued (0,1) probability measures).

The Kochen-Specker and Bell theorems can be formulated (following Pitowsky) as follows:

The Kochen-Specker Theorem. *There is a set of observables $\mathcal{E}$ such that for all ρ the classical probability measure P_ρ does not exist.*

Bell's Theorem. *There is a set of local observables $\mathcal{E}$ on $\mathcal{H} \otimes \mathcal{H}$ and a state $\rho \in \mathcal{H} \otimes \mathcal{H}$ such that the classical probability measure P_ρ does not exist.*

The Bub–Clifton is the positive result:

The Bub–Clifton Theorem. *For every pure state $\rho = |\psi\rangle\langle\psi|$ and every observable R, there is a maximal extension $\mathcal{E}$ of $\{R\}$ for which there exists a classical probability measure P_ρ. The extension $\mathcal{E}$ is unique if we require invariance with respect to automorphisms of the subspace structure of $\mathcal{H}$ (the projective geometry of $\mathcal{H}$) that preserve ρ and R.*

The pure state ρ can be expressed as a linear ► superposition of orthogonal 1-dimensional projection operators (► projection) ρ_r onto the non-null eigenspaces $\{V_r\}$ of R: $\rho = \bigvee_r \rho_r = \sum_r \rho_r$. The theorem shows that the set of observables $\mathcal{E}$ contains all the maximal observables whose spectral measures comprise:

(i) The 1-dimensional projection operators ρ_r,
(ii) The 1-dimensional projection operators onto any orthogonal basis in the orthocomplement of the subspace spanned by the projections ρ_r, i.e., the 'null space' V_{null} that is the range of the projection operator $I - \sum_r \rho_r$,

and all the non-maximal observables which are functions of these maximal observables.

Equivalently, $\mathcal{E}$ consists of all the observables whose eigenspaces are spanned by the rays defined by (i) and (ii) above.

According to the theorem, even though the set $\mathcal{E}$ contains non-commuting observables, there exists a classical probability measure P_ρ for the observables in $\mathcal{E}$, i.e., a measure space $(X, \mathfrak{F}, P_\rho)$, where the elements of the space X are the projection operators ρ_r, which are in 1-1 correspondence with the 2-valued homomorphisms—representing bivalent truth-value assignments—on the lattice of subspaces generated by the 1-dimensional projectors in (i) and (ii) above, and hence in 1-1 correspondence with the 2-valued homomorphisms on the ranges of values of the observables in $\mathcal{E}$.

Nakayama [5] has constructed a topos-theoretic extension of the theorem.

A quantum measurement interaction can be represented schematically as follows:

$$|s\rangle|r\rangle \xrightarrow{U(t)} \sum_i c_i|s_i\rangle|r_i\rangle$$

where $|s\rangle = \sum_i c_i|s_i\rangle$ is the initial state of the measured system expressed as a linear superposition of the eigenstates $|s_i\rangle$ of the measured observable S, $|r\rangle$ is the initial state of the measuring instrument with indicator or 'pointer' observable R, and $U(t)$ is the unitary transformation implementing the measurement interaction between the system and the measuring instrument that sets up a correlation between eigenvalues of S and pointer positions. (Note that for the systems we use as measuring instruments, the pointer observable R commutes with the instrument-environment interaction Hamiltonian, so the correlation between eigenvalues of S and pointer positions R induced by the system-instrument Hamiltonian is preserved under the instrument-environment interaction.) If we take the pointer observable R as 'preferred,' in the sense that it always has a definite (determinate) value, then the set of definite-valued observables $\mathcal{E}$ for the state $|\psi\rangle = \sum_i c_r |s_i\rangle|r_i\rangle$ after the measurement interaction includes the observables whose spectral measures contain the projection operators onto the states $|\psi_i\rangle = |s_i\rangle|r_i\rangle$. It follows that $\mathcal{E}$ contains the measured observable S and the pointer observable R. For this state $\rho = |\psi\rangle\langle\psi|$, there exists a classical measure space $(X, \mathfrak{F}, P_\rho)$, where the elements of X are the projection operators $\rho_i = |\psi_i\rangle\langle\psi_i|$, in 1-1 correspondence with the 2-valued homomorphisms on the ranges of values of the observables in $\mathcal{E}$. So the elements of X can be identified with the alternative possible states of affairs that are the outcomes of the quantum measurement process.

This observation underlies the demonstration in [2, 3] that various 'no collapse' interpretations, including Bohr's ► complementarity principle interpretation, ► modal interpretations, and ► Bohm's hidden variable theory, can all be represented as 'preferred observable' interpretations, for different choices of the preferred observable (e.g., in the case of Bohm's theory, the preferred observable is position in the configuration space of all the Bohmian particles).

Literature

1. J. S. Bell: *On the Einstein-Podolsky-Rosen Paradox.* Physics. 1:195–200, 1964. Reprinted in John Stuart Bell, *Speakable and Unspeakable in Quantum Mechanics*, Cambridge University Press, Cambridge, 1989.
2. J. Bub: *Interpreting the Quantum World.* Cambridge University Press, Cambridge, 1997.
3. J. Bub, R. Clifton: *A uniqueness theorem for no collapse interpretations of quantum mechanics.* Studies in the History and Philosophy of Modern Physics. 27:181–219, 1996.
4. S. Kochen, E. P. Specker: *On the problem of hidden variables in quantum mechanics.* Journal of Mathematics and Mechanics. 17:59–87, 1967.
5. K. Nakayama: *Topos-theoretic extension of a modal interpretation of quantum mechanics.* arXiv e-print quant-ph/0711.2200, 2007.
6. I. Pitowsky: *Quantum Probability, Quantum Logic.* Springer, Berlin, 1989.
7. I. Pitowsky: *George Boole's 'conditions of possible experience' and the quantum puzzle.* British Journal for the Philosophy of Science. 45:95–125, 1994.

C

Casimir Effect

Peter W. Milonni and Umar Mohideen

The Casimir effect is a force associated with the ► zero-point energy of a field. The effect originally considered by Hendrik B. G. Casimir (1909–2000) is the attraction between two uncharged, perfectly conducting plates (Fig. 1). According to quantum theory, there is energy in the electromagnetic field even at the absolute zero of temperature. For a field of frequency ν, this energy is $\frac{1}{2}h\nu$, identical to the zero-point energy of a harmonic oscillator having the same frequency. The total zero-point energy is then $\frac{1}{2}h$ times the sum over all the field frequencies, these being determined by Maxwell's equations and the boundary conditions. In the example of Fig. 1, Maxwell's equations allow field modes of arbitrarily large frequency both between the plates and outside them, and therefore the zero-point field energy is infinite when the plates are separated by a finite distance d as well as when they are infinitely far apart. However, the difference in zero-point energy for the two cases is finite, and its dependence on the plate separation d implies a force $F = -\pi hc/480d^4$ per unit area.

The force between conducting plates is the most widely cited Casimir effect, but such effects can be derived – usually with considerable difficulty – for more complicated geometries as well as for dielectric media, and more generally they appear whenever topological constraints are imposed on quantum fields. Because of their close association with zero-point energy in empty space, Casimir effects are often cited as evidence of the nontrivial nature of the vacuum in quantum field theory.

Casimir effects are generally rather weak. However, due to its inverse fourth-order distance dependence it is a dominant effect at the nanometer scale and impacts experimental searches for extra dimensions, new forces outside the standard model and the design of micromachines. The first experimental searches for the Casimir effect were constrained by the available technology and understanding of systematic errors. Sparnaay, and later Overbeek and von Blokland, qualitatively showed the attractive Casimir force using a spring balance technique but they were limited due to large experimental errors. Experimental progress accelerated in 1997 with Lamoreaux's demonstration of the Casimir effect using the torsion pendulum. Increasing precision has been demonstrated with techniques using the Atomic Force Microscope and microelectromechanical oscillators. Presently precision of the order of a percent has been reported, restricted by both theoretical and experimental uncertainties. Experiments with simple periodic non-planar surfaces have also been reported. The extraordinary theoretical and experimental activity of the last few years should lead to measurements of increased precision and demonstrations of some of the fascinating nontrivial geometry dependences of the Casimir force.

D. Greenberger et al. (eds.), *Compendium of Quantum Physics: Concepts, Experiments, History and Philosophy*, © Springer-Verlag Berlin Heidelberg 2009

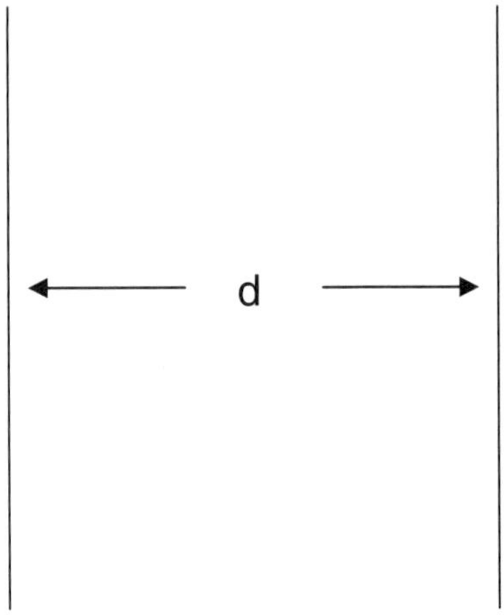

Fig. 1 Two parallel, perfectly conducting plates experience an attractive Casimir force

Primary Literature

1. Casimir, H. B. G.: *On the attraction between two perfectly conducting plates.* Proc. K. Ned. Akad. Wet. 51, 793 (1948)

Secondary Literature

2. Sparnaay, M. J.: *Measurement of attractive forces between flat plates.* Physica 24, 751 (1958)
3. Van Blokland, P. H. G. M., J. T. G. Overbeek: *Van der Waals forces between objects covered with a chromium layer.* J. Chem. Soc. Faraday Trans. 74, 2637 (1978)
4. Milonni, P. W.: *The Quantum Vacuum: An Introduction to Quantum Electrodynamics.* San Diego: Academic (1994)
5. Mostepanenko, V. M., N. N. Trunov: *The Casimir Effect and its Applications.* Oxford: Clarendon Press (1997)
6. Lamoreaux, S. K.: *Demonstration of the Casimir force in the 0.6 to 6 mm range.* Phys. Rev. Lett. 78, 5 (1997)
7. Bordag, M., U. Mohideen, V. M. Mostepanenko: *New developments in the Casimir effect.* Phys. Rep. 353, 1 (2001)
8. Milton, K. A.: *The Casimir Effect. Physical Manifestations of Zero-Point Energy.* Singapore: World Scientific (2001)
9. Chan, H. B., V. A. Aksyuk, R. N. Kleiman, D. J. Bishop, F. Capasso: *Quantum mechanical actuation of microelectromechanical systems by the Casimir force.* Science 291, 1941 (2001)
10. Chen, F., U. Mohideen, G. L. Klimchitskaya, V. M. Mostepanenko: *Demonstration of the lateral Casimir force.* Phys. Rev. Lett. 88, 101801–1 (2002)
11. Lamoreaux, S. K.: *The Casimir force: background, experiments, applications.* Rep. Prog. Phys. 68, 201 (2005)

Cathode Rays

Theodore Arabatzis

C

The detection of cathode rays was a by-product of the investigation of the discharge of electricity through rarefied gases. The latter phenomenon had been studied since the early eighteenth century. By the middle of the nineteenth century it was known that the passage of electricity through a partly evacuated tube produced a glow in the gas, whose color depended on its chemical composition and its pressure. Below a certain pressure the glow assumed a stratified pattern of bright and dark bands.

During the second half of the nineteenth century the discharge of electricity through gases became a topic of intense exploratory experimentation, primarily in Germany [21]. In 1855 the German instrument maker Heinrich Geißler (1815–1879) manufactured improved vacuum tubes, which made possible the isolation and investigation of cathode rays [23]. In 1857 Geissler's tubes were employed by Julius Plücker (1801–1868) to study the influence of a magnet on the electrical discharge. He observed various complex and striking phenomena associated with the discharge. Among those phenomena were a "light which appears about the negative electrode" and a fluorescence in the glass of the tube ([9], pp. 122, 130).

The understanding of those phenomena was advanced by Plücker's student and collaborator, Johann Wilhelm Hittorf (1824–1914), who observed that "if any object is interposed in the space filled with glow-light [emanating from the negative electrode], it throws a sharp shadow on the fluorescent side" ([5], p. 117). This effect implied that the "rays" emanating from the cathode followed a straight path. Furthermore, Hittorf showed that those rays could be deflected by the action of a magnet. In 1876 they were dubbed cathode rays (Kathodenstrahlen) by Eugen Goldstein (1850–1930) [2, 24]. Thus, by the late 1870s cathode rays had been identified and some of their main observable properties had been established.

The nature of cathode rays remained a controversial subject for some years to come. There were two opposing views concerning their constitution. The first view was maintained by British and French scientists, who identified cathode rays with streams of charged particles. A well-known advocate of that view was the British experimentalist William Crookes (1832–1919). Crookes studied electrical discharges through highly rarefied gases: "[T]he exhaustion carried out [is so high] that the dark space around the negative pole ... entirely fills the tube." ([1], p. 6) Under those conditions the behavior of cathode rays could be studied in isolation, without interference from other discharge phenomena. Thus, Crookes determined, in a particularly clear manner, several properties of cathode rays: their "power of exciting phosphorescence" (p. 7), their propagation in straight lines (p. 12), their power to cast shadows (p. 15), their capacity to "exert strong mechanical action where they strike" (p. 17) and to "produce heat when their motion is arrested" (p. 24), and their deflection by a magnet (p. 20). He put forward the hypothesis that cathode rays were charged molecules, "molecular bullets", which he justified on the basis of their magnetic deflection and their capacity to perform mechanical work. Furthermore, from the direction of their magnetic deflection he inferred that they were negatively charged. Several years later, in 1895, Jean B. Perrin (1870–1942) would arrive at the same conclusion by means of a different experiment [8].

Another eminent scientist who defended the particulate interpretation of cathode rays was Arthur Schuster (1851–1934). In 1884 he suggested that they were negatively charged atoms [10]. In 1890 he calculated the upper and lower bounds of their charge to mass ratio (e/m), based on measurements of their magnetic deflection and an estimate of their velocity. The lower limit was close to the charge to mass ratio of electrolytic ions. The upper limit was three orders of magnitude higher ([11], pp. 546–547).

The second view concerning the nature of cathode rays was advocated by some German physicists, who identified them with processes in the ether. Their main argument was that cathode rays have some of the properties of light-waves. For instance, they both travel in straight lines and produce fluorescence. The ethereal interpretation of cathode rays received additional support in 1883, when Heinrich Hertz (1857–1894) failed to deflect them by an electric field [3,22]. In the following years, new experimental facts were discovered which seemed to undermine further the interpretation of cathode rays as charged particles. In 1892 Hertz showed that they could penetrate thin sheets of metal (e.g., gold, silver, aluminum) [4]. In 1893 his student, Philipp Lenard (1862–1947), built upon Hertz's work to investigate the behavior of cathode rays outside the vacuum tube. He devised a tube with a thin metallic "window" facing the cathode. The cathode rays passed through that window and, thus, Lenard could measure their mean free path outside the tube. As it turned out, it was much longer than that of atoms and molecules. Furthermore, he showed that their absorption depended only on the density of the absorbing substance [7].

Thus, different experimental results supported different accounts of the nature of cathode rays. Furthermore, the evidential import of some of those results was ambiguous. On the one hand, the magnetic deflection of cathode rays, which indicated that they were charged particles, was compatible with an ethereal interpretation of their nature. It was conceivable that the magnetic field altered the state of the ether so as to produce a deflection of the rays ([17], p. 285). On the other hand, the capacity of cathode rays to pass through thin metallic sheets, which suggested that they were waves in the ether, could be accommodated by the hypothesis that cathode rays were charged particles. In 1893 J. J. Thomson (1856–1940) argued that the capacity in question was only apparent: what really happened, according to Thomson, was that the material bombarded by cathode rays turned into a source of cathode rays itself.

The cathode ray controversy was resolved by Thomson in 1897. He had studied electrical discharges in gases since 1883 and the discovery of ► X-rays by Wilhelm Conrad Röntgen (1845–1923) rekindled his interest in cathode rays. In a lecture to the Royal Institution on 30 April 1897, Thomson argued that cathode rays were composed of minute, sub-atomic particles that he named "corpuscles". Their small size followed, according to Thomson, from Lenard's results concerning their mean free path outside the cathode ray tube. A further indication of their small size was provided by Thomson measurements of their mass to charge ratio, which turned out to be very small in comparison to the corresponding ratio of hydrogen ions [12].

A few months later, in October 1897, Thomson presented his case for the particulate interpretation of cathode rays in more detail [13]. He reported a novel result favoring that interpretation: the deflection of cathode rays by an electric field. Furthermore, he reported a series of measurements of the mass to charge ratio (m/e) of cathode ray particles, whose purpose was to enable him to figure out their identity. He obtained those measurements by means of two different approaches. The first one was based on measurements of the charge carried by cathode rays, the heat produced by their impact on a target, and the effect of a magnetic field on their trajectory. A combination of those data led to an estimate of m/e. The guiding idea behind the second approach was to place cathode rays under the influence of an electric and a magnetic field and to adjust the intensity of the latter "so that the electrostatic deflexion [sic] was the same as the magnetic" ([13], p. 309). It was then possible to calculate m/e on the basis of directly measurable parameters. Thomson obtained the following value: $m/e = H^2 l / F\Theta$, where H and F were, respectively, the intensities of the magnetic and the electric fields, l the length of the region under the influence of the field, and Θ the angle of electric (or magnetic) deflection. Both methods indicated that the value of m/e was three orders of magnitude smaller than "the smallest value of this quantity previously known, and which is the value for the hydrogen ion in electrolysis" ([13], p. 310). Furthermore, the value of m/e was independent of the material of the cathode and the chemical composition of the gas within the cathode ray tube. This independence suggested to Thomson that the "corpuscles" were universal constituents of all material substances.

In the early months of 1897 analogous results of the charge to mass ratio of cathode rays were reported by Emil Wiechert (1861–1928) and Walter Kaufmann (1871–1947). Those physicists, however, drew different conclusions from their experiments. Wiechert identified the constituents of cathode rays with disembodied charges [14, 15]; and Kaufmann suggested that the unexpectedly large ratio of e/m refuted the particulate interpretation of cathode rays [6]. According to our knowledge today, the cathode rays are nothing but swiftly moving ▶ electrons.

Primary Literature

1. W. Crookes, On Radiant Matter, Nature **20**, 419–423, 436–440 (1879).
2. E. Goldstein, Vorläufige Mittheilungen über elektrische Entladungen in verdünnten Gasen, Königliche Preussische Akademie der Wissenschaften zu Berlin. Monatsberichte, 279–295 (1876).
3. H. Hertz, Versuche über die Glimmentladung, Annalen der Physik **19**, 782–816 (1883); Engl. transl. Experiments on the Cathode Discharge, in H. Hertz, *Miscellaneous Papers* (London, 1896), pp. 224–254.
4. H. Hertz, Ueber den Durchgang der Kathodenstrahlen durch dünne Metallschichten, Annalen der Physik und Chemie **45**, 28–32 (1892); Engl. transl. On the Passage of Cathode Rays Through Thin Metallic Layers, in H. Hertz, *Miscellaneous Papers* (London, 1896), pp. 328–331.

5. W. Hittorf, Ueber die Elektricitätsleitung der Gase, Annalen der Physik **136**, 1–31, 197–234 (1869); Engl. transl. On the Conduction of Electricity in Gases, in *Physical Memoirs, Selected and Translated from Foreign Sources Under the Direction of the Physical Society of London*, vol. 1 (London, 1891), pp. 111–166.
6. W. Kaufmann, Die magnetische Ablenkbarkeit der Kathodenstrahlen und ihre Abhängigkeit vom Entladungspotential, Annalen der Physik und Chemie **61**, 544–552 (1897).
7. P. Lenard, Ueber Kathodenstrahlen in Gasen von atmosphärischem Druck und im äussersten Vacuum, Annalen der Physik und Chemie **51**, 225–267 (1894).
8. J. Perrin, Nouvelles propriétés des rayons cathodiques, Comptes Rendus **121**, 1130 (1895); Engl. transl. New Experiments on the Kathode Rays, Nature **53**, 298–299 (1896).
9. J. Plücker, Ueber die Einwirkung des Magneten auf die elektrischen Entladungen in verdünnten Gasen, Annalen der Physik **103**, 88–106 (1858); Engl. transl. On the Action of the Magnet upon the Electrical Discharge in Rarefied Gases, Philosophical Magazine (4th ser.) **16**, 119–132 (1858).
10. A. Schuster, The Bakerian Lecture: Experiments on the Discharge of Electricity through Gases. Sketch of a Theory, Proceedings of the Royal Society of London **37**, 317–339 (1884).
11. A. Schuster, The Bakerian Lecture: The Discharge of Electricity through Gases. (Preliminary Communication), Proceedings of the Royal Society of London **47**, 526–561 (1889–1890).
12. J. J. Thomson, Cathode Rays, Proceedings of the Royal Institution **15**, 419–432 (1897).
13. J. J. Thomson, Cathode Rays, Philosophical Magazine (5th ser.) **44**, 293–316 (1897).
14. E. Wiechert, Ueber das Wesen der Elektrizität, Schriften der Physikalisch-Ökonomischen Gesellschaft zu Königsberg **38**, 3–12 (1897).
15. E. Wiechert, Experimentelles über die Kathodenstrahlen, Schriften der Physikalisch-Ökonomischen Gesellschaft zu Königsberg **38**, 12–16 (1897).

Secondary Literature

16. P. F. Dahl, *Flash of the Cathode Rays: A History of J J Thomson's Electron* (Bristol: Institute of Physics Publishing, 1997).
17. O. Darrigol, *Electrodynamics from Ampère to Einstein* (Oxford: Oxford University Press, 2000).
18. E. A. Davis and I. J. Falconer, *J. J. Thomson and the Discovery of the Electron* (London: Taylor & Francis, 1997).
19. I. Falconer, Corpuscles, Electrons and Cathode Rays: J. J. Thomson and the 'Discovery of the Electron', British Journal for the History of Science **20**, 241–276 (1987).
20. S. M. Feffer, Arthur Schuster, J.J. Thomson, and the discovery of the electron, Historical Studies in the Physical Sciences **20**, 33–61 (1989).
21. E. Hiebert, Electric Discharge in Rarefied Gases: The Dominion of Experiment: Faraday, Plücker, Hittorf, in A. J. Kox and D. M. Siegel (eds.), *No Truth Except in the Details. Essays in Honor of Martin Klein* (Dordrecht: Kluwer, 1995), pp. 95–134.
22. G. Hon, H. Hertz, 'The electrostatic and electromagnetic properties of the cathode rays are either *nil* or very feeble.' A case-study of an experimental error, Studies in History and Philosophy of Science **18**, 367–382 (1987).
23. F. Müller, *Gasentladungsforschung im 19. Jahrhundert* (Berlin: GNT, 2004).
24. M. Hedenus, *Der Komet in der Entladungsröhre. Eugen Goldstein, Wilhelm Foerster und die Elektrizität im Weltraum* (Stuttgart: GNT 2007).

Causal Inference and EPR

Mauricio Suárez

The status of causality in the EPR experiment has always been a source of controversy. A condition of local *causality* is implicit in the original EPR criterion of reality: "If, without in any way disturbing the system, we can predict with certainty (i.e., with probability equal to unity) the value of a physical quantity, then there exists an element of physical reality corresponding to this physical quantity." In the EPR set-up both systems have separated and are no longer interacting so it is assumed that "no real change can take place in the second system in consequence of anything that may be done to the first system" [1, p. 779]. The non-disturbance clause in the antecedent is hence satisfied, and we may predict with certainty the values of properties in the distant wing. In other words: although the theory does not represent causal influences, there seems *prima facie* to be physical determination of values across a spatial gap. This notoriously led EPR to draw the conclusion that the theory is incomplete; but in the aftermath of ► Bell's theorem it is customary to draw the alternative conclusion – that there is non-local causation in nature. Indeed Bell's theorem has been the driving force of scepticism regarding local causality in the literature. In the last two decades the scepticism has linked up to a more general worry concerning the inference of causal hypotheses from statistical correlations in quantum mechanics. For physicists these issues matter to the evaluation of the compatibility of quantum mechanics with special relativity theory, and the prospects of a unified quantum gravitational theory. For philosophers these issues are key to a thorough assessment of the philosophical implications of quantum mechanics; and in addition EPR has become one benchmark against which all methodologies of causal inference are routinely tested.

The EPR Experiment Briefly Reviewed

Recall that in Bohm's version of the EPR experiment two particles ("1" and "2") are simultaneously created at some event "*e*" in the singlet state Ψ and move in opposite directions. In a Minkowski space-time diagram, both particles describe symmetric paths along the time axis (see Fig. 1). The ► Stern–Gerlach apparati that measure these particles' ► spin at each wing of the experiment are at rest in the laboratory frame so their world lines are represented by vertical lines "A_1" and "A_2" in that frame. Each time the experiment is repeated, laboratory technicians can freely select a particular orientation of the measurement apparatus in each wing, and we denote such events as "*a*" and "*b*". Each particle's spin is measured on interaction

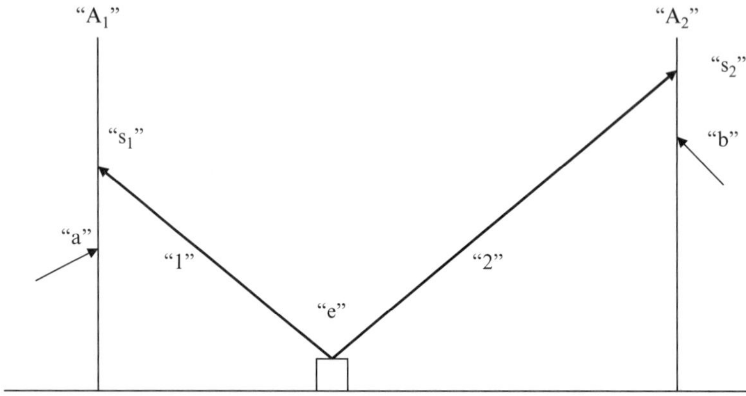

Fig. 1 EPR in space-time setting

with the associated measuring device on the corresponding wing. The outcomes that are produced are denoted by "s_1" and "s_2", respectively, and are known as the "outcome-events":

The Argument Against Causality in EPR

An essay by Bas van Fraassen [2] has been particularly influential in setting a default view against causality in EPR among philosophers of physics and foundational physicists alike. Van Fraassen's argument tracks Bell's own reasoning, with the notorious factorizability condition playing a key role. But there is a significant difference: whereas Bell was concerned with factorizability as a condition of physical ► locality, Van Fraassen takes it to be a condition of causality, in the tradition of Reichenbach's *Principle of the Common Cause*. The putative conclusion of this influential argument is that the principle of the common cause fails in quantum mechanics: there are quantum phenomena that have no causal explanation.

Let us briefly review the argument. Van Fraassen rules out a direct causal link between the wings by appeal to special relativity theory. I will not discuss this assumption here, although it is controversial (see e.g. [10] for an extended critique). The main statistical condition at the heart of Bell's theorem (the notorious "factorizability" condition) is:

$$\text{prob}\left(s_1 \,\&\, s_2/a \,\&\, b \,\&\, \Psi\right) = \text{prob}\left(s_1/a \,\&\, \Psi\right)\text{prob}\left(s_2/b \,\&\, \Psi\right) \text{ (FACT)}$$

The condition can be further analysed into three Reichenbachian screening-off conditions, which in different versions have received the names "causality" or "outcome independence"; "hidden locality" or "parameter independence"; and "hidden autonomy":

$$\begin{aligned} \text{prob}\left(s_1/s_2 \;\&\; a \;\&\; b \;\&\; \Psi\right) &= \text{prob}\left(s_1/a \;\&\; b \;\&\; \Psi\right) \\ \text{prob}\left(s_2/s_1 \;\&\; a \;\&\; b \;\&\; \Psi\right) &= \text{prob}\left(s_2/a \;\&\; b \;\&\; \Psi\right) \end{aligned} \qquad \text{(Causality)}$$

$$\begin{aligned} \text{prob}\left(s_1/a \;\&\; b \;\&\; \Psi\right) &= \text{prob}\left(s_1/a \;\&\; \Psi\right) \\ \text{prob}\left(s_2/a \;\&\; b \;\&\; \Psi\right) &= \text{prob}\left(s_2/b \;\&\; \Psi\right) \end{aligned} \qquad \text{(Hidden Locality)}$$

$$\text{prob}\left(\Psi/a \;\&\; b\right) = \text{prob}\left(\Psi\right) \qquad \text{(Hidden Autonomy)}$$

C

However, in the ► Aspect Experiment a violation of (Hidden Locality) would be as much in conflict with relativity as a direct causal link; while a violation of (Hidden Autonomy) would entail backwards-in-time causation. Hence (Causality) must bear the blame for the violation of factorizability, and indeed it is easy to show that in an EPR experiment with parallel settings and perfect anticorrelation, (Causality) is false. This seems to imply that no causal model is viable for the EPR correlations, and that Reichenbach's principle of the common cause is false as a matter of fact: not all well established correlations admit of a screening-off causal model.

Arguments in Favour of Causality in EPR

However influential, the above argument is not conclusive, and several authors explicitly or implicitly take issue with it. Maudlin [10] argues that direct causation between the wings remains compatible with relativity, and objects to the analysis of factorizability in terms of the three conditions above. Healey [8] and Cartwright and Jones [4] object to the screening-off condition on common causes more generally. Fine [6] accepts the argument but claims that no causal explanation was required in the first place. Bohmian mechanics is widely believed to reject "hidden locality". Price [11] rejects "hidden autonomy", and builds "backwards in time" models following Costa de Beauregard [5]. Höfer-Szabo et al. [9] argue that Van Fraassen's proof assumes not just common causes, but what they term *common common causes*; without this assumption, they claim, Reichenbach's Principle may be rescued (their claim has also been recently contested – see Butterfield [3]). Some of the various options are mapped out in detail in [12]. (See also ► Bohm's approach to EPR paradox; EPR problem; Indeterminism).

Primary Literature

1. A. Einstein, B. Podolski, N. Rosen: *Can Quantum-Mechanical Description of the World Be Considered Complete?* Physical Review, **47**, 777–780 (1935)
2. B. Van Fraassen: *The Charybdis of Realism: Epistemological Implications of Bell's Inequality.* Synthese, **52**, 25–38 (1982)

Secondary Literature

3. J. Butterfield: *Stochastic Einstein Locality Revisited*. The British Journal for the Philosophy of Science **58**, 805–67 (2007)
4. N. Cartwright, M. Jones: *How to Hunt Quantum Causes*. Erkenntnis, **35**, 205–231 (1991)
5. O. Costa de Beauregard: *Time Symmetry and the Einstein Paradox*. Nuovo Cimento, **42B**, 41–64 (1977)
6. A. Fine: *Do Correlations Need to Be Explained?*, in J. T. Cushing, E. McMullin (eds.), Philosophical Consequences of Quantum Theory (Notre Dame University Press 1989, 175–194)
7. Grasshoff et al.: *Minimal Assumption Derivation of a Bell-type Inequality*. The British Journal for the Philosophy of Science, **56**, 663–680 (2005)
8. R. Healey: *Chasing Quantum Causes: How Wild is the Goose*. Philosophical Topics, **20**, 181–204 (1992)
9. G. Höfer-Szabo, M. Redei, L. Szabo: *On Reichenbach's Common Cause Principle and Reichenbach's Notion of Common Cause*. The British Journal for the Philosophy of Science, **50**, 569–589 (1999)
10. T. Maudlin: *Quantum Non-Locality and Relativity* (Blackwell, Oxford 1995)
11. H. Price: *Time's Arrow and Archimedes' Point* (Oxford University Press, Oxford 1997)
12. M. Suárez: *Causal Inference in Quantum Mechanics: A Reassessment*, in F. Russo, J. Williamson (eds.), Causality and Probability in the Sciences (London College Texts 2007, 65–106)

Clauser-Horne-Shimony-Holt (CHSH) – Theorem

See ► Bell's theorem; Loopholes in Experiments.

Cluster States

Hans J. Briegel

1 Introduction

Cluster states [1] form a class of multiparty entangled quantum states with surprising and useful properties. The main interest in these states draws from their role as a *universal resource* in the one-way quantum computer [2, 3]: Given a collection of sufficiently many particles that are prepared in a cluster state, one can realize any ► quantum computation by simply measuring the particles, one by one, in a specific order and basis (see Fig. 1). By the measurements, one exploits ► correlations in quantum mechanics which are rich enough to allow for universal logical processing.

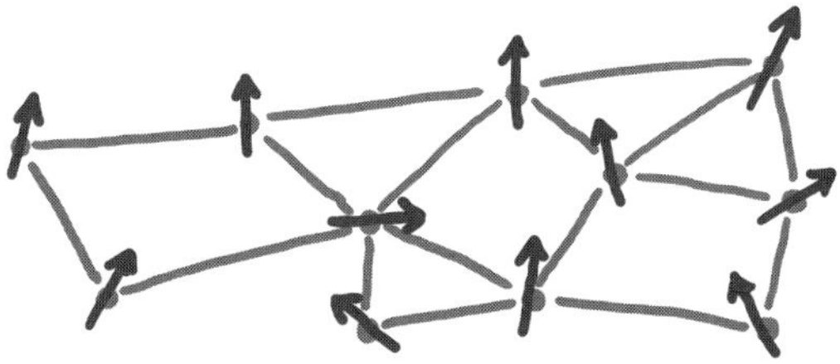

Fig. 1 A collection of N particles in an entangled cluster state can serve as a quantum computer: A quantum computation can be realized by simply measuring the particles, one by one, in a specific order and basis [2]

Owing to this property, the cluster state has attracted considerable attention in recent foundational studies on the power of quantum computation. It has also become an interesting object from the perspective of multipartite entanglement theory and the study of quantum mechanical ▶ nonlocality.

Cluster states belong to the larger set of so-called graph states [4], which comprise many of the entangled states known in quantum information theory and foundations. Examples include the Bell or Einstein–Podolsky–Rosen (EPR) states, the Greenberger–Horne–Zeilinger (▶ GHZ) states, and states that appear in quantum error correction. Graph states have been providing a playground for the study of multipartite ▶ entanglement, and investigating their role as resources in measurement-based quantum computation has revealed connections to other fields, including quantum many-body physics, graph theory, topological codes, statistical physics, and even mathematical logic. For a recent review of these developments see e.g., [5].

2 Entangled Clusters of Qubits in Randomly Occupied Lattices

The study of cluster states was initially motivated by experimental developments with cold atomic gases in optical lattices. In optical lattices, standing laser fields are used to create a periodic potential, in which atoms can be trapped by electric dipole forces. It had been shown theoretically in [6] that, by tuning the laser parameters, one could realize collision-type interactions between neighboring atoms to create entanglement with respect to their internal atomic states. This meant that simple lattice manipulations would allow one to entangle entire arrays of atoms with the control of a few laser parameters. However, it was not clear what kind of entangled states would be created and what they could be used for.

On the experimental side, a realization of these ideas required, first, to fill the lattice with exactly one atom per lattice site and, second, to entangle them, both of which were a formidable challenge. Regarding the first point, early experiments [7] achieved a situation where about 44% of all sites were occupied with exactly one atom, with the other sites empty. In such *randomly occupied lattices*, one obtains

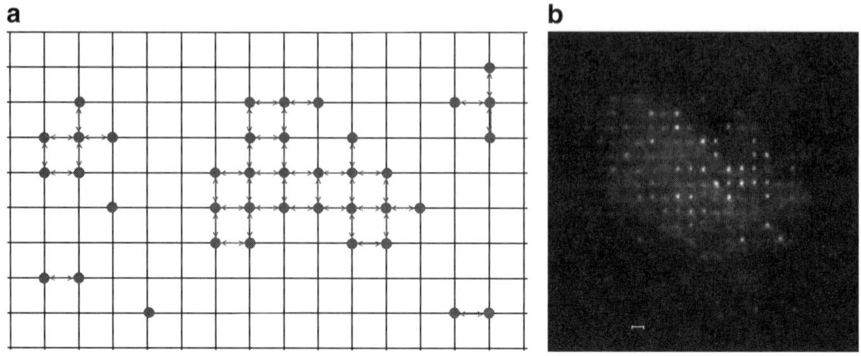

Fig. 2 (**a**) Quantum mechanical particles (red filled circles) are trapped at different sites of a 2-dimensional lattice. If each site is either occupied with one particle or left empty, with site-occupation probability p, $(0 \leqslant p \leqslant 1)$, then one will observe clusters $C \subset Z^2$ of neighboring particles, as indicated in the figure. Suppose that each particle has two internal states $|0\rangle$ and $|1\rangle$ defining a spin or a qubit, and that each particle is initially brought in a superposition $|0\rangle + |1\rangle$ of the two states. A simple Ising-type interaction (red arrows between the circles), switched on for a certain time, will then create entanglement between all particles that belong to one and the same cluster. The resulting entangled state is called a "cluster state" [1]. (**b**) Images of individual atoms (not entangled) trapped in an optical lattice. The figure is taken from [9] (courtesy D.S. Weiss). Reprinted by permission from Macmillan Publishers Ltd: Nature Physics 3, 556–560; copyright (2007)

clusters of neighboring atoms, as shown in Fig. 2b. Randomly occupied lattices play an important role in statistical physics, e.g., in the context of percolation theory and phase transitions.

In [1], we studied the quantum states that can be generated in such clusters of two-level atoms (qubits) with an Ising-type interaction. Such an interaction can be realized by simple interferometric lattice manipulations, inducing state-dependent collisions as proposed in [6]. An entangled state can be created (see Fig. 2a) by first bringing first each atom into a ▶ superposition state – which can be achieved with a simple laser pulse – and subsequently switching on the Ising interaction for a certain time span. By this operation, each cluster of atoms can be brought into a joint entangled state, which we called *"cluster state"* in [1]. We showed that such entangled clusters of atoms would have distinct entanglement properties: Their entanglement would be remarkably robust and, furthermore, any pair of atoms in such a cluster could be brought into a maximally entangled Bell state by simple measurements on other particles of the cluster (see Fig. 4a). Simple operations would thus allow one to maximally entangle arbitrary two atoms in a cluster. This could thus be used to establish arbitrary "teleportation channels" within a cluster, and it also provided perspectives for encoding and quantum error correction in random clusters [8].

Later experiments [10] used different techniques, based on the superfluid-to-Mott phase transition of a ▶ Bose–Einstein condensate as predicted in [11]. By using this method, it was possible to realize, in effect, large "atomic crystals" where each of the lattice sites was occupied by exactly one atom. This corresponded to a situation where a single cluster of atoms would extend over the entire lattice (or at least over large parts thereof). In these experiments, it was also possible for the first time to *entangle* the atoms using cold controlled collisions [6, 12] and to create the cluster states as predicted in [1]. Today, the word "cluster state" is often used referring to the completely filled lattice in 1, 2, or 3 dimensions.

For a detailed exposition of the ideas how to generate cluster states and to study multiparty entanglement in the specific context of ultracold atoms in optical lattices, see [8]. For a review of recent experimental work in this area, see [13]. More recently, there has also been significant progress in the experimental realization of cluster states using photons (▶ light quanta) [28–30].

The notion of cluster states can be straightforwardly generalized to so-called *graph states*, which will be described in the next section.

3 Mathematical Description of Cluster and Graph States

For the clusters shown in Fig. 2a, the entangling interactions occur between neighboring particles, whereby the neighborhood relation is defined by the underlying lattice. These clusters can be regarded as special instances of graphs with a more general interaction pattern, giving rise to a larger class of states.

Let $G = (V, E)$ by a simple mathematical graph, where $V = \{1, ..., N\}$ denotes a set of vertices, and $E \subseteq [V]^2$ denotes the set of edges that connect pairs of vertices. In the present context, each vertex is associated with a qubit, while each edge is associated with an interaction. The *graph state* $|G\rangle$ associated with graph G is a (pure) quantum mechanical state of N qubits, i.e. $|G\rangle \in \left[\mathbb{C}^2\right]^{\otimes N}$, described by the following set of linear equations:

$$K^{(a)}\,|G\rangle = |G\rangle\,, \quad \forall a = 1, ..., N \tag{1}$$

where $K^{(a)} = \sigma_x^{(a)} \bigotimes_{\{a,b\}\in E} \sigma_z^{(b)}$ denotes a *correlation operator* that acts nontrivially on qubit a and all of its neighboring qubits, see Fig. 3a. In the language of quantum physics, the N (hermitean) correlation operators $K^{(a)}$, $a = 1, ..., N$ form a complete set of commuting observables (CSCO) [14]. The graph state $|G\rangle$ associated with the graph G is then, by convention, the common eigenstate of these observables with all eigenvalues equal to $+1$. In general, the CSCO describes properties of a system that can be observed simultaneously without disturbing the state [14]. Here, they give rise to *strict quantum correlations*, such as $\left\langle K^{(a)}\right\rangle = +1$.

In Fig. 3, we see examples of graphs representing specific quantum states. These pictures can be interpreted in two ways. First, they indicate the interaction pattern under which the graph state can be created, if the vertices represent qubits and the edges represent Ising-type interactions. Second, they give a concise graphical encoding of the correlation operators (CSCO) that stabilize the state according to eq. (1).

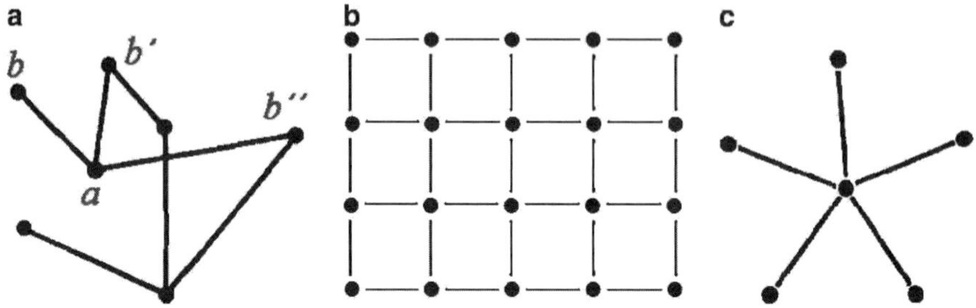

Fig. 3 (**a**) Graph $G = (V, E)$ with a selected vertex $a \in V$ and its neighbors $b, b', b'' \in V$ highlighted. Vertices correspond to qubits and edges to specific (interactions or) quantum correlations (see text). (**b**) Graph corresponding to the 2D cluster state $|C_{N \times M}\rangle$. (**c**) Star graph, corresponding to the N-particle Greenberger–Horne–Zeilinger state $|GHZ_N\rangle$

An explicit formula for the graph state $|G\rangle$ is given by an expansion into the computational basis

$$|G\rangle = 2^{-N/2} \sum_{s \in \{0,1\}^N} (-1)^{s^T \Gamma s} |s\rangle$$

wherein Γ is the adjacency (or neighborhood) matrix of the graph G, the summation multiindex $s = s_1 s_2 s_3 \ldots s_N \in \{0, 1\}^N$ runs over all binary strings of length N, and $s^T \Gamma s$ is understood as the matrix multiplication of matrix Γ with a column s and a row s^T. The summation as written above involves exponentially many terms. For some families of graph states a more compact basis expansion – with a fewer number of terms – can be found by a suitable choice of local basis for each qubit. For example, for the family of N-qubit GHZ states, represented by the star graph in Fig. 3c, a more compact representation is given by $|G_{\text{star}}\rangle \equiv |GHZ_N\rangle = 2^{-1/2} \left[|0\rangle_z |0\rangle_x^{\otimes N-1} + |1\rangle_z |1\rangle_x^{\otimes N-1} \right]$, where by $|0\rangle_z = \sigma_z |0\rangle_z$ and $|0\rangle_x = \sigma_x |0\rangle_x$ are eigenstates of the Pauli spin operators σ_z and σ_x ("spin up" in z and x direction), respectively. For most families, however, including the cluster states, even the most compact expansion still requires an exponential number of terms, which makes calculations with explicit expansions often cumbersome and inefficient.

Graph and in particular cluster states exhibit a number of remarkable properties, some of which we are going to describe in the following. For example, all graph states *violate local realism*, that is, they exhibit certain correlations that cannot be explained by any local hidden variable model (see ► nonlocality). At the same time, the fragility of their entanglement under the influence of *decoherence* depends strongly on the specific state. It is quite different for the GHZ and the cluster state, which can be considered as two extreme representatives of graph states.

Graph states were first mentioned in [3] as a natural generalization of cluster states, which were introduced in [1]. The similar notion of "graph codes" was studied independently by Schlingemann and Werner in the context quantum error correction [15]. A classification of graph states in terms of their entanglement was first systematically investigated in [16]. A comprehensive exposition of the theory of graph states can be found in the Varenna Lecture Notes [4].

4 Physical Properties

Measurements play an important role, both in studies of quantum correlations and nonlocality, and in studies of ▶ decoherence, (▶ experimental observation of decoherence). A question that turned out to be very fruitful in the studies of cluster states was this: What is the effect of measurements on individual particles, regarding the state of the remaining, unmeasured particles? How robust is the entanglement under measurements? It is, for example, known that a measurement (in a suitable basis) on a single qubit of an N-qubit GHZ state is sufficient to destroy all entanglement in the state. Is the same true for the cluster state? Suppose that each qubit of a 1D cluster state of N qubits is held by a different party, and each party can measure its qubit in an arbitrary basis. How many measurements are needed if the parties seek to *destroy* all entanglement in the state? How *persistent* is its entanglement against such destructive attempts?

In [1] we showed that for the cluster state this number grows linearly with N (more precisely, it is given by the largest integer smaller or equal to $N/2$): About half of all parties have to measure their qubit to completely destroy all entanglement in the state. Thus, compared to the GHZ state, the entanglement in a cluster state is rather robust.

Taking a more constructive point of view, one may ask which type of states can be created out of the cluster state by local measurements. As was shown in [1], the cluster state is not only highly persistent, it is also entangled in such a way that a *Bell state* can be created *between any two qubits* belonging to the same cluster, by measurements on other qubits of the cluster (see Fig 4a). These two properties make the cluster state an interesting candidate for applications in quantum computation and communication. The first property (high persistency) indicates the possibility of applying various measurements on the state, without immediately destroying its entanglement. The second property tells us that cluster states are useful to establish "teleportation channels" between arbitrary qubits that belong to a given cluster.

Entanglement resource: These properties carry over to higher dimensions and provide the two dimensional cluster state with a large degree of flexibility; they also indicate that the cluster state can be seen an entanglement *resource* which can be used to create other interesting entangled states. In Fig. 4, examples of this resource property are shown. The cluster is there used as a resource from which one can obtain collections of Bell states or GHZ states. An interesting question is, which other states can be obtained from this resource? Are the Bell and the GHZ states special? The answer is both surprising and profound, namely *any other state* can be obtained from this resource – if it is sufficiently large – by single-particle

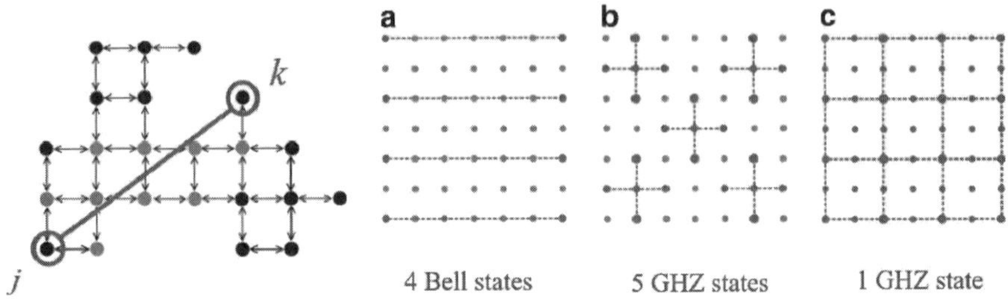

Fig. 4 Cluster states serve as a resource for other highly entangled states, such as the Bell or GHZ states. Left: Measurements on the gray qubits allow one to project qubits labeled j and k into a maximally entangled Bell state. Right: Measurements on the gray qubits allow one to project the unmeasured (red) qubits into (**a**) 4 Bell states (**b**) 5 GHZ states of 5 qubits (**c**) one GHZ state of 16 qubits. (Figures adapted from [1])

measurements, as was shown in [2]. The resource is indeed *universal* for arbitrary state preparation! This last property makes the cluster state the basis of an entirely new concept of quantum computation, the *one-way quantum computer*, which will be described in the last section.

Decoherence: How fast quantum do states become classical? The persistency of the entanglement in a cluster state under measurements is closely related to another fundamental property, namely its robustness under decoherence. Decoherence is an effect that arises from interactions of the system's degrees of freedom with (usually a vast number of) uncontrolled degrees of freedom of the environment. One can view decoherence as the result of "environment-induced measurements" [17] on the system. If the cluster state has a high persistency of entanglement, it should therefore also be robust against decoherence. This is indeed the case, as was shown in the papers [19, 20]. These investigations have also led to a new perspective on the *lifetime of entanglement* in macroscopic bodies and thus the concept of the Schrödinger cat [18]. The lifetime of entanglement in a state of a system, given a specific interaction with its environment, can be defined as the time it takes until the resulting (mixed) state of the system becomes completely separable, that is, all correlations are classical [19]. A lower bound on this lifetime is found by showing that after certain time, given many copies of the ► mixed state, one can distil again the original state. (► Entanglement, purification and distillation) This analysis was carried out in [19,20]. It was found that, under a wide class of system–environment interactions, the lifetime of (multipartite) entanglement of the cluster state is largely independent of the number N of particles, i.e., the size of the system. For the GHZ state, in contrast, this time goes to zero for $N \to \infty$. One conclusion of these investigations is that, given any generic interaction between the system and the environment, a macroscopic GHZ state cannot exist. In contrast, the lifetime of the entanglement in a cluster state is largely independent of its size and does not vanish in the macroscopic limit. Besides implications for the idea of "macroscopic entanglement" and ► Schrödinger cat, this has also practical implications regarding the realization of a fault-tolerant quantum computation based on cluster states [26, 27].

In the next section, we will review this most striking and important application of cluster states for universal measurement-based quantum computation.

5 One-Way Quantum Computer

Quantum computers are devices that use quantum mechanical properties of their information carriers for enhanced ways of information processing [21]. It has been shown [22] that certain mathematical problems, such as *factoring* a large integer into primes, which plays an important role in modern data encryption schemes, can be solved much faster on a quantum computer (▶ quantum communication) than with any known algorithm on a classical computer. A standard model of a quantum computer resembles the Boolean circuit representation of a classical computation, and consists of a sequence of quantum gates – a quantum circuit – applied to a few qubits at a time. Such quantum gates are elementary unitary operations that take over the role of Boolean gates, e.g., AND, OR, NOT (or their reversible counterparts). The computation is thereby a coherent process which creates entangled superpositions of different states of the quantum register.

In the one-way quantum computer, introduced by Raussendorf and Briegel in [2], a quantum computation is instead realized by a sequence of simple measurements on an entangled resource state of many qubits. A universal resource state is the *cluster state* in two (or three) dimensions, and the measurements act on single qubits at a time. In contrast to the quantum circuit model, here the elementary building blocks are not quantum gates but single-qubit measurements. A quantum algorithm then corresponds to a *pattern of measurement directions* on the cluster (see Fig. 5), together with the classical processing of the measurement results. While the result of an individual measurement is random – owing to the entanglement of the resource – the quantum computation is nevertheless deterministic, and the random outcome of a measurement is compensated by the choice of basis for subsequent measurements. The measurements are thus adaptive, introducing a *temporal order* into the scheme, which determines the run time of a quantum algorithm. It has been shown in [2, 3] that any quantum algorithm, for which an efficient quantum circuit description exist, can also be run efficiently on the one-way quantum computer, i.e., by single-qubit measurements on a cluster state of sufficient size.

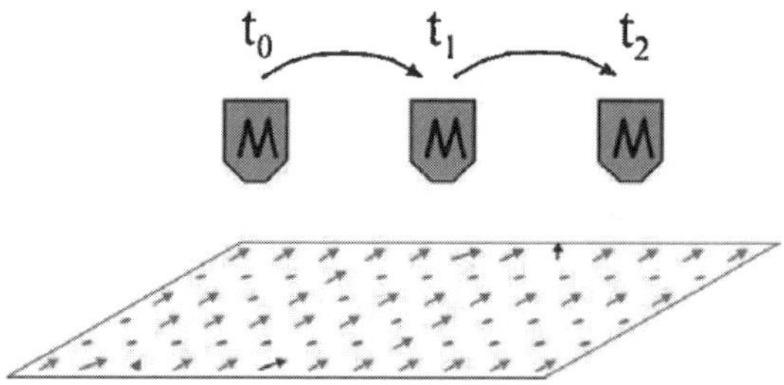

Fig. 5 Scheme of the one-way quantum computer. A quantum computation is realized by a sequence (t_0, t_1, ...) of adaptive measurements $\mathcal{M}$ on single qubits, here arranged on a lattice, exploiting the entanglement of the cluster state. Any quantum algorithm corresponds to a specific (spatial and temporal) pattern of measurement directions

One of the main features of the one-way model lies in the clear separation between the preparation of the quantum resource (the cluster state) and its processing. This has practical advantages, in that the resource can be prepared off-line, independent of the quantum algorithm one wants to perform, which is preferential for certain implementations. It is also conceptual appealing and has opened new perspectives in the study of fundamental questions e.g., regarding the *computational power* of a quantum computer: Whatever the computational power of a (one-way) quantum computer is, it must originate in the entanglement of the resource state! From this perspective, a quantum computation may also be regarded as a *processing of quantum correlations*. Starting with a set of basic, but universal, correlations carried by the initial cluster state, it is an unfolding of more and more complicated correlations in the course of the computation. This viewpoint puts the fundamental notion of quantum correlations at the focus of the theory.

While the one-way quantum computer is an abstract concept, a variety of proposals exist for its concrete physical implementation (for a review see e.g., [5]). Laboratory experiments using cold atoms in optical lattices have a great practical potential, even though up-to-date the addressing of individual atoms remains a challenge. More recently, experiments using polarization-entangled photons have been reported [31–34], demonstrating the principles of one-way quantum computation.

For a detailed exposition of the one-way quantum computer, the reader is referred to [3]. Reviews that treat the one-way model and other formats of measurement-based computation from various perspectives are given in [23–25]. Recent investigations have related the computational power of resource states to their entanglement and revealed interesting connections of measurement-based quantum computation to other fields such as graph theory, classical statistical physics, and even mathematical logic. For a review of these developments, and for further references, see [5].

Literature

1. H.J. Briegel, R. Raussendorf: Persistent entanglement in arrays of interacting particles. Physical Review Letters 86, 910–913 (2001).
2. R. Raussendorf, H.J. Briegel: A one-way quantum computer. Physical Review Letters 86, 5188–5191 (2001).
3. R. Raussendorf, D.E. Browne, H.J. Briegel: Measurement-based quantum computation using cluster states. Physical Review A 68, 022312 (2003).
4. M. Hein, W. Dür, J. Eisert, R. Raussendorf, M. Van den Nest, H.J. Briegel: Entanglement in graph states and its applications. In International School of Physics Enrico Fermi (Varenna, Italy), Course CLXII, Quantum computers, algorithms and chaos, eds. G. Casati et al. (2006). E-print available at http://arxiv.org/abs/quant-ph/0602096.
5. H.J. Briegel, D.E. Browne, W. Dür, R. Raussendorf, M. Van den Nest: Measurement-based quantum computation, Nature Physics 5, 19–26 (2009).
6. D. Jaksch, H.J. Briegel, J.I. Cirac, C.W. Gardiner, P. Zoller: Entanglement of atoms via cold controlled collisions. Physical Review Letters 82, 1975–1978 (1999).
7. M.T. De Pue, C. McCormick, S. Lukman Winoto, S. Oliver, D.S. Weiss: Physical Review Letters 82, 2262–2265 (1999).

8. H.J. Briegel, R. Raussendorf, A. Schenzle: Optical lattices as a playground for studying multiparticle entanglement. In Laserphysics at the limit, ed. by H. Figger et al. (Springer, Berlin, 2002, 236–261).
9. K. Nelson, X. Li, D.S. Weiss: Imaging single atoms in a three-dimensional array. Nature Physics 3, 556–560 (2007).
10. M. Greiner, O. Mandel, T. Esslinger, T.W. Hänsch, I. Bloch: Quantum phase transition from a superfluid to a Mott insulator in a gas of ultracold atoms. Nature 415, 39–44 (2002).
11. D. Jaksch, C. Bruder, J.I. Cirac, C.W. Gardiner, P. Zoller: Cold Bosonic atoms in optical lattices. Physical Review Letters 81, 3108–3111 (1998).
12. O. Mandel, M. Greiner, A. Widera, T. Rom, T.W. Hänsch, I. Bloch: Controlled collisions for multi-particle entanglement of optically trapped atoms. Nature 425, 937–940 (2003).
13. I. Bloch: Ultracold quantum gases in optical lattices. Nature Physics 1, 23–30 (2005).
14. C. Cohen-Tannoudji, B. Diu, F. Laloë, Quantum Mechanics, Vol. I (Wiley, 1977).
15. D. Schlingemann, R.F. Werner: Quantum error-correcting codes associated with graphs. Physical Review A 65, 012308–012315 (2002).
16. M. Hein, J. Eisert, H.J. Briegel: Multi-party entanglement in graph states. Physical Review A 69, 062311–062330 (2004).
17. W. Zurek: Decoherence, einselection, and the quantum origins of the classical. Reviews of Modern Physics 75, 715–775 (2003).
18. E. Schrödinger, Die gegenwärtige Situation in der Quantenmechanik, Naturwissenschaften 23 (1935).
19. W. Dür, H.J. Briegel: Stability of macroscopic entanglement under decoherence. Physical Review Letters 92, 180403—180406 (2004).
20. M. Hein, W. Dür, H.J. Briegel: Entanglement properties of multipartite entangled states under decoherence. Physical Review A 71, 032350–032375 (2005).
21. M.A. Nielsen, I.L. Chuang: Quantum computation and quantum information (Cambridge University Press, Cambridge, 2000).
22. P.W. Shor: Algorithms for quantum computation: discrete logarithms and factoring. In Proceedings, 35th Annual Symposium on Fundamentals of Computer Science (IEEE Press, Los Alamitos, California, 1994.
23. D.E. Browne, H.J. Briegel: One-way quantum computation - a tutorial introduction. Preprint available at http://arxiv.org/abs/quant-ph/0603226
24. M.A. Nielsen: Cluster-state quantum computation. Preprint available at http://arxiv.org/abs/quant-ph/0504097.
25. R. Jozsa: An introduction to measurement-based quantum computation. Preprint available at http://arxiv.org/abs/quant-ph/0508124.
26. R. Raussendorf, S. Bravyi, J. Harrington: Long-range quantum entanglement in noisy cluster states. Physical Review A 71, 062313 (2005).
27. R. Raussendorf, J. Harrington, K. Goyal, Topological fault-tolerance in cluster state quantum computation, New Journal of Physics 9, 199 (2007).
28. N. Kiesel, C. Schmid, U. Weber, G. Toth, O. Gühne, R. Ursin, H. Weinfurter: Experimental analysis of a four-qubit photon cluster state. Physical Review Letters 95, 210502 (2005).
29. C.Y. Lu, X.Q. Zhou, O. Gühne, W.B. Gao, J. Zhang, Z.S. Yuan, A. Goebel, T. Yang, J.W. Pan: Experimental entanglement of six photons in graph states. Nature Physics 3, 91–95 (2007).
30. G. Vallone, E. Pomarico, F. De Martini, P. Mataloni, V. Berardi: Realization and characterization of a 2-photon 4-qubit linear cluster state. Physical Review Letters 98, 180502 (2007).
31. P. Walther, K.J. Resch, T. Rudolph, E. Schenck, H. Weinfurter, V. Vedral, M. Aspelmeyer, A. Zeilinger: Experimental one-way quantum computing. Nature 434, 169–176 (2005).
32. R. Prevedel, P. Walther, F. Tiefenbacher, P. Bohl, R. Kaltenbaek, T. Jennewein, A. Zeilinger: High-speed linear optics quantum computing using active feed-forward. Nature 445, 65–69 (2007)
33. K. Chen, C.-M. Li, Q. Zhang, Y.A. Chen, A. Goebel, S. Chen, A. Mair, J.-W. Pan: Experimental realization of one-way quantum computing with two-photon four-qubit cluster states. Physical Review Letters 99, 120503 (2007).
34. G. Vallone, E. Pomarico, F. De Martini, P. Mataloni: Active one-way quantum computation with 2-photon 4-qubit cluster states. Physical Review Letters 100, 160502 (2008).

Coherent States

Peter W. Milonni and Michael Martin Nieto

Coherent states (of the harmonic oscillator) were introduced by Erwin Schrödinger (1887-1961) at the very beginning of quantum mechanics in response to a complaint by Lorentz that Schrödinger's ► wave function did not display classical motion. Schrödinger obtained solutions that were Gaussians having the width of the ground state. The expectation values of the coordinate and momentum for these Gaussian solutions oscillate in time in just the same way as the coordinate and momentum in the classical theory of the harmonic oscillator.

In modern parlance Schrödinger's solutions are the 2-parameter $(\langle x\rangle, \langle p\rangle)$ states

$$\psi_{\mathrm{cs}} = [2\pi(\Delta x)^2]^{-1/4} \exp\left[-\left(\frac{x-\langle x\rangle}{2\Delta x}\right)^2 + \mathrm{i}\frac{\langle p\rangle x}{\hbar}\right] \tag{1}$$

satisfying equality in the uncertainty relation

$$(\Delta x)^2(\Delta p)^2 \geqslant \frac{\hbar^2}{4} \tag{2}$$

and having "widths" equal to those of the ground state, $(\sqrt{2}\Delta x) \equiv (\hbar/m\omega)^{1/2}$.[1] These can be called *minimum uncertainty coherent states.*

In the 1960s there was a reawakening of interest in these states in terms of the boson operator formalism. Two other, equivalent formulations of coherent states were obtained. The first yields the *annihilation operator coherent states*, $|\alpha\rangle$, defined by

$$a|\alpha\rangle = \alpha|\alpha\rangle, \tag{3}$$

where a ($a^\dagger$) is the annihilation (creation) operator (► creation and annihilation operator). The second yields the *displacement operator coherent states*

$$|\alpha\rangle \equiv D(\alpha)|0\rangle = \exp[\alpha a^\dagger - \alpha^* a]|0\rangle. \tag{4}$$

The real and imaginary parts of the complex number α are the two parameters which give the solution as

$$|\alpha\rangle = \exp\left[-\frac{1}{2}|\alpha|^2\right] \sum_{n=0}^{n=\infty} \frac{\alpha^n}{\sqrt{n!}}|n\rangle, \tag{5}$$

[1] ► Squeezed states, whose width oscillates with time, were introduced in 1927 by E. H. Kennard. They are a 3-parameter set of Gaussians whose widths are *not* that of the ground state.

where $|n\rangle$ are the number states, i.e., the energy eigenstates of the harmonic oscillator. From the Hermite polynomial generating function these can be shown to be identical to the Gaussians of the minimum-uncertainty coherent states, where

$$\operatorname{Re}\alpha = \langle x\rangle \left(\frac{m\omega}{2\hbar}\right)^{1/2}, \qquad \operatorname{Im}\alpha = \langle p\rangle \left(\frac{1}{2m\omega\hbar}\right)^{1/2}. \tag{6}$$

These ideas have been applied to non-harmonic systems, involving different symmetries and/or potentials. There the coherence properties are not as strong in general, since it is the equally-spaced levels of the harmonic oscillator which allow the system never to decohere if there is no damping or excitation.

An especially interesting system is described by the even- and odd-coherent states ("cat" states). They are higher-power states, eigenvalues of aa. They are given by

$$|\alpha;+\rangle = [\cosh|\alpha|^2]^{-1/2}\sum_{n=0}^{\infty}\frac{\alpha^{2n}}{\sqrt{(2n)!}}|2n\rangle \rightarrow \psi_+(x), \tag{7}$$

$$|\alpha;-\rangle = [\sinh|\alpha|^2]^{-1/2}\sum_{n=0}^{\infty}\frac{\alpha^{2n+1}}{\sqrt{(2n+1)!}}|2n+1\rangle \rightarrow \psi_-(x). \tag{8}$$

$$\psi_\pm(x) = \frac{\mathrm{e}^{-\mathrm{i}2x_0p_0}\left[\exp[-\frac{1}{2}(x-x_0)^2]\mathrm{e}^{\mathrm{i}p_0x} \pm \exp[-\frac{1}{2}(x+x_0)^2]\mathrm{e}^{-\mathrm{i}p_0x}\right]}{2^{1/2}\pi^{1/4}\left[1\pm\exp[-(x_0^2+p_0^2)]\right]^{1/2}}. \tag{9}$$

where we have set $\hbar$ and $m = 1$.

The ► wave packet of these states are two Gaussians, at positions π apart in the phase-space circle. The Gaussians keep their shapes as they move as a normal coherent state would in time evolution, until they overlap. When the even states, composed of $n = 0, 2, 4, \ldots$ number states. interfere, they have a maximum central peak. (See the left graph in Fig. 1.) The odd states are composed of $n = 1, 3, 5, \ldots$

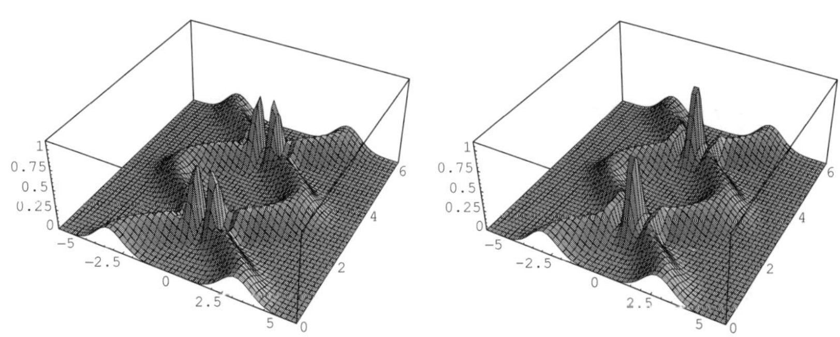

Fig. 1 The time evolution of the even- and odd-coherent states $\rho_\pm(x,t)$. The initial conditions are $x_0 = 2^{3/2}$ and $p_0 = 0$. The position is along the x-axis, time is along the y axis, and the Z-axis displays the probability density

number states. When the odd Gaussians interfere there is a central minimum and two slightly smaller peaks on each side. (See the right graph in Fig. 1.)

These states have been observed experimentally (Monroe et al.).

The coherent states have been especially useful in quantum optics. Each mode of the electromagnetic field may be described formally as a harmonic oscillator, and different quantum states of the oscillator correspond to different states of the field. The field from a single-mode laser operating far enough above threshold can be described for many purposes as a coherent state; it differs from a coherent state in that its phase drifts randomly. But its photon counting statistics and other properties make the light from a single-mode laser practically indistinguishable from a coherent state.

The quantum theory of optical coherence is based on "normally ordered" products of lowering and raising operators a and $a^{\dagger}$ which act, respectively, as photon annihilation and creation operators. The fact that coherent states are eigenstates of lowering operators implies that the expectation value of a normally ordered field operator product $f(a, a^{\dagger})$ reduces to the deterministic function $f(\alpha, \alpha^{*})$ for a coherent state. A coherent state of the field therefore comes closest to the idealized classical stable wave in which there are no random field fluctuations. Thus a coherent-state field exhibits maximal fringe visibility or "coherence" in a Michelson interferometer, for instance, and it is maximally coherent as well when more complicated interference effects involving higher orders of field products are considered.

Primary Literature

1. Schrödinger, E.: Der stetige Übergang von der Mikro- zur Makromechanik. Naturwiss. 14, 664 (1926). Translated into English in Schrödinger, E.: *Collected Papers in Wave Mechanics*. London, Blackie & Son (1928), p. 41
2. Glauber, R. J.: *Coherent and Incoherent States of the Radiation Field*. Phys. Rev. 131, 2766 (1963)
3. Klauder, J. R.: *Continuous-Representation Theory*. J. Math. Phys. 4, 1055, 1058 (1963)
4. Sudarshan, E. C. G.: *Equivalence of Semiclassical and Quantum Mechanical Descriptions of Statistical Light Beams*. Phys. Rev. Lett. 10, 277 (1963)

Secondary Literature

5. Klauder, J. R., B.-S. Skagerstam: *Coherent States – Applications in Physics and Mathematical Physics*. Singapore, World Scientific (1985)
6. Mandel, L., E. Wolf: *Optical Coherence and Quantum Optics*. Cambridge, Cambridge University Press (1995)
7. Nieto, M. M., D. R. Truax: *Higher-Power Coherent and Squeezed States*. Opt. Commun. 179, 197 (2000)
8. Monroe, C., D. M. Meekhof, B. E. King, D. J. Wineland: *A "Schrödinger Cat" Superposition State of an Atom*. Science 272, 1131 (1996)

Color Charge Degree of Freedom in Particle Physics

C

O.W. Greenberg

Color has two facets in ▶ particle physics. One is as a three-valued charge degree of freedom, analogous to electric charge as a degree of freedom in electromagnetism. The other is as a ▶ gauge symmetry, analogous to the $U(1)$ gauge theory of electromagnetism. Color as a three-valued charge degree of freedom was introduced by Oscar W. Greenberg [1] in 1964. Color as a gauge symmetry was introduced by Yoichiro Nambu [2] and by Moo Young Han and Yoichiro Nambu [3] in 1965. The union of the two contains the essential ingredients of ▶ Quantum Chromodynamics, QCD. The word "color" in this context is purely colloquial and has no connection with the the color that we see with our eyes in everyday life.

The theoretical and experimental background to the discovery of color centers around events in 1964. In 1964 Murray Gell-Mann [4] and George Zweig [5] independently proposed what are now called "quarks," particles that are constituents of the observed strongly interacting particles, "hadrons," such as protons and neutrons. Quarks gave a simple way to account for the ▶ quantum numbers of the hadrons. However quarks were paradoxical in that they had fractional values of their electric charges, but no such fractionally charged particles had been observed. Three "flavors" of quarks, up, down, and strange, were known at that time. The group $SU(3)_{\text{flavor}}$, acting on these three flavors, gave an approximate symmetry that led to mass formulas for the hadrons constructed with these quarks. However the spin 1/2 of the quarks was not included in the model. (Quarks, see also ▶ Mixing and Oscillations of Particles; Particle Physics; Parton Model; QCD; QFT.)

The quark spin 1/2 and the symmetry $SU(2)_{\text{spin}}$ acting on the two states of spin 1/2 were introduced in the model by Feza Gürsey and Luigi Radicati [6]. They combined $SU(2)_{\text{spin}}$ with $SU(3)_{\text{flavor}}$ into a larger $SU(6)_{\text{spin-flavor}}$ symmetry. This larger symmetry unified the previously known mass formulas for the octet of spin-1/2 baryons and the decuplet of spin-3/2 baryons. Using this $SU(6)$ theory Mirza A.B. Bég, Benjamin W. Lee and Abraham Pais [7] calculated the ratio of the magnetic moments of the proton and neutron to be -3/2, which agrees with experiment to within 3%. However the successful $SU(6)$ theory required that the configuration of the quarks that gave the correct lowlying baryons must be in a symmetric state under permutations. This contradicts the ▶ spin statistics theorem of Wolfgang Pauli [8], according to which quarks as spin-1/2 particles have ▶ Fermi statistics and must be in an antisymmetric state under permutations.

In the same year 1964 Oscar W. Greenberg [1] recognized that this contradiction could be resolved by allowing quarks to have a new hidden three-valued charge, expressed in terms of parafermi statistics of order three. This was the discovery of color. The antisymmetrization of the hidden degree of freedom allows the quarks in baryons to be in the observed symmetric configuration of the visible degrees of freedom: space, spin and flavor. Greenberg called this model the "symmetric quark model" for baryons. As an observable test of this model, Greenberg constructed a table of the spin, ▶ parity, isospin and strangeness of the orbital excitations of the ground-state quark configurations in this model.

In 1964 the hidden color charge on top of the fractionally charged quarks seemed unduly speculative to some. Independent evidence for the existence of color came when measurements of the properties of excited baryons confirmed the predictions of the symmetric quark model. It was only in 1968 that Haim Harari [9], as rapporteur for baryon spectrocopy, adopted the symmetric quark model as the correct model of baryons.

Additional evidence for color came from the ratio of the annihilation cross section for $e^+e^- \rightarrow hadrons$ to that for $e^+e^- \rightarrow \mu^+\mu^-$ and from the decay rate for $\pi^0 \rightarrow \gamma\gamma$. Both of these follow from the gauge theory and the parastatistics version of color. Further consequences of color require the gauged theory of color, quantum chromodynamics, ▶ QCD, described below.

In 1965 Yoichiro Nambu [2] and, in a separate paper, Moo Young Han and Yoichiro Nambu [3] proposed a model with three sets of quark triplets. Their model has two different $SU(3)$ symmetries. One called $SU(3)'$ has the original $SU(3)_{\text{flavor}}$ symmetry of the quark model and the other, called $SU(3)''$, makes explicit the hidden three-valued color charge degree of freedom that had been introduced in the parastatistics model of Greenberg. This model allows the $SU(3)''$, which can be identified with the present $SU(3)_{\text{color}}$ if the quark charges are chosen fractional, to be gauged. Indeed Nambu [2] and Han and Nambu [3] introduced an octet of what we now call "gluons" as the mediator of the force between the quarks. The gauging of the three-valued color charge carried by quarks with fractional electric charges is the present QCD, the accepted theory of the strong interactions.

The model of Han and Nambu assigned integer charges to their three triplets to avoid the fractional electric charges of the original quark model. This aspect of the Han-Nambu model conflicts both with experiment and with exact color symmetry and is not part of QCD. Greenberg and Daniel Zwanziger [10] made the identity of the 3 of parafermi statistics of order 3 and the 3 of $SU(3)_{\text{color}}$ with fractionally-charged quarks explicit in 1966.

In addition to the consequences of the parastatistics model, QCD leads to other important results. These include (a) permanent confinement of quarks and color, (b) asymptotic freedom ▶ QCD; QFT, discovered by David J. Gross [11], H. David Politzer [12] and Frank Wilczek [11] in 1973, which reconciles the low energy behavior of quarks confined in hadrons with the quasi-free behavior of quarks that interact at high energy and momentum transfer in the ▶ parton model, (c) running of coupling constants and high-precision tests of QCD at high energy, and (d) jets in high energy collisions.

Note: References [1] through [12] are primary references. References [13] through [18] are secondary references.

Literature

1. O.W. Greenberg: Spin and Unitary Spin Independence in a Paraquark Model of Baryons and Mesons. Phys. Rev. Lett. **13**, 598-602 (1964)
2. Y. Nambu: A Systematics of Hadrons in Subnuclear Physics. In *Preludes in Theoretical Physics*, ed. A. de Shalit, H. Feshbach and L. Van Hove (North Holland, Amsterdam, 1966)
3. M.Y. Han and Y. Nambu: Three-Triplet Model with Double $SU(3)$ Symmetry. Phys. Rev. **139**, B1006-B1010 (1965)
4. M. Gell-Mann: A Schematic Model of Baryons and Mesons. Phys. Lett. **8**, 214-215 (1964)
5. G. Zweig, CERN Reports 8182/TH.401 and 8419/TH.412 (1964). The latter is reprinted as An $SU(3)$ Model for Strong Interaction Symmetry and Its Breaking. In *Developments in the Quark Theory of Hadrons*, ed. D.B. Lichtenberg and S.P. Rosen (Hadronic Press, Nonamtum, Mass., 1980)
6. F. Gürsey and L. Radicati: Spin and Unitary Spin Independence of Strong Interactions. Phys. Rev. Lett. **13**, 173-175 (1964)
7. M.A.B. Bég, B.W. Lee and A. Pais: SU(6) and Electromagnetic Interactions. Phys. Rev. Lett. **13**, 514-517, erratum 650 (1964) The magnetic moment ratio was found independently by B. Sakita: Electromagnetic Properties of Baryons in the Supermultiplet Scheme of Elementary Particles. Phys. Rev. Lett. **13**, 643-646 (1964)
8. W. Pauli: The Connection between Spin and Statistics. Phys. Rev. **58**, 716-722 (1940)
9. H. Harari, in *14th Int. Conf. on High-Energy Physics*, ed. J. Prentki and J. Steinberger (CERN, Geneva, 1968)
10. O.W. Greenberg and D. Zwanziger: Saturation in Triplet Models of Hadrons. Phys. Rev. **150**, 1177-1180 (1966)
11. D. Gross and F. Wilczek: Ultraviolet Behavior of Nonabelian Gauge Theories. Phys. Rev. Lett. **30**, 1343-1346 (1973)
12. H.D. Politzer: Reliable Perturbative Results for Strong Interactions? Phys. Rev. Lett. **30**, 1346-1349 (1973)
13. J.S. Trefil: *From Atoms to Quarks* (Scribner's, New York, 1980)
14. D. Griffiths: *Introduction to Elementary Particles* (Harper & Row, New York, 1987)
15. A. Watson: *The Quantum Quark* (Cambridge University Press, Cambridge, 2004)
16. A. Pais: *Inward Bound: Of Matter and Forces in the Physical World*, (Oxford, Oxford, 1986)
17. T.D. Lee: *Particle Physics and Introduction to Field Theory* (Harwood Academic Publishers, Chur, 1981)
18. O.W. Greenberg: *Quarks*. Ann. Rev. Nucl. Part. Sci. **28**, 327-386 (1978)

Complementarity Principle

Henry Stapp

Niels Bohr introduced and explained his concept of "complementarity" in his famous 1927 Como Lecture (reproduced in [1]. He recognized the need for the mathematical formalism of quantum mechanics to be imbedded in a rationally coherent conceptual framework if it were to serve as the core of an acceptable scientific theory. Yet the applications of the formalism were based upon the integration of two logically incompatible conceptual structures, the mathematical formalisms of classical and quantum physics. The applications that we normally make of quantum theory involve three physical systems: (1), the system being examined; (2), the measuring devices by means of which we probe its properties; and (3), our own physical bodies. All three systems are composed of atoms, and hence must be describable in terms of the mathematical concepts of quantum theory. Yet our observations are described in terms of the contents of our sense experiences, which, for the phenomena under consideration, are described in terms of the concepts of classical physics.

Classical physics postulates that, at each instant of time, each elementary particle is located at some definite point in space, and has a definite velocity, and hence also a definite momentum. On the other hand, in quantum mechanics an elementary particle is represented by a distribution of possibilities, where the distributions in position and in momentum are related by Fourier transformation. This entails that localization at a point in position space demands a complete lack of localization in momentum space, and vice versa. Bohr associates "causation" with the law of conservation of momentum and energy, and hence is able to say that:

The very nature of quantum theory thus forces us to regard the claim of space-time co-ordination and the claim of causality, the union of which characterizes the classical theories, as complementary but exclusive features of the description, symbolizing the idealization of observation and definition respectively. ([1], p. 54)

Bohr explains that:

> The quantum theory is characterized by the acknowledgement of a fundamental limitation in the classical physical ideas when applied to atomic phenomena. ... its essence may be expressed in the so-called quantum postulate, which attributes to any atomic process an essential discontinuity, or rather individuality, completely foreign to classical theories and symbolized by Planck's quantum of action. ... the quantum postulate implies that any observation of atomic phenomena will involve an interaction with the agency of observation not to be neglected. Accordingly, an independent reality in the ordinary physical sense can neither be ascribed to the phenomena nor to the agencies of observation. After all, the concept of observation is in so far arbitrary as it depends upon which objects are included in the system to be observed. Ultimately, every observation can, of course, be reduced to our sense perceptions." ([1], p. 53)

These passages gives a glimpse of the range and complexity of the ideas that Bohr wants to integrate into his rationally coherent foundation for the application and use of quantum theory.

The elaboration that he provides in the remainder of the Como lecture is lengthy, but its essence is summarized and updated in his 1958 paper "Quantum physics and Philosophy: Causality and Complementarity", in which he says:

> Within the scope of classical physics, all characteristic properties of a given object can in principle be ascertained by a single experimental arrangement, although in practice various arrangements are often convenient for the study of different aspects of the phenomena. In fact, data obtained in such a way simply supplement each other and can be combined into a consistent picture of the behaviour of the object under investigation. In quantum mechanics, however, evidence about atomic objects obtained by different experimental arrangements exhibits a novel kind of complementary relationship. Indeed, it must be recognized that such evidence which appears contradictory when combination into a single picture is attempted, exhaust all conceivable knowledge about the object. Far from restricting our efforts to put questions to nature in the form of experiments, the notion of *complementarity* simply characterizes the answers we can receive by such inquiry, whenever the interaction between the measuring instruments and the objects form an integral part of the phenomena. ([2], p.4)

Compactly stated, the essential idea here is that in quantum theory the information provided by different experimental procedures that in principle cannot, because of the physical character of the needed apparatus, be performed simultaneously, cannot be represented by any mathematically allowed quantum state of the system being examined. The elements of information obtainable from incompatible measurements are said to be complementary: taken together they exhaust the information obtainable about the state. On the other hand, any preparation protocol that is maximally complete, in the sense that all the procedures are mutually compatible and are such that no further procedure can add any more information, can be represented by a quantum state, and that state represents in a mathematical form all the conceivable knowledge about the object that experiments can reveal to us.

As regards the closely connected issue of causality, Bohr says:

> In the treatment of atomic problems, actual calculations are most conveniently carried out with the help of a Schrödinger state function, from which the statistical laws governing observations obtainable under specified conditions can be deduced by definite mathematical operations. It must be recognized, however, that we are dealing here with a purely symbolic procedure, the unambiguous physical interpretation of which in the last resort requires reference to the complete experimental arrangement. ([2], p. 5)

This relegation of the Schrödinger state function, which gives the space-time representation of the atomic substrate of all systems, to a purely symbolic status, might seem to be denigrating this Schrödinger representation of the state relative to others. But the point is rather that it puts the Schrödinger space-time representation on a par with the others:

> In fact, wave mechanics, just as the matrix mechanics, represents on this view a symbolic transcription of the problem of motion of classical mechanics adapted to the requirements of quantum theory and only to be interpreted by an explicit use of the quantum postulate. ([1], p.75)

All of this must be understood within the basic pragmatic premise of Bohr's approach:

> In our description of nature the purpose is not to disclose the real essence of phenomena but only to track down as far as possible relations between the multifold aspects of our experience. ([1], p. 18)

Literature

1. N. Bohr: *Atomic theory and the description of nature* (Cambridge University Press, Cambridge 1934)
2. N. Bohr: *Essays 1958–1962 on atomic physics and human knowledge* (Wiley, New York 1963)
3. K. Camilleri: *Bohr, Heisenberg and the divergent views of complementarity*, Studies in History and Philosophy of Modern Physics **38** (2007), 514–528
4. H. Folse: *The philosophy of Niels Bohr. The framework of complementarity* (Dordrecht: North Holland Publ. 1985)
5. D. Murdoch: *Niels Bohr's philosophy of physics* (Cambridge: Cambridge University Press 1987)

Complex-Conjugate Number

Roderich Tumulka

The complex-conjugate number, or conjugate number, of a complex number $z = x + \mathrm{i}y$ with real part x and imaginary part y is the number $x - \mathrm{i}y$, usually denoted $\overline{z}$ or z^*. (The notation z^* is more frequent in quantum physics.)

The definition implies the following properties. Every complex number is the conjugate of its conjugate:

$$\overline{\overline{z}} = z\,, \quad \text{or} \quad (z^*)^* = z. \tag{1}$$

That is, conjugate numbers come in pairs, except for the cases in which a number is conjugate to itself; the latter case occurs if and only if the number $z = x + \mathrm{i}y$ has vanishing imaginary part y, that is if and only if z is real:

$$z^* = z \Leftrightarrow z \in \mathbb{R}. \tag{2}$$

Conjugation, i.e., the operation of taking the conjugate, defines a mapping $^* : \mathbb{C} \to \mathbb{C}$. This mapping is real-linear, i.e.,

$$(z + w)^* = z^* + w^* \quad \text{and} \quad (\lambda z)^* = \lambda(z^*) \tag{3}$$

for all $z, w \in \mathbb{C}$ and $\lambda \in \mathbb{R}$. It is not complex-linear, as there exist $z, w \in \mathbb{C}$ for which $(zw)^* \neq z(w^*)$, but instead conjugation is multiplicative, i.e.,

$$(zw)^* = z^* w^*. \tag{4}$$

If the set of complex numbers is represented as a plane then conjugation corresponds to reflection across the real axis (see Fig. 1). Complex-conjugate numbers have equal modulus (absolute value), $r = |z| = |z^*|$, and opposite phase angles (arguments) $\varphi(z) = -\varphi(z^*)$. As a related fact, for all $\varphi \in \mathbb{R}$ and $z \in \mathbb{C}$,

$$(\mathrm{e}^{\mathrm{i}\varphi})^* = \mathrm{e}^{-\mathrm{i}\varphi} \quad \text{and} \quad (e^z)^* = e^{z^*}. \tag{5}$$

Moreover,

$$z^* z = |z|^2. \tag{6}$$

The real and imaginary part of a complex number z can be expressed using z and z^*:

$$\operatorname{Re} z = \tfrac{1}{2}(z + z^*)\,, \quad \operatorname{Im} z = \tfrac{1}{2\mathrm{i}}(z - z^*). \tag{7}$$

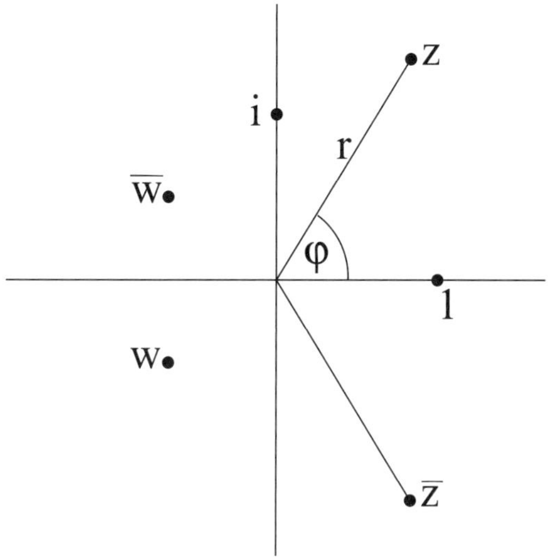

Fig. 1 The complex plane, with example numbers z and w and their complex conjugate numbers $\overline{z}$ and $\overline{w}$

For a function $f(z)$ of a complex variable z one defines the *Wirtinger derivatives*

$$\frac{\partial f}{\partial z} = \frac{1}{2}\left(\frac{\partial u}{\partial x} + \frac{\partial v}{\partial y}\right) + \frac{i}{2}\left(\frac{\partial v}{\partial x} - \frac{\partial u}{\partial y}\right) \tag{8}$$

$$\frac{\partial f}{\partial z^*} = \frac{1}{2}\left(\frac{\partial u}{\partial x} - \frac{\partial v}{\partial y}\right) + \frac{i}{2}\left(\frac{\partial v}{\partial x} + \frac{\partial u}{\partial y}\right), \tag{9}$$

where $x = \mathrm{Re}\, z$, $y = \mathrm{Im}\, z$, $u = \mathrm{Re}\, f$, and $v = \mathrm{Im}\, f$.

Compton Experiment (or Compton Effect)

Friedel Weinert

The famous Compton experiment concentrates on the wave rather than the particle aspect of quantum phenomena. It had been observed that the wavelength of ► X-rays is increased when they are scattered off matter. Arthur Compton (1892–1962) showed that this behaviour could be explained by assuming that the X-rays were photons (► light quantum). When photons are scattered off ► electrons, part of their energy is transferred to the electrons. The loss of energy is translated into a reduction of frequency, which in turn leads to a lengthening of the wavelength of the scattered photons. This happens because the relation $E = h\nu = hc/\lambda$ holds. In these experiments, first carried out between 1919 and 1922, the scattering of X-rays is treated as a collision of photons with electrons (Fig. 1).

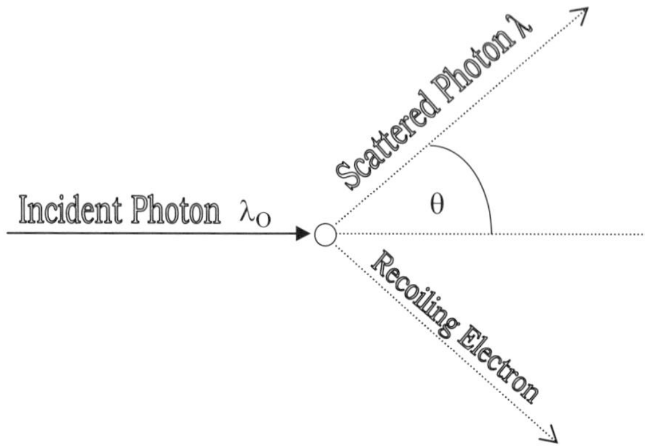

Fig. 1 Compton's model of the scattering process

The wavelength of the scattered photon, λ, can be related to its *initial* wavelength, λ_0, to the electron mass, m_e, and the scattering angle, θ, by the relation $\lambda - \lambda_0 = h/m_e c\,(1 - \cos\theta)$. We should note that Compton was not content with stating the equation. He also sought an explanation. Compton's description of his model conveys the flavour of a mechanistic explanation.

From the point of view of the quantum theory, we may suppose that any particular quantum of X-rays is not scattered by all the electrons in the radiator, but spends all of its energy upon some particular electron. This electron will in turn scatter the ray in some definite direction, at an angle with the incident beam. This bending of the path of the quantum of radiation results in a change in its momentum. As a consequence, the scattering electron will recoil with a momentum equal to the change in momentum of the X-ray. The energy in the scattered ray will be equal to that in the incident ray minus the kinetic energy of the recoil of the scattering electron; and since the scattered ray must be a complete quantum, the frequency will be reduced in the same ratio as is the energy. Thus on the quantum theory we should expect the wavelength of the scattered X-rays to be greater than that of the incident rays.

In terms of a causal account, the effect is the increase in wavelength of the scattered photon, caused by a collision with an electron. Note that Compton's explanation dispenses with the above-stated Compton scattering formula, i.e. the precise numerical determination of the wavelength, λ, of the scattered photon.

Primary Literature

1. A. H. Compton: A Quantum Theory of the Scattering of X-Rays by Light Elements. The Physical Review **21**, 483–502 (1923)
2. A. H. Compton: The Spectrum of Scattered X-Rays. The Physical Review **22**, 409–413 (1923)
3. W. Bothe, H. Geiger: Über das Wesen des Comptoneffekts. Zeitschrift für Physik **32**, 639 (1925)

Secondary Literature

4. M. H. Shamos: *Great Experiments in Physics* (Dover Publications, New York, 1987, 348–58)
5. R. Stuewer: *The Compton Effect – Turning Point in Physics* (Science History, New York, 1975)
6. B. Falkenburg: *Teilchenmetaphysik*. (Spektrum, Heidelberg, 1995, 102–5)

Consistent Histories

Robert B. Griffiths

The consistent histories interpretation of quantum mechanics was introduced by Griffiths in 1984 [1], and further developed by Omnès in 1987 [2]. It is essentially identical to the decoherent histories approach of Gell–Mann and Hartle that first appeared in 1989 [3]. See the monographs [4] and [5] for a detailed treatment and more extensive bibliographies.

In essence, what the consistent histories approach does is to introduce probabilities into quantum mechanics in a fully consistent and physically meaningful way. In Copenhagen quantum mechanics (i.e., the version in most current textbooks) probabilities are introduced with reference to *measurements* and refer (if one is careful) only to measurement *outcomes*, macroscopic states of the measurement apparatus ("pointer positions") after the measurement is over. (► Born rule; Metaphysics in Quantum Mechanics; Nonlocality; Orthodox Interpretation; Schrödinger's Cat; Transactional Interpretation). How these probabilities are related to the microscopic quantum properties supposedly measured is obscure, due to the infamous measurement problem. (► Bohmian mechanics; Measurement theory; Metaphysics in Quantum Mechanics; Modal Interpretation; Objectification; Projection Postulate.) By contrast, the consistent histories approach assigns probabilities to both microscopic and macroscopic states of affairs, using the same formalism for both, without any reference to measurements. Actual laboratory measurements can then be discussed in purely quantum terms using the same principles that apply to any quantum process. ► Hidden variables play no role in the consistent histories approach, which employs the standard quantum ► Hilbert space. And there is no such thing as a classical world or classical measuring apparatus lying outside the quantum domain. Instead, classical physics is an approximation to quantum mechanics, one that works very well in certain situations.

Copenhagen quantum mechanics is a "black box" description in which a macroscopic preparation procedure is followed by a macroscopic measurement outcome, and what happens in between cannot be discussed in terms of microscopic physics if one wants to avoid paradoxes. The consistent histories approach opens the box without generating paradoxes (► errors and paradoxes in quantum mechanics), and thus extends Copenhagen to allow a consistent discussion of microscopic (or macroscopic) quantum physics in probabilistic terms.

Let us see how this works for a spin-half particle whose z component of angular momentum S_z can take on only two values, $+1/2$ and $-1/2$ in units of $\hbar$. These correspond to orthogonal vectors (or rays) in a two-dimensional complex Hilbert space. Each vector can be interpreted as the logical negation of the other, so $+1/2$ and $-1/2$ are mutually exclusive possibilities, one of which must be true. The actual value can be determined by carrying out a ▶ Stern–Gerlach measurement; see Spin; Vector model.

As there are no preferred directions in space, the preceding comments apply equally to the x component of angular momentum, S_x, which is either $+1/2$ or $-1/2$. In classical physics the conjunction of two descriptions of a physical system is always a meaningful description; thus "$L_x = 0.002$ J s AND $L_z = -0.002$ J s" makes perfect sense when referring to two components of angular momentum of a spinning top. But "$S_x = +1/2$ AND $S_z = -1/2$" for a spin-half particle cannot be associated with any vector in the quantum Hilbert space, and in the consistent histories approach it is considered a *meaningless* statement: quantum mechanics can assign it no meaning. Similarly, "$S_x = +1/2$ OR $S_z = -1/2$" is meaningless. Note that "meaningless" is very different from "false," since the logical negation of a false statement is a true statement, whereas the negation of a meaningless statement is equally meaningless. For more details, see Sect. 4.6 of [5].

The *single framework rule* of consistent histories states that two (or more) *incompatible* quantum descriptions – such as $S_x = +1/2$ and $S_z = -1/2$, or other properties represented by noncommuting projectors – cannot be combined to form a meaningful quantum description. Quantum incompatibility is a concept difficult to grasp and easily misunderstood, so the following analogy may be helpful. A photographer taking pictures of Mt. Rainier may do so from a variety of different directions or perspectives: north, south, east, etc. The perspective is chosen by the photographer and has no effect on the reality represented by the mountain. The chosen perspective makes it possible to answer certain questions but not others on the basis of the resulting photograph: a view from the south will not indicate what is happening on the northern slopes. Now replace the photographer with a physicist, the mountain with a spin-half particle, and the choice of perspective with a decision to measure a particular component of its angular momentum. The physicist's choice is free and has no influence on the physical reality associated with the particle before it is measured. However, several photographs of a mountain taken from different perspectives can be combined to provide a more complete description, whereas this is not possible for measurements of different components of spin-half angular momentum. The issue is not that the apparatus will perturb the particle – it certainly will, but we are interested in the particle's state *before* the measurement. The point is that there is *no physical reality* associated with simultaneous values of S_x and S_z, and what is not real cannot be measured.

The consistent histories approach treats the time development of a quantum system as probabilistic, rather than deterministic, and uses ▶ Schrödinger's equation to calculate the requisite probabilities. In the simplest case the ▶ Born rule gives

$$\Pr(\phi_j \mid \psi) = |\langle\phi_j|T(t_1, t_0)|\psi\rangle|^2 \tag{1}$$

for the conditional probability that the quantum system is in the state $|\phi_j\rangle$, belonging to the ► orthonormal basis $\{|\phi_j\rangle\}$, at time t_1, given the state $|\psi\rangle$ at time t_0. Here $T(t', t)$ is the unitary time development operator that results from solving Schrödinger's equation; it is $\exp[-i(t' - t)H/\hbar]$ if the Hamiltonian H is independent of time.

Several comments are in order. First, (1) applies to a closed or isolated quantum system, as Schrödinger's equation only works for this case. Second, unlike Copenhagen, the probability (1) refers not to outcomes of some external measurement, but to physical states inside the closed system, independent of whether or not it is being measured. (These could be pointer states if the measurement apparatus is itself part of the closed quantum system, i.e., inside the box.) Third, the states $\{|\phi_j\rangle\}$ must be orthogonal, for only then do they represent mutually exclusive possibilities appropriate for a quantum sample space. Nonorthogonal states are incompatible (unless multiples of each other), and hence it is meaningless to ask whether one or the other occurred. Fourth, one need not assume that t_0 precedes t_1. The ► Born rule and its consistent extensions (see below) work equally well for both senses of time, so that introducing probabilities into quantum mechanics does not in and of itself single out a direction of time.

The right side of (1) is often written as $|\langle\phi_j|\hat{\psi}\rangle|^2$, where $|\hat{\psi}\rangle = T(t_1, t_0)|\psi\rangle$ is obtained from $|\psi\rangle$ by integrating Schrödinger's equation from t_0 to t_1. When used in this way $|\hat{\psi}\rangle$, which is typically incompatible with the basis states $\{|\phi_j\rangle\}$, does not represent the physical reality of the quantum system at time t_1. It is instead a mathematical construct, a pre-probability in the terminology of [5], used for computing probabilities. One could equally well compute these probabilities by starting with each of the $|\phi_j\rangle$ and integrating Schrödinger's equation in the reverse direction from t_1 to t_0, making no reference whatsoever to $|\hat{\psi}\rangle$. For further discussion, see Sect. 9.4 of [5].

Indeed, $|\hat{\psi}\rangle$ could be the infamous ► Schrödinger's cat state. To discuss whether the cat is dead or alive, the consistent historian adopts an orthonormal basis (or a decomposition of the identity, see [5]) for which these terms make sense, and computes probabilities. As $|\hat{\psi}\rangle$ is a computational tool, it requires no physical interpretation. One could instead adopt an orthonormal basis that includes $|\hat{\psi}\rangle$ as one of its elements, in which case it occurs with probability 1. But then it makes no sense to ask whether the cat is dead or alive, since the corresponding quantum properties are incompatible with $|\hat{\psi}\rangle$.

In order to describe a quantum system at more than two times it is necessary to extend the Born rule to families of quantum histories. A *history* is simply a sequence of quantum events represented by vectors – or, more generally, subspaces – of the quantum Hilbert space at successive times. A family is a collection of mutually exclusive histories, the quantum counterpart of the sample space of a stochastic process in ordinary probability theory. Extending the Born rule is nontrivial because assigning probabilities in a meaningful way requires a *consistent family* or *framework* in which appropriate consistency (or ► decoherence) conditions are satisfied. Different consistent families may be incompatible with each other, in which case they cannot be combined (single-framework rule), even though each one provides a

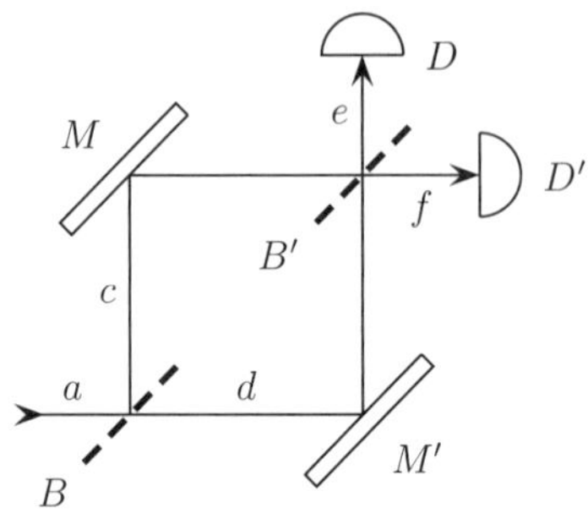

Fig. 1 Mach-Zehnder interferometer

valid set of possibilities for describing the time development of the quantum system. Rather than discussing the details, found in Chaps. 10 and 11 of [5], let us consider a particular application.

The figure shows a Mach–Zehnder interferometer: B and B' are beam splitters, M and M' mirrors, D and D' detectors. Suppose the unitary time development of a photon ▶ wave packet passing through the interferometer has the (schematic) form $|a\rangle \to (|c\rangle + |d\rangle)/\sqrt{2} \to |f\rangle$. This history can be embodied in a family $\mathcal{F}_1$, which remains consistent when extended to include the event that D' is, and D is not, triggered by the arrival of the photon. Within this family it makes no sense to ask whether the photon passes through the c or d arm of the interferometer, for those properties are incompatible with $(|c\rangle + |d\rangle)/\sqrt{2}$. There is a second consistent family $\mathcal{F}_2$ in which the photon while inside the interferometer is either in the c arm or in the d arm, two mutually exclusive possibilities. One can extend $\mathcal{F}_2$ to a consistent family including later states of D and D', but only by using macroscopic quantum ▶ superposition (Schrödinger cat states). Thus a "which arm?" description ($\mathcal{F}_2$) precludes a "which detector?" description ($\mathcal{F}_1$), and vice versa. No fundamental quantum principle singles out one of the two incompatible families $\mathcal{F}_1$ or $\mathcal{F}_2$ as "the correct" description, just as there is no "correct" perspective from which to photograph Mt. Rainier. Instead, certain descriptions are useful when addressing certain physical questions. The same sort of analysis can be applied to the famous ▶ double-slit interference paradox; see Sect. 13.1 of [5].

Quantum measurements pose no difficulty in the consistent histories approach. By adopting an appropriate framework one can show that the measurement *outcome* (pointer position) for a properly constructed quantum-mechanical apparatus is appropriately correlated with, and thus reveals, a property the microscopic system possessed *before* the measurement took place. In brief, measurements actually measure something, as has long been believed by experimental physicists. See Chaps. 17 and 18 of [5] for details. In Chaps. 23 and 24 of [5] it is shown explicitly, by applying appropriate quantum principles, that the nonlocal influences sometimes thought to arise in the Einstein–Podolsky–Rosen gedanken (▶ EPR) experiment are completely spurious: they come about from improperly assuming that "▶ wave function collapse" is a physical process, rather than a mathematical technique for computing conditional probabilities that can be obtained by completely different methods.

This removes an apparent conflict with relativity theory. Indeed, the consistent histories approach, unlike some other interpretations of quantum theory, is perfectly compatible with special relativity [6]. A number of other quantum paradoxes can be resolved or "tamed" in the sense that a consistent analysis is possible using quantum principles, and one is able to identify the point(s) at which an improper use of classical reasoning has led to an apparent contradiction. See Chaps. 19–25 of [5].

C

Here are brief comments on the relationship of consistent histories with some other approaches to quantum interpretation. The connection with Copenhagen (current textbooks) was discussed above. The Everett or ► many-worlds interpretation regards the ► wave function of a closed system ("universe") as representing physical reality, whereas in consistent histories it is a mathematical tool, $|\hat{\psi}\rangle$ in the preceding discussion, useful for computing some but not all of the probabilities of real histories. ► Bohmian mechanics and consistent histories contradict each other about what happens inside the box [7]. Because it solves the Schrödinger cat problem in a completely different way, consistent histories has no need of the nonunitary dynamics employed in spontaneous localization. Unlike Bohmian mechanics and spontaneous localization, there is no conflict between consistent histories and special relativity. Since it employs rules to delineate meaningful descriptions, consistent histories is (or employs) a form of "► quantum logic" in the sense of specifying rules for correct reasoning in the quantum domain. These rules are, however, different from those employed in what is usually called ► quantum logic. See [8] for the relationship between consistent histories and the ► Ithaca interpretation of Mermin.

Primary Literature

1. R. B. Griffiths: Consistent histories and the interpretation of quantum mechanics. J. Stat. Phys. **36**, 219 (1984); Quantum interpretation using consistent histories, in *Fundamental Questions in Quantum Mechanics*, edited by L. M. Roth, and A. Inomata, (Gordon and Breach Science Publishers, New York, 1986, p. 211); Making consistent inferences from quantum measurements, in *New Techniques and Ideas in Quantum Measurement Theory*, edited by D. M. Greenberger, (New York Academy of Sciences, New York, 1986, p. 512); Am. J. Phys. **55**, 11 (1987).
2. R. Omnès: Interpretation of quantum mechanics. Phys. Lett. A **125**, 169 (1987); Logical reformulation of quantum mechanics I, II, III. J. Stat. Phys. **53**, 893, 933, 957 (1988).
3. M. Gell-Mann, J. B. Hartle: Quantum Mechanics in the Light of Quantum Cosmology, in *Complexity, Entropy, and the Physics of Information*, edited by W. Zurek (Addison Wesley, Reading, 1990, p. 425); Classical equations for quantum systems, Phys. Rev. D **47**, 3345 (1993)

Secondary Literature

4. R. Omnès: *Understanding Quantum Mechanics* (Princeton University Press, Princeton, 1999)
5. R. B. Griffiths: *Consistent Quantum Theory* (Cambridge University Press, 2002) and http://quantum.phys.cmu.edu.

6. R. B. Griffiths: Consistent resolution of some relativistic quantum paradoxes. Phys. Rev. A **66**, 062101 (2002)
7. R. B. Griffiths: Bohmian mechanics and consistent histories. Phys. Lett. A **261**, 227 (1999)
8. R. B. Griffiths: Probabilities and quantum reality: are there correlata? Found. Phys. **33**, 1423 (2003).

Copenhagen Interpretation

See ▸ Born rule; Consistent Histories; Metaphysics in Quantum Mechanics; Nonlocality; Orthodox Interpretation; Schrödinger's Cat; Transactional Interpretation.

Correlations in Quantum Mechanics

Richard Healey

The statistical algorithm of quantum mechanics predicts that measurements will reveal correlations among the values of magnitudes ("▸ observables"). Whenever such measurements have been performed, they have borne out the predictions. But the patterns exhibited by these correlations can be difficult to square with classical intuitions – about probability, about the nature and properties of quantum systems, and about causal connections between systems.

In a ▸ Hilbert space formulation, an observable is represented by a ▸ self-adjoint operator, while the state of a system is represented by a normalized vector (perhaps a ▸ wave function) or more generally a ▸ density operator $\hat{W}$ (a self-adjoint operator with unit trace). If $\{O_1, ..., O_n\}$ is a set of observables on a system represented by pairwise commuting operators $\{\hat{O}_1, ..., \hat{O}_n\}$, then quantum mechanics predicts that measured values of all these observables in state $\hat{W}$ will conform to a joint probability distribution $pr(O_1 \in \Delta_1, ..., O_n \in \Delta_n)$ given by

$$pr(O_1 \in \Delta_1, ..., O_n \in \Delta_n) = Tr\left[\hat{W}\hat{O}_1(\Delta_1).....\hat{O}_n(\Delta_n)\right] \tag{1}$$

where $\hat{O}_i(\Delta_i)$ is the element of the spectral resolution of $\hat{O}_i$ corresponding to Borel set Δ_i of possible values ($i = 1, ..., n$). If any two operators $\hat{O}_i$, $\hat{O}_j$ in such a set fail to commute, then no joint distribution is predicted.

For example, a simple quantum mechanical model of a Hydrogen atom ▸ Bohr's atom model will predict a joint probability distribution for energy, total angular momentum, and z-component of angular momentum in any state; but it will never predict a joint probability distribution for energy, position and momentum, nor for z-component and x-component of angular momentum.

The orthodox view of this reticence takes non-commuting operators to represent incompatible observables–pairs of observables that can never be jointly measured with arbitrary precision because at most one of each pair may have a precise value in any state. In general, there are no theoretical restrictions on the precision with which any single observable may be measured. So measurement cannot be taken always faithfully to reveal the value of the measured observable.

A number of "no-go" theorems may be cited in support of this orthodox view [6, 11]. But when are the joint distributions that quantum mechanics *does* predict compatible with an underlying joint distribution for *all* observables? Fine [4] shows that the necessary and sufficient condition for four two-valued observables A_i, B_j $(i, j = 1, 2)$ to have a joint distribution compatible with the given joints is that the following system of (BCH) inequalities be satisfied, for $i \neq i'$ and $j \neq j'$:

$$\begin{aligned} -1 \leqslant\ & \mathrm{pr}(A_i, B_j) + \mathrm{pr}(A_i, B_{j'}) + \mathrm{pr}(A_{i'}, B_{j'}) - \mathrm{pr}(A_{i'}, B_j) \\ & -\mathrm{pr}(A_i) - \mathrm{pr}(B_{j'}) \leqslant 0 \end{aligned}$$

As we shall see, for some observables and quantum states quantum mechanics predicts values for the terms in this expression that violate the inequalities: these predictions have been verified. Such observables then have no joint distribution.

The state of a non-relativistic particle may be represented in a tensor product Hilbert space $\mathcal{H} = \mathcal{H}_1 \otimes \mathcal{H}_2$, where $\mathcal{H}_2$ is used to represent its ► spin. But not every vector in a product space is itself expressible as a tensor product of vectors, one from each space. A vector state of the form $|\ \psi_1\rangle \otimes \cdots \otimes |\ \psi_n\rangle$ is said to be *separable*. The state of a pair of particles may also be represented in a tensor product of the spaces used to represent their individual states. When their joint state is nonseparable between these component spaces, the particles are said to be *entangled*, and their state exhibits state holism (► Holism in Quantum Mechanics). The total spin space for a pair of spin-1/2 particles is a tensor product of two-dimensional spin spaces that includes nonseparable spin states, including the singlet spin state

$$|\ \psi_{\mathrm{s}}\rangle = \frac{1}{\sqrt{2}}\,(|\uparrow\rangle \otimes |\downarrow\rangle - |\downarrow\rangle \otimes |\uparrow\rangle) \tag{2}$$

Any spin component A_i on one particle is compatible with any spin component B_j on the other, so quantum mechanics predicts a joint distribution for every such pair. There are many choices of four such observables for which these violate the (BCH) inequalities in the singlet state and other entangled states.

Quantum mechanics predicts that measurements of the *same* spin-component on each particle in the singlet state will yield *different* results with probability 1. Einstein believed that if particles in such a pair are widely separated, then each must have its own real state, and any influence on the state of one can have no direct influence on the state of the other [3]. On that basis his argument would conclude that each particle in the singlet state *has* a definite value of spin-component in every direction. But every way of distributing such values among many pairs will yield a statistical distribution conforming to the (BCH) inequalities [2, 13]. So unless the statistics systematically differ between measured and unmeasured pairs, measurements confirming (quantum mechanically-predicted) statistics in violation of (BCH) inequalities refute this conclusion. These measurements have been successfully performed in circumstances where the event of choice and execution of a measurement on one particle is spacelike separated from the analogous event on the other [1]. Not only is there no known mechanism by which the measurement on one particle could influence the result of the other measurement: any such influence would have to be superluminal, undetectable and unpreventable, and extraordinarily selective. Although Einstein dismissed this possibility as "spooky" action at a distance, the observed violations of (BCH) inequalities show we may have to live with just such a novel kind of non-local "causal" connection [10], [► Causal Inference and EPR].

But causation is a relation between distinct events. Perhaps it is wrong to regard each particle, or measurement event, as a distinct entity, each with its own properties. If a pair together constitute an indivisible whole, then the question of causal relations among its parts doesn't arise. The clearest violations of (BCH) inequalities involve the polarization states of pairs of photons (► light quantum). A two-photon state of the quantized electromagnetic field is perhaps best not thought to consist of two distinct particles–certainly not if each were considered to have its own trajectory. From this perspective, violation of (BCH) inequalities only seems strange if one fails to acknowledge the fundamental *holism* underlying quantum mechanics. It is neither the properties of quantum objects nor their probabilistic relations that strain our non-classical intuitions, but the objects themselves. Such ontological holism is also suggested by the fact that violations of (BCH)-type inequalities occur even in the vacuum state of a quantum field [14].

Leggett [9] has proposed a test of macroscopic realism that relies on an unusual application of (BCH)-type inequalities involving measurements on a single system. Here the quantum correlations that cause problems for a classical world-view concern measurements at different times of the current circulating in an RF SQUID. There are quantum mechanical states that are ► superpositions of different directions of current circulation. Assuming these are measurable without disturbance, then measurements of the current at carefully chosen times will reveal correlations that are incompatible with the assumption that the current is always circulating either one way or the other.

Investigations of the nature of light have uncovered correlations that seemed surprising on the assumption that light is "composed" of photons. Hanbury, Brown and Twiss [5] investigated correlations between the responses of two separated detectors to a weak light source. They expected the responses of the detectors to be *uncorrelated*, on the grounds that each photon could activate only one detector at a time. Instead they found strong correlations. These could be explained by a ► semi-classical model in which light is treated classically but the detectors are treated quantum-mechanically. The anticorrelations expected on the photon hypothesis only showed up much later after the incoherent light source was replaced by a source to which single excited atoms made independent coherent contributions [7, 8].

Correlations play a starring role in some proposed interpretations of quantum mechanics. Mermin [12] claims that while correlations have physical reality, that which they correlate does not. This view of correlations without correlata has produced philosophical debate but little consensus.

See Consistent histories, Ignorance interpretation, Ithaca Interpretation, Many Worlds Interpretation, Modal Interpretation, Orthodox Interpretation, Transactional Interpretation.

Primary Literature

1. A. Aspect, J. Dalibard, G. Roger, Experimental test of Bell's Inequalities using time-varying analyzers, Physical Review Letters 49: 1804–7, 1982.
2. J.S. Bell, On the Einstein-Podolsky-Rosen paradox, Physics 1: 195–200, 1964.
3. A. Einstein, Quantentheorie und Wirklichkeit, Dialectica 2: 320–4, 1948.
4. A. Fine, Joint distributions, quantum correlations, and commuting observables, Journal of Mathematical Physics 23: 1306–10, 1982.
5. R. Hanbury-Brown and R.Q. Twiss, Correlations between photons in two coherent beams of light, Nature 177: 27–9, 1956.

Secondary Literature

6. J. Bub, Interpreting the Quantum World, Cambridge University Press, Cambridge 1997.
7. P. Grangier, G. Roger and A. Aspect, Experimental evidence for a photon anticorrelation effect on a beamsplitter, Europhysics Letters 1: 173–9, 1986.
8. M. Hennrich, A. Kuhn, and G. Rempe, Transition from antibunching to bunching in cavity QED, Phys. Rev. Lett. 94: 53604-1–4, 2005.
9. A.J. Leggett and A. Garg, Quantum mechanics versus macroscopic realism: Is the flux there when nobody looks?, Phys. Rev. Lett. 54: 857–60, 1985.
10. T. Maudlin, Quantum Non-locality and Relativity, Blackwell, Oxford 1994.
11. N.D. Mermin, Simple unified form for the major no-hidden-variables theorems, Phys. Rev. Lett. 65: 3373–7, 1990
12. N.D. Mermin, What is quantum mechanics trying to tell us?, American Journal of Physics 66: 753–67, 1998.
13. A. Shimony, Bell's Theorem at http://plato.stanford.edu/entries/bell-theorem/.
14. S.J. Summers and R.F. Werner, The vacuum violates Bell's inequalities, Phys. Lett. A 110: 257–9, 1985.

Correspondence Principle

Brigitte Falkenburg

The correspondence principle is due to Niels Bohr (1885–1962). According to Bohr, the principle justifies the use of formal classical expressions in quantum theory and a physical interpretation of quantum theory in terms of classical concepts. The principle emerged from his use of classical concepts and formal analogies in ► *Bohr's atomic model* of 1913. Before the rise of quantum mechanics (i.e., in "old" quantum theory), Bohr employed the principle in order to establish inter-theoretical relations between the classical theory of radiation and the quantum theory of atomic spectra. After the rise of quantum mechanics, he justified his ► *complementarity* view of quantum mechanics in terms of the correspondence between mutually exclusive quantum phenomena on the one hand and the classical concepts of wave or particle (*particle picture, wave picture*) (► Franck–Hertz experiment; Davisson–Germer experiment; Stern–Gerlach experiment; Schrödinger equation) on the other hand.

Werner Heisenberg (1901–1976) made heuristic use of Bohr's correspondence principle when he developed his ► matrix mechanics. In 1930, he developed a generalized version of the correspondence principle which emphasized the heuristic and interpretative aspects of the correspondence principle.

See also ► Bohmian mechanics; Measurement theory; Metaphysics in Quantum Mechanics; Modal Interpretation; Objectification; Projection Postulate.

In view of the quantum measurement problem, a generalized correspondence principle is indispensable up to the present day. In particular, it underlies the ► semi-classical models of atomic and nuclear physics, condensed matter physics etc.

Classical Concepts in "Old" Quantum Theory

► Bohr's *atomic model* of 1913 was based on quantum postulates which violate the classical laws of radiation. The model raised the question of how the quantized transitions between the stationary electron states relate to the classical theory of radiation. In order to explain this, Bohr postulated a formal analogy between the harmonics of classical radiation and the various quantum jumps from a given stationary state. This analogy warranted the asymptotic agreement between the classical and quantum-theoretical radiations in the limit of large ► quantum numbers (when the quantum jumps become very small) [1, 9, 10]. Together with Ehrenfest's "adiabatic hypothesis" (which concerned the energy of the permitted electron motions [2]), the analogy justified a limited use of the classical concepts of energy and frequency in quantum theory. In particular, it made it possible to interpret the quantum law $\Delta E = h\nu$ in terms of the classical concepts of energy and frequency. This was the germ of the correspondence principle. 1914–1918, Bohr elaborated the analogy for periodic systems and extended it to multi-periodic systems and more general cases [10]. He managed to derive ► selection rules for the line splitting of the hydrogen spectrum in an electric or magnetic field, i.e., the ► Stark and ► Zeeman effects. After Einstein had introduced transition probability coefficients [3], Bohr expected that the limited use of classical electrodynamics should also give correct intensities and polarizations for the spectral lines. The calculations were performed by Hendrik Anthony Kramers (1894–1952) [4], who applied the correspondence principle to the Fourier analysis of the classical stationary motions and derived in this way the intensities and polarizations of the hydrogen lines, including the fine structure, Stark and Zeeman effects.

Finally, in 1920 Bohr gave the following explicit formulation of the correspondence principle [5, p. 23–24; quoted in 10, p. 137–138]:

> [...] there is found [...] to exist a far-reaching correspondence between the various types of possible transitions between the stationary states on the one hand and the various harmonic components of the motion on the other hand. This correspondence is of such a nature that the present theory of spectra is in a certain sense to be regarded as a rational generalization of the ordinary theory of radiation.

Here, the correspondence principle comes in two steps. First, it states the actual correspondence of the possible quantum transitions to components of the classical motion. Second, it claims that the quantum theory of atomic spectra should be regarded as a "rational generalization" of the classical theory of radiation. The first point justified the use of classical concepts in quantum theory. The second point justified the heuristic use of the correspondence principle for the derivation of quantum laws.

To regard the quantum theory of atomic spectra as a "rational generalization" of the classical theory of radiation has two aspects, a formal and an interpretative one [10, p. 82; 12]. The classical orbit is merely formal since it can by no means be measured and is only related to the quantum radiation in a formal, indirect manner. At the same time, the correspondence principle associates the symbol ν in the formal expression $\Delta E = h\nu$ with the familiar quantity of a light frequency measured by a spectrometer, in accordance with the laws of classical wave optics.

In old quantum theory, the correspondence principle had a hybrid theoretical status. On the one hand, it was a meta-theoretical principle. It established inter-theoretical relations between classical radiation theory and the laws of old quantum theory. On the other hand, it put inner-theoretical constraints on the formulation of quantum laws, thus making the extension of old quantum theory possible. Hence, Bohr's correspondence principle should not be confused with an empirical rule of correspondence in the sense of empiricist philosophy of science. It does much more than only assigning the empirical concept of a "line in the spectrum" to the formal law of radiation $\Delta E = h\nu$, as Ernest Nagel (1901–1985) suggested [14]. In particular, it does not relate theoretical concepts directly to an observational language. Rather, it is an inter-theoretical relation that establishes a formal (numerical) and interpretative (physical) analogy between classical radiation theory and quantum theory. This two-fold analogy allows for the continued use of the classical concepts of 'frequency', 'wavelength', 'energy', 'polarization', etc. in the quantum theory of atoms and line spectra. Even taken as an internal principle of old quantum theory, the correspondence principle only expresses constraints that derive from an inter-theoretical relation.

Correspondence and Complementarity

Quantum mechanics emerged from the crisis of old quantum theory confronted by the anomalous Zeeman effect and other problems with which the correspondence principle could not cope. Nevertheless, Bohr's correspondence principle played a crucial heuristic role for Heisenberg when he developed his matrix mechanics. After the rise of quantum mechanics, Heisenberg emphasized that the correspondence principle helps to obtain a quantum theory from quantizing the corresponding classical theory (see below).

In view of quantum mechanics, Bohr employed the correspondence principle in order to interpret the formal quantum concepts. He considered Schrödinger's ► wave function Ψ as a mere symbol, as a formal tool that lacks any direct physical meaning [9,15]. His ► *complementarity* view of quantum mechanics aimed at interpreting quantum phenomena in terms of the corresponding classical concepts. According to his famous Como lecture, ► Heisenberg's *uncertainty relations* describe quantum phenomena which correspond to mutually exclusive classical descriptions and appear under mutually exclusive experimental conditions [6]. Bohr's examples of complementary quantum phenomena are ► *particle tracks* and ► *scattering events* such as the ► *Compton effect*, on the one hand, and interference fringes, on the other hand. The physical magnitudes attributed to these phenomena (i.e., either momentum-energy, or spatio-temporal magnitudes) are classical. According to Bohr's writings of 1927 and later, any physical magnitude attributed to a quantum phenomenon represents the outcome of a measurement, and all measurement results have to be expressed in classical terms. Bohr thought that a full understanding of quantum phenomena is only possible in terms of the corresponding classical concepts (i.e., either momentum-energy or spatio-temporal location) and classical models (i.e., the complementary *wave* and *particle picture* ► Franck–Hertz experiment; Davisson–Germer experiment; Stern–Gerlach experiment; Schrödinger equation [9–11,13,15].

The Generalized Correspondence Principle

In 1930, Heisenberg generalized Bohr's correspondence principle. His generalized principle deals explicitly with inter-theoretical relations, extending Bohr's original analogy between classical and quantized radiation frequencies to many more physical quantities. Heisenberg emphasizes three features of the general correspondence principle [7, p. 70]:

1. It postulates a detailed analogy between the quantum theory and the appropriate "mental picture", i.e., the classical *wave* or *particle picture*.
2. This analogy is a "guide to the discovery of formal laws", i.e., it has heuristic meaning in the formation of a quantum theory. Here, Heisenberg means the well-known ► quantization of a classical theory.
3. In addition, it "furnishes the interpretation of the formal laws in terms of the mental picture used", i.e., the analogy tells us that we may attribute to the quantized ► observables the physical properties of the corresponding classical wave or particle picture.

Like Bohr's original version, Heisenberg's generalized correspondence principle is a principle of semantic continuity [10, p. 133–137; 11; 12, p. 188–194]. It guarantees that the predicates for the classical physical properties of 'position', 'momentum', 'mass', 'energy', etc. can also be defined in the domain of quantum mechanics, and that one may interpret them operationally in accordance with classical measurement methods. It provides many inter-theoretical relations by means of which the formal concepts and models of quantum mechanics can be filled with physical meaning. Bohr and Heisenberg both called this physical meaning "intuitive", even though in quite a different sense [6,11].

In modern textbooks of quantum mechanics, the generalized correspondence principle shows up for example in ► Ehrenfest theorem.

Correspondence in Semi-Classical Models

Often, the general correspondence principle helps to interpret the abstract formalism of a quantum theory in such a way that it can be applied against the background of classical physics and on semi-classical conditions. In the semi-classical models of quantum physics, the correspondence principle is tacitly employed up to the present day. Important examples stem from condensed matter physics, atomic and nuclear physics, as well as ► *particle physics*.

In condensed matter physics, the macroscopic state of a solid is necessarily presupposed. As a macroscopic state, it has obviously to be described in classical terms. As Philip K. Anderson (*1923) emphasized, the existence of a solid (or the regularity of the ground states of most assemblages of atoms, respectively) can not be explained by quantum theory [16, p. 3]. In addition, the quantum behavior of a complex many-particle system cannot be calculated *ab initio*. Therefore, semi-classical approximations are indispensable in condensed matter physics or atomic physics. Many ► *scattering experiments* of atomic, nuclear, and ► *particle physics* are based on ► semi-classical models, too. The models of the scattering of subatomic particles off the atoms inside macroscopic measuring devices are based on several semi-classical conditions. In these models, a generalized correspondence principle is employed in the following ways [12, pp. 125–160]:

1. The simplest models of quantum mechanical scattering theory correspond to classical Rutherford scattering. Exact correspondence between the classical and quantum mechanical differential scattering cross sections (► *scattering experiments*) is given in the case of the Rutherford formula, that is, for the Coulomb potential, for non-relativistic probe particles, and in the absence of quantum mechanical ► spin or exchange effects.
2. In the domain of ► relativistic quantum mechanics and ► quantum field theory, there is a chain of models of quantum mechanical scattering theory, namely Mott scattering and Dirac scattering, that approximately correspond to Rutherford

scattering under well-defined conditions. Here, the tacit use of a generalized correspondence principle is extended to the inter-theoretical relation between relativistic and non-relativistic concepts.

3. To describe the charge distribution inside the atom by a classical form factor (▶ *nuclear models*) is based on the correspondence between the quantum mechanical many-particle ▶ wave function $|\Psi(\mathbf{r})|^2$ of charged subatomic particles and the classical charge distribution $\rho(r)$, which is the Fourier transform of a classical form factor $F(q)$.
4. In the domain of relativistic quantum field theory, the above correspondence assumptions (1)–(3) come together in the definition of *structure functions*, which express (via correspondence to the classical case, again) the momentum distributions of the partons (▶ parton model) or quark constituents of the nucleons, the proton and neutron (▶ *large angle scattering*).
5. The data analysis of the *particle tracks* taken in such (▶ *scattering experiments*) is based on a similar chain of models, which relate the quantum mechanics of scattering to the corresponding classical case.

In all ▶ semi-classical models, the generalized correspondence principle bridges the semantic gaps between quantum theory and the classical theories, which are due to the unresolved problems of the ▶ *measurement process*. Hence, the correspondence principle connects the languages of classical physics and quantum theory. In a further common generalization, it bridges the languages of non-relativistic and relativistic theories.

Limitations of Correspondence

Obviously, the correspondence principle does not exhaust the domain of the current quantum theories. Indeed quantum mechanics emerged from its limitations in old quantum theory. These early limitations were due to the spin-orbit coupling effects in the spectra of complex atoms. Later, the ▶ nonlocality of quantum mechanics predicted in the famous ▶ *EPR* paper showed up. Today, in addition to the EPR correlations many non-local quantum phenomena without any classical correspondence are known, such as, e.g., super conductivity, the Bohm–Aharanov effect, etc.

However, the semi-classical models of quantum physics are affected by the limitations of the correspondence principle, too. In particular, such limitations are relevant for the data analysis of ▶ *particle tracks*. According to the classical *particle picture*, a particle loses energy along its track due to dissipation, where the energy loss is due to the ionization of atoms (e.g., in Wilson's cloud chamber). There is indeed a classical model of the process, namely Bohr's classical calculation of energy loss by ionization [8]. However, for charged particles that pass the cloud chamber with non-vanishing energy, the results of this model are wrong by a factor of 2. In addition, the non-relativistic model of the energy loss via ionization no longer corresponds to the relativistic description of the scattering processes along the track of a particle of high energy. In particular, the process of pair creation, which becomes the more probable the higher the particle energy is, does not have any classical analogue [12, p. 174–187].

C

Primary Literature

1. N. Bohr: *On the Constitution of Atoms and Molecules* (1913). In: Bohr's Collected Works (BCW). Amsterdam: North-Holland (Elsevier) 2, 161–233.
2. P. Ehrenfest: Adiabatische Invarianten und Quantentheorie. Annalen der Physik 51 (1916), 327–52.
3. A. Einstein: Zur Quantentheorie der Strahlung. Physikalische Zeitschrift 18 (1917), 121–28.
4. H. A. Kramers: *Intensities of spectral lines*. Der Kongelige Danske Videnskabernes Selskab, Mathematisk-fysiske Meddelser 8/3/3 (1919), 284–388.
5. N. Bohr: *The Theory of Spectra and Atomic Constitution*. (Cambridge: Cambridge University Press 1922).
6. W. Heisenberg: Über den anschaulichen Inhalt der quantentheoretischen Kinematik und Mechanik. Z. für Physik 43 (1927), 172–98. N. Bohr: *The Quantum Postulate and the Recent Development of Atomic Theory*. Como Lecture (1927). Modified version: Nature 121 (1928), 580–90. Both versions in: BCW 6, 109–58.
7. W. Heisenberg: Physikalische Prinzipien der Quantentheorie. Leipzig: Hirzel 1930. Engl. Transl.: *The Physical Principles of Quantum Theory.* Transl. by C. Eckart and F.C. Hoyt. (Chicago: University of Chicago Press, 1930).
8. N. Bohr: *On the Theory of the Decrease of Velocity of Moving Electrified Particles on Passing through Matter* (1913). In: BCW Vol. 2, 18–39. On the Decrease of Velocity of Swiftly Moving Electrified Particles on Passing through Matter. In: BCW 8, 127–60.

Secondary Literature

9. C. Chevalley: *Introduction and Glossaire*. In: Niels Bohr: Physique atomique et connaissance humaine. (Paris: Gallimard, 1991), 17–144 and 345–567.
10. O. Darrigol 1992, *From c-Numbers to q-Numbers*. (Berkeley: University of California Press 1992), Part B.
11. B. Falkenburg: *Bohr's Principles of Unifying Quantum Disunities*. Phil. naturalis 35/1 (1998), 95–120. Functions of Intuition in Quantum Physics. In: Carson and Huber (eds.): Intuition and the Axiomatic Method. (Dordrecht: Springer 2006), 267–92.
12. B. Falkenburg: *Particle Metaphysics. A Critical Account of Subatomic Reality*. (Heidelberg: Springer 2007).
13. K. M. Meyer-Abich: Korrespondenz, Individualität und Komplementarität. (Wiesbaden: Steiner 1965).
14. E. Nagel: *The Structure of Science. Problems in the Logic of Scientific Explanation*. (London: Routledge & Kegan Paul 1961).
15. H. Pringe: *Critique of the Quantum Power of Judgement*. (Berlin: De Gruyter 2007).
16. P. W. Anderson: *Concepts in Solids*. (Singapore: World Scientific 1997).

Counterfactuals in Quantum Mechanics

Lev Vaidman

Counterfactuals in quantum mechanics appear in discussions of (a) ► nonlocality, (b) pre- and post-selected systems, and (c) ► interaction-free measurement; Quantum interrogation. Only the first two issues are related to counterfactuals as they are considered in the general philosophical literature:

If it were that $\mathcal{A}$, then it would be that $\mathcal{B}$.

The truth value of a counterfactual is decided by the analysis of similarities between the actual and possible counterfactual worlds [1].

The difference between a counterfactual (or counterfactual conditional) and a simple conditional: *If $\mathcal{A}$, then $\mathcal{B}$,* is that in the actual world $\mathcal{A}$ is not true and we need some "miracle" in the counterfactual world to make it true. In the analysis of counterfactuals out of the scope of physics, this miracle is crucial for deciding whether $\mathcal{B}$ is true. In physics, however, miracles are not involved. Typically:

$\mathcal{A}$: *A measurement $\mathcal{M}$ is performed*

$\mathcal{B}$: *The outcome of $\mathcal{M}$ has property $\mathcal{P}$.*

Physical theory does not deal with the questions of which measurement and whether a particular measurement is performed? Physics yields conditionals: "If $\mathcal{A}_i$, then $\mathcal{B}_i$". The reason why in some cases these conditionals are considered to be counterfactual is that several conditionals with incompatible premises $\mathcal{A}_i$ are considered with regard to a single system.

The most celebrated example is the Einstein–Podolsky–Rosen (► EPR problem) argument in which incompatible measurements of the position or, instead, the momentum of a particle are considered. Stapp has applied a formal calculus of counterfactuals to various EPR-type proofs [2,3] and in spite of extensive criticism [4–9], continues to claim that the nonlocality of quantum mechanics can be proved without the assumption "reality" [10].

Let me give here just the main point of this controversy. Stapp provides elaborate arguments in which an *a priori uncertain* outcome of a measurement of O in one location might depend on the measurements performed on an entangled quantum particle in another location. But if *anything* is different in a counterfactual world, the outcome of the measurement of O need not be the same as in the actual world. The core of the difficulty is this randomness of the outcomes of quantum measurements. The formal philosophical analysis of counterfactuals which uses similarity criteria, presupposes that in a counterfactual world which is identical to the actual world in all relevant aspects up until the measurement of O, the outcome has to be the same. Thus, Stapp's analysis tacitly adopts the *counterfactual definiteness* [4, 5] which is essentially equivalent to "reality" or ► hidden variables and which is absent in the conventional quantum theory.

Important examples of quantum counterfactuals are *elements of reality*. Consider the following *definition* [11]:

> If we can *infer* with certainty that the result of measuring at time t of an observable O is o, then, at time t, there exists an element of reality $O = o$.

If we consider several elements of reality which cannot be verified together, we obtain counterfactuals. A celebrated example is the Greenberger–Horne–Zeilinger (▶ GHZ) entangled state of three spin-$\frac{1}{2}$ particles [4, 13]:

$$|\Psi\rangle = \frac{1}{\sqrt{2}}(|{\uparrow}\rangle_A|{\uparrow}\rangle_B|{\uparrow}\rangle_C - |{\downarrow}\rangle_A|{\downarrow}\rangle_B|{\downarrow}\rangle_C). \tag{1}$$

We consider spin component measurements of these three particles in the x and y directions. The counterfactuals (the elements of reality) have a more general form than merely "the value of O is o", they are properties of a set of three measurements:

$$\begin{aligned}
\{\sigma_{Ax}\}\{\sigma_{Bx}\}\{\sigma_{Cx}\} &= -1,\\
\{\sigma_{Ax}\}\{\sigma_{By}\}\{\sigma_{Cy}\} &= 1,\\
\{\sigma_{Ay}\}\{\sigma_{Bx}\}\{\sigma_{Cy}\} &= 1,\\
\{\sigma_{Ay}\}\{\sigma_{By}\}\{\sigma_{Cx}\} &= 1.
\end{aligned} \tag{2}$$

Here $\{\sigma_{Ax}\}$ signifies the outcome of a measurement of σ_x of particle A, etc. Since one cannot measure for the same particle both σ_x and σ_y at the same time, this is a set of counterfactuals. It is a very important set because no local hidden variable theory can ensure such outcomes with certainty; there is no solution for the set of equations (2).

Lewis's theory of counterfactuals is asymmetric in time [14]. The counterfactual worlds have to be identical to the actual world during the whole time before $\mathcal{A}$, but not after. This creates difficulty in applications of counterfactuals to physics and especially to quantum mechanics because "before" and "after" are not absolute concepts. Different Lorentz observers might see different time ordering of measurements performed at different places. Finkelstein [15] and Bigaj [16] have attempted to define time asymmetric counterfactuals to overcome this difficulty. But in my view, the time asymmetry of quantum counterfactuals is an unnecessary burden [17]. We *can* consider a time symmetric (or time neutral) definition of quantum counterfactuals.

The general strategy of counterfactual theory is to find counterfactual worlds closest to the actual world. In the standard approach, the worlds must be close only before the measurement. In the time-symmetric approach, the counterfactual worlds should be close to the actual world both before and after the measurement at time t. Quantum theory allows for a natural and non-trivial definition of "close" worlds as follows: *all outcomes of all measurements performed before and after the measurement of O at time t are the same in the actual and counterfactual worlds.*

A peculiar example of time symmetric counterfactuals is the *three box paradox* [18]. Consider a single particle prepared at time t_1 in a ▶ superposition of being in three separate boxes:

$$|\Psi_1\rangle = \frac{1}{\sqrt{3}}(|A\rangle + |B\rangle + |C\rangle). \tag{3}$$

At a later time t_2 the particle is found in another superposition:

$$|\Psi_2\rangle = \frac{1}{\sqrt{3}}(|A\rangle + |B\rangle - |C\rangle). \tag{4}$$

For this pre- and post-selected particle, a set of counterfactual statements, which are *elements of reality* according to the above definition, is:

$$\begin{aligned} \mathbf{P}_A &= 1, \\ \mathbf{P}_B &= 1. \end{aligned} \tag{5}$$

Or, in words: if we open box A, we find the particle there for sure; if we open box B (instead), we also find the particle there for sure. Indeed, not finding the particle in box A (or B) collapses the pre-selected state (3) to a state which is orthogonal to the post-selected state (4).

Beyond these counterfactual statements, there are numerous manifestations of the claim that in some sense, this single particle is indeed in two boxes simultaneously. A single photon which interacts with this particle scatters as if there are two particles: one in A and one in B, but two or more photons (▶ light quantum) do not "see" two particles. Many photons see this single particle as two particles if the photons interact weakly with the particle. Indeed, there is a useful theorem which says that if a strong measurement of an observable O yields a particular outcome with probability 1, (i.e. there is an element of reality) then a weak measurement yields the same outcome. Sometime this is called a *weak-measurement element of reality* [19]. The outcomes of weak measurements are *weak values* (▶ weak value and weak measurements):

$$\begin{aligned} (\mathbf{P}_A)_w &= 1, \\ (\mathbf{P}_B)_w &= 1. \end{aligned} \tag{6}$$

Contrary to the set of counterfactuals above, the weak measurements can be performed simultaneously both in box A and box B. Thus, the existence of counterfactuals helps us to know the outcome of real (weak) measurement.

The three-box paradox and other time-symmetric quantum counterfactuals have raised a significant controversy [11, 20, 21, 21–28]. It seems that the core of the controversy is that quantum counterfactuals about the results of measurements of ▶ observables, and especially "elements of reality" are understood as attributing values to observables which are not observed. But this is completely foreign to quantum mechanics. Unperformed experiments have no results! "Element of reality" is just a shorthand for describing a situation in which we know with certainty the outcome of a measurement *if* it is to be performed, which in turn helps us to know how weakly coupled particles are influenced by the system. Having "elements of reality" does not mean having values for observables. The semantics are misleading since "elements of reality" are not "real" in the ontological sense.

An attempt to give counterfactuals some ontological sense, at the cost of placing artificial constraints on the context in which counterfactuals are considered, was made by Griffiths [29]. He showed that counterfactuals have no paradoxical features when only ► *consistent histories* are considered. Another recent step in this direction are quantum counterfactuals in very restrictive "measurement-ready" situations [30].

Penrose [31] used the term "counterfactuals" in a very different sense:

> *Counterfactuals* are things that might have happened, although they did not in fact happen.

In interaction-free measurements [32], an object is found because it might have absorbed a photon, although actually it did not. This idea has been applied to "counterfactual computation" [33], a setup in which the outcome of a computation becomes known in spite of the fact that the computer did not run the algorithm (in case of one particular outcome [34]).

In the framework of the ► Many-Worlds Interpretation, Penrose's "counterfactuals" are counterfactual only in one world. The physical Universe incorporates all worlds, and, in particular, the world in which Penrose's "counterfactual" is actual, the world in which the "counterfactual" computer actually performed the computation.

This work has been supported in part by the European Commission under the Integrated Project Qubit Applications (QAP) funded by the IST directorate as Contract Number 015848 and by grant 990/06 of the Israel Science Foundation.

Literature

1. D. Lewis: *Counterfactuals*. Oxford, Blackwell (1973).
2. H.P. Stapp: *S-Matrix interpretation of quantum theory*. Phys. Rev. D 3, 1303 (1971).
3. H.P. Stapp: *Nonlocal character of quantum theory*. Am. J. Phys. 65, 300 (1997).
4. B. Skyrms: *Counterfactual definiteness and local causation*. Phil. Sci. 49, 43 (1982).
5. M. Redhead: *Incompleteness, Nonlocality, and Realism: A Prolegomenon to the Philosophy of Quantum Mechanics*. New York, Oxford University Press (1987).
6. R.K. Clifton, J.N. Butterfield, M. Redhead: *Nonlocal influences and possible worlds – a Stapp in the wrong direction*. Br. J. Philos. Sci. 41, 5 (1990).
7. D. Mermin: *Nonlocal character of quantum theory?* Am. J. Phys. 66, 920 (1998).
8. W. Unruh: *Nonlocality, counterfactuals, and quantum mechanics*. Phys. Rev. A 59, 126 (1999).
9. A. Shimony, H. Stein: *Comment on Nonlocal character of quantum theory*, Am. J. Phys. 69, 848 (2001).
10. H.P. Stapp: *Comments on Shimony's an analysis of Stapp's 'a Bell-type theorem without hidden variables'*, Found. Phys. 36, 73 (2006).
11. L. Vaidman: *The meaning of elements of reality and quantum counterfactuals*: Reply to Kastner. Found. Phys. 29, 856 (1999).

12. D.M. Greenberger, M.A. Horne, A. Zeilinger: *Going beyond Bell's theorem.* In Bell Theorem, Quantum Theory and Conceptions of the Universe, M. Kafatos, ed., p. 69, Dordrecht, Kluwer, (1989).
13. N.D. Mermin: *Quantum mysteries revisited.* Am. J. Phys. 58, 731 (1990).
14. D. Lewis: *Counterfactual dependence and time's arrow.* Nous 13, 455 (1979).
15. J. Finkelstein: *Space-time counterfactuals.* Synthese 119, 287 (1999).
16. T. Bigaj: *Counterfactuals and spatiotemporal events.* Synthese 142, 1 (2004).
17. L. Vaidman: *Time-symmetrized counterfactuals in quantum theory.* Found. Phys. 29, 755 (1999).
18. Y. Aharonov, L. Vaidman: *Complete description of a quantum system at a given time.* J. Phys. A 24, 2315 (1991).
19. L. Vaidman: *Weak-measurement elements of reality.* Found. Phys. 26, 895 (1996).
20. W.D. Sharp, N. Shanks: *The rise and fall of time-symmetrized quantum mechanics.* Philos. Sci. 60, 488 (1993).
21. R.E. Kastner: *Time-symmetrised quantum theory, counterfactuals and 'advanced action'.* Stud. Hist. Philos. Mod. Phy. 30 B, 237 (1999).
22. L. Vaidman: *Defending time-symmetrised quantum counterfactuals.* Stud. Hist. Philos. Mod. Phy. 30 B, 337 (1999).
23. R.E. Kastner: *The three-box paradox and other reasons to reject the counterfactual usage of the ABL rule.* Found. Phys. 29, 851 (1999).
24. R.E. Kastner: *The nature of the controversy over time-symmetric quantum counterfactuals.* Phil. Sci. 70, 145 (2003).
25. L. Vaidman: (2003) *Discussion: Time-Symmetric Quantum Counterfactuals.* e-print: PITT-PHIL-SCI00001108 (2003).
26. U. Mohrhoff: *Objective probabilities, quantum counterfactuals, and the ABL rule* A response to R. E. Kastner. Am. J. Phys. 69, 864 (2001).
27. K.A. Kirkpatrick: *Classical three-box 'paradox'.* J. Phys. A 36, 4891 (2003).
28. T. Ravon, L. Vaidman: *The three-box paradox revisited.* J. Phys. A 40, 2882 (2007).
29. R.B. Griffiths: *Consistent quantum counterfactuals.* Phys. Rev. A 60, R5 (1999).
30. D.J. Miller: *Counterfactual reasoning in time-symmetric quantum mechanics.* Found. Phys. Lett. 19, 321 (2006).
31. R. Penrose: *Shadows of the Mind.* Oxford, Oxford University Press (1994).
32. A.C. Elitzur, L. Vaidman: *Quantum mechanical interaction-free measurements.* Found. Phys. 23, 987 (1993).
33. G. Mitchison, R. Jozsa: *Counterfactual Computation.* Proc. R. Soc. Lond. A 457, 1175 (2001).
34. L. Vaidman: *Impossibility of the counterfactual computation for all possible outcomes.* Phys. Rev. Lett. 98, 160403 (2007).

Covariance

K. Mainzer

Covariance means form invariance, i.e. the form of a physical law is unchanged (invariant) with respect to transformations of reference systems. Covariance can be distinguished from ▶ invariance which refers to quantities and objects [2]. The covariant formulation of laws implies that the form of laws is independent of the state of motion in a reference system that an observer takes. In that sense, all fundamental laws of classical and relativistic physics are covariant [3, 4]. According to the definition of covariance, the gauge principle (▶ gauge symmetry; symmetry) can also be considered a principle of gauge covariance [5].

In quantum mechanics, measurable quantities (eigenvalues, probabilities, expectation values) are invariants (▶ invariance) with respect to unitary transformations (▶ symmetry). But the form of laws changes in a ▶ Heisenberg picture or ▶ Schrödinger picture. The fundamental laws of quantum mechanics can also be formulated in a covariant form with respect to arbitrary unitary transformations [1]. In this case the fundamental laws are represented by the following schemes:

1. Heisenberg's commutation relation:

$$[Q_K, P_L] = \mathrm{i}\eta\, \delta_{KL}, [Q_K, Q_L] = 0, [P_K, P_L] = 0$$

2. Heisenberg's equation of motion for operators:

$$\frac{\mathrm{d}F}{\mathrm{d}t} = \frac{\partial F}{\partial t} + \frac{1}{\mathrm{i}h}[F, H]\,(F{=}F(Q_K, P_K, t))$$

3. Equation of movement for a general state and eigenvalues:

$$\frac{\mathrm{d}\,|\psi\rangle}{\mathrm{d}t} = \frac{\partial\,|\psi\rangle}{\partial t} - \frac{1}{\mathrm{i}h} H\,|\psi\rangle\,, \frac{\mathrm{d}\,|f_\Gamma\rangle}{\mathrm{d}t} = \frac{\partial\,|f_\Gamma\rangle}{\partial t} - \frac{1}{\mathrm{i}h} H\,|f_\Gamma\rangle$$

   The concept of state $|\psi\rangle = |\psi\,(t)\rangle$ resp. $|f_\Gamma\rangle = |f_\Gamma\,(t)\rangle$ is generalized as $|\psi\rangle = |\psi\,(Q_K\,(t)\,, P_L\,(z)\,, t)\rangle$ resp. $|f_\Gamma\rangle = |f_\Gamma\,(Q_K\,(t)\,, P_L\,(z)\,, t)\rangle$ which allows the partial time-depending derivation of states. This formulation yields a maximal symmetry between the equations of motion between operators and states.
4. Eigenvalue equation:

$$F\,|f_\Gamma\rangle = f_\Gamma\,|f_\Gamma\rangle$$

These equations can be considered a picture-free formulation of quantum mechanics, because they are covariant with respect to arbitrary unitary transformations.

Primary Literature

1. E. Schmutzer: Eine neue Grundlegung der Quantenmechanik. Nova Acta Leopoldina 44(8), 79–158 (1976)

Secondary Literature

2. F. Friedman: *Foundations of Space-Time Theories* (Princeton University Press, Princeton, 1983)
3. K. Mainzer: *Symmetrien der Natur* (De Gruyter, Berlin, 1988; English trans. *Symmetries of Nature*, De Gruyter, New York, 1996)

4. K. Mainzer: *Symmetry and Complexity. The Spirit and Beauty of Nonlinear Science* (World Scientific, Singapore, 2005)
5. T.Y. Cao: Gauge Theory and the Geometrization of Fundamental Physics. In: H.R. Brown, R. Harré (eds.): *Philosophical Foundations of Quantum Field Theory* (Clarendon Press, Oxford, 1988)

CPT Theorem

Claus Kiefer

The CPT theorem is a theorem for local relativistic quantum field theories in Minkowski space-time. Here, C means 'charge conjugation', P 'parity transformation' ('space inversion'), and T 'time inversion'; while C and P are implemented by ► unitary operators, T is implemented by an antiunitary operator.

The CPT theorem states that these field theories are invariant under the combined combination of C, P, and T; one therefore speaks of CPT symmetry. The original proof by Gerhart Lüders [1] and Wolfgang Pauli [2] was performed within Lagrangian field theories; Res Jost then presented a more general proof using axiomatic quantum field theory [3].

The importance of the CPT theorem stems from the fact that the assumptions for this theorems are very general; in fact, they are believed to be universally valid for field theories in flat space-time. The main assumption is Lorentz ► invariance, which implements the principle of special relativity; in addition, one has to assume that the fields obey the standard commutation relations. The proof in [3], besides being more general, has also the advantage that it provides a simple method to calculate the CPT transform of a field directly, without having to calculate C, P, and T separately and to multiply them.

The Standard Model of elementary particles ► quantum field theory; particle physics describes the strong and the electroweak interaction by a local relativistic field theory and therefore implements the CPT symmetry; however, it violates CP symmetry (and therefore T symmetry), as has been confirmed by many experimental tests.

CPT symmetry entails in particular that the masses of particles and antiparticles must be equal. This, in turn, provides the most precise test of this symmetry. The current experimental bounds result mainly from the limit of the mass difference between the neutral K-meson K^0 and its antiparticle, $\bar{K}^0$ [4]:

$$\left| \frac{m_{\bar{K}^0} - m_{K^0}}{m_{K^0}} \right| \leqslant 10^{-18}.$$

The CPT symmetry also entails equal lifetimes for particles and antiparticles. More details on the CPT theorem can be found in references [5, 6].

It is clear from its proof that the CPT theorem is not expected to hold if the main assumption – Lorentz symmetry – is violated. This should apply, in particular, to a fundamental theory of ► quantum gravity, since already the classical theory of gravity (Einstein's theory of general relativity) is not a Lorentz-invariant theory (it possesses instead ► diffeomorphism invariance). Since, moreover, time seems to be absent in quantum gravity, the theorem cannot even be formulated at the most fundamental level.

Literature

1. G. Lüders: On the Equivalence of Invariance under Time Reversal and under Particle-Anti-Particle Conjugation for Relativistic Field Theories. K. Dan. Vidensk. Selsk. Mat.-fys. Medd. **28**(5), (1954)
2. W. Pauli: Exclusion Principle, Lorentz Group and Reflection of Space-Time and Charge. In *Niels Bohr and the Development of Physics*, ed. by W. Pauli (Pergamon Press, New York 1955, 30–51)
3. R. Jost: Eine Bemerkung zum CPT-Theorem. Hel. Phys. Acta **30**, 409–416 (1957)
4. W.-M. Yao et al.: The Review of Particle Physics. J. Phys. G: Nucl. Part. Phys. **33**, 1 (2006)
5. R.F. Streater, A.S. Wightman: *PCT, Spin and Statistics, and All That* (Princeton University Press, Princeton 2000).
6. O.W. Greenberg: Why is CPT fundamental? Found. Phys. **36**, 1535–1553 (2006). A related online version is http://arxiv.org/abs/hep-ph/0309309.

Creation and Annihilation Operators

Christopher Witte

Creation and annihilation operators are linear ► operators on a so called *Fock space* associated to a complex ► Hilbert space. The interpretation of creation and annihilation operators in multi-particle quantum system is that they *increase* and *lower*, respectively, the number of particles of the system by one. Some of many applications of these operators can be found in the study of oscillations in solids, quantum optical systems, spin systems and general free quantum fields.

Fock Space. Let $\mathcal{H}$ be a complex Hilbert space and $\mathcal{H}^{\otimes n}$ the n-fold tensor product of $\mathcal{H}$. The orthogonal direct ℓ_2-sum of Hilbert spaces $\mathscr{F}(\mathcal{H}) := \bigoplus_{n=0}^{\infty} \mathcal{H}^{\otimes n}$ (with $\mathcal{H}^{\otimes 0} := \mathbb{C}$) is called the *Fock space* over $\mathcal{H}$.

The n-particle *symmetrization operator* $S_+^{(n)}$ and *antisymmetrization operator* $S_-^{(n)}$ are defined by linear extension of

$$S_\pm^{(n)}(f_1 \otimes \cdots \otimes f_n) = \frac{1}{n!} \sum_{\sigma \in \mathcal{S}_n} (\pm 1)^\sigma f_{\sigma(1)} \otimes \cdots \otimes f_{\sigma(n)}$$

(sum over all permutations σ, with $(-1)^\sigma$ the *signature* of σ) and for the sake of completeness $S_\pm^{(0)} := 1$. They are orthogonal projectors onto the *bosonic* ($S_+^{(n)}$) and *fermionic* ($S_-^{(n)}$) n-particle space $\mathcal{H}_\pm^{\otimes n} := S_\pm^{(n)}\mathcal{H}^{\otimes n}$. The orthogonal direct ℓ_2-sum of Hilbert spaces $\mathscr{F}_\pm(\mathcal{H}) := \bigoplus_{n=0}^\infty \mathcal{H}_\pm^{\otimes n}$ is called the *symmetric* or *bosonic Fock space* ($\mathscr{F}_+$) and the *antisymmetric* or *fermionic Fock space* ($\mathscr{F}_-$) over $\mathcal{H}$. These spaces are used as state spaces for systems with identical particles of variable number. The element $1 \in \mathcal{H}_\pm^{\otimes 0}$ will be denoted by Ω, when embedded into a Fock space, and called the *vacuum* or no-particle state.

By linear extension the following sets of operators are defined on the Fock spaces: For any $f \in \mathcal{H}$ i.) the *creation operator* $a^*(f)$ is defined by

$$a^*(f)S_\pm^{(n)}(f_1 \otimes \cdots \otimes f_n) := \sqrt{n+1}S_\pm^{(n+1)}(f \otimes f_1 \otimes \cdots \otimes f_n),$$

thus mapping n-particle states to $(n+1)$-particle states, and ii.) the *annihilation operator* $a(f)$ is defined by

$$a(f)S_\pm^{(n)}(f_1 \otimes \cdots \otimes f_n) := \frac{1}{\sqrt{n}}\sum_j (\pm 1)^{j-1}\langle f, f_j\rangle S_\pm^{(n-1)}(f_1 \otimes \cdots \hat{f}_j \cdots \otimes f_n),$$

where $\hat{f}_j$ denotes the omission of the j-th factor such that this operator maps n-particle states to $(n-1)$-particle states. On the vacuum Ω the action of the operator is defined to be $a(f)\Omega = 0$.

Given any ► orthonormal basis $\{e_i\}$ of the one-particle Hilbert space $\mathcal{H}$ the sum of operators $\sum_i a^*(e_i)a(e_i)$ converges on each n-particle space to the n-fold of the identity operator. Therefore it is common to write the formal sum $N := \sum_i a^*(e_i)a(e_i)$, where N denotes the self-adjoint *number operator* with discrete spectrum and eigenspaces $\mathcal{H}_\pm^{\otimes n}$ for eigenvalue $n \in \mathbb{N}_0$. The eigenvectors of the number operator, i.e., the elements of $\mathcal{H}_\pm^{\otimes n}$ embedded into Fock space, are also called *Fock states*.

Another important class of vectors especially in bosonic systems are the eigenvectors of the annihilation operator, obeying $a(f)\psi_\alpha^f = \alpha\psi_\alpha^f$, with generally complex eigenvalue α. Contrary to the Fock states, the statistical distribution of the results in a *number measurement* in these states is a *Poisson distribution*. These states are usually called ► coherent states and are of great importance in the study of quantum optical systems (see, e.g., [4]).

Occupation-Numbers. In the *bosonic* n-particle space $\mathcal{H}_+^{\otimes n}$ an orthonormal basis related to a one-particle basis $\{e_i\}$ is given by

$$e(n_1, n_2, \ldots) := \sqrt{\frac{n!}{n_1!n_2!\ldots}}\, S_+^n(e_{i_1} \otimes \ldots \otimes e_{i_n}),$$

where n_i is the number of indices among $i_1, \ldots, i_n$ which are equal to i. Evidentally $\sum_i n_i = n$, and $e(0, 0, \ldots) = \Omega$. Considering the vectors $e(n_1, n_2, \ldots)$ for all values of $n \in \mathbb{N}_0$, a basis of the symmetric Fock space $\mathcal{F}_+(\mathcal{H})$ consisting of Fock states is induced. The representation of vectors and operators of $\mathcal{H}_+^{\otimes n}$ and $\mathcal{F}_+(\mathcal{H})$ with respect to the basis $\{e(n_1, n_2, \ldots)\}$ is called the *occupation-number representation associated with* $\{e_i\}$.

C

The bosonic creation and annihilation operators can be replaced by the discrete set of operators $a_i^* := a^*(e_i)$ and $a_i := a(e_i)$. The action of these operators on the basis is given by

$$a_i^* e(n_1, \ldots, n_i, \ldots) = \sqrt{n_i + 1}\, e(n_1, \ldots, n_i + 1, \ldots)$$
$$a_i e(n_1, \ldots, n_i, \ldots) = \begin{cases} \sqrt{n_i}\, e(n_1, \ldots, n_i - 1, \ldots) & \text{if } n_i \neq 0 \\ 0 & \text{if } n_i = 0. \end{cases}$$

An orthonormal basis in the *fermionic* n-particle space $\mathcal{H}_-^{\otimes n}$ is given by

$$e(n_1, n_2, \ldots) := \sqrt{n!}\, S_-^n (e_{i_1} \otimes \ldots \otimes e_{i_n}),$$

where $i_1 < i_2 < \ldots < i_n$, $n_i = 1$ or $n_i = 0$ depending on whether the vector e_i is among $e_{i_1}, \ldots, e_{i_n}$ or not, and $\sum_i n_i = n$; the basis vectors define the *occupation-number representation* for fermions. The creation and annihilation operators $a_i^* := a^*(e_i)$ and $a_i := a(e_i)$ act according to

$$a_i^* e(n_1, \ldots, n_i, \ldots) = \begin{cases} (-1)^{s_i}\, e(n_1, \ldots, n_i + 1, \ldots) & \text{if } n_i = 0 \\ 0 & \text{if } n_i = 1 \end{cases}$$
$$a_i e(n_1, \ldots, n_i, \ldots) = \begin{cases} 0 & \text{if } n_i = 0 \\ (-1)^{s_i}\, e(n_1, \ldots, n_i - 1, \ldots) & \text{if } n_i = 1 \end{cases}$$

where $s_i = \sum_{j=1}^{i-1} n_j$ (i.e., s_i is the number of indices i_j satisfying $i_j < i$).

Any self-adjoint one-particle operator A acting on $\mathcal{H}$ gives rise to a self-adjoint operator on Fock space (as well bosonic as fermionic) acting on all particles identically, sometimes called the "second ▶ quantization" of the operator. It is defined by the formal sum

$$d\Gamma(A) := \sum_{n=0}^{\infty} \sum_{\nu=1}^{n} \mathbf{1} \otimes \ldots \otimes A \otimes \ldots \otimes \mathbf{1},$$

where in the inner sum A is at the ν-th position. This can be written in an easy way using creation and annihilation operators:

$$d\Gamma(A) = \sum_{i,j} A_{ij} a_i^* a_j,$$

with matrix elements $A_{ij} = \langle e_i, A e_j \rangle$. Translated to the occupation-number representation one finds

$$Ae(n_1, \ldots, n_i, \ldots) = \sum_i n_i A_{ii} e(n_1, n_2, \ldots)$$
$$+ \sum_{i \neq j} \sqrt{(n_i + 1) n_j} A_{ij} e(n_1, \ldots, n_i + 1, \ldots, n_j - 1, \ldots).$$

The easiest example of such an operator is the above seen *number operator*: $N = d\Gamma(\mathbf{1})$.

It is worth noting at this point that the "second quantization" of unitary operators is defined differently, namely by $\Gamma(U) := \sum_{n=0}^{\infty} U \otimes \ldots \otimes U$. In this way the useful relation $\exp(\mathrm{i}td\Gamma(H)) = \Gamma(\exp(\mathrm{i}tH))$ in the realm of unitary one-parameter groups holds true (see ► Hamiltonian operator).

Canonical Commutation and Anticommutation Relations. The bosonic annihilation and creation operators are *unbounded* linear operators and can be defined on the dense subset D_+ of the bosonic Fock space $\mathscr{F}_+(\mathcal{H})$ constituted by finite sums of n-particle vectors [1]. On this subset they are formal adjoints of each other in the way the notation suggest: $a(f)^*|_{D_+} = a^*(f)|_{D_+}$. Furthermore they fulfil on D_+ the following relations:

$$[a(f), a^*(g)] = \langle f, g \rangle; \quad [a(f), a(g)] = [a^*(f), a^*(g)] = 0,$$

called *canonical commutation relations* (CCRs). Together with the property $a(f)\Omega = 0$ the CCRs define the action of the bosonic creation and annihilation operators, justifying the term "canonical" [2, 5]. The operators $A(f) := (a(f) + a^*(f))/\sqrt{2}$ are *essentially self-adjoint* and thus one can form unitary operators $W(f) = \exp(\mathrm{i}A(f))$ with these. The CCRs can expressed equivalently by these so called *Weyl operators*:

$$W(f)W(g) = W(f + g)\mathrm{e}^{-\mathrm{i}\,\mathrm{Im}\langle f,g \rangle /2}.$$

In the study of coherent states it is worth noting that the Weyl operators map the vacuum to coherent states: $a(f)W(f)\Omega = (i\langle f, f\rangle/\sqrt{2})\, W(f)\Omega$. The C*-algebra generated by the Weyl operators is called the *CCR algebra*.

The fermionic annihilation and creation operators are *bounded* linear operators with norm $\|a(f)\| = \|a^*(f)\| = \|f\|$. Indeed the mapping $f \mapsto a^*(f)$ is an isometric embedding of Banach spaces, whereas the mapping $f \mapsto a(f)$ is antilinear, i.e., $a(\lambda f) = \bar{\lambda} a(f)$ for $\lambda \in \mathbb{C}$, and isometric. Thus both sets of operators are defined on the whole fermionic Fock space $\mathscr{F}_-(\mathcal{H})$ and are adjoints of each other: $a(f)^* = a^*(f)$. By defining the *anticommutator* $[A, B]_+ = AB + BA$, one finds

$$[a(f), a^*(g)]_+ = \langle f, g \rangle; \quad [a(f), a(g)]_+ = [a^*(f), a^*(g)]_+ = 0,$$

called *canonical anticommutation relations* (CARs). The basic consequence $a^*(f)^2 = 0$ is a demonstration of the *Pauli* ► *exclusion principle* in fermionic systems. Together with the property $a(f)\Omega = 0$ the CARs define the action of the fermionic creation and annihilation operators. The norm closure of polynomials in the $a(f)$ and $a^*(f)$ form a C*-Algebra, called the *CAR algebra*. A detailed description of CCRs and CARs can be found in [3].

Continuous Representations. If the Hilbert space $\mathcal{H}$ is represented in form of a function space $L^2(\mathbb{R}^n)$, it is common to introduce creation and annihilation operators *in a point* $a^*(x)$ and $a(x)$. Mathematically these are *operator-valued distributions*, defined by

$$a^*(f) = \int a^*(x) f(x) \mathrm{d}^n x; \quad a(f) = \int a(x) \overline{f(x)} \mathrm{d}^n x.$$

While $a(x)$ can still be interpreted as a densely defined, but not *closable* operator, $a^*(x)$ is not an operator at all. Formally the operator-valued distributions fulfil the *continuous CARs and CCRs*

$$[a(x), a^*(y)]_\pm = \delta(x - y); \quad [a(x), a(y)]_\pm = [a^*(x), a^*(y)]_\pm = 0.$$

Examples. The most basic bosonic Fock space is $\mathscr{F}_+(\mathbb{C}) = \bigoplus_{n=0}^\infty \mathbb{C}$, which is canonically isomorphic to the sequence space ℓ_2. Each n-particle space is one-dimensional and spanned by the sequence $e(n) = (\delta_{nk})_k$, with the Kronecker delta being different from zero only at the n-th position. These vectors form an orthonormal basis of ℓ_2, and define the occupation-number representation in this case. The action of creation and annihilation operator is $a_1^* e(n) = \sqrt{n+1}\, e(n+1)$ and $a_1 e(n) = \sqrt{n}\, e(n-1)$ (the indices of the operators can be omitted due to one-dimensionality of $\mathcal{H}$).

This example is relevant in the study of the one-dimensional *quantum mechanical harmonic oscillator*, modeled on the Hilbert space $L^2(\mathbb{R})$. By defining annihilation and creation operators on this space, one can find a suitable isomorphism to $\mathscr{F}_+(\mathbb{C})$. On $L^2(\mathbb{R})$ we set $a := \sqrt{m\omega/(2\hbar)}\,(x + \mathrm{i}p/(m\omega))$ and $a^* := \sqrt{m\omega/(2\hbar)}\,(x - \mathrm{i}p/(m\omega))$, where x and p denote position and momentum operators and m and ω are positive constants. The two operators obey the CCRs (with f set to unity) and the operator a has a one-dimensional kernel, from which we choose a normed representative $\Omega = |0\rangle = (m\omega/(\pi\hbar))^{1/4} \exp\left(-m\omega x^2/(2\hbar)\right)$. By defining $|n\rangle := (a^*)^n |0\rangle/\sqrt{n}$ one finds an orthonormal basis and thus the isomorphism onto $\mathscr{F}_+(\mathbb{C})$ by $|n\rangle \mapsto e(n)$. The Hamiltonian operator of an oscillator of mass m and frequency ω can be expressed in the simple form $\mathrm{H} = \hbar\omega(a^* a + 1/2)$. The operator $N = a^* a$ is the number operator in the one dimensional setting with $N|n\rangle = n|n\rangle$. Thus the n-particle states are the eigenstates of the Hamiltonian operator, with $\mathrm{H}\,|n\rangle = (n + 1/2)\hbar\omega|n\rangle$. The term "particle" is somewhat misleading in this context, since it does not refer to the single oscillating particle, but to so called *phonons*, which is a name for each "quantum" of oscillation energy, numbered by n. The "vacuum" state refers to the absence of any such oscillation quantum and defines the ground state of the system. *Coherent states* of the oscillator, given by $a\psi_\alpha = \alpha\psi_\alpha$, can be derived by the Weyl operator from the vacuum $\psi_\alpha = W(-\mathrm{i}\sqrt{2}\alpha)|0\rangle$. The Weyl operator can be expressed by position and momentum operators, leading to an interpretation as *displacement operator* in phase space. Coherent states can thus be seen as elongated ground states with a certain momentum. They are not stationary, but stay coherent with only the phase of the eigenvalue α changing in time.

The typical Hamiltonian of a bosonic many-particle system with constant particle number reads

$$H = d\Gamma(H_1) + \frac{1}{2}\sum_{\mu\neq\nu} V_{\mu\nu}.$$

Here the one-particle Hamiltonian H_1 (kinetic energy and potential energy in an exterior field) is used in "second quantization" $d\Gamma(H_1)$, and

$$V_{\mu\nu}(e_{k_1}\otimes\ldots\otimes e_{k_\mu}\otimes\ldots\otimes e_{k_\nu}\otimes\ldots\otimes e_{k_n}) := \sum_{i,j} V_{ijk_\mu k_\nu} e_{k_1}\otimes\ldots\otimes e_i\otimes\ldots e_j\otimes\ldots\otimes e_{k_n},$$

e_i being at position μ, e_j at position ν, acts only on the μ-th and ν-th tensor factor nontrivially and $V_{ijk_\mu k_\nu}$ is the matrix element of some two-body interaction operator V.

Due to the special form of H, acting on each particle identically, it makes sense to write the Hamiltonian H in occupation-number representation. H can be represented in terms of creation and annihilation operators according to

$$H = \sum_{i,j} H_{ij} a_i^* a_j + \frac{1}{2}\sum_{i,j,k,l} V_{ijkl} a_i^* a_j^* a_k a_l$$

where the matrix elements of H_1 are $H_{ij} := \langle e_i, H_1 e_j\rangle$. In particular, if the basis vectors e_i are eigenvectors of H_0, $H_0 e_i = E_i e_i$, then $d\Gamma(H_1) = \sum_i E_i a_i^* a_i$, i.e.,

$$d\Gamma(H_1)e(n_1, n_2, \ldots) = \sum_i n_i E_i e(n_1, n_2, \ldots).$$

The most basic fermionic Fock space is $\mathscr{F}_-(\mathbb{C}) = \mathbb{C}\oplus\mathbb{C} = \mathbb{C}^2$, since the antisymmetrization operator reduces all n-particle spaces for $n \geqslant 2$ to $\{0\}$ in this case. The vectors Ω and $a^*\Omega$ can be identified with the canonical basis of $\mathbb{C}^2$ and span the vacuum and the 1-particle space, respectively. The annihilation and creation operator can be represented by matrices:

$$a = \begin{pmatrix} 0 & 1 \\ 0 & 0 \end{pmatrix}; \quad a^* = \begin{pmatrix} 0 & 0 \\ 1 & 0 \end{pmatrix}.$$

This system can be taken as model for a single locally fixed electron with ► spin in a magnetic field. The Hamiltonian operator of such a system is basically given by a multiple of the number operator a^*a, i.e.,

$$\mathrm{H} = 2\mu_{\mathrm{S}} B a^* a = 2\mu_{\mathrm{S}} B \begin{pmatrix} 0 & 0 \\ 0 & 1 \end{pmatrix},$$

with μ_{S} the spin magnetic moment of the electron and B the magnetic field.

Likewise fermionic Fock spaces over finite dimensional Hilbert spaces $\mathbb{C}^n$ have dimension 2^n and are isomorphic to $(\mathbb{C}^2)^{\otimes n}$. Therefore they can be used to model n-electron spin systems (see, e.g., [2]). The general formalism to write a fermionic system in occupation-number representation is analogous to the bosonic case seen above.

C

Primary Literature

1. K. Ito (editor): *Encyclopedic Dictionary of Mathematics (2nd edn.)*. MIT Press, Cambridge (1993)
2. W. Thirring: *Quantum Mathematical Physics: Atoms, Molecules and Large Systems*. Springer, Berlin (2002)
3. O. Bratteli, D.W. Robinson: *Operator Algebras and Quantum Statistical Mechanics 2*. Springer, New York (1979)

Secondary Literature

4. M.O. Scully, M.S. Zubairy: *Quantum Optics*. Cambridge University Press, Cambridge (1997)
5. J. Glimm, A. Jaffe: *Quantum Physics*. Springer, New York (1981)

Creation and Detection of Entanglement

Dagmar Bruß

The fundamental equation of non-relativistic quantum mechanics, the ▶ Schrödinger equation, is linear. Thus, superpositions of its solutions (quantum states) constitute solutions as well. This is the famous ▶ superposition principle. Given a composite quantum system, i.e. a quantum system that consists of two or more subsystems, superpositions of its states can be either separable or entangled [1]. The quantum state of a bipartite system, i.e. a system consisting of two subsystems A (located at Alice's lab) and B (located at Bob's lab), is an element of the tensored Hilbert space $\mathcal{H} = \mathcal{H}_A \otimes \mathcal{H}_B$. A pure bipartite state $|\psi\rangle \in \mathcal{H}_A \otimes \mathcal{H}_B$ is called *separable* if and only if $|\psi\rangle = |a\rangle \otimes |b\rangle$, where $|a\rangle \in \mathcal{H}_A$ and $|b\rangle \in \mathcal{H}_B$. It is *entangled* otherwise.

A mixed bipartite density matrix ϱ, acting on $\mathcal{H}_A \otimes \mathcal{H}_B$, is called separable if and only if it can be written as [2] $\varrho = \sum_i p_i |a_i\rangle\langle a_i| \otimes |b_i\rangle\langle b_i|$, with $|a_i\rangle \in \mathcal{H}_A$ and $|b_i\rangle \in \mathcal{H}_B$. It is entangled otherwise. Here the coefficients p_i are probabilities, i.e. $0 \leqslant p_i \leqslant 1$ and $\sum_i p_i = 1$. In general $\langle a_i | a_j\rangle \neq \delta_{ij}$, and also Bob's states need not be orthogonal. This decomposition is not unique. Note that a mixed separable state may contain classical correlations, but no quantum correlations (entanglement), see the reviews [24–26] and general textbooks on quantum information, e.g. [27–29].

The definition of a separable state can be interpreted as follows: as a separable state is a statistical mixture of projectors onto product states, Alice and Bob can create a separable state locally in their corresponding laboratories, with the help of communication over a classical channel (e.g. a telephone). In other words, any state that can be prepared without interaction of the subsystems is not entangled. In order to create entanglement, the subsystems have to interact via some entangling (non-local) Hamiltonian [3]. When a Hamiltonian acts for a certain time, one can consider its action as a quantum gate. The most simple quantum gate that allows to entangle two qubits (any two-level system can be considered as a qubit) is the CNOT gate, with the truth table $|00\rangle \rightarrow |00\rangle, |01\rangle \rightarrow |01\rangle, |10\rangle \rightarrow |11\rangle, |11\rangle \rightarrow |10\rangle$, i.e. the second qubit (target) is flipped if the first qubit (control) is in state $|1\rangle$. A simple quantum network, consisting of a Hadamard gate, with the truth table $|0\rangle \rightarrow \frac{1}{\sqrt{2}}(|0\rangle + |1\rangle), |1\rangle \rightarrow \frac{1}{\sqrt{2}}(|0\rangle - |1\rangle)$, applied to the first qubit, and a subsequent CNOT gate acting on both qubits, creates from the four possible inputs $|00\rangle, |01\rangle, |10\rangle, |11\rangle$ the four (maximally entangled) Bell states $|\Phi^+\rangle = \frac{1}{\sqrt{2}}(|00\rangle + |11\rangle), |\Psi^+\rangle = \frac{1}{\sqrt{2}}(|01\rangle + |10\rangle), |\Phi^-\rangle = \frac{1}{\sqrt{2}}(|00\rangle - |11\rangle), |\Psi^-\rangle = \frac{1}{\sqrt{2}}(|01\rangle - |10\rangle)$, respectively. All quantum networks can be built from a certain set of one- and two-qubit gates (universality theorem, see, e.g. [27]). Thus, the main experimental challenge for the creation of entanglement lies in the realisation of two-qubit quantum gates with low noise. Nowadays it is routine to entangle two qubits, represented by photons (▶ light quantum), atoms or ions, so the experimental attention moved towards creation of entanglement between more than two subsystems.

The above general definition of separability vs. entanglement holds for bipartite quantum states, but can be generalized to multipartite quantum states (states of composite systems with more than two subsystems). However, for multipartite states it is not sufficient to distinguish only between separable and entangled states, as the structure of the set of states is much richer than that: already for quantum systems composed of three qubits there are four different types of states: separable states, biseparable states (i.e. two of the three subsystems are entangled with each other, while the third one is separable from the others), and two classes of genuinely tripartite entangled states (each subsystem is entangled with both others): the GHZ-class [4] and the W-class [5]. A typical ▶ GHZ state consists of a superposition of two product states, where each of the three qubits in the first term is orthogonal to the corresponding one in the second term, e.g. $|GHZ\rangle = \frac{1}{\sqrt{2}}(|000\rangle + |111\rangle)$. A typical W state consists of a superposition of three terms that are permutations of each other and have one excitation each, i.e. $|W\rangle = \frac{1}{\sqrt{3}}(|001\rangle + |010\rangle + |100\rangle)$. The entanglement of a GHZ state is more fragile (with respect to the loss of one subsystem) than that of a W state: tracing out one of the three particles leads to a separable state of the remaining two particles for a GHZ state, but to an entangled state for a W state. Mixed states of three qubits can be classified according to their decomposition into projectors onto pure states [6]. For more than three subsystems the number of entanglement classes grows accordingly. When creating multipartite entanglement, one is mainly interested in that type of entanglement where all subsystems are entangled with each other (genuine multipartite entanglement).

The task of controlled creation of multipartite entanglement is very challenging, due to the impediment of ▶ decoherence. At present quantum optical methods provide the most advanced experimental tools to engineer and control entanglement. Entanglement between atoms and photons has been created in a cavity [7, 8]. Here, a 3-particle GHZ state was produced by first creating a Bell state of an atom and a cavity mode (photon), and then entangling this Bell state with another atom. Photons (▶ light quantum) can be entangled with each other via the non-linear process of parametric downconversion. Interference of independent photon pairs and conditional detection allowed to create a 3-photon GHZ state [9] and a 4-photon GHZ state [10]. Recently, even a 5-photon GHZ state has been realised in the laboratory [11]. Another method to entangle polarised photons consists of using a strong pump power in parametric downconversion, and thus reaching a reasonable probability for simultaneous emission of four entangled photons. In this way, a 4-photon singlet state [12] (which is invariant under simultaneous basis rotations) and a 3-photon W state [13] were produced. - The record in the number of entangled particles is held by the implementation with ion traps. Here, the ions are entangled via a collective excitation mode (phonon bus) [14]. Already in 2000 it was possible to create a 4-particle GHZ state [15]. Meanwhile even a GHZ state of 6 ions has been achieved [16]. The class of W states has first been produced with 3 ions [17], and recently even an 8-qubit W-state has been created [18].

In any experiment that aims at creating entanglement one also has to take into account the existence of noise, and thus one needs a method to prove that the produced state is indeed entangled. Here, three methods are of importance: first, one can perform *state tomography*, i.e. one measures every element in the ▶ density matrix and then uses theoretical tools to determine whether the density matrix is entangled. Second, one can perform a *Bell inequality* test: if a Bell inequality (▶ Bell's

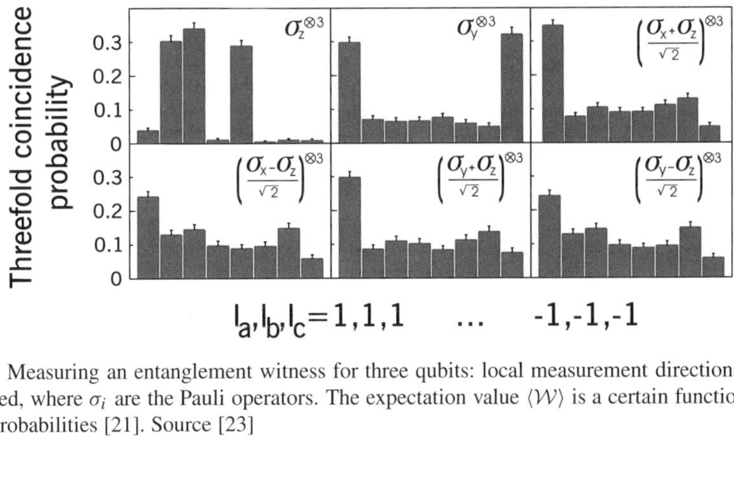

Fig. 1 Measuring an entanglement witness for three qubits: local measurement directions are as indicated, where σ_i are the Pauli operators. The expectation value $\langle \mathcal{W} \rangle$ is a certain function of all these probabilities [21]. Source [23]

theorem) is violated, the state is entangled. Note, however, that this is not an optimal criterion for the detection of entanglement, because there exist states (even states of two qubits) that do not violate any Bell inequality [2]. Third, one can use the tool of so-called *entanglement witnesses*. Entanglement witnesses are Hermitean operators that are constructed such that they detect entanglement: they lead to a positive expectation value for any separable state, but have a negative expectation value for some entangled states [19, 20, 26]. An entanglement witness is an observable and can be decomposed into local measurements [21]. Therefore witnesses provide a simple tool for entanglement detection: a negative expectation value of a witness implies the existence of entanglement [22]. Regarding multipartite quantum systems, witnesses have been constructed that prove the existence of genuine multipartite entanglement [6]. For example, for 3 qubits $\mathcal{W} = 2/3 \cdot \mathbf{1} - | W\rangle\langle W |$ is a witness that detects noisy W-states. Here, 2/3 is the maximal squared overlap of a W state with any pure biseparable state, and therefore the witncss $\mathcal{W}$ has a positive expectation value for all biseparable states.

As an example for the creation of entanglement with polarised photons, and the detection of entanglement via witnesses we show data from [23]. Here, a 3-partite W state was produced, and the witness $\mathcal{W}$ given above was measured, by collecting results from local coincidence measurements in different polarisation directions, as indicated in the figure. The expectation value of $\mathcal{W}$ was determined as $\langle\mathcal{W}\rangle = -0.197 \pm 0.018$. This value is higher than the theoretically expected one of -0.333, but this can be explained by noise that systematically increases the expectation value. The negative expectation value clearly proves the existence of genuine 3-partite entanglement.

See also entanglement; entanglement purification and distillation; entropy of entanglement.

Primary Literature

1. E. Schrödinger: Naturwissenschaften 23, 807 (1935).
2. R. Werner: *Quantum states with Einstein-Podolsky-Rosen correlations admitting a hidden-variable model.* Phys. Rev. A 40, 4277 (1989).
3. W. Dür, G. Vidal, J. I. Cirac, N. Linden, S. Popescu: *Entanglement Capabilities of Nonlocal Hamiltonians*, Phys. Rev. Lett. 87, 137901 (2001).
4. D. Greenberger, M. Horne, A. Zeilinger: *Bell's theorem, Quantum Theory and Conceptions of the Universe*, ed. M. Kafatos, Kluwer, Dordrecht (1989).
5. W. Dür, G. Vidal, J. I. Cirac: *Three qubits can be entangled in two inequivalent ways*, Phys. Rev. A 62, 062314 (2000).
6. A. Acín, D. Bruß, M. Lewenstein, A. Sanpera: *Classification of Mixed Three-Qubit States.* Phys. Rev. Lett. 87, 040401 (2001).
7. A. Rauschenbeutel et al: *Step-by-step engineered multiparticle entanglement.* Science 288, 2024 (2000).
8. J. Raimond, M. Brune, S. Haroche: *Manipulating quantum entanglement with atoms and photons in a cavity.* Rev. Mod. Phys. 73, 565 (2001).
9. D. Bouwmeester et al: *Observation of Three-Photon Greenberger-Horne-Zeilinger Entanglement.* Phys. Rev. Lett. 82, 1345 (1999).

10. J.-W. Pan et al: *Experimental Demonstration of Four-Photon Entanglement and High-Fidelity Teleportation*. Phys. Rev. Lett. 86, 4435 (2001).
11. Z. Zhao et al: *Experimental Demonstration of Five-photon Entanglement and Open-destination Teleportation*. Nature 430, 54 (2004).
12. M. Eibl et al: *Experimental Observation of Four-Photon Entanglement from Parametric Down-Conversion*. Phys. Rev. Lett. 90, 200403 (2003).
13. M. Eibl et al: *Experimental Realization of a Three-Qubit Entangled W State*. Phys. Rev. Lett. 92, 077901 (2004).
14. I. Cirac, P. Zoller, *Quantum Computations with Cold Trapped Ions*. Phys. Rev. Lett. 74, 4091 (1995).
15. C. Sackett et al: *Experimental entanglement of four particles*. Nature 404, 256 (2000).
16. D. Leibfried et al: *Creation of a six-atom "Schrodinger cat" state*. Nature 438, 639 (2005).
17. C. Roos et al: *Control and measurement of three-qubit entangled states*. Science 304, 1478 (2004).
18. H. Häffner et al: *Scalable multiparticle entanglement of trapped ions*. Nature 438, 643 (2005).
19. M. Horodecki: P. Horodecki, R. Horodecki, *Separability of Mixed States: Necessary and Sufficient Conditions*. Phys. Lett. A 223, 1 (1996).
20. B. Terhal: *Bell Inequalities and the Separability Criterion*. Phys. Lett. A 271, 319 (2000).
21. O. Gühne et al: *Detection of entanglement with few local measurements*. Phys. Rev. A 66, 062305 (2002).
22. M. Barbieri et al: *Detection of Entanglement with Polarized Photons: Experimental Realization of an Entanglement Witness*. Phys. Rev. Lett. 91, 227901 (2003).
23. M. Bourennane et al: *Experimental Detection of Multipartite Entanglement using Witness Operators*. Phys. Rev. Lett. 92, 087902 (2004).

Secondary Literature

24. M. Lewenstein, D. Bruß, J. I. Cirac, B. Kraus, M. Kuś, J. Samsonowicz, A. Sanpera, R. Tarrach: *Separability and distillability in composite quantum systems – a primer*. J. Mod. Opt. 47, 2841 (2000).
25. D. Bruß: *Characterizing entanglement*. J. Math. Phys. 43, 4237 (2002).
26. R. Horodecki, P. Horodecki, M. Horodecki, K. Horodecki: *Quantum entanglement*. arXiv: quant-ph/0702225, subm. to Rev. Mod. Phys.
27. M. Nielsen, I. Chuang: *Quantum Computation and Information*. Cambridge University Press (2000).
28. *Quantum Information: An Introduction to Basic Theoretical Concepts and Experiments* (Springer Tracts in Modern Physics, 173). Eds. G. Alber, T. Beth, M. Horodecki, P. Horodecki, R. Horodecki, M. Rötteler, H. Weinfurter, R. Werner, A. Zeilinger. Springer-Verlag (April 2001).
29. *Lectures on Quantum Information*. Eds. D. Bruß, G. Leuchs. WILEY-VCH Weinheim (2007).

C

D

Davisson–Germer Experiment

Friedel Weinert

The Davisson–Germer experiment (1927) was the first measurement of the wavelengths of ► electrons. C. J. Davisson, who worked in the Bell Research Laboratories, received the Nobel Prize in Physics for the year 1937 together with George P. Thomson from the University of Aberdeen in Scotland, who independently also found experimental indications of electron diffraction. According to the Copenhagen Interpretation of Quantum Mechanics, ► wave-particle duality leads to particles also exhibiting wave-like properties like extension in space and interference.

Clinton J. Davisson (1881–1958) and Lester H. Germer (1896–1971) investigated the reflection of electron beams on the surface of nickel crystals. When the beam strikes the crystal, the nickel atoms in the crystal scatter the electrons in all directions. Their detector measured the intensity of the scattered electrons with respect to the incident electron beam. Their normal polycrystalline samples exhibited a very smooth angular distribution of scattered electrons. In early 1925, one of their samples was inadvertently recrystallized in a laboratory accident that changed its structure into nearly monocrystalline form. As a result, the angular distribution manifested sharp peaks at certain angles. As Davisson and Germer soon found out, other monocrystalline samples also exhibited such anomalous patterns, which differ with chemical constitution, angle of incidence and orientation of the sample. Only in late 1926 did they understand what was going on, when Davisson attended the meeting of the British Association for the Advancement of Science in Oxford. There Born spoke about de Broglie's ► matter-waves and Schrödinger's ► wave mechanics. Their later measurements completely confirmed the quantum mechanical predictions for electron wavelength λ as a function of momentum p: $\lambda = h/p$. But their initial experiments (unlike G.P. Thomson's) were conducted in the context of industrial materials research on filaments for vacuum tubes, not under any specific theoretical guidance.

The phenomenon of electron diffraction is quite general and can be explained by the wave nature of atomic particles. Planes of atoms in the crystal (Bragg planes) are regularly spaced and can produce a constructive interference pattern, if the so-called Bragg condition ($n\lambda = 2\,d\ \sin\theta = D\sin\phi$, where d is the spacing of atomic planes and D is the spacing of the atoms in the crystal) is satisfied. This condition basically states that the reflected beams from the planes of atoms in the crystal will give an intensity maximum, or interfere constructively, if the distance, which the wave travels between two successive planes ($2\,d\sin\theta$), amounts to a whole number of wavelengths ($n\lambda,\ n = 1, 2, 3\ldots$). This is illustrated in Fig. 1.

D. Greenberger et al. (eds.), *Compendium of Quantum Physics: Concepts, Experiments, History and Philosophy*, © Springer-Verlag Berlin Heidelberg 2009

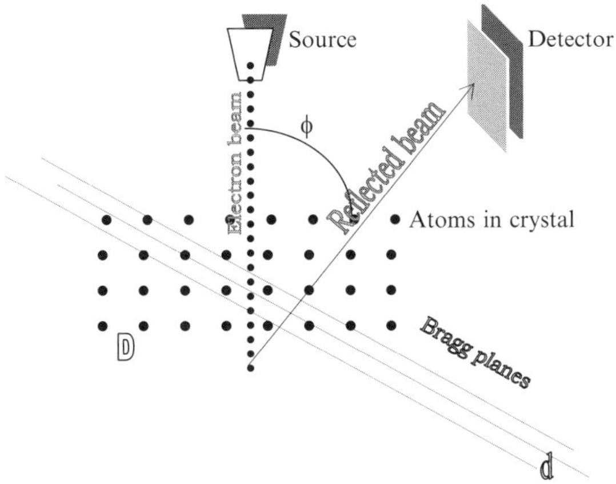

Fig. 1 *Davisson–Germer Experiment*: Scattering of electrons by a crystal for 54 eV electrons

In their experiment, Davisson and Germer found that the intensity reached a maximum at $\phi = 50°$ (for an initial kinetic energy of the electrons of 54 eV, normal incidence as indicated and ϕ as the scattering angle). From a *philosophical* point of view this experiment reveals a striking feature. It demonstrates the existence of de Broglie waves (► de Broglie wavelength). Yet we can speak of causation, not in a deterministic but in a *probabilistic* sense. There is clearly, on the observational level, a conditional dependence of the intensity of the reflected beam on the set of antecedent conditions. These antecedent conditions are also conditionally prior to their respective effects. There is of course no local causal mechanism, as the causal situation covers a stream of particles. There is only a certain likelihood that one particular particle in these experiments will be scattered in a particular direction.

But sufficiently much is known about scattering of atomic particles to establish a causal dependence between the antecedent and consequent conditions. In the Davisson–Germer experiment the wavelength of the electron beam, scattered at 50°, is 0.165 nm. This is the effect to which specific antecedent conditions correspond: the electron beam has initial kinetic energy of 54 eV; the lattice spacing of the nickel atoms is known, from which the spacing of the Bragg planes can be calculated; the condition for constructive interference is also known. There is quite a general dependence of the interference effects on the regular spacing of the atom planes in the crystal. It is used regularly in the study of atomic properties and is completely analogous to the use of X-ray diffraction by Max von Laue, Paul Knipping and Walter Friedrich in 1912. Under certain conditions, particles such as electrons thus exhibit wave-like characteristics like electromagnetic radiation.

Primary Literature

1. C. Davisson, L. H. Germer: Diffraction of Electrons by A Crystal of Nickel, Physical Review **30/6**, 705–40 (1927)

Secondary Literature

2. A. P. French, E. F. Taylor: *An Introduction to Quantum Physics* (Stanley Thornes Publishers, Cheltenham 1998, 64–72, Chapman & Hall 1979)
3. K. Krane: *Modern Physics* (Wiley, New York 1983, 89–93)
4. A. Russo: Fundamental research at Bell laboratories: The discovery of electron diffraction. Historical Studies in the Physical Sciences **12**, 117–160 (1981)
5. P. A. Tipler: *Modern Physics* (Worth Publishers, New York 1978, 169–73)

De Broglie Wavelength ($\lambda = h/p$)

Bruce R. Wheaton

Initially a thought in the thesis of young Louis de Broglie in 1923 for his doctorate from the Sorbonne. Attempting to reconcile special relativity with the quantum transformation relation (QTR), de Broglie assumed a hypothetical "phase wave" traveling faster than light that guides the physical displacement of an ▶ electron (see ▶ matter waves). In the thesis he derived its putative wavelength in the degenerate case of dipole oscillation, equal to ▶ Planck's constant divided by the momentum of the linearly oscillating particle; at the same time deriving the action-integral representation of the ▶ Bohr atom's orbital states by forcing every elliptical orbit to contain an integral number of phase wavelengths, as in Fig. 1.

With de Broglie, others (Einstein ▶ light-quantum, Schrödinger ▶ wave mechanics and Dirac ▶ QED) recognized the generality of the de Broglie wave

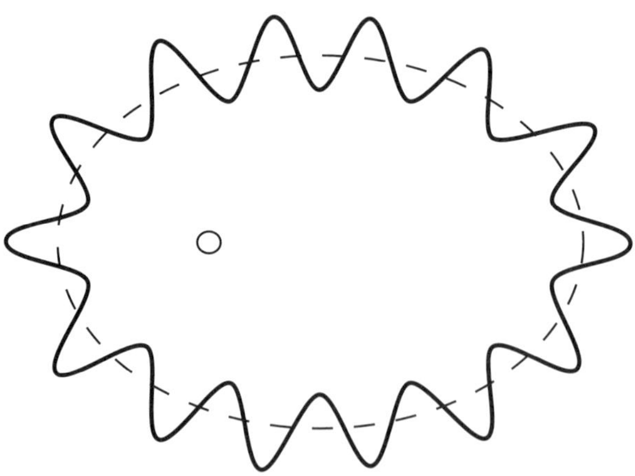

Fig. 1 Louis de Broglie's "beautiful result" of 1923 imagining a sinewave. Figure (c) 2009 TAPSHA, with thanks to Lauren Zimmermann

representation to all micro-processes of matter, confirming a missing permutation of matter with light in ► wave-particle duality. Some of the most important precursors to $\lambda = h/p$ were, in fact, concerns about the apparent ► light-quantum behavior of high-frequency X- and γ-rays.

Although the de Broglie wavelength, which predicted the electron diffraction found in 1927, applies only on the most microscopic level, it lately has come to have practical consequences. At extremely low temperatures ($<10^{-9}$ °K achieved by evaporative cooling) the ► wave packet of particles increase in wavelength, spread, and combine with others producing a sea of undifferentiated bosons (► Bose–Einstein statistics) (rather than the non-fungible fermions (► Fermi-Dirac statistics) they may have started as) in what is called a ► "Bose–Einstein condensate" or BEC. It has a macroscopic de Broglie wavelength (up to 30 μm so it can actually be photographed with visible light) because the entire assemblage of millions of atoms functions as a single ► wave function. See Fig. 2.

On the down-slope approach to this transition from atomic to Bosonic hierarchy lie ► superconductivity, ► superfluidity, the lowest temperatures yet attained and a demonstrated matter-wave "laser" (masem?) One of the most remarkable characteristics of a BEC is its phenomenally large effective group index of refraction ($v_g \approx di/dv$ so slows by as much as 10^{-8}) which, in almost stopping an incident light beam, may lead to information storage in un-heard of density albeit

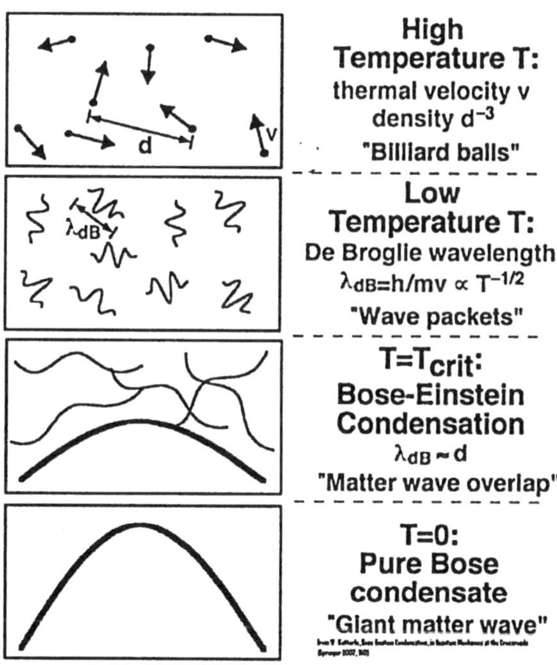

Fig. 2 How the de Broglie wave behaves on the downslope of temperature. From W. Ketterle, Bose–Einstein Condensation: Identity Crisis for Indistinguishable Particles. *Quantum Mechanics at the Crossroads* (Berlin: Springer 2007). p. 160

only in a momentary BEC. Other properties may lead to unprecedentedly fast multi-processing super-conducting computers, inter alia,from this quite literally "quintessential" new state of matter.

Primary Literature

1. De Broglie, M., La relation $h\nu = \varepsilon$ dans les phénomènes photoélectriques, Atomes et electrons (Paris: Gauthier, 1923), 80–100
2. Ellis, C. D., β-ray spectra and their meaning, R. Soc. Lond. Proc, 101A (1922), 1–17
3. De Broglie, L., A tentative theory of light quanta, Phil. Mag., 47 (1924), 446–458
4. De Broglie, L., Thèses. Recherches sur la théorie des quanta (Paris: Masson, 1924 and 1963)
5. Bose, S. N., Plancks Gesetz und Lichtquantenhypothese, Zeits. Physik, 26 (1924), 178–181
6. Schrödinger, E., Quantisierung als Eigenwertproblem (erste Mitteilung), Ann. Phys., 79 (1926), 361–376
7. Dirac, P., The quantum theory of the electron, I, Royal Soc. London, Proc., A117 (1928), 610–624
8. Anderson, M., J. Ensher, M. Matthews, C. Wieman, E. Cornell, Observation of Bose-Einstein condensation in a dilute atomic vapor, Science, 269 (1995), 198–201
9. Inouye, S., T. Pfau, S. Gupta, A. Chikkatur, A. Görlitz, D. Pritchard, W. Ketterle, Observation of phase-coherent amplification of atomic matter waves, Nature, 402 (1999), 641–643

Secondary Literature

10. De Broglie, L., Le dualisme des ondes et des corpuscules dans l'oeuvre de Albert Einstein (Paris: Hendaye, 1955)
11. Wheaton, B., The tiger and the shark: Empirical roots of wave-particle dualism (Cambridge: CUP, 1983), part 5
12. Darrigol, O., Strangeness and soundness in Louis de Broglie's early works, Physis, 30 (1993), 303–372
13. Ketterle, W., When atoms behave as waves: Bose-Einstein condensation and the atom laser, Rev. modern physics, 74 (2002), 1131–1151
14. Wheaton, B., Atomic waves in private practice, Quantum mechanics at the crossroads (Berlin: Springer, 2007), 39–71

Decay

See ► Mixing and Oscillations of Particles, Nuclear Fission, Particle Physics, Quantum Zeno Effect, Rigged Hilbert Spaces and Time Asymmetric Quantum Theory, Rigged Hilbert Spaces in Quantum Physics, Symmetry, Radioactive Decay Law.

Decoherence

E. Joos

The term *decoherence* is used in many fields of (quantum) physics to describe the disappearance or absence of certain superpositions of quantum states. Decoherence is a consequence of the unavoidable interaction of virtually all physical systems with their environment. In particular, macroscopic objects must be strongly entangled if quantum theory is universally valid [1,2]. Decoherence then explains within quantum theory why macroscopic objects seem to possess their familiar classical properties. No additional classical concepts are required for a consistent quantum description. Decoherence explains, for example, why particles appear localized in space (hence there is no need for an additional particle concept). Contradictory levels of description (classical and quantum) are no longer needed, instead a consistent description in terms of a universal ► wave function can be pursued.

The basic mechanism of decoherence is the unavoidable and generally irreversible disappearance of certain phase relations from the states of (local) systems by interaction with their environment according to the ► Schrödinger equation. Equivalently, decoherence describes irreversibly *increasing* entanglement as a consequence of a unitary global dynamics. Phase relations between certain states of a system are preserved globally (because of the assumed unitarity), but are no longer locally accessible, thus leading to apparent non-unitarity – or, in other words – to an apparent violation of the quantum ► superposition principle. This non-unitarity can be described as a disappearance of non-diagonal (in a certain basis) elements of the ► density matrix characterizing the local system. The two most important consequences of decoherence are suppression of interference and the selection of a set of preferred (dynamically stable) states.

The mechanisms underlying decoherence phenomena have much in common with quantum measurements. In the paradigmatic example of a macroscopic mass point scattering photons (► light quantum), and molecules, recoil is negligible like in an "ideal" measurement. This scheme also represents the case of "pure" decoherence: only the state of the environment changes, depending on the state of the "measured" object (here the position of the mass point).

Different components $|n\rangle$ of the state of the considered system may influence the environment Φ in different ways,

$$\left(\sum_n c_n|n\rangle\right)|\Phi_0\rangle \stackrel{t}{\longrightarrow} \sum_n c_n|n\rangle|\Phi_n(t)\rangle.$$

The resulting global superposition still contains phase relations connecting all components, but these are now a property of the total state and no longer relevant locally. Generically, phase relations originating from the initial superposition are distributed over an increasing number of degrees of freedom, rendering this process effectively irreversible. Local observations are operationally characterized by the system's density matrix ρ_S which changes according to

$$\rho_S = \sum_{n,m} c_m^* c_n |m\rangle\langle n| \xrightarrow{t} \sum_{n,m} c_m^* c_n \langle\Phi_m|\Phi_n\rangle |m\rangle\langle n| \,.$$

Non-diagonal terms are reduced by a factor $|\langle\Phi_m|\Phi_n\rangle| \leqslant 1$, which represents the overlap of corresponding environmental states. If these are approaching orthogonality,

$$\langle\Phi_m|\Phi_n\rangle \approx \delta_{mn},$$

the density matrix becomes approximately diagonal in this basis,

$$\rho_S \approx \sum_n |c_n|^2 |n\rangle\langle n|.$$

The result of this interaction is a density matrix which seems to describe an *ensemble* of states $|n\rangle$ with the respective probabilities [3]. However, this density matrix only represents an *apparent* (non-statistical) ensemble ("improper mixture"), not a genuine ensemble of quantum states (► ensembles in quantum mechanics). Coherence is not lost but is only *delocalized* into the larger system. The basis $\{|n\rangle\}$ characterizing dynamically stable states is defined solely by the properties of the interaction. These states are inert against further decoherence (with respect to the same basis). A complete treatment of realistic cases has to include the Hamiltonian governing the evolution of the system itself (as well as that of the environment), leading to a large variety of consequences [11,12,13].

Some fundamental examples of decoherence are the following.

- Localization and trajectories
  Coherence between macroscopically different positions of macroscopic objects disappears *very* rapidly because of the strong influence of scattering processes [2]. Trajectories thus emerge just as ► particle tracks in a bubble chamber as a consequence of the locality of interactions.
  In this way the equations of reversible classical mechanics are derivable from irreversible decoherence processes. In the macroscopic domain, decoherence is a much faster process than dissipation.
- Molecular configurations and robust states
  Most molecules appear to have a shape. Obvious examples are chiral molecules such as sugar – in contrast to small molecules (such as ammonia) appearing mostly in energy eigenstates. Parity (energy) eigenstates of a symmetric molecular Hamiltonian would immediately decohere (into local mixtures) because the shape of the molecule is monitored by the environment. Additional stabilization may be achieved by the ► *Zeno effect*. The robustness of these molecules resembles a classical ("macroscopic") state. Again, in this way classical properties are *created* by decoherence.

- ► Superselection rules
  Local charges are always accompanied by their Coulomb field (Gauss law). This may explain the charge superselection rule (usually derived from a kinematical constraint), if viewed as caused by dynamical coupling between local charge and Coulomb field. If a charge is decohered by its own field, a charge superselection rule does not need to be postulated separately. In quantum gravity, superpositions of different masses should be decohered by coupling to the spatial curvature.
- Quantum and classical fields
  Fields are decohered by coupling to matter (charges). ► *Coherent states* are usually the most stable states [7, 8] under decoherence, therefore they represent classical fields.
- Quantum gravity and space-time
  Entangled superpositions of space-time curvature and matter necessarily emerge in all versions of ► *quantum gravity*. Even if the precise form of a theory of quantum gravity is not known, decoherence should explain the classical structure of spacetime [9,14].
- ► Quantum jumps
  Exponential decay represents the textbook example for quantum "randomness", but an exactly exponential decay law is incompatible with the Schrödinger equation (this is related to the ► *Zeno effect*). Instead, the Schrödinger equation leads to superposition of different decay times (as observed in cavities). As soon as decay fragments interact with the environment, decay becomes irreversible (and usually exponential). The appearance of "quantum jumps" thus has its origin in very small, but finite decoherence times.
- Classical and ► quantum chaos
  According to the ► *correspondence principle* there should exist quantum states which mimic the behavior found for classically chaotic systems. Already the breakdown of ► *Ehrenfest theorems* shows that this is not the case. Instead, *open* systems show a behavior resembling classical chaos. Omission of decoherence has been shown to lead to unacceptable ► Schrödinger cat like states for large objects (such as the chaotically tumbling moon Hyperion).
- Quantum Computers
  Quantum computing schemes depend decisively on controllable unitary evolution of certain states ("qubits"). Since decoherence irreversibly delocalizes the required phase relations, it represents a major challenge to the practical realization of quantum computers. Error correction schemes try to reconstruct the lost coherences by scaling up the system with redundant bits, thereby possibly causing even larger sensitivity to decoherence.
- Decoherence in the brain
  The quantum superposition principle would allow "non-classical" states, like that of a superposition of a neuron firing and not firing. Quantum coherence effects in the brain have been repeatedly suggested. Quantitative estimates [10] showed, however, that the brain is such a "hot" environment that any non-classical states would decohere on a very small timescale. This dynamical selection of certain states is important for defining observers (which play a crucial role for some interpretations) in a quantum framework.

Decoherence represents a straightforward application of quantum concepts (in particular, wave function(al)s) to *all* physical objects. The essential new feature of quantum states, namely (kinematical) quantum non-locality, is responsible for all *local* consequences of ► entanglement. Therefore, decoherence does not have any classical analogue, while it is also based on an arrow of time in the form of a special (cosmological) initial condition.

Decoherence can explain why and how *within* quantum theory certain objects (including fields) *appear* classical to "local" observers. It can, of course, not explain conscious observers.

In many situations decoherence leads to a selection of a special set of dynamically stable (robust) states, which are relatively stable, thereby representing "classical" states (in a quantum framework). Classical properties are then *not* an a priori attribute of objects, but only come into being through the irreversible interaction with the environment. If all physical states are expressed in terms of quantum states, all the well-known paradoxes (► errors and paradoxes in quantum mechanics) which arise from intermingling incompatible notions can be avoided. Secondary concepts, such as "observable" can be derived from the dynamics of quantum states. Traditional, but ill-defined concepts, such as dualism, ► Heisenberg uncertain relations, or ► complementarity principle appear obsolete from this point of view.

Because decoherence acts, for macroscopic systems, on an extremely short time scale, it appears to work discontinuously, although decoherence is a smooth process. This is why "events", "particles", or "quantum jumps" seem to be observed. Only in the special arrangement of experiments, where systems are used that lie at the border between microscopic and macroscopic, can this smooth nature of decoherence be observed [4, 5, 6].

There are some common misinterpretations of decoherence. First, decoherence does *not* mean a disturbance of the system by the environment ("noise"). Quite to the contrary, in the case of "pure" decoherence, the system disturbs the environment. The local consequences result solely from quantum ► nonlocality.

Phenomena which mimic decoherence also arise in a statistical description using either an ensemble of differently prepared initial states or different Hamiltonians. This may lead to similar effects (e.g. disappearance of interference fringes), but has nothing to do with decoherence proper [11].

Decoherence leads to only an apparent collapse, in contrast to what would be traditionally expected in a quantum measurement. This apparent collapse is, however, operationally indistinguishable from a real collapse because of the irreversibility of decoherence [15]. See also ► Experimental Observation of Decoherence.

Primary Literature

1. H.D. Zeh: On the interpretation of measurement in quantum theory. Found. Phys. 1, 69 (1970)
2. E. Joos, H.D. Zeh: The emergence of classical properties through interaction with the environment. Z. Phys. B 59, 223 (1985)
3. W.H. Zurek: Pointer basis of quantum apparatus: Into what mixture does the wave packet collapse? Phys. Rev. D 24, 1516 (1981)

4. M. Brune, E. Hagley, J. Dreyer, X. Maître, A. Maali, C. Wunderlich, J.M. Raimond, S. Haroche: Observing the progressive decoherence of the "Meter" in a quantum measurement. Phys. Rev. Lett. 77, 4887 (1996)
5. L. Hackermüller, K. Hornberger, B. Brezger, A. Zeilinger, M. Arndt: Decoherence of matter waves by thermal emission of radiation. Nature 427, 711 (2004)
6. P. Sonnentag, F. Hasselbach: Measurement of decoherence of electron waves and visualization of the quantum-classical transition. Phys. Rev. Lett. 98, 200402 (2007)
7. O. Kübler, H.D. Zeh: Dynamics of quantum correlations. Ann. Phys. (N.Y.) 76, 405 (1973)
8. W.H. Zurek, S. Habib, J.P. Paz: Coherent states via decoherence. Phys. Rev. Lett 70, 1187 (1993)
9. E. Joos: Why do we observe a classical spacetime? Phys. Lett. A116, 6–8 (1986)
10. M. Tegmark: Importance of quantum decoherence in brain processes. Phys. Rev. E61, 4194 (2000) – also quant/ph-9907009

D

Secondary Literature

11. E. Joos, H.D. Zeh, C. Kiefer, D. Giulini, J. Kupsch, I.-O. Stamatescu: Decoherence and the Appearance of a Classical World in Quantum Theory: 2nd edn. Springer, Berlin (2003)
12. W.H. Zurek: Decoherence, Einselection, and the Quantum Origin of the Classical. Rev. Mod. Phys. 75, 715 (2003)
13. M. Schlosshauer: Decoherence and the Quantum-to-Classical Transition. Springer 2007
14. C. Kiefer: Quantum Gravity. Oxford University Press, 2nd ed. (2007)
15. H.D. Zeh: Roots and Fruits of Decoherence. In: Quantum Decoherence, B. Duplantier, J.M. Raimond, V. Rivasseau, eds. (Birkhäuser 2006) – also quant-ph/0512078

Degeneracy

Daniel M. Greenberger

In quantum mechanics, when there is more than one solution to the ► Schrödinger equation for a given energy, the energy level is said to be degenerate. In one dimension, if $V(x)$ is even, i.e., $V(-x) = V(x)$ (and $V(\pm\infty) \to 0$), then for bound states ($E < 0$, $\psi(\infty) \to 0$), there will generally be one solution. For unbound states, ($E > 0$, $\psi(\infty)$ finite), there are two solutions for a given E, one an even function of x, and one an odd function of x ($\psi(-x) = -\psi(x)$), or any linearly independent combination of the two, so that for unbound solutions there is a two-fold degeneracy.

In more general circumstances, such as in 3-D problems, if there are several degenerate solutions, and one makes a unitary transformation between any of them, $\psi_i = R_{ij}\psi_j$, then R will commute with the Hamiltonian, $[\mathrm{R}, \mathrm{H}] = 0$, and so R, which usually generates some symmetry group, will be a constant of the motion. For example, if the (3-D) potential is spherically symmetric, $V = V(r)$, the angular part of the solution to the Schrödinger equation will be the spherical harmonics, $Y_{\ell m}(\vartheta, \varphi)$, which are degenerate, and one can transform between them with the different components of $\boldsymbol{L}$, the angular momentum, ► Spin; Stern–Gerlach experiment; Vector model which is a constant of the motion, and is also the operator which generates rotations, and mixes up the $Y_{\ell m}$.

Occasionally the symmetry is non-existent, or more usually, not apparent, and the degeneracy is called "accidental". A famous example is the Kepler (Coulomb) problem, with the potential $V = -\alpha/r$, whose energies are $E_n = -E_0/n^2$, which are independent of ℓ. This contrasts with the case for any other potential $V = (\alpha r^n)$, for which $E = E_{n\ell}$. But for this special potential there is a hidden symmetry that explains this, and there is another constant of the motion, the Runge-Lenz vector, $\boldsymbol{R}$,

$$\boldsymbol{R}_{\text{class.}} = \frac{1}{m\alpha}\boldsymbol{p} \times \boldsymbol{L} - \hat{\boldsymbol{r}},$$
$$\boldsymbol{R}_{\text{quant.}} = \frac{1}{2m\alpha}(\boldsymbol{p} \times \boldsymbol{L} - \boldsymbol{L} \times \boldsymbol{p}) - \hat{\boldsymbol{r}},$$

where $\hat{\boldsymbol{r}}$ is the unit vector $\boldsymbol{r}$/r. The quantum form differs from the classical one by having been symmetrized, so as to be Hermitian. (An even deeper connection exists, in that if the system is imbedded in a 4-D Euclidean space, then $\boldsymbol{L}$ and $\boldsymbol{R}$ are the generators of rotations.)

The connection between the degeneracy of the Hamiltonian and the existence of ► symmetry groups is very profound, and leads, e.g., to the classification and representations of crystal symmetries.

Also, when one adds a perturbation to a symmetrical system, the perturbation generally has a lesser symmetry than the original Hamiltonian, and this leads to the splitting of the degeneracy. In the unperturbed Hamiltonian, any independent, orthogonal combination of the degenerate solutions is an equally good basis for describing the system. But under the lesser symmetry of the perturbation, only a single combination, or subset of combinations of the solutions will still be proper to describe the system with the perturbation (i.e., will make the perturbation matrix V_{ij} diagonal).

Furthermore, if there is a symmetry operator A that commutes with both the unperturbed Hamiltonian, and the perturbation, so that

$$H_0 |n, a\rangle = E_n |n, a\rangle, \quad A |n, a\rangle = a |n, a\rangle,$$
$$[A, H_0] = [A, V] = 0,$$

then for the perturbation,

$$\langle n, a' | V | n, a\rangle = \delta_{a,a'} f(n),$$

so that symmetries dictate whether or not the perturbation can split the degeneracy.

So, as a general rule, it is the symmetries of the system that determine the structure of the Hamiltonian, and they are revealed in the degeneracy of the solutions. For a detailed analysis of the relation between symmetry and degeneracy, see Elliot and Dawber, below.

Literature

A detailed discussion of the relation between symmetry and degeneracy, via a discussion of group theory can be found in:

J. P. Elliott and P. G. Dawber: *Symmetry in Physics*, Volume 1 and 2 (Cambridge University Press, Cambridge 1979)

and many other books devoted to group theory in physics. They also have a brief discussion of the Kepler (Coulomb) problem.
A more complete discussion of the Kepler problem is given in:

M. J. Englefield: *Group Theory and the Coulomb Problem* (Wiley, New York 1972)

Quick discussions of the symmetry aspects are given in some quantum theory texts.
See e.g., L. D. Landau and E. M. Lifshitz: *Quantum Mechanics: Non-Relativistic Theory*, revised third edition (Pergamon Press, Oxford 1977)

Delayed-Choice Experiments

A.J. Leggett

The phenomenon of ► "wave-particle duality" is at the heart of quantum mechanics, indeed has been described as "the one real mystery" of the subject. If we consider the standard Young's slits setup shown in Fig. 1, we may imagine for definiteness that the experiment is done with electrons (► Double-slit Experiment), then in the absence of "inspection" the probability of arrival of an electron on the final screen

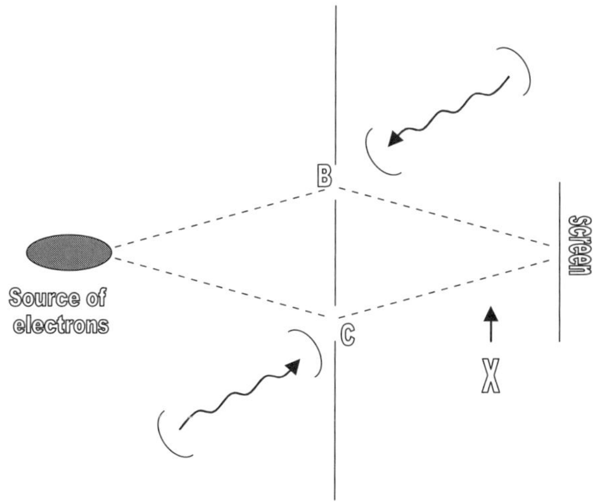

Fig. 1 The standard Young's slit setup. We may or may not choose to 'inspect' whether a given electron passes through slit B or slit C; the brackets indicate the optionality of the observation

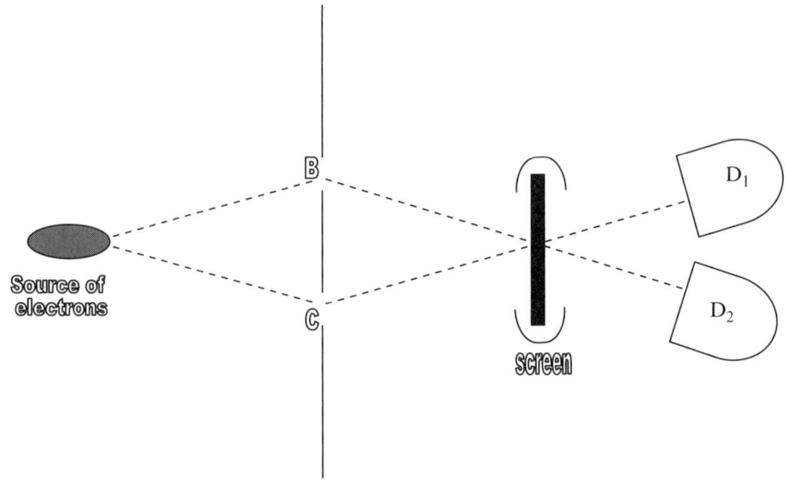

Fig. 2 An experiment illustrating "wave-particle duality" for photons. The brackets around the screen indicate that it may be either left in place (to indicate the "wave" aspect) or removed (to indicate the "particle" aspect)

shows the usual interference pattern – the electron appears to behave as a wave. If on the other hand we arrange to inspect which path is followed (e.g. by shining light on the intermediate slits as in the Heisenberg "gamma ray microscope" thought experiment ► Heisenberg microscope; which-way experiments), then the electron is always found, like a classical particle, to take one route or the other, and under these conditions no interference occurs at the final screen. If we replace the ► electrons with photons (► light quantum), we expect a similar duality to manifest itself; however, in this case, since it is very difficult to detect a photon without destroying it, it is more convenient to try to display the "particle" aspect by removing the final screen and replacing it by a pair of detectors as indicated in Fig. 2; again we will find that one detector or the other clicks, never both.

If D_1 clicks we can infer that the photon in question came through slit C, if D_2 clicks that it came through B. As is well known, Bohr interpreted experiments of this type to indicate that the very nature ("wave" or "particle") of elementary objects such as electrons or photons depends on the arrangement of the macroscopic experimental apparatus used to examine them; the arrangements needed to see wavelike behavior on the one hand and particle-like behavior on the other are always mutually exclusive ("complementarity"). This is particularly obvious in the example of the photon, and for definiteness I will from now on restrict myself to this case, although an entirely parallel discussion could be given for the case of an electron.

(See Consistent histories, Ignorance Interpretation, Ithaca Interpretation, Many Worlds Interpretation, Modal Interpretation, Orthodox Interpretation, Transactional Interpretation).

Is it necessary that the photon should as it were know in advance of entering the apparatus whether the latter has been set up in the "wave" configuration (Fig. 2) with the screen S in place or the "particle" one (S removed)? This question was already raised by implication [1] within a few years of the birth of quantum mechanics, and in 1978 John Archibald Wheeler (1911–2008) [2] pointed out that it can be answered, at least in principle, by an experiment in which we leave the decision as to which configuration to use until after the ▶ wave packet representing the photon is well within the apparatus (let us say to the right of point X in Fig. 1). Such an experiment is called a "delayed-choice" experiment, and several have been done over the last 30 years, not only on photons but also on hydrogen atoms ▶ Bohr's atomic model and neutrons; without exception they have indicated that it does not matter whether the choice of configuration is made well in advance or only at the "last moment", the counting statistics are quite independent of this.

In the case of photons, if the dimensions of the apparatus are of the order of 3 m (a fairly typical value), the transit time is about 10 ns, and it is therefore essential, in conducting a meaningful delayed-choice experiment, that the time needed to make the "choice" should be substantially smaller than this. (For atoms and neutrons the requirement is somewhat less stringent). This obviously rules out the possibility of physically inserting or removing a screen as in Fig. 2; however, it turns out that one can get around this difficulty by exploiting the polarization degree of freedom. (For a different technique which does not rely on this, see below). The basic idea is to correlate (or decline to correlate) the path taken by the photon with its polarization, a choice which can be realized over a few nanoseconds with the help of a device such as a Pockels cell (which can rotate the plane of polarization by 90°).

A possible schematic realization is shown in Fig. 3: The photons emitted by the source are polarized (for example) in the plane of the paper, and in the absence of the Pockels cell (or if it is in place but not activated) this polarization is maintained throughout the experiment for both beams, so that they interfere at BS_2 with a relative phase which is controlled by the phase shifter. Thus, under these conditions the output of the detector D_1 (for example) is a periodic function of the phase difference introduced by the shifter ("wave" behavior). If on the other hand the Pockels cell is activated, the polarization of a photon in the lower beam is rotated out of

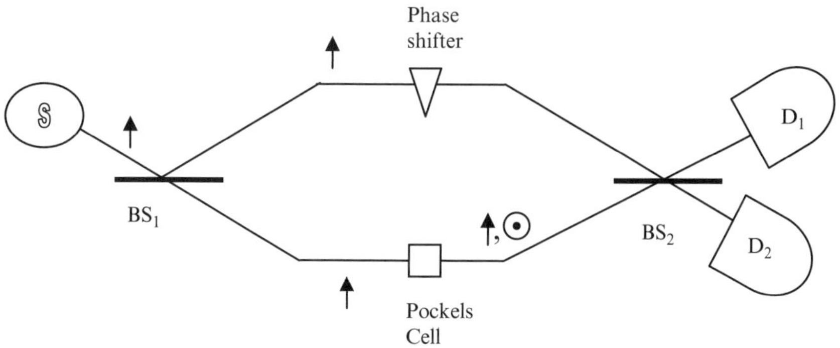

Fig. 3 Schematic realization of a polarization-mediated delayed-choice experiment. The notation to the right of the Pockels cell indicates that the polarization may, depending on our choice, be either in-plane or out-of-plane

the plane of the paper, so is perpendicular to that of the upper beam; the path taken by a given photon is now effectively "labelled" by its polarization. Under these conditions there can be no interference at BS_2 (which we assume is polarization-insensitive), and the output of detector D_1 is exactly the sum of what it would be for each of the two beams separately; since for each beam alone the output is independent of the position of the phase shifter, the total output of D_1 when the Pockels cell is activated is similarly insensitive to the latter ("particle" behavior). The crucial point is that the cell can be activated *after* the incoming photon wave packet has split at BS_1.

Over the last twenty years a number of experiments along these general lines have been done; the one closest to Wheeler's original proposal is probably that of ref. [3], which uses a setup similar though not identical[1] to that of Fig. 3. In this experiment the length of the interferometer was 48 m, and the choice as to whether or not to activate the switching cell was made by a quantum random number generator (QRNG) close to the far end; with this geometry the photon enters the future light cone of the random choice event long after it has passed the initial beam splitter. The use of the QRNG is designed to ensure that the photon has no way of "knowing" the choice ahead of time. The results are clear-cut: If one selects those photons for which the "wave" configuration was realized and plots the dependence of the output of one of the detectors on the phase shift between the two beams, one finds a well-defined sinusoidal pattern with visibility of 94%. If on the other hand one selects those photons which experienced the "particle" configuration, the corresponding plot is flat within experimental error.

An interesting variant of the "delayed-choice" experiment was reported in ref. [4]. The schematic setup is shown in Fig. 4: the "source" is prepared in such a way that there are nonzero mutually coherent amplitudes for a *pair* of photons to be emitted back-to-back by either of two regions A and B. Photon no.1 is registered by the screen S long before photon no.2 hits BS_1 or BS_2. The point of this arrangement is that any photon detected by D_3 (D_4) could only have come from source A(B); on the other hand, a photon arriving in D_1 or D_2 could have come from either source. Under these conditions, if we select only those photons 1 whose partners 2 were detected in (say) D_4 (let's call this the "D_4-correlated subensemble" of photons 1), we find that the distribution on the screen S is flat; on the other hand, if we select only those whose partners were detected in (say) D_1 ("D_1-correlated" subensemble), we obtain a well-defined fringe pattern (with a complementary pattern for those whose partners were detected in D_2). At first sight this is puzzling, since the detection of photon 1 on screen S took place well before the corresponding photon 2 "knew" whether it would be transmitted or reflected by BS1/2 and thus whether it will be detected by D_3/D_4 or by D_1/D_2.

In fact, there is no real paradox here (or in any of the other delayed-choice experiments); a consistent application of the quantum measurement axioms predicts

[1] Note in particular that in contrast to the setup of Fig. 3, in ref. [4] the activation of the electro-optical cell corresponds to the "wave" configuration and its non-activation to the "particle" configuration.

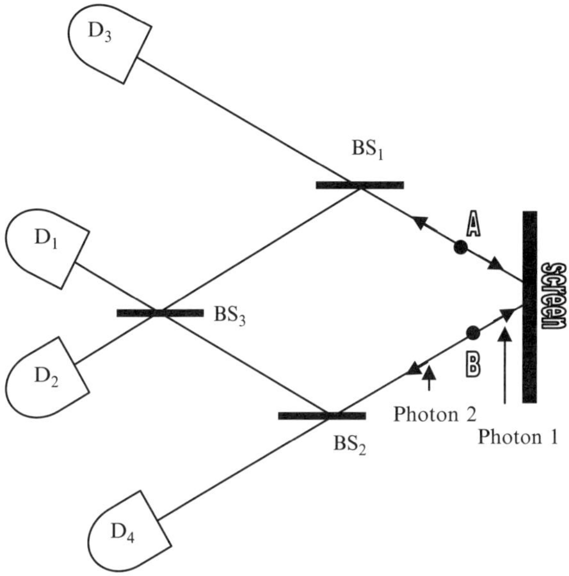

Fig. 4 The experimental arrangement of Kim et al. [4]

precisely the experimentally observed results. In particular, let us consider a case in which photon no.1 is detected at a point where the pattern corresponding to (say) the D_1-correlated subensemble has a node. When we say that the photon is "detected", we imply that it has induced a (quasi-) macroscopic event and thus satisfied what is usually considered the criterion for having undergone a "measurement". If at this point we apply the standard ▶ projection postulate to the two-photon system, we find that following the projection the ▶ wave function of photon 2 is automatically such that its amplitude to arrive in D_1 is zero, so everything is consistent. What the "delayed-choice" experiments really illustrate, in a spectacular way, is the pitfalls of applying the projection postulate at too early a stage in the game, while nothing has been registered at the macroscopic level and there is still a possibility of mutual interference of the possible alternatives.[2]

Primary Literature

1. K.F. von Weizsäcker: Ortsbestimmung eines Elektrons durch ein Mikroskop. Z. Phys. **70**, 114 (1931)
2. J.A. Wheeler: The "past" and the "delayed-choice" double-slit experiment, in A.R. Marlow, ed., *Mathematical Foundations of Quantum Theory* (Academic, New York 1978, pp. 9–48)

[2] This work was supported by the National Science Foundation through grant no.NSF-EIA-01-21568.

3. V.J. Jacques et al.: Experimental realization of Wheeler's delayed-choice gedanken experiment. Science **315**, 966 (2007)
4. Y.H. Kim et al.: Delayed "choice" quantum eraser. Phys. Rev. Lett. **84**, 1 (2000)

Secondary Literature

5. R.P. Feynman: *The Character of Physical Law* (MIT, Cambridge 1965)
6. N. Bohr: *Atomic Physics and Human Knowledge* (Wiley, New York 1958)

Density Matrix

Leslie Ballentine

A matrix representation of the ► *state operator*. So named because in the *position* basis its diagonal elements are equal to the position probability density. This name is older than the modern term *state operator*, and is still frequently used in its place, especially in many-electron theory and ► quantum chemistry. The name *density matrix* is not entirely accurate, since in the position basis it is not really a matrix, but rather a function of two continuous variables. If a discrete basis is chosen (such as the *spin* basis), then it becomes a genuine matrix, but its diagonal elements are *probabilities* rather than densities. ► States, pure and mixed, and their representation.

Density Operator

Werner Stulpe

Density operator, an operator used to describe (mixed) quantum states. A *density operator* [1–6], also called *statistical operator* or – somehow misleading – density matrix, is a positive trace-class ► operator ρ of trace 1 acting in some separable complex ► Hilbert space $\mathcal{H}$; i.e., ρ is a linear operator defined on $\mathcal{H}$ with values in $\mathcal{H}$ that satisfies $\rho = \rho^*$, $\langle\phi|\rho\phi\rangle \geqslant 0$ for all $\phi \in \mathcal{H}$, and $\operatorname{tr}\rho = \sum_i \langle\phi_i|\rho\phi_i\rangle = 1$, $\phi_1, \phi_2, \ldots$ being a complete orthonormal system in $\mathcal{H}$. In particular, ρ is a compact self-adjoint ► operator; in consequence, a density operator has the spectral decomposition $\rho = \sum_i \lambda_i P_{\chi_i}$ (► self-adjoint operator) where $\lambda_1, \lambda_2, \ldots$ are the nonzero eigenvalues of ρ, counted according to their multiplicity and arranged according to $\lambda_1 \geqslant \lambda_2 \geqslant \ldots > 0$, $\sum_i \lambda_i = 1$, $\chi_1, \chi_2, \ldots$ is an orthonormal system of corresponding eigenvectors (supplemented by the eigenvectors belonging to the possible eigenvalue 0, the system $\chi_1, \chi_2, \ldots$ is complete), and $P_{\chi_i} = |\chi_i\rangle\langle\chi_i|$ are the corresponding one-dimensional orthogonal projections (▶ projection). Moreover, for a bounded linear operator A in $\mathcal{H}$, $\operatorname{tr} \rho A$ exists (▶ operator) and $\operatorname{tr} \rho A = \sum_i \lambda_i \langle\chi_i|A\chi_i\rangle$; if in addition A is self-adjoint, then $\operatorname{tr} \rho A$ is a real number.

The set $\mathcal{S}(\mathcal{H})$ of all density operators is *convex*, i.e., the *convex linear combination* $\rho = \alpha\rho_1 + (1-\alpha)\rho_2$ of any $\rho_1, \rho_2 \in \mathcal{S}(\mathcal{H})$, $0 \leqslant \alpha \leqslant 1$, belongs to $\mathcal{S}(\mathcal{H})$. The set $\mathcal{S}(\mathcal{H})$ is even σ-*convex*, i.e., for any sequence $\rho_1, \rho_2, \ldots$ of density operators and any sequence of numbers satisfying $0 \leqslant \alpha_i \leqslant 1$ and $\sum_i \alpha_i = 1$, $\rho = \sum_{i=1}^{\infty} \alpha_i \rho_i \in \mathcal{S}(\mathcal{H})$ where the sum converges in the operator norm and even in the trace norm (▶ operator). An *extreme point* of the convex set $\mathcal{S}(\mathcal{H})$ is a density operator ρ that admits only trivial *convex decompositions*, i.e., $\rho = \alpha\rho_1 + (1-\alpha)\rho_2$, $\rho_1, \rho_2 \in \mathcal{S}(\mathcal{H})$, and $0 < \alpha < 1$ imply $\rho_1 = \rho_2 = \rho$. The extreme points of $\mathcal{S}(\mathcal{H})$ are the one-dimensional orthogonal projections $P_\psi = |\psi\rangle\langle\psi|$, $\|\psi\| = 1$. Physically, the extreme points P_ψ describe the *pure states* of conventional Hilbert-space quantum mechanics (equivalently, a pure ▶ state can be described by the unit vector ψ which is uniquely determined up to a phase factor $e^{i\alpha}$, $\alpha \in \mathbb{R}$). A ▶ *mixed state* is described by a density operator that is not an extreme point. So $\mathcal{S}(\mathcal{H})$ can be considered as the set of all quantum states and the set $\operatorname{ex} \mathcal{S}(\mathcal{H})$ of the extreme points of $\mathcal{S}(\mathcal{H})$ as the set of all pure states. For $\rho \in \mathcal{S}(\mathcal{H})$, the statement $\rho \in \operatorname{ex} \mathcal{S}(\mathcal{H})$ is equivalent to $\rho = \rho^2$.

For instance, if $\psi_1, \psi_2, \ldots$ is a nonorthogonal sequence of unit vectors and $\alpha_1, \alpha_2, \ldots$ a sequence of numbers satisfying $0 < \alpha_i < 1$ and $\sum_i \alpha_i = 1$, then $\rho = \sum_i \alpha_i P_{\psi_i}$, $P_{\psi_i} = |\psi_i\rangle\langle\psi_i|$, is a density operator with a spectral decomposition $\rho = \sum_i \lambda_i P_{\chi_i}$ into mutually orthogonal states P_{χ_i}. That is physically, the state ρ can be prepared both as the ▶ *mixture* of the states $P_{\psi_1}, P_{\psi_2}, \ldots$ in ratio $\alpha_1 : \alpha_2 : \ldots$ and as the mixture of the states $P_{\chi_1}, P_{\chi_2}, \ldots$ in ratio $\lambda_1, \lambda_2, \ldots$. Even the decomposition of a density operator into orthogonal states is in general not unique, as the example $\rho = \frac{1}{2}(P_{\phi_1} + P_{\phi_2}) = \frac{1}{2}(P_{\chi_1} + P_{\chi_2}) = \frac{1}{2}P$ shows where ϕ_1, ϕ_2 and χ_1, χ_2 are two different orthonormal bases of a two-dimensional subspace $\mathcal{X}$ of $\mathcal{H}$ and P is the orthogonal projection onto $\mathcal{X}$. In particular, for spin-$\frac{1}{2}$ systems, ϕ_1 and ϕ_2 can be the eigenstates (eigenvectors) of the operator S_z of the z-component of spin whereas χ_1 and χ_2 can be the eigenstates of S_x. The decomposition of a density operator $\rho \in \mathcal{S}(\mathcal{H})$ into mutually orthogonal pure states P_{χ_i} corresponds to the spectral decomposition $\rho = \sum_i \lambda_i P_{\chi_i}$, under the condition $\lambda_1 \geqslant \lambda_2 \geqslant \ldots > 0$ the spectral decomposition is unique if and only if the nonzero eigenvalues λ_i of ρ are nondegenerate, i.e., of multiplicity 1. Besides the decomposition into orthogonal pure states, every ▶ mixed state $\rho \in \mathcal{S}(\mathcal{H})$ can be decomposed in many ways into pure states P_{ψ_i} not being mutually orthogonal [3], so $\rho = \sum_i \lambda_i P_{\chi_i} = \sum_i \alpha_i P_{\psi_i}$ where $0 < \alpha_i < 1$ and $\sum_i \alpha_i = 1$.

(Spectral decomposition, see ▶ Ignorance interpretation; Measurement theory; Objectification; Operator; Probabilistic Interpretation; Propensities in Quantum Mechanics; Self-adjoint operator; Wave Mechanics).

For a density operator $\rho \in \mathcal{S}(\mathcal{H})$ and a bounded self-adjoint operator A satisfying $0 \leqslant A \leqslant I$, $0 \leqslant \operatorname{tr} \rho A \leqslant 1$ holds; in particular, if Q is an orthogonal ▶ projection, then $0 \leqslant \operatorname{tr} \rho Q \leqslant 1$. The orthogonal projections can be interpreted as *ideal (sharp) yes-no measurements* performed on quantum systems (▶ effect), and $\operatorname{tr}\rho Q$ is interpreted to be the *probability for the outcome 'yes' of the measurement Q in the state ρ*. If ρ is a pure state, i.e., $\rho = P_\psi$, then $\operatorname{tr}\rho Q = \langle\psi|Q\psi\rangle$. Moreover, *quantum observables* (▶ observable) are traditionally described by (in general unbounded) operators A; if E is the spectral measure of A (▶ self-adjoint operator) and B a Borel set of the real line (e.g., an interval), then $\operatorname{tr}\rho E(B)$ is the probability that a measurement of A in the state ρ yields a value in B. The mapping μ_ρ defined by $\mu_\rho(B) = \operatorname{tr}\rho E(B)$ is a probability measure on the Borel sets of $\mathbb{R}$, called the *probability distribution of the observable A in the state ρ*. Furthermore, if $\operatorname{tr}\rho A$ exists, it is the *expectation value of A* (for a definition of $\operatorname{tr}\rho A$ in the case of an unbounded operator A, see [7]).

A mixed state $\rho = \sum_i \alpha_i P_{\psi_i}$ can be established by preparing the pure states $P_{\psi_1}, P_{\psi_2}, \ldots$ with respective probabilities $\alpha_1, \alpha_2, \ldots$. This preparation procedure can be generalized. If a preparation device produces pure states $P = P_\psi$ whose occurrence is subject to a probability distribution μ on the set $\operatorname{ex}\mathcal{S}(\mathcal{H})$ of all pure states (i.e., μ is a probability measure on the Borel sets of the one-dimensional orthogonal projections), then the probability for the outcome 'yes' of a measurement Q, $Q = Q^2 = Q^*$, is given by $l(Q) = \int_{\operatorname{ex}\mathcal{S}(\mathcal{H})} \operatorname{tr} PQ\ \mu(dP)$. Replacing Q in this equality by a general bounded self-adjoint operator $A \in \mathcal{B}_s(\mathcal{H})$ (▶ operator), l becomes a bounded linear functional on $\mathcal{B}_s(\mathcal{H})$. Moreover, l is *positive*, i.e., $l(A) \geqslant 0$ for all $A \geqslant 0$, and l is *normal*, i.e., for every sequence of operators $A_n \in \mathcal{B}_s(\mathcal{H})$ such that $A_n \leqslant A_{n+1}$ and $\|A_n\phi - A\phi\| \to 0$ for all $\phi \in \mathcal{H}$ as $n \to \infty$ where $A \in \mathcal{B}_s(\mathcal{H})$, l satisfies $l(A_n) \to l(A)$ as $n \to \infty$. Just the normal positive linear functionals on $\mathcal{B}_s(\mathcal{H})$ can be represented by positive trace-class operators [6], that is, $l(A) = \operatorname{tr}\rho A$ where $\rho \geqslant 0$. Since $l(A) = \int_{\operatorname{ex}\mathcal{S}(\mathcal{H})} \operatorname{tr} PA\ \mu(dP)$ and μ is a probability measure, ρ is of trace 1, i.e., ρ is a density operator. Hence, the probability considered above reads $l(Q) = \int_{\operatorname{ex}\mathcal{S}(\mathcal{H})} \operatorname{tr} PQ\ \mu(dP) = \operatorname{tr}\rho Q$ where ρ describes the underlying preparation procedure which is determined by μ; formally, one can write $\rho = \int_{\operatorname{ex}\mathcal{S}(\mathcal{H})} P\ \mu(dP)$. In general, many different probability distributions on $\operatorname{ex}\mathcal{S}(\mathcal{H})$ give rise to the same quantum state ρ.

The states of quantum systems consisting of two subsystems with the respective Hilbert spaces $\mathcal{H}_1$ and $\mathcal{H}_2$ are described by the density operators acting in the tensor product $\mathcal{H}_1 \otimes \mathcal{H}_2$ [3, 4, 8]. For every density operator $\rho \in \mathcal{S}(\mathcal{H}_1 \otimes \mathcal{H}_2)$, there exist a uniquely determined density operator $\rho_1 \in \mathcal{S}(\mathcal{H}_1)$ such that, for all $A \in \mathcal{B}_s(\mathcal{H}_1)$, $\operatorname{tr}\rho(A \otimes I) = \operatorname{tr}\rho_1 A$ where I is the unit operator of $\mathcal{H}_2$; $A \otimes I$ are those observables of the composite systems that concern only their first components. The operator ρ_1 is called the *reduced state of ρ w.r.t. $\mathcal{H}_1$* or the *partial trace of ρ w.r.t. $\mathcal{H}_2$*. The latter name is related to the explicit representation $\rho_1 = \sum_{i,j,k} \langle\phi_i \otimes \chi_k|\,\phi_j \otimes \chi_k\rangle |\phi_i\rangle\langle\phi_j|$ where $\phi_1, \phi_2, \ldots$ and $\chi_1, \chi_2, \ldots$ are complete orthonormal systems in $\mathcal{H}_1$ and $\mathcal{H}_2$, respectively. Analogously, the reduced state of ρ *w.r.t.* $\mathcal{H}_2$ (the partial trace *w.r.t.* $\mathcal{H}_1$) is defined. The reduced states of a pure state $\rho = P_\psi \in \operatorname{ex}\mathcal{S}(\mathcal{H}_1 \otimes \mathcal{H}_2)$ are pure if and only if ψ is of the form $\phi \otimes \chi$ in which case $\rho_1 = P_\phi$ and $\rho_2 = P_\psi$. If $\rho = P_\psi$ where $\psi \in \mathcal{H}_1 \otimes \mathcal{H}_2$ is not of the form $\phi \otimes \chi$, i.e., if ρ is an *entangled* pure state (▶ entanglement), then both the reduced states are mixed. In fact, for every vector $\psi \in \mathcal{H}_1 \otimes \mathcal{H}_2$ there exist orthogonal systems $\phi_1, \phi_2, \ldots$ in $\mathcal{H}_1$ and

$\chi_1, \chi_2, \ldots$ in $\mathcal{H}_2$ such that $\psi = \sum_i \alpha_i \phi_i \otimes \chi_i$ where $\alpha_i > 0$ [3, 4, 2]. If $\|\psi\| = 1$ and $\rho = P_\psi$, then $\rho_1 = \sum_i |\alpha_i|^2 P_{\phi_i}$ and $\rho_2 = \sum_i |\alpha_i|^2 P_{\chi_i}$. So the pure states of $\mathcal{S}(\mathcal{H}_1 \otimes \mathcal{H}_2)$ yield in general mixed reduced states. More generally, for a state $\rho = \rho_1 \otimes \rho_2$, the partial traces are just ρ_1 and ρ_2; for an *entangled* state ρ (i.e., for a state $\rho \in \mathcal{S}(\mathcal{H}_1 \otimes \mathcal{H}_2)$ that is not of the form $\rho_1 \otimes \rho_2$), both the partial traces are mixed states.

D

A *face* $\mathcal{F}$ of the convex set $\mathcal{S}(\mathcal{H})$ is a subset of $\mathcal{S}(\mathcal{H})$ being closed under convex linear combinations as well as under convex decompositions, that is, $\mathcal{F} \subseteq \mathcal{S}(\mathcal{H})$ is a convex set such that $\rho \in \mathcal{F}$, $\rho = \alpha\rho_1 + (1-\alpha)\rho_2$, $\rho_1, \rho_2 \in \mathcal{S}(\mathcal{H})$, and $0 < \alpha < 1$ imply that $\rho_1, \rho_2 \in \mathcal{F}$. The empty set and the whole set $\mathcal{S}(\mathcal{H})$ are the trivial faces of $\mathcal{S}(\mathcal{H})$, and the extreme points of $\mathcal{S}(\mathcal{H})$ correspond to the one-element faces of $\mathcal{S}(\mathcal{H})$. The set $\Phi(\mathcal{S}(\mathcal{H}))$ of all faces of $\mathcal{S}(\mathcal{H})$ can be ordered by inclusion; it is obvious that the partially ordered set $\Phi(\mathcal{S}(\mathcal{H}))$ is a complete lattice. The same holds true for the set $\Phi_n(\mathcal{S}(\mathcal{H}))$ of all faces of $\mathcal{S}(\mathcal{H})$ that are closed w.r.t. the trace norm. For every orthogonal projection Q, $\mathcal{F}_Q = \{\rho \in \mathcal{S}(\mathcal{H}) \mid \operatorname{tr} \rho Q = 1\}$ is such a trace-norm closed face. Moreover, the mapping assigning the face $\mathcal{F}_Q$ to every Q, is an order isomorphism between the orthocomplemented lattice $\mathcal{P}(\mathcal{H})$ of all orthogonal projections (▶ projection, quantum logic) and the lattice $\Phi_n(\mathcal{S}(\mathcal{H}))$ [3]; so $\Phi_n(\mathcal{S}(\mathcal{H}))$ is, as $\mathcal{P}(\mathcal{H})$, an atomic complete orthomodular lattice.

Primary Literature

1. H. Weyl: *Gruppentheorie und Quantenmechanik* (Hirzel, Leipzig 1928, second edition 1931); English *The Theory of Groups and Quantum Mechanics* (Dutten, New York 1931)
2. J. von Neumann: *Mathematische Grundlagen der Quantenmechanik* (Springer, Berlin 1932, reprint 1968); English *Mathematical Foundations of Quantum Mechanics* (Princeton University Press, Princeton 1955)

Secondary Literature

3. G. Ludwig: *Foundations of Quantum Mechanics I* (Springer, New York 1983)
4. G. Ludwig: *Einführung in die Grundlagen der Theoretischen Physik, Band 3* (Vieweg, Braunschweig 1976)
5. E. Prugovečki: *Quantum Mechanics in Hilbert Space* (Academic, New York 1971, second edition 1981; reprint Dover Publications, Mineola 2006)
6. E. B. Davies: *Quantum Theory of Open Systems* (Academic, London 1976)
7. W. Stulpe: *Classical Representations of Quantum Mechanics Related to Statistically Complete Observables* (Wissenschaft und Technik Verlag, Berlin 1997; quant-ph/0610122), p. 91
8. M. Reed and B. Simon: *Methods of Modern Mathematical Physics I: Functional Analysis* (Academic, New York 1972, second edition 1980)

Diffeomorphism Invariance

Christian Heinicke

Diffeomorphism invariance refers to the form invariance of tensor(-equations)s under diffeomorphisms ([5], see also ▶ covariance).

A diffeomorphism Φ is a one-to-one mapping of a differentiable manifold M (or an open subset) onto another differentiable manifold N (or an open subset). Moreover, Φ (and its inverse Φ^{-1}) is differentiable. The concept of a diffeomorphism is intrinsically tied to the concept of a differentiable manifold. Here, we are mainly concerned with the four-dimensional spacetime manifold. The curves in Fig. 1 correspond to coordinate lines. There are two interpretations of the action of a diffeomorphism. A *passive diffeomorphism* changes one coordinate system to another one, like a cartesian to a polar coordinate system. Thus, one just changes the description of one and the same manifold ($\mathsf{M} = \mathsf{N}$). An *active diffeomorphism* corresponds to a transformation of the manifold which may be visualized as a smooth deformation of a continuous medium.

Now let a (tensor) field $\mathbf{T}$ be a solution of a *diffeomorphism invariant field equation*. By applying a diffeomorphism we obtain a transformed field $\tilde{\mathbf{T}}$ which still is a solution to the field equation.

Passively interpreted, $\mathbf{T}$ and $\tilde{\mathbf{T}}$ describe one and the same field in different coordinate systems. Passive diffeomorphism invariance is achievable by formulating the fundamental differential equations of a theory in a coordinate free way. One may argue that this is a purely mathematical task and involves no physics, i.e. means no restriction to a theory (Kretschmann, 1917 [2]). But even if the "decoordinatization" may seem quite "harmless" the interpretation of the basic terms of the theory is modified. Moreover, in specific cases, such as in the development of general relativity, there can emerge substantial generalizations.

Interpreted as active transformation $\mathbf{T}$ and $\tilde{\mathbf{T}}$ describe two distinct fields in the same coordinate system. "Distinct" here means that the field is "redistributed" (or "spread differently") over the manifold. From this point of view one would say that the field equation has the property to allow for (local) symmetry or gauge transformations of the field (▶ symmetries). Such local symmetries are not ensured automatically by a coordinate free formulation but have to be enforced dynamically (▶ gauge theories). Invariance under active diffeomorphisms raises important

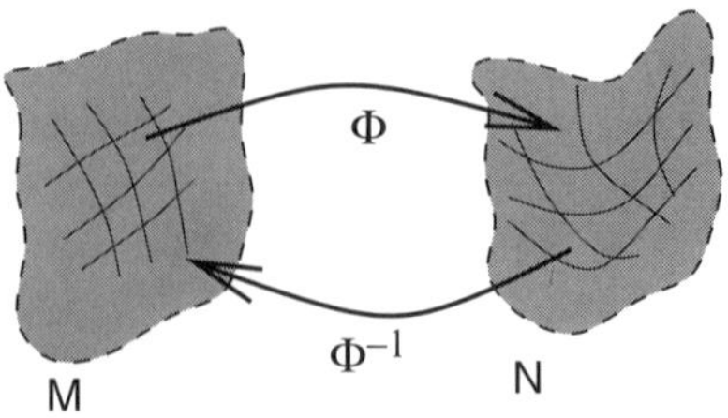

Fig. 1 Passive vs. active diffeomorphism: re-coordinatization vs. deformation

interpretational questions. Do the ("gauge equivalent") fields **T** and $\tilde{\mathbf{T}}$ represent distinct physical situations? If so, does the (diffeomorphism invariant) theory fail to prescribe the dynamics of the field uniquely? These questions are addressed in the famous hole-argument, originally put forward by Einstein in 1913 in the context of his search for the theory of general relativity [1]. Later, these difficulties were circumvented by focusing on (gauge-) invariant observables. Nevertheless, the values of fields alone can not be used to individuate points of the manifold. This makes a realistic interpretation of the manifold as spacetime less tenable. Therefore, diffeomorphism invariance (general covariance) plays an important role in the context of the spacetime structuralism-realism debate [3].

Earman, Stachel, Norton revived the hole argument in view of modern developments in spacetime and gauge theories. The discussion still continues [4].

Literature

1. A. Einstein: Die formale Grundlage der allgemeinen Relativitätstheorie.... Sitzungsber. Preuss. Akad. Wiss. Phys.-math. Klasse, **2** (1914) 1030–1085
2. E. Kretschmann: Über den physikalischen Sinn der Relativitätspostulate ... Ann. Phys. 358, 575–614 (1918)
3. J. Earman: *World Enough and Spacetime* (MIT, Cambridge, MA 1989)
4. J.D. Norton: The hole argument. In *The Stanford Encyclopedia of Philosophy*. http://plato.stanford.edu/entries/spacetime-holearg/ (2004)
5. T. Frankel: *The Geometry of Physics* (Cambridge University Press, Cambridge 1997)

Dirac Equation

Helge Kragh

The Dirac equation is a fundamental wave equation that satisfies the requirements of the special theory of relativity. Shortly after the appearance of the ▶ Schrödinger equation, several physicists attempted to extend it to the relativistic domain. The result – known as the Klein-Gordon-equation ▶ relativistic quantum mechanics – was however unable to describe ▶ electrons correctly. Paul A.M. Dirac realized that the formal structure of the Schrödinger equation, the form $H\psi = \mathrm{i}\hbar\partial\psi/\partial t$, had to be retained also in a relativistic theory, implying that the ▶ Hamilton operator must be of the first order in the space derivatives. By "playing around with mathematics" he derived in late 1927 a wave equation which was linear in both space and time derivatives. For a free electron he wrote it as $(W/c + \boldsymbol{\alpha}\cdot\boldsymbol{p} + \beta m_0 c)\psi = 0$, where the quantities $\boldsymbol{\alpha}$ and β were 4×4 matrices. In later literature the matrices were often designated $\gamma_\mu(\mu = 1, 2, 3, 4)$.

As Dirac showed in his paper of 1928, the operators or matrices have the mathematical properties that $\gamma_\mu^2 = 1$ and, for $\mu \neq \nu$, $\gamma_\mu \gamma_\nu + \gamma_\nu \gamma_\mu = 0$. In fact, it were these relations that led him to the equation. Dirac had not originally thought of ► spin, but discovered that his equation was able to account for the electron's magnetic moment, hence its spin. When it turned out that the equation provided a full explanation of the hydrogen spectrum (► spectroscopy), including the fine-structure components, it was quickly accepted by the physics community as the fundamental equation for the electron and presumably also the proton. Only after World War II, with the discovery of the Lamb shift, was it shown that the predictions from Dirac's theory disagree slightly with the measured spectrum.

Dirac's relativistic equation led to serious conceptual difficulties, principally because the wave function has four components rather than the two corresponding to the electron's spin states. Its solutions seemingly referred to electrons with negative energy – entities with no physical meaning. The so-called "±-difficulty" was turned into a success with Dirac's theory of the anti-electron (and other anti-particles) which he developed 1929–31. According to Dirac's theory of 1931, two of the four components of the ► wave function referred to an electron with positive electrical charge, soon to be known as a positron. When the positron was detected in cosmic-ray experiments 1932–33, it was considered a great triumph of the Dirac equation. In 1995 a plaque was unveiled in Westminster Abbey, commemorating Dirac. It contains a version of the Dirac wave equation in the compact form $i\gamma \cdot \partial\psi = m\psi$.

Primary Literature

1. P. A. M. Dirac: The Quantum Theory of the Electron. Proceedings of the Royal Society of London A **117**, 610–624; **A 118**, 351–361 (1928)
2. P. A. M. Dirac: *The Principles of Quantum Mechanics* (Clarendon, Oxford 1930)

Secondary Literature

3. H. Kragh: *Dirac. A Scientific Biography* (Cambridge University Press, Cambridge 1990)
4. J. Mehra, H. Rechenberg: *The Historical Development of Quantum Theory*. (Springer, New York 2000, Vol. 6, Part 1)

Dirac Notation

Roderich Tumulka

The "bra"-and-"ket" notation (introduced by Dirac) uses the symbols $|\psi\rangle$ and $\langle\psi|$ for vectors in and linear forms on ► Hilbert space.

In this notation, if ψ is a vector in Hilbert space $\mathscr{H}$ then $|\psi\rangle$ is just another notation for ψ, and $\langle\psi|$ means the mapping $\phi \mapsto \langle\psi|\phi\rangle$, a *linear form* $\mathscr{H} \to \mathbb{C}$ defined using the scalar product $\langle\cdot|\cdot\rangle$ of $\mathscr{H}$. Turning $|\psi\rangle$ into $\langle\psi|$ is a conjugate-linear operation: $\langle\phi+\psi| = \langle\phi| + \langle\psi|$ and $\langle z\psi| = z^*\langle\psi|$ for $z \in \mathbb{C}$.

Linear forms are also called *co-vectors*, and the set of all linear forms is called the *dual space*. Thus, $\langle\psi|$ is the co-vector naturally associated with the vector ψ. The difference between vectors and co-vectors is basically the same as the difference between a column and a row in matrix theory (linear algebra), or between the contravariant components u^μ and the covariant components u_μ of a 4-vector in relativity theory. The *Riesz lemma* of functional analysis implies that every continuous linear form $\mathscr{H} \to \mathbb{C}$ (only the continuous ones are usually considered) is of the form $\phi \mapsto \langle\psi|\phi\rangle$ for a suitable $\psi \in \mathscr{H}$; as a consequence, there is a one-to-one correspondence between vectors and (continuous) co-vectors, and $\mathscr{H}$ is, up to complex conjugation, its own *continuous dual space*.

D

As the notation suggests, the scalar product $\langle\phi|\psi\rangle$ is the same as the linear form $\langle\phi|$ applied to the vector $|\psi\rangle$. That is why Dirac called $\langle\phi|$ a "bra" vector and $|\psi\rangle$ a "ket" vector: bra + ket = bracket; that is, when written one after the other, they form the scalar product. When written in the opposite order, $|\psi\rangle\langle\phi|$, they form not a number but an operator $|\chi\rangle \mapsto |\psi\rangle\langle\phi|\chi\rangle$. In particular, if $\|\psi\| = 1$ then $|\psi\rangle\langle\psi|$ is the projection to the 1-dimensional subspace spanned by ψ. Moreover, if T is an operator then $\langle\phi|T|\psi\rangle$ means the same as $\langle\phi|T\psi\rangle$ or $\langle T^*\phi|\psi\rangle$.

The Dirac notation has another advantage: If some vectors ψ_n are indexed by some index n then one can write $|n\rangle$ instead of $|\psi_n\rangle$, provided there is no danger of misunderstanding. For example, an ▶ orthonormal basis can be denoted $|1\rangle, |2\rangle, |3\rangle, \ldots$, so that the matrix elements of an operator T can be written as $T_{nm} = \langle n|T|m\rangle$, the identity operator as

$$I = \sum_n |n\rangle\langle n| \,, \tag{1}$$

and the orthonormality relation as

$$\langle n|m\rangle = \delta_{nm} \,. \tag{2}$$

An extension of the ▶ Dirac equation concerns *generalized orthonormal bases* (such as the position basis in quantum mechanics), which consist of a unitary isomorphism $\mathscr{H} \to L^2(\Omega)$ and thus permits us to write every vector $\psi \in \mathscr{H}$ as a square-integrable function $\psi(q)$ on some set Ω (such as $\Omega = \mathbb{R}^{3N}$), whereas an ▶ orthonormal basis in the ordinary sense permits us to write a vector $\psi \in \mathscr{H}$ as a sequence $\langle 1|\psi\rangle, \langle 2|\psi\rangle, \ldots$ of numbers, the components of ψ. The extended ▶ Dirac notation introduces the symbol $|q\rangle$ as if the generalized basis was an ordinary basis, and to treat this symbol as if it denoted a vector in $\mathscr{H}$. (In quantum mechanics, in fact, $|q\rangle$ of the position basis represents the Dirac delta function $\delta(\cdot - q)$, which is not a square-integrable function and thus does not belong to $\mathscr{H}$; similarly, the kets of the momentum basis $|k\rangle$ represent the non-normalizable functions $x \mapsto e^{ikx}$.)

Thus, one writes

$$\psi(q) = \langle q|\psi\rangle\,, \tag{3}$$

while the orthonormality relation can be expressed as

$$\langle q|q'\rangle = \delta(q - q')\,, \tag{4}$$

and the identity operator as

$$I = \int |q\rangle\langle q|\,\mathrm{d}q\,. \tag{5}$$

See also the contributions on ► Rigged Hilbert Spaces.

Double-Slit Experiment (or Two-Slit Experiment)

Gregg Jaeger

The phenomenon of interference arises in both classical and quantum physics. In everyday life, more general interference effects can be seen, for example, patterns formed on the surface of a body of water when the wakes of two passing ships merge and pass through each other. Mathematically, this effect is due to the addition of corresponding physical quantities, such as wave height in the case of surface waves on water, to produce modulated patterns. These patterns can be made to exhibit clear regularities, particularly in simple situations. This effect has most often been studied by passing light through a pair of slits in a diaphragm, due in particular to an influential experiment in the early nineteenth century performed by Thomas Young [4] in which a double-slitted screen was used to produce an interference pattern. This pattern was readily explained in terms of classical light beams as waves traveling in the classical electromagnetic field. However, there are important differences between quantum interference and the more familiar effect of interference in classical physics. In particular, in quantum mechanical situations there are complex amplitudes, which therefore mathematically involve a phase contribution, that add, giving rise to characteristically quantum behavior, rather than real-valued *intensities* which are sometimes also referred to as amplitudes which add as in the case of water waves. It is important, from the ontological perspective, to recognize that quantum mechanical quantities do not directly describe substances, unlike in the classical ether theory of Christiaan Huygens, for example.

At the time of its appearance, the double-slit experiment of Young was understood to resolve a long-running debate regarding the nature of light as to whether light is best understood as composed of waves or composed of particles. Robert Hooke, in his book *Micrographia* [1] of 1665, had initially suggested that light propagation may involve “very short” vibratory motions in some underlying mechanical medium, making reference to the mechanical properties of diamond in particular. However, because Hooke provided no specific experimental evidence supporting this view, it was not particularly influential in his scientific environment in which, by then, empirical evidence had already become paramount. At the time, various phenomena were in need of explanation by making use of one or the other of these two ontologies, including the observation of rays and shadows, diffraction, reflection, refraction, the polarization of light, and rainbows. Huygens later emerged as the primary advocate of what is now identified as the wave ontology, which was used in his 1690 book *Traité de la lumière* [2], whereas Isaac Newton was the primary advocate of the particle ontology, which was used in his *Opticks* [3] of 1704. (► Wave-Particle Duality)

D

Huygens was able to explain the appearance of linearly propagating patterns of light by considering the net effect of locally originating radial propagation of finite-speed influences. Mechanically, Huygens described light as a solitary longitudinal pulse moving at a uniform rate, in contrast to water wave motion, through homogeneous material through an elastic ether medium determined by its composition. He was able within this limited wavelike picture to make headway by explaining both reflection and refraction. Importantly, however, this picture left no room for a mathematical description involving a *phase*. As a result, there were difficulties in explaining other of the above-mentioned phenomena, rainbows in particular, using this picture. By contrast, Newton's corpuscular theory was able to explain rainbows, as well as reflection and refraction. Famously, Newton first explained the production of colored light from white light by prisms. The theory was referred to as the *corpuscular theory* because, in it, light beams are represented as many localized individual bodies of colored matter, which could be variously combined and separated by media. The separation of variously colored corpuscles by a glass prism provided an adequate explanation of rainbows.

Newton's conception of light then held sway for nearly a century, until the appearance of Thomas Young's [4] article “Experiments and Calculations Relative to Physical Optics” in the *Philosophical Transactions* of the Royal Society of London, in which the double-slit experiment was reported. In Young's experiment, light was allowed to pass through a slit in a diaphragm, after which it then encountered a second diaphragm horizontally distanced from the first with two slits equally spaced vertically about the vertical location of the first slit, and finally impinged on a detection screen in a pattern of light and dark fringes. This sort of apparatus is now referred to as a *Young interferometer*. Because, by Huygens' principle, light continually expands radially from every point where it is present, it will do so from each of the three slits; first, the single slit feeds equally the remaining two slits, after which emanations from these two slits are able to encounter each other. As a result, light from each of the two slits meets on the detection screen, producing a distinctive pattern of illuminated and dark points. In this way, the pattern at the detection screen, particularly the dark regions thereof, can be understood as due to the addition of contributions from each of the pair of slits. By contrast, when only one of the two slits was unblocked, no such pattern was seen but only illumination symmetrically fading vertically about the position horizontally located directly in front of the unblocked slit.

At the very turn of the twentieth century, due to the influence of Young's experimental results and the further development of classical electromagnetic field theory by James Clerk Maxwell and others, light was believed to be fundamentally wave-like whereas matter was continued to be understood as fundamentally particulate. With the advent of quantum mechanics, the understanding of the fundamental nature of both light and matter changed again. This was due equally to the success of Albert Einstein's light-particle or *photon* hypothesis [6], which explained the then surprising ► photoelectric effect, and to Louis de Broglie's hypothesis [5] that both light and matter exhibit wavelike behavior in accordance with the relation $\lambda = h/p$, where h is ► Planck's constant and p is momentum; X-ray diffraction experiments of von Laue [7] and ► Davisson–Germer experiment [8] electron diffraction experiment confirmed the latter hypothesis.

Now, after the formal completion of modern quantum theory, quantum interference as observed in double-slit experiments is understood to arise due to the ► superposition of quantum states, which occurs when there is ► indistinguishability in principle by a precise measurement of alternative sequences of quantum states that originate with a common initial preparation. In the quantum mechanical double-slit experiment (for an instructive, more detailed and yet elementary discussion, see [16]), elementary systems such as ► electrons impinge precisely in one direction on a double-slit diaphragm and strike a detection screen, much as in the last stages of Young's original arrangement (Fig. 1). Take $a_i(x)$ to be the quantum probability amplitude corresponding to the passage through slit i $(i = 1, 2)$ of a diaphragm toward the vertical spatial point x on the measurement screen oriented precisely perpendicularly to the direction of the initial horizontal beam. The probability density of later finding these systems at x upon measurement is then $p_i(x) = |a_i(x)|^2$. The normalized quantum amplitude for systems being found at x when both slits are passable, so that either slit might be entered on the way to the screen, is $a_{12}(x) = \frac{1}{\sqrt{2}}\big(a_1(x) + a_2(x)\big)$, according to the amplitude superposition principle. The probability density of arrival at a point x of the detection screen upon measurement is

$$p_{12}(x) = \frac{1}{2}\Big[|a_1(x)|^2 + |a_2(x)|^2 \\ \qquad + |a_1(x)a_2(x)|\big(\exp\left[\mathrm{i}(\theta_2(x) - \theta_1(x))\right] + \exp\left[\mathrm{i}(\theta_1(x) - \theta_2(x))\right]\big)\Big],$$

the complex square of $a_{12}(x)$, where the $\{\theta_i(x)\}$ are the phases of the complex numbers $\{a_i(x)\}$ in the polar representation. Integrating $p_{12}(x)$ provides the detection rates observed in realizations of this ideal experiment.

The important difference between this quantum-mechanical experiment and the analogous one in which particles are described by classical mechanics is that the probability density $p_{12}(x) \not\propto p_1(x) + p_2(x)$ in the quantum case: the density is not additive, as it is in the classical experiment. The quantum-mechanical predictions

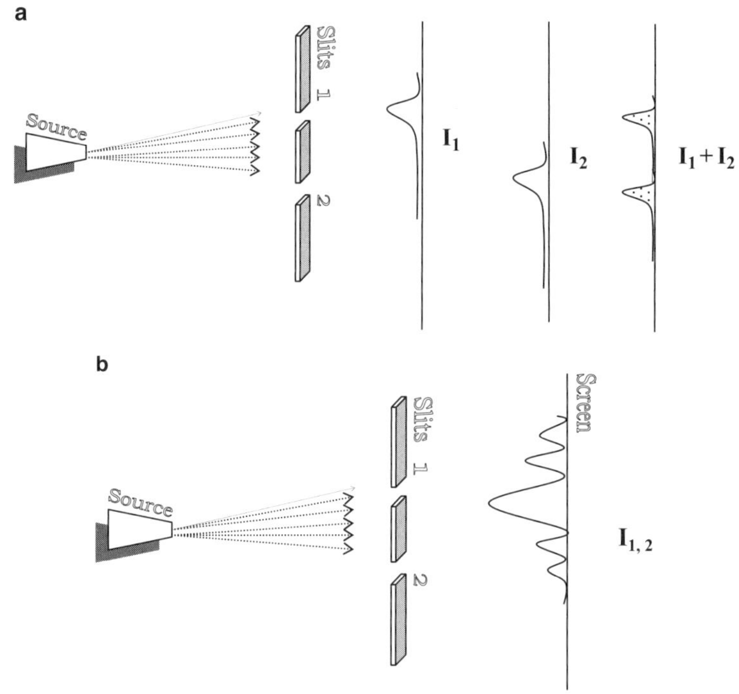

Fig. 1 (**a**) Particle-like behaviour of particles from a sand blast aimed at two slits. Depending on whether slit 1 or slit 2 is open, patterns I_1 or I_2 will form respectively. (**b**) Wavelike behaviour of electrons, when both slits are open. Adapted from F. Weinert, The Scientist as Philosopher (Springer 2004, 58)

are confirmed by observation, even in the case that the systems are sent into this apparatus only one at a time. Such independency from intensity was first clearly observed in a related 'feeble' light diffraction experiment by G. I. Taylor [9], and is also exhibited in the interference of massive electrically neutral particles. The analogue of Young's experiment was carried out by Jönsson and Möllenstedt [10, 11], and a conclusive demonstration with individual electrons was achieved by Tonomura et al. [12]. Further suggested reading regarding historical and conceptual issues involving the nature of light and the double-slit experiment are [13–15]. More detail of the very interesting history of the experiment with references to realizations with atoms and molecules can be found in the Physics World Editorial of 1 September, 2002 [17].

Primary Literature

1. R. Hooke: Micrographia (London 1665)
2. C. Huygens: Traité de la lumière (Leiden 1690)
3. I. Newton: Opticks (London 1704)

4. Young, T.: On the theory of light and colours (The 1801 Bakerian Lecture). Philosophical Transactions of the Royal Society of London **92**, 12–48 (1802)
5. L. de Broglie: A tentative theory of light quanta. Philosophical Magazine **47**, 446 (1924); Thesis, Recherche sur la théorie des quanta, Annales de Physique, 10th Series, III (1925)
6. A. Einstein: Über einen die Erzeugung und Verwandlung des Lichtes betreffenden heuristischen Gesichtspunkt. Annalen der Physik **17**, 132–148 (1905).
7. Knipping, W.P., von Laue, M.: Interferenzerscheinungen bei Röntgenstrahlen. Sitzungsberichte der Bayerischen Akademie der Wissenschaften, Math.-Phys. Klasse, 303–322 (1912)
8. Davisson, C., Germer, L.H.: The scattering of electrons by a single crystal of nickel. Nature **119**, 558–560 (1927); Diffraction of electrons by a crystal of nickel, Physical Review **30**, 705–740 (1927)
9. Taylor, G.I.: Interference fringes with feeble light. Proceedings of the Cambridge Philosophical Society **15**, 114–115 (1909)
10. Möllenstedt, G., Jönsson, C.: Elektronen-Mehrfachinterferenzen an regelmäßig hergestellten Feinspalten. Zeitschrift für Physik **155**, 472–474 (1959)
11. Jönsson, C.: Elektroneninterferenzen an mehreren künstlich hergestellten Feinspalten. Zeitschrift für Physik **161**, 454–474 (1961); English translation by D. Brandt and S. Hirschi: Electron Diffraction at Multiple Slits, American Journal of Physics **42**, 4–11 (1974)
12. Tonomura, A., Endo, J., Matsuda, T., Kawasaki, T., Ezawa, H.: Demonstration of single-electron build-up of an interference pattern. American Journal of Physics **57**, 117–120 (1989)

Secondary Literature

13. E. Scott Barr: Men and milestones in optics II. Thomas Young. Applied Optics **2**, 639–647 (1963).
14. A. H. Compton: The corpuscular properties of light, Reviews of Modern Physics **1**, 74–89 (1929).
15. H. Crew (Ed.): The wave-theory of light: Memoirs by Huygens, Young and Fresnel (New York 1900).
16. Feynman, R.P., Leighton, R.B., Sands, M.: The Feynman Lecture on Physics, Vol. 3, Ch. 37 (Reading MA 1963).
17. Rodgers, P. (Editor, Physics World): The double-slit experiment (Editorial), http://physicsworld.com/cws/article/print/9745 (2002).

E

Effect

Paul Busch

The term *effect* was introduced by G. Ludwig [1] as a technical term in his axiomatic reconstruction of quantum mechanics. Intuitively, this term refers to the "effect" of a physical object on a measuring device. Every experiment is understood to be carried out on a particular ensemble ("Gesamtheit") of objects (▶ ensembles in quantum mechanics), all of which are subjected to the same *preparation procedure*; each object interacting with the measuring device triggers one of the different possible measurement outcomes. Technically, *preparation procedures* and *effects* are used as primitive concepts to postulate the existence of probability assignments: each measurement outcome, identified by its *effect*, and each preparation procedure are assumed to determine a unique probability which represents the probability of the occurrence of that particular outcome. Thus, an *effect* can be taken to be the probability assignment, associated with a given outcome, to an ensemble of objects, or the preparation procedure applied to this ensemble [3].

In Hilbert space quantum mechanics, an effect is defined as an affine map from the set of states to the interval [0,1], or equivalently, as a linear operator E whose expectation value $\mathrm{tr}[\rho E]$ for any state (▶ density operator) ρ lies within [0,1]. From this it follows that E is a positive bounded, hence selfadjoint, ▶ operator.

Two selfadjoint bounded linear operators are said to be ordered as $A \leqslant B$ (A is less than B) if $\mathrm{tr}[\rho A] \leqslant \mathrm{tr}[\rho B]$ for all states ρ. Thus, an effect E is a positive bounded operator with the property that $O \leqslant E \leqslant I$, where O and I are the null and identity operators, respectively.

Among the effects are the projection operators (▶ projection) , P, with the idempotency property $P^2 = P$. They are singled out as those effects for which the generalized Lüders operation $\rho \mapsto E^{1/2}\rho E^{1/2}$ is repeatable, that is, $\mathrm{tr}[E\rho E] = \mathrm{tr}[E^{1/2}\rho E^{1/2}]$ for all states ρ. The condition $E = E^2$ can be expressed as $EE' = O$, where $E' := I - E$ is the *complement* effect of E. It is thus seen that for an effect that is not a projection, there is in general a nonzero probability, in a repeated Lüders measurement, of obtaining complementary outcomes. By contrast, two complementary projections P and $P' = I - P$ satisfy $PP' = O$, they are mutually orthogonal. If projections are interpreted as *properties*, then effects which are not projections are sometimes called *unsharp properties*, in an operational sense made precise in [2].

Another characterization of the set of projections is given by the fact that the set of effects is convex and the extreme elements are exactly the projections. Further details on mathematical and physical aspects of effects and their application can be found in [4–6].

D. Greenberger et al. (eds.), *Compendium of Quantum Physics: Concepts, Experiments, History and Philosophy*,

Primary Literature

1. Ludwig, G.: Versuch einer axiomatischen Grundlegung der Quantenmechanik und allgemeinerer physikalischer Theorien (Attempt at an axiomatic foundation of quantum mechanics and more general physical theories), Zeitschrift für Physik **181**, 233–260 (1964)
2. Busch, P.: Unsharp reality and joint measurements for spin observables, Physical Review D **33**, 2253–2261 (1986)

Secondary Literature

3. Ludwig, G.: *Foundations of Quantum Mechanics, Vol. 1* (Springer, Berlin, 1983)
4. Davies, E.B.: *Quantum Theory of Open Systems* (Academic, London, 1976)
5. Kraus, K.: *States, Effects and Operations* (Springer, Berlin, 1983)
6. Busch, P., Grabowski, M., Lahti, P.: *Operational Quantum Physics* (Springer, Berlin, 1995; 2nd corrected printing 1997)

Ehrenfest Theorems

Erich Joos

The Ehrenfest theorems establish a formal connection between the time dependence of quantum mechanical expectation values of ► observables and the corresponding classical equations of motion. Although mean values alone are insufficient to derive classical behavior from quantum mechanics, the validity of the Ehrenfest relations is an important requirement for a partial derivation of classical physics.

If the system (here a particle in one dimension, with obvious generalization to more complex systems) is governed by a ► Schrödinger equation with Hamiltonian

$$H = \frac{p^2}{2m} + V(x),$$

the mean values for position, momentum and energy obey the relations

$$\frac{\mathrm{d}}{\mathrm{d}t}\langle x\rangle = \frac{\langle p\rangle}{m},$$

$$\frac{\mathrm{d}}{\mathrm{d}t}\langle p\rangle = -\left\langle \frac{\mathrm{d}}{\mathrm{d}x}V(x)\right\rangle,$$

and

$$\frac{\mathrm{d}}{\mathrm{d}t}\langle H\rangle = 0.$$

The mean value of position therefore follows a law of motion *similar* to Newton's:

$$m\frac{\mathrm{d}^2}{\mathrm{d}t^2}\langle x\rangle = -\left\langle \frac{\mathrm{d}V}{\mathrm{d}x}\right\rangle = \langle F\rangle .$$

These relations are a special case of the general time-dependence of expectation values of an observable $A(t)$,

$$\frac{\mathrm{d}}{\mathrm{d}t}\langle A\rangle = \frac{\mathrm{i}}{\hbar}\langle\Psi|\,[H,A]\,|\Psi\rangle + \langle\Psi|\,\frac{\partial A}{\partial t}\,|\Psi\rangle\,,$$

which follows immediately from the definition of the expectation value $\langle A\rangle = \langle\Psi|\,A\,|\Psi\rangle$ and Schrödinger's equation $\mathrm{i}\hbar\partial_t\,|\Psi\rangle = H\,|\Psi\rangle$.

E

Further considerations:

1. Quite independent of the chosen interpretation of quantum states, the mean value $\left\langle \frac{\mathrm{d}}{\mathrm{d}x}V(x)\right\rangle$ is different from $\frac{\mathrm{d}}{\mathrm{d}x}V(\langle x\rangle)$. Only if $V(x)$ is a polynomial of degree 2 – that is, for a free particle, motion in a homogenous field and the harmonic oscillator – does the mean value follow the classical law of motion. For all other cases, a strongly localized ▶ wave packet is required, a condition which is rapidly violated for classically chaotic systems. The range of validity of classical equations is sometimes called "Ehrenfest time". Beyond this time-scale wave packet dispersion becomes essential.
2. Historically, Ehrenfest's theorem played an important role in establishing the "correspondence limit" of quantum mechanics, that is, the hope (or the requirement) that classical mechanics be contained in quantum mechanics as a limiting case. This "▶ correspondence principle" fails, however, for at least two reasons: As already mentioned, mean values for general wave packets and potentials do not follow classical laws, second, macroscopic systems do not obey a Schrödinger equation, since they are manifestly open systems.

   A spectacular example of failure of the "correspondence principle" is provided by the rotation of Hyperion, a moon of Saturn. Hyperion's rotation is chaotic with an estimated Ehrenfest time of only 20 years.
3. Extension to open systems. For some important classes of open systems, relations similar to that shown by Ehrenfest can be derived. Mean values are then calculated from dynamical equations for the density matrix ρ describing the open system according to $\frac{\mathrm{d}}{\mathrm{d}t}\langle A\rangle = \frac{\mathrm{d}}{\mathrm{d}t}\mathrm{tr}\,(A\rho) = \mathrm{tr}\left(A\frac{\mathrm{d}\rho}{\mathrm{d}t}\right)$ for a time-independent observable A. For example, from the equation for "Quantum Brownian motion" (a particle immersed in a heat bath of temperature T),

$$\mathrm{i}\frac{\partial}{\partial t}\rho = \left[\frac{p^2}{2m} + V(x),\rho\right] + \frac{\eta}{2m}\left[x,\{p,\rho\}\right] - \mathrm{i}\eta k_\mathrm{B}T\left[x,[x,\rho]\right],$$

one finds

$$\frac{\mathrm{d}}{\mathrm{d}t}\langle x\rangle = \frac{\langle p\rangle}{m},$$
$$\frac{\mathrm{d}}{\mathrm{d}t}\langle p\rangle = -\left\langle\frac{\mathrm{d}}{\mathrm{d}x}V(x)\right\rangle - \frac{\eta}{m}\langle p\rangle$$

and

$$\frac{\mathrm{d}}{\mathrm{d}t}\langle H\rangle = \frac{2\eta}{m}\left[\frac{k_{\mathrm{B}}T}{2} - \left\langle\frac{p^2}{2m}\right\rangle\right].$$

In this case, motion is damped (with friction constant η), while energy approaches its equilibrium value. Re-evaluations of the Ehrenfest theorem for open quantum systems (often described by Lindblad equations derived from a Schrödinger equation that includes the environment (see ► decoherence)) are important for a proper understanding of the relation between classical and quantum physics.

Primary Literature

1. P. Ehrenfest: Bemerkung über die angenäherte Gultigkeit der klassischen Mechanik innerhalb der Quantenmechanik. Z. Phys **45**, 455–457 (1927)
2. E. Joos, H. D. Zeh: The emergence of classical properties through interaction with the environment. Z. Phys. **B59**, 223–243 (1985)
3. W.H. Zurek: Decoherence, chaos, quantum-classical correspondence, and the algorithmic arrow of time. Phys. Scripta **T76**, 186–1988 (1998)

Secondary Literature

4. F. Haake: *Quantum signatures of chaos* (Springer, Berlin, 2001)
5. E. Joos, H. D. Zeh, C. Kiefer, D. Giulini, J. Kupsch, I.-O. Stamatescu: *Decoherence and the appearance of a classical World in quantum theory* (Springer, Berlin, 2003)
6. Z. P. Karkuszewski, J. Zakrewski, W. H. Zurek: Breakdown of correspondence in chaotic systems: Ehrenfest versus localization times. Phys. Rev. **A65**, 042113 (2002)

Eigenstates, Eigenvalues

See ► States pure and mixed, and their Representations.

Einstein Locality

Henry P. Stapp

In 1935 Albert Einstein, in collaboration with Boris Podolsky and Nathan Rosen, published a landmark paper entitled "Can quantum mechanical description of physical reality be considered complete?" [1] Einstein had already been engaged for several years in a discussion with Niels Bohr about the completeness of quantum theory. In the1935 paper Einstein did not challenge the claim of the quantum theorists that their theory was complete in the pragmatic/epistemological sense that it gives all possible empirically testable predictions about connections between the various aspects of "our knowledge." In the 1935 paper Einstein et. al. effectively accepted this claim of epistemological completeness, but defined the question they were addressing to be the completeness of quantum mechanics as a description of physical reality.

"Physical reality" is a slippery concept for scientists, when it becomes separated from empirically testable predictions. Hence Einstein and his colleagues were faced with the difficult task of introducing this term into the discussion in a way that could not easily be dismissed as vague metaphysics by a physics community which, greatly impressed by the empirical successes of quantum mechanics, was in no mood to be sucked into abstruse philosophical dialectics. Yet Einstein and his colleagues did succeed in coming up with a formulation that shook the complacency of physicists in a way that continues to reverberate to this day.

The key to their approach was to tie the needed characterization of physical reality to a peculiar *nonlocal* feature of the quantum mechanical treatment of two-particle systems.

The mathematical rules of quantum theory permit the generation of a state of two particles that has predicted properties that appear, at least at first sight, to violate a basic precept of the special theory of relativity, namely the exclusion of instantaneous (i.e., faster-than-light) action at a distance. (► Locality)

Quantum theory generally allows any one of several alternative possible measurements to be performed on a particle that lies in some experimental region R. The choice of the measurement to be performed in R is treated in quantum mechanics as a boundary condition that can be "freely chosen" by the experimenter. According to the Copenhagen interpretation, performing the measurement is supposed to affect the particle being measured in a way such that the observed outcome specifies the measured property of the state of the particle *after* the measuring process is complete. (See ► Born rule; Consistent Histories; Metaphysics in Quantum Mechanics; Nonlocality; Orthodox Interpretation; Schrödinger's Cat; Transactional Interpretation). But then if two alternative possible measurements are *mutually incompatible*, in the sense that either one or the other can be performed, but not both at the same time, then there is no logical reason why the particle should have at the same time well defined values of *both* of the two properties.

The mathematical structure of quantum theory does in fact involve various properties of a particle that cannot, within that theoretical structure, have simultaneously well defined values. Potential inconsistencies are evaded by claiming that any two such theoretically incompatible properties are also empirically incompatible, in the sense that they cannot be measured simultaneously. But Einstein *et. al.* constructed an argument designed to show that the values of certain of these properties are, nevertheless, simultaneous elements of physical reality. Such a demonstration would render quantum mechanical account incomplete, as a description of physical reality!

To bring "physical reality" into the discussion, in conjunction with the question of completeness, Einstein et. al, noted that the basic precepts of quantum theory ensure that there is a state (▶ wave function) of two particles that has the following properties:

1. The two particles lie at the time of a measurement performed on particle 1, in two large regions that lie very far apart.
2. There is a pair of measurable properties, X_1 and P_1, which are the location and the momentum of particle 1, respectively, that are neither simultaneously representable nor simultaneously measurable; and also a pair of measurable properties, X_2 and P_2, of particle 2 that are, likewise, neither simultaneously representable nor simultaneously measurable.
3. The prepared state of the two particle system, before the measurement is performed on particle 1, is such that measuring the value of X_1 determines the value of X_2, whereas measuring the value of P_1 determines the value of P_2.

These properties entail that the experimenter in the region where the first particle lies can come to know either X_2 or P_2, depending upon which measurement he chooses to perform. This choice controls physical measuring actions that are confined to the region where particle 1 is located, and this region is very far from the region where particle 2 is located. Consequently, any physically real property of the faraway particle 2 should, according to the precepts of the theory of relativity, be left undisturbed by the nearby measurement process: the distance between the two regions can be made so great that the physical consequences of performing the measurement on particle 1 cannot reach the region where particle 2 is located without traveling superluminally: faster than the speed of light ▶ superluminal communication.

These considerations permit Einstein *et. al.* to introduce "physical reality" by means of their famous "criterion of physical reality":

> If, without in any way disturbing a system, we can predict with certainty (i.e., with probability unity) the value of a physical property, then there exists an element of physical reality corresponding to this physical property.

If a measurement were to be performed in the region where particle 2 is located then the quantum theorist could argue that this measurement could disturb the particle, and hence there would be no reason why properties X_2 and P_2 should exist simultaneously. But the situation under consideration allows either of the two (simultaneously incompatible) properties of particle 2 to be determined (predicted with certainty) without anything at all being done in the region where that particle 2 is located, and hence, according to the ideas of the theory of relativity, "without in any way disturbing that system." Thus Einstein and his colleagues infer, on the basis of their criterion of physical reality, that *both* properties are physically real. However, these two properties cannot be represented simultaneously by any quantum mechanical wave function. Hence Einstein et.al. "conclude that the quantum mechanical description of physical reality given by wave functions is not complete."

Anticipating an objection, Einstein *et. al.* complete their argument by saying:

> One could object to this conclusion on the grounds that our criterion of reality is not sufficiently restrictive. Indeed, one would not arrive at our conclusion if one insisted that two or more physical quantities can be regarded as simultaneous elements of reality *only when they can be simultaneously measured or predicted.* On this point of view, since either or the other, but not both simultaneously, of the quantities P [here P_2] or Q [here X_2] can be predicted they are not simultaneously real. This makes the reality of P and Q depend upon which measurement is made of the first system, which does not disturb the second system in any way. No reasonable definition of reality can be expected to permit this.

E

If one examines the situation considered by Einstein *et. al.* in the explicit formulation of relativistic quantum field theory given by Tomonaga [2] and Schwinger [3] one finds that the quantum state (wave function) of particle 2 after the measurement is performed on particle 1 depends not simply on which measurement is performed on particle 1, but jointly upon which measurement is performed and what its outcome is.

In a general context it is neither problematic nor surprising that what a person can *predict* should depend not only upon which measurement he performs, but also upon what he learns by experiencing the outcome of that experiment, and hence upon both which measurement is chosen and performed, and which outcome then appears.

In *classical* relativistic physics an *outcome* in one region can be *correlated* to an *outcome* in a faraway region – that is space-like separated from the first – without their being any hint or suggestion of any faster-than-light transfer of information. Such correlations can arise from a common cause lying in the earlier (preparation) region from which each of the two later experimental regions can be reached by things traveling at the speed of light or less.

In relativistic quantum field theory, as in relativistic classical theory, merely performing the measurement action on particle 1 does not affect any measurable or predictable property of particle 2. In both the classical and quantum versions the subsequent *outcome* pertaining to particle 1 is *correlated* (through the earlier initial preparation) to a predictable and measurable outcome pertaining to the faraway particle 2. Thus, although this experimenter's choice and his consequent action on particle 1 have, by themselves, no *direct* faraway effects, this choice and action-by determining the physical significance (X_1 or P_1) of the local outcome, and thereby also the physical significance (X_2 or P_2) of the correlated faraway outcome-do influence the *nature* of the particular property of the faraway property of particle 2 that is revealed to the experimenter who is performing the measurement on particle 1, by his experiencing the outcome of the experiment that he has chosen and performed. But this sort of "influence" would, as in the classical case, fall far short of any indication of the need for any superluminal action at a distance, or of any superluminal transfer of information about the nearby free choice to the faraway region. All that has happened, in both the classical and quantum cases, is that the nearby experimenter has learned the value of an outcome that is *correlated* to the value of the outcome that a particular faraway experiment would have if the faraway experimenter were to choose to perform that particular experiment.

To identify what makes the quantum case different from classical case suppose one has two balls, one red and one green, and one hot the other cold. Suppose they are shot in opposite directions into two far-apart labs. Simply measuring the color of the ball reaching the first lab does not immediately disturb in any way anything in the other lab. But knowing the outcome of this color measurement allows one to know something about what will be found if color is measured also in the second lab. *But in the classical case this real property of the system that arrives in the second lab would not be nullified or eradicated if one had chosen to measure temperature instead of color*. It is the claimed *nullification* of one *kind* of property of particle 2 or another, on the basis of which kind of experiment is performed on particle 1, that distinguishes the quantum case from the classical one. It entails the need for some sort of leaping of the information about which action was chosen and performed on particle 1 to the region where particle 2 is being measured. The need for this nullification arises from the fact that no wave function can represent a well defined value of both X_2 and P_2.

In spite of this apparent violation of the notion that no information about the free choice made in region 1 can get to region 2, relativistic quantum field theory is compatible with the basic requirement of relativity theory that no "signal" can be transmitted faster than light. A *signal* is a carrier of information that allows a receiving observer to know which action was taken by a distant sender. Because the receiver does not know, superluminally, which *outcome* was observed by the sender, she, the receiver, cannot know, superluminally, which action was taken by the sender. Hence no signal can be sent.

The sender, who knows both which experiment he has freely chosen and performed, and which outcome has appeared, knows, on the basis of his knowledge of both the theory and this outcome, more about what the receiver will experience than the receiver herself can know.

Quantum theory, by focusing on knowledge and prediction, is able neatly to sort out these observer dependent features. The theory carries one step further Einstein's idea that science needs to focus on what actual observers can know and deduce on the basis of their own observations. But quantum theory places a crucial restriction on definability that classical relativistic theory lacks: a person by his choice of probing action performed in one region can cause one *type* of property in a faraway region to become *undefined in principle*, within the theory, because an incompatible *type* of property becomes defined there.

In the book *Albert Einstein: Philosopher–Physicist* Einstein [4, p. 85] gives a short statement of his locality condition:

> The real factual situation of the system S_2 is independent of what is done with the system S_1, which is spatially separated from S_2.

The problem of reconciling this condition with quantum theory is that quantum theory is a theory of predictions (about outcomes of observations) not a theory of reality. The probing action performed on system S_1 by the experimenter does not, by itself, *disturb* in any way the real factual system S_2. This action, by itself, does not allow any new prediction to be made about any outcome of any measurement made on S_2. Hence one may quite reasonably claim that "the real factual situation of the system S_2" is not disturbed by the mere action of performing the faraway measurement. And it is in no way surprising that what kind of predictions one can make about the faraway correlated system depends upon what kind of nearby measurement is chosen. Einstein's challenge is to the quantum theoretical claim that if the quantum state, which pertains to predictions, allows no predictions about a property then that property is in reality ill-defined.

If one accepts the quantum claim that the property itself is ill-defined if the property is ill-defined in the quantum theoretic state then the argument of Einstein et al. shows that the condition of no-faster-than-light action is violated in quantum theory. It is violated because the choice made in one region determines, no matter which outcome occurs, which *kind* of properties of the faraway particle becomes, within the quantum framework, ill defined.

The conclusion is that Einstein's argument leads, *within the quantum theoretical framework*, not to a proof of some incompleteness of quantum theory, but rather to a proof of the existence within theory of a faster-than-light transfer to a faraway region of the information about which measurement is performed in the nearby region.

This conclusion depends, however, on accepting the basic precept of quantum theory that if two properties of a system cannot be simultaneously represented by a wave function and one of these two properties is defined then the other cannot exist. Einstein rejected that premise. The question thus arises: Can the requirement of no superluminal transfer of information be upheld if one rejects the quantum precept that properties that cannot be simultaneously represented by any quantum state cannot be considered to be simultaneously definite.

This question has been studied by John Bell [5] and others within the special context of theories that postulate the existence of pertinent real hidden-variables. (▶ Bell's Theorem) Those arguments show that, within this hidden-variable context, the answer to the question posed at the end of the preceding paragraph is 'No'! Once the notion is accepted that decisions as to which measurements are performed are controlled by free choices that can go either way, it is impossible to reconcile *even merely the predictions* of quantum theory for all of the then-allowed alternative possible measurements with the demand that there be no superluminal transfer of information about which measurements are freely chosen. (▶ Nonlocality)

Primary Literature

1. A. Einstein, B. Podolsky, N. Rosen: Can quantum mechanical description of physical reality be considered complete? Phys. Rev. **47**, 177–180 (1935)
2. J. Schwinger: The theory of quantized fields I. Phys. Rev. **82**, 914–927 (1951)
3. S. Tomonaga: On a relativistically invariant formulation of the quantum theory of wave fields. Prog. Theor. Phys. **1**, 27–42 (1946)
4. A. Einstein: Autobiographical Notes, in P. A. Schilpp (ed.), *Albert Einstein: Philosopher-Scientist*. (Open Court, La Salle, Ill. 1949, 2–94)
5. J. S. Bell: On the Einstein-Podolsky-Rosen paradox. Physics **1**, 195–200 (1964); reprinted in J. S. Bell: *Speakable and unspeakable in quantum mechanics* (Cambridge University Press, Cambridge 1987, 14–21)

Secondary Literature

6. J. Cushing, E. McMullin eds.: *Philosophical Consequences of Quantum Theory*. (University of Notre Dame Press, Notre Dame, Indiana 1989)
7. D. Howard: Holism, Separability, and the Metaphysical Implications of the Bell Experiments, in Cushing/McMullin eds. (1989), 224–53
8. D. Howard: Einstein's Philosophy of Science. *The Stanford Encyclopedia of Philosophy (Spring 2004 Edition)*, Edward N. Zalta ed., URL =<http://plato.stanford.edu/archives/spr2004/entries/einstein-philscience/>.
9. M. Lange: *The Philosophy of Physics*. (Blackwell, London 2002, Ch. 9)

Electron Interferometry

J.C.H. Spence

Massive-particle interferometry can provide tests of fundamental ideas in quantum mechanics, due to the presence of mass and charge, not possible with the more familiar optical interferometry. Most importantly, since the first observation of electron diffraction in 1927 by Davisson, Germer and Thomson [1] (and the observation of electron Fresnel edge fringes by Boersch in 1940 [2]), it has been clear that matter diffracts, according to de Broglie's 1924 hypothesis. (▶ Davisson–Germer Experiment) The subsequent demonstration of Young's pinhole and biprism experiments (discussed below) with ▶ electrons about fifty years ago has since led to astonishing demonstrations of, for example, the diffraction of beams of buckyballs by a grating [3] and effects of gravity on neutron interferometry [4]. For neutrons and electrons, both Fermions, new effects due to ▶ spin and the ▶ exclusion principle might also be expected, not seen with photons (▶ light quantum). Perhaps the most famous experiments to date have been tests of the ▶ Aharonov–Bohm effect using electrons, and those using neutrons to see the effects of gravity on interference, but there have been many more (including an electron Sagnac interferometer and experiments on ▶ decoherence). The separate but closely related field of electron holography has come to prominence in recent decades, with applications in materials science and superconducting vortex imaging. Here we briefly review work on electron interferometry, first reviewed at an early stage by Denis Gabor [5], and also provide some guidance to the rapidly growing contemporary electron holography literature. Historically, it is of interest to note that the analysis of multiple scattering, and the role of the mean inner potential, in the experiments of Davisson and Germer by H. Bethe in his thesis work introduced Floquet's theorem into condensed matter physics for periodic structures, leading to the review article which founded modern condensed matter physics [6]. Bethe and Bloch were both students of A. Sommerfeld in 1928.

The construction of an electron interferometer requires a beam-splitter and a small, bright source of electrons. This should be of sufficiently small size d_s to produce a spatial coherence width L_c which spans the beam-splitter. ($L_c \sim \lambda/\Theta_c$ for a source at distance $L = d_s/(2\Theta_c)$ from the beamsplitter). Prior to the development of the field-emission electron source in 1968 [7] the use of heated tungsten wire pointed filaments produced values of $L_c < 1$ micrometer, so that early workers understood the need for an extremely small beamsplitting device, which limited development of the field. But even before the peak of interest in the Aharonov–Bohm effect in the 1960s, both amplitude and wavefront dividing beamsplitters had been demonstrated for electron beams. The first, using Bragg scattering [8], has since been abandoned in favor of the Mollenstedt and Duker electrostatic biprism, which may be said to have founded the field of electron interferometry [9].

(The convenient ability to adjust fringe spacing with a biprism using the applied voltage, and lack of inelastic scattering background favored it over the Bragg beamsplitter). The biprism uses a micron-sized wire (originally spider's web, then quartz fibers) held at a small potential running across the beam (normal to the page at B) as shown in Fig. 1. The charge on this wire creates a field which deflects rays from the source S around it such that they appear to come from virtual sources S′ and S″. In fact a cone of rays is deflected, so that S′ and S″, being images of S, are coherent if S is small. These act as Young's pin-holes to produce the interference fringes at F by exact analogy with an optical biprism. For these experiments it was natural to use the recently developed electron microscope, which produced a very high quality beam of electrons at a kinetic energy of about $E = 100\,\text{keV}$, corresponding to a relativistically corrected ► de Broglie wavelength of about $\lambda = 0.004\,\text{nm} = |\boldsymbol{k}|^{-1}$. (The longitudinal coherence length of an electron beam, $L_z \sim \lambda\, E/(2\Delta E)$ is maximized by reducing electronic fluctuations ΔE in the accelerating voltage E. The largest possible values of L_c and L_z are needed by modern transmission electron microscopes to produce high resolution phase-contrast images of atoms; they therefore provide the highest quality electron beams for interferometry, together with high mechanical and thermal stability. Low-energy biprism instruments are discussed below). The earliest pioneering work on the development of the electron biprism was undertaken at the University of Tübingen and used to measure L_c and L_z. Soon after, it became clear that by placing an electron-transparent sample in one arm of the interferometer at D, an off-axis electron hologram could be formed. (The in-line geometry was being investigated at the same time by Mulvey, Gabor and Haine in the UK – Gabor's original Noble-prize winning proposal for holography was devoted to electron interference, not light. The history of electron interferometry is therefore inextricably linked with that of electron holography). Modern work uses electron microscopes fitted with a field-emission electron source. This emits electrons from a source size of about $d_s = 2\,\text{nm}$ diameter with a brightness (measured in particles per unit solid angle per unit area) which exceeds that of current generation synchrotrons [10]. The dramatic success of electron interferometry is due primarily to these two inventions – the biprism and the field-emission electron gun.

Using an electron biprism, Feynman's "only one mystery" of quantum mechanics can immediately be demonstrated. Figure 2 shows Young's fringes obtained using

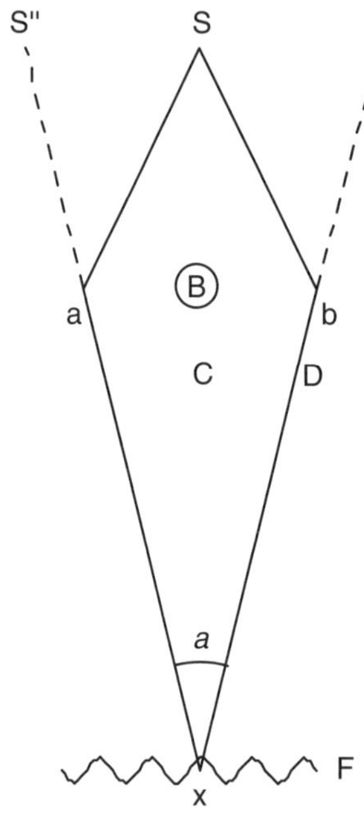

Fig. 1 The electron biprism

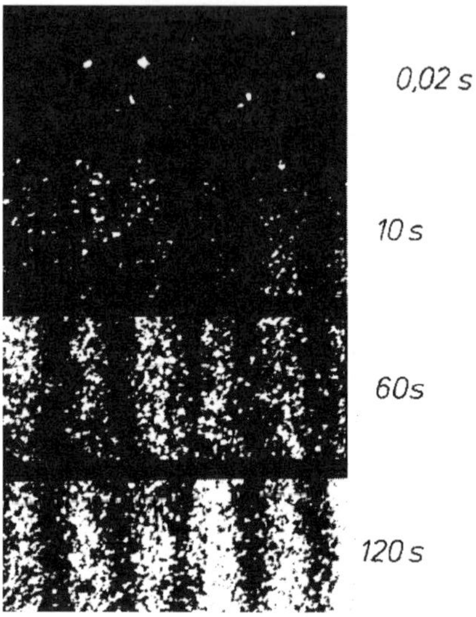

Fig. 2 Young's fringes formed using coherent electrons of very low intensity, recorded as a function of increasing exposure time. There is only one electron in the interferometer at any instant, yet an interference pattern develops with time [11]

coherent electrons and a biprism [11]. The important point is that the intensity has been reduced to such a low value that the electrons arrive one at a time, and the flight time of the electrons is much shorter than the time between their arrival at the detector. Nevertheless, the statistical buildup of an interference pattern is observed. (A similar experiment was undertaken for light by G.I.Taylor in 1909 [12]).

Despite the brightness of field-emission sources, if intense focussing by lenses is avoided, electron–electron interactions can normally be neglected in an electron microscope beam, and each electron reaches the detector before the next leaves the source. Then spin interactions can be neglected and the scalar theory of first-order optical coherence [29] (for bosons) can be applied to electron interferometry (fermions). If each of the beams in Fig. 1 are of unity amplitude, the fringe intensity recorded on the screen at F is then

$$I(x) = 2 + 2|\mu| \cos(2\pi q x + \varphi_c + \Delta\varphi(x)) \tag{1}$$

where the complex degree of coherence is $\mu = |\mu| \exp(\mathrm{i}\varphi_c)$, $q = |k|\ a$ (a is the angle between beams arriving at the detector, controlled by the voltage on the biprism wire, and setting the period of the fringes) and $\Delta\phi$ is the phase difference along the two optical paths a and b from source to detector point x. The complex degree of coherence may be expressed as a product of factors describing spatial and temporal coherence. These factors are proportional to the Fourier transform of the source intensity distribution (spatial coherence) and the distribution of wave numbers (temporal coherence). The biprism therefore offers a method of measuring both types of coherence. (Temporal coherence measurement requires a variable time delay to be introduced, by passing one beam along the axis of a cylinder held at a fixed potential [11]).

To understand the effect of the addition of fields into one or both arms of the interferometer, we require an expression for the refractive index of a medium with finite permeability traversed by an electron beam. For the ► Aharonov–Bohm effect we might imagine a solenoid at C in Fig. 1, with axis normal to the page, and return flux at infinity. (A clear description of the Aharonov–Bohm effect is given in the undergraduate lectures of R. Feynman [30]). For electron holography, an electron-transparent thin sample with internal fields might be placed at D. The refractive index expression was first given by Ehrenberg and Siday in 1949 [13], however the implications of this paper were not fully appreciated until the work of Aharonov and Bohm [14] a decade later. The precise form of the interaction had been controversial at that time. These papers showed that an electron would experience a measurable phase-shift even in the absence of a magnetic field $\boldsymbol{B} = \text{curl}\,\boldsymbol{A}$, (or resulting classical force), provided the vector potential $\boldsymbol{A}$ was non-zero. (This emphasis on the fundamental nature of potentials coincided with Maxwell's original formulation of electrodynamics, and differs from the standard modern form of his equations in terms of fields, first published by Heaviside long after Maxwell's death). For potentials weak compared with the accelerating potential, the phase shift is given by

$$\Delta\varphi = \sigma \int_{a-b} V(\boldsymbol{r})\mathrm{d}z - \frac{2\pi e}{h} \oint_{a+b} \boldsymbol{A}(\boldsymbol{r})\mathrm{d}\boldsymbol{s} \qquad (2)$$

for electrostatic potential V, interaction constant $\sigma = 2\pi\,|e|/hv$ and electron velocity v with charge e. The optical paths a (SaX) and b (SbX) are indicated in Fig. 1.

Since the first test of equation 2 with $\mathrm{V} = \mathrm{B} = 0$ at the electron trajectory in 1960, many experimental tests of the Aharnonov–Bohm effect have been published (see [15] for a review). All confirm the existence of a measurable phase-shift according to equation 2 if $\boldsymbol{A}$ is finite. Early objections regarding leakage of fields and the proximity of the return flux were met in the most sophisticated experiment, in which a torroidal magnet, coated with superconductor, was inserted into one arm of an electron interferometer, with the beam passing along its axis. The Meissner effect in the coating then confines the flux below T_c to within the torroid, and the field on its axis is zero [15].

The effects of inelastic scattering in one arm of the interferometer have been analysed in several papers, and the results have important implications for electron holography. An energy change as small as 4×10^{-15} eV results in a beat frequency of 1 Hz in the observed fringes, and fringe motion (consistent with the ► Heisenberg uncertainty relations). This effect has been observed [16] using the doppler shift from a moving electron mirror, or ramped electric or magnetic fields in one path. (Related effects are observed in the interference fringes observed very briefly due to interference between different lasers, if the recording time is less than the beat period). For electron holography, this has the remarkable effect that, for long recording times, we may consider that images reconstructed from off-axis electron holograms are formed from purely elastic scattering in the sample, since electrons loosing more than 4×10^{-15} eV while traversing the sample (e.g. due to phonon excitation) cannot produce stable time-independent fringes by interference with the reference wave (which has not lost energy). Electron holography therefore acts as a very efficient elastic energy filter [16]. There has been considerable discussion in the literature regarding ▶ "which way" experiments, in which a small energy loss in one arm might be used to signal the path taken by an electron [11].

For some purposes a low-energy table-top electron interferometer has advantages. Typical values of $\Delta E/E$ (which controls the temporal coherence) for electron microscopes operating at hundreds of kilovolts are 10^{-6}, whereas the spatial coherence width is proportional to λ, which increases at low energy. But stray fields and potentials, to which low-energy instruments are extremely susceptible, make their design very challenging. (The effect of time-dependent stray magnetic fields, for example, may result in enlargement of the virtual electron source size within a field-emission tip, resulting in loss of coherence [17]). Such a small instrument of 30 cm length with high performance has been constructed at the University of Tübingen [18]. This instrument includes a Wien filter, which imparts a different group velocity to the ▶ wave packet in one arm of the interferometer, without introducing a phase difference (the wavepackets in each arm are thus shifted longitudinally). The instrument operates at 150eV – 3 keV using a field-emission source, includes three biprisms, quadrupole lenses (to magnify the fringes) and extensive magnetic shielding. The fringes are detected on a channel plate, viewed by a charge-coupled device. Since it is powered by batteries, it may readily be rotated, and so has been used to form the electron equivalent of a Sagnac interferometer, with the path SaXbS taking the place of the loop in the Sagnac optical interferometer. The observation of an electron Sagnac effect [19] demonstrates that the coupling of inertial potentials and fields is independent of charge.

Most recently, this instrument has been used to demonstrate the electron antibunching effect [20]. Unlike the bunching of photons observed in the Hanbury Brown and Twiss experiment, the Pauli ▶ exclusive principle for electrons prevents overlapping wavetrains due to antisymmetrization of the ▶ wave function [21]. The result is a reduced probability (compared with classical particles) of detecting two electrons within a coherence time $\tau = L_c/v$. The electron arrival times are more uniformly distributed than Boltzman classical particles, and fluctuations reduced. A strong antibunching effect requires crowding of electrons in phase space, yet the degeneracy of a field-emitter is only about 10^{-4} (electrons per cell in phase space – maximum two, with opposite spins), unlike the values of 10^{15} for lasers (unrestricted Bosons). The degeneracy (and coherence parameters) may be measured from observations of Fresnel edge fringes [22]. In addition, electron detectors with time resolution $\tau \sim 10^{-14}$ s do not exist. Nevertheless, by detecting the arrival times at two detectors of an electron beam whose coherence patch spanned both detectors it has been possible recently to detect electron antibunching by comparing the results of coherent and incoherent illumination [20]. Finally, a variant of this instrument has been used to observe decoherence effects directly [23] (▶ decoherence, experimental observation of decoherence), as discussed above for inelastic scattering in electron holography [16]. The transition to classical behaviour of a quantum system is supposed to occur as a result of ▶ entanglement of its wave function with the environment, resulting in an incoherent mixture of states and loss of interference effects. Under these conditions of classical behaviour it should be possible to determine which path the electron took. Anglin and Zurek [24] proposed an interferometric experiment to test this idea, which has recently been implemented by electron interferometry. Both beams of the biprism interferometer pass over a resistive plate (tens of microns above it), in which they may induce polarization charges and Joule heating. The fringes are observed as a function of the height of the beam above the plate. The fading of the fringes with decreasing gap is clearly seen as coupling with phonon excitations in the plate increases [23]. A variety of more exotic electron interference experiments have been proposed by M. Silverman [21], such as those which test many-particle, multivalued wavefunction, and spin effects. These require a more subtle interpretation of Dirac's famous dictum that "each electron interferes only with itself". The simplest directly observable many-body effect in electron beams is the Boersch effect, in which Coulomb interactions along the direction of travel broaden the energy distribution. Lateral coulomb repulsion causes an angular divergence, which degrades the spatial resolution in time-resolved electron microscopy. At present, as a result of this effect, resolution is limited to a few nanometers, unlike the Angstrom level of resolution possible in CW mode.

Gabor's original proposal for electron holography in 1948 had the aim of eliminating the aberrations of electron lenses. This aim was finally achieved in 1995, when, for the first time, atomic-resolution images were reconstructed from an off-axis electron hologram whose resolution (about one Angstrom) exceeded that of the same state-of-the-art instrument in its conventional (Scherzer) imaging mode [25]. Since that time, aberration-correction devices have provided a simpler approach to this resolution, and electron holography has undergone a recent renaissance for other reasons – including the ability to map out electric and magnetic fields inside materials and nanostructures, from semiconductor devices to magnetic bacteria, ferroelectrics [26] and computer memory elements [27]. Other applications include the ability to image vortices and their quantization in superconductors at low temperature, and the ability to image magnetic domain structures in nanoparticles (see [28] for a review). Most recently, three-dimensional electron holography of internal fields has been developed, with important implications for semiconductor devices. At the same time, new solutions to the phase problem have been developed, which allow "interferometry without an interferometer" by extracting the phase difference information which is encoded within scattered intensities. It has recently been shown that this phase information may be extracted if scattering is sampled at the Shannon sampling interval (for a review of this field, see [31]).

Primary Literature

1. C.J. Davidson, L.H. Germer: Diffraction of electrons by a crystal of nickel. Phys.Rev.**30**, 705–15 (1927)
2. H. Boersch: Fresnelsche Beugeung im Elecktrononmikroskop. Naturwissenschaften **28**, 709–12 (1940)
3. L. Hackermuller, S. Uttenthaler, K. Hornberger, E. Reiger, B. Brezger, A. Zeilinger, M. Arndt: Wave nature of biomolecules and fluorofullerenes. Phys. Rev. Lett. **91**, 090408 (2003)
4. R. Colella, A.W. Overhauser, S.A. Werner: Observation of gravitationally induced quantum interference" Phys.Rev. Lett. **34**, 1472–74 (1975)
5. D. Gabor: Theory of electron interference experiments. Rev Mod Phys. **28**, 260–76 (1956)
6. H. Bethe, A. Sommerfeld in: *Handbuch der Physik*. Ed H. Geiger Vol **14** (1928, Ch 3, p.333)
7. A.V. Crewe: Electron gun using a field-emission source. Rev. Sci. Instrum **39**, 576–88 (1968)
8. L. Marton: Electron beam interferometer. Phys. Rev. **90**, 490–91 (1953)
9. G. Mollenstedt, H. Ducker: Fresnelscher Interferenzversuch mit einem Biprism fur Electronenwellen. Naturwissenschaften **42**, 41–41 (1955)
10. J.C.H. Spence, M. Howells: Synchrotron soft X-ray sources and electron sources: a comparison. Ultramic **93**, 213–22 (2002)
11. H. Lichte: Electron interference: mystery and reality". Phil. Trans. R. Soc. Lond. **A360**, 897–20 (2002)
12. G.I. Taylor: Interference fringes with feeble light. Proc. Camb. Philos. Soc. **15**, 114–15 (1909)
13. W. Ehrenberg, R. Siday: The refractive index in electron optics and the principles of dynamics. Proc. Phys. Soc. **B62**, 8–21 (1949)
14. Y. Aharonov, D. Bohm: Significance of electromagnetic potentials in quantum theory. Phys Rev. **115**, 485–91 (1959)
15. M. Peshkin, A. Tonomura: The Aharonov-Bohm effect". Lecture notes in physics Vol. 340 (Springer, Berlin 1989). See also Phys. Rev. Lett. **56**, 792–94 (1986)
16. D. van Dyck, H. Lichte, J.C.H. Spence. Ultramic **81**, 187–99 (2000). See also Ultramic **106**, 1012–18 (2006) and references therein.
17. M. Scheinfein, W. Qian, J.C.H. Spence. "Aberrations of emission cathodes: Nanometer diameter field-emission electron sources". J. Appl. Phys. **73**, 2057–68 (1993)
18. F. Hasselbach: A ruggedized miniature UHV electron biprism interferometer for fundamental experiments". Z. Phys. B (Cond. Matter.) **71**, 443–48 (1988)
19. F. Hasselbach, M. Nicklaus: Sagnac experiment with electrons. Phys. Rev. **A48**, 143–51 (1993)
20. H. Kiesel, A. Renz, F. Hasselbach: Observation of Hanbury-Brown Twiss anticorrelations for free electrons. Nature **418**, 392–94 (2002)
21. M.P. Silverman: On the feasibility of observing electron antibunching in a field-emission beam. Phys Lett. **A120**, 442–46 (1987)
22. J.C.H. Spence, W. Qian, M.P. Silverman: Electron source brightness and degeneracy from Fresnel fringes. J. Vac. Sci. Tech. **A12**(2), 542–47 (1994)
23. P. Sonnentag, F. Hasselbach: Decoherence of electron waves due to induced charges moving through a nearby resistive material. Braz. J. Phys. **35**, 385–90 (2005)
24. J.R. Anglin, J.P. Paz, W.H. Zurek: Deconstructing decoherence. Phys Rev. **A55**, 4041–45 (1997)
25. A. Orchowski, W. Rau, H. Lichte: Electron holography surmounts resolution limit of electron microscopy". Phys. Rev Lett. **74**, 399–02 (1995)
26. M. Lehmann, H. Lichte: Electron holographic materials analysis at atomic dimensions. Cryst. Res. Technol. **40**, 149–60 (2005)
27. H. Hu, H. Wang, M. McCartney, D. Smith: Switching mechanisms for nanoscale Co elements studied by electron holography. Phys. Rev. **B73**, 153401 (2006)
28. A. Tonomura. Electron holography and the A-B effect. Nucl. Instrum Meth. **A314**, 297–05 (1992)

Secondary Literature

29. R. Loudon: *The quantum theory of light* (Oxford University Press, Oxford 2000)
30. *The Feynman Lectures on Physics* (Addison-Wesley 2005, Vol. 3)
31. J.C.H. Spence: Lensless (Diffractive) Imaging, in: *Science of Microscopy*. P. Hawkes, J.C.H. Spence Eds. (Springer, Berlin 2007, Ch. 19)

E

Electrons

Theodore Arabatzis

The discovery of the electron was a complex and extended process, stretching from Faraday's investigation of electrolysis to Millikan's oil-drop experiments [18]. The results of four different fields (electrochemistry, electromagnetic theory, ► spectroscopy, and ► cathode rays) converged to support the existence of a novel subatomic constituent of matter. Faraday's experiments on electrolysis, interpreted from the perspective of the atomic theory of matter, implied that electricity has an atomic structure [4]. That is, electricity appears in naturally occurring units. In 1891 George Johnstone Stoney (1826–1911) named those units "electrons" ([13], p. 583, [30]).

In 1894 Stoney's electrons were appropriated by Joseph Larmor (1857–1942) to overcome certain empirical and conceptual problems faced by Maxwell's electromagnetic theory ([6], pp. 806 ff.). Larmor's electrons were supposed to be universal constituents of matter and were represented as structures in the all-pervading ether. On the continent a similar electromagnetic theory had been proposed by Hendrik Antoon Lorentz (1853–1928), who developed a synthesis of British and Continental traditions in electromagnetism [7]. Lorentz's theory incorporated Maxwell's suggestion that electromagnetic phenomena are wave processes in the ether and the suggestion of continental theorists (e.g., Wilhelm Weber) that these phenomena are due to the action of charged particles. Lorentz named those particles "ions", in analogy with the ions of electrolysis.

A crucial event for the development of Larmor's and Lorentz's theories was an experimentally discovery by Pieter Zeeman (1865–1943). In 1896 Zeeman observed that the spectral lines of sodium widen under the influence of a magnetic field (► Zeeman effect). Drawing on Lorentz's theory, he attributed the modification of the sodium spectrum to the influence of magnetism on the mode of vibration of the "ions". From the observed widening he was able to calculate their charge to mass ratio, which to everyone's surprise turned out to be three orders of magnitude larger than that of the electrolytic ions [17]. That was the first indication that Lorentz's ions, as well as Larmor's electrons, were much smaller than ordinary ions. In 1899 Lorentz changed the name of his "ions" to "electrons" [18].

Electron theories received additional support by the theoretical and experimental investigation of ► cathode rays. The nature of those rays had been the subject of considerable debate. The controversy subsided in 1897, when J. J. Thomson (1856–1940) showed that they were composed of "corpuscles", minute charged particles. From the electric and magnetic deflections of those particles he calculated their mass to charge ratio (m/e). It turned out that the value of m/e was three orders of magnitude smaller than "the smallest value of this quantity previously known, and which is the value for the hydrogen ion in electrolysis" ([15], p. 310).

In 1899 Thomson reported measurements of the mass to charge ratio of the particles produced in the ► photoelectric effect as well as by thermionic emission. Those measurements indicated that the particles in question were identical with the constituents of cathode rays [16]. Henri Becquerel (1852–1908) reached a similar conclusion about the identity of the recently discovered β-rays, which were shown to be "entirely comparable to . . . cathode rays, or masses of negative electricity transported with great speed" ([1], p. 210). Thus, by the end of the nineteenth century the electron had surfaced in a variety of theoretical and experimental contexts.

In the beginning of the twentieth century, β-rays were employed as a tool to adjudicate between contemporary electromagnetic theories, which gave different accounts of the electron's shape and structure. First, the theory developed by Max Abraham (1875–1922) implied that the electron was a rigid sphere with a uniform (surface or volume) distribution of charge, whose shape was not affected by its motion through the ether. Second, according to H. A. Lorentz's theory of electrons and Albert Einstein's relativity theory, the electron was deformable and contracted in the direction of its motion. Third, Alfred Bucherer (1863–1927) and Paul Langevin (1872–1946) suggested that a moving electron would be deformed but its volume would remain constant. All of those theories implied that the mass of the electron depended on its velocity. However, their quantitative predictions about that dependence differed. Walter Kaufmann (1871–1947) undertook an experimental research program that aimed at elucidating the nature of the electron's mass and its variation with velocity. He determined the velocity dependence of the charge to mass ratio of β-rays, on the basis of their electric and magnetic deflections. His results seemed to contradict the predictions of the "Lorentz–Einstein" theory and to favor the theories of Abraham, Bucherer, and Langevin [5]. Lorentz, for one, thought "very likely that we shall have to relinquish this idea [of a deformable electron] altogether" ([8], p. 213). His pessimism, however, was not vindicated by subsequent developments. By the mid-1910s the combined efforts of theoreticians and experimentalists had shown that Kaufmann's results were erroneous [20, 24–26].

The 1910s saw the culmination of a research program that aimed at measuring the charge of the electron. Its origins go back to the late nineteenth century and the experimental method devised by C. T. R. Wilson (1869–1959) to obtain artificial clouds and raindrops. J. J. Thomson employed Wilson's method to measure the charge of the "ions" (i.e., electrons) liberated "when a negatively electrified metal plate . . . is illuminated by ultra-violet light" ([16], p. 548). Thomson's work, as well as subsequent efforts along similar lines, were beset by many uncertainties (e.g., due to the evaporation of cloud droplets). Their main limitation was that they provided information about the statistical average of a great number of individual charges. Those difficulties were met by Robert Millikan (1868–1953). From 1909 onwards Millikan was able to get a grip on individual electrons. His meticulous observations of charged oil drops, moving under the simultaneous action of gravity and an electric field, enabled him to measure the charge of individual electrons [9]. Those measurements established that electricity has an atomic structure and eliminated the possibility of the electron being "a *statistical* mean of charges which are themselves greatly divergent" ([11], p. 58; cf. [23]). Thus, they provided "[t]he most direct and unambiguous proof of the existence of the electron" ([10], p. 55].

The electron also played a key role in the development of ▶ atomic models [22]. From 1913 to 1928 a quantum physics of the electron was gradually developed. Niels Bohr (1885–1962) and Arnold Sommerfeld (1868–1951) imposed restrictive conditions on the size, shape, and direction in space of the orbit of electrons bound within the atom. Those conditions were expressed as ▶ quantum numbers, which "denote the state of the electron in question" ([12], p. 150). In 1924 Wolfgang Pauli (1900–1958) attributed a fourth quantum number to the electron in an attempt to come to terms with the complexities of the anomalous Zeeman effect and the regularities of the periodic table. Furthermore, Pauli formulated an ▶ exclusion principle, which prohibited the coexistence of identical electrons (i.e., with the same quantum numbers) in the same atom. In 1925 Samuel Goudsmit (1902–1978) and George Uhlenbeck (1900–1988) proposed a semi-classical interpretation of the fourth quantum number as a manifestation of ▶ spin, that is, as a self-rotation of the electron. This interpretation led to several paradoxes (▶ errors and paradoxes in quantum mechanics) and was subsequently abandoned [18]. Spin was reconceptualized as a quantum mechanical property with no classical correlate. However, the incorporation of spin into the new quantum mechanics encountered difficulties, until P. A. M. Dirac (1902–1984) showed in 1928 that spin could be derived from his relativistic wave equation [27].

During the 1920s the wave character of the electron was also established. In 1923 Louis de Broglie (1892–1987) developed a synthesis of particle and wave conceptions of matter. The wave properties of matter implied that "[a] group of electrons that traverses a sufficiently small aperture will exhibit diffraction effects" ([2], p. 549; transl. in [29], p. 263; ▶ matter waves; ▶ de Broglie wavelength). De Broglie's suggestion was confirmed in 1927–28, when Clinton Davisson (1881–1958) and Lester Germer (1896–1971) in the US and George Paget Thomson (1892–1975) in England discovered experimentally electron diffraction [3, 14, 28] ▶ Davisson–Germer experiment.

Primary Literature

1. H. Becquerel, Contribution à l'ètude du rayonnement du radium, Comptes Rendus **130**, 206–211 (1900).
2. L. de Broglie, Quanta de lumière, diffraction et interférences, Comptes Rendus **177**, 548–550 (1923).

3. C. J. Davisson and L. Germer, Diffraction of electrons by a crystal of nickel, Physical Review **30**, 705–740 (1927).
4. H. Helmholtz, On the modern development of Faraday's conception of electricity, Journal of the Chemical Society **39**, 277–304 (1881).
5. W. Kaufmann, Über die Konstitution des Elektrons, Annalen der Physik **19**, 487–553 (1906).
6. J. Larmor, A dynamical theory of the electric and luminiferous medium, Philosophical Transactions of the Royal Society of London A **185**, 719–822 (1895).
7. H. A. Lorentz, *Versuch einer Theorie der electrischen und optischen Erscheinungen in bewegten Körpern* (Leiden: E. J. Brill, 1895).
8. H. A. Lorentz, *The Theory of Electrons* (New York: The Columbia University Press, 1909).
9. R. Millikan, On the elementary electrical charge and the Avogadro constant, Physical Review **2**, 109–143 (1913).
10. R. A. Millikan, The electron and the light-quant from the experimental point of view, Nobel Lecture, May 23, 1924, in *Nobel Lectures: Physics, 1922–1941* (Amsterdam: Elsevier, 1965), pp. 54–66.
11. R. A. Millikan, *The Electron: Its Isolation and Measurement and the Determination of some of its Properties* (Chicago. The University of Chicago Press, 2nd ed., 1924; 1st ed., 1917).
12. A. Sommerfeld, *Atomic Structure and Spectral Lines* (London: Methuen, 3rd rev. ed., 1934).
13. G. J. Stoney, On the Cause of Double Lines and of Equidistant Satellites in the Spectra of Gases, The Scientific Transactions of the Royal Dublin Society (2nd ser.) **4**, 563–608 (1888–1892).
14. G. P. Thomson, Experiments on the diffraction of cathode rays, *Proceedings of the Royal Society* **117A**, 600–609 (1928). This article was published in February 1928.
15. J. J. Thomson, Cathode Rays, Philosophical Magazine (5th ser.) **44**, 293–316 (1897).
16. J. J. Thomson, On the Masses of the Ions in Gases at Low Pressures, Philosophical Magazine (5th ser.) **48**, 547–567 (1899).
17. P. Zeeman, Doublets and Triplets in the Spectrum produced by External Magnetic Forces, Philosophical Magazine (5th ser.) **44**, 55–60, 255–259 (1897).

Secondary Literature

18. T. Arabatzis, *Representing Electrons: A Biographical Approach to Theoretical Entities* (Chicago: The University of Chicago Press, 2006).
19. J. Z. Buchwald and A. Warwick (eds.), *Histories of the Electron: The Birth of Microphysics* (Cambridge, MA: MIT, 2001).
20. J. T. Cushing, Electromagnetic Mass, Relativity, and the Kaufmann Experiments, American Journal of Physics **49**, 1133–1149 (1981).
21. S. Goldberg, The Abraham Theory of the Electron: The Symbiosis of Experiment and Theory, Archive for History of Exact Sciences **7**, 7–25 (1970).
22. J. L. Heilbron, *Historical Studies in the Theory of Atomic Structure* (New York: Arno Press, 1981).
23. G. Holton, Subelectrons, Presuppositions and the Millikan-Ehrenhaft Dispute, in Holton, *Scientific Imagination: Case Studies* (Cambridge: Cambridge University Press, 1978), pp. 25–83.
24. G. Hon, Is the Identification of Experimental Error Contextually Dependent? The Case of Kaufmann's Experiment and Its Varied Reception, in J. Z. Buchwald (ed.), *Scientific Practice: Theories and Stories of Doing Physics* (Chicago: The University of Chicago Press, 1995), pp. 170–223.
25. M. Janssen and M. Mecklenburg, From Classical to Relativistic Mechanics: Electromagnetic Models of the Electron, in V. F. Hendricks, K. F. Jørgensen, J. Lützen and S.A. Pedersen (eds.), *Interactions: Mathematics, Physics and Philosophy, 1860–1930* (Dordrecht: Springer, 2006), pp. 65–134.
26. A. I. Miller, *Albert Einstein's Special Theory of Relativity, Emergence (1905) and Early Interpretation (1905–1911)* (New York: Springer, 1998).

27. M. Morrison, Spin: All is not What it Seems, Studies in History and Philosophy of Modern Physics **38**, 529–557 (2007).
28. A. Russo, Fundamental Research at Bell Laboratories: The Discovery of Electron Diffraction, Historical Studies in the Physical Sciences **12**, 117–160 (1981).
29. B. R. Wheaton, *The Tiger and the Shark: Empirical Roots of Wave-Particle Dualism* (Cambridge: Cambridge University Press, 1983).
30. J. G. O'Hara: George Johnstone Stoney, F.R.S., and the Concept of the Electron, Notes and Records of the Royal Society of London **29** (2), 265–276 (1975).

Ensembles in Quantum Mechanics

Leslie E. Ballentine

> The attempt to conceive the quantum-theoretical description as the complete description of the individual systems leads to unnatural theoretical interpretations, which immediately become unnecessary if one accepts the interpretation that the description refers to *ensembles* of systems and not to individual systems.
> – Albert Einstein (1879–1955) [1], p. 671.

This quotation is perhaps the most famous statement of the *ensemble* interpretation of quantum mechanics. The role of the ensemble in quantum mechanics ranges from innocuous to profound, and even controversial.

The innocuous role of the ensemble stems from the fact that quantum mechanics does not predict the actual events, but only the probabilities of the various possible outcomes (► *probability in quantum mechanics*) of the various possible events. In order to compare the predictions of quantum mechanics with experiment, one must prepare a ► state and measure some dynamical variable, and repeat this preparation–measurement sequence many times. The relative frequencies of the various outcomes in this ensemble of results can then be compared with the theoretical probabilities predicted by quantum mechanics. Thus it is natural to say that quantum mechanics describes the statistics of an ensemble of similarly prepared systems.

Here, as in classical statistical mechanics, one should not confuse the *ensemble* of systems with an *assembly* of systems into a composite. For example, if the system is a single particle, then the ensemble is a conceptual set of replicas of it, each in its own environment, whereas the assembly would be a many-particle system. The role of the ensemble is to enable statistical analysis; its members do not interact with or influence each other.

The more significant role of the ensemble interpretation is exemplified by ► *Schrödinger's cat paradox* [2], which involves an unstable atom, a cat, and a mechanism that releases a poison to kill the cat when the atom decays. The initial state vector of the system, $|\phi_1\rangle|live\rangle$, describes an atom in an excited state and a live cat. The final state vector, after the atom has decayed and the cat is dead, will be $|\phi_0\rangle|dead\rangle$. At an intermediate time equal to one half-life of the unstable atomic state, the normalized state vector will be

$$|\Psi\rangle = (|\phi_1\rangle|live\rangle + |\phi_0\rangle|dead\rangle)/\sqrt{2} \quad (1)$$

Now how are we to interpret the state vector $|\Psi\rangle$, which apparently describes a coherent superposition of macroscopically distinct components, namely a live cat and a dead cat? It makes no sense as a realistic description of an *individual* system. The paradox is not changed at all if we include the effect of the *environment*, i.e. ► decoherence. In place of (1), we will have

$$|\Psi\rangle = (|\phi_1\rangle|live\rangle|e_1\rangle + |\phi_0\rangle|dead\rangle|e_2\rangle)/\sqrt{2} \quad (2)$$

where $|e_1\rangle$ and $|e_2\rangle$ are states of the environment. But (2) is still a coherent superposition of macroscopically distinct components; indeed, the paradox is even worse, since we now have a superposition of two environmental states, which is an even more macroscopic superposition than that in (1).

But if the state vector is regarded only as the generator of probability distributions for the ► observables of an *ensemble* of similarly prepared systems, then $|\Psi\rangle$ makes perfectly good sense. If the experiment is repeated many times, in one half of the cases the cat will be found to be alive, and in the other half of the cases it will be found to be dead [4, 5].

The limitations of the *ensemble* interpretation can be expressed by the question, "Is that all there is?" The world is made up of individual systems and individual events, not ensembles and probabilities, so the description of the world by quantum mechanics seems somewhat incomplete. An extention of the theory to describe individual events, not merely their probabilities, would, indeed, be desirable, but it would appear to require new fundamental developments that go beyond those of present day quantum mechanics. A broad review of ensemble interpretations is given in [6].

Primary Literature

1. A. Einstein: in *Albert Einstein: Philosopher-Scientist*, ed. P. A. Schilpp (Harper & Row, New York), 665–88 (1949).
2. E. Schrödinger: Die gegenwärtige Situation in der Quantenmechanik, Naturwissenschaften. 23, 807-12, 823–28, 844–49 (1935); english translation, *The Present Situation in Quantum Mechanics*, in Wheeler and Zurek (1983), 152–67.
3. J. A. Wheeler and W. H. Zurek: *Quantum Theory and Measurement*. (Princeton University Press, Princeton, NJ) (1983).
4. L. E. Ballentine: *The Statistical Interpretation of Quantum Mechanics*. Rev. Mod. Phys. 42, 358–81 (1970).
5. L. E. Ballentine: *Quantum Mechanics: A Modern Development*. (World Scientific, Singapore) (1998).

Secondary Literature

6. D. Home, M. A. B. Whitaker: *Ensemble Interpretations of Quantum Mechanics*. Phys. Rep. 210(4), 223–317 (1992).

Entanglement

E

Peter Mittelstaedt

Consider two proper quantum systems S_1 and S_2 with ► Hilbert spaces $\mathcal{H}_1$ and $\mathcal{H}_2$, respectively. If S_1 and S_2 are independently prepared in the pure states $\varphi_1 \in \mathcal{H}_1$ and $\varphi_2 \in \mathcal{H}_2$, then the compound system $S_1 + S_2$ is correctly described in the tensor-product Hilbert space $\mathcal{H}_1 \otimes \mathcal{H}_2$ by the product state $\psi_0 = \varphi_1 \otimes \varphi_2$. In this case, the state $\psi_0(S_1 + S_2)$ determines uniquely the states φ_1 and φ_2 of subsystems S_1 and S_2, respectively.

In the general case, the state $\psi(S_1 + S_2)$ of the compound system cannot be written as a product of states referring to S_1 and S_2. (A state of this general kind can be prepared by a convenient interaction between the two systems for a limited period of time, as in a scattering process of S_1 and S_2.) However, even if the state $\psi(S_1 + S_2)$ (after the interaction) cannot be written as a product, it can be decomposed with respect to two orthonormal systems in $\mathcal{H}_1$ and $\mathcal{H}_2$ into a weighted sum of products. In particular, for any pure state $\psi(S_1 + S_2)$ there exist orthonormal systems ${\xi^{(1)}}_i \in \mathcal{H}_1$ and ${\eta^{(2)}}_k \in \mathcal{H}_2$ that allow for a *biorthogonal decomposition* [1,2]

$$\psi(S_1 + S_2) = \sum_i c_i \xi_i^{(1)}(S_1) \otimes \eta_i^{(2)}(S_2)$$

with *one* summation index i and complex numbers c_i. In this state, the two systems are called *entangled* provided that the sum consists of more than one term. The *entanglement* [3] of S_1 and S_2 means in particular, that the compound state $\psi(S_1 + S_2)$ does not provide definite information about pure states of S_1 and S_2. We can only say, that the probability p_n for finding S_1 in the state ${\xi^{(1)}}_n(S_1)$ and S_2 in the state ${\eta^{(2)}}_n(S_2)$ is given by the value $p_n = |c_n|^2$.

Primary Literature

1. J. Von Neumann: *Mathematische Grundlagen der Quantenmechanik* (Springer, Berlin, 1932, 231–232)
2. J. M. Jauch: *Foundations of Quantum Mechanics* (Addison-Wesley, Reading, MA, 1968, 179–182)

3. 'Entanglement' is the English translation of the German word "Verschränkung", first introduced by E. Schrödinger: Die gegenwärtige Situation in der Quantenmechanik. Die Naturwissenschaften **23**, 807–49 (1935)

Entanglement Purification and Distillation

Dagmar Bruß

In quantum mechanics, subsystems of a composite system can exhibit correlations (▶ correlations in quantum mechanics) that are stronger than any classical correlations. Quantum correlations are also called *entanglement* [1]. A mixed quantum state ϱ consisting of two subsystems (i.e. a bipartite state) can be either separable or entangled. It is separable [2] if $\varrho = \sum_i p_i | a_i \rangle\langle a_i | \otimes | b_i \rangle\langle b_i |$, with p_i being probabilities, and entangled otherwise. Entanglement can be quantified via entanglement measures. Maximally entangled states are pure, and mixing generally decreases entanglement. For further reading on entanglement, see [18–20] and general textbooks on quantum information, e.g. [21–23].

In quantum information entanglement is viewed as a resource, see protocols such as quantum teleportation [3], superdense coding [4] or entanglement-based quantum cryptography (▶ quantum communication) [5]. Therefore, one is interested in maximally entangled (pure) quantum states. In a realistic scenario, noise due to interaction with the environment (▶ decoherence) or imperfect gate operations generally reduces both purity and entanglement of a given state. However, if one has several copies of some less than maximally entangled state available, it is possible that the two parties Alice (A) and Bob (B) concentrate or *distill* the entanglement, by acting locally on their parts of the states (in their corresponding laboratories) and exchanging classical information via a telephone. Thus, by using so-called *local operations and classical communication* (LOCC) they can create fewer pairs with higher entanglement and higher degree of purity. This process is called *entanglement purification* or *entanglement distillation*.

In this context, two topics are of interest: First, one wants to find *distillation protocols* that are as efficient as possible. Second, one studies the possibility of distillation. The "distillability problem" is phrased as: given a certain density matrix ϱ, is it *distillable* or not?

For pure, but not maximally entangled states, it is possible to increase the entanglement by "local filtering" [6]. Here Alice and Bob apply certain local operators, and with some probability p arrive at a state with higher entanglement. However, as it is not possible to increase entanglement on average by local operations, with probability $1 - p$ the resulting state is less entangled than before. The first purification and distillation protocols for mixed states were suggested in [7, 8]. In [7] the given state ϱ is first brought by random local rotations into a standard form, namely a Bell-diagonal state (a mixture of the four maximally entangled Bell states). Then, Alice and Bob apply local CNOT-gates to two copies of ϱ, and each of them measures his or her second qubit. If their measurement outcomes agree, the singlet fidelity of the first pair, i.e. its overlap with $|\psi_-\rangle = \frac{1}{\sqrt{2}}(|01\rangle - |10\rangle)$, and thus its entanglement, has increased. Otherwise this pair has to be thrown away. This procedure is repeated in an iterative way, thus gradually increasing the entanglement. Note that this protocol can also be generalised to higher dimensions. However, it is a very wasteful protocol, concerning the resource of entangled states.

E

The efficiency of distillation for qubits can be improved by replacing the CNOT operation by a permutation on more than two qubits. For details on improvements of distillation protocols, for the link between entanglement distillation and error correction that led to security proofs in quantum key distribution, and for multipartite distillation protocols, see the literature given in [20].

A quantum state ϱ is called *n-distillable* if there exists a number n of copies such that Alice and Bob can create with LOCC a state that is arbitrarily close to a maximally entangled state. A quantum state ϱ is called *distillable* if there exists a number n for which ϱ is n-distillable. Which quantum states are distillable? At the moment, this question has been only partially answered. It was found in [9] that *all* entangled two-qubit states are distillable. This statement does not hold for higher dimensions. Clearly, a necessary condition for a quantum state to be distillable is that it is entangled. It has been shown [9] that a further necessary condition for distillability of ϱ is the non-positivity of the partial transpose of ϱ. The partial transpose [10] of a composite density matrix is given by transposing only one of the subsystems. As the definition of a separable state is $\varrho_{\text{sep}} = \sum_i p_i |a_i\rangle\langle a_i| \otimes |b_i\rangle\langle b_i|$, the partial transpose of a separable state reads $\varrho_{\text{sep}}^{\text{T}_\text{A}} = \sum_i p_i\,(|a_i\rangle\langle a_i|)^{\text{T}} \otimes |b_i\rangle\langle b_i|$, where the index T denotes the transpose, and T_A denotes the partial transpose with respect to Alice. As $(|a_i\rangle\langle a_i|)^{\text{T}}$ is some quantum state of Alice, $\varrho_{\text{sep}}^{\text{T}_\text{A}}$ describes a positive semidefinite density matrix. (A Hermitian matrix σ is called "positive semidefinite" if $\langle\psi|\sigma|\psi\rangle \geqslant 0$ for all vectors $|\psi\rangle$, or, equivalently, if all eigenvalues are greater or equal zero.) The property $\varrho^{\text{T}_\text{A}} \geqslant 0$ is called *positive partial transpose* (PPT) of ϱ. For bipartite systems with low dimensions, namely for composite states of dimension 2×2 and 2×3, positivity of the partial transpose is a necessary and sufficient condition for separability [11]. For higher dimensions, however, there exist entangled PPT states [12]. They are called *bound entangled states*, as their entanglement cannot be distilled. The concept of bound entanglement can be generalised also to multipartite quantum states.

A necessary and sufficient criterion for distillability of a given bipartite state ϱ was derived in [13]: "The state ϱ is distillable if and only if there exists $|\psi^{(2)}\rangle = c_1|e_1\rangle|f_1\rangle + c_2|e_2\rangle|f_2\rangle$ such that $\langle\psi^{(2)}|(\varrho^{\text{T}_\text{A}})^{\otimes n}|\psi^{(2)}\rangle < 0$ for some n." Here, $|\psi^{(2)}\rangle$ is written in the bi-orthogonal *Schmidt decomposition*, with $\langle e_1|e_2\rangle = 0 = \langle f_1|f_2\rangle$. Thus, $|\psi^{(2)}\rangle$ denotes a state with Schmidt rank 2 (i.e. the Schmidt decomposition has two terms). The matrix $(\varrho^{\text{T}_\text{A}})^{\otimes n}$ denotes the n-fold tensor product of $\varrho^{\text{T}_\text{A}}$. The above criterion implies that a state with a positive partial transpose is undistillable: if $\varrho^{\text{T}_\text{A}} \geqslant 0$, then $(\varrho^{\text{T}_\text{A}})^{\otimes n} \geqslant 0$.

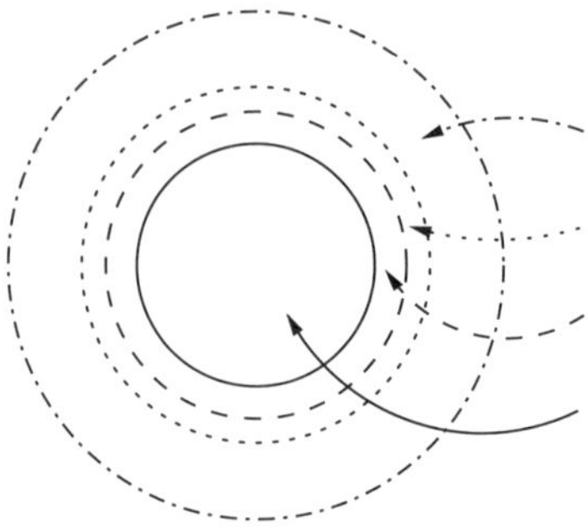

Fig. 1 The set of bipartite quantum states and their distillability properties

It is an open question whether non-positivity of the partial transpose (NPT) is also a sufficient criterion for distillability. Based on a family of states introduced in [14, 15], there is the (unproven) conjecture that NPT-undistillable states exist. Somewhat surprising, many copies may be needed for entanglement distillation: it has been shown [16] that for every n there exists a state that is distillable, but not n-distillable. This fact illustrates the difficulty of proving the mentioned conjecture, as one has to take into account the limit $n \to \infty$. Our present understanding of how the set of all bipartite quantum states is decomposed into separable, entangled undistillable and distillable states is summarized in Fig. 1. Experimentally, distillation of a two-qubit mixed state via local filtering has been achieved [17].

See also creation and detection of entanglement; entropy of entanglement.

Primary Literature

1. E. Schrödinger: Naturwissenschaften 23, 807 (1935).
2. R. Werner: Quantum states with Einstein-Podolsky-Rosen correlations admitting a hidden-variable model. Phys. Rev. A 40, 4277 (1989).
3. C. Bennett et al: Teleporting an unknown quantum state via dual classical and Einstein-Podolsky-Rosen channels, Phys. Rev. Lett. 70, 1895 (1993).
4. C. Bennett, S. Wiesner: Communication via one- and two-particle operators on Einstein-Podolsky-Rosen states, Phys. Rev. Lett. 69, 2881 (1992).
5. A. Ekert: Quantum cryptography based on Bell's theorem. Phys. Rev. Lett. 67, 661 (1991).
6. N. Gisin: Hidden quantum nonlocality revealed by local filters. Phys. Lett. A 210, 151 (1996).
7. C. Bennett, G. Brassard, S. Popescu, B. Schumacher, J. Smolin, W. Wootters: Purification of Noisy Entanglement and Faithful Teleportation via Noisy Channels. Phys. Rev. Lett. 76, 722 (1996).
8. D. Deutsch, A. Ekert, R. Jozsa, C. Macchiavello, S. Popescu, A. Sanpera: Quantum Privacy Amplification and the Security of Quantum Cryptography over Noisy Channels. Phys. Rev. Lett. 77, 2818 (1996).
9. M. Horodecki, P. Horodecki, R. Horodecki: Inseparable Two Spin-1/2 Density Matrices Can Be Distilled to a Singlet Form. Phys. Rev. Lett. 78, 574 (1997).
10. A. Peres: Separability Criterion for Density Matrices. Phys. Rev. Lett. 77, 1413 (1996).
11. M. Horodecki, P. Horodecki, R. Horodecki: Separability of Mixed States: Necessary and Sufficient Conditions. Phys. Lett. A 223, 1 (1996).

12. P. Horodecki: Separability criterion and inseparable mixed states with positive partial transposition. Phys. Lett. A 232, 333 (1997).
13. M. Horodecki, P. Horodecki, R. Horodecki: Mixed-State Entanglement and Distillation: Is there a Bound Entanglement in Nature? Phys. Rev. Lett. 80, 5239 (1998).
14. W. Dür, J.I. Cirac, M. Lewenstein, D. Bruß: Distillability and partial transposition in bipartite systems. Phys. Rev. A 61, 062313 (2000).
15. D. DiVincenzo, P. Shor, J. Smolin, B. Terhal, A. Thapliyal: Evidence for bound entangled states with negative partial transpose. Phys. Rev. A 61, 062312 (2000).
16. J. Watrous: Many Copies May Be Required for Entanglement Distillation. Phys. Rev. Lett. 93, 010502 (2004).
17. Z.-W. Wang et al: Experimental Entanglement Distillation of Two-Qubit Mixed States under Local Operations, Phys. Rev. Lett. 96, 220505 (2006).

E

Secondary Literature

18. M. Lewenstein, D. Bruß, J. I. Cirac, B. Kraus, M. Kuś, J. Samsonowicz, A. Sanpera, R. Tarrach: Separability and distillability in composite quantum systems – a primer. Journ. Mod. Opt. 47, 2841 (2000).
19. D. Bruß: Characterizing entanglement. J. Math. Phys. 43, 4237 (2002).
20. R. Horodecki, P. Horodecki, M. Horodecki, K. Horodecki: Quantum entanglement. arXiv: quant-ph/0702225, subm. to Rev. Mod. Phys.
21. M. Nielsen, I. Chuang: *Quantum Computation and Information*. Cambridge University Press (2000).
22. Quantum Information: *An Introduction to Basic Theoretical Concepts and Experiments* (Springer Tracts in Modern Physics, 173). Eds. G. Alber, T. Beth, M. Horodecki, P. Horodecki, R. Horodecki, M. Rötteler, H. Weinfurter, R. Werner, A. Zeilinger, Springer-Verlag (April 2001).
23. *Lectures on Quantum Information*. Eds. D. Bruß G. Leuchs: WILEY-VCH Weinheim (2007).

Entropy of Entanglement

Dominik Janzing

An essential feature of an entangled joint state (▶ entanglement) of two physical systems A, B is that the state of each subsystem is always *mixed* even though the joint state of the bipartite system may be *pure*. The entropy of the subsystems can therefore be used to quantify the entanglement of pure bipartite quantum states. For simplicity, we restrict ourselves to finite dimensions. Every pure state on $\mathbb{C}^\ell \otimes \mathbb{C}^d$ (with $d \leqslant \ell$) can be written as

$$|\gamma\rangle = \sum_{j=1}^{d} c_j |\phi_j\rangle \otimes |\psi_j\rangle$$

where $|\phi_j\rangle$ and $|\psi_j\rangle$ are *orthonormal* vectors defined in the ► Hilbert spaces of system A and B, respectively. The number of summands in this so-called *Schmidt decomposition* [3–5] is at most the dimension of the smaller subsystem. The state is *entangled* if the number of terms is at least 2. When restricting our attention to one of the subsystems we no longer can describe its quantum state by a ► wave function. Instead, the "reduced states" of A and B are given by the ► density operators

$$\rho_A = \sum_{j=1}^{d} |c_j|^2 |\phi_j\rangle\langle\phi_j| \qquad \text{and} \qquad \rho_B = \sum_{j=1}^{d} |c_j|^2 |\psi_j\rangle\langle\psi_j| \, .$$

The following argument, which uses basically the von Neumann projection postulate, shows why this is the case. A measurement on system B cannot change the mixed state of A as long as the measurement result is ignored.[1] Consider a von Neumann measurement corresponding to a self-adjoint observable $\mathcal{B}$ having the states $|\psi_j\rangle$ as (non-degenerate) eigenvectors. A possible choice is

$$\mathcal{B} := \sum_{j=1}^{d} j |\psi_j\rangle\langle\psi_j| .$$

Given that the measurement result is j, which happens with probability

$$p_j := |c_j|^2, \tag{1}$$

the wave function of the joint system has been "collapsed" to the state

$$|\phi_j\rangle \otimes |\psi_j\rangle .$$

The state of A is then given by $|\phi_j\rangle$. When ignoring the measurement result we thus obtain

$$\rho_A := \sum_{j=1}^{d} p_j |\phi_j\rangle\langle\phi_j| .$$

Using similar measurements on system A we conclude that the state of the right hand system reads

$$\rho_B := \sum_{j=1}^{d} p_j |\psi_j\rangle\langle\psi_j| ,$$

[1] Since this fact is sometimes blurred by incorrect descriptions of the phenomenon of *entanglement*, it should be stressed that such a locality principle still remains true in quantum theory: For distant subsystems, measurements on B can only change the statistics of experiments performed on A if the result is communicated to A, where an operation is performed that depends on the result.

The key observation to quantify entanglement is that the eigenvalues of both density operators are the same. Hence their von-Neumann entropies (► quantum entropy) coincide, i.e.,

$$S(\rho_B) = S(\rho_A) = \mathcal{H}(p), \tag{2}$$

where $\mathcal{H}(p)$ denotes the Shannon entropy [6] of the probability distribution $(p_1, \ldots, p_d)$ defined in (1). The entropy thus can considered as a property of the bipartite state, the *entropy of entanglement.*

The interpretation of the entropy of entanglement is not obvious. Note that the state of the joint system is *completely known* in the sense of being a pure state and the wave functions of the subsystems are *not defined.* It would therefore not be justified to consider the entropy as "missing knowledge" on the states of A and B.

In order to describe an information-theoretic interpretation, we show that the entropy of entanglement is the maximal amount of *classical* information that measuring one system can provide about the results of measurements performed on the second. First consider observables $\mathcal{A}$ and $\mathcal{B}$ having the vectors $|\phi_j\rangle$ and $|\psi_j\rangle$ that appear in the Schmidt decomposition as non-degenerate eigenvectors. The uncertainty of the measurement results of A is given by the entropy $\mathcal{H}(p)$. However, given the measurement result of B, the entropy is 0 since both results will always coincide. Hence the result of B provides the information $\mathcal{H}(p)$ about the result of A. The following argument shows that there cannot exist any pair of measurements for which the mutual information exceeds the entropy of entanglement. Label the results of an arbitrary measurement performed on B by i in some index set I (for simplicity we assume I to be countable) and denote the probability to obtain i by q_i. Let σ_i denote the state of A given that the result of B was i. Due to the so-called Holevo-bound [1], measurements performed on an unknown quantum state taken from a set of states $\{\sigma_i \mid i \in I\}$, each occurring with probability q_i, can never provide more information than

$$\chi := S\Big(\sum_{i\in I} q_i\sigma_i\Big) - \sum_{i\in I} q_i S(\sigma_i).$$

According to our locality arguments above, the mixture $\sum_{i\in I} q_i\sigma_i$ coincides with ρ_A. Hence we get $\chi \leqslant S(\rho_A)$. This shows that the classical information about the measurement outcomes of A obtained by measurements on B can never exceed $S(\rho_A)$. Hence the entropy of entanglement is the maximal *classical mutual information* between measurement results performed on both systems separately. It should be emphasized that this amount of classical information does not coincide with the *quantum mutual information*

$$I(A:B) = S(\rho_A) + S(\rho_B) - S(\rho),$$

which is for pure states twice the entropy of entanglement.

An alternative interpretation of entropy of entanglement is that it quantifies the amount of quantum information that has to be transferred if one party wants to send his/her part of the entangled state to a third party. To sketch this idea, we consider a scenario where two parties A and B share n copies of the entangled state in (2) and B

wants to forward his part of the entangled states to a third party C such that the state $|\gamma\rangle^{\otimes n}$ is shared by A and C instead of being shared by A and B. A straightforward way to achieve this would be to transfer the d^n-dimensional quantum system from B to C. However, this is not the most economical method. The key observation for saving communication resources is the following. The Schmidt decomposition of $|\gamma\rangle$ defines in a straightforward way a Schmidt decomposition of the n-fold copy:

$$|\gamma\rangle^{\otimes n} = \sum_{1\leqslant j_1,\ldots,j_n\leqslant d} c_{j_1}\cdots c_{j_n}|\phi_{j_1}\rangle\otimes\cdots\otimes|\phi_{j_n}\rangle\otimes|\psi_{j_1}\rangle\otimes\cdots\otimes|\psi_{j_n}\rangle. \quad (3)$$

This sum may contain d^n non-zero coefficients, but often many of them can be ignored since their total contribution is small. Roughly speaking, we drop those n-tuples $j_1,\ldots,j_n$ for which the numbers n_j of occurrences of index j do not satisfy

$$\sum_j \frac{n_j}{n}\log\frac{n_j}{n} \approx \sum_j |c_j|^2\log|c_j|^2.$$

After formalizing this condition appropriately,[2] one can show that the contribution of such "untypical terms" is negligible for $n\to\infty$. The numbers $N(n)$ of remaining terms satisfy

$$\lim_{n\to\infty}\frac{\log N(n)}{n} = S(\rho_B).$$

The entanglement thus can be transferred from B to C using $N(n)$-dimensional quantum systems in such a way that the resulting state coincides with the desired one in the asymptotics $n\to\infty$. One can furthermore show that $\lim_{n\to\infty}(\log N(n))/n < S(\rho_B)$ would not work. Hence $S(\rho_B)$ quantifies the asymptotic number of qubits per copy required to transfer the entanglement to C (provided that the entropy is measured in terms of bits, i.e. is defined using the logarithm to the basis 2.)

The fact that the restriction of pure entangled states to subsystems have non-vanishing entropy has important implications for *quantum* thermodynamics as opposed to *classical* thermodynamics. If a quantum system couples to an environment the joint dynamics can generate entanglement between the two systems. Hence the entropy of the system can increase. The decisive difference to classical physics is that this can happen even though the state of the environment is *perfectly known*. For this reason, models of the transition of a physical system to its thermal equilibrium do not necessarily require the assumption of *incomplete knowledge* about the state of the environment [7, 8]. Assuming that system and environment is in a pure joint state, strong entanglement is (for an environment being a high-dimensional quantum system) the typical situation rather than being the exception. To be more specific we consider B and A as models for the system and its environment, respectively. If $d \ll \ell$ the overwhelming majority of pure states (see [8] for

---

[2] Compare ► quantum entropy and the definition of "typical sequences" in classical coding theory [6] as well as the definition of "typical subspaces" in quantum coding theory [2].

details) have the property that the restriction to system B is close to the maximally mixed state

$$\rho_B := \frac{1}{d}\mathbf{1}.$$

Imposing some physically natural assumptions on the Hamiltonians of A and B and their interactions, Ref. [8] derives furthermore a statement that makes the thermodynamical relevance of entanglement even more obvious: almost every pure joint state lying in the subspace corresponding to some small interval of energy values has the property that its restriction to B is close to the thermodynamical Gibbs state.

E

See also creation and detection of entanglement; entanglement purification and distillation.

Literature

1. M. Nielsen and I. Chuang: *Quantum Information and Computation.* (Cambridge University Press, Cambridge, 2000)
2. B. Schumacher: Phys. Rev. A **51**(4), 2738–2747 (1995)
3. E. Schmidt: Math. Ann. **63** (1907)
4. A. Ekert, P. Knight: Am. J. Phys. **63**, 415 (1995)
5. A. Peres: *Quantum Theory: Concepts and Methods.* (Kluwer, Dordrecht, 1995)
6. T. Cover, J. Thomas: *Elements of Information Theory.* (Wiley, New York, 1991)
7. J. Gemmer, M. Michel, G. Mahler: *Quantum Thermodynamics: Emergence of Thermodynamic Behavior Within Composite Quantum Systems.* (Springer, Berlin, 2005)
8. S. Popescu, A. Short, A. Winter: The foundations of statistical mechanics from entanglement: Individual states vs. averages. *Nature Physics* **2**(11), 754–758 (2006). http://arxiv.org/abs/quant-ph/0511225

EPR-Problem (Einstein-Podolsky-Rosen Problem)

Peter Mittelstaedt

In 1935, Einstein, Podolsky, and Rosen published a paper [1] in which they tried to show that the quantum-mechanical description of physical reality is not complete. For the demonstration of this result, the authors made use of two assumptions, the principle of reality (R) and the principle of ▸ locality (L). These assumptions read:

(R): If, without in any way disturbing a system, we can predict with certainty (i.e. with probability equal to unity) the value of a physical quantity, then there exists an element of physical reality corresponding to this physical quantity.

(L): If two systems can not interact with each other, then a measurement of one system can not change the state of the other system.

Based on the principles (R) and (L), the authors of the EPR-article tried to show that quantum mechanics is incomplete. For the demonstration of this result, they made use of a thought experiment, which was later simplified by Bohm and Aharonov [2]. The argument reads as follows:

Consider two ▶ spin $1/2$ particles S_1 and S_2 prepared in a $^1S_0$ state $\Psi(S_1 + S_2)$ (with total spin 0) but separated such that they no longer interact. If a measurement of spin $\sigma_1(\boldsymbol{n})$ of S_1 in direction $\boldsymbol{n}$ results in the value $s_1 = +1/2$, then a subsequent measurement of spin $\sigma_2(\boldsymbol{n})$ of S_2 in the same direction leads with certainty to the value $s_2 = -1/2$.

For demonstrating the incompleteness of quantum mechanics on the basis of this thought experiment, we refer to the principles (R) and (L), which from a logical point of view are both implications. Since systems S_1 and S_2 are assumed to have a sufficiently large distance, they can no longer interact. Then, the premise of (L) is satisfied and thus the conclusion holds that a measurement of $\sigma_1(\boldsymbol{n})$ at S_1 cannot change S_2. Furthermore, since the outcome s_1 of a $\sigma_1(\boldsymbol{n})$-measurement determines the value $s_2 = -s_1$ of the observable $\sigma_2(\boldsymbol{n})$, the premise of (R) is satisfied. Hence, the conclusion of (R) holds too, that is the value s_2 of $\sigma_2(\boldsymbol{n})$ is an objective property of the system S_2 (after preparing the compound system in the state Ψ). Because this argument may be applied to the spin observables for any direction $\boldsymbol{n}$, we conclude that the value s_2 of $\sigma_2(\boldsymbol{n})$ for any direction $\boldsymbol{n}$ objectively pertains to the system S_2 after preparing the state Ψ. Hence, on the one hand the value s_2 of $\sigma_2(\boldsymbol{n})$ in S_2 is objectively determined, even if the observer subjectively does not know it. However, on the other hand, quantum mechanics does not allow to determining this value but only its probability. Therefore, quantum mechanics is not complete.

Neither the authors of the EPR paper nor their opponents recognised, that the incompleteness argument is not correct. Formally, this can be seen in the following way: Consider the last step of the argument that led to the conclusion of (R) which states, that for every $\boldsymbol{n}$ the system S_2 has an objective value $\{+1/2, -1/2\}$ of $\sigma_2(\boldsymbol{n})$ with probability $1/2$. Hence, the subsystem S_2 is in a ▶ mixed state $W_2(S_2) = 1/2\ P[\phi_n{}^{(2)}] + 1/2\ P[\phi_{-n}{}^{(2)}]$ admitting an ▶ ignorance interpretation, i.e. S_2 is in a "proper mixture" [3]. This means that the compound system $S_1 + S_2$ with the preparation Ψ is in a mixed state

$$W_\Psi(S_1 + S_2) = 1/2 P[\phi_n^{(1)} \otimes \phi_{-n}^{(2)}] + 1/2 P[\phi_{-n}^{(1)} \otimes \phi_n^{(2)}] \tag{1}$$

Therefore, for the calculation of the expectation values of the compound system, the states Ψ and W_Ψ are equivalent. This claim can easily be checked. For the special observable

$$\mathrm{B}\left(\boldsymbol{n}', \boldsymbol{n}''\right) := \sigma_1\left(\boldsymbol{n}'\right) \otimes \sigma_2\left(\boldsymbol{n}''\right) \tag{2}$$

the expectation values with regard to Ψ and W_Ψ must be identical.

After a short calculation from this derives

$$\boldsymbol{n}' \cdot \boldsymbol{n}'' - (\boldsymbol{n} \cdot \boldsymbol{n}')(\boldsymbol{n} \cdot \boldsymbol{n}'') = 0 \text{ (VO)} \tag{3}$$

as the condition of value objectification (VO) of σ_1 and σ_2. Except for a few special triples $(\boldsymbol{n},\ \boldsymbol{n}',\ \boldsymbol{n}'')$ this equation is violated in quantum mechanics. Hence, the EPR-argument does not result in the incompleteness of quantum mechanics, but in a contradiction. In addition, an elementary calculation shows that from (VO) we can derive Bell's inequalities ▶ Bell's theorem [4]

$$|\boldsymbol{n}' \cdot \boldsymbol{n} - \boldsymbol{n}''| \leqslant \boldsymbol{n} \cdot \left(\boldsymbol{n} - \boldsymbol{n}''\right), \quad |\boldsymbol{n}' \cdot \boldsymbol{n} + \boldsymbol{n}''| \leqslant \boldsymbol{n} \cdot \left(\boldsymbol{n} + \boldsymbol{n}''\right) \tag{4}$$

E

which are known to contradict quantum mechanics for appropriate triples of values.

Of course, the contradiction must be eliminated. Since the reality principle is fulfilled in quantum mechanics [5], the principle of locality must be abandoned. The resulting ▶ nonlocality of quantum mechanics has been confirmed in quantum mechanics since 1980 by a great number of experiments. We mention here the experiments from Aspect [6] to Weihs [7]. See also ▶ Bohm's approach to EPR; Causal Inference and EPR.

Literature

1. A. Einstein, B. Podolsky, N. Rosen: Can Quantum-Mechanical Description of Physical Reality Be Considered Complete? Phys. Rev. **47**, 777–780 (1935)
2. D. Bohm, Y. Aharonov: Discussion of Experimental Proof for the Paradox of Einstein, Rosen and Podolsky. Phys. Rev. **108**, 1070–1076 (1957)
3. B. d'Espagnat: *Veiled Reality* (Westview Press, Cambridge 2003, 105)
4. J. S. Bell: On the Einstein-Podolsky-Rosen Paradox. Physics **1**, 195–200 (1964)
5. P. Mittelstaedt: Einsteins Objections Against Quantum Mechanics. Phys. Philos. Open Access Online Journal, http://physphil.uni-dortmund.de, 1–15 (2006)
6. A. Aspect et al.: Experimetal Tests of Bell's Inequalities Using Time-Varying Analysers. Phys. Rev. Lett. **49**, 1804–1807 (1982)
7. G. Weihs: Violation of Bell's Inequality Under Strict Einstein Locality Conditions. Phys. Rev. Lett. **81**, 5039–5043 (1998)

Errors and Paradoxes in Quantum Mechanics

Daniel Rohrlich

According to one definition, a paradox is a statement that seems self-contradictory or absurd but may be true; according to another, a paradox is a true self-contradiction and therefore false. Let us define paradox to be an apparent contradiction that follows from apparently acceptable assumptions via apparently valid deductions. Since logic admits no contradictions, either the apparent contradiction is not a contradiction, or the apparently acceptable assumptions are not acceptable, or the apparently valid deductions are not valid. A paradox can be useful in developing a physical theory; it can show that something is wrong even when everything appears to be right.

Paradoxes in physics often arise as thought experiments. For example, to refute Aristotle's statement that a heavy body falls faster than a light one, Galileo [1] invented a paradox: Suppose, with Aristotle, that a large stone falls faster than a small stone. If the stones are tied together, the smaller stone will then retard the large one. But the two stones tied together are heavier than either of them. "Thus you see how, from your assumption that the heavier body moves more rapidly than the lighter one, I infer that the heavier body moves more slowly." Such free invention of paradoxes as thought experiments marks especially the development of twentieth century physics, i.e. of the relativity and quantum theories.

Both relativity theory and quantum theory are well supplied with paradoxes. In relativity theory, however, well known paradoxes such as the twin paradox have accepted resolutions. These paradoxes arise from intuitions, typically about simultaneity, that relativity theory rendered obsolete. By contrast, not all well known paradoxes of quantum theory have accepted resolutions, even today. Below we briefly review seven quantum paradoxes.

In keeping with our definition above, we do not distinguish between "apparent" and "true" paradoxes. But we distinguish between apparent and true contradictions. A true contradiction is a fatal flaw showing that a physical theory is wrong. By contrast, apparent contradictions may arise from errors; they may also arise from a conceptual gap in a theory, i.e. some ambiguity or incompleteness that is not fatal but can be removed by further development of the theory. Thus we can classify [2] physics paradoxes into three classes: Contradictions, Errors and Gaps. The first three paradoxes below are examples of a Contradiction, an Error and a Gap, respectively.

1. By 1911, Rutherford and his co-workers had presented striking experimental evidence (back-scattering of alpha particles, ► large-angle scattering; scattering experiments) that neutral atoms of gold have cores of concentrated positive charge. According to classical electrodynamics, an atom made of ► electrons surrounding a positive nucleus would immediately collapse; but the gold foil in Rutherford's experiment evidently did not collapse. This contradiction between experimental evidence and classical theory was not merely apparent: it showed that atoms do not obey classical electrodynamics. Faced with this evidence, Bohr broke with classical theory and explained the stability of matter by associating ► quantum numbers $n = 1, 2, 3, \dots$ with the allowed orbits of electrons in atoms. Although ► Bohr's model described well only the hydrogen atom, quantum numbers characterize all atoms.

2. Einstein invented thought experiments to challenge Bohr's [3] principle of ► complementarity. One thought experiment involved two-slit interference (► double-slit experiment). (See Fig. 1.) Let a wave of (say) electrons of wavelength λ, collimated by a screen with a single slit, impinge on a screen with two slits with separation d. An electron interference pattern – dark lines with separation $D = \lambda L/d$ – emerges on a third screen a distance L beyond the second. In Fig. 1, however, the experiment is modified to measure also the transverse recoil of the second screen (the screen with the two slits). Why the modification? According

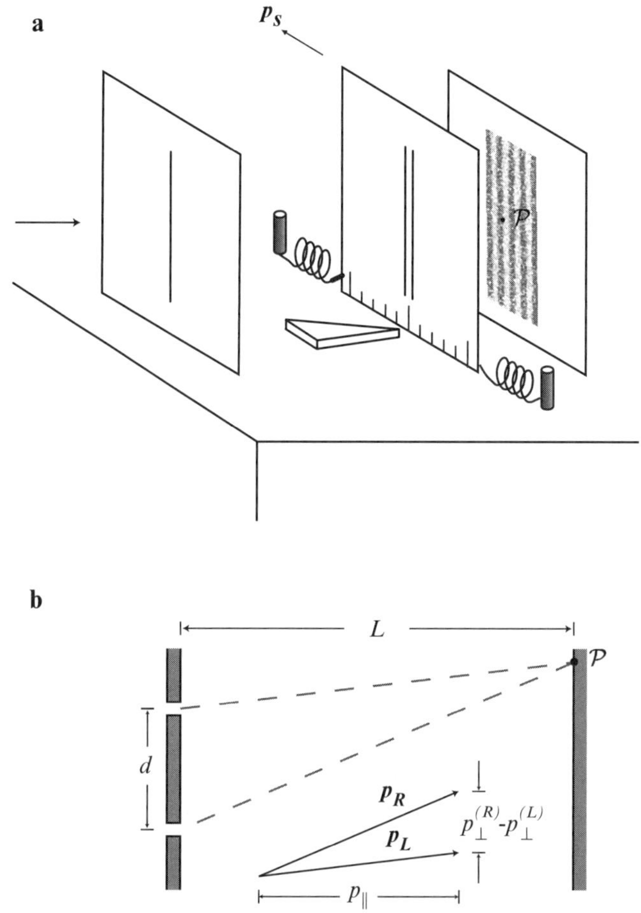

Fig. 1 (**a**) A two-slit interference experiment adapted for measuring the transverse momentum of the middle screen. (**b**) The second and third screens seen from above, with interfering electron paths and corresponding momenta

to Bohr, a setup can demonstrate *either* wave behavior (e.g. interference) of electrons *or* particle behavior (e.g. passage through a single slit), but not *simultaneous* wave and particle behavior; these two behaviors are complementary (▶ "wave-particle duality") and no setup can simultaneously reveal complementary behaviors. Einstein's modified experiment apparently shows electron interference while also revealing through which slit each electron passes (e.g. an electron passing through the right slit makes the screen recoil more strongly to the right) and thus contradicts the principle of complementarity.

To analyze the modified experiment, let $\boldsymbol{p}^{(L)}$ and $\boldsymbol{p}^{(R)}$ denote the momentum of an electron if it arrives at $\mathcal{P}$ via the left and right slits, respectively, and let $p_\perp^{(L)}$ and $p_\perp^{(R)}$ denote the respective transverse components. From a measurement of the change in transverse momentum p_s of the screen with accuracy $\Delta p_s \leqslant p_\perp^{(R)} - p_\perp^{(L)}$, we can infer through which slit an electron passed. But now apply ▶ Heisenberg's uncertainty principle to the second screen:

$$\Delta x_s \geqslant h/\Delta p_s \geqslant h/[p_\perp^{(R)} - p_\perp^{(L)}],$$

where x_s is the transverse position of the second screen. Similarity of triangles in Fig. 1(b) implies that $|\boldsymbol{p}^{(R)} - \boldsymbol{p}^{(L)}|$ (which equals $|p_\perp^{(R)} - p_\perp^{(L)}|$), divided by the electron's longitudinal momentum $p_\|$, equals d/L. The longitudinal momentum $p_\|$ is h/λ (assuming $p_\|$ large compared to the transverse momentum). Thus

$$\Delta p_s < \frac{d}{L}(h/\lambda).$$

We obtain $\Delta p_s < h/D$ and thus $\Delta x_s > D$. The uncertainty in the transverse *position* x_s of the screen, arising from an accurate enough measurement of its transverse *momentum* p_s, is the distance D between successive dark bands in the interference pattern, and so the interference pattern is completely washed out. Precisely when Einstein's thought experiment succeeds in showing through which slit each electron passes, it fails to show electron interference; that is, it *obeys* the principle of complementarity after all.

3. In 1931, Landau and Peierls [4] considered the following model measurement of the electric field $\boldsymbol{E}$ in a region. Send a charged test particle through the region; the electric field deflects the particle, and the change in the momentum $\boldsymbol{p}$ of the test particle is a measure of $\boldsymbol{E}$. But an accelerated, charged particle radiates, losing an unknown fraction of its momentum to the electromagnetic field. Reducing the charge on the test particle reduces radiation losses but then $\boldsymbol{p}$ changes more slowly and the measurement lasts longer (or is less accurate). On the basis of their model, Landau and Peierls concluded that an instantaneous, accurate measurement of $\boldsymbol{E}$ is impossible. They obtained a lower bound $\Delta|\boldsymbol{E}| \geqslant \sqrt{\hbar c}/(cT)^2$ as the minimum uncertainty in a measurement of $|\boldsymbol{E}|$ lasting a time T. Their conclusion is paradoxical because it leaves the instantaneous electric field $\boldsymbol{E}$ with no theoretical or experimental definition. However, the Landau–Peierls model measurement is too restrictive. Bohr and Rosenfeld [5] found it necessary to modify the model in many ways; one modification was to replace the (point) test particles of Landau and Peierls with extended test bodies. In their modified model, they showed how to measure electric (and magnetic) fields instantaneously. Note that the electric field is not a *canonical* variable, i.e. it is not one of the generalized coordinates and momenta appearing in the associated Hamiltonian. (It depends on the time derivative of $\mathbf{A}$, the electromagnetic vector potential, which *is* a canonical variable.) The resolution of this sort of paradox is that quantum measurements of canonical and noncanonical variables differ systematically [6].

4. *Zeno's paradoxes* are named for the Greek philosopher who tried to understand motion over shorter and shorter time intervals and found himself proving that motion is impossible. The quantum Zeno paradox [7] (▶ *quantum* Zeno effect) seems

to prove that quantum evolution is impossible. Consider the evolution of a simple quantum system: a ▶ spin-1/2 atom precesses in a constant magnetic field. If we neglect all but the spin degree of freedom, represented by the ▶ Pauli spin matrices σ_x, σ_y and σ_z, the Hamiltonian is

$$H = \mu B \sigma_z$$

where the direction of the magnetic field defines the z-axis and μ is the Bohr magneton. Suppose that at time $t = 0$ the state is

E

$$|\psi(0)\rangle = \frac{1}{\sqrt{2}}[|\uparrow\rangle + |\downarrow\rangle]$$

(where $\sigma_z|\uparrow\rangle = |\uparrow\rangle$ and $\sigma_z|\downarrow\rangle = -|\downarrow\rangle$). Solving ▶ Schrödinger's equation

$$\mathrm{i}\hbar\frac{\mathrm{d}}{\mathrm{d}t}|\psi(t)\rangle = H|\psi\rangle,$$

we obtain the time evolution:

$$\begin{aligned}|\psi(t)\rangle &= \mathrm{e}^{-\mathrm{i}Ht/\hbar}|\psi(0)\rangle \\ &= \frac{1}{\sqrt{2}}\left[\mathrm{e}^{-\mathrm{i}\mu Bt/\hbar}|\uparrow\rangle + \mathrm{e}^{\mathrm{i}\mu Bt/\hbar}|\downarrow\rangle\right].\end{aligned}$$

At $t = 0$, a measurement of σ_x is sure to yield 1; at time $t = T \equiv h/4\mu B$, the σ_x measurement is sure to yield -1; at intermediate times, a measurement may yield either result.

At no time does a measurement of σ_x yield a value other than 1 and -1; the spin component σ_x jumps discontinuously from 1 to -1 (▶ quantum jumps) and defines a moment in time by jumping. *When* does the spin jump? We cannot predict when it will jump, but we can make many measurements of σ_x between $t = 0$ and $t = T$. The jump in σ_x must occur between two successive measurements. When it does, we will know when the jump occurred, to an accuracy Δt equal to the time between the measurements. But now we apparently violate the uncertainty relation for energy and time:

$$\Delta E \Delta t \geqslant \hbar/2.$$

Here E is the energy of the measured system and t is time as defined *by* the system. (Although t is not an ▶ operator, we can define t via an operator that changes smoothly in time, and then derive $\Delta E \Delta t \geqslant \hbar/2$ indirectly [8].) The problem is that the uncertainty ΔE in the energy cannot be greater than the difference $2\mu B$ between the two eigenvalues of H; but the measurements can be arbitrarily dense, i.e. Δt can be arbitrarily small.

Since quantum mechanics will not allow a violation of the uncertainty principle, we may guess that the atomic spin will simply refuse to jump! A short calculation verifies this guess. Consider N measurements of σ_x, at equal time intervals, over a period of time T. The interval between measurements is T/N. What is the probability of finding the spin unchanged after the first measurement? The state at time $t = T/N$ is

$$\frac{1}{\sqrt{2}}\left[\mathrm{e}^{-\mathrm{i}\mu BT/N\hbar}|\uparrow\rangle + \mathrm{e}^{\mathrm{i}\mu BT/N\hbar}|\downarrow\rangle\right],$$

so the probability of finding the spin unchanged is $\cos^2(\mu BT/N\hbar)$. Hence the probability of finding the spin unchanged at time T, after N measurements, is $\cos^{2N}(\mu BT/N\hbar)$. As N approaches infinity, $\cos^{2N}(\mu BT/N\hbar)$ approaches 1: the spin never jumps. Here quantum evolution is impossible. But consider a dual experiment: instead of N measurements of σ_x on an atom in a magnetic field, consider N measurements of $\sigma_x \cos(2\mu Bt/\hbar) + \sigma_y \sin(2\mu Bt/\hbar)$, at equal time intervals, on an atom in no magnetic field ($H = 0$). In the limit $N \to \infty$, the atom precesses: each measurement of $\sigma_x \cos(2\mu Bt/\hbar)+\sigma_y \sin(2\mu Bt/\hbar)$ yields 1. Experiments from 1990 on have progressively demonstrated such ▶ quantum Zeno effect.

5. A thought experiment due to Einstein, Podolsky and Rosen [9] (▶ EPR problem) shows how to measure precisely the position $\boldsymbol{x}_\mathrm{A}(T)$ *or* the momentum $\boldsymbol{p}_\mathrm{A}(T)$ of a particle A at a given time T, *indirectly* via a measurement on a particle B that once interacted with A. The measurement on B is spacelike separated from $\boldsymbol{x}_\mathrm{A}(T)$, and so it cannot have any measurable effect on $\boldsymbol{x}_\mathrm{A}(T)$ or $\boldsymbol{p}_\mathrm{A}(T)$ (no superluminal signalling). It is indeed reasonable to assume (▶ Einstein locality; superluminal communication) that the measurement on B has no effect whatsoever on $\boldsymbol{x}_\mathrm{A}(T)$ or $\boldsymbol{p}_\mathrm{A}(T)$; thus $\boldsymbol{x}_\mathrm{A}(T)$ and $\boldsymbol{p}_\mathrm{A}(T)$ are simultaneously defined (in the sense that either is measurable without any effect on the other) and a particle has a precise position and momentum simultaneously. Since quantum mechanics does not define the precise position and momentum of a particle simultaneously, quantum mechanics does not completely describe particles. EPR envisioned a theory that would be *consistent* with quantum mechanics but more complete, just as statistical mechanics is consistent with thermodynamics but more complete.

Almost 30 years after the EPR paper, Bell [10] proved a startling, and – to Bell himself – disappointing theorem: Any more complete theory of the sort envisioned by EPR would contradict quantum mechanics! Namely, the correlations of any such theory must obey ▶ Bell's inequality; but according to quantum mechanics, some correlations of entangled states (▶ entanglement) of particles A and B violate Bell's inequality. If quantum mechanics is correct, then there can be no theory of the sort envisioned by EPR. Experiments have, with increasing precision and rigor, demonstrated violations of Bell's inequality and ruled out any theory of the sort envisioned by EPR.

6. In 1927, at the fifth Solvay congress, Einstein presented "a very simple objection" to the ▶ probability interpretation of quantum mechanics. According to quantum mechanics, the state of an electron approaching a photographic plate is an extended object; the probability density for the electron to hit varies smoothly over the plate. Once the electron hits somewhere on the plate, however, the probability for the electron to hit anywhere else drops to zero, and the state of the electron collapses instantaneously. (▶ Wave function collapse). But instantaneous collapse of an extended object is not compatible with relativity. A related paradox is the following.

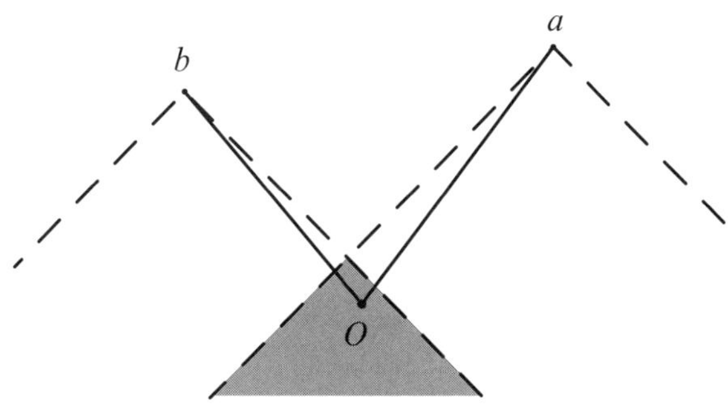

Fig. 2 Two atoms, produced in an entangled state at O, fly off in opposite directions (*solid lines*) in this spacetime figure. Alice measures a spin component of one atom at a; Bob measures a spin component of the other atom at b. Collapse cannot occur anywhere outside the past light cones of a and b (*dotted lines*), hence it cannot occur anywhere outside the intersection of their past light cones (*shaded region*)

Figure 2 shows two atoms, prepared in an entangled state at O, flying off in different directions. (For simplicity, assume that they separate at nonrelativistic speeds.) One atom enters the laboratory of Alice, who measures a component of its spin at a; the other enters the laboratory of Bob, who measures a component of its spin at b. After Alice's measurement, the atoms are not in an entangled state anymore, hence collapse cannot occur anywhere outside the past light cone of a. Likewise, collapse cannot occur anywhere outside the past light cone of b. Hence collapse cannot occur anywhere outside the *intersection* of the past light cones of a and b. But then, in the inertial reference frame of Fig. 2, the state of the atoms just before either measurement is a product (collapsed) state, not an entangled state. Now this conclusion contradicts the fact that, by repeating this experiment on many pairs of atoms, Alice and Bob can obtain violations of Bell's inequality, i.e. can demonstrate that the atomic spins were in an entangled state until Bob's measurement. This paradox shows that there can be no Lorentz-invariant account of the collapse. In general, observers in different inertial reference frames will disagree about collapse. They will not disagree about the results of local measurements, because local measurements are spacetime events, hence Lorentz invariant; but they will have different accounts of the collapse of nonlocal states. Collapse is Lorentz *co*variant [11].

7. ► Schrödinger's Cat is a paradox of quantum evolution and measurement. For simplicity, let us consider just the σ_z degree of freedom of spin-1/2 atoms and define a superposition of the two normalized eigenstates $|\uparrow\rangle$ and $|\downarrow\rangle$ of σ_z:

$$|\Psi_{\alpha\beta}\rangle = \alpha|\uparrow\rangle + \beta|\downarrow\rangle;$$

we assume $|\alpha|^2 + |\beta|^2 = 1$. The ► Born probability rule states that a measurement of σ_z on many atoms prepared in the state $|\Psi_{\alpha\beta}\rangle$ will yield a fraction approaching $|\alpha|^2$ of atoms in the state $|\uparrow\rangle$ and a fraction approaching $|\beta|^2$ of atoms in the state $|\downarrow\rangle$. If quantum mechanics is a complete theory, it should be possible to describe these measurements themselves using Schrödinger's equation. We can describe a measurement on an atom abstractly by letting $|\Phi_0\rangle$ represent the initial state of a measuring device, and letting $|\Phi_\uparrow\rangle$ or $|\Phi_\downarrow\rangle$ represent the final state of the measuring device if the state of the atom was $|\uparrow\rangle$ or $|\downarrow\rangle$, respectively. If the Hamiltonian for the measuring device and atom together is H, during a time interval $0 \leqslant t \leqslant T$ that includes the measurement, then the Schrödinger equation implies

$$\mathrm{e}^{-\mathrm{i}\int_0^T H\mathrm{d}t/\hbar}|\uparrow\rangle \otimes |\Phi_0\rangle = |\uparrow\rangle \otimes |\Phi_\uparrow\rangle,$$
$$\mathrm{e}^{-\mathrm{i}\int_0^T H\mathrm{d}t/\hbar}|\downarrow\rangle \otimes |\Phi_0\rangle = |\downarrow\rangle \otimes |\Phi_\downarrow\rangle.$$

(The spin states do not change as they are eigenstates of the measured observable σ_z.) If the initial spin state is neither $|\uparrow\rangle$ nor $|\downarrow\rangle$ but the superposition $|\Psi_{\alpha\beta}\rangle$, the evolution of the superposition is the superposition of the evolutions:

$$\mathrm{e}^{-\mathrm{i}\int_0^T H\mathrm{d}t/\hbar}|\Psi_{\alpha\beta}\rangle \otimes |\Phi_0\rangle = \alpha|\uparrow\rangle \otimes |\Phi_\uparrow\rangle + \beta|\downarrow\rangle \otimes |\Phi_\downarrow\rangle.$$

The right side of this equation, however, does not describe a completed measurement at all: the measuring device remains entangled with the atom in a superposition of incompatible measurement results. It does not help to couple additional measuring devices to this device or to the atom; since the Schrödinger equation dictates linear, unitary evolution, additional devices will simply participate in the superposition rather than collapse it. Even a cat coupled to the measurement will participate in the superposition. Suppose the measuring device is triggered to release poison gas into a chamber containing a cat, *only* if the spin state of the measured atom is $|\uparrow\rangle$. The state of the atom, measuring device and cat at time $t = T$ will be a superposition of $|\uparrow\rangle \otimes |\Phi_\uparrow\rangle \otimes |\text{dead}\rangle$ and $|\downarrow\rangle \otimes |\Phi_\downarrow\rangle \otimes |\text{live}\rangle$ with coefficients α and β, respectively. So we do not know how to describe even one measurement using Schrödinger's equation.

Paradoxes 1–4 and 6 and their resolutions are not controversial. Paradoxes 5 and 7, however, do excite controversy. For many physicists, the EPR paradox and Bell's theorem remain unresolved because, for them, renouncing the "reasonable" assumption of EPR is just not a resolution. As one distinguished physicist put it [12], "Anybody who's not bothered by Bell's theorem has to have rocks in his head." (No such statement would apply to any well known paradox in relativity theory.)

The Schrödinger Cat paradox has been resolved several times over – with spontaneous "collapse" of quantum states [13], nonlocal ► "hidden variables" [14], ► "many (parallel) worlds" [15] and future boundary conditions [16] (conditions on the future state in a ► "two-state" vector formalism [17]) – but since experiments are consistent with all these resolutions, there is no one accepted resolution, at least within nonrelativistic quantum mechanics. The predictions of quantum mechanics with and without collapse differ, but the differences are (so far) not accessible to experiment. (There is even a proof [18] that if quantum mechanics is correct and an experiment could verify that a cat is in the superposition $\alpha|\text{dead}\rangle + \beta|\text{live}\rangle$, i.e. if it could verify that collapse has not occurred, the same experiment could transform the state |dead⟩ into the state |live⟩, i.e. it could revive a dead cat.) However, it is doubtful whether all these resolutions can be made relativistic.

Literature

E

1. G. Galilei, Dialogues Concerning Two New Sciences, trans. H. Crew and A. de Salvio (Dover, New York, 1954), p. 63.
2. Y. Aharonov and D. Rohrlich, Quantum Paradoxes: Quantum Theory for the Perplexed (Wiley-VCH, Weinheim, 2005), Chap. 1.
3. N. Bohr, Discussion with Einstein on epistemological problems in atomic physics, in Albert Einstein: Philosopher–Scientist, ed. Paul A. Schilpp (Tudor Publishing Company, New York, 1951), pp. 201–41.
4. L. Landau and R. Peierls, Erweiterung des Unbestimmtheitsprinzips für die relativistische Quantentheorie, Z. Phys. **69**, 56–69 (1931); trans. Extension of the uncertainty principle to the relativistic quantum theory, in Collected Papers of Landau, ed. D. ter Haar (Gordon and Breach, New York, 1965), pp. 40–51; also in J. A. Wheeler and W. H. Zurek, Quantum Theory and Measurement (Princeton University Press, Princeton, 1983), pp. 465–476.
5. N. Bohr and L. Rosenfeld, Zur Frage der Messbarkeit der elektromagnetischen Feldgrössen, Det Kgl. Danske Vid. Selsk. Mat. fys. Medd. **XII**(8), p. 65 (1933); trans. by A. Petersen, On the question of the measurability of electromagnetic field quantities, in Selected Papers of Léon Rosenfeld, eds. R. S. Cohen and J. Stachel (Reidel, Dordrecht, 1979), pp. 357–400; also in J. A. Wheeler and W. H. Zurek, op. cit. pp. 479–522.
6. Y. Aharonov and J. L. Safko, Measurement of noncanonical variables, Ann. Phys. **91**, 279–294 (1975); see also Y. Aharonov and D. Rohrlich op. cit., Chap. 8.
7. B. Misra and E. C. G. Sudarshan, The Zeno's paradox in quantum theory, J. Math. Phys. **18**, 756–763 (1977).
8. Aharonov and Rohrlich, op. cit., Sect. 8.5.
9. A. Einstein, B. Podolsky and N. Rosen, Can quantum-mechanical description of physical reality be considered complete? Phys. Rev. **47**, 777–780 (1935).
10. J. S. Bell, On the Einstein–Podolsky–Rosen paradox, Physics **1**, 195–200 (1964).
11. Y. Aharonov and D. Z. Albert, Is the usual notion of time evolution adequate for quantum-mechanical systems? II. Relativistic considerations, Phys. Rev. **D29**, 228–234 (1984); Aharonov and Rohrlich op. cit., Chap. 14.
12. Quoted in N. D. Mermin, Is the moon there when nobody looks? Reality and the quantum theory, Physics Today **38**(4), 38–47 (1985).
13. G. C. Ghirardi, A. Rimini and T. Weber, Unified dynamics for microscopic and macroscopic systems, Phys. Rev. **D34**, 470–491 (1986); Disentanglement of quantum wave functions: answer to 'Comment on Unified dynamics for microscopic and macroscopic systems', Phys. Rev. **D36**, 3287–3289 (1987); P. Pearle, Combining stochastic dynamical state-vector reduction with spontaneous localization, Phys. Rev. **A39**, 2277–2289 (1989); G. C. Ghirardi, P. Pearle and A. Rimini, Markov processes in Hilbert space and continuous spontaneous localization of systems of identical particles Phys. Rev. **A42**, 78–89 (1990); P. Pearle, Relativistic collapse model with tachyonic features, Phys. Rev. **A59**, 80–101 (1999).
14. D. Bohm, A suggested interpretation of the quantum theory in terms of 'hidden' variables, parts I and II, Phys. Rev. **85** 166–179 and 180–193 (1952), reprinted in Wheeler and Zurek, op. cit., pp. 369–82 and 383–96.
15. H. Everett, III, 'Relative state' formulation of quantum mechanics, Rev. Mod. Phys. **29**, 454–462 (1957), reprinted in Wheeler and Zurek, op. cit., pp. 315–23; J. A. Wheeler, Assessment of Everett's 'relative state' formulation of quantum theory, Rev. Mod. Phys. **29**, 463–465 (1957).

16. Y. Aharonov and E. Gruss, eprint quant-ph/0507269 (2005); Aharonov and Rohrlich, op. cit., Sect. 18.3.
17. Y. Aharonov and L. Vaidman, On the two-state vector reformulation of quantum mechanics, Phys. Scrip. **T76**, 85–92 (1998).
18. Aharonov and Rohrlich op. cit., p. 131.

Exclusion Principle (or Pauli Exclusion Principle)

Michela Massimi

The exclusion principle, introduced by Wolfgang Pauli in 1925 [1], is a fundamental scientific principle in quantum mechanics. It explains a wide range of phenomena, from the stability of matter at the level of stars and galaxies to the inner constitution of particles at the level of coloured quarks. The exclusion principle states that there cannot be in nature two ► electrons, or two protons, or two coloured quarks, or, more in general, any two fermions (i.e. spin-$1/2$ particles obeying the ► Fermi–Dirac statistics) in the same dynamic state. Formally, this means that any system consisting of two or more indistinguishable fermions is expressed by antisymmetric functions as opposed to symmetric functions. Symmetric functions for two indistinguishable particles are such that the state vector of the composite system does not change sign under permutation of space and spin coordinates of the two particles, i.e.

$$1/\sqrt{2}\left(|a_1^r\rangle \otimes |a_2^s\rangle + |a_1^s\rangle \otimes |a_2^r\rangle\right)$$

whereas in antisymmetric functions the state vector does change sign under permutation of the space and spin coordinates of the two particles

$$1/\sqrt{2}\left(|a_1^r\rangle \otimes |a_2^s\rangle - |a_1^s\rangle \otimes |a_2^r\rangle\right)$$

The exclusion principle then prescribes the mathematical nature of quantum states allowed for fermions: it excludes all classes of mathematically possible states different from the antisymmetric ones. To say that the state vector of the composite system is antisymmetric is mathematically equivalent to saying that the dynamic states of the two particles are different. Although the exclusion principle is normally associated with the above formulation in terms of antisymmetrization of the state vector of a composite system, this was not Pauli's original formulation of the principle. In fact, the actual origins of the exclusion principle can be traced back to the Bohr–Sommerfeld old ► quantum theory before 1925.

The principle was indeed introduced by Pauli at the end of 1924 as an "extremely natural" empirical rule in the attempt to provide an explanation for some spectroscopic anomalies that had vexed physicists such as Alfred Landé (1888–1976), Werner Heisenberg (1901–76), Niels Bohr (1885–1962) and Wolfgang Pauli (1900–1958) in the early 1920s. According to the Bohr–Sommerfeld old quantum theory, each bound electron in an atom should be characterised in terms of a set of ► quantum numbers describing the energy state, n, the angular momentum, l, and the orientation with respect to a magnetic field, m_l, respectively ► Spin; Stern–Gerlach experiment; Vector model. The Bohr–Sommerfeld theory (► Bohr's atomic model) was used to explain the closure of electronic shells in atoms according to the periodic table, as well as to account for atomic spectra. But by 1921 it became clear that there were some serious problems with Bohr's schema for the closure of electronic shells; nor were the quantum numbers sufficient to account for the complex spectral lines observed in some chemical elements, such as alkali metals and alkaline earths, among others. Even more puzzling were some spectroscopic anomalies observed when chemical elements were placed in a weak or strong external magnetic field: these spectroscopic anomalies were known as anomalous ► Zeeman effect and ► Paschen–Back effect, respectively. An understanding of both spectroscopic anomalies and closure of electronic shells required some drastic changes in the old quantum theory, and between 1921 and 1924 Alfred Landé, Werner Heisenberg, Niels Bohr and Wolfgang Pauli all tried to tackle these problems and put forward different theoretical proposals (see [4, 6]). A conclusive understanding came only in 1924, when in his three-year long struggle to understand the anomalous Zeeman effect, Pauli abandoned the previous theoretical models and came up with the bold idea of introducing a fourth degree of freedom for electrons in atoms, which he referred to as the electron's *Zweideutigkeit* (the "twofold", or, as is more frequently translated, "two-valued" intrinsic angular momentum of electron). A year later, Ralph Kronig (1904–1995) and, independently, George E. Uhlenbeck (1900–1988) and S. Goudsmit (1902–78) reinterpreted this fourth degree of freedom as the electron ► spin, s. In conjunction with the introduction of a fourth degree of freedom for the electron, Pauli introduced also a new empirical rule for the closure of electronic shells.

> "I can trace back the closure of groups (...) to a single prescription that seems to me extremely natural. I am thinking of a so strong magnetic field that all electrons can be characterised through the symbol n_{k_1,m_1,m_2}. Then it should be forbidden that more than one electron with the same (equivalent) n belongs to the same values of the three quantum numbers k_1, m_1, m_2. When an electron corresponds to a given n_{k_1,m_1,m_2}–state, this state is occupied."[1]

Thus, the exclusion principle was born as an empirical rule for the closure of electronic shells that Pauli called *Ausschließungsregel* or *meine Ausschlußregel*

[1] Pauli's letter to Alfred Landé, 24 November 1924. In [3], p. 180. Note here that n refers to the so-called principal quantum number defining the energy state of the electron; k is the azimuthal quantum number (in modern notation l) defining the orbital angular momentum, and m_1, m_2 are two magnetic quantum numbers representing the interaction energy with a strong magnetic field.

(exclusion rule), while Heisenberg teasingly referred to it Pauli's *Verbot der äquivalenten Bahnen* (Pauli's prohibition of equivalent orbits). Pauli admitted that 'we cannot give a closer foundation to this rule, yet it seems to present itself in a very natural way'. [1, p. 776] There was a long way to go for this empirical rule to be promoted to the rank of a scientific principle in the new quantum mechanics after 1925. The history of the exclusion principle is entwined with the development of quantum mechanics after 1925 as a new theoretical framework into which Pauli's rule was built from the ground up. When fifteen years later, in 1940, Pauli proved the ► spin–statistics theorem [2], it became clear that not only electrons but *any* half-integral spin particle obeyed the Fermi–Dirac statistics and hence the exclusion principle. The impact of this result for subsequent scientific developments is striking: for instance, when quarks were introduced in the 1960s, they were automatically taken as particles obeying the exclusion principle, given their half-integral spin and the consequent spin-statistics connection established by Pauli's theorem. This was the beginning of a research programme that led to quarks (see ► Color Charge Degree of Freedom in Particle Physics; Mixing and Oscillations of Particles; Particle Physics; Parton Model; QCD; QFT) and hence to ► quantum chromodynamics (QCD).

The history of the exclusion principle raises an important philosophical issue: why and how could Pauli's empirical rule – tentatively introduced in the context of the old quantum theory to solve some puzzling spectroscopic phenomena – become a building-block of quantum mechanics? Answering this question means addressing the challenging philosophical issue of what a scientific principle is, how it originates and how it can possibly be experimentally tested and verified. For a philosophical analysis of these questions in relation to the history of Pauli's exclusion principle, see [5].

Primary Literature

1. W. Pauli: Über den Zusammenhang des Abschlusses der Elektronengruppen im Atom mit der Komplexstruktur der Spektren. Zeitschrift für Physik **31**, 765–83 (1925)
2. W. Pauli: The Connection Between Spin and Statistics. Physical Review **58**, 716–22 (1940)
3. W. Pauli: *Wissenschaftlicher Briefwechsel mit Bohr, Einstein, Heisenberg u.a., Band* 1: *1919–1929*, edited by A. Hermann, K. von Meyenn, V. F. Weisskopf (Berlin: Springer 1979)

Secondary Literature

4. J. L. Heilbron: The Origins of the Exclusion Principle. Historical Studies in the Physical Sciences **13**, 261–310 (1982)
5. M. Massimi: *Pauli's exclusion principle* - The origin and validation of a scientific principle (Cambridge: Cambridge University Press 2005)
6. D. Serwer: Unmechanischer Zwang – Pauli, Heisenberg, and the Rejection of the Mechanical Atom 1923–1925. Historical Studies in the Physical Science **8**, 189–256 (1977)

Experimental Observation of Decoherence

Maximilian Schlosshauer

In the 1980s, theoretical estimates showed that on macroscopic scales decoherence occurs extremely rapidly, thus effectively precluding the observation of nonclassical ► superposition states [21–23]. This immediately led to the question of how we may experimentally observe the continuous action of ► decoherence and thus the smooth transition from quantum to classical. Several challenges have to be overcome in the design of such experiments. The system is to be prepared in a nonclassical superposition of mesoscopically or even macroscopically distinguishable states (► Schrödinger-cat state) with a sufficiently long decoherence time such that the gradual action of decoherence can be resolved. The existence of the superposition must be verified, and a scheme for monitoring decoherence must be devised that introduces a minimal amount of additional decoherence. Starting in the mid-1990s, several such experiments have been successfully performed, using physical systems such as:

- Cavity QED (atom–photon interactions) [1];
- Fullerenes (C_{60}, C_{70}) and other mesoscopic molecules [2];
- Superconducting systems (SQUIDs, Cooper-pair boxes) [3].

Other experimental domains are promising candidates for the observation of decoherence; however, the necessary superposition states have not yet been realized:

- Bose–Einstein condensates [24];
- Nano-electromechanical systems [4].

These five classes of experiments are described below (for a more detailed account, see, e.g., Chap. 6 of [21]). Such experiments are important for several reasons. They are impressive demonstrations of the possibility of generating nonclassical states of mesoscopic and macroscopic objects. They show that the boundary between quantum and classical is smooth and can be moved by varying the relevant experimental parameters. For example, by engineering different strengths and types of environmental interactions, wide ranges of decoherence rates can be obtained and the system can be driven into different preferred ("environment-superselected") bases [5]. The experiments also allow us to test and improve decoherence models. Finally, they may reveal deviations from unitary quantum mechanics and thus may be used to test quantum mechanics itself [3]. This would require sufficient shielding of the system from decoherence so that an observed (full or partial) ► wave function collapse could be unambiguously attributed to some novel nonunitary mechanism in nature, such as that proposed by the ► GRW theory. However, this shielding would be extremely difficult to implement in practice: The large number of atoms required for the collapse mechanism to be effective also leads to strong decoherence [6]. None of the superpositions realized in current experiments disprove existing collapse theories [7].

Cavity QED

In 1996 Brune et al. at *Ecole Normale Supérieure* in Paris generated a superposition of radiation fields with classically distinguishable phases involving several photons (► light quantum) [1, 8, 24]. This experiment was the first to realize a mesoscopic ► Schrödinger-cat state and to observe and manipulate its decoherence in a controlled way.

The experimental procedure is as follows. A rubidium atom is prepared in a superposition of distinct energy eigenstates $|g\rangle$ and $|e\rangle$ corresponding to two circular Rydberg states. The atom enters a cavity C containing a radiation field containing a few photons. The field effectively measures the state of the atom: If the atom is in the state $|g\rangle$, the field remains unchanged, whereas if the state is $|e\rangle$, the ► coherent state $|\alpha\rangle$ of the field undergoes a phase shift ϕ, $|\alpha\rangle \longrightarrow |e^{i\phi}\alpha\rangle$. The experiment achieved $\phi \approx \pi$. The linearity of the evolution implies that the initial superposition of the atom is amplified into an entangled atom-field state of the form $\frac{1}{\sqrt{2}}(|g\rangle|\alpha\rangle + |e\rangle|-\alpha\rangle)$. The atom then passes through an additional cavity, further transforming the superposition. Finally, the energy state of the atom is measured. This disentangles the atom and the field and leaves the latter in a superposition of the mesoscopically distinct states $|\alpha\rangle$ and $|-\alpha\rangle$.

To monitor the decoherence of this superposition, a second rubidium atom is sent through the apparatus. One can show that, after interacting with the field superposition state in cavity C, the atom will always be found in the same energy state as the first atom if the ► superposition has not been decohered. This correlation rapidly decays with increasing decoherence. Thus, by recording the measurement correlation as a function of the wait time τ between sending the first and second atom through the apparatus, the decoherence of the field state can be monitored. Experimental results were in excellent agreement with theoretical predictions. The influence of different degrees of "nonclassicality" of the field superposition state was also investigated. It was found that decoherence became faster as the phase shift ϕ and the mean number $\bar{n} = |\alpha|^2$ of photons in the cavity C was increased. Both results are expected, since an increase in ϕ and $\bar{n}$ means that the components in the superposition become more distinguishable. Recent experiments have realized superposition states involving several tens of photons [9].

Fullerenes and Other Mesoscopic Molecules

These experiments were carried out by the group of Anton Zeilinger and Markus Arndt at the University of Vienna [2] and are also described in ► Mesoscopic Quantum Phenomena. Basically, they represent sophisticated versions of the ► double-slit experiment. Spatial interference patterns are here demonstrated for mesoscopic molecules such as the fullerenes C_{60} and C_{70} (containing $O(1{,}000)$ microscopic constituents), the fluorinated fullerene $C_{60}F_{48}$ (mass $m = 1632$ amu), and the biomolecule $C_{44}H_{30}N_4$ ($m = 614$ amu, width over 2 nm). Since the ▶ de Broglie wavelength of these rather massive molecules is on the order of picometers and since it is impossible to manufacture slits of such small width, standard double-slit interferometry is out of reach. Instead the experiments make use of the Talbot–Lau effect, a true interference phenomenon in which a plane wave incident on a diffraction grating creates an "image" of the grating at multiples of a distance L behind the grating. In the experiment, the molecular density (at a macroscopic distance L) is scanned along the direction perpendicular to the molecular beam. An oscillatory density pattern (the image of the slits in the grating) is observed, confirming the existence of coherence and interference between the different paths of each individual molecule through the grating.

Decoherence is measured as a decrease of the visibility of this pattern. Such decoherence can be understood as a process in which the environment obtains information about the path of the molecule (see also ▶ Which-way experiment). This leads to a decay of spatial coherence at the level of the molecule. As described under ▶ Mesoscopic Quantum Phenomena, controlled decoherence induced by collisions with background gas particles and by emission of thermal radiation from heated molecules has been observed, showing a smooth decay of visibility in agreement with theoretical predictions. These successes have led to speculations that one could perform similar experiments using even larger particles such as proteins, viruses, and carbonaceous aerosols. Such experiments will be limited by collisional and thermal decoherence and by noise due to inertial forces and vibrations [10].

Superconducting Systems

See also ▶ Superconductivity. The idea of using superconducting quantum two-state ("qubit") systems for the generation of macroscopic superposition states goes back to the 1980s [11]. The main systems of interest are superconducting quantum interference devices (SQUIDs) and Cooper-pair boxes.

SQUIDs A SQUID consists of a ring of superconducting material interrupted by thin insulating barriers, called Josephson junctions (Fig. 1a). At sufficiently

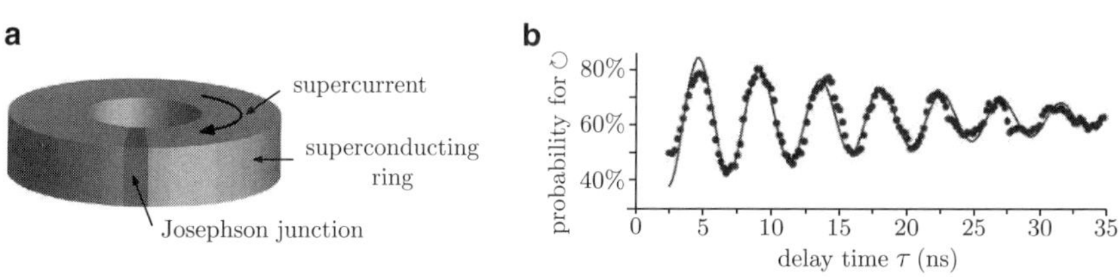

Fig. 1 (**a**) Schematic illustration of a SQUID. A ring of superconducting material is interrupted by Josephson junctions, which induce the flow of a dissipationless supercurrent. (**b**) Decoherence in a superconducting qubit. The damping of the oscillation amplitude corresponds to the gradual loss of coherence from the system. Figure adapted with permission from [14]. Copyright 2003 by AAAS

low temperatures, electrons of opposite spin condense into bosonic *Cooper pairs* (► BKS theory). Quantum-mechanical tunneling of Cooper pairs through the junctions leads to the flow of a persistent resistance-free "supercurrent" around the loop (Josephson effect), which creates a magnetic flux threading the loop. The collective center-of-mass motion of a macroscopic number ($\sim 10^9$) of Cooper pairs can then be represented by a ► wave function labelled by a single macroscopic variable, namely, the total trapped flux Φ through the loop. The two possible directions of the supercurrent define a quantum-mechanical two-state system with basis states $\{|\circlearrowright\rangle, |\circlearrowleft\rangle\}$. By adjusting an external magnetic field, the SQUID can be biased such that the two lowest-lying energy eigenstates $|0\rangle$ and $|1\rangle$ are equal-weight superpositions of the persistent-current states $|\circlearrowright\rangle$ and $|\circlearrowleft\rangle$. Such superposition states involving μA currents flowing in opposite directions were first experimentally observed in 2000 by Friedman et al. [12] and van der Wal [13] using spectroscopic measurements.

The decoherence of these superpositions was first measured by Chiorescu et al. [14] using Ramsey interferometry [24]. Two consecutive microwave pulses are applied to the system. During the delay time τ between the pulses, the system evolves freely. After application of the second pulse, the system is left in a superposition of the persistent-current states $|\circlearrowright\rangle$ and $|\circlearrowleft\rangle$ with the relative amplitudes exhibiting an oscillatory dependence on τ. A series of measurements in the basis $\{|\circlearrowright\rangle, |\circlearrowleft\rangle\}$ over a range of delay times τ then allows one to trace out an oscillation of the occupation probabilities for $|\circlearrowright\rangle$ and $|\circlearrowleft\rangle$ as a function of τ (Fig. 1b). The envelope of the oscillation is damped as a consequence of decoherence acting on the system during the free evolution of duration τ. From the decay of the envelope we can thus infer the decoherence timescale. Chiorescu et al. [14] measured a characteristic decoherence timescale of 20 ns. Recent experiment have achieved decoherence times of up to 4 μs [15].

Cooper-pair boxes Superpositions states and their decoherence have also been observed in superconducting devices whose key variable is charge (or phase), instead of the flux variable Φ used in SQUIDs. Cooper-pair boxes consist of a tiny superconducting "island" onto which Cooper pairs can tunnel from a reservoir through a Josephson junction. Two different charge states of the island, differing by at least one Cooper pair, define the basis states. Coherent oscillations between such charge states were first observed in 1999 [16]. In 2002, Vion et al. [17] reported thousands of coherent oscillations with a decoherence time of 0.5 μs. Similar results have been obtained for phase qubits.

Prospective Experimental Domains

Bose–Einstein condensates (BECs) In ► Bose-Einstein condensation, a macroscopic number of atoms undergoes a quantum phase transition into a condensate in which the atoms lose their individuality and occupy the same quantum state [24]. While quantum effects such as interference patterns – created by the overlap of different condensates or by coherently splitting and recombining a single condensate – have been experimentally observed, the preparation of superposition states involving macroscopically distinguishable numbers of particles have to date been unsuccessful. Theoretical studies of decoherence in BECs have played an important role in qualitatively and quantitatively understanding the challenges and conditions for the generation of such superpositions (see, e.g., [18]). The dominant source of decoherence was found to be collisions between condensate and noncondensate atoms. Decoherence models have suggested improved experimental procedures that may soon enable production of the desired superposition states. Existing proposals include: Modified condensate traps for faster evaporation of the decoherence-inducing thermal cloud of noncondensate atoms; creation of superpositions of relative-phase (instead of number-difference) states; environment engineering to shrink the thermal cloud; and faster generation of the superposition.

Nano-electromechanical systems (NEMS) NEMS are nanometer-to-micrometer-sized crystalline mechanical resonators, such as a cantilever or beam, coupled to nanoscale electronic transducers that detect the high-frequency vibrational motion of the resonator (Fig. 2a) [4]. Despite their macroscopic size, the resonators can be effectively treated as one-dimensional quantum harmonic oscillators (representing the lowest, fundamental flexural mode). NEMS are interesting systems from both applied and fundamental points of view and offer many opportunities for a study of quantum behavior at the level of macroscopic mechanical systems. In particular, Armour, Blencowe, and Schwab [19] have proposed a scheme for the experimental generation of superpositions of two well-separated displacements of the resonator and a measurement of the decoherence of this superposition (Fig. 2b). Here, a Cooper-pair box (prepared in a superposition of two charge states $|0\rangle$ and $|1\rangle$) is electrostatically coupled to the displacement of the resonator. This creates an entangled box-resonator state of the form $\frac{1}{\sqrt{2}}(|0\rangle|P_0\rangle + |1\rangle|P_1\rangle)$, where $|P_0\rangle$ and $|P_1\rangle$ are distinct center-of-mass states of the resonator. Existence of the superposition may subsequently be confirmed through interferometric techniques. Due to strong decoherence, no such superpositions have yet been experimentally realized.

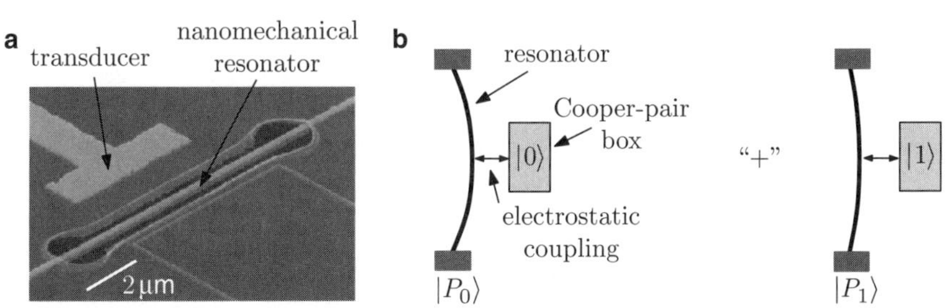

Fig. 2 (**a**) Nano-electromechanical system built by the Schwab group at Cornell University. (**b**) Proposed scheme for creating a superposition of two displacements of the resonator (see text). Figure reprinted with permission from [20]. Copyright 2004 by AAAS

Theoretical models of decoherence in NEMS are currently being developed to suggest improvements to experimental structures that could lead to sufficiently long-lived spatial superposition states.

Primary Literature

1. J. M. Raimond, M. Brune, S. Haroche: Manipulating quantum entanglement with atoms and photons in a cavity. Rev. Mod. Phys. **73**, 565–582 (2001).
2. M. Arndt, K. Hornberger, A. Zeilinger: Probing the limits of the quantum world. Phys. World **18**, 35–40 (2005).
3. A. J. Leggett: Testing the limits of quantum mechanics: motivation, state of play, prospects. J. Phys. Condens. Matter **14**, R415–R451 (2002).
4. M. Blencowe: Quantum electromechanical systems. Phys. Rep. **395**, 159–222 (2004).
5. J. F. Poyatos, J. I. Cirac, P. Zoller: Quantum reservoir engineering with laser cooled trapped ions. Phys. Rev. Lett. **77**, 4728–4731 (1996).
6. M. Tegmark: Apparent wave function collapse caused by scattering. Found. Phys. Lett. **6**, 571–590 (1993).
7. A. Bassi, G. C. Ghirardi: Dynamical reduction models. Phys. Rep. **379**, 257–426 (2003).
8. M. Brune, E. Hagley, J. Dreyer, X. Maître, A. Maali, C. Wunderlich, J. M. Raimond, S. Haroche: Observing the progressive decoherence of the "meter" in a quantum measurement. Phys. Rev. Lett. **77**, 4887–4890 (1996).
9. A. Auffeves, P. Maioli, T. Meunier, S. Gleyzes, G. Nogues, M. Brune, J. M. Raimond, S. Haroche: Entanglement of a mesoscopic field with an atom induced by photon graininess in a cavity. Phys. Rev. Lett. **91**, 230405 (2003).
10. K. Hornberger, Thermal limitation of far-field matter-wave interference. Phys. Rev. A **73**, 052102 (2006).
11. A. J. Leggett: Macroscopic quantum systems and the quantum theory of measurement. Suppl. Prog. Theor. Phys. **69**, 80–100 (1980).
12. J. R. Friedman, V. Patel, W. Chen, S. K. Yolpygo, J. E. Lukens: Quantum superposition of distinct macroscopic states. Nature **406**, 43–46 (2000).
13. C. H. van der Wal, A. C. J. ter Haar, F. K. Wilhelm, R. N. Schouten, C. J. P. M. Harmans, T. P. Orlando, S. Lloyd, J. E. Mooij: Quantum superposition of macroscopic persistent-current states. Science **290**, 773–777 (2000).
14. I. Chiorescu, Y. Nakamura, C. J. P. M. Harmans, J. E. Mooij: Coherent quantum dynamics of a superconducting flux qubit. Science **21**, 1869–1871 (2003).
15. P. Bertet, I. Chiorescu, G. Burkard, K. Semba, C. J. P. M. Harmans, D. P. DiVincenzo, J. E. Mooij: Dephasing of a superconducting qubit induced by photon noise. Phys. Rev. Lett. **95**, 257002 (2005).
16. Y. Nakamura, Y. A. Pashkin, J. S. Tsai: Coherent control of macroscopic quantum states in a single-Cooper-pair box. Nature **398**, 786–788 (1999).
17. D. Vion, A. Aassime, A. Cottet, P. Joyez, H. Pothier, C. Urbina, D. Esteve, M. H. Devoret: Manipulating the quantum state of an electrical circuit. Science **296**, 886–889 (2002).
18. D. A. R. Dalvit, J. Dziarmaga, W. H. Zurek: Decoherence in Bose–Einstein condensates: Towards bigger and better Schrödinger cats. Phys. Rev. A **62**, 013607 (2000).
19. A. D. Armour, M. P. Blencowe, K. C. Schwab: Entanglement and decoherence of a micromechanical resonator via coupling to a Cooper-pair box. Phys. Rev. Lett. **88**, 148301 (2002).
20. M.D. LaHaye, O. Buu, B. Camarota, K.C. Schwab: Approaching the quantum limit of a nonmechanical resonator. Science 304, 74–77 (2004)

Secondary Literature

21. M. Schlosshauer: Decoherence and the Quantum-to-Classical Transition (Springer, Berlin 2007).
22. E. Joos, H. D. Zeh, C. Kiefer, D. Giulini, J. Kupsch, I.-O. Stamatescu: Decoherence and the Appearance of a Classical World in Quantum Theory (Springer, New York 2003).
23. W. H. Zurek: Decoherence, einselection, and the quantum origins of the classical. Rev. Mod. Phys. **75**, 715–775 (2003).
24. R. Kaiser, C. Westbrook, F. David (Eds.): Coherent Atomic Matter Waves, Les Houches Session LXXII, Les Houches Summer School Series (Springer, Berlin 2001).

F

Fermi–Dirac Statistics

Simon Saunders

Fermi–Dirac statistics are one of two kinds of statistics exhibited by ▶ identical quantum particles, the other being ▶ *Bose-Einstein statistics.* Such particles are called *fermions* and *bosons* respectively (the terminology is due to Paul Adrien Maurice Dirac (1902–84) [1]). In the light of the ▶ spin-statistics theorem, and consistent with observation, fermions are invariably spinors (of half-integral spin), whilst bosons are invariably scalar or vector particles (of integral spin). See ▶ spin.

In general, in quantum mechanics, the available states of a homogeneous many-particle system in thermal equilibrium, for given total energy, are counted as *equiprobable*. For systems of *exactly similar* ('identical') fermions or bosons, states which differ only in the permutation of two or more particles are not only counted as equiprobable – they are *identified* (call this *permutivity*). Fermions differ from bosons in that no two fermions can be in exactly the same 1-particle state. This further restriction follows from the *Pauli* ▶ *exclusion principle*. The thermodynamic properties of gases of such particles were first worked out by Enrico Fermi (1901–54) in 1925 [2], and, independently, by Dirac in 1926 [3].

To understand the consequences of these two restrictions, consider a system of N weakly-interacting identical particles, with states given by the various 1-particle energies ϵ_s together with their degeneracies – the number C_s of distinct 1-particle states of each energy ϵ_s. From permutivity, the total state of a gas is fully specified by giving the number of particles with energy ϵ_s in each of the C_s possible states, i.e. by giving the *occupation numbers* n_k^s for each s, $k = 0, 1, , , , C_s$. We suppose all possible states of the same total energy E and, supposing particle number is conserved, of the same total number N, are available to the N particles when in thermal equilibrium, i.e. all sets of occupations numbers that satisfy:

$$\sum_{s,k=1}^{k=C_s} n_k^s = \sum_s N_s = N; \ \sum_s N_s \epsilon_s = E. \tag{1}$$

Since this is quantum mechanics, we suppose that ▶ superpositions of such states are available to the system as well.

Imposing Pauli's restriction that no two particles can be in the same 1-particle state, it follows that the occupation numbers are all zeros and ones and that $C_s \geqslant N_s$.

The number of distinct sets of occupation numbers $n_0^s, n_1^s, .., n_{C_s}^s$ that sum to N_s satisfying this condition is:

$$\frac{C_s!}{N_s!(C_s - N_s)!}.$$

Since the occupation number states span the subspace of the total ► Hilbert space to which the N_s particles are confined, this is the dimensionality – the 'volume'– of the available spate space for fermions of energy s.

For comparison, if the exclusion principle is not obeyed, the number of distinct sets of $\{n_k^s\}$ that sum to N_s is rather:

F

$$\frac{(C_s + N_s - 1)!}{N_s!(C_s - 1)!}$$

the state-space measure that applies to bosons of energy s. The total number of distinct sets of occupation numbers for $N = \sum_s N_s$ particles is then for fermions:

$$P_- = \prod_s \frac{C_s!}{N_s!(C_s - N_s)!}$$

and for bosons:

$$P_+ = \prod_s \frac{(C_s + N_s - 1)!}{N_s!(C_s - 1)!}.$$

By conventional reasoning, the equilibrium coarse-grained distribution is that for which $P_\pm$ is a maximum. The equilibrium entropy is proportional to the logarithm of this number, $S_\pm = k \log P_\pm$, where k is Boltzmann's constant. Using the Stirling approximation for $x \gg 1$, $\log x! \approx x \log x - x$, the two entropy functions are:

$$S_\pm = k \log P_\pm \approx k \sum_s [\mp C_s \log C_s - N_s \log N_s - (\mp C_s - N_s) \log(C_s \pm N_s)].$$

If this is to be stationary under independent variation of the numbers $N_s \to N_s + \delta N_s$, subject to the constraints (1), then

$$0 = \delta \log P_\pm = \sum_s [-\delta N_s \log N_s - \delta N_s \log(C_s \pm N_s)].$$

Were the variations δN_s completely independent each term in this summand would have to vanish. Introducing undetermined Lagrange multipliers a, β, for each of the constraints (1), conclude rather that for each s:

$$-\delta N_s \log N_s - \delta N_s \log(C_s \pm N_s) - \alpha - \beta \epsilon_s = 0.$$

Rearranging:

$$N_s = C_s(\mathrm{e}^{\alpha+\beta\epsilon_s} \pm 1)^{-1}. \tag{2}$$

In the case of light quanta, there is no constraint on particle number and the multiplier α does not occur. The multiplier β meanwhile has its usual meaning, $\beta = 1/kT$, where T is the absolute temperature. C_s is the number of distinct 1-quanta states in the energy range $[\epsilon_s, \epsilon_s + \mathrm{d}\epsilon_s]$, where $\epsilon_s = h\nu_s$. It is given by:

$$C_s = 8\pi V \nu_s^2 \mathrm{d}\nu_s / c^3 \tag{3}$$

(obtained either classically, from the wave theory, or by Bose's method). From (2) and (3) the Planck ► black-body radiation law follows immediately. The numbers N_s, of (2) are proportional to the radiation energy density in the frequency range $[\nu_s, \nu_s + \mathrm{d}\nu_s]$, which can be directly measured.

The contrast with the statistics of non-identical particles is that in the latter case (failing permutivity) there is the further question of which of the N_s particles is in which of the C_s one-particle states ($C_s^{N_s}$ possible distributions in all). There is also the question of how the N particles are partitioned into the occupation numbers N_1,N_2, ...N_s, ... Taking both into account, the total number of distinct states P_0 with occupation numbers N_1, ..., N_s, .. is:

$$\frac{N!}{N_1!..N_s!....}\prod_s C_s^{N_s}. \tag{4}$$

By a similar calculation as before, this yields:

$$N_s = C_s \mathrm{e}^{-\alpha-\beta\epsilon_s} \tag{5}$$

Evidently (2) (for either sign) and (5) are approximately the same for $C_s \gg N_s$ (equivalently, when $\alpha + \beta\epsilon_s \gg 1$), and the difference in the statistics for identical and non-identical particles disappears.

At the other extreme, for bosons for which $C_s \ll N_s$, from (2) it follows:

$$N_s = C_s(\alpha + \beta\epsilon_s). \tag{6}$$

For $\alpha = 0$, and C_s as given by (3), (6) is the *Rayleigh–Jeans black-body distribution*; (5) is the *Wien distribution*. The discovery of ► Planck's constant began with the puzzle of how to understand these distributions, which yielded the observed long ($C_s \ll N_s$) and short ($C_s \gg N_s$) wavelength behaviour respectively, and with Planck's black body formula (2) (with negative sign), obtained by interpolating between them [10]. The method of counting (4) is associated with *Maxwell–Boltzmann* or *classical* statistics. It was derived, using specifically quantum-mechanical methods, by Paul Ehrenfest (1880–1933) and George Uhlenbeck (1900–88) immediately after the discovery of Fermi's statistics. They concluded that '► wave mechanics does not yet *per se* imply the refutation of Boltzmann's method' [4, p. 24]. The difference, in quantum mechanics, resides *solely* in the assumption of permutivity. It is an easy slide to think, since classical statistical mechanics delivers the same statistics as quantum mechanics for non-identical particles, that classical particles likewise are non-identical (and do not satisfy permutivity), i.e. that the *correct* classical count of states P_0 is (4). But Josiah Williard Gibbs (1839–1903) had argued for the permutivity of classical particles long before [6], and for a non-quantized classical phase space, permutivity makes no difference to the statistics [11]. That is, computing the *volume* of classical phase space, subject to permutivity, rather than a count of equiprobable states, one should use:

F

$$P_0 = \prod_s \frac{C_s^{N_s}}{N_s!} \tag{7}$$

rather than (4). The logarithm of P_0 as given by (7) yields an extensive entropy function, as required [12].

Fermi in 1924 was led to assume that no two ► electrons could occupy the same elementary volume in phase space, because only thereby could he obtain agreement with the Sarkur–Stern expressions for the chemical potential and absolute entropy [5]. That was enough, the following year, to get out a new equation of state, but little more. Dirac, a few months later, had many more fragments of the nascent theory of quantum mechanics to hand. He considered the question of how to formulate permutivity in terms of ► matrix mechanics directly. He was led to the question by Heisenberg's dictum: the new mechanics was to be restricted to *observable* quantities. In matrix mechanics the observable quantities were the matrix elements, corresponding to the intensities of the various transition processes giving rise to line spectra. In the still unresolved problem of the helium atom, the question arose of how to treat a transition involving both electrons in one-particle states ψ_n, ψ_m, of the form $(mn) \to (m'n')$, and its relation to the transition $(mn) \to (n'm')$. Only the sum of the two, Dirac noted, was observable. 'Hence, in order to keep the essential characteristic of the theory that it shall enable one to calculate only observable quantities, one must adopt the second alternative that (mn) and (nm) count as only one state.' [3, p. 667].

Incorporating this into the matrix mechanics (and in particular in terms of his theory of uniformizing variables) presented certain technical difficulties, whereas in wave mechanics the way forward was much easier (an early indicator for Dirac that Schrodinger's wave theory may have definite advantages over the matrix mechanics). In the two particle case the state (mn) of the composite system of electrons, labelled 1 and 2, must be of the form

$$\psi_{mn} = a_{mn}\psi_m(1)\psi_n(2) + b_{mn}\psi_n(1)\psi_m(2) \tag{8}$$

where $a_{nm} = \pm b_{nm}$ (and superpositions of such). Dirac observed that the antisymmetric case ($a_{nm} = -b_{nm}$) leads to Pauli's principle and the symmetric case to the Bose–Einstein statistical mechanics. He went on to deduce the theory just sketched;

he thought, as had Fermi, that the new statistics, applying as it did to electrons in the atom, was likely to apply to material gases as well.

Dirac shortly after remarked on the possibility of alternative ('more complicated') representations of the permutation group, other than the completely symmetrized (boson) and antisymmetrized (fermion) representations (in 1930, in the first edition of his *Principles*). These alternatives lead to a variety of different statistics – *parastatistics* – that are not realized in nature (or not in 3+1 dimensions; special considerations apply to particles effectively restricted to two spatial dimensions). It was thought, for a time, that they might offer an alternative to the quark model of deep inelastic scattering, but without success [13].

Werner Heisenberg (1901–76) as well as Dirac had been preoccupied with the helium problem. His earlier papers in 1926 on the helium and related 2-electron spectra had made use of the Pauli exclusion principle and, for the first time, the Schrödinger wave mechanics (albeit only as a calculational tool). He too arrived at the two classes of states (8), but under a somewhat different interpretation from Dirac's, and with no understanding of the fact that they gave rise to different statistics. He was led, rather, to an idea absent from Dirac's paper – that a two-electron system, each with identical allowed energies $E_m(1) = E_m(2)$, $E_n(1) = E_n(2)$ (with $E_n > E_m$), would in wave-theoretic terms be subject to *resonance*, with energy $E_n - E_m$ passing from one electron to the other under the transition $(mn) \rightarrow (nm)$ (states that Dirac had *identified*). Likewise the perturbation due to the electron charge 'will in general contain terms corresponding to transitions in which the systems 1 and 2 switch places ('den Platz tauschen')' [7, p. 417].

Thus did the idea of *exchange forces* first arise. A similar interpretation was advanced by Walter Heitler (1904–81) and Fritz London (1900–54) the following year in their treatment of the homopolar bond [8]. But by this time, as Heitler went on to remark, this question of interpretation had become closely wed to disputes over other interpretative issues in quantum mechanics, notably over Schrödinger's continuous beat picture of emission and absorption processes as compared to Born's statistical interpretation [14]. What was being exchanged, Heitler concluded, 'remained completely unclear.' ([9, p. 48]).

What *was* clear was that in any of the symmetric, triplet states of spin, for which the spatial ▶ wave function must be antisymmetric, the norm of the wave-function for electron coordinates close together is extremely small (and for coinciding coordinates, vanishes). In this sense electrons in bound states with correlated spins effectively repel one another. Those with anticorrelated spins, in the antisymmetric singlet state, have greater amplitudes for small relative distances, for their spatial wavefunction must then be symmetric – the amplitude is much greater than if there were no overall symmetry requirement on the state (the case of non-identical fermions). This effect is independent of the Coulomb force altogether, and plays a key role in ferromagnetism as well as in the chemical bond, as Heisenberg was shortly to show, again with reference to 'electron exchange', and 'exchange forces'.

Whether interpreted as an exchange force involving the ▶ identity of quantum particles over time, or as a consequence of permutivity and the Pauli exclusion principle, Fermi–Dirac statistics is fundamental to the whole of quantum chemistry and throughout the physics of the solid state.

Primary Literature

1. P.A.M. Dirac: The Principles of Quantum Mechanics, 3rd edition, Oxford: Oxford University Press (1945)
2. E. Fermi: Aur Quantelung des idealen einatomigen Gases, Zeitschrift für Physik **36**, 902–912 (1926)
3. P.A.M. Dirac: On the theory of quantum mechanics, Proceedings of the Royal Society (London) **A112**, 281–305 (1926)
4. P. Ehrenfest and G. Uhlenbeck: Die wellenmechanische Interpretation der Botzmannschen Statistik neben der neueren Statistiken, Zeitschrift für Physik, **41**, 24–26 (1927)
5. E. Fermi: Considerazioni sulla quantizzazione dei sistemi che contegno degli elementi identici, Nouvo Cimento **1**, 145–152 (1924)
6. J. Gibbs: Elementary Principles in Statistical Mechanics, Woodbridge, Connecticut: Ox Bow Press (1981)
7. W. Heisenberg: Merhrkorperproblem und Resonanz in der Quantenmechanik, Zeitschrift für Physik, **38**, 411–426 (1926)
8. W. Heitler and F. London: Wechselwirkung neutraler Atome und homöopolare Bindung nach der Quantenmechanik, Zeitschrift für Physik **44**, 455–472 (1927)
9. W. Heitler: Störungsenergie und Austausch beim Merhkörprobleme, Zeitschrift für Physik **46,** 47–72 (1927)

Secondary Literature

10. F. Hund: Geschichte der Quantentheorie (Manheim: Bibliographisches Institut. Translated as The History of Quantum Theory, G. Reece (trans.), London, Harrap 1974)
11. N. Huggett: Atomic metaphysics, Journal of Philosophy, **96**, 5–24 (1999)
12. S. Saunders: On the explanation for quantum statistics, Studies in the History and Philosophy of Modern Physics **37**, 192–211 (2006)
13. S. French: The separable uberty of quantum chromodynamics, Studies in the History and Philosophy of Modern Physics **26**, 87–105 (1995)
14. C. Carson: The peculiar notion of exchange forces I: origins in quantum mechanics, 1926–28, Studies in the History and Philosophy of Modern Physics **27**, 23–45 (1996)

F

Feynman Diagrams

David Kaiser

Feynman diagrams are a powerful pictorial tool for making calculations in quantum theory. They were invented by the American theoretical physicist Richard Feynman (1918–88) during the late 1940s, in the context of ► quantum electrodynamics (QED), physicists' quantum-mechanical theory of electric and magnetic forces. The diagrams were intended to provide a shorthand for the famously unwieldy mathematics of QED calculations, in which it had become common, since the 1930s, for physicists to mistakenly conflate or omit terms within long series of expressions. Feynman unveiled his new techniques at a private conference in 1948. He also coached a young protégé, Freeman Dyson (born 1923, at that time a graduate student at Cornell University in upstate New York, where Feynman taught), in how to use the diagrams. Feynman and Dyson each published a pair of articles on the new techniques during 1949 [1].

Feynman's own route to the diagrams involved a major re-thinking of quantum mechanics, based on his notion of ▶ path integrals, which he developed for his dissertation at Princeton University in 1942. Dyson, on the other hand, recognized that the diagrams could be useful for calculations in ▶ quantum field theory independent of Feynman's particular ideas about path integrals. Well into the 1960s, most applications of Feynman diagrams, and most discussion of them in textbooks, followed Dyson's prescriptions, until Feynman's path integrals entered the mainstream [4, 5].

As in any quantum-mechanical calculation, the main item of interest is a complex number, or "amplitude," whose absolute square yields a probability. For example, $A(t, \boldsymbol{x})$ might represent the amplitude that a particle will be found at a point $\boldsymbol{x}$ at time t. Then the probability of finding the particle there at that time will be $|A(t, \boldsymbol{x})|^2$. (See ▶ Born rule)

In QED, amplitudes are composed from a few basic ingredients, each of which has an associated mathematical expression. Most often, the basic ingredients refer to the behavior of virtual particles (see ▶ QED) – particles that pop into existence by "borrowing" energy from the vacuum, as long as they pay that energy back sufficiently quickly, on timescales set by the ▶ Heisenberg uncertainty principle. To illustrate how the diagrams work, we may write, schematically:

- Amplitude for a virtual electron to travel undisturbed from spacetime point x to spacetime point y: $B(x, y)$;
- Amplitude for a virtual photon to travel undisturbed from spacetime point x to spacetime point y: $C(x, y)$;
- Amplitude for an electron and photon to scatter: eD.

Here e is the charge of the electron, which governs how strongly electrons and photons will interact, and we label coordinates as $x = (t, \boldsymbol{x})$.

Feynman introduced his diagrams to keep track of all the different ways that electrons and photons (▶ light quantum) could interact. The rules for using the diagrams are fairly straightforward: at every "vertex," draw two electron lines meeting one photon line. Draw all of the topologically distinct ways that electrons and photons can scatter (subject to this rule of always having two electron lines meet one photon line). Then build an equation: substitute factors of $B(x, y)$ for every virtual electron line, $C(x, y)$ for every virtual photon line, and eD for every vertex. Lastly, because these vertices can occur anywhere in space and time, integrate over all the spacetime points involving virtual particles.

The diagrammatic accounting scheme is so useful because e is so small: $e^2 \sim 1/137$, in appropriate units. That means that diagrams that involve fewer vertices – and hence fewer factors of this small number, e – tend to contribute more to the overall amplitude than complicated diagrams, which contain many vertices and hence many factors of e. Thus physicists can approximate an amplitude, A, by expanding it in a series of progressively complicated terms, known as a "perturbation-series expansion." In principle the series includes an infinite number of distinct contributions – there are an infinite number of different ways in which virtual electrons and photons can scatter– but as a practical matter, physicists can truncate the series at a desired level of accuracy.

For example, consider how an electron is scattered by an electromagnetic field. Quantum-mechanically, the field can be described as a collection of photons. In the simplest case, the electron (*straight line*) will scatter just once from a single photon (*dotted line*) at just one vertex (circle at the point x_0):

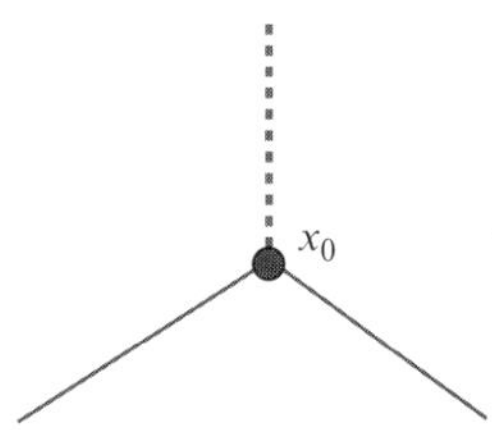

$$A_{(1)} = eD$$

In this case the electron is real, not virtual, and hence the only contribution comes from the vertex.

Many more things can happen to the hapless electron. At the next level of complexity, the incoming electron might shoot out a virtual photon before scattering from the electromagnetic field, reabsorbing the virtual photon at some other point:

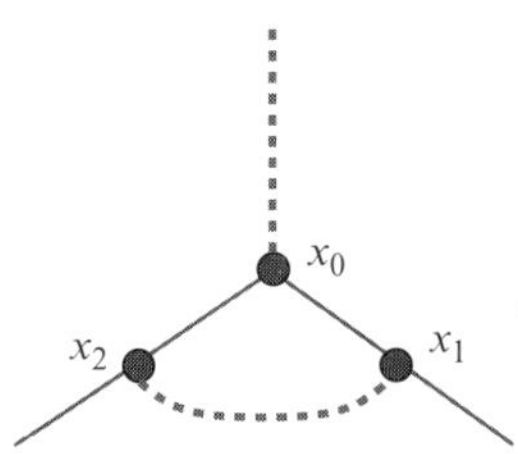

$$A_{(2)} = e^3 \int DB(1,0)\ DB(0,2)\ DC(1,2)$$

In this more complicated diagram, electron lines and photon lines meet in three places, and hence the contribution to the overall amplitude from this diagram is proportional to e^3. Thus it is roughly one hundred times smaller in magnitude that the contribution from the simplest diagram.

Still more complicated things can happen. At the next level of complexity, seven distinct Feynman diagrams enter:

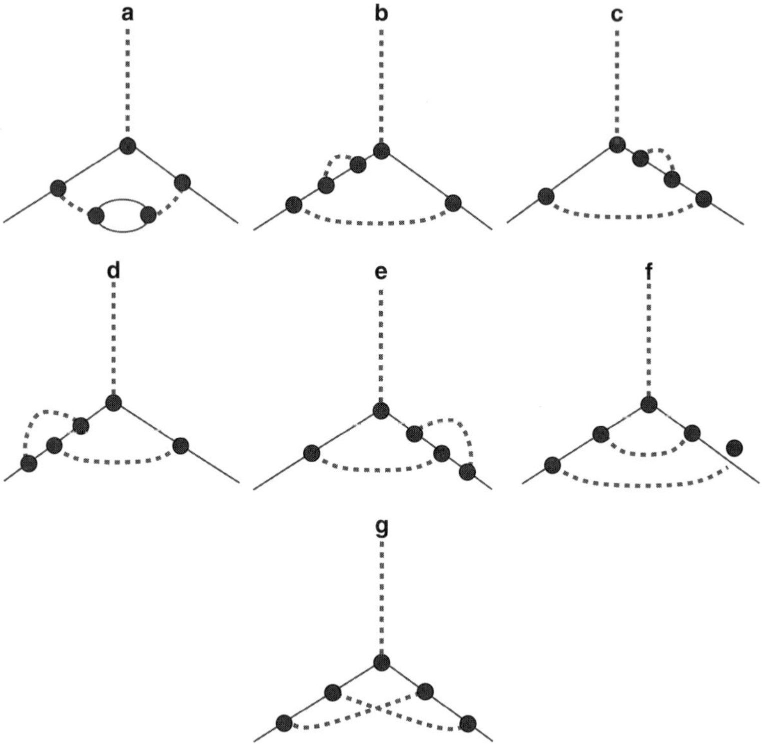

Each diagram – labeled *a*, *b*, *c*, and so on – will contribute a distinct term to the overall amplitude. All seven of these contributions, deriving from diagrams that contain five vertices, will be proportional to e^5. As an example, consider the contribution from the diagram labeled *a* at the upper left. We may label its contribution $A_{(3)}{}^a$, meaning that this term enters at the third level of approximation, and stems from diagram *a*:

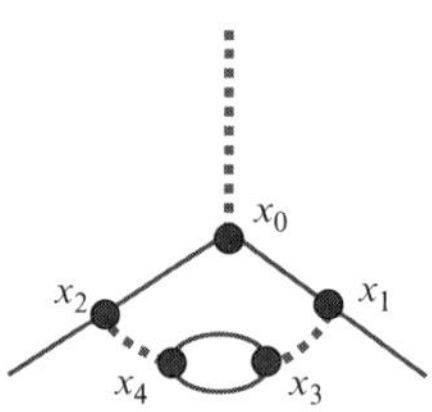

$$A_{(3)}{}^a = e^5 \int D B(1,0)\, D B(0,2)\, D C(1,3)\, D \times B(3,4)\, D B(4,3)\, C(4,2)$$

Similar terms can be written for each of the remaining diagrams at this level of approximation, leading to terms such as $A_{(3)}{}^b$, $A_{(3)}{}^c$, right through $A_{(3)}{}^g$. The total amplitude for an electron to scatter from the electromagnetic field may then be written as the sum of all these terms:

$$A = A_{(1)} + A_{(2)} + A_{(3)}{}^a + A_{(3)}{}^b + A_{(3)}{}^c + \ldots$$

and the probability for this interaction is $|A|^2$.

Robert Karplus and Norman Kroll first attempted this type of calculation using Feynman's diagrams soon after learning the new techniques from Dyson [2]. Eight years later, several other physicists found a few algebraic errors in the calculation, whose correction only affected the fifth decimal place of the original answer. Since the 1980s, Tom Kinoshita of Cornell University has gone all the way to diagrams containing *eight* vertices – a calculation involving 891 distinct Feynman diagrams, accurate to *ten* decimal places [3].

Although Feynman diagrams were developed as a tool for calculating the effects of weakly-interacting forces (such as electromagnetism), the diagrams were quickly adapted during the 1950s and 1960s to treat all kinds of other interactions, from the strong nuclear force, to many-body interactions in condensed-matter physics, to gravitation, and beyond [5]. They have become a ubiquitous element of the physicist's toolkit.

Primary Literature

1. F. J. Dyson: The radiation theories of Tomonaga, Schwinger, and Feynman. Phys. Rev. **75**, 486 (1949); F. J. Dyson: The S matrix in quantum electrodynamics. Phys. Rev. **75**, 1736 (1949); R. P. Feynman: The theory of positrons. Phys. Rev. **76**, 749 (1949); R. P. Feynman: Space-time approach to quantum electrodynamics. Phys. Rev. **76**, 769 (1949)
2. R. Karplus, N. M. Kroll: Fourth-order corrections in quantum electrodynamics and the magnetic moment of the electron. Phys. Rev. **77**, 536 (1950)
3. T. Kinoshita, M. Nio: Improved α^4 term of the electron anomalous magnetic moment. Phys. Rev. D **73**, 013003 (2006)

Secondary Literature

4. S. S. Schweber: *QED and the Men Who Made it: Dyson, Feynman, Schwinger, and Tomonaga* (Princeton University Press, Princeton 1994)
5. D. Kaiser: *Drawing Theories Apart: The Dispersion of Feynman Diagrams in Postwar Physics* (University of Chicago Press, Chicago 2005)

Fine-Structure Constant

Helge Kragh

The fine-structure constant is a dimensionless constant of nature, given by $\alpha = e^2/\hbar c$, in electrostatic cgs units, where e is the elementary charge, $\hbar$ ► Planck's constant ($=h/2\pi$), and c the velocity of light. The number is a measure of the strength of electromagnetic interactions. The numerical value of α is known with great accuracy:

$$\alpha^{-1} = 137.035 \cdot 999 \cdot 76 \pm 0.000 \cdot 000 \cdot 50$$

The name "fine-structure constant" relates to ▶ spectroscopy, but even before it was revealed in spectra it was realized that the ratio $e^2/\hbar c$ might be of theoretical significance. In 1905 Max Planck pointed out that e^2 and $\hbar c$ have the same dimensions and the same order of magnitude. However, if α was ever "discovered" the honour must go to Arnold Sommerfeld, who in 1915–16 extended Niels Bohr's theory of the hydrogen atom (▶ Bohr's atomic model) to the domain of special relativity. He derived the energy levels in the relativistic case and found that the H_α line would appear as a doublet with a "fine-structure separation" given by α. Measurements made by Friedrich Paschen confirmed the theory and resulted in α^{-1}=137.9.

With the emergence of quantum mechanics in 1925–26, it turned out that α was intimately connected with the electron's ▶ spin, a relationship fully explained by Paul A.M. Dirac's relativistic wave equation of 1928 (▶ Dirac equation and ▶ relativistic quantum mechanics). Inspired by Dirac's theory, Arthur S. Eddington suggested that α was a fundamental quantity connected also to cosmological quantities such as the number of particles in the universe. Moreover, he believed that the numerical value of α could be derived *a priori*, and that the result must be an integer: α^{-1}=137. Although experiments disagreed with Eddington's claim, and his theory was generally rejected, it led to many attempts to relate α to pure numbers or other constants of nature. This kind of "alpharology" was particularly popular in the 1930s and has continued until the present. Although numerology à la Eddington has today a low reputation, some physicists still believe that it should be possible to calculate the value of α purely deductively. So far, all attempts have failed.

Because α can be determined from the spectra of distant luminating objects, such as quasars, it is possible to check if the quantity has varied over cosmological time. Speculations of a time-varying α go back to the 1930s and in 2001 measurements from absorption lines in quasars indicated that α might have been smaller in the past. However, more recent and accurate data suggest that the fine-structure constant is indeed constant: it had the same value billions of years ago as it has today.

Primary Literature

1. A. Sommerfeld: Zur Quantentheorie der Spektrallinien. Annalen der Physik **51**, 1–94 (1916)
2. A. S. Eddington: *Relativity Theory of Protons and Electrons* (Cambridge University Press, Cambridge 1936)

Secondary Literature

3. J. D. Barrow: *The Constants of Nature – From Alpha to Omega* (Jonathan Cape, London 2002)
4. H. Kragh: Magic Number: A Partial History of the Fine-Structure Constant. Archive for History of Exact Sciences **57**, 395–431 (2003)

Franck–Hertz Experiment

Friedel Weinert

In 1913 Bohr took Rutherford's nucleus model of the hydrogen atom as the basis for his quantized atom model (► Bohr's atomic model; Rutherford atom). Although it was not the first, it was the first successful atom model. A year later, two Berlin experimenters, James Franck (1882–1964) and Gustav Hertz (1887–1975), unaware of Bohr's model and its implications, performed an experiment which later turned out to be one of its strongest corroborations. For the so-called *Franck–Hertz experiment*, they were awarded the Nobel Prize for Physics in 1925. In this experiment ► electrons are ejected from a cathode, **C**, into a tube filled with mercury gas (see Fig. 1). The energy of the electrons can be increased in a controllable manner by accelerating them towards the positively charged grid, **G**, through the potential difference V_a. Electrons fly through the grid towards anode **A**. Between **G** and **A**, a small retarding voltage, V_r, decelerates the electrons. They will only reach the anode **A**, if their energies V exceed V_r, where they will be recorded by the ammeter A.

Collisions between the atoms and the electrons will occur. Only electrons with sufficient energy will cause the mercury atoms to make transitions to higher states of energy. The electrons will lose their energy to the atoms. When $V_a = 4.9$ V, the curve drops very sharply.

The two experimenters initially thought they had measured mercury's ionization potential.

As Bohr pointed out in August 1915 but Franck and Hertz only realized in 1917, the Bohr atomic model provides a perfect explanation for this behaviour. The electrons near the grid lose all their energy to the mercury atoms and are unable to overcome the small retarding potential, V_r, to reach the anode. A drop in the current, I_a, is observed. When $V_a = 9.8$ V, another drop in the curve occurs. The electrons either excite the atoms to higher energy levels or lose 4.9 V more than once. The excited mercury atoms in turn will return to their ground energy state and emit photons with energies corresponding to the energy intake. The experiment displayed

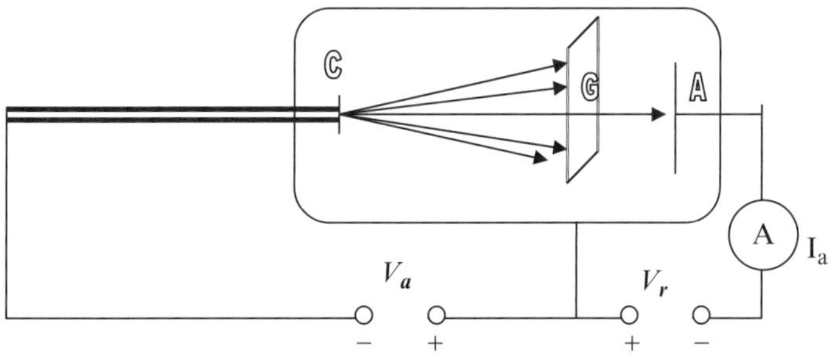

Fig. 1 Franck–Hertz experiment (1914)

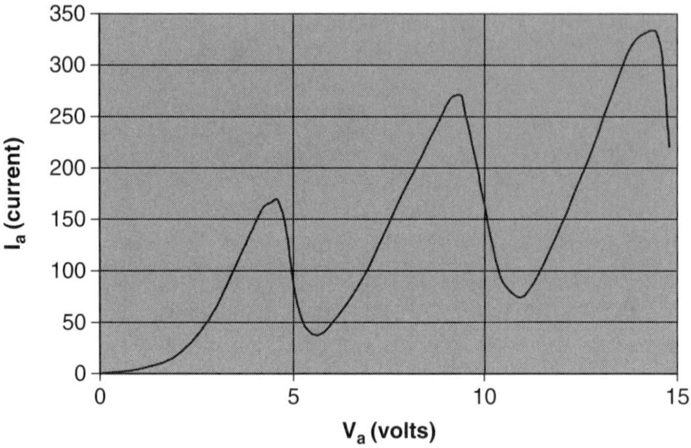

Fig. 2 Frank–Hertz experiment (1914): Dependence of current (I_a) on accelerating potential (V_a)

the loss of the electronic energy at discrete levels. Later, more precise experiments confirmed that the higher states of energy of the atoms corresponded to the discrete energy levels calculated from the Bohr model. The observable results are shown in Fig. 2.

As in quantum mechanics there is a traditional distinction between the *wave* and the *particle* picture, we should note that the Franck–Hertz experiment illustrates the particle picture of quantum mechanical processes. (For the wave-picture ► Stern–Gerlach experiment and ► Davisson–Germer experiment) In this experiment the particle picture gives rise to a *probabilistic* notion of causality, since we are not in a position to predict which electron will collide with which mercury atom and how much energy it will transfer.

Primary Literature

1. J. Franck, G. Hertz: Über Zusammenstöße zwischen Elektronen und den Molekülen des Quecksilberdampfes und die Ionisierungsspannung desselben. Verhandlungen der Deutschen Physikalischen Gesellschaft **16**, 457–67 (1914). With supplementary texts and commentary by Armin Hermann in: *Die Elektronenstoßversuche* (Battenberg, München 1967, Dokumente der Naturwissenschaft 9)
2. J. Franck, P. Jordan: *Anregung von Quantenspüngen durch Stöße* (Springer, Berlin 1926)

Secondary Literature

3. A. P. French, E. F. Taylor: *An Introduction to Quantum Physics* (Stanley Thornes Publishers, Cheltenham 1998, 30–1, Chapman & Hall 1979)
4. K. Krane: *Modern Physics* (Wiley, New York 1983, 169–70)
5. P. A. Tipler: *Modern Physics* (Worth Publishers, New York 1978, 155–7)

6. H. Geiger, K. Scheel eds.: *Quanten* (Springer, Berlin 1926, Handbuch der Physik 23)
7. J. Lemmerich: *Max Born, James Franck: Physiker in ihrer Zeit* (Reichert, Wiesbaden 1982)
8. J. Lemmerich: *Aufrecht im Sturm* (Spektrum, Berlin 2005) (Biography of J. Franck)
9. J. Kuczera: *Gustav Hertz* (Teubner, Leipzig 1985)

Functional Integration; Path Integrals

F

Cecile DeWitt-Morette

Definitions

Functional integral is, by definition, an integral over a space of functions. The functions are the variables of integration. When the variables are paths, the functional integral is usually called a "path integral". For example, let x be a path parameterized by time $t \in T$, taking its values in a D-dimensional manifold $\mathbb{M}^D$, i.e.

$$x : T \to \mathbb{M}^D \quad \text{by} \quad t \mapsto x(t), \tag{1}$$

a sum over all paths x is a path integral.

To compute a path integral

$$\int_{\mathbb{X}} \mathrm{D}x \; F(x), \qquad x \in \mathbb{X}, \tag{2}$$

one needs to define the domain of integration $\mathbb{X}$, a norm on $\mathbb{X}$, a volume element $\mathrm{D}x$ on $\mathbb{X}$, and choose an integrable functional F on $\mathbb{X}$.

If the variable of integration is a field, a functional integral is sometimes called "a sum over histories".

Functional integration is a rich and powerful mathematical technique because the domain of integration $\mathbb{X}$ is an infinite dimensional space. Short of having intuitive understanding of infinite dimensional spaces of functions, we have extensive studies of such spaces developed during the last century.

Path Integrals, A Modern Approach to ► Quantization

Functional integration entered physics in 1942 in the doctoral dissertation of Richard P. Feynman, "The Principle of Least Action in Quantum Mechanics" [1]. The goal was a formulation of ► quantum electrodynamics beginning with quantum mechanics formulated in terms of the classical action functional S of a given system.

Schematically, the path integral constructed by Feynman that gives the probability amplitude for a particle, known to be at a at a time t_a, be found at b at time t_b is

$$< b, t_b \mid a, t_a > = \int_{\mathbb{X}_{a,b}} \mathrm{D}x \; \exp(\frac{\mathrm{i}}{\hbar} S(x)) \tag{3}$$

where $\mathbb{X}_{a,b}$ is the space of all paths x from a to b. The paths x were replaced by n of their values

$$\{x(t_1), x(t_2), \ldots, x(t_n)\}\,, \qquad x(t) \in \mathbb{R}^D$$

for n ordered values of t_i in the interval $[t_a,\ t_b]$.

The path integral is approximated by an integral over $(\mathbb{R}^D)^n$.

This crude approximation was both beneficial and detrimental.

- It lead Feynman to a powerful formulation of Quantum Electrodynamics in terms of diagrams, and to the award of the 1965 Nobel prize. The diagrams corresponding to a particular matrix element are both an aid to its calculation and a picture of its physical process. They rapidly became popular ▶ Feynman Diagrams.
- Unfortunately, the time-slicing approximation is fundamentally deficient because it ignores the domain of integration. Indeed a functional space is rarely the limit of $\mathbb{R}^{Dn}$ when n goes to infinity. It also ignores the topological properties of the range $\mathbb{M}^D$ of the paths. In addition, it makes it extremely awkward, not to say impossible, to implement the two basic techniques for computing integrals: integration by parts and change of variable of integration.

Gaussian Integrals, Semi-classical Approximations

Gaussian integrals are easily defined by their Fourier transforms.

In one-dimension the Fourier transform of a real gaussian is:

$$\int_{\mathbb{R}} \frac{\mathrm{d}x}{\sqrt{a}} \exp(-\pi a x^2) \; \exp(-2\pi \mathrm{i}\, x' x) \; := \; \exp(-\pi \frac{x'^2}{a});$$

the right hand side defines the gaussian on the left.

In D-dimensions the Fourier transform of a real or complex gaussian is:

$$\int_{\mathbb{R}^D} \mathrm{D}x \; \exp(-\frac{\pi}{s} Q(x)) \; \exp(-2\pi \mathrm{i} \langle x', x \rangle) \; := \; \exp(-\pi s\, W(x')) \tag{4}$$

where

$$\begin{aligned} \mathrm{D}x &= \mathrm{d}x^1\,\mathrm{d}x^2\ldots\mathrm{d}x^D\,(\det Q_{ij})^{-1/2}\,, \qquad s \in \{1, i\}, \\ Q(x) &= \sum_{ij} Q_{ij}x^i x^j\,, \\ W(x') &= W^{ij}x'_i x'_j, \quad \sum Q_{ij}W^{jk} = \delta_i^k, \quad x' \in \mathbb{R}_D(\text{dual of } \mathbb{R}^D). \end{aligned}$$

A gaussian *functional* integral has the same structure. Given a quadratic form W on the dual space $\mathbb{X}'$ of the domain of integration $\mathbb{X}$, one defines a gaussian volume element $\mathrm{D}x\ \exp(-\frac{\pi}{s}\ Q(x))$ by its Fourier transform $\exp(-\pi s W(x'))$. Integrating polynomials with respect to gaussian volume elements follow the same rules in finite and infinite dimensions. The Feynman diagrams are the graphic representation of integrals of polynomials with respect to a gaussian volume element.

F

In order to use gaussian techniques in the integral (3), one expands the action functional $S(x)$ around its value at a fiducial choice x_0 often chosen to be a classical solution x_{cl} of the Euler–Lagrange equation:

$$S(x) = S(x_{\mathrm{cl}}) + \frac{1}{2}\,S''(x_{\mathrm{cl}})\ J.J + \frac{1}{3!}\,S'''(x_{\mathrm{cl}})\ J.J.J + \ldots \tag{5}$$

The second variation $S''(x_{\mathrm{cl}})\ J.J$ of the action functional S is a quadratic form on the space of vector fields J on $T_{x_{\mathrm{cl}}}\mathbb{X}$ (tangent space to $\mathbb{X}$ at x_{cl}). The calculus of variation provides powerful techniques [2] for computing gaussian integrals defined by the second variation of the classical action functional.

If one terminates the expansion (5) at the second variation, the integral (3) is the semi-classical approximation of the matrix element on the left-hand side.

The second variation is degenerate in many interesting situations: conservation laws, caustics, etc...; then the contributions of the first, third...variations come into play and provide explicit results. For example, explicit cross sections of glory scattering of waves (scalar, electromagnetic and gravitation) by black holes can be obtained from gaussian integrals where the second variation is degenerate. The explicit result is given in terms of Bessel functions [2].

Geometrical and Topological Applications

From a small seed in 1942, functional integration in Quantum Physics has grown into a large and widespread tree [2]. Just a few examples corresponding to a variety of paths and a variety of action functionals:

1. A path

$$x : \mathbb{R} \to \mathbb{M}^D \qquad \text{by} \qquad s \mapsto x(s)$$

is characterized by

- Its analytic properties: it is an element of a space of continuous paths, or a Sobolev space when the action functional contains a kinetic energy term, or a space of Poisson paths for solutions of the telegrapher equation and the Dirac equation.
- Its domain: s can be the time in a fixed time interval, or the time in a path-dependent time interval (e.g. in an interval terminating at a first-exit time), or the intrinsic time of a given process, etc... The parameter s need not be a time variable, it can be any ordering parameter, e.g. a scale variable in coarse-graining problems.
- Its range: $\mathbb{M}^D$ can be a (pseudo) riemannian manifold, and/or a multiply connected space, or a fibre bundle.
  Detailed calculations of all these cases can be found in [2].

2. An action functional

- If S is a Chern–Simons action [3], functional integration provides an intrinsic definition of Jones polynomials of knot theory in 3-dimensions, explicit evaluations of topological invariants and applications to physics.
- S maybe defined on supervariables (commuting and anticommuting variables). Functional integrals in supersymmetric quantum mechanical systems can be used for proving the Atiyah–Singer index theorem, for computing the index [4,5], and for related results.

Conclusion

From a heuristic tool, functional integration is gradually becoming a mathematical tool. Path integrals are by now a well-defined, robust tool. A number of explicit path integrals can be found in [6]. A number of functional integrals in Quantum Field Theory are mathematically reliable.

The power of functional integrals stems from the fact that function spaces are infinite dimensional. For example, a linear change of variable of integration for $x \in \mathbb{R}^D$ can be useful but it is not spectacular; a linear change of variable of integration for $x \in \mathbb{X}$ offers a great variety of possibilities, and uses concepts and techniques from several areas of analysis [2].

Primary Literature

1. R. P. Feynman: *The Principle of Least Action in Quantum Mechanics*. Princeton University Press No 2948 (1942) (University Microfilms, Ann Arbor Michigan). R. P. Feynman, with L. Brown: Feynman's Thesis: *A New Approach to Quantum Theory*. (World Scientific, Singapore 2005). R. P. Feynman: *Space-time approach to non-relativistic quantum mechanics*. Reviews of Modern Physics. 20, 367–87 (1948).

2. P. Cartier, C. DeWitt-Morette: *Functional Integration; Action and Symmetries*. (Cambridge University Press, Cambridge 2006). This book has an extensive bibliography.

Secondary Literature

3. E. Witten: *Quantum field theory and the Jones polynomial Communications in Mathematical Physics*. 121, 351–99 (1989).
4. L. Alvarez-Gaume: *Supersymmetry and Index Theorem in Supersymmetry*. Proceedings of the 1984 NATO School in Bonn, ed. by K. Dietz, R. Flume, G. U. Gehlen, V. Rittenberg (1984).
5. B. DeWitt: *Supermanifolds*. 2nd Edition, 385–89. (Cambridge University Press, Cambridge, 1992). B. DeWitt: *The Global Approach to Quantum Field Theory*. Revised edition (Oxford University Press, Oxford, 2004).
6. C. Grosche, F. Steiner: *Handbook of Feynman Path Integrals*. (Springer, Berlin, 1998).

F

G

Gauge Symmetry

Holger Lyre

Gauge symmetries characterize a class of physical theories, so-called *gauge theories* or *gauge field theories*, based on the requirement of the invariance under a group of transformations, so-called *gauge transformations*, which occur in a theory's framework if the theory comprises more variables than there are physically independent degrees of freedom. Gauge ► symmetry was firstly acknowledged in Maxwell's electrodynamics, where the vector potential shows a freedom of transformation in the sense that it is not uniquely determined by the Maxwell field equations, but only up to adding the derivative of a scalar function. Since all three fundamental quantum field theoretic interactions as well as gravity can be reconstructed within a gauge theoretic framework, gauge field theories represent the backbone of modern physics today, that is, the physics of the Standard Model and beyond. ► Quantum field theory; particle physics.

Short History and Core Idea

In modern notation based on four-tensor-valued fields, classical Maxwellian electrodynamics is captured by the Lagrangian $\mathcal{L}_E = -\frac{1}{4}F_{\mu\nu}F^{\mu\nu} - j^{\mu}A_{\mu}$ with a tensor $F_{\mu\nu} = \partial_{\mu}A_{\nu} - \partial_{\nu}A_{\mu}$ comprising the electric and magnetic field strengths and the vector potential A_{μ}. The Maxwell field equations follow from the variation of $\mathcal{L}_E$ according to A_{μ} as a dynamic field variable as $\partial_{\mu}F^{\mu\nu} = j^{\nu}$ and $\epsilon^{\mu\nu\lambda\sigma}\partial_{\nu}F_{\lambda\sigma} = 0$ (Bianchi identity of $F^{\mu\nu}$). While it seems natural to consider A_{μ} as a basic variable, the true observable quantity of the theory, the field $F^{\mu\nu}$, remains unchanged under gauge transformations of the potential

$$A_{\mu}(x) \rightarrow A'_{\mu}(x) = A_{\mu}(x) - \partial_{\mu}\alpha(x). \tag{1}$$

Here α may be any differentiable scalar function, either constant or dependent on the spacetime variable x. This amounts to saying that Maxwellian electrodynamics shows a gauge freedom under both global and local gauge transformations.

The gauge freedom of classical electrodynamics went largely unrecognized. In 1918, however, Hermann Weyl conjectured a unified theory of gravitation and electrodynamics by extending Einstein's idea of his thus completed general theory of relativity. Here the Riemannian geometry of spacetime itself becomes dynamical. However, while in Riemannian geometry the comparison of directions at two points

D. Greenberger et al. (eds.), *Compendium of Quantum Physics: Concepts, Experiments, History and Philosophy*, © Springer-Verlag Berlin Heidelberg 2009

depends on the paths connecting these points, the comparison of lengths does not. Weyl argued that, in a true infinitesimal geometry, the scale of length should also undergo a change, such that, under parallel transport, a "gauge measure of length" (in German: "Eichmaßstab") should undergo a change $d\ell = A_\mu(x)dx^\mu \, \ell$, where the function A_μ should be identified with the Maxwell potential. This latter suggestion is established by the formalism within which the above formulae of Maxwellian electrodynamics can be derived including the gauge transformations (1) of the potential, thus leading to a "geometrization" of electromagnetics. Einstein applauded to the admirable depth and boldness of Weyl's mathematical invention, but at the same time recognized the physical failure of the theory, since in Weylian spacetime the length of a rod and speed of a clock would depend on the history, in contrast to the observed uniquely defined frequencies of the spectral lines of chemical elements (the second clock effect in Weylian spacetime can also be considered as a classical analogue of the ► Aharonov-Bohm effect).

G

In 1929, however, and based on earlier work of Fock and London, Weyl found the correct way to establish the idea of "gauging." He realized that one must gauge the internal phase factor of the quantum ► wave function in order to get a recipe to combine a free matter field theory with the theory of electromagnetic interaction. This recipe is nowadays widely known as the "gauge principle" (see next section). In fact, his 1929 paper "Electron and gravitation" must in retrospect be considered one of the cornerstone papers of twentieth century physics, because in it Weyl not only invented the gauge principle, but also developed the first systematic formulation of the spinor and tetrad formalism (cf. [36–40] for the history of gauge theories).

Gauge Principle and Yang–Mills Theories

The Lagrangian of the free Dirac matter field $\mathcal{L}_D = \bar{\psi}(i\gamma^\mu\partial_\mu - m)\psi$ admits global gauge symmetry transformations $\psi' = e^{iq\alpha}\psi$ which form the unitary group $U(1)$. From Noether's first theorem, $j^\mu = q\bar{\psi}\gamma^\mu\psi$ follows as the conserved charge density current. To construct a $U(1)$ gauge theory the invariance of $\mathcal{L}_D$ under local phase transformations

$$\psi'(x) = e^{iq\alpha(x)}\psi(x), \qquad (2)$$

also known as gauge transformations of the first kind, is postulated. This postulate can be fulfilled under the replacement of the usual derivative in $\mathcal{L}_D$ by the *covariant derivative*

$$\partial_\mu \;\rightarrow\; D_\mu = \partial_\mu + iqA_\mu(x) \qquad (3)$$

with a vector field A_μ which itself obeys the local gauge transformations (1) of the second kind. It seems obvious to identify A_μ with the electromagnetic gauge potential and to end up with the total Lagrangian

$$\mathcal{L}_{DM} = \bar{\psi}(i\gamma^\mu\partial_\mu - m)\psi - j_\mu A^\mu - \frac{1}{4}F_{\mu\nu}F^{\mu\nu} \qquad (4)$$

of the combined Dirac-Maxwell matter and interaction field theory obeying full local gauge invariance. This is the idea of the *gauge principle*, its simplest field-theoretic application leads to an abelian $U(1)$ gauge theory, on which quantum electrodynamics is based.

In 1954, Yang and Mills extended the idea to non-abelian gauge groups $SU(n)$. In the Standard Model the theory of electroweak interaction is considered an $SU_I(2) \times U_Y(1)$ of flavor and hypercharge and the theory of strong interaction an $SU_C(3)$ gauge theory of nucleonic color charge. The most important difference to the abelian case is the appearance of an additional term in the potentials B^a_μ in the field strength $F^a_{\mu\nu} = (\partial_\mu B^a_\nu - \partial_\nu B^a_\mu - g_n f^{abc} B^b_\mu B^c_\nu)$ (with couplings g_n and SU(n) structure functions f^{abc}) due to the non-commutativity of the SU(n) generators $\hat{t}^a$ such that only the product $F^a_{\mu\nu}\hat{t}^a$ transforms homogeneously under local gauge transformations and that the Lagrangian $\mathcal{L}_{YM} = -\frac{1}{4} F^{a\mu\nu} F^a_{\mu\nu}$ includes self-interacting terms proportional to $g_n \partial B B^2$ and $g_n^2 B^4$. Hence, the gauge bosons themselves carry charge (cf. [30–35] for modern textbook presentations of gauge theories).

Fibre Bundles and Constrained Systems

The appropriate mathematical description of gauge theories is given within the enlarged geometrical arena of principal fibre bundles and their associated vector bundles [2, 31, 32, 34]. A fibre bundle is a structure $\langle \mathbf{E}, \mathcal{M}, \pi, \mathbb{F}, G \rangle$ with bundle space $\mathbf{E}$, base manifold $\mathcal{M}$, projection map $\pi : \mathbf{E} \to \mathcal{M}$, fibre space $\mathbb{F}$ and structure group G. Fibre bundles can be considered as generalizations of the Cartesian product in the sense that they look locally like $\mathcal{M} \times \mathbb{F}$ (all fibres $\mathbb{F}_p = \pi^{-1}(p)$ at $p \in \mathcal{M}$ being homeomorphic to the typical fibre $\mathbb{F}$). A local trivialisation is given by a ▶ diffeomorphism $\phi_i : \mathcal{U}_i \times \mathbb{F} \to \pi^{-1}(\mathcal{U}_i)$ for some open set $\mathcal{U}_i \subset \mathcal{M}$. In order to obtain the global bundle structure the local chart domains $\mathcal{U}_i$ must be glued together with transition functions $t_{ij}(p) = \left(\phi^{-1} \circ \phi_j\right)(p)$. If the fibre is given by an n dimensional linear vector space $\mathbb{V}^n$ the bundle is called a vector bundle. For a principal bundle $\mathbb{P}(\mathcal{M}, G)$ the fibre $\mathbb{F}$ is identical to the structure group G. To any principal bundle there exists a totality of associated vector bundles with the same structure group and transition functions.

In the Lagrangian view of gauge theories one usually considers fibre bundles over spacetime $\mathcal{M}$ as base space with a continuous Lie group, the gauge group G, as structure group. The connection of the principal bundle $\mathbb{P}(\mathcal{M}, G)$ is physically interpreted as the gauge potential, which takes values in the Lie algebra $\mathfrak{g}$ of G. The generators represent the gauge bosons. The derivative of the connection, the bundle curvature, encodes the interaction field strength. The connection can be thought of as a rule which decomposes the tangent of $\mathbb{P}$ into a horizontal and a vertical part $\mathbb{T}_u\mathbb{P} = \mathbb{V}_u\mathbb{P} \oplus \mathbb{H}_u\mathbb{P}$ for every $u \in \mathbb{P}$, it is defined as a $\mathfrak{g}$-valued one-form projecting $\mathbb{T}_u\mathbb{P}$ to $\mathbb{V}_u\mathbb{P} \cong \mathfrak{g}$. This idea is also expressed by the covariant derivative (2). Matter fields are defined as (local) sections in some associated bundle $\mathbf{E}$ of $\mathbb{P}$, usually a vector bundle. A fibre bundle section is defined as a mapping $\sigma : \mathcal{M} \rightarrow \mathbf{E}$ and can be thought of as a generalization of a tangent vector field. With $\pi(\sigma(p)) = p$ the section $\sigma(p) \in \mathbb{F}_p$ is local. A principal bundle is trivial, if it admits a global section.

Phenomenological high energy physics mostly uses the Lagrangian formulation of gauge theories, but for certain purposes, in particular for the formulation of canonical general relativity, the Hamiltonian approach seems better suited (cf. [33]). Earman [7, 8] has argued at length for the appropriateness of the Hamiltonian view also for the purposes of philosophy of physics because of its mathematical rigor. The transition from the Lagrangian velocity phase space $\mathcal{V}(\mathbf{q}, \dot{\mathbf{q}})$ to the Hamiltonian phase space $\Gamma(\mathbf{q}, \mathbf{p})$ is mediated by a Legendre transformation. If Noether's second theorem applies, the canonical momenta $\mathbf{p} = \frac{\partial L}{\partial \dot{\mathbf{q}}}$ are not independent and primary constraints $\varphi(\mathbf{q}, \mathbf{p}) = 0$ exist. These constraints generate gauge transformations (elements of G) which form gauge orbits $[\mathbf{p}]$ (equivalence classes of $\mathbf{p}$ under G), such that one ends up with a reduced phase space $\tilde{\Gamma} = \Gamma/G$. For instance, in Maxwellian electrodynamics in vacuo the canonical variable $\vec{E}$ is subject to the constraint $div\ E = 0$.

The Interpretation of Gauge Symmetry

A first point of interest is whether local gauge transformations are observable. Textbooks sometimes give the false impression that this could indeed be the case, since it is for instance possible to change the interference pattern in a ► double-slit experiment by inserting a phase shifter. Such a device, however, does not instantiate a local phase transformation, but rather a relative phase change between the two parts corresponding to the two slits of the total wave function $\psi = \psi_I + \psi_{II}$. In particular, as Brading and Brown [4] have pointed out, the phase of ψ_I at some point on the interference screen will be changed under a local gauge transformation by the same amount as the phase of ψ_{II} at that same point.

Philosophy of physics has especially focussed on the logic of the gauge principle. There is a certain consensus (cf. [5, 11, 17, 20, 25] that in a wide variety of the textbook literature the gauge principle is overstated, since it is sometimes said to "dictate" the interaction from the mere requirement of local gauge invariance. However, let $|x\rangle$ be the position representation of a wave function $\Psi(x) = \langle x|\phi\rangle$, where $\{|\phi\rangle\}$ span an abstract ► Hilbert space, then local gauge transformations $|x'\rangle = e^{i\chi(x)}|x\rangle = \hat{U}|x\rangle$ must properly be seen as mere changes in $|x\rangle$. Such a change of representation affects the ► operators as well, which generally transform as $\hat{O}' = \hat{U}\hat{O}\hat{U}^{+}$. In the particular case of the derivative or momentum operator one gets the covariant derivative as a result, which is thus uncovered as a mere change in the position representation. In fibre bundle terminology, this amounts to saying that the inhomogeneous term in the covariant derivative (2) includes a flat connection only, where the corresponding curvature or gauge field strength is still zero. Hence, no non-vanishing gauge field is enforced by the requirement of local gauge symmetry.

Gauge symmetry structure is, as Redhead [24] has dubbed it, mere surplus structure. Only the gauge-invariant quantities figure as candidates for observable entities. ► Quantum Electrodynamics, for instance, is quite aptly characterized as a $U(1)$ gauge theory, insofar as the $U(1)$-invariant tensor $F^{\mu\nu}$ can found to be realized in nature as the electrodynamic field strength. Unfortunately, up to now there seems to exist no straightforward procedure to identify the symmetries which are gauge as opposed to other, empirically significant symmetries in a given theoretical framework.

Because of the gauge freedom of constrained Hamiltonian systems there exists no unique system evolution in phase space, but rather an indeterministic time-evolution where a unique phase space point p_t must be replaced by a gauge orbit $[p_t]$. Earman [7,8] has pointed out that this breakdown of determinism is a general feature of the gauge freedom of constraint Hamiltonian systems (in analogy to the notorious "hole argument" based on the Leibniz equivalence of diffeomorphic models of spacetime theories). The real conceptual problem here is to develop general rules for deciding whether certain transformations in the mathematical apparatus of physical theories are gauge transformations or not.

Another philosophical debate concerns the question about the genuine entities in gauge theories. Here the variety of answers spans a whole spectrum. The genuine candidate for the basic entity in field theories is the field strength as a more or less directly measurable quantity. In view of the typical gauge-theoretic non-local effects such as the Aharonov-Bohm effect, many authors favor the gauge potential as the basic entity (► Aharonov-Bohm effect), which is, however, gauge-dependent and not directly observable. A third option concerns holonomies or Wilson-loops as non-separable but gauge-invariant entities ([3], particularly Healey [11–13]). Further proposals consider the whole fibre bundle structure [23] or the retarded Greens function representation of the charge distribution [21] up to the view that we are dealing with a genuine case of ontological indeterminacy and that we should direct our ontological commitment only at the group theoretic, structural content of gauge theories in the sense of structural realism [18]. Obviously, the debate about the ontology of gauge theories has not been settled.

Gauge Theories of Gravity

General relativity can in fact be considered a gauge theory proper, not in the above sense of a quantum but a classical gauge field theory. An informal application of the gauge principle starts from spinless matter following trajectories described by the geodesic equation $\frac{\mathrm{d}}{\mathrm{d}\tau} v^{\mu}(\tau) = 0$ with four-velocity v^{μ} in flat Minkowski spacetime. The formal transition from special to general relativity basically amounts to replacing partial by covariant derivatives. Geodesic trajectories on curved spacetime are thus described by $\frac{\mathrm{d}}{\mathrm{d}\tau} v^{\mu}(\tau) + \{\}^{\mu}_{\nu\rho}\, v^{\nu}(\tau)\, v^{\rho}(\tau) = 0$, where the Christoffel symbols of the connection are derived from the metric according to $\{\}^{\lambda}_{\mu\nu} = \frac{1}{2}\, g^{\lambda\sigma}(\partial_{\mu} g_{\sigma\nu} + \partial_{\nu} g_{\sigma\mu} - \partial_{\sigma} g_{\mu\nu})$. Under coordinate transformations

$x \to x'$ the connection transforms inhomogeneously as $\{\}^{\lambda}_{\mu\nu} \to \{\}'^{\lambda}_{\mu\nu} = \frac{\partial x'^{\lambda}}{\partial x^{\rho}} \frac{\partial x^{\sigma}}{\partial x'^{\mu}} \frac{\partial x^{\tau}}{\partial x'^{\nu}} \{\}^{\rho}_{\sigma\tau} + \frac{\partial x'^{\lambda}}{\partial x^{\rho}} \frac{\partial^2 x^{\rho}}{\partial x'^{\mu} \partial x'^{\nu}}$, whereas the Riemann curvature tensor $R^{\kappa}{}_{\lambda\mu\nu} = \partial_{\mu}\{\}^{\kappa}_{\lambda\nu} - \partial_{\nu}\{\}^{\kappa}_{\lambda\mu} + \{\}^{\rho}_{\lambda\nu}\{\}^{\kappa}_{\rho\mu} - \{\}^{\rho}_{\lambda\mu}\{\}^{\kappa}_{\rho\nu}$ as well as the Ricci tensor $R_{\mu\nu} = R^{\lambda}{}_{\mu\lambda\nu}$ and the Ricci scalar $R = R^{\mu}{}_{\mu}$ all transform homogeneously. They can be used in the Einstein-Hilbert Lagrangian $\mathcal{L}_{GR} \sim \sqrt{-g}\, R$, where g is the determinant of the metric, leading to the Einstein equations.

Here again the mere appearance of Christoffel symbols in the geodesic equation cannot enforce spacetime to be curved, but rather ensures a covariant, i.e. coordinate independent, representation. Historically, the first attempt to gauge gravity goes back to Utiyama [26], who considered a gauge theory of the (homogeneous) Lorentz group. It is a remarkable feature of gravitational gauge theories that the choice of the kinetic term for the gauge fields and the corresponding gauge group is far less restricted than in the Yang-Mills case. Cho [6] has developed first a pure translational gauge field theory with a particular choice of a quadratic Lagrangian. In such a gauge theory of the four-dimensional translation group $\mathbb{R}^{1,3}$ one does not end up with a curved Riemann space but rather a flat teleparallel Weitzenböck space, where the gravitational field strength is represented by the torsion instead of the curvature tensor. There is an ongoing debate whether both approaches can in fact shown to be empirically equivalent [10, 22], rendering the ontology of gravity – curvature or torsion – indetermined [16, 19]. In search of a more fundamental physics various accounts of extended gauge groups such as for instance affine or super groups have been considered (cf. [15] and [14] for overviews).

The issue of gauging gravity is also intimately connected to the longstanding debate about the status of the requirement of general covariance and the distinction between ► covariance and ► symmetry groups, whether gauge or not. While any sensible physical theory should allow for a generally covariant formulation, in general relativity the diffeomorphism group seems to play a double role as covariance and gauge group. A recent discussion, following Anderson's [1] classic distinction between dynamic and absolute objects, has been given by Guilini [9].

G

Primary Literature

1. J. L. Anderson: *Principles of Relativity Physics.* Academic, New York (1967).
2. S. Y. Auyang: *How is Quantum Field Theory Possible?* Oxford University Press, New York (1995).
3. G. Belot: *Understanding Electromagnetism.* The British Journal for the Philosophy of Science. 49: 531–55 (1998).
4. K. Brading, H. R. Brown: *Are Gauge Symmetry Transformations Observable?* The British Journal for the Philosophy of Science. 55 (4): 645–65 (2004).
5. H. R. Brown: *Aspects of Objectivity in Quantum Mechanics.* In J. Butterfield, Pagonis, C., editors. From Physics to Philosophy. Cambridge University Press, Cambridge (1999).
6. Y. M. Cho: *Einstein Lagrangian as the translational Yang-Mills Lagrangian.* Physical Review. D 14 (10): 2521–525 (1976).
7. J. Earman: *Gauge Matters.* Philosophy of Science. 69 (3), Suppl.: S209–S220 (2002).

8. J. Earman: *Tracking Down Gauge: An Ode to the Constrained Hamiltonian Formalism.* In: K. Brading, E. Castellani, editors. Symmetries in Physics: Philosophical Reflections. Cambridge University Press, Cambridge (2003).
9. D. Giulini: *Remarks on the Notions of General Covariance and Background Independence.* In: E. Seiler, I.-O. Stamatescu, editors. Approaches to Fundamental Physics. Springer, Berlin (2007).
10. K. Hayashi, T. Shirafuji: *New General Relativity*. Physical Review. D 19 (12): 3524–553 (1979).
11. R. Healey: *On the Reality of Gauge Potentials*. Philosophy of Science. 68 (4): 432–55 (2001).
12. R. Healey: *Gauge Theories and Holisms*. Studies in History and Philosophy of Modern Physics. 35 (4): 619–42 (2004).
13. R. Healey: *Gauging What's Real*. Oxford University Press, New York (2007).
14. F. W. Hehl, J. D. McCrea, E. W. Mielke, Y. Ne'eman: *Metric-Affine Gauge Theory of Gravity: Field Equations, Noether Identities, World Spinors, and Breaking of Dilation Invariance*. Physics Reports. 258: 1–171 (1995).
15. D. Ivanenko, G. Sardanashvily: *The Gauge Treatment of Gravity*. Physics Reports. 94 (1): 1–45 (1983).
16. C. Liu: *Gauge Gravity and the Unification of Natural Forces*. International Studies in the Philosophy of Science. 17 (2): 143–59 (2003).
17. H. Lyre: *The Principles of Gauging. Philosophy of Science*. 68 (3) Suppl.: S371–S381 (2001).
18. H. Lyre: *Holism and Structuralism in U(1) Gauge Theory*. Studies in History and Philosophy of Modern Physics. 35 (4): 643–70 (2004).
19. H. Lyre, T. O. Eynck: *Curve It, Gauge It, or Leave It? Underdetermination in Gravitational Theories*. Journal for General Philosophy of Science. 34 (2): 277–303 (2003).
20. C. Martin: *Gauge Principles, Gauge Arguments and the Logic of Nature*. Philosophy of Science. 69 (3), Suppl.: S221–S234 (2002).
21. J. Mattingly: *Which Gauge Matters*. Studies in History and Philosophy of Modern Physics. 37 (2): 243–62 (2006).
22. J. M. Nester: *Is there Really a Problem with the Teleparallel Theory?* Classical and Quantum Gravity. 5: 1003–010 (1988).
23. A. Nounou: *A Fourth Way to the Aharonov-Bohm Effect*. In: K. Brading, E. Castellani, editors. Symmetries in Physics: Philosophical Reflections. Cambridge University Press, Cambridge (2003).
24. M. Redhead: *The Interpretation of Gauge Symmetry*. In: M. Kuhlmann, H. Lyre, A. Wayne, editors. Ontological Aspects of Quantum Field Theory. World Scientific, Singapore (2002).
25. P. Teller: *The Gauge Argument*. Philosophy of Science. 67 (3), Suppl.: S466–S481 (2000).
26. R. Utiyama: *Invariant Theoretical Interpretation of Interaction*. Physical Review. 101 (5): 1597–1607 (1956).
27. T. T. Wu, C. N. Yang: *Concept of Nonintegrable Phase Factors and Global Formulation of Gauge Fields*. Physical Review. D 12 (12): 3845–857 (1975).
28. C. N. Yang: *Integral Formalism for Gauge Fields*. Physical Review Letters. 33 (7): 445–447 (1974).
29. H. Weyl: *Gravitation und Elektrizität. Sitzungsberichte der Preußischen Akademie der Wissenschaften*. p. 465–480 (1918).

Secondary Literature

30. T.-P. Cheng, L.-F. Li: *Gauge Theory of Elementary Particle Physics: Problems and Solutions*. Oxford University Press, New York (2000).
31. M. Göckeler, T. Schücker: *Differential Geometry, Gauge Theories and Gravity*. Cambridge University Press, Cambridge (1987).

32. R. Gambini, J. Pullin: *Loops, Knots, Gauge Theories and Quantum Gravity*. Cambridge University Press, Cambridge (1996).
33. M. Henneaux, C. Teitelboim: *Quantization of Gauge Systems*. Princeton University Press, Princeton (1992).
34. M. Nakahara: *Geometry, Topology and Physics*. IOP Publishing, Bristol (1990).
35. L. H. Ryder: *Quantum Field Theory*. Cambridge University Press, Cambridge (1985, ^{2}1996).
36. T. Y. Cao: *Conceptual Developments of 20th Century Field Theories*. Cambridge University Press, Cambridge (1997).
37. L. O'Raifeartaigh: *The Dawning of Gauge Theory*. Princeton University Press, Princeton (1995).
38. L. O'Raifeartaigh, N. Straumann: *Gauge Theory: Historical Origins and Some Modern Developments*. Reviews of Modern Physics. 72 (1): 1–23 (2000).
39. J. C. Taylor: *Gauge Theories in the Twentieth Century*. World Scientific, Singapore (2001).
40. V. P. Vizgin: *Unified Field Theories*. Birkhäuser, Basel (1994).

Generalizations of Quantum Statistics

O.W. Greenberg

The general principles of quantum theory allow statistics more general than bosons or fermions. (▸ Bose-Einstein statistics and ▸ Fermi statistics are discussed in separate articles.) The restriction to Bosons or Fermions requires the symmetrization postulate, "the states of a system containing N identical particles are necessarily either all symmetric or all antisymmetric under permutations of the N particles," or, equivalently, "all states of identical particles are in one-dimensional representations of the symmetric group [1]." Messiah and Greenberg discussed quantum mechanics without the symmetrization postulate [2]. The spin-statistics connection, that integer ▸ spin particles are bosons and odd-half-integer spin particles are fermions [3], is an independent statement. Identical particles in 2 space dimensions are a special case, "▸ anyons." Braid group statistics, a nonabelian analog of anyons, are also special to 2 space dimensions.

All ▸ observables must be symmetric in the dynamical variables associated with identical particles. Observables can not change the permutation symmetry type of the wave function; i.e. there is a superselection rule separating states in inequivalent representations of the symmetric group and when identical particles can occur in states that violate the ▸ spin statistics theorem their transitions must occur in the same representation of the symmetric group. One can not introduce a small violation of statistics by assuming the Hamiltonian is the sum of a statistics-conserving and a small statistics-violating term, $H = H_S + \epsilon H_V$, as one can for violations of ▸ parity, charge conjugation, etc. Violation of statistics must be introduced in a more subtle way.

Doplicher et al. [4, 5] classified identical particle statistics in 3 or more space dimensions. They found parabose and parafermi statistics of positive integer orders, which had been introduced by Green [6], and infinite statistics, which had been introduced by Greenberg [7, 8]. Parabose (parafermi) statistics allows up to p identical particles in an antisymmetric state (symmetric) state. Infinite statistics allows any number of identical particles in a symmetric or antisymmetric state.

Trilinear commutation relations,

$$[[a_k^\dagger, a_l]_\pm, a_m^\dagger]_- = 2\delta_{lm} a_k^\dagger \tag{1}$$

with the vacuum condition, $a_k|0\rangle = 0$, and single-particle condition, $a_k a_l^\dagger|0\rangle = p\delta_{kl}|0\rangle$, define the Fock representation of order p parabose (parafermi) statistics. Green found two infinite sets of solutions of these commutation rules, one set for each positive integer p, by the ansatz,

$$a_k^\dagger = \sum_{\alpha=1}^{p} b_k^{(\alpha)\dagger}, \quad a_k = \sum_{\alpha=1}^{p} b_k^{(\alpha)}, \tag{2}$$

where the $b_k^{(\alpha)}$ and $b_k^{(\beta)\dagger}$ are bose (fermi) operators for $\alpha = \beta$ but anticommute (commute) for $\alpha \neq \beta$ for the parabose (parafermi) cases. The integer p is the order of the parastatistics. For parabosons (parafermions) p is the maximum number of particles that can occupy an antisymmetric (symmetric) state. The case $p = 1$ corresponds to the usual Bose or Fermi statistics. Greenberg and Messiah [9] proved that Green's ansatz gives all Fock-like solutions of Green's commutation rules. Local observables in parastatistics have a form analogous to the usual ones; for example, the local current for a spin-1/2 theory is $j_\mu = (1/2)[\bar{\psi}(x), \psi(x)]_-$. From Green's ansatz, it is clear that the squares of all norms of states are positive; thus parastatistics [10] gives a set of orthodox positive metric theories. Parabose or parafermi statistics for $p > 1$ give gross violations of Bose or Fermi statistics so that parastatistics theories are not useful to parametrize small violations of statistics.

The bilinear commutation relation

$$a(k)a^\dagger(l) - qa^\dagger(l)a(k) = \delta(k, l), \tag{3}$$

with the vacuum condition, $a(k)|0\rangle = 0$, define the Fock representation of quon statistics. Positivity of norms requires $-1 \leqslant q \leqslant 1$ [11, 12]. Outside this range the squared norms become negative. There is no commutation relation involving two a's or two $a^\dagger$'s. There are $n!$ linearly independent n-particle states in ► Hilbert space if all ► quantum numbers are distinct; these states differ only by permutations of the order of the ► creation operators.

For $q \approx \pm 1$, quons provide a formalism that can parametrize small violations of statistics so that quons are useful for quantitative tests of statistics. At $q = 1(-1)$ only the symmetric (antisymmetric) representation of $\mathcal{S}_n$ occurs. The quon operators interpolate smoothly between fermi and bose statistics in the sense that as $q \to \mp 1$ the antisymmetric (symmetric) representations smoothly become more heavily weighted.

Although there are $n!$ linearly independent vectors in Fock space associated with a degree n monomial in ► creation operators that carry disjoint quantum numbers acting on the vacuum, there are fewer than $n!$ observables associated with such vectors. The general observable is a linear combination of projectors on the irreducibles of the symmetric group.

A convenient way to parametrize violations or bounds on violations of statistics uses the two-particle density matrix. For fermions, $\rho_2 = (1 - v_F)\rho_a + v_F\rho_s$; for bosons, $\rho_2 = (1-v_B)\rho_s + v_B\rho_a$. In each case the violation parameter varies between zero if the statistics is not violated and one if the statistics is completely violated. R.C. Hilborn [13] pointed out that the transition matrix elements between symmetric (antisymmetric) states are proportional to $(1 \pm q)$ so that the transition probabilities are proportional to $(1 \pm q)^2$ rather than to $(1 \pm q)$.

G

Several properties of kinematically relativistic quon theories hold, including a generalization of Wick's theorem, cluster decomposition theorems and (at least for free quon fields) the ► CPT theorem; however ► locality in the sense of the commutativity of ► observables at spacelike separation fails [7]. The nonrelativistic form of locality

$$[\rho(x), \psi^\dagger(y)]_- = \delta(x-y)\psi^\dagger(y), \tag{4}$$

where ρ is the charge density, does hold.

Greenberg and Hilborn [14] derived the generalization of the result due to Wigner [15] and to Ehrenfest and Oppenheimer [16] that a bound state of bosons and fermions is a boson unless it has an odd number of fermions, in which case it is a fermion generalizes for quons: A bound state of n identical quons with parameter $q_{\text{constituent}}$ has parameter $q_{\text{bound}} = q_{\text{constituent}}^{n^2}$ [14].

Primary Literature

1. A.M.L. Messiah: *Quantum Mechanics, Vol. II* (North-Holland, Amsterdam, 1962), p 595
2. A.M.L. Messiah and O.W. Greenberg: Symmetrization Postulate and Its Experimental Foundation. Phys. Rev. **136**, B248–B267 (1964)
3. W. Pauli: The Connection Between Spin and Statistics. Phys. Rev. **58**, 716–722 (1940)
4. S. Doplicher, R. Haag, J. Roberts: Local Observables and Particle Statistics I. Commun. Math. Phys. **23**, 199–230 (1971)
5. S. Doplicher, R. Haag, J. Roberts: Local Observables and Particle Statistics I. Commun. Math. Phys. **35**, 49–85 (1974)
6. H.S. Green: A Generalized method of field quantization. Phys. Rev **90**, 270–273 (1952)
7. O.W. Greenberg: Example of Infinite Statistics. Phys. Rev. Lett. **64**, 705–708 (1990)
8. O.W. Greenberg: Particles with Small Violations of Fermi or Bose Statistics. Phys. Rev. D **43**, 4111–4120 (1991)
9. O.W. Greenberg and A.M.L. Messiah: Selection Rules for Parafields and the Absence of Para Particles in Nature. Phys. Rev. **138**, B1155–B1167 (1965)
10. G.F. Dell' Antonio, O.W. Greenberg and E.C.G. Sudarshan: Parastatistics: Axiomatic Formulation, Connection with Spin and Statistics and TCP Theorem for the General Field Theory. In *Group Theoretical Concepts and Methods in Elementary Particle Physics*, ed. F. Gürsey, Gordon and Breach, New York, 1964, p 403, introduced the word "parastatistics"
11. D. Zagier: Realizability of a Model in Infinite statistics. Commun. Math. Phys. **147**, 199–210 (1992)

12. R. Speicher: Generalized Statistics of Macroscopic Fields. Lett. Math. Phys. **27**, 97–104 (1993)
13. R.C. Hilborn: Connecting q-Mutator Theory with Experimental Tests of the Spin-Statistics Connection. in *Spin-Statistics Connection and Commutation Relations*, ed. R.C. Hilborn and G.M. Tino (American Institute of Physics, Melville, NY, 2000), pp 128–141
14. O.W. Greenberg and R.C. Hilborn: Quon Statistics for Composite Systems and a Limit on the Violation of the Pauli Principle for Nucleons and Quarks. Phys. Rev. Lett. **83**, 4460–4463 (1999)
15. E.P. Wigner: Math. und Naturwiss. Anzeiger der Ungar. Ak. der Wiss. **46**, 576 (1929); translated in *Collected Works of E.P. Wigner*
16. P. Ehrenfest and J.R. Oppenheimer: Note on the Statistics' of Nuclei. Phys. Rev. **37**, 333–338 (1931)

Secondary Literature

17. O.W. Greenberg, D.M. Greenberger and T.V. Greenbergest: (Para)bosons, (Para)fermions, Quons and other Beasts in the Managerie of Particle Statistics, in *Quantum Coherence and Reality*, ed. J.S. Anandan and J.L. Safko, (World Scientific, Singapore, 1994), pp 301–312
18. Y. Ohnuki and S. Kamefuchi: *Quantum Field Theory and Parastatistics*, (University of Tokyo, Berlin, 1982)
19. R. Haag: *Local Quantum Physics*, (Springer, Berlin, 1992)
20. R.C. Hilborn and G.M. Tino, editors: *Spin-Statistics Connection and Commutation Relations*, (American Institute of Physics, Melville, NY, 2000)

GHZ (Greenberger–Horne–Zeilinger) Theorem and GHZ States

Daniel M. Greenberger

The GHZ states (Greenberger–Horne–Zeilinger states) are a set of *entangled states* that can be used to prove the GHZ theorem, which is a significant improvement over ► *Bell's Theorem* as a way to disprove the concept of "elements of reality", a concept introduced by ► *EPR* problem (Einstein–Podolsky–Rosen) in their attempt to prove that quantum theory is incomplete. Conceding that they did not quite know what "reality" is, EPR nonetheless said that it had to contain an "element of reality" as one of its properties. This was that if one could discover a property of a system (i.e., predict it with 100% certainty) by making an experiment elsewhere, that in no way interacted with the system, then this property was an element of reality. The argument was that since one had not in any way interacted with the system, then one could not have affected this property, and so the property must have existed before one performed one's experiment. Thus the property is an intrinsic part of the system, and not an artifact of the measurement one made.

From a common-sense point of view, this proposition seems unassailable, and yet quantum theory denies it. For example, in the Bohm form of the EPR experiment, one has a particle that decays into two, that go off in opposite directions. If the original particle had ▶ spin 0, while each of the two daughters has spin 1/2, then if the one going to the right has its spin up, the one going to the left will have its spin down, and vice-versa. So the spin of each of the daughters is an element of reality, because if one measures the spin of the particle on the right as up, one can predict with 100% certainty that the other will be spin down, etc. EPR would conclude from this that, since we did not interfere with the particle on the left in any way, then we could not have changed its spin, and so it had to have been spin down from the moment the original particle decayed.

How can quantum theory deny this? By pointing out that since the original spin was 0, we did not have to measure the spin of the particle on the right as up or down, but we could have measured it at 90° from the vertical. Then the particle on the left would be 90° from the vertical in the opposite direction. In fact we could have measured the spin of the particle on the right in any direction, and the one on the left would be opposite it. (This is because in quantum theory, there are only two possibilities for the spin in any direction, up along that direction, or down, opposite it.) So how could the particle on the left know in which direction we were going to measure the particle on the right? Therefore the direction of its spin can not be said to exist until after the spin direction of the particle on the right is measured. Now this argument also seems unassailable, although it leads to the exact opposite conclusion from that of EPR, namely that the state of a particle cannot be defined until a measurement is made on it. And so the EPR argument has fascinated physicists since it was first given, in 1935.

G

Until Bell's theorem in 1964, it did not seem that the conflict here was experimentally decidable. But Bell took the EPR argument seriously, and saw that together with completeness, another postulate of EPR (all elements of reality must have some counterpart in a complete theory), it implied that there must exist some function $A(\alpha, \lambda)$, where α represents the angle along which the spin of the particle on the right is measured, and λ represents any other parameters that must be set to determine the outcome of the measurement. (These are now called ▶ hidden variables). The result of the experiment, the possible values for A, can only be ± 1, representing the two possible outcomes, up or down. There is a similar function representing the particle on the left, $B(\beta, \lambda)$, where β is the angle along which its spin will be measured, and the value of λ is set by nature when the particle decays.

In any given decay, one can measure the spin of the two particles along any two directions, α and β, and one will obtain the product $A(\alpha, \lambda)B(\beta, \lambda)$, as the result of the measurement. Then when one takes the result of many measurements, one will obtain an average of this product as

$$E(\alpha, \beta) = \int \mathrm{d}\lambda \rho(\lambda) A(\alpha, \lambda) B(\beta, \lambda),$$

where $\rho(\lambda)$ is some positive weighting function over the λ's, since we cannot know how often each value of λ will occur. The only limitation on this average is that when $\beta = \alpha$, then $E = -1$, since this is the condition imposed by the fact that the original particle has spin 0, and if you measure the two daughters along the same direction, the spins will be opposite each other. Equivalently, if $\beta = \alpha \pm \pi$, then $E = +1$. These two cases are known as the "perfect correlation" cases, since they represent the case where an element of reality exists, and one can predict the outcome for the product with 100% certainty. (That the function A depends only on α, and not on β, is known as the ► locality, which we have also taken to be true.)

From this form for $E(\alpha, \beta)$, as a weighted product over A and B, Bell was able to prove an inequality that the average function, E, had to satisfy, which has come to be known as the Bell inequality. Any realistic description based on the EPR elements of reality must obey this inequality. But the quantum theory expectation value violates this inequality for most sets of angles (α, β), and thus the Bell inequality established an experimental test to determine whether the EPR postulates were correct or not. The long experimental history of making the inequality experimentally useful, and the subsequent confirming of quantum theory is a fascinating tale, but it is not our concern here. Here we merely note that it is ironical that when $\beta = \alpha$, the perfect correlation case that inspired the controversy, the Bell inequality is not violated. This is because in this case it is easy to make a realistic model that explains the result. The violation occurs when one takes arbitrary angles.

The GHZ theorem concerns three particles. It considers only perfect correlations, so one does not have to take an average over many experiments. In theory one could use only a single event to prove a contradiction with the EPR result, although in practice one always needs statistics in an experiment. The GHZ theorem shows that one can construct three-particle situations in which there are perfect correlations (meaning that by measuring two particles, one can make a prediction with 100% certainty what a measurement of the third particle will yield), in which a classical, realistic interpretation will yield a particular result, while quantum mechanics predicts the exactly opposite result.

We will give a very clever version of the experiment, due to David Mermin. Consider three spin 1/2 particles. Now look at the four Hermitian operators A, B, C, D, which represent ► observables, and which are defined as

$$A = \sigma_x^1 \sigma_y^2 \sigma_y^3, \quad B = \sigma_y^1 \sigma_x^2 \sigma_y^3, \quad C = \sigma_y^1 \sigma_y^2 \sigma_x^3, \quad D = \sigma_x^1 \sigma_x^2 \sigma_x^3.$$

Here the σ's are the ► Pauli spin matrices, and the superscripts tell which particle the matrix operates on, while the subscripts define the component of the spin. All these ► operators commute with each other:

$$\begin{aligned}
AB &= \sigma_x^1 \sigma_y^2 \sigma_y^3 \sigma_y^1 \sigma_x^2 \sigma_y^3 = (\sigma_x^1 \sigma_y^1)(\sigma_y^2 \sigma_x^2)(\sigma_y^3 \sigma_y^3) = (\mathrm{i}\sigma_z^1)(-\mathrm{i}\sigma_z^2)1^3 = \sigma_z^1 \sigma_z^2, \\
BA &= \sigma_y^1 \sigma_x^2 \sigma_y^3 \sigma_x^1 \sigma_y^2 \sigma_y^3 = (\sigma_y^1 \sigma_x^1)(\sigma_x^2 \sigma_y^2)(\sigma_y^3 \sigma_y^3) = (-\mathrm{i}\sigma_z^1)(\mathrm{i}\sigma_z^2)1^3 = \sigma_z^1 \sigma_z^2, \\
&[A, B] = 0 = [A, C] = [B, C], \\
AD &= \sigma_x^1 \sigma_y^2 \sigma_y^3 \sigma_x^1 \sigma_x^2 \sigma_x^3 = 1^1(-\mathrm{i}\sigma_z^2)(-\mathrm{i}\sigma_z^3) = -\sigma_z^2 \sigma_z^3, \\
DA &= \sigma_x^1 \sigma_x^2 \sigma_x^3 \sigma_x^1 \sigma_y^2 \sigma_y^3 = 1^1(\mathrm{i}\sigma_z^2)(\mathrm{i}\sigma_z^3) = -\sigma_z^2 \sigma_z^3, \\
&[A, D] = 0 = [B, D] = [C, D].
\end{aligned}$$

(Here 1^3 means unity for particle 3, which is unity.) Thus all the operators commute and they can all be measured at the same time, and simultaneously diagonalized. Finally, their product satisfies the relation

$$
\begin{aligned}
ABCD &= \sigma_x^1\sigma_y^2\sigma_y^3\sigma_y^1\sigma_x^2\sigma_y^3\sigma_y^1\sigma_y^2\sigma_x^3\sigma_x^1\sigma_x^2\sigma_x^3 \\
&= (\sigma_x^1\sigma_y^1\sigma_y^1\sigma_x^1)(\sigma_y^2\sigma_x^2\sigma_y^2\sigma_x^2)(\sigma_y^3\sigma_y^3\sigma_x^3\sigma_x^3) = 1^1(-\mathrm{i}\sigma_z^2)(-\mathrm{i}\sigma_z^2)1^3 \\
&= -1^1 1^2 1^3 = -1 .
\end{aligned}
$$

The above is a quantum calculation. From the point of view of a classical, realistic theory, if one measures σ_x^1, the x component of the spin for particle 1, one will get m_x^1, which $= \pm 1$. Thus, if one measures the operator ABCD, one will get

$$
ABCD = (m_x^1 m_y^2 m_y^3)(m_y^1 m_x^2 m_y^3)(m_y^1 m_y^2 m_x^3)(m_x^1 m_x^2 m_x^3) = +1 .
$$

The product must be $= +1$, because every term appears in the product twice. But quantum mechanically, the product is -1. The reason for the difference between this and the quantum result, -1, is that even though one can make all the measurements at the same time quantum mechanically, all the spin components do not commute. (One must measure the operators A, B, C, and D, not the individual particle spins.) Thus in principle, we could make this one measurement of ABCD, and distinguish between the EPR view of reality and the quantum-mechanical one.

What are the quantum mechanical states that simultaneously diagonalize the operators A, B, C, and D? The particles cannot be in one of the states of say σ_x^1, because then one could not at the same time measure σ_y^1. So the particle cannot be in any one state, but must be in a state that is not a simple product of the states of each of the particles. In other words, it must be in an an entangled state (▶ entanglement). We call the spin states $|m_z^1 = +1\rangle \equiv |\uparrow^1\rangle$, $|m_z^1 = -1\rangle \equiv |\downarrow^1\rangle$, and simplify further by leaving out the superscripts for the particles, so that we merely denote the state $|\uparrow^1\rangle|\downarrow^2\rangle|\uparrow^3\rangle \equiv |\uparrow\downarrow\uparrow\rangle$.

Then we can use the properties of the spin states, namely

$$
\begin{aligned}
\sigma_x|\uparrow\rangle &= |\downarrow\rangle , & \sigma_x|\downarrow\rangle &= |\uparrow\rangle , \\
\sigma_y|\uparrow\rangle &= \mathrm{i}|\downarrow\rangle , & \sigma_y|\downarrow\rangle &= -\mathrm{i}|\uparrow\rangle , \\
\sigma_z|\uparrow\rangle &= |\uparrow\rangle , & \sigma_z|\downarrow\rangle &= -|\downarrow\rangle ,
\end{aligned}
$$

to verify that the state $|\psi_1\rangle = \frac{1}{\sqrt{2}}(|\uparrow\uparrow\uparrow\rangle + |\downarrow\downarrow\downarrow\rangle)$ satisfies

$$
\begin{aligned}
A|\psi_1\rangle &= \tfrac{1}{\sqrt{2}}\sigma_x^1\sigma_y^2\sigma_y^3(|\uparrow\uparrow\uparrow\rangle + (\downarrow\downarrow\downarrow\rangle) = \tfrac{1}{\sqrt{2}}(-|\downarrow\downarrow\downarrow\rangle - |\uparrow\uparrow\uparrow\rangle) = -|\psi_1\rangle , \\
B|\psi_1\rangle &= \tfrac{1}{\sqrt{2}}\sigma_y^1\sigma_x^2\sigma_y^3(|\uparrow\uparrow\uparrow\rangle + (\downarrow\downarrow\downarrow\rangle) = -|\psi_1\rangle , \\
C|\psi_1\rangle &= \sigma_y^1\sigma_y^2\sigma_x^3|\psi_1\rangle = -|\psi_1\rangle , \\
D|\psi_1\rangle &= \sigma_x^1\sigma_x^2\sigma_x^3|\psi_1\rangle = +|\psi_1\rangle .
\end{aligned}
$$

So this state diagonalizes each of the operators A, B, C, and D. So does the state $|\psi_2\rangle = \frac{1}{\sqrt{2}}(|\uparrow\uparrow\uparrow\rangle - |\downarrow\downarrow\downarrow\rangle)$. And in fact so do all the eight states

$$
\begin{aligned}
|\psi_1\rangle &= \tfrac{1}{\sqrt{2}}(|\uparrow\uparrow\uparrow\rangle + |\downarrow\downarrow\downarrow\rangle)\,, & |\psi_2\rangle &= \tfrac{1}{\sqrt{2}}(|\uparrow\uparrow\uparrow\rangle - |\downarrow\downarrow\downarrow\rangle)\,,\\
|\psi_3\rangle &= \tfrac{1}{\sqrt{2}}(|\uparrow\uparrow\downarrow\rangle + |\downarrow\downarrow\uparrow\rangle)\,, & |\psi_4\rangle &= \tfrac{1}{\sqrt{2}}(|\uparrow\uparrow\downarrow\rangle - |\downarrow\downarrow\uparrow\rangle)\,,\\
|\psi_5\rangle &= \tfrac{1}{\sqrt{2}}(|\uparrow\downarrow\uparrow\rangle + |\downarrow\uparrow\downarrow\rangle)\,, & |\psi_6\rangle &= \tfrac{1}{\sqrt{2}}(|\uparrow\downarrow\uparrow\rangle - |\downarrow\uparrow\downarrow\rangle)\,,\\
|\psi_7\rangle &= \tfrac{1}{\sqrt{2}}(|\downarrow\uparrow\uparrow\rangle + |\uparrow\downarrow\downarrow\rangle)\,, & |\psi_8\rangle &= \tfrac{1}{\sqrt{2}}(|\downarrow\uparrow\uparrow\rangle - |\uparrow\downarrow\downarrow\rangle)\,.
\end{aligned}
$$

These eight entangled states are called the GHZ states, and the concept can be generalized to many particles.

The operators A, B, and C form what is called a completely commuting set of operators, and we could label the states by the eigenvalues of these operators, acting on the states, so that

$$
\begin{aligned}
&A\,|\psi_i\rangle = a_i\,|\psi_i\rangle\,, \quad B\,|\psi_i\rangle = b_i\,|\psi_i\rangle\,, \quad C\,|\psi_i\rangle = c_i\,|\psi_i\rangle\,, \quad a_i, b_i, c_i = \pm 1,\\
&|\psi_i\rangle \equiv |a_i, b_i, c_i\rangle\,.
\end{aligned}
$$

(The operator D is redundant, since D $= -$ABC, and $d_i = -a_i b_i c_i$.) Then

$$
\begin{aligned}
&|\psi_1\rangle = |- - -\rangle\,, \quad |\psi_2\rangle = |+ + +\rangle\,, \quad |\psi_3\rangle = |+ + -\rangle\,, \quad |\psi_4\rangle = |- - +\rangle\,,\\
&|\psi_5\rangle = |+ - +\rangle\,, \quad |\psi_6\rangle = |- + -\rangle\,, \quad |\psi_7\rangle = |- + +\rangle\,, \quad |\psi_8\rangle = |+ - -\rangle\,.
\end{aligned}
$$

The GHZ states are entangled, non-local, and from a realistic point of view, acausal, and as we have seen, even their perfect correlations cannot be explained as elements of reality. They have been created in the laboratory, not as particles with spin 1/2, but rather as photon states, where their degrees of freedom, rather than being spin up or spin down, have been their polarization states, H or V, for horizontal or vertical, or equivalently, + or −, for circular polarization, and in some cases their position, rather than polarization, meaning, for example, whether they were transmitted or reflected by a beam splitter. The Mermin experiment above has been performed, using photons (▶ light quantum) by the group of A. Zeilinger in Vienna (see the bibliography).

Literature

The EPR argument was first made by

1. A. Einstein, B. Podolsky, N. Rosen: Can Quantum-Mechanical Description of Physical Reality Be Considered Complete? Phys. Rev. **47**, 777–780 (1935)
   Bohr's answer was given in

2. N. Bohr: Can Quantum-Mechanical Description of Physical Reality be Considered Complete? Phys. Rev. **48**, 696–702 (1935)
Bell's Theorem was first stated in
3. J. S. Bell: On the Einstein-Podolsky-Rosen Paradox. Physics **1**, 195–200 (1964)
These and many other papers relevant to the discussion, up to 1982, are contained in the book
4. J. A. Wheeler, W. H. Zurek (eds.): *Quantum Theory and Measurement* (Princeton University Press, Princeton 1983)
All John Bell's papers relevant to the discussion are contained in the book
5. J. S. Bell: *Speakable and Unspeakable in Quantum Mechanics* (Cambridge University Press, Cambridge 1987). They are also included in the book M. Bell, K. Gottfried, M. Veltman (eds.): *John S. Bell on the Foundations of Quantum Mechanics* (World Scientific, Singapore 2001).
GHZ states were first introduced in a very short paper by
6. D. M. Greenberger, M. A. Horne, A. Zeilinger in M. Kafatos (ed.): *Bell's Theorem, Quantum Theory, and Conceptions of the Universe* (Kluwer, Dordrecht 1989, 73–76). An expanded, and much more complete version is given in D. M. Greenberger, M. A. Horne, A. Zeilinger, A. Shimony: Bell's Theorem without Inequalities. Am. J. Phys. **58**, 1131–43 (1990)
The Mermin experiment we discuss was first described by
7. D. Mermin, Physics Today **43/6**, 9 (June 1990); Quantum Mysteries Revisited. Am. J. Phys. **58**, 731–34 (1990). See also Asher Peres: *Quantum Theory: Concepts and Methods* (Kluwer, Dordrecht, the Netherlands 1995)
The Mermin experiment has been essentially performed with a multi-photon system by
8. J.-W. Pan, D. Bouwmeester, M. Daniell, H. Weinfurter, A. Zeilinger: Experimental Test of quantum nonlocality in three-photon Greenberger–Horne–Zeilinger entanglement. Nature **403**, 515–9 (2000). See also the review by J.-W. Pan and A. Zeilinger, in R. A. Bertlmann, A. Zeilinger (eds.): *Quantum (Un)speakables – From Bell to Quantum Information* (Springer, Berlin 2002, 225–240).

Gleason's Theorem

Carsten Held

On a ► Hilbert space H, the quantum-mechanical trace formula provides a probability measure. Let {P} be the set of projection operators (► projection) on H and let, for a given statistical operator W, μ be a function from {P} into [0, 1] defined by $\mu(\mathrm{P}) = \mathrm{Tr}\,(\mathrm{P} \cdot \mathrm{W})$. Let $\{\mathrm{P}_i\} \subset \{\mathrm{P}\}$ be a countable set of mutually orthogonal projection operators. Then, $\mu(\sum_i \mathrm{P}_{|\mathrm{q}_i\rangle}) = \sum_i \mu(\mathrm{P}_{|\mathrm{q}_i\rangle})$ (countable additivity), $\mu(\mathrm{I}) = 1$, if $\sum_i \mathrm{P}_{|\mathrm{q}_i\rangle} = \mathrm{I}$, where I is the identity operator (probability of the certain event), and $\mu(\mathrm{P}_0) = 0$, where P_0 is the operator projecting on the zero space (probability of the impossible event). Hence for every particular set $\{\mathrm{P}_i\}$, μ fulfils the familiar probability axioms, i.e. is a probability measure. Obviously, μ is not a probability measure defined on the whole set {P}, since countable additivity is fulfilled, not for arbitrary, but only for mutually orthogonal elements of {P}. We have, in effect, defined a generalised probability function, a function on the lattice of projection operators such that every restriction to a Boolean sublattice is a probability measure.

Are there generalised probability functions besides the quantum-mechanical ones? Given that dim (H)>2, the answer is no. In other words, on a Hilbert space of dimension greater than two the quantum-mechanical probability measures are the only ones forming generalised probability functions. This remarkable claim is the content of Gleason's Theorem [1]. The theorem is important for the question whether quantum mechanics is complete in the following sense. Assume that, if a quantum-mechanical system S is in a pure state $|a_1\rangle$ such that $\mu(\mathrm{P}_{|a_1\rangle}) = 1$ (i.e. the probability that S is found, upon an A-measurement, to have a_1 equals 1), then it has the physical property represented by a_1 (the eigenvalue of A pertaining to $|a_1\rangle$). Completeness can be characterized as the idea that the properties ascribed to S in this way are the only ones (i.e. the "if" in the previous sentence should be replaced by "if and only if") and incompleteness as the idea that there are more. Explicitly, let A, B, ... be pairwise non-commuting operators (complementary observables) on S which are non-degenerate and have eigenvalues (possible S-properties) $a_1, a_2, \ldots, b_1, b_2, \ldots$ (Non-degeneracy of A means that if dim (H) $= n$, then A has n distinct values.) Quantum mechanics prescribes that S can be in only one of the states $|a_1\rangle, |a_2\rangle, \ldots, |b_1\rangle, |b_2\rangle, \ldots$ E.g., if S is in $|a_1\rangle$, then completeness means that it does not have any value of B and incompleteness that it does. The latter idea now can be expressed as follows. Every one of the observables A, B, ... has one of its values or, equivalently: one of the $|a_1\rangle, |a_2\rangle, \ldots$ gets assigned the number 1, the others the number 0, one of the $|b_1\rangle, |b_2\rangle, \ldots$ gets assigned the number 1, the others the number 0, and so on. Since each of the sets $\{|a_1\rangle, |a_2\rangle, \ldots\}, \{|b_1\rangle, |b_2\rangle, \ldots\}, \ldots$ is an ► orthonormal basis of H, incompleteness becomes the task of assigning 1 to one vector in such a basis and 0 to all others and doing this for all bases of H. Is such an assignment possible or not? It is easy to see that if it is impossible for a space H with dim (H) $= n$, then it is impossible for all spaces H with dim (H) $= m \geqslant n$, all defined over the same field. And it is comparatively easy to see that if such an assignment is impossible for a space R^n, an n-dimensional space over the real numbers, then it is impossible over C^n, a space of identical dimension over the complex numbers (see [6], p. 124, [2], pp. 323–25). So, an impossibility proof of the incompleteness assumption reduces to showing that in R^3 it is impossible to assign the number 1 to exactly one vector – in any orthonormal basis – (the number 0 to the two others) and do so consistently for all bases under the conditions that (a) vectors of different bases but lying in the same ray get assigned the same number and (b) any vector gets assigned a unique number, although it can belong to many bases.

The assignment described is indeed impossible, but there are two different ways to prove this. First, one can show the impossibility directly (i.e. constructively) by writing out a set of bases that make the assignment impossible. This is the route taken by the Kochen–Specker Theorem (► Kochen–Specker Theorem). Or one can exploit Gleason's Theorem for an indirect proof. It follows immediately from the theorem that all probability measures on H, with dim (H)>2, are continuous. Especially, every μ on C^3 is continuous and induces a map μ' on R^3 that is also continuous. Every such μ' can be visualized as an assignment of values from [0, 1] to all points on the surface of the unit sphere in R^3 such that the values vary continuously. On the other hand, the map required for realizing the above incompleteness assumption must be discontinuous. Intuitively, when all points on the surface of the unit sphere in R^3 are assigned numbers 1 and 0 only and both values occur, the map must be discontinuous. So the incompleteness assumption is refuted.

What should we think about conditions (a) and (b)? Condition (a) is unproblematic and plays no substantial role in the argument. It just reminds us that the space R^3, though intuitively accessible, is not a direct representation of physical space. In the full quantum-mechanical Hilbert space C^3, vectors $|a\rangle$ and $-|a\rangle$ represent the same state and the map μ' on R^3 is defined to respect this constraint. A possible assignment of 0 and 1 values to basis vectors in R^3 will likewise have to respect (a) because R^3 is a stand-in for quantum-mechanical C^3 where it is respected automatically. Condition (b) seems to explicate a trivial premise of the assignment task. The task of assigning 1 and 0 to all R^3 basis vectors would be trivially possible if we did not look for an assignment to all vectors, at once. However, this implies that any vector gets assigned a unique number, although it can belong to many bases and this condition can be interpreted in terms of the corresponding physics. It is called the assumption of non-contextuality. Assume that we wish to assign values to observables beyond the quantum-mechanical allowances. These values might not be ontologically independent from each other, but it seems reasonable to require that they are epistemologically independent in the following sense: The value of an ► observables does not depend on which other observables are measured in conjunction with it. In particular, consider a non-degenerate observable $A = \sum_i a_i P_{|a_i\rangle}$ on $H = C^n$. Ascribing some value to A implies ascribing values to all the $P_{|a_i\rangle}$. (Ascribing, e.g., a_k will ascribe 1 to $P_{|a_k\rangle}$ and 0 to each $P_{|a_i\rangle}$ with $i \neq k$.) But, given $n > 2$, there is for an arbitrary eigenvector $|a_m\rangle$ of A a non-degenerate A′ sharing this eigenvector, but no others, with A. Does the value of $P_{|a_m\rangle}$ depend on whether it is measured as a function of A or of A′? Answering no means to endorse non-contextuality, answering yes to reject it. (If $n = 3$ the eigenvectors of A and A′ can be directly represented as two orthogonal triples in R^3 sharing just $|a_m\rangle$.) So, denying condition (b) (i.e. assuming hidden S properties to be contextual) opens a loophole in the no-hidden-variables argument from Gleason's Theorem. An exactly parallel reply can of course be made in connection with the Kochen–Specker Theorem (► Kochen–Specker Theorem for more discussion).

Gleason's original proof of his theorem is mathematically involved. An elementary proof was given by Cooke et al. in 1985 [2]. It is reproduced and extensively commented by Hughes [4].

Primary Literature

1. A.M. Gleason: Measures on the Closed Subspaces of a Hilbert Space, J. Math. Mech. **6**, 885–93 (1957); reprinted in [3], 123–34

G

Secondary Literature

2. R.M. Cooke, M, Keane, and W. Moran: An Elementary Proof of Gleason's Theorem, Math. Proc. Camb. Phil. Soc. **98**, 117–28 (1985); reprinted in [4], 321–46
3. C. Hooker (ed.): The Logico-Algebraic Approach to Quantum Mechanics (Reidel, Dordrecht, 1975)
4. R.I.G. Hughes: The Structure and Interpretation of Quantum Mechanics (Harvard University Press, Cambridge, MA 1989)
5. S. Kochen and E. Specker: The Problem of Hidden Variables in Quantum Mechanics, J. Math. Mech. **17**, 59–87 (1967); reprinted in [3], 293–328
6. M. Redhead: Incompleteness, Nonlocality, and Realism: A Prolegomenon to the Philosophy of Quantum Mechanics. (Clarendon Press, Oxford, 1987)

Grover's Algorithm

See ► quantum computation.

GRW Theory (Ghirardi, Rimini, Weber Model of Quantum Mechanics)

Roman Frigg

Consider a toy system consisting of a marble and box. The marble has two states, $|\Psi_{\text{in}}\rangle$ and $|\Psi_{\text{out}}\rangle$, corresponding to the marble being inside or outside the box. These states are eigenvectors of the operator $\hat{B}$, measuring whether the marble is inside or outside the box. The formalism of quantum mechanics (QM) has it that not only $|\Psi_{\text{in}}\rangle$ and $|\Psi_{\text{out}}\rangle$ themselves, but any ► superposition $|\Psi_m\rangle = a\,|\Psi_{\text{in}}\rangle + b\,|\Psi_{\text{out}}\rangle$ where a and b are complex numbers such that $|a|^2 + |b|^2 = 1$, can be the state of the marble. What are the properties of the marble in such a state? This question is commonly answered by appeal to the so-called Eigenstate-Eigenvalue Rule (EER): An observable $\hat{O}$ has a well-defined value for a quantum a system S in state $|\Psi\rangle$ if, and only if, $|\Psi\rangle$ is an eigenstate of $\hat{O}$. Since $|\Psi_{\text{in}}\rangle$ and $|\Psi_{\text{out}}\rangle$ are eigenstates of $\hat{B}$, EER yields that the marble is either inside (or outside) the box if its state is $|\Psi_{\text{in}}\rangle$ (or $|\Psi_{\text{out}}\rangle$). However, states like $|\Psi_m\rangle$ defy interpretation on the basis of EER and we have to conclude that if the marble is in such a state then it is neither inside nor outside the box. This is unacceptable because we know from experience that marbles are always either inside or outside boxes. Reconciling this fact of everyday experience with the quantum formalism is the infamous measurement problem. See also ► Bohmian mechanics; Measurement theory; Metaphysics in Quantum Mechanics; Modal Interpretation; Objectification; Projection Postulate.

Standard quantum mechanics solves this problem, following a suggestion of von Neumann's, by postulating that upon measurement the system's state is instantaneously reduced to one of the eigenstates of the measured observable, which leaves the system in a state that can be interpreted on the basis of EER (▶ Measurement Theory). However, it is generally accepted that this proposal is ultimately unacceptable. What defines a measurement? At what stage of the measurement process does the ▶ wave function collapse take place (trigger problem)? And why should the properties of a system depend on actions of observers?

GRW Theory (sometimes also 'GRW model') is a suggestion to overcome these difficulties (the theory has been introduced in Ghirardi et al. [1]; Bell [6] and Ghirardi [10] provide short and non-technical presentations of the theory; for a comprehensive discussion of the entire research programme to which GRW Theory belongs see Bassi and Ghirardi [5]). The leading idea of the theory is to eradicate observers from the picture and view state reduction as a process that occurs as a consequence of the basic laws of nature. The theory achieves this by adding to the fundamental equation of QM, the ▶ Schrödinger equation, a stochastic term which describes the state reduction occurring in the system. (For this reason GRW theory is not, strictly speaking, an interpretation of QM; it is a quantum theory in its own right).

A system governed by GRW theory evolves according to the Schrödinger equation all the time except when a state reduction, a so-called hit, occurs (hits are also referred to as 'hittings', 'perturbations', 'spontaneous localisations', 'collapses', and 'jumps'). A crucial assumption of the theory is that hits occur at the level of the micro constituents of a system (in the above example at the level of the atoms that make up the marble). The crucial question then is: when do hits occur and what exactly happens when they occur?

GRW Theory posits that the occurrence of hits constitutes a Poisson process. Generally speaking, Poisson processes are processes characterised in terms of the number of occurrences of a particular type of event in a certain interval of time τ, for instance the number of people passing through a certain street during time τ. These events are Poisson distributed if the probability that the number of events occurring during τ, n, takes value m is given by $p(n = m) = e^{-\lambda\tau}(\lambda\tau)^m/m!$, where λ is the parameter of the distribution. One can show that λ is also the mean value of the distribution and hence it can be interpreted as the average number of events occurring per unit time. GRW theory sets $\lambda = 10^{-16}\,\text{s}^{-1}$ and posits that this is a new constant of nature. Hence, in a macroscopic system that is made up of about 10^{23} atoms there are on average 10^7 hits per second.

A hit transforms the system's state into another state according to a probabilistic algorithm that takes the position basis as the privileged basis (in that the reduction process leads to a localisation of the system's state in the position basis). Let $|\Psi_S\rangle$ be state of the entire system (e.g. the marble) before the hit occurs. When the kth particle, say, is hit the state is instantaneously transformed into another, more localised state:

$$|\Psi_S\rangle \rightarrow \left|\Psi_S^{\text{hit}}\right\rangle = \frac{L_{k,c}\,|\Psi_S\rangle}{\left\|L_{k,c}\,|\Psi_S\rangle\right\|}.$$

$L_{k,c}$, the localisation operator, that has the shape of a Gaussian (a bell-shaped curve) centred around c, which is chosen at random according to the distribution $p_k(c) = \left\| L_{k,c} \left|\Psi_S\right\rangle \right\|^2$; the width σ of the Gaussian is also a new constant of nature, and it is of the magnitude 10^{-7} m. The choice of this distribution assures that the predictions of GRW Theory coincide almost always with those of standard QM (there are domains in which the two theories do not yield the same predictions, but these are (so far) beyond the reach of experimental test; see Rimini [15]).

Due to the mathematical structure of QM (more specifically, due to the fact that $|\Psi_S\rangle$ is the tensor product of the states of all its micro constituents) the hits at the micro level 'amplify': if the marble is in state $|\Psi_m\rangle$ and kth particle gets hit, then the entire state is transformed into a highly localised state, i.e. all terms except one in the superposition are suppressed. This is GRW's solution of the measurement problem. A macro system gets hit 10^7 times per second and hence superpositions are suppressed almost immediately; micro systems are not hit very often and hence retain their 'quantum properties' for a very long time.

This proposal faces two important formal problems. First, the ▶ wave function of systems of identical particles has to be either symmetrical (in the case of Bosons) or antisymmetrical (in the case of Fermions), and remain so over the course of time. GRW theory violates this requirement in that wave functions that are symmetric (or antisymmetric) at some time need not be (and generally are not) symmetric (or antisymmetric) at later times. Second, although hits occur at the level of the system's wave function, the fundamental equation of the theory is expressed in terms of the density matrix. This strikes physicists as odd and one would like to have an equation governing the evolution of the wave function itself. Both difficulties are overcome within the so-called CSL model (for 'continuous spontaneous localization') introduced in Pearle [3] and Ghirardi et al. [2]. The model belongs to the same family of proposals as GRW theory in that it proposes to solve the measurement problem by an appeal to a spontaneous localisation processes. The essential difference is that the discontinuous hits of GRW theory are replaced by a continuous stochastic evolution of the state vector in ▶ Hilbert space (similar to a diffusion process).

Another serious problem concerns the nature of GRW hits. Unlike the state reduction that von Neumann introduced into standard QM, the hits of GRW theory do not leave the system's state in an *exact* position eigenstate; the post-hit state is highly peaked, but nevertheless fails to be a precise position eigenstate. This is illustrated schematically in Fig. 1. Hence, strictly speaking the post-hit states are not interpretable on the basis of EER and we are back where we started; this problem is also know as the 'tails problem' (see Albert and Loewer [4]). Common wisdom avoids this conclusion by pointing out that GRW post-hit states are close to eigenstates and positing that being close to an eigenstate is as good as being an eigenstate. This has been challenged by Lewis [12], who presents an argument for the conclusion that this move has the undesirable consequence that arithmetic does not apply to ordinary macroscopic objects. For a critical discussion of this argument see Frigg [12].

What is the correct interpretation of the theory? That is, what, if anything, does the theory describe? The answer to this question is less obvious than it might

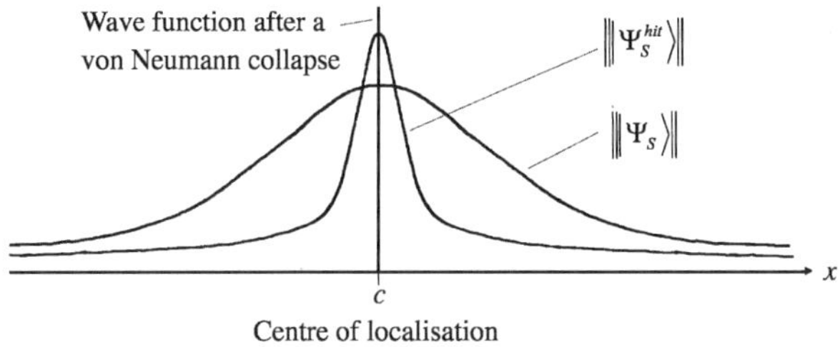

Fig. 1 GRW hit

seem. Clifton and Monton [7] regard it as a 'wave function only theory' according to which the world literally is just the wave function that the theory describes. Monton [14] later criticises this view as mistaken and suggests a variation of the mass density interpretation originally proposed by Ghirardi et al. [11]. Lewis [12] points out that all versions of the mass density interpretation lead to a violation of common sense and should hence not be regarded as a problem-free alternative.

How should we interpret the probabilities that the theory postulates in its hit mechanism? Are they best interpreted as propensities, frequencies, Humean chances, or yet something else? Or should the quest for such an interpretation be rejected as ill-conceived? This question is discussed in Frigg and Hoefer [9] who come to the conclusion that GRW probabilies can be understood either as single case propensities or as Humean chances, while all other options are ruled out by GRW Theory itself. See also ► Metaphysics in Quantum Mechanics; Quantum State Diffusion Theory.

Literature

1. G. C. Ghirardi, A. Rimini, T. Weber: Unified Dynamics for Microscopic and Macroscopic Systems. Physical Review **34D**, 470–91 (1986)
2. G. C. Ghirardi, P. Pearle, A. Rimini: Markov Processes in Hilbert Space and Spontaneous Localization of Systems of Identical Particles. Physical Review **42A**, 78–89 (1990)
3. P. Pearle: Combining Stochastic Dynamical State-Vector Reduction with Spontaneous Localization. Physical Review **39A**, 2277–2289 (1989)
4. D. Z. Albert, B. Loewer: Tails of Schrödinger's Cat. In *Perspectives on Quantum Reality: Non-Relativistic, Relativistic, and Field-Theoretic*, ed. by R. Clifton (Kluwer, Dordrecht 1995, 81–92)
5. A. Bassi, G. C. Ghirardi: Dynamical Reduction Models. Physics Reports **379**, 257–426 (2003)
6. J. S. Bell: Are There Quantum Jumps? In *Speakable and unspeakable in quantum mechanics*, ed. J. S. Bell (Cambridge University Press, Cambridge 1987, 201–212)
7. R. Clifton, B. Monton: Losing Your Marbles in Wavefunction Collapse Theories. British Journal for the Philosophy of Science **50**, 697–717 (1999)
8. R. Frigg: On the Property Structure of Realist Collapse Interpretations of Quantum Mechanics and the So-Called 'Counting Anomaly'. International Studies in the Philosophy of Science **17**, 43–57 (2003)
9. R. Frigg, C. Hoefer: Probability in GRW Theory. Forthcoming in Studies in History and Philosophy of Science **38** (2007)

10. G. C. Ghirardi: Collapse Theories. *The Stanford encyclopedia of philosophy* (Spring 2002 Edition), URL = http://plato.stanford.edu/archives/spr2002/contents.html
11. G. C. Ghirardi, R. Grassi, F. Benatti: Describing the Macroscopic World: Closing the Circle Within the Dynamic Reduction Program. Foundations of Physics **25**, 5–38 (1995)
12. P. J. Lewis: Quantum Mechanics, Orthogonality, and Counting. British Journal for the Philosophy of Science **48**, 313–328 (1997)
13. P. J. Lewis: Interpreting Spontaneous Collapse Theories. Studies in History and Philosophy of Modern Physics **36**, 165–180 (2005)
14. B. Monton: The Problem of Ontology for Spontaneous Collapse Theories. Studies in History and Philosophy of Modern Physics **35**, 407–421 (2004)
15. A. Rimini: Spontaneous Localization and Superconductivity. In *Advances in quantumPhenomena*, ed. by E. Beltrametti et al. (Plenum, New York 1995, 321–333)

H

Hamiltonian Operator

Christopher Witte

Hamiltonian operator, a term used in a quantum theory for the linear operator on a complex ▶ Hilbert space associated with the generator of the dynamics of a given quantum system. Under most circumstances this operator is assumed to be *self-adjoint*, thus having real spectrum. The spectral values are in such a case interpreted as possible resulting values of an energy measurement performed on the system. The Hamiltonian operator can then be seen as synonymous with the *energy operator*, which serves as a model for the energy *observable* of the quantum system.

In these two aspects of (a) generating the dynamics and (b) representing the energy observable, the *Hamiltonian operator* in quantum theory plays a rôle very much analogous to that of the *Hamiltonian function* in classical theories. Historically this fact became obvious as soon as modern quantum mechanics was constituted by Heisenberg, Schrödinger, Dirac and others. Schrödinger himself used a term for this mathematical object that translates to *"the wave operator analogous to the Hamiltonian function"* [5] in comparing his ▶ wave mechanics to Heisenberg's ▶ matrix mechanics. Due to this obvious similarity to the Hamiltonian function of classical mechanics the symbol H and the names *energy operator* or *Hamiltonian operator* came into use (see, e.g., [1] for a relatively early example).

The concept of a Hamiltonian operator is useful in almost any quantum theory, be it quantum mechanics or a quantum field theory. Nevertheless, since quantum field theories are usually considered in a relativistic setting, the meaning of dynamics is more complicated due to the lack of an absolute time parameter. This problem can be dealt with in an elegant way by an algebraic approach to such theories (see, e.g., [4]). In much the same way the measurement process and the concept of *energy of the system* need refinement. Especially in approaches to a theory of *quantum gravity* the significance of the Hamiltonian operator becomes much different, since such an operator should rather be seen as a *constraint operator* than a generator of dynamics. The concept of an energy operator fails completely to be applicable [6]. To avoid these complications in this encyclopedic overview, we will restrict the detailed description to the realm of non-relativistic quantum mechanics.

Generator of dynamics. The most simple quantum mechanical systems are closed, conservative systems. For such systems the homogeneity of time suggests that their dynamics is induced by a *symmetry of the system* [2]. By Wigner's theorem such a symmetry can be either a *unitary* or an *anti-unitary* operation on the underlying Hilbert space $\mathcal{H}$. Since the effects of the dynamics should tend to identity in a measurement context when time steps become small, the operations must form a weakly continuous one-parameter-group of unitary operators

$U: \mathbb{R} \to \boldsymbol{U}(\mathcal{H}), t \mapsto U(t)$, where $\boldsymbol{U}(\mathcal{H})$ denotes the group of ▶ unitary operators on $\mathcal{H}$. The unitary operators $U(t)$ are called *evolution operators*, since they describe evolution of a pure state at time t_0 to the state at time t_1 by $\psi(t_0) \mapsto \psi(t_1) = U(t_1 - t_0)\psi(t_0)$. The term "one-parameter-group" actually means, that the mapping U is a *group homomorphism*, such that $U(t_1)U(t_0) = U(t_0 + t_1)$ and $U(t)^{-1} = U(-t)$. The notion of "weak continuity" refers to the claim that the mapping $t \to \langle \phi, U(t)\psi \rangle$ for arbitrary $\phi, \psi \in \mathcal{H}$ be continuous. This kind of continuity ensures that statistical distributions of arbitrary measurements (more specifically their moments) vary continuously with time. Stone's theorem (see, e.g., [7]) states that such one-parameter-groups are exactly those, which are generated by a ▶ self-adjoint operator. Explicitly, there is a self-adjoint operator H, the *Hamiltonian operator* such that $U(t) = \exp(-\mathrm{i}t\,\mathrm{H}/\hbar)$ for all $t \in \mathbb{R}$ (exp denoting the operator exponential function). The Hamiltonian operator can be found from the evolution operator by differentiation

$$\langle \phi, \mathrm{H}\,\psi \rangle = \mathrm{i}\hbar \frac{\mathrm{d}}{\mathrm{d}t} \langle \phi, U(t)\psi \rangle|_{t=0}.$$

This equation defines the self-adjoint operator H on its domain $\mathscr{D}(\mathrm{H})$, which is *dense in*, but generally not *equal to* the Hilbert space $\mathcal{H}$. The domain of the Hamiltonian is invariant under the action of the evolution operators and the equation above describes the derivative of the curve $\psi(t) = U(t)\psi(0)$, which a pure quantum state passes in time:

$$\mathrm{i}\hbar \frac{\mathrm{d}}{\mathrm{d}t} \psi(t) = \mathrm{H}\,\psi(t).$$

This equation, which describes the infinitesimal generation of the quantum dynamics, is the famous ▶ *Schrödinger equation*. The dynamics of *mixed states* follows according to the definition of mixing directly from the dynamics of pure states: $\rho(t) = U(t)\rho(0)U^*(t)$. By differentiation this leads to

$$\mathrm{i}\hbar \frac{\mathrm{d}}{\mathrm{d}t} \rho(t) = [\mathrm{H}, \rho(t)],$$

an equation usually called *von Neumann equation*. Sometimes this equation is also called *quantum Liouville equation*, in analogy to the dynamical equation for density distributions in classical Hamiltonian mechanics.

The ▶ *Heisenberg picture* of a quantum system models the dynamics in a different but equivalent way to the above seen so called ▶ *Schrödinger picture*. States are seen as time-independent in the Heisenberg picture, whereas ▶ observables carry the time dependence of the system. Since the statistical outcome of any measurement performed on the system must not depend on the picture chosen, one must have for any observable A the identity $\mathrm{Tr}(\rho(t)A) = \mathrm{Tr}(\rho(0)A^H(t))$ and thus $A^H(t) = U^*(t)AU(t)$ for the time dependent observable in the Heisenberg picture. From this dynamics one gets the differential equation of motion

$$\mathrm{i}\hbar \frac{\mathrm{d}}{\mathrm{d}t} A^H(t) = [A^H(t), \mathrm{H}]$$

for observables, which is correspondingly called the *von Neumann equation* in the Heisenberg picture.

The dynamics of nonconservative systems is more complicated in general. In some important cases the dynamics of the system is still unitary, but the evolution operators do not form a one parameter group. Instead the more general case is that of a *two-parameter-groupoid* $U\colon \mathbb{R} \times \mathbb{R} \to \boldsymbol{U}(\mathcal{H}), t \mapsto U(t_1, t_2)$ with properties $U(t_3, t_2)U(t_2, t_1) = U(t_3, t_1)$ and $U(t_2, t_1)^{-1} = U(t_1, t_2)$. Such systems with a time dependent Hamiltonian can be seen as analogous to holonomic-rheonomous classical systems. The infinitesimal generator of such a groupoid is a *time dependent* self-adjoint operator $\mathrm{H}(t)$ and can be calculated by

$$\langle \phi, \mathrm{H}(t)\psi \rangle = \mathrm{i}\hbar \frac{\mathrm{d}}{\mathrm{d}\tilde{t}} \langle \phi, U(\tilde{t}, t)\psi \rangle|_{\tilde{t}=t}.$$

H

Nevertheless *integration* of a time dependent Hamiltonian operator to get back the evolution operators is non-trivial and must not be done by simply taking the operator exponential function.

The dynamics of general *open* quantum systems cannot be modeled in the same way as seen above. The dynamical mapping $\rho(t_0) \mapsto \rho(t_1) = V(t_1, t_0)(\rho(t_0))$ can only be given in the ► mixed state context, or equivalently as a quantum stochastical process of pure states, since time evolution doesn't conserve purity of states (details can be found, e.g., in [3]).

Energy operator: In classical mechanics the generator of the dynamical group of a holonomic-scleronomous system is the generator of a symmetry operation. By Noether's theorem the generator of such a symmetry is a constant of motion with the physical interpretation of the total *energy* of the system. In much the same way the time-independent Hamiltonian operator of a conservative quantum system can be seen as a constant of motion, as the statistical distribution of the *observable* H is constant, i.e., for any natural number n the expectation value of H^n is constant in time: $\mathrm{i}\hbar\frac{\mathrm{d}}{\mathrm{d}t}\,\mathrm{Tr}(\rho(t)\,\mathrm{H}^n) = \mathrm{Tr}([\mathrm{H}, \rho(t)]\,\mathrm{H}^n) = 0$. By analogy it is justified to call this observable *energy*, and the spectrum of H is to be interpreted as possible outcomes of an energy measurement. Eigenstates of this observable correspond to preparations of the system with sharp energy values. They are solutions to the eigenvalue equation $E\psi = \mathrm{H}\,\psi$, where E is a certain discrete spectral value. This equation is sometimes called the *time-independent Schrödinger equation*, and its solutions show an especially simple time dependence: $\psi(t) = \psi(0)\exp(\mathrm{i}Et/\hbar)$, i.e., only a time dependent phase factor is changed. These states are called *stationary*, since statistical distributions of (time-independent) observables in such states are invariant in time. For energy values E in the continuous spectrum there are no solutions to the time-independent Schrödinger equation in Hilbert space. Nevertheless, in the context of ► *rigged Hilbert space* one can find weak solutions called *improper* eigenstates. These have a physical interpretation as scattering states and are stationary as well.

The spectrum of the Hamiltonian operator is usually *bounded from below*, i.e., there is a lower limit for the energy of the system. This condition is a necessity for systems, which can *in principle* interact with the environment, for otherwise such systems could act as an infinite source of energy, since they do not admit a *ground state*. Also, no thermodynamical equilibrium is possible for systems with Hamiltonian operator not bounded from below, since the usual expression for the *equilibrium state* $\rho_\beta = \exp(-\beta\,\mathrm{H})/\,\mathrm{Tr}(\exp(-\beta\,\mathrm{H}))$ would in that case not yield a well-defined operator for any inverse temperature $\beta > 0$.

The analogy to the classical energy function becomes most obvious, if one chooses the *Schrödinger representation* for the Hilbert space $\mathcal{H} = L^2(\mathbb{R}^{3n}) \otimes \mathcal{H}_{\mathrm{int}}$, where $\mathbb{R}^{3n}$ is the n-particle position space and $\mathcal{H}_{\mathrm{int}}$ the space of internal degrees of freedom (usually ▶ spin). For systems without internal degrees of freedom the Hamiltonian operator takes the form of a certain kind of partial differential operator, called *Schrödinger operator* [8]:

$$\mathrm{H} = \sum_k \left(\frac{1}{2m_k} \left(\frac{\hbar}{i} \nabla_k - q_k \boldsymbol{A}(\boldsymbol{x}_k) \right)^2 \right) + V,$$

with functions $\boldsymbol{A}(\boldsymbol{x})$ (exterior magnetic vector potential) and $V(\boldsymbol{x}_1, \ldots, \boldsymbol{x}_n)$ (common potential) of suitable integrability and differentiability. Since $\boldsymbol{p}_k = \frac{\hbar}{\mathrm{i}} \nabla_k$ is the momentum operator of the k-th particle, this can be seen as the formal translation of the classical Hamiltonian function of n charged particles with magnetic terms. Some important special cases of Schrödinger operators will be listed below.

(a) For a single particle in the absence of a magnetic field, one gets the standard Hamiltonian operator

$$\mathrm{H} = -\frac{\hbar^2}{2m} \Delta + V,$$

with *Laplace operator* Δ and single particle potential $V(\boldsymbol{x})$.

(b) A single *electron* in an electromagnetic field, taking spin into account, can be described by a Hamiltonian of the form

$$\mathrm{H} = \frac{1}{2m} \left(\frac{\hbar}{\mathrm{i}} \nabla + e\boldsymbol{A}\right)^2 - e\phi + \frac{e}{m} \boldsymbol{S} \cdot \boldsymbol{B},$$

where the electromagnetic potentials ϕ and $\boldsymbol{A}$ as well as the magnetic field $\boldsymbol{B}$ are functions of position $\boldsymbol{x}$ and possibly time t. This operator acts on the Hilbert space $\mathcal{H} = L^2(\mathbb{R}^3) \otimes \mathbb{C}^2$, where the space of internal degrees of freedom of the spin-1/2 electron can be taken to be $\mathbb{C}^2$. The formal scalar product $\boldsymbol{S} \cdot \boldsymbol{B}$ of a matrix-valued vector and a vector-valued function yields a matrix-valued function, which admits a natural action as an operator on the tensor product Hilbert space.

(c) The Hamiltonian operator of an ion or atom of N electrons and a nucleus of atomic number Z is given by

$$\mathrm{H} = \sum_{k=1}^{N} \left(-\frac{\hbar^2}{2m} \Delta_k - \frac{Ze^2}{4\pi\epsilon_0 |\boldsymbol{x}_k|} \right) + \sum_{k<l} \frac{e^2}{4\pi\epsilon_0 |\boldsymbol{x}_k - \boldsymbol{x}_l|},$$

with nucleus located in the origin of the co-ordinate system.

(d) The Hamiltonian operator of an electron in an atom, using the central-field model, but taking into account the spin-orbit coupling, reads

$$H = -\frac{\hbar^2}{2m}\Delta + V(|\boldsymbol{x}|) + \frac{1}{2m^2c^2|\boldsymbol{x}|}\frac{\mathrm{d}}{\mathrm{d}|\boldsymbol{x}|}V(|\boldsymbol{x}|)\,\boldsymbol{L}\cdot\boldsymbol{S}$$

where V is some effective potential.

Primary Literature

1. J. von Neumann: *Grundlagen der Quantenmechanik.* Springer, Berlin (1932)
2. J.M. Jauch: *Foundations of Quantum Mechanics.* Addison-Wesley, Reading (1968)
3. H.P. Breuer, F. Petruccione, *The Theory of Open Quantum Systems.* Oxford University Press, Oxford (2002)
4. O. Bratteli, D.W. Robinson: *Operator Algebras and Quantum Statistical Mechanics 1.* Springer, New York (1979)

H

Secondary Literature

5. E. Schrödinger: *Über das Verhältnis der Heisenberg-Born-Jordanschen Quantenmechanik zu der meinen.* Annalen der Physik (4), 79, 734–56 (1926)
6. T. Thiemann: *Lectures on Loop Quantum Gravity.* Lecture Notes in Physics 631, 41–135, (2003) arXiv:gr-qc/0210094v1
7. S. Reed, B. Simon: *Methods of Modern Mathematical Physics.* Volume 1: Functional Analysis. Academic, New York (1972)
8. H.L. Cycon et al.: *Schrödinger Operators.* Springer, Berlin (1987)

Hardy Paradox

Antonio Acín

Since the seminal work by Bell [1], it is known that the results obtained when measuring a quantum state in space separated regions can display some counter-intuitive form of correlations, often named as quantum ► nonlocality. The standard Bell scenario consists of a source emitting a pair of particles to two distant observers, Alice and Bob, who can choose between m different measurements of n possible outcomes. The choice of the measurement by Alice and Bob is denoted by x and y, while a and b label the corresponding measurement outcome, see Fig. 1. By measuring the particles, the parties can estimate the correlations between the measurement

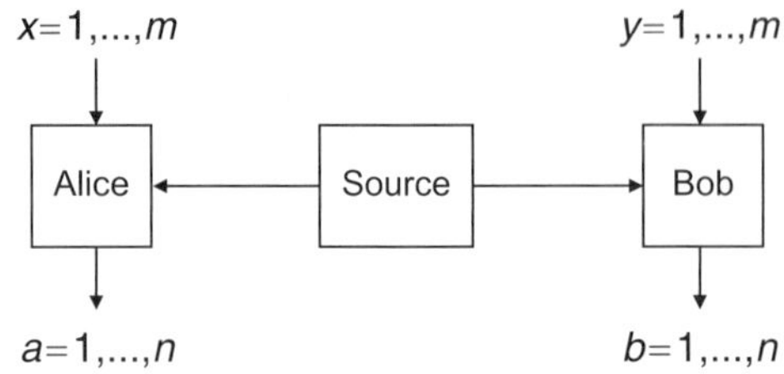

Fig. 1 Standard Bell scenario: two distant parties receive correlated quantum particles from a source. Alice and Bob choose between m possible measurements of n outcomes. The choice of measurement is labeled by x and y and the obtained outcome by a and b

outcomes, described by a conditional probability distribution $p\,(a, b|x, y)$. The timing is such that the particles are emitted at the source before Alice and Bob decide which measurement to perform. It is also assumed that the parties are situated in distant labs, so there does not exist any form of communication between them. This can be guaranteed, for instance, if Alice's measurement is outside the light-cone defined by Bob's measurement, and viceversa: Einstein's special relativity implies that there cannot be any causal influence between the measurements. Under these conditions, any possible correlation between Alice and Bob's measurement outcomes should have been defined at the source.

In what follows, we consider the simplest case where Alice and Bob have to perform two different Stern Gerlach measurements on two spin-one-half particles. The measurements are defined by two directions, corresponding to the directions of the Stern–Gerlach measurement apparatuses for each party, namely $\hat{a}_1$ and $\hat{a}_2$ for Alice and $\hat{b}_1$ and $\hat{b}_2$ for Bob, while the outcomes of these measurements are $a_1, a_2, b_1, b_2 = \pm 1$. Note that here we replace the previous general notation, that is a, b, x and y, by the more physical ► spin notation given by the direction of the spin measurements and the ± 1 outcomes.

In 1993 Lucien Hardy showed that in this scenario, it is possible to choose a quantum state of two spin-one-half particles and measurements by Alice and Bob such that:

1. If Alice measures along the first direction and obtains the result $+1$, Bob, when measuring along the first direction also gets $+1$. This means that $p\,(b_1 = -1|a_1 = +1) = 0$ which implies $p\,(a_1 = +1, b_1 = -1) = 0$.
2. If Bob measures along the first direction and obtains the result $+1$, Alice, when measuring along the second direction also gets $+1$. This means that $p\,(a_2 = -1|b_1 = +1) = 0$ which implies $p\,(a_2 = -1, b_1 = +1) = 0$.
3. If Alice measures along the second direction and obtains the result $+1$, Bob, when measuring along the second direction also gets $+1$. This means that $p\,(b_2 = -1|a_2 = +1) = 0$ which implies $p\,(a_2 = +1, b_2 = -1) = 0$.
4. If Alice measures along the first direction and Bob along the second direction, sometime their outcomes are $a_1 = +1$ and $b_2 = -1$. This means that $p\,(a_1 = +1, b_2 = -1) \neq 0$.

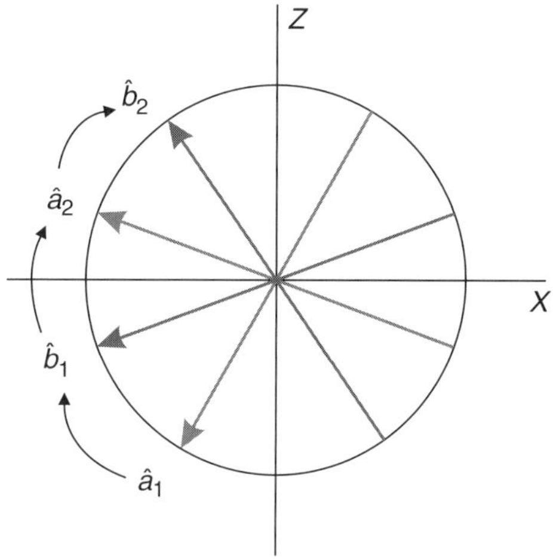

H

Fig. 2 Spin measurements in Hardy's paradox. The measurements are defined by the direction of the Stern–Gerlach apparatus. The arrow of each measurement indicates the outcomes associated to a positive result. If Alice measures $\hat{a}_1$ and gets the positive outcome, she is projecting Bob's state into $+\hat{b}_1$. All these implications are shown by *curved arrows* in the figure

Using standard quantum concepts and the pictorial representation of spin measurements, it is possible to get an intuition about these measurements. Recall that, according to Quantum Mechanics, the measurement by one of the parties, say Alice, on her quantum particle projects the other particle, Bob's particle, into a quantum state that depends on Alice's result. Hardy's choice of measurements and states is such that, for instance, when Alice measures along the direction $\hat{a}_1$ her quantum particle and obtains the result $+1$, she is aligning (projecting) the spin of Bob's particle along the positive direction defined by $\hat{b}_1$ (see also Fig. 2). Thus, if Bob measures along this direction, he will always obtain the result $+1$. The same reasoning applies to the remaining directions.

Let's now apply our classical intuition to this situation. As discussed above, since it is assumed that there is no communication between the particles when measured, all observed ► correlations in quantum mechanics should have been established at the source. That is, before leaving the source, the parties get some instructions about which result, $+1$ or -1, corresponds to the each of the two measurements. Remember that the choice of measurement by Alice and Bob is made after the particles leave the source. This is why the particles should carry information about the two possible measurements by each party. These instructions are nothing but a list specifying the outcomes for each measurement, for example $\{a_1 = +1, a_2 = +1, b_1 = -1, b_2 = +1\}$. Since the scope of these instructions is to reproduce the observed quantum correlations, they cannot be in contradiction with properties 1–4 listed above. This means that in all the cases where $a_1 = +1$, b_1 should also be equal to +1 because of property 1. If this was not the case, $p\,(a_1 = +1, b_1 = -1)$ could not be zero. Now, a_2 has to be $+1$ as well, because of property 2. But, then, because of property 3, $b_2 = +1$. That is, the chain of implications $a_1 = +1 \to b_1 = +1 \to a_2 = +1 \to b_2 = +1$, see also Figure 2, implies that the probability of observing $a_2 = +1, b_2 = -1$ has to be zero. However, Hardy's paradox, and in particular property 4 above, shows that this reasoning is wrong in the quantum case! More precisely, it is not that the reasoning is wrong but it is just another manifestation of the fact that quantum ▶ nonlocality cannot be explained using classical correlations, as first shown by ▶ Bell's Theorem.

Once the paradox is presented, one can try to "optimize the surprise", in the sense of preparing the quantum state and measurements such that $p\,(a_1 = +1, b_2 = -1)$ is maximized. As shown by Hardy [2], the solution to this problem gives $p\,(a_1 = +1, b_2 = -1) = 0.09$. Interestingly, Hardy's paradox does not work for the singlet state, which in many senses can be considered as the most correlated quantum state of two ▶ spin-one-half particles. When the two distant observes share a singlet state, if Alice measures along $\hat{a}_1$ and gets the result $+1$, she knows that Bob's particle is projected onto the orthogonal state and, therefore, he will get the opposite result when measuring along the same direction. In this sense the singlet state has the strongest form of anti-correlations. Note however that these perfect anti-correlations appear when Alice and Bob measure along the same direction. Therefore, it is impossible to derive the chain of implications that was crucial in the construction of Hardy's paradox. The proof, however, works for any other quantum state of two spin-one-half particles, provided it is not product.

To conclude, Hardy's paradox provides an alternative and elegant proof of Bell's theorem. It is worth mentioning here that, from an experimental point of view, it does not provide any advantage over other existing versions of this Theorem. In particular, it is based on combination of events that have zero probability, which is impossible as soon as we introduce some reasonable form of noise in the system [3]. However, it is perhaps one of the simplest demonstrations of the weirdness and beauty of quantum correlations.

Literature

1. J. S. Bell, On the Einstein Podolsky Rosen paradox, Physics **1**, 195 (1964).
2. L. Hardy, Nonlocality for two particles without inequalities for almost all entangled states, Phys. Rev. Lett. **71**, 1665 (1993).
3. Nevertheless, it is possible to construct experimenally testable proofs of Bell's theorem based on Hardy's construction. See for instance, D. Boschi, S. Branca, F. De Martini and L. Hardy, Ladder proof of nonlocality without inequalities: theoretical and experimental results, Phys. Rev. Lett. **79**, 2755 (1997); W. T. Irvine, J. F. Hodelin, C. Simon and D. Bouwmeester, Realization of Hardys thought experiment with photons, Phys. Rev. Lett. **95**, 030401 (2005).

Heisenberg Microscope

Marianne Breinig

In 1925 Werner Heisenberg published the first coherent mathematical formulation of quantum theory, now referred to as ► matrix mechanics. One year later, Erwin Schrödinger presented an alternative theory which became known as wave mechanics. ► Wave mechanics was considered the more intuitive theory and was favored by many physicists of that time. In 1927 Heisenberg published a paper to show that the predictions of matrix mechanics, which lead to the ► Heisenberg uncertainty relations, should not be considered counterintuitive, but should be viewed as being built into every measurement.

In the paper Heisenberg introduced a thought experiment to measure the position of an electron with a microscope which uses high-energy gamma rays for illumination. By reducing the wavelength of the gamma rays and by increasing the diameter of the microscope objective, the position of the electron can be measured as accurately as desired. Assuming diffraction-limited optics, the uncertainty in the position measurement is on the order of $\Delta x \cong \lambda/(2\sin\theta)$ (Fig. 1).

However, as a gamma ray scatters off the electron whose position is being measured, into the solid angle subtended by the microscope objective at the position of this electron, energy and momentum conservation require that the electron recoils. This ► Compton scattering process produces an uncertainty in the momentum of the scattered electron, since the gamma ray can be scattered into any angle within the acceptance cone of the objective. The uncertainty in the x-component of the

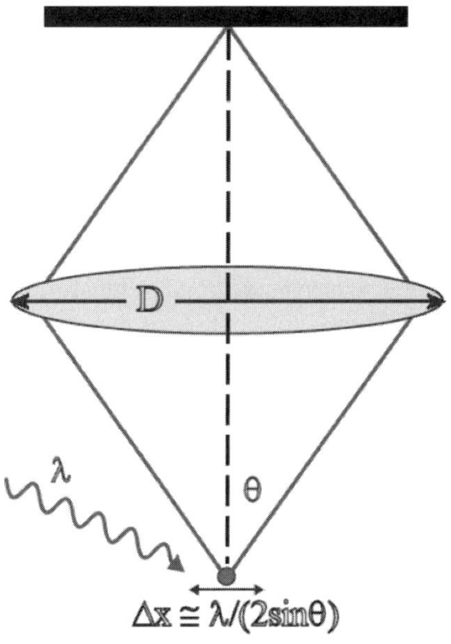

Fig. 1 Diffraction-limited optics

momentum is on the order of $\Delta p_x \cong (2h/\lambda)\sin\theta$. The Heisenberg microscope thought experiment therefore leads to a product of uncertainties $\Delta x\ \Delta p_x \cong h$.

The thought experiment contains the notion that the uncertainty relation is a result of a disturbance of the electron by the measurement process. This may lead to the assumption that without this disturbance the electron could have a well defined position and momentum, which conflicts with our current understanding of quantum mechanics. The uncertainty principle applies to all quantum objects and should not be viewed as only the result of us not being able to make an accurate measurement.

Literature

1. W. Heisenberg: Über quantentheoretische Umdeutung kinematischer und mechanischer Beziehungen. Zeitschrift für Physik **33**, 879–893 (1925)
2. W. Heisenberg: Über den anschaulichen Inhalt der quantentheoretischen Kinematik und Mechanik. Zeitschrift für Physik **43**, 172–198 (1927); English translation in (Wheeler and Zurek, 1983, pp. 62–84)
3. J.A. Wheeler, W.H. Zurek (eds): *Quantum Theory and Measurement* (Princeton NJ, Princeton University Press 1983)

Heisenberg Picture

Marianne Breinig

In non-relativistic quantum mechanics, the state of a physical system at a fixed time t_0 is defined by specifying a ket $|\psi(t_0)\rangle$ belonging to the space ε. ε is a complex, separable ► Hilbert space, a complex linear vector space in which an inner product is defined and which possesses a countable ► orthonormal basis. The vectors in such a space have the properties mathematical objects must have in order to be capable of describing a quantum system.

In the Heisenberg picture the time evolution of a physical system is described as a continuous, passive unitary transformation. Passive unitary transformations change the basis vectors but leave the state vectors unchanged. ► Operators are defined through their action on the basis vectors and therefore change under a passive unitary transformation.

Let the state vector in the Heisenberg picture be $|\psi_H\rangle$ at $t = t_0$. As the system evolves, the state vector will not change. The ► Schrödinger equation is replaced by an equation describing the time evolution of any operator Ω_H in the Heisenberg picture. If the operator does not depend explicitly on time, then

$$\frac{d\Omega_H}{dt} = \frac{1}{i\hbar}\left[\Omega_H, H_H\right].$$

The Heisenberg picture leads to equations similar to the classical equations of motion and is often used to explore general properties of quantum systems and the formal analogy between classical and quantum theory.

One can switch from the ► Schrödinger picture to the Heisenberg picture at any time t by applying a unitary transformation. The transformation

$$|\psi_H\rangle = U(t_0, t)|\psi_S(t)\rangle = U^T(t, t_0)|\psi_S(t)\rangle = |\psi_S(t_0)\rangle$$

yields the state vectors $|\psi_H\rangle$ in the Heisenberg picture given the state vector $|\psi_S(t)\rangle$ in the Schrödinger picture, and the transformation

$$\Omega_H(t) = U^T(t, t_0)\Omega_S U(t, t_0)$$

yields the operator $\Omega_H(t)$ in the Heisenberg picture given the operator Ω_S in the Schrödinger picture. This is a change of representation. The matrix elements of any operator, Ω, are independent of the representation.

H

Primary Literature

1. C. Cohen-Tannoudji, B. Diu, F. Laloë: *Quantum Mechanics*, Volume 1 (Wiley, New York 1977, 312–314)

Heisenberg Uncertainty Relation (Indeterminacy Relations)

Paul Busch and Brigitte Falkenburg

The term Heisenberg uncertainty relation is a name for not one but three distinct trade-off relations which are all formulated in a more or less intuitive and vague way in Heisenberg's seminal paper of 1927 [1]. These relations are expressions and quantifications of three fundamental limitations of the operational possibilities of preparing and measuring quantum mechanical systems which are stated here informally with reference to position and momentum as a paradigmatic example of canonically conjugate pairs of quantities:

(A) *It is impossible to prepare states in which position and momentum are simultaneously arbitrarily well localized. In every state, the probability distributions of these ► observables have widths that obey an uncertainty relation.*

(B) *It is impossible to make joint measurements of position and momentum. But it is possible to make approximate joint measurements of these observables, with inaccuracies that obey an uncertainty relation.*

(C) *It is impossible to measure position without disturbing momentum, and vice versa. The inaccuracy of the position measurement and the disturbance of the momentum distribution obey an uncertainty relation.*

Of these three statements, only (A) was immediately given a precise formulation. Heisenberg only proved $\Delta(Q,\varphi)\Delta(P,\varphi) = \hbar/2$ for the standard deviations of position Q and momentum P in a Gaussian state φ; this was successively generalized soon afterwards by Weyl, Kennard, Robertson and Schrödinger, and the most general form for two observables represented as ▶ selfadjoint operators A, B is given by

$$\Delta(A,T)^2\Delta(B,T)^2 \geqslant \tfrac{1}{4}|\langle[A,B]\rangle_T|^2 + \tfrac{1}{4}[\langle\{A,B\}_+\rangle_T - 2\langle A\rangle_T\langle B\rangle_T]^2. \quad (1)$$

Here the notation $\langle X\rangle_T := \mathrm{tr}[TX]$ is used for the expectation value of an operator X in a state T, and $\Delta(X,T)^2 := \langle X^2\rangle_T - \langle X\rangle_T^2$; further, $[A,B] = AB - BA$ and $\{A,B\}_+ = AB + BA$. Relation (1) holds for all states T for which all expectation values involved are well-defined and finite. For an account of the early formal and conceptual developments of the uncertainty relation the reader is referred to the monograph [9].

It should be noted that uncertainty relations can be formulated in terms of other measures of the widths of the relevant probability distributions; these are sometimes more stringent than the above, particularly in cases where the standard deviation is infinite or otherwise an inadequate representation of the width.

The Heisenberg uncertainty relation (1) is commonly called indeterminacy relation, reflecting the interpretation that this relation expresses an objective limitation on the definition of the values of noncommuting quantities and not just a limitation to accessing knowledge about these values. Successful tests of the uncertainty relation in single-slit and interferometric experiments with neutrons and recently with fullerenes have been reported in [2–5].

The other two uncertainty relations, (B) and (C), have proved significantly harder to make precise and prove. Heisenberg only illustrated their validity by means of idealized thought experiments, such as the ▶ γ-ray microscope experiment and the single- or ▶ double-slit experiment. Other authors, notably Einstein, Margenau and Popper, proposed experiments which were intended to demonstrate that the uncertainty relations are only statistically relevant and have no bearing on the properties of the individual quantum system.

In recent quantum optics, a *which way* thought experiment was proposed in order to show that Niels Bohr's ▶ *complementarity* principle is more fundamental than the uncertainty relation (C) [6]. A polemic debate arose about this question [7]. Finally, a ▶ *which way* experiment with single atoms showed that for the “complementary” observables D (path distinguishability) and V (visibility of interference fringes) a duality relation holds, which is indeed a generalized type (A) uncertainty relation [7, 8, 11]. Hence, a debate on (C) could be settled in terms of (A).

A general proof of both (B) and (C) (assuming that these relations are in fact valid) requires the development of a theory of approximate joint measurements (► *observable*) of noncommuting observables, which has become possible on the basis of the generalized notion of an ► *observable* represented as a positive operator measure (POM) and the corresponding extended measurement theory. The quality of the approximation of one observable by means of another can be assessed and quantified by comparing the associated probability distributions. Similarly, the disturbance of one observable, B, due to the measurement of another one, A, can be quantified by a comparison of the probability distributions of B immediately before and after the measurement of A. In the case of position and momentum, the theory of approximate joint measurements is well developed and has led to rigorous formulations of trade-off relations in the spirit of (B) and (C). The conceptual development that has led to this result is reviewed in [10]. Work on obtaining formalizations of (B) and (C) for general pairs of noncommuting observables is still under way.

H

Primary Literature

1. Heisenberg, W.: Über den anschaulichen Inhalt der quantentheoretischen Kinematik und Mechanik. Zeitschrift für Physik **43**, 172–198 (1927).
2. Shull, C.: Single-slit diffraction of neutrons, Physical Review **179**, 752–754 (1969).
3. Kaiser, H., S.A. Werner, E.A. George: Direct measurement of the longitudinal coherence length of a thermal neutron beam. Physical Review Letters **50**, 560–563 (1983).
4. Klein, A.G., G.I. Opat, W.A. Hamilton: Longitudinal coherence in neutron interferometry. Physical Review Letters **50**, 563–565 (1983).
5. Nairz, O., M. Arndt, A. Zeilinger: Experimental verification of the Heisenberg uncertainty principle for fullerene molecules. Physical Review A **65**, 032109/1-4 (2002).
6. Scully, M.O., B.G. Englert, H. Walther: Quantum optical tests of complementarity. Nature **351**, 111–116 (1991).
7. Storey, E.P., S.M. Tan, M.J. Collet, D.F. Walls: Path detection and the uncertainty principle. Nature **367**, 626–628 (1994). Complementarity and Uncertainty. Nature **375**, 368 (1995).
8. Dürr, S., T. Nonn, G. Rempe: Origin of quantum-mechanical complementarity probed by 'which-way' experiment in an atom interferometer. Nature **395**, 33–37 (1998). Dürr, S., G. Rempe: Can wave-particle duality be based on the uncertainty relation? American Journal of Physics **68**, 1021–1024 (2000).

Secondary Literature

9. Jammer, M.: The Philosophy of Quantum Mechanics (Wiley, New York, 1974).
10. Busch, P., T. Heinonen, P.J. Lahti: Heisenberg's uncertainty principle. arXiv:quant-ph/0609185.
11. Falkenburg, B.: Particle Metaphysics (Springer, Berlin, 2007) 296–311.

Hermitian Operator

See ► Hilbert space, Indistinguishability, Operator, Propensities in Quantum Mechanics, Rigged Hilbert Space in Quantum Mechanics, Self-adjoint operator, Superposition Principle, Wave Mechanics.

Hidden Variables

B. J. Hiley

Standard quantum mechanics, in the hands of von Neumann, makes the assumption that the ► wave function, $\psi(\boldsymbol{r}, t)$, provides the most complete description of state of an evolving system. It then uses the Born probability postulate (► Born rule) and assumes that the probability of finding the system at position $\boldsymbol{r}$ at time t is given by $P = |\psi(\boldsymbol{r}, t)|^2$. This gives an essentially statistical theory, ► probability interpretation but a statistical theory unlike those found in classical situations where all the dynamical variables such as position, momentum, angular momentum etc., are well defined but unknown.

The dynamical variables of a quantum system are determined by the eigenvalues of operators called ► 'observables'. Given a quantum state, not all the dynamical variables have simultaneous values. For example, if the position is sharply defined, then the momentum is undefined and *vice-versa*. In other words there exist sets of complementary variables such that if one set are well defined, the other set are completely undefined. This is the feature that underlies the ► Heisenberg uncertainty principle.

Furthermore it is assumed that the complementary set of variables cannot even be postulated to exist with unknown numerical values. Thus ► quantum statistics do not emerge from averaging over a set of unknown parameters. This means that quantum statistics must have a very different origin from classical statistics and these statistics are totally different from the statistics that arise, for example, in statistical mechanics. This surprising result was already noticed by Born when he first introduced the ► probability interpretation. He wrote "But, of course, anybody dissatisfied with these ideas may feel free to assume that there are additional parameters not yet introduced into the theory which determine the individual event" [1]. These new variables could then be regarded as *hidden*. This then is one of the ideas lying behind the search for a hidden variable interpretation of quantum theory.

This point of view was strongly opposed by Bohr on what today would be regarded as a philosophical argument. For Bohr the ► Heisenberg uncertainty relations implied an *indivisibility of the quantum of action*, which in turn implied that it was not possible to make a sharp separation between the properties of the observed system and those of the observing apparatus. In other words, quantum phenomena introduced a radically novel notion wholeness, where it is impossible to make a sharp separation between what is being observed and the means used for its observation. If this proposition is correct then it is, in principle, not possible to introduce other, unknown variables belonging to the observed system which could be integrated over to obtain the required statistics. Technically this is summarised with the statement that no *dispersion free* ensembles exist. (▶ Ensembles in Quantum Mechanics). The existence of such ensembles would imply that it is possible to make a sharp separation between the observed and the means of observation.

Mathematical support for "no dispersion free ensembles" came from von Neumann, who in his classic book *Mathematical Foundations of Quantum Mechanics* claimed to have proved that no dispersion free ensembles could exist without destroying the predictions of the formalism. Von Neumann writes "Nor would it help if there existed other, as yet undiscovered, physical quantities, in addition to those represented by the operators in quantum mechanics, because the relations assumed in quantum mechanics would have to fail already for the by now known quantities discussed above [His postulates I and II]. It is therefore not, as is often assumed, a question of a re-interpretation of quantum mechanics – the present system of quantum mechanics would have to be objectively false, in order that another description of the elementary processes than the statistical one be possible." [2]

Although there were some objections raised against the precise nature of the proof, there was a consensus view that von Neumann was right and it was, in fact, not possible to reproduce the results of the quantum formalism using hidden variables [19]. In other words it was generally believed that von Neumann's theorem had carried the day. Indeed Wiener sums up the situation very nicely. He writes "One might suppose that it is still possible to maintain that a particle such as an electron still has a definite momentum and a definite position, whether we can measure them simultaneously or not, and that there are precise laws of motion into which this position and momentum enter. Von Neumann has shown that this is not the case, and that the indeterminacy of the world is genuine and fundamental." [3]

However in 1952 Bohm [4] produced a counter example to the von Neumann theorem showing that it was, in fact, possible after all to reproduce exactly all the results of the quantum formalism by attributing definite values to all the dynamical variables such as position, momentum, angular momentum, etc. To carry this through consistently and in agreement with the uncertainty principle, it was necessary to assume the values of the complementary set to be definite but unknown. In other words the complementary set could be assumed to be the 'hidden variables'. In this way it was not necessary to add any new exotic variables but merely to assume that a particle had all its dynamical variables well-defined and having definite values. It was simply that we could not measure all the values simultaneously so that the complementary set must remain unknown. Some features of the Bohm model had been anticipated years before by de Broglie [5, 6] but he had not been able to counter the objections raised by Pauli [7]. One of the important features of Bohm's approach was to answer these objections and show the model provided a consistent account of quantum phenomena [8, 9].

The appearance of this counter example led to a revival of interest, not only in hidden variable theories themselves [18], but also in attempts to generalise von Neumann's theorem which clearly did not lead to the type of general conclusions claimed for it by Wiener [3]. It was not until 1966 that Bell [10] pointed out exactly where the limitations of the von Neumann proof and its subsequent generalisations [11–13] lay. Although these authors attempted to assume as little as possible about quantum mechanics, what they did assume did not apply to a whole raft of possible hidden variable theories including the model proposed by Bohm.

Specifically they made the restrictive assumption that the dynamical variables of a system must be simultaneously eigenvalues of *all* the dynamical ► operators whether they commuted or not. However as we have seen, we can only measure, and therefore know, the values of a commuting subset of operators in a given situation so why make that particular assumption? Why not attribute eigenvalues only to one set of variables, while the values of the complementary set were not necessarily eigenvalues? This complementary set only become eigenvalues when measurements corresponding to their operators are actually made. This is what the ► Bohm model does.

In this model these new measurements can actually change the values of the dynamical variables so that they are, in general, no longer eigenvalues of the first set of operators. In this sense measurement is "participatory" and is not passively revealing what is already there. Thus the values attributed to the particle depend on a given context defined by the given experimental arrangement. This supports Bohr's view of "the impossibility of a sharp separation between the behaviour of atomic objects and the interaction with the measuring instruments which serve to define the conditions under which the phenomena appear." [14].

Although there is no mathematical way to exclude the type of hidden variable theory introduced by Bohm, there is still a considerable debate as to whether such theories are physically viable. For example, in the Bohm approach particles in ► entangled states are non-locally connected [15]. Indeed it was the Bohm model that led Bell [10] to ask if all theories that attributed simultaneous well defined values to all dynamical variables were non-local. What Bell [16] was able to show was that all local theories must satisfy an inequality (Bell inequalities, ► Bell's theorem), which was not satisfied by the quantum formalism and, more importantly, experiments were shown to violate the inequality. Even though the Bohm approach accounts for this ► non-locality, there is still a general reluctance to accept such approaches even when extended to field theories [17].

An excellent review of the history of the evolution of hidden variable theories will be found in Belinfante [18] and Jammer [19]. For a critical appraisal of hidden variable theories and their relation to non-locality see Bell [20]. See also Bohm Interpretation; Bohmian Mechanics.

Primary Literature

1. M. Born: Zur Quantenmechanik der Strossvorgänge. Z. Phys. **37**, 863–867 (1927)
2. J. von Neumann: *Mathematical Foundations of Quantum Mechanics* (Princeton University Press, Princeton 1955, p. 324)

3. N. Wiener: The Role of the Observer. Phil. Sci. **3**, 307–319 (1936)
4. D. Bohm: A Suggested Interpretation of the Quantum Theory in Terms of Hidden Variables, I and II. Phys. Rev. **85**, 166–179; 180–193 (1952)
5. L. de Broglie: La méchanique ondulatoire et la structure atomique de la matiére et du rayonnement. J. Phys. Radium, 6e series **8**, 225–241 (1927)
6. L. de Broglie: *Non-linear Wave Mechanics: a Causal Interpretation* (Elsevier, Amsterdam 1960)
7. W. Pauli: *Reports on the 1927 Solvey Congress* (Gauthiers-Villars et Cie, Paris 1928, p. 280)
8. D. Bohm, B. J. Hiley: *The Undivided Universe: an Ontological Interpretation of Quantum Theory* (Routledge, London 1993)
9. P. R. Holland: *The Quantum Theory of Motion* (Cambridge University Press, Cambridge 1993)
10. J. S. Bell: On the Problem of Hidden Variables in Quantum Theory. Rev. Mod. Phys. **38**, 447–452 (1966)
11. A. M. Gleason: Measures on the Closed Sub-spaces of Hilbert Space. J. Math. Mechs. **6**, 885–893 (1957)
12. J. M. Jauch, C. Piron: Can Hidden Variables be Excluded from Quantum Mechanics. Helv. Phys. Acta **36**, 827–837 (1963)
13. S. Kochen, E. P. Specker: The Problem of Hidden Variables in Quantum Mechanics. J. Math. Mech. **17**, 59–87 (1967)
14. N. Bohr: *Atomic Physics and Human Knowledge* (Science Editions, New York 1961, p. 39)
15. D. Bohm, B. J. Hiley: On the Intuitive Understanding of Nonlocality as Implied by Qauntum Theory. Found. Phys. **5**, 93–109 (1975)
16. J. S. Bell: On the Einstein-Podolsky-Rosen paradox. Physics **1**, 195–200 (1964)
17. D. Bohm, B. J. Hiley, P. N. Kaloyerou: An Ontological Basis for the Quantum Theory: II - A Causal Interpretation of Quantum Fields. Phys. Reports **144**, 349–375 (1987)

H

Secondary Literature

18. F. J. Belinfante: *A Survey of Hidden-Variables Theories* (Pergamon Press, Oxford 1973)
19. M. Jammer: *The Philosophy of Quantum Mechanics* (Wiley-Interscience, New York 1974)
20. J. S. Bell: *Speakable and Unspeakable in Quantum Mechanics* (Cambridge University Press, Cambridge 1987)

Hidden-Variables Models of Quantum Mechanics (Noncontextual and Contextual)

Abner Shimony

In the following discussion of hidden variables models of quantum mechanics the ► Hilbert space formulation of quantum mechanics and the standard interpretation of its notation and concepts will be taken to be initially understood, even though challenges to the standard interpretation are implicit in the proposals of ► hidden variables.

Very soon after the formulation of the new quantum mechanics by Werner Heisenberg (1901–76) and Erwin Schrödinger (1887–1961) its advocates, notably Niels Bohr (1885–1962) [1], made strong claims that the new theory provided a complete framework for physics. Nevertheless, conjectures that quantum mechanics does not provide a complete description of physical reality materialized in each of the two competing (but equivalent, as was eventually recognized) formulations of the theory by Heisenberg and Schrödinger. The ► Heisenberg Uncertainty Principle – asserting a limitation on the precision of simultaneous determinations of position and linear momentum – suggested to Albert Einstein (1879–1955) [2] that the uncertainty was due to limitations of customary experimentation, and that two quantum mechanically incompatible quantities could in principle be shown to have simultaneous precise values by more sophisticated measuring procedures. Max Born's (1882–1970) ► probabilistic interpretation [3] of Schrödinger's wave function – that the ► wave function $\Psi(\boldsymbol{r},\ t)$, where $\boldsymbol{r}$ is position of a particle in three-space and t is the time coordinate, is connected with a physically observable quantity by the rule

$$|\psi(\boldsymbol{r},t)|^2 \mathrm{d}\boldsymbol{r} = \text{probability that at time } t \text{ the particle is found in the interval } (\boldsymbol{r}, \boldsymbol{r}+\mathrm{d}\boldsymbol{r}) \tag{1}$$

– suggested to Louis de Broglie (1892–1987) [4] and Einstein [2] that quantum mechanics, despite its predictive power is an incomplete physical theory in a manner analogous to the relation between classical statistical mechanics and classical mechanics. This appeal to an analogy was greatly strengthened by Einstein's paper (► EPR) with Podolsky and Rosen [5] in 1935, studying a wave function in which the positions of particles 1 and 2 are strictly correlated when ψ is expressed in the position representation, and their linear momenta are strictly correlated when it is expressed in the momentum representation. They postulate a sufficient condition for the existence of an element of physical reality: "If, without in any way disturbing a system, we can predict with certainty (i.e., with probability equal to unity) the value of a physical quantity, then there exists an element of physical reality corresponding to this physical quantity" [5, p. 777]. When this sufficient condition is applied to the pair of correlated particles 1 and 2, with the tacit assumption that the outcome of a measurement on one of the particles cannot causally affect the outcome of a measurement on the other – a consequence of relativistic causality if the two measurements are events with space-like separation – they inferred that both position and linear momentum are elements of physical reality of both 1 and 2. This conclusion suggested models in which the quantum state was regarded as an incomplete description of physical reality, in need of supplementation by "hidden variables." In spite of John von Neumann's [6] argument of 1932 (influential but later shown to inconclusive), that a hidden variables model cannot agree with all of the experimental predictions of standard quantum mechanics, and Bohr's widely accepted epistemological critique [7] in1935 of EPR's argument, the early attraction of hidden variables survived (undoubtedly because of Einstein's prestige) at least as a heterodox curiosity, but it finally was seriously investigated with greater subtlety in the latter half of the twentieth century.

Important new subtleties were the distinction between "noncontextual" and "contextual" hidden-variable models, first articulated explicitly (though without these names) by Bell [8] in 1966, and the recognition that these two kinds of models required different analyses. A noncontextual hidden variables model postulated that an isolated physical system is characterized by a complete state λ, which is the compendium of the real properties of the system at a definite time – the prototype being a point in the Gibbsian phase space of a classical mechanical system. When λ is given, then the result of measuring any property A of the system at the given time by an ideal measuring apparatus (one that introduces no distortions due to its own imperfections) is a function $A(\lambda)$. The outcome of the measurement is assumed to be independent of other properties $B, C, \ldots$ that may be measured simultaneously with A, and indeed such an independence may be tacitly assumed to be intrinsic to the ideal character of the measurement process.

H

The program of noncontextual hidden variables models was demonstrated in various ways to be incompatible with the predictions of quantum mechanics for a system associated with a Hilbert space of dimension 3 or greater – by ► Gleason [9], Bell [10], ► Kochen and Specker [11], Belinfante [12], Mermin [13], and others. A particularly simple proof was given by Belinfante and followers concerning a system of spin unity (neglecting the configuration space variables of this system), for which quantum mechanics predicts the following constraint: the measurement of two of the squared components of spin $s_x{}^2$, $s_y{}^2$, $s_z{}^2$ – where $\boldsymbol{x}$, $\boldsymbol{y}$, and z are three orthogonal directions – will yield value 1 (in units of ► Planck's constant h divided by 2π) and one of them will yield value 0. In a noncontextual model the complete state λ will ascribe values to each component of ► spin, regardless of what other components are measured with it. The proof of incompatibility of the noncontextual hidden variables assignment of definite values to all spin components proceeds by cleverly choosing an appropriate set of directions $\boldsymbol{n}$, most belonging to more than one orthogonal triad of directions in the set, and then showing that the quantum mechanical constraint on values of $s_n{}^2$ can be satisfied only if for some $\boldsymbol{n}$ this value is 1 when $\boldsymbol{n}$ is measured along with $\boldsymbol{r}$ and $\boldsymbol{s}$ in one orthogonal triad and 0 when it is measured along with $\boldsymbol{r}'$ and $\boldsymbol{s}'$ in another orthogonal triad. (The number of directions considered in this proof is 138. In other proofs fewer directions suffice but the argumentation is more complex.)

John Stewart Bell (1928–90) gave a new lease on life to the program of hidden variables by proposing *contextuality*. In the physical example just considered the complete state λ in a contextual hidden variables model would indeed ascribe an antecedent element of physical reality to each squared spin component $s_n{}^2$ but in a complex manner: the outcome of the measurement of $s_n{}^2$ is a function $s_n{}^2$ (λ, C) of the hidden variable λ and the *context* C, which is the set of quantities measured along with $s_n{}^2$. If the context C is the pair $(s_u{}^2, s_v{}^2)$, then $s_n{}^2$ (λ, C) is 1, and if C is $(s_{u'}{}^2, s_{v'}{}^2)$ the value is 0. In other words, the demonstration by Belinfante and his followers of the impossibility of a noncontextual hidden variables theory for quantum mechanics is converted into a demonstration of the compatibility of a contextual theory. Bell argues practically that "the result of an observation may reasonably depend not only upon the state of the system (including the hidden variables) but also on the complete disposition of the apparatus" [14].

Two important questions remain concerning contextual hidden variables models: how do they account for the probabilistic character of quantum mechanical predictions, and what are the constraints on the context C?

The first question is answered by assuming an appropriate probability distribution ρ over the space Λ of hidden variables. The specification of λ is determined by the physical circumstances which determine the quantum state of the system – viz. the mode of preparation of the state and interactions with the environment of the system – and when these circumstances are not sufficiently precise to fix λ exactly they may suffice to determine a distribution ρ over the space Λ. The integral $\int A(\lambda, C)\mathrm{d}\rho$ over the space Λ will recover the quantum mechanical expectation value of the quantity A if the contextual hidden variables theory is properly constructed.

As to the second question, a minimum constraint on the context C is that it consist of quantities that are quantum mechanically compatible, that is represented by ► self-adjoint operators which commute with each other. If A is a projection operator P (► projection) of interest, a natural context with this property is a maximal Boolean algebra of projection operators containing P, studied intensively by Stanley P. Gudder [15].

Another reasonable constraint on C of great conceptual importance was proposed by Bell when the system of interest consists of two or more spatially separated parts, and the physical quantity of interest A concerns one of these parts. C should not include quantities whose measurements are events with space-like separation from the measurement of A, since there would be a violation of relativistic locality if those measurements affected the outcome of the measurement of A. This ► locality constraint on the context has been studied intensively by Bell and his followers. When the context C satisfies the locality constraint, Bell and his followers derived inequalities which are violated by the quantum mechanical predictions of a large class of systems [16]. Consequently, even though contextual hidden variables models may agree with the predictions of quantum mechanics when the locality constraint is not imposed on C, no local contextual hidden variables model can recover all the quantum mechanical predictions. Very briefly, without providing details, we can assert that experimental tests of local contextual hidden variables models against quantum mechanics have strongly supported the latter [17]. See also ► Bohm Interpretation, Bohmian Mechanics.

Literature

1. N. Bohr: *Atomic Physics and Human Knowledge* (Wiley, New York, 1958, 60–61)
2. A. Einstein: In *Electrons et Photons – Rapports et Discussions du Cinquième Conseil de Physique Tenu à Bruxelles... sous les Auspices de l'Institut International de Physique Solvay* (Gauthier-Villars, Paris 1928, 256)
3. M. Born: Quantenmechanik der Stossvorgänge. Zeitschrift fur Physik **38**, 803 (1926)

4. L. de Broglie: La mécanique ondulatoire et la structure atomique de la matiêre et du rayonnement. Journal de Physique et du Radium **8**, 225 (1927)
5. A. Einstein, B. Podolsky, N. Rosen: Can quantum-mechanical description of physical reality be considered complete? Physical Review **47**, 777 (1935)
6. J. von Neumann: *Mathematical Foundations of Quantum Mechanics* (Princeton University Press, Princeton, 1955, 305–324)
7. N. Bohr: Can quantum-mechanical description of physical reality be considered complete? Physical Review **48**, 696 (1935)
8. J.S. Bell: On the problem of hidden variables in quantum theory. Reviews of Modern Physics **38**, 450 (1966). Reprinted with numerous related articles in *Speakable and Unspeakable in Quantum Mechanics* (Cambridge University Press, Cambridge, 2004). The name "contextualistic" was introduced by A. Shimony: Experimental test of local hidden variable theories, in B. d'Espagnat ed.: *Foundations of Quantum Mechanics* (Academic, New York 1971), and a shortening to "contextual" was performed by E.G. Beltrametti and C. Cassinelli: *The Logic of Quantum Mechanics* (SIAM, 1983)
9. A. Gleason: Measures on the closed subspaces of a Hilbert space. Journal of Mathematics and Mechanics **6**, 885 (1957)
10. J.S. Bell: Ref. [8], sect. 5.
11. S. Kochen, E. Specker: The problem of hidden variables in quantum Mechanics. Journal of Mathematics and Mechanics **17**, 59 (1967)
12. F. Belinfante: *A Survey of Hidden-Variables Theories* (Pergamon, Oxford, 1973). This sketch of a proof is filled out in E. de Obaldia, A. Shimony, and F. Wittel: Amplification of Belinfante's demonstration of the non-existence of a dispersion free state. Foundations of Physics **18**, 1013 (1988)
13. N.D. Mermin: Hidden variables and the two theorems of John Bell. Reviews of Modern Physics **65**, 803 (1993)
14. J.S. Bell: Ref. [8] (1966) p. 451 and (2004), p. 9.
15. S. Gudder: On hidden-variable theories. Journal of Mathematical Physics **11**, 431 (1970)
16. J.S. Bell: On the Einstein-Podolsky-Rosen Paradox. Physics **1**, 195 (1964); reprinted in Bell, Ref. [8] (2004), 14.
17. A. Aspect: Introduction: John Bell and the second quantum revolution, in Bell: Ref. [8] (2004), xvii–xxxix

Hilbert Space

Erhard Scholz and Werner Stulpe

Hilbert space, a generalization of the concept of Euclidean vector space, i.e., of a finite-dimensional real vector space equipped with a scalar product. A *Hilbert space* $\mathcal{H}$ [7–12] is a vector space over the real or complex numbers (sometimes over the quaternions) in which a scalar product is defined and which is complete w.r.t. the norm induced by the scalar product.

The *scalar product* in a complex Hilbert space $\mathcal{H}$ associates any two vectors $\phi, \psi \in \mathcal{H}$ with a complex number $\langle\phi|\psi\rangle$ such that (i) $\langle\phi|\psi\rangle$ is *linear* in ψ, i.e., $\langle\phi|\chi + \psi\rangle = \langle\phi|\chi\rangle + \langle\phi|\psi\rangle$ and $\langle\phi|\lambda\psi\rangle = \lambda\langle\phi|\psi\rangle$ where $\phi, \chi, \psi \in \mathcal{H}$ and $\lambda \in \mathbb{C}$, (ii) $\langle\phi|\psi\rangle = \overline{\langle\psi|\phi\rangle}$ where the bar denotes complex conjugation, (iii) $\langle\phi|\phi\rangle \geqslant 0$ for all $\phi \in \mathcal{H}$, and (iv) $\langle\phi|\phi\rangle = 0$ if and only if $\phi = 0$; as a consequence of (i) and (ii), the scalar product is *antilinear* in the first argument, i.e.,

$\langle\phi+\chi|\psi\rangle = \langle\phi|\psi\rangle + \langle\chi|\psi\rangle$ and $\langle\lambda\phi|\psi\rangle = \overline{\lambda}\langle\phi|\psi\rangle$. Property (i) refers to the physicists' convention, according to the mathematicians' convention the scalar product is linear in the first argument and antilinear in the second.—For a real Hilbert space, $\lambda \in \mathbb{C}$ in (i) is replaced by $\lambda \in \mathbb{R}$, (ii) reads $\langle\phi|\psi\rangle = \langle\psi|\phi\rangle$, and the scalar product is linear in both arguments.

An important consequence of the properties (i)–(iv) of the scalar product is the *Cauchy-Schwarz inequality*, stating that $|\langle\phi|\psi\rangle| \leqslant \|\phi\|\,\|\psi\|$ where $\|\phi\| = \sqrt{\langle\phi|\phi\rangle}$. This inequality becomes an equality if and only if the vectors ϕ and ψ are linearly dependent. The properties (i)–(iv) and the Cauchy-Schwarz inequality entail that the association of every $\phi \in \mathcal{H}$ with the real number $\|\phi\|$ is a *norm*, i.e., (i) $\|\phi\| \geqslant 0$ and $\|\phi\| = 0$ if and only if $\phi = 0$, (ii) $\|\lambda\phi\| = |\lambda|\,\|\phi\|$ where $\phi \in \mathcal{H}$ and $\lambda \in \mathbb{C}$ ($\lambda \in \mathbb{R}$ in case of a real Hilbert space), and (iii) the triangle inequality holds, i.e., $\|\phi + \psi\| \leqslant \|\phi\| + \|\psi\|$ where $\phi, \psi \in \mathcal{H}$. In the triangle inequality of a norm that is induced by a scalar product, equality holds if and only if the vectors ϕ and ψ are linearly dependent.

The Hilbert-space norm enables one to define some analytical and topological concepts in $\mathcal{H}$. In particular, a sequence of vectors $\phi_n \in \mathcal{H}$ *converges to the limit* $\psi \in \mathcal{H}$ if $\|\phi_n - \psi\| \to 0$ as $n \to \infty$, i.e., for every $\epsilon > 0$ there exists a positive integer $N(\epsilon)$ such that $\|\phi_n - \psi\| < \epsilon$ for $n \geqslant N(\epsilon)$. A sequence of vectors ϕ_n is called a *Cauchy sequence* if, for every $\epsilon > 0$, there exists an $N(\epsilon)$ such that $\|\phi_n - \phi_m\| < \epsilon$ for all $m, n \geqslant N(\epsilon)$. Every convergent sequence is a Cauchy sequence; conversely, in the general case of a vector space equipped with a norm, a Cauchy sequence need not have a limit. By definition, a Hilbert space is *complete*, i.e., every Cauchy sequence in $\mathcal{H}$ is convergent.

A subset S of a Hilbert space $\mathcal{H}$ is called *dense in* $\mathcal{H}$ if every ϵ-neighborhood of any $\psi \in \mathcal{H}$ contains an element $\phi \in S$, i.e., for any $\psi \in \mathcal{H}$ and every $\epsilon > 0$ there exists a vector $\phi \in S$ such that $\|\phi - \psi\| < \epsilon$. A Hilbert space is called *separable* if there exists a sequence of vectors $\phi_n \in \mathcal{H}$ being dense in $\mathcal{H}$.

A subset $S \subseteq \mathcal{H}$ is called *closed* if the limit of every in $\mathcal{H}$ convergent sequence of vectors $\phi_n \in S$ belongs to S, briefly, if from $\phi_n \in S$ and $\|\phi_n - \psi\| \to 0$ as $n \to \infty$, $\psi \in \mathcal{H}$, it follows that $\psi \in S$. A linear submanifold $\mathcal{S}$ of a Hilbert space can be closed (in which case $\mathcal{S}$ is often called a *subspace of* $\mathcal{H}$), but need not (a finite-dimensional submanifold is closed); a subspace is, with the scalar product inherited from $\mathcal{H}$, a Hilbert space itself. A linear submanifold can be dense in $\mathcal{H}$; dense submanifolds play an important role as domains of linear operators (► operator).

A real or complex vector space equipped with a norm is called a *normed space*. The concepts limit of a sequence, Cauchy sequence, completeness, dense subset or dense linear submanifold, closed subset or submanifold, and separability apply more generally to normed spaces. A complete normed space is called a *Banach space*. Hilbert spaces are particular Banach spaces, namely those whose norm is induced by a scalar product.

Two vectors ϕ, ψ of a Hilbert space $\mathcal{H}$ (two subsets S_1, S_2 of $\mathcal{H}$) are called *orthogonal to each other* if $\langle\phi|\psi\rangle = 0$ (if $\langle\phi|\psi\rangle = 0$ for all $\phi \in S_1$ and all $\psi \in S_2$). For a subset $S \subseteq \mathcal{H}$, the *orthocomplement* $S^\perp$ consists of all vectors $\chi \in \mathcal{H}$ satisfying $\langle\chi|\phi\rangle = 0$ for all $\phi \in S$; $S^\perp$ is a subspace, i.e., a closed linear submanifold. If

$\mathcal{X}$ is a subspace of $\mathcal{H}$, every vector $\psi \in \mathcal{H}$ can, according to $\psi = \phi + \chi$, uniquely be decomposed into a vector $\phi \in \mathcal{X}$ and a vector $\chi \in \mathcal{X}^{\perp}$; that is, the Hilbert space can be represented as the direct sum $\mathcal{H} = \mathcal{X} \oplus \mathcal{X}^{\perp}$. The latter decomposition of $\mathcal{H}$ entails that, for every subspace $\mathcal{X}$, there exists the *orthogonal projection onto* $\mathcal{X}$ (► projection).

A family of vectors $\phi_i \in \mathcal{H}$ where i belongs to some index set I, is called an *orthonormal system* if $\langle\phi_i|\phi_j\rangle = \delta_{ij}$. A maximal orthonormal system is called a *complete orthonormal system in* $\mathcal{H}$, a *Hilbert basis of* $\mathcal{H}$, or an ► *orthonormal basis*. In every Hilbert space, there exists a complete orthonormal system, and different such systems have the same cardinality, the latter being called the *Hilbert-space dimension of* $\mathcal{H}$. Given a Hilbert basis $\phi_i, i \in I$, every vector $\psi \in \mathcal{H}$ can be expanded into the series $\psi = \sum_{i\in I} \alpha_i\phi_i$ where $\alpha_i = \langle\phi_i|\psi\rangle$ and only countably many α_i are not zero. Different Hilbert spaces are *isomorphic*, i.e., there exists a one-to-one correspondence between the spaces that preserves linearity and the scalar products (► unitary operator), if and only if their bases have the same cardinality.—A Hilbert space is separable if and only if it has a countable Hilbert basis $\phi_1, \phi_2, \ldots$. All infinite-dimensional separable Hilbert spaces are isomorphic. Although in a separable infinite-dimensional Hilbert space there exist only countably many mutually orthogonal vectors, there always exist uncountably many linearly independent vectors.

The standard realization of a finite-dimensional complex (real) Hilbert space is the space $\mathbb{C}^n$ ($\mathbb{R}^n$). The straightforward infinite-dimensional generalization of $\mathbb{C}^n$ is the separable Hilbert space l^2 of the *square-summable* complex sequences $u = (\xi_1, \xi_2, \ldots)$, $\xi_i \in \mathbb{C}$, $\sum_{i=1}^{\infty} |\xi_i|^2 < \infty$, with the scalar product $\langle u|v\rangle = \sum_{i=1}^{\infty} \overline{\xi}_i\eta_i$. The other typical example of a separable infinite-dimensional Hilbert space is the space $L^2(M, \mathrm{d}x)$ of the (equivalence classes of the) *square-integrable* complex-valued functions on M where M is $\mathbb{R}$, $\mathbb{R}^n$, or a measurable subset of $\mathbb{R}^n$ of nonzero Lebesgue measure and $\mathrm{d}x$ indicates the Lebesgue measure; $\phi \in L^2(M, \mathrm{d}x)$ satisfies $\int_M |\phi(x)|^2\mathrm{d}x < \infty$, the scalar product is defined by $\langle\phi|\psi\rangle = \int_M \overline{\phi(x)}\psi(x)\,\mathrm{d}x$, and functions differing only on a set of measure zero are considered to be equal. More generally, if (Ω, Σ, μ) is any measure space, the space $L^2(\Omega, \Sigma, \mu)$ of the w.r.t. μ square-integrable functions on Ω is a (possibly nonseparable) Hilbert space. Besides $L^2(M, \mathrm{d}x)$, an important particular case is the separable Hilbert space $L^2(\mathbb{R}, \mu)$ where Σ is the σ-algebra of the Borel sets of $\mathbb{R}$ and μ a finite Borel measure.

Hilbert spaces are useful in functional analysis, in classical physics, and in quantum physics where they serve as state spaces of quantum systems. Their study was initiated in analytical terms by David Hilbert (1862–1943). His student Erhard Schmidt (1876–1959) introduced the geometric language of function spaces to the field. An axiomatic definition of infinite-dimensional separable Hilbert spaces was given by Johann von Neumann (1903–1957) in one of his first papers on the foundations of quantum mechanics [1].

The space l^2 was introduced by Hilbert in a famous series of publications on integral operators (1904–1910). He proved that integral operators with a symmetric kernel (those being particular compact self-adjoint operators, ► operator) can be diagonalized by a suitable change of the basis [2]. Moreover, l^2 was shown to be isomorphic to the Hilbert space $L^2(M, \mathrm{d}x)$ where M is the real line or any of its intervals (*Riesz-Fischer theorem*, 1906). David Hilbert and Erhard Schmidt further showed that every completely continuous ("vollstetig") Hermitian operator in a separable Hilbert space can be diagonalized, i.e., in modern language, every compact symmetric (self-adjoint) operator has a complete orthonormal system of eigenvectors (*Hilbert-Schmidt theorem*). This result can be generalized to the *spectral theorem* for all bounded Hermitian (self-adjoint) operators and even for unbounded self-adjoint ones (▶ self-adjoint operator). These insights became important for quantum theory.

In order to characterize states of a physical system which do not behave pointlike but appear in some way or other as "spread out" (probabilistically or in the sense of a classical continuous field), the quantum theorists of the 1920s found sufficiently useful infinite-dimensional linear spaces for describing and understanding basic quantum properties of matter. Werner Heisenberg (1901–1976), supported by Pascual Jordan (1902–1980) and Max Born (1882–1970), introduced infinite matrices (comparable to Hilbert's matrices in l^2, ▶ matrix mechanics) as essential symbolic representatives for the new quantum mechanics, whereas Erwin Schrödinger (1887–1961) introduced function spaces (similar to $L^2(M, \mathrm{d}x)$) for his wave functions (▶ wave function, ▶ wave mechanics) and linear operators as symbolic representatives. The seemingly different approaches of Heisenberg and Schrödinger were subsumed in a common formal framework by Paul A. M. Dirac (1902–1984) in his "bra-ket" formalism to express the duality structure of the underlying normed vector spaces. On the other hand they could be conceptually unified in the language of Hilbert spaces. The latter approach, at that time mathematically better founded, was initiated by David Hilbert, Lothar Nordheim (1899–1985), Johann von Neumann, and Hermann Weyl (1885–1955) between 1926 and 1928. It was spelt out by von Neumann in the late 1920s in a series of path-breaking publications.

Central to the usefulness of Hilbert spaces in quantum physics is the peculiar ▶ superposition of quantum probabilities which allows successfully to characterize pure (▶ states, pure & mixed) of quantum systems by normed Hilbert-space vectors or, more precisely, by rays in Hilbert space (a *ray* is a vector up to any complex nonzero factor). For mixed states, density matrices (▶ density operator) in Hilbert space have to be used [3, 4]. Physical quantities (▶ observable) can be encoded by self-adjoint operators and their spectrum, time evolution and symmetries by unitary group representations (▶ unitary operator, ▶ symmetry).

The most important operators used by Schrödinger are unbounded and are not defined on the entire Hilbert space $L^2(M, \mathrm{d}x)$. Thus the main challenge for von Neumann was to develop a whole new field of mathematical properties of unbounded operators acting in Hilbert space. In particular he succeeded in finding a convincing generalization of the spectral theorem for bounded Hermitian operators to the case of unbounded self-adjoint ones. On this basis he concluded Hilbert's attempts at a (first) axiomatization of quantum mechanics [4, 5]. His later researches on a quantum logical interpretation (▶ quantum logic) of the orthomodular lattice of the closed subspaces of a separable Hilbert space were less successful in achieving their original goals. They contributed, however, to a highly consequential research program for the study and classification of C^*-algebras (▶ algebraic quantum mechanics) [5].

Challenging questions remain open in the theory of nonseparable Hilbert spaces. These arise mathematically from infinite tensor products of separable Hilbert spaces and physically from the study of quantum fields.

Primary Literature

1. J. von Neumann: Mathematische Begründung der Quantenmechanik. Nach- richten Göttinger Gesellschaft der Wissenschaften, 1–57 (1927). In Collected Works, Vol. I (Pergamon, New York 1961, 151–207)
2. D. Hilbert: *Grundzüge einer allgemeinen Theorie der Integralgleichungen* (Teubner, Leipzig 1912)
3. H. Weyl: *Gruppentheorie und Quantenmechanik* (Hirzel, Leipzig 1928, second edition 1931); English *The Theory of Groups and Quantum Mechanics* (Dutten, New York 1931)
4. J. von Neumann: *Mathematische Grundlagen der Quantenmechanik* (Springer, Berlin 1932, reprint 1968); English *Mathematical Foundations of Quantum Mechanics* (Princeton University Press, Princeton 1955)
5. J. Lacki: The early axiomatization of quantum mechanics: Jordan, von Neumann and the continuation of Hilbert's program. Archive for History of Exact Sciences **54**, 279–318 (2000)
6. M. Rédei: Von Neumann's concept of quantum logic and quantum probability. In John von Neumann and the Foundations of Quantum Mechanics, ed. by M. Rédei and M. Stöltzner (Kluwer, Dordrecht 2005, 153–172)

Secondary Literature

7. F. Riesz and B. Sz.-Nagy: *Functional Analysis* (Frederick Ungar Publishing, New York 1955; reprint Dover Publications, Mineola 1991)
8. M. Reed and B. Simon: *Methods of Modern Mathematical Physics I: Functional Analysis* (Academic, New York 1972, second edition 1980)
9. W. Mlak: *Hilbert Spaces and Operator Theory* (Springer, Dordrecht 1991)
10. D. Werner: *Funktionalanalysis* (Springer, Berlin 1995, third edition 2000)
11. G. Ludwig: *Foundations of Quantum Mechanics I* (Springer, New York 1983), Appendix IV
12. E. Prugovečki: *Quantum Mechanics in Hilbert Space* (Academic, New York 1971, second edition 1981; reprint Dover Publications, Mineola 2006)

H

Holism in Quantum Mechanics

Richard Healey

In slogan form, holism is the thesis that the whole is more than the sum of its parts. *Explanatory holism* is the view that a satisfactory explanation of the behavior of a system cannot be given by explaining the behavior of its parts. *Property holism* is the view that the properties of a whole are not wholly determined by those of its parts. *Ontological holism* denies that some supposedly composite object has (proper) parts. Quantum phenomena exhibit holism of at least the first two kinds.

Quantum mechanics is often applied to a system as a whole, even though it is known to be composed of many subsystems. Such applications supply many instances of explanatory holism. Interference has been experimentally demonstrated between beams of sodium atoms and of fullerenes (C_{60} molecules ► mesoscopic quantum phenomena) [11]. The result of these experiments is readily explained by direct application of quantum mechanics to such composite objects. It would be futile to try to explain their behavior by applying quantum mechanics to their quark and lepton components. Many phenomena in condensed matter physics are explained by applying quantum mechanics directly to systems composed of very large numbers of atomic or subatomic particles: only in special cases can the theory be applied at the level of these components [9].

Even when classical physics is applied to the behavior of the solar system by treating planets as wholes, the planetary motions and interplanetary gravitational forces are readily understood to be constituted by the motions and gravitational interactions of their constituent particles in a way that permits a simple summation. But any attempt to analyze the behavior of a compound quantum system into the behavior of its components encounters a barrier: In quantum mechanics, the state of a compound system is not always determined by the states of its components: each such failure of determination in quantum mechanics is an example of *state holism*. Schrödinger called the subsystems in such a compound state 'entangled' [5]. Assuming a system's state specifies its properties, state holism implies property holism.

Consider two ► spin 1/2 particles that emerge from an interaction in the singlet state

$$| \psi_s \rangle = \frac{1}{\sqrt{2}} (|\uparrow\rangle \otimes |\downarrow\rangle - |\downarrow\rangle \otimes |\uparrow\rangle) \tag{1}$$

Suppose that before the interaction, the state of the ith particle was represented by a vector $| \psi_i \rangle \in \mathcal{H}_i$ $(i = 1, 2)$, where $\mathcal{H}_i$ is a 2-dimensional complex vector space. The state of the pair was then represented by the vector $| \psi_1 \rangle \otimes | \psi_2 \rangle$, an element of the 4-dimensional tensor product space $\mathcal{H} = \mathcal{H}_1 \otimes \mathcal{H}_2$. But while the vector $| \psi_s \rangle$ is also an element of $\mathcal{H}_1 \otimes \mathcal{H}_2$, there is no pair of vectors $| \varphi_1 \rangle \in \mathcal{H}_1$, $| \varphi_2 \rangle \in \mathcal{H}_2$ such that $| \psi_s \rangle = | \varphi_1 \rangle \otimes | \varphi_2 \rangle$. The singlet spin state is entangled: the state of neither particle may be represented by a vector in its own state space. There is a sense in which a typical state of a compound system is entangled: the set of entangled vectors in a tensor product Hilbert space is dense. Moreover, because it must be totally antisymmetric under particle exchange, *every* state-vector representing a system composed of more than one electron is entangled, whether or not these ► electrons have previously interacted.

It is still possible to represent the state of a component of an entangled state, not by a vector but by a density operator. Consider the more general entangled spin state

$$| \psi \rangle = \alpha \, (|\uparrow\rangle \otimes |\downarrow\rangle + \beta \, |\downarrow\rangle \otimes |\uparrow\rangle) \quad : \quad |\alpha|^2 + |\beta|^2 = 1 \tag{2}$$

Assignment of the reduced density operator $W_1 = |\alpha|^2 \, |\uparrow\rangle\langle\uparrow| + |\beta|^2 \, |\downarrow\rangle\langle\downarrow|$ to the first particle and $W_2 = |\beta|^2 \, |\uparrow\rangle\langle\uparrow| + |\alpha|^2 \, |\downarrow\rangle\langle\downarrow|$ to the second particle will predict the same statistics as $|\psi\rangle$ for the measurement of any spin magnitude on either particle alone. (These reduced states are arrived at by "tracing over" the ► Hilbert space of the rest of the system: see e.g. [7].) But note that if one does take W_i to be the state of the ith particle, then these states do *not* determine $|\psi\rangle$ as the state of the pair: many other states of the pair are equally compatible with individual states $\{W_1, W_2\}$, including $W_1 \otimes W_2$ and

$$| \psi\rangle = \alpha \, (|\uparrow\rangle\otimes |\downarrow\rangle - \beta \, |\downarrow\rangle\otimes |\uparrow\rangle) \tag{3}$$

If the state of an entangled component is represented by its reduced density operator, then these states fail to determine the state of the whole system.

H

A third option is to assign a *relative* state to each component in an entangled state. The first particle in (2) would be assigned state $|\uparrow\rangle$ relative to state $|\downarrow\rangle$ for the second particle, but state $|\downarrow\rangle$ relative to state $|\uparrow\rangle$ for the second particle. This option is favored by the so-called relational interpretation of quantum mechanics [8].

On each of these three options, quantum mechanics implies state holism. All three conflict with Einstein's view that, for a pair of separated systems *AB*

> The real state of the pair *AB* consists precisely of the real state of *A* and the real state of *B*, which states have nothing to do with one another. [4]

Bohm proposed an interpretation of quantum mechanics that seems to accord better with Einstein's view [1]. In its "minimal" version this takes the real state of a system of particles to be completely specified by the positions of all the particles. Each particle has a determinate trajectory, with velocity determined by the gradient of the phase of the particles' ► wave function, evaluated at the positions of all the particles. But this interpretation conflicts with property holism to the extent that the wave-function (or the resultant velocity field, or "quantum potential") must itself be included in the whole system to which quantum mechanics is applied. Bohm himself stressed the holism of the quantum world [2]. This is in keeping with the fact that on his interpretation the wave-function never "collapses" on measurement ► wave function collapse. Such "collapse" provided Schrödinger with a mechanism for periodically disentangling quantum states.

The indivisibility of a quantum field manifests a kind of holism. Their indiscernibility, superposability and failure of localization makes field quanta like photons (► light quantum) poor candidates for distinct parts of the field, suggesting ontological holism. If one insists on breaking the field into parts by covering space-time by open regions (as one does in algebraic quantum field theory), then one has a case of state holism: states on the local algebras of ► observables typically fail to determine a state on a global space-time algebra.

In the early days of quantum mechanics, Bohr advocated a different kind of holism. He took the essence of quantum theory to be expressed in

> ...the so-called quantum postulate, which attributes to any atomic process an essential discontinuity, or rather individuality...symbolized by Planck's quantum of action. [3]

He took this to imply that

> any observation of atomic phenomena will involve an interaction with the agency of observation not to be neglected. Accordingly, an independent reality in the ordinary physical sense can neither be ascribed to the phenomena nor to the agencies of observation. (ibid.)

The entire experimental arrangement, including both "atomic" system and measuring device must therefore be treated as an indivisible whole. Neither has a state independent of the other. The former may be ascribed a quantum state while the latter must be described classically. But the choice of the experimenter on how to divide the entire experimental arrangement into these two parts is to an extent arbitrary. Any ascription of quantum state is therefore doubly relative – to a choice of experimental arrangement, and to a subsidiary choice as to how to analyze the entire arrangement into parts. Only in this doubly relativized sense do quantum systems or measuring devices have properties. Such properties arc not independent of the arrangement and its division, and cannot therefore be taken to determine the properties of the whole experimental arrangement. This, too, is incompatible with property holism.

Primary Literature

1. D. Bohm: *A suggested interpretation of the quantum theory in terms of 'hidden' variables, I and II.* Physical Review 85: 166–93, 1952.
2. D. Bohm: *Wholeness and the Implicate Order* (New York: Routledge, Kegan Paul, 1980).
3. N. Bohr: *The quantum postulate and the recent development of atomic theory.* Nature 121: 580–90, 1928.
4. A. Einstein: *Letter to E. Schrödinger of June 19th, 1935*; quoted and translated by D. Howard in Einstein on Locality and Separability. Studies in History and Philosophy of Science 16: 171–201, 1985.
5. E. Schrödinger: *Discussion of probability relations between separated systems.* Proceedings of the Cambridge Philosophical Society 31: 555–63, 1935.

Secondary Literature

6. R. Healey: *Holism and nonseparability in physics* at http://plato.stanford.edu/entries/physics-holism/.
7. J.M. Jauch: *Foundations of Quantum Mechanics* (Reading, MA: Addison Wesley, 1968).
8. F. Laudisa, C. Rovelli: *Relational quantum mechanics* at http://plato.stanford.edu/entries/qm-relational/.
9. A.J. Leggett: *The Problems of Physics* (New York: Oxford University Press, 1987), p. 113.
10. T. Maudlin: *Part and whole in quantum mechanics*, in E. Castellani, ed. Interpreting Bodies (Princeton: Princeton University Press: 1998), 46–60.
11. O. Nairz, M. Arndt, A. Zeilinger: *Quantum interference experiments with large molecules.* American Journal of Physics 71: 319–25, 2003.

I

Identity of Quanta

Simon Saunders

Identity. From very early days of quantum theory it was recognized that quanta were statistically strange (see ▶ Bose–Einstein statistics). Suspicion fell on the *identity* of quanta, of how they are to be counted [1, 2]. It was not until Paul A. Dirac's (1902–1984) work of 1926 (and his discovery of ▶ Fermi–Dirac statistics [3]) that the nature of the novelty was clear: the quantum state of exactly similar particles of the same mass, charge, and ▶ spin must be *symmetrized*, yielding states either symmetric or antisymmetric under permutations. This is the *symmetry postulate* (SP).

The SP further implies that expectation values of particle ▶ observables are invariant under permutations. The latter looks temptingly like the sort of principle on which one might hope to found the theory of quantum identity. It is called the *indistinguishability postulate* (IP) – see ▶ indistinguishability. But it turns out to be weaker than the SP, the principle we are interested in.

The question we shall pose is this: what does the SP tell us about quantum ontology? By a large margin, the consensus today is that the founding fathers were on to something, and that the SP implies or otherwise reflects a *failure of particle identity* in quantum mechanics, whether identity over time, or identity at a time (or identity *simpliciter*, without regard to time). For quantum mechanics itself, even for exactly similar particles, does not require the SP; such particles can perfectly well be described by unsymmetrized states and their superpositions.

Identity over time. It is common to most interpretations of quantum mechanics that the underlying ontology need not be localized – that particles have no trajectories. In which case, there may be no good criterion of particle identity over time. (▶ See Consistent histories, Ignorance interpretation, Ithaca Interpretation, Many Worlds Interpretation, Modal Interpretation, Orthodox Interpretation, Transactional Interpretation).

Of course that cannot be the whole story: unsymmetrized quantum mechanical systems also lack trajectories, but obey Maxwell–Boltzmann statistics [4]. In fact, it is already over-simplistic: the existence or otherwise of trajectories is not an all or nothing affair. It is true that no continuous sequence of 1−particle states defines a curve in configuration space (or momentum space or any other sub-manifold of the classical phase space), but there are certainly evolutions under which symmetric and antisymmetric states define smooth curves ('orbits') of 1-particle states in *quantum* state space (▶ Hilbert space) – see ▶ indistinguishability. In terms of these the SP appears to have only a humble role, as ruling out any further fact as to which particle is attached to which orbit. The same can be said of the analogous symmetrization postulate as applied to classical particle trajectories [5].

D. Greenberger et al. (eds.), *Compendium of Quantum Physics: Concepts, Experiments, History and Philosophy*,

This point has appeared puzzling to some. Doesn't the SP imply the IP? If particles can be associated with 1-particle states, or orbits of such, why can't they be individuated accordingly, in violation of the IP? Surely in the classical case we can always distinguish the particle by the trajectory, in violation of the IP? [6, p.7-8]. But this is to confuse the question of which particle is in which state, or sequence of states, or trajectory, which cannot be determined by any observation according to the IP, with the question of what distinguishes the states, or sequences of states or trajectories from each other, which in principle is perfectly observable [7]. The atoms (1−particle states) in the bottle of helium by the door are distinguishable from those (1−particle states) in the laser trap in the corner.

The SP then, blocks the question of which particle is in which state, or sequence of states. Classically, by mean of the trajectories, one can still say of two particles at two different times if they are the same or different – whether or not they lie on the same trajectory. In quantum mechanics, where orbits of 1−particle states may not be defined at all, there can be no such guarantee (this independent of symmetrization). This and the SP now lead to something new. For the SP implies that given two exactly similar particles with momenta in directions a and b, the state (a, b) (to use ► Dirac notation [3]) is the same as (b, a); we should read these states as unordered pairs; but now given two particles initially in the state $(1, 2)$, and finally in the state (a, b), understood as unordered pairs, there will in general be *two* ways of linking them - by a transition $1 \to a$, $2 \to b$, and the 'exchange' transition $2 \to a$, $1 \to b$. If both transition amplitudes are appreciable, they may interfere with each other, and their relative phase will make a difference to the total transition probability. The relative phase is in turn different for symmetric states than for antisymmetric ones [8].

This point was in Richard Feynman's (1918–1988) view the key to understanding ► quantum statistics. The rule is:

Bosons (Amplitude direct) + (Amplitude exchanged)
Fermions (Amplitude direct) – (Amplitude exchanged).

In Feynman's notation [9], $\langle a|1\rangle = a_1$ is the amplitude for particle 1 to scatter in direction a, and similarly $\langle a|2\rangle = a_2$, etc. The total amplitude is the sum (bosons) or difference (fermions) of the amplitudes for the two ► Feynman diagrams shown in Fig. 1:

$$\langle a|1\rangle\langle b|2\rangle \pm \langle b|1\rangle\langle a|2\rangle = a_1 b_2 \pm b_1 a_2.$$

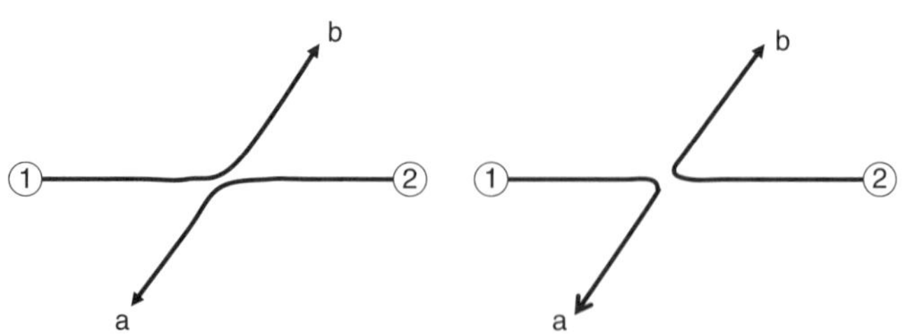

Fig. 1 Feynman diagrams for direct and exchange transition amplitudes

The probability for bosons as $a \to b$ is then $\lim_{a \to b} |a_1 b_2 + b_1 a_2|^2 = 4|b_1 b_2|^2$; for fermions it vanishes. In the case of unsymmetrized particles, one of the processes $\langle a|1\rangle\langle b|2\rangle$, $\langle b|1\rangle\langle a|2\rangle$ results, with probability $|a_1 b_2|^2$ and $|b_1 a_2|^2$, respectively; in the limit $a \to b$ one cannot tell which has occurred, and the probabilities should be summed to obtain $2|b_1 b_2|^2$, exactly half the cross-section for bosons. Bosons, relative to unsymmetrized particles, act as though they attract one another, whilst fermions repel.

The point dovetails neatly with the Copenhagen interpretation ► Born rule; Consistent Histories; Metaphysics in Quantum Mechanics; Nonlocality; Orthodox Interpretation; Schrödinger's Cat; Transactional Interpretation. According to this, if the experimental set-up permits the determination of the path (trajectory, orbit), taken by the particle – as would be possible if the particles differed in their state-independent properties (but which could also be ensured by other means) – there could be no interference effects (think of the two-slit experiment). This is reflected in the formalism by rules for using the measurement postulates: whether we should first take the absolute square of the amplitudes and then add, or add the amplitudes and then take the absolute square.

One might wonder if such a close link to the problem of measurement is a virtue of Feynman's approach. On the other hand, one could say the link was obvious from the beginning, purely on the basis of ► Bohmian mechanics. In that theory trajectories are introduced explicitly, but one can still derive the same transition probabilities, consistent with quantum statistics.

Identity at a Time or Identity Simpliciter. Does the SP pose a still deeper challenge to the concept of identity? Many think it does, and point to the apparent failure in quantum mechanics of Gottfried W. Leibniz's (1646–1716) theory of identity, in particular his *principle of identity of indiscernibles* (PII).

Yet the history to this suggestion is curious, for when the PII was first brought up in the context of the SP, by Hermann Weyl (1885–1955), the principle was supposed to be vindicated, not undermined:

> The upshot of it all is that the electrons satisfy Leibniz's principium identitatis indiscernibilium, or that the electronic gas is a 'monomial aggregate' (Fermi–Dirac statistics). In a profound and precise sense physics corroborates the Mutakallimûn: neither to the photon nor to the (positive and negative) electron can one ascribe individuality. As to the Leibniz–Pauli Exclusion Principle, it is found to hold for electrons but not for photons. [10, p.247].

Quantum mechanics, for Weyl, posed no special problem for Leibniz's philosophy, at least as goes fermions.

For those focused on the question of quantities assigned to particles on the basis of their place in the N–fold tensor product of 1–particle states, these comments made no sense. They are determined as expectation values of operators of the form

$$\langle \Psi, I \otimes \ldots \otimes I \otimes A \otimes I \otimes \ldots \otimes I \Psi \rangle$$

(where A is a 1–particle observable). Include by all means other statistical properties, and marginal probability distributions, likewise attributed to particles or particle pairs of k−tuples on the basis of their place in the tensor product structure; if Ψ is symmetrized, every particle (or particle pair or k−tuple) has exactly the same 1−particle expectation value for A, and the same statistical properties and marginal probability distributions. It seems, then, that the PII must *comprehensively* fail in quantum mechanics, for fermions as well as bosons, as claimed by Henry Margenau (1901–1997) [11]. Similar conclusions were reached by others in subsequent studies [12, 13].

There is, however, a rather obvious rejoinder to this argument, namely that by particles we really mean 1−particle states and properties. Our concern is not with which particle has which state or property, but with what those states and properties are. At least in *some* circumstances, particles may be identified with 1−particle states. Thus in 2−particle case, for $\{\phi_i\}$ an ▶ orthonormal basis for the 1−particle space, consider states of the form:

$$\Psi_{\pm}^{ij} = \frac{1}{\sqrt{2}}(\phi_i \otimes \phi_j \pm \phi_j \otimes \phi_i),\ i \neq j. \tag{1}$$

Ψ_{+}^{ij} is symmetric; Ψ_{-}^{ij} is antisymmetric. In Dirac's notation, they are states (i, j), understood as an unordered pair. As such they *manifestly* describe two particles, one being state ϕ_i, one being state ϕ_j; one having property P_{ϕ_i}, the other property P_{ϕ_j} (where P_ϕ is the projection on the state ϕ). It was understandable for Weyl to speak of the 'Leibniz–Pauli Exclusion Principle', at least in the case of electrons, in certain circumstances – in atoms subject to sufficiently strong external fields, so as to completely remove every energy ▶ degeneracy. In that case each electron is uniquely identified by its four ▶ quantum numbers.

But these are special cases. In the case of superpositions of vectors $\Psi_{\pm}^{ij}$, more than two 1-particle states are involved; there may be no pair of distinguished properties, one for each particle. And of course even if there are definite 1-particle states or properties for each particle, in the case of bosons there could spell trouble: they may be precisely the same (as with product states $\phi_j \otimes \phi_j$). Even for a state of the form (1) there may be a difficulty, as with the spherically symmetric singlet state of spin of two spin-$\frac{1}{2}$ particles. This state can be written in many ways:

$$\Psi_{-}^{0} = \frac{1}{\sqrt{2}}(\phi_{+}^{x}\phi_{-}^{x} - \phi_{-}^{x}\phi_{+}^{x}) = \frac{1}{\sqrt{2}}(\phi_{+}^{y}\phi_{-}^{y} - \phi_{-}^{y}\phi_{+}^{y}) = \frac{1}{\sqrt{2}}(\phi_{+}^{z}\phi_{-}^{z} - \phi_{-}^{z}\phi_{+}^{z}) \tag{2}$$

where $\phi_{\pm}^{x}$ are eigenstates of the x-component of spin, etc., as exploited by Bohm (1917–1992) in his formulation of the ▶ EPR thought experiment. It seems each particle must have every component of spin, or none.

We should be clearer on what the PII actually says. It is usually stated as the principle "it is not possible for there to exist two individuals possessing all their properties (relational and non-relational) in common" [14, p. 9] (where the principle is the stronger the fewer the admissible properties and relations). Traditionally, philosophical debates on this principle have centered on what is to count as admissible: surely not relations involving identity and proper names, which threaten to trivialize the PII altogether. But there has been less interest in questions of logical form, and the meaning of 'relational properties'. If indeed *properties*, then they correspond to complex monadic predicates, presumably involving relations with other things only through bound quantification. But this is not the only, or the most important way in which relations are used in predication. Restricted to these, the PII is unnecessarily stringent. Why not allow that things may be discerned by relations as well as relational properties? But take *this* step and it is not obvious that the PII fails in quantum mechanics.

For the sake of clarity, the point is worth formalizing. Let L be a first-order language with a finite primitive vocabulary. Let s and t be L-terms (variables or proper names). Then the principle stated in terms of relational properties has the form:

$$s = t \underset{\text{def}}{=} \bigwedge_{\text{all primitive } L\text{-predicates } F} [\forall\forall...\forall F(...s...) \leftrightarrow \forall\forall...\forall F(...t....)] \tag{3}$$

where, if F is an n-ary predicate, there are $n - 1$ quantifiers $\forall$ (so that $\forall\forall...\forall F$ is 1-ary). This clearly fails to capture the full generality of relational predication: on the RHS of (3) should be conjoined conditions of the form:

$$\forall\forall...\forall[F(...s...) \leftrightarrow F(...t...)] \tag{4}$$

Proceeding in this way, one arrives at a definition of identity that, unlike (3), satisfies the formal axioms of identity and is essentially unique. As such it was championed by Willard van Orman Quine (1908–2000) [15].

Given this, if s and t are exactly similar, but $s \neq t$, they need not differ in any relational property, but only if for some F (4) is false. (4) would fail, for example, if for some dyadic F, $F(st)$ is true and F is irreflexive. F may even be symmetric too, thus incorporating permutation symmetry [5, 7].

As applied to quantum mechanics, it would then be enough, to discern ▶ electrons in the singlet state of ▶ spin, that they satisfy an irreflexive relation. And so they do: in the state (2), the relation 's has opposite $x-$component of spin to t' is clearly irreflexive and clearly true. Indeed, analogous statements hold for every component of spin, as (2) shows. But this does not imply the electrons each have any definite component of spin; compare 's is one mile apart from t', which may be true, for the space-time relationist, even though neither s nor t has any particular position in space.

A similar relation of anticorrelation for any state of the form (1) is easily specified:

$$(P_{\phi_i} - P_{\phi_j}) \otimes (P_{\phi_i} - P_{\phi_j})\Psi^{ij}_{\pm} = -\Psi^{ij}_{\pm}. \tag{5}$$

The generalization to superpositions of finitely-many such states is

$$\frac{1}{d}\sum_{i.j=1}^{d}(P_{\phi_i} - P_{\phi_j}) \otimes (P_{\phi_i} - P_{\phi_j}) \sum_{i\neq j=1}^{d} c_{ij}\Psi^{ij}_{\pm} = -\sum_{i\neq j=1}^{d} c_{ij}\Psi^{ij}_{\pm} \tag{6}$$

where $c_{ij} = c_{ji}$. Since for fermions the RHS of (6) is the most general state possible, fermions, at least in finite dimensions, are always discernible. Evidently the same cannot be said of bosons; symmetric product states, such as $\phi_j\phi_j$, can be discerned by these methods only if subject to an evolution which leaves them entangled [16].

The upshot is that violation of the PII is neither sufficient nor necessary for the SP. But it would be wrong to conclude that the two principles are completely unrelated. There is, indeed, a very simple sense in which the PII together with exact similarity implies the SP, for they imply that states of affairs that differ only by permutations of particles should be identified – in Dirac's notation, that (a, b) and (b, a) be identified. But then the same principles should apply to classical statistical mechanics as well (for classical particles may surely be exactly similar); the explanation of quantum statistics cannot be traced to these – or not in isolation from other features of quantum mechanics, whether to do with identity over time, or the discrete nature of probability measures on Hilbert space [5], in line with early suggestions by Max Planck (1858–1947) and Hendrik A. Lorentz (1853–1928) [17].

Identity Operator

See ► Dirac notation; POVM.

Literature

1. P. Ehrenfest: Welche Züge der Lichtquantuenhypothese spielen in der Theorie der Wärmestrahlung eine wesentliche Rolle, Annalen der Physik **36**, 91–118 (1911).
2. L. Natanson, Über die statistische Theorie der Strahlung, Physikalische Zeitschrift **12**, 659–666 (1911).
3. P.A.M. Dirac: On the theory of quantum mechanics, Proceedings of the Royal Society (London) **A112**, 281–305 (1926).
4. P. Ehrenfest and G. Uhlenbeck: Die wellenmechanische Interpretation der Botzmannschen Statistik neben der neueren Statistiken, Zeitschrift für Physik **41**, 24–26 (1927).
5. S. Saunders, On the explanation for quantum statistics, Studies in the History and Philosophy of Modern Physics **37**, 192–211 (2006).
6. A. Bach, Indistinguishable Classical Particles, Berlin: Springer (1995).
7. S. Saunders, Are quantum particles objects? Analysis **66,** 52–63 (2006).
8. R. Oppenheimer: On the quantum theory of electronic impacts, Physical Review **32,** 361–376 (1928).
9. R. Feynman: The Feynman Lectures on Physics, Vol.III, R. Feynman, R. Leighton and M. Sands (eds.), Reading, MA: Addison-Wesley (1965).
10. H. Weyl, Philosophy of Mathematics and Natural Science, Princeton: Princeton University Press (1949).
11. H. Margenau, The exclusion principle and its philosophical importance, Philosophy of Science **11,** 187–208 (1944).
12. S. French and M. Redhead, Quantum Physics and the Identity of Indiscernibles, British Journal for the Philosophy of Science **39**, 233–246 (1988).

13. D. Dieks, Quantum Statistics, Identical Particles and Correlations, Synthese **82**, 127–155, (1990).
14. S. French and D. Krause, Identity in Physics: A Historical, Philosophical and Formal Analysis, Oxford: Clarendon Press, (2006).
15. W. V. Quine, Set Theory and Its Logic, Cambridge, MA: The Belnap Press of Harvard University Press (1969).
16. F. Muller and S. Saunders, Discerning fermions, British Journal for the Philosophy of Science **59** (2008).
17. O. Darrigol, Statistics and cominbatorics in early quantum theory, II: Early symptoma of indistinghability and holism, Historical Studies in the Physical and Biological Sciences **21**, 237–298 (1991).

Ignorance Interpretation of Quantum Mechanics

I

Peter Mittelstaedt

Let S be a proper quantum system with ▶ Hilbert space $\mathcal{H}_S$ that is prepared in a ▶ *mixed state* given by the self-adjoint operator $W_S = W_S^+$ with $tr\{W_S\} = 1$. Here, we assume that W_S is not a pure state, i.e. $W_S \neq W_S{}^2$.

Two kinds of mixed states can be distinguished by their preparation.

(a) A "mixture of states" [1], a "real mixture" [2], or a "Gemenge" [3] is an ensemble $\Gamma_S(p_k, \varphi_k)$ of pure states φ_k with probabilities p_k.
(b) System S is a subsystem of a compound system $S^* = S + S'$ with Hilbert space $\mathcal{H}^* = \mathcal{H}_S \otimes \mathcal{H}_{S'}$ that is prepared in a pure state $\Psi^*(S + S')$.

In case (a) the mixed state $W_S = \sum_i p_i \, P[\varphi_I]$ may be considered as a formal description of the "Gemenge" $\Gamma_S(p_k, \varphi_k)$ in terms of Hilbert space quantum mechanics. Hence, there are obviously no difficulties for interpreting the state $W_S = \sum_i p_i \, P[\varphi_i]$ as a description of a system S that is objectively in one of the states φ_i, which is, however subjectively unknown to the observer who knows only the probability p_i. In this situation, we say that the state W_S admits "ignorance interpretation".

In case (b) the mixed state of the subsystem S of S^* is given by the partial trace $W_S = tr' \, P[\Psi^*]$ where tr' denotes the summation over the degrees of freedom of S'. It is easy to demonstrate that $W_S = W_S{}^+$ with $tr\,\{W_S\} = 1$. However, nothing is known about the decomposition of W_S into weighted components corresponding to pure states. If the spectrum of W_S is not degenerate, then there is a uniquely defined spectral decomposition $W_S = \sum p_i \, P[\psi_i]$ of W_S into orthogonal, i.e. mutually exclusive states ψ_i. The states ψ_i are eigenstates of the operator W_S and the coefficients p_i are the eigenvalues. Hence, for any $i \in \mathbf{N}$ the eigenvalue equation $W_S \, \psi_i = p_i \, \psi_i$ holds. However, the decomposition of the state W_S is by no means unique since there are infinitely many decompositions of W_S into nonorthogonal states $\psi_i{}'$. Hence, the operator W_S would represent formally an infinite number of ensembles (▶ ensembles in quantum mechanics) $\Gamma^{(n)}$ $(W_S) := \Gamma(p_i{}^{(n)}, \ \psi_i{}^{(n)})$. This means that the state W_S is not sufficient to determine a particular mixture $\Gamma^{(n)}(W_S)$ of states that is actually realized. Even if W_S admits "ignorance interpretation" and can be interpreted as the description of some Gemenge $\Gamma(W_S)$, new arguments must be added for a complete determination of the Gemenge $\Gamma(W_S)$ in question.

Spectral decomposition, see ▶ Density operator; Measurement theory; Objectification; Operator; Probabilistic Interpretation; Propensities in Quantum Mechanics; Self-adjoint operator; Wave Mechanics.

However, the main question is still open. Does a given mixed state admit at all "ignorance interpretation"? In other words, is it allowed to assume that a system S with the mixed state $W_S = \sum p_i P[\varphi_i]$ is actually in one of the pure states φ_i, which is, however, unknown to the observer who knows only the probabilities p_i. If this interpretation of W_S were correct, then the mixed state would express the observers "ignorance" of the actual pure state but not the objective indeterminacy of this state. It is one of the most fundamental results of quantum mechanics that a mixed state in general does not admit "ignorance interpretation". The reason for this result is that the assumption of a objectively decided pure state leads in general to contradictions with well established results in quantum mechanics. This can be shown in various ways and on different levels of generality [4, 5]. See also States, pure and mixed, and their representations.

Literature

1. P. Mittelstaedt: *The Interpretation of Quantum Mechanics and the Measurement Process* (Cambridge University Press, Cambridge 1998, 35–37)
2. B. d'Espagnat: *Reality and the Physicist* (Cambridge University Press, Cambridge 1989, 85)
3. W. Heisenberg: *Physikalische Prinzipien der Quantentheorie* (S. Hirzel, Leipzig 1930, 43)
4. P. Busch et al.: *The Quantum Theory of Measurement* (Springer, Heidelberg 1991/96)
5. P. Mittelstaedt: *The Interpretation of Quantum Mechanics and the Measurement Process* (Cambridge University Press, Cambridge 1998, 79–82)

Indeterminacy Relations

See ▶ Heisenberg Uncertainty Relation.

Indeterminism and Determinism in Quantum Mechanics

Brigitte Falkenburg and Friedel Weinert

It is an often repeated claim in the literature that quantum mechanics is indeterministic and that it has put an end to the classical notion of causality. From the impossibility of determining the exact spatio-temporal trajectory of an atomic system, for instance, Heisenberg infers 'the invalidity of the causal law' in quantum mechanics [1]. What is tacitly assumed in such views is a chain of reasoning, which leads from determinism to causality. One form of determinism – predictive determinism – is the view that a sufficient knowledge of the laws of nature and appropriate boundary conditions will enable a superior intelligence to predict the future states of the physical world and to retrodict its past states with infinite precision. Laplace attributes this capacity to his famous demon: for the demon the physical world stretches out like the frames of a filmstrip. Each frame is caused by an earlier frame and in its turn causes a later frame. From the present frame the Laplacean demon is capable of predicting and retrodicting all other frames. Hence the demon identifies determinism and causality. 'We ought to regard the present state of the universe as the effect of its antecedent state and as the cause of the state that is to follow' [9]. Laplace assumes that these states are unique and can be determined with mathematical precision such that prediction and retrodiction become possible. The laws of physics are typically expressed in differential equations which describe the evolution of some physical parameter, P, as a function of time, t. As one state of a system, S_1, evolves to another state, S_2, where this temporal evolution is made precise by the employment of differential equations, it becomes easy to think of differential equations as precise mathematical representations of causal laws [10]. This is indeed how Einstein presented the matter: 'The differential law is the only form which completely satisfies the modern physicist's demand for causality' [2]. Although Russell [11] had argued that the 'law of causality (...) is the product of a bygone age' he nevertheless admitted causal laws in the form of functional relations and differential equations into physics.

This functional model of causality enjoyed great popularity amongst physicists. But the experimental results from quantum mechanics – like the ► double-slit experiments – seemed to threaten the Laplacean identification of determinism and causality. Physicists reacted to this threat in three different ways.

1. An older generation of physicists (Einstein, von Laue, Planck) wished to retain the notion of causality and its identification with determinism. 'An event is causally determined when it can be predicted with certainty.' [3] They never abandoned the hope of a causal-deterministic understanding of quantum mechanics.

2. A second group of physicists (Bohr, Heisenberg, Pauli) concluded that quantum mechanics had become both indeterministic and acausal. Let us neglect for the moment that the ▶ Schrödinger equation is a deterministic differential equation in an abstract ▶ Hilbert space and concentrate instead on the decay law and on ▶ Heisenberg's uncertainty relations (or indeterminacy relations). Rutherford's ▶ radioactive decay law is statistical in nature; it expresses the probability of the disintegration rate of an ensemble of atoms (▶ ensembles in quantum mechanics) rather than the disintegration rate of an individual atom. The latter is unpredictable in the sense that it can only be expressed by the whole range of the decay curve of the ensemble. James Jeans therefore concluded that causality had disappeared from the physical world picture. [4] Due to the discovery of his indeterminacy relations, Heisenberg arrived at a similar conclusion. The indeterminacy principle shows that neither the antecedent nor the consequent conditions of the causality principle, as Heisenberg sees it, can be satisfied: 'If we know exactly the determinable properties of a closed system at a given point in time, we can calculate precisely the future behaviour of the properties of this system.' [1] The indeterminacy principle excludes the simultaneous knowledge of the antecedent conditions of an atomic system by non-commuting ▶ operators $[x]$, $[p_x]$, $[x]$, $[E]$, $[t]$; but it also excludes the precise knowledge of the future behaviour of the individual system. Bohr [5] agreed with Heisenberg that the indeterminacy relations spelt the end of the classical notion of causality. He argued that his notion of ▶ complementarity should be regarded as a generalization of the notion of causality. Complementarity means that quantum mechanics must employ both the particle picture ▶ Franck–Hertz experiment and the wave picture ▶ Davisson–Germer experiment; Stern–Gerlach experiment; Schrödinger equation to describe the behaviour of atomic systems. But the indeterminacy relations:

$$\Delta x \Delta p > \hbar \tag{1a}$$

$$\Delta E \Delta t > \hbar \tag{1b}$$

produce, according to Bohr, the following dilemma:

(i) The determination of the spatio-temporal location, x, of atomic particles, say in a double-slit experiment, leads to an unavoidable disturbance of dynamic variables, like momentum p.

(ii) The determination of the value of dynamic variables, like energy, E, or momentum, p, leads to an unavoidable loss of precise coordination regarding the spatio-temporal location of the particles, i.e. t, x.

Quantum mechanics must employ both the particle and the wave picture but each leads to a loss of information, as relations (1a,b) show, which prevents the precise spatio-temporal determination known from classical particles. Physicists like Bohr, Heisenberg and Pauli were content to conclude that the indeterminacy relations implied the acausal nature of quantum mechanical systems. Their argument went through on the assumption of an identification of determinism with causality.

(3) This traditional identification, however, harboured the conceptual possibility of a third response. The philosopher Ernst Cassirer [6] maintained a functional view of causality, claiming that causality (or determinism) is preserved at the level of Schrödinger's ▸ wave function, whereas the individual quantum events or measurement results were indeterministic. Physicists agree that the ▸ Schrödinger equation is a deterministic equation in Hilbert space. Max Born [7] and Louis de Broglie [8], however, argued, unlike Cassirer that, the notion of causality could be retained in quantum mechanics, despite its observable indeterminism, even if the functional view of causality was abandoned. The Born-de Broglie move had two consequences:

(i) The notions of determinism and causality became disentangled; it was possible to accept the indeterminism of quantum mechanics without giving up the notion of causality.
(ii) The notion of causality needed to be modified in order to speak of causal relations in quantum mechanics.

I

To illustrate these consequences, consider a schematic representation of the Davisson–Germer experiment, i.e. de Broglie's thought experiment. A beam of ▸ electrons is targeted at a crystal; call this phenomenon A. The encounter of the beam with the surface of the crystal will lead to diffraction effects, B_1, B_2, B_3, which will be recorded at different points on a recording screen (Fig. 1).

As is well-known the rules of quantum mechanics do not permit a precise prediction of the diffraction effects, i.e. their precise spatio-temporal location. Yet it is possible to speak of a causal situation in this case for the experiments show that the observable consequent effects, B_1, B_2, B_3, are dependent on the antecedent condition A. We can speak of a 'conditional dependence' because (a) the experimental situation leads to the identification of a cluster of relevant antecedent and consequent conditions and (b) the distribution of the occurrence of the consequent conditions is statistically dependent on the anterior conditions. Such a conditional dependence of the consequent conditions, B, on the antecedent condition, A, is further emphasized by the absence of B in the absence of A (indicated in Fig. 1). A conditional dependence, indicated in de Broglie's thought experiment, is clearly observable in many of the classic experiments in quantum mechanics: ▸ Davisson–Germer experiment, ▸ Frank–Hertz experiment, ▸ Stern–Gerlach experiment, ▸ large-angle scattering; ▸ scattering experiments; ▸ which-way experiments.

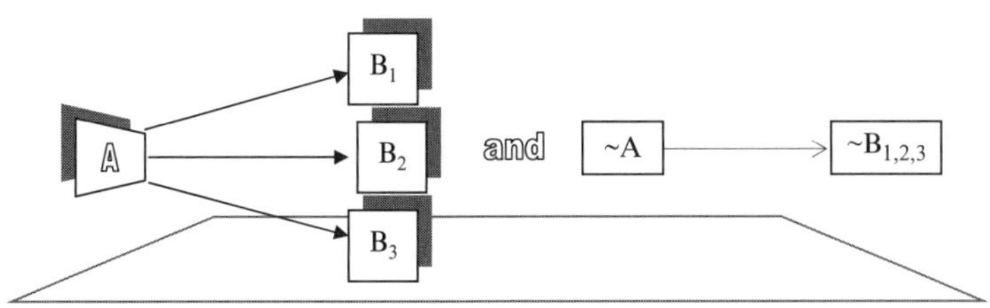

Fig. 1 De Broglie's causal thought experiment

A consequence of the acceptance of both indeterministic and causal relations in quantum mechanics is a revised view of these relations: a conditional model of causality [12]. According to such a conditional model, it is possible (here in the context of quantum mechanical experiments) to specify a cluster of antecedent conditions (further specified in terms of necessary and sufficient conditions) and a cluster of consequent conditions (observable effects in quantum mechanical experiments). It is observed that between the antecedent and consequent conditions lawlike statistical relations obtain, which specify the probability with which the consequent conditions may be expected to occur. For instance in the Stern–Gerlach experiments, when the silver atoms are in the ground state, there is a 50% chance for the atoms to be deflected either upward or downward, a deflection which, under these conditions, is due to the spin or the intrinsic angular momentum of the spinning electron in the outer shell of the silver atoms in the atom beam. ▶ Spin; Stern–Gerlach experiment; Vector model. Hence, given the lawlike statistical dependence between antecedent and consequent conditions, the distribution of the observable events is specified. On such a conditional model of causality, experiments in quantum mechanics reveal causal relations in the absence of deterministic predictability of individual events and a traceable mechanism linking particular causes and effects.

In thc famous EPR argument [13], Einstein raised a further concept of causality. (▶ Causal Inference and EPR) According to it, there is no causal relation between two space-like separated events. Hence, the wave function of a compound system (functional causality) or the predictions obtained from it (probabilistic causality) come together with the a-causal correlation of events at a space-like distance (Einstein causality or ▶ Einstein locality). Therefore, quantum mechanics raises the conceptual problem that there is no longer an unambiguous concept of causality [14, pp. 316–319].

Primary Literature

1. W. Heisenberg: Über den anschaulichen Inhalt der quantentheoretischen Kinematik und Mechanik. Zeitschrift für Physik **43**, (1927), 174–79, 197; W. Heisenberg: Kausalgesetz und Quantenmechanik. Annalen der Philosophie **2**, (1931), 179–81; reprinted in W. Heisenberg: *Gesammelte Werke/Collected Works* Abteilung C, Band I: Physik und Erkenntnis 1927–1955; eds. W. Blum, H.P. Dürr/H. Rechenberg, München/Zürich: Piper (1984–5), 29–39
2. A. Einstein: Newtons Mechanik und ihr Einfluβ auf die Gestaltung der theoretischen Physik. Die Naturwissenschaften **15**, (1927), 273–6; reprinted in A. Einstein: *Ideas and Opinions* (Alvin Redman, London 1954, 253–61)
3. M. Planck: Die Kausalität in der Natur (1932), in Planck, *Vorträge und Erinnerungen* (Wissenschaftliche Buchgesellschaft, Darmstadt 1975, 252).
4. J. Jeans: *Physics and Philosophy* (Cambridge University Press, Cambridge 1943, 149–50). Earlier Franz Exner, *Vorlesungen über die Physikalischen Grundlagen der Naturwissenschaften* (1919) had already expressed the view that the micro-world is indeterministic. Historically, the roots of indeterminism seem to reach beyond the emergence of quantum mechanics to the 19th work on the 2nd law of thermodynamics and the problem of irreversibility. See St. G. Brush, Irreversibility and Indeterminism: Fourier to Heisenberg, Journal of the History of Ideas **37**, (1976), 603–30.

5. N. Bohr: The Quantum Postulate and the Recent Development of Atomic Theory. Supplement to Nature (April 14 1928), 580–90, reprinted in N. Bohr: *Atomic Theory and the Description of Nature* (Cambridge University Press, Cambridge 1934, 52–91); N. Bohr: The Atomic Theory and the Fundamental Principles underlying the Description of Nature (1929), in N. Bohr: *Atomic Theory and the Description of Nature* (Cambridge University Press, Cambridge 1934, 103–119); N. Bohr: Wirkungsquantum und Naturbeschreibung. Die Naturwissenschaften **17** (1929), 483–6; N. Bohr: Kausalität und Komplimentarität. Erkenntnis **6**, (1936), 293–302; N. Bohr: Quantum Physics and Philosophy – Causality and Complementarity, reprinted in N. Bohr, *Essays on Atomic Physics and Human Knowledge, 1958–1962* (Interscience Publishers, New York 1958, 1–7)
6. E. Cassirer: Determinismus und Indeterminismus in der Modern Physik (1936), in: E. Cassirer, *Gesammelte Werke* 19. Ed. by B. Recki (Felix Meiner Verlag, Hamburg 2008); English translation: *Determinism and Indeterminism in Modern Physics* (Yale University Press, New Haven 1956)
7. M. Born: *Natural Philosophy of Cause and Chance* (Clarendon, Oxford 1949, 102, 124)
8. L. de Broglie: *Continu et Discontinu en Physique moderne* (Albin Michel, Paris 1941, 64–6)

I

Secondary Literature

9. P. Laplace: *Théorie analytique des probabilities* (Couvier, Paris 1820, Introduction VI–VII). Translation quoted from E. Nagel: The Structure of Science (Routledge & Kegal Paul, London 1961, 281–2)
10. Ph. Frank: *Das Kausalgesetz und seine Grenzen* (Springer, Wien 1932, 142–46). Note that the functional view was already held by Mach and Boltzmann, see M. Stöltzner, Vienna Indeterminism: Mach, Boltzmann, Exner, Synthese **119**, (1999), 85–111
11. B. Russell: *On the notion of cause*. Proceedings of the Aristotelian Society **XIII**, (1912–13), 1–26; reprinted in B. Russell: On the Philosophy of Science, edited, with an Introduction, by Charles A. Fritz (The Bobbs–Merrill Company, Indianapolis 1965, 163–186)
12. D. Bohm: *Causality and Chance in Modern Physics* (Routledge and Kegan Paul, London 1957) J.L. Mackie: The Cement of the Universe (Clarendon Press, Oxford 1980); F. Weinert: *The Scientist as Philosopher* (Springer, Berlin 2004, Chap. 5)
13. A. Einstein, B. Podolsky, N. Rosen: *Can quantum-mechanical description of physical reality be considered complete?* Physical Review **47**, (1935), 777–80
14. B. Falkenburg: *Particle Metaphysics. A Critical Account of Subatomic Reality* (Springer, Heidelberg 2007)

Indistinguishability

Nick Huggett and Tom Imbo

In the considerable physical and philosophical literature,[1] ‘indistinguishability’, and the related concept of ‘identicality’, are used in many ways, and in the resulting

[1] See [1] for a cross section of the philosophical literature, and a comprehensive bibliography of the subject.

confusion the logical relations between the various notions are often obscured, with unfortunate consequences. This article will use them in the following senses, which are most useful and (likely) common:

> Particles are *identical* if they share in common all their constant properties, such as mass, charge, spin and so on: that is, if they agree in all their state-independent or *intrinsic* properties. Particles are *indistinguishable* if they satisfy the indistinguishability postulate (*IP*). This postulate states that all observables O must commute with all particle permutations P: $[O, P] = 0$. Put informally, the IP is the requirement that no expectation value of any observable is affected by particle permutations.

The IP presupposes the following formal structure: assume that we have a system of n identical quantum particles, and that if n were equal to 1 then the state space of the system would be $\mathcal{H}_1$. The natural assumption for $n > 1$ is that the state space $\mathcal{H}$ describing the system is a subspace of the tensor product, $\mathcal{H}_n$, of n copies of $\mathcal{H}_1$. That is,

$$\mathcal{H} \subseteq \mathcal{H}_n \equiv \bigotimes_{i=1}^{n} \mathcal{H}_1. \tag{1}$$

We assume that $\mathcal{H}$ is closed under the action of arbitrary permutations, P, which permute the n factors of $\mathcal{H}_n$. Any such operator is a product of 'particle exchange operators' P_{ij} ($1 \leqslant i, j \leqslant n$). P_{ij} interchanges the ith and jth copies of $\mathcal{H}_1$ in $\mathcal{H}_n$: for instance (for $n = 2$),

$$P_{12}(|\phi\rangle \otimes |\psi\rangle) = |\psi\rangle \otimes |\phi\rangle. \tag{2}$$

For example, if the particles are either bosons or fermions then the appropriate state spaces are the symmetric ($P_{ij}|\Psi\rangle = |\Psi\rangle$) and antisymmetric ($P_{ij}|\Psi\rangle = -|\Psi\rangle$) subspaces of $\mathcal{H}_n$ respectively. Operators that commute with all permutations are called *symmetric*. The IP says that only symmetric Hermitian operators are observables; any non-symmetric Hermitian operators on $\mathcal{H}$ do not correspond to observable quantities if the IP holds.

Logical Relations

Oftentimes (e.g., [2], 275–6) an attempt is made to connect identicality and indistinguishability by appeal to the fact that in quantum mechanics (QM), unlike classical mechanics, particles cannot have varying continuous trajectories. Even if a particle has a definite location at some times, its position will be indefinite at times in between. Why? States of definite position – eigenstates of position – are necessarily orthogonal, and it is impossible for a system to occupy a continuous series of orthogonal states. (Any unitary evolution between such states will take a finite time, and under measurement the probability of collapse to an orthogonal state is zero.) And of course there is nothing special about position in this regard: even if the spectrum of an operator is continuous, no quantum evolution corresponds to a continuous trajectory through the spectrum.

This line of thought is supposed to lead directly to the conclusion that identical quantum particles (unlike classical particles) cannot be distinguished by continuous trajectories (through space or the spectrum of any observable). So there are two questions: (*i*) Does this conclusion – call it *trajectory indistinguishability* – actually follow? (*ii*) What do these considerations have to do with indistinguishability as defined earlier?

Trajectory Indistinguishability

First (*i*). This argument is supposed to show that quantum particles are trajectory indistinguishable, *without appeal to the IP* (from which it follows immediately, as discussed below). The idea behind the argument is that quantum particles can only be distinguished by continuous trajectories that are constant – because, as we just saw, varying continuous trajectories are impossible. But the identicality of the particles is supposed to preclude their being distinguished by constant properties. However, there is a fallacy in this line of thought. A property is 'intrinsic' if it is independent of any *possible* state of the system, not simply if it is a constant of some particular evolution; so identical particles *can* be distinguished by constant trajectories.

For instance, let $\mathcal{H}_1$ be a 2-dimensional ► Hilbert space spanned by $\{|\lambda_1\rangle, |\lambda_2\rangle\}$, eigenstates of the time-independent observable A with eigenvalues λ_1 and λ_2, respectively. Further suppose that $\mathcal{H} = \mathcal{H}_1 \otimes \mathcal{H}_1$, and that all Hermitian operators are observables and indeed allowed ► Hamiltonian operators. Then one possible evolution of the system is $\Psi(t) = |\lambda_1\rangle \otimes |\lambda_2\rangle$ (for all t), in which the particles are distinguished by their constant 'trajectories' – the first always has the value λ_1 for A and the other λ_2.[2] But the values of A are not state independent: there are states in $\mathcal{H}$ in which the value of A for the first particle is not λ_1, for instance $(|\lambda_2\rangle \otimes |\lambda_1\rangle)$, and states in which the particles have no definite A value, for instance $(a|\lambda_1\rangle \otimes |\lambda_2\rangle + b|\lambda_2\rangle \otimes |\lambda_1\rangle)$. So A is not intrinsic, and indeed (supposing the particles do share their truly intrinsic properties) the example shows that identical particles can, after all, be trajectory distinguishable.

Note that in this example, the ► operators corresponding to the value of A for the two particles violate the IP, and hence their values would not constitute physical trajectories if the IP held. Indeed, although identical quantum particles are not necessarily trajectory indistinguishable, they will be if they are indistinguishable.[3]

---

[2] A is not an operator on $\mathcal{H}$, so what is meant here is that $\Psi(t)$ is an eigenstate of $A \otimes I$ with eigenvalue λ_1, and of $I \otimes A$ with eigenvalue λ_2. That is, following the standard understanding, the operator 'corresponding' to A for the first particle is $A \otimes I$, and so on.

[3] It is often assumed that all single particle observables have the form $I \otimes \ldots \otimes I \otimes A \otimes I \ldots \otimes I$ (which violates the IP), but one might imagine a more general conception. What is essential, how-

Indistinguishability

In answer to (ii), the impossibility of continuously varying trajectories does not support indistinguishability in the sense of the IP. The IP is a constraint on which operators can be observables, but the impossibility of continuously varying trajectories is a fact about all Hermitian operators, whether or not they satisfy the IP. Hence this impossibility places absolutely no restriction on observables at all, once we adopt the quantum formalism.

Indeed, there are consistent (though hypothetical) quantum systems of identical particles that violate the IP: for instance, a collection of identical 'quantum Maxwell–Boltzmann' particles. For n such particles the state space is the full Hilbert space $\mathcal{H}_n$ of (1) – i.e., $\mathcal{H} = \mathcal{H}_n$ – and every (sufficiently well-behaved) Hermitian operator is an observable (as in the example above). Note that while this formalism is commonly used for non-identical particles, a system of n identical particles can also have $\mathcal{H}_n$ as its state space. Such particles are said to obey quantum Maxwell–Boltzmann or 'infinite' statistics.[4] This system clearly violates the IP, because some observables are non-symmetric: $[O, P] \neq 0$. In this sense then, the particles are 'distinguishable'.

While it is widely known, at least implicitly, that identicality does not imply the indistinguishability postulate, it seems rarely to be explicitly acknowledged, with certain resultant confusions about the nature of identical particles.[5] For example, it seems that the 'problem of identical particles' is often taken to be the problem of understanding how the symmetrization postulate (SP) – that all particles are either bosons or fermions – can be shown to follow from the indistinguishability postulate, as if the latter were more secure than the former (e.g., [4]). But there are no first principle grounds for holding indistinguishability either; certainly not as a logical consequence of quantum identicality. Thus both the symmetrization and indistinguishability postulates are on a very similar footing. As a matter of empirical fact, all known particles satisfy both, but no purely logical grounds exist for either. Indeed, the situation is that the SP entails the IP, but not the converse.[6] Thus, if

---

ever, is that an observable representing a property of one particle be related by permutation to the observable representing the same property of another particle: as $A \otimes I$ and $I \otimes A$ are. But the IP means that permutations leave observables unchanged, in which case there cannot be a pair of *distinct* observables representing the same property for a pair of particles: hence no such pair of particles can have distinct trajectories.

[4] Such a system has second-quantized realizations whose particles are known as 'quons'. See [3] and references therein.

[5] Part of the confusion arises because 'identicality' is often used to mean indistinguishability. Although logically unproblematic, this usage obscures the possibility of particles that share their intrinsic properties, but violate the IP.

[6] An operator on $\mathcal{H}_n$ leaves the subspace of bosonic states, $\mathcal{H}_+$, invariant iff its action on $\mathcal{H}_+$ is the same as that of its projection onto $\mathcal{H}_+$; this latter operator necessarily satisfies the IP. Now, observables for a system of identical bosons must leave $\mathcal{H}_+$ invariant, else measurement collapses will not be well-defined. So not all Hermitian operators on $\mathcal{H}_n$ can be bosonic observables, only those whose action on $\mathcal{H}_+$ is the same as that of a symmetric operator; similarly for fermions, hence SP implies IP.

one principle explains the other (and if entailment is a form of explanation), it is symmetrization that explains indistinguishability, not the other way around!

Of course, the fact that all known species of elementary particles are either bosons or fermions suggests that there may be some reason, some important principle, explaining why nature does not explore the many other options. Much work has been devoted to showing which additional principles are necessary to prove the IP or SP; but none of these principles seem more natural or secure than what is meant to be shown.

To summarize: It is important to keep clear the relations between the concepts of identicality, trajectory indistinguishability and indistinguishability (and symmetrization). First, identicality entails neither trajectory indistinguishability nor indistinguishability (though the former follows from the latter); the impossibility of continuously varying trajectories in QM is nothing but a red herring. Second, the SP implies the IP, but not the converse. So, to summarize the summary,

$$\text{SP} \Rightarrow \text{IP} \Rightarrow \text{Trajectory Indistinguishability}$$

but none of these follow from identicality.

Approximate Distinguishability

It is important to note that one can sometimes treat indistinguishable particles as 'approximately' distinguishable.

First, which properties are to count as intrinsic is a system-relative matter. Consider a system of two ► electrons that are in distinct *constant* spin-z eigenstates, one ► spin up and the other spin down, so that the spins function as intrinsic distinguishing properties for the particles. Now, this may seem surprising since the particles in question are identical fermions at a fundamental level, and hence their states are antisymmetric under the exchange operator P_{12}. Antisymmetrization (and, similarly, symmetrization for identical bosons) implies that the z-spins can never distinguish particle 1 – that is, the particle associated with the first 'slot' in the tensor product space – from particle 2 – the one associated with the second slot. For example, their state cannot be something like $|\uparrow\rangle \otimes |\downarrow\rangle \otimes |\psi\rangle$, in which particle 1 is the spin-up electron and particle 2 the spin-down electron, and $|\psi\rangle$ represents the non-spin portion of the two particle state. Suppose, however, that the Hilbert space of the system in question is spanned by states of the form

$$(|\uparrow\rangle \otimes |\alpha\rangle) \otimes (|\downarrow\rangle \otimes |\beta\rangle) - (|\downarrow\rangle \otimes |\beta\rangle) \otimes (|\uparrow\rangle \otimes |\alpha\rangle), \qquad (3)$$

The SP can be derived from the conjunction of the IP and the assumption that the representation of the permutation group is 1-dimensional on $\mathcal{H}$: $P|\Psi\rangle = \lambda|\Psi\rangle$. The point is that there is no independent justification for the latter conjunct, which can be consistently relaxed, as we shall see the final section.

(in which, for example, the first term assigns spin-up and the non-spin state $|\alpha\rangle$ to particle 1, and spin-down and the non-spin state $|\beta\rangle$ to particle 2). Then we can 'distinguish' a spin-up particle from a spin-down particle in the following sense. In a state such as (3), $|\alpha\rangle$ ($|\beta\rangle$) is associated with spin-up (spin-down) in both terms. Hence we can simply denote the state by $|\alpha\rangle \otimes |\beta\rangle$ in which it is understood that the 'new' particle 1 – that associated with the first slot in the new notation – is spin-up and the new particle 2 – that associated with the second slot – is spin-down. So although the state is antisymmetric at a fundamental level, in this effective description we have two particles that are distinguished by their spins. Since the electrons are identical in the fundamental sense, and distinguished by constant properties in the effective description of this system, it would perhaps be more accurate to say, not that the electrons are approximately distinguishable, but that they are approximately non-identical.[7]

Second, while particles cannot be distinguished by continuously varying, exact positions, they can by continuously varying *approximate* positions. In the classical limit, identical particles have ► wave function that are peaked in space with little overlap for some period; they are approximately trajectory distinguishable. Quantum mechanics does allow such states to evolve in a continuous way, with the peaks moving through space – as the existence of the classical limit demands. (And of course similar points hold for other observables.) If the particles in question are identical bosons or fermions, then these approximately distinct trajectories will serve to distinguish in just the way that spins did for the two electrons: we will be able to give an effective description of states in which the new ith slot is associated with the ith spatial trajectory. This is exactly what goes on for instance when we refer to an electron localized in a particular region of space, distinct from all other electrons.[8]

Why It Matters

Carefully distinguishing the concepts discussed in this article reveals a wider range of possibilities for multi-particle quantum systems, as is now briefly explained.

Messiah and Greenberg [6] were the first to exploit systematically the fact that the IP (which they called 'identicality'!) was not sufficient for the symmetrization postulate. Specifically, they relaxed the latter postulate and considered more general state spaces. Building on this work, Hartle, Stolt and Taylor (e.g., [7]) showed how to classify all types of identical, indistinguishable quantum particle statistics (compatible with a principle of 'cluster decomposition') according to the transformation properties of their state spaces under the action of particle permutations. However, they considered only observables satisfying the IP, which we have just seen to be

[7] Although the particles in the example of p. 313 are not fermions, they are – just for the evolution described – non-identical in a similar sense.

[8] Related issues in both classical and quantum mechanics are discussed in [5].

an *ad hoc* restriction on observables. Thus, one may ask: 'Does also relaxing the IP allow an even richer classification of statistics by the transformation properties of states *and observables* under the action of particle permutations?'

And indeed it does, as Espinoza et al. [8] have recently shown. Bose and Fermi particles – what are usually called 'quanta' – are of course still examples of the types now classified, as are parastatistical particles and quantum Maxwell–Boltzmann particles, and a countable infinity of others. In every case categorized by Hartle, Stolt and Taylor (except for bosons and fermions which necessarily satisfy the indistinguishability postulate) there is an associated distinguishable case now possible in which non-symmetric observables are allowed. Any two systems with different statistics – whether they differ in the transformation properties of their states or observables or both – will have different partition functions and hence different thermodynamic behaviors. In particular, whether the indistinguishability postulate holds makes a real physical difference for a system of identical particles – or at least it would were we to discover identical yet distinguishable particles in nature.

Literature

1. E. Castellani (ed.): *Interpreting Bodies: Classical and Quantum Objects in Modern Physics* (Princeton University Press, Princeton, NJ 1998).
2. J.M. Jauch: *Foundations of Quantum Mechanics* (Addison-Wesley, New York 1968).
3. O.W. Greenberg: Particles with Small Violations of Fermi or Bose Statistics. Phys. Rev. D 43, 4111-20 (1991).
4. Bros, A.M.L. Messiah and D.N. Williams: Theoretical Basis of the Symmetrization Postulate. Phys. Rev. 149, 1008–9 (1966).
5. S. Saunders: On the Explanation for Quantum Statistics. Stud. Hist. Philos. Mod. Phys. 37, 192-211 (2006).
6. A.M.L. Messiah and O.W. Greenberg: Symmetrization Postulate and Its Experimental Foundation. Phys. Rev. 136, B248-67 (1964).
7. J.B. Hartle, R.H. Stolt and J.R. Taylor: Paraparticles of Infinite Order. Phys. Rev. D 2, 1759-60 (1970).
8. R. Espinoza, T.D. Imbo and M. Satriawan: Identicality, (In)distinguishability and Quantum Statistics. In preparation and UIC preprint (2003).

Interaction-Free Measurements (Elitzur–Vaidman, EV IFM)

Lev Vaidman

The interaction-free measurements proposed by Elitzur and Vaidman [1] (EV IFM) is a quantum mechanical method to find an object that interacts with other systems solely via its explosion without exploding it. In this method, an object can be found without "touching it", i.e. without any particle being at its vicinity.

The basic idea of the method is as follows. A quantum test particle is being split into a ► superposition of two separated states. One of these states is being split again into a superposition of two output states while the other is being split into a (different) superposition of the same output states. The phases of the various parts are tuned in such a way that there is a destructive interference at one of the outputs. At this output there is a detector. This is the EV device ready for action.

The simplest EV device is the Mach–Zehnder interferometer, Fig. 1. To use it, the device should be placed in such a way that only one of the intermediate states

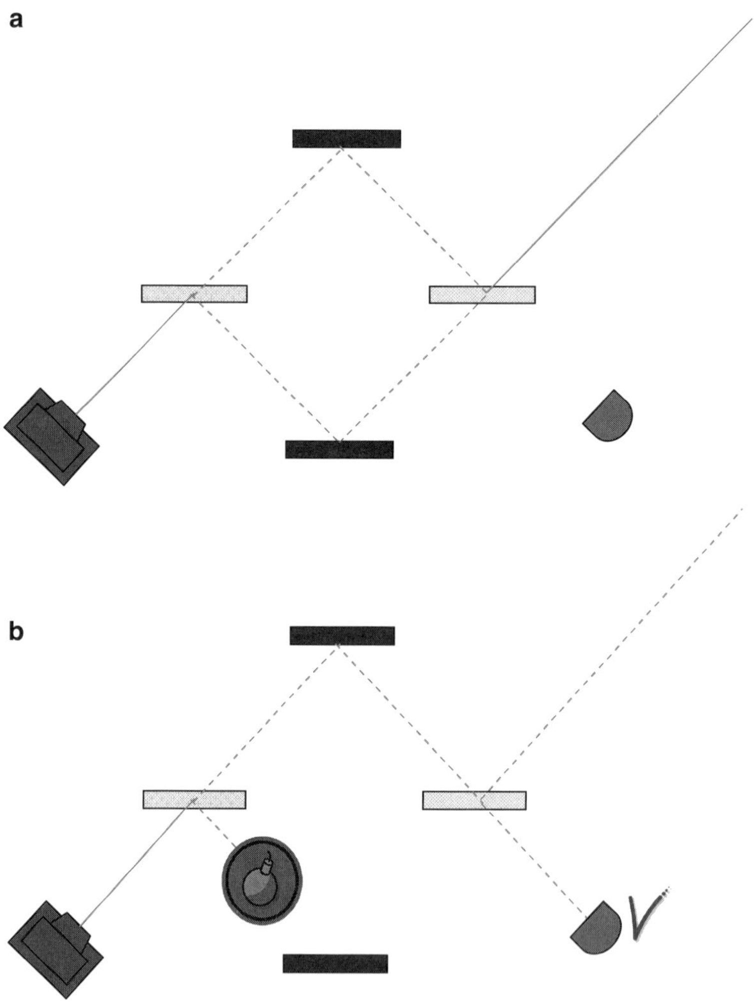

Fig. 1 The Elitzur–Vaidman scheme (**a**) When the interferometer is empty and properly tuned, photons do not reach the detector. (**b**) If the exploding object is present, the detector has the probability 25% to detect the photon sent through the interferometer, and in this case we know that the object is inside the interferometer without exploding it

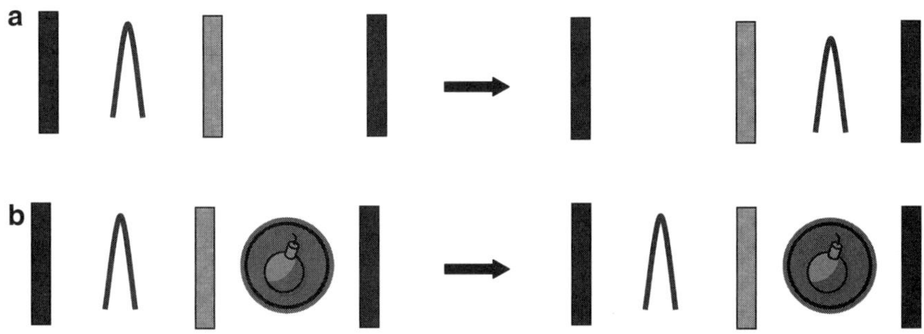

Fig. 2 The Kwiat et al. scheme. (**a**) If the cavities are empty, the photon after N bouncing moves completely from the left cavity to the right cavity. (**b**) If the object is present in the second cavity, after the same N bounces it will remain in the first cavity with probability close to 1 for large N

interacts with the object. If the object is present, the destructive interference is spoiled and the detector might click announcing that the object is present. In this case, no explosion has occurred, since the particle can be found only in one place. The particle can also be "found" by the object, so in half of the cases the object explodes. The probability of finding the object on the first run is just one quarter, so the efficiency of the method is low, but given that the detector clicks, the object is present with certainty.

The EV method was improved using the ► quantum Zeno effect [2] and the probability of the explosion could be made arbitrary small. This, however, requires more time: the quantum test particle has to traverse the interaction region many times. Conceptually, the simplest implementation of this improvement is a device consisting of two identical cavities A and B connected by a highly reflective wall, see Fig. 2. If we place a photon in one cavity, the evolution brings it to another cavity after N bounces in one cavity. At this moment, a detector tests for the presence of the photon in cavity A. This is the device which is ready for action. We place it in such a way that the interaction region of possible explosive object is cavity B. The detector will click with probability close to 1. (The probability for the failure, which is an explosion of the object, is of the order of $1/N$). It will not click for sure if the object is absent.

Setups similar to the EV device were considered before by Renninger [3] and Dicke [4]. However, they did not realize the effect because in their analysis the object and the test particle were reversed: they pointed out the peculiar property that the EV test particle changes its state while the EV explosive object (their measuring device) has not changed at all, it was a *negative result experiment*.

The EV method can find in an interaction-free manner not only exploding objects, but any opaque object. This experiment, however, is somewhat more difficult to implement. For finding an explosive device we could use, instead of a single particle source, a weak laser beam. If the click happens before the explosion, we know that the object is there. For an opaque object, we need a single particle source: if we get a click sending only one photon, we know that there is a opaque object somewhere inside the interferometer and that it did not absorb any photon.

One of the most paradoxical features of the EV IFM is that the test particle in some sense never passes in the vicinity of the interaction region. How can we get information about a region when nothing passed through it and nothing came out of it? Indeed, when we hear the click announcing the presence of the object, there is no record of any kind in our world showing that the test particle was near the object.

A way to resolve this paradox is to note that of our intuition regarding causality in our world is based on physical laws. These laws, however, describe our Universe which includes many worlds, including the one in which the test particle visited the interaction region (and there was an explosion). In this picture it is easy to understand why there is no interaction free method for finding out that the interaction region is empty. Since there is no parallel world in which an explosion occurs, we cannot verify that the region is empty without passing through it.

Let us consider now what happens when the EV IFM device is used for finding a quantum object. If the ▶ wave function of the quantum object spreads over space such that only part of it overlaps with the interaction region, the successful EV IFM localizes the object to the interaction region without changing its internal state (without exploding it). The momentum of the object is changed in this procedure. In this respect it is no different from any other nondemolition measurement of the projection on the interaction region. The name "energy exchange free measurement" frequently associated with the EV proposal, thus does not reflect the unique features of the EV IFM [5, 6].

Energy exchange is relevant for the Penrose modification of the EV IFM [5], in which the goal is different: We are to distinguish between objects which explode whenever their trigger is touched and duds where the trigger is locked to the object which do not explode. The dud serves as a mirror in the Mach–Zehnder interferometer which produces a destructive interference in its detector. A good exploding device cannot serve as a mirror and thus the detector might click announcing that the object is not a dud. Penrose's explanation of the core of the IFM is *counterfactual* [7, p. 135]: the object caused the detector to click because it could have exploded, although it did not. This is the origin of the name *counterfactual computation* [8, 9] for a quantum computer which yields the outcome without "running" the algorithm. Note, however, that as we cannot establish the *absence* of an object in an interaction-free manner, we cannot have a counterfactual computation for all possible outcomes [10].

In Penrose's IFM, when the detector clicks, we can claim, as before, that the quantum test particle was not at the vicinity of the exploding object. However, when the EV IFM device is used for finding a quantum object, the click of the detector does not ensure that the quantum test particle was not present in the interaction region. It might that the whole quantum wave of the test particle passes the interaction region. This happens when the observed quantum object is the "test particle" of the EV IFM measuring the presence of the original test particle. This setup is known as Hardy's paradox. This consideration shows that the claim that the EV IFM localizes quantum objects to the interaction region is strictly speaking incorrect. But limitation is minor: anyone observing the location of the object (and not a superposition of localized states) after the EV IFM announcement about its location, will find that EV IFM method is not mistaken.

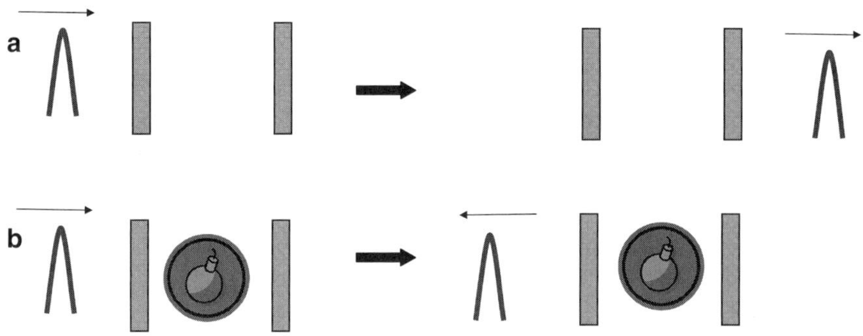

Fig. 3 The Paul and Pavičić scheme. (**a**) If the cavity is empty, the photon passes through it with very high probability. (**b**) If the object is present in the cavity, the photon is reflected with very high probability

There have been numerous experiments performing the EV IFM. The original EV scheme was first implemented in laboratory by Kwiat et al. [2]. (► Quantum Interrogation) Later, Kwiat et al. also performed an experiment of their improved scheme which combines the EV setup with the Zeno Effect [12] reaching efficiency of about 70%. Technical problems make further improvement difficult. It is not easy to tune the optical cavities and it is very difficult to put the photon into the first cavity at a particular moment for starting the process.

When the goal is a practical application of the EV IFM, the best approach is the Paul and Pavičić setup [13] which is, essentially a Fabry Perot interferometer, Fig. 3. There is only one cavity build with almost 100% reflecting mirrors, which is tuned to be transparent when empty. If, however, there is an object inside the cavity, it becomes almost 100% reflective mirror which allows finding the object without exploding it. The method has a conceptual drawback that in principle the photon can be reflected even if the cavity is empty, thus, detecting reflected photon cannot ensure presence of the object with 100% certainty. But this drawback has no meaning for actual experiment because noise in an ideal setup is usually larger. This method was first implemented in a laboratory by Tsegaye et al. [14] and recent experiment reached the efficiency of 88% [15]. The method has a potential to improve controlled-not gate for quantum information processing [16].

Applying the EV device for imaging semitransparent objects [17–19] hardly pass the strict definition of the IFM in the sense that the photons (► light quantum) do not pass in the vicinity of the object, but they achieve a very important practical goal, since we "see" the object significantly reducing the irradiation of the object: this can allow measurements on fragile objects.

The EV IFM is one of the quantum paradoxes (► Errors and Paradoxes in Quantum Mechanics. It is a task which cannot be performed in the realm of classical physics, but can be done in the framework of quantum theory. Progress in experimental demonstrations of the method shows that it has a potential for practical applications. See also ► Quantum Interrogation.

Primary Literature

1. A. C. Elitzur, and L. Vaidman, Found. Phys. **23**, 987 (1993).
2. P. Kwiat, H. Weinfurter, T. Herzog, A. Zeilinger, and M. Kasevich, Phys. Rev. Lett. **74**, 4763 (1995).

Secondary Literature

3. M. Renninger, Z. Phys. **136**, 251 (1953).
4. R. H. Dicke, Am. J. Phys. **49**, 925 (1981).
5. L. Vaidman, Found. Phys. **33**, 491 (2003)
6. R. Penrose, *Shadows of the Mind*, p. 240 (Oxford University Press, Oxford, 1994).
7. See Counterfactuals in Quantum Mechanics, in this volume, pp. 132–6.
8. G. Mitchison, and R. Jozsa, Proc. R. Soc. Lond. A **457**, 1175 (2001).
9. O. Hosten, M. T. Rakher, J. T. Barreiro, N. A. Peters, and P. G. Kwiat, Nature, **439**, 949 (2006).
10. L. Vaidman, Phys. Rev. Lett. **98**, 160403 (2007).
11. L. Hardy, Phys. Rev. Lett. **68**, 2981 (1992).
12. P. G. Kwiat, A. G. White, J. R. Mitchell, O. Nairz, G. Weihs, H. Weinfurter, and A. Zeilinger, Phys. Rev. Lett. **83**, 4725 (1999).
13. H. Paul, and M. Pavičić, J. Opt. Soc. Am. B **14**, 1275 (1997).
14. T. K. Tsegaye, E. Goobar, A. Karlsson, G. Bjork, M. Y. Loh, and K. H. Lim, Phys. Rev. A **57**, 3987 (1998).
15. N. Namekata, and S. Inoue, J. Phys. **B 39**, 3177 (2006).
16. M. Pavičić, Phys. Rev. A **75**, 032342 (2007).
17. J. S. Jang, Phys. Rev. A **59**, 2322 (1999).
18. G. Mitchison, and S. Massar, Phys. Rev. A **63**, 032105 (2001).
19. H. Azuma, Phys. Rev. A **74**, 054301 (2006).

Interpretations of Quantum Mechanics

See ► Consistent histories, Ignorance interpretation, Ithaca Interpretation, Many Worlds Interpretation, Modal Interpetation, Orthodox Interpretation, Transactional Interpretation.

Invariance

K. Mainzer

Invariance, in general, means that quantities or objects do not change with respect to transformations [7]. Invariance of quantities and objects can be distinguished from ► covariance which refers to form invariance of laws and equations [8]. In mathematics, a function of coordinates is called invariant with respect to a transformation

T, if the function remains unchanged by application of T to the coordinates. In geometry, for example, lengths and angles are invariants with respect to orthogonal transformations of Cartesian coordinates. Double proportions are invariants of projective transformations. In physics, basic quantities like energy, linear momentum, or angular momentum are invariants, because their conservation results from the ► symmetry properties of the interactions under global space and time continuous transformations.

Examples of continuous transformations are the translation in space, the rotation around a given axis, and the translation in time. For a particle of mass m moving in a one-dimensional space, its classical motion is governed by Newton's equation

$$m\ddot{x} = F.$$

If the interaction force F derives from an energy potential $U(x)$, that is $F = -\frac{dU}{dx}$, and if the potential is constant, i.e. independent of x, then $m\ddot{x} = 0$. Integration gives $m\dot{x} = C$, where C is a constant. Therefore, the invariance of $U(x)$ under the space translation

$$T_a : x' \rightarrow x = x + a$$

leads to the conservation of the linear momentum $m\dot{x}$. The parameter a can take any real value, hence T_a is a continuous transformation. In a similar way, one can show that the invariance of a potential under continuous rotations in space leads to the conservation of the angular momentum and the invariance under translation in time leads to the principle of energy conservation. These crucial connections between the symmetries of a system and the conservation laws are the consequences of a general theorem, Emmy Noether's theorem: If a Lagrangian theory is invariant under a N-parameter continuous transformation (in the sense that the Lagrangian function is invariant) then the theory possesses N conserved quantities [1]. Noether's theorem is not only true in classical and relativistic physics [9]. According to the ► correspondence principle, it also holds in quantum physics.

Historically, Noether's theorem from 1918 did not come immediately into the view of quantum physicists. The reason is that early quantum mechanics emphasized the Hamiltonian frame work in mechanics and the new formulation of symmetries being associated with unitary or antiunitary representations of groups in the Hilbert spaces of states (► symmetry). All three classical text books on group theory and quantum mechanics, namely those by Hermann Weyl (1928) [2], Eugene Paul Wigner (1931) [3], and Bartel Laendert van der Waerden (1932) [4] did not deal with Lagrangian equations of action integrals and their invariance properties. In non-relativistic quantum mechanics the fundamental ► observables are position operators Q_A and momentum operators P_A. The Hamilton ► operator $H = H(Q_A, P_A, t)$ depends on them. Time t is only a parameter. (► Time in quantum mechanics). The Lagrangian framework was rediscovered with the rise of ► quantum field theory and elementary ► particle physics. For the thirties of the last century there is only a paper of Moisei A. Markov [5] which explicitly and systematically applied Noether's theorem to the currents of a Dirac particle in an external electromagnetic field. After some textbooks on classical field theory quoting Noether's paper the breakthrough came with Edward L. Hill's exposition of Noether's results in 1951 [6] which was quoted in textbooks on quantized fields in the fifties of the last century.

In general, quantum field theory refers to independent field operators $u_A(x^i)$ $(A = 1,\ 2,\ \ldots)$ as fundamental quantities of the theory. The Galilean coordinates x^i are parameters. There is a formal correspondence

$$\begin{aligned} t &\to x^i\,, \\ Q_A(t) &\to u_A\left(x^i\right) = u_A\left(x^\mu, t\right). \end{aligned}$$

The coordinates x^μ describe the continuum of the space of position. Therefore, quantum field theory can be considered a quantum mechanical system with non-countably-infinite many degrees of freedom. In the Lagrangian theory of fields the operator of Lagrangian density plays a central role. It has the same external form like the classical Lagrange density. A classical Lagrange density function can be differentiated in the usual way with respect to field functions and their derivatives. The derivation of an operator with respect to operators is in general problematic because of the non-commutativity of the operators. But with appropriate rules of partial differentiation, the results of Noether's theorem can be transferred to quantum field theory [10].

Primary Literature

1. E. Noether: Invariante Variationsprobleme. Nachr. Ges. Wiss. Göttingen, Math. Phys. Kl. (1918) 235–257
2. H. Weyl: *Gruppentheorie und Quantenmechanik* (Hirzel, Leipzig 1928; English trans. *The Theory of Groups and Quantum Mechanics*, Dover, New York 1950, 147)
3. E.P. Wigner: *Gruppentheorie und ihre Anwendung auf die Quantenmechanik der Atomspektren* (Vieweg, Braunschweig 1931; English trans. *Group Theory*, Academic, New York 1959)
4. B.L. van der Waerden: Die gruppentheoretische Methode in der Quantenmechanik (Springer, Berlin 1932)
5. M. Markov: Zur Diracschen Theorie des Elektrons, I. Teil. Physikal. Zeitschr. D. Sowjetunion **10**, 773–808 (1936)
6. E.L. Hill: Hamilton's Principle and the Conservation Theorems of Mathematical Physics. Rev. Mod. Phys. **23**, 253–260 (1951)

Secondary Literature

7. K. Mainzer: *Symmetrien der Natur* (De Gruyter, Berlin 1988, English trans. *Symmetries of Nature*, De Gruyter, New York 1996)
8. F. Friedman: *Foundations of Space-Time Theories* (Princeton University Press, Princeton 1983)
9. H.A. Kastrup: The Contributions of Emmy Noether, Felix Klein and Sophus Lie to the Modern Concept of Symmetries in Physical Systems. In: M.G. Doncel, A. Hermann, L. Michel, A. Pais (Eds.): *Symmetries in Physics (1600–1980)* (Servei de Publicacions, UAB, Barcelona 1987,113–163)
10. E. Schmutzer: *Symmetrien und Erhaltungssätze der Physik* (Akademie-Verlag, Berlin/Ost 1972)

Ithaca Interpretation of Quantum Mechanics

Jeffrey A. Barrett

The Ithaca Interpretation of quantum mechanics was proposed by the physicist N. David Mermin (*1935) as an attempt to understand quantum mechanics by supposing that the only proper subject of physics is correlations between ► observables. Further, while correlations are taken to have physical reality, that which they correlate is not. Quantum mechanics with no dynamical collapse is then taken to be an entirely adequate physical theory since it can be understood as describing correlations without correlate.

Mermin's presentation of the Ithaca Interpretation starts by taking quantum mechanics without the collapse postulate as given. (► Wave function collapse). He then seeks to infer what physical reality must be in order for this theory to be taken as providing a complete and accurate physical description. What Mermin refers to as the Theorem of the Sufficiency of Subsystem Correlations, the SSC Theorem, plays a central role in characterizing what he takes to be the essential structure of quantum-mechanical states – Mermin cites Wootters [3] for an earlier proof of the theorem.

The SSC Theorem says that the mean values of the products of subsystem observables, over a particular resolution of a system into subsystems, suffice to uniquely determine the quantum-mechanical state of a given system. Mermin understands this to mean that the quantum-mechanical state of a complex system is nothing more than a coding of the correlations, or joint probabilities, between the observables of its subsystems. And since the quantum-mechanical state of a system determines the correlations between observables of its subsystems and nothing more, he concludes that the quantum-mechanical description of the reality extends only to such ► correlations. On the assumption that the quantum-mechanical state of a system provides a complete description of physical reality, since the state determines the joint probabilities for observables of its subsystems but not the probabilities of physical properties in fact obtaining, Mermin concludes that physical reality consists in correlations without there being any physical correlata described by the correlations. Once one recognizes that physics, properly conceived, concerns correlations without correlata, he argues, one recognizes that the quantum-mechanical description is entirely adequate as a complete and accurate description of the physical world since it fully characterizes precisely these correlations.

Using quantum mechanics to determine the proper subject of physics, then judging the adequacy of quantum mechanics as a physical theory using the standard of adequacy derived from the theory itself is clearly circular, but Mermin argues that there are historical precedents for such an argument. Just as electrodynamics taught us that it is possible to have physical fields without there being any physical medium to support them, quantum mechanics teaches us that it is possible to have physical correlations without there being any physical correlata to support them. The argument is that one should listen to what quantum mechanics is trying to tell us rather than to try to impose our intuitions concerning the interpretation of joint probabilities and the nature of probabilistic explanation on quantum mechanics. The central question then in judging the adequacy of the Ithaca Interpretation concerns the degree to which one should be willing to allow a physical theory to determine the explanatory standards for its own assessment.

Part of the puzzle here is that Mermin recognizes that the quantum measurement problem remains a problem in the context of the Ithaca Interpretation. He concedes that "When *I* look at the scale of the apparatus *I know* what it reads. Those absurdly delicate, hopelessly inaccessible, global system correlations *obviously* vanish completely when they connect up with *me*" He insists, however, that explaining the particular outcome of a measurement when there are no physical correlata (and, for that matter, explaining why we have to update our probability calculations after performing a measurement) "is a puzzle about consciousness which we should not get mixed up with the efforts to understand quantum mechanics as a theory of subsystem correlations in the non-conscious world" ([1], 759).

One way to understand the argument would be to suppose that while Mermin takes quantum mechanics to provide a complete and accurate description of physical reality, he does not take physical reality to determine the mental states of observers. Indeed, on the Ithaca Interpretation of quantum mechanics, *reality* is explicitly defined to be "physical reality plus that on which physics is silent, its conscious perception" ([1], 766). This distinction allows one, if one wished, to locate correlata as features of the nonphysical conscious world and thus to explain how it is possible to know the result of a measurement when physical reality consists in only correlations without correlata. The cost of this line of explanation would, it seems, be a commitment to a strong variety of mind-body dualism.

Primary Literature

1. D. N. Mermin: What is Quantum Mechanics Trying to Tell Us? American Journal of Physics **66/9**, 753–767 (1998)
2. D. N. Mermin: The Ithaca Interpretation of Quantum Mechanics. Notes for a lecture given at the Golden Jubilee Workshop on Foundations of Quantum Theory, Tata Institute, Bombay, September 9–12, 1996. arXiv:quant-ph/9609013 v 1: 17 Sep 1996

Secondary Literature

3. W. K. Wootters in *Complexity, Entropy, and the Physics of Information*, W.H. Zurek ed. (Addison-Wesley, Redwood City, CA 1990, 39–46)

J

jj-Coupling

Klaus Hentschel

The ► vector model provides various ways of calculating the vectorial sum of all the contributing angular momenta l_i (► Spin; Stern–Gerlach experiment; Vector model) and ► spins $s_i = 1/2$ for atoms with more than one ► electron. Either all the l_i are first summed up to one L, and then combined with $S = \sum_i s_i$, or all the l_i and s_i are first summed up separately to j_i with $J = \sum_i j_i$. The noncommutativity of ► operators makes these two procedures in general non-equivalent, yielding different combinatorics, and thus different energy levels and transitions. The first possibility is called ► Russell–Saunders coupling (valid for the lighter, hydrogen-like atoms ► Bohr's atom model). The latter is called jj-coupling, yielding the better approximation for heavier atoms and for the energetically higher terms. jj-coupling assumes a strong interaction between each l_i and the corresponding s_i of each electron. There is thus no definite L and S, but only a well-defined J which also implies that the prohibition of intercombinations with $\Delta S \pm 1$ is no longer in place, and the only ► selection rules applying for jj-coupling are $\Delta J = 0$ or ± 1, and similar for the individual j_i. The ► Landé g-formulae also have to be revised for this case; see [1].

Primary Literature

1. G. Herzberg: *Atomic Spectra and Atomic Structure* (New York: Prentice-Hall 1937, 2nd ed. New York: Dover Publications 1944, 174–5)

Secondary Literature

2. C. Candler: *Atomic Spectra and the Vector Model* (Princeton: Van Nostrand 1937, esp. chaps. 11, 16–18)
3. http://wwwchem.uwimona.edu.jm:1104/courses/RScoupling.html (accessed July 20, 2006)

K

Kaluza–Klein Theory

Daniela Wuensch

Theodor Kaluza (1885–1954) set forth his idea of unifying gravitation and electromagnetism within five-dimensional space-time in a paper from the beginning of 1919. It was presented by Albert Einstein (1879–1955) to the Berlin Academy of Science and published in its *Sitzungsberichte* in 1921 under the title "Zum Unitätsproblem der Physik" [4]. Having received the manuscript from the author in April 1919, Einstein was so impressed with the idea of unifying the basic forces in a five-dimensional space that he used it himself up to the mid-1940s in eight of his own papers.

Kaluza's idea of unifying gravitation and electromagnetism goes back to David Hilbert's (1862–1943) unification program and to the pioneering work of two of his pupils: Hilbert sought unification within a four-dimensional space by having electromagnetism come from gravitation in 1915 [2]. His pupil Gunnar Nordström (1881–1923) explored unification within a five-dimensional unwarped (Minkowskian) space in the foregoing year [7]. The unification attempt by Hermann Weyl (1885–1955) in 1918, finally, was to apply a gauge transformation within a four-dimensional space with a generalized non-Riemannian metric [10]. Although Nordström was the first to introduce a five-dimensional space, it was Kaluza's theory from 1919 that proposed a realistic unification of the two interactions. Nordström's theory predated the general theory of relativity (1915) so the gravitation was derived from electromagnetism within a space described by a Minkowskian flat metric. As a consequence, it could not explain phenomena like light deflection and was therefore condemned as a prerelativistic theory.

Kaluza's idea, which was to serve as the model for the design of all unified theories in higher-dimensional spaces, was as follows: Within a five-dimensional space (with a Riemannian metric) there exists a unique five-dimensional gravitational force that upon projection onto the four-dimensional space of our experience splits into two phenomena: our familiar natural forces, being four-dimensional Einsteinian gravitation (known from the general theory of relativity), and Maxwellian electromagnetism. Thus in Kaluza's theory – as in all modern higher-dimensional unified theories, and unlike Nordström's – the fundamental force is gravitation. It, according to Kaluza, is the originator of electromagnetism. (Modern-day higher-dimensional unified theories attribute all the other forces to gravitation as well.) Electromagnetism is thus an effect of the fifth dimension. The fifth components of the metric tensor $g_{\mu 5} (\mu = 1, 2, 3, 4)$ are identical to the Maxwellian electromagnetic field A_μ. Kaluza was able to show that the five-dimensional momentum p_5

D. Greenberger et al. (eds.), *Compendium of Quantum Physics: Concepts, Experiments, History and Philosophy*, © Springer-Verlag Berlin Heidelberg 2009

is proportional to the electric charge ρ_0, which offers a possible explanation for electric charge as a five-dimensional effect.

Kaluza applied his "cylinder condition" in order to explain why the fifth dimension is not perceptible in any phenomena of our experience. It states that the first derivative of all physical quantities after the fifth dimension must be null:

$$\frac{\partial f(x)}{\partial x^5} = 0$$

This condition determines the structure of the five-dimensional Kaluza space, in that the fifth dimension forms the cylinder's central axis. The points on the cylinder's surface correspondingly make up our familiar four-dimensional space. Einstein found fault with this preference for the fifth dimension because it limits the covariance within five-dimensional space. He also argued for a then radical conception of field theory that makes all particles interpretable as condensations of a field. Kaluza's theory should, he thought, yield the ► electron as a product of its unified field, which was not the case.

In 1926 Oskar Klein (1894–1977) succeeded in linking Kaluza's theory [4] with quantum mechanics [5,6]. He quantified the fifth components of momentum according to the rule:

K

$$p_5 = n \times \frac{h}{l}$$

(n = quantum number, h = ► Planck's constant, l = period of the fifth dimension, i.e., the circumference of a tiny circle).

It differed from Kaluza's theory in the following way [12]: Instead of having the fifth dimension form the central axis of a cylinder of infinite extension, Klein had it curled up ("compactified") into a tiny circle of magnitude $l = 10^{-30}$ cm.

$$l = \frac{hc\sqrt{2k}}{e}$$

(c = velocity of light, e = electron charge, k = Einstein's gravitational constant)

The term "Kaluza–Klein theory" was first used in 1933 by Oswald Veblen (1880–1960) [9], who together with Banesh Hoffmann (1906–1986) had given the theory its projective form in 1930 [3]. The theory became a purely formal construct in which the five-dimensional space is no longer attributed any physical reality. It serves instead as a mathematical space from which the real four-dimensional space emerges as a projection. Pascual Jordan (1902–1980) and André Lichnerowicz (1915–1998) were among the proponents of this construct from 1945 on.

Wolfgang Pauli (1900–1958) and others working on quantum mechanics rejected the five-dimensional Kaluza–Klein theory in the mid-1930s, however, because it offered no way to quantify field theories [13].

Two new interactions were discovered during the 1930s, the weak and the strong interactions. Gauge theories – an idea Hermann Weyl originally developed in 1918 and generalized in 1929 – were the first to prove successful in unifying the three natural forces: electromagnetism, the weak and strong forces, by means of common symmetry properties.

In the continued search for a new theory to unify all four natural forces, the Kaluza–Klein theory was rediscovered in the 1960s. Bryce S. DeWitt (1923–2004) came up with the idea of combining the ► symmetry transformation contained in gauge theories with the higher-dimensionality of physical space in the Kaluza–Klein theory in 1963 [1].

Three advocates of superstring theory: In 1975, J. Scherk (1946–1979), together with H. J. Schwarz (born in 1941), E. Crammer (born in 1942) and J. Scherk (in 1976) and in 1976 E. Crammer (born in 1942) together with J. Scherk introduced the idea that the higher dimensions should be regarded as true physical dimensions "on a par with the four observed dimensions." [11, p. 412] They suggested that the obvious differences between the four observed dimensions and the extra microscopic ones could arise from a spontaneous breakdown in the vacuum symmetry, i.e., from a process of 'spontaneous compactification' of the extra dimensions (curling up as the universe cooled).

At the beginning of the 1980s the initiator of the superstring revolution, Edward Witten (born in 1951), explored in his article "Search for a Realistic Kaluza–Klein-Theory" (1981) [11] whether the theory could serve as a conceptual basis for the unification of all the natural forces. "This theory," he exclaimed, "is surely one of the most remarkable ideas ever advanced for unification of electromagnetism and gravitation" [11, p. 415]. Thus Kaluza–Klein theories began to be considered as the potential beginning of a paradigm shift.

Kaluza–Klein theories still serve as a model for superstring theory as well as for the M-theory propounded by Edward Witten in 1995 which endows space with eleven dimensions. It is based on Kaluza's idea that apparently different natural forces may be unified by introducing additional spatial dimensions, with the unifying force being higher-dimensional gravitation. It also takes up Klein's idea of compactifying additional dimensions and explains why the additional dimensions are not perceptible: Their extremely small size makes them technically immeasurable.

Lisa Randall (born in 1962) and Raman Sundrum (born in 1964) developed a new form of unification in 1999 that does not use the Kaluza–Klein model but reverts back to Kaluza's original idea of a fifth dimension of infinite extension [6].

Primary Literature

1. B. S. DeWitt: *Dynamical Theory of Groups and Fields*. Relativity, Groups and Topology, C. DeWitt and B.S. DeWitt (Eds.) 725, (Gordon & Breach: New York and London 1964)
2. D. Hilbert: *Die Grundlagen der Physik*. Erste Mitteilung am 20. November 1915 Nachrichten von der Königlichen Gesellschaft der Wissenschaften zu Göttingen, Mathematisch-physikalische Klasse, 395–407 (1915)
3. B. Hoffmann and O. Veblen: *Projective Relativity*. The Physical Review **36**, 810–822 (1930)
4. T. Kaluza: *Zum Unitätsproblem der Physik*. Sitzungsberichte der Königlich Preußischen Akademie der Wissenschaften zu Berlin 966–972 (1921)
5. O. Klein: *Quantentheorie und fünfdimensionale Relativitätstheorie*. Zeitschrift für Physik **37**, 895–906 (1926a)
6. O. Klein: *The Atomicity of Electricity as a Quantum Theory Law*. Nature **118**, 516 (1926b)

7. G. Nordström: *Über die Möglichkeit, das elektromagnetische Feld und das Gravitationsfeld zu vereinigen*. Physikalische Zeitschrift **15**, 504–506 (1914)
8. L. Randall: *Warped Passages: Unraveling the Mysteries of the Universe's Hidden Dimensions*. (Ecco: New York 2005)
9. O. Veblen: *Projektive Relativitätstheorie*. (Springer: Berlin 1933)
10. H. Weyl: *Gravitation und Elektrizität*. Sitzungsberichte der Königlich Preußischen Akademie der Wissenschaften zu Berlin 465–478 (1918)
11. E. Witten: *Search for a Realistic Kaluza–Klein Theory*. Nuclear Physics **B 186**, 412–428 (1981)

Secondary Literature

12. D. Wuensch and H. Goenner: *Kaluza's and Klein's Contributions to Kaluza-Klein theory*. Proceedings of the Tenth Marcel Grossmann Meeting July 20 to 26, 2003, Rio de Janeiro, Brasil, 2041–2047 (World Scientific: Singapore, 2005)
13. D. Wuensch: *Der Erfinder der fünften Dimension. Theodor Kaluza, Leben und Werk*. (Termessos: Göttingen 2007)

Kochen–Specker Theorem

Carsten Held

Quantum mechanics generates, for chosen ► observables and state assignments, measurement outcome predictions. What does it mean to ask whether the theory *completely* describes the systems it in fact describes? Assume that, if a quantum-mechanical system S is in a pure state $|a_1\rangle$ such that prob $(a_1) = 1$ (i.e., the probability that S is found, upon an A-measurement, to have a_1 equals 1), then it has the physical property represented by a_1 (the eigenvalue of A pertaining to $|a_1\rangle$). Completeness then can be characterized as the idea that the properties ascribed to S in this way are the only ones and incompleteness as the idea that there are more. The possible S properties not derivable from S's quantum-mechanical state are usually called ► hidden variables.

Incompleteness (i.e., the presence of hidden variables) can be related to S's description in ► Hilbert space H as follows. In every orthogonal set of vectors spanning H and thus representing a non-degenerate observable, there is one vector representing a possessed property and thus being ascribed the number 1, the others the number 0. We now ask the question whether such an assignment (representing incompleteness) is possible. This is a mathematical problem that turns out to be reducible to the Hilbert space R^3 (the familiar three-dimensional space over the real numbers) and for this space to the task of assigning the number 1 to exactly one vector, in any ► orthonormal basis (the number 0 to the two others), under the condition that vectors of different bases but lying in the same ray get assigned the same number. Call such an assignment a 0–1 valuation.

Indeed, a 0–1 valuation is impossible on R^3. This can be shown either, by reductio ad absurdum, from Gleason's Theorem ([11]; ► *Gleason's Theorem*), or constructively, by finding finitely many R^3 vectors such that no 0–1 valuation is possible. In 1967, Kochen and Specker ([1]) explicitly presented such a set for the first time, whence finite vector sets without a 0–1 valuation are generally called Kochen–Specker (KS) sets. It is immediately obvious that a KS set must contain vectors not only from many bases, but many interlocking ones, i.e., bases sharing one vector. The decisive first step of Kochen and Specker's result then is to prove that a set of 10 vectors can form a certain structure of orthogonality relations (see [1], p. 68, [22], p. 126, [12] Fig. 1) only if two of these vectors, v_1 and v_2, make an angle smaller than $\sin^{-1}(1/3)$. Now, it turns out that there is no 0–1 valuation of this set where v_1 is assigned the number 1 and v_2 the number 0, so in any larger set containing this one, if v_1 is assigned the number 1, then so must be v_2. This is the heart of Kochen and Specker's argument (since in a hidden-variables construction vectors assigned 1 and 0 should be allowed to be arbitrarily close). Indeed, this initial step of the proof had been established independently by John Bell, a year earlier ([3], pp. 7–8), drawing directly on Gleason's Theorem. For this reason, some researchers refer to the result as the Bell–Kochen–Specker Theorem ([8], [18]). Kochen and Specker's argument involves a quite complicated structure consisting of 15 (partly interlocking) copies of the 10-vector set just described (see [1], p. 69, [22], p. 130, [12] Fig. 2). Finally, the original KS set contains 117 vectors. In later proofs, inconsistency has been achieved using KS sets with only 33 (Bub, [6]) or 31 (Conway and Kochen, described in [21], p. 114) vectors. Moving up to R^4, we can find a KS set with only 18 vectors (Cabello et al., [7]). It has recently been argued ([16, 20]) that all these arguments, except Cabello et al., tacitly refer to many more vectors so that the KS sets in question are actually larger. What is at issue here is that a traditional KS set contains only those vectors necessary to show the impossibility of a 0–1 valuation, but by choosing these we have tacitly chosen more. E.g., the original KS 10-vector set is a subset of a set of five interlocking bases, i.e. a set of 15 vectors ([20] Fig. 6 (i), (ii)), but five of these vectors can be ignored in the argument. Now, if we really *construct* these sets starting from one basis and rotating it stepwise into the others, we will inevitably drag along vectors we do not explicitly need to show a 0–1 valuation to be impossible. This question of the actual size of a concrete Kochen–Specker set is important not so much for determining the record of the smallest such set, but for an experimental realisation, which actually involves procedures equivalent to basis rotations.

It is crucial to analyse in what sense an advocate for incompleteness is committed to the impossible task of producing a 0–1 valuation for a KS set. There are two very different ways in which this question may be taken. Consider first the observation that a KS set contains many triples (or higher n-tuples) of vectors, but that (identifying a tuple with a set of projection operators (► projection) corresponding to one maximal, i.e. non-degenerate, observable) in any quantum-mechanical experiment only one of these triples can be measured. Initially, this seems an irrelevant point. The hidden-variables program essentially is about whether quantum system S can possess properties prior to measurement such that a faithful measurement procedure would reveal just the properties predicted by quantum mechanics. KS sets are finite sets of vectors such that S cannot simultaneously possess the pertaining properties, be they jointly measurable or not. It has been observed, however, that any R^3 basis vector, corresponding to a possible S property, can be arbitrarily well approximated by another vector with only rational coordinates. So for any finite measurement resolution, there is a rational approximation to a KS set vector that is indistinguishable from it and might have been measured instead. Now, sets of rational vectors, approximating any KS set, have a 0–1 valuation [15, 17]. Indeed, the whole set of R^3 basis vectors with purely rational coordinates possesses such a valuation [10]. So, one might defend an incompleteness interpretation of quantum mechanics assuming that only rational vectors have values. In the attempt to measure certain two vectors one would be measuring (perhaps faithfully) not the two intended real vectors but, unwittingly, two of their rational approximations. These vectors would not all stand in the strict orthogonality relations imposed for a KS set, hence would not be the members of such a set. However, due to the ineluctably finite measurement precision the situation would seem empirically indistinguishable from the one described in quantum mechanics. On a closer look, however, this impression dissolves. Quantum mechanics makes exact statistical predictions for vectors standing at specific angles (like v_1 and v_2 in Kochen and Specker's 10-vector set) also when such vectors have only rational coordinates. A 0–1 valuation for a set of only rational vectors, in order to meet these predictions in one place, must violate them in another (see [7]). So, even if it were reasonable for a hidden-variables interpretation to assume that we live in a "toy universe" ([19], p. 3), where only rational vectors have values, the fixation of such values would lead to predictions at odds with quantum mechanics. There is no evidence that quantum mechanics fails in these cases, so these artificial constructions ultimately do not diminish the Kochen–Specker Theorem's force of ruling out hidden-variables interpretations of the theory (see also [2, 5]).

There are sets of purely rational basis vectors allowing a 0–1 valuation such that KS sets are arbitrarily well approximated, which have an interesting property: Every basis vector in such a set belongs to only one basis, so there are in fact no strictly interlocking bases in these sets. On the other hand, the original Kochen–Specker Theorem and all simplified versions make crucial use of interlocking bases. It is crucial to all these arguments that a vector gets assigned a unique number, regardless of the fact that it can be (and, for the Kochen–Specker arguments to get started, always is) a member of several different bases. This opens a second way in which the hidden-variables proponent might reject the 0–1 valuation task. The assumption that every vector is assigned a unique number is generally called non-contextuality because then the value is considered to be independent of which basis the vector belongs to. In physical terms this means that whether S possesses a certain property is independent of the context of other observables considered to have certain values. It turns out that any vector in any one basis corresponds to an observable, say $C = f(A)$, that is the function of one maximal observable A, but C can also be the function of another maximal observable B (i.e., $C = g(B)$), with A and B not being jointly observable. Non-contextuality is the idea that the value of C does not depend on whether we determine it via a measurement suitable to measure A or another incompatible one, suitable to measure B.

It is sometimes said that the arguments from Bell's Theorem ([4]; ► *Bell's Theorem*) and the Kochen–Specker Theorem prove the unconditional completeness of quantum mechanics. As we have seen, there is one way to deny this: one has to deny noncontextuality, i.e., one must subscribe to contextual hidden variables. This is not a well-researched possibility. However, it can be shown that quantum mechanics and completeness, both reasonably formalized, are in a fundamental conceptual conflict and in a sense inconsistent [13]. So, contextual hidden variables interpretations deserve serious interest, after all.

There are two main ways to think about contextual hidden variables (► hidden variable models). The value of an observable might be contextual because it changes depending on which other observables are measured in conjunction with it (causal contextuality; see [22] p. 133–34, [12], Sect. 5.3). This idea is directly opposed to a basic motivating idea of the hidden-variables program, namely the idea that measurement faithfully reveals existing values, and accordingly it has not drawn much interest, in the literature. Alternatively, f(A) and g(B) might simply be taken to be different observables, although they are represented by the same mathematical object, operator C (ontological contextuality; see [22] p. 135, [12], Sect. 5.3). What we would reasonably require of such a position is that it physically motivates or explains in which sense these observables, though represented by the same operator, are different and no promising proposal has hitherto been made.

Primary Literature

1. S. Kochen and E. Specker: *The Problem of Hidden Variables in Quantum Mechanics*, J. Math. Mech. **17**, 59–87 (1967); reprinted in [12], 293–328

Secondary Literature

2. D.M. Appleby: *Existential Contextuality and the Models of Meyer, Kent and Clifton*, Phys. Rev. A **65**, 022105 (2002), arXiv: quant-ph/0005056
3. J.S. Bell: *On the Problem of Hidden Variables in Quantum Mechanics*, Rev. Mod. Phys. **38**, 447–452 (1966); reprinted in J.S. Bell: *Speakable and Unspeakable in Quantum Mechanics* (Cambridge University Press; Cambridge 1987), 1–13
4. J.S. Bell: *On the Einstein-Podolsky-Rosen Paradox*, Physics **1**, 195–200 (1964); reprinted in J.S. Bell: *Speakable and Unspeakable in Quantum Mechanics* (Cambridge University Press; Cambridge 1987), 14–21
5. Th. Breuer: *Kochen–Specker Theorem for Finite Precision Spin One Measurements*, Phys. Rev. Lett. **88**, 240402 (2002), arXiv: quant-ph/0206035
6. J. Bub: *Schütte's Tautology and the Kochen–Specker Theorem*, Found. Phys. **26**, 787–806 (1996)
7. A. Cabello: *Finite-precision measurement does not nullify the Kochen–Specker theorem*, Phys. Rev. A **65**, 052101 (2002), arXiv: quant-ph/0104024

8. A. Cabello, J. Estebaranz, and G. García-Alcaine, *Bell–Kochen–Specker Theorem: A Proof with 18 vectors*, Phys. Lett. A **212**, 183 (1996), arXiv: quant-ph/9706009
9. A. Cabello and G. Garcìa-Alcaine: *Proposed Experimental Test of the Bell–Kochen–Specker Theorem*, Phys. Rev. Lett. **80**, 1797–1799 (1998), arXiv: quant-ph/9709047
10. R.K. Clifton and A. Kent: *'Simulating Quantum Mechanics by Non-Contextual Hidden Variables'*, Proc. Roy. Soc. Lond. A **456**, 2101–2114 (2000), arXiv: quant-ph/9908031
11. A.M. Gleason: *Measures on the Closed Subspaces of a Hilbert Space*, J. Math. Mech. **6**, 885–93 (1957); reprinted in [14], 123–134
12. C. Held: *The Kochen–Specker Theorem*, at http://plato.stanford.edu/entries/kochen-specker/.
13. C. Held: *Axiomatic Quantum Mechanics and Completeness*, Found. Phys. **38**, 707–732 (2008), arxiv: quant-ph/0705.2763
14. C. Hooker (ed.): *The Logico-Algebraic Approach to Quantum Mechanics* (Reidel, Dordrecht, 1975)
15. A. Kent: *'Non-Contextual Hidden Variables and Physical Measurements'*, Phys. Rev. Lett. **83**, 3755–3757 (1995), arXiv: quant-ph/9906006
16. J.-Å. Larsson: *A Kochen–Specker Inequality*, Europhys. Lett. **58**, 799–805 (2002), arXiv: quant-ph/0006134
17. D.A. Meyer: *Finite Precision Measurement Nullifies the Kochen–Specker Theorem* Phys. Rev. Lett. **83** 3751–3754 (1999), arXiv: quant-ph/9905080
18. N.D. Mermin: *Simple Unified Form of the Major No-Hidden Variables Theorems*, Phys. Rev. Lett. **65**, 3373–3376 (1990)
19. N.D. Mermin: *'A Kochen-Specker Theorem for Imprecisely Specified Measurements'* (1999), arXiv: quant-ph/9912081
20. M. Pavičić, J.-P. Merlet, B. McKay, and N.D. Megill, *'Kochen-Specker Vectors'* J. Phys. A **38**, 1577–1592 (2005), arXiv: quant-ph/0409014
21. A. Peres: *Quantum Theory: Concepts and Methods* (Kluwer, Dordrecht 1993)
22. M. Redhead: *Incompleteness, Nonlocality, and Realism: A Prolegomenon to the Philosophy of Quantum Mechanics*. (Clarendon, Oxford 1987)

L

Landé's g-factor and g-formula

Klaus Hentschel

In 1919, the young theoretician Alfred Landé (1888–1976) in Frankfurt am Main showed in his habilitation thesis that satisfactory agreement could be reached between observed splittings of spectral lines in the ► Zeeman effect if one assumed that, in general, ► electrons contribute more to the total energy of the system than had been expected according to classical electron theory.

Instead of $\boldsymbol{\mu} \cdot \boldsymbol{B} = \boldsymbol{\mu_0} \boldsymbol{m_J}$, set $\boldsymbol{\mu} \cdot \boldsymbol{B} = g\, \mu_0\, m_J$ with μ the electron's magnetic moment, μ_0 Bohr's magneton: $\mu_0 = -e\hbar/2m$ and m_J the magnetic quantum number. The so-called Landé g-factor thus describes deviations of experimentally observed magnetic moments from the classical case with $g = 1$. According to Landé, in general

$$g = (\boldsymbol{L} + 2\boldsymbol{S})\mathrm{g} \cdot \boldsymbol{J}/J^2 = 1 + \frac{J(J+1)S(S+1) - L(L+1)}{2J(J+1)}$$

Under the assumption of what later came to be called ► Russell-Saunders coupling, Landé could also derive the ratio of the intervals in a Zeeman multiplet. A physical explanation of the foregoing has to make use of the then widely popular ► vector model.

In the ► vector model (more fully described in [1] or [2]), the total angular momentum $\boldsymbol{J}$ is the vectorial sum $\boldsymbol{J} = \boldsymbol{L} + \boldsymbol{S}$, with $\boldsymbol{L}$ angular momentum of the electrons, and $\boldsymbol{S}$ the spin. ► Spin; Stern–Gerlach experiment; Vector model.

Then the total magnetic moment of the atomic system is given by $\boldsymbol{\mu} = \boldsymbol{\mu_0}\ (\boldsymbol{L} + 2\boldsymbol{S})$. Because the spin contributes twice as much to the total magnetic moment as does the orbit, $\boldsymbol{\mu}$ is not parallel to $\boldsymbol{J}$, but precesses around $\boldsymbol{J}$. In an external magnetic field $\boldsymbol{B}$, the component of magnetic moment $\boldsymbol{\mu}$ in the direction of $\boldsymbol{J}$ yields a contribution of $-\boldsymbol{\mu}_J \cdot \boldsymbol{B}$. Now, after a short calculation, Landé's g-factor as defined by $\mathrm{g} = \boldsymbol{\mu} \cdot \boldsymbol{B}/\mu_0\, m_J$ results:

$$g = (\boldsymbol{L} + 2\boldsymbol{S})\mathrm{g} \cdot \boldsymbol{J}/J^2 = 1 + \frac{J(J+1)S(S+1) - L(L+1)}{2J(J+1)}$$

Thus, retrospectively, Landés g-formula appears to be a straightforward consequence of quantum mechanics. But Landé arrived at this formula without that later knowledge, in a single-handed effort to come to grips with observed regularities in the splitting of spectral lines, emitted in a magnetic field, the so-called ► Zeeman effect.

D. Greenberger et al. (eds.), *Compendium of Quantum Physics: Concepts, Experiments, History and Philosophy*,

According to Landé's own reminiscences: "Thus, working quite alone in Frankfurt am Main without encouragement from colleagues, I found the key, the g-factor, which then opened the drawer with the g-formula in it, while whole groups of older physicists, even the great atomist Sommerfeld, remained in the dark" . . . "I cracked the magnetic code of atomic structure by the g-factor, followed its applications in the g-formula."

Primary Literature

1. G. Herzberg: *Atomic Spectra and Atomic Structure* (New York: Prentice-Hall 1937, 2nd edn. New York: Dover Publications 1944)
2. C. Candler: *Atomic Spectra and the Vector Model* (Princeton: Van Nostrand 1937)

Secondary Literature

3. A. Barut: Alfred Landé, in: *Physiker und Astronomen in Frankfurt* (Neuwied: Metzner, 1989), 38–45, also available online as http://www.physik.uni-frankfurt.de/paf/paf38.html
4. P. Forman: Alfred Landé and the anomalous Zeeman Effect, 1919–1921. Historical Studies in the Physical Sciences **2**, 153–261 (1970)

Large-Angle Scattering

Brigitte Falkenburg

In the ► *scattering experiments* of ► particle physics, large-angle scattering indicates the recoil of the scattered "probe" particles at an impenetrable small or point-like scattering center. In the history of subatomic physics, it happened twice that unexpected large-angle scattering was observed in a crucial experiment. Both discoveries are based on a classical or ► semi-classical model of the atomic nucleus (► Rutherford atom).

Rutherford Scattering

In order to investigate subatomic structure, Ernest Rutherford (1871–1937) scattered α particles from radioactive radiation sources off thin gold foil. In 1909, Rutherford's assistants Hans Geiger (1882–1945) and Ernest Marsden (1889–1970) performed scattering experiments with low energy α-particles of around 5 MeV. They observed unexpected backward scattering at an angle of >90°.

Rutherford spent two years calculating the probability of multiple backward scattering in several atomic models ► Rutherford atom. The homogeneous charge distribution of Thomson's plum pudding model of the atom (► *atomic models*) could not explain Geiger's and Marsden's discovery. Finally Rutherford derived his famous formula for the Coulomb scattering, i.e., the scattering of a charged particle at a point-like positive charge described by a Coulomb potential [1]. Rutherford's atomic model with a point-like nucleus explained the backward scattering in terms of the differential cross section (► *scattering experiments*)

$$\mathrm{d}\sigma/\mathrm{d}\theta = (\hbar c/4E)^2 (ZZ'\alpha)^2 \sin^{-4}\frac{\theta}{2},$$

with the scattering angle θ, where E is the kinetic energy of the probe particles, Z, Z' are the charge numbers of the probe particles and the atomic nucleus, and α is the fine structure constant. The formula predicts a non-negligible probability of large-angle scattering.

The prediction of the formula was confirmed in subsequent scattering experiments which measured the angular distribution of the scattered α-particles [5, 6]. Rutherford's model included an additional term for the shielding by the electrons which turned out to be negligible. The experiments were neither sensitive to deviations from Rutherford's formula due to strong interactions between the α-particles and the gold nucleus, nor to quantum mechanical effects. For the Coulomb potential, the quantum mechanics of scattering results in Rutherford's formula, too.

Pointlike Nucleon Constituents

In 1968, a similar discovery recurred in a high-energy scattering experiment at the SLAC (Stanford Linear Accelerator). Large angle scattering was observed for inelastic electron-nucleon scattering [2, 8]. The measured total cross section (► *scattering experiments*) turned out to be scale invariant, i.e., the crucial dimensionless quantity obtained from it did not depend on the scattering energy of the probe particles (► *nucleus models*). In a far-reaching formal analogy to the Rutherford scattering, James Bjorken (*1934) and Richard P. Feynman (1918–1988) interpreted this scale invariance as evidence for pointlike scattering centers within the protons and neutrons that constitute the atomic nucleus [3, 4, 7]. Their interpretation was based on the heuristic idea that the higher the energy of the probe particles is, the smaller structures can be measured in a ► *scattering experiment*. Bjorken and Feynman concluded that the scale invariance of the measured cross section indicated pointlike partons within the proton and neutron, i.e., particles that carry fractional charges and certain fractions of the proton or neutron momentum (► parton model). After carrying out other scattering experiments of a similar type and after accumulating much more additional experimental evidence, these "partons" were identified with the quarks of the quark model established in 1963 (▶ Rutherford atom; Quarks, see ▶ Color Charge Degree of Freedom in Particle Physics; Mixing and Oscillations of Particles; Particle Physics; Parton Model; QCD; QFT) [8].

Primary Literature

1. E. Rutherford: The scattering of α and β particles by matter and the structure of the atom. Phil. Mag. 21 (1911), 669–688.
2. W.H.K. Panofsky: Rapporteur talk. In: J. Prentki and J. Steinberger (eds.):, *Proceedings of the Fourteenth International Conferences on High-Energy Physics*. Vienna 1968. Geneva: CERN, 23–39.
3. J.D. Bjorken: Asymptotic sum rules at infinite momentum. Phys. Rev. 179 (1969), 1547–1553. J.D. Bjorken and E.A. Paschos: Inelastic electron and γ-proton scattering and the structure of the nucleon. Phys. Rev. 185 (1969), 1975–1982. J.D. Bjorken: High-energy inelastic neutrino-nucleon interactions. Phys. Rev. D 1 (1970), 3151–3160.
4. R.P. Feynman: Very high-energy collisions of hadrons. Phys. Rev. Lett. 23 (1969), 1415–1417.

Secondary Literature

5. A. Pais: *Inward Bound*. Oxford: Clarendon Press 1986, 190–192.
6. G.L. Trigg: *Crucial Experiments in Modern Physics*. New York: Crane Russah & Company 1971, Chap. 2.
7. B. Falkenburg: Probing Subatomic Structure. In: *Particle Metaphysics. A Critical Account of Subatomic Reality*. Heidelberg: Springer 2007, 125–160.
8. M. Riordan: *The Hunting of the Quark*. New York: Simon & Schuster 1987, Chaps. 6–8.

Light Quantum

Klaus Hentschel

The light quantum concept comprises 12 layers of meaning which matured at very different times, thus refuting the simplistic legend that Albert Einstein (1879–1955) singlehandedly discovered them all in 1905. Einstein's "heuristic point of view" was actually regarded with extreme skepticism until 1922. Today's understanding of the subject takes for granted that light quanta:

- Are particle-like and localized
- Propagate at a finite velocity
- Have equal velocity for all colors (i.e., frequencies)
- Transmit energy E
- Transmit momentum $p = E/c$ (giving rise to radiation pressure)

- Have energy E correlated with their frequency ν ($E \sim \cdot\nu$)
- Obey strict ► quantization in their energy E (i.e., $E = h\nu$ with ► Planck's constant)
- Are emitted and absorbed by matter
- Exhibit ► wave-particle duality
- Transmit angular momentum with ► spin 1
- Exhibit ► indistinguishability with other light quanta of same E and spin orientation
- Obey the ► Bose–Einstein statistics.

The term 'photon' was introduced in late 1926 by the American physical chemist Gilbert Lewis (1875–1946), that is roughly 20 years after Einstein's famous paper from 1905 and 1 year after the discovery of electron spin in 1925. The other layers of meaning of the word 'light quanta' have complex histories of their own, extending variously back in time and tightly intertwined with other strands of the history of ► quantum theory to 1925.

Corpuscularity or Particle Characteristics

We find particle theories of light, in the broadest sense of the word, as far back as the atomists of Ancient Greece, but Sir Isaac Newton (1643–1727) first conceived a more developed model of this type. His early papers in the Royal Society's *Philosophical Transactions* conceal his basic conception of light as a corpuscle. Nevertheless, his *Principia* from 1687 as well as the queries in his *Opticks* from 1704 provide clear hints at this projectile model. His *Mathematical Principles of Natural Philosophy*, for instance, derive light diffraction from a stronger attraction of light particles to the denser medium, and in query 29 of *Opticks* he asks: "Are not the Rays of Light very small Bodies emitted from shining Substances?" ([9], p. 370; cf. also *Principia*, book I, Sect. XIV § 141ff.).

When critics tried to nail him down on this projectile model of light, Newton replied with his distinction between facts and hypotheses. "*that light is a body* [. . .], it seems, is taken for my *Hypothesis*. 'Tis true, that from my Theory I argue the *Corporeity* of Light; but I do it without any absolute positiveness, as the word *perhaps* intimates; and make it at most but a very plausible *consequence* of the Doctrine, and not a fundamental *Supposition*."[1] Newton knew perfectly well that he could not prove without an element of doubt that the corpuscular model of light was right. Unlike the Cartesians, he was adverse to hypothesizing out of the blue, but that did not stop him from frequently making *heuristic* use of such hypotheses and models.

[1] Newton's reply to Hooke, 1672, reprinted with Hooke's attacks in I.B. Cohen (ed.) *Isaac Newton's Papers & Letters on Natural Philosophy* (Cambridge, Mass.: Harvard Univ. Press, 1958), quotes from pp. 118f.

Newton's cautious wording in his essays on light are remarkably similar to Einstein's in his paper from 1905 on 'A heuristic point of view concerning the production and transformation of light'. Albert Einstein (1879–1955) writes: "monochromatic radiation of low density... behaves *as if* it were composed of mutually independent energy quanta." ([5], **2**, p. 161) This fictionalistic as-if conjunctive reveals the same intellectual reserve with which Newton enveloped his projectile model. Just like Newton, Einstein also had a more urgent statement to defend, a statement that was likewise more phenomenological than the light-quantum model: namely, the equation $E = h \cdot \nu$. The underlying *model* of light was left in the background.

Just two years before Einstein's 1905 paper, the director of the Cavendish laboratory in Cambridge, Joseph John Thomson (1856–1940), had also speculated about corpuscular localized field quanta in an effort to explain two anomalies in the propagation of ► x-rays, which Röntgen had discovered in late 1895: (1) the extremely directed and point-like effects of these hard rays, then referred to as "needle" radiation; and (2) the fact that its intensity does not diminish as $1/r^2$ but remains almost the same even over longer distances, if occasional ionization of directly hit gas molecules is disregarded. In his *Autobiography*, Robert Millikan still refers to the "Thomson–Planck–Einstein conception of localized radiant energy (i.e., the corpuscular or photon conception of light)" rather than 'Einstein's light quanta'. Speculations about the corpuscularity of specific types of radiation are thus older than Einstein's" heuristic point of view" from 1905.

Constancy of the Velocity of Light

Like Newton, Einstein also considered the *corpuscularity* of light in connection with its *propagation velocity*. The constancy of its propagation velocity was, as we know, one of the axioms of his paper which appeared three months later in the *Annalen der Physik*: 'On the Electrodynamics of Moving Bodies' [2]. Before Einstein arrived at his postulate of the constancy of light velocity in a vacuum, he carefully considered its dependence on the velocity of its emitter, as suggested in the projectile theory of light. We know this from his correspondence with Paul Ehrenfest as well as from his comments on contemporary papers by Walter Ritz (1878–1909), who was working on exactly such types of emission theories. Einstein's postulate of a constant velocity of light in all inertial systems was a direct consequence of the failure of emission theories. This is a concealed but interesting link between the famous papers from 1905. "Turn the problem into a postulate, that's how you get by", Einstein later joked.

Energy and Momentum Transfer (Radiant Pressure)

The insight that light can transfer energy and momentum also has a long history extending back into the early modern period [22]. In 1905, the existence of radiation pressure had just recently been established experimentally by Pjotr Lebedev (1866–1912) and confirmed to an accuracy of 1% by Ernest Fox Nichols (1869–1924) and Gordon Ferrie Hull (1870–1956). The decisive papers fall exactly within the period when Einstein was studying articles in the *Annalen der Physik*, among other physics journals, during his free time as an examiner at the Swiss Patent Office. Remarks in his papers show that he knew about the "just recently experimentally confirmed light pressure, which plays such an important role in the theory of radiation" ([5], **2**, pp. 300, 483, 565).

Proportionality Between Energy and Frequency

If Einstein had relied on the literature, he would have missed the correlation between the energy and frequency of light. Both Lebedev and Nichols & Hull assumed from classical electrodynamics that the energy of light was always proportional to its intensity: $E \sim I \sim H^2 + D^2$. Lebedev explicitly writes in 1901: "These pressure forces of light are directly proportional to the impinging amount of energy and *independent* of *the color of light*." Nichols and Hull thought they were able to confirm this two years later (1903), because their measurements of the light pressure initially suggested (independently of the filters chosen) a frequency-independent energy proportional to the light's intensity. This false conclusion is generally concealed in the professional folklore. Einstein's extraordinary sense for the validity of experimental results saved him from being led astray. Instead of just relying on this one experimental strand, he linked experimental results from the most disparate areas of scientific inquiry. Each of these individual strands might have led to a dead end, but woven together they yielded a dense fabric: Einstein realized that "the observations on black-body radiation, photoluminescence, the generation of cathode rays from ultraviolet light and other *groups of phenomena* concerning the generation or transformation of light would appear better comprehensible under the assumption that the energy of light was discontinuously distributed." The third of these experimental strands was the ► photoelectric effect. Experimentalist Philipp Lenard (1862–1947) had assumed that UV radiation acts only as a trigger to release charges (see [18, 19]). Einstein's interpretation suggested" that the excited light is composed of energy quanta [...]. The generation of cathode rays by light can be understood in the following way. Energy quanta penetrate into the surface layer of the body and their energy is transformed at least in part into the kinetic energy of electrons. [...] Furthermore, it has to be assumed that upon leaving the body each electron must expend work P (characteristic of the body)" ([1], p. 145f.). According to Einstein, the maximum kinetic energy of these 'electricity quanta' was therefore $h\nu - P$.

Lenard had not sought this frequency dependence according to his own model. He had found a slight dependence of the limiting potential on the type of light used but had not followed up on this hint. Ten years had to go by before Robert A. Millikan (1868–1953) verified Einstein's prediction experimentally beyond doubt. He had expressly set out to *refute* Einstein's prediction: "I spent ten years of my life testing that 1905 equation of Einstein's and, *contrary to all my expectations*, I was compelled in 1915 to assert its unambiguous verification in spite of its unreasonableness since it seemed to violate everything we knew about the interference of light." This shows that contrary to the claims of certain sociologists of science, experimenters do *not* always confirm what they anticipate. Even after publishing his findings in 1916, Millikan continued to have qualms about Einstein's light quantum, this "bold, not to say reckless hypothesis".

Quantization

For Max Planck (1858–1947), energy ► quantization only served as an emergency solution to prevent that the interaction between radiation and resonator lead to an increasing dominance of oscillations of ever diminishing frequency in the radiation field. Planck conceived the energy of electromagnetic radiation as *continuous* because Maxwellian electrodynamics is a continuum theory. In Planck's ► quantum theory, discontinuity is only at play during the process of energy transmission from the radiation field to the oscillator.

L

This is where Einstein found fault. In a frequently quoted letter to his friend Conrad Habicht (1876–1958) from May 1905, Einstein announced a "very revolutionary" paper. For the first time, quantization was explicitly *not* limited to resonators or the interaction between matter and the field, but also was required of the energy of the electromagnetic field itself: "the energy of a propagating ray of light emitted from one point [is] not continuously distributed over an augmenting space but is composed of a finite number of energy quanta localized in points in space, which move without dividing and can only be absorbed and generated as a whole" ([1], p. 133).

A terminological and conceptual broadening soon followed: 'light energy quanta' (partitioning into packets of energy) became 'light quanta' (light as a particle-like phenomenon). Just as with Planck's energy quantization in 1900 and later with the so-called ► Bose-Einstein statistics in 1924/25, here also we see a *gradual* realization of the radical implications of this step. While in 1905 Einstein's emphasis lay on energy considerations, a particle-like conception emerges in Einstein's letter to Sommerfeld from Sept. 29, 1909, where he speaks of "the ordering of the energy of light around discrete points which move with light velocity" ([5], **5**, doc 179). So by then the first seven levels have been spelled out. The momentum of light quanta only came into play in Einstein's Salzburg talk of 1909, and even more explicitly so in his paper on induced emission in 1916. According to Einstein's mental model, the interaction between matter and the field would consist of the emission and subsequent absorption of such quantized packets of energy: This idea reappears in Bohr's model of the atom. Unlike ► Bohr's atomic model of a later date, Einstein's paper of 1905 offers no specific model of this process.

How did Einstein argue for the existence of light quanta of energy or at least for their plausibility? He resorted to his typical strategy of following two separate derivations at the same time. He analysed a single system according to two different theoretical methods as far as he could. In a second step, he sought to equate the physical expressions obtained by these two different paths. This is only possible if $E = h\nu$ is true. q.e.d. (cf. [13, 16] for details).

By juxtaposing an ideal gas obeying Boltzmann statistics with radiation in the Wien limit, Einstein thus arrived at the light quantum hypothesis: "monochromatic radiation of low density [at the Wien limit] acts *as if it were* composed of mutually independent quanta of energy of the magnitude $(R\beta\nu)/N \cdot \nu[= h \cdot \nu]$". ([1], p. 143) As is typical of Einstein's thinking, the originality of this consideration lay in the new way of linking different chains of reasoning; here, classical combinatorics with statistical mechanics of Boltzmann and Gibbs and radiation theory of Wien and Planck. This derivation also reveals another characteristic of Einstein's thinking: the constant vacillation between micro- and macro-physics as encapsulated in $S = k \ln W$, which Einstein termed the Boltzmann formula and used to its fullest in both directions.

Einstein's correspondence with Lorentz and his Salzburg lecture of 1909 show that he certainly had a quite fully developed model of light quanta: "For the time being the most natural interpretation seems to me to be that the occurrence of electromagnetic fields of light is associated with singular points just like the occurrence of electrostatic fields according to the electron theory. It is not out of the question that in such a theory the entire energy of the electromagnetic field might be viewed as localized in these singularities, exactly like in the old theory of action at a distance. I more or less imagine each such singular point as being surrounded by a field of force which has essentially the character of a plane wave and whose amplitude decreases with the distance from the singular point." ([4], p. 581).

Einstein shied away from explicitly discussing this conceptual model because he had encountered three profound problems in its development:

1. How to explain interference, implying deviations from a point-like structure.
2. How to interpret partial reflection: the splitting of photons is impossible!
3. Problems with particle characteristics of light quanta: if they transmit energy, then they do have mass according to $E = mc^2$, but no massive particle can have the velocity of light.

While the solution to the third enigma was, of course, to assume a vanishing rest mass of the photon, the other two problems proved to be much harder, as they were intimately linked with the thorny issue of ► wave-particle duality.

Reception of the Light Quantum

Strangely enough, one of the first advocates of the light quantum hypothesis was the later antirelativist and Nazi proponent Johannes Stark (1874–1957). His arguments were foremost experimentally based (see, e.g., [14, 20]):

1. Photoelectric effect
2. Shortwave limit of X-ray ► *bremsstrahlung*
3. Intensity minimum of the Doppler effect
4. (generally:) discrete excitation energy of atoms
5. (personally:) his tendency to go against generally accepted opinions

But Stark had to swallow criticism for his support of the light quantum. Arnold Sommerfeld (1868–1951) and many others remained skeptical. In a letter dated 4 Dec. 1909, Sommerfeld reflected on: "the really very hypothetical and uncertain light quantum theory [. . .] Not as if I were doubting the significance of the quantum of action. But the form in which you present it (light quantum) appears, not just to me but also to Planck, very daring." Max Planck was similarly skeptical. In the *Annalen der Physik* of January 1910 he wrote: "I cannot at the moment acknowledge compelling proof in favor of the corpuscular theory of light any more for J. Stark's experiments on X-rays than for A. Einstein's deductions."

The great majority of physicists at that time were even more reluctant, particularly Planck. He saw "no compelling reason" for abandoning Maxwell's equations along with its continuum physics. His skepticism of the light quantum hypothesis was shared by many others.

Conclusion

A complex concept like 'light quantum' does not emerge at once. Some of its layers of meaning are very old. Others only became evident in Einstein's paper of 1905; the full-fledged concept of photons only emerged at the end of 1926. Some physicists had already realized some of these layers on their own. But this does not diminish the profundity of Einstein's insight that the energy in a field of radiation is strictly quantized (1905) and that light quanta also carry momentum (1909). No one else had the courage or the far-reaching intellectual perspicuity for these two bold steps. Furthermore, Einstein's Salzburg talk was a first step towards ► wave-particle duality, later further elaborated by Louis and Maurice de Broglie, Niels Bohr and others (► Born rule; Consistent Histories; Metaphysics in Quantum Mechanics; Nonlocality; Orthodox Interpretation; Schrödinger's Cat; Transactional Interpretation of quantum mechanics). But Einstein's most important achievement was drawing together all these individual insights into a first quantum theory of radiation. As with his theory of relativity, his greatest strength lay in *tracking down* heuristically fruitful ideas from the large reservoir of then conceivable options, consistently shedding elements that did not agree and weaving these previously separate strands into theories that were not just consistent but also empirically adequate.

Primary Literature

1. A. Einstein: Über einen die Erzeugung und Verwandlung des Lichtes betreffenden heuristischen Gesichtspunkt, Annalen der Physik (4th ser.) **17**, 132–148 (1905), reprinted in [5], Vol. 2, 149–169
2. A. Einstein: Zur Elektrodynamik bewegter Körper, Annalen der Physik (4th ser.) **17**, 891–921 (1905), reprinted in [5], Vol. 2, 275–310
3. A. Einstein: Zum gegenwärtigen Stand des Strahlungsproblems, Physikalische Zeitschrift **10**, 185–193 and the discussion, 224f., 323f. (1909), reprinted in [5], Vol. 2, 541–555
4. A. Einstein: Über die Entwicklung unserer Anschauungen über das Wesen und die Konstitution der Strahlung, Physikalische Zeitschrift **10**, 817–825 (1909), reprinted in [5], Vol. 2, 563–583
5. A. Einstein: *Collected Papers of Albert Einstein*, Vol. 2: *The Swiss Years: Writings, 1900–1909*, Vol. 5: *Correspondence, 1902–1914*, ed. by J. Stachel, R. Schulmann et al. (Princeton: Princeton University Press 1989)
6. P. Lenard: Ueber die lichtelektrische Wirkung, Annalen der Physik (4th ser.) **8**, 149–198 (1902)
7. G.N. Lewis: The conservation of photons, Nature **118**, 874–875 (1926)
8. R. Millikan: *The Autobiography* (New York: Prentice-Hall 1950, reprint New York: Arno Press 1980)
9. I. Newton: *Opticks* (reprint of 4th ed. 1730 with Foreword by Einstein New York: Dover Publications 1952)
10. J.J. Thomson: *Electricity and Matter*, Silliman Lectures as guest lecturer for Yale University in 1903 (Westminster: Constable 1904), trans. into German: *Elektrizität und Materie* (Braunschweig: Vieweg 1904)

Secondary Literature

11. S. Bergia, C. Ferrario, V. Monzoni: Side paths in the history of physics: the idea of light molecule from Ishiwara to de Broglie, Rivista di Storia della Scienze **2**, 71–97 (1983)
12. R. McCormmach: J.J. Thomson and the structure of light, British Journal for the History of Science **3**, 362–387 (1967)
13. J. Dorling: Einstein's introduction of photons, British Journal for the Philosophy of Science **22**, 1–8 (1971)
14. K. Hentschel: Einstein über Lichtquanten. Zur stufenweisen Herausbildung eines mentalen Modells, Naturwissenschaftliche Rundschau **58,6**, 311–319, **58,7**, 363–371 (2005)
15. G. Holton: Millikan's struggle with theory, Europhysics News **31,3** (2000) also online as www.europhysicsnews.com/full/03/article3/article3.html
16. J. Norton: Atoms, entropy, quanta: Einstein's miraculous argument of 1905, Studies in the History and Philosophy of Modern Physics **37**, 71–100 (2006)
17. J. Stachel: Einstein's light quantum hypothesis, in: *Einstein from B to Z* (Basel: Birkhäuser 2002) 427–444
18. R. Stuewer: Non-Einsteinian interpretations of the photoelectric effect, *Minnesota Studies in the Philosophy of Science* (Minneapolis: University of Minnesota 1970) 246–263
19. B.R. Wheaton: Philipp Lenard and the photoelectric effect, 1889–1911, Historical Studies in the Physical Sciences **9**, 299–322 (1978)
20. B.R. Wheaton: *The Tiger and the Shark. Empirical Roots of Wave-Particle Dualism* (Cambridge: Cambridge University Press 1983)
21. K.-H. Wiederkehr: Zur Geschichte der lichtelektrischen Wirkungen, Technikgeschichte **40**, 93–103 (1973)
22. J. Worrall: The pressure of light, Studies in the History and Philosophy of Science **13**, 133–171 (1982)

Locality

Henry P. Stapp

Locality: The locality assumption is sometimes called "local causes". It is the requirement that each physical event or change has a physical cause, and that this cause can be localized in the immediate space-time neighborhood of its effects. A collision of two billiard balls or the mechanical connections between the parts of a steam engine are clear examples. A more subtle example is the feature of classical electromagnetism that any change in the velocity of a moving charged particle can be regarded as being caused by the action upon this particle of the electric and magnetic fields existing in the immediate space-time neighborhood of that particle at the moment at which the change in velocity occurs, and that any change in the electric and magnetic fields are likewise caused by physically describable properties that are located very close to where that change occurs.

This idea that all physical effects are consequences of essentially "contact" interactions was part of the intellectual milieu, stemming from the ideas of Rene Descartes, in which Isaac Newton worked while creating the foundations of modern physics. However, his universal law of gravitational attraction was stated as a law of instantaneous action over astronomical distances, a clear violation of the idea that all physical effects have local causes. Newton tried unsuccessfully to devise some local mechanical idea of how gravity worked, but in the end asserted his famous "hypothesis non fingo" (I feign [pretend to make] no hypothesis [about how gravity works]) [1, p. 671]. He relied, instead, on the empirical success of his simple inverse-square-law postulate to account for a huge amount of empirical data. Yet as regards basic metaphysics he wrote: "That one body can act upon another at a distance through the vacuum, without the mediation of anything else, by and through which their action and force my be conveyed from one to another, is to me so great an absurdity that I believe that no man who has in philosophical matters a competent faculty of thinking can ever fall into it." [1, p. 636]. This statement is a trenchant formulation of the notion of locality. It took more than two centuries of development before Einstein came up with an explanation, in terms of the idea distortions of space-time that allowed the requirement of locality to be met for gravity. Einstein's special theory of relativity imposes the condition that no localized measurable output can depend upon the character of a localized physical input before a point moving at the speed of light can travel from the smallest region in which the input is localized to the smallest region in which the output is located. This locality condition is required to hold in any classical physical theory that is called "relativistic".

This idea of locality is fairly simple and straightforward in classical physics, because in that setting everything has a material basis and all causal effect are associated with transfers of momentum or energy, which moves about in a continuous way. In quantum theory the fundamental substrate of change is more ephemeral, having the character of information expressed as changing potentialities for observable events to occur. These potentialities normally change in a continuous way, but, in conventional quantum mechanics, they change abruptly in association with the occurrence of an observable (or actually observed) event. And a "cause", such as the performance of a freely chosen measurement in one region, can have an instant far-away effect without any energy or momentum *traveling* from the region of the cause to the region of the effect.

In the quantum context a suitable definition of locality pertains to information: Locality requires that no information about which measurement is freely chosen and performed in one space-time region can be present in another space-time region unless a point traveling at the speed of light or less can get from the first region to the second. Or in terms of outcomes: no statement whose truth is determined solely by which outcomes appear in one space-time, under conditions freely chosen in that region, can be true if one experiment is freely performed in a region that is space-like-separated from the first region, but be false if another experiment is freely chosen there. The term "freely chosen" means only that in the argumentation this choice is not to be constrained in any way. Locality defined in either of these ways appears to be violated in relativistic quantum field theory. These violations are discussed the entries ► Nonlocality and ► Einstein locality.

Literature

1. I. Newton: *Principia Mathematica*. Ed. F. Cajori (University of California Press Berkeley 1964/1687)
2. A. Einstein, L. Infeld: *The Evolution of Physics*: The growth of ideas from early concepts to relativity and quanta (Simon and Schuster, New York 1938, Chapter 3)
3. A. Einstein: Quantenmechanik und Wirklichkeit. Dialectica **2**, 32–24 (1948)
4. J. Cushing, E. McMullin: *Consequences of Quantum Theory* (Notre Dame Press, Notre Dame, Indiana, 1989)

Loopholes in Experiments

Gregor Weihs

Introduction

Shortly after John S. Bell's proof of his celebrated theorem (► Bell's Theorem) in 1964 [6] experiments started [13] that tried to check whether nature actually was as counterintuitive as the theorem implied. At the same time it became clear that it would be very difficult to carry out an experiment that tested Bell's original version of the inequality, because it had been derived using very stringent assumptions.

The first difficulty was with Bell's assumption of perfect correlations. That is, if the measurement functions of the hidden variable model are $A(a, \lambda)$ and $B(b, \lambda)$, where λ denotes the hidden variable, a and b the analyzer directions, Bell had assumed that they obey $A(a, \lambda) = -B(a, \lambda)$. This assumption, however, is difficult to justify, because no real experiment will ever live up to it.

Upon realizing this Clauser, Horne, Shimony, and Holt (CHSH) (► Bell's theorem) [11] were able to derive an inequality without assuming perfect correlations. This version of the inequality is the best known one and it reads

$$|E(a, b) + E(a, b')| + |E(a', b) - E(a', b')| \leqslant 2, \tag{1}$$

where a, a', b, b' are two choices of a measurement parameter on each side and $E(a, b) = p_{++}(a, b) + p_{--}(a, b) - p_{+-}(a, b) - p_{-+}(a, b)$ is a correlation between the measurement results obtained on the two sides of the experiment. The quantities p are either theoretically predicted or experimentally determined probabilities of the binary outcomes $+1$ and -1. Entangled quantum systems can violate this inequality with the l.h.s. attaining values of up to $2\sqrt{2}$. In the same work, CHSH realized that there was another problem. The detection efficiencies for visible photons (► light quantum) were too small and one wouldn't be able to violate the inequality experimentally.

This was the first discovery of what has since been called loopholes in attempted experimental refutations of objective local theories. In the following we will see that there are two main loopholes, *efficiency* and *locality*. Besides these, there is a range of other, lesser known issues. To date, no experiment was able to achieve closure of all loopholes.

John S. Bell had his own view of a generic experiment to test the inequality that avoids the use of any microscopic description. It is shown in Fig. 1, which is drawn following his Fig. 7 in Ref. [7]. In this picture, all we have is an elongated apparatus with a central "go" trigger signal input and an "experiment ready" indicator, as well as a signal input and a result output at each of the two ends. The parameters of the measurements a and b are injected a short time before the results are expected to occur.

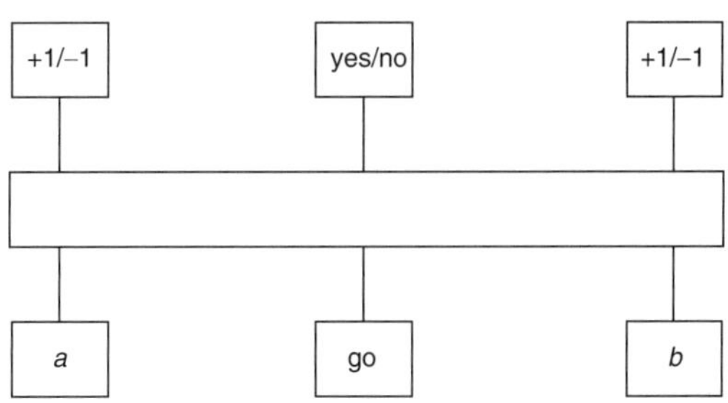

Fig. 1 Adapted version of J. S. Bell's schematic of a general EPR set-up [7]

L

Efficiency

While this idealized picture requires no microscopic description, it leaves no room for the cases where either one or both outputs do not yield an result. The effect of particle loss in real experiments [16] is usually treated by adding a hypothetical third (inconclusive) outcome "0" in addition to the ± 1. Then, if η is the conditional probability of getting a ± 1 on the one side when we detected ± 1 on the other side, Eq. (1) is modified to read

$$|E(a,b) + E(a,b')| + |E(a',b) - E(a',b')| \leqslant \frac{4}{\eta} - 2, \qquad (2)$$

where the correlations E now include the 0 outcome. For values of η smaller than $2(\sqrt{2}-1) \approx 83\%$ the r.h.s. of inequality (2) becomes bigger than $2\sqrt{2}$, the maximal value attainable by measurements on entangled quantum systems and a violation of the inequality is impossible.

Since 83% is still very difficult to achieve, experiments with lower efficiencies are often interpreted with the help of auxiliary assumptions. Events with conclusive results on both sides are called *coincidences*. The *fair sampling assumption* stipulates that the coincidences represent an unbiased (fair) sample of the underlying distribution in question. Using this assumption all the quantities required for Eq. (1) are then derived from the set of coincidences only and they can violate the inequality, regardless of the efficiency.

The fair sampling assumption is not the only way of treating inefficiency and the somewhat weaker hypothesis of "*no enhancement*" introduced by Clauser and Horne [10] uses additional measurements in which the analyzers (filters) are removed in order to bound the possible dependence of the detection probability on the analyzer direction. This bound is then a limitation for any objective local theory that tries to explain the experimental results. While we won't delve further into this particular assumption, it should be noted that in the same Ref. [10] Clauser and Horne also introduced a version of Bell's inequality, the CH inequality, which turned out to admit a lower detection efficiency threshold. Eberhard [12] showed that non-maximally entangled quantum systems could violate this inequality even at efficiencies as low as 2/3 when they are the same on both sides. Recent studies [8,9] of the asymmetric case, where one side may detect their particle with close to 100% efficiency reduces the requirement for detection efficiency at the other side to 1/2. These results are interesting, because in experiments with one atom and one photon the atomic state may be measured with close to 100% efficiency.

Pearle [23] was the first to show that objective local theories can exploit the efficiency loophole by making the local hidden variable determine the detection probability dependent on the analyzer setting. Obviously this is not compatible with the fair sampling assumption. But even experimentally one has to be careful not to introduce analyzer dependent bias. Such a bias can even lead to "superviolations", in particular when doing binary outcome measurements on higher dimensional systems [22].

Setting the analyzer direction in an experiment usually involves mechanical rotation, electro-optical or acousto-optical switching or phase shifting. All these processes tend to have side effects, such as beam deviation, distortion or attenuation. Therefore it is experimentally difficult to have perfectly unbiased detection efficiency. To allow for some variation, a relaxed version of the fair-sampling assumption [2] allows a local variation of the detection efficiency with the analyzer setting, but excludes nonlocal influences.

While various researchers have been trying to construct a loophole-free experiment [14, 15], to date only the experiments by Rowe et al. [24] and Matsukevich et al. [21] closed the detection efficiency loophole. Instead of the more common optical experiments they used entangled pairs of ions in a traps. Since the ions could be stored for days, the efficiency in the traditional sense is 100%. In these experiments, the limiting factors for the violation were the finite state preparation fidelity and the measurement errors, both of which were good enough to yield a clean result that refutes objective local theories. While in Ref. [23] the two ions were only separated by about 3 μm, too close even to measure each ion's state separately, Ref. [24] extended this distance to about 1 m by storing the two ions in separate traps and entangling them using emitted photons to project the ions into an entangled state.

Since the measurements on the ions are slow for locality one would need a large separation of several kilometers between the two sides of a Bell experiment. It seems unlikely that two optical fiber-connected ion traps separated by such a distance could be built very soon. Therefore efforts are still underway to improve the detection efficiency of optical photons [25]. So far, the highest reported experimental efficiencies for optical experiments were about 30% [3, 19].

Locality

In Ref. [7], Bell expressed his view that more important than detection efficiency would be to implement a dynamic experiment, in which the analyzers were switched just before the measurement. More precisely, the time interval of the series of events in which **a.** a decision is made on a setting, **b.** that setting is implemented, and **c.** the particle is detected (an irreversible process with macroscopic consequences happens) needs to be much smaller than L/c, where L is the length of the apparatus (see Fig. 2). In this way, one can be sure that information about the setting on the one side cannot influence the measurement of a particle on the other, since the two series of events **a-c** on either side are spacelike separated. Since the source of the particles will always be timelike separated from the events **a-c** it does not matter where the source is placed, or how fast the particles fly [29]. Yet, in order to enforce the independence of the random number generators (or the freedom of choice) one has to place them outside the forward lightcones of the source, which was implemented for the first time in a recent experiment [25].

The first experiment to attempt this was Aspect's [4] (► Aspect Experiment), in which he employed fast acousto-optic switches to choose an analyzer direction

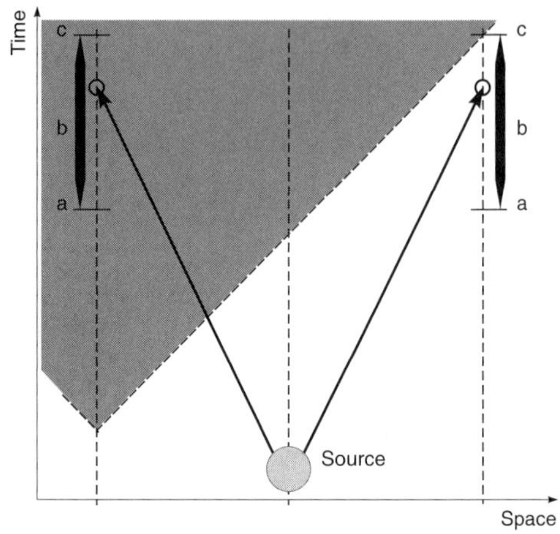

Fig. 2 Spacetime diagram of a Bell experiment. The whole measurement process (indicated by the bold black double arrows) including **a.** the random decision on a setting, **b.** the implementation of the setting, and **c.** the macroscopic registration of the event, must be spacelike separated from the corresponding process on the other side

on both sides. The switches were controlled by periodic signals, because it wasn't reasonably possible at the time to build fast enough random number generators. Two experiments in the 1990s went beyond Aspect's by including fast and random switching [29] and very large (10 km) separation [27].

In connection with the proposals for completely loophole-free experiments, it seems that it would be very difficult to achieve spacelike separation for any length L that is less than 10 m. This is because we don't only have to consider the rates at which we can generate random numbers (1 GHz seems to be quite a challenge here), but also the various delays and latencies that occur in signal generators and detectors. The sum of these delays is unlikely to be less than a few nanoseconds, corresponding to a length L of a few meters.

Other Loopholes

Randomness and Free Will

Closely related to ► locality is the question of randomness. Bell's theorem only makes sense, if we believe that regions of spacetime or subsystems can in fact be isolated, so that they can be truly independent of what is going on elsewhere. Proponents of objective local theories frequently deny the existence of randomness that is independent from the Bell experiment in question. This constitutes a loophole but at the cost of serious consequences for the ways in which we can describe the world altogether. Since one should apply the same logic to all situations this reasoning brings us very close to an all-encompassing determinism, in which there are no independent events in the whole universe.

To take this to the extreme, nothing in the experiments prevents us from replacing the random number generators with humans who decide on the analyzer setting [18].

If the humans aren't allowed to be independent then there is no free will. Amazingly, this could actually be testable in a few years from now, if a source of entangled photon pairs can be put into an earth orbit, so that the separation between the two observer stations L corresponds to a signalling time of the order of seconds, i.e. within human response times. The current distance record for Bell experiments is 144 km [28].

Coincidences

In any experiment that doesn't have perfect efficiency, it will be necessary to determine which detections on one side belong to the detections on the other side. This can open up a further loophole, albeit one that is closely connected to the efficiency one. The decision on whether a certain event has a partner event on the other side is customarily made by imposing a coincidence time window. The size of this window is usually fitted to the relative timing spread between events on either side of the experiment, caused by the finite timing resolution of the detectors and circuits.

Difficulties can arise when the relative timing is different for different experimental channels on one side. In this case a fixed coincidence window can lead to a bias, because it may reject more events in one channel than another. This effect and the fact that objective local theories can exploit it, has been called the *coincidence loophole* [20]. The bias may even introduce an apparent nonlocal influence, caused by the coincidence post-selection based on settings on the far side [2].

Remedies for this loophole include pulsed experiments where the pairs are produced in narrow pulses with long spaces in between. Then, coincidences can be counted naturally, without an artificial window as long as all the timing errors are small compared to the pulse repetition period.

Accidentals

In the earlier tests of Bell's inequality it was customary to subtract background rates — so called accidental coincidences. Accidentals occur in a situation of low detection/collection efficiency, high detector noise and poor timing resolution. In such a situation, there is a chance that two events, of different origin are registered simultaneously. For example, one detector could observe a noise click, whereas the other one receives an actual signal. Since these events are typically independent of parameter settings they form a more or less uniform background rate. Frequently, these rates have been measured by recording event pairs that occur with a large time delay between the two sides in addition to the simultaneous events (coincidences). One would then subtract from every rate the accidental rate and calculate the correlations from the corrected rates. With the advent of experiments based on spontaneous parametric down-conversion sources and better detectors, this practice has become obsolete.

Double Detections

In an ideal Bell experiment there is always one and only one answer to a measurement. In the discussion of detection efficiency we have seen that events are frequently missed. A lesser known fact is that multiple detections can occur in the case of an experiment that has detectors in both output channels of an analyzer. In optical experiments these events stem from detector noise and from double pair emissions, as a consequence of the thermal emission statistics of the usual photon pair sources. Various treatments have been suggested, such as removal of double events or performing a random choice. Since double detections are usually negligible, any procedure will work and hardly change the result.

Memory

Another potential loophole [1] is the so-called memory loophole. It claims that because experiments are done by averaging over repetitions in time rather than simultaneous measurements on an ► ensemble, an objective local theory could exploit the results of previous measurements to achieve a violation of a Bell inequality test. However, it was shown [5, 17] that even for relatively small numbers of repetitions the achievable violations are very small.

Literature

1. Accardi, L., Regoli, M.: Locality and Bell's inequality (2000). URL http://arxiv.org/abs/quant-ph/0007005. Quant-ph/0007005
2. Adenier, G., Khrennikov, A.: Is the fair sampling assumption supported by EPR experiments? J. Phys. B (2006)
3. Altepeter, J., Jeffrey, E., Kwiat, P.: Phase-compensated ultra-bright source of entangled photons. Opt. Exp. **13**, 8951–8959 (2005). URL http://www.opticsinfobase.org/abstract.cfm?URI=oe-13-22-8951
4. Aspect, A., Dalibard, J., Roger, G.: Experimental test of Bell's inequalities using time-varying analyzers. Phys. Rev. Lett. **49**, 1804–1807 (1982). DOI 10.1103/PhysRevLett.49.1804
5. Barrett, J., Collins, D., Hardy, L., Kent, A., Popescu, S.: Quantum nonlocality, bell inequalities, and the memory loophole. Phys. Rev. A **66**, 042111 (2002). DOI 10.1103/PhysRevA.66.042111
6. Bell, J.S.: On the Einstein Podolsky Rosen paradox. Physics **1**, 195–200 (1964)
7. Bell, J.S.: Bertlmann's socks and the nature of reality. J. Phys. **42**, C2 41–61 (1981)
8. Brunner, N., Gisin, N., Scarani, V., Simon, C.: Detection loophole in asymmetric bell experiments. Phys. Rev. Lett. **98**, 220403 (2007). DOI 10.1103/PhysRevLett.98.220403. k URL http://link.aps.org/abstract/PRL/v98/e220403
9. Cabello, A., Larsson, J.A.: Minimum detection efficiency for a loophole-free atom-photon bell experiment. Phys. Rev. Lett. **98**, 220402 (2007). DOI 10.1103/PhysRevLett.98.220402. URL http://link.aps.org/abstract/PRL/v98/e220402
10. Clauser, J.F., Horne, M.A.: Experimental consequences of objective local theories. Phys. Rev. D **10**, 526–535 (1974). DOI 10.1103/PhysRevD.10.526

11. Clauser, J.F., Horne, M.A., Shimony, A., Holt, R.A.: Proposed experiment to test local hidden-variable theories. Phys. Rev. Lett. **23**, 880–884 (1969). DOI 10.1103/PhysRevLett.23.880
12. Eberhard, P.H.: Background level and counter efficiencies required for a loophole-free Einstein-Podolsky-Rosen experiment. Phys. Rev. A **47**, R747–R750 (1993). DOI 10.1103/PhysRevA.47.R747
13. Freedman, S.J., Clauser, J.F.: Experimental test of local hidden-variable theories. Phys. Rev. Lett. **28**, 938–941 (1972). DOI 10.1103/PhysRevLett.28.938
14. Freyberger, M., Aravind, P., Horne, M., Shimony, A.: Proposed test of Bell's inequality without a detection loophole using entangled Rydberg atoms. Phys. Rev. A **53**, 1232–1244 (1996). DOI 10.1103/PhysRevA.53.1232
15. Fry, E.S., Walther, T., Li, S.: Proposal for a loophole-free test of the Bell inequalities. Phys. Rev. A **52**, 4381–4395 (1995). DOI 10.1103/PhysRevA.52.4381
16. Garg, A., Mermin, N.: Detector inefficiencies in the Einstein-Podolsky-Rosen experiment. Phys. Rev. D **35**, 3831–3835 (1987). DOI 10.1103/PhysRevD.35.3831
17. Gill, R.D.: Accardi contra Bell (cum mundi): The impossible coupling. In: M. Moore, S. Froda, L. C. (eds.) Mathematical Statistics and Applications: Festschrift for Constance van Eeden, *IMS Lecture Notes - Monograph Series*, vol. 42, pp. 133–154. Institute of Mathematical Statistics, Beachwood, Ohio (2003). URL http://arxiv.org/abs/quant-ph/0110137
18. Gill, R.D., Weihs, G., Zeilinger, A., Żukowski, M.: Comment on "Exclusion of time in the theorem of Bell" by K. Hess and W. Philipp. Europhys. Lett. **61**, 282–283 (2003)
19. Kurtsiefer, C., Oberparleiter, M., Weinfurter, H.: High efficiency entangled photon pair collection in type II parametric fluorescence. Phys. Rev. A **64**, 023802 (2001). DOI 10.1103/PhysRevA.64.023802
20. Larsson, J.A., Gill, R.D.: Bell's inequality and the coincidence-time loophole. Europhys. Lett. **67**(5), 707–713 (2004). DOI 10.1209/epl/i2004-10124-7. URL http://stacks.iop.org/0295-5075/67/707
21. Matsukevich, D.N., Maunz, P., Moehring, D.L., Olmschenk, S., Monroe, C.: Bell Inequality Violation with Two Remote Atomic Qubits. Phys. Rev. Lett. **100**, 150404 (2008). DOI 10.1103/PhysRevLett.100.150404
22. Oemrawsingh, S.S.R., Aiello, A., Eliel, E.R., Nienhuis, G., P. Woerdman, J.: How to observe high-dimensional two-photon entanglement with only two detectors. Phys. Rev. Lett. **92**, 217901 (2004). DOI 10.1103/PhysRevLett.92.217901
23. Pearle, P.: Hidden variable example based upon data rejection. Phys. Rev. D **2**, 1418–1425 (1970). DOI 10.1103/PhysRevD.2.1418
24. Rowe, M.A., Kielpinski, D., Meyer, V., Sackett, C.A., Itano, W.M., Monroe, C., Wineland, D.J.: Experimental violation of a bell's inequality with efficient detection. Nature **409**, 791–794 (2001)
25. Scheidl, T., Ursin, R., Kofler, J., Ramelow, S., Ma, X.-S., Herbst, T., Ratschbacher, L., Fedrizzi, A., Langford, N., Jennewein, T., Zeilinger, A.: Violation of local realism with freedom of choice. arXiv:0811.3129v1 [quant-ph]. URL http://arxiv.org/abs/0811.3129v1
26. Takeuchi, S., Kim, J., Yamamoto, Y., Hogue, H.H.: Development of a high-quantum-efficiency single-photon counting system. Appl. Phys. Lett. **74**, 1063–1065 (1999). DOI 10.1063/1.123482
27. Tittel, W., Brendel, J., Zbinden, H., Gisin, N.: Violation of Bell inequalities by photons more than 10 km apart. Phys. Rev. Lett. **81**, 3563–3566 (1998). DOI 10.1103/PhysRevLett.81.3563
28. Ursin, R., Tiefenbacher, F., Schmitt-Manderbach, T., Weier, H., Scheidl, T., Lindenthal, M., Blauensteiner, B., Jennewein, T., Perdigues, J., Trojek, P., Omer, B., Furst, M., Meyenburg, M., Rarity, J., Sodnik, Z., Barbieri, C., Weinfurter, H., Zeilinger, A.: Entanglement-based quantum communication over 144 km. Nat. Phys. **3**, 481–486 (2007). DOI 10.1038/nphys629
29. Weihs, G., Jennewein, T., Simon, C., Weinfurter, H., Zeilinger, A.: Violation of Bell's inequality under strict Einstein locality conditions. Phys. Rev. Lett. **81**, 5039–5043 (1998). DOI 10.1103/PhysRevLett.81.5039

Lüders Rule

Paul Busch and Pekka Lahti

The *Lüders rule* describes a change of the state of a quantum system under a selective measurement: if an ▶ *observable* A, with eigenvalues a_i and associated eigenprojections P_i, $i = 1, 2, \ldots$, is measured on the system in a ▶ *state* T, then the state transforms to $\widetilde{T}_k := P_k T P_k / \mathrm{tr}\,[T P_k]$ on the condition that the result a_k was obtained. This rule was formulated by Gerhart Lüders (1920–95) [1] as an elaboration of the work of John von Neumann (1903–57) [2] on the measurement process and it is an expression of the ▶ *projection postulate*, or the collapse of the wave function (▶ wave function collapse).

From the perspective of quantum ▶ *measurement theory*, the Lüders rule characterizes just one (albeit distinguished) form of state change that may occur in appropriately designed measurements of a given observable with a discrete spectrum. In general, the notion of *instrument* is used to describe the state changes of a system under a measurement, whether selective or not. The Lüders instrument $\mathcal{I}^L$ consists of the *operations* $\mathcal{I}^L_X$ of the form $\mathcal{I}^L_X(T) = \sum_{a_i \in X} P_i T P_i$, and it is characterized as a repeatable, ideal, nondegenerate measurement [3, Theorem IV.3.2], see also [8, Theorem 4.7.2]. In such a measurement, with no selection or reading of the result, the state of the system undergoes the transformation $T \mapsto \mathcal{I}^L_{\mathbb{R}}(T) = \sum_i P_i T P_i = \sum_i \mathrm{tr}\,[T P_i] \widetilde{T}_i$, the projection postulate then saying that if a_k is the actual measurement result, this state collapses to $\widetilde{T}_k$.

Lüders measurements offer an important characterization of the compatibility of ▶ observables A, B with discrete spectra: A and B commute if and only if the expectation value of B is not changed by a nonselective Lüders operation of A in any state T [1]. This result is the basis for the axiom of local commutativity in relativistic quantum field theory: the mutual commutativity of observables from local algebras associated with two spacelike separated regions of spacetime ensures, and is necessitated by, the impossibility of influencing the outcomes of measurements in one region through nonselective measurements performed in the other region.

The Lüders rule is directly related to the notion of conditional probability in quantum mechanics, conditioning with respect to a single event. According to ▶ *Gleason's theorem* [4], the generalized probability measures μ on the projection lattice $\mathcal{P}(\mathcal{H})$ of a complex ▶ Hilbert space $\mathcal{H}$ with dimension $\dim(\mathcal{H}) \geqslant 3$ are uniquely determined by the state operators through the formula $\mu(P) = \mathrm{tr}\,[T P]$ for all $P \in \mathcal{P}(\mathcal{H})$. For any μ and for any P such that $\mu(P) \neq 0$ there is a unique generalized probability measure μ_P with the property: for all $R \in \mathcal{P}(\mathcal{H})$, $R \leqslant P$, $\mu_P(R) = \mu(R)/\mu(P)$. The state operator defining μ_P is given by the Lüders form: if μ is determined by the state T, then μ_P is determined by the state $PTP/\mathrm{tr}\,[T P]$ [5].

The Lüders rule is also an essential structural element in axiomatic reconstructions of quantum mechanics. As shown in [6], it occurs in various disguised forms as an axiom in ► *quantum logic*; for example, it plays a role in the formulation of the covering law; see also [9, Chapter 16], [10].

The Lüders rule has a natural generalization to measurements with a discrete set of outcomes $a_1, a_2, \ldots$, represented by a positive operator measure such that each a_i is associated with a positive operator A_i. The generalized Lüders instrument, defined via the operations $T \mapsto \mathcal{I}_X^L(T) = A_i^{1/2} T A_i^{1/2}$, is known to have approximate repeatability and ideality properties [7]. The Lüders theorem extends to generalized measurements under certain additional assumptions [11] but is not valid in general [12].

The Lüders rule is widely used as a practical tool for the effective modeling of experiments with quantum systems undergoing periods of free evolution separated by iterated measurements. It is successfully applied in the ► quantum jumps approach [13]. The single- and ► *double-slit experiments* with individual quantum objects are the classic illustrations of the physical relevance of the Lüders rule.

L

Primary Literature

1. G. Lüders: Über die Zustandsänderung durch den Meßprozeß, Annalen der Physik **8**, 322–328 (1951), available on-line at http://www.physik.uni-augsburg.de/annalen/history/historic-papers/1951_443_322-328.pdf English translation by K.A. Kirkpatrick: Concerning the state-change due to the measurement process, Ann. Phys. (Leipzig) **15**, 663–670 (2006)
2. J. von Neumann: *Mathematische Grundlagen der Quantenmechanik* (Springer, Berlin 1932, 2nd edn. 1996); English translation (by R. Beyer): Mathematical Foundations of Quantum Mechanics (Princeton University Press, Princeton 1955)
3. E. B. Davies: *Quantum Theory of Open Systems* (Academic, London 1976)
4. A. M. Gleason: Measures on the Closed Subspaces of a Hilbert Space, Journal of Mathematics and Mechanics **6**, 885–893 (1957)
5. G. Cassinelli, N. Zanghi: Conditional Probabilities in Quantum Mechanics. I. Conditioning with Respect to a Single Event, Il Nuovo Cimento **73**, 237–245 (1983)
6. K. Bugajska, S. Bugajski: The Projection Postulate in Quantum Logic, Bulletin de L'Academie Polonaise des Sciences, Ser. sci. math., astr. et phys. Bull. Acad. Pol. Sci., Série des sc. math. astr. et phys. **21**, 873–887 (1973)
7. P. Busch: Unsharp reality and joint measurements for spin observables, Physical Review D **33**, 2253–2261 (1986)

Secondary Literature

8. P. Busch, P. Lahti, P. Mittelstaedt: *The Quantum Theory of Measurement* (Springer, Berlin 1991, 2nd revised edn. 1996)
9. E. Beltrametti, G. Cassinelli: *The Logic of Quantum Mechanics* (Addison-Wesley, Reading, MA 1981)

10. P. J. Lahti, S. Bugajski: Fundamental Principles of Quantum Theory. II. From a Convexity Scheme to the DHB Theory, International Journal of Theoretical Physics **24**, 1051–1080 (1985)
11. P. Busch, J. Singh: Lüders theorem for unsharp quantum measurements, Physics Letters A **249**, 10–12 (1998)
12. A, Aries, S. Gudder: Fixed points of quantum operations, Journal of Mathematical Physics **43**, 5872–5881 (2002)
13. M. B. Plenio, P.L. Knight: The quantum-jump approach to dissipative dynamics in quantum optics, Reviews of Modern Physics **70**, 101–144 (1998)

M

Mach–Zehnder Interferometer

See ▶ Consistent Histories; Interaction-free measurements (EV); One and Two Photon Interference.

Magnetic Resonance

Antoine Weis

Magnetic resonance (MR) is an important experimental technique by which the ▶ spin orientation of an isolated particle (atom, ion, nucleus, electron, neutron, ...) or the macroscopic polarization of an ensemble of particles (▶ ensembles in quantum mechanics) can be manipulated in a controlled way.

In general, a particle with a spin $\boldsymbol{S}$ has a magnetic moment $\boldsymbol{\mu}$ oriented either parallel or antiparallel to $\boldsymbol{S}$. The spin manipulation in MR relies on the coupling of $\boldsymbol{\mu}$ to one or more external magnetic fields $\boldsymbol{B}_0$ via the interaction Hamiltonian $H = -\boldsymbol{\mu} \cdot \boldsymbol{B}_0$. If $\boldsymbol{\mu}$ is not along $\boldsymbol{B}_0$ then the interaction induces a precession of $\boldsymbol{S}$, and hence of $\boldsymbol{\mu}$, around $\boldsymbol{B}_0$ at the Larmor frequency defined by

$$\omega_0 = \frac{\boldsymbol{\mu} \cdot \boldsymbol{B}}{\hbar} \equiv \gamma \left|\boldsymbol{B}_0\right| , \qquad (1)$$

where the gyromagnetic ratio γ connects the magnetic field to the associated precession frequency. For an ensemble of particles, the spin polarization $\boldsymbol{P}$ is defined as the quantum mechanical expectation value $\boldsymbol{P} = \langle \boldsymbol{S} \rangle = \left\langle S_x, S_y, S_z \right\rangle$ of the spin operators S_i, and the (macroscopic) ensemble magnetization correspondingly as $\boldsymbol{M} = \langle \boldsymbol{\mu} \rangle$.

Figure 1 shows a typical arrangement of a magnetic resonance experiment involving a static magnetic field $\boldsymbol{B}_0$ and a much weaker magnetic field $\boldsymbol{B}_1(t)$, perpendicular to $\boldsymbol{B}_0$ and rotating around $\boldsymbol{B}_0$ at a frequency ω_{rf}. The index rf stands for radio-frequency, as many of the original magnetic resonance experiments were carried out in that frequency range. The apparatus consists of a polarizer which orients the spins of the particles in an initially unpolarized sample. The magnetic resonance proper takes place in the central part in which the spin orientation is flipped and finally the analyzer measures the number of particles whose spin has undergone a reversal (spin flip). The insert on the upper left shows the geometry of the MR process in a frame rotating around the field $\boldsymbol{B}_0$ at the frequency ω_{rf}. In that frame the $\boldsymbol{B}_1$ field becomes static. At the same time a rotating observer experiences, as

D. Greenberger et al. (eds.), *Compendium of Quantum Physics: Concepts, Experiments, History and Philosophy*, © Springer-Verlag Berlin Heidelberg 2009

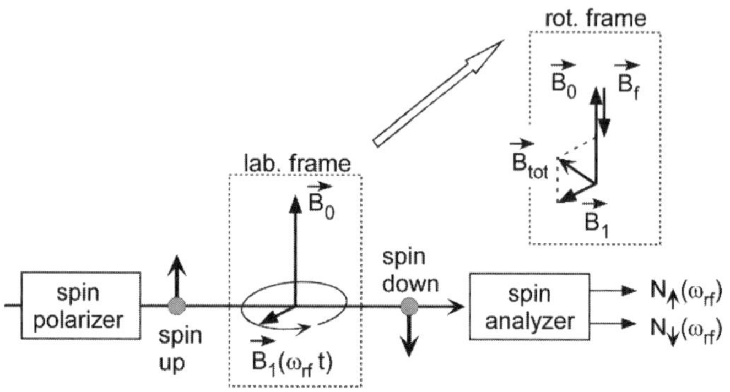

Fig. 1 Typical components of an apparatus for performing and detecting magnetic resonance

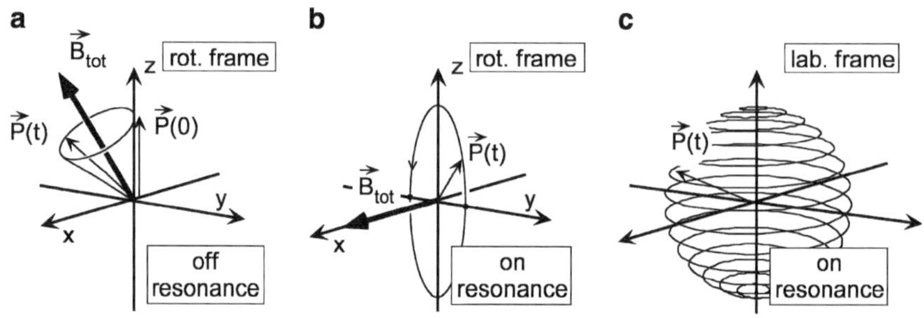

Fig. 2 Spin dynamics in the rotating (**a**,**b**) and laboratory (**c**) frames: off resonance case $\omega_{rf} \neq \omega_0$ (**a**) and on-resonance case $\omega_{rf} = \omega_0$ (**b**,**c**)

a consequence of the Larmor theorem, a fictitious magnetic field $B_f = -\omega_{rf}/\gamma$ which partially compensates $\boldsymbol{B}_0$. The dynamics of the spin flip process consists in the precession of the polarization, initially oriented along $+\hat{z}$, around the field $\boldsymbol{B}_{tot}$ as shown in Figs. 2a, b.

When the resonance condition $\omega_0 = \omega_{rf}$ is met the fictitious field $\boldsymbol{B}_f$ compensates the external field $\boldsymbol{B}_0$ exactly and $\boldsymbol{B}_{tot} = \boldsymbol{B}_1$ (Fig. 2b). In this case the spin flip probability, i.e., the probability to find a negative value of S_z becomes maximal.

In the rotating frame the precession around the total field occurs at the effective Rabi frequency $\Omega_{eff} = \sqrt{\omega_1^2 + (\omega_0 - \omega_{rf})^2}$, where $\omega_1 = \gamma B_1$ is the Rabi frequency associated with the field $\boldsymbol{B}_1$. On resonance, the polarization precesses at the frequency ω_1 around $\boldsymbol{B}_1$, a motion referred to as Rabi nutation or Rabi flopping. If one transforms back to the laboratory frame by rotating the (static) rot. frame at the frequency $-\omega_{rf}$ around $\boldsymbol{B}_0$ the polarization will follow the trajectory shown in Fig. 2c, in which one recognizes the fast precession, at ω_{rf}, and the slow nutation, at ω_1.

In pulsed MR experiments the $\boldsymbol{B}_1$ field is applied as a pulse of a duration τ. If the duration is chosen such that $\omega_1 \tau = \pi/2\,(= \pi)$ one speaks of a $\pi/2 - (\pi-)$ pulse respectively. In the former case the spin is flipped from the $+\hat{z}$ direction to the x-y

plane, while in the latter case the spin makes one half of a Rabi nuation cycle moving it from $+\hat{z}$ to $-\hat{z}$. In 1949 N. Ramsey introduced a variant of MR spectroscopy in which the rf field is applied as two spatially (or temporally) separated phase–coherent $\pi/2$-pulses. This so-called method of separated oscillatory fields yields a considerable of the resonance linewidths.

Remarks

1. The apparatus shown in Fig. 1 is close to the original set-up used by I. I. Rabi to observe magnetic resonance. Here, preparation, MR, and detection occur as three spatially separated steps. Other variants use a static sample and apply the three steps in a time sequential order (pulsed MR).
2. The external magnetic field $\boldsymbol{B}_0$ lifts the ► degeneracy of the atomic levels coupled by the MR transition. In atoms, level degeneracies can also be lifted by internal magnetic fields, leading, e.g., to fine structure and hyperfine structure splittings, between whose multiplet components one can drive MR transitions. In that case no external field $\boldsymbol{B}_0$ is needed.
3. The first (polarizing stage) can be realized in different ways. In an atomic beam one can use a Stern–Gerlach magnet (► Stern–Gerlach experiment) to select a given polarization state. Alternatively, the Boltzmann factor $\exp(-\boldsymbol{\mu}\cdot\boldsymbol{B}_0/kT)$ in a large field and/or at low temperature yields a small, but finite polarization, used, e.g., in nuclear magnetic resonance imaging or NMR/ESR spectroscopy. In dilute samples, such as gases of paramagnetic atoms, the process of optical pumping with spin polarized light can be used to achieve a large degree of spin polarization.
4. The dynamics of the magnetic resonance process is described by the Bloch equations

$$\begin{pmatrix}\dot{P}_x\\ \dot{P}_y\\ \dot{P}_z\end{pmatrix} = \begin{pmatrix}P_x\\ P_y\\ P_z\end{pmatrix} \times \begin{pmatrix}\omega_1\\ 0\\ \omega_0-\omega_{\mathrm{rf}}\end{pmatrix} - \begin{pmatrix}\gamma_2 P_x\\ \gamma_2 P_y\\ \gamma_1(P_z-P_0)\end{pmatrix}, \tag{2}$$

where γ_1 and γ_2 are the longitudinal and transverse spin relaxation rates respectively, and where P_0 is the equilibrium polarization achieved in the polarizing stage. The steady state polarization $\boldsymbol{P}$ has a longitudinal component P_z given by

$$\begin{aligned}P_z/P_0 &= \frac{\gamma_2^2+(\omega_0-\omega_{\mathrm{rf}})^2}{(\omega_0-\omega_{\mathrm{rf}})^2+\gamma_2^2(1+\omega_1^2/\gamma_1\gamma_2)}\\ &= 1-\frac{\gamma_2^2}{(\omega_0-\omega_{\mathrm{rf}})^2+\gamma_2^2(1+\omega_1^2/\gamma_1\gamma_2)}\frac{\omega_1^2}{\gamma_1\gamma_2}.\end{aligned} \tag{3}$$

5. For practical reasons the rotating field $\boldsymbol{B}_1(t)$ is often realized as a linearly oscillating field. The counter-rotating component of that field leads to a small shift of the resonance frequency, known as Bloch–Siegert shift.
6. The detection of the spin flip can occur as in Fig. 1 by measuring the number of particles in the specific spin states, or alternatively by detecting the magnetic field radiated during the magnetic resonance transition by pick-up coils. In dilute gases the same light beam used to polarize the medium can be used to detect the magnetic resonance (ODMR = optically detected magnetic resonance).
7. One speaks of nuclear magnetic resonance (NMR) or electron spin resonance (ESR), also called electron paramagnetic resonance (EPR), when the magnetic moments involved in the MR are of nuclear or electronic origin, respectively.
8. The treatment given above is a purely classical treatment, valid for an ensemble of spins (or a single spin) interacting with a classical radiation field. The fully quantum treatment of the problem, i.e., the interaction of a single two-level system with a single mode of the radiation field is treated by the Jaynes–Cummings model.

Applications

Equation 1 points to the possible applications of MR. If $\boldsymbol{\mu}$ is known the measurement of ω_0 is equivalent to a measurement of B_0 (magnetometry). Conversely, if B_0 is known, MR allows one to determine $\boldsymbol{\mu}$. This is used for the precision measurement of the magnetic moments of elementary particles, nuclei, atoms, and molecules (metrology) or their possible alterations by fundamental interactions (electric dipole moments of elementary particles). A spatial variation of the field B_0 leads to a corresponding spatial encoding of the resonance frequency ω_0. In medicine this is used in magnetic resonance imaging (MRI), where controlled field gradients yield spatially resolved MR signals from the body tissue (actually from the protons' magnetic moments), which allows one to infer the proton density, and hence the hydrogen content of the tissue. MR also plays an important role in analytical chemistry, where one uses the fact that the local field seen, e.g., by protons of large organic molecules depends on their position within the molecular structure (chemical shift). Atomic clocks, presently the most precise timekeepers, are based on a MR transition between the two hyperfine levels of the ^{133}Cs ground state. The clock mechanism consists in locking a microwave oscillator to the hyperfine frequency of the atom (metrology). MR plays an important role in recent developments such as the evaporative cooling of atoms on the way to a ► Bose–Einstein condensate or the selective manipulation of q-bits in ► quantum computation. The physics of MR is common for all two-level quantum systems interacting with a time dependent perturbation. The equivalent equations in the case of an optical transition in a two level atom are known as the optical Bloch equations or Maxwell–Bloch equations.

Primary Literature

1. I. Rabi, J. R. Zacharias, S. Millman, P. Kusch: A New Method of Measuring Nuclear Magnetic Moment. Phys. Rev. **55**, 526 (1939)
2. F. Bloch, A. Siegert: Magnetic Resonance for Nonrotating Fields. Phys. Rev. **57**, 522 (1940)
3. F. Bloch: Nuclear Induction. Phys. Rev. **70**, 460 (1946)
4. N. F. Ramsey: A New Molecular Beam Resonance Method. Phys. Rev. 996 (1949)
5. W. F. Bell, A. L. Bloom: Optical Detection of Magnetic Resonance in Alkali Metal Vapor. Phys. Rev. **107**, 1559 (1957)

Secondary Literature

6. A. Abragam: *Principles of Nuclear Magnetism* (Oxford University Press, USA; Reprint edition October 27, 1983)

Many Worlds Interpretation of Quantum Mechanics

M

Jeffrey A. Barrett

Hugh Everett III developed his relative-state formulation of quantum mechanics while a graduate student in physics at Princeton University [5–7]. It was a reaction to his belief that the standard von Neumann–Dirac collapse formulation of quantum mechanics could not be consistently applied to systems which, like the universe, contained observers. Everett proposed solving the quantum measurement problem by dropping the collapse postulate from the standard formulation of quantum mechanics then deducing the empirical predictions of the standard collapse theory as the subjective experiences of observers who were themselves treated as physical systems described by the theory. While it remains unclear precisely how Everett intended for this to work, the relative-state formulation of quantum mechanics is often taken to be identical to Bryce DeWitt's popular many-worlds interpretation of Everett [1,2,4]. (See also ► Bohmian mechanics; Measurement theory; Metaphysics in Quantum Mechanics; Modal Interpretation; Objectification; Projection Postulate).

On Everett's relative state formulation of quantum mechanics observers were to be thought of as automatically functioning machines possessing recording devices that could be correlated with their environments. Everett's goal then was to deduce the appearance of the statistical predictions of quantum mechanics with the collapse postulate, as physical records in the memory of the observer, from pure wave mechanics without the collapse postulate: "We are then led to the novel situation in which the formal theory is objectively continuous and causal, while subjectively discontinuous and probabilistic. While this point of view thus shall ultimately justify our use of the statistical assertions of the orthodox view, it enables us to do so in a logically consistent manner, allowing for the existence of other observers" [7, p. 9].

Consider an observer M measuring a system S initially in a ► superposition of states corresponding to different values ϕ_i of the observable being measured. The initial state of the composite system is

$$|\text{“ready to measure”}\rangle_M \otimes \sum_{i=1}^{n} \alpha_i |\phi_i\rangle_S. \tag{1}$$

Here M is determinately ready to make a measurement, but, given the standard eigenvalue-eigenstate link, the object system S, has no determinate value for the observable being measured.

If we assume that M has the disposition to perfectly correlate its memory with the state of the system being observed, then it follows from the linearity of the deterministic dynamics that the state of the composite system after M's interaction with S will be

$$\sum_{i=1}^{n} \alpha_i |\text{“the result is } \phi_i\text{”}\rangle_M \otimes |\phi_i\rangle_S. \tag{2}$$

Everett confesses that this post-measurement state is puzzling: "As a result of the interaction the state of the measuring apparatus is no longer capable of independent definition. It can be defined only relative to the state of the object system. In other words, there exists only a correlation between the states of the two systems. It seems as if nothing can ever be settled by such a measurement" [6, p. 318]. And he describes the problem one faces in interpreting pure ► wave mechanics: "This indefinite behavior seems to be quite at variance with our observations, since physical objects always appear to us to have definite positions. Can we reconcile this feature of wave mechanical theory built purely on [the deterministic linear dynamics] with experience, or must the theory be abandoned as untenable?" [6, p. 318].

Everett then presents his solution to this problem of indeterminate measurement records in pure wave mechanics:

> It is ... an inescapable consequence that after the interaction has taken place there will not, generally, exist a single observer state. There will, however, be a superposition of the composite system states, each element of which contains a definite observer state and a definite relative object-system state. Furthermore ... each of these relative object system states will be, approximately, the eigenstates of the observation corresponding to the value obtained by the observer which is described by the same element of the superposition. Thus, each element of the resulting superposition describes an observer who perceived a definite and generally different result, and to whom it appears that the object-system state has been transformed into the corresponding eigenstate. In this sense the usual assertions of [the collapse postulate] appear to hold on a subjective level to each observer described by an element of the superposition" (1973, p. 10).

The fundamental relativity of quantum-mechanical states is the central principle of Everett's formulation of quantum mechanics. On this principle there are typically no simple state or property attributions to subsystems of a composite system in an entangled state; rather, property attributions to one subsystem must typically be made only relative to property attributions to the other subsystems of a composite system. In the post-measurement state above, M recorded "the result is ϕ_i" relative to S having property ϕ_1 but recorded "the result is ϕ_2" relative to S having property ϕ_2, etc. Similarly, there is no simple matter of fact concerning which property S has. S has property ϕ_1 relative to M recording "the result is ϕ_i" but S has property ϕ_2 relative to M recording "the result is ϕ_2", etc.

While the notion of a relative state is clear enough; it remains unclear how Everett meant for the principle of the relativity of states to explain an observer's apparent determinate measurement records and how the statistical predictions of the standard formulation of quantum mechanics were to be derived from pure wave mechanics together with this principle. An observer will have different, but correlated, records relative to different properties of the measured system, but this is not by itself sufficient to derive the standard predictions of quantum mechanics as appearances to the observer insofar as it does not explain why it seems to the observer that she has recorded a single, fully determinate measurement result.

Bryce DeWitt's [3] popular interpretation of Everett seeks to explain just this. On the most straightforward version of DeWitt's many-worlds or, perhaps better, splitting-worlds interpretation, there is one world corresponding to each term in the expansion of the post-measurement state when written in a specified preferred basis, and the preferred basis is chosen so that each term in the expansion of the post-measurement state describes a world where there is in fact a determinate measurement record (Fig. 1). Given the preferred basis presupposed above, the post-measurement state describes n worlds, since there are n terms in the expansion of the state in this basis: one world where M determinately records "the result is ϕ_i", another where M determinately records "the result is ϕ_2", etc.

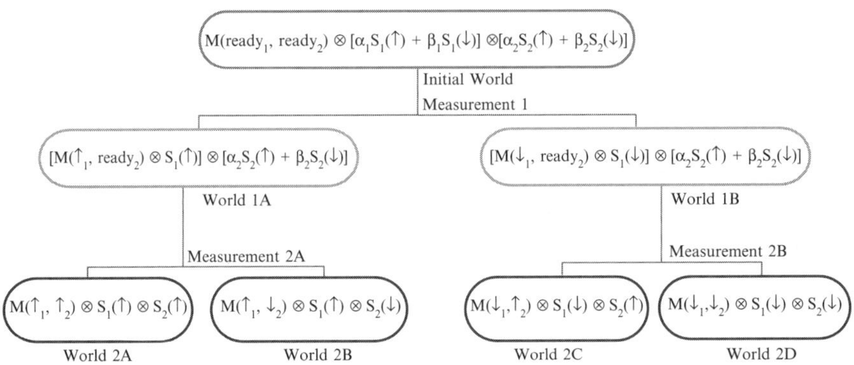

Fig. 1 Sequential measurements in the splitting worlds interpretation. On DeWitt and Graham's interpretation of probability, coefficients are represented in the proportion of each type of world.

In the introduction to their anthology on Everett's theory, DeWitt and Graham explain that Everett's interpretation of quantum mechanics

> denies the existence of a separate classical realm and asserts that it makes sense to talk about a state vector for the whole universe. This state vector never collapses and hence reality as a whole is rigorously deterministic. This reality, which is described jointly by the dynamical variables and the state vector, is not the reality we customarily think of, but is a reality composed of many worlds. By virtue of the temporal development of the dynamical variables the state vector decomposes naturally into orthogonal vectors, reflecting a continual splitting of the universe into a multitude of mutually unobservable but equally real worlds, in each of which every good measurement has yielded a definite result and in most of which the familiar statistical quantum laws hold (1973, p. v).

DeWitt admits that the constant splitting of worlds whenever the states of systems become correlated is counterintuitive: "I still recall vividly the shock I experienced on first encountering this multiworld concept. The idea of 10^{100} slightly imperfect copies of oneself all constantly spitting into further copies, which ultimately become unrecognizable, is not easy to reconcile with common sense. Here is schizophrenia with a vengeance" (1973, p. 161). But while the theory is counterintuitive, it does provide a direct explanation for why it seems to an observer that she has record a particular determinate measurement result, something that was unclear in Everett's original account. The explanation here is because each copy of the observer does in fact have a determinate record: in the post-measurement state above there are n observers, each occupying a different world and each with a perfectly determinate measurement record.

A standard complaint against such many-worlds formulations of quantum mechanics is that they are ontologically extravagant. One would presumably only ever need one physical world, our world, to account for our experiences. On the other hand, postulating the actual existence of a different physical world corresponding to each term in the quantum-mechanical state may allow one to explain our determinate measurement records while taking the standard deterministically-evolving quantum state to be in some sense a complete and accurate description of the physical facts. The explanatory tradeoff here is between the theoretical elegance of the linear dynamics alone and the metaphysical extravagance of branching worlds.

A more serious problem for many-worlds formulations is that, in order to explain determinate measurement records, the theory requires one to choose a preferred basis so that observers can be thought to have determinate measurement records in each term of the quantum-mechanical state as expressed in this basis in order to account for their determinate experiences. The problem is that not just any basis will do this – one needs to select a preferred basis that makes records determinate *given how observers have in fact chosen to record their measurement results,* but it is unclear what basis would make our most immediately accessible physical records, those records that determine our experiences and beliefs, determinate in every Everett world.

It has been suggested that ► decoherence considerations might resolve the preferred basis problem. On this proposal, rather than stipulating an ad hoc preferred basis, one would seek to explain how the interactions between measuring devices and their environments serve to select the basis that determines what worlds there are. One might, for example, argue that the environments of measuring devices will quickly become correlated to their pointer variables, then stipulate that such correlations select a preferred basis that guarantees that the values of the pointer variables will determinate in each Everett world. Note that decoherence considerations alone do not explain the determinate measurement records; rather, since an observer gets a determinate result in each Everett world, it is whatever stipulation one adopts concerning how environmental interactions determine what worlds there are that ultimately explains the determinate measurement records. General decoherence considerations are then to provide justification for the particular stipulation one adopts.

Perhaps the most difficult problem for many-world formulations concerns the statistical predictions of quantum mechanics and how probability is understood in the theory. The standard collapse theory predicts that M will record "the result is ϕ_i" with probability $|\alpha_i|^2$, but it is unclear how one is to make sense of this when M in fact gets every possible measurement result in some world. (▶ Wave function collapse). It will not do to simply claim that our world is *typical* since, if there is one world for each term in the preferred basis expansion of the post-measurement state, the standard ▶ quantum statistics will typically fail to hold in most worlds. There are several proposals for solving such problems, but it remains unclear whether any of the current proposals will ultimately prove satisfactory [1, 2, 8].

Literature

1. Albert, D. Z.: 1992, Quantum Mechanics and Experience (Harvard University Press, Cambridge, MA).
2. Barrett, J.: 1999, The Quantum Mechanics of Minds and Worlds (Oxford University Press, Oxford).
3. DeWitt, B. S.: 1971, The Many-Universes Interpretation of Quantum Mechanics, in Foundations of Quantum Mechanics (Academic, New York). Reprinted in DeWitt and Graham (1973), pp. 167–218.
4. DeWitt, B. S., and Graham, N. (eds): 1973, The Many- Worlds Interpretation of Quantum Mechanics (Princeton University Press, Princeton).
5. Everett, H.: 1957a, On the Foundations of Quantum Mechanics, thesis submitted to Princeton University, March 1, 1957, in partial fulfillment of the requirements for the Ph.D. degree.
6. Everett, H.: 1957b, 'Relative State' Formulation of Quantum Mechanics, Reviews of Modern Physics 29:454–462. Reprinted in Wheeler, J. A., and Zurek, W. H. (eds), Quantum Theory and Measurement (Princeton University Press, Princeton, 1983, pp. 315–323).
7. Everett, H.: 1973, The Theory of the Universal Wave Function, in DeWitt and Graham (1973).
8. Wallace, D.: 2002, Worlds in the Everett Interpretation, Studies in History and Philosophy of Modern Physics, 33B(4):637–661.

Matrix Mechanics

Henry Stapp

The theoretical ideas formulated in the seventeenth century by Isaac Newton (1643–1727) and Galileo (1564–1642) reigned as the fundamental scientific precepts until the year 1900, when Max Planck's (1858–1947) work on the emission of light from a hole in a heated hollow sphere showed that something was fundamentally amiss. Planck's work identified a new constant of nature, called Planck's quantum of action (► Planck's constant), that was alien to classical physics, and that evidently needed to be integrated into a revised physics, to be called quantum mechanics. A big step toward this new physics seemed to be model of the atom devised in 1913 by Niels Bohr (1885–1962) (see ► Bohr's atomic model). It was a space-time picture of the atom in which the ► electrons instead of spiraling inward and gradually radiating away their energies, as demanded by classical physics, were usually confined to stable orbits, which were specified in terms of Planck's quantum of action.

The very strange thing about this model was that no light was emitted by the circling electron when it was in one of these orbits. Light was emitted, instead, when an electron jumped from one orbit to another. However, its frequency was not some average of the frequencies of the light that classical physics predicted should be emitted from the electron of each of the two orbits: it was, instead, the *difference* of these two frequencies.

A large amount of experimental data was being collected at that time about the energy levels of various atoms, and about the rates at which the transitions between different levels occurred (► spectroscopy, ► quantum jumps). The excitations of atoms from various states to more excited states could be induced by the absorption of light, and the theory of this absorption and re-emission of light was called dispersion theory.

Intensive efforts to construct a rationally coherent quantum mechanics were being pursued by many groups, including most prominently those led by Niels Bohr in Copenhagen, Max Born (1882–1970) in Göttingen, and Arnold Sommerfeld (1868–1951) in Munich. But the key breakthrough was made by Werner Heisenberg (1901–1976).

Heisenberg was a prodigy. He entered the University of Munich in 1920 at age 18, and received his Ph. D 3 years later. In 1921 he published with Sommerfeld's approval a bold and original paper on the anamolous ► Zeeman effect, and in 1922 had co-authored two papers with Sommerfeld, and had closely collaborated on another with Max Born. In September of 1924 he began a stay in Copenhagen where he collaborated with Bohr and co-authored a paper on dispersion theory with Bohr's assistant Hendrik Kramers (1894–1952). Thus when he returned to Göttingen in April of 1925 he was only 23, but had spent the better part of 5 years working intensively in close collaboration with the leaders of the field.

The state of affairs was at that point extremely muddled, with the Copenhagen-based Bohr–Kramers–Slater dispersion theory recently falsified by data. Also, a recent closely reasoned paper by Heisenberg's close colleague, Wolfgang Pauli (1900–1958), argued that the entire program of basing the theory on space-time pictures akin to Bohr's model was a "swindle", and called for a new mathematical foundation: "It seems to me. . .without doubt that not only the dynamical concept of force, but also the kinematic concept of motion of classical theory, will have to experience profound modification. . . .I believe that the energy and momentum values of the stationary states are much more real than "orbits". (Pauli to Bohr, 12 December 1924).

Armed with all this deep knowledge and wise council, and influenced by Einstein's 1905 success in shedding unhelpful intuitions and biases concerning space and time by focusing on observable properties, Heisenberg tried to find a new foundation for atomic physics based not on a space-time picture of what was going on, but rather on mathematical connections between observable quantities. The ► observables in the abundant and accurate data pertaining to the dispersion of light were energy levels of the "stationary states", whatever they were, and transition amplitudes between these states. The transition amplitudes refer to *two* states and thus form a square array. In order to establish some sort of correspondence with the classical idea of an atom Heisenberg needed arrays corresponding to the variable of classical physics, such as momentum, position, acceleration, etc. and needed to form the analogs of products of these "quantities". He constructed what seemed to be the needed rules, by comparing to some apparently valid rules of dispersion theory, and discovered that, for certain quantities X and Y, XY was different from YX. This troubled Heisenberg, but did not deter him.

M

Because atomic systems are complicated, Heisenberg considered first a one-dimensional anharmonic oscillator, obtained by adding an extra force term.

The results for that case, and in particular his proof that energy was strictly conserved – it was a violation of strict energy conservation that had doomed the Bohr–Kramers–Slater theory ► BKS theory – convinced him that he had found the basic structure he needed. Its subsequent successful applications to innumerable physical situations by thousands of physicists, with no proven failures. has borne out his optimism.

Born was quick to point out that the arrays of numbers, with their rule of multiplication, were objects already well studied by mathematicians. They are called "matrices", and the quantum theory based on them was, for a time, called "matrix mechanics", particularly to distinguish it from what appeared at first to be an alternative quantum mechanics devised by Erwin Schrödinger, and called "wave mechanics". The two theories were eventually shown to be formally equivalent by Schrödinger, whose approach did seem to provide a space-time description of the kind that Heisenberg and Pauli had deemed impossible. However, Heisenberg, Pauli, and Bohr held that the Schrödinger wave was an abstract formal structure that could be used to compute observable quantities, because of the proved formal equivalence, but that it could not be regarded as describing an actually existing space-time structure, because of the "► quantum jumps" that the wave needs to undergo in order to keep it in line with human experience.

Because of the formal equivalence of the two forms, the two names "matrix mechanics" and "wave mechanics" have largely fallen out of use now, being replaced by the more inclusive name "quantum mechanics".

The see how these ideas work in actual practice one may consider the simplest case, in which the quantum system being examined has just two states, labeled by an index i that can take two alternative possible values, 1 or 2. Then the relevant arrays are sets of four (complex) numbers z_{ij} where the two indices i and j each can take, independently, the value 1 or 2. If one has two such sets z_{ij} and w_{ij} then an array called $(zw)_{ij}$ is defined by the rule $(zw)_{ij} = z_{i1}w_{1j} + z_{i2}w_{2j}$.This is the standard rule of matrix multiplication, in this two dimensional case.

Pauli defined four 2-by-2 matrices of interest:

(σ_0) defined by $((\sigma_0)_{11} = 1,\ (\sigma_0)_{12} = 0,\ (\sigma_0)_{21} = 0,\ (\sigma_0)_{22} = 1)$,
(σ_1) defined by $((\sigma_1)_{11} = 0,\ (\sigma_1)_{12} = 1,\ (\sigma_1)_{21} = 1,\ (\sigma_1)_{22} = 0)$,
(σ_2) defined by $((\sigma_2)_{11} = 0,\ (\sigma_2)_{12} = -i,\ (\sigma_2)_{21} = i,\ (\sigma_2)_{22} = 0)$,
(σ_3) defined by $((\sigma_3)_{11} = 1,\ (\sigma_3)_{12} = 0,\ (\sigma_3)_{21} = 0,\ (\sigma_3)_{22} = -1)$.

Laborious computations can then be simplified by writing matrices of interest as linear combination: $a = a_0\sigma_0 + a_1\sigma_1 + a_2\sigma_2 + a_3\sigma_3$, and using the following results:

$$\text{for any } i,\ \sigma_i\sigma_i = \sigma_0;\ \sigma_0\sigma_i = \sigma_i\sigma_0 = \sigma_i;\ \sigma_1\sigma_2 = i\sigma_3;\ \sigma_2\sigma_1 = -i\sigma_3.$$

These results follow directly from the definitions and multiplication rules specified above. Notice that in the last two equations the order in which the matrices are multiplied matters.

The rule that connects the mathematical symbols to our observations is this:

Each elementary observation upon the system is associated with a "projection operator" P. (Projection operators P must satisfy PP = P). (► Projection).

Let P_1 be the projection operator that corresponds in the mathematics to our knowledge that an associated set of preparation conditions have been met.

Let P_2 be the projection operator that corresponds in the mathematics to the condition that a subsequent observation fulfills an associated set of conditions. Then the predicted probability that a system known to be prepared in accordance with the conditions corresponding to P_1 at time $t = 0$ will be observed at time $t > 0$ to fulfill the conditions corresponding to P_2 is

$$\text{Trace } P_2 \ (\exp - iHt)\ P_1\ (\exp iHt)\,,$$

where H is the matrix that corresponds to energy, here assumed to have no explicit dependence on time, and for any X, Trace $X = X_{11} + X_{22}$, for this 2-by-2 case. (I use units in which Planck's constant of action is 2π.)

Suppose, for example, that $P_1 = (1 + \sigma_3)/2$, which corresponds the prepared system's being in the state $i = 1$, and that $P_2 = (1 - \sigma_3)/2$, which corresponds to the system's being observed to be in the state $i = 2$. Suppose $H = e\,\sigma_1$.

Using the fact (deducible from the power series expansion of exp x) that $\exp(-iet\sigma_1) = (\cos et - i\sigma_1 \sin et)$ one can easily deduce *just from the rules given above* that the probability identified above is $(\sin et)^2$. The calculation is carried out without referring to any space-time picture of what is going on.

Literature

1. Beller, M: *Matrix Theory before Schrödinger*, Isis **74** (1983) 469–491.
2. Heisenberg, W: *Über quantentheoretische Umdeutung kinematischer und mechanischer Berechnungen*, Zeitschrift für Physik **33** (1925), 879–893. (English translation in: B. L. van der Waerden: Sources of Quantum Mechanics, New York: Dover 1983, pp. 261–276)
3. MacKinnon, E: *Heisenberg, Models, and the Rise of Matrix Mechanics*, Historical Studies in the Physical Sciences **8** (1977), 137–188.
4. Mehra, J & Rechenberg, H: *The Historical Development of Quantum Mechanics*, Volume 2. Chapters III, IV, & V (2001).
5. Pauli, W: *Wissenschaftlicher Briefwechsel*, Volume 1, Berlin: Springer 1979.

Matter Waves

Bruce R. Wheaton

Among the audacious proposals in the evolution of natural philosophy, Louis de Broglie's (1892–1987) claim in 1923 that atoms possess a wave-property sits at top rank. Substantial matter had from ancient times been ascribed to particles like those we encounter everyday. While there was always doubt whether light is material or a disturbance in a medium (see ► wave-particle duality) there had never been much doubt about matter. A noteworthy, late nineteenth century exception in the wake of Maxwellian success in field theory came to be called "the electromagnetic world-view," based on Kantian idealism, that described ponderable matter as secondary properties of the primary æther.

However, Albert Einstein's (1879–1955) tri-partite recasting of matter, light, and time in 1905 gave a molecular explanation in accord with that of Jean Perrin (1870–1942) to long-observed ► Brownian motion, and atoms prevailed. In the 1920s, practical concerns of physicists in France led to de Broglie's recognition of a paradox, particularly in the domain of ► x-rays, when he tried to bring coherence to both new theories: of the quantum and of relativity.

France may seem an unlikely locale and 31-year old Louis de Broglie an even more unlikely source for so earth-shaking an inspiration. But under the tutelage of elder brother Maurice (1875–1960), Louis and a cadre of young physicists tried to apply the new *fin-de-siècle* discoveries in physics to improve French industrial process control. Entirely privately funded, and virtually independent of academic

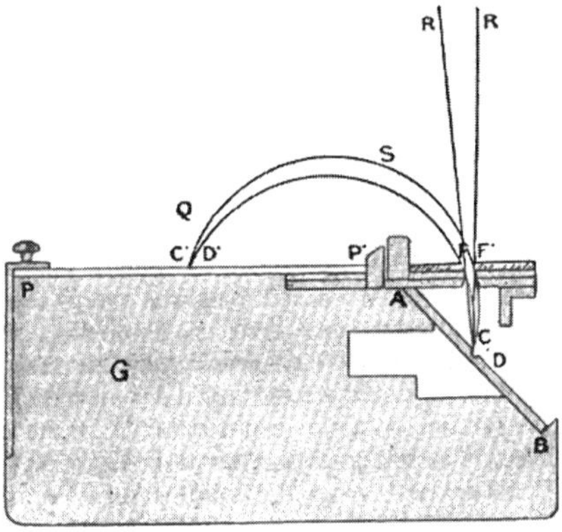

Fig. 1 M. de Broglie & A. Dauvillier's sample at C, irradiated with x-rays, emits electrons into the normal magnetic field that are sorted by velocity onto photoplate PP'. From M. de Broglie, ***Les rayons x*** (Paris, 1922), 142

physics, Maurice's "Laboratoire française des rayons-x" drew talent from all of Europe. For our purposes the most important was Alexandre Dauvillier (1892–1979), whose passion was the x-ray ► photoelectric effect. Together with Maurice, he showed experimentally by the 1921 Solvay congress, that x-rays must be absorbed by matter in discrete quantum units, using β-ray spectroscopy to measure velocities of emitted ► electrons (Fig. 1). In all cases the corpuscular behavior of e-m radiation prevailed. Charles Ellis (1895–1980) presented equivalent results for nuclear ► γ-rays at the same session.

"Little Louis" heard all of the Solvay discussions and tried to bring coherence to what he called the "dual wave-particle nature of radiation." He turned to Einstein's other two remarkable products of 1905, the ► light-quantum and relativity theory. In brief, relativity predicts that time intervals on a moving particle will appear lengthened to a stationary observer: that makes an observed frequency *lower*. But quantum theory predicts a moving particle possesses more energy and exhibits a *higher* frequency. Louis found a clever, most perplexing, way to reconcile this conundrum. "We debated the most pressing and baffling issues of the time," Louis recalled to his elder brother, "particularly the interpretation of results in your experiments on the x-ray photoeffect."

Louis' inspiration in 1923 was to posit a *virtual* wave that accompanies (actually precedes) every particle of matter. He had turned Einstein's light-quantum on its head: if light can be corpuscular, matter can be undulatory. Every particle of matter, he posited, has a guiding "phase wave" that travels faster than the particle such that $\boldsymbol{v}_{\mathrm{p}}\boldsymbol{v}_{\mathrm{w}} = \boldsymbol{c}^2$. The advantage is that these two oscillations maintain constructive interference at a moving point in space that essentially defines the observed trajectory of the particle. His hypothesis owed much to prior work by Vito Volterra (1860–1940), Marcel Brillouin (1854–1948) and Erwin Schrödinger (1887–1961) on theories of "retarded potentials." Louis' phase wave travels faster than the velocity of light, has wavelength $\lambda = \boldsymbol{h/p}$, carries no energy, and so he referred to it as an *onde fictiv*.

But this wave has a physical significance beyond a mere calculating device. On its basis he derived the action-integral representation of stable electron orbits in the Bohr atom (each revolution a standing wave-like band), and explained the contemporaneous Compton–Debye effect. This influential experiment on generalized x-ray scattering also confirmed the corpuscular nature of x-rays. The audacious proposal of an inescapable wave-property of atoms "stuck the issue right under the nose" of Erwin Schrödinger who clarified the concept into the new ► wave mechanics in 1926. Louis' phase wave of 1923 also predicted diffraction of an electron beam, experiments corroborated by 1929 leading to his Nobel Prize.

Primary Literature

1. L. de Broglie: Recherches sur la théorie des quanta. Annales de physique, *10th ser.*, **3**, 22–128 (1925). Reprinted by Masson, 1925 and in facsimile by Fondation Louis de Broglie, 1992, with useful appended matter.
2. L. de Broglie: A tentative theory of light quanta. Philosophical Magazine, *6th ser.*, **47**, 446–58 (1924)
3. M. de Broglie: La relation $h\nu = \varepsilon$ dans les phénomènes photoélectriques, in *Atomes et electrons* (Paris, 1923), 80–100 (Solvay III.)
4. E. Schrödinger: Über eine bemerkenwerte Eigenschaft der Quantenbahnen eines einzelnen Elektrons. Zeitschrift fur Physik **12**, 13–23 (1922)
5. E. Schrödinger: Quantisierung als Eigenwertproblem (erste Mitteilung). Annalen der Physik, *4th Ser.*, **79**, 361–76 (1926)

M

Secondary Literature

6. O. Darrigol: Strangeness and soundness in Louis de Broglie's early works. Physis **30**, 303–72 (1993)
7. W. Heisenberg: Die Entwicklung der Quantentheorie, 1918–1928. Die Naturwissenschaften **17**, 490–96 (1929)
8. A. Russo: Fundamental research at Bell Laboratories: The discovery of electron-diffraction. Historical studies in the physical sciences **12**, 117–60 (1981)
9. B. R. Wheaton: Atomic waves in private practice, in *Quantum Mechanics at the crossroads*, eds. J. Evans, A. Thorndike (Springer, Heidelberg 2006), 39–71
10. B. R. Wheaton: *The tiger and the shark*: Empirical roots of wave-particle dualism (Cambridge University Press, Cambridge 1983)

Measurement Problem

See ► Bohmian mechanics; Measurement theory; Objectification; Projection Postulate.

Measurement Theory

Paul Busch and Pekka Lahti

The term *measurement theory* refers to that part of a physical theory in which the empirical and operational content of the concepts of the theory is determined. Measurements are analyzed both as operational procedures defining the ► *observables* of the theory and as physical processes which are themselves subject to the laws of physics.

In classical physics, measurements are performed in order to determine the values of one or several observables of the physical system under consideration. Classical physics allowed the idealized notion that every physical quantity has a definite value at any time, and that this value can be determined with certainty by measurement without influencing the object system in a significant way. By contrast, in quantum mechanics both features fail to hold without strong qualifications. Accordingly, in their seminal paper of 1935 [1], Einstein, Podolsky and Rosen used elements of this description as a sufficient criterion of physical reality, applicable both in classical and quantum mechanics:

> "If, without in any way disturbing a system, we can predict with certainty (i.e., with probability equal to unity) the value of a physical quantity, then there exists an element of physical reality corresponding to that physical quantity."

As far as *observable* elements of reality represented by quantum mechanics are concerned, this condition must also be regarded as necessary. Hence, an observable is understood to have a definite value if the probability that a measurement indicates a particular value of the observable is equal to one. In quantum mechanics, this can only be satisfied if the system is in an eigenstate of the observable associated with the value in question. Moreover, it turns out that in quantum mechanics the interaction between a measuring apparatus and the measured system is generally not negligible. This leads to the necessity of reconsidering what it means that a measurement determines the value of an observable. Here this question is discussed for the case of an observable represented by a selfadjoint operator A (acting on a complex separable ► Hilbert space $\mathcal{H}$) with nondegenerate discrete spectrum $\{a_1, a_2, \dots\}$, associated ► orthonormal basis of eigenvectors $\{\varphi_1, \varphi_2, \dots\}$, and *spectral decomposition* $A = \sum_i a_i P_i$, where $P_i = |\varphi_i\rangle\langle\varphi_i|$ denotes the projection onto the one-dimensional subspace spanned by φ_i. (Spectral decomposition, see ► Density operator; Ignorance interpretation; Objectification; Operator; Probabilistic Interpretation; Propensities in Quantum Mechanics; Self-adjoint operator; Wave mechanics).

A minimal requirement for a physical interaction process between an object system and an apparatus to qualify as a measurement of A is the so-called *calibration condition*: whenever the system is in an eigenstate, the apparatus should indicate the corresponding eigenvalue unambiguously after the interaction has ceased. In quantum mechanics, a measurement is modeled by representing the apparatus by a Hilbert space $\mathcal{H}_A$, the pointer observable as a selfadjoint operator Z acting on $\mathcal{H}_A$ and the coupling between object and apparatus as a unitary operator U acting on the tensor product Hilbert space $\mathcal{H} \otimes \mathcal{H}_A$ of the total system. Together with the initial apparatus *state* T_A, these elements, collected into a quadruple $\langle \mathcal{H}_A, T_A, U, Z \rangle$, constitute a *measurement scheme*.

Assuming, for simplicity, that the apparatus initially is in a pure state, described by a unit vector ϕ, the calibration condition can be formalized as follows: the measurement scheme has to be such that for any eigenstate φ_i of A there is an associated (normalized) eigenstate ϕ_i of the pointer Z so that U effects the following transition:

$$\varphi_i \otimes \phi \rightarrow U(\varphi_i \otimes \phi) = \psi_i \otimes \phi_i . \tag{1}$$

Here ψ_i is some normalized vector state in $\mathcal{H}$, and the ϕ_i are mutually orthogonal. Thus, if the observable A initially has a definite value a_i, the pointer observable of the apparatus will indicate this value with probability equal to one, in accordance with the ► *Born probability rule*. If condition (1) is satisfied for all φ_i, the given measurement scheme is called a *premeasurement* of A.

If the system is initially in a vector state φ which is not an eigenstate of A, then φ is a ► superposition of eigenstates of A, that is, $\varphi = \sum_i c_i \varphi_i$ with more than one of the c_i nonzero. Together with the linearity of U, the rule (1) still determines unambiguously the final state of the total system:

M

$$\varphi \otimes \phi = \sum_i c_i \varphi_i \otimes \phi \rightarrow U\varphi \otimes \phi = \sum_i c_i \psi_i \otimes \phi_i . \tag{2}$$

The final state is a superposition of mutually orthogonal states, and the probability for the pointer to indicate a value a_i is equal to $|c_i|^2 = |\langle \varphi | P_i \varphi \rangle|^2$, thus justifying the Born probability interpretation of the latter expression.

This simplified description also highlights the fundamental dilemma of quantum measurement theory known as the quantum measurement problem, the problem of objectification, or the collapse problem: if an observable A does not have a definite value, then according to quantum mechanics, a premeasurement of A will leave the object-plus-apparatus system in an *entangled* state in which the pointer observable does not have a definite value – in stark contrast to the fact that every real measurement ends with a definite pointer position. This leaves one with the following alternative: on the one hand, if one requires that quantum mechanics should include an account of its measuring processes – that is, this theory should be semantically complete – then it turns out that the occurrence of definite measurement outcomes contradicts the quantum mechanical account of the measurement dynamics – that is, this theory is semantically inconsistent; on the other hand, if one requires semantical consistency, then quantum mechanics cannot be semantically complete [8]. In the first case, a modification of the axioms of quantum mechanics is required. In the second case, there is no consistent quantum measurement theory, unless an appropriate reinterpretation of what it means for an observable to have a definite value can be found.

There is an enormous amount of literature dealing with the quantum measurement problem, and as yet there is no generally accepted resolution. Rigorous technical presentations of the problem and the spectrum of interpretational options are found, for example, in [9] and [10], whereas philosophical aspects are discussed in [11]. A valuable cross-section of the older literature until 1980 is reprinted in the volume [12]. Interestingly, the founders of quantum mechanics (e.g., [2, 3]) identified the reality of the collapse of the wave function (▶ wave function collapse) or state vector but did not regard it as a conceptual problem. It was von Neumann in 1932 [4] who pointed out the tension between the collapse process as a random event and the deterministic (unitary, linear) Schrödinger dynamics of a closed system. Somewhat later, Schrödinger [5] conceived his infamous ▶ Schrödinger cat paradox to highlight the apparent absurdity of the possibility, suggested by quantum mechanics, of observing macroscopic systems in superpositions of states corresponding to such discernible situations as a cat being dead or alive.

Adopting the collapse postulate has since been taken by many as a pragmatic way of suspending the measurement problem. Following this route, there remains the task for quantum measurement theory to show that quantum mechanics entails the possibility in principle of measuring any of its observables. For an observable represented as a POVM (▶ *observable*), the above calibration condition is generally not applicable. However, whenever that condition does apply, it implies the reproduction of probabilities for the object observable in terms of the pointer statistics. This latter condition, called *probability reproducibility condition* [9], can always be taken as the defining criterion for a measurement scheme to constitute a measurement of a given observable. This characterization of the measurements of an observable implements the Born interpretation (▶ *Born rule*) of the quantum mechanical probabilities and the idea that any observable is identified by the totality of its statistics. The formal implementation of these ideas, which constitute the mathematical framework of quantum measurement theory, are briefly summarized in the text box below.

Tools of Quantum Measurement Theory

Every measurement scheme $\langle \mathcal{H}_A, T_A, U, Z\rangle$ defines a unique observable of the object system. If the pointer observable Z is represented as a POVM on the (Borel) sets of $\mathbb{R}$ (say), then for each state T of the object system, the following defines a probability measure on the real line (X denotes any Borel subset of $\mathbb{R}$ and I is the identity operator):

$$X \mapsto \mathrm{tr}[UT \otimes T_A U^* I \otimes Z(X)] \equiv \mathrm{tr}[T E(X)]. \tag{3}$$

This equation, valid for all states T, entails the existence of a positive operator $E(X)$ associated with each set X; moreover, the fact that $X \mapsto \mathrm{tr}[T E(X)]$ is a probability measure for each T ensures that $E : X \mapsto E(X)$ is a POVM on the (Borel) subsets of $\mathbb{R}$.

It is a fundamental theorem of the quantum theory of measurement that for every observable there are measurement schemes (in fact, infinitely many) such that (3) is fulfilled for all object states T [6].

With the existence of premeasurements for any observable thus secured, another task of quantum measurement theory is the description of the effect of a measurement on the object system. Given a measurement scheme for an observable E, one can ask for the probabilities of the outcomes of any subsequent measurement. If F is another POVM on the (Borel) subsets of $\mathbb{R}$, to be measured immediately after the E measurement, the sequential joint probability for obtaining a value of E in a set X and a value of F in a set Y is

$$\mathrm{tr}[UT \otimes T_A U^* F(Y) \otimes Z(X)] \equiv \mathrm{tr}[\mathcal{I}_X(T) F(Y)]. \tag{4}$$

This relation, valid for all states T, all observables F and all X, Y, determines a unique non-normalized object state $\mathcal{I}_X(T)$; substituting for $F(Y)$ the identity operator, it is seen that $\mathrm{tr}[\mathcal{I}_X(T)] = \mathrm{tr}[TE(X)]$. Dividing the joint probability in (4) by the latter probability gives the conditional probability for the occurrence of an outcome in Y given that the first measurement led to an outcome in X. Thus $\mathcal{I}_X(T)$ can be taken to play the role of the final object state in accordance with the collapse postulate. The map $T \mapsto \mathcal{I}_X(T)$ is known as a *(quantum) operation*, and $X \mapsto \mathcal{I}_X$ is an operation-valued measure called the *instrument* induced by the given measurement scheme [7].

Any instrument arising from a measurement scheme has the property of *complete positivity*: that is, for any operation $\mathcal{I}_X$, if extended to a linear map $I_n \otimes \mathcal{I}_X$ acting on the trace class operators of the Hilbert spaces $\mathbb{C}^n \otimes \mathcal{H}$, the extended map is positive for each n. It is another fundamental theorem of quantum measurement theory that every completely positive instrument can be realized by some (in fact, infinitely many) measurement schemes [6].

M

With the conceptual tools of measurement theory outlined in the above box, it has become possible to eliminate some long-standing myths and corroborate a number of equally long-standing folk truths. For example, it has long been held without questioning that any measurement collapses the object system into an eigenstate of the measured observable. Measurements with that property are called repeatable. In the example leading to (1), repeatability is achieved by putting $\psi_i = \varphi_i$; but it is by no means necessary to assume that every measurement has this property. Moreover, according to a theorem due to Ozawa [6], in order for an observable to admit a repeatable measurement, this observable must be discrete, that is, have a countable set of values.

The realization that measurements necessarily disturb the object system was made early on in the history of quantum mechanics. However, the nature of that "disturbance" and its quantification have remained the subject of much debate until recently, when it was realized that the notion of instrument allows a rigorous and effective description of the state changes due to measurements. Yet another fundamental theorem of quantum measurement theory is given by the statement that there is no measurement which does not change at least some of the states of the system under investigation: a measurement scheme that leaves unchanged all states of the object defines a trivial observable, that is one whose probability measures do not depend on the state. Thus, there is no information gain in quantum measurements without some disturbance.

The trade-off between information gain and disturbance in quantum measurements has been recognized as a resource for novel applications of quantum measurements, particularly in quantum cryptography, ► quantum communication a sub-field of the new area of *quantum information* science. This is one example for the importance of quantum measurement theory as an applied discipline besides its foundational role.

Applications of quantum measurement theory ranging from nondemolition measurements and analyses of basic experiments to open quantum systems and quantum tomography are covered, for instance, by the monographs [13–17].

Primary Literature

1. Einstein, A., B. Podolsky, N. Rosen: Can quantum-mechanical description of physical reality be considered complete? Physical Review **47**, 777–780 (1935).
2. Heisenberg, W.: Über den anschaulichen Inhalt der quantentheoretischen Kinematik und Mechanik. Zeitschrift für Physik **43**, 172–198 (1927).
3. Bohr, N.: The quantum postulate and the recent development of atomic theory. Nature **121**, 580–590 (1928).
4. Neumann, J. von: *Mathematische Grundlagen der Quantenmechanik* (Springer, Berlin 1932, 2nd ed. 1996); English translation (by R. Beyer): Mathematical Foundations of Quantum Mechanics (Princeton University Press, Princeton 1955).
5. Schrödinger, E.: Die gegenwärtige Situation in der Quantenmechanik. Die Naturwissenschaften **23**, 807–812; 824–828; 844–849 (1935). English translation in [12].
6. Ozawa, M.: Quantum measuring processes of continuous observables, Journal of Mathematicsl Physics **25**, 79–87 (1984).
7. Davies, E.B., J.T. Lewis: An operational approach to quantum Probability, Communications in Mathematical Physics **17**, 239–260 (1970).

Secondary Literature

8. Mittelstaedt, P.: *Philosophical Problems of Modern Physics* (D. Reidel, Dordrecht 1976).
9. Busch, P., P. Lahti, P. Mittelstaedt: *The Quantum Theory of Measurement* (Springer, Berlin 1991, 2nd revised edn. 1996).
10. Bub, J.: *Interpreting the Quantum World* (Cambridge University Press, Cambridge 1997).
11. Mittelstaedt, P.: *The Interpretation of Quantum Mechanics and the Measurement Process* (Cambridge University Press, Cambridge 1997).
12. Wheeler, J.A., W.H. Zurek.: *Quantum Theory and Measurement* (Princeton University Press, Princeton 1983).
13. Braginsky, V.B., F.Ya. Khalili: *Quantum Measurement*, edited by K.S. Thorne, (Cambridge University Press, Cambridge 1992).
14. Busch, P., M. Grabowski, P. Lahti: *Operational Quantum Physics* (Springer, Berlin 1995, 2nd corrected printing 1997).

15. Muynck, W.M. de: *Foundations of Quantum Mechanics : An Empiricist Approach* (Kluwer, Dordrecht, 2002).
16. Breuer, H.-B., F. Petruccione: *The Theory of Open Quantum Systems*, (Oxford University Press, Oxford 2002).
17. Paris, M.G.A., J. Řeháček (Eds.): *Quantum State Estimation*, (Springer, Berlin, 2004).

Mesoscopic Quantum Phenomena

Markus Arndt

Quantum physics was first developed to understand the properties of small individual objects such as photons (► light quantum), atoms and molecules. And many features of quantum physics, such as the discreteness of energy levels, the ► superposition of mutually exclusive states, quantum interference or ► entanglement are usually not directly accessible to our human senses. Colloquially we therefore often separate between microscopic and macroscopic in the sense of 'being observable or unobservable by the unaided eye' rather than in the more physical sense where microscopic would refer to objects in the micrometer size range. In physics, the notion of mesoscopic quantum phenomena is generally used for systems with dimensions somewhere in the middle (in Greek: meso = middle) between the microscopic and the macroscopic world. In practice, mesoscopic systems mostly range between a few and a few hundred nanometers. They are large enough to contain many particles and can therefore be described by average properties, such as density or conductivity. On the other hand they are small enough for their lateral extensions to match characteristic lengths, such as the coherence length or the mean free path. Mesoscopic quantum systems therefore often exhibit unique physical properties such as size-dependent electronic properties, transport phenomena and more. The following examples select some of the most quoted mesoscopic quantum phenomena [6,9–11].

M

Mesoscopic Quantum Confinement

Quantum dots are zero-dimensional nanostructures in the sense that they confine the quantum wave function in all three directions [1]. This has to be contrasted with for instance one-dimensional quantum wires, two-dimensional electron gases atomic ensembles (► ensembles in quantum mechanics) or three-dimensional bulk solids. Quantum dots are often referred to as artificial, ultra-cold trapped atoms, since they exhibit a size-dependent discrete energy spectrum. Optical transition lines in small dots are blue-shifted with respect to those in larger dots. Q-dots realize the textbook example of a particle in the box: strong confinement leads to strong wavefunction curvature, high momentum and large energy splittings. Q-dots can be realized lithographically, or with a suitable arrangement of interfaces between different materials. Colloidal semiconductor nanocrystals may measure up to about 10 nm. Self-assembled quantum dots on surfaces range between 10 and 50 nm. Lithographically patterned or self-assembled semiconductor dots may extend to 100 nm. Quantum dots are for instance the basis for blue lasers, single-photon emitters, fluorescent markers in biology and many other applications.

Mesoscopic Quantum Conductance

Singe electron capacitors and single electron transistors When two conductors are separated by a thin insulating barrier, current flow is forbidden classically, while ► tunnelling is still allowed quantum mechanically. Mesoscopic devices with lateral extensions around 100 nm and a barrier thickness of about 1 nm exhibit interesting conductance properties as their electric capacity gets as small as 1 Femtofarad.

A single electron transistor can then be formed by sandwiching a conducting island between two such junctions and by capacitively connecting it to a third gate electrode. A positive voltage to the gate electrode will lower the energy levels of the island and an electron can tunnel first onto the island and then further on to the drain electrode. The charging of the island with a single electron can already suffice to raise the voltage ($U = e/C$) such that a second electron cannot enter the same transistor at the same time. In order to observe such a *Coulomb blockade* the device temperatures has to be about 1 K, sufficient to suppress thermal excitations.

Josephson Effects In a Josephson device two superconducting leads are separated by a thin insulator material. The appearance of an electric DC current across the tunnelling junction in the absence of any external electromagnetic field is known as the DC Josephson effect [2]. This current is a genuine quantum phenomenon, and uniquely determined by the phase difference of the quantum ► wave functions on either side of the insulator. By adding a fixed voltage, the quantum phase will start oscillating in time and the applied DC voltage therefore induces an alternating current (*AC Josephson effect*).

Mesoscopic Electron Interference

Diffraction of free ► electrons has been known since the experiments by Davisson and Germer (► Davisson–Germer experiment) in 1927. More recent experiments have proven that ► *electron interferometry in mesoscopic systems* is equally feasible, interesting and sometimes unavoidable. In order to maintain coherence, perturbations have to be minimized and such experiments are done in low-dimensional electron systems with semi-conductor wave guides or in strong external magnetic fields. These demonstrations show that electron coherence can extend up to one micrometer in cold solids and mesoscopic electron interferometers have for instance been applied to explore the ► Aharonov–Bohm effect, ► Berry-phase or ► decoherence. (Cf. ► environmental observation of decoherence). Natural interference of electron wave functions is also at the basis of *universal conductance fluctuations* [9–11]: mesoscopic systems exhibit ballistic electron transport when their impurity content is sufficiently low and the elastic mean free path of the charge carriers at least comparable to the size of the system. The terminal conductance may then exhibit reproducible fluctuations on the order of the quantum of conductance $e^2 h^{-1}$ when the chemical potential, magnetic field or impurity configuration is varied. These fluctuations arise from quantum-interference effects due to the phase-coherent electron transport.

Anderson Localization was also first established in the context of mesoscopic disordered media: it describes the observation that the diffusive spreading of waves can be suppressed in randomly disordered media, because of interference between multiple scattering path-ways. When applied to microwaves in chaotic potentials, this is a classical wave phenomenon. For electrons in solids this is a genuine mesoscopic quantum phenomenon [3, 9–11].

The integer and fractional ► *Quantum Hall effects* also fall into the category of mesoscopic quantum transport phenomena. They are observed in two-dimensional electron systems at low temperatures and in strong magnetic fields. The Hall conductance in such a configuration is quantized in integer or fractional unities of $e^2 h^{-1}$, with the electron charge e and the ► Planck's constant h [4].

New Directions in Mesoscopic Quantum Physics

Quantum 'Mechanics': With the improvements of nanotechnologies and cooling technologies it has recently become possible to cool nanomechanical cantilevers with masses in the nanogram regime close to their quantummechanical ground state [5]. Cold cantilevers are also promising for new schemes heading towards mesoscopic entanglement [13].

For a long time, mesoscopic quantum phenomena counted generally as a subfield of condensed matter physics. Over recent decades, however, photonic, atomic and molecular systems have been extended to truly mesoscopic dimensions:

Atomic Bose–Einstein condensates [6] (► Bose–Einstein condensation) can be composed of more than one million atoms and exhibit coherence lengths well beyond the micrometer scale. Many studies with ultra-cold degenerate atomic ensembles are concerned with the classification of quantum phenomena according to their dimensionality. Long-range order can be observed in three-dimensional systems at low temperature (BEC). In two-dimensional systems long-range order is destroyed by thermal fluctuations at any finite temperature. But superfluid quasi-condensates can still be observed, which are related to a short-range topological order. Also in one dimension, mesoscopic atom clouds exhibit a quantum phenomenon: strongly interacting bosons may form a Tonks–Girardeau gas.

A mesoscopic superposition of photonic field states can be created by sending Rydberg atoms through a coherent field trapped in a microwave cavity. The interaction between atoms and microwave photons can be designed such that the phase of the photon field can simultaneously point into two different directions after the interaction. With several dozens of photons in the cavity this is a mesoscopic realization of a ► Schrödinger cat. The fragility of such large superposition states can be traced by monitoring their decay as a function of time and as a function of the 'distance' between the mutually exclusive states in the superposition. These experiments [7] beautifully illustrate many aspects of decoherence theory [12].

In macromolecule interferometry, complex many-body systems can be shown to exhibit the behaviour of delocalized matter waves with transverse coherence widths of the order of a micrometer [8]. Massive molecules, such as the fullerenes C_{60} and C_{70} or even biomolecules still show this phenomenon. They are composed of several dozens of atoms and exhibit quantum motion even though they may attain internal temperatures as high as 1,000 K. A major interest in such experiments is the understanding of the transition between quantum and classical behaviour.

Fullerenes are mesoscopic quantum objects in the sense that they exhibit many bulk properties of classical objects and still behave quantum mechanically when appropriately prepared. The bulk behavior manifests itself in collective excitations, such as plasmons, excitons or the large number of vibrational modes which are statistically excited according to a microcanonical temperature. But also the thermal emission of photons, electrons and molecular fragments at elevated temperatures have similarities with thermal radiation, glow emission and evaporation of bulk media. The ► de Broglie wavelength and coherence length of fullerenes in a thermal beam at 900 K amounts to only a few picometers, which is a few hundred times smaller than the molecule itself. Because of all that one might be tempted to identify a fullerene with a classical body. And yet it can be shown that C_{60} can delocalize over several micrometers and exhibit de Broglie quantum interference when diffracted at mechanical gratings.

It is interesting to explore how quantum coherence is destroyed on the way towards complex and larger bodies. In particular the interaction between the molecules and their environment has raised a lot of interest: Collisions with residual gas molecules but also photons emitted by the hot fullerenes themselves can reveal which-path information inside the interferometer. This also leads to decoherence via entanglement between the fullerene and the colliding or emitted particles. Figure 1 shows the experimental setup of a near-field matter wave interferometer for C_{70} as recently realized in Vienna. And it demonstrates the mesoscopic quantum nature of the experiment: Under high-vacuum conditions and at sufficiently low internal temperature the visibility of the molecular interference fringes is high and demonstrate the quantum nature of the fullerene. At increasing pressure of the residual gas or high internal temperature, the coupling to the environment becomes so strong that the intrinsic quantumness becomes effectively unobservable.

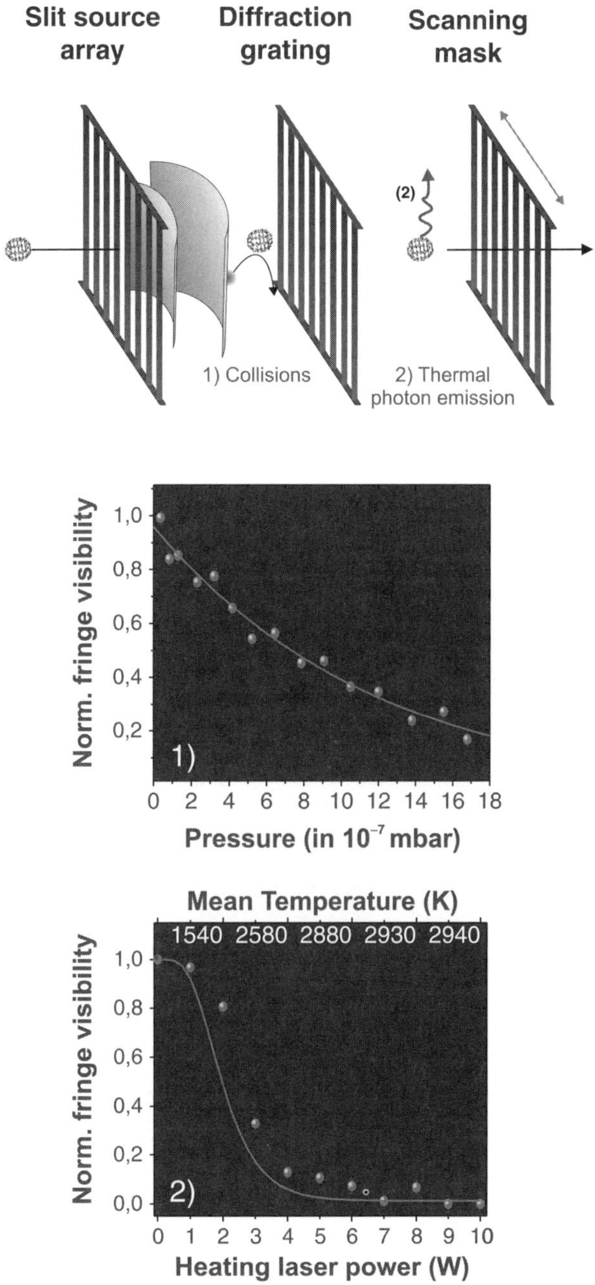

Fig. 1 Interference of hot complex molecules is a mesoscopic quantum phenomenon that serves in the exploration of the quantum-classical transition

Primary Literature

1. P.M. Petroff, A. Lorke, A. Imamoglu: Epitaxially self-assembled quantum dots. Phys. Today 54, (5), 46–52 (2001).
2. B.D. Josephson: The discovery of tunnelling supercurrents. Rev. Mod. Phys. 46, 251–54 (1974).
3. P.W. Anderson: Absence of diffusion in certain random lattices. Phys. Rev. 109, 1492 (1958).
4. K. von Klitzing: Nobel lecture: The quantized Hall effect. Rev. Mod. Phys. 58, 519–31 (1986).
5. K.C. Schwab, M.L. Roukes: Putting mechanics into quantum mechanics. Phys. Today 7, 36 (2005).
6. E. Cornell, C.E. Wieman: Nobel Lecture: Bose–Einstein condensation in a dilute gas, the first 70 years and some recent experiments. Rev. Mod. Phys. 74, 875 (2002).
7. S. Haroche: Entanglement, decoherence and the quantum/classical boundary. Phys. Today 51, (7), 36–42 (1998).
8. M. Arndt, K. Hornberger, A. Zeilinger: Probing the limits of the quantum world. Phys. World 18, 35 (2005).

Secondary Literature

9. J. Imry: *Introduction to Mesoscopic Physics*. Oxford University Press, Oxford (1997).
10. Y. Yamamoto: *Mesoscopic Quantum Optics*. Wiley-Interscience, New York (1999).
11. I. O. Kulik: *Quantum Mesoscopic Phenomena and Mesoscopic Devices in Microelectronics*. Springer, Amsterdam (2000).
12. E. Joos, H.D. Zeh, C. Kiefer, D. Giulini, J. Kupsch, I.O. Stamatescu: *Decoherence and the Appearance of a Classical World in Quantum Theory*. Springer, Berlin (2003).
13. D. Vitali, S. Gigan, A. Ferreira, H. R. Böhm, P. Tombesi, A. Guerreiro, V. Vedral, A. Zeilinger, M. Aspelmeyer: *Optomechanical entanglement between a movable mirror and a cavity field.* Phys. Rev. Lett. 98, 030405 (2007).

Metaphysics of Quantum Mechanics

Craig Callender

Quantum mechanics, like any physical theory, comes equipped with many metaphysical assumptions and implications. The line between metaphysics and physics is often blurry, but as a rough guide, one can think of a theory's metaphysics as those foundational assumptions made in its interpretation that are not usually directly tested in experiment. In classical mechanics some examples of metaphysical assumptions are the claims that forces are real, that inertial mass is primitive, and that space is substantival. The distinctive feature of these claims is that they are all rather far removed from ordinary tests of the theory. Newton defended all three of the above claims at one time or other, whereas Mach attacked each one; however, both scientists agreed on enough of the formalism and its connection to experiment to predict (e.g.) the same periods for given pendulums. What they disagreed about were the ingredients necessary to use classical mechanics to explain and understand the world.

Controversy engulfed the metaphysics of classical mechanics soon after its origin. Newton's idea of forces proved extremely contentious among the scientists of his time. Although metaphysical assumptions need not be controversial, quantum mechanics is also no stranger to metaphysical dispute. If anything, here the situation is more undecided because the theory was born with two different formalisms (Heisenberg's ► matrix mechanics, wave functions) and no clear interpretation. Heisenberg [1] originally offered a merely instrumental understanding of his formalism (later he opted for an interpretation employing discontinuous quantum jumps), whereas Schrödinger [2] viewed his theory as having physical content: it described, he thought, the evolution of continuous matter waves. The formalisms subsequently proved to be equivalent, but the metaphysical pictures could hardly have been more different. Soon thereafter, Bohr's ► complementarity thesis took shape, ► Heisenberg's uncertainty principle was discovered, and Born provided a ► probabilistic interpretation of the wavefunction. The combination of these three theses formed the essential core of the so-called Copenhagen interpretation. Associated especially with Bohr [3], the Copenhagen interpretation is itself the subject of active interpretation [10], and few advocates of the theory agree on all of the theses commonly associated with it. (See ► Born rule; Consistent Histories; Nonlocality; Orthodox Interpretation; Schrödinger's Cat; Transactional Interpretation). Nevertheless, if correct, it makes dramatic metaphysical assumptions. These include the ideas that measurement brings into being the measured property as opposed to revealing it, that there is a "complementarity" between dynamic and kinematic aspects of the world, and that all properties of atoms are inherently contextual – that is, irreducibly relative to a measuring apparatus.

Stepping back from its history, we see that the basic ontology of the quantum world is very much undetermined. Thanks to the infamous measurement problem [7,8] we have an extra layer of assumptions that might be called metaphysical – although in another sense these assumptions are simply the ordinary claims of any physical theory. The reason for this extra layer is that one must first solve the measurement problem and then provide the best interpretation of that solution. Experiment cannot yet decide among these theories, and in some cases, never will. Thus the choice of solution is not directly tested in experiment, nor are some of assumptions made by any given solution. The metaphysics of quantum mechanics thus hangs on both a particular solution to the measurement problem and then the best interpretation of that solution. (For measurement problem, see ► Bohmian mechanics; Measurement theory; Modal Interpretation; Objectification; Projection Postulate).

Working in the Schrödinger formalism, the measurement problem arises from the (1) linearity of the equation evolving the wave function, and (2) the claim that the ► wave function or quantum state is representationally complete – that is, that there are properties of kind A in the world if and only if the quantum state is in an eigenstate of the operator $\mathfrak{R}$ believed to represent that property. If linear dynamical evolution of the quantum state is uninterrupted, then the ► superpositions of microscopic states necessary for quantum predictions will evolve into superpositions of macroscopic states. And if the quantum state offers a complete representation of what there is, then the systems described by these macroscopic superpositions do not have any definite measurable properties. Since measurements seem to have determinate outcomes, we appear to have an inconsistency between the theory and experience.

Putative solutions to this problem fall naturally into three classes. The first class consists of theories (sometimes dubbed "hidden variable theories") denying that the quantum state is representationally complete. In addition to the wavefunction evolving according to some linear equation, there are posited what J.S. Bell [7] calls "beables" (as opposed to ► observables) and a dynamics for these beables. Beables are the basic ontology of the theory. In classical electromagnetism, they are the electric and magnetic fields; in Newtonian mechanics, the beables are the particles. In quantum mechanics, typically particle or field ontologies are posited. The ontology is dualistic: interpreted realistically, there are both beables and wavefunctions in the world. The best-known version of this kind of reaction was first discovered by de Broglie but later developed by Bohm [5]. According to this theory, there are in addition to wavefunctions particles with always-determinate trajectories evolving in three-dimensional space, governed by an equation that is a function of the system's wavefunction. Even within a solution in this class one finds varying metaphysical pictures [7, 12]. One can find deterministic and indeterministic Bohm theories, particle and field-based theories, theories that treat ► spin as a beable and ones that do not – even theories that do not treat fermions as beables. Some believe the wavefunction is part of reality, others that it is nomological, and still others treat it instrumentally.

The second class of solutions are unified in their claim that the evolution of the quantum state is not always linear. So-called "collapse" theories state that upon measurement there is an instantaneous ► wavefunction collapse from a superposition to an eigenstate (when the state is expanded in the relevant basis for the observable being measured). Proposals for what triggers this collapse include the "classicality" of the device (some Bohrians – although perhaps not Bohr [10]), non-physical minds (Wigner) [11], and in more recent theories, such as GRW [4] (after Ghirardi, Rimini and Weber), certain thresholds being reached in the system's mass density or particle number.

Again, even within one class of putative solutions, we find a diverse array of possible metaphysical assumptions. In some theories the wavefunction represents an objective part of reality, in others our state of knowledge. Even within a particular solution, say, ► GRW [4], there are a variety of metaphysical pictures available. In one especially radical interpretation of GRW, there is nothing but a sometimes-collapsing wavefunction evolving in 3N-dimensional state space, where N is the number of "constituents" of the system. On this view, 3-dimensional objects like us are aspects of the universal wavefunction that have grown "clumpy" in $3N$-dimensional configuration space. According to the "mass density" theory, there is a continuous distribution of mass throughout spacetime, and the mass density at a point is a function of the wavefunction. Yet according to the "flash ontology" theory, the basic ontology is one of primitive spacetime events that are the loci of GRW collapses [11].

The third class of solutions tries to explain away the mismatch between macroscopic superpositions and experience by neither supplementing the wavefunction description of the world nor interrupting its linear evolution. Originally developed by Everett [6], advocates of the so-called relative-state interpretation claim that our experience supervenes upon macroscopic superpositions in a way that is more complicated than one normally thinks. According to the "many worlds" version, quantum measurements literally split the world into two or more mini-worlds – one corresponding to each possible measurement outcome. The most interesting versions of Everettian theories, however, do not add anything to the wavefunction but instead discover different observers as emergent from complex relations encoded in the wavefunction of the world [13]. It is hardly necessary to say that the metaphysical implications of this view for our conception of ourselves, the external world and probabilities—to name just three topics – are quite dramatic.

Finally, it is worth mentioning that there is a very different group (e.g. [9]), inspired by Bohr that treats quantum mechanics instrumentally. These thinkers consider the wavefunction to be solely an epistemic device that gives observers information about the probabilities of finding various outcomes. Collapse of the wavefunction is viewed as merely the modification of one's subjective credence in light of new information. Because the wavefunction does not represent a genuine state of a real physical system, and these theorists are silent about what the information is information about, the theory offers no physical picture of the world.

In general, no matter the solution to the measurement problem, we expect any non-instrumental version of quantum mechanics to provide answers to various metaphysical questions. Is the wavefunction epistemic or ontological? What is the basic ontology (i.e. beables) of the theory? Do we live in ▶ Hilbert space or four-dimensional spacetime? What is the mechanism responsible for the non-local quantum correlations? What is the interpretation of the probabilities given to us by Born's rule? Do measurements create or reveal the measured properties? Answers to these questions will hang on both the best solution to the measurement problem and the best interpretation of that solution. It is important not to confuse these two issues. For instance, it is commonly said that quantum mechanics implies that atoms don't have determinate trajectories; but strictly speaking, these conclusions follow only from some versions of some interpretations. The original Bohm theory is an empirically adequate (for non-relativistic phenomena) counterexample to this claim, for instance.

The same warning applies to what is one of the most vexed metaphysical questions surrounding quantum mechanics, the question of determinism. (▶ Indeterminism and determinism in quantum mechanics) A physical theory is deterministic if, roughly, given a complete state of the universe at any one time, a unique past and future follow. With suitable assumptions classical mechanics is deterministic. With the advent of quantum mechanics, many of the theory's founders famously declared that determinism was "dead". The Schrödinger evolution of the wavefunction is deterministic; however, the collapse of the wavefunction is stochastic, so the full theory is indeterministic. Quantum mechanics proved, they thought, that "God plays dice". However, as we have just seen, this claim is interpretation-dependent. There are plenty of no-collapse interpretations of quantum mechanics, e.g. Everett, Bohm, and some versions of these are deterministic. The question of whether "God plays dice" is still open.

(See Consistent histories, Ignorance interpretation, Ithaca Interpretation, Many Worlds Interpretation, Modal Interpetation, Orthodox Interpretation, Transactional Interpretation).

Interestingly, the many interpretations of quantum mechanics illustrate why the line between metaphysics and physics is sometimes blurry. Given current technology, there is no way to experimentally decide between, say, a Wignerian collapse theory ("human consciousness causes collapse ► Wigner's friend") and one or more versions of GRW ("reaching a threshold of particle number in the system makes collapse likely"). But in principle these theories do issue different predictions for some observables. In this sense, the metaphysics of today may be the physics of tomorrow. In addition, even before any crucial experiment is performed—and it is not clear that there ever will be such between certain pairs of interpretations—we see that science can have a real bearing on these metaphysical disputes. Scientists value more than good predictions. They also prize simplicity, unification, consilience and other theoretical virtues. Even if there is no test between two given interpretations, there may be good reasons to adopt one over another. One interpretation may possess a symmetry others do not, resolve a problem others cannot, or uniquely extend to a promising new theory (say, some version of ► quantum gravity).

Primary Literature

1. W. Heisenberg: The Physical Content of Quantum Kinematics and Mechanics. Zeitschrift fur Physik **43**, 172 (1927)
2. E. Schrödinger: *Collected Papers on Wave Mechanics* (Chelsea, New York, 1982)
3. N. Bohr: *Atomic Theory and the Description of Nature* (Cambridge University Press, Cambridge, 1934)
4. G.C. Ghirardi, A. Rimini, T. Weber: Unified Dynamics for Microscopic and Macroscopic Systems. Physical Review **D 34**, 470 (1986)
5. D. Bohm: A Suggested Interpretation of the Quantum Theory in Terms of Hidden Variables. I & II.' Physical Review **85**, 166, Physical Review, **85**, 180 (1952)
6. H. Everett, H.: Relative State Formulation of Quantum Mechanics. Review of Modern Physics **29**, 454–462 (1957)

Secondary Literature

7. J.S. Bell: *Speakable and Unspeakable in Quantum Mechanics* (Cambridge University Press, New York, 1987)
8. D.Z. Albert: *Quantum Mechanics and Experience* (Harvard University Press, Cambridge, MA, 1992)

9. J. A. Wheeler: Information, Physics, Quantum: The Search for Links, In *Complexity, Entropy and the Physics of Information* (Addison-Wesley, Redwood City, 1990), edited by W. Zurek.
10. J. Faye: Copenhagen Interpretation of Quantum Mechanics, In *The Stanford Encyclopedia of Philosophy (Summer 2002 Edition)*, edited by E. Zalta, URL = http://plato.stanford.edu/archives/sum2002/entries/qm-copenhagen/
11. G.C. Ghirardi: Collapse Theories, In *The Stanford Encyclopedia of Philosophy (Fall 2007 Edition)*, edited by E. Zalta, URL=<http://plato.stanford.edu/archives/fall2007/entries/qm-collapse/>
12. S. Goldstein: Bohmian Mechanics, In *The Stanford Encyclopedia of Philosophy (Fall 2007 Edition)*, edited by E. Zalta, URL=<http://plato.stanford.edu/archives/fall2007/entries/qm-bohm/>
13. J. Barrett: Everett's Relative-State Formulation of Quantum Mechanics, In *The Stanford Encyclopedia of Philosophy (Spring 2003 Edition)*, edited by E. Zalta, URL=<http://plato.stanford.edu/archives/spr2003/entries/qm-everett/>

Mixed State

Peter Mittelstaedt

The most general state of a proper quantum system S with ▶ Hilbert space $\mathcal{H}_S$ is given by a self-adjoint positive operator with trace 1, i.e. by an operator

$$\mathrm{W_S} = \mathrm{W_S}^{+} \geqslant 0 \text{ with tr } \{\mathrm{W_S}\} = 1.$$

It can be shown that these positive trace class 1 operators form a convex set $\mathcal{T}_1{}^{+}(\mathcal{H}_\mathrm{S})$ [1].

Two kinds of states must be distinguished. If $\mathrm{W_S}$ is idempotent, i.e. $\mathrm{W_S} = \mathrm{W_S}^2$, then $\mathrm{W_S}$ is a *pure state* given by an projection operator $\mathrm{P}[\varphi]$ where $\varphi \in \mathcal{H}_\mathrm{S}$ is an element of $\mathcal{H}_\mathrm{S}$. If, however, $\mathrm{W_S} \neq \mathrm{W_S}^2$, then $\mathrm{W_S}$ describes a *mixed state*. As any self-adjoint operator, a *mixed state* $\mathrm{W_S}$ can be decomposed according to its spectral decomposition

$$\mathrm{W_S} = \sum_i w_i \, \mathrm{P}[\mathcal{M}_i]$$

with real numbers w_i such that $0 \leqslant w_i \leqslant 1$ and projection operators P (▶ projection), which project on subspaces $\mathcal{M}_\mathrm{i}$ of $\mathcal{H}_\mathrm{S}$. It must be emphasised, however, that the decomposition is not uniquely defined, since there are many other, non-orthogonal decompositions of $\mathrm{W_S}$. If, in addition, the operator $\mathrm{W_S}$ has a degenerate spectrum, there are also infinitely many orthogonal decompositions.

There are two kinds of mixed states of S given by an operator $\mathrm{W_S} = \sum_i w_i \, P[\varphi_i]$ with $0 \leqslant w_i \leqslant 1$, which are distinguished by their preparation.

(a) Mixture of states

Assume that a preparation apparatus does not work completely accurately and prepares systems with states $\varphi_1, \varphi_2, \varphi_3 \ldots$, say, with *a priori* probabilities $p_1, p_2, p_3 \ldots$, which depend on the construction of the apparatus. In this case, any single system is actually in one of the states φ_i, which one, however, is not known to the observer who knows only the probabilities. This very special kind of a mixed stated is called a "mixture of states" [2], or "real mixture" [3] or a "Gemenge" [4]. A "Gemenge" $\Gamma_S(p_k, \varphi_k)$ is a classical mixture of states φ_k with weights p_k. Formally, it can be described by the state operator $W_S = \sum_k w_k \, P[\varphi_k]$, since this mixed state operator leads to the same statistical predictions as the "Gemenge" $\Gamma_S(p_k, \varphi_k)$.

(b) ► Mixed state (in general)

Let $S = S_1 + S_2$ be a compound system with ► Hilbert space $\mathcal{H}$ that consists of two subsystems S_1 and S_2 with Hilbert spaces $\mathcal{H}_1$ and $\mathcal{H}_2$, such that $\mathcal{H} = \mathcal{H}_1 \otimes \mathcal{H}_2$ is the tensor product Hilbert space. If S is prepared in a pure state $\Phi(S)$, then the subsystems S_1 and S_2 are in the reduced mixed states $W(S_1) = tr_2 \{P[\Phi(S)]\}$ and $W(S_2) = tr_1\{P[\Phi(S)]\}$, where "$tr_k$" denotes the partial trace with respect to system S_k. To say that the subsystem S_1 is in a mixed state $W(S_1)$ means, that we consider only those properties of the total system S that are concerned with the degrees of freedom of system S_1, neglecting in this way all possible correlations between S_1 and S_2. (► Entanglement). The state $W(S_1)$ is a genuine mixed state except when $\Phi(S)$ is a product state $\Phi = \varphi(S_1) \otimes \psi(S_2)$. In this special situation $W(S_1)$ is the pure state $P[\varphi(S_1)]$. In general, $W(S_1)$ does not admit an "► ignorance interpretation". The mixed state $W(S_1)$ is also called – somewhat misleadingly – "improper mixture" [3]. See also ► density operator; objectification; states in quantum mechanics; states, pure and mixed and their representation.

Literature

1. J. M. Jauch: *Foundations of Quantum Mechanics* (Addison-Wesley, Reading, MA, 1968, 96)
2. P. Mittelstaedt: *The Interpretation of Quantum Mechanics and the Measurement Process* (Cambridge University Press, Cambridge, 1998, 35–37)
3. B. d'Espagnat: *Reality and the Physicist* (Cambridge University Press, Cambridge, 1989, 85)
4. W. Heisenberg: *Physikalische Prinzipien der Quantentheorie* (S. Hirzel, Leipzig, 1930, 43)

Mixing and Oscillations of Particles

Andrzej K. Wróblewski

In 1955 Murray Gell-Mann and Abraham Pais analyzed the behaviour of neutral particles under the operation C of charge conjugation which changes every particle into its anti-particle [1]. According to the proposed scheme of classification of K mesons, the neutral kaon K^0 was assumed to possess an anti-particle $\overline{K}^0$ distinct from itself (at that time these particles were called θ^0 and $\overline{\theta}^0$, respectively).

Gell-Mann and Pais were able to show that in that case the neutral kaon must be considered to be a "particle mixture", exhibiting two distinct lifetimes and different decay modes. The two mesons K^0 and $\overline{K}^0$, are states of definite strangeness $S = +1$ and $S = -1$, and they are produced as such in the strong interactions which conserve strangeness. However, when these neutral particles then propagate through empty space both can decay to pions by the weak interactions, with $|\Delta S| = 1$. Their mixing can occur *via* virtual intermediate pion states, e.g. $K^0 \leftrightarrows 2\pi \leftrightarrows \overline{K}^0$. These are second-order $\Delta S = 2$ weak transitions. In the modern language of quarks and intermediate bosons, the transitions occur between valence quarks, as shown in Fig. 1. (Quarks, see ▶ Color Charge Degree of Freedom in Particle Physics; Particle Physics; Parton Model; QCD; QFT).

At that time it was believed that the particles which decay by the weak interactions were eigenstates of combined parity CP. These eigenstates are quantum mechanical linear superpositions of the K^0 and $\overline{K}^0$,

$$| K_1\rangle = [| K^0\rangle + | \overline{K}^0\rangle]/\sqrt{2} \text{ of } CP = +1,$$
$$| K_2\rangle = [| K^0\rangle - | \overline{K}^0\rangle]/\sqrt{2} \text{ of } CP = -1.$$

Conservation of CP required the K_1 to decay into two pions and the K_2 into three pions. Because of the large difference in available kinetic energy in two-pion and three-pion decays, the K_2 was expected to have much longer lifetime. In essence Gell-Mann and Pais predicted that only half of the neutral kaons underwent the decay into two pions which was well known at that time, while the other half remained undetected. These bold predictions of Gell-Mann and Pais were soon confirmed experimentally. In 1957 Leon Lederman and his group discovered a long-lived neutral kaon decaying into three pions [2]. The mean lifetime of K_2 was about 500 times

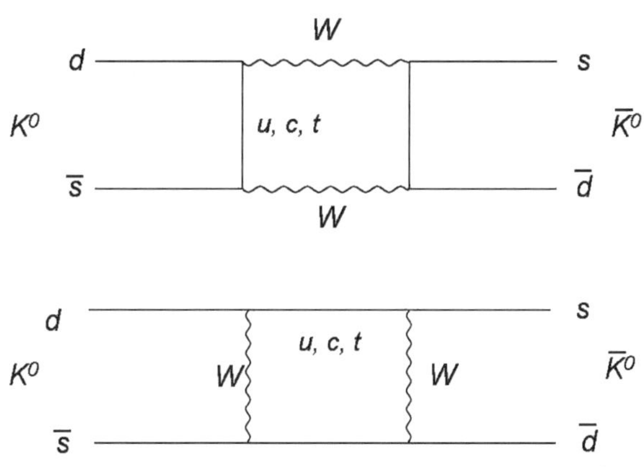

Fig. 1 Feynman diagrams explaining the oscillations between K^0 and $\overline{K}^0$. Similar "*box*" diagrams account for the oscillations of neutral charm mesons $D^0 \leftrightarrows \overline{D}^0$ and neutral bottom mesons $B^0 \leftrightarrows \overline{B}^0$

longer than that of K_1. In the following year the change in time of the nature of the neutral particle produced in association with the Λ^0 hyperon was detected [3]. The particle, initially the K^0 of strangeness $+1$, was observed to interact with matter to produce another hyperon, thus proving to be a strangeness -1 particle. Yet another experimental confirmation of the particle-mixture theory was the observation of regeneration of the short-lived neutral K meson [4].

Thus, an initially pure beam of K^0 will turn into its anti-particle $\overline{K}^0$ while propagating, which will turn back into the original particle, and so on. This is called particle oscillation (strangeness oscillation, or more generally, flavour oscillation). On observing the weak decay into leptons, it was found that a K^0 always decayed into the electron, whereas the anti-particle $\overline{K}^0$ decayed into the positron. Analysis of the time dependence of this semileptonic decay also showed the phenomenon of flavour oscillation and allowed the extraction of the mass splitting between the K_1 and K_2. In 1964 Jim Christenson, James Cronin, Val Fitch, and René Turlay discovered that CP invariance was violated in the decays of long-lived neutral kaons [5]. Thus, the short-lived neutral kaon K_S and the long-lived neutral kaon K_L had to be redefined as $\mid K_S\rangle = [(1+\varepsilon) \mid K^0\rangle + (1-\varepsilon) \mid \overline{K}^0\rangle]/\sqrt{2(1+\varepsilon^2)}$ and $\mid K_L\rangle = [(1+\varepsilon) \mid K^0\rangle - (1-\varepsilon) \mid \overline{K}^0\rangle]/\sqrt{2(1+\varepsilon^2)}$, where ε is a small parameter responsible for CP symmetry breaking.

After the discovery of the charm quark and the bottom quark, physicists have been searching for the flavour oscillations of neutral charm mesons $D^0 \leftrightarrows \overline{D}^0$ and neutral bottom mesons $B^0 \leftrightarrows \overline{B}^0$. The lifetimes of these mesona are of order of a picosecond which makes the experiments much more difficult than those with neutral kaons. The mixing of neutral B mesons was first studied in 1987 and that of neutral D mesons was discovered only in 2007.

The mixing of quarks was first considered by Nicola Cabibbo in 1963 [6]. At that time only three quarks, u, d, and s were known. In order to explain observed differences in branching ratios of semileptonic decays of strange particles Cabibbo proposed that the d and s quarks are mixed and it is the mixture $d\prime = d \cos\theta_C + s \sin\theta_C$ which takes part in the weak interactions. The mixing angle $\theta_C = 12.7°$ is called the Cabibbo angle. Later Makoto Kobayashi and Toshihide Maskawa [7] generalized this idea to the three families of quarks. In the Standard Model (► Quantum field theory, particle physics) the mixing of quarks is described by a 3 × 3 matrix called the CKM matrix after its proponents Cabibbo, Kobayashi and Maskawa. It is written as $\begin{pmatrix} d' \\ s' \\ b' \end{pmatrix} . = V_{\text{CKM}} \begin{pmatrix} d \\ s \\ b \end{pmatrix}$. The elements of the CKM matrix have been determined in a large number of experiments.

In 1957 Bruno Pontecorvo, inspired by the paper of Gell-Mann and Pais [1], pointed out that if lepton number is not absolutely conserved and neutrinos have finite masses, then mixing may occur between neutrino ν and its anti-particle, anti-neutrino $\overline{\nu}$, so that neutrino could be a "mixed" particle [8]. At that time only one neutrino was known. In 1962 Ziro Maki, Masami Nakagawa, and Shoichi Sakata generalized Pontecorvo's idea to the case of three families of leptons [9]. We know now that the neutrino oscillation data can consistently be described within a three-neutrino mixing scheme with massive neutrinos, in which the flavor states $\nu_\alpha(\alpha = e, \mu, \tau)$ are mixed with the mass states $\nu_i(i = 1, 2, 3)$ *via* the unitary 3×3 Pontecorvo-Maki-Nagakawa-Sakata lepton mixing matrix (PMNS matrix). The mass states $\nu_i(i = 1, 2, 3)$ propagate with slightly different frequencies because of their mass differences. If at the start there is a pure ν_e beam, oscillations would occur and at subsequent times one would have admixtures of ν_e with ν_μ and ν_τ (► Particle physics). The oscillations of neutrinos originating from interactions of high energy cosmic ray particles in earth's atmosphere were discovered in 1998 by the Super-Kamiokande Collaboration [10]. Neutrino oscillations also provided the explanation of the deficit of neutrinos coming to earth from the sun as observed in several experiments which were sensitive only to ν_e produced in thermonuclear reactions in the sun's interior. It was experimentally confirmed that in the passage to the earth some of these electron neutrinos changed into muon neutrinos which could be detected by the Solar Neutrino Observatory in Canada [11].

Primary Literature

1. M. Gell-Mann, A. Pais, Behavior of Neutral Particles Under Charge Conjugation, The Physical Review **97**, 1387 (1955)
2. K. Lande, E. T. Booth, J. Impeduglia, L. Lederman, Observation of Long-Lived Neutral V Particles, The Physical Review **103**, 1901 (1957)
3. E. Boldt, D. O. Caldwell, Y. Pal, θ_1^0- θ_2^0 Mass Difference, Physical Review Letters **1**, 150 (1958)
4. R. H. Good, R. P. Matsen, F. Muller, O. Piccioni, W. M. Powell, H. S. White, W. B. Fowler, R, W. Birge, Regeneration of Neutral K Mesons and Their Mass Difference, The Physicial Review **124**, 1223 (1961)
5. J. H. Christenson, J. W. Cronin, V. L. Fitch, R. Turlay, Evidence for the 2π Decay of the K_2^0 Meson, Physical Review Letters **13**, 138 (1964).
6. N. Cabibbo, Unitary Symmetry and Leptonic Decays, Physical Review Letters **10**, 531 (1963)
7. M. Kobayashi, T. Maskawa, CP-Violation in the Renormalizable Theory of Weak Interactions, Progress in Theoretical Physics **49**, 652 (1973)
8. B. Pontecorvo, Mesonium and Anti-Mesonium (in Russian), Soviet Journal Experimental and Theoretical Physics **33**, 549 (1957); Inverse β−Processes and Non-Conservation of Lepton Charge (in Russian), ibidem **34**, 247 (1958)
9. Z. Maki, M. Nakagawa, S. Sakata, Remarks on the Unified Model of Elementary Particles, Progress in Theoretical Physics **28**, 870 (1962).
10. Y. Fukuda et al., (Super-Kamiokande Collaboration), Evidence for Oscillations of Atmospheric Neutrinos, Physical Review Letters **81**, 1562 (1998).
11. Q. R. Ahmad et al., (SNO Collaboration), Direct Evidence for Neutrino Flavor Transformation from Neutral-Current Interactions in the Sudbury Neutrino Observatory, Physical Review Letters **89**, 011301 (2002)

Secondary Literature

12. H. Frauenfelder, E. M. Henley: *Nuclear and Particle Physics* (Benjamin, Reading, Mass. 1975, 507–50)
13. D. H. Perkins, *Introduction to High Energy Physics*, 4th Edition, Cambridge University Press 2000.

Modal Interpretations of Quantum Mechanics

Meir Hemmo

Modal interpretations seek to solve the measurement problem within no collapse quantum mechanics and to account for the nonlocal ▶ correlations in quantum mechanics in EPR- and Bell-type scenarios in a way that might be compatible with special relativity. Various modal interpretations have been proposed, from the mid 1970s onwards, by Van Fraassen [1, 2], Kochen [3], Healey [4], Dieks [5], Bub [6], and others. These versions are quite different from each other. We present below some of their main, and in some cases shared, features.

Consider the scheme of a generic measurement of the z-spin of a spin half particle. Suppose that the composite system, particle plus pointer, is initially prepared at $t = 0$ in the state

$$|\Psi_0\rangle = \Big(\alpha|-_z\rangle + \beta|+_z\rangle\Big) \otimes |\psi_0\rangle, \tag{1}$$

where $|\alpha|^2 + |\beta|^2 = 1$ and we assume that $\alpha \neq \beta$. Here the $|\pm_z\rangle$ are the z-spin eigenstates and $|\psi_0\rangle$ is the ready state of the pointer. Suppose that the interaction correlates, respectively, the $|\pm_z\rangle$ states with the eigenstates $|\psi_\pm\rangle$ of the pointer observable. We assume that the time evolution is described by the ▶ Schrödinger equation alone, i. e. there is no collapse of the state, as modal interpretations require. This means that the interaction maps the initial state at $t = 0$ to the superposition at the final time $t = 1$:

$$|\Psi_1\rangle = \alpha|+_z\rangle \otimes |\psi_+\rangle + \beta|-_z\rangle \otimes |\psi_-\rangle, \tag{2}$$

in which there is a one-to-one correlation between the ▶ spin states $|\pm_z\rangle$ and the pointer states $|\psi_\pm\rangle$. But due to the entanglement in (2) one can only assign reduced states to the particle and to the pointer which are quantum mechanically *mixed*:

$$\begin{aligned} \rho_1 &= \alpha^2|+_z\rangle\langle+_z| + \beta^2|-_z\rangle\langle-_z| \\ \rho_2 &= \alpha^2|\psi_+\rangle\langle\psi_+| + \beta^2|\psi_-\rangle\langle\psi_-|. \end{aligned} \tag{3}$$

This is the scenario in which the measurement problem (or ▶ Schrödinger's cat paradox) arises in standard quantum mechanics. See also ▶ Bohmian mechanics; Measurement theory; Metaphysics in Quantum Mechanics; Objectification; Projection Postulate. On the standard theory, an observable is assigned one of its eigenvalues if and only if the system is in the corresponding eigenstate (this is sometimes called the eigenstate-eigenvalue link). And so, if (2) were the final state after the measurement, the pointer observable (and also the z-spin) would have no definite value, and so the measurement would have no definite outcome. To solve this problem, the so-called ▶ projection postulate or the collapse of the state in measurement is introduced in the standard theory: that is, the state (2) collapses onto one of its components $|+_z\rangle \otimes |\psi_+\rangle$ or $|-_z\rangle \otimes |\psi_-\rangle$ with respective probabilities $|\alpha|^2$ or $|\beta|^2$ as given by the ► Born rule.

Assuming that quantum states don't collapse in measurement, how could we understand the state (2) and the quantum mechanical probabilities for collapses in a way that is consistent with our experience of definite pointer readings? Van Fraassen [1,2] observed that *any* decomposition of the ► mixed state of, say, the pointer in the post-measurement state (2) can be interpreted as describing a set of what he calls possible *value* states of the pointer. The quantum mechanical (Born) probability can then be understood, not as describing the effects of collapses as in the standard theory, but rather as describing our *ignorance* with respect to the actual value state of the pointer when it is in the mixed state (3) generated by (2). By this Van Fraassen in fact rejects the standard interpretation of quantum states via the eigenstate-eigenvalue link (in fact, only its 'only if' direction). On his proposal, the quantum state doesn't fix (with probability one) the value state of the pointer, nor does it completely determine the set of the possible value states. The quantum state has a dynamical role (and is called *dynamical* state) in generating the *probabilities* over the possible value states and in restricting the possible sets of values states (in future interactions). But of course this latter restriction is not enough since a mixed state is not uniquely decomposable as a mixture of pure states (► states, pure and mixed) (with an ignorance interpretation of the probabilities) and moreover not all decompositions of, say the pointer's mixed state can be possible at the same time, on pain of a Kochen-Specker contradiction. And so the question in Van Fraassen's approach is *which* amongst all the *possible* sets of value states allowed by the quantum state correspond to the *actual* circumstances in our world (this is the origin of the term *modal* interpretation.) Van Fraassen's proposes various conditions to this effect in what he calls the *Copenhagen Variant* of the modal interpretation (see [2]).

M

Kochen [3] proposed an interpretation which can be seen as a more restrictive modal interpretation than Van Fraassen's (Kochen doesn't refer to his view as modal). On his proposal the sets of the possible properties of the particle and of the pointer in our example are *determined* by the quantum state (2) as follows. According to the *biorthogonal* decomposition theorem (for proof see [7,8]), the expansion in which the state (2) is written in terms of the *biorthogonal* bases states, $|\pm_z\rangle$ and $|\psi_\pm\rangle$ on the factor spaces, always *exists* and is *unique* whenever the coefficients are not equal. So we can consider the biorthogonal expansion in (2) as depicting *uniquely* the sets of the possible properties (or value states) of the pointer and of the particle together with the quantum mechanical probability distribution over these properties. (Degenerate cases of equal probabilities might be treated as unphysical having 'measure zero'.) And as in Van Fraassen's approach we can interpret the quantum probabilities as reflecting ignorance about the values actually possessed by the particle and the pointer *without* collapsing the state (2). Kochen developed a relational view which is meant to justify the choice of the biorthogonal expansion of (2) as somehow preferred by relying on the symmetry of this expansion. He calls this symmetric relation *witnessing*.

The idea that the biorthogonal expansion of states like (2) has a distinguished physical role in depicting the actual value states of quantum systems has been developed in great detail with an explicit realistic interpretation of quantum mechanics by Healey [4] and Dieks [5] (with interesting insights and differences). In Healey's approach the biorthogonal expansion plays an important role in assigning properties that are in general *holistic* (i.e. properties that are not inherited from the properties of the subsystems; see below) to composite systems. For example, if we add the description of the pointer's interaction with the environment in our measurement scheme above, the final state will be:

$$|\Psi_2\rangle = \alpha|+_z\rangle \otimes |\psi_+\rangle \otimes |E_+\rangle + \beta|-_z\rangle \otimes |\psi_-\rangle \otimes |E_-\rangle, \qquad (4)$$

where the environment states $|E_\pm\rangle$ relative to the pointer states $|\psi_\pm\rangle$ become very quickly approximately orthogonal for almost any initial state of the environment (this is one feature of environmental ► decoherence, see [9]). And now Healey assigns properties via Kochen's prescription to any bi-partition of the three subsystems, e. g. particle + pointer and environment, pointer and particle + environment, etc., where the holistic properties of composite systems are assigned to the composites *independently* of the properties of the subsystems that make them up. For example, in the state (4) the composite properties of, say the particle + pointer turn out to be *close* (in inner product) to the products of the properties of the particle and of the pointer alone (this is due to the decoherence of the pointer), whereas the property of the total system particle+pointer+environment which is just their quantum state isn't even nearly a product property. In Healey's approach such properties play an essential role in accounting for EPR- and Bell-type ► nonlocality.

Vermaas and Dieks [10] generalised Kochen's prescription by adopting a rule that prefers the *spectral* (or diagonal) decomposition of the reduced density operators corresponding to quantum mechanically mixed states. The spectral resolution of a ► density operator always exists and is unique by the spectral theorem (because density operators are self-adjoint; see any textbook on functional analysis). And this means that every system can be assigned value states directly via its quantum state, so that one need not rely on the quite restrictive bi-partition form of the biorthogonal expansion. And moreover, Kochen's prescription turns out to be a special case of the spectral theorem for a composite of two systems in a pure state. This can be seen in our example above, where the reduced states in (3) of the pointer and of the particle are already written in their spectral form. But the Vermaas–Dieks prescription can be applied also in the triple case above in state (4) in order to assign properties directly to the three subsystems. Under certain idealized assumptions about the interaction with the environment, the properties assigned to the pointer via the spectral resolution of its reduced state will be close to the pointer states in (3) in correspondence with our experience. And in this sense the Vermaas–Dieks prescription (as well as Kochen's) turns out to be empirically adequate. But again, this is not the most general case (see below).

In standard quantum mechanics spectral and biorthogonal decompositions don't seem to have the special role assigned to them in modal approaches (as 'markers' of properties). And so it is natural to ask in this context what is special from a physical point of view about these choices. Of course, as we just mentioned, the question might turn out to bear on empirical considerations, and indeed we shall come back to it shortly. (For various attempts to justify these rules, see Bacciagaluppi [8], Dieks and Vermaas [11], Healey and Hellman [12], Bub [6], Vermaas [13] Ruetsche [14] and Bub and Clifton [15], and references therein.) But it is important to note that what characterises modal approaches is not some particular choice of value states but rather that some such choice is made (sometimes under certain conditions), and that the quantum mechanical probability distribution has nothing to do with collapses but rather expresses ignorance about the actual value states. Indeed, there are other modal approaches with entirely different ways of defining the value states. For example, in Bub's approach [6] the value states are assigned only to macroscopic systems that interact with their environment, and they correspond to ► observables that commute with the decoherence Hamiltonian. In our example above, this means that only the pointer is assigned extra value states, and these will be, by construction, the $|\psi_{\pm}\rangle$ (since the pointer observable commutes with the decoherence Hamiltonian). This idea has been also developed by Hemmo [16,17] and applied to the ► consistent histories approach. In yet other versions the value states are selected by entropy minimisation (Spekkens and Sipe [18]), or in various relational ways (Bene and Dieks [19], Berkovitz and Hemmo [20]).

It is clear that modal interpretations solve the measurement problem for ideal measurements which have final states like (2), since for example, the reduced state of the pointer (taken by partial tracing) is diagonal in the pointer basis as can be seen from (3). However, the measurement problem immediately re-appears if we relax idealizations and allow for imperfect correlations and disturbances in the measurement interaction. It has been noticed by Albert and Loewer [21] that for nearly degenerate initial states (e. g. states in which α and β in (1) are almost equal) slight imperfections in the measurement are enough to make the final state of the pointer *not* even nearly diagonal in the pointer basis. And this just means in modal interpretations that the measurement has *no* determinate pointer readings. Bacciagaluppi and Hemmo [22] showed that the problem might be avoided if one takes into account the decoherence interaction of the pointer with the environment as in (4), but, again, only under certain idealizations, this time on the decoherence interaction. It has been shown by Bacciagaluppi [23] that in continuous models of decoherence (with position being the pointer observable) it is the continuous nature of the interaction with the environment itself which seems to result in extreme near ► degeneracy. And under these circumstances the modal recipe seems to break down, since it picks out *delocalised* ► wave functions for the pointer. Obviously, this result strongly undermines modal interpretations in the versions sketched above. For more details on this problem, see Bacciagaluppi [8], Hemmo [16], Bub [6], Dieks and Vermaas [11], Healey and Hellman [12] and Vermaas [13] and references therein. Similar problems arise in the attempts to generalise these versions to quantum field theory (see Dieks [24], Butterfield and Halvorson [25] and for criticism Earman and Reutche [26]). Other versions of the modal interpretation, for example, versions relying on decoherence (Bub [6], Hemmo [17]) and the relational versions (Bene and Dieks [19], Berkovitz and Hemmo [20]) are unaffected by this problem.

Another consequence of modal interpretations is that composite systems do not inherit their properties from their subsystems (this is sometimes called failure of property *composition*). Although, as we said, for macroscopic systems in decoherence situations (as in (4)) property composition can be recovered, in general the properties assigned to a composite system are not products of the properties of its subsystems, in fact they do not generally have the form of product properties at all. It has been shown by Bacciagaluppi [27] and Clifton [28] that the introduction of property composition (together with the fact that the ► Hilbert space of composite systems can be factorised into factor spaces in many different ways) leads to a Kochen-Specker contradiction. Therefore, properties in different factorisations cannot be pasted together (see also Butterfield and Halvorson [25]). This problem prompted the so-called *atomic* modal interpretation (Bacciagaluppi and Dickson [29]) in which the above rules are applied only to a class of fundamental atomic systems, whereas composites of atomic systems inherit their properties from their subsystems by composition.

We saw up to now that in modal interpretations the *complete* physical state of a system is given by a *pair* of states at each time: the generally mixed quantum state and the actual value state of the system. The time evolution of the quantum state of a system is fixed deterministically by the Schrödinger evolution of the state of the total system. And this evolution is supposed to generate an ignorance probability distribution over the value states at all times. But, there seems to be no connection between the evolution of the quantum state of the system and the value states that actually obtain at a time. The problem arises already in our simple example above. The particle has a spin $+$ value in some direction at $t = 0$ in state (1), and by the modal recipe, it has a $+_z$ or $-_z$ value at $t = 1$ in state (2). We know that state (1) evolves to state (2) by the Schrödinger equation. But this evolution doesn't explain what is it that brings about the $+_z$ or $-_z$ value at $t = 1$. In standard quantum mechanics the connection between the quantum state and the outcomes we observe is made by the collapse postulate and the Born rule. But here we don't know whether and how the value state at $t = 1$ depends on the value state at $t = 0$. It seems that in the modal recipe some connection of this sort is missing. And obviously the fact that the probability distribution over the value states is given by the quantum probabilities is in equal need of explanation: given that the probabilities reflect ignorance, why are they distributed in accordance with Born's rule?

Following Bell's [30] stochastic dynamics for hidden variable theories, Bacciagaluppi and Dickson [29] proposed a class of general dynamics for the value states that answers these questions. According to their proposal modal interpretations are in fact hidden variable theories where the dynamics of the value states is in general *stochastic*, and it yields the quantum probability distribution over the value states at any given time, just as desired. Bacciagaluppi, Donald and Vermaas [31] have further shown that the evolution of value states can be naturally defined to follow a *continuous* path in Hilbert space. These two results have more or less solved the problem of dynamics for modal interpretations in the spectral resolution versions. An alternative view which relies on sets of decoherent histories and their probabilities has been proposed by Hemmo [17]. An explicitly nonlocal dynamics which depends on the measure of ► entanglement of the state of a system has been proposed by Berkovitz and Hemmo [20] in the context of relativity theory.

Modal interpretations reproduce the quantum mechanical correlations in EPR and Bell-type experiments, and so they are nonlocal in Bell's sense, just like standard quantum mechanics. But are they consistent with relativity theory, in the sense that they satisfy relativistic (Lorentz) invariance? In this context no-go theorems have been proved by Dickson and Clifton [32], Arntzenius [33] and Myrvold [34] given some locality conditions on the dynamics of the properties and certain meshing conditions on their assignment by all Lorentz frames. Dickson and Clifton require local properties of spacelike separated systems in Bell-type situations to evolve under local dynamical laws. If no measurements are carried out, a condition they call *stability* requires the properties to evolve deterministically. If measurements are carried out the local transition probabilities are determined by the local reduced state of each system, such that all Lorentz frames agree on the local transition probabilities (this is called invariant transition probabilities). They show that modal interpretations with such local dynamics are committed to Bell-type inequalities, and therefore cannot reproduce the quantum mechanical predictions.

Myrvold arrived at a similar result by considering four intersecting hyperplanes in Minkowsky spacetime, also in a Bell-type situation. The joint probabilities of the properties of spatially separated systems at the regions of intersection of the hyperplanes are just the Born probabilities, as determined by the quantum state on each hyperplane. Myrvold then argues that relativistic invariance requires that these joint probabilities be mutually consistent. And he shows, on the assumption that the dynamics of the properties satisfies a certain locality condition (roughly, that local properties remain invariant under transformations that leave the reduced state of the system unchanged), that this is impossible for some quantum states and evolutions. This is again, a Bell-type scenario: given a locality condition (on the dynamics), there is no joint probability distribution over the properties, which yields as marginals the quantum mechanical predictions on all hyperplanes.

The dynamics by Bacciagaluppi and Dickson [29] is local in the above sense, and therefore seems to be ruled out by the no-go theorems. Berokovitz and Hemmo [20] proposed a nonlocal dynamics which gets around the no-go theorems, but in which the value states and the transition probabilities turn out to be hyperplane-dependent. Versions in which properties are assigned to systems only under certain decoherence conditions also seem to get around these theorems (see e. g. Dieks [35]). But the crucial and persisting and still open question is whether these or other modal interpretations can be extended to a genuine relativistic theory.

Literature

1. B. Van Fraassen: *Quantum Mechanics: An Empiricist View* (Clarendon Press, Oxford 1991, Chap. 9)
2. B. Van Fraassen: *The Einstein-Podolsky-Rosen Paradox*. Synthese **29**, 291–309 (1974)

3. S. Kochen: *A New Interpretation of Quantum Mechanics*. In Symposium on the Foundations of Modern Physics, ed. by P. Lahti, and P. Mittelstaedt (World Scientific, Singapore 1985, 151–69)
4. R. Healey. *The Philosophy of Quantum Mechanics: An Interactive Interpretation* (Cambridge University Press, Cambridge 1989)
5. D. Dieks: *Resolution of the Measurement Problem Through Decoherence of the Quantum State*. Phys. Lett. A **142**, 439–46 (1989)
6. J. Bub: *Interpreting the Quantum World* (Cambridge University Press, Cambridge 1997)
7. E. Schrödinger: *Discussion of Probability Relations Between Separated Systems*. Proc. Cam. Philol. Soc. **31**, 555–563 (1936)
8. G. Bacciagaluppi: *Topics in the Modal Interpretation of Quantum Mechanics* (Cambridge University, PhD thesis, 1996)
9. E. Joos, H.D. Zeh, C. Kiefer, D. Giulini, J. Kupsch, I.-O. Stamatescu: *Decoherence and the Appearance of a Classical World in Quantum Theory* (Springer, Berlin 2003)
10. P. Vermaas, D. Dieks: *The Modal Interpretation of Quantum Mechanics and Its Generalization to Density Operators*. Found. Phys. **25**, 145–58 (1995)
11. D. Dieks, P. Vermaas (eds.): *The Modal Interpretation of Quantum Mechanics* (Kluwer, Dordrecht 1998)
12. R. Healey, G. Hellman (eds.): *Quantum Measurement, Decoherence and Modal Interpretations: Minnesota Studies in the Philosophy of Science* **17** (University of Minnesota Press, Minneapolis 1998)
13. P. Vermaas: *A Philosopher's Look at Quantum Mechanics: Possibilities and Impossibilities of a Modal Interpretation* (Cambridge University Press, Cambridge 1999)
14. L. Ruetsche: *On the Verge of Collapse: Modal Interpretation of Quantum Mechanics* (University of Pittsburgh, PhD thesis, 1995)
15. J. Bub, R. Clifton: *A Uniqueness Theorem for 'No Collapse' Interpretations of Quantum Mechanics*. Stud. Hist. Philos. Mod. Phys. **27**, 181–219 (1996)
16. M. Hemmo: *Quantum Mechanics Without Collapse: Modal Interpretations, Histories and Many Worlds* (Cambridge University, PhD thesis, 1996)
17. M. Hemmo: *Quantum Histories in the Modal Interpretation*. In Dieks and Vermaas [11], 253–77 (1998)
18. R. Spekkens, J. Sipe: *A Modal Interpretation of Quantum Mechanics Based on a Principle of Entropy Minimization*. Found. Phys. **31**, 1431–64 (2001)
19. G. Bene, D. Dieks: *A Perspectival Version of the Modal Interpretation of Quantum Mechanics and the Origin of Macroscopic Behavior*. Found. Phys. **32**, 645–71 (2002)
20. J. Berkovitz, M. Hemmo: *Can Modal Interpretations of Quantum Mechanics Be Reconciled with Relativity*. Philos. Sci. **72**, 789–801 (2005)
21. D. Albert, B. Loewer: *Wanted Dead Or Alive: Two Attempts to Solve Schrödinger's Paradox*. In A Fine, M. Forbes, L. Wessels (eds.), PSA 1990, vol 1, 277–85 (East Lansing, Michigan, 1990)
22. G. Bacciagaluppi, M. Hemmo: *Modal Interpretations, Decoherence and Measurements*. Stud. Hist. Philos. Mod. Phys. **27**, 239–77 (1996)
23. G. Bacciagaluppi: *Delocalised Properties in the Modal Interpreration of a Continuous Model of Decoherence*. Found. Phys. **30**, 1431–1444 (2000)
24. D. Dieks: *Events and Covariance in the Interpretation of Quantum Field Theory*. In Ontological Aspects of Quantum Field Theory, ed. by M. Kuhlmann, H. Lyre and A. Wayne (World Scientific, Singapore 2002, Chap. 11)
25. J. Butterfield, H. Halvorson (eds.): *Quantum Entanglement: Selected Papers, by Rob Clifton* (Clarendon Press, Oxford 2004)
26. J. Earman and L. Ruetsche: *Relativistic Invariance and Nodal Interpretations*. Philos. Sci. **72** (4), 557–83 (2005)
27. G. Bacciagaluppi: *A Kochen-Specker Theorem in the Modal Interpretation of Quantum Mechanics*. Int. J. Theor. Phys. **34**, 1205–16 (1995)
28. R. Clifton: *Why Modal Interpretations of Quantum Mechanics Must Abandon Classical Reasoning About the Values of Observables*. Int. J. Theor. Phys. **34**, 1302–12 (1995)

29. G. Bacciagaluppi, M. Dickson: *Dynamics for Modal Interpretations*. Found. Phys. **29**, 1165–201 (1999)
30. J. S. Bell: *Beables for Quantum Field Theory. Speakable and Unspeakable in Quantum Mechanics* (Cambridge University Press, Cambridge 1987, pp. 173–80)
31. G. Bacciagaluppi, M. Donald, P. Vermaas: *Continuity and Discontinuity of Definite Properties in the Modal Interpretation*. Helv. Phys. Acta **68**, 679–704 (1995)
32. M. Dickson, R. Clifton: *Lorentz-Invariance in Modal Interpretations*. In Dieks and Vermaas [11] 9–48 (1998)
33. F. Arntzenius: Curiouser and Curiouser: *A Personal Evaluation of Modal Interpretations*. In Dieks and Vermaas [11], 337–77 (1998)
34. W. Myrvold: *Modal Interpretations and Relativity*. Found. Phys. **32**, 1773–84 (2002)
35. D. Dieks: *Consistent Histories and Relativistic Invariance in the Modal Interpretation of Quantum Mechanics*, Phys. Lett. A **265**, 317–325 (2000)

N

Neutron Interferometry

Helmut Rauch

Neutrons are elementary massive particles consisting of one "up" and two "down" *quarks*; but in neutron interference experiments they exhibit wave features only. In this case, the ▶ wave function describing thermal neutrons can be split, reflected and superposed coherently by means of dynamical Bragg diffraction from a perfect silicon single crystal. The coherent beam parts are widely separated, and they can be influenced individually by nuclear, magnetic or gravitational interaction. This technique has first been tested 1974 at a small 250 kW TRIGA reactor in Vienna [1]. The monolithic design of such interferometers guarantees the parallelism of the reflecting lattice planes up to a fraction of their lattice distance, which is a necessary condition for coherent beam splitting. This experimental method has been adapted from X-ray interferometry developed earlier [2]. The figure shows various types of such interferometers as they are used now at several neutron sources around the world.

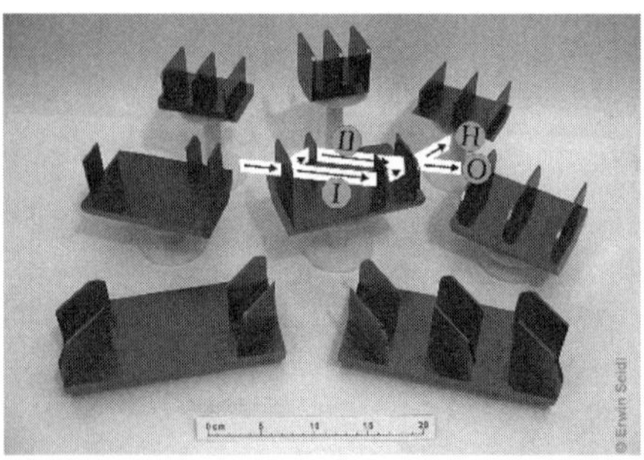

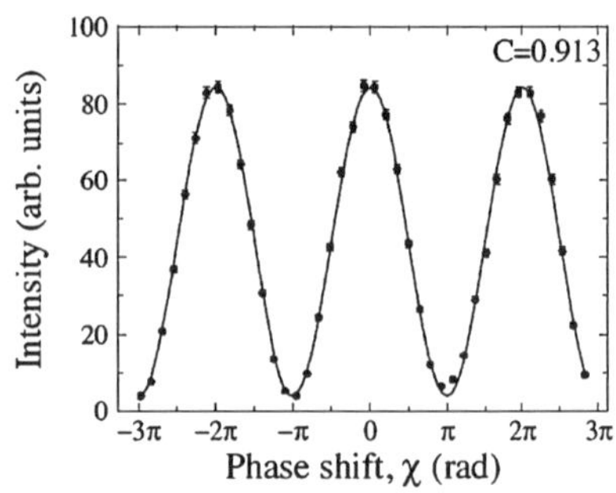

A well balanced and insulated interferometer can provide interference fringes with a contrast higher than 90% (see figure). The intensity modulation due to relative phase shifts between the coherent beams can be achieved by any material or magnetic or gravitational field. The related interaction for neutrons with wavelength λ can be described by an index of refraction n which causes a phase shift $\chi = (1-n)\,k\,D = -Nb_c\lambda D$ where $k = 2\pi/\lambda$ denotes the k-number, N the particle density, b_c the coherent scattering length and D the thickness of the material

$$I \propto \left|\psi^{\mathrm{I}} + \psi^{\mathrm{II}}\right|^2 \propto 1 + \cos\chi$$

D. Greenberger et al. (eds.), *Compendium of Quantum Physics: Concepts, Experiments, History and Philosophy*, © Springer-Verlag Berlin Heidelberg 2009

Neutron interferometry always takes place in the regime of self interference since the phase space density of existing neutron sources is rather small, with the consequence that during a certain time interval there is only one neutron within the interferometer while the following one is still in a uranium nucleus of the reactor fuel.

The main scientific achievements during the years of applying and developing this neutron interferometric technique were:

- The verification of the 4π-symmetry of spinor wavefunctions (▶ *Berry's Phase*)
- The observation of the Earth gravitational and rotational effect
- The observation of coherent spin superposition
- The observation of the neutron Fizeau effect
- The observation of the magnetic Josephson effect
- The observation of the topological ▶ *Aharonov–Casher* and the scalar ▶ Aharonov–Bohm effect
- The observation of single and multiple photon exchange within time-dependent magnetic fields
- The experimental separation of the geometric and dynamical phases

A detailed description of these experiments can be found in the book "Neutron Interferometry", [3].

More recently, *quantum contextuality* could be verified which implies an entanglement of external (beam path) and internal (▶ spin) degrees of freedom for a single particle system. In this connection, the ▶ *Kochen–Specker Theorem* has been tested indicating that a measurement of commuting ▶ observables depends on the order in which the measurements have been done [4]. Several recent investigations have also dealt with *non-adiabatic and non-cyclic phases* and they show that nowadays the complete quantum state can be measured. Neutron phase tomography has been developed as well, providing a kind of non-interaction imaging technique. Broad interest have found investigations directed towards *decoherencing and dephasing* effects (▶ decoherence) since the separated beams can be exposed to various fluctuating conditions (magnetic noise fields, etc.). The transition from a pure to a ▶ mixed state and several state retrieval methods have also been investigated. The sensitivity against fluctuating and dissipative forces of coherent and non-classical ▶ Schrödinger cat-like states is an important topic in order to understand how a classical world emerges from the quantum mechanical properties of nature.

Perfect crystal neutron interferometers can be seen as relatively robust macroscopic quantum devices since the whole system operates under ordinary atmospheric conditions and environmental effects have to become rather strong to destroy the typical quantum behaviour. Neutron interferometry can be considered as a pioneering step preparing the path towards interferometry and quantum optics with even heavier particles like *atoms, molecules, fullerenes, etc.* (▶ Mesoscopic quantum phenomena). Nowadays neutron interferometry has been established as a laboratory tool for basic quantum phenomena.

Literature

1. H. Rauch, W. Treimer, U. Bonse: Test of a single-crystal neutron interferometer. Phys. Lett. A47 (1974) 369–371
2. U. Bonse, M. Hart: An X-ray interferometer. Appl. Phys. Lett. 6 (1965) 155–156
3. H. Rauch, S.A. Werner: *Neutron Interferometry*, Clarendon Press, Oxford 2000
4. Y. Hasegawa, R. Loidl, G. Badurek, S. Filipp, J. Klepp, H. Rauch: Evidence for entanglement and full tomographic analysis of Bell states in a single-neutron system. Phys. Rev. A76 (2007) 052108

No-Cloning Theorem

Stefan Weigert

There is no quantum-mechanical device which outputs a *perfect* copy of an *arbitrary* pure quantum state $|\psi\rangle$ while leaving the original intact. Such an apparatus would be described by a unitary operator $\hat{U}$ acting as

$$\hat{U}|\psi\rangle \otimes |0\rangle = |\psi\rangle \otimes |\psi\rangle \, ,$$

where $|0\rangle$ is a fixed 'blank' input state. However, due to the linearity of the operator $\hat{U}$ this equation is consistent only if the input states $|\psi\rangle$ are pairwise orthogonal. A contradiction arises if one requires that the device work correctly for non-orthogonal states as well. It is also impossible to duplicate (or *broadcast*) non-commuting *mixed* states.

Two proofs of the No-Cloning theorem [1, 2] have been published in 1982, both triggered by a claim that the use of entangled states (► entanglement) would allow one to transmit information with supraluminal speed. However, the proposed scheme cannot be implemented since it relies on the perfect cloning of quantum states. Considering the elementary nature of its proof, the No-Cloning theorem and its generalization to mixed states [3] have been discovered surprisingly late.

The No-Cloning theorem captures a fundamental aspect of the structure of quantum mechanics. Its limiting character plays an important role in the theory of quantum information. For example, the theorem forbids to copy the information carried by a state $|\psi\rangle$ at the end of a ► quantum computation. Thus, although desirable, no safety copies of the result embodied in the state $|\psi\rangle$ can be made, it cannot be distributed to other parties or multiplied for ► quantum state reconstruction. At the same time, the security of quantum cryptography (► quantum communication) relies on the No-Cloning theorem: if two parties establish a secret key by exchanging quantum states through a quantum channel, eavesdroppers are not able to reliably copy the states unknown to them. The theorem is consistent with quantum teleportation (► quantum communication) since the unknown input state is destroyed irretrievably once the process has been completed.

Quantum cloning machines have been devised to produce one or more *approximate* copies of an unknown quantum state [4]. To achieve optimal cloning the devices take into account the number N of identically prepared (unknown) input states, the number M of desired output copies, whether pure or mixed states are to be duplicated, and whether the cloner is required to work for arbitrary input states, i.e. *universally*, or for a limited set of input states only. Optimal cloning machines are conceptually linked to ► quantum state reconstruction and the impossibility to use quantum correlations (► correlations in quantum mechanics) for signaling.

Primary Literature

1. W. K. Wootters and W. H. Zurek: A single quantum cannot be cloned. Nature 299, 802–803 (1982)
2. D. Dieks: Communication by EPR devices. Phys. Lett. A 92, 271 (1982) 271–272
3. H. Barnum, C. Caves, C. Fuchs, R. Jozsa, and B. Schumacher: Noncommuting mixed states cannot be broadcast. Phys. Rev. Lett. 76, 2818–2821 (1996)
4. V. Bužek and M. Hillery: Quantum copying: Beyond the no-cloning theorem. Phys. Rev. A 54, 1844–1852 (1996)

Secondary Literature

5. A. Peres: How the no-cloning theorem got its name. Fortschr. Phys. 51, 458–461 (2003)
6. N. D. Mermin: *Quantum Computer Science*. (Cambridge University Press, Cambridge 2007, 39–40)
7. V. Scarani, S. Iblisdir, N. Gisin, and A Acín: Quantum cloning. Rev. Mod. Phys. 77, 1225–1256 (2005)

Nonlocality

Henry Stapp

Nonlocality: In quantum mechanics the term "nonlocality" refers to an apparent failure of a certain relativity-theory-based ► locality assumption. This assumption is that no information about which experiment is freely chosen and performed in one space-time region can be present in a second space-time region unless a point traveling at the speed of light (or less) can reach the second region from the first. This assumption is valid in relativistic *classical* physics. Yet quantum theory permits the existence of certain experiments in which this locality assumption seems to fail. Einstein called the faster-than-light effect evidently entailed by conventional (Copenhagen) quantum theory "spooky action at a distance". (For Copenhagen interpretation, see ► Born rule; Consistent Histories; Metaphysics in Quantum Mechanics; Orthodox Interpretation; Schrödinger's Cat; Transactional Interpretation.)

The simplest of the experiments pertinent to this issue involve two measurements performed in two space-time regions that lie so far apart that nothing traveling at the speed of light or less can pass from either of these two regions to the other. The experimental arrangements are such that an experimenter in each region – or perhaps some device that he has set up – is able to choose between two alternative possible measurements. The locality assumption then demands, for each region, that the truth of statements exclusively about the outcomes of the possible measurements performed in that region be independent of which experiment is "freely chosen" in the other (faraway) region.

The first actual experiment exhibiting these features was carried out by Aspect, Grangier, and Roger [1] ► Aspect Experiment. Dozens of other such experiments have been carried out since, and the validity of the quantum predictions appears to be borne out.

The significance of this nonlocality property of quantum theory is clouded by several considerations. The first is that although the conventional quantum precepts do appear to entail the need for some sort of *sub rose*, behind-the-scenes, faster-than-light transfer of information (► Einstein Locality), this effect cannot be used to send a superluminal *signal*: no one can use this effect to transfer, superluminally, information that he or she possesses to a faraway colleague ► superluminal communication. This limitation on *signal* velocity, together with other relativistic features of the actually verifiable predictions of the theory, allows relativistic quantum field theory to be called "relativistic" in spite of the apparently entailed faster-than-light transfer of information.

It might seem contradictory to assert first that locality fails, and hence that information about which experiment is freely chosen and performed in a first region is present in a second region, yet then to assert that the experimenter in the first region cannot use this feature to send information to a colleague in the second region. The resolution of the puzzle is that the dependence of faraway measurable properties on the choice made by the nearby experimenter arises only via nature's choice of the *outcome* of the nearby experiment. The faraway colleague, lacking all knowledge about which outcome occurs in the sender's region, must treat that outcome as unknown. This leads to a quantum theoretical averaging over these outcomes that exactly eliminates all dependence upon the sender's free choice of anything that the receiving colleague can observe.

A second clouding consideration is this: in order to analyze the consequences of the non-dependence of some property upon a free choice one must consider, theoretically, or logically, within one argument, the consequences of various alternative choices. But, in the cases of interest, only one of the alternative possibilities can actually occur in any one existing empirical/experimental situation. Thus the argument needed to demonstrate the existence of faster-than-light transfer of information requires some sort of counterfactual reasoning that involves considering in one argument the predictions about outcomes of several experiments that cannot all be actually performed.

A logical opening to counterfactual argumentation is provided by the precepts of quantum theory themselves. Bohr often emphasized the freedom of experimenters to choose which experiment is actually performed. This *freedom to choose* is important in quantum theory for the following reason: the quantum state (▶ wave function) of a physical system provides the basis for predictions about outcomes of *whichever* experiment is freely chosen and performed: predictions for various alternative possible choices are given by the theory, even though only one of the alternatives can be realized physically. On the other hand, the structure of the quantum mathematics entails that the *outcomes* of certain pairs of measurements, between which the experimenter is considered free to choose, cannot be simultaneously represented within this mathematics. This theoretical limitation upon the theoretically representable outcomes is reconciled with the claim of the pragmatic or epistemological completeness of quantum theory by noting that whenever the outcomes of the two measurements cannot be theoretically represented simultaneously then the two experiments also cannot be physically performed simultaneously. Hence the theoretical and physical limitations match, and completeness can be claimed.

The validity of this way of arguing for the completeness of the theory was brought into question by a 1935 paper by Einstein, Podolsky, and Rosen ▶ EPR Problem. Because these authors were endeavoring to prove an internal inconsistency of the quantum precepts, they were careful not to *assume* that, contrary to the precepts of quantum theory, the outcomes of mutually incompatible measurements were simultaneously well defined. On the contrary, they used the quantum prohibition on well defined values of mutually incompatible properties to deduce that they could influence by their nearby choice which of two faraway mutually incompatible properties was real. Thus what they actually proved was that Copenhagen precepts entailed the existence of faster-than-light transfer of information, though not faster-than-light signaling.

N

In 1964 John Bell published a follow-up to the 1935 paper of Einstein et al. Because it was, specifically, the Copenhagen prohibition against well defined values for the outcomes of mutually incompatible measurements that allowed Einstein *et al.* to deduce the need for faster-than-light transfer of information, Bell [2] inquired whether dropping that Copenhagen precept could extinguish the need for faster-than-light information transfer. Bell forthrightly contravened the Copenhagen ban on determinate outcomes of mutually incompatible measurements by introducing "deterministic hidden variables". These ▶ hidden variables specify, simultaneously, the outcomes of *all* of the alternative possible experiments under consideration. Bell then showed [▶ Bell's Theorem] that, within this deterministic hidden variable structure, one cannot reconcile the validity of the predictions of quantum theory (in these experiments) with the locality assumption that the outcomes in each region be independent of which experiment is performed in the other (faraway) region.

The hidden-variable machinery introduced by Bell is actually superfluous: all that is really needed is the assumption that in any given empirical *instance*, prior to the independent choices made by the experimenters in the two far-apart region, any one of the allowed pairs of choices *could* occur, and that for each such pair of choice (of which pair of measurement is performed) some long sequence of N pairs of numbers represent outcomes that *could* occur in the pair of regions if N repetitions of the selected pair of measurements were performed. The existence of such sequences of pairs of numbers specifying possible outcomes follows from Bell's hidden-variable machinery. But they refer only to *performable actions* and *observable outcomes*. Thus they can be stated without bringing in any notions of "microscopic", "invisible", or other "hidden" variables. The assumption that such a set of pairs of numbers specifying outcomes exists is called "counterfactual definiteness". This assumption cannot be consistently reconciled with the assumed validity of the predictions of quantum theory for each of the measurement possibilities available to the experimenters, if one demands also that outcomes in each region be independent of which experiment is chosen and performed in the faraway region [3].

Bell [4] and others [5] went on to consider, instead of *deterministic* local hidden-variable theories, rather *probabilistic* local hidden variable theories. But, as shown by Stapp [6], and independently by Fine [7], this change does not substantially change the situation, because the two detailed formulations are, from a logical point of view, essentially equivalent.

The locality assumption fails, therefore, under either of these two opposing conditions on outcomes: either the Copenhagen prohibition of well defined values of outcomes of mutually incompatible measurements, or the counterfactual definiteness assumption that for each of the two times two, or four, possible combinations of measurements available to the experimenters, some set numbers represents outcomes that could occur if that pair of measurement were to be selected by the experimenters.

In both of these two cases some special conditions pertaining to outcomes are imposed.

The question thus naturally arises whether locality fails also under the weaker assumptions that, for some selected experimental situation, the predictions of quantum theory are valid and the two choices (one made in each of two very far apart regions, and determining which measurement will be performed in that region) can be treated as two independent free variables.

The answer is affirmative! Under experimental conditions described by Hardy [8] there are again two far apart experimental space-time regions, labeled R and L, and in each region an experimenter chooses between a first or second possible measurement and he observes and records there whether the first or second possible outcome of the *single measurement* that he performs occurs. In some specific frame of reference the space-time region L will be earlier than the space-time region R. Quantum theory makes four pertinent predictions. The first two prediction combine with the locality condition that "the outcome observed and recorded in the earlier space-time region does not depend upon which measurement is chosen and performed later" to prove, under the condition that the *first* of the two alternative possible measurements is chosen in the earlier region, the truth of the following statement [9]:

SR: If performing the first measurement in the later region gives the first of the two possible outcomes, then performing, instead, the second measurement would (necessarily) give the first of the two possible outcomes of that second experiment.

Under the condition that the first measurement is performed in the earlier region, the first two predictions of quantum theory in the Hardy case are:

1. If the first measurement is performed in the later region and the first possible outcome appears there, then the first possible outcome must have appeared in the earlier region.
2. If the second measurement is performed in the later region and the first possible outcome appeared in the earlier region, then the first possible outcome must appear in the later region.

Notice that the first of these two predictions is analogous in form to the predictions used by Einstein, Podolsky, and Rosen, in their argument, except that here the possible outcomes are just two in number, rather that a continuum. But the second prediction, which is again a prediction with certainty (probability unity), in the idealized limit that is being considered here, pertains to the case in which the pairing of measurements in the two regions is different from what it was for the first prediction. This crossing of the pairings creates a potent new logical situation.

Combining these two predictions with the assumption that changing the choice of which experiment was performed in the later region cannot affect what already happened earlier in the faraway region entails the truth of SR.

The second two predictions hold under the condition that the second measurement is performed in the earlier region. They are:

3. If the first possible outcome appears in the earlier region and the first measurement is performed in the later region, then the first possible outcome will appear in the later region
4. If the first possible outcome appears in the earlier region and the second measurement is performed in the later region, and then the second possible outcome will sometimes occur in the later region,

Quantum theory predicts that no matter which of the measurements under consideration is performed, each possible outcome will occur half the time. Thus the common premise of (3) and (4) is sometimes satisfied. Combining these two predictions with the assumption that changing the choice of which experiment was performed in the later region cannot affect what already happened earlier in the faraway region entails that SR. sometimes fails: the assertion SR is false.

The fact that statement SR about outcomes of measurements performable in the later region is true if the first possible measurement is chosen and performed in the earlier region but is false if the second possible measurement is chosen and performed in that earlier region means that information about which experiment is performed in the earlier region must be present in the later region. This conclusion contradicts the locality condition that information about which choice is freely made by an experimenter in one region cannot be present in a second region unless the second can be reached from the first by traveling no faster than light.

David Mermin [10] gives a rather compelling argument that the predictions of quantum theory are very mysterious if one tries to deny the existence of superluminal information transfer. Shimony [11] and Jarrett [12], like most other contributors to the nonlocality issue, tie their analyses to Bell's theorem, and hence to hidden-variable reality" assumptions that conflict with the precepts of quantum theory. Hence it is not clear that it is the locality assumption, rather than the reality assumption, that fails.

Jarrett and Shimony call by the names "locality" and "parameter independence", respectively, a certain property that is satisfied by the predictions of quantum theory, and that is entailed by the requirement of no superluminal signaling. Using Jarrett's weak definition (i.e., weak locality requirement) one would call quantum theory "local". However, Shimony emphasizes that because entangled states of well separated bodies exist "there is a peculiar kind of *quantum nonlocality* in nature. To get to the crux of the matter I have defined locality to be the requirement of no superluminal transfer of information about which measurements are chosen and performed by experimenters, and taken nonlocality to be the failure of that condition. According to this definition, conventional (Copenhagen) quantum theory and relativistic quantum field theory are nonlocal, though in a way that does not allow superluminal signaling.

Primary Literature

1. A. Aspect, P. Grangier, G. Roger: Experimental tests of realistic local theories via Bell's Theorem. Phys. Rev. Lett. **47**, 460–463 (1981)
2. J. S. Bell: On the Einstein Podolsky Rosen Paradox. Physics **1**, 195–200 (1964)
3. H. P. Stapp: Whiteheadian approach to quantum theory and generalized Bell's theorem. Found. Phys. **9**, 1–25 (1979)
4. J. S. Bell: Introduction to the hidden variable question. *Foundations of Quantum Mechanics*, Proceedings of the International School of Physics 'Enrico Fermi', Course II (Academic, New York 1971, 279–81)
5. J. F. Clauser, M. A. Horne, A. Shimony, R. A. Holt: Proposed experiment to test local hidden-variable theories. Phys. Rev. Lett. **23**, 880–884 (1969)
6. H. P. Stapp: Non-local character of quantum mechanics. Epistemological letters (June 1978) (Association F. Gonseth, Case Postal 1081, Bienne, Switzerland)
7. A. Fine: Hidden variables, joint probabilities, and Bell inequalities. Phys. Rev. Lett. **48**, 291–295 (1982)
8. J. Hardy: Nonlocality for two particles without inequalities for almost all entangled states. Phys. Rev. Lett. **71**, 1665–1668 (1993)
9. H. P. Stapp: A Bell-type theorem without hidden variables. Am. J. Phys. **72**, 30–33 (2003)
10. N. D. Mermin: Quantum mysteries for anyone, in *Philosophical consequences of quantum theory*, eds. J.T. Cushing, E. McMullin (Notre Dame University Press, Notre Dame 1987, 49–59)
11. A. Shimony: Our world view and microphysics, in *Philosophical consequences of quantum theory*, eds. J.T. Cushing, E. McMullin (Notre Dame University Press, Notre Dame 1987, 25–37)
12. J. P. Jarrett: Bell's theorem: a guide to the implications, in *Philosophical consequences of quantum theory*, eds. J.T. Cushing and E. McMullin (Notre Dame University Press, Notre Dame 1987, 60–79)

Secondary Literature

13. J.T. Cushing, E. McMullin eds: *Philosophical consequences of quantum theory* (Notre Dame University Press, Notre Dame 1987)

Nuclear Fission

Hanne Andersen

Nuclear fission is a process in which a heavy nucleus splits into two much lighter nuclei. For some very unstable nuclei fission can happen spontaneously, but that is a very rare event. Usually, the process is induced by the excitation of the nuclei by bombarding them with particles or with gamma rays. Heavy nuclei have a greater neutron/proton ratio than the lighter nuclei, and the fragments will therefore contain too many neutrons. To reduce the excess of neutrons, two or three neutrons will be emitted by the fragments immediately, and the fragments will then decay by β-decay until stable isotopes are reached.

Nuclear fission was discovered in the 1930s when nuclear physics was still a young research field. At this time, a completely new realm of phenomena opened up when researchers discovered that radioactivity could be induced in heavy elements when bombarding them with neutrons. Initially, it had been discovered by Irene Curie (1897–1956) and her husband Fréderic Joliot (1900–1958) in 1934 that when bombarding light elements with alpha particles, these would transmute into radioactive isotopes of near-by elements. Because of the positive charge of the alpha particles, Curie and Joliot could only induce radioactivity in light elements. However, Enrico Fermi (1901–1954) soon realized that neutron bombardment could be used to induce radioactivity in heavy elements. After a series of experiments, Fermi and his collaborators reported that for a large number of elements of any atomic weight, neutron bombardment would produce unstable elements which emitted β-particles. Fermi's team therefore concentrated on the heavy nuclei thorium and uranium, since their general instability might give rise to successive disintegrations. For uranium, the last element in the periodic table as it was then known, such a series of β-emissions would lead to elements that did not exist in nature, and it attracted the attention of scientists around the world when Fermi's group reported that they had identified the first such transuranic element by chemical analysis of one of the decay products.

The German chemist Ida Noddack (1896–1978) objected that no conclusion of this sort could be drawn on the basis of the chemical analyses conducted by Fermi's team. She imagined that maybe a nucleus could break apart into several light elements, but the chemical analyses that the Fermi group had made were based on the assumption that the element had an atomic number close to that of uranium and did not take the possibility of light elements into account. However, Noddack's suggestion that the nucleus could split did not comply with the physical model of the nucleus that was accepted among her contemporaries. In his quantum mechanical theory of α-decay from the late 1920s, George Gamow (1904–1968) had shown that if nuclear disintegration was treated as a ► tunneling phenomenon, only particles up to the size of the α-particle were energetically capable of tunneling through the potential barrier. This result had been tacitly accepted among nuclear physicists to such an extent that the possibility of larger decay products were never even mentioned.

Diagrams used to illustrate disintegration were only suited for illustrating the transformation of one nucleus into another nucleus of almost the same size. Similarly, most notations could only represent the idea that a projectile hit a nucleus which, as a result, transformed into another nucleus by the emission of a particle. Noddack's suggestion did not fit with this way of thinking, and it remained ignored by other scientists in the field.

Other groups of scientists soon began pursuing Fermi's line of research. Not only Curie and Joliot in Paris started similar experiments, also a group in Berlin consisting of the physicist Lise Meitner (1878–1968) and the two chemists Otto Hahn (1879–1968) and Fritz Strassmann (1902–1980) went into the race of discovering new transuranic elements. This research was based on two assumptions. Nuclear physics dictated that the nuclear changes would always be very small and that the chemical analyses of the decay products could therefore be focused on just a few heavy elements. Further, although it was at the time disputed whether there would be a second series like the lanthanides in the periodic table, it was still assumed that the transuranic elements would chemically resemble the transition elements.

Based on these assumptions several new transuranic elements were identified, but most results were complex and required a variety of new hypothesis to be explained. Some transmutations led to extraordinarily long beta decay series which were difficult to understand. Other processes were not supposed to be energetically possible. Likewise, too many decay series seemed to originate from the same isotope. As these anomalies accumulated, it became increasingly difficult to integrate them all into a picture that made sense, and it was reported in several publications that the results were troubling and difficult to reconcile with standard concepts of the nucleus.

Finally came the anomaly that led to the discovery of nuclear fission. Hahn and Strassmann had identified a particular daughter element as radium in a precipitation process where it behaved like the alkaline earth element barium. However, on December 19th, 1938 Hahn and Strassmann discovered that they could not separate the product that they assumed to be radium from its barium carrier. The element they had produced did not just behave chemically *like* barium, it *was* barium. But then the original nucleus had not just transmutated into another heavy nucleus, instead it had simply split into much lighter elements. In a series of letters to Meitner, who had had to flee from Germany, Hahn described that although he knew that it was ruled out by the laws of physics, as a chemist he had to conclude that the nucleus had been divided.

Meitner discussed the results of Hahn and Strassmann with her nephew, the physicist Otto Frisch (1904–1979). On the basis of another model of the nucleus which had been advanced by the Danish physicist Niels Bohr (1885–1962) in 1936 and which treated the nucleus as an oscillating droplet, Meitner and Frisch conceived the explanation that adding energy by neutron bombardment, these oscillations could become so violent that the drop would divide into two smaller drops. Further, they pointed out that for heavy nuclei the surface tension produced by the short range nuclear forces was so effectively reduced by the increased nuclear charge that only relatively little energy was required to produce such critical deformations.

Thus, instead of considering quantum-mechanical tunnel-effects that would necessarily be extremely small for the large masses involved, Meitner and Frisch offered an explanation that was essentially classical. This explanation was consolidated further a few months later when Bohr and Wheeler offered quantitative computations of the qualitative ideas suggested by Meitner and Frisch.

However, this new discovery had far-reaching consequences for all the previous results on transuranic elements. New categorizations of all the previously examined processes had to be made, now distinguishing transuranic elements from fission products by their lack of recoil. Thus, Fréderic Joliot in Paris and Edwin M. McMillan (1907–1991) at Berkeley both developed experiments in which they measured the energy of the fission fragments by observing the distances they travelled from each other as a result of their mutual recoil.

Once fission had been discovered, a number of new research questions immediately suggested themselves. Most importantly, the splitting of a heavy nucleus into two light nuclei would produce a few free neutrons. If the released neutrons could cause new nuclei to split, a continuous chain reaction might occur. How to sustain such a chain reaction became another new research question. Due to the difference between the binding energy of a heavy nucleus and that of its fission products energy is also produced in the process. With the world at the edge of war, this research question gradually became more and more important and eventually gave rise to what has later become one of the prime examples of modern big science: the Manhattan project's creation of the first atomic bomb.

Primary Literature

1. N. Bohr, J.A. Wheeler: The Mechanism of Nuclear Fission. Physical Review (2nd ser.) **56**, 426–450 (1939)
2. S. Flügge: Kann der Energiegehalt der Atomkerne technisch nutzbar gemacht werden. Die Naturwissenschaften **27**, 402–410 (1939)
3. I. Noddack: Über das Element 93. Angewandte Chemie **47**, 653–655 (1934)
4. O. Hahn, F. Strassmann: Über den Nachweis und das Verhalten der bei der Bestrahlung des Urans mittels Neutronen entstehenden Erdalkalimetalle. Die Naturwissenschaften **27**, 11–15 (1939)
5. L. Meitner, O. Frisch: Disintegration of Uranium by Neutrons: A New Type of Nuclear Reaction. Nature **143**, 239–240 (1939)

Secondary Literature

6. H. Andersen: Categorization, Anomalies and the Discovery of Nuclear Fission. Studies in the History and Philosophy of Modern Physics **27**, 463–492 (1996)
7. R. Rhodes: *The Making of the Atomic Bomb* (Simon & Shuster, New York 1986)
8. R. L. Sime: The Search for Transuranium Elements and the Discovery of Nuclear Fission. Physics in Perspective **2**, 48–62 (2000)
9. R. Stuewer: The Origin of the Liquid-Drop Model and the Interpretation of Nuclear Fission. Perspectives on Science **2**, 76–129 (1994)

Nuclear Models

Brigitte Falkenburg

The atomic nucleus is made up of protons and neutrons, where the latter are made up of quarks (► *particle physics*). It is a complex compound system which is held together by the strong interaction and may change its charge by radioactive processes due to the (electro)weak interaction, giving rise to ► nuclear fission and fusion. Due to the complexity of the nuclei and their constituents (the nucleons, the proton and neutron), there are several nuclear models. It is remarkable that quantum mechanical and ► semi-classical models co-exist with the quark ► parton model of ► *quantum field theory*. Quarks, see ► Color Charge Degree of Freedom in Particle Physics; Mixing and Oscillations of Particles; Particle Physics; Parton Model; QCD; QFT.

History

In the classical Rutherford model of the atom (► Rutherford atom; Bohr's atom model), the atomic nucleus is a classical point charge which generates a Coulomb potential. Ernest Rutherford (1871–1937) first found deviations from his scattering formula (► *large angle scattering*) in 1909, when he made scattering experiments with α particles and hydrogen. He interpreted them in his classical model as indications of nuclear force effects. At that time it was already clear that the atomic nucleus must have a complex structure. In 1932, James Chadwick (1891–1974) discovered the neutron. In the same year, Werner Heisenberg (1901–1976) proposed a dynamic symmetry of the neutron and proton in view of the charge independence of the nuclear forces, giving rise to the concept of "isospin" [4]. In the 1930s, Carl Friedrich von Weizsäcker (1912–2007) developed the liquid droplet model. In the late 1940s, Maria Göppert–Mayer (1906–1972), Hans D. Jensen (1907–1973) and Eugene P. Wigner (1902–1995) developed the nuclear shell model [1, 2]. In the 1950s, Robert Hofstadter (1915–1990) investigated the structure of heavy and light nuclei by measuring their electromagnetic form factors in ► *scattering experiments* [3]. In the 1960s and 1970s, the quark model of the proton and neutron was developed in terms of group theory (► *symmetry; particle physics*, the quark–parton model was developed on the basis of electron-nucleon scattering, and the quark model was established (► *large angle scattering, parton model, scattering experiments*).

Liquid Droplet Model and Shell Model

The liquid droplet model and the shell model are based on the quantum mechanics of a many-particle system. According to the liquid droplet model, a heavy nucleus behaves like a Fermi gas. As ► spin 1/2 particles, the protons and neutrons obey Pauli's principle, i.e., they are in different quantum states and behave independently. According to the nuclear shell model, the nuclei form a periodic system of stable and unstable energy states. In both models, there is a sum rule for the mass and energy of the nucleus and its constituent parts. The nucleus mass differs from the mass of its protons and neutrons by the binding energy.

Form Factors

The Rutherford model of the atom and Rutherford's scattering formula are the basis for describing the nucleus as a non-pointlike structure in terms of *form factors* [5]. In the classical model of scattering, an internal structure of the scattering center is described by an extended charge distribution $\rho(\boldsymbol{r})$ rather than a point charge. In the non-relativistic case, the form factor is the Fourier transform of the charge distribution. For the Coulomb potential, the classical description of the scattering gives exactly the same result as the quantum mechanics of scattering. Based on this exact ► *correspondence*, the classical concept of the form factor could be extended to quantum mechanics. In this way, a semi-classical model of the nucleus is obtained, according to which the charge distribution generated by a quantum mechanical many-particle system corresponds to a classical charge distribution. The classical form factor describing the nucleus is then combined with the quantum mechanics of scattering. According to this semi-classical model, a pointlike scattering center has a form factor 1 which does not depend on the momentum transfer of the scattering. In ► *scattering experiments*, pointlike particles give rise to "scaling" behaviour, i.e., to a dimensionless effective cross section that does not depend on the energy of the scattered probe particles, while non-pointlike structures or extended charge distribution give rise to "scaling" violations, i.e., an energy dependence of the dimensionless quantity extracted from a measured cross section.

The Quark–Parton Model

In *high energy physics*, the above semi-classical model was extended to the relativistic domain, giving rise to the "structure functions" of the proton and neutron. The unexpected discovery of ► *large-angle scattering* and "scaling behaviour" of electron-nucleon scattering in 1968 gave rise to the quark–parton model of the proton and neutron [6]. The quark–parton model is a constituent model of the nucleons proton and neutron. It gives sum rules for the mass-energy, momentum and spin of the quarks and the proton or neutron. Scaling violations in certain kinematic domains indicate that there are further nucleon constituents, namely quark-antiquark pairs generated by virtual processes of quantum field theory and gluons, i.e., the exchange particles of the strong interaction or quanta of ► quantum chromodynamics.

Primary Literature

1. E. Fermi: *Nuclear Physics*. University of Chicago Press, Chicago 1950.
2. M. Goeppert-Mayer, J.H.D. Jensen: *Elementary Theory of Nuclear Shell Structure*. Wiley, New York 1955.
3. R. Hofstadter: *Electron Scattering and Nuclear Structure*. Rev. Mod. Phys. 28 (1956), 214–59. R. Hofstadter, F. Bumiller, M.R. Yearian: *Electromagnetic Structure of the Proton and Neutron*. Rev. Mod. Phys. 30 (1958), 482–97.

Secondary Literature

4. A. Pais: *Inward Bound*. Clarendon Press, Oxford 1986, Chaps. 15 and 17.
5. B. Falkenburg: *Probing Subatomic Structure and The Parts of Matter, in: Particle Metaphysics*. A Critical Account of Subatomic Reality. Springer, Heidelberg 2007, 125–61 and 246–56.
6. M. Riordan: *The Hunting of the Quark*. Simon & Schuster, New York 1987, Chaps. 6–8.

O

Objectification

Peter Mittelstaedt

The Concept of Objectification

In quantum mechanics, the term "objectification" is used for the attribution of a state or of the value of an observable to a quantum system. Correspondingly, the concepts of strong and weak objectification are used by some authors for *state attribution* and *value attribution*, respectively. Objectification may refer to the situation before the measurement (the *preparation*) and to the situation after a measurement (the *reading*). In particular, the so-called "problem of objectification" is concerned with the situation after the measuring process. It is also called the "measurement problem". See also ► Bohmian mechanics; Measurement theory; Metaphysics in Quantum Mechanics; Modal Interpretation; Projection Postulate.

This problem has a long history. Already in his book "Mathematische Grundlagen der Quantenmechanik" of 1932, J. von Neumann [1] observed, that a first and preliminary theory of the quantum measurement process does not lead to the objectification of the measurement result such, that the object system possesses the measured value of the observable in question after the measurement. For correcting this obvious deficiency of quantum mechanics, von Neumann introduced the "projection postulate" as a new and additional requirement for quantum mechanics. In contrast, Heisenberg [2] argued that the indispensability of the separation between the quantum object and the apparatus after the measurement is the real origin for the impossibility to objectifying the values of the system and of the apparatus – but not a deficiency of quantum theory.

Objectification in the Quantum Measurement Process

The quantum theory of ► measurement, first conceived by J. von Neumann (ref [1], 233–238) and further developed my many authors [3, 4, 7] considers the object system S as well as the measurement apparatus M as proper quantum systems. The measurement of an observable A, with discrete and nondegenerate values A_i and eigenstates $\varphi\ (A_i)$ is treated in three steps.[1] In the first step, the *preparation*, the systems S and M are dynamically independent and prepared in pure states φ and Φ,

[1] For sake of simplicity we mention here the only simplest version of the measurement process. Generalisations can be found in the literature, e.g. in ref. [3] and [4].

respectively. In the second step, the *premeasurement*, the interaction Hamiltonian $H_{\text{int}}(A)$ between systems S and M is turned on for some time interval Δt. If the interaction $H_{\text{int}}(A)$ is suited for a measurement of the observable A, then the preparation state $\Psi(\text{S}+\text{M}) = \varphi \otimes \Phi$ of the compound system S + M will be changed within the time interval Δt to the state after the premeasurement

$$\Psi'(\text{S}+\text{M}) = \exp\{-(i/\hbar)H_{\text{int}}(A)\Delta t\}\Psi(\text{S}+\text{M}) = \sum c_i \varphi(A_i) \otimes \Phi_i,$$

where Φ_i are eigenstates of the pointer observable that refer to pointer values Z_i. The coefficients c_i are given by the scalar products $c_i = (\varphi(A_i), \varphi)$. It can be shown that for any observable A there exists an interaction $H_{\text{int}}(A)$ that provides a state $\Psi'(\text{S}+\text{M})$ after the premeasurement with the bi-orthogonal decomposition as shown here.

In the third step of the measurement, *objectification* and *reading*, the systems S and M are again dynamically independent but still correlated. Considered as subsystems of S+M in the entangled state $\Psi'(\text{S}+\text{M})$, S and M can be described by the correlated ► mixed state $W'_{\text{S}} = \sum |c_i|^2 \, \text{P}[\varphi(A_i)]$, $W'_{\text{M}} = \sum |c_i|^2 \, \text{P}[\Phi_i]$, respectively. There are two kinds of mixed states. Formally, a mixed state is a self-adjoint positive operator W with trace 1. As any self-adjoint operator it can be decomposed according to its spectral decomposition as $\text{W} = \sum w_i \, \text{P}[\varphi_{\text{I}}]$ with $0 \leqslant w_i \leqslant 1$. (The states W'_{S} and W'_{M} discussed here are already written in their spectral decomposition). The two kinds of mixed states are distinguished by their preparation. a) If object systems are prepared in states φ_i, say, with a priori probabilities w_i, then any single system is said to be in a mixed state $\text{W} = \sum w_i \, \text{P}[\varphi_{\text{I}}]$, i.e. it is in one of the states φ_i with probability w_i. This very special kind of a mixed state is called a "mixture of states" [4], a "real mixture" [8], or a Gemenge [2]. It is a classical mixture which can, however, formally be described by the operator $\text{W} = \sum w_i \, \text{P}[\varphi_{\text{I}}]$. b) If a compound system $\text{S} = \text{S}_1 + \text{S}_2$ of subsystems S_1 and S_2 is prepared in a pure state $\Phi(\text{S})$, then the subsystem S_1, say, is in the reduced mixed state $\text{W}(\text{S}_1) = \text{tr}_2\{\text{P}[\Phi(\text{S})]\}$ where "tr_2" denotes the trace with respect to system S_2. The state $\text{W}(\text{S}_1)$ is a mixed state which does, however, in general not admit an "► ignorance interpretation". It is also called "improper mixture" [8]. Spectral decomposition, see ► Density operator; Ignorance interpretation; Measurement theory; Operator; Probabilistic Interpretation; Propensities in Quantum Mechanics; Self-adjoint operator; Wave Mechanics.

The two mixed states W'_{S} and W'_{M} that appear in the third step of the measuring process are improper mixtures of this kind, which do not admit ignorance interpretation. This means, that it is not allowed to say that the system S with the state W'_{S} is actually in one of the states $\varphi(A_i)$, but the observer does not know the state. In other words, neither the state $\varphi(A_i)$, nor the value A_i can be attributed to the system, which means that the measuring result cannot be objectified. The opposite assumption, that the system S were in a state $\varphi(A_i)$ and would possess the value A_i, leads to a contradiction with the statistical predictions of quantum mechanics. – The same conclusions hold, mutatis mutandis, for the state W'_{M} of the apparatus M, and for the state Φ_{i} and the value Z_i of the pointer.

The impossibility to objectify the values A_i of the measured observable and even the values Z_i of the pointer observable, is called the "measurement problem". There are many attempts to solve this problem, either within the framework of quantum mechanics or by convenient generalisations or modifications of this theory.[2]

Objectification of Unsharp Observables and Unsharp Objectification

The most promising attempt to solve the problem of objectification within the well known quantum mechanics in ► Hilbert space consists of a generalisation of the concept of an observable to unsharp ► observables. Formally, the projection-valued or (PV) measures, which correspond to ► self-adjoint operators, are replaced in this attempt by the more general positive operator valued (POV) measures. Originally, the expectation of the advocates of this attempt was, that in spite of the non-objectification theorems for (PV) observables[3] at least unsharp (POV) observables can be objectified.

However, within the quantum theory of measurement that is formulated in terms of (POV) observables, it could be shown that neither system objectification nor pointer objectification can be obtained. There was only a small chance to achieve pointer objectification, if even for the pointer observable an unsharp (POV) observable is used. This situation is also called "unsharp objectification" [5,6]. However, reading of an unsharp pointer observable corresponds to a situation, where pointer states which belong to different pointer values, are no longer strictly orthogonal. Thus, "one cannot claim with certainty, that the reading one means to have taken is reproducible on a 'second look' at the pointer".[4] Hence, even if unsharp objectification could be achieved in this way, we would loose the reliability of the results of our reading.

O

Primary Literature

1. von Neumann, J.: *Mathematische Grundlagen der Quantenmechanik*. Springer, Berlin (1932).
2. Heisenberg, W.: *Die PhysikalischenPrinzipien der Quantentheorie*. S. Hirzel, Leipzig (1930).
3. Busch P., P. J. Lahti, P. Mittelstaedt: *The Quantum Theory of Measurement*. Springer, Heidelberg (1991/1996).
4. Mittelstaedt, P.: *The Interpretation of Quantum Mechanics and the Measurement Process*. Cambridge University Press, Cambridge (1998).
5. Busch, P., A. Shimony: *Insolubility of Quantum Measurement Problem for Unsharp Observables*. Stud. Hist. Philos. Mod. Phys. Vol. 27, 397–404 (1996).
6. Busch, P.: *Can 'Unsharp Objectification' Solve the Quantum Measurement Problem?* Int. J. Theor. Phys. Vol. 37, 241–47 (1998).

---

[2] Reports about these attempts can be found in the literature, e.g. [3], chapter IV and [4], chapters 4 and 5.

[3] The non-objectification theorems can be found in ref. [4], pp. 82–88.

[4] Ref. [6], p. 246.

Secondary Literature

7. Wheeler, J. A., W. H. Zurek (eds.): *Quantum Theory and Measurement.* Princeton University Press, Princeton NJ (1983).
8. d'Espagnat, B.: *Reality and the Physicist.* Cambridge University Press, Cambridge, p. 85 (1989).

Objective Quantum Probabilities

Storrs McCall

Objective quantum probabilities represent the polar opposite to the Bayesian approach to quantum probabilities, which assumes probabilities to be subjective degrees of belief. In the objective theory, probabilities of quantum events are part of the physical world, and take their values independently of what human beings believe. The first objective theory was Karl Popper's propensity theory of probabilities, which identified propensities as the dispositional properties of particles to assume certain states under given conditions [1]. The propensity theory placed Popper squarely on the "particle" side of de Broglie's and Bohr's ▶ wave-particle duality. Propensities, however, suffered from the defect that Popper was unable to specify where in the physical world the values of his propensities lay. The present theory deals with this problem in locating precise quantum probability values in space-time structure.

Imagine that a spin-1/2 particle with direction of ▶ spin at an angle of 60° to the vertical is passed through an "HV apparatus", a vertically-oriented Stern–Gerlach magnet with two exit channels which separates particles into a "spin-up" stream (direction of spin v or vertical) and a "spin-down" stream (direction of spin h or horizontal). The spin-60° particle has a probability of $\cos^2 30° = 3/4$ of emerging in the spin-up channel. In the objective theory, this value is encoded in space-time structure in the following way. Imagine that at the time the particle enters the apparatus the 4-dimensional manifold divides into non-mutually-accessible future branches, and that on 75% of these branches the particle is measured spin-up and that on 25% it is measured spin-down. Figure 1, part (i), depicts a simple instance of this branching in space-time.

The future branches represent *possible outcomes* of the experiment, and the relative proportionality of sets of branches containing different kinds of outcome represent the *probabilities* of each outcome. But when the particle has exited from the apparatus there is only one *actual* outcome, and this "passage from potentiality to actuality" (Heisenberg [2]) is represented by the progressive vanishing of all branches but one in space-time structure (Fig. 1 part (ii)). There will, of course, always be many more branching surfaces and future branches higher up on the selected branch.

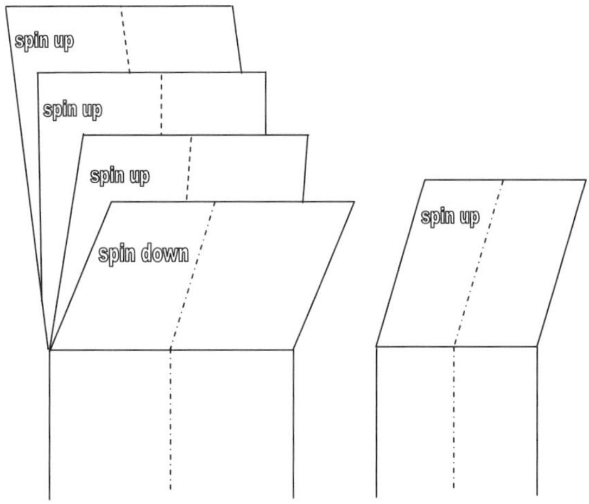

Part (i): time t Part (ii): time t+ t'

Fig. 1 At time t, 75% are spin-up branches and 25% spin-down branches. Hence the probability of a spin-up branch being randomly selected as the sole surviving "actual" branch is 0.75

The emergence of actuality, and the progressive vanishing of all but one future branch, is one of the two principal differences between the present theory and the ► many worlds interpretation of quantum mechanics. The other is that in the many-worlds theory the probabilities of the different future outcomes are "put in by hand", whereas in the objective theory probability values are represented by *branch-proportionality*. The probability of a spin-up or a spin-down outcome is determined by the proportionality of the spin-up and spin-down branch subsets relative to the set of all branches above a given branching surface. (In the example above, this is the totality of spin-up and spin-down branches when the particle measured by the HV apparatus has spin-orientation 60°). The latter set is *symmetric* in the sense that each branch has an equal chance of being selected as the actual branch. The breaking of this symmetry and the selection of the actual branch models the *collapse of the superposition*, i.e. the superposition of vertically-oriented and horizontally-oriented spin-states which describes the state of the particle when it enters the apparatus. Collapse in branching space-time is constituted by random branch selection of the actual branch.

In the example given of the particle with spin-orientation 60° the probabilities of the different future outcomes were 3/4 and 1/4, and it might be asked whether only rational probability values, corresponding to proportions among *finite* sets of different outcomes, can be objectively represented. The answer is no. Although Georg Cantor has shown that there can be no fixed proportions among subsets of a denumerably infinite set, there exist non-denumerably infinite sets of branches with a tree-like structure which possess subsets with proportionality corresponding to any real number between 0 and 1 [3]. Under appropriate initial conditions, the proportion of spin-up branches in some experiment will be precisely $\cos^2 20°$, an irrational number.

In relativistic 4-dimensional Minkowski space-time, the surfaces along which branches split are 2-dimensional spacelike hypersurfaces. These are constant-time hyperplanes in different frames of reference, and since the number of different inertial frames is unlimited, so will be the number of families of parallel hypersurfaces along which space-time branching occurs. Each of these families partitions space-time. The hypersurfaces in them criss-cross one another, and make the overall branching structure very complex. The complexity is necessary if we are to have a way of relativistically transforming the description of a quantum process in one frame of reference to a description of the same process in another frame [4]. The fact that branching is along spacelike hypersurfaces greatly increases the number of branches, since in one and the same set of branches there may be found, for example, proportionalities (and hence probabilities) for the outcomes of a ► Stern–Gerlach experiment in Montreal, for the possible transition from one energy state to another of a hydrogen atom in Alpha Centauri, and for the pending death of a mosquito in Mexico. The probability values of all these different events are Lorentz-invariant, remaining the same no matter which hypersurface they sit upon.

An important consequence of the space-time modelling of objective quantum probabilities, and in particular the splitting of branches along spacelike hypersurfaces, is the light shed by this approach on the nonlocal correlations and influences seen in the EPR ► Aspect experiment. If two entangled photons (► light quantum) with parallel polarization emitted by an atomic cascade are sent through a pair of aligned two-channel HV analyzers, either both photons will pass h or both will pass v. If the analyzers are misaligned, the left analyzer being HV and the right one oriented at an angle φ to the vertical, as in Fig. 2, the probabilities of the joint measured outcomes (v, φ^+), (v, φ^-), (h, φ^+), and (h, φ^-) are respectively $\mathrm{p}(v, \varphi^+) = \mathrm{p}(h, \varphi^-) = 1/2 \cos^2 \varphi$, $\mathrm{p}(v, \varphi^-) = \mathrm{p}(h, \varphi^+) = 1/2 \sin^2 \varphi$ [5].

When $\varphi = 30°$, $1/2 \sin^2 \varphi = 1/8$ and $1/2 \cos^2 \varphi = 3/8$. Let A and B denote the polarization measurement events on the left and right photons respectively. A branching space-time diagram yielding the probability values for the joint outcomes (v, φ^+), (v, φ^-), (h, φ^+), and (h, φ^-) is given in Fig. 3.

From the diagram, $\mathrm{p}(v) = \mathrm{p}(v, \varphi^+) + \mathrm{p}(v, \varphi^-) = 3/8 + 1/8 = 1/2$, and $\mathrm{p}(\varphi^+) = \mathrm{p}(v, \varphi^+) + \mathrm{p}(h, \varphi^+) = 3/8 + 1/8 = 1/2$. Consequently $\mathrm{p}(v, \varphi^+) \neq \mathrm{p}(v) \times \mathrm{p}(\varphi^+)$, which is to say that the outcome v on the left is not independent of the outcome φ^+ on the right. The EPR experiment provides an instance of the "distant correlations" of observed outcomes that have intrigued and baffled students of quantum physics for the last 70 years.

Since the two photons in the entangled quantum state are flying apart from each other at the speed of light, the two measurements at A and B will be spacelike separated events. Their outcomes are correlated, but the correlation cannot be explained

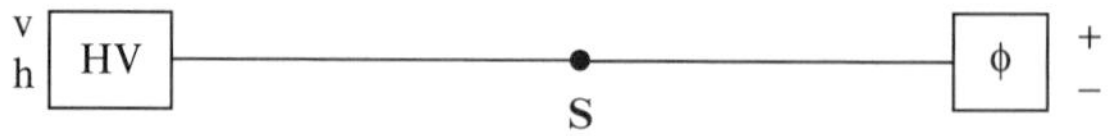

Fig. 2 Two entangled photons leave a source S and enter left and right polarization analyzers

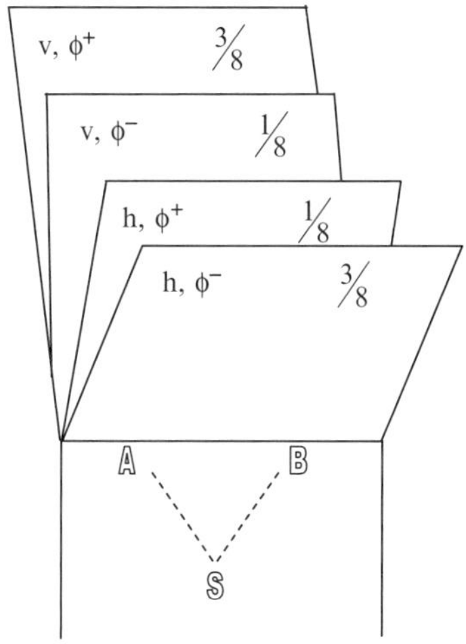

Fig. 3 The relative proportions of possible joint measurement outcomes when $\phi = 30°$

in terms of "► hidden variables", or instruction sets which travel with the photons. The problem becomes particularly acute when a frame of reference is chosen in which A occurs before B, or vice versa. If the stochastic outcome of the left measurement is v, then the probability of the right photon being measured φ^+ is 3/4. But if the left outcome happens to be h, then the probability for φ^+ on the right is 1/4. How does the information about the outcome on the left get communicated to the photon on the right, so that it "knows" its probability of being measured φ^+? Barring superluminal signalling (► superluminal communication), which would require causal influences travelling faster than light, there exists no apparent answer this question.

That being said, an explanation of the distant EPR correlations based on branching space-time structures is possible, when splitting takes place along spacelike hypersurfaces. Figure 3 is a picture of such a structure relative to a frame of reference in which the left and right measurement events are simultaneous. Figure 4 pictures the same experiment in a frame in which A occurs before B. Since A and B are spacelike separated events, such a frame always exists.

In Fig. 4, splitting occurs along a constant-time hypersurface on which A occurs, but relative to which B is future. The photon at A has a 50% probability of being measured v or h. If it is measured v, the branches on which it is measured h vanish instantaneously, along the whole length of the hypersurface. On the sole remaining v-branch, the probability of the right photon being measured φ^+ is 3/4. If, however, the stochastic outcome of the left measurement had been h, then all v-branches would have vanished, and the probability of the right photon passing φ^+ would have been 1/4 instead of 3/4.

Figure 5 is a picture of the same EPR experiment in a frame in which the right measurement B occurs before A. As before, branch attrition explains how the right outcome φ^+ or φ^- instantaneously affects the probabilities for the outcome on the left. The conclusion is that branch attrition along spacelike hypersurfaces or hyperplanes provides an objective, realistic explanation of the instantaneous

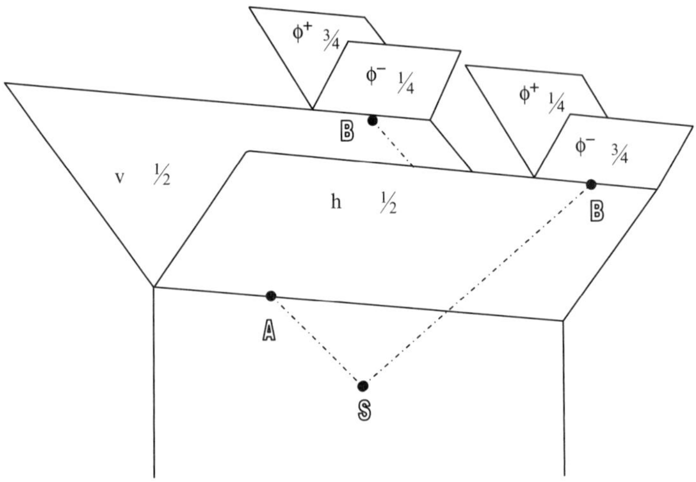

Fig. 4 In a frame in which A occurs before B, the probabilities for the right outcome depend upon the outcome of the left measurement

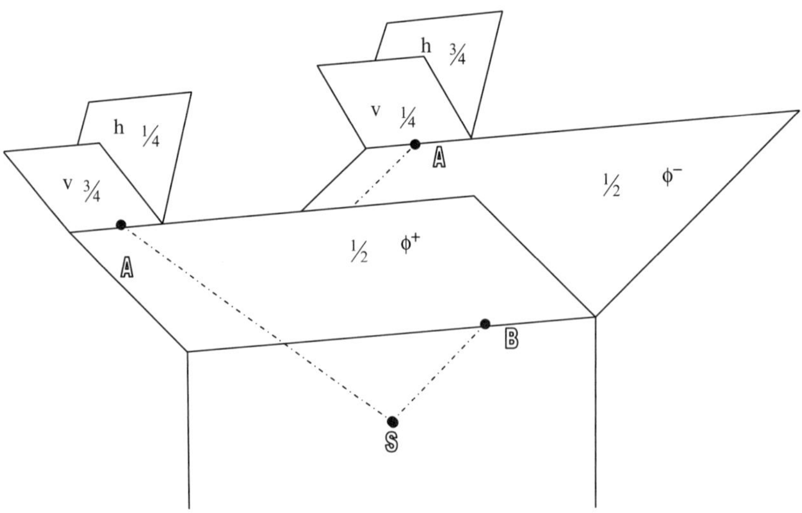

Fig. 5 When B occurs before A, the dependence is reversed. The probability of the left outcome depends on the right outcome

information transfers which underlie the distant correlations of the EPR experiment. These information transfers do not involve superluminal signalling, since nothing travels from B to A or vice versa. Nevertheless, information is effectively transferred by the instantaneous vanishing of the non-actual branches along hypersurfaces. See also ► Probability in Quantum Mechanics; Propensities in Quantum Mechanics.

Primary Literature

1. K. Popper: *Quantum Theory and the Schism in Physics* (Routledge, London 1982, 126)
2. W. Heisenberg: *Physics and Philosophy* – The Revolution in Modern Science (Harper, New York 1958, 54–5)
3. S. McCall: *A Model of the Universe* (Clarendon, Oxford 1994, 86–92)

Secondary Literature

4. S. McCall: QM and STR. Philosophy of Science (Proceedings) **67**, S535–48 (2000)
5. See, for example, J. Baggott: *The Meaning of Quantum Theory* (Oxford University Press, Oxford 1992, 120–27, 141)

Observable

Paul Busch and Pekka Lahti

The term *observable* has become the standard name in quantum mechanics for what used to be called physical quantity or measurable quantity in classical physics. This term derives from *observable quantity* ("beobachtbare Grösse"), which was used by Werner Heisenberg in his groundbreaking work on ▶ matrix mechanics [1] to emphasize that the meaning of a physical quantity must be specified by means of an operational definition. Together with a state (▶ *states in quantum mechanics*), an observable determines the probabilities of the possible outcomes of a measurement of that observable on the quantum system prepared in the given state. Conversely, observables are identified by the totalities of their measurement outcome probabilities. Examples of observables in quantum mechanics are position, velocity, momentum, angular momentum, spin, and energy. ▶ Spin; Stern–Gerlach experiment; Vector model.

O

In elementary quantum mechanics, the observables of a physical system are represented by, and identified with, *selfadjoint operators* A acting in the ▶ Hilbert space $\mathcal{H}$ associated with the system. For any pure states of the system (▶ *states, pure and mixed*), represented by a unit vector $\psi \in \mathcal{H}$, the probability $p_{\psi}^{A}(X)$ that a measurement of A leads to a result in a (Borel) set $X \subset \mathbb{R}$ is given by the inner product of ψ with $E^{A}(X)\psi$, that is, $p_{\psi}^{A}(X) = \langle \psi | E^{A}(X)\psi \rangle$; here $E^{A}(X)$ is the spectral *projection* of A associated with the set X, and the map $X \mapsto E^{A}(X)$ is called the *spectral measure* of A. The probability measures p_{ψ}^{A}, with ψ varying over all possible pure states of the system, determine the observable A. The expectation, or average $\int x \, dp_{\psi}^{A}(x)$, of the measurement outcome distribution of an observable A in a state ψ can be expressed as $\langle \psi | A\psi \rangle$ whenever ψ is in the domain of the operator A.

The statistical meaning of quantum observables was first recognized by Max Born [2] who proposed that, in the position representation, the absolute square $|\psi|^2$ of the '▶ wave function' ψ gives the probability density of observing a quantum object at a given point (▶ Born rule). This idea was systematically elaborated by John von Neumann [3] who formulated and proved the spectral theorem for selfadjoint (hypermaximal hermitian) operators and applied it to obtain the interpretation of expectations as statistical averages given above.

In his seminal paper on the uncertainty relations [4] Werner Heisenberg argued, among other things, that

> all concepts which can be used in classical theory for the description of a mechanical system can also be defined exactly for atomic processes in analogy to classical concepts.

This statement can be substantiated in precise form by virtue of the mathematical fact that for any value x in the *spectrum* of a selfadjoint operator A and for each $\epsilon > 0$ there is a state ψ such that $p_\psi^A((x-\epsilon, x+\epsilon)) = 1$. In particular, if A has an eigenvalue a, that is, there is a state ψ such that $A\psi = a\psi$, then in such an eigenstate of A a measurement of A is certain to yield the value a. Such a situation is commonly described by saying that observable A has a *definite value* if the state of the system is an eigenstate of A. The generic situation in quantum mechanics, however, is that most observables have no definite value in a given pure state.

It is a basic feature of quantum mechanics that there are pairs of observables, such as position and momentum, which do not commute. This fact, which lies at the heart of the ▶ *complementarity principle* and ▶ *Heisenberg uncertainty relation*, reflects a fundamental limitation on the possibilities of assigning definite values to observables and to the possibilities of measurements in the quantum world. For example, among the pairs of observables with discrete spectra there are those that do not commute, and this implies that they do not share a complete system of eigenvectors. Then A has eigenstates that are not eigenstates of B. Moreover, according to a theorem due to von Neumann [5], observables A, B are jointly measurable, that is, they have a *joint observable* (see below), if and only if they commute.

The idea of identifying an observable (with real values) with the totality of the outcome probabilities in a measurement does not single out spectral measures, but is exhausted by the wider class of (real) *positive operator (valued) measures*, or *semispectral measures*. A positive operator measure is a map $E : X \mapsto E(X)$ that assigns to every (Borel) subset X of $\mathbb{R}$ a *positive operator* $E(X)$ in such a way that for every pure state ψ the map $X \mapsto p_\psi^E(X) := \langle\psi|E(X)\psi\rangle$ is a probability measure. This definition extends readily to cases where the measurement outcomes are represented as elements of $\mathbb{R}^n$ or more general sets. Excellent expositions of the definition and properties of positive operator measures can be found, e.g., in [8,9].

Observables represented by positive operator measures which are not projection valued are referred to as *generalized observables*, or *unsharp observables*, while spectral measures and generally all projection valued measures are called *standard*, or *sharp observables*. Commonly used acronyms for positive operator measures are ▶ POVM or POM.

The generalized representation of observables as positive operator measures was discovered by several authors in the 1960s (e.g., [6, 7, 10–13]) and has by now become a standard element of quantum mechanics. It has greatly advanced the mathematical coherence and conceptual clarity of the theory. For instance, the problem of the (approximate) joint measurability of noncommuting observables such as position and momentum and the relevance of the ► *Heisenberg uncertainty relations* to this question is now fully understood; for a survey, see, e.g. [14].

Two (real) POMs E, F are jointly measurable if and only if there is a third POM, G, defined on the (Borel) subsets of $\mathbb{R}^2$, which has E and F as marginals, that is, $E(X) = G(X \times \mathbb{R})$ and $F(Y) = G(\mathbb{R} \times Y)$ for all (Borel) subsets X, Y of $\mathbb{R}$. For the joint measurability of two unsharp observables E, F, their mutual commutativity is sufficient but not necessary. If one of the observables is sharp, then the joint measurability implies commutativity.

As two noncommuting standard observables are never jointly measurable, one can only try to approximate them (in a suitable sense) by some other observables which in turn may be jointly measurable. This turns out indeed to be possible as has been well demonstrated in the cases of position and momentum or spin components.

Finally, the introduction of POMs has widely increased the applicability of quantum mechanics in the description of realistic experiments (see, e.g., [15, 16]), and POMs are now in full use also in the relatively new fields of ► quantum computation and information, see, e.g., [17, 18]. See also ► PVOM; Rigged Hilbert Spaces; Superselection Rules; Wave function collapse.

Primary Literature

O

1. Heisenberg, W.: *Über quantentheoretische Umdeutung kinematischer und mechanischer Beziehungen*. Zeitschrift für Physik **33** 879-893 (1925).
2. Born, M.: *Zur Quantenmechanik der Stoßvorgänge*. Zeitschrift für Physik **37**, 863-867 (1926).
3. von Neumann, J.: *Mathematische Begründung der Quantenmechanik*. Göttinger Nachrichten 1–57 (1927). Wahrscheinlichkeitstheoretischer Aufbau der Quantenmechanik. *Göttinger Nachrichten* 245–272 (1927). Allgemeine Eigenwerttheorie Hermitescher Funktionaloperatoren, Mathematische Annalen **102**, 49–131 (1929). This all is summarized in his book Mathematische Grundlagen der Quantenmechanik (Springer, Berlin 1932, 2nd ed. 1996); English translation (by R. Beyer): Mathematical Foundations of Quantum Mechanics (Princeton University Press, Princeton 1955).
4. Heisenberg, W.: *Über den anschaulichen Inhalt der quantentheoretischen Kinematik und Mechanik*. Zeitschrift für Physik **43**, 172–198 (1927).
5. von Neumann, J.: *Über Funktionen von Funktionaloperatoren*. Annals of Mathematics **32**, 191–226 (1931).
6. Ludwig, G.: *Versuch einer axiomatischen Grundlegung der Quantenmechanik und allgemeinerer physikalischer Theorien*. Zeitschrift für Physik **181** 233–260 (1964).
7. Davies, E.B., J.T. Lewis: *An Operational Approach to Quantum Probability*. Communication in Mathematical Physics **17**, 239–260 (1970).

Secondary Literature

8. Davies, E.B.: *Quantum Theory of Opens Systems* (Academic, New York 1976).
9. Berberian, S.K.: *Notes on Spectral Theory* (D. van Nostrand, Princeton 1966).
10. Helstrom, C.W.: *Quantum Detection and Estimation Theory* (Academic, New York 1976).
11. Holevo, A.S.: *Probabilistic and Statistical Aspects of Quantum Theory* (North Holland, Amsterdam 1982).
12. Ludwig, G.: *Foundations of Quantum Mechanics* (Springer, Berlin 1983).
13. Kraus, K.: *States, Effects, and Operations* (Springer, Berlin 1983).
14. Busch, P., T. Heinonen, P. Lahti: *Heisenberg's Uncertainty Principle*, Physics Reports 452 (2007) 155–176.
15. Busch, P., M. Grabowski, P. Lahti: *Operational Quantum Physics*, Springer, LNP m31, 1995, 2nd Corrected Printing 1997.
16. Peres, A.: *Quantum Theory: Concepts and Methods* (Kluwer, Dordrecht 1993).
17. Nielsen, M.A., I.L.Chuang: *Quantum Computation and Quantum Information* (Cambridge University Press, Cambridge 2000).
18. Stenholm, S., K.-A. Suominen: *Quantum Approach to Information* (Wiley, New York 2005).

One- and Two-Photon Interference

Paul G. Kwiat

Taylor's version of Young's double-slit experiment with an attenuated light source is often hailed as one of the key experiments demonstrating quantum mechanical interference [1]. Although it would be incorrect to say it is not quantum – all optical interference effects have their origin at the quantum level – it is now generally accepted that such experiments are not *non-classical*. They usually allow a semi-classical description in which the detector is treated quantum mechanically, but the field is treated classically. In fact, such descriptions can also account for a host of other "quantum" phenomena, such as resonance fluorescence and the photoelectric effect [2,3]. For single-photon interference, one can readily convert from the classical field to the quantum mechanical description simply by relating the probability of a photon being detected at a given location and time to the intensity of the classical field.

The need for a quantum description of the light – the need for "photons" (▶ light quantum) – arises when one considers higher-order photon statistics, e.g., involving coincidences between 2 or more detectors. In fact, this is now the method of choice for characterizing would-be single-photon sources [4]: send the light onto a beam-splitter and measure the coincidence rate between the detectors in each output. For a true single-photon input – formally described as an $n = 1$ Fock number state – the coincidence rate will fall to zero (at equal detection times), a very non-classical effect (a classical field would necessarily cause detections in *both* output ports).[1]

[1] More precisely, one measures $g^{(2)}(0)$, the second-order correlation function, equal to the number of coincidence counts in a given time interval, divided by the product of the singles counts (at the two detectors) in that interval. For an n-photon Fock state, $g^{(2)}(0) = 1 - 1/n$.

Currently there is great interest in developing such single-photon sources, for applications in metrology and quantum information processing (▶ quantum communication). For example, the original quantum cryptography protocols assumed the key material was transmitted using single-photon states [5], so as to deny any potential eavesdropper the possibility of "tapping" the line. More recently, sources of single photons "on-demand" are a critical resource for realizing scalable optical quantum computing [6]. At present a number of physical systems are being explored as single-photon sources. In the first category, a single quantum emitter – e.g., an atom, ion, or quantum dot – is excited, and consequently decays, either spontaneously or in a driven transition, emitting precisely one photon in the process. Much effort is directed to using cavities to tailor the mode into which the photon is preferentially emitted [7]. A second strategy is to employ systems that always emit *pairs* of photons: using one photon as a "trigger" then heralds the presence of the other photon. Examples include 2-photon transitions in atoms, or most prevalent, pair sources from nonlinear optics, e.g., spontaneous parametric down-conversion [8,9]. In the down-conversion process, a high-energy pump photon splits – via the interaction in a non-linear crystal – into two daughter photons, traditionally called the "signal" and "idler" (Fig. 1).

Following earlier experiments (by Clauser et al. [10]), Grangier et al., performed the first interference experiment using a light field in a single-photon Fock state [11], based on a two-photon atom cascade as mentioned above. One photon was used as a trigger to condition the presence of the other photon, which was then directed to a Mach-Zehnder interferometer. The resulting interference fringes, built up one photon at a time, displayed a visibility $>98\%$, verifying Dirac's statement that a single photon interferes with itself [12]. The same technique has been routinely adopted to down-conversion sources, and used to demonstrate, e.g., ▶ Berry's phase at the single-photon level [13].

O

Once one has single photons, the concept of the *trajectory* of the photon inside an interferometer becomes well defined. One finds that the existence of any which way information,[2] labeling which path a photon took, will reduce the contrast of

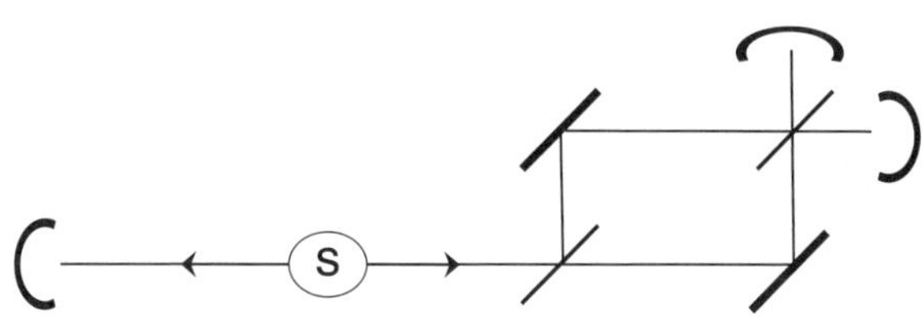

Fig. 1 In a basic demonstration of single-photon interference [11], a two-photon cascade source S is used to conditionally prepare a single photon, which is directed into a Mach-Zehnder interferometer. Even though at most one of the two interferometer detectors fires at a given time, high-visibility interference fringes are observed, supporting Dirac's dictum that in such experiments "each photon then interferes only with itself" [12]. (Fig. based on [11])

[2] This information may be due to entanglement to another photon or atom, or simply an entanglement between the path and some other degree of freedom, e.g., polarization of the single photon.

interference fringes,[3] quantitatively described by $V^2+D^2 \leqslant 1$, where V is the fringe visibility, and D is the distinguishability of the paths [14]; the inequality holds if the which-way quantum marker (▶ which-way experiment) is initially in a mixed (i.e., uncertain) state [15, 16].

Curiously, the distinguishing information can sometimes be subsequently removed, by making a suitable measurement on the which-path system. This phenomenon, in which interference can be recovered, is known as a ▶ "quantum eraser" [17], and has now been demonstrated in many experiments (e.g., [16,18,19]). Note that the interference is only revealed by correlating the detections of the photon with particular measurement results on the which-way marker, thereby preventing any superluminal signalling (▶ superluminal communication).

One recent experiment of this sort directed photons emitted from a single excited nitrogen-vacancy color center (in a diamond nanocrystal) into a Mach–Zehnder interferometer [20]. Waveplates were used to set the photon polarization in the two paths to horizontal and vertical. The output of the interferometer was directed through a rapidly switchable polarization analyzer, and then to a single-photon detector. Results showed that the measurement could either reveal which-way information (by analyzing in the horizontal-vertical basis) or could recover fringes or anti-fringes (by analyzing in the $\pm 45°$ basis). Moreover, the experiment had a ▶ delayed-choice aspect [21] – the choice in which basis to measure the photon was made after the photon ▶ wave packet had already passed the initial beamsplitter of the interferometer; however, this did not affect the results.

Another interesting series of experiments arises when the interfering photon can originate in more than one source. In the first of these experiments [22] light beams from two independent single-mode lasers demonstrated interference fringes, even when the intensities were so low that only a single photon was in the interferometer at any given time. From a wave perspective this is hardly surprising – e.g., one has no trouble accepting that signals from two radio towers can interfere. The understanding at the quantum level is that there is no way, even in principle, to distinguish from which laser a given photon originated, due to the fact that the quantum state of the laser itself is negligibly altered by emission of a photon.

However, a quite different result can be obtained for a more 'quantum' light source. Consider, for example, trying to interfere the signal photons from two independent down-conversion crystals, by superposing the photons' spatial modes on a beamsplitter (see Fig. 2a). In this case there is no interference, because the simultaneous emission of the twin idler photon from one of the crystals labels which source a given signal photon came from; even if one does not measure the idler photon, the mere possibility that one could in principle make such a measurement is enough to prevent interference. However, it is possible to arrange the crystals in such a way that this information is not available (Fig. 2b): by directing the idler mode of the first crystal to pass through the second crystal and completely overlap the second

[3] Following Feynman, to calculate the probability of any outcome, we must add the probability amplitudes of indistinguishable processes that lead to this outcome, and then take the absolute square. If the processes are *in principle* distinguishable, then we simply add the probabilities directly.

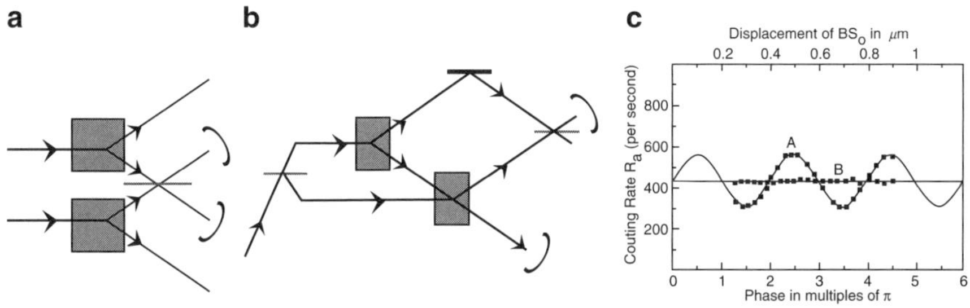

Fig. 2 (**a**) One does normally not observe single-photon interference when the signal modes from two down-conversion crystals are combined, because the idler photons carry distinguishing information about which crystal produced a given signal photon. (**b**). However, if the idler modes are made to overlap, this information is in principle unobtainable. (**c**). Interference in the signal singles rate is observable as any of the phases in the overall experiment are adjusted (A), unless the idler mode between the crystals is blocked (B). Data reprinted with permission from Fig. 2 in Ref. [23]: Copyright (1991) by the American Physical Society

crystal's idler mode, any process-labeling by these photons can be eliminated. The consequence is that single-photon interference fringes are once again observable in the output of the beamsplitter combining the two signal modes (Fig. 2c); this interference occurs as *any* of the path lengths in the experiment are varied [23]. Experiments have also demonstrated that if a time-dependent gate is introduced in the idler arm between the two crystals, the observation of interference of the signal photons depends on the state of the gate at the time when the idler photon amplitude was passing through it [24]: A closed gate – allowing which-path information – destroys the interference.

Allowing for more than one photon opens the way for a multitude of purely quantum multi-photon interference effects. Here we will only discuss two of the main 2-photon interference phenomena. The most well-known and arguably the most important example is the Hong–Ou–Mandel interferometer [25]. Two identical photons are directed to opposite sides of a 50–50 beamsplitter, so each individually has a 50% likelihood to be transmitted or reflected[4] (Fig. 3). If these were classical light fields, then sometimes both of the detectors at the outputs would fire simultaneously, corresponding to the possibility that both fields were transmitted or both reflected. However, following Feynman, we must add the probability amplitudes of indistinguishable processes. In the Hong–Ou–Mandel interferometer, the two indistinguishable processes that could lead to a coincidence detection (both photons being transmitted, with net probability amplitude $\frac{1}{\sqrt{2}}\frac{1}{\sqrt{2}} = \frac{1}{2}$, and both being reflected, with net probability amplitude $\frac{i}{\sqrt{2}}\frac{i}{\sqrt{2}} = -\frac{1}{2}$) completely destructively interfere[5]. Hence, if the photons arrive at the beamsplitter simultaneously, there is a

[4] There is no single-photon interference, because each photon is *not* in a superposition of being in the upper and lower path; also, there is no definite phase relationship between the two photons.

[5] The extra factors of "i"—$\pi/2$ phase shifts—are required to satisfy unitarity and conservation of probability/energy.

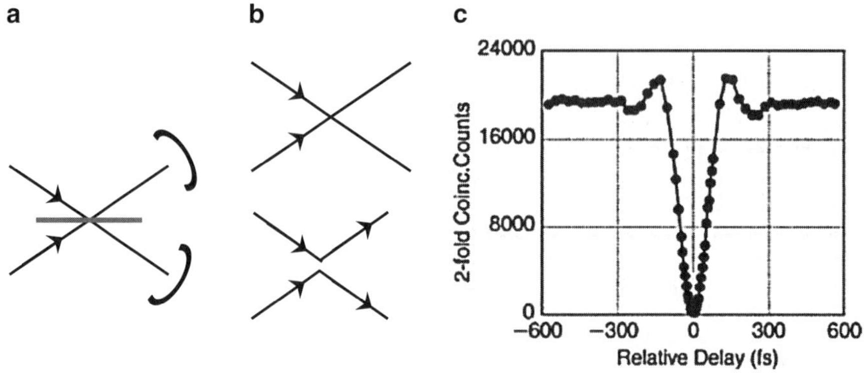

Fig. 3 (**a**) In the Hong–Ou–Mandel interferometer [25], two identical photons are directed onto opposite sides of a 50–50 beamsplitter, aligned so that the reflected and transmitted modes completely overlap. (**b**) If the photons arrive at the beamsplitter simultaneously, the 'transmitted-transmitted' and 'reflected-reflected' processes destructively interfere with each other. (**c**) A dip is observed in the observed coincidence rate (data from [9])

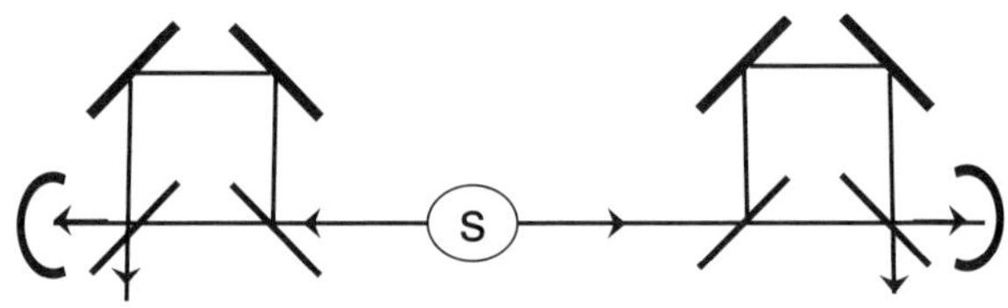

Fig. 4 Schematic of a two-photon interference effect [29], in which each of the down-conversion photons is directed into an unbalanced Mach-Zehnder interferometer. Although the path imbalance precludes any single-photon interference, two-photon interference fringes (depending on the sum of the relative phases in the interferometers) may be observed, due to the indistinguishability of the processes in which both photons take the short paths and both take the long paths in their respective interferometers

dip in the coincidence rate (Fig. 3c) as both photons then take the same output port. The Hong–Ou–Mandel interference effect has now been used to enable precision relative timing measurements [25, 26], and is the central technique to enable Bell-state analysis for quantum teleportation [27, 28] (▶ quantum communication) and various quantum logic gates [6].

As a final example of 2-photon interference, each of the signal and idler photons can be directed into its own, quite imbalanced, Mach–Zehnder interferometer (Fig. 4). In this case, no interference is observable in any of the singles rates because the interferometer imbalance is much larger than the coherence length of the photons. However, if the two interferometers are matched to each other, interference fringes can be observed in the coincidence rates between detectors at the outputs of each interferometer: For continuous-wave pumping, processes in which both photons take the short paths or both take the long paths in their respective interferometers – corresponding to two different emission times for the pair – are in principle indistinguishable, and thus interfere. One observes coincidence interference fringes which depend nonlocally on the sum of the phases in both interferometers [29]. This 2-photon interference effect has been used to demonstrate the ► nonlocality of quantum mechanics (i.e., producing violations of a suitable Bell's inequality; ► Bell's theorem) [30, 31], and forms the basis of some entanglement-based on quantum cryptography implementations [32, 33].

Primary Literature

1. G. I. Taylor: Interference Fringes with Feeble Light. Proc. Camb. Philos. Soc. **15,** 114 (1909)
2. P. A. Franken: Collisions of light with atoms. Atomic Physics, edited by B. Bederson et al. (Plenum, NY), 377–387 (1969)
3. P. W. Milonni: What (If Anything) Does the Photoelectric Effect Teach Us? Am. J. Phys. **65,** 11–12 (1997)
4. C. Santori, D. Fattal, J. Vuckovlc, G. S. Solomon, Y. Yamamoto: Indistinguishable Photons from a Single-Photon Device. Nature **419,** 594–597 (2002)
5. N. Gisin, G. Ribordy, W. Tittel, H. Zbinden: Quantum Cryptography. Rev. Mod. Phys. **74,** 145–195 (2002)
6. J. L. O'Brien: Optical Quantum Computing. Science **318,** 1567–1570 (2007)
7. P. Grangier, B. Sanders, J. Vuckovic (Eds.): Focus on Single Photons on Demand. New J. Phys. 6 (2004)
8. C. K. Hong, L. Mandel: Experimental Realization of a Localized One-Photon State. Phys. Rev. Lett. **56,** 58–60 (1986)
9. T. Pittman, B. Jacobs, J. Franson: Heralding single photons from pulsed parametric down-conversion. Opt. Commun. **246,** 545–550 (2005)
10. J. F. Clauser: Experimental Limitations to the Validity of Semiclassical Radiation Theories. Phys. Rev. A **6,** 49–54 (1972)
11. P. Grangier, G. Roger: A. Aspect: Experimental Evidence for a Photon Anticorrelation Effect on a Beam Splitter: A New Light on Single-Photon Interferences. Europhys. Lett. **1,** 173–179 (1986)
12. P. A. M. Dirac: *The Principles of Quantum Mechanics,* 4th edn. (Oxford University Press, London 1958)
13. P. G. Kwiat, R. Y. Chiao: Observation of a nonclassical Berry's phase for the photon. Phys. Rev. Lett. **66,** 588–591 (1991)
14. B. -G. Englert: Fringe Visibility and Which-Way Information: An Inequality. Phys. Rev. Lett. 77, 2154–2157 (1996)
15. P. D. D. Schwindt, P. G. Kwiat, B. -G. Englert: Quantitative Wave-Particle Duality and Non-erasing Quantum Erasure. Phys. Rev. A **60,** 4285–4290 (1999)
16. A. Trifonov, G. Bjork, J. Soderholm, T. Tsegaye: Comprehensive Experimental Test of Quantum Erasure. Euro. Phys. J. D **18,** 251–258 (2002)
17. M. O. Scully, B. -G. Englert, H. Walther: Quantum Optical Tests of Complementarity. Nature **351,** 111–116 (1991)
18. T. Herzog, P. G. Kwiat, H. Weinfurter, A. Zeilinger: Complementarity and the Quantum Eraser. Phys. Rev. Lett. **75,** 3034 (1995).
19. Y. -H. Kim, R. Yu, S. P. Kulik, Y. Shih, M. O. Scully: Delayed 'Choice' Quantum Eraser. Phys. Rev. Lett. **84,** 1 (2000)
20. V. Jacques, et al.: Experimental Realization of Wheeler's Delayed-Choice Gedanken Experiment. Science **315,** 966–968 (2007)
21. J. A. Wheeler: Frontiers of Time. In: *Proceedings in the formulation of physics,* ed. by G. T. diFrancia (North-Holland, Amsterdam 1979)
22. R. L. Pfleegor, L. Mandel: Interference of Independent Photon Beams. Phys. Rev. **159,** 1084–1088 (1967)

23. X. Y. Zou, L. J. Wang, L. Mandel: Induced coherence and indistinguishability in optical interference. Phys. Rev. Lett. **67,** 318–321 (1991)
24. L. J. Wang, J. -K. Rhee: Propagation of Transient Quantum Coherence. Phys. Rev. A **59,** 1654–1657 (1999)
25. C. K. Hong, Z. Y. Ou, L: Mandel: Measurement of Subpicosecond Time Intervals Between Two Photons by Interference. Phys. Rev. Lett. **59,** 2044–2046 (1987)
26. A. M. Steinberg, P. G. Kwiat, R. Y. Chiao: Measurement of the Single-Photon Tunneling Time. Phys. Rev. Lett. **71,** 708 (1993)
27. C. H. Bennett, et al.: Teleporting an Unknown Quantum State Via Dual Classical and Einstein–Podolsky–Rosen channels. Phys. Rev. Lett. **70,** 1895–1899 (1993)
28. D. Bouwmeester, J. -W. Pan, K. Mattle, M. Eibl, H. Weinfurther, A. Zeilinger: Experimental Quantum Teleportation. Nature **390,** 576–579 (1997)
29. J. D. Franson: Bell Inequality for Position and Time. Phys. Rev. Lett. **62,** 2205–2208 (1989)
30. J. Brendel, E. Mohler, W. Martienssen: Experimental Test of Bell's Inequality for Energy and Time. Europhys. Lett. **20,** 575–580 (1992)
31. P. G. Kwiat, A. M. Steinberg, R. Y. Chiao: High-Visibility Interference in a Bell-Inequality Experiment for Energy and Time. Phys. Rev. A, **47,** R2472–R2475 (1993)
32. W. Tittel, J. Brendel, H. Zbinden, N. Gisin: Quantum Cryptography Using Entangled Photons in Energy-Time Bell States. Phys. Rev. Lett. **84,** 4737–4740 (2000)
33. I. Marcikic, et al.: Distribution of Time-Bin Entangled Qubits over 50 km of Optical Fiber. Phys. Rev. Lett. **93,**180502 (2004)

Secondary Literature

34. D. M. Greenberger, M. A. Horne, A. Zeilinger: Multiparticle interferometry and the superposition principle. Phys. Today **46,** 22–29 (1993)
35. L. Mandel: Quantum effects in one-photon and two-photon interference. Rev. Mod. Phys. **71,** S274–S282 (1999)
36. A. Zeilinger, G. Weihs, T. Jennewein, M. Aspelmeyer: Happy Centenary, Photon. Nature **433,** 230–238 (2005)
37. A. M. Steinberg, P. G. Kwiat, R. Y. Chiao: Quantum Optical Tests of the Foundations of Physics. In Springer Handbook of Atomic, Molecular, and Optical Physics, ed by G. W. F. Drake (Springer, Berlin 2006, 1185–1207)
38. R. Hillmer, P. Kwiat: A Do-It-Yourself Quantum eraser. Sci. Am. **296,** 90 (2007)

Operational Quantum Mechanics, Quantum Axiomatics and Quantum Structures

Diederik Aerts

Operational quantum mechanics and quantum axiomatics have their roots in a work of John von Neumann in collaboration with Garett Birkhoff, that is almost as old as quantum mechanics itself [1]. Indeed already during the beginning years of quantum mechanics, the formalism that is now referred to as standard quantum mechanics [5], was thought to be too specific by the founding fathers themselves. One of the questions that obviously was at the origin of this early dissatisfaction is: 'Why would a complex ► Hilbert space deliver the unique mathematical structure for a complete description of the microworld? Would that not be amazing? What is so special about a complex Hilbert space that its mathematical structure would play such a fundamental role?'

Let us turn for a moment to the other great theory of physics, namely general relativity, to raise more suspicion towards the fundamental role of the complex Hilbert space for quantum mechanics. General relativity is founded on the mathematical structure of Riemann geometry. In this case however it is much more plausible that indeed the right fundamental mathematical structure has been taken. Riemann developed his theory as a synthesis of the work of Gauss, Lobatsjevski and Bolyai on non-Euclidean geometry, and his aim was to work out a theory for the description of the geometrical structure of the world in all its generality. Hence Einstein took recourse to the work of Riemann to express his ideas and intuitions on space time and its geometry and this lead to general relativity. General relativity could be called in this respect 'the geometrization of a part of the world including gravitation'.

There is, of course, a definite reason why von Neumann used the mathematical structure of a complex Hilbert space for the formalization of quantum mechanics, but this reason is much less profound than it is for Riemann geometry and general relativity. The reason that Heisenberg's matrix mechanics and Schrödinger's ► wave mechanics turned out to be equivalent is that the first made use of l_2, the set of all square summable complex sequences, and the second of $L_2(\mathbb{R}^3)$, the set of all square integrable function of three variables, and the two spaces l_2 and $L_2(\mathbb{R}^3)$ are canonical examples of a complex Hilbert space. This means that Heisenberg and Schrödinger were working already in a complex Hilbert space, when they formulated matrix mechanics and wave mechanics, without being aware of it. This made it a straightforward choice for von Neumann to propose a formulation of quantum mechanics in an abstract complex Hilbert space, reducing ► matrix mechanics and wave mechanics to two possible specific representations.

One problem with the Hilbert space representation was known from the start. A (pure) state of a quantum entity is represented by a unit vector or ray of the complex Hilbert space, and not by a vector. Indeed vectors contained in the same ray represent the same state or one has to renormalize the vector that represents the state after it has been changed in one way or another. It is well known that if rays of a vector space are called points and two dimensional subspaces of this vector space are called lines, the set of points and lines corresponding in this way to a vector space, forms a projective geometry. What we just remarked about the unit vector or ray representing the state of the quantum entity means that in some way the projective geometry corresponding to the complex Hilbert space represents more intrinsically the physics of the quantum world as does the Hilbert space itself. This state of affairs is revealed explicitly in the dynamics of quantum entities, that is built by using group representations, and one has to consider projective representations, which are representations in the corresponding projective geometry, and not vector representations [6].

The title of the article by John von Neumann and Garett Birkhoff [1] that we mentioned as the founding article for operational quantum axiomatics is 'The logic of quantum mechanics'. Let us explain shortly what Birkhoff and von Neumann do in this article. First of all they remark that an operational proposition of a quantum entity is represented in the standard quantum formalism by an orthogonal projection operator or by the corresponding closed subspace of the Hilbert space $\mathcal{H}$. Let us denote the set of all closed subspaces of $\mathcal{H}$ by $\mathcal{L}(\mathcal{H})$. Next Birkhoff and von Neumann show that the structure of $\mathcal{L}(\mathcal{H})$ is not that of a Boolean algebra, the archetypical structure of the set of propositions in classical logic. More specifically it is the distributive law between conjunction and disjunction

$$(a \vee b) \wedge c = (a \wedge c) \vee (b \wedge c) \tag{1}$$

that is not necessarily valid for the case of quantum propositions $a, b, c \in \mathcal{L}(\mathcal{H})$. A whole line of research, called ► quantum logic, was born as a consequence of the Birkhoff and von Neumann article. The underlying philosophical idea is that, in the same manner as general relativity has introduced non-Euclidean geometry into the reality of the physical world, quantum mechanics introduces non-Boolean logic. The quantum paradoxes (► errors and paradoxes in quantum mechanics) would be due to the fact that we reason with Boolean logic about situations with quantum entities, while these situations should be reasoned about with non-Boolean logic.

Although fascinating as an approach [7], it is not this idea that is at the origin of quantum axiomatics. Another aspect of what Birkhoff and von Neumann did in their article is that they shifted the attention on the mathematical structure of the set of operational propositions $\mathcal{L}(\mathcal{H})$ instead of the Hilbert space $\mathcal{H}$ itself. In this sense it is important to pay attention to the fact that $\mathcal{L}(\mathcal{H})$ is the set of all operational propositions, i.e. the set of yes/no experiments on a quantum entity. They opened a way to connect abstract mathematical concepts of the quantum formalism, namely the orthogonal projection operators (► projection) or closed subspaces of the Hilbert space, directly with physical operations in the laboratory, namely the yes/no experiments.

George Mackey followed in on this idea when he wrote his book on the mathematical foundations of quantum mechanics [2]. He starts the other way around and considers as a basis the set $\mathcal{L}$ of all operational propositions, meaning propositions being testable by yes/no experiments on a physical entity. Then he introduces as an axiom that this set $\mathcal{L}$ has to have a structure isomorphic to the set of all closed subspaces $\mathcal{L}(\mathcal{H})$ of a complex Hilbert space in the case of a quantum entity. He states that it would be interesting to invent a set of axioms on $\mathcal{L}$ that gradually would make $\mathcal{L}$ more and more alike to $\mathcal{L}(\mathcal{H})$ to finally arrive at an isomorphism when all the axioms are satisfied. While Mackey wrote his book results as such were underway. A year later Constantin Piron proved a fundamental representation theorem. Starting from the set $\mathcal{L}$ of all operational propositions of a physical entity and introducing five axioms on $\mathcal{L}$ he proved that $\mathcal{L}$ is isomorphic to the set of closed subspaces $\mathcal{L}(V)$ of a generalized Hilbert space V whenever these five axioms are satisfied [3]. Let us elaborate on some of the aspects of this representation theorem to be able to explain further what operational quantum axiomatics is about.

We mentioned already that Birkhoff and von Neumann had noticed that the set of closed subspaces $\mathcal{L}(\mathcal{H})$ of a complex Hilbert space $\mathcal{H}$ is not a Boolean algebra, because distributivity between conjunction and disjunction, like expressed in (1), is not satisfied. The set of closed subspaces of a complex Hilbert space forms however a lattice, which is a more general mathematical structure than a Boolean algebra, moreover, a lattice where the distributivity rule (1) is satisfied is a Boolean algebra, which indicates that the lattice structure is the one to consider for the quantum mechanical situation. To make again a reference to general relativity, the lattice structure is indeed to a Boolean algebra what general Riemann geometry is to Euclidean geometry. And moreover, meanwhile it has been understood why the structure of operational propositions of the world is not a Boolean algebra but a lattice. This is due to the fact that measurements can have an uncontrollable influence on the state of the physical entity under consideration [4]. Hence the intuition of Birkhoff and von Neumann, and later Mackey, Piron and others, although only mathematical intuition at that time, was correct.

Axiomatic quantum mechanics is more than just an axiomatization of quantum mechanics. Because of the operational nature of the axiomatization, it holds the potential for 'more general theories than standard quantum mechanics' which however are 'quantum like theories'. In this sense, we believe that it is one of the candidates to generate the framework for the new theory to be developed generalizing quantum mechanics and relativity theory [4]. Let us explain why we believe that operational quantum axiomatics has the potential to deliver such a generalization of relativity theory and quantum mechanics. General relativity is a theory that brings part of the world that in earlier Newtonian mechanics was classified within dynamics to the geometrical realm of reality, and more specifically confronting us with the prescientific and naive realistic vision on space, time, matter and gravitation. It teaches us in a deep and new way, compared to Newtonian physics, 'what are the things that exists and how they exist and are related and how they influence each other'. But there is one deep lack in relativity theory: it does not take into account the influence of the observer, the effect that the measuring apparatus has on the thing observed. It does not confront the subject-object problem and its influence on how reality is. It cannot do this because its mathematical apparatus is based on the Riemann geometry of time-space, hence prejudicing that time-space is there, filled up with fields and matter, that are also there, independent of the observer. There is no fundamental role for the creation of 'new' within relativity theory, everything just 'is' and we are only there to 'detect' how this everything 'is'. That is also the reason why general relativity can easily be interpreted as delivering a model for the whole universe, whatever this would mean. We know that quantum mechanics takes into account in an essential way the effect of the observer through the measuring apparatus on the state of the physical entity under study. In a theory generalizing quantum mechanics and relativity, such that both appear as special cases, this effect should certainly also appear in a fundamental way. We believe that general relativity has explored to great depth the question 'how can things **be** in the world'. Quantum axiomatics explores in great depth the question 'how can things be **acted** in the world'. And it does explore this question of 'action in the world' in a very similar manner as general relativity theory does with its question of 'being of the world'. This means that operational quantum axiomatics can be seen as the development of a general theory of 'actions in the world' in the same manner that Riemann geometry can be seen as a general theory of 'geometrical forms existing in the world'. Of course Riemann is not equivalent to general relativity, a lot of detailed physics had to be known to apply Riemann resulting in general relativity. This is the same with operational quantum axiomatics, it has the potential to deliver the framework for the theory generalizing quantum mechanics and relativity theory.

We want to remark that in principle a theory that describes the possible actions in the world, and a theory that delivers a model for the whole universe, should not be incompatible. It should even be so that the theory that delivers a model of the whole universe should incorporate the theory of actions in the world, which would mean for the situation that exists now, general relativity should contain quantum mechanics, if it really delivers a model for the whole universe. That is why we believe that Einstein's attitude, trying to incorporate the other forces and interactions within general relativity, contrary to common believe, was the right one, globally speaking. What Einstein did not know at that time was 'the reality of ▶ nonlocality in the micro-world'. Nonlocality means non-spatiality, which means that the reality of the micro-world, and hence the reality of the universe as a whole, is not time-space like. Time-space is not the global theatre of reality, but rather a crystallization and structuration of the macro-world. Time-space has come into existence together with the macroscopic material entities, and hence it is 'their' time and space, but it is not the theatre of the microscopic quantum entities. This fact is the fundamental reason why general relativity, built on the mathematical geometrical Riemannian structure of time-space, cannot be the canvas for the new theory to be developed. A way to express this technically would be to say that the set of events cannot be identified with the set of time-space points as is done in relativity theory. Recourse will have to be taken to a theory that describes reality as a kind of pre-geometry, and where the geometrical structure arises as a consequence of interactions that collapse into the time-space context. We believe that operational quantum axiomatics can deliver the framework as well as the methodology to construct and elaborate such a theory.

Mackey and Piron introduced the set of yes/no experiments but then immediately shifted to an attempt to axiomatize mathematically the lattice of (operational) propositions of a quantum entity, Mackey postulating right away an isomorphism with $\mathcal{L}(\mathcal{H})$ and Piron giving five axioms to come as close as possible to $\mathcal{L}(\mathcal{H})$. Also Piron's axioms are however mostly motivated by mimicking mathematically the structure of $\mathcal{L}(\mathcal{H})$. In later work Piron made a stronger attempt to found operationally part of the axioms [8], and this attempt was worked out further in [9], to arrive at a full operational foundation only recently [4].

Also mathematically the circle was closed only recently. There do exist a lot of finite dimensional generalized Hilbert spaces that are different from the three standard examples, real, complex and quaternionic Hilbert space. But since a physical entity has to have at least a position observable, it follows that the generalized Hilbert space must be infinite dimensional. At the time when Piron gave his five axioms that lead to the representation within a generalized Hilbert space, there only existed three examples of generalized Hilbert spaces that fitted all the axioms, namely real, complex and quaternionic Hilbert space. Years later Hans Keller constructed the first counterexample, more specifically an example of an infinite dimensional generalized Hilbert space that is not isomorphic to one of the three standard Hilbert spaces [10]. The study of generalized Hilbert spaces, nowadays also called orthomodular spaces, developed into a research subject of its own, and recently Maria Pia Solèr proved a groundbreaking theorem in this field. She proved that an infinite dimensional generalized Hilbert space that contains an orthonormal base is isomorphic with one of the three standard Hilbert spaces [11]. It has meanwhile also been possible to formulate an operational axiom, called 'plane transitivity' on the set of operational propositions that implies Solèr's condition [12], which completes the axiomatics for standard quantum mechanics by means of six axioms, the original five axioms of Piron and plane transitivity as sixth axiom.

An interesting and rather recent evolution is taking place, where quantum structures, as developed within this operational approach to quantum axiomatics, are used to model entities in regions of reality different of the micro-world [13–20]. We believe that also this is a promising evolution in the way to understand deeper and more clearly the meaning of quantum mechanics in all of its aspects. See also ► algebraic quantum mechanics; relativistic quantum mechanics.

Primary Literature

1. Birkhoff, G., von Neumann, J.: *The logic of quantum mechanics*. Annals of Mathematics. 37, 823–43 (1936).
2. Mackey, G.: *Mathematical Foundations of Quantum Mechanics*. Benjamin, New York (1963).
3. Piron, C.: *Axiomatique quantique*. Helvetica Physica Acta. 37, 439–68 (1964).
4. Aerts, D., Aerts, S.: *Towards a general operational and realistic framework for quantum mechanics and relativity theory*. In A. C. Elitzur, S. Dolev, N. Kolenda (Eds.), Quo Vadis Quantum Mechanics? Possible Developments in Quantum Theory in the 21st Century. 153–208. Springer, New York (2004).

O

Secondary Literature

5. von Neumann, J. V.: *Mathematische Grundlagen der Quantenmechanik*. Springer, Berlin (1932).
6. Wigner, E. P.: *Group Theory and its Applications to Quantum Mechanics of Atomic Spectra*. Academic, New York (1959).
7. Mittelstaedt, P.: *Philosophische Probleme der Modernen Physik*. Bibliographisches Institut, Manheim (1963).
8. Piron, C.: *Foundations of Quantum Physics*. Benjamin, Massachusetts (1976).
9. Aerts, D.: *Description of many physical entities without the paradoxes encountered in quantum mechanics*. Foundations of Physics. 12, 1131–70 (1982).
10. Keller, H.: *Ein nicht-klassischer Hilbertscher Raum*. Mathematische Zeitschrift. 172, 1432–1823 (1980).
11. Solèr, M. P.: *Characterization of Hilbert spaces by orthomodular spaces*. Communications in Algebra. 23, 219–43 (1995).

12. Aerts, D., Van Steirteghem, B.: *Quantum axiomatics and a theorem of M. P. Solèr*. International Journal of Theoretical Physics. 39, 497–502, archive ref and link: quant-ph/0105107 (2000).
13. Eisert, J., Wilkens, M., Lewenstein, M.: *Quantum games and quantum strategies*. Physical Review Letters. 83, 3077–080 (1999).
14. Schaden, M.: Quantum finance: *A quantum approach to stock price fluctuations*. Physica A. 316, 511–38 (2002).
15. Widdows, D., Peters, S.: *Word vectors and quantum logic: Experiments with negation and disjunction*. In Mathematics of Language 8. 141–54. Bloomington, Indiana (2003).
16. Aerts, D., Czachor, M.: *Quantum aspects of semantic analysis and symbolic artificial intelligence*. Journal of Physics A, Mathematical and Theoretical. 37, L123–L132 (2004).
17. Aerts, D., Gabora, L.: *A theory of concepts and their combinations I & II. Kybernetes*. 34, 167–91; 192–221 (2005).
18. Bruza, P. D., Cole, R. J.: *Quantum logic of semantic space: An exploratory investigation of context effects in practical reasoning*. In S. Artemov, H. Barringer, A. S. d'Avila Garcez, L.C. Lamb, J. Woods (Eds.): We Will Show Them: Essays in Honour of Dov Gabbay. College Publications (2005).
19. Bagarello, F.: *An operatorial approach to stock markets*. Journal of Physics A. 39, 6823–840 (2006).
20. Busemeyer, J. R., Wang, Z., Townsend, J. T.: *Quantum dynamics of human decision making*. Journal of Mathematical Psychology. 50, 220–41 (2006).

Operator

Werner Stulpe

Operator, a technical term that is used for a mapping associating elements of some more or less abstract space uniquely with elements of the same or some other abstract space. If $\mathcal{X}$ and $\mathcal{Y}$ are such spaces (e.g., vector spaces or spaces of functions), an *operator from* $\mathcal{X}$ *to* $\mathcal{Y}$ (or *from* $\mathcal{X}$ *into* $\mathcal{Y}$) assigns exactly one element $\psi \in \mathcal{Y}$ to every element ϕ belonging to some specified subset D_A of $\mathcal{X}$; one writes $\psi = A\phi$. The set D_A is called the *domain of* A, the set of all elements $\psi \in \mathcal{Y}$ of the form $\psi = A\phi$, $\phi \in D_A$, is called the *range of* A. If B is a second operator from $\mathcal{Y}$ to $\mathcal{Z}$ such that $R_A \subseteq D_B$, then the *product* BA is defined by the successive application of A and B, i.e., $BA\phi = B(A\phi)$ where $D_{BA} = D_A$ and $R_{BA} \subseteq R_B$. An operator A from $\mathcal{X}$ into $\mathcal{Y}$ is called *invertible* if $A\phi_1 = A\phi_2$, $\phi_1, \phi_2 \in D_A$, implies $\phi_1 = \phi_2$; in this case the *inverse operator* A^{-1} is defined to be that operator that takes $\psi \in R_A$, $\psi = A\phi$, back to the uniquely determined $\phi \in D_A$. So $D_{A^{-1}} = R_A$, $R_{A^{-1}} = D_A$, and $A^{-1}\psi = \phi$ for $\psi = A\phi$; furthermore, $A^{-1}A\phi = \phi$ and $AA^{-1}\psi = \psi$.

In quantum physics, linear operators acting in a complex ▶ Hilbert space $\mathcal{H}$ play a dominant role [1–7]. An *operator* A *in* $\mathcal{H}$, i.e., from $\mathcal{H}$ to $\mathcal{H}$, is called *linear* if (i) D_A is a linear submanifold of $\mathcal{H}$, (ii) $A(\phi + \chi) = A\phi + A\chi$ for all $\phi, \chi \in D_A$, and (iii) $A(\lambda\phi) = \lambda A\phi$ for all complex numbers $\lambda \in \mathbb{C}$ and all $\phi \in D_A$. As a consequence, the range R_A is also a linear submanifold of $\mathcal{H}$. An operator in

$\mathcal{H}$ satisfying conditions (i), (ii), and (iii') $A(\lambda\phi) = \overline{\lambda}A\phi$ where $\overline{\lambda}$ is the complex conjugate of λ, is called *antilinear*.

A linear operator A acting in $\mathcal{H}$ is called *bounded* if $\|A\phi\| \leqslant c\|\phi\|$ for some real number $c \geqslant 0$ and all $\phi \in D_A$ (for the definition of the norm $\|\phi\|$ of a vector $\phi \in \mathcal{H}$, ► Hilbert space). A (not necessarily linear) operator A is *continuous* if from $\|\phi_n - \phi\| \to 0$ as $n \to \infty$, $\phi_n \in D_A$, and $\phi \in D_A$ it follows that $\|A\phi_n - A\phi\| \to 0$ as $n \to \infty$. A linear operator is continuous if and only if it is bounded. A (not necessarily linear) operator A is said to be *closed* if from $\|\phi_n - \phi\| \to 0$ as $n \to \infty$, $\phi_n \in D_A$, and $\|A\phi_n - \psi\| \to 0$ as $n \to \infty$ it follows that $\phi \in D_A$ and $\psi = A\phi$. Since the closure of the linear submanifold D_A is a Hilbert space itself, we can assume that either $D_A = \mathcal{H}$ or that $D_A \neq \mathcal{H}$ is dense in $\mathcal{H}$ (► Hilbert space). So, for linear operators, the following cases can be distinguished:

1. $D_A = \mathcal{H}$ and A is bounded. Then A is continuous and closed.
2. $D_A = \mathcal{H}$ and A is closed. Then, according to the so-called *closed-graph theorem*, A is bounded and continuous.
3. $D_A = \mathcal{H}$ and A is not bounded (equivalently, not continuous, resp., not closed). This possible case is only of pathological interest.
4. $D_A \neq \mathcal{H}$, D_A is dense in $\mathcal{H}$, and A is bounded. Then A is continuous, not closed, but can uniquely be extended to a bounded linear operator defined on $\mathcal{H}$.
5. $D_A \neq \mathcal{H}$, D_A is dense in $\mathcal{H}$, and A is not bounded (resp., not continuous), but closed.
6. $D_A \neq \mathcal{H}$, D_A is dense in $\mathcal{H}$, and A is not bounded and not closed. Such an operator can be *closable*, i.e., A can have a closed extension. A closable operator A always has a smallest closed extension, called its *closure* $\overline{A}$.

For a bounded linear operator A in $\mathcal{H}$, the smallest number c such that $\|A\phi\| \leqslant c\|\phi\|$ holds for all $\phi \in D_A$, is called its *operator norm* $\|A\|$. Let $\mathcal{B}(\mathcal{H})$ be the set of all bounded linear operators defined *on* $\mathcal{H}$ (i.e., $D_A = \mathcal{H}$). According to $(A+B)\phi = A\phi + B\phi$ and $(\lambda A)\phi = \lambda A\phi$ where $A, B \in \mathcal{B}(\mathcal{H})$, $\lambda \in \mathbb{C}$, and $\phi \in \mathcal{H}$, an addition of the operators of $\mathcal{B}(\mathcal{H})$ and a multiplication by numbers is defined. So $\mathcal{B}(\mathcal{H})$ becomes a complex vector space and, equipped with the operator norm, a complex Banach space (► Hilbert space). Moreover, since operators $A, B \in \mathcal{B}(\mathcal{H})$ can be multiplied, the product AB being an element of $\mathcal{B}(\mathcal{H})$ satisfying $\|AB\| \leqslant \|A\| \, \|B\|$, $\mathcal{B}(\mathcal{H})$ is a Banach algebra with some additional structure (► algebraic quantum mechanics).

An operator $A \in \mathcal{B}(\mathcal{H})$ is called *compact* if, for a bounded sequence of vectors $\phi_n \in \mathcal{H}$, the sequence $A\phi_n$ contains a convergent subsequence. The set $\mathcal{C}(\mathcal{H})$ of all compact operators is a norm-closed subspace of $\mathcal{B}(\mathcal{H})$ and an ideal of $\mathcal{B}(\mathcal{H})$, i.e., $A \in \mathcal{C}(\mathcal{H})$ and $B \in \mathcal{B}(\mathcal{H})$ implies $AB, BA \in \mathcal{C}(\mathcal{H})$.

A linear operator A in a Hilbert space $\mathcal{H}$ with domain D_A dense in $\mathcal{H}$ (including the case $D_A = \mathcal{H}$) is called *symmetric* or *Hermitian* if $\langle\phi|A\psi\rangle = \langle A\phi|\psi\rangle$ for all $\phi, \psi \in D_A$. A densely defined linear operator in $\mathcal{H}$ is symmetric if and only if the scalar products $\langle\phi|A\phi\rangle$, $\phi \in D_A$, are real. A symmetric operator defined on $\mathcal{H}$ is necessarily bounded. The concept of the symmetry of a linear operator can be sharpened to that of self-adjointness which is defined below.

A real or complex number λ is said to be an *eigenvalue* of a linear operator A acting in $\mathcal{H}$ if there is a nonzero vector $\phi \in D_A$ such that $A\phi = \lambda\phi$, ϕ is called an *eigenvector*. The set of all eigenvectors belonging to the same eigenvalue is, together with the zero vector, a linear submanifold, the *eigenspace*; the eigenspaces of A are closed if $A \in \mathcal{B}(\mathcal{H})$ or if A is closed. Finitely many eigenvectors belonging to different eigenvalues are linearly independent. It is possible that a linear operator has no eigenvalues; however, it can also happen that an operator even in a separable ▶ Hilbert space has a continuum of eigenvalues (in this context, eigenvalues are understood precisely as defined here, the so-called *improper eigenvalues* are not considered). A compact operator $A \in \mathcal{C}(\mathcal{H})$ has at most countably many eigenvalues with zero as only possible accumulation point where the eigenspaces belonging to nonzero eigenvalues are finite-dimensional. A symmetric or ▶ self-adjoint operator in a separable Hilbert space also has at most countably many eigenvalues, these are real and the eigenspaces are orthogonal (▶ Hilbert space) to each other. In general, such an operator does not have a complete orthonormal system of eigenvectors, instead a self-adjoint operator has a so-called spectral decomposition which is a generalization of the case of a complete orthonormal system of eigenvectors and which is essential for quantum mechanics.

(Spectral decomposition, see ▶ Density operator; Ignorance interpretation; Measurement theory; Objectification;Probabilistic Interpretation; Propensities in Quantum Mechanics; Self-adjoint operator; Wave mechanics.)

Most of the concepts and statements mentioned until now are also valid for operators acting in a complex or real Banach space $\mathcal{X}$ (▶ Hilbert space) or even for operators from one Banach space $\mathcal{X}$ to some other Banach space $\mathcal{Y}$ (in the case of a real Banach or Hilbert space, the condition $\lambda \in \mathbb{C}$ must be replaced by $\lambda \in \mathbb{R}$, and there are no antilinear operators). The eigenvalue problem, of course, makes sense only for linear operators acting in $\mathcal{X}$, and symmetric or ▶ self-adjoint operators exist only in a Hilbert space $\mathcal{H}$ (in the case of a real Hilbert space, the criterion $\langle\phi|A\phi\rangle \in \mathbb{R}$ for the symmetry of a densely defined linear operator A does not apply). Furthermore, the set $\mathcal{B}(\mathcal{X})$ of all bounded linear operators defined on a Banach space $\mathcal{X}$, with values in $\mathcal{X}$, is a Banach algebra with a less rich structure than $\mathcal{B}(\mathcal{H})$. In the more general context of operators between different Banach spaces, the set $\mathcal{B}(\mathcal{X}, \mathcal{Y})$ of all bounded linear operators defined on $\mathcal{X}$, with values in $\mathcal{Y}$, is again a Banach space, but no longer an algebra, and the subspace $\mathcal{C}(\mathcal{X}, \mathcal{Y})$ of the compact operators is not an ideal.

For a linear operator A in a Hilbert space $\mathcal{H}$ with dense domain D_A, the *adjoint operator* A^* is defined as follows. The domain D_{A^*} of A^* consists of all vectors $\phi \in \mathcal{H}$ for which there exists a vector χ_ϕ such that $\langle\phi|A\psi\rangle = \langle\chi_\phi|\psi\rangle$ holds for all $\psi \in D_A$; since D_A is dense in $\mathcal{H}$, χ_ϕ is uniquely determined, and $A^*\phi = \chi_\phi$, $\phi \in D_{A^*}$, concludes the definition of A^*. In particular, $\langle\phi|A\psi\rangle = \langle A^*\phi|\psi\rangle$ for all $\psi \in D_A$ and all $\phi \in D_{A^*}$. The adjoint A^* is a closed linear operator, but the linear submanifold D_{A^*} is in general not dense in $\mathcal{H}$; in fact, D_{A^*} is dense if and only if A is closable in which case $\overline{A} = A^{**}$ (by definition, $A^{**} = (A^*)^*$). For $A \in \mathcal{B}(\mathcal{H})$, A^* is also bounded with domain $D_{A^*} = \mathcal{H}$. A densely defined linear operator in $\mathcal{H}$ is symmetric if and only if A^* is an extension of A (briefly written as $A \subseteq A^*$), i.e., A^* coincides with A on D_A, but possibly has a larger domain.

A densely defined linear operator in $\mathcal{H}$ is called *self-adjoint* (► self-adjoint operator) if $A = A^*$, i.e., $\langle\phi|A\psi\rangle = \langle A\phi|\psi\rangle$ for all $\phi, \psi \in D_A = D_{A^*}$. A bounded self-adjoint operator must necessarily be defined on $\mathcal{H}$. For a linear operator defined on $\mathcal{H}$, the concepts of symmetry and self-adjointness are equivalent; a self-adjoint operator defined on $\mathcal{H}$ is bounded. A self-adjoint operator $A \in \mathcal{B}(\mathcal{H})$ is said to be *positive*, briefly written as $A \geqslant 0$, if $\langle\phi|A\phi\rangle \geqslant 0$ for all $\phi \in \mathcal{H}$. If $A \in \mathcal{B}(\mathcal{H})$ is positive, the equation $B^2 = BB = A$ has a unique positive solution $B \in \mathcal{B}(\mathcal{H})$; B is called the *square root of* A and is denoted by $B = A^{\frac{1}{2}}$. The set $\mathcal{B}_s(\mathcal{H})$ of all bounded self-adjoint operators on $\mathcal{H}$ form a real Banach space; defining $A \leqslant B$ for $A, B \in \mathcal{B}_s(\mathcal{H})$ by $B - A \geqslant 0$, $\mathcal{B}_s(\mathcal{H})$ becomes partially ordered, in fact, $\mathcal{B}_s(\mathcal{H})$ is an ordered Banach space.

For a positive operator $A \in \mathcal{B}_s(\mathcal{H})$, the ► *trace* $\operatorname{tr} A = \sum_{i=1}^{\infty}\langle\phi_i|A\phi_i\rangle$ is well-defined, i.e., independent of the complete orthonormal system $\phi_1, \phi_2, \ldots$ of $\mathcal{H}$ (in this context, assume that $\mathcal{H}$ is an infinite-dimensional separable complex Hilbert space, the finite-dimensional case is trivial); $\operatorname{tr} A$ can be infinite. An arbitrary operator $A \in \mathcal{B}(\mathcal{H})$ is called a *trace-class operator* if $\operatorname{tr}(A^*A)^{\frac{1}{2}} < \infty$ (observe that $A^*A \geqslant 0$). For a trace-class operator, the *trace* $\operatorname{tr} A = \sum_{i=1}^{\infty}\langle\phi_i|A\phi_i\rangle$ exists and is well-defined. The set $\mathcal{C}^1(\mathcal{H})$ of all trace-class operators is a linear submanifold of $\mathcal{B}(\mathcal{H})$ and, equipped with the *trace norm* $\|A\|_1 = \operatorname{tr}(A^*A)^{\frac{1}{2}}$, a complex Banach space. The trace defines a linear functional on $\mathcal{C}^1(\mathcal{H})$ (i.e., a linear operator with range $\mathbb{C}$) which satisfies $\operatorname{tr} A^* = \overline{\operatorname{tr} A}$. If $A \in \mathcal{C}^1(\mathcal{H})$ and $B \in \mathcal{B}(\mathcal{H})$, then $AB, BA \in \mathcal{C}^1(\mathcal{H})$ where $\operatorname{tr} AB = \operatorname{tr} BA$ and $|\operatorname{tr} AB| \leqslant \|A\|_1\|B\|$. Moreover, according to $\Lambda_B(A) = \operatorname{tr} AB$ a bounded linear functional Λ_B on the Banach space $\mathcal{C}^1(\mathcal{H})$, i.e., an element of the dual space $(\mathcal{C}^1(\mathcal{H}))^*$, is defined, and by virtue of the association $B \mapsto \Lambda_B$ the spaces $\mathcal{B}(\mathcal{H})$ and $(\mathcal{C}^1(\mathcal{H}))^*$ are norm-isomorphic.— The space $\mathcal{C}_s^1(\mathcal{H})$ of the self-adjoint trace-class operators is, by means of the trace norm and the partial order inherited from $\mathcal{B}_s(\mathcal{H})$, an ordered real Banach space. If $A \in \mathcal{C}_s^1(\mathcal{H})$ and $B \in \mathcal{B}_s(\mathcal{H})$, then $\operatorname{tr} AB$ is real and the dual space $(\mathcal{C}_s^1(\mathcal{H}))^*$ is norm-isomorphic to $\mathcal{B}_s(\mathcal{H})$.

A *Hilbert–Schmidt operator* is an element $A \in \mathcal{B}(\mathcal{H})$ for which $\operatorname{tr} A^*A < \infty$. The set $\mathcal{C}^2(\mathcal{H})$ of all Hilbert-Schmidt operators is a Hilbert space where the scalar product is given by $\langle A|B\rangle = \operatorname{tr} A^*B$, $A, B \in \mathcal{C}^2(\mathcal{H})$; so the *Hilbert-Schmidt norm* reads $\|A\|_2 = (\operatorname{tr} A^*A)^{\frac{1}{2}}$. The following statements hold: $\mathcal{C}^1(\mathcal{H}) \subseteq \mathcal{C}^2(\mathcal{H}) \subseteq \mathcal{C}(\mathcal{H}) \subseteq \mathcal{B}(\mathcal{H})$; $\mathcal{C}^1(\mathcal{H})$, $\mathcal{C}^2(\mathcal{H})$, and $\mathcal{C}(\mathcal{H})$ are linear submanifolds as well as ideals of the algebra $\mathcal{B}(\mathcal{H})$; whereas $\mathcal{C}(\mathcal{H})$ is closed w.r.t. the operator norm, $\mathcal{C}^1(\mathcal{H})$ and $\mathcal{C}^2(\mathcal{H})$ are not closed (provided that $\dim\mathcal{H} = \infty$), but dense in $\mathcal{C}(\mathcal{H})$; for the Banach spaces $(\mathcal{C}^1(\mathcal{H}), \|\cdot\|_1)$, $(\mathcal{C}^2(\mathcal{H}), \|\cdot\|_2)$, $(\mathcal{C}(\mathcal{H}), \|\cdot\|)$, and $(\mathcal{B}(\mathcal{H}), \|\cdot\|)$ the dualities $(\mathcal{C}(\mathcal{H}))^* \cong \mathcal{C}^1(\mathcal{H})$, $(\mathcal{C}^1(\mathcal{H}))^* \cong \mathcal{B}(\mathcal{H})$, and $(\mathcal{C}^2(\mathcal{H}))^* \cong \mathcal{C}^2(\mathcal{H})$ are valid.

Like linear operators acting in a finite-dimensional vector space, operators $A \in \mathcal{B}(\mathcal{H})$ have matrix representations. Assume that $\mathcal{H}$ is an infinite-dimensional separable Hilbert space; let $\phi_1, \phi_2, \ldots$ be a complete orthonormal system in $\mathcal{H}$. Then, for $\chi \in \mathcal{H}$, $\chi = \sum_{i=1}^{\infty}\alpha_i\phi_i$ and $\psi = A\chi = \sum_{i=1}^{\infty}\beta_i\phi_i$ where $\alpha_i = \langle\phi_i|\chi\rangle$ and $\beta_i = \langle\phi_i|A\chi\rangle$. Moreover, $\beta_i = \left\langle\phi_i\middle|A\left(\sum_{j=1}^{\infty}\alpha_j\phi_j\right)\right\rangle = \sum_{j=1}^{\infty}\langle\phi_i|A\phi_j\rangle\alpha_j$. The complex numbers $a_{ij} = \langle\phi_i|A\phi_j\rangle$ are called the *matrix elements of* A *w.r.t.*

$\phi_1, \phi_2, \ldots$, and $\beta_i = \sum_{j=1}^{\infty} a_{ij}\alpha_j$, $i = 1, 2, \ldots$, is called the *matrix representation of* $\psi = A\chi$. One can write $\begin{pmatrix} \beta_1 \\ \beta_2 \\ \vdots \end{pmatrix} = \begin{pmatrix} a_{11} & a_{12} & \cdots \\ a_{21} & a_{22} & \cdots \\ \vdots & \vdots & \ddots \end{pmatrix} \begin{pmatrix} \alpha_1 \\ \alpha_2 \\ \vdots \end{pmatrix}$ where the column vectors are elements of the Hilbert space l^2 since $\|\psi\|^2 = \sum_{i=1}^{\infty} |\alpha_i|^2 < \infty$ and $\|A\psi\|^2 = \sum_{i=1}^{\infty} |\beta_i|^2 < \infty$.—If A is an unbounded operator with domain D_A dense in $\mathcal{H}$, then $\psi = A\chi$, $\chi \in D_A$, has a matrix representation w.r.t. $\phi_1, \phi_2, \ldots$ whenever $\phi_1, \phi_2, \ldots$ belongs to D_A as well as to D_{A^*}. Moreover, $\phi_1, \phi_2, \ldots \in D_{A^*}$ and $\phi_1, \phi_2, \ldots \in D_A$ entail that D_{A^*} is dense in $\mathcal{H}$, $A^{**} = \overline{A}$ exists, and $\phi_1, \phi_2, \ldots \in D_{A^{**}}$. So the action of A^* can also be represented in matrix form, the matrix elements a^*_{ij} of A^* satisfy $a^*_{ij} = \overline{a}_{ji}$. In particular, every symmetric or self-adjoint operator enables a matrix representation of $\psi = A\chi$ if $\chi, \phi_1, \phi_2, \ldots \in D_A$. The matrix elements of a symmetric or self-adjoint operator satisfy $a_{ij} = \overline{a}_{ji}$, i.e., the matrix elements constitute a *Hermitian matrix*.

Literature

1. F. Riesz and B. Sz.-Nagy: *Functional Analysis* (Frederick Ungar Publishing, New York 1955; reprint Dover Publications, Mineola 1991)
2. M. Reed and B. Simon: *Methods of Modern Mathematical Physics I: Functional Analysis* (Academic, New York 1972, second edition 1980)
3. N. Dunford and J. T. Schwartz: *Linear Operators* (Interscience Publishers, New York), *Part I: General Theory* (1958), *Part II: Spectral Theory* (1963)
4. W. Młak: *Hilbert Spaces and Operator Theory* (Springer, Dordrecht 1991)
5. D. Werner: *Funktionalanalysis* (Springer, Berlin 1995, third edition 2000)
6. G. Ludwig: *Foundations of Quantum Mechanics I* (Springer, New York 1983), Appendix IV
7. E. Prugovečki: *Quantum Mechanics in Hilbert Space* (Academic, New York 1971, second edition 1981; reprint Dover Publications, Mineola 2006)

See also ► Creation and Annihilation Operators
► Density Operator
► Projection
► Self-Adjoint Operator
► State Operator
► Unitary Operator

Orthodox Interpretation of Quantum Mechanics

Henry Stapp

Eugene Wigner, in a paper entitled *The Problem of Measurement* [1], used the term "orthodox interpretation" to identify the interpretation spelled out in mathematical detail by John von Neumann in his book *Mathematische Grundlagen der*

Quantenmechanik [2]. Von Neumann, in the chapter on the measuring process, shows how to expand the quantum mechanical description of a system to include the physical variables of the measuring device, or, more generally, the physical variables of any system that interacts with an original system of interest. He then gives a detailed analysis of the process of measurement.

Von Neumann calls the unitary evolution of the quantum state (or wave function) generated by the ► Schrödinger equation by the name "process 2". The process-2 quantum mechanical evolution is a mathematical generalization of the deterministic evolution of a dynamically closed system in classical physical theory. The quantum mechanical process 2, like its classical counterpart, is deterministic: given the quantum state at any time, the state into which will evolve at any later time via process 2 is completely fixed.

Von Neumann considers an (idealized) situation involving a sequence of physically described measuring devices each performing a good measurement on the outcome variables of the preceding device, leading eventually to the retina, then to the optical nerves, and finally to the higher brain centers directly associated with the consciousness of the observer. There is no apparent reason for the process 2 to fail at any point, provided the full environment (essentially the entire physically described universe) is included in the physical system. But in general the process 2 evolution will lead to a state in which the higher brain centers directly associated with consciousness will have non-negligible components corresponding to different incompatible experiences, such as seeing the pointer of a measuring device simultaneously at several distinct positions.

Von Neumann notes that "It is entirely correct that the measurement or the related process of subjective perception is a new entity relative to the physical environment and is not reducible to the latter. Indeed, it leads into the intellectual inner life of the individual, which is extra-observational by its very nature (since it must be taken for granted by any conceivable observation or experiment)."

To tie the quantum mathematics usefully to human experience von Neumann invokes another process, which he called "process 1". Process 1 partitions the state into a particular collection of components each corresponding to a distinct possible experience, but only one of which will survive the "► wave function collapse" or the "reduction of the ► wave packet" associated with process of measurement or observation.

Wigner proves that process 1 can never be a consequence of process 2 alone: some other process, not the quantum analog of the deterministic classical law of evolution, must come in. As in the classical case, one must of course respect the condition that the quantum system be dynamically closed. This means that if any macroscopic element is included in the quantum mechanically described system then one must effectively include the whole universe, due to the non-negligible effects of the environment upon a macroscopic system.

Von Neumann notes that, in line with the precepts of the Copenhagen interpretation, "we must always divide the world into two parts, the one being the observed system, the other the observer", and that "quantum mechanics describes the events which occur in the observed portion of the world, so long as they do not interact with the observing portion, with the aid of process 2, but as soon as such an interaction occurs, i.e., a measurement, it requires an application of process 1." (For Copenhagen interpretation see ▶ Born rule; Consistent Histories; Metaphysics in Quantum Mechanics; Nonlocality; Schrödinger's Cat; Transactional Interpretation.)

The von Neumann/Wigner approach is, in this regard, not identical to the Copenhagen interpretation specified by Bohr and Heisenberg, who, in keeping with their pragmatic epistemological stance, resist treating the entire physical universe as a quantum system obeying the linear deterministic unitary law. Bohr ties this limitation in the applicability of the normal quantum rules to the fact that any attempt to obtain sufficient knowledge about any living organism, in order to enable us to make useful predictions, would probably kill the organism. Hence "the strict application of those concepts adapted to our description of inanimate nature might stand in a relationship of exclusion to the consideration of the laws of the phenomena of life" [3]. This argument is effectively a cautious suggestion that the breakdown of process 2 might be associated with biological systems: i.e., with life. But von Neumann says "there arises the frequent necessity of localizing some of these processes at points which lie within the portion of space occupied by our own bodies. But this does not alter the fact of their belonging to the 'world about us', the objective environment referred to above."

Wigner's suggestion for dealing with this gross mismatch between the process-2 generated activities of our brains and the contents of our streams of conscious experiences, evidently stems from a desire to have a rationally coherent *ontological* understanding of nature herself; an understanding of the reality that actually exists. Noting that process 1 is associated with the occurrence of *observable* events, and hence the implied need for an observer, Wigner suggest that the breakdown of process 2 is due to the interaction of the physically described aspects of nature with the consciousness of a conscious being [4]. (▶ Wigner's Friend) This physically efficacious consciousness stands outside the physically described aspects of nature controlled by process 2. Von Neumann calls it the observer's "abstract ego".

Conscious experiences are certainly real, and real things normally have real effects. The most straightforward conclusion would seem to be that process 1 specifies features of the interaction between the brain activities that are directly associated with conscious experiences and the conscious experiences with which those activities are associated.

This solution is in line with Descartes' idea of two "substances", that can interact in our brains, provided "substance" means merely a carrier of "essences". The essence of the inhabitants of *res cogitans* is "felt experience". They are thoughts, ideas, and feelings: the realities that hang together to form our streams of conscious experiences. But the essence of the inhabitants of *res extensa* is not at all that of the sort of persisting stuff that classical physicists imagined the physical world to be made of.

They are indeed represented in terms of mathematically described properties assigned to space-time points, but their essential nature is that of "potentialities for the psycho-physical events to occur". These events occur at the interface between the psychologically and physically described aspects of nature, and the laws governing their interaction are given by von Neumann. The causal connections between "potentialities for psychologically described events to occur" and such events themselves are easier to comprehend and describe than causal connections between the corresponding features of classical physics. For, both sides of the duality are conceptually more like "ideas" than like "rocks".

Primary Literature

1. E. Wigner: The Problem of Measurement. American Journal of Physics **31**, 6–15 (1963)
2. J. Von Neumann: *Mathematische Grundlagen der Quantnmechanik* (Springer, Berlin 1932) (Translated as *Mathematical Foundations of Quantum Mechanics*, Princeton University Press, 1955)
3. N. Bohr: *Atomic Theory and the Description of Nature* (Cambridge University Press, Cambridge UK 1961, 22–23)
4. E. Wigner: Remarks on the Mind-Body Problem, in *The Scientist Speculates*, I. J. Good, ed. (Heinemann, London; Basic Books, New York 1962, 284–302)

Secondary Literature

5. N. Bohr: *Atomic Physics and Human Knowledge* (Wiley, New York 1958)

Orthonormal Basis

Roderich Tumulka

Orthonormal basis (plural orthonormal bases): a set B of vectors in Euclidean or Hilbert space such that every vector can be written as a (finite or infinite) linear combination of vectors from B, while all vectors from B have length 1 and any two of them are orthogonal. The number of vectors in B then equals the dimension of the space, which can be finite or infinite.

In the infinite-dimensional ► Hilbert spaces considered in quantum physics, the appropriate sense of linear combination is that of a *convergent series*

$$\psi = \sum_{n=1}^{\infty} c_n \, \phi_n \,, \tag{1}$$

where $B = \{\phi_1, \phi_2, \ldots\}$ and c_n are complex coefficients, called the *expansion coefficients* of ψ relative to B. (A basis in the sense that linear combinations are convergent series is called a *Schauder basis*, whereas a basis in the sense that linear combinations can only involve finitely many terms is a called a *Hamel basis*.) Thus, an orthonormal basis in (separable) Hilbert space is a set $B = \{\phi_1, \phi_2, \ldots\}$ of vectors such that every vector ψ can be written in the form (1), and

$$\langle \phi_n | \phi_m \rangle = \delta_{nm}, \tag{2}$$

where $\langle \cdot | \cdot \rangle$ is the scalar product in Hilbert space, $\delta_{nm} = 1$ if $n = m$ and $\delta_{nm} = 0$ otherwise.

A set of vectors that satisfies (2) but does not permit us to represent every vector in the form (1) is called an *orthonormal set* or *orthonormal sequence*; it is an orthonormal basis of a closed subspace. The word "orthonormal" means pairwise orthogonal ($\langle \phi_n | \phi_m \rangle = 0$ for all $n \neq m$) and normalized ($\langle \phi_n | \phi_n \rangle = 1$ for all n). A set of vectors that permits us to represent every vector ψ in the form (1) is called a *complete set*; if for every ψ the coefficients c_n are unique then the set is called a *basis*, but not orthonormal if it does not satisfy (2). A *complete orthonormal set* is the same as an orthonormal basis.

If, relative to an orthonormal basis $\{\phi_1, \phi_2, \ldots\}$, ψ has expansion coefficients c_n – as expressed in (1) – and ψ' has expansion coefficients c'_n then

$$\langle \psi | \psi' \rangle = \sum_{n=1}^{\infty} c_n^* c'_n, \tag{3}$$

where * denotes the ► complex conjugate. The coefficients can be computed according to

$$c_n = \langle \phi_n | \psi \rangle. \tag{4}$$

Just as a vector ψ is represented, relative to an orthonormal basis, by a sequence of numbers c_n, an ► operator T is represented by an (infinite) matrix $T_{nm} = \langle \phi_n | T \phi_m \rangle$. An operator T is *diagonal* in an orthonormal basis if $T_{nm} = 0$ for $n \neq m$. A self-adjoint operator T can be diagonalized (i.e., an orthonormal basis can be found in which T is diagonal) if and only if T has pure point spectrum. To diagonalize a self-adjoint operator T with continuous spectrum, one needs the concept of a *generalized orthonormal basis*: in this case, the basis elements are themselves not contained in the Hilbert space. For example, the generalized orthonormal basis diagonalizing the quantum-mechanical position operator on the Hilbert space $L^2(\mathbb{R})$ of square-integrable functions consists of Dirac delta functions, not contained in $L^2(\mathbb{R})$, and the generalized orthonormal basis diagonalizing the momentum operator consists of plane waves e^{ikx}, which are not square-integrable either. A generalized orthonormal basis can be defined rigorously as a unitary isomorphism between the given Hilbert space and $L^2(\Omega, \mu)$, where Ω is the index set of the generalized basis ($\Omega = \mathbb{R}$ in the examples above), and μ is a measure on Ω (the Lebesgue volume measure in the examples above).

Less frequently in quantum physics, one has to deal with Hilbert spaces of uncountable dimension, so-called *non-separable* Hilbert spaces. For such spaces, an orthonormal basis should be understood as a set B of vectors that is orthonormal (i.e., $\langle\phi|\phi\rangle = 1$ and $\langle\phi|\chi\rangle = 0$ for every $\phi, \chi \in B$ with $\phi \neq \chi$) and that is complete in the sense that for every vector ψ there exist $\phi_1, \phi_2, \ldots \in B$ such that ψ can be written as a countable linear combination of $\phi_1, \phi_2, \ldots$ as in (1). One should distinguish the concept of an orthonormal basis in a non-separable Hilbert space from that of a generalized orthonormal basis in a separable Hilbert space.

P

Parity

Andrzej K. Wróblewski

The term is used in two ways, first, as the operation P of spatial inversion, and the second as a numerical quantity associated with the system. Parity in the second sense is a multiplicative quantum number (▶ Quantum numbers) which could be $+1$ or -1. In quantum mechanics the operation of spatial inversion is described by equation $\mathbf{P}\ \Psi(\vec{r}) = \mathrm{P}\ \Psi(-\vec{r})$, where the unitary parity operator $\mathbf{P}$ acting on a ▶ wave function Ψ has only two eigenvalues $\mathrm{P} = +1$ or $\mathrm{P} = -1$ which correspond to even and odd parity, respectively.

By convention, protons and neutrons have been assigned the same positive intrinsic parity. The intrinsic parity of the pion has been established experimentally to be negative. The total parity of the system of particles is the product of their intrinsic parities and the spatial parity given by $(-1)^l$, where l denotes the angular momentum of the wave function of the system. Thus the parity of a particle of spin l decaying into two pions is just $(-1)^l$ and that of a particle of spin l decaying into three pions equals $(-1)^{l+1}$.

History of parity began in 1924, when Otto Laporte (1902–1971), and independently Henry Norris Russell (1877–1957), analyzed the structure of the spectrum of iron and titanium and found that there were two kinds of energy levels, such that the transitions never occurred between levels of the same kind but always between levels of the first and the second kind. No convincing explanation of the existence of two types of levels was found within the framework of the old quantum theory. Then, in 1927, Eugene Wigner (1902–1995) analyzed Laporte's finding and showed that the two types of levels and the selection rule followed from the invariance of the Schrödinger equation (▶ Schrödinger equation) under the operation of inversion of coordinates $x \longrightarrow -x$, $y \longrightarrow -y$, $z \longrightarrow -z$. This property was originally called "*Spiegelung*", at least until 1933, when the term was still used by Pauli. The name "parity" appeared later. In 1935, Condon and Shortley used the term "parity operator" in their book on atomic spectra.

In modern language the two types of energy levels found by Laporte and Russell are states of positive and negative parity. The electric dipole transitions between states of the same parity are forbidden by parity conservation in electromagnetic interactions. The intrinsic parity of the emitted photons (▶ light quantum) is negative and in order for the total parity of the system to be conserved the parity of the atomic state must change.

The concept of parity conservation was quickly accepted by physicists. As the authors of a well-known textbook [13] put it: 'Since invariance under space reflection

D. Greenberger et al. (eds.), *Compendium of Quantum Physics: Concepts, Experiments, History and Philosophy*, © Springer-Verlag Berlin Heidelberg 2009

is intuitively so appealing (why should a left- and a right-handed system be different?), conservation of parity quickly became a sacred cow'.

Complications appeared in the early 1950s. Several new "mesons", i.e. particles with mass intermediate between the electron and the proton, were discovered (► Particle physics). When more precise data became available, the two particles $K_{\pi 3} \equiv \tau^{\pm} \longrightarrow \pi^{\pm} + \pi^{+} + \pi^{-}$ and $K_{\pi 2} \equiv \theta^{\pm} \longrightarrow \pi^{\pm} + \pi^{0}$ appeared to have almost identical masses and lifetimes, although their parities seemed to be different. The decay properties of the θ were simple. The decay $\theta^{0} \longrightarrow \pi^{0} + \pi^{0}$ has been observed. The Bose–Einstein statistics (► Quantum statistics) requires the system of two neutral pions to have even parity and therefore even orbital momentum l. The intrinsic parity of the θ must be even and its spin (► Spin) must be zero. The spin of the τ meson was established to be even. Because of the decay into three pions the τ parity was found to be negative. This became known as the tau-theta puzzle. There were several attempts to solve it. Of course it could have been just a coincidence: two different particles of almost identical mass and lifetime but different parities. But usually physicists are wary when they encounter coincidences.

In August, 1955, Tsung Dao Lee (b. 1926) and Jay Orear [1] proposed to explain the tau-theta puzzle by assuming that there are two different particles; the heavier one decays rapidly into the lighter: $\tau \longrightarrow \theta + \gamma$ or $\theta \longrightarrow \tau + \gamma$. This hypothesis had soon to be rejected because of negative results of the search for the supposed γ rays. In December, 1955, Lee and Chen Ning Yang (b. 1922) came forward with another explanation [2]. All particles with odd strangeness S were assumed to be "parity doublets", that is, two particles with opposite parity. The θ^{+} and τ^{+} were assumed to have the same spin but opposite parity (such as, e.g. 0^{+} and 0^{-}). Thus, in particular, for the reaction $\pi^{+} + n \longrightarrow \Lambda_{1}^{0} + \theta^{+}$, one obtained a reaction of equal amplitude by taking the parity conjugation of all the particles $\pi^{+} + n \longrightarrow \Lambda_{2}^{0} + \tau^{+}$. Here Λ_{2}^{0} was the parity conjugated state of Λ_{1}^{0}.

P

One of its main topics of discussion during the Sixth Annual Rochester Conference in April, 1956, was the rapidly growing field of the new elementary particles, in particular, the tau-theta puzzle. However, no convincing solution was found. A few weeks later Lee and Yang discussed the possibility that parity could be violated in weak processes. After consultations with Chien Shiung Wu (1912–1997) from Columbia, an expert in beta decay, they soon discovered that nobody has ever proved that parity conservation was valid for weak interactions. They presented analysis of the problem in the paper submitted on June 22, 1956 [3]. Several possible experimental tests of parity conservation in β decay were listed in this paper. Lee and Yang suggested to measure the angular distribution of the ► electrons coming from β decays of oriented nuclei, e.g. Co^{60}. If θ is the angle between the orientation of the parent nucleus and the momentum of the electron, an asymmetry of distribution between θ and $180^{0} - \theta$ constitutes an unequivocal proof that parity is not conserved in β decay. The angular distribution of the β radiation was assumed to be of the form $I(\theta)d\theta = (\text{constant})(1 + a\cos\theta)\,\sin\theta\,d\theta$. If $a \neq 0$, one would then have a positive proof of parity nonconservation in β decay. Lee and Yang also proposed to study the distribution of the angle θ between the μ momentum and the electron momentum

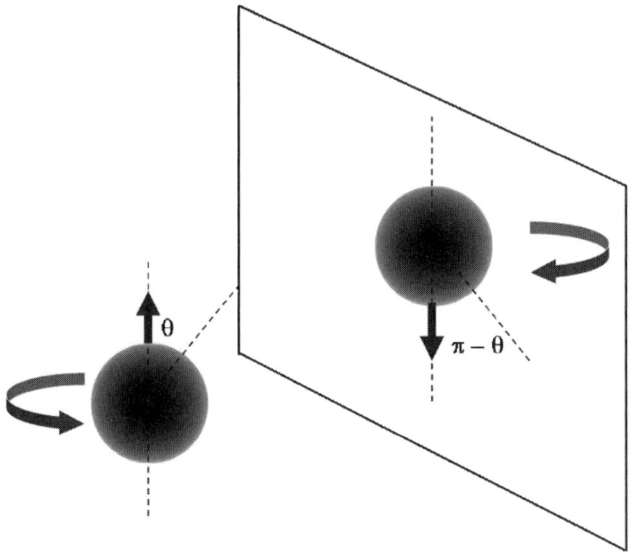

Fig. 1 The direction of rotation and the spin of rotating object are reversed by mirror reflection. Thus, if parity is conserved, the emission of electrons at angles θ and $\pi - \theta$ must be the same

in the decay processes $\pi \longrightarrow \mu + \nu$, $\mu \longrightarrow e + \nu + \nu$, starting from a π meson at rest. If parity is not conserved the distribution would not in general be identical for θ and $\pi - \theta$.

Chien Shiung Wu resolved to try an experiment even before Lee and Yang submitted their paper for publication. The idea of an experiment with Co^{60} was simple only in theory (Fig. 1). In order to make the measurement possible the radioactive nuclei must be aligned (polarized) so that their spins pointed in the same direction. It required very low temperatures, otherwise the thermal motion of the nuclei would destroy the alignment. Wu combined forces with Ernest Ambler (b. 1923) whose group at the National Bureau of Standards in Washington was involved in a nuclear orientation work.

The Co^{60} nucleus emits both β and γ rays. The degree of polarization can be measured by the anisotropy of the γ radiation, which is emitted more in the polar direction than in the equatorial plane. The β particles from ^{60}Co could not penetrate any substantial thickness of matter. For this reason Wu and her collaborators had to locate the radioactive nuclei in a very thin layer of only 0.002 inch on a surface of cerium magnesium (cobalt) nitrate. The β counter had to be placed inside the demagnetization cryostat. The β particles emitted by ^{60}Co nuclei were detected by scintillations in a thin anthracene crystal located inside the vacuum chamber about 2 cm above the ^{60}Co source. The scintillations were transmitted through a glass window and a Lucite light pipe 4 feet long to a photomultiplier located at the top of the cryostat.

The paper [3] by Lee and Yang was published only on October 1, 1956, but its contents was known earlier because of a circulated preprint. Most physicists rejected the idea of parity nonconservation as too fantastic and adverse to universally accepted notions on symmetries in physics.

First readings confirming parity violation were obtained by Wu's team on December 27, but the results were not consistently reproducible in the following days. They announced success only on January 9, 1957, after everything had been checked and rechecked. A few days earlier, during a discussion among Columbia physicists over a meal in a cafe on Friday, January 4, Leon Lederman (b. 1922) learned about Wu et al. preliminary results. He quickly realized that it was possible to check Lee and Yang's ideas about decay processes $\pi \longrightarrow \mu + \nu$, $\mu \longrightarrow e + \nu + \nu$, by using the muon beam from the cyclotron at the Nevis Laboratory of Columbia University. He explained the idea over the phone to his colleague, Richard Garwin (b. 1928). It took Garwin, Lederman, and Lederman's graduate student Marcel Weinrich, just little over 48 hours to prepare and carry out the experiment with a muon beam from the university cyclotron. The two papers [4, 5] from Columbia University were submitted for publication on January 15.

The chain of decays $\pi \longrightarrow \mu + \nu$, $\mu \longrightarrow e + \nu + \nu$ was studied also at the University of Chicago. Valentine Telegdi (1922–2006) read a preprint of Lee and Yang paper in August and, not knowing about Wu et al. effort, began an experiment similar in many respects to that of Lederman. With his postdoctoral researcher, Jerome Friedman (b. 1930), he exposed nuclear emulsion to a π^+ beam of the University of Chicago synchrocyclotron. They scanned the emulsions for characteristic $\pi \longrightarrow \mu + \nu$ events. In each case the scanner followed the muon to the end of its range and measured the angle of the positron emission. Their paper [6] was submitted for publication on January 17, two days after the two papers from Columbia. With 2000 $\pi \longrightarrow \mu \longrightarrow e$ events Telegdi and Friedman were able to determine that the electron emission indeed followed the linear law of the form $1+a \cos\theta$, postulated by Lee and Yang, and determined $a = 0.174 \pm 0.038$.

At the beginning of 1957 an experiment similar to that of Wu et al. has also been done in Leyden with ^{58}Co, which is a positron emitter [7]. It decays into ^{58}Fe and emits a positron and a neutrino $^{58}\text{Co} \longrightarrow {}^{58}\text{Fe} + e^{+} + \nu$. In this case the positron was found to be preferentially emitted along the direction of the nuclear spin (magnetic field) (Fig. 2).

There were numerous experiments checking parity nonconservation in various circumstances. Good review of these works can be found in [14], whereas a popular account of the theory is given in [15]. Parity nonconservation effects have been well explained by the two-component theory of the neutrino proposed independently by Landau (1908–1968) [8], Salam (1926–1996) [9], and Lee and Yang [10]. Massless neutrinos were assumed to possess a "handedness" to their spin. All neutrinos in nature were found to spin in a left-handed sense relative to their direction of flight, whereas antineutrinos were right-handed.

The discovery that parity is not conserved in weak interactions increased interest in the discrete symmetry operations, the charge conjugation C and time reversal T. It was shown that relativistic locality required invariance of the Lagrangian of any

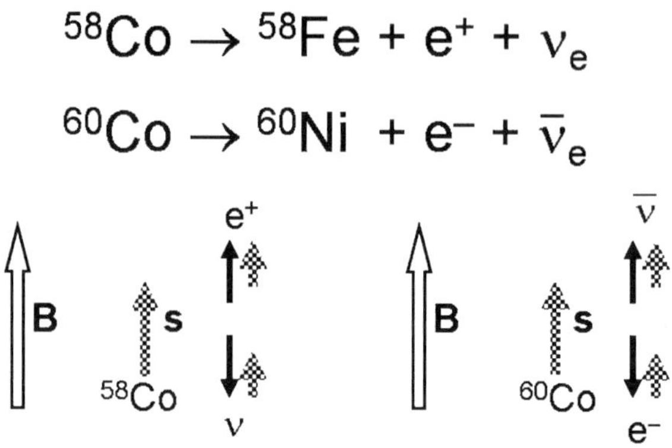

Fig. 2 Comparison of beta decays of ^{60}Co and ^{58}Co. The electrons from the ^{60}Co decay are emitted preferentially into the hemisphere opposite to the nuclear spin **s**, whereas the positrons from the ^{58}Co are emitted preferentially along the spin of the nucleus. It illustrates the left-handedness of neutrinos and right-handedness of antineutrinos

system under the combined operation CPT, irrespective of order of the three operations (▶ CPT theorem). The two-component theory of the neutrino allowed a natural formulation of a CP-conserving, but P- and C- violating, weak interaction. Then, in 1964, the unexpected discovery of CP nonconservation in kaon decay [11] took the physics community by surprise. It followed from the CPT theorem that time reversal symmetry must also be violated. It was indeed confirmed in 1998 by experiments at CERN [12].

Primary Literature

1. T. D. Lee, J. Orear, Speculations on Heavy Mesons, The Physical Review **100**, 932 (1955)
2. T. D. Lee, C. N. Yang, Mass Degeneracy of the Heavy Mesons, The Physical Review **102**, 290 (1956); Possible Interference Phenomena Between Parity Doublets, The Physical Review **104**, 822 (1956)
3. T. D. Lee, C. N. Yang, Question of Parity Conservation in Weak Interactions, The Physical Review **104**, 254 (1956)
4. C. S. Wu, E. Ambler, R. W. Hayward, D. D. Hoppes, R. P. Hudson, Experimental Test of Parity Conservation in Beta Decay, The Physical Review **105**, 1413 (1957)
5. R. L. Garwin, L. M. Lederman, M. Weinrich, Observations of the Failure of Conservation of Parity and Charge Conjugation in Meson Decays: The Magnetic Moment of a Free Muon, The Physical Review **105**, 1415 (1957)
6. J. I. Friedman, V. L. Telegdi, Nuclear Emulsion Evidence for Parity Nonconservation in the Decay Chain $\pi^+ \longrightarrow \mu^+ \longrightarrow e^+$, The Physical Review **105**, 1681 (1957)
7. H. Postma, W. J. Huiskamp, A. R. Miedema, M. J. Steenland, H. A. Tolhoek, C. J. Gorter, Asymmetry of the Positron Emission by Polarized ^{58}Co Nuclei, *Physica* **23**, 259 (1957); Asymmetry of the Positron Emission by Polarized ^{58}Co and ^{58}Mn Nuclei, Physica **24**, 157 (1958)
8. L. Landau, On the Conservation Laws for Weak Interactions, Nuclear Physics **3**, 127 (1957)
9. A. Salam, On Parity Conservation and Neutrino Mass, Il Nuovo Cimento, **5**, 299 (1957)
10. T. D. Lee, C. N. Yang, Parity Nonconservation and the Two-Component Theory of the Neutrino, The Physical Review **105**, 1671 (1957)

11. J. Christenson, J. W. Cronin, V. L. Fitch, R. Turlay, Evidence for the 2π Decay of the K_2^0 Meson, Physical Review Letters **13**, 138 (1964)
12. A. Angelopoulos et al., First Direct Observation of Time-reversal Non-invariance in the Neutral Kaon System, Physics Letters B **444**, 43–51 (1998)

Secondary Literature

13. H. Frauenfelder, E. M. Henley, *Nuclear and Particle Physics* (Benjamin, Reading, MA 1975, 258–416)
14. C. S. Wu, Parity Experiments in Beta Decay, Reviews of Modern Physics **31**, 783 (1959); V. L. Telegdi, The Early Experiments Leading to the $V - A$ Interaction, in: *Pions to Quarks. Particle Physics in the 1950s*, ed. Laurie M. Brown, Max Dresden, Lillian Hoddeson (Cambridge University Press, Cambridge 1989)
15. T. D. Lee, Space Inversion, Time Reversal and Particle-Antiparticle Conjugation, Physics Today **19**(3), 23 (1966)

Particle Physics

Kim Milton

P

The first discovered of what we would now call an elementary particle is the electron; although its discovery was a long and complicated process, J. J. Thomson's experiments of 1897 played a decisive role, since he was the first to obtain a quantitative value for e. Remarkably, precision experiments conducted last year (2006) [1] show that the electron still possesses no structure other than that demanded by quantum mechanics and relativity – it is a point particle. The proton, as the nucleus of the hydrogen atom, was identified as soon as the Rutherford scattering experiment demonstrated the ► model of the atom (► large-angle scattering); its partner in the nucleus, the neutron, was discovered by Chadwick in 1932. (For a review of the history of particle physics told in words of some of its creators, see Ref. [13]. See also Refs. [14, 15].)

Antiparticles were theoretically predicted by P.A.M. Dirac in 1928 on the basis of his famous ► Dirac equation describing the relativistic electron, or more generally, any particle carrying ► spin $\hbar/2$ [2]. At first he thought the positive proton was the antiparticle to the negative electron, but then he was convinced that the antiparticle had to have the same mass as the particle (this is now seen as a consequence of the famous ► CPT theorem). The positron was actually discovered in 1933 by Anderson, Blackett, and Occhialini [3]. The antiproton was found in 1955 [4].

Nuclear forces were studied extensively in the 1930s, aided immeasurably by the use of Lawrence's cyclotron. It was clear that new forces beyond those known since ancient times, gravity and electromagnetism, had to come into play in order to hold the nucleus together, overcoming the strong Coulomb repulsion of its positive protons. Yukawa in 1935 proposed the existence of a mesotron (now meson); the exchange of which between protons and neutrons could explain the strong nuclear force [5]. (This was analogous to the explanation of electromagnetism through the exchange of the massless photon between charged particles ▶ QED.) However, unlike electromagnetism strong nuclear forces have a very short range ($\sim 10^{-15}$ m), and so, by the ▶ Heisenberg uncertainty principle, must correspond to the exchange of a massive particle some 200 times heavier than the mass of an electron. Indeed, in 1938 Neddermeyer and Anderson discovered a particle of mass ~100 MeV,[1] which we now call the muon. However, these particles turned out to be not strongly interacting, and resulted in a period of confusion, which was only resolved in 1947, when what we now call the pion, indeed Yukawa's mesotron, was discovered by Lattes, Occhialini, and Powell at a mass of about 140 MeV. The pion (π) could decay into a muon (μ) plus a neutrino (ν),

$$\pi^{\pm} \rightarrow \mu^{\pm} + \nu,$$

where the superscripts denote the charges of the particles. The neutral, massless, neutrino had been proposed by Pauli in 1930 to explain the apparent failure of the conservation of energy in the so-called β-decay of the neutron,

$$n \rightarrow p + e^{-} + \bar{\nu},$$

where the overbar signifies that is actually an antineutrino that is produced here. (This is called β-decay because the electron was earlier called a β-ray.)

So the muon was the first "unwanted" particle discovered. (I.I. Rabi once said, "who ordered that?") It turned out to be the first member discovered of the second generation or family. As new accelerators were built after the Second World War, such as Berkeley's Bevatron and Brookhaven National Laboratory's Cosmotron, a proliferation of new particles, mostly very strongly interacting and very unstable, living only maybe 10^{-23} s, were discovered. Many of these particles carried a new quantum number called "strangeness," conserved by the strong interactions – therefore the lightest of these lived much longer. By the late 1960s hundreds of strongly interacting particles, dubbed hadrons, had been discovered. Some were fermions, like the electron, having spin equal to an integer plus one-half times $\hbar$; these were called baryons. Those whose spins where integers times $\hbar$, bosons, were called mesons. (The ▶ spin statistics theorem is reflected here: Only one fermion can occupy a given quantum state, while any number of bosons can do so. The latter allows for the phenomenon of ▶ Bose condensation, which is responsible for ▶ superconductivity and ▶ superfluidity.) This proliferation of particles represented a crisis for particle physics, for not all these states could be elementary constituents of matter.

---

[1] In particle physics, it is customary to adopt "natural units" in which $c = \hbar = 1$.

Many efforts were made to bring order out of this chaos. The first great success came to Gell-Mann in 1961 [6] (there were of course precursors and competitors) who proposed, not too seriously, the quark model as a mathematical way to organize the various particles under a symmetry group called SU(3), the group of 3×3 unitary matrices having determinant one. The reality of quarks was not taken seriously until the late 1960s, when high-energy scattering experiments at Stanford ("deep-inelastic scattering") suggested, somewhat like the Rutherford model of the atom, that point-like constituents existed inside the proton and neutron, which were first called partons, but are now recognized as quarks [7]. Quarks, see ► Color Charge Degree of Freedom in Particle Physics; Mixing and Oscillations of Particles; Parton Model; QCD; QFT.

The next step was taken by Schwinger [8], Glashow [9], Weinberg [10], and Salam [11], who discovered (1957–71) that electromagnetic and weak nuclear forces (those responsible for β decay) could be "unified" into a single theory, the so-called electroweak unification. It is represented by the product of two groups, SU(2)×U(1). To understand the strong nuclear force, Greenberg introduced the idea of "color," a new quantum number carried by quarks, and shortly after the success of the electroweak theory, Gell-Mann and others proposed that color SU(3) (not to be confused with the flavor SU(3) mentioned in the previous paragraph) would be the underlying symmetry of the strong interactions between the quarks, and thus was born quantum chromodynamics or ► QCD.

The resulting picture is called the Standard Model (SM) of particle physics. (► Quantum field theory). Matter is composed of fermions, quarks and leptons, the latter being particles that feel the electroweak forces but not the strong ones. The leptons consist of charged particles, like the electron and muon, and neutral particles of very small mass, the neutrinos. The forces are carried by bosons: the photon, and its weak partners, $W^{\pm}$ and Z^0, and gluons, which come in eight color states. The quarks and lepton occur in pairs, grouped in three families:

$$\begin{pmatrix} u \\ d \end{pmatrix} \quad \begin{pmatrix} c \\ s \end{pmatrix} \quad \begin{pmatrix} t \\ b \end{pmatrix}$$

$$\begin{pmatrix} e \\ \nu_e \end{pmatrix} \quad \begin{pmatrix} \mu \\ \nu_\mu \end{pmatrix} \quad \begin{pmatrix} \tau \\ \nu_\tau \end{pmatrix}$$

The masses of the quarks and leptons are given in Table 1. Neutrino masses are very small, but now known to be nonzero. The neutrino flavor eigenstates, which couple

Table 1 Approximate masses of quarks and charged leptons in millions of electron volts, MeV. Masses for the quarks are the so-called current algebra masses, not constituent masses.

$m_u \sim 2$	$m_c \sim 1200$	$m_t = 174{,}000$
$m_d \sim 6$	$m_s \sim 100$	$m_b = 4200$
$m_e = .511$	$m_\mu = 106$	$m_\tau = 1777$
$m_{\nu_e} \neq 0$	$m_{\nu_\mu} \neq 0$	$m_{\nu_\tau} \neq 0$

P

to the weak interactions, are not the same as the mass eigenstates. This leads to the phenomenon of neutrino mixing. This is a bit complicated to describe for three kinds of neutrinos. If we make the approximation of two-state mixing, the probability of a neutrino of type α turning into a neutrino of type β is [16]

$$P(\nu_\alpha \to \nu_\beta) = \sin^2 2\theta_{\alpha\beta} \sin^2 \left(1.27\Delta m^2_{\alpha\beta}(\mathrm{eV}^2)\frac{L(\mathrm{km})}{E(\mathrm{GeV})}\right)$$

Recent observations appear to give for the values of the parameters here, the mixing angles $\theta_{\alpha\beta}$ and the mass differences $\Delta m^2_{\alpha\beta} = m^2_\alpha - m^2_\beta$,

$$\Delta m^2_{21} \approx 8 \times 10^{-5}\,\mathrm{eV}^2, \quad \Delta m^2_{32} \approx 2 \times 10^{-3}\,\mathrm{eV}^2,$$

$$\sin^2 2\theta_{12} \sim 0.86, \ \sin^2 2\theta_{23} > 0.92, \ \sin^2 2\theta_{13} < 0.19$$

Interactions are mediated by gauge bosons, which have the following properties (m is the mass, and S the spin):

8 Gluons: $g \quad m_g = 0, \ S = 1$

3 Electroweak bosons: $W^\pm, \ Z^0$

$$m_W = 80.4\,\mathrm{GeV}, \ m_Z = 91.2\,\mathrm{GeV}, \ S = 1$$

1 Photon: $\gamma, \quad m_\gamma = 0, \ S = 1$

1 Graviton: $\mathfrak{g}, \quad m_\mathfrak{g} = 0, \ S = 2$

(Here, for completeness, we make reference to gravity, which is not actually described by the Standard Model.) The group-theory structure of the interactions within the Standard Model are given by the product of three unitary groups:

$$SU(3) \times SU(2) \times U(1)$$

The mathematics of this group gives reaction rates that are completely in accord with experiment.

We do not know where the masses of the elementary particles come from. In the Standard Model, the masses are accommodated by another particle, the *Higgs boson*. The Higgs boson is the only element of the Standard Model not yet discovered: Since it has yet to be seen, $m_H > 115\,\mathrm{GeV}$. The expectation is that the Higgs boson will be discovered at the Large Hadron Collider (LHC).

Although there is no evidence that the Standard Model breaks down even at the highest energies, and in fact, QED is valid to fantastic precision, and Newtonian gravity holds to $\sim 50\,\mu\mathrm{m}$ [12] (both these limits were greatly extended during the past year), parameters (masses and couplings) in the Standard Model are unexplained. Therefore many physicists speculate that new physics lies beyond the

Standard Model. The most popular extension is supersymmetry (SUSY), which is the hypothesis that for every fermion there is a partner boson, and vice versa. However, at present, there is no evidence for SUSY particles, and in fact strong evidence against SUSY (coming from limits on the electric dipole moment of the electron and neutron). It is hoped that supersymmetric partners to SM particles will be found at the LHC. Other more exotic possibilities, such as signatures for large extra dimensions (also rendered less likely by the precision gravity tests), will be searched for there as well. See also Color Charge Degree in Particle Physics.

Primary Literature

1. B. Odom, D. Hanneke, B. D'Urso, and G. Gabrielse: New Measurement of the Electron Magnetic Moment Using a One-Electron Quantum Cyclotron. Phys. Rev. Lett. **97**, 030801 (2006); G. Gabrielse, D. Hanneke, T. Kinoshita, M. Nio, and B. Odom: New Determination of the Fine Structure Constant from the Electron g Value and QED. Phys. Rev. Lett. **97**, 030802 (2006)
2. P. A. M. Dirac: The Quantum Theory of the Electron. Proc. R. Soc. Lond. A **117**, 610 (1928)
3. C. D. Anderson: The Positive Electron. Phys. Rev. **43**, 491 (1933). P. M. S. Blackett and G. P. S. Occhialini: Some Photographs of the Tracks of Penetrating Radiation. Proc. R. Soc. Lond. A **139**, 699 (1933)
4. O. Chamberlain, E. Segre, C. Wiegand and T. Ypsilantis: Observation of Anti-Protons. Phys. Rev. **100**, 947 (1955)
5. H. Yukawa: On the Interaction of Elementary Particles. Proc. Phys. Math. Soc. Jpn. **17**, 48 (1935)
6. M. Gell-Mann: Symmetries of Baryons and Mesons. Phys. Rev. **125**, 1067 (1962)
7. J. I. Friedman and H. W. Kendall: Deep Inelastic Electron Scattering. Ann. Rev. Nucl. Sci. **22**, 203 (1972)
8. J. Schwinger: A Theory of Fundamental Interactions. Ann. Phys. (N.Y.) **2**, 407 (1957)
9. S. L. Glashow. Partial Symmetries of Weak Interactions. Nucl. Phys. **22**, 579 (1961)
10. S. Weinberg: A Model of Leptons. Phys. Rev. Lett. **19**, 1264 (1967)
11. A. Salam: Weak and Electromagnetic Interactions, in N. Svartholm, ed. Elementary Particle Theory, Relativistic Groups and Analyticity. (Stockholm: Almqvist and Wiksell, 1968, pp. 367–377)
12. D. J. Kapner, T. S. Cook, E. G. Adelberger, J. H. Gundlach, B. R. Heckel, C. D. Hoyle and H. E. Swanson: Tests of the Gravitational Inverse-Square Law Below the Dark-Energy Length Scale. Phys. Rev. Lett. **98**, 021101 (2007) [arXiv:hep-ph/0611184]

Secondary Literature

13. *Colloque International sur l'Histoire de la Physique des Particules*. Journal de Physique, Colloque C8, supplément au no. 12, Tome 43 (1982)
14. Gordon Fraser, ed: *The Particle Century*. (London: Institute of Physics, 1998)
15. Val Fitch and Jonathan Rosner: *Elementary particle physics in the second half of the 20th century, in Twentieth Century Physics*, ed. Laurie Brown, A. Pais, and B. Pippard (Philadelphia: Institute of Physics, 1995), vol. 2, pp. 635–794
16. Particle Data Group, W.-M. Yao et al.: *Review of Particle Physics*. J. Phys. G: Nucl. Part. Phys. **33**, 1 (2006) [frequently updated online: http://pdg.lbl.gov/pdg.html]. (Contains both experimental data and short theoretical reviews.)

Particle Tracks

Brigitte Falkenburg

Particle tracks are sequences of adjacent position measurements caused by subatomic particles. As quantum phenomena to which the particle picture applies (► Franck–Hertz experiment), they constitute the empirical basis of ► particle physics. The dynamic properties of the underlying particles are measured by means of a semi-classical measurement theory. The generation of particle tracks, however, is explained in the wave picture by the quantum mechanics of scattering. ► Davisson–Germer experiment; Stern–Gerlach experiment; Schrödinger equation.

History

In 1912 particle tracks were first observed and photographed in Wilson's cloud chamber. They stemmed from radioactive radiation sources. For α-rays only a continuous track was visible, while for β-rays (► electrons) the individual measurement points could be clearly distinguished [1]. Since the 1920s, Wilson's cloud chamber helped to investigate particle tracks from cosmic rays, the most famous being the positron track observed by Anderson in 1932 [2]. Since the 1950s, particle tracks are also generated in the accelerator experiments of high energy physics.

Measuring Devices

The first decades of particle physics were based on various methods of taking photographs [3]. The cloud chamber developed by Charles T.R. Wilson (1869–1959) was filled with over-saturated steam. Charged particles ionize the hydrogen atoms ► Bohr's atom model of the steam, giving rise to observable condensation droplets. In the 1940s, nuclear emulsions made it possible to record the tracks of charged particles from cosmic rays and to develop their pictures photographically with a very high spatial resolution (of 1 μm). In the bubble chamber, developed in the 1950s by Donald A. Glaser (*1926) for the ► *scattering experiments* performed in particle accelerators, the ionization gives rise to gas bubbles in liquid hydrogen instead of condensation droplets in steam. The bubble chamber made it possible to detect and photograph a variety of particle tracks at the same time.

In modern electronic particle detectors, the particle tracks are no longer observable on a photograph. They are recorded electronically and reconstructed by computer programs. For example, a drift chamber detects and amplifies the electric currents caused by the passage of charged particles through a grid of wires. In this way, the observable particle tracks of the first decades of particle physics have been replaced by electronic data and their reconstruction. Only after a lengthy process of data analysis by means of reconstruction programs do they become visible on a computer display.

Measurement Theory

Particle tracks have characteristic phenomenological features, above all, the density of the measurement points, the curvature in a magnetic field, the track length, and the temporal order of the single position measurements (i.e., the flight direction). They give important hints for particle identification. In the first decades of particle physics, they made it possible to estimate the mass and charge of unknown particles. The flight direction can be inferred from the energy loss along a track which results in a characteristic increase of the track curvature. (In this way, Anderson identified the positive electrons as a particle with the electron's mass and charge, of opposite sign.)

In order to measure the dynamic properties of the underlying particle, the points of a particle track in space-time are connected or "fitted" by the trajectory of a massive charged particle. The trajectory is the data model [10] of a particle track. This data model is based on the classical model of a massive charged particle of mass m, charge q, and momentum $\vec{\boldsymbol{p}} = m\vec{\boldsymbol{v}}$, which loses energy along the track due to subsequent inelastic collisions with the detector atoms. In the model, the track ends when the particle is stopped. The track length indicates the kinetic energy lost during the passage of the particle through the detector. An empirical law, the so-called *energy-range* relation, connects the kinetic energy of a massive charged particle to its range (or track length) in different materials.

P

Based on this model, the particle tracks taken in an experiment are analyzed by means of a *semi-classical* measurement theory. This measurement theory contains [12]:

1. The classical Lorentz force $\boldsymbol{F} = q/c\ (\boldsymbol{E} + \vec{\boldsymbol{v}} \times \boldsymbol{B})$. It describes the momentum change of a massive charged particle in an external electric field $\boldsymbol{E}$ or magnetic field $\boldsymbol{B}$. According to the Lorentz force, the momentum of a particle of known mass and charge can be determined from the track curvature.
2. The laws of relativistic kinematics for particles of high energy, in particular, the law of energy-momentum conservation.
3. A dissipation term $\Delta E/\Delta x$ for the average energy loss ΔE per finite detector length Δx along the track [5]. The differential energy loss $\mathrm{d}E$ per path length $\mathrm{d}x$ obtained from $\mathrm{d}E/\mathrm{d}x \approx \Delta E/\Delta x$ is combined with the Lorentz force, giving rise to a differential equation for the mean momentum decrease due to energy dissipation along the track of a charged particle.
4. The empirical energy-range relation for charged particles in a given material, giving rise to a rough estimate of the average energy loss $\Delta E/\Delta x$. The average

range in a given material was measured for charged particles of known energy, for many materials [1, 23].

5. Quantum electrodynamic predictions for the dissipation of energy and the deflection of charged particles by subatomic scattering processes. They are based on the quantum mechanics of scattering [4] and the ► *quantum electrodynamic* description of ionization, ► *bremsstrahlung*, pair creation, and multiple scattering [5, 6].
6. Quantum mechanical conservation laws for ► spin, ► parity, isospin and other internal dynamic properties of subatomic particles, associated with the group theoretical definition of particles as the irreducible representations of ► *symmetry* groups.

The quantum electrodynamic laws which enter the measurements are supported by empirical laws. These empirical laws make it possible to test the quantum electrodynamic formulae independently. During the phase of consolidation of quantum electrodynamics, experimenters like Anderson exerted substantial effort to determine the mass and charge of particles by improvements in such independent semi-empirical measurement procedures. After the consolidation of quantum electrodynamics, the semi-empirical methods remained in the measurement theory. To the present day, they make it possible to perform several consistency checks on the measurements.

Mott's Idealized Quantum Mechanical Model

Strictly speaking, however, quantum mechanics is *incompatible* with the classical trajectories of the above measurement theory. So, how do they fit together?

Shortly after the development of quantum mechanics it was shown that the generation of particle tracks in a Wilson chamber is perfectly compatible with quantum mechanical scattering theory. As Werner Heisenberg (1901–1976) stressed in his 1930 book on quantum mechanics, the probability of α-particle deflection due to repeated ionization of molecules in the vapour is non-zero only if the connecting line of the two molecules runs parallel to the velocity direction of the α-particles [7]. The calculation was first carried out by Nevill F. Mott (1905–1996) in 1929 [8], based on Born's 1926 quantum mechanics of scattering which gave rise to the ► probabilistic interpretation of quantum mechanics [4]. According to quantum mechanics, the scattering is not due to the impact of a particle but to the diffraction of a wave, lacking the classical trajectory of a deflected particle and the corresponding classical impact parameter. The squared ► wave function predicts the probability of particle detections at a certain scattering angle.

Mott calculated the probability for two subsequent collisions of an α-particle and a hydrogen atom ► Rutherford atom with the effect of the ionization of both atoms. The ionized atoms give rise to observable measurement points, where the observation of a droplet is a *position measurement*. But the observation of the particle deflection given by straight lines drawn between the adjacent droplets is a *momentum measurement*. Heisenberg showed in 1930 that the uncertainty relation for position and momentum holds for any ionization process along the track. Due to the finite size of the water molecules in the Wilson chamber, the position and momentum measurement cannot both be sharp [7]. The inaccuracy of the position measurement for individual measurement points of a particle track and the measurement error of the particle momentum obtained from a curved particle track using the expression for the Lorentz force are typically more than 12 (!) orders of magnitude larger than ▶ Heisenberg's *uncertainty relation*.

Thus, the quantum mechanical explanation of single measurement points of a particle track is in perfect correspondence to the classical particle picture, the only difference being the unobservable classical path *between* the position measurements.

Mott's and Heisenberg's calculations neglect the energy loss associated with the ionization processes that give rise to the observable measurement points. The particle is described as if it did not transfer a definite amount of energy to the hydrogen atom when ionizing it. The calculations deal with the amplitudes of inelastic collisions, but they are performed as if the momentum state of the charged particle remained unaffected by the energy transfer that gives rise to ionization. This 'unrealistic' neglect of the momentum transfer is reasonable, since the energy loss of an α-particle due to ionization of hydrogen atoms is *very* small compared to the kinetic energy of the α-particle. Therefore the momentum of an α-particle remains practically unchanged along its track in the Wilson chamber.

Under such idealized conditions, the classical and quantum descriptions of a track agree for any sequence of measurement points. This ▶ *correspondence* between the classical and the quantum cases holds not only for the straight particle tracks calculated by Mott and Heisenberg but also for the curved tracks in a magnetic field. For a weak external field, the ▶ Schrödinger equation for a stationary beam of particles predicts approximately the classical beam deflection which is described by the Lorentz force [11].

Realistic Tracks with Energy Loss

In the case of a substantial amount of energy loss along a particle track, the agreement between the classical and the quantum descriptions vanishes. Nevertheless, the ▶ semi-classical model has to be maintained for the data analysis of individual particle tracks.

The first quantum mechanical calculation of non-negligible energy loss was given by Hans Bethe (1906–2005) in 1930 [5, 12]. Bethe's ▶ semi-classical model adds classical assumptions about the individual scattering processes along a particle track to Born's quantum mechanics of scattering. The calculation is performed in time-dependent perturbation theory. It results in a formula for the quantum mechanical expectation value $\langle E \rangle$, the mean energy loss per atom and per incoming particle (in the limit of infinitely many incoming particles, $N_{\text{in}} \rightarrow \infty$). Then the result is applied to the scattering processes along an individual particle track, giving rise to an expression for the mean energy loss ΔE per length Δx of matter. Hence, the ► semi-classical model assumes that the expectation value $\langle E \rangle$ means the *average energy loss of a charged particle by successive scatterings from many detector atoms along an individual track*, normalized to the number of atoms per path Δx.

Mott's and Bethe's calculations hold for the non-relativistic domain. According to Bethe's 1930 results, energy loss due to ionization is small and the shape of particle tracks is smooth. For relativistic particles, however, the semi-classical picture breaks down. ► Quantum Electrodynamics predicts that a particle does not lose its energy smoothly. Due to quantum fluctuations, the energy loss along a particle track may become completely irregular and extreme deviations from the classical path may occur. Several kinds of processes may give rise to large fluctuations in the energy loss. In addition to energy loss due to ionization, quantum electrodynamics predicts processes of ► *bremsstrahlung* and pair creation, that is, the emission of a photon or an electron–positron pair, respectively. These processes are associated with large fluctuations in the energy loss along a particle track. They give rise to irregular deflections which violate the classical shape of a track predicted by Mott in 1929 and presupposed also by Bethe in his 1930 energy loss calculation. In the data analysis of modern high energy ► *scattering experiments*, these fluctuations have to be corrected at the probabilistic level.

Primary Literature

1. E. Rutherford, J. Chadwick, C.D. Ellis: *Radiations from Radioactive Substances.* Cambridge: Cambridge University Press 1930. Repr.: Cambridge: Cambridge University Press 1951.
2. C.D. Anderson: *The Apparent Existence of Easily Deflectable Positives.* Science 76 (1932), 238. The Positive Electron. Phys. Rev. 43 (1933), 491–94.
3. C.F. Powell, P.H. Fowler, D.H. Perkins: *The Study of Elementary Particles by the Photographic Method.* London: Pergamon Press 1959.
4. M. Born: *Zur Quantenmechanik der Stoßvorgänge.* Z. F. Phys. 37 (1926), 863–67 Engl. transl. in [9]. M. Born, *Quantenmechanik der Stoßvorgänge.* Z. f. Phys. 38 (1926), 803–27.
5. H. Bethe: *Zur Theorie des Durchgangs schneller Korpuskularstrahlen durch Materie.* Ann. Phys. 53 (1930), 25–400.
6. H. Bethe: *Bremsformel für Elektronen relativistischer Geschwindigkeit.* Z. f. Phys. 76 (1932), 293–99. W. Heitler and F. Sauter: *Stopping of Fast Particles with Emission of Radiation and the Birth of Positive Electrons.* Nature 132 (1933), 892. F. Bloch: *Bremsvermögen von Atomen mit mehreren Elektronen.* Z. f. Phys. 81 (1933), 363–76. H. Bethe and W. Heitler: *On the Stopping of Fast Particles and on the Creation of Positive Electrons.* Proc. R. Soc. Lond. A 146 (1934), 83–112.
7. W. Heisenberg: *Physikalische Prinzipien der Quantentheorie.* Leipzig: Hirzel 1930. Engl. Transl.: *The Physical Principles of Quantum Theory.* Transl. by C. Eckart and F.C. Hoyt. Chicago: University of Chicago Press 1930, 70.
8. N.F. Mott: *The Wave Mechanics of α-Rays.* Proc. R. Soc. A 126 (1929), 79–84. Repr. in [9].
9. J.A. Wheeler and W.H. Zurek: *Quantum Theory and Measurement.* Princeton (N. J.), Guildford: Princeton University Press 1983.

Secondary Literature

10. P. Suppes: *Models of Data*. In: E. Nagel, P. Suppes, and A. Tarski (eds.): *Logic, Methodology and Philosophy of Science*. Proceedings of the 1960 International Congress. Stanford: Stanford University Press 1962, 252–61.
11. N.F. Mott, H.S. Massey: *The Theory of Atomic Collisions*. Oxford: Clarendon Press 1965.
12. B. Falkenburg: *Particle Metaphysics. A Critical Account of Subatomic Reality.* Springer: Heidelberg 2007, p. 169–87.

Parton Model

O.W. Greenberg

The parton model pictures hadrons as a collection of pointlike quasi-free particles. The model describes the cross section for high-energy scattering of hadrons with another particle as an incoherent sum of the cross sections of the pointlike partons in the hadron with the other particle. The hadronic factors in the cross sections are parametrized by "structure functions." The parton model expresses the structure functions in terms of parton distribution functions that give the longitudinal momentum distribution of the partons in the given hadron. The parton distribution functions are found from experimental data in a given process and are used in the description of other processes (Fig. 1).

The prototype process for the parton model is $eN \to e'X$, where e and e' are the incident and scattered electron, N is the target nucleon, and X is the set of final state hadrons. The particles in the final state X are not measured, so the cross section is for the sum over all hadronic final states, an "inclusive" cross section. This contrasts with an "exclusive" cross section in which the final states are restricted to a specific subset. In the prototype process, $eN \to e'X$, the kinematics of the inclusive scattering depends on the momentum transfer $q = k - k'$ from the electron to the hadrons and the invariant mass, W, of the hadronic final state, where $W^2 = (p+q)^2 = M^2+2M\nu+q^2$, and M is the mass of the target nucleon or other hadron. Here k and k' are the energy-momentum 4-vectors of the incident and scattered electron, p is the energy-momentum 4-vector of the target hadron, and $\nu = E - E'$ is the energy transfer to the target hadron in its rest frame.

J.D. Bjorken [1] predicted that the hadronic factor in the cross section would depend only on the ratio $x = (-q^2)/(2p \cdot q) = (-q^2)/(2M\nu)$, rather than on ν and $-q^2$ separately, on the basis of an algebra of local currents. This property, called "scaling," was expected to hold in the "deep inelastic" limit in which the energy transfer and momentum transfer are much larger than the target hadron mass. R.P. Feynman [2] interpreted scaling in terms of constituents of the nucleon that he called "partons." Bjorken and E.A. Paschos [3, 4] gave early discussions of electron-nucleon and neutrino-nucleon scattering in the deep inelastic limit. The

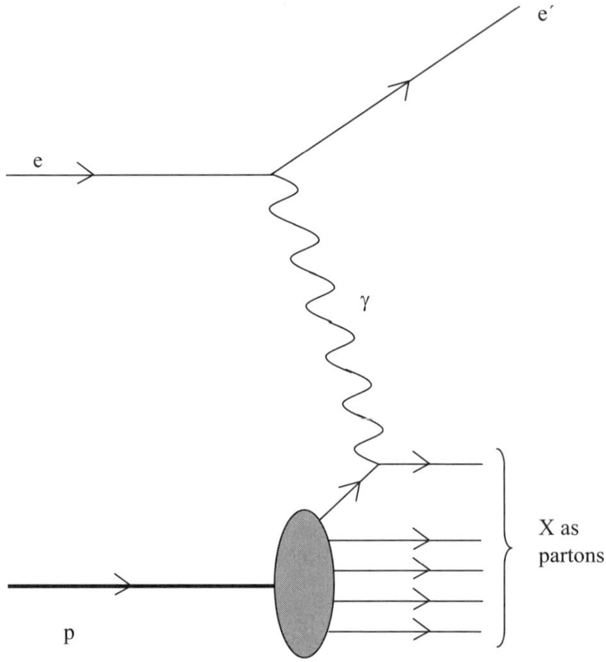

Fig. 1 Parton Model; $ep \to e'X$. $e(e')$ is the incident (scattered) electron. γ is the exchanged photon. p is the incident proton. X is the final hadronic state

Bjorken x can be identified with the fraction of the longitudinal hadron momentum carried by a given parton.

C.G. Callan and D.J. Gross [5] showed that the commutators of the electric current give information about the carriers of electric charge. Subsequent data on deep inelastic scattering showed that the carriers of charge have ► spin 1/2 and can be identified with quarks, ► Color Charge Degree of Freedom in Particle Physics; Mixing and Oscillations of Particles; Particle Physics; QCD; QFT (see [6] for early data and [7,8] for recent data in the references). Other sum rules together with data show that the charged partons carry only about 1/2 of the energy-momentum of the nucleon. The other half is carried by gluons and other neutral particles. Several reviews discuss sum rules below (see A.J. Buras [9], C. Bourrely and J. Soffer [10] and F. Close [22] in the references).

Surprisingly, scaling sets in at rather low energy and momentum transfer, so-called "precocious" scaling. [11] The paper of Bloom and Gilman also pointed out a duality between resonances and smooth scaling behavior which later led to the dual resonance model and even later to string theory. The partons are identified with the "valence" quarks that account for the electric charge, isospin and strangeness of the hadron, and with "sea" quarks that correspond to extra quark-antiquark pairs as well as with "gluons," which are quanta of the color gauge group that mediate quark interactions and have zero electric charge, isospin and strangeness. S.D. Drell,

D.J. Levy and T.-M. Yan extended the parton model to hadron-hadron scattering and gave the celebrated Drell-Yan mechanism for the production of lepton pairs (see [12] in the references for a review).

More detailed processes, such as semi-inclusive processes in which some of the final state hadrons are measured, require parton fragmentation functions [13], as well as parton distribution functions, for their description. The fragmentation functions account for the conversion of partons into hadrons in the final states. Gross and Wilczek [14] and H. Georgi and Politzer [15] showed that quantum chromodynamics predicts logarithmic corrections to scaling. The DGLAP formalism [16] expresses these corrections in parton language.

Scattering experiments with polarized beams and targets give information that cannot be obtained from unpolarized experiments. The EMC experiment with polarized muons scattering on polarized protons [17] led to the "spin crisis," that only about 1/4 of the spin of the proton is carried by quarks [18] (see reviews in [19]).

Feynman gave arguments that partons don't interact with each other in first approximation because in the limiting infinite momentum frame there is a separation of scales between the (slow) parton-parton interactions and the (fast) interaction with the scattered lepton. [13] The running of coupling constants that follows from asymptotic freedom ► Color Charge Degree of Freedom in Particles Physics; QCD; QFT provides further understanding of the mystery that quarks are permanently confined in hadrons viewed at low energy, but are quasi-free viewed as partons at high energy. [20, 21]

R.E. Taylor, H.W. Kendall and J.I. Friedman describe the pathbreaking experimental discoveries that stimulated the invention of the parton model [6]. P.M. Nadolsky et al. [7] and J. Blumlein et al. [8] analyse recent data on parton distributions ► nuclear models.

Primary Literature

1. J.D. Bjorken: Asymptotic Sum Rules at Infinite Momentum. Phys. Rev. **179**, 1527–1553 (1969).
2. R.P. Feynman: Very High-Energy Collisions of Hadrons. Phys. Rev. Lett. **23**, 1415–1417 (1969); The behavior of hadron collisions at extreme energies, in *Stony Brook 1969, Proceedings, Conference On High Energy Collisions*, ed. C.N. Yang, et al, (Gordon and Breach, New York 1969), 237–258. The word parton first appears here.
3. J.D. Bjorken and E.A. Paschos: Inelastic Electron-Proton and γ-Proton Scattering and the Structure of the Nucleon. Phys. Rev. **185**, 1975–1982 (1969).
4. J.D. Bjorken and E.A. Paschos: High-Energy Inelastic Neutrino-Nucleon Interactions. Phys. Rev. D **1**, 3151–3160 (1970).
5. C.G. Callan, Jr. and D.J. Gross: High-Energy Electroproduction and the Constitution of the Electric Current. Phys. Rev. Lett. **22**, 156–159 (1969).
6. R.E. Taylor: Deep inelastic scattering: The early years. Rev. Mod. Phys. **63**, 573–595 (1991); H.W. Kendall: Deep inelastic scattering: Experiments on the proton and the observation of scaling. Rev. Mod. Phys. **63**, 597–614 (1991); J.I. Friedman: Deep inelastic scattering: Comparisons with the quark model. Rev. Mod. Phys. **63**, 615–627 (1991).
7. P.M. Nadolsky, et. al.: Implications of CTEQ global analysis for collider observables. arXiv 0802.0007.

8. J. Blumlein, H. Bottcher and A. Guffanti: Non-Singlet QCD Analysis of Deep Inelastic World Data at $O(\alpha(s)^3)$. Nucl. Phys. B **774**, 182–207 (2007).
9. A.J. Buras: Asymptotic Freedom in Deep Inelastic Processes in the Leading Order and Beyond. Rev. Mod. Phys. **52**, 199–276 (1980).
10. C. Bourrely and J. Soffer: Deep inelastic scattering of leptons and hadrons in the QCD parton model and experimental tests. in *Scattering and Inverse Scattering in Pure and Applied Science*. ed. E.R. Pike and Pierre Sabatier. (Academic, New York 2001) pp. 1565–1581.
11. E.D. Bloom and F. Gilman: Scaling, Duality, and the Behavior of Resonances in Inelastic Electron-Proton Scattering. Phys. Rev. Lett. **25**, 1140 (1970).
12. S.D. Drell and T.-M. Yan: Partons and Their Applications at High Energies. Ann. Phys. (NY) **66**, 578 (1971).
13. R.P. Feynman, *Photon-Hadron Interactions*, (Benjamin, Reading 1972)
14. D.J. Gross and F. Wilczek: Asymptotically Free Gauge Theories. 1. Phys. Rev. D **8**, 3633–3652 (1973); and 2. Phys. Rev. D **9**, 980–993 (1974).
15. H. Georgi and H.D. Politzer: Electroproduction Scaling in an Asymptotically Free Theory of Strong Interactions. Phys. Rev. D **9**, 416–420 (1974).
16. Yu.L. Dokshitzer: Calculation of the Structure Functions for Deep Inelastic Scattering and e+ e− Annihilation by Perturbation Theory in Quantum Chromodynamics. Sov. J. JETP **46**, 641–653 (1977); V.N. Gribov and L.N. Lipatov: e+ e- Pair Annihilation and Deep Inelastic e p Scattering in Perturbation Theory. Sov. J. Nucl. Phys. **15**, 438–450 (1972) and Deep inelastic e p scattering in perturbation theory. Sov. J. Nucl. Phys. **15**, 438–450 (1972); and G. Alterelli and G. Parisi: Asymptotic Freedom in Parton Language. Nucl. Phys. B **126**, 298–318 (1977).
17. J. Ashman, et al: An investigation of the spin structure of the proton in deep inelastic scattering of polarised muons on polarised protons. Nucl. Phys. B **328**, 1–35 (1989).
18. S.J. Brodsky, J. Ellis and M. Karliner: Chiral Symmetry and the Spin of the Proton. Phys. Lett. B **206**, 309–315 (1988) and J. Ellis and M. Karliner: An Analysis of the Angular Momentum of the Proton. Phys. Lett. B **213**, 73–80 (1988).
19. X. Ji: The Spin of the Proton: Looking Back and Looking Forward. AIP Conf Proc. **915**, 16–21 (2007); and Generalized Parton Distributions. Annu. Rev. Nucl. Part. Sci. **54**, 413–450 (2004).
20. D.J. Gross and F. Wilczek: Ultraviolet Behavior of Non-Abelian Gauge Theories. Phys. Rev. Lett. **30**, 1343–1346 (1973).

Secondary Literature

21. H.D. Politzer: Reliable Perturbative Results for Strong Interactions? Phys. Rev. Lett. **30**, 1346–1349 (1973).
22. F.E. Close, *An Introduction to Quarks and Partons*, (Academic, London, 1980).
23. F. Halzen and A.D. Martin, *Quarks & Leptons*, (Wiley, New York, 1984).
24. D.H. Perkins, *Introduction to High Energy Physics*, (Addison-Wesley, London, 1987).
25. M. Riordan: *The Hunting of the Quark*, (Simon & Schuster, New York, 1987).

Paschen–Back Effect

Klaus Hentschel

In 1921, two experimental physicists in Tübingen, Friedrich Paschen (1865–1947) and Ernst Back (1881–1959), observed that with strongly increasing magnetic field

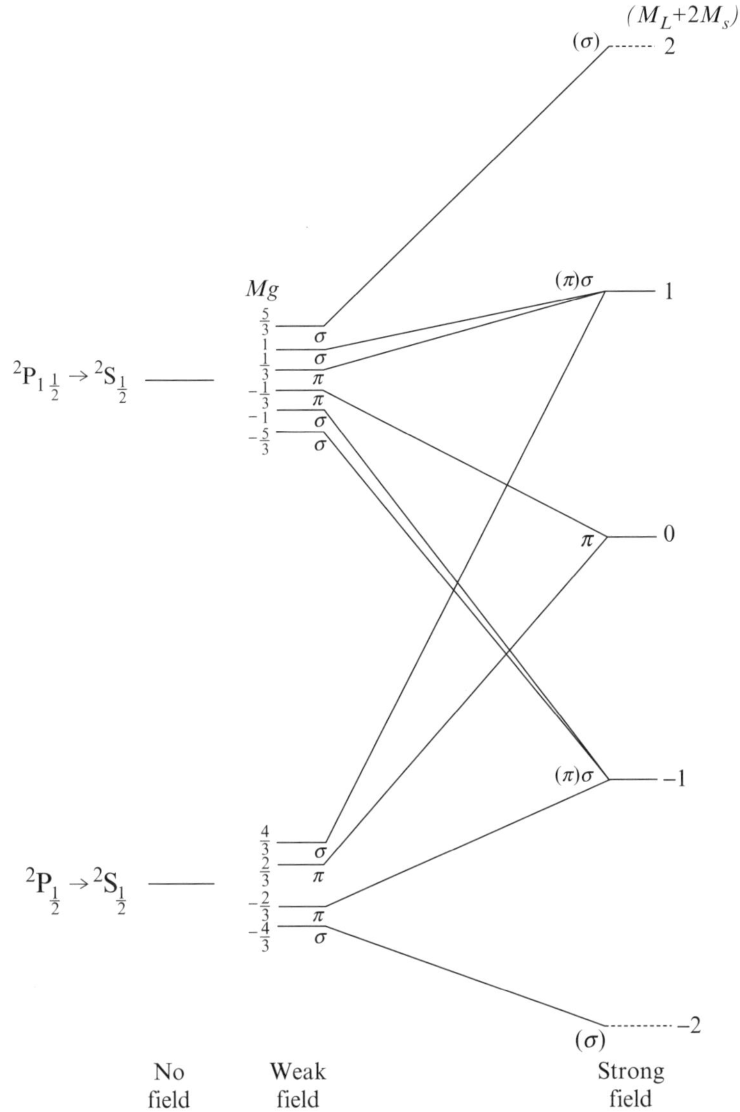

Fig. 1 Diagrammatic sketch of the changes occurring in a principal doublet as the field is increased; where π or σ is enclosed in brackets, this component fades in a strong field
Source: Chris Candler, *Atomic Spectra* (Cambridge, Cambridge Univ. Press 1937; London, Hilger & Watts [2]1964, 86)

strength, the complicated multiplets of the anomalous ► Zeeman effect change into the simpler patterns typical of the normal Zeeman effect (see Fig. 1). Initially, this observation remained inexplicable. With the discovery of ► spin in late 1925, however, and the realization that the anomalous Zeeman effect is characteristic of systems with spin $S > 0$, whereas the normal Zeeman effect governs atoms with a total $S = 0$, the Paschen–Back effect could be understood as a decoupling of S and orbital angular momentum L, since the influence of the total spin becomes neglectable for diminishing spin-orbit coupling. (See also ► Russell–Saunders coupling, ► jj-coupling, Stern–Gerlach experiment and ► vector model).

Literature

1. F. Paschen, E. Back: Normale und anomale Zeemaneffekte. Annalen der Physik (4th ser.), **39**, 897–932; **40**, 960–970 (1912)
2. F. Paschen, E. Back: Liniengruppen magnetisch vervollständigt. Physica **1**, 261–273 (1921)
3. G. Herzberg: *Atomic Spectra and Atomic Structure* (Dover, New York 1944, 112ff)

Pauli Exclusion Principle

See ► exclusion principle.

Pauli Spin Matrices

Roderich Tumulka

The Pauli spin matrices are the following 3 complex 2×2 matrices:

$$\sigma_x = \begin{pmatrix} 0 & 1 \\ 1 & 0 \end{pmatrix}, \quad \sigma_y = \begin{pmatrix} 0 & -i \\ i & 0 \end{pmatrix}, \quad \sigma_z = \begin{pmatrix} 1 & 0 \\ 0 & -1 \end{pmatrix}. \tag{1}$$

These matrices represent the spin observables along the x- (respective y- and z-)axis of physical 3-space for a spin-$\frac{1}{2}$ particle, relative to an ► orthonormal basis of spin space consisting of eigenvectors of σ_z. (Spin observables are measured, e.g., in the ► Stern–Gerlach experiment.) The spin observable along any direction in physical 3-space defined by the unit vector $\boldsymbol{n} = (n_x, n_y, n_z)$ is given, relative to the same basis, by

$$\sigma_{\boldsymbol{n}} = n_x\,\sigma_x + n_y\,\sigma_y + n_z\,\sigma_z = \boldsymbol{n} \cdot \boldsymbol{\sigma} \tag{2}$$

with $\boldsymbol{\sigma} = (\sigma_x, \sigma_y, \sigma_z)$. The spin observable is related to the angular momentum observable $J_{\boldsymbol{n}}$ along $\boldsymbol{n}$ according to

$$J_{\boldsymbol{n}} = \tfrac{\hbar}{2}\sigma_{\boldsymbol{n}} + L_{\boldsymbol{n}}, \tag{3}$$

where $L_{\boldsymbol{n}} = \boldsymbol{n} \cdot \boldsymbol{L}$ is the $\boldsymbol{n}$-component of the orbit angular momentum operator $\boldsymbol{L} = \boldsymbol{q} \times \boldsymbol{p}$. ▶ Spin; Stern–Gerlach experiment; Vector model.

The Pauli spin matrices, named after Wolfgang Pauli (1900–1958), are self-adjoint (= Hermitian) and unitary. Each of them (as well as $\sigma_{\boldsymbol{n}}$ for every unit vector $\boldsymbol{n}$) has trace equal to zero, determinant equal to -1, and eigenvalues 1 and -1.

The Pauli matrices belong to the fundamental structure of spin space, as spin space is defined to be a 2-dimensional complex vector space $\mathcal{H}_{\text{spin}} \cong \mathbb{C}^2$ coupled to physical 3-space by a law specifying how the elements of spin space transform under rotations. The law involves the Pauli matrices and asserts that the rotation through the angle $\varphi \in \mathbb{R}$ about the axis spanned by the unit vector $\boldsymbol{n} \in \mathbb{R}^3$ transforms the vector $\psi \in \mathcal{H}_{\text{spin}}$ from spin space into

$$\psi' = \pm \mathrm{e}^{-(\mathrm{i}/2)\varphi\sigma_{\boldsymbol{n}}}\psi \,. \tag{4}$$

(Exponentiation of a matrix can be defined by means of the power series $\mathrm{e}^x = \sum x^k/k!$.) As a consequence, for the rotation through an infinitesimal angle $\delta\varphi$ one can write, neglecting higher order terms in $\delta\varphi$,

$$\psi' = \psi - \tfrac{\mathrm{i}}{2}\delta\varphi\,\sigma_{\boldsymbol{n}}\,\psi \,. \tag{5}$$

From this equation one can read off that the matrix $-(\mathrm{i}/2)\sigma_{\boldsymbol{n}}$ (acting on ψ) represents the rate of change of ψ per angle when rotating around $\boldsymbol{n}$.

Expressing these facts in a technical way, spin space is endowed with an irreducible projective representation of the rotation group $SO(3)$ (the set of all orthogonal real 3×3 matrices with determinant 1), called the "spin-$\frac{1}{2}$ representation." Using the fact that $SO(3)$ can be "unfolded" yielding the group $SU(2)$ (the set of all unitary complex 2×2 matrices with determinant 1), the irreducible projective representation of $SO(3)$ can be translated into an irreducible representation of $SU(2)$, in fact the natural representation on $\mathbb{C}^2$ defined by matrix multiplication. In this translation, the rotation by φ about $\boldsymbol{n}$ corresponds to the matrix $\pm\mathrm{e}^{-(\mathrm{i}/2)\varphi\sigma_{\boldsymbol{n}}} \in SU(2)$, where the sign ambiguity arises from the "unfolding." The Lie algebra $su(2)$ associated with the Lie group $SU(2)$ consists of the infinitesimal generators of $SU(2)$, and thus of all matrices of the form $-(\mathrm{i}/2)\varphi\sigma_{\boldsymbol{n}}$, and that is the 3-dimensional real vector space of all traceless skew-adjoint 2×2 matrices, of which $\mathrm{i}\sigma_x, \mathrm{i}\sigma_y, \mathrm{i}\sigma_z$ form a basis.

The Pauli matrices satisfy the commutation relations

$$[\sigma_i, \sigma_j] = 2\mathrm{i}\sigma_k \tag{6}$$

if ijk is any cyclic permutation of xyz. Except for the factor 2, these relations are the same as those of any angular momentum operators; the reason is that these are the defining relations of the Lie algebra $su(2)$, which is also the Lie algebra of the rotation group $SO(3)$, and thus are relations characteristic of rotations in physical 3-space.

Higher spins: For spin-s particles, $s \in \frac{1}{2}\mathbb{Z}$, the matrices analogous to the Pauli spin matrices are 3 complex $(2s + 1) \times (2s + 1)$ matrices. Higher dimensions: If physical space had dimension d instead of 3, there would be $d(d - 1)/2$ Pauli spin matrices, as that number is the dimension of the rotation group $SO(d)$.

Primary Literature

1. W. Pauli: Zur Quantenmechanik des magnetischen Elektrons. Zeitschrift für Physik 43, 601–625 (1927)

Secondary Literature

2. R. U. Sexl, H. K. Urbantke: *Relativity, Groups, Particles.* (Springer-Verlag, New York, 2001)

Photoelectric Effect

Bruce R. Wheaton

When electromagnetic radiation strikes a metal, ► electrons are released. This simple statement hides a considerable history stretching back to Galvani and not plumbed entirely to this day.

In its initial form, the effect was discovered by Heinrich Hertz (1857–94) during his path-breaking corroboration of Maxwell's laws in 1887. He was using spark-discharges in one part of his laboratory in Karlsruhe to stimulate other, much weaker ones, in another. To see the weaker ones he began to shield his eyes from the bright primary spark, then, inspired, realized that the length of the weak ones diminished when the blue primary spark light failed to reach the secondary electrodes. He called it "a peculiar and surprising property of the spark," showed by elimination that the ultra-violet light of the primary eased the secondary sparks from the metal electrodes, and put the matter out for others to investigate because it deterred him from his Maxwellian objective.

Many took up the challenge with telling results. Wilhelm Hallwachs (1859–1922) in Dresden gave it its modern form when he found that ultra-violet light from almost any source will discharge a negatively-charged zinc plate. Augusto Righi (1850–1920) in Padua named the device a "coppia fotoelettrica." By 1889 both Hallwachs and Righi showed that a neutral plate will acquire a positive charge from the action. One must note here that the concept of the "electron" did not exist except in a few prescient minds at the time, so the active mechanism remained unclear.

That circumstance changed in the mid-1890s with the pioneering investigations of ion-currents by Joseph John Thomson (1856–1940) at the famous Cavendish Laboratory in Cambridge. He studied ► cathode-rays in the newly possible vacuum: streams of negative electricity visible and accessible to quantitative study within those glass vacuum tubes. Convinced that there was a negatively-charged "corpuscle" constituting the beam, he sought all means to measure its properties. In 1898, after proclaiming its existence by a clever determination of its charge/mass ratio using crossed electric and magnetic fields, he eagerly sought its charge; the photoelectric effect made it possible.

If his electrons were emitted from the plate AB in Fig. 1, passing them through a magnetic field would bend them into cyclodial trajectories.[1] Were he then to probe the region of the plate with an electrical collector CD, the height of their cycloid

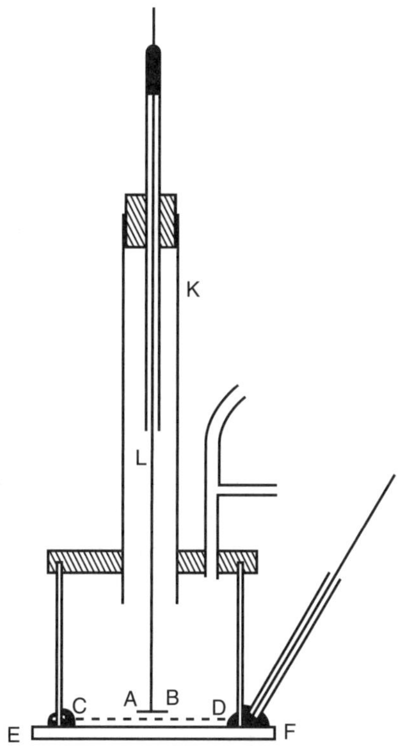

Fig. 1 Thomson shone uv light though a quartz plate EF at the bottom of the device, irradiating plate AB. He then moved AB closer to grid CD until it first collected charge. From Thomson (1899), p. 550

[1] This is true for electrons emitted on one side of normal to the plate, those emitted on the other side describe tortured paths not pictured but that never reach the full excursion from plate AB.

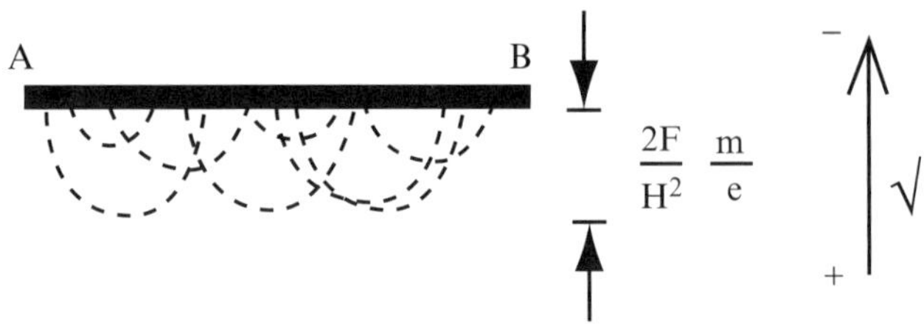

Fig. 2 Cycloidal paths of corpuscles emitted from plate AB on one side of the normal. From [11, p. 88]

(hence their velocity) was easily measured, as in Fig. 2. So the photoeffect gave the first accurate determination of the charge *e* in 1899. This was 7 years before Millikan's oil-drop experiment.

Philipp Lenard (1862–1947) was convinced, like his mentor Hertz, that the cathode rays were ætherial disturbances. So in 1902 he tried and failed to disprove Thomson's results. In the most far-reaching study of photoelectric emission to the time, he found that the velocity of emitted cathode rays seemed entirely independent of the intensity of radiation, but only depended on the *type* of light used. He did not say it depended on the frequency or color of the light and concluded that there was therefore no conversion of radiant energy to electron kinetic energy occurring in the effect, but that some sort of resonant action of the light would "trigger the release of electrons" from metal atoms with the energy they had possessed *within* the atom. Until he finally rejected this "triggering" action in 1911, his views formed the majority opinion amongst physicists because the energy of released photo-electrons seemed entirely too great to have collected from a wave-front of radiation in the very short time ($<10^{-3}$ s.) which Alexandr Stolyetov (1839–96) in Moscow had found it to occur in 1889.

Far in the background lay the heretical proposal in 1905 by the unknown Albert Einstein (1879–1955) that there must be a particulate nature to ultra-violet light. In 1905, as part of his recasting of physics, he derived a linear law for the electric potential that stopped the fastest released electrons as proportional to the *frequency,* not the intensity of the incident light. This "quantum transformation relation" or (QTR) side-stepped the æther altogether in favor of a ▶ "light-quantum" interpretation of ultra-violet (and visible) radiation. In reaction to Planck's statistical "quantum" of 1900, Einstein's physical light-quantum carried energy proportional to frequency, and was absorbed in quantum units. Einstein was well aware of Lenard's findings but explained them in an entirely different (he said "truly revolutionary") manner. Why "revolutionary"? Were light a continuous wave, how did the atom know when enough energy had been absorbed?

In 1913, Einstein's light-quantum was judged "erroneous" by leading German physicists. Even in 1916, when Millikan showed Einstein's linear photoeffect law to be entirely accurate empirically, the idea was almost universally rejected (even by Millikan.) But Einstein received the 1921 Nobel Prize for the idea when tides began to turn. The ▶ Compton Effect and Louis de Broglie's hypothesis of ▶ matter waves

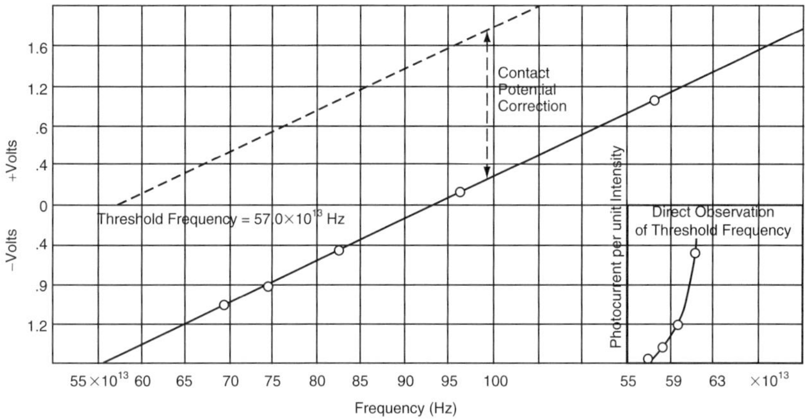

Fig. 3 Millikan's unambiguous 1916 demonstration of Einstein's predicted linear law for the photoelectric effect in lithium. From [9, p. 240], by permission

fairly convinced the next generation that Einstein had been right all along. See also ► "light-quantum", ► "wave-particle duality" and ► "quantum theory".

Primary Literature

1. H. Hertz: Über einen Einfluß des ultravioletten Lichtes auf die elektrische Entladung. Ann. Phys. (3rd ser.) **31**, 983–1000 (1887)
2. W. Hallwachs: Über den Einfluß des Lichtes auf elektrostatisch geladene Körper. Ann. Phys. (3rd ser.) **33**, 301–12 (1887)
3. J. J. Thomson: On the masses of the ions at low pressures. Phil. Mag. (5th ser.) **48**, 547–67 (1899)
4. Ph. Lenard: Über die lichtelektrische Wirkung. Ann. Phys. (4th ser.), **8**, 149–98 (1902)
5. A. Einstein: Über einen Erzeugung und Verwandlung des Lichtes betreffenden heuristischen Gesichtspunkt. Ann. Physik (4th ser.) **17**, 132–48 (1905)
6. R. Millikan: A direct photoelectric determination of Planck's '***h***'. Phys. Rev. (2nd Series) **4**, 355–88 (1916)

Secondary Literature

7. M. Klein: Einstein's first paper on quanta. Natural philosopher, **2**, 59–86 (1963)
8. B. R. Wheaton: Philipp Lenard and the photoelectric effect, 1889–1911. Hist. Stud. Phys. Sci. **9**, 299–323 (1978)
9. B. R. Wheaton: *The Tiger and the Shark: Empirical roots of wave- particle dualism* (Cambridge University Press, Cambridge 1983)
10. B. R. Wheaton: Was the photoelectric effect discovered? A spatial nomograph of micro-ontology. BJHS in press since 2003
11. B. Wheaton: *The photoelectric effect and the origin of the quantum theory of free radiation*, M.A. thesis (University of California, Berkeley, 1971; available through WorldCat)

Photon

See ► Light Quantum.

Pilot Waves

Basil James Hiley

The notion of a pilot wave was first proposed by Louis de Broglie (1892–1987) in his doctorate thesis in 1924 [1] and eventually published in 1927 [2]. Earlier experiments on the ► photoelectric effect showed the need to introduce the notion of a 'packet' of electromagnetic energy, the photon, into what had till then been thought to be a purely wave phenomena. How then was it possible to bring together the particle and the wave, two apparently contradictory physical notions into one theory?

De Broglie summarised his ideas in what he called "the theory of the double solution". In this approach he proposed that the equations of ► wave mechanics would admit two kinds of solution. One solution would be a continuous wave solution, Ψ, and the other would be a singular solution which would represent the physical "particle". This singular solution would be localised and incorporated within the extended wave phenomena. De Broglie's brilliant perception [3] was that this idea could be applied, not only to photons (► light quantum), but to quantum particles in general, namely those with non-zero rest mass. What was missing was the general non-linear wave equation which would unite wave behaviour with particle behaviour in one comprehensive theory.

To a first approximation, de Broglie [10] argued that we can treat the two solutions separately provided we find some way of locking the particle to the Ψ wave, which he assumed would satisfy the ► Schrödinger equation. To achieve this de Broglie first noticed that a particle has an internal energy, $m_0c^2 = h\nu_0$, so that it can be compared with a small clock of proper frequency ν_0. When the particle is in motion with a velocity v, relativity tells us that its frequency would be $\nu = \nu_0(1 - v^2/c^2)^{1/2}$. This is different from the frequency of a wave which transforms as $\nu_1 = \nu_0/(1 - v^2/c^2)^{1/2}$. However combining these two results gives $\nu_1 = \nu(1 - v^2/c^2)$, a relation which we will now exploit.

How does this result 'lock' the wave and particle aspects together? Notice that an observer will see the moving particle represented by a wave $\psi = \sin(2\pi\nu_1 t)$. If at time $t = 0$ there is agreement between the internal phase of the particle described by ψ and the phase of the wave Ψ, then we want this agreement in phase to persist throughout the movement of the particle.

At time t, the particle will have moved a distance $x = vt$ from its original position. Its internal motion will then be represented by $\psi = \sin[2\pi(x/v)\nu_1]$. Now the Ψ-wave at this point will be given by

$$\Psi = \sin[2\pi\nu(t - xv/c)] = \sin[2\pi(x/v)\nu(1 - v^2/c^2)].$$

Using the relation $\nu_1 = \nu(1 - v^2/c^2)$, we find the Ψ-wave is given by $\Psi = \sin[2\pi(x/v)\nu_1]$, which is exactly the same as the internal motion represented by ψ. In this way the particle is locked to the wave, so that the wave can be regarded as "piloting" the particle.

In pursuing the idea, de Broglie [10] then analysed the singularity further and found that the velocity of the particle could be given by $\boldsymbol{v} = \nabla\phi/m$ where ϕ was the phase of the wave. He regarded this as a fundamental formula and called it "the guidance formula". Furthermore he immediately recognised the similarity with the classical Hamilton-Jacobi theory of classical mechanics in which there appears a canonical relation $\boldsymbol{p} = \nabla S$, where S was the classical action. It is through this relation that de Broglie had anticipated the 1952 ► Bohm model [4].

De Broglie was invited to present these ideas at the 1927 Solvay Congress held in Brussels, which he did under the title "The Pilot-Wave Theory". The paper was not well received and the alternative ► probabilistic interpretation of Bohr and others was preferred by most of those present. During the course of the conference Pauli [5] raised detailed objections to the work, which de Broglie was unable to answer at the time and he was disappointed that Einstein did not support his ideas. As a consequence de Broglie stopped working on this approach.

However de Broglie did take up his ideas again [6, 10] after David Bohm (1917–92) [4] published his papers containing an analysis of the ► Schrödinger equation that exploited formulae similar to those presented in the pilot-wave theory. The significant feature of Bohm's work for de Broglie was that Pauli's specific objections had been answered. Furthermore the papers also outlined how the ideas could be extended to deal with, not only many of the troubling paradoxes of the standard interpretation (► errors and paradoxes in quantum mechanics), but also how to extend these ideas to ► quantum field theory.

More recently Dürr, Goldstein and Zangí [7] have proposed a new way of deriving the guidance condition. They begin by assuming the velocity of the particle is determined by the ► wave function, ψ so that $\boldsymbol{v} = \boldsymbol{v}^{\psi}$. Then by also assuming Galilean invariance, together with $\boldsymbol{v}^{a\psi} = \boldsymbol{v}^{\psi}$ and time-reversal symmetry, $\boldsymbol{v}^{\psi *} = \boldsymbol{v}^{\psi}$, they were able to derive the de Broglie guidance condition,

$$\boldsymbol{v}^{\psi} = \frac{\hbar}{m}\Im\frac{\nabla\psi}{\psi} = \frac{\nabla\phi}{m}.$$

where ϕ is again the phase of the wave. Dürr *et al.* called their approach "Bohmian mechanics", a rather unfortunate terminology as Bohm himself had argued against the notion of "mechanics" as underlying quantum phenomena, arguing that his preferred term was "quantum non-mechanics" [8], a position he maintained throughout his life [9].

However the possibility of a mechanical explanation of quantum phenomena is a legitimate area for exploration and shows how far one can take these ideas without the need to follow the more exotic interpretations of the formalism discussed elsewhere in this compendium. In fact approaches based on such considerations do provide a consistent and coherent account of quantum phenomena, removing many of the paradoxes thrown up by the even more conventional approaches. Nevertheless there has been a general reluctance amongst the majority of physicists to embrace the approach based on the notion of a pilot wave.

A comprehensive survey of the pilot wave theory can be found in de Broglie [10, 11].

Primary Literature

1. L. de Broglie: *Thèses de Doctorat* (Masson, Paris 1924)
2. L. de Broglie: La méchanique ondulatoire et la structure atomique de la matiére et du rayonnement. J. Phys. Radium, 6e series **8**, 225–41 (1927)
3. L. de Broglie: Ondes et quanta. Compt. Rendus **177**, 507–510 (1923)
4. D. Bohm: A Suggested Interpretation of the Quantum Theory in Terms of Hidden Variables I. Phys. Rev. **85**, 166–193 (1952)
5. W. Pauli: *Électrons et photons. Comptes Rendus du Ve Conseil de Physique Solvay* (Gauthier-Villars, Paris 1928, pp. 282–4)
6. L. de Broglie: The Reinterpretation of Wave Mechanics. Found. Phys. **1**, 5–15 (1970)
7. D. Dürr, S. Goldstein, N. Zangí: Quantum Equilibrium and the Origin of Absolute Uncertainty. J. Stat. Phys. **67**, 843–907 (1992)
8. D. Bohm: *Quantum Theory* (Prentice-Hall, Englewood Cliffs, NJ, 1951, p. 167)
9. D. Bohm: Hidden variables and the Implicate Order, in *Quantum Implications: Essays in Honour of David Bohm,* ed. B. J. Hiley, F. D. Peat (Routledge, London, 1987, pp. 33–44)

Secondary Literature

10. L. de Broglie: *Non-linear Wave Mechanics: A Causal Interpretation* (Elsevier, Amsterdam, 1960)
11. L. de Broglie: *The Current Interpretation of Wave Mechanics: A Critical Study* (Elsevier, Amsterdam, 1964)

Planck's Constant *h*

Dieter Hoffmann

Planck's constant *h* is one of the fundamental constants of nature and crucial for our physical understanding of atomic and subatomic processes. It was introduced in 1899 by Max Planck (1858–1947) in the context of his investigations of heat radiation. While trying to derive Wien's radiation law and to examine the thermal equilibrium between matter and radiation in a cavity, Planck used a model now called Planck's resonator. Its entropy was defined as

$$S = -\frac{E}{av} \ln \frac{E}{ebv},$$

where v is the frequency of the radiation and "a and b stand for two universal positive constants." (Planck 1899, p. 465) Planck had already calculated the value for constant b (now designated h) in "thermodynamic fashion" as $b = 6.885 \times 10^{-27}$erg sec. The current best value for h is 6.27×10^{-27} erg sec or 6.626×10^{-34} J s.

By the way, Planck also showed that the second constant a is defined by h/k, where k is Boltzmann's constant and depends on the definition of temperature. With h and k one can calculate very precisely the values for Loschmidt's constant (L) and the electric elementary quantum (e) from heat radiation measurements.

In the same paper from May 1899 Planck also pointed out that this new fundamental constant of nature opens up the possibility of combining the gravitational constant (G) and the velocity of light (c) "to define units for length, mass, time and temperature which keep their meaning for any time and any civilization, even extraterrestrial and unhuman ones. Therefore one can designate them as 'natural units'." ([1], p. 480; [2], p. 121)

Soon thereafter, during the fall of 1900, Planck noticed that the meaning of b resp. h was not restricted to metrology or the foundation of natural units. By way of explanation by a new radiation formula – the so-called ► Planck's radiation law, which replaced Wien's law – the constant h again plays a central role. For the energy of Planck's resonators, which regulate the exchange of energy between matter and radiation in a cavity, one had to postulate:

P

$$E = hv$$

This introduction of discrete levels of energy and its revolutionary character for the physical understanding of nature was not yet fully understood at that time. Initially, it merely agreed with the available measurement data. Planck himself first spoke of discrete energy levels of his resonators in 1908. That is why the beginning of our modern understanding of the quantum character of atomic processes and the crucial role of h is signified less by Planck than by Albert Einstein and his hypothesis of ► light quanta from 1905 as well as his and Paul Ehrenfest's analysis of Planck's radiation law in 1905/06. It took an additional decade for the revolutionary character of Planck's quantum hypothesis and Planck's constant to become fully clear and quantum physics to become a central part of modern physical research. This was not the work of Planck and his generation but of a younger one, the founders of quantum mechanics during the 1920. With this theory and the ► Heisenberg uncertainty principle, the fundamental role of h for our understanding of the atomic world was fully elucidated.

Primary Literature

1. M. Planck: Über irreversible Strahlungsvorgänge, 5th communication. Sitzungsberichte der Preußischen Akademie der Wissenschaften 1899, pp. 440–480.
2. M. Planck. Über irreversible Strahlungsvorgänge. Annalen der Physik 1 (1900) 69–122.
3. M. Planck. Über die Elementarquanta der Materie und der Elektrizität. Annalen der Physik 4 (1901) 564–566.

Secondary Literature

4. R. Ladenburg: Die Methoden zur *h*-Bestimmung und ihre Ergebnisse. In Handbuch der Physik. Edited by H. Geiger and K. Scheel. Berlin 1993, vol. XXIII, pp. 1–22.
5. M. Planck. Die Entdeckung des Wirkungsquantums. Edited by A. Hermann. Dokumente der Naturwissenschaft, Abteilung Physik, vol. 11, Munich 1969.

POVM (Positive Operator Value Measure)

Roderich Tumulka

POVM: *positive-operator-valued measure*, also called *generalized observable*. A mathematical object, consisting of a family of operators on ► Hilbert space, that occurs in quantum theoretical formulas for the probability distribution of the random outcome of a quantum mechanical experiment. The concept of POVM contains, as a special case, that of ► observables represented by ► self-adjoint operators.

Overview

Outline of Definition. The word "measure" in "positive-operator-valued measure" is understood in the sense of mathematical probability and measure theory [5], where it means "additive set function". A *set function* $E(\cdot)$ is a function whose argument is a set (rather than a number, or a point in space). Possible arguments are subsets Δ of a basic set Ω. Typical relevant examples of Ω include the real line $\mathbb{R}$, n-space $\mathbb{R}^n$, or finite sets. A set function is called *additive* if for any two *disjoint* sets Δ_1, Δ_2 it is true that

$$E(\Delta_1 \cup \Delta_2) = E(\Delta_1) + E(\Delta_2). \tag{1}$$

(The full mathematical definition, see below, requires slightly more.)

Examples of measures include probability measures, for which $E(\Delta)$ is a number between zero and one, giving the probability that a given random variable assumes a value in the set Δ. For a POVM, $E(\Delta)$ is a (bounded) positive operator on a Hilbert space $\mathscr{H}$. An ▶ operator T is called *positive* if $\langle\phi|T\phi\rangle \geqslant 0$ for all $\phi \in \mathscr{H}$; this is also sometimes called *positive semi-definite* in the literature; every (bounded) positive operator is self-adjoint. Finally, it is part of the definition of a POVM that it is normalized in the sense that $E(\Omega) = I$, where I is the identity operator on $\mathscr{H}$, $I\psi = \psi$. In case Ω is a finite (or countable) set, $E(\Delta)$ can be expressed by singletons:

$$E(\Delta) = \sum_{\omega\in\Delta} E(\{\omega\}). \tag{2}$$

(Below we write $E\{\omega\}$ instead of $E(\{\omega\})$.)

Probabilities from POVMs. From a POVM $E(\cdot)$ on a set Ω one can create probability measures on Ω in the following way: Given any vector $\psi \in \mathscr{H}$ with $\|\psi\| = 1$, then

$$\mathbb{P}_\psi(\Delta) = \langle\psi|E(\Delta)|\psi\rangle \tag{3}$$

defines a probability measure $\mathbb{P}_\psi(\cdot)$ on Ω. To see this, note that $\langle\psi|E(\Delta)|\psi\rangle$ is a nonnegative real number since $E(\Delta)$ is a positive operator, and

$$\mathbb{P}_\psi(\Omega) = \langle\psi|E(\Omega)|\psi\rangle = \langle\psi|I|\psi\rangle = \|\psi\|^2 = 1. \tag{4}$$

Physical Role. The physical relevance of POVMs is based on the following *main theorem about POVMs*: *For every quantum physical experiment $\mathcal{E}$ whose possible outcomes lie in a space Ω, there exists a POVM $E(\cdot)$ on Ω such that, whenever the experiment $\mathcal{E}$ is carried out on a quantum system with state vector ψ, the random outcome Z has probability distribution given by*

P

$$\mathbb{P}(Z \in \Delta) = \langle\psi|E(\Delta)|\psi\rangle\,. \tag{5}$$

Observables. When all operators $E(\Delta)$ are *projection operators* (▶ projection) then $E(\cdot)$ is also called a *PVM* or *projection-valued measure*. The widespread concept of ▶ observables as represented by self-adjoint operators is contained in the concept of POVM as the special case of a PVM on $\Omega = \mathbb{R}$. The self-adjoint operator A usually called the "observable" is obtained from $E(\cdot)$ by setting

$$A = \int_{\mathbb{R}} E(\mathrm{d}\lambda)\,\lambda\,. \tag{6}$$

Conversely, given A, the spectral theorem for self-adjoint operators provides the right hand side of this equation, that is, provides the unique PVM $E(\cdot)$ on $\mathbb{R}$ that makes this equation true. Thus, the self-adjoint operator A summarizes the entire information encoded in the PVM $E(\cdot)$ in just one operator.

Examples

Observables as represented by self-adjoint operators correspond to the simplest cases of quantum experiments, usually connected with "ideal measurements." POVMs are necessary for more complex experiments.

Time of Arrival. Send a particle towards a detector and measure the time at which the detector clicks. As a consequence of the main theorem about POVMs, the statistics of the random result, depending on the initial wave function of the particle, is given by a POVM, i.e., is of the form (5). Since this POVM is a "proper POVM", i.e., not a PVM, there is no self-adjoint operator summarizing it; in other words, there is no "time operator". (► Time in quantum mechanics).

Sequence of Ideal Measurements. Readers familiar with the formalism of ideal quantum measurement of an observable (self-adjoint operator) A may consider a sequence of such measurements, first one corresponding to A_1, then another corresponding to A_2, and so on, up to A_n. Suppose that these measurements are carried out one immediately after another, so that we can neglect the unitary time evolution in between. Suppose further that the A_i have purely discrete spectrum. Note that the operators A_i need not commute with each other, as they are not measured simultaneously, but in a specified order. The sequence of outcomes forms a vector in $\mathbb{R}^n$, whose distribution is given by a POVM $E(\cdot)$ that can be constructed from the PVMs $E_i(\cdot)$ associated by (6) with A_i as follows:

$$E\big\{(\lambda_1, \ldots, \lambda_n)\big\} = E_1\{\lambda_1\}^{1/2} \cdots E_n\{\lambda_n\}^{1/2} E_n\{\lambda_n\}^{1/2} \cdots E_1\{\lambda_1\}^{1/2}. \quad (7)$$

(The powers $1/2$ can be omitted as $P^{1/2} = P$ for every projection P; however, in the above form the equation still defines a POVM $E(\cdot)$ when the $E_i(\cdot)$ are themselves proper POVMs.)

In case the A_i commute with each other, $E(\cdot)$ is a PVM on $\mathbb{R}^n$. In this sense, a PVM can represent a family of commuting observables. In particular, the three position operators Q_x, Q_y, Q_z of non-relativistic quantum mechanics of a single particle together give rise to the following PVM $P(\cdot) = E(\cdot)$ on $\mathbb{R}^3$:

$$P(\Delta)\psi(x, y, z) = \begin{cases} \psi(x, y, z) & \text{if } (x, y, z) \in \Delta, \\ 0 & \text{otherwise.} \end{cases} \quad (8)$$

However, when the A_i do not commute then $E(\cdot)$ is not a PVM but a proper POVM.

To make the setting more general, we can allow that the choice of second observable A_2 depends on the outcome of the first measurement. To take this into account, replace $E_i\{\lambda_i\}$ in (7) by $E_{i,\lambda_1,\ldots,\lambda_{i-1}}\{\lambda_i\}$.

Position Measurements with Constraints. In some cases, not all square-integrable functions on $\mathbb{R}^3$ are possible as physical wave functions of a single particle, but only those from a suitable subspace $\mathscr{H}_{\text{phys}}$ [2,4]. For example, photon ► wave function are functions $\Psi : \mathbb{R}^3 \to \mathbb{C}^3$ obeying the constraint $\nabla \cdot \Psi = 0$. As another example, Dirac wave functions $\psi : \mathbb{R}^3 \to \mathbb{C}^4$ are usually regarded as physical only if they consist exclusively of Fourier components with positive energy, in other words, if they lie in the positive spectral subspace $\mathscr{H}_{\text{phys}}$ of the Dirac Hamiltonian. In this case, the usual position operators and the associated PVM as in (8) often map physical state vectors into unphysical ones, and are thus not defined as operators on the physical Hilbert space $\mathscr{H}_{\text{phys}}$. The problem is solved by replacing the "generalized position observable" $P(\cdot)$ with $\tilde{P}(\cdot)$ defined by

$$\tilde{P}(\Delta) := P_{\text{phys}}\, P(\Delta)\, P_{\text{phys}}, \tag{9}$$

where P_{phys} denotes the projection to $\mathscr{H}_{\text{phys}}$. Then $\tilde{P}(\Delta)$ is an operator on $\mathscr{H}_{\text{phys}}$, and $\tilde{P}(\cdot)$ is a proper POVM on $\mathbb{R}^3$.

Fuzzy Measurements. An ideal detector, when detecting the particle in the region $\Delta \subseteq \mathbb{R}^3$, would collapse the wave function $\psi(x, y, z)$ to the function in (8). Real detectors, however, might, for example, cut off the wave function in an unsharp way, corresponding to a proper POVM $\tilde{P}(\cdot)$ that arises from the PVM $P(\cdot)$ of (8) by smearing out (convolving) with a "bump function" f (for example a Gaussian):

$$\tilde{P}(\Delta) = \int_{\Delta} \mathrm{d}^3x \int_{\mathbb{R}^3} P(\mathrm{d}^3y)\, f(y - x). \tag{10}$$

The Main Theorem About POVMs

It is not difficult to understand the main theorem; here is a simple argument [3]. Suppose the experiment $\mathcal{E}$ begins at time t_1 and ends at time t_2, and suppose the quantum state of system and apparatus at time t_1 is $\Psi(t_1) = \psi \otimes \phi$. We make three assumptions: (1) The time evolution from t_1 to t_2 is given by a unitary operator U. (2) The ► Born rule, according to which the probability distribution of the configuration Q at time t_2 is given by $\langle \Psi(t_2)|P(\cdot)|\Psi(t_2)\rangle$ with $P(\cdot)$ the position PVM as in (8). (3) The outcome Z is a function f of the configuration Q at time t_2. Then, for $\Delta \subseteq \Omega$,

$$\mathbb{P}(Z \in \Delta) = \mathbb{P}\big(Q \in f^{-1}(\Delta)\big) = \langle \Psi(t_2)|P(f^{-1}(\Delta))|\Psi(t_2)\rangle \tag{11}$$

$$= \langle \psi \otimes \phi|U^* P(f^{-1}(\Delta))U|\psi \otimes \phi\rangle = \langle \psi|E(\Delta)|\psi\rangle \tag{12}$$

with

$$E(\Delta) = \langle \phi|U^* P(f^{-1}(\Delta))U|\phi\rangle, \tag{13}$$

where the scalar product in (13) is a *partial scalar product* in the Hilbert space of the apparatus. It can be shown that (13) defines a POVM.

P

Mathematical Aspects

Definition. The mathematical definition of POVM contains some details we have omitted above. The family of sets Δ for which $E(\Delta)$ is defined is required to be a σ*-algebra*, i.e., closed under the complement operation $\Delta \mapsto \Omega \setminus \Delta$ and under forming countable intersections. A POVM $E(\cdot)$ is further supposed to be σ*-additive*, i.e., additive for any countable union of pairwise disjoint sets $\Delta_1, \Delta_2, \ldots,$

$$E\Big(\bigcup_{i=1}^{\infty} \Delta_i\Big) = \sum_{i=1}^{\infty} E(\Delta_i), \tag{14}$$

where the series on the right hand side is required to converge weakly, i.e., $\sum \langle\psi|E(\Delta_i)\psi\rangle$ converges for every $\psi \in \mathscr{H}$. (Then it automatically also converges strongly, i.e., $\sum E(\Delta_i)\psi$ converges for every $\psi \in \mathscr{H}$.)

Integration. Just as integrals can be defined relative to a probability measure $\mathbb{P}$, $\int \mathbb{P}(\mathrm{d}\omega)\, f(\omega)$, one can define integrals relative to a POVM. Such integrals have occurred above in (6) and (10). One can define them by

$$\Big\langle \psi \Big| \int E(\mathrm{d}\omega)\, f(\omega) \Big| \psi \Big\rangle = \int \langle\psi|E(\mathrm{d}\omega)|\psi\rangle\, f(\omega)\,. \tag{15}$$

Primary Literature

1. E. B. Davies, J. T. Lewis: An Operational Approach to Quantum Probability. Commun. Math. Phys. **17**, 239–60 (1970)
2. S. T. Ali, G. G. Emch: Fuzzy Observables in Quantum Mechanics. J. Math. Phys. **15**, 176–82 (1974)
3. D. Dürr, S. Goldstein, N. Zanghì: Quantum Equilibrium and the Role of Operators as Observables in Quantum Theory. J. Stat. Phys. **116**, 959–1055 (2004).
4. K. Kraus: Position Observables of the Photon. In *The Uncertainty Principle and Foundations of Quantum Mechanics* ed. by W.C. Price and S.S. Chissick (Wiley, New York 1977, 293–320)

Secondary Literature

5. P. Billingsley: *Probability and Measure* (Wiley, New York 1986)
6. E. B. Davies: *Quantum Theory of Open Systems* (Academic, London 1976)

Probabilistic Interpretation of Quantum Mechanics

Brigitte Falkenburg and Peter Mittelstaedt

The probabilistic interpretation of quantum mechanics is based on Born's 1926 papers and von Neumann's formal account of quantum mechanics in ► Hilbert space. According to Max Born (1882–1970), the quantum mechanical ► wave function Ψ does not have any direct physical meaning, whereas its square $|\Psi|^2$ is a probability [1] ► Born rule, probability in quantum mechanics. According to Johann von Neumann (1903–1957), the scalar product $(\Psi, O\Psi)$ of the pure states Ψ and $O\Psi$ is the expectation value of the observable O, with spectral decomposition $O = \sum O_i \mathrm{P}(O_i)$, in the state Ψ. The products $(\Psi, \mathrm{P}(O_i)\Psi)$ give the probabilities of the possible measurement outcomes O_i [2].

(Spectral decomposition, see ► Density operator; Ignorance interpretation; Measurement theory; Objectification; Operator; Propensities in Quantum Mechanics; Self-adjoint operator; Wave mechanics).

The probabilistic interpretation holds for all quantum theories, i.e., for non-relativistic or ► relativistic quantum mechanics as well as for quantum field theory. In general, the probabilities for transitions between two quantum states are calculated from the density matrix of a quantum system. In ► *quantum field theory*, this is the S-matrix. The squared S-matrix element or scattering amplitude gives the transition probability of a certain type of particle interaction. In this way, the S-matrix is directly related to the effective cross section of particle reactions in ► *scattering experiments*.

In view of the probabilistic interpretation, it has been discussed for decades whether quantum theory refers to individual quantum systems or only to ensembles of identically prepared systems. The laws provided by the theory are statistical and they are experimentally confirmed to a very high degree of accuracy. But our scientific language is concerned with individual systems: with the properties of a system, its preparation, its development in time, and the measurement of its objective ► observables. The difficulties in understanding the physical behavior of an individual system on the basis of an essentially statistical theory gave rise to von Neumann's quantum theory of ► *measurement* [2], to the ► *hidden variable theories*, and to the ► *many worlds interpretation*. The latter had an enormous impact [3] on recent attempts to make the relation between individual quantum systems and their probabilistic behavior more precise, in the quantum theory of measurement [5–7].

Born's Derivation

In order to interpret the wave function, Born generalized the ► Schrödinger equation from bound states inside the atom to a scattering problem, laying the grounds for the quantum mechanics of scattering, too [1]. He applied the Schrödinger equation to the stationary wave of an asymptotically free quantum state, in ▶ *correspondence* to the scattering of classical particles at the Coulomb potential. Born's model employs ▶ *wave-particle duality* in the following sense. The scattering process is calculated in the *wave picture* ▶ Davisson–Germer experiment; Stern–Gerlach experiment; Schrödinger equation, whereas the scattering outcomes are interpreted in the *particle picture* ▶ Franck–Hertz experiment.

In the *wave picture*, a plane wave φ_{in} is diffracted by a hydrogen atom ▶ Bohr's atom model in the ground state Ψ_0. In first approximation, the diffraction results in a superposition of spherical waves $\varphi_{nm}(q)$ with amplitudes $f_{nm}(\theta)$ (which today are called the scattering amplitudes, in the quantum theory of scattering):

$$\varphi_{\text{out}} = \sum_{nm} c_{nm} \int f_{nm}(\theta)\, \varphi_{nm}(q)\, \mathrm{d}\Omega .$$

Here, q is the momentum transfer to the atom, nm indicates the state of the atom after a transition from an electron from state n to state m, θ is the propagation angle, and the integration is taken over the solid angle Ω. Due to the momentum transfer, the superposition φ_{out} is entangled with the ground state Ψ_0 and the excited states $\Psi_{nm}(q)$ of the atom. A momentum transfer q may give rise to the excitation of an electron in state n to the mth level. Accordingly, the momentum transfer is quantized. This explains the results of the ▶ *Franck–Hertz* collision experiment.

In the *particle picture* which applies to the detection of scattered particles, θ is the scattering angle related to the momentum transfer q by $q = |\boldsymbol{p} - \boldsymbol{p}'| \cos\theta$. Here, $\boldsymbol{p}, \boldsymbol{p}'$ is the particle momentum before and after the scattering. Born stated the scattering outcomes in terms of an *Ausbeutefunktion* $\Phi_{nm}(q)$ which is identical to the differential cross section (▶ *scattering experiments*) of the scattering:

$$\Phi_{nm}(q) = c_{nm}^2 |f_{nm}(\theta)|^2 .$$

Finally, Born argued that the calculation is only consistent with the empirical scattering results if the squared amplitudes $|\Phi_{nm}(q)|^2$ of the outgoing waves are interpreted as the probabilities for the scattering of particles of momentum $\boldsymbol{p}' = \boldsymbol{p} - \boldsymbol{q}$ into direction θ. This interpretation relates the squared amplitudes of the partial waves $\varphi_{nm}(q)$ to the relative frequencies of the particles measured in direction θ. The relation between both quantities is a correspondence rule in an empiricist sense [8], i.e., a rule for relating a theoretical concept to an observable quantity (in contradistinction to Bohr's ▶ *correspondence principle*, which establishes inter-theoretic relations).

Problems of Born's Approach

Born established the probabilistic interpretation by plausible reasoning, but obviously he did not give any proof. The probabilistic interpretation is merely operational, without being anchored in the axiomatic foundations of quantum mechanics. (In von Neumann's approach, it is established by the problematic ▶ projection postulate [2].) Born's model relates the squared amplitude of the scattered wave to the observed particle detections. Here, "scattered wave" means the asymptotic behavior of the diffracted wave φ_{out}, considered a long time after the interaction between the incoming plane wave φ_{in} and the atom, as if after the interaction there was no ▶ entanglement between φ_{out} and the atom wave function Ψ.

Following Born, the quantum mechanical wave function determines the probabilistic ensemble (▶ ensembles in quantum mechanics), whereas the measurement outcomes are the individual events. This gave rise to a widespread pragmatic view of ▶ *wave-particle duality*, according to which the waves show up in the quantum probabilities and the particles in the individual events [9].

In order to say more about the obscure relation between the wave and particle pictures employed in the above model, Born's papers [1] suggested also some philosophical ideas *beyond* the probabilistic interpretation. He suspected whether there might be quantities that causally fix the measurement outcomes, giving, however, a tentative answer in the negative. And he characterized Schrödinger's wave function Ψ in terms of a ghost-like particle-guiding field or pilot wave and the transfer of energy or momentum in terms of corpuscle propagation. These ideas, which stem from Albert Einstein (1879–1955) [10], were later taken up in the ▶ *hidden variables* approach [4].

Born's derivation of the probabilistic interpretation has crucial gaps. First, (theoretical) probabilities and (empirical) relative frequencies are different. Probabilities may only be identified with relative frequencies in the limit of infinitely many events or measurement outcomes. Born neglected this point. Second, there is no explanation of how a statistical law may emerge from an interaction of individual quantum systems. Third, the quantum mechanics of scattering cannot explain why finally individual particles are detected. The quantum theory of measurement addressed these questions, providing answers for the first and second (see below), whereas the third gap, the notorious measurement or objectification problem, remains (▶ *ignorance interpretation, measurement theory*) [6,7].

P

The Split-Beam Experiment

Let us consider the split-beam experiment of Fig. 1, which has been realized both with photons (▶ light quantum) and with neutrons. The state φ of the incoming photon is split by a beam splitter BS_1 into two orthogonal components described by orthonormal states $\varphi(B)$ and $\varphi(\neg B)$. The two parts of the split beam are reflected at two (fully reflecting) mirrors M_1 and M_2 and recombined with a phase shift δ at a second beam splitter BS_2. In the experiment there are two mutually exclusive measuring arrangements: If the detectors D_1 and D_2 are in the positions (D_1^B, D_2^B) one observes which way (B or $\neg$B) the photon came. If the detectors are in the position (D_1^A, D_2^A) one observes the interference pattern, i.e., the intensities which

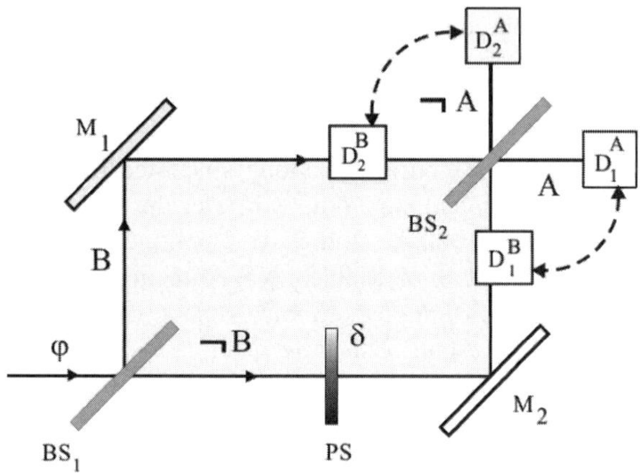

Fig. 1 Photon split-beam experiment with beam splitters BS_1 and BS_2, two fully reflecting mirrors M_1 and M_2, a phase shifter PS providing a phase shift δ, and two detectors D_1 and D_2 in mutually exclusive positions (D_1^A, D_2^A) and (D_1^B, D_2^B)

depend on the phase δ. In this experiment the object system S is prepared in the state φ (which belongs to the two-dimensional Hilbert space $H_2 = C^2$). There are two non-commuting observables, the path observable B with eigenstates $\varphi(B)$, $\varphi(\neg B)$ and the interference observable A with eigenstates $\varphi(A)$ and $\varphi(\neg A)$. The probability for B (to register a photon in D_2^B) and for $\neg B$ (to register a photon in D_1^B) reads

$$p(\varphi, B) = p(\varphi, \neg B) = 1/2.$$

The probability for A (to register a photon in D_1) and for $\neg A$ (to register a photon in D_2) reads

$$p(\varphi, A) = \cos^2(\delta/2) \quad \text{and} \quad p(\varphi, \neg A) = \sin^2(\delta/2),$$

respectively. [7] This means that the relative frequency of photons detected at D_1 is approximately given by $\cos^2(\delta/2)$ and the relative frequency of photons detected at D_2 by $\sin^2(\delta/2)$.

The Measurement Process

The quantum mechanical probabilities in the split-beam experiment refer to the state of the system *after* the measurement. According to the ► *measurement theory*, we consider both the object system S and the apparatus M as proper quantum systems with Hilbert spaces H_S and H_M, respectively. Let us further assume that the systems S and M are prepared in pure states $\varphi \in H_S$ and $\Phi \in H_M$. A measurement process of the observable A and in particular of the observable P(A), which is given by the projection operator $P[\varphi(A)]$ of the eigenstate $\varphi(A)$, can be described by a unitary operator U_A acting on the tensor product state $\varphi \otimes \Phi$ of the compound system $S+M$. After the measurement process the compound system is in the pure state $U_A(\varphi \otimes \Phi)$, whereas the object system is given in the reduced ▶ *mixed state*

$$W_S(\varphi, A) = \cos^2 \delta/2\, P[(\varphi(A)] + \sin^2 \delta/2\, P[(\varphi(\neg A)],$$

i.e., by a weighted sum of projection operators $P[\varphi(A_i)]$ of states $\varphi(A_i)$ with $A_i \in \{A, \neg A\}$.

The Probability Reproducibility Condition

The interpretation of the ▶ *mixed state* $W_S(\varphi, A)$ of the object system after the premeasurement is usually given by the following *probability reproducibility condition.* The (formal) probability distribution $p(\varphi, A_i)$, $A_i \in \{A, \neg A\}$ induced by the preparation φ and the measured observable $P(A)$ is reproduced in the statistics of the post-measurement values $(Z_A, Z_{\neg A})$ and states $(\Phi_A, \Phi_{\neg A})$ of the pointer. In case of repeatable measurements, i.e., when a realistic interpretation of the observables is possible, this means that $p(\varphi, A_i)$ is also reproduced in the statistics of the states $(\varphi(A), \varphi(\neg A))$.

On the basis of these arguments we can now formulate the main problem. Let an ensemble of systems S be given, which before the measurement are identically prepared and after the premeasurement of A in the reduced ▶ *mixed state* $W_S(\varphi, A)$. Is it then possible to justify that the (formal) probability $p(\varphi, A_i)$ is reproduced in the statistics of the measurement results A and ¬A, respectively?

P

In order to answer this question, consider a large number of identically prepared systems S_i in states φ^i which are not eigenstates of the observable $P(A)$. Let us further assume that the unitary operator U_A used for a measurement of the observable $P(A)$ fulfills the calibration postulate for repeatable measurements. Then we know that a measurement of the observable $P(A)$ in case of the particular preparation $\varphi(A)$ leads with certainty to the states Φ_A and $\varphi(A)$ showing the result A. Are we able to show, on the basis of this *probability free interpretation* of quantum mechanics, that for arbitrary preparations $\varphi \neq \varphi(A)$, $\varphi \neq \varphi(\neg A)$, the formal probability $p(\varphi, A_i)$ induced by φ and $P(A)$ is reproduced in the statistics of the measuring outcomes A_i? If this is the case, then the probability reproducibility condition is a theorem of the probability free theory and no longer an additional postulate.

The Emergence of Statistical Laws in Quantum Mechanics

Let us consider N independent systems S_i with identical preparation φ_i as a compound system S^N in the tensor product state

$$(\varphi)^N = \varphi^1 \otimes \varphi^2 \otimes \ldots \otimes \varphi^N,\ (\varphi)^N \in H(S^N),$$

where $H(S^N)$ is the tensor product of N Hilbert spaces $H(S_i)$. A premeasurement of A transforms the initial state φ^i of each system S^i into the mixed state

$$W^i = p(\varphi^i, A)\, P[(\varphi^i, A)] + p(\varphi^i, \neg A)\, P[(\varphi^i, \neg A)]$$

with eigenstates $\varphi^i(A_k)$ of A corresponding to results A_k. If A is measured on each system S_i, then the measurement result is given by a sequence $\{A_{l(1)}, ... A_{l(N)}\}$ of system properties $A_{l(i)}$ and states $\varphi(A_{l(i)})$, respectively, with an index sequence $l := \{l(1), l(2), ... l(N)\}$ such that $A_{l(i)} \in \{A_k\} = \{A, \neg A\}$.

In the N-fold tensor product Hilbert space $H(S^N)$ of the compound system S^N, the special states $(\varphi)_l^N = \varphi^{(1)}(A_{l(1)}) \otimes ... \otimes \varphi^{(N)}(A_{l(N)})$ with $\varphi^{(i)}(A_{l(i)}) \in H(S_i)$ form an ► orthonormal basis. The relative frequency $f^N(k, l)$ of outcomes $A_k \in \{A, \neg A\}$ in the state $(\varphi)_l^N$ is then given by $f^N(k, l) = 1/N \sum \delta_{l(i),k}$. We can now define in $H(S^N)$ an operator *"relative frequency of systems with properties* A_k*"* by

$$f_k^N := \Sigma f^N(k, l)\, P[(\varphi)_l^N],$$

where the sum is taken over all sequences l. The eigenvalue equation of this operator

$$f_k^N (\varphi)_l^N = f^N(k, l)(\varphi)_l^N$$

then shows that the relative frequency of the measurement result A_k is an objective property of S^N in the state $(\varphi)_l^N$ and given by $f^N(k, l)$. The eigenvalue equation can also be written in the equivalent form

$$\mathrm{tr}\{P[(\varphi)_l^N](f_k^N - f^N(k, l))^2\} = 0.$$

After a premeasurement of P(A) a system S_i is in a mixed state W_i. If N premeasurements of P(A) are performed, then the state of the compound system S^N is given by the N-fold tensor product state

$$(W)^N = W^1 \otimes W^2 \otimes ... \otimes W^N$$

of these mixed states W^i. One easily verifies that the expectation value of f_k^N in this product state is given by $p(\varphi, A_k)$. However, in general the state $(W)^N$ is not an eigenstate of the relative frequency operator f_k^N with eigenvalue $p(\varphi, A_k)$. This means that

$$T_k^N := \mathrm{tr}\{(W)^N (f_k^N - p(\varphi, A_k))^2\} \neq 0$$

and that the relative frequency of outcomes A_k is not an objective property of the system S^N in the state $(W)^N$. In contrast to this somewhat unsatisfactory result one finds that for large values of N the post-measurement product state $(W)^N$ of the compound system S^N becomes an eigenstate of the operator f_k^N and the value of the relative frequency of results A_k approaches the probability $p(\varphi, A_k)$. Indeed, one finds after some tedious calculations [3, 5, 7]

$$\mathrm{T}_k^N := 1/N\,\mathrm{p}(\varphi, \mathrm{A}_k)(1 - (\varphi, \mathrm{A}_k))$$

and thus one finally obtains the desired result

$$\lim_{N\to\infty} \mathrm{tr}\{(W)^N (f_k^N - \mathrm{p}(\varphi, \mathrm{A}_k))^2\} = 0.$$

This means that in the limit of an infinite number N of systems the state $(W)^N$ is an eigenstate of the operator f_k^N of the relative frequency of results A_k and that the compound system S^N possesses the relative frequency $\mathrm{p}(\varphi,\mathrm{A}_k)$ of A_k as an objective property. In order to ensure this way of reasoning against mathematical objections one has to guarantee that the overwhelming majority of index sequences $l = \{l_i\}$ are random sequences and that the contribution of the non-random sequences can be neglected [5,7].

Primary Literature

1. M. Born: Zur Quantenmechanik der Stoßvorgänge. Z. f. Phys. **37**, 863–867 (1926). M. Born: Quantenmechanik der Stoßvorgänge. Z. f. Phys. **38**, 803–827 (1926)
2. J. von Neumann: *Mathematische Grundlagen der Quantenmechanik* (Springer, Berlin 1932). Engl. Transl. by R. Beyer: *Mathematical Foundation of Quantum Mechanics* (Princeton University Press, Princeton 1955)
3. B. S. DeWitt: The Many-Universes Interpretation of Quantum Mechanics, in: *Foundations of Quantum Mechanics*. IL Corso, B. d'Espagnat (ed.) (Academic, New York 1971, 167–218)
4. D. Bohm: A Suggested Interpretation of the Quantum Theory in Terms of "Hidden" Variables, I and II. Phys. Rev. *85*, 166–179 (1952)
5. P. Mittelstaedt: The objectification in the measuring process and the many worlds interpretation. In: *Symposium on the Foundations of Modern Physics* (World Scientific, Singapore 1990, 261–279). P. Mittelstaedt: Quantum Measurement I. The Measuring Process and the Interpretation of Quantum Mechanics. Int. J. Theor. Phys. **32**, 1763–1775 (1993). P. Mittelstaedt: The Emergence of Statistical Laws in Quantum Mechanics. In: M. Ferrero and A. van der Merwe (eds.): *New Developments on Fundamental Problems in Quantum Physics* (Kluwer, Dordrecht 1997, 265–274). P. Mittelstaedt: Quantum Mechanics without Probabilities. In: *John von Neumann and the Foundations of Physics.* M. Rédei and M. Stöltzner (eds.) (Kluwer, Dordrecht 2001, 189–200)

P

Secondary Literature

6. P. Busch, P. J. Lahti, and P. Mittelstaedt: *The Quantum Theory of Measurement* (Springer, Berlin 1991, 2nd edition 1996)
7. P. Mittelstaedt: *The Interpretation of Quantum Mechanics and the Measurement Process* (Cambridge University, Cambridge Press 1998)
8. E. Nagel: *The Structure of Science. Problems in the Logic of Scientific Explanation* (Routledge & Kegan Paul, London 1961)
9. B. Falkenburg: *Particle Metaphysics. A Critical Account of Subatomic Reality* (Heidelberg, Berlin, New York, Springer 2007), p. 269–272, 278–289.
10. M. Jammer: *The Conceptual Development of Quantum Mechanics* (McGraw-Hill, New York 1966)

Probability in Quantum Mechanics

Abner Shimony

The concept of probability played an important role in the very beginning of ► quantum theory, when Max Planck (1858–1947) postulated the discrete emission and absorption of radiation in a ► black body radiation. The quantum statistical mechanics developed by Planck and his successors has extraordinary consequences treated elsewhere in this Compendium. Here, however, the emphasis will be upon the unprecedented role of probability in the quantum mechanical treatment of the state of a physical system, which will be discussed first in the ► wave mechanical formulation of Schrödinger, and then in the more abstract and general ► Hilbert space formulation.

The thesis of Louis de Broglie (1892–1987) of 1924 [1] postulated that waves are associated with ► electrons and that a discrete atomic state is determined by the occurrence of an integral number of wave lengths in a circular trajectory about the atomic nucleus. Erwin Schrödinger (1887–1961) [2] systematized and generalized de Broglie's idea in the first of his series of papers on ► wave mechanics.

Schrödinger assumed that the wave associated with a system of electrons is a complex-valued function ψ whose argument $\mathbf{r} = (\mathrm{r}_1, \ldots, \mathrm{r}_N)$ is positioned in an N-dimensional configuration space $\mathbf{R}$ ($N = 3$ in the case of a single electron). If the ► state ψ of the system of electrons is stationary with energy E, then ψ was assumed to satisfy the time-independent differential equation

$$\left[\sum\nolimits_n (2m_n)^{-1}(-i\hbar\partial/\partial \mathrm{r}_n)^2 + V(\mathrm{r}_1, \ldots, \mathrm{r}_N)\right]\psi(\mathbf{r}) == E\psi, \tag{1}$$

where V is the potential energy as a function of position in configuration space. In later papers in the series Schrödinger [3] wrote a time-dependent equation for ψ $(\mathbf{r}, t)$, explored analogies to Hamilton's comparison of appropriately formulated classical mechanics and optics, and developed methods for solving his wave mechanical equations. He also examined [4] the conceptual relations between his wave mechanics and Heisenberg's [5] formulation of quantum mechanics (► matrix formulation).

The fourth of Schrödinger's pioneering papers [6] on wave mechanics suggested that $\psi^*\psi$ be interpreted as a "weight function" of a charge distribution. In particular, if ψ is the wave associated with an electron, then $e[\psi(\mathbf{r})]^*\psi(\mathbf{r})$, where e is the electric charge of the electron, was interpreted as the electron's charge density at $\mathbf{r}$. But this interpretation was difficult to reconcile with the evidence for the indivisibility of the electron and the quantum mechanical predictions of the spatial spreading of an unbound electron. In a quantum mechanical analysis of collision phenomena Max Born (1882–1970) [7] proposed an alternative interpretation of the wave function which was almost universally accepted: that $[\psi(\mathbf{r})]^*\psi(\mathbf{r})d\mathbf{r}$ is the probability that the system be found in the infinitesimal region $d\mathbf{r}$ about $\mathbf{r}$; in other words $[\psi(\mathbf{r})]^*\psi$

(**r**) is a probability density of position in configuration space. To ensure that the probability of finding the system anywhere in configuration space is unity, which is the conventional representation of certainty in probability theory, it suffices to multiply $\psi(\mathbf{r})$ by a scalar independent of **r** such that the integral of this density over **R** is unity.

Various pioneers, among them London [8] and Dirac [9], gave prescriptions for extracting information about the probability distributions of other quantities than position from the wave function ψ. Typically in their prescriptions ψ is expanded in a complete orthonormal set of functions $\phi_n(\mathbf{r})$, each square integrable over **R**,

$$\psi(\mathbf{r}) = \sum_n c_n \phi_n(\mathbf{r}), \sum_n |c_n|^2 = 1 \tag{2}$$

where

$$\int_R [\phi_n(\mathbf{r})]^* \phi_m(\mathbf{r}) d\mathbf{r} = \delta_{mn} (\text{orthonormality}) \tag{3a}$$

and

$$A\phi_n(\mathbf{r}) = a_n \phi_n(\mathbf{r}), \tag{3b}$$

with A a self-adjoint linear operator on function space. Physically A represents an observable quantity A, the eigenvalues a_n being the values of A in the physical states represented respectively by the $\phi_n(\mathbf{r})$. The physical interpretation of the expansion (2) is that the probability of finding the quantity A to have value a when the particle is prepared in the state represented by ψ is

$\text{Prob}_\psi(A = a) = \Sigma' |c_{n|}|^2$, with the sum Σ' taken over all n such that

$$a_n = a. \tag{4}$$

Lacking in the foregoing generalized probabilistic interpretation of the ► wave function is the procedure for associating self-adjoint linear operators A with physical quantities A. The pioneers treated this problem by an intuitive combination of analogies to classical mechanics with Heisenberg's analysis of the relation of momentum to position. A rigorous treatment, notably by George Whitelaw Mackey (1916–2006) [10], applies the theory of induced representations of groups.

Although Schrödinger's expression of the quantum mechanical state as a function of position in configuration space – the wave function – was extraordinarily valuable both intuitively and practically, it lacked mathematical generality and rigor. A series of investigations by David Hilbert, Lothar Nordheim, and John von Neumann [11], most notably the last [12], used the theory of Hilbert space to achieve a more general and more abstract formulation of quantum mechanics than Schrödinger's.

A Hilbert space is a vector space endowed with an inner product (for quantum mechanical purposes usually taken to be complex), with a norm, and complete in this norm. A vector space is a set of elements closed under the operation of vector addition + and multiplication of vectors by scalars, which are elements of a field F, and with standard behavior of the null vector **0** and of the scalars 0 and 1. (In standard quantum mechanics F is taken to be the set C of complex numbers.) A complex inner product on the vector space V is a mapping of ordered pairs ϕ, χ of vectors in V into C, denoted by $\langle\phi|\chi\rangle$, satisfying the following conditions:

1. $\langle\phi|\phi\rangle \geqslant 0$ and equals 0 only if $\phi = \boldsymbol{0}$
2. $\langle\phi|\chi\rangle = \langle\chi|\phi\rangle^{*}$, where * represents complex conjugation.
3. $\langle\phi|\chi\rangle = a.\langle\phi|\chi\rangle$ for any scalar a.
4. $\langle\phi|\chi + \sigma\rangle = \langle\phi|\chi\rangle + \langle\phi|\sigma\rangle$.

A *norm* on V is introduced without further postulation by the definition

$$\phi \rightarrow ||\phi|| = |\langle\phi\,|\phi\rangle|^{1/2}. \tag{5}$$

A *Cauchy sequence* in V relative to this norm is a sequence $\{\phi_n\}$ with the property that for any positive ε there is an integer M such that

$$||\phi_n - \phi_m|| < \varepsilon \tag{6}$$

for n and m greater than M. The space V is *complete* if *every Cauchy sequence* $\phi_{\rm n}$ converges in the norm to some vector ϕ in *V.* In common applications of quantum mechanics the Hilbert space associated with a system is assumed to be separable, that is, to have a denumerable basis from which a sequence can be constructed by addition and scalar multiplication to converge in the norm to any given vector in V.

An idealized but illuminating bridge between the quantum mechanical view of physical systems and the Hilbert space formulation is the concept of a "logic of questions", discussed by various authors including Birkhoff and von Neumann [13], Piron [14], Mackey [15], and Varadarajan [16]. A physical system can be characterized by systematically answering *yes-no questions* about its properties, whose answers assert or deny attributions to the system. According to empiricist science the answers to these questions are determined by measurements, but the entire set of questions can be endowed with a rigorous structure only if the measurements are ideally accurate and error free. The idealized set of questions, which will be called the *logic of questions*, is assumed to be a *complete orthocomplemented lattice*. A lattice L is (1) a partially ordered set of elements: i.e., there is a binary relation such that for all elements $q < r$ and $r < q$ imply $q = r$; $q < r$ and $r < s$ imply $q < s$; and $q < q$; (2) there is unique element $\boldsymbol{0}$ such that $\boldsymbol{0} < q$ for all elements q, and a unique element $\boldsymbol{1}$ such that $q < \boldsymbol{0}$ for all q; and (3) for any nonempty finite subset F of L there exist in L a unique least upper bound and a unique greatest lower bound of F with respect to the ordering relation <, denoted by $\vee_{q\varepsilon F}$ and $\wedge_{q\varepsilon F}$ respectively. L is *orthocomplemented* if there is a one-one mapping q $\rightarrow q^{\perp}$ such that $q^{\perp\perp} = q$, q < r implies $r^{\perp} < q^{\perp}$, the least upper bound of q and $q^{\perp}$ is $\boldsymbol{1}$, and the greatest lower bound of q and $q^{\perp}$ is $\boldsymbol{1}$. A lattice is *complete* if the restriction of finiteness of F in condition (3) is replaced by denumerable infinity. When the abstract structure of a complete orthocomplemented lattice is applied to the logic of questions the element **1** is interpreted as the question whose answer is necessarily 'yes' and the element **0** is interpreted as the question whose answer is necessarily 'no'; the orthocomplementation operation generates from the question q the question $q^{\perp}$ whose answer under any circumstances is the opposite of the answer to q in the same circumstances.

With these preliminaries, the Hilbert space formulation of quantum mechanics (if one sets aside simplifications like particles of finite spin whose properties in configuration space are deliberately neglected) can be compactly formulated: (I) the logic of questions of a quantum mechanical system is isomorphic to the lattice of closed linear subspaces of a Hilbert space H [17].

Equivalently, the logic of questions is isomorphic to the lattice of projection operators on H, where P is a projection operator if it is a linear operator on H, self-adjoint in the sense that for all pairs of vectors ϕ, χ $\langle \phi | P\chi \rangle = \langle P\phi | \chi \rangle$, and idempotent in the sense that $P^2 = P$.

The formulation can be strengthened by inserting the adjectives "separable infinite-dimensional" before "Hilbert space" for all cases in which the system is explicitly located in a configuration space, omitting these adjectives only when the ► spin aspects of the system are studied as a convenient simplification in abstraction from configuration aspects.

It should be noted that the explicit construction of the isomorphism asserted in Axiom (I) is mathematically intricate, using the theory of induced representations of groups (see Mackey [10]), as indicated in the paragraph after Eq. (4).

In usual textbook expositions of quantum mechanics there is not only an axiom relating the logic of questions to the lattice of projections on the Hilbert space but another axiom giving a Hilbert space characterization of the *states* which assign probabilities to the questions: it is assumed that a *pure state* S (later to be contrasted with a ► *mixed state*) is represented many-one by non-null vectors in the Hilbert space H, or more elegantly one-one by a one-dimensional subspace $E(\phi) = \{\phi\}$, which consists of all scalar multiples of an arbitrary non-null vector ϕ associated with the state. Then *(II) the probability that question Q has answer 'yes' when the state is represented by ϕ, or equivalently by the one-dimensional subspace $E\{\phi\}$, is $\langle \phi | \mathbf{Q} | \phi \rangle / \langle \phi | \phi \rangle$, where* **Q** *is the projection operator associated with the question Q; and of course this expression is simplified when ϕ has norm unity.*

A remarkable theorem of Andrew Mattei Gleason (*1921) (► Gleason's theorem) [18] shows that assumption (II) of the preceding paragraph is almost superfluous. If the Hilbert space has dimension greater than 2, then the only *states* S in the sense of probability measures on the lattice of questions which satisfy the standard axioms of probability – non-negativity, ascription of probability unity to the identity operator I, and additivity of the probability assigned to the least upper bound of mutually orthogonal questions – have the form

$$\text{Prob(Q has answer yes/S)} = \sum\nolimits_n p_n \langle \phi_n | \mathbf{Q} | \phi_n \rangle, \tag{7}$$

where the p_n is a sequence of non-negative real numbers summing to unity, and the ϕ_n are mutually orthogonal vectors each of unit norm. If there is only one term ϕ_1 in the right-hand side of (7),

$$\text{Prob(Q has answer yes/S)} = (\phi_1 | \mathbf{Q} | \phi_1), \tag{8}$$

then the state S is a *pure state*, represented in H in only one way except for the trivial recourse to scalar multiples of ϕ_1; if there is more than one term in the right-hand side of (7), the state S is *mixed* and can be represented non-trivially in different ways.

Heisenberg [19] emphasizes a great conceptual difference between the probabilities expressed by the coefficients p_n in (7) and that expressed by the inner product $(\phi_1|\mathbf{Q}|\phi_1)$ in (8). He calls the former "subjective" because they do not express an intrinsic indefiniteness of the constitution of the system, but rather a kind of partial knowledge and partial ignorance on the part of the scientist, as is the case with the probabilities occurring in classical statistical mechanics. The inner product in (8) he calls "objective", because it does not stem from ignorance on the part of the scientist. Indeed, the vector ϕ_1 represents the state of the system *an sich, maximally characterized.* The fact that some of the questions Q have neither 'yes' nor 'no' as answers but probabilities intermediate between these extremes, characterizes the system itself and only derivatively the scientist's knowledge of the system. Heisenberg suggests the name "potentiality" for this modality of objective reality, which is intermediate between full actuality and mere logical possibility. Although he borrowed this name from Aristotle, he actually generalized Aristotle's embryological sense of "potentiality" and introduced a radically new philosophical concept, which may very well be the most profound contribution of quantum mechanics to philosophy. See also ► Objective Quantum Probabilities; Propensities in Quantum Mechanics.

Primary Literature

1. L. de Broglie: *Thèses présentées à la Faculté des Sciences de l'Université de Paris* (Masson et Cie., Paris 1924)
2. E. Schrödinger: Quantisierung als Eigenwertproblem. Annalen d. Physik **79**, 361 (1926). Reprinted, as are [3] and [4], in E. Schrödinger, *Abhandlungen zur Wellenmechanik* (J.A. Barth. Leipzig. 1926), translated as *Collected Papers on Wave Mechanics* by J.F. Shearer and W.M. Deans (Blackie & Son, London/Glasgow, 1928)
3. E. Schrödinger: Quantisierung als Eigenwertproblem. Second communication, Annalen d. Physik **79**, 489 (1926); third communication, Annalen d. Physik **80**, 437 (1926); fourth communication, Annalen d. Physik **81**, 109 (1926)
4. E. Schrödinger: Über das Verhältnis der Heisenberg-Born-Jordanschen Quantenmechanik zu der meinen. Annalen d. Physik **79**, 734 (1926)
5. W. Heisenberg: Über quantentheoretische Umdeutung kinematischer und mechanischer Beziehungen. Zeitschrift f. Physik **33**, 879 (1925).
6. E. Schrödinger: Quantisierung als Eigenwertproblem. Fourth communication, reference [3].
7. M. Born: Quantenmechanik der Stossvorgänge. Zeitschrift f. Physik **38**, 803 (1926)
8. F. London: Winkelvariable und kanonische Transformationen in der Undulationsmechanik. Zeitchrifts f. Physik **40**, 193 (1926)
9. P.A.M. Dirac: The physical interpretation of the quantum dynamics. Proceedings of the Royal Society of London A, **113**, 621 (1926). Also *The Principles of Quantum Mechanics* (Clarendon Press, Oxford, 1930) and three subsequent editions.
10. G. Mackey: *Induced Representations of Groups and Quantum Mechanics* (W.A, Benjamin, New York, 1968, chapters 2–3)
11. D. Hilbert, J. von Neumann, and L. Nordheim: Über die Grundlagen der Quantenmechanik. Mathematische Annalen **98**, 1 (1927)

12. J. von Neumann: *Mathematische Grundlagen der Quantenmechanik* (Springer, Berlin 1932); transl. as *Mathematical Foundations of Quantum Mechanics* by R.T. Beyer (Princeton University Press, Princeton 1955)
13. G. Birkhoff and J. von Neumann: The logic of quantum mechanics. Annals of Mathematics **37**, 823 (1936)
14. C. Piron: *Foundations of Quantum Physics* (W.A. Benjamin, Reading MA, 1976, 6–9)
15. G. Mackey: *Mathematical Foundations of Quantum Mechanics* (W.A. Benjamin, New York 1964, 61–73)
16. V.S. Varadarajan: *Geometry of Quantum Theory* (van Nostrand, Princeton 1968, 8–11)
17. G. Mackey: Ref. 15, p. 71.
18. A.M. Gleason: Measures on the closed subspaces of a Hilbert space. J. Mathematics and Mechanics **6**, 885 (1957)
19. W. Heisenberg: *Physics and Philosophy* (Harper, New York, 1962, 54, 185)

Secondary Literature

20. M. Jammer: *The Conceptual Development of Quantum Mechanics* (McGraw-Hill, New York 1966, 293–322)
21. S.-I. Tomonaga: *Quantum Mechanics*, vol. II (North-Holland, Amsterdam 1966, chap. 8, pp. 163–200; chap. 9, pp. 276–287)

Projection

Werner Stulpe

P

Projection, an *idempotent* linear ► operator defined on a vector space with values in that vector space. That is, a linear operator P acting in a real or complex vector space $\mathcal{V}$ is called a *projection* if (i) $D_P = \mathcal{V}$, D_P being the domain of P, and (ii) $P^2 = P$. For a projection P, $I - P$ is also a projection, I being the unit operator. Defining $\mathcal{X} = R_P$ and $\mathcal{Y} = R_{I-P}$ where R_P and R_{I-P} denote the ranges of P and $I - P$, a projection P induces the decomposition of the vector space $\mathcal{V}$ into the linear submanifolds $\mathcal{X}$ and $\mathcal{Y}$ according to the direct sum $\mathcal{V} = \mathcal{X} \oplus \mathcal{Y}$. That is, every vector $\psi \in \mathcal{V}$ can be written as a sum $\psi = \phi + \chi$ where $\phi \in \mathcal{X}$, $\chi \in \mathcal{Y}$, and ϕ, χ are uniquely determined (in particular, the zero vector is the only vector contained in both, $\mathcal{X}$ and $\mathcal{Y}$). Conversely, given a decomposition of $\mathcal{V}$ into the direct sum of any two complementary submanifolds $\mathcal{X}$ and $\mathcal{Y}$, $\mathcal{V} = \mathcal{X} \oplus \mathcal{Y}$, then a projection P is defined according to $P\psi = \phi$ where $\psi = \phi + \chi$, $\phi \in \mathcal{X}$, $\chi \in \mathcal{Y}$; P is the *projection onto* $\mathcal{X}$ *w.r.t.* $\mathcal{Y}$. A projection P in a Banach space $\mathcal{V}$ need not be continuous, i.e., need not be bounded (► operator); a projection onto $\mathcal{X}$ w.r.t. $\mathcal{Y}$ is continuous if and only if $\mathcal{X}$ and $\mathcal{Y}$ are closed submanifolds of the Banach space $\mathcal{V}$ [1].

In the context of Hilbert spaces (► Hilbert space), the concept of a projection can be sharpened to that of an orthogonal projection [1–5], the latter being a projection that is self-adjoint (▶ operator, ▶ self-adjoint operator). That is, a linear operator P in a complex (or real) Hilbert space is an *orthogonal projection* if (i) $D_P = \mathcal{H}$, (ii) $P = P^2$, and (iii) $P = P^*$. An orthogonal projection is a positive (▶ operator) bounded operator with norm (▶ operator) $\|P\| = 1$ for $P \neq 0$; $\mathcal{X} = R_P$ and $\mathcal{Y} = R_{I-P}$ are subspaces of $\mathcal{H}$ (i.e., closed linear submanifolds), and $\mathcal{Y}$ is the orthocomplement of $\mathcal{X}$ (▶ Hilbert space). Thus, an orthogonal projection induces the *orthogonal decomposition* $\mathcal{H} = \mathcal{X} \oplus \mathcal{X}^\perp$. Conversely, every subspace $\mathcal{X}$ of $\mathcal{H}$ induces the orthogonal decomposition $\mathcal{H} = \mathcal{X} \oplus \mathcal{X}^\perp$ of the Hilbert space $\mathcal{H}$ and in consequence the definition of an orthogonal projection P according to $P\psi = \phi$ where $\psi = \phi + \chi$, $\phi \in \mathcal{X}$, $\chi \in \mathcal{X}^\perp$; P is the *orthogonal projection onto* $\mathcal{X}$. Hence, there is a one-one correspondence between the orthogonal projections in $\mathcal{H}$ and the subspaces of $\mathcal{H}$. In the sequel, 'projection' means 'orthogonal projection' in $\mathcal{H}$.

Let P_1 and P_2 be projections projecting onto the subspaces $\mathcal{X}_1$ and $\mathcal{X}_2$, respectively. The product P_1P_2 is zero if and only if $\mathcal{X}_1$ and $\mathcal{X}_2$ are orthogonal to each other, i.e., $\mathcal{X}_2 \subseteq \mathcal{X}_1^\perp$; equivalently, the sum $P_1 + P_2$ is a projection, $P_1 + P_2$ projects onto $\mathcal{X}_1 \oplus \mathcal{X}_2$. The product P_1P_2 is a projection if and only if P_1 and P_2 commute, i.e., $P_1P_2 = P_2P_1$, P_1P_2 projects onto $\mathcal{X}_1 \cap \mathcal{X}_2$; $P_1P_2 = P_2P_1$ is equivalent to the existence of three mutually orthogonal projections E_1, E_2, and F such that $P_1 = E_1 + F$ and $P_2 = E_2 + F$. The difference $P_1 - P_2$ is a projection if and only if $\mathcal{X}_1 \supseteq \mathcal{X}_2$, $P_1 - P_2$ projects onto $\mathcal{X}_1 \ominus \mathcal{X}_2 = \mathcal{X}_1 \cap \mathcal{X}_2^\perp$; $\mathcal{X}_1 \supseteq \mathcal{X}_2$ is equivalent to $P_2 = P_1P_2$. If $P_1, \ldots, P_n$ are projections onto mutually orthogonal subspaces, say, $\mathcal{X}_i$, then the sum $\sum_{i=1}^n P_i$ is the projection onto the direct sum $\bigoplus_{i=1}^n \mathcal{X}_i$. If $P_1, P_2, \ldots$ is an infinite sequence of projections onto mutually orthogonal subspaces $\mathcal{X}_i$, then, by $P\phi = \sum_{i=1}^\infty P_i\phi$ where $\phi \in \mathcal{H}$, a projection P is defined which projects onto $\bigoplus_{i=1}^\infty \mathcal{X}_i$. In the latter case, the infinite sum $\sum_{i=1}^\infty P_i$ does not converge in the operator norm (unless in the trivial case that $P_i = 0$ for all $i > N$), instead it converges *strongly*, i.e., $(\sum_{i=1}^\infty P_i)\phi = \sum_{i=1}^\infty P_i\phi$ for all $\phi \in \mathcal{H}$.

As a subset of the ordered real Banach space $\mathcal{B}_s(\mathcal{H})$ (▶ operator) of the bounded ▶ self-adjoint operators in $\mathcal{H}$, the set $\mathcal{P}(\mathcal{H})$ of all projections inherits the partial order of $\mathcal{B}_s(\mathcal{H})$. That is, $P_1 \leqslant P_2$, $P_1, P_2 \in \mathcal{P}(\mathcal{H})$, if and only if $\langle\phi|P_1\phi\rangle \leqslant \langle\phi|P_2\phi\rangle$ for all $\phi \in \mathcal{H}$. The statement $P_1 \leqslant P_2$ is equivalent to $\mathcal{X}_1 \subseteq \mathcal{X}_2$ where $\mathcal{X}_1 = R_{P_1}$ and $\mathcal{X}_2 = R_{P_2}$. The partially ordered set $\mathcal{P}(\mathcal{H})$ is a complete lattice with the zero operator as its smallest element and the unit operator as its greatest element, and the association of every element $P \in \mathcal{P}(\mathcal{H})$ with $P^\perp = I - P$ is an orthocomplementation of $\mathcal{P}(\mathcal{H})$. Thus, $(\mathcal{P}(\mathcal{H}), \leqslant, \perp)$ is a complete orthocomplemented lattice which, in addition, is orthomodular and atomic (▶ quantum logic); $\mathcal{P}(\mathcal{H})$ is isomorphic to the orthocomplemented lattice of the subspaces of the Hilbert space $\mathcal{H}$ where the set of the subspaces is ordered by the set-theoretic inclusion.

If P is a projection onto $\mathcal{X}$ and $\phi_1, \phi_2, \ldots$ a complete orthonormal system in $\mathcal{X}$ (▶ Hilbert space, ▶ orthonormal basis), then $P\psi = \sum_i \langle\phi_i|\psi\rangle\phi_i$, $\psi \in \mathcal{H}$. If $\mathcal{X}$ is one-dimensional and ϕ a unit vector in $\mathcal{X}$, one writes $P = |\phi\rangle\langle\phi|$ (▶ Dirac notation); so, if the dimension of $\mathcal{X}$ is greater than one, $P = \sum_i |\phi_i\rangle\langle\phi_i|$ where the sum, if it is infinite, converges strongly. In particular, one writes $I = \sum_i |\phi_i\rangle\langle\phi_i|$ where $\phi_1, \phi_2, \ldots$ is a complete orthonormal system of $\mathcal{H}$.

Literature

1. D. Werner: *Funktionalanalysis* (Springer, Berlin 1995, third edition 2000)
2. F. Riesz and B. Sz.-Nagy: *Functional Analysis* (Frederick Ungar Publishing, New York 1955; reprint Dover Publications, Mineola 1991)
3. M. Reed and B. Simon: *Methods of Modern Mathematical Physics I: Functional Analysis* (Academic, New York 1972, second edition 1980)
4. G. Ludwig: *Foundations of Quantum Mechanics I* (Springer, New York 1983), Appendix IV
5. E. Prugovečki: *Quantum Mechanics in Hilbert Space* (Academic, New York 1971, second edition 1981; reprint Dover Publications, Mineola 2006)

Projection Postulate

Sheldon Goldstein

In quantum mechanics, the state of a system is given by its ► wave function, a vector ψ in the ► Hilbert space of the system. The behavior of the wave function is governed by two dynamical laws [1, 2]: (1) When the system is closed, i.e., when it does not interact with its environment, its wave function evolves according to ► Schrödinger's equation

$$i\hbar\partial\psi/\partial t = H\psi, \tag{1}$$

where H is the Hamiltonian of the system. (2) When a measurement is performed on the system in state ψ, its wave function changes in a different way; it "collapses,"

P

$$\psi \mapsto P\psi/\|P\psi\|, \tag{2}$$

to its (normalized) projection onto the subspace of its Hilbert space associated with the result of the measurement. Here P is the corresponding ► projection operator, and the denominator provides the normalization, with $\|\cdot\|$ the Hilbert space norm, $\|\psi\|^2 = \langle\psi|\psi\rangle$, given by the inner product $\langle\cdot|\cdot\rangle$ with which the Hilbert space is equipped. This transition occurs with probability $\|P\psi\|^2$, the probability of the corresponding result. This rule is called the *projection postulate;* the associated change of quantum state (2) is usually referred to as the ► *wave function collapse* or as the *reduction of the state vector.*

Strictly speaking, the projection postulate governs, not any measurement, but only the most basic sort of measurement, called an *ideal measurement*, one which changes the wave function as little as possible consistent with obtaining the relevant information. In the simplest case, of an ideal measurement of a quantum observable A – a self-adjoint operator on the Hilbert space of the system – with non-degenerate spectrum λ_α and corresponding ► orthonormal basis of eigenvectors $|A = \lambda_\alpha\rangle \equiv |\lambda_\alpha\rangle$,

$$A\,|\lambda_\alpha\rangle = \lambda_\alpha\,|\lambda_\alpha\rangle, \tag{3}$$

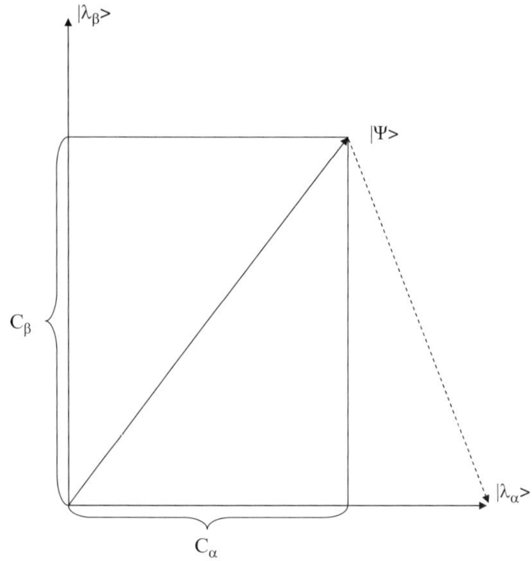

Fig. 1 Illustration of the Projection Postulate – the transition (4) from $C_\alpha|\lambda_\alpha\rangle + C_\beta|\lambda_\beta\rangle$ (with $C_\alpha > 0$ and $C_\beta > 0$) to $|\lambda_\alpha\rangle$

we have that $P = |\lambda_\alpha\rangle\langle\lambda_\alpha|$, so that (2) becomes

$$\psi \mapsto |\lambda_\alpha\rangle \tag{4}$$

(up to an irrelevant phase factor) when the result of the measurement is λ_α, and this occurs with probability $|\langle\lambda_\alpha|\psi\rangle|^2$ (Fig. 1). In other words, if

$$\psi = \sum c_\beta\,|\lambda_\beta\rangle \quad (\text{with} \sum |c_\beta|^2 = 1), \tag{5}$$

then according to the projection postulate, an ideal measurement of A in the state ψ will yield the result λ_α and wave function $|\lambda_\alpha\rangle$ with probability $|c_\alpha|^2$.

It should perhaps be stressed that what is intended by "result" in the projection postulate is the fine-grained result, corresponding to a single eigenvalue λ_α. For example, if the measurement yields the result that A is in the interval (a, b) (and this interval contains more than one eigenvalue λ_β), the after-measurement wave function will of course not be given by (2) with P the projection operator corresponding to (a, b),

$$P = \sum_{a<\lambda_\beta<b} |\lambda_\beta\rangle\langle\lambda_\beta|, \tag{6}$$

but rather will be the eigenstate $|\lambda_\alpha\rangle$ belonging to the specific eigenvalue λ_α found in the measurement – the fine-grained result. Nonetheless, if instead of an ideal measurement of A itself an ideal measurement or determination of whether or not A is in (a, b) were performed, the use of (6) would indeed be appropriate.

In standard quantum theory, the projection postulate plays a crucial but controversial role: crucial, because standard quantum theory makes contact with physics and the results of experiments via the measurement axioms of quantum theory, the most important of which is the projection postulate; and controversial, because the projection postulate appears to conflict with Schrödinger's equation. This apparent conflict is the notorious *measurement problem* of quantum mechanics, or, what amounts to the same thing, the paradox of ▶ *Schrödinger's cat*. See also ▶ Bohmian mechanics; Measurement theory; Metaphysics in Quantum Mechanics; Modal Interpretation; Objectification.

A variety of proposals have been put forward for resolving the measurement problem. For many of these, whether in fact they do solve the problem remains highly controversial. Two proposals that clearly resolve the measurement problem are the ▶ *GRW theory* and the ▶ *pilot-wave* formulation of quantum mechanics (▶ *Bohmian mechanics*). In the former, collapse of the wave function during measurement is achieved, and the projection postulate recovered, by a stochastic modification of the Schrödinger dynamics on the microscopic level [3]. (See Consistent histories, Ignorance interpretation, Ithaca Interpretation, Many Worlds Interpretation, Modal Interpretation, Orthodox Interpretation, Transactional Interpretation).

The cleanest resolution is provided by Bohmian mechanics. In Bohmian mechanics, arguably the simplest version of quantum mechanics, the projection postulate emerges in a straightforward manner as a consequence of the measurement-like interactions between system and apparatus that are present when an ideal measurement occurs [4]. A critical ingredient in this derivation is the notion of the wave function of a subsystem of a larger system, a notion made possible by the additional structure, beyond the wave function, present in Bohmian mechanics, namely the actual configuration Q of the larger system.

Primary Literature

1. P. A. M. Dirac: *The Principles of Quantum Mechanics* (Oxford University Press, Oxford 1930)
2. J. von Neumann: *Mathematische Grundlagen der Quantenmechanik* (Springer, Berlin 1932); English translation by R. T. Beyer: *Mathematical Foundations of Quantum Mechanics* (Princeton University Press, Princeton 1955)
3. G. C. Ghirardi, A. Rimini, T. Weber: Unified Dynamics for Microscopic and Macroscopic Systems, Physical Review **D 34**, 470–491 (1986)
4. D. Dürr, S. Goldstein, N. Zanghì: Quantum Equilibrium and the Role of Operators as Observables in Quantum Theory, Journal of Statistical Physics **116**, 959–1055 (2004), quant-ph/0308038

Secondary Literature

5. D. J. Griffiths: *Introduction to Quantum Mechanics* (Benjamin Cummings, San Francisco 2004)
6. R. Shankar: *Principles of Quantum Mechanics* (Springer, Berlin 1994)

Propensities in Quantum Mechanics

Mauricio Suárez

Propensities are probabilistic dispositions, and there is a long history of informal appeals to dispositional terms in connection with quantum mechanics, going all the way back to the founding of the discipline. A dispositional account of quantum properties is, for instance, arguably implicit in the early quantum theory in Bohr's model of the atom, since transitions between quantum orbitals can be described as stochastic processes that *bring about* certain values of quantum properties with certain probabilities. Similarly, on the orthodox Copenhagen interpretation, measurements do not reveal pre-existent values of physical quantities, but *bring about* values with some well-defined probability. (See ► Born rule; Consistent Histories; Metaphysics in Quantum Mechanics; Nonlocality; Orthodox Interpretation; Schrödinger's Cat; Transactional Interpretation). Then, in addition, starting in the 1950s there has been a succession of attempts to employ explicit dispositional notions, such as propensities, in order to resolve the paradoxes of quantum mechanics (► errors and paradoxes in quantum mechanics). Two stand out: Henry Margenau's *latency* interpretation, and Karl Popper's *propensity interpretation* of quantum probability.

Margenau's *Latency* Interpretation

Different interpretations of quantum mechanics can be in general fruitfully distinguished in terms of the answers they provide to the paradigmatic question concerning the general interpretation of superposed states. Suppose that the state of a quantum system is ψ, a ► superposition of eigenstates of the Hermitian operator that represents the observable Q. The standard interpretational rule within orthodox quantum mechanics, the *eigenstate/eigenvalue link* (e/e link) states that a system in state ψ can be said to have a value of a property Q if and only if ψ is an eigenstate of the Hermitian operator that represents the property. The paradigmatic question regarding these states is then the following: *What does it mean – with respect to the property represented by the observable Q – for a quantum system to be in state ψ which is not an eigenstate of the Hermitian operator that represents Q?* Propensity views of quantum mechanics vary greatly in their details but they all coincide in their answer to the paradigmatic interpretational question: *It means that the system possesses the* propensity *to exhibit a particular value of Q if Q is measured on this system in state ψ.*

In an excellent pioneering article Henry Margenau [1] argued in favour of latent quantities, or *latencies*. Margenau's key contribution was the basic *template* for propensity views. Suppose that state ψ can be written as a linear combination $\psi = \Sigma_n c_n |v_n\rangle$ of the eigenstates v_n of the *latent* observable represented by Q with spectral decomposition given by $Q = \Sigma_n a_n |v_n\rangle\langle v_n|$. Margenau then answered the paradigmatic interpretational question very precisely as follows: *a system in state* ψ *has a latent property Q if and only if it possesses a propensity to manifest eigenvalue* a_i *with probability* $|c_i|^2$ *in a measurement of Q.*

(Spectral decomposition, see ► Density operator; Ignorance interpretation; Measurement theory; Objectification; Operator; Probabilistic Interpretation; Self-adjoint operator; Wave mechanics).

However, Margenau went beyond the basic template in some unhelpful ways. For instance he conflated the possession of a property with the manifestation of a value of the property – a distinction that makes no sense for categorical properties, but is essential in order to understand dispositional property ascriptions in general. A failure to draw this distinction led Margenau to inappropriately link the actualisation of latent properties with their existence. So in the absence of a measurement of position, for instance, an electron has no value of position, and as a consequence it has no position at all. This conflation renders Margenau's attempt to solve the quantum paradoxes largely unsuccessful, and brings about additional difficult issues related to the ► identity of quantum objects.

The conflation is unfortunately present also in Heisenberg's well known appeal to Aristotelian potentialities [2], but can be avoided by distinguishing carefully the possession of a propensity from its manifestation. To be coherent a propensity view must deny a common presupposition behind the (e/e link), namely that it is legitimate to ascribe a property to a system if and only if the system takes a value of the property. It would then follow in accordance with the (e/e link) that a system possesses a property if and only if the system's state is an eigenstate of the operator that represents the property. But any coherent propensity (or more generally dispositional) account must ascribe a property without manifestation.

Popper's Propensity Interpretation of Quantum Probability

Karl Popper's propensity interpretation of quantum mechanics is surely his most important contribution to the philosophy of physics. Popper conceived the propensity interpretation of quantum mechanics as both a milestone of his philosophical career, and a key to his philosophical system. He defended it in a large number of his writings, and over a very large period of time (for instance Popper [3, 4]). It was a milestone since it was a consideration of the nature of quantum phenomena that led him to abandon the frequency theory of probability, and adopt instead a propensity interpretation for objective probabilities in general. And it was a key to Popper's philosophical system because the propensity interpretation of probability i) resolved the paradoxes of quantum mechanics; ii) re-established the possibility of a thoroughly realist interpretation of the quantum theory, of physics, and of science in general; and iii) provided strong empirical confirmation in favour of the propensity interpretation of the calculus of probability.

However, Popper's account is subject to three lethal objections that render it untenable. The first criticism was raised by Neal Grossman [5], and shows that

Popper's account confuses quantum mixtures and superpositions. In essence the problem is that for any observable Q every superposed state $\psi = \sum_n c_n|\nu_n\rangle$ can be shown to be statistically indistinguishable from an appropriate mixture $W_{n=1}^{\infty} = \Sigma_n|c_n|^2|\nu_n\rangle\langle\nu_n \mid$ over the eigenstates $\{|\nu_n\rangle\}$ of the operator that corresponds to Q. Since Popper identifies propensities with probability distributions, he has no option but to identify the propensities generated by both states. Yet both states are different, as is shown in any experiment that measures any observable other than Q on systems in these states.

The second difficulty was first raised by Peter Milne [6], and is related to the notion of interference of propensity waves invoked by Popper in order to account for the ► double-slit experiment. Popper's identification of propensities with whole experimental set-ups entails that any small change in the experimental set-up, such as the closing of a slit, essentially brings about a change in the propensity ascribed. Milne employed this fact to refute Popper's account of interference experiments, such as the two slit experiment. Popper's account entails that in each of the experiments *A* and *B* with one or the other slit open a different propensity ascription "A" and "B" is in order. The interference pattern that results in the experiment with both slits open is then just the result of the interference of both propensities "A" and "B". But Milne shows that there is no reason on Popper's account to expect propensities "A" and "B" to be co-present in the interference experimental set-up, since this is distinct from both *A* and *B*.

The final objection to Popper's propensity account is Humphrey's notorious paradox [7], which shows that propensities are not in general probabilities, and vice versa, since propensities are time-asymmetric but conditional probabilities are not. Together these three objections essentially refute Popper's propensity interpretation of quantum probabilities.

New Prospects for Propensities

The failure of propensity accounts in the past sometimes gives all propensity interpretations a bad name in the philosophy of physics. But this is essentially unfair since, as we have seen, it is not propensities *per se* that have been shown to be inapplicable to quantum mechanics, but rather *particular uses* of them. It remains possible to apply propensities to quantum mechanics in more appropriate ways. In particular propensity accounts could abandon the ideal of interpreting probabilities in general. Instead propensities can be used to *explain* certain probabilities. Some of the presuppositions underlying the (e/e link) will also need to be confronted. Finally, it must be possible to ascribe propensities to quantum systems in the absence of any experimental set-up. Three recent accounts that go some way towards meeting these goals are Maxwell [8], Thompson [9] and Suárez [10]. See also ► Objective Quantum Probabilities; Probability in Quantum Mechanics.

Primary Literature

1. H. Margenau: Advantages and disadvantages of various interpretations of the quantum theory. Physics Today **7**(10), 6–13 (1954)
2. W. Heisenberg: *Physics and Philosophy* (George Allen and Unwin, London 1958)
3. K. Popper: The Propensity interpretation of calculus of probability, and the quantum theory, in S. Körner ed.: *Observation and Interpretation*: A Symposium of Philosophers and Physicists (Butterworths, London 1957, 65–70)
4. K. Popper: *Quantum Theory and the Schism in Physics*. Volume III of the Postscript to the Logic of Scientific Discovery (Hutchinson, London 1982)

Secondary Literature

5. N. Grossman: Quantum mechanics and interpretations of probability theory. Philosophy of Science **39**, 451–460 (1972)
6. P. Milne: A note on Popper, propensities and the two slit experiment. British Journal for the Philosophy of Science **36**, 66–70 (1985)
7. P. Humphreys: Why propensities can not be probabilities. Philosophical Review **94**, 557–570 (1985)
8. N. Maxwell: Quantum propensiton theory: A testable resolution to the wave/particle dilemma. British Journal for the Philosophy of Science **39**, 1–50 (1988)
9. I. Thompson: Real dispositions in the physical world. British Journal for the Philosophy of Science **39**, 67–79 (1988)
10. M. Suárez: Quantum selections, propensities and the problem of measurement. British Journal for the Philosophy of Science **55**, 219–255 (2004)

P

Protective Measurements

Lev Vaidman

Protective measurement [1] is a method for measuring an expectation value of an observable on a single quantum system. The quantum state of the system can be protected by a potential, when the state is a nondegenerate energy eigenstate with a known gap to neighboring states, or via ▶ quantum Zeno effect by frequent projection measurements.

Apart from protection, the procedure consists of a standard von Neumann measurement with weak coupling which is switched on and, after a long time, switched off, adiabatically. The interaction Hamiltonian for protective measurement of O is:

$$H_{\text{int}} = g(t)PO, \tag{1}$$

where P is a momentum conjugate to Q, the pointer variable of the measuring device. The interaction Hamiltonian is small as in *weak measurements*, [2, p. 845]. In both cases the initial state of the pointer is such that $\langle Q\rangle_{\text{in}} = 0, \ \langle P\rangle_{\text{in}} = 0$. In weak measurement, the weakness is due to small uncertainty in P which requires a large uncertainty of the pointer variable Q. Thus, although for the final ▶ wave function of the pointer, $\langle Q\rangle_{\text{fin}} = \langle\Psi|O|\Psi\rangle$, a single measurement does not allow obtaining significant information about $\langle\Psi|O|\Psi\rangle$. In protective measurement, the pointer is well localized at zero, which requires large uncertainty in P and the weakness is due to a small value of the coupling $g(t)$. The coupling to the measurement device is weak, yet long enough so that we still have $\int g(t)\mathrm{d}t = 1$. The result is again $\langle Q\rangle_{\text{fin}} = \langle\Psi|O|\Psi\rangle$, but this time, the pointer is well localized, so we can learn the value of the expectation value from a single experiment. This is so if during the measurement, the quantum state of the system remains close to $|\Psi\rangle$. Given the adiabatic switching of the measurement interaction, its small value, and the protection of the state, this is indeed the case.

One of the basic results of quantum mechanics is that when a measurement of a variable O with eigenvalues o_i is performed on a quantum system described by the state $|\Psi\rangle$, the probabilities p_i for obtaining outcome o_i satisfy:

$$\langle\Psi|O|\Psi\rangle = \sum p_i o_i \,. \tag{2}$$

This is why the expression $\langle\Psi|O|\Psi\rangle$ is called the expectation value of O. In protective measurements we obtain this value not as a statistical average, but as a reading of a measuring device coupled to a *single* system.

A sufficient number of protective measurements performed on a single system allow measuring its quantum wave function. This provides an argument against the claim that the quantum wave function has a physical meaning only for an ensemble of identical systems. Therefore, protective measurements have some merit even when the protection is achieved via frequent projection measurements on the state $|\Psi\rangle$ with no new information obtained during the whole procedure. If the protection of the state is via a known energy gap to any orthogonal state, then the protection measurement provides new information: we can find the whole wave function. Thus, protective measurement of the quantum wave function of an ion in a trap can yield the the trap's potential.

Numerous objections to the validity and meaning of protective measurements have been raised [4–8]. The validity of the result was questioned due to misunderstanding of what the protective measurement is [9–11]. The issue of meaning: "Is the wave function of a single particle an ontological entity?" [3] is open to various interpretations. Some will say 'yes' even before hearing about protective measurement, others say 'no' just because protective measurements are never 100% reliable. The protective measurement procedure is not a proof that we should adopt one interpretation instead of the other, but it is a good testbed which shows advantages and disadvantages of various interpretations. For example, the Bohmian interpretation does not provide a natural explanation of how a protective measurement can "draw" the whole wave function of an ion in a ground state of a trap, since the Bohmian position of the ion hardly changes during the measurement [12, 13].

The protective measurements method can be extended to pre- and post-selected systems described by a ► two-state vector formalism $\langle\Phi| \quad |\Psi\rangle$ [14]. It requires separate different protections for the forward and backward evolving quantum states which are achieved by pre- and post-selection of quantum states of systems which provide the protection [15]. The outcome of such protective measurements is not the expectation value, but the ► *weak value*, $\frac{\langle\Phi|O|\Psi\rangle}{\langle\Phi|\Psi\rangle}$ [2, p. 845]. A realistic setup for such protective measurement is a weak coupling to a variable of a decaying system which is post-selected not to decay [16].

Theoretical analysis of protective measurements leads to deeper understanding of quantum reality while its experimental realization (which seems feasible in a near future) might be useful for more effective gathering of information about quantum systems [17].

Primary Literature

1. Y. Aharonov and L. Vaidman, Measurement of the Schrodinger wave of a single particle, *Phys. Lett.* **A 178**, 38 (1993).

Secondary Literature

2. Weak Values and Weak Measurements, in this volume, pp. 840–2.
3. Y. Aharonov, J. Anandan, and L. Vaidman, Meaning of the wave function, Phys. Rev. **A 47**, 4616 (1993).
4. W.G. Unruh, Reality and measurement of the wave function, Phys. Rev. **A 50**, 882 (1994).
5. C. Rovelli, Meaning of the wave function - Comment, Phys. Rev. **A 50**, 2788 (1994).
6. P. Ghose and D. Home, An analysisi of Aharonov-Anandan-Vaidman model, Found. Phys. **25**, 1105 (1995).
7. O. Alter and Y. Yamamoto, Protective measurement of the wave function of a single squeezed harmonic-oscillator state, Phys. Rev. **A 53**, R2911 (1996); **A 56**, 1057 (1997).
8. J. Uffink, How to protect the interpretation of the wave function against protective measurements, Phys. Rev. **A 60**, 3474 (1999).
9. Y. Aharonov, J. Anandan and L. Vaidman, The meaning of protective measurements, Found. Phys. **26**, 117 (1996).
10. Y. Aharonov and L. Vaidman, Protective measurement of the wave function of a single squeezed harmonic-oscillator state – Comment, Phys. Rev. **A 56**, 1055 (1997).
11. N.D.H. Dass and T. Qureshi, Critique of protective measurements, Phys. Rev. **A 59**, 2590 (1999).
12. Y. Aharonov, and L. Vaidman, About position measurements which do not show the bohmian particle position, in *Bohmian Mechanics and Quantum Theory: An Appraisal*, J.T. Cushing, A. Fine, S. Goldstein, eds. (Kluwer, Dordrecht 1996).
13. Y. Aharonov, B.G. Englert, and M.O. Scully, Protective measurements and Bohm trajectories, Phys. Lett. **A 263**, 137 (1999).
14. The Two-State Vector Formalism, this volume, pp. 802–6.
15. Y. Aharonov, and L. Vaidman, Protective Measurements of Two-State Vectors, in *Entanglement and Passion-at-a-Distance*, R.S. Cohen, M. Horne and J. Stachel (eds.), BSPS 1-8, (Kluwer, Dordrecht 1997).

16. Y. Aharonov, et al., Adiabatic measurements on metastable systems, Phys. Rev. Lett. **77**, 983 (1996).
17. S. Nussinov, Scattering experiments for measuring the wave function of a single system, Phys. Lett. **B 413**, 382 (1997).

Pure States

See ► Density operator; Ignorance interpretation; Kochen–Specker theorem; Mixed states; Objectification; Observable; Probability in Quantum Mechanics; Quantum entropy; States in Quantum Mechanics; States, pure and mixed and their Representation; Superselection Rules; Wave function collapse.

Q

Quantization (First, Second)

Helge Kragh

If there is a second quantization, presumably there is also a first quantization. The latter term refers to the ordinary application of the ► Schrödinger equation to physical objects characterized by ► wave functions, while the surrounding environment (such as an electromagnetic field) is treated classically. In second quantization the environment is treated quantum-mechanically – the field is quantized – and the wave function is considered as a dynamical system subject to quantization. To put it differently, one takes the wave function of an already quantized system and turns it into an ► operator.

The method of second quantization goes back to works of Paul A.M. Dirac and Pascual Jordan in 1927. Dirac used a kind of second quantization to the electromagnetic field by identifying the coefficients of the Fourier expansion of the field as photon ► creation and annihilation operators. He showed that there is a close connection between quantum fields and statistics, and derived in this way that photons obey ► Bose-Einstein statistics. Jordan went considerably further, in part alone and in part in works together with coauthors. Whereas Dirac restricted his approach to photons (► light quantum), Jordan quantized ► matter waves given by the Schrödinger equation, first non-relativistically and, with Eugene Paul Wigner in 1928, relativistically. Jordan's quantization could be performed in two ways, leading either to ► Bose-Einstein or ► Fermi-Dirac statistics. In the latter case it gave a quantum-mechanical justification of Pauli's ► exclusion principle.

It was Jordan's field-quantization method that was taken up by other physicists and used in quantum field theory. It is also in Jordan's paper of 1927 that the name "second quantization" first appears. Dirac, who did not appreciate Jordan's method of second quantization, did not consider the discreteness of matter a property deducible from quantum mechanics. Jordan, on the other hand, claimed ambitiously to have derived from quantization of matter fields the very existence of particles. "The basic fact of electron theory, the existence of discrete electric particles, appears. . . as a characteristic quantum phenomenon," he wrote in 1927; "indeed, it means exactly that matter waves occur only in discrete quantum states."

Second quantization was discussed by Pauli at the 1927 Solvay congress. Einstein did not like the idea and later told Oskar Klein that "second quantization, that is sin squared." In spite of some opposition, Jordan's method was developed by several physicists in the years around 1930. It was applied by Wolfgang Pauli and Werner Heisenberg in their relativistic quantum theory of wave fields 1929–30 and given a new formulation by V. Fock in 1932. Fock's version allowed the translation

D. Greenberger et al. (eds.), *Compendium of Quantum Physics: Concepts, Experiments, History and Philosophy*,

of the formalism of second quantization into the language of conventional quantum mechanics, which helped making the method more acceptable.

Primary Literature

1. P. A. M. Dirac: The Quantum Theory of the Emission and Absorption of Radiation. Proceedings of the Royal Society of London **A 114,** 243–265 (1927)
2. P. Jordan: Über eine neue Begründung der Quantenmechanik. Zeitschrift für Physik **44,** 1–25 (1927)
3. V. Fock: Konfigurationsraum und zweite Quantelung. Zeitschrift für Physik **75,** 622–647 (1932)

Secondary Literature

4. O. Darrigol: The origin of quantized matter waves. Historical Studies in the Physical Sciences **16,** 197–253 (1986)
5. H. D. Zeh: There Is no "first" quantization. Physics Letters **A 309,** 329–334 (2003)

Quantization (Systematic)

N.P. Landsman

The term *quantization* (in the sense described here) refers to attempts to construct a mathematical description of a quantum system from its formulation as a classical system (which is supposed to be known). Such attempts go back to the pioneers of the old quantum theory (Planck, Einstein, Bohr, Sommerfeld); see [16] and ► Quantization: (First, Second). (The opposite procedure is the subject of the ► quasi-classical limit.)

The modern era of quantization theory started with Heisenberg's famous paper [5] from 1925, in which he proposed the idea of a 'quantum-theoretical reinterpretation (*Umdeutung*) of classical observables.' All later work on quantization may be said to consist of various different implementations of this idea.

The first successful such implementation consisted of the position and momentum ► operators introduced by Schrödinger [9], i.e. $\hat{q}^j = x^j$ and $\hat{p}_j = -i\hbar\partial/\partial x^j$, seen (in modern parlance) as unbounded operators on the ► Hilbert space $L^2(\mathbb{R}^3)$. Substituting these expressions into the classical Hamiltonian yields the left-hand side of the ► Schrödinger equation. These operators satisfy the so-called *canonical commutation relations*

$$[\hat{p}_j, \hat{q}^k] = -i\hbar\delta_j^k, \tag{1}$$

along with $[\hat{p}_j, \hat{p}_k] = 0$ and $[\hat{q}^j, \hat{q}^k] = 0$. This fact formed the basis of the various equivalence proofs of ► matrix mechanics and ► wave mechanics that were given at the time by Schrödinger, Dirac, and Pauli; the first genuine mathematical proof of this equivalence is due to von Neumann [8].

Approaches to quantization that are based on the canonical commutation relations are usually called *canonical quantization*. Dirac [3, 4] made the important observation that the canonical commutation relations resemble the Poisson brackets in classical mechanics. He suggested that a quantization map $f \mapsto Q(f)$ (in which a function f on phase space, seen as a classical observable, is replaced by some operator on a Hilbert space interpreted as the corresponding quantum observable) should satisfy the condition

$$\frac{\mathrm{i}}{\hbar}[Q(f), Q(g)] = Q(\{f, g\}). \tag{2}$$

This is indeed the case for $f(p,q) = p_j$ or q^k and $g(p,q)$ likewise, provided we follow Schrödinger in putting $Q(p_j) = \hat{p}_j$ and $Q(q^j) = \hat{q}^j$. For more complicated observables, however, Dirac's condition turns out to hold only asymptotically as $\hbar \to 0$. For example, in the first systematic account of the quantization of a particle moving in flat space, Weyl [11] proposed that a function f on classical phase space $\mathbb{R}^{2n}$ corresponds to the operator

$$Q(f)\Psi(x) = \int_{\mathbb{R}^{2n}} \frac{\mathrm{d}^n p\, \mathrm{d}^n q}{(2\pi\hbar)^n}\, \mathrm{e}^{\mathrm{i}p(x-q)/\hbar} f\left(p, \tfrac{1}{2}(x+q)\right) \Psi(q). \tag{3}$$

on $L^2(\mathbb{R}^n)$. This reproduces Schrödinger's position and momentum operators, but satisfies (2) only if f and g are at most quadratic in p and q (and according to the so-called Groenewold–van Hove theorem prescriptions different from Weyl's will not fare better). This violation of Dirac's condition is well understood now, since is has been recognized that the essence of the process of quantization is that it yields a *deformation* of the classical algebra of ► observables [1, 2]. The idea of deformation quantization is particularly relevant to physics in the framework of ► algebraic quantum theory [14, 17] (see also [13] for other aspects of Weyl quantization).

The quantization problem on phase spaces other than $\mathbb{R}^{2n}$ (or, more generally, cotangent bundles of Riemannian manifolds, to which Weyl's quantization method is easily generalized [14]) has to be treated by different means. In fact, even on flat space one can sympathize with Mackey's lamentation that 'Simple and elegant as this model [i.e. canonical quantization] is, it appears at first sight to be quite arbitrary and ad hoc. It is difficult to understand how anyone could have guessed it and by no means obvious how to modify it to fit a model for space different from $\mathbb{R}^n$.' ([15], p. 283). Mackey himself explained and generalized canonical quantization on the basis of symmetry arguments that apply whenever a symmetry group G acts on configuration space Q (with associated phase space T^*Q); for flat space $Q = \mathbb{R}^3$ ones takes $G = E(3) = SO(3) \ltimes \mathbb{R}^3$, the Euclidean symmetry group of rigid translations and rotations. Mackey's generalization of the canonical commutation relations (1) consists of his notion of a *system of imprimitivity*. Given an action of a group G on a space Q, such a system consists of a Hilbert space H, a unitary representation U of G on H, and a projection-valued measure $E \mapsto P(E)$ on Q with values in H, such that

$$U(x)P(E)U(x)^{-1} = P(xE), \tag{4}$$

for all $x \in G$ and all (Borel) sets $E \subset Q$. One notices that position and momentum are assigned a quite different role in this procedure: the former are replaced by the projection-valued measure $E \mapsto P(E)$, whereas the latter are treated as the (infinitesimal) generators of symmetries. Each irreducible system of imprimitivity provides a valid quantization of a particle moving on Q. Mackey's imprimitivity theorem classifies all possibilities; for example, for $Q = \mathbb{R}^3$and $G = E(3)$ one finds that each irreducible representation of $SO(3)$ yields a possible quantization. This is Mackey's explanation of ▶ spin. More generally, if $Q = G/K$ is a homogeneous G-space with stability group K, then each irreducible representation of K induces a system of imprimitivity and hence a quantization of the system (and *vice versa*). Let us note that the modern way of understanding this method involves groupoids and their C^*-algebras, which not only lead to a vast generalization of Mackey's approach but in addition put it under the umbrella of deformation quantization [14].

Geometric quantization is a method that starts from the symplectic (or, in old-fashioned language, 'canonical') structure of phase space. This method was independently introduced by Kostant [6] and Souriau [10] and is still being developed; cf. [12, 18]. Although its formalism is quite general, geometric quantization is most effective in the presence of a Lie group acting canonically and transitively on phase space. If successful, the method then yields a representation of the Lie algebra of this group, whose elements play the role of quantum observables.

The procedure starts with a phase space M (i.e. a symplectic manifold), and as a first step towards a quantum theory one constructs a map $f \mapsto Q^{\mathrm{pre}}(f)$ from functions on M to operators on the Hilbert space $L^2(M)$. This map turns out to satisfy Dirac's condition (2) exactly. In the special case $M = \mathbb{R}^{2n}$, it is given by

$$Q^{\mathrm{pre}}(f)\Phi = -\mathrm{i}\hbar\{f, \Phi\} + \left(f - \sum_j p_j \frac{\partial f}{\partial p_j}\right)\Phi, \tag{5}$$

where $\{f, \Phi\}$ is the Poisson bracket (which makes sense if $\Phi \in L^2(\mathbb{R}^{2n})$ is assumed differentiable). Unfortunately, the Hilbert space is wrong and the ensuing representation of the canonical commutation relations $Q^{\mathrm{pre}}(q^k) = q^k + \mathrm{i}\hbar\partial/\partial p_k$ and $Q^{\mathrm{pre}}(p_j) = -\mathrm{i}\hbar\partial/\partial q^j$ is highly reducible: it contains an infinite number of copies of the Schrödinger representation on $L^2(\mathbb{R}^n)$. The second step of the method therefore involves a procedure to cut down the size of the Hilbert space $L^2(M)$ by a certain geometric technique, but through this step only some of the operators (5) remain well defined. Those that are still satisfy (2), however, which fact lies at the basis of the construction of Lie algebra representations from geometric quantization. Despite some successes in that direction, with considerable impact on mathematics, the method of geometric quantization remains unfinished and somewhat unsatisfactory for physics.

Like geometric quantization, *phase space quantization* starts with the Hilbert space $L^2(M)$, but instead of (5) one constructs a quantization map $f \mapsto Q^{\mathrm{p}}(f)$ by

$$Q^{\mathrm{p}}(f) = pfp, \tag{6}$$

where p is a suitable projection operator on $L^2(M)$ (so that the operator $Q^{\mathrm{p}}(f)$ effectively acts on $pL^2(M)$). This projection is constructed from a so-called reproducing kernel K on $L^2(M)$, and has the form $p\Phi(z) = \int_M \mathrm{d}w\, K(z,w)\Phi(w)$. This kernel, in turn, comes from a family of ▶ coherent states - here construed as maps $z \mapsto \Psi_z$ from M to the set of unit vectors in an auxiliary Hilbert space H - by means of $K(z,w) = (\Psi_z, \Psi_w)$ (the inner product in H). See [12, 14]. The best-known example is $M = \mathbb{R}^{2n}$ with coherent states $\Psi^{\hbar}_{(p,q)}(x) = (\pi\hbar)^{-n/4}\exp((-(x-q)^2 + ip(2x-q))/2\hbar)$ in $H = L^2(\mathbb{R}^n)$, yielding what is often called *Berezin quantization* Q^{B} on $\mathbb{R}^{2n}$. It has the advantage over Weyl quantization and geometric quantization of being positive (in the sense that $(\Phi, Q^{\mathrm{B}}(f)\Phi) \geqslant 0$ for all Φ whenever $f \geqslant 0$) and bounded (i.e. $Q^{\mathrm{B}}(f)$ is a bounded operator if f is a bounded function on M).

Quantization theory remains a very active area of research in physics and mathematics [12]. See also ▶ Functional integration; path integrals.

Primary Literature

1. F. Bayen, M. Flato, C. Fronsdal, A. Lichnerowicz & D. Sternheimer: Deformation theory and quantization I, II. Ann. Phys. (N.Y.) **110**, 61–110, 111–151 (1978).
2. F.A. Berezin: General concept of quantization. Commun. Math. Phys. **40**, 153–174 (1975).
3. P.A.M.Dirac: The fundamental equations of quantum mechanics. Proc. R. Soc. Lond. **A109**, 642–653 (1926).
4. P.A.M. Dirac: *The Principles of Quantum Mechanics* (The Clarendon Press, Oxford 1930).
5. W. Heisenberg: Über die quantentheoretische Umdeutung kinematischer und mechanischer Beziehungen. Z. Phys. **33**, 879–893 (1925).
6. B. Kostant: Quantization and unitary representations. *Lect. Notes Math.* **170**, 87–208 (1970).
7. G.W. Mackey: *Induced Representations of Groups and Quantum Mechanics* (W. A. Benjamin, New York 1968).
8. J. von Neumann: Die Eindeutigkeit der Schrödingerschen Operatoren. Math. Ann. **104**, 570–578 (1931).
9. E. Schrödinger: Über das Verhältnis der Heisenberg-Born-Jordanschen Quantenmechanik zu der meinen. Ann. Phys. **79**, 734–756 (1926).
10. J.-M. Souriau: *Structure des systèmes dynamiques* (Dunod, Paris (1969). Translated as: *Structure of Dynamical Systems: A Symplectic View of Physics* (Birkhäuser, Boston 1997).
11. H. Weyl: *Gruppentheorie und Quantenmechanik* (S. Hirzel, Leipzig 1928). Translated as: *The Theory of Groups and Quantum Mechanics* (Dover, New York 1931).

Secondary Literature

12. S.T. Ali & M. Englis: Quantization methods: a guide for physicists and analysts. Rev. Math. Phys. **17**, 391–490 (2005).
13. D.A. Dubin, M.A. Hennings & T.B. Smith: *Mathematical Aspects of Weyl Quantization and Phase* (World Scientific, Singapore 2000).
14. N.P. Landsman: *Mathematical Topics Between Classical and Quantum Mechanics* (Springer, New York 1998).
15. G.W. Mackey: *The Scope and History of Commutative and Noncommutative Harmonic Analysis* (American Mathematical Society, Providence 1992).
16. J. Mehra & H. Rechenberg: *The Historical Development of Quantum Theory. Vol. 1: The Quantum Theory of Planck, Einstein, Bohr, and Sommerfeld: Its Foundation and the Rise of Its Difficulties* (Springer, New York 1982).
17. M.A. Rieffel: Quantization and C^*-algebras. *Contemp. Math.* **167**, 66–97 (1994).
18. J. Śniatycki: *Geometric Quantization and Quantum Mechanics* (Springer, Berlin 1980).

Quantum Chaos

Stefan Weigert

The term Quantum Chaos designates a body of knowledge which has been established in an attempt to understand the implications of Classical Chaos for quantum systems. Classical Mechanics successfully describes many aspects of the macroscopic world in a phenomenological way. Chaotic behaviour being ubiquitous, its presence begs for an explanation in terms of (non-relativistic) quantum mechanics, the fundamental theory to describe matter. Only the *deterministic* part of the quantum time evolution generated by ► Schrödinger's equation is of interest here while the probabilistic element introduced by quantum ► measurements is ignored.

An autonomous classical Hamiltonian system with $N \geqslant 2$ degrees of freedom is either integrable or non-integrable. The time evolution of integrable systems is *quasi-periodic*, hence simple: N global constants of motion exist which force trajectories in phase space to evolve on tori of dimension N. The distance between initially close trajectories increases at most *linearly* with time; the Lyapunov exponent, a measure for the rate of divergence of nearby trajectories, is equal to zero. In the vast majority of cases, however, fewer than N constants of motion exist and the system is non-integrable. A typical trajectory now may explore a larger part of phase space while still evolving deterministically. Due to their highly complicated – apparently chaotic – time evolutions, trajectories with similar initial conditions tend to diverge at an *exponential* rate. This property makes long-term predictions of the system's dynamics unreliable if not effectively impossible.

A considerable amount of studies relevant to Quantum Chaos revolve around three questions: (1) Is it possible to (approximately) quantize classically chaotic systems by exploiting their phase-space structure? (2) What are quantum mechanical manifestations–also known as *precursors* or *signatures*–of Classical Chaos? (3) Does a rigorous distinction between regular and chaotic quantum systems exist?

To answer these questions, quantum systems from many branches of physics and chemistry have been studied afresh from a new perspective. They include nuclei, atoms and molecules in the presence of strong electromagnetic fields, and microwaves in cavities, for example. The approaches to explore the properties of these systems range from experimental and numerical to rigorously mathematical.

For a long time, complicated dynamical behaviour has been assumed (tacitly) to require *many* interacting constituents such as the molecules of a gas. Their large number justifies the use of powerful statistical methods. Dynamical chaos, however, results from non-linear interactions between only a few degrees of freedom. This fundamental property of Classical Mechanics has been widely recognized only in the second half of the 20th century, when it became one of the driving forces to study quantum mechanical counterparts of classical systems with effectively unpredictable time evolution.

Widely studied models include quantum particles restricted to move in two-dimensional regions known as *billiards*, pairs of coupled spins or a single periodically driven spin. Reducing the continuous time evolution of a classically chaotic system to an iterated *map* has proved advantageous in many cases. Maps are simple to formulate but capture essential features of the dynamics. A thoroughly studied example is the (classical or quantum) *standard map* describing a kicked rotor. Many other systems such as an electron in a one-dimensional hydrogen atom in the presence of a periodically modulated electric field give rise to the same or structurally similar maps. (► Bohr's atom model).

(1) If a quantum system has a classically chaotic limit, it is usually hard to extract useful information from its ► Schrödinger equation. Often, extensive numerical calculations are the only means to determine (the spatial structure, say, of) excited states and the corresponding energy levels. A substantial amount of work has thus been devoted to generalize the *torus quantization*, an early method to 'quantize' classical systems which precedes and thus bypasses ► Schrödinger's equation. Its original formulation relies on the phase space of the system being foliated entirely by tori. This structure, however, only exists if the system is *integrable*, i.e. it must possesses as many global constants of motion as it has degrees of freedom. The foliation is destroyed if a perturbation is added to the system, and only a skeleton of closed trajectories known as *periodic orbits* continues to exist. Einstein realized in 1917 that the quantization conditions are not generally applicable [1]. The new approach, initiated in the early 1970s, relies on the fact that, even in a non-integrable system, isolated periodic orbits survive and continue to determine the quantum properties of the system to a large extent. To see this, one uses the ► path-integral formulation of quantum mechanics. The resulting *trace formula* provides an alternative and often efficient road to (approximately) quantize a classically chaotic system [2].

(2) The statistics of energy levels exhibit striking differences for different quantum systems. After appropriate normalization, the spacings between the energy eigenvalues of systems with a classically *regular* limit are described well by a Poisson distribution: small spacings dominate. The small spacings are suppressed for systems with a chaotic classical limit, resulting in a distribution derived by Wigner in 1951 to statistically describe observed energy spectra of nuclei [3]. The overall shapes of the distributions are universal in the sense that they only depend on symmetry properties such as the presence or absence of time reversal invariance of the system. It turns out that the spectra of *random matrices*, with matrix elements drawn from specific distributions determined by the symmetries, have very similar spectral properties [4]. This confirms the intuitively appealing picture that a Hamiltonian describing a quantum system with a classically chaotic limit correspond to a matrix with 'random' entries.

The spatial structure of energy eigenstates of a quantum system may also anticipate whether it has a classically chaotic counterpart or not [5], as do scattering amplitudes. It is the classical periodic orbits which, to a large extent, determine the properties of both bounded and open quantum systems in the ► quasi-classical regime defined by $S/\hbar \ll 1$, where S is the value of the classical action associated with a typical periodic orbit.

The Anderson model of conduction in a one-dimensional disordered solid predicts that its energy eigenstates are confined to only small parts of the available space. Mathematically, the quantum standard map is structurally identical to the Anderson Hamiltonian if discrete time is thought to label lattice sites [6]. The resulting *dynamical localization* is used to explain that electron diffusion in a driven hydrogen atom [7] deviates from classically expected behaviour: the atom is ultimately not ionized since the diffusion is suppressed quantum mechanically.

(3) Ideally, a concept such as Quantum Chaos should rest upon a definition which is inherently quantum mechanical: it should not depend on properties of quantum systems which emerge only in the classical limit. The challenge is to put each (non-relativistic) quantum system with only a few degrees of freedom, say, in one of two disjoint classes using quantum mechanical concepts only. So far, no such division entailing sets of systems with *provably* different properties has been agreed upon [8].

Another fundamental aspect is the question to what degree ► Schrödinger's equation, as a *linear* equation, is capable to generate complicated time evolutions. Is it conceivable that the evolution of a quantum state is as difficult to predict as a trajectory of a classically chaotic system, typically resulting from coupled non-linear differential equations? An appropriate Fourier transform of such a trajectory will reveal a *continuous* spectrum of frequencies, an unmistakable sign for the trajectory being highly irregular. If a similar approach is taken within a time-independent quantum system, the resulting spectrum will be determined by the energy eigenvalues of the system which are a *discrete* set if the quantum system has bound states only. This observation explains why externally driven quantum systems and scattering processes are promising candidates when searching for chaotic behaviour in quantum mechanics.

The tendency of quantum mechanics to suppress chaos is supported by a phase space perspective: quantization can be thought of as introducing a 'granular' structure ► quantization. Its scale relates to the non-commutativity of position and momentum operators measured by the value of ► Planck's constant $\hbar$. Thus, the evolution of arbitrarily fine structures in phase space, a hallmark of Classical Chaos, appears forbidden. Nevertheless, the time evolution of a quantum system may be as difficult to predict as a classical irregular trajectory if *commuting* observables such as two (or more) position operators undergo a complicated dynamics in configuration space [9].

Primary Literature

1. A. Einstein: Zum Quantensatz von Sommerfeld und Epstein. Verh. Deutsch. Phys. Ges. 19, 82–92 (1917)
2. M.C. Gutzwiller: Periodic Orbits and Classical Quantization Conditions. J. Math. Phys. 12, 343–358 (1971)
3. E.P. Wigner: On the Statistical Distribution of the Widths and Spacings of Nuclear Resonance Levels. Proc. Camb. Philos. Soc. 47, 790–798 (1951)
4. Bohigas, O., Giannoni, M.J., Schmit, C.: Characterization of Chaotic Quantum Spectra and Universality of Level Fluctuation Laws. Phys. Rev. Lett. 52, 1–4 (1984)
5. M.V. Berry: Regular and Irregular Semiclassical Wavefunctions. J. Phys. A 10, 2083–2091 (1977)
6. S. Fishman, Grempel, D.R., Prange, R.E.: Chaos, Quantum Recurrences, and Anderson Localization. Phys. Rev. Lett. 49, 509–512 (1984)
7. J.E. Bayfield, P.M. Koch: Multiphoton Ionization of Highly Excited Hydrogen Atoms. Phys. Rev. Lett. 33, 258–261 (1974)
8. S. Weigert: The Problem of Quantum Integrability. Physica D 56, 107–119 (1992)
9. S. Weigert: Quantum Chaos in the Configurational Quantum Cat Map. Phys. Rev. A 48, 1780–1798 (1993)

Q

Secondary Literature

10. M. V. Berry: Quantum Chaology. Proc. R. Soc. A 413, 183–198 (1987)
11. F. Haake: *Quantum Signatures of Chaos*. (Springer, Berlin 2001)
12. M. C. Gutzwiller: *Chaos in Classical and Quantum Mechanics*. (Springer, Berlin 1990)
13. *Chaos and Quantum Physics*, (Les Houches 1989, Session LII), ed. by M.J. Giannoni, A. Voros, and J. Zinn-Justin (North-Holland, Amsterdam 1991)
14. M. C. Gutzwiller: Resource Letter ICQM-1: The Interplay between Classical and Quantum Mechanics. Am. J. Phys. 66, 304–324 (esp. Part VIII) (1998)
15. H.-J. Stöckmann: *Quantum Chaos*. (Cambridge University Press, Cambridge 1999)

Quantum Chemistry

Ana Simões

When introducing the *International Journal of Quantum Chemistry* in 1967, the Swedish quantum chemist Per-Olov Löwdin (1916–2000) defined the then forty-year old discipline in the following manner:

> Quantum chemistry deals with the theory of the electronic structure of matter: atoms, molecules, and crystals. It describes this structure in terms of wave patterns, and it uses physical and chemical experience, deep-going mathematical analysis, and high-speed electronic computers to achieve its results. Quantum mechanics has rendered a new conceptual framework for physics and chemistry, and it has led to a unification of the natural sciences which was previously inconceivable; the recent development of molecular biology shows also that the life sciences are now approaching the same basis.
>
> Quantum chemistry is a young field which falls between the historically developed areas of mathematics, physics, chemistry, and biology.

Written at a time in which quantum chemistry was experiencing intense networking and growing internationalization and was exploring the potential of a promising instrument – the electronic digital computer – at the same time as extending its domain to molecules of biological interest, the definition bears witness to the challenges posed by this recent juncture when contrasted with the previous state of things. It calls attention to the subject-matter of quantum chemistry – the elucidation of the electronic make-up of atoms, molecules and aggregates of molecules; the interplay of inputs from theory, experiment, mathematics and computation in building the methodological apparatus of quantum chemistry; its relationship with the neighboring disciplines of mathematics, physics, and biology; and finally the assessment of the role of quantum mechanics in providing a unifying framework for the natural sciences and eventually for the life sciences. The influence of quantum chemistry was to extend to all branches of chemistry, from physical, organic, analytical, and inorganic chemistry to biochemistry.

Evidence of the difficulties encountered in positioning the new field in relation to neighboring areas such as chemistry, physics and mathematics lies in the multiplicity of names attributed to the field extending well into the period when Löwdin founded the journal. Extra evidence includes the different names assigned to chairs occupied by its practitioners, the titles of journals used as outlets for their publications or the descriptions of courses taught on the subject. The new field has been called mathematical chemistry, quantum theory of valence, molecular quantum mechanics, theoretical chemistry, chemical physics as well as the now standard quantum chemistry. Although hard to ascertain, the first appearance of the designation 'quantum chemistry' in the literature is probably due to Arthur Erich Haas (1884–1941), the professor of physics at the University of Vienna who published in 1929 *Die Grundlagen der Quantenchemie*, a collection of four lectures delivered to the Physico-Chemical Society in Vienna.

Löwdin wrote this passage forty years since the German physicists Walter Heitler (1904–1981) and Fritz London (1900–1954) published their 1927 joint paper usually considered as marking the birthday of quantum chemistry. Heitler and London extended Heisenberg's quantum-mechanical treatment of the two indistinguishable ► electrons in the helium atom (1926) to the quantum-mechanical explanation of the formation of the hydrogen molecule. They started with a ► wave function that took into consideration the ► indistinguishability of the two electrons and minimized the system's energy by using perturbation theory. They obtained two values for the energy expressed as a function of three integrals – Coulomb integral, exchange integral and overlap integral – and showed that attraction between the two atoms occurred only when electrons had opposite spins ('electron pairing'), giving rise to a covalent bond. Covalent bonds were thus shown to be pure quantum-mechanical effects and ► spin became one of the most significant indicators of valence behavior. Despite a selection of the simplest of all molecules, the rationale behind this first successful attempt to solve an intrinsically chemical problem – understanding why and how atoms combine to form molecules – was to treat it as a many-body problem, which they handled by means of the integration of ► Schrödinger's equation. The difficulty in solving Schrödinger's equation for molecular systems exactly lay at the heart of quantum chemistry.

This state of affairs was soon encapsulated in Paul A.M. Dirac's 1929 dictum to the effect that 'the underlying physical laws necessary for the mathematical theory of a large part of physics and the whole of chemistry are [now] completely known, and the difficulty is only that the exact application of these laws leads to equations much too complicated to be soluble'. This statement has been cited frequently by historians and philosophers of science in the context of discussions on the hypothetical reduction of chemistry to physics. Chemists, however, took it as a historical prediction (not a philosophical claim) proven wrong due to the inability to foresee the importance of exact computations for chemistry. Extending this argument, one may well claim that Dirac was unable to foresee that a new breed of chemists would emerge sharing a culture very different from the reductionist culture of physicists but taking seriously the perspectives opened up by the use of quantum mechanics. By embracing different methodological and ontological commitments, they successfully devised semi-empirical approximate methods which became a constitutive feature of quantum chemistry in its first decades, and which had to face the challenge of an era of wholly theoretical (*ab initio*) computations following the extensive use of electronic digital computers after World War II.

While Heitler and London attempted unsuccessfully to extend their pioneering work to polyelectronic molecules using group theory to help generalize results derived by perturbation methods, other German physicists tried to understand quantum-mechanically the nature of the chemical bond. Friedrich Hund (1896–1997) classified the electronic quantum states of diatomic molecules and Erich Hückel (1896–1980) built a theoretical model for the benzene molecule. By the late 1930s, they all had abandoned the field as it proved impossible to treat in an analytical manner Schrödinger's equation for molecules including more than three electrons.

In the meantime, the Americans Linus Pauling (1901–1992) (Nobel Prize 1954), John Clarke Slater (1900–1976) and Robert Sanderson Mulliken (1896–1986) (Nobel Prize 1966) developed a different perspective for the quantum-mechanical explanation of the chemical bond. While German physicists thought the theories of the chemical bond should be derived from first principles firmly grounded on the postulates of quantum mechanics, Americans acknowledged the importance of quantum mechanics and, at the same time, aimed at developing semi-empirical methods dependent on the formulation of short-cut rules based on a sort of induction from available data (which in many instances they gathered themselves) together with the introduction of concepts which facilitated the making of approximations.

Pauling's valence bond approach, envisioning molecules as aggregates of atoms bonded together along privileged directions, was meant to extend classical structure theory. Both Slater (1931) at M.I.T. and Pauling (1931–1933) at Caltech built on Heitler and London's 1927 valence bond paper, but outlined a semi-empirical approach based on the idea of hybridization of atomic orbitals to form bond orbitals possessing directional character. In this way they explained the formation of molecules such as water and methane. Pauling subsequently attempted to understand the formation of more complex molecules, dealing with the stability of aromatic and conjugated compounds. In molecules such as benzene, for which no single structure seemed to represent adequately all its properties, Pauling suggested that the molecule could be represented as a hybrid of two or more conventional forms, a situation he dubbed 'resonance among several valence-bond structures'. Introduced in 'The Nature of the Chemical Bond' series, the 'theory of resonance' was further developed and presented to a wider audience in the famous book *The Nature of the Chemical Bond* (1939).

Clarification of the relations between electronic states and the structure of molecular spectra (1928–1932) was the basis on which Mulliken grounded his rejection of the ontological foundation of classical valence theory. He refused to reduce a molecule to an aggregate of atoms, and built it instead from nuclei and electrons. Reasoning by analogy with Bohr's building-up principle, Mulliken considered that molecules were formed by feeding electrons into orbitals encircling two or more nuclei. Electrons were delocalized in the sense that there was a non-zero probability of finding them near more than one nucleus. The assignment of quantum numbers to electrons in molecules was achieved by exploring the relations to the united-atom description and the separated-atom description put forward by Hund, and the classification of molecular orbitals in polyatomic molecules applied group theory (1932–1935). New auxiliary concepts were introduced such as promoted and unpromoted electrons, bonding, non-bonding and anti-bonding electrons, and varying bonding power of electrons. Mulliken's approach was semi-empirical in the sense that the relative order of energy states was obtained from quantum mechanics but energy levels were dependent on spectroscopic and thermochemical data.

To highlight the choice of opposite methodological stances, noted already by John H. van Vleck (1899–1980) and Albert Sherman (1907–38) in their 1935 review paper, historians of science have suggested that the usual division appearing in the chemical literature and in textbooks between the Heitler–London–Slater–Pauling valence bond method (VB) and the Hund–Mulliken method of molecular orbitals (MO) should be replaced by another dichotomy – Mulliken–Pauling–Slater versus the Heitler–London-Hund.

In the meantime, John Lennard-Jones (1894–1954), the British physicist from Cambridge University who was soon to hold the Plummer Chair in Theoretical Chemistry (1932), had introduced the physical simplification of representing molecular orbitals as linear combinations of atomic orbitals (LCAO) (1929), a step that proved crucial to the subsequent mathematization of MO theory. Together with Douglas R. Hartree (1897–1958) and Charles Alfred Coulson (1910–1974), who was to become the first holder of the Chair of Theoretical Chemistry at the University of Oxford (1972), these British theoreticians played a decisive role in the further development of quantum chemistry. All strongly influenced by Mulliken's legacy, they perceived the problems of quantum chemistry first and foremost as calculational problems; and by devising novel calculational methods they tried to bring quantum chemistry within the realm of applied mathematics.

Until the 1950s VB theory dominated quantum chemistry for reasons that were not due to its empirical adequacy, explanatory power or predictive ability when compared with MO theory; they rather depended on contrasting rhetorical skills and personal characteristics of the advocates of both theories. The ascendancy of the MO theory was largely associated with the contributions of Coulson, its advocate who rivalled with Pauling in rhetorical and pedagogical skills. His textbook *Valence* (1952) counterbalanced the approach set previously in *The Nature of the Chemical Bond*. Furthermore, MO theory profited from being easily adapted to the classification of the excited states of molecules – one of the realms of molecular spectroscopy – and, above all, was suitable for computer programs. In fact, in the period right after the end of World War II, quantum chemists were eager to take advantage of electronic digital computers in the computation of molecular wave functions and energy levels.

Considered a 'watershed', the international program outlined at the Shelter Island Conference (1951) clarified chemical concepts such as electron pairs, bond energies and bond orders, hybridization and chemical reactivity. But, above all, it aimed at obtaining formulas for the troublesome multi-central integrals which acted as 'bottlenecks' to the integration of Schrödinger's equation in the *ab initio* manner. These formulas thus became available to the community of quantum chemists in standardized tables. While at first dependent on human computers aided by desk calculators, the program soon evolved to articulate an efficient cooperative network that took advantage of the slowly increasing number of electronic digital computers available to the international community. Computers turned into an essential tool to calculate the time-consuming integrals of the increasingly sophisticated versions of the MO method (Pariser–Parr–Pople, Self Consistent Field, Hartree–Fock, Configuration Interaction, etc.) and in many instances replaced laboratory experiments as sources of new data, especially in the investigation of molecules otherwise inaccessible to experimentation.

By 1959 a conference convened in Boulder, Colorado, debated the impact of computers in quantum chemistry. In an after-dinner speech delivered at the end of the conference, Coulson announced the splitting of the community into two distinct groups – those interested in exact calculations in molecules including up to 20 electrons (*ab-initionists*) and those still faithful to semi-empirical methods, those loath to abandon conventional chemical concepts and those claiming that chemistry was still an experimental science built around quite elementary concepts. The split resulted from diverging views concerning the use of large-scale electronic computers, and pointed to deep, perhaps irreconcilable, divisions among the practitioners of quantum chemistry (Group I included the *ab-initionists*, those who explored the potentialities of electronic computers, while Group II included the *a posteriorists* who did not bet on the importance of electronic computers for quantum chemistry). In 1965, John A. Pople (1925–2004) (Nobel Prize 1988) illustrated these divisions with a chart, later known as the 'hyperbola of quantum chemistry', which depicted the inverse relationship between the size of the molecules under study and the sophistication of computational methods.

Also reflecting on the impact of computers, the French quantum chemist Alberte Pullman (1920-) whose group was extending MO theory to biological molecules, predicted the merging of the two groups into a single group of '*ab initio* for everybody' (1970). Sensing that in the near future ever more powerful but also cheaper computers would become available to increasingly large fractions of the quantum chemical community, she pressed theoreticians to abandon their 'ivory tower of abstractions' to venture into the exploration of real problems of chemistry, ranging from the hydrogen molecule to biological macromolecules. In fact, by 1990 Martin Karplus (1930-) suggested replacing the two-dimensional Pople diagram by a three-dimensional one including as an extra dimension the estimated accuracy of calculation for the system under consideration. At the same time, he changed the linear scale of the axis in Pople's diagram representing the size of the molecule (which covered 1–100 electrons) by a logarithmic scale going up to 10^6 electrons. This change highlighted the possibility of conducting *ab initio* computations at a satisfactory accuracy for reasonably complex molecules and their reactions. Furthermore, Karplus recognized that density functional methods appeared to violate the 'hyperbola of quantum chemistry' in the sense that they fall within the range of accuracy and sophistication of Hartree-Fock type calculations but handle molecules with a larger number of electrons within available computer time.

Having these recent developments in mind, one wonders whether Dirac's 1929 prediction has been fulfilled to a significant degree. One wonders further whether the divorce in the quantum chemical community that haunted the perceptive Coulson in time converged into a peaceful cohabitation and eventually into a successful marriage of the two different cultures of practitioners.

Primary Literature

1. C. A. Coulson: *Valence* (Oxford University Press, Oxford 1952)
2. C. A. Coulson: Recent developments in valence theory – Symposium fifty years of valence. Pure and Applied Chemistry **24**, 257–87 (1970)

3. P. A. M. Dirac: Quantum mechanics of many-electron systems. Proceedings of the Royal Society of London **A123,** 714–33 (1929)
4. W. Heitler, F. London: Wechselwirkung neutraler Atome und homöopolare Bindung nach der Quantenmechanik. Zeitschrift für Physik **44,** 455–72 (1927)
5. M. Karplus: Three-Dimensional 'Pople Diagram'. Journal of Physical Chemistry **94**, 5435–6 (1990)
6. J. A. Pople: Two-Dimensional Chart of Quantum Chemistry. Journal of Chemical Physics **43,** S229–30 (1965)
7. D. A. Ramsay, J. Hinze, eds.: *Selected Papers of R.S. Mulliken* (University Press of Chicago, Chicago 1975)
8. J. Van Vleck, A. Sherman: The Quantum Theory of Valence. Reviews of Modern Physics **7,** 167–228 (1935)
9. L. Pauling: *The Nature of the Chemical Bond and the Structure of Molecules and Crystals* (Cornell University Press, New York 1939)

Secondary Literature

10. K. Gavroglu: Philosophical Issues in the History of Chemistry. Synthese **111,** 283–304 (1997)
11. K. Gavroglu, ed.: Studies in the History and Philosophy of Modern Physics **31** (2000). Special Issue: Theoretical Chemistry in the Making: Appropriating Concepts and Legitimizing Techniques.
12. K. Gavroglu, A. Simões: The Americans, the Germans and the Beginnings of Quantum Chemistry: The Confluence of Diverging Traditions. Historical Studies in the Physical and Biological Sciences **25,** 47–110 (1994)
13. A. Karachalios: On the making of quantum chemistry in Germany. Studies in the History and Philosophy of Modern Physics **31,** 493–510 (2000)
14. I. N. Levine: *Quantum Chemistry* (Allyn and Bacon, Inc., ^{3}1983)
15. B. S. Park: The 'Hyperbola of Quantum Chemistry': the Changing Practice and Identity of a Scientific Discipline in the Early Years of Electronic Digital Computers, 1945–1965. Annals of Science **60,** 219–47 (2003)
16. B. S. Park: In the 'context of pedagogy': Teaching strategy and theory change in quantum chemistry, in D. Kaiser, ed.: *Pedagogy and the Practice of Science. Historical and Contemporary Perspectives* (MIT, Cambridge 2005, 287–319)
17. R. Parr: *The Quantum Theory of Molecular Electronic Structure* (W.A. Benjamin, New York 1963)
18. A. Simões: Dirac's claim and the chemists. Physics in Perspective **4,** 253–66 (2002)
19. A. Simões: Chemical Physics and Quantum Chemistry in the Twentieth-Century, in *Modern Physical and Mathematical Sciences*, M.J. Nye, ed. (Cambridge University Press, Cambridge 2003, Vol.5, 394–412)
20. A. Simões, K. Gavroglu: Quantum Chemistry *qua* Applied Mathematics. The Contributions of Charles Alfred Coulson (1910–1974). Historical Studies in the Physical Sciences **29,** 363–406 (1999)
21. A. Simões, K. Gavroglu: Quantum Chemistry in Great Britain: Developing a Mathematical Framework for Quantum Chemistry. Studies in the History and Philosophy of Modern Physics **31,** 511–48 (2000)

Quantum Chromodynamics (QCD)

Kim Milton

Quantum chromodynamics (QCD) is the theory of the strong interaction between the quarks that constitute the strongly interacting particles, the hadrons, consisting of baryons and mesons. (▶ Particle Physics. Quarks, see ▶ Color Charge Degree of Freedom in Particle Physics; Mixing and Oscillations of Particles; Particle Physics; Parton Model; QFT). It is modeled on the extremely successful theory of ▶ electrons and photons, quantum electrodynamics (▶ QED). However, unlike the latter, which is tested now to 10th order in the strength of the electric charge of the electron, it is not easy to compare QCD with experiment. This is not only because the coupling is strong, not weak as in electrodynamics, but also because of the related fact that the fundamental components of the theory, the quarks and the force-carrying gluons, have never been directly seen, and are generally believed to be unattainable because of the phenomenon of confinement.

Yukawa [1] was the first to start to understand the strong force in terms of his posited "mesotron," what we now call the pion. He believed that the strong nuclear force between protons and neutrons in the nucleus could be understood in terms of the exchange of a mesotron between these particles, the short range of the nuclear force reflecting the fact that the mass of the mesotron was around $100\,\mathrm{MeV}\,c^{-2}$. However, by the end of the 1950s dozens of strongly interacting particles, most rapidly decaying, had been discovered in cosmic rays and accelerators, and these could not all be fundamental. Physicists searched for various schemes to unite the zoo of particles, and Gell-Mann [2] and Zweig [3] independently came up with the quark model, which was first not taken very seriously except as a way to describe the group theory that organized the hadrons. This group was called by Gell-Mann the eightfold way, but in fact it was simply SU(3), the group of three by three unitary matrices with determinant one. The fundamental representation of the group was realized by three quarks, what we now call up, down, and strange. (Now we know there are six "flavors" of quarks, up, down; charm, strange; and top, bottom; grouped in three families or generations of two each.) The quarks had fractional charge; up had charge $+2/3$, down had charge $-1/3$ in units of the electron charge, and each carried ▶ spin $\hbar/2$.

The quark model could be used to classify all the observed strongly interacting particles: baryons, like the proton and neutron, were composed of three quarks, and mesons, like the pion, were composed of a quark and an antiquark. No other combinations seemed then, or now, to appear in nature. (The recent flap over pentaquarks has ended with no believable evidence for exotic states.) However, it supplied no dynamics, and it left open the question of why quarks were not seen.

The next major step was supplied by Greenberg [4], who noted that to consistently describe baryons in the quark model required a new quantum number, called color, since three "charges" were required, called, say, red, green, and blue. The simplest example is the famous Δ baryon resonance, which comes in four charge states, −1, 0, 1, 2. The Δ^{++} state should be composed of three up quarks, each having charge +2/3. The spin of the Δ was 3/2, precisely what one would expect by adding the spins of the three up quarks in a symmetrically aligned state. Correspondingly, the spatial wave function should have zero orbital angular momentum, and should therefore be symmetric as well. But the ► wave function of a fermion must be totally antisymmetric under interchange of the constituent coordinates, so if it is symmetric in space, and symmetric in spin, it must be antisymmetric in something else, color. Such a state is one with no net color, the antisymmetrical combination of red, green, and blue. So the rule became, only those states are allowed which are color singlets; these are just the mesons and baryons described above.

The mathematics of this is that the color group is also SU(3) (no relation to the flavor SU(3) group introduced in the eightfold way); the quarks are triplets under color, and in terms of irreducible representations of SU(3), labeled by their dimensionality, baryons and mesons are described by

$$3 \otimes 3 \otimes 3 = 10 \oplus 8 \oplus 8 \oplus 1; \quad 3 \otimes \bar{3} = 8 \oplus 1,$$

the singlet in each case corresponding to the color part of a hadron wavefunction.

Still there was no dynamics. That came from earlier work in the 1950s, when Yang and Mills [5] discovered non Abelian gauge theories (quantum electrodynamics is an Abelian gauge theory). The great success of these theories came with the electroweak synthesis (► Particle Physics), carried out by Schwinger [6], Glashow [7], Weinberg [8], and Salam [9]. As soon as that approach was seen to be successful and consistent in 1971 [10], it was natural to apply it to the quark model. However, how could that theory be consistent with the confinement property that only color singlet states appear in nature? The answer came with the work of Politzer [11], Gross, and Wilczek [12], who showed that a non Abelian gauge theory like that based on SU(3) would have the property of *asymptotic freedom*: ► Color Charge Degree of Freedom in Particles Physics; QFT. That is, the force becomes strong at large distance (low energies) but weak at short distance (high energies). This is just what is needed to explain the quark model, where inside the nucleons (neutrons and protons) the quarks are nearly free, but they can never get more than about 10^{-15} m away from each other. This was also quite consistent with the deep-inelastic experiments which had appeared by 1970 which showed nearly free point-like constituents within the nucleons [13].

Gell-Mann is usually attributed as author of QCD [14]. The theory is governed by a Lagrangian density very similar to that of QED,

$$\mathcal{L}_{\mathrm{QCD}} = -\frac{1}{4} F^{a}_{\mu\nu} F^{a\mu\nu} - \sum_{f} \bar{q}_f \left[\gamma^{\mu} \frac{1}{\mathrm{i}} \left(\partial_{\mu} - \mathrm{i} g \frac{\lambda^{a}}{2} A^{a}_{\mu} \right) + m_f \right] q_f .$$

The difference between QCD and QED is that in the former there are eight colors of gluon fields, which are represented by the index a (repeated indices are to be summed over). The sum over f represents summing over the different flavors of

Q

quarks; each flavor of quark has three components in color space, and the eight matrices λ^a live in that space:

$$\lambda_1 = \begin{pmatrix} 0 & 1 & 0 \\ 1 & 0 & 0 \\ 0 & 0 & 0 \end{pmatrix}, \quad \lambda_2 = \begin{pmatrix} 0 & -\mathrm{i} & 0 \\ \mathrm{i} & 0 & 0 \\ 0 & 0 & 0 \end{pmatrix}, \quad \lambda_3 = \begin{pmatrix} 1 & 0 & 0 \\ 0 & -1 & 0 \\ 0 & 0 & 0 \end{pmatrix},$$

$$\lambda_4 = \begin{pmatrix} 0 & 0 & 1 \\ 0 & 0 & 0 \\ 1 & 0 & 0 \end{pmatrix}, \quad \lambda_5 = \begin{pmatrix} 0 & 0 & -\mathrm{i} \\ 0 & 0 & 0 \\ \mathrm{i} & 0 & 0 \end{pmatrix}, \quad \lambda_6 = \begin{pmatrix} 0 & 0 & 0 \\ 0 & 0 & 1 \\ 0 & 1 & 0 \end{pmatrix},$$

$$\lambda_7 = \begin{pmatrix} 0 & 0 & 0 \\ 0 & 0 & -\mathrm{i} \\ 0 & \mathrm{i} & 0 \end{pmatrix}, \quad \lambda_8 = \frac{1}{\sqrt{3}} \begin{pmatrix} 1 & 0 & 0 \\ 0 & 1 & 0 \\ 0 & 0 & -2 \end{pmatrix}.$$

The matrices satisfy the group property

$$\left[\frac{\lambda^a}{2}, \frac{\lambda^b}{2}\right] = \mathrm{i} f^{abc} \frac{\lambda^c}{2},$$

where the f^{abc} are the structure constants of the SU(3) group. The non Abelian field strength is constructed in terms of potentials as follows,

$$F^a_{\mu\nu} = \partial_\mu A^a_\nu - \partial_\nu A^a_\mu + g f^{abc} A^b_\mu A^c_\nu.$$

Note that the theory states that, unlike photons (▶ light quantum), gluons carry color, and hence couple to each other.

As in QED, Feynman rules can be readily written down to describe how to carry out perturbative calculations in powers of the coupling constant g. (A readable discussion is in [17]). These are, however, somewhat more complex than those in QED, and by themselves, of somewhat limited utility. Perturbation theory does not capture the confinement property, and in any case we do not want to calculate scattering amplitudes for free quarks, but for observable particles, the hadrons. To do this, semiempirical models are used to construct form factors and structure functions, so there are rather few direct tests of QCD itself. The structure functions encode our ignorance about the real wavefunction of hadrons. Moreover, because g is rather large, $g^2/4\pi\hbar c$ ranging from 0.1 to 2 depending on the process (remember that the strength of the coupling decreases as the energy of the process increases), higher corrections may in fact turn out to be larger that the leading terms, so perturbation theory is intrinsically unreliable. There are various methods to reduce this unreliability (for example, what is called analytic perturbation theory [15]), and lattice gauge theory [16] is a viable approach to transcend perturbation theory, but it may be fair to say that QCD, although nearly universally believed true, is not yet a quantitative model of strong interactions.

Primary Literature

1. H. Yukawa: On the Interaction of Elementary Particles. Proc. Phys. Math. Soc. Jpn. **17**, 48 (1935)
2. M. Gell-Mann: Symmetries of Baryons and Mesons. Phys. Rev. **125**, 1067 (1962)
3. G. Zweig: An SU(3) Model for Strong Interaction Symmetry and its Breaking, CERN-TH-401, and An SU(3) Model for Strong Interaction Symmetry and its Breaking. 2, CERN-TH-412, NP-14146. Published in 'Developments in the Quark Theory of Hadrons.' Volume 1. Edited by D. Lichtenberg and S. Rosen. (Nonantum, MA, Hadronic, 1980)
4. O. W. Greenberg: Spin and Unitary Spin Independence in a Paraquark Model of Baryons and Mesons. Phys. Rev. Lett. **13**, 598 (1964)
5. C. N. Yang and R. L. Mills: Conservation of Isotopic Spin and Isotopic Gauge Invariance. Phys. Rev. **96**, 191 (1954)
6. J. Schwinger: A Theory of Fundamental Interactions. Ann. Phys. (N.Y.) **2**, 407 (1957)
7. S. L. Glashow: Partial Symmetries of Weak Interactions. Nucl. Phys. **22**, 579 (1961)
8. S. Weinberg: A Model of Leptons. Phys. Rev. Lett. **19**, 1264 (1967)
9. A. Salam: Weak and Electromagnetic Interactions. Published in 'Elementary Particle Theory, Relativistic Groups and Analyticity.' Edited by N. Svartholm. (Stockholm, Almqvist and Wiksell, 1968, pp. 367–377)
10. G. 't Hooft: Renormalizable Lagrangians for Massive Yang-Mills Fields. Nucl. Phys. B **35**, 167 (1971)
11. H. D. Politzer: Reliable Perturbative Results for Strong Interactions? Phys. Rev. Lett. **30**, 1346 (1973)
12. D. J. Gross and F. Wilczek: Ultraviolet Behavior of Non-Abelian Gauge Theories. Phys. Rev. Lett. **30**, 1343 (1973)
13. J. I. Friedman and H. W. Kendall: Deep Inelastic Electron Scattering. Annu. Rev. Nucl. Part. Sci. **22**, 203 (1972)
14. H. Fritzsch, M. Gell-Mann, and H. Leutwyler: Advantages of the Color Octet Gluon Picture. Phys. Lett. B **47**, 365 (1973)
15. K. A. Milton and I. L. Solovtsov: Analytic Perturbation Theory in QCD and Schwinger's Connection Between the Beta Function and the Spectral Density. Phys. Rev. D **55**, 5295 (1997) [arXiv:hep-ph/9611438]
16. K. G. Wilson: Confinement of Quarks. Phys. Rev. D **10**, 2445 (1974)

Secondary Literature

17. I. J. R. Aitchinson and A. J. G. Hey: Gauge Theories in Particle Physics, 2nd Ed. (Adam Hilger, Bristol, 1989)

Quantum Communication

Michel LeBellac

In our everyday world, almost all the information exchanged, stored and processed is encoded in the form of discrete entities called bits, which take by convention the values zero or one. In the computers and optical fibers of today's information and communication technology, the bits are carried by electric currents or light beams, corresponding to macroscopic fluxes of ▶ electrons or photons (▶ light quantum) respectively, and they are stored in memories of various kinds, for example, magnetic. Although the basic physics which underlies the operation of a transistor or a laser is quantum physics, each elementary bit corresponds to a large number of elementary quantum systems, and its behavior can be described classically due to the strong coupling to the environment.

In the past twenty years, physicists have been able to manipulate with an increasing accuracy individual quantum objects, such as photons, atoms, neutrons... This opens the way for using quantum two-state systems to exchange, store and process information, by selecting two orthogonal states spanning the ▶ Hilbert space of states: using ▶ Dirac's notation for the state vectors, one of the states, $|0\rangle$, encodes the value zero of the bit, the other one, $|1\rangle$, the value one. In this article, we shall discuss three aspects of quantum communication: quantum cryptography, quantum dense coding and quantum teleportation.

An elementary example of quantum two-state system is given by photon polarization, where one may choose a basis of linearly polarized states and associate, by convention, the vertical polarization ($\updownarrow$) with the value zero of the bit and the horizontal polarization ($\leftrightarrow$) with the value one. Storing information with individual photons is still far beyond present technical capabilities, but transmission of information is easy to implement. The two people exchanging information being conventionally called Alice and Bob, Alice may send Bob individual photons which are either vertically polarized, or horizontally polarized. Any message written in binary language is a series of 0s and 1s, and the message 0110101 will be encoded in the sequence of photon polarizations $\updownarrow \leftrightarrow \leftrightarrow \updownarrow \leftrightarrow \updownarrow \leftrightarrow$, which will be sent via, for example, an optical fiber. To read the message (see Fig. 1), Bob uses a polarizing beamsplitter to separate the photons of vertical and horizontal polarization, and two detectors tell him whether the photon was horizontally or vertically polarized: each photon carries one bit of information. Although this technique has a rather poor efficiency compared to standard bit transmission via photon pulses in optical fibers, a

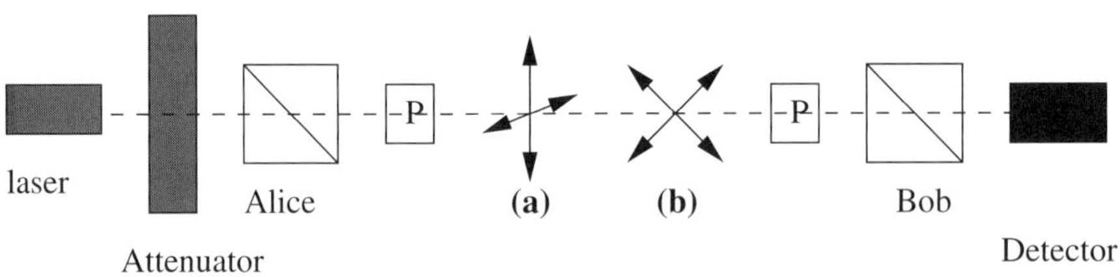

Fig. 1 Schematical depiction of the BB84 protocol. A laser beam is attenuated such that it simulates individual photons. A laser is used for practical reasons: it would be safer to use a single photon source, but these sources are not yet available commercially. A birefringent plate selects the polarization, which can be rotated by means of Pockels cells P. The photons are either vertically/horizontally polarized **(a)** or polarized at $\pm 45^\circ$ **(b)**

few hundreds of kbit s^{-1} as compared to tens of Gbit s^{-1}, we shall see later on that it can be modified in order to ensure the security of the transmission.

A quantum two-state system which can be used to store, process or transmit bits of information is called a quantum bit, or qubit: see [4]–[6] for an overview. We may hope for some gain by going from bits to qubits, because, in contrast to classical bits, we can build linear ► superposition $|\varphi\rangle$ of states $|0\rangle$ and $|1\rangle$, for example vertically and horizontally polarized states, $|\updownarrow\rangle \equiv |0\rangle$ and $|\leftrightarrow\rangle \equiv |1\rangle$

$$|\varphi\rangle = \cos\frac{\theta}{2}\,|0\rangle + \mathrm{e}^{\mathrm{i}\phi}\,\sin\frac{\theta}{2}\,|1\rangle \tag{1}$$

Since the angles θ and ϕ in (1) can vary continuously, it may seem that a qubit contains much more information than a classical one (in fact an infinite amount of information!). However, we must use an orthogonal basis for measurement, and the result of the measurement will always be zero or one, whatever the basis, so that our hopes of getting more from a qubit than from a classical bit look unfounded. This pessimistic observation is confirmed by Holevo's theorem [1]: N qubits may transmit at most N bits of information. Fortunately, ► entanglement will allow us to bypass this theorem.

The simplest application of quantum communication is quantum cryptography, as it uses only single qubits, at least in its most elementary version. Moreover, it is the only application which is now coming on the market. 'Quantum cryptography' is a catchy phrase, but it is somewhat inaccurate. A better terminology is quantum key distribution (QKD). In fact, there is no encryption of a message using quantum physics; the latter is used only to ensure that the key needed in secret key systems of encryption is not intercepted by a spy, so that quantum cryptography solves the problem of secure key distribution. Of course, this problem does not exist in public key systems, such as RSA (Rivest, Shamir and Adleman) encryption, whose security relies on the difficulty of finding the prime factors of a large integer. As we have seen, a message, encrypted or not, can be transmitted using the two orthogonal linear polarization states of a photon, but, in addition, we shall make use of the basic laws of quantum physics in order to be sure that the message has not been intercepted. Two complementary (incompatible) bases are chosen at random by Alice, for example $\{|\updownarrow\rangle, |\leftrightarrow\rangle\}$ and $\{|\nearrow\rangle, |\nwarrow\rangle\}$, where

$$|\nearrow\rangle \equiv |+\rangle = \frac{1}{\sqrt{2}}(|\updownarrow\rangle + |\leftrightarrow\rangle) \qquad |\nwarrow\rangle \equiv |-\rangle = \frac{1}{\sqrt{2}}(|\updownarrow\rangle - |\leftrightarrow\rangle) \tag{2}$$

to send Bob photons of four types, either polarized vertically ($\updownarrow$) or horizontally ($\leftrightarrow$) in the first basis, or polarized along axes rotated by $\pm 45^{\circ}$ in the second basis: ($\nearrow$) or ($\nwarrow$), corresponding to the values zero and one of the bit respectively. Similarly, Bob analyzes the photons sent by Alice using the same orthogonal bases chosen at random. After recording a sufficient number of photons, Bob publicly announces the sequence of bases he has used, but not his results. Alice compares her sequence of bases to Bob's and publicly gives him the list of bases identical with his. About half of the bits, those corresponding to a different choice of bases, are rejected, and then Alice and Bob are certain that the values of the other bits are the same. These are the bits which will be used to construct the key, and they are known only to Bob and Alice, because an eavesdropper only knows the list of bases and not the results. The protocol we have described is called BB84, from the names of its inventors Bennett and Brassard [2].

We still need to be sure that the message has not been intercepted and that the key it contains can be used without risk. Alice and Bob choose at random a subset of their key and compare publicly not only their choice of bases, but also the bit values. The consequence of interception of the photons by a spy would be a reduction of the correlation between the values of their bits. The security of the protocol depends on the fact that a spy cannot find out the polarization state of a photon unless he knows beforehand the basis in which it was prepared. Of course, the raw process which we have just described does not take into account the possibility of errors, which must be corrected thanks to a classical error correcting code, while a second classical process, called privacy amplification, ensures the secrecy of the key, even if an eavesdropper has been able to correctly guess some of the bits. As optical fibers do not allow one to control the direction of polarization over large distances, in practice qubits are encoded in the phase of the photon ► wave function, and Mach-Zehnder interferometers are used to fix the phase at one end of the line and to measure it at the other end. Rates of transmission of 50 kbit s^{-1} have been reached over distances up to 100 km. Other quantum cryptography protocols have been proposed, which use either three incompatible bases, or entangled states.

Let us now turn to multi-qubit systems, which will be used for dense coding and teleportation. Unlike the classical case, most of the information contained in a generic quantum mechanical system is stored in the form of ► correlations between its subsystems. Dense coding and teleportation make essential use of these correlations. Let us recall that a two-qubit state which cannot be written as a tensor product is called an entangled state. A convenient orthogonal basis in the Hilbert space $\mathcal{H}_A \otimes \mathcal{H}_B$ of two qubits A and B is the so-called Bell's basis, made of the four Bell states

$$|\Psi_0\rangle = \frac{1}{\sqrt{2}}(|0_A \otimes 0_B\rangle + |1_A \otimes 1_B\rangle) = (\sigma_{0A} \otimes I_B)|\Psi_0\rangle \quad (3)$$

$$|\Psi_1\rangle = \frac{1}{\sqrt{2}}(|1_A \otimes 0_B\rangle + |0_A \otimes 1_B\rangle) = (\sigma_{1A} \otimes I_B)|\Psi_0\rangle \quad (4)$$

$$|\Psi_2\rangle = \frac{1}{\sqrt{2}}(-|1_A \otimes 0_B\rangle + |0_A \otimes 1_B\rangle) = (\mathrm{i}\sigma_{2A} \otimes I_B)|\Psi_0\rangle \quad (5)$$

$$|\Psi_3\rangle = \frac{1}{\sqrt{2}}(|0_A \otimes 0_B\rangle - |1_A \otimes 1_B\rangle) = (\sigma_{3A} \otimes I_B)|\Psi_0\rangle \quad (6)$$

where $\sigma_0 = I$ and the σ_is are the ► Pauli spin matrices.

Dense coding and teleportation rely on the use of measurements in the Bell basis. How to perform such a measurement is not a priori obvious because one is limited to measuring individual qubits. The first step consists of disentangling the Bell states thanks to a quantum logic gate called a control-NOT, or cNOT gate. Quantum logic gates are unitary operations acting in the Hilbert space of one or several qubits. The cNOT gate is a two-qubit quantum gate, acting in $\mathcal{H}_A \otimes \mathcal{H}_B$ which has the following action on a two-qubit state

$$\mathrm{cNOT}|x_A \otimes y_B\rangle = |x_A \otimes (x_A \oplus y_B)\rangle \tag{7}$$

where $x_A, y_B = 0, 1$ and $\oplus$ is addition modulo 2; x is the control bit and y the target bit. It is important to observe that the cNOT gate is not a tensor product $M_A \otimes M_B$ of two operators M_A and M_B: this is precisely the reason why this gate may transform a tensor product $|\varphi_A \otimes \varphi_B\rangle$ into an entangled state, or vice-versa.

We also need the Hadamard gate H, which is a unitary transformation on individual qubits; when H is applied to an eigenstate $|0\rangle$ or $|1\rangle$ of σ_3, the result is an eigenstate $|\pm\rangle$ (2) of σ_1, $\sigma_1|\pm\rangle = \pm|\pm\rangle$

$$\mathrm{H}|0\rangle = \frac{1}{\sqrt{2}}\big(|0\rangle + |1\rangle\big) = |+\rangle, \qquad H|1\rangle = \frac{1}{\sqrt{2}}\big(|0\rangle - |1\rangle\big) = |-\rangle \tag{8}$$

and conversely, since $\mathrm{H}^2 = I$, $\mathrm{H}|+\rangle = |0\rangle$, $\mathrm{H}|-\rangle = |1\rangle$.

To measure in the Bell basis, we first apply a cNOT gate, followed by a Hadamard gate on qubit A. A measurement of the two qubits sketched in Fig. 2 (a) will give a result in the form (x_A, y_B) and the four possible results will be in one-to-one correspondence with the Bell states

$$|\Psi_0\rangle \Longleftrightarrow (0_A 0_B) \quad |\Psi_1\rangle \Longleftrightarrow (0_A 1_B) \quad |\Psi_2\rangle \Longleftrightarrow (1_A 1_B) \quad |\Psi_3\rangle \Longleftrightarrow (1_A 0_B) \tag{9}$$

Dense coding works as follows (Fig. 2 (b)): Alice and Bob share an entangled pair AB of qubits, for example in the state $|\Psi_0\rangle$. Alice wants to send Bob two bits of

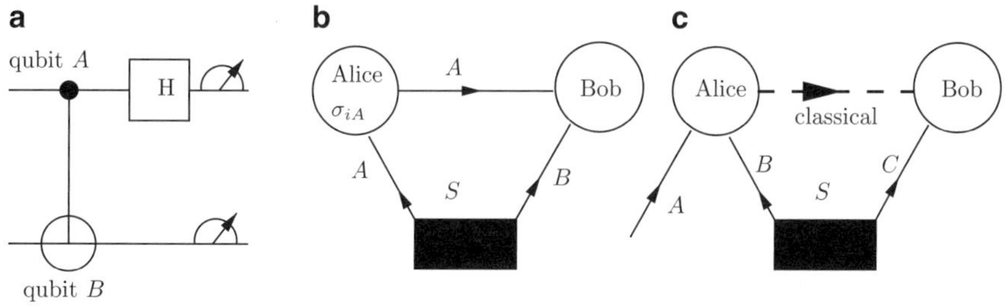

Fig. 2 **(a)** Measurement in the Bell basis: a cNOT gate, where A is the control bit and B the target bit, is followed by a Hadamard gate applied on qubit A. The diagrams are read from *left to right*, in the direction opposite to that of the operator products. **(b)** Dense coding: Alice applies σ_{iA} on qubit A and Bob performs a Bell measurement on the AB pair. S is a source of entangled particles. **(c)** Quantum teleportation: Alice makes a Bell measurement on the AB pair. The classical communication channel is represented by a *dashed line*

information by exchanging only one qubit, the two bits being encoded in a number i, $i = 0, 1, 2, 3$. She applies to qubit A the operator σ_{iA}, $i = 0, 1, 2, 3$, see (3)-(6). Then Bob receives one of the four states (3)–(6) and he measures the AB pair in the Bell basis, as explained above. From the measurement result, he will be able to find the value of i. Dense coding seems to bypass Holevo's theorem, but there is no contradiction, because the assumptions needed in the proof of the theorem explicitly exclude that Alice and Bob share an entangled pair.

Quantum teleportation [3] allows one to transport quantum information from one location to another, without any physical transfer of the associated quantum-information carrier. To give an example of another physical realization of qubits, let us assume that the qubits are now carried by the ▶ spin states of spin 1/2 particles, and that Alice wishes to transfer to Bob the information about the spin state $|\varphi_A\rangle$ of a particle A (Fig. 2(c))

$$|\varphi_A\rangle = \lambda|0_A\rangle + \mu|1_A\rangle \qquad |\lambda|^2 + |\mu|^2 = 1 \tag{10}$$

which is unknown to both partners, without sending him this particle directly. The principle of information transfer consists of using an auxiliary pair of entangled particles B and C of spin 1/2 shared between Alice and Bob. Particle B is used by Alice and particle C is sent to Bob (Fig. 2 (c)). Particles B and C may be, for example, in the entangled spin state $|\Psi_0^{BC}\rangle$. The initial three-particle state $|\Phi_{ABC}\rangle$ can be written in terms of the Bell states of the AB pair

$$\begin{aligned}|\Phi_{ABC}\rangle =& \frac{1}{2}|\Psi_0^{AB}\rangle \otimes (\lambda|0_C\rangle + \mu|1_C\rangle) + \frac{1}{2}|\Psi_1^{AB}\rangle \otimes (\lambda|1_C\rangle + \mu|0_C\rangle) \\ &+ \frac{1}{2}|\Psi_2^{AB}\rangle \otimes (\lambda|1_C\rangle - \mu|0_C\rangle) + \frac{1}{2}|\Psi_3^{AB}\rangle \otimes (\lambda|0_C\rangle - \mu|1_C\rangle)\end{aligned} \tag{11}$$

Alice measures the previously unentangled AB pair in the Bell basis: the measurement projects particle C in a state which is directly linked to its result. If, for example, Alice finds the state $|\Psi_1^{AB}\rangle$, then she knows that Bob is going to receive particle C in the state

$$|\varphi_C\rangle = \lambda|1_C\rangle + \mu|0_C\rangle$$

and she will be able to inform Bob by a classical channel (for example, a telephone) of the quantum state of qubit C. If necessary, Bob can apply a suitable rotation in order to recover the original state (10). Notice that Bob 'knows' the spin state of particle C only once he has received the result of Alice's measurement. This information must be sent by a classical channel, at a speed at most equal to that of light. There is therefore no instantaneous transmission of information at a distance. As possible applications, quantum teleportation could be used to build quantum relays for long distance quantum cryptography, or to provide a way for distant qubits in a quantum computer to interact without the requirement of physical proximity.

Primary Literature

1. A. H. Holevo: Some estimates of the information content transmitted by a quantum communication channel, Probl. Pered. Inform. **9**, 3 (1973)
2. C. Bennett and G. Brassard: Quantum cryptography: Public key distribution and coin tossing, Proceedings of IEEE conference on computers, systems and signal processing, Bangalore, India, p. 175 (1984)
3. C. Bennett, G. Brassard, C. Crépeau, R.Jozsa, A. Peres and W. Wootters: Teleporting an unknown quantum state via dual classical and EPR channels, *Phys. Rev. Lett.* **70**, 1895 (1993)

Secondary Literature

4. M. A. Nielsen and I. L. Chuang: *Quantum computing and quantum information* (Cambridge University Press, Cambridge 2000)
5. M. Le Bellac: *A short introduction to quantum computation and quantum information* (Cambridge University Press, Cambridge 2006)
6. G. Jaegger: *Quantum information: an overview* (Springer 2007)

Quantum Computation

Michel LeBellac

By using specific properties of quantum mechanics: ► superposition principle and ► entanglement, quantum computers can outperform classical ones when carrying out certain type of computation, see [5]–[7] for an overview. The basic information unit processed by a quantum computer is the qubit, a two-state quantum system living in a ► Hilbert space where one can choose an ► orthonormal basis of two vectors $|0\rangle$ and $1\rangle$. If we wish to store in a qubit register an integer $\mathbf{x}$, $0 \leqslant \mathbf{x} \leqslant 2^n - 1$

$$\mathbf{x} = 2^{n-1}x_{n-1} + 2^{n-2}x_{n-2} + \cdots + 2x_1 + x_0 \tag{1}$$

with $x_i = 0$ or $x_i = 1$, we need n qubits from which we construct the tensor product state

$$|\mathbf{x}\rangle = |x_{n-1} \otimes x_{n-2} \otimes \cdots \otimes x_1 \otimes x_0\rangle \tag{2}$$

State vectors of the form (2) form a basis of the 2^n-dimensional space $\mathcal{H}^{\otimes n}$ called the computational basis, and it might be concluded, because of the superposition principle, that an n-qubit register is able to encode 2^n states at the same time. However, a measurement of the n-qubits will give only one result corresponding to one of the states (2), and the challenge of quantum computation is to use interference and entanglement in order to exploit this exponentially growing information.

Q

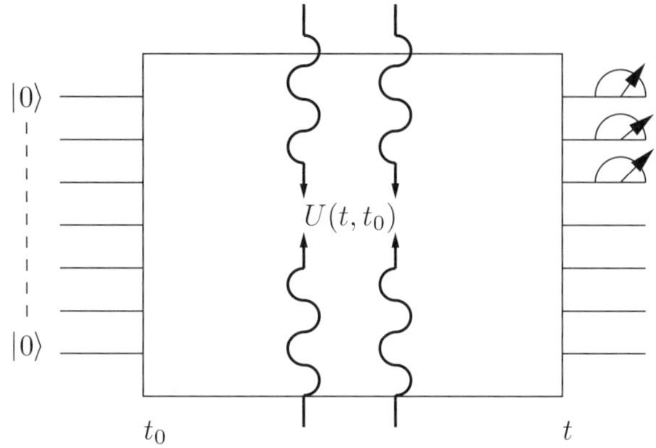

Fig. 1 Schematic depiction of the basic principle of a quantum calculation. n qubits are prepared in the state $|0\rangle$. They undergo a unitary and deterministic evolution in the space $\mathcal{H}^{\otimes n}$ from time $t = t_0$ to time t described by a unitary operator $U(t, t_0)$ acting in $\mathcal{H}^{\otimes n}$. The wiggly arrows represent interactions with external classical fields. A measurement of the qubits (or a subset thereof, the first three in this figure) is made at time t

A calculation performed on a quantum computer is shown schematically in Fig. 1, where n qubits are all prepared in the state $|0\rangle$ at time $t = t_0$: this is the preparation stage of the quantum system. The qubits then undergo a unitary quantum evolution described by a unitary operator $U(t, t_0)$ acting in $\mathcal{H}^{\otimes n}$ which performs the desired operations, for example, the calculation of a function. The experimental difficulty is to avoid unwanted interactions with the environment, otherwise ▶ decoherence would make the evolution nonunitary: if the qubits interact with the environment, the unitary evolution occurs in a Hilbert space which is larger than $\mathcal{H}^{\otimes n}$, because it includes the degrees of freedom of the environment along with those of the qubits. Interactions with external classical fields are compatible with unitary evolution and they are indeed needed to manipulate qubits by Rabi oscillations, which is the most common way of acting on computational qubits. Once the quantum evolution has been completed, a measurement is made on the qubits (or on a subset thereof) at time t in order to obtain the result of the calculation. An important point is that intermediary states of the calculation cannot be observed between t_0 and t, because any measurement would modify the unitary evolution: the qubits can be measured at the entrance and at the exit of the box of Fig. 1, but not inside it. Another essential point is that the unitary evolution is reversible: if we know the state vector at time t, we can recover the state vector at time t_0 using $U^{-1}(t, t_0) = U(t_0, t)$. As a consequence, classical algorithms, which contain irreversible logic gates, cannot be directly transposed to quantum ones. One needs first to transform these algorithms into reversible (classical) ones, which can be done with little reduction in efficiency.

The most general quantum evolution is a unitary transformation in $\mathcal{H}^{\otimes n}$, and the most general quantum logic gate is a $2^n \times 2^n$ unitary matrix operating in $\mathcal{H}^{\otimes n}$. A theorem of linear algebra states that any unitary transformation in $\mathcal{H}^{\otimes n}$ can be decomposed into a product of cNOT gates and unitary transformations on one qubit. In practice, in addition to the Hadamard and cNOT gates, one-qubit gates called the phase gate U_{ph} and the $\pi/8$ gate are also frequently encountered

$$U_{\text{ph}} = \begin{pmatrix} 1 & 0 \\ 0 & \mathrm{i} \end{pmatrix} \qquad U_{\pi/8} = \begin{pmatrix} 1 & 0 \\ 0 & \mathrm{e}^{\mathrm{i}\pi/4} \end{pmatrix} \tag{3}$$

Any unitary operation in $\mathcal{H}^{\otimes n}$ can be approximated with arbitrary accuracy by a combination of cNOT gates and a small number of one-qubit gates, for example the Hadamard gate H and the two one-qubit gates (3). Schematically, a quantum algorithm works as follows: an input register of n-qubits stores an integer $\mathbf{x}$, $0 \leqslant \mathbf{x} \leqslant 2^n - 1$, and an output register stores m-qubits, $0 \leqslant \mathbf{y} \leqslant 2^m - 1$. An elementary example of a quantum circuit is drawn in Fig. 2: this circuit has the following action on the initial state vector $|x_1 \otimes x_0 \otimes y_1 \otimes y_0\rangle$, where x_1 and x_0 are stored in the input register, y_1 and y_0 in the output register

$$|x_1 \otimes x_0 \otimes y_1 \otimes y_0\rangle \to |x_1 \otimes x_0 \otimes (y_1 \oplus x_0 \oplus 1) \otimes (y_0 \oplus x_1 \oplus x_0)\rangle \tag{4}$$

where $\oplus$ is addition modulo 2. If the function $f(\mathbf{x})$ is given by

$$f(0) = 2 \qquad f(1) = 3 \qquad f(2) = 1 \qquad f(3) = 0 \tag{5}$$

then the action of the circuit can be summarized by

$$|\mathbf{x} \otimes \mathbf{y}\rangle \to U_f|\mathbf{x} \otimes \mathbf{y}\rangle = |x \otimes [\mathbf{y} \oplus f(\mathbf{x})]\rangle \tag{6}$$

where $\oplus$ is now addition modulo 2 without carry over. The transformation U_f is clearly a unitary operation, since $U_f^2 = I$. For a generic function $f(\mathbf{x})$, U_f will be built in analogy to (6).

Quantum parallelism relies on using linear combinations of vectors of the computational basis, obtained by application of the Hadamard gate H. Indeed, if we

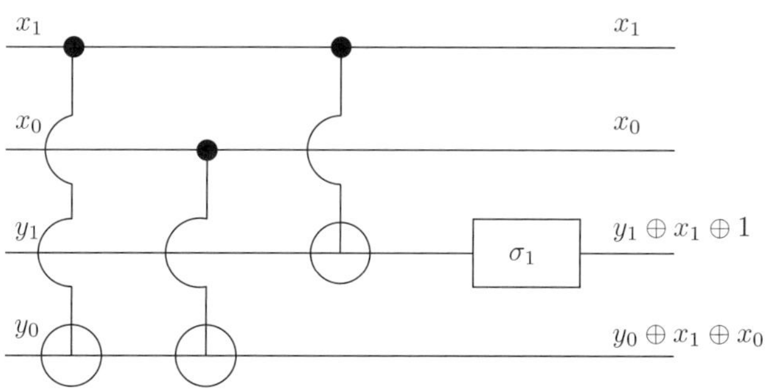

Fig. 2 An elementary quantum circuit with three cNOT gates (x_1 and x_0 = control bits, y_1 and y_0 = target bits) and a one-qubit gate σ_1 which computes the function $f(\mathbf{x})$ (5)

apply H to the input register in the state $|0^{\otimes n}\rangle$ before U_f, the state vector of the final state will be, by linearity,

$$|\Psi_{\text{fin}}\rangle = U_f|(\text{H}^{\otimes n}\, 0^{\otimes n}) \otimes 0^{\otimes m}\rangle = \frac{1}{2^{n/2}} \sum_{\mathbf{x}=0}^{2^n-1} |\mathbf{x} \otimes f(\mathbf{x})\rangle \qquad (7)$$

In principle, this state vector contains the 2^n values of the function $f(\mathbf{x})$ (not necessarily all of them different). For example, if $n = 100$, it contains the $\sim 10^{30}$ values of $f(\mathbf{x})$: it is this exponential growth of states which allows quantum parallelism to deal efficiently with some exponentially complex problems. A measurement can of course give only one of these values, but it is nevertheless possible to extract useful information about the relations between the values of $f(\mathbf{x})$ for an ensemble (▶ ensembles in quantum mechanics) of different values of $\mathbf{x}$, of course at the price of losing the individual values. A classical computer, on the other hand, would have to evaluate $f(\mathbf{x})$ for all these values of $\mathbf{x}$ independently. The art of quantum computing is to construct an interference pattern in which the desired result stands out with a reasonable probability against a small background.

Two broad classes of quantum algorithm have been identified so far. The first class, to which belongs Grover's algorithm, allows quadratic speed up with respect to classical algorithms. Grover's algorithm [1], for example, is able to find an entry in an unstructured data base of N elements in $\sim\sqrt{N}$ steps, while a classical algorithm needs an average of $N/2$ steps. This class of algorithm exploits the superposition principle, but not entanglement. Shor's algorithm [2] belongs to the second class, and makes essential use of entanglement. Its purpose is to find the prime factors of an integer N. If implemented some day (in a distant future) on an actual quantum computer, this algorithm would be able to break the widely used RSA encryption, whose security relies on the difficulty of factoring large numbers. As of today, the best algorithm running on a classical computer needs $\sim\exp[1.9 \ln^{1/3} N \ln\ln^{2/3} N]$ computational steps to find the prime factors. Since the number of steps grows faster than any polynomial in $\ln N$, the number of bits which specify the size of the problem, it has been conjectured that factorization is an exponentially complex problem, also called an "intractable" problem. On the contrary, the problem becomes of polynomial complexity with Shor's algorithm, where the number of computational steps is $\sim(\ln N)^3$. Finally one should also mention that quantum computers could be used to simulate efficiently quantum systems, but it is somewhat frustrating that no new really interesting quantum algorithm has been discovered in the past ten years, which could be added to Grover's and Shor's algorithms.

As in classical computers, errors may arise in processing or storing information, and it is necessary to develop error correcting codes. Classical error correcting codes are based on redundancy: for example, one makes three copies of each bit and retrieves the correct value by a majority rule. Classical error correcting codes cannot be directly transposed to qubits, first because the ▶ no-cloning theorem forbids reproducing an unknown qubit state, and second because errors may affect continuous variables. For example, in the general qubit state

$$|\varphi\rangle = \cos\frac{\theta}{2}\,|0\rangle + \mathrm{e}^{\mathrm{i}\phi}\,\sin\frac{\theta}{2}\,|1\rangle \tag{8}$$

noise could lead to continuous variations of the angles θ and ϕ. Fortunately, these errors can be kept under control by taking care of a finite, and in fact small set of errors. Error correcting codes have been developed, which are based on seven qubits (Steane) or nine qubits (Shor) and are able to deal with all kinds of error.

Quantum algorithms have had an important impact on the theory of algorithmic complexity. We assume, as is usually done, the validity of the strong version of the Church-Turing thesis: any computational model can be simulated efficiently, that is with at most a polynomial increase in the number of computational steps, by a universal probabilistic Turing machine. Then, it is possible to define two main classes of algorithmic complexity. The first class is the polynomial class **P**, that of problems which can be solved with a number of computational steps that is polynomial in the number of bits characterizing the size of the problem: these problems are called "tractable". The second class is the **NP** class, that of problems in which a trial solution can be checked in a polynomial number of steps. Clearly, **P**⊂**NP**, and a celebrated conjecture, which to this day remains unproven, states that $\mathbf{P} \neq \mathbf{NP}$, which means that there exist problems that are termed intractable. Numerous complexity classes have been identified, such as that of **NP** complete problems: finding a polynomial algorithm to solve one **NP** complete problem, for example the "traveling salesman problem", would automatically imply a polynomial solution for any **NP** problem. Quantum computers are important because they make the strong version of the Church-Turing thesis questionable. In fact, if factorization is an intractable problem (as suggested by experience but is still unproven), then Shor's algorithm contradicts this strong version. Using a quantum computer it is possible to find the prime factors of an integer N by a number of steps which is a polynomial in $\ln N$, whereas a classical computer can only do this in an exponential number of steps. However, it must be acknowledged that factorization is not **NP** complete.

A final important issue is that of physical realizations of quantum computers. The storage and processing of quantum information requires physical implementations of qubits possessing the following properties (di Vincenzo criteria [3]):

(i) they must be scalable, that is, capable of being extended to a sufficient number of qubits, with well defined qubits
(ii) they must have qubits which can be initialized in the state $|0\rangle$
(iii) they must have qubits which are carried by physical states of sufficiently long lifetime, so as to ensure that the quantum states remain coherent throughout the calculation
(iv) they must possess a set of universal quantum gates: unitary transformations on individual qubits and a cNOT gate, which are obtained by controlled manipulations

(v) there must be an efficient procedure for measuring the state of the qubits at the end of the calculation (readout of the results).

The main enemy of quantum computers is interaction with the environment leading to decoherence, a consequence of which is the loss of the phase in the ► superposition of qubits. The calculations must be performed in a time less than the decoherence time τ_{dec}. If a quantum gate takes a time τ_{op}, the figure of merit for a quantum computer is the ratio

$$n_{\text{op}} = \frac{\tau_{\text{dec}}}{\tau_{\text{op}}}.$$

This is the maximum number of operations that the quantum computer can perform.

There are at present two main avenues of research: realizations using as qubits degrees of freedom carried by individual atoms or ions, which are "clean" systems, at least in principle, but not easily scalable, and realizations based on solid state technology, using as qubits collective degrees of freedom such as superconducting circuits or quantum dots, which are "dirty" systems, but more easily scalable because one can adapt conventional microchip technology. The present state of the art does not allow experimenters to manipulate more than three of four qubits in a fully controlled way (seven with a NMR based quantum computer [4], which however is not scalable), and, barring an unexpected technological breakthrough, it will take many years before a reasonably powerful quantum computer sees the light of the day.

Primary Literature

1. L. K. Grover: A fast quantum mechanical algorithm for data base search, Proceedings of 28th Annual ACM Symp. Theory Comp., p. 212 (1996)
2. P. W. Shor: Algorithms for quantum computation: Discrete logarithm and factoring, Proceedings of 35th annual symposium on foundations of computer science, IEE, p. 124 (1994)
3. D. di Vincenzo: The physical implementation of quantum computation, *Fortshcr. Phys.* **48**, 771 (2000)
4. L. Vandersypen, M. Steffen, G. Breyta, C. Yannoni, M. Sherwood, and I. Chuang: Experimental realization of quantum Shor's factoring algorithm using nuclear magnetic resonance, *Nature* **414**, 883 (2001)

Secondary Literature

5. M. A. Nielsen and I. L. Chuang: *Quantum computing and quantum information* (Cambridge University Press, Cambridge 2000)
6. M. Le Bellac: *A short introduction to quantum computation and quantum information* (Cambridge University Press, Cambridge 2006)
7. G. Jaegger: *Quantum information: an overview* (Springer 2007)

Quantum Electrodynamics (QED)

Kim Milton

The theory of quantum electrodynamics was born immediately following the formulation of quantum mechanics. In 1927 Dirac put Maxwell's classical theory of electromagnetism together with Planck's and Einstein's ideas of quanta [1]. The following year he came up with his famous equation describing a relativistic electron [2], and with that all the ingredients for a quantum field theory of an electron interacting with photons (► light quantum) were present. In two decisive papers in 1929 Heisenberg and Pauli [3, 4] developed a consistent theory of quantum electrodynamics. (For a detailed history of the development of quantum electrodynamics, see [17]. For scientific biographies of Julian Seymour Schwinger (1918–94) and Richard Feynman (1918–88), who solved the problems of QED, see [18, 19].)

Thus the equations governing quantum electrodynamics were formulated throughout the 1930s, which followed from the following Lagrangian density:

$$\mathcal{L} = -\frac{1}{4}F^{\mu\nu}F_{\mu\nu} - \bar{\psi}\left[m + \gamma^\mu\left(\frac{1}{\mathrm{i}}\partial_\mu - eA_\mu\right)\right]\psi,$$

where A_μ is the four-vector potential describing the photon, in terms of which the electromagnetic field strength is constructed, $F_{\mu\nu} = \partial_\mu A_\nu - \partial_\nu A_\mu$, and ψ is the electron field, $\bar{\psi} = \psi^\dagger\gamma^0$. Here appear the 4×4 Dirac matrices, which satisfy the anticommutation relation

$$\{\gamma^\mu, \gamma^\nu\} = -2g^{\mu\nu},$$

in the metric $g^{\mu\nu} = \mathrm{diag}\,(-1, 1, 1, 1)$. In the canonical ► quantization scheme, we regard the fields as operator-valued, satisfying the canonical equal-time commutation relations in the radiation gauge where $\nabla \cdot \boldsymbol{A} = 0$:

$$[\boldsymbol{A}_{\perp i}(\boldsymbol{x}, t), \boldsymbol{E}_{\perp j}(\boldsymbol{y}, t)] = -\mathrm{i}\left(\delta_{ij} - \frac{\nabla_i\nabla_j}{\nabla^2}\right)\delta(\boldsymbol{x} - \boldsymbol{y}),$$

$$\{\psi_\alpha^\dagger(\boldsymbol{x}, t), \psi_\beta(\boldsymbol{y}, t)\} = \delta_{\alpha\beta}\delta(\boldsymbol{x} - \boldsymbol{y}).$$

Here, $\boldsymbol{E}_i = F^{0i}$, and $\perp$ denotes the transverse part, $\nabla \cdot \boldsymbol{E}_\perp = 0$, while α, β are Dirac indices. The second relation is an anticommutation relation for the electron field, reflecting the fact that it is a Fermion.

However, when people tried to calculate using this theory, assuming an expansion in the small parameter called the fine structure constant, $\alpha = e^2/4\pi\hbar c = 1/137$, all but the most trivial processes turned out to be divergent. There were some notable successes during this period, perhaps most important being the Euler-Heisenberg

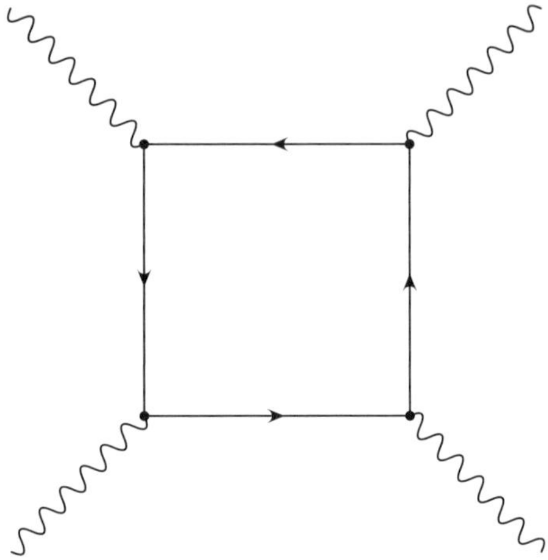

Fig. 1 The light-by-light scattering graph, where the *solid line* represents an electron. The *wavy lines* represent photons

Lagrangian that describes exactly the quantum motion of an electron in a constant background electromagnetic field [5]. Among other processes, this represents the scattering of light by light, a phenomenon not directly yet observed, although present as an internal process in the well-tested theory of the anomalous magnetic moment of the electron. This scattering process can be represented pictorially by what we would now call a ▸ Feynman diagram, see Fig. 1. Here the loop represents an electron, as a virtual particle, one that does not satisfy the ordinary balance between energy and momentum,

$$E^2 \neq m^2c^4 + p^2c^2.$$

Thus, it can only propagate for a short distance and for a short period of time.

Oppenheimer and many others struggled with the theory of quantum electrodynamics, but little progress was made until after the second world war, when using techniques developed during the war experimentalists established that two predictions of the Dirac theory of the electron were invalid. One was that the $2s_{1/2}$ and $2p_{1/2}$ states of the hydrogen atom should be degenerate, that is, have equal energy; the nondegeneracy is called the Lamb shift, after it was conclusively established by Willis Lamb [6]. The second turned out to be an deviation from the Dirac g-factor of the electron, its anomalous magnetic moment, unexpectedly discovered by Nafe, Nelson, and Rabi [7], and by Kusch and Foley [8]. This set the stage for solving the theory, and in Schwinger's words, showed that "electrodynamic effects were neither infinite nor zero, but finite and small, and demanded understanding."

So after these results were announced at the Shelter Island conference in June 1947, theoretical developments rapidly followed. Based on discussions at the meeting, Bethe published a nonrelativistic calculation of the Lamb Shift [9]. By December, Schwinger had a relativistic calculation of this effect (with some incorrect details), and most importantly had calculated the anomalous magnetic moment of the electron [10],

$$\boldsymbol{\mu} = \frac{e}{2mc} g \mathbf{S},$$

where $\mathbf{S}$ is the spin operator for the electron, and the g-factor, to first order in α, was

$$\frac{g}{2} = 1 + \frac{\alpha}{2\pi}.$$

The correction to the Dirac value $g_D = 2$ was in perfect agreement with experiment. A quantitative theory of quantum electrodynamics had been achieved. What Schwinger had done in this famous 1-page paper was to isolate the infinities that occurred in the calculations into redefinitions, or ▶ *renormalization*, of the mass and charge of the electron.

Feynman rapidly caught up, and based on the propagator methods he had begun to develop in his Ph.D. thesis at Princeton, derived a pictorial method of calculating processes in QED. Although initially meeting with disbelief, the method turned out to be simpler than Schwinger's earlier methods, and is now the universal formulation of perturbative quantum field theory: the famous Feynman diagrams [11,12]. Before Feynman's papers appeared, Dyson had established that the methods of Schwinger and Feynman, although appearing so different, were actually mathematically equivalent [13, 14], although Feynman had already demonstrated that equivalence to his own satisfaction.

The Feynman rules for quantum electrodynamics are exhibited in Fig. 2. The lines represent the particle propagators, and the vertices interactions. In addition, for external on-shell lines one must supply an appropriate ▶ wave function: for the photon a polarization vector $e^{\mu}_{p\lambda}$, and for the electron, a spinor $u_{p\sigma}$ or $u^{*}_{p\sigma}\gamma^{0}$. Furthermore, a factor of -1 must be supplied for each closed Fermion loop, reflecting the statistics of Fermions. By putting these components together in all possible ways, one arrives at the quantum-mechanical amplitude for a process. Thus, for example, the ▶ Feynman diagram that corresponds to the famous Schwinger correction to the magnetic moment of the electron is shown in Fig. 3. By using the

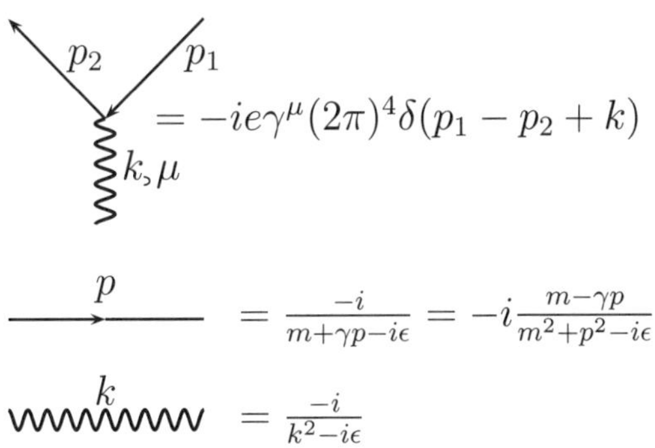

Fig. 2 Feynman rules for QED

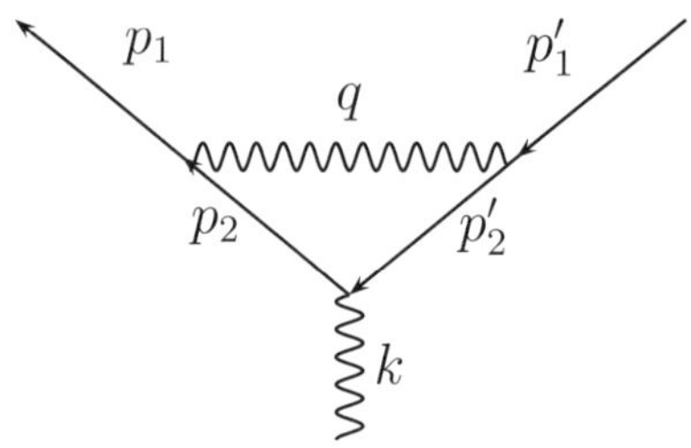

Fig. 3 Feynman diagram giving rise to the electric and magnetic form factors of the electron in order α

above rules the amplitude corresponding to this graph is given by (in momentum space with $k = p_1 + p'_1 = p_2 + p'_2$)

$$-e^3 \int \frac{d^4 p_2}{(2\pi)^4} \bar{\psi}(-p_1)\gamma^\mu \frac{m - \gamma \cdot p_2}{m^2 + p_2^2} \gamma^\lambda A_\lambda(k) \frac{m + \gamma \cdot p'_2}{m^2 + p_2'^2} \gamma_\mu \frac{1}{(p_1 - p_2)^2} \psi(-p'_1).$$

A calculation of a few pages yields the result for the g-factor of the electron:

$$\frac{g-2}{2} = \frac{\alpha}{2\pi}.$$

Since 1949, progress in QED has been considerable. A great many processes have been calculated, and agreement with experiment is spectacular. With the aid of computers, even the $O(\alpha^5)$ corrections to $g - 2$ have been computed. Last year [15, 16] a new precision experiment has yielded the most exquisite test of quantum electrodynamics to date. Because the experiment is more accurate than other measurements of the fine structure constant α, it can be used to determine that constant most precisely. The experiment is consistent with the statement that the electron is a point particle down to an incredible distance of 6×10^{-24} m.

Primary Literature

1. P. A. M. Dirac: Quantum Theory of Emission and Absorption of Radiation. Proc. R. Soc. Lond. A **114**, 243 (1927)
2. P. A. M. Dirac: The Quantum Theory of the Electron. Proc. R. Soc. Lond. A **117**, 610 (1928)
3. W. Heisenberg and W. Pauli: Zur Quantenmechanik der Wellenfelder. Z. Phys. **56**, 1 (1929)
4. W. Heisenberg and W. Pauli: Zur Quantenmechanik der Wellenfelder II. Z. Phys. **59**, 168 (1930)
5. W. Heisenberg and H. Euler: Consequences of Dirac's Theory of Positrons. Z. Phys. **98**, 714 (1936)
6. Willis E. Lamb, Jr. and Robert C. Retherford: Fine Structure of the Hydrogen Atom by a Microwave Method. Phys. Rev. **72**, 241 (1947)
7. J. E. Nafe, E. B. Nelson, and I. I. Rabi: The Hyperfine Structure of Atomic Hydrogen and Deuterium. Phys. Rev. **71**, 914 (1947)
8. P. Kusch and H. M. Foley: Precision Measurement of the Ratio of the Atomic 'g Values' in the $^2P_{3/2}$ and $^2P_{1/2}$ States of Gallium. Phys. Rev. **72**, 1256 (1947)
9. H. A. Bethe: The Electromagnetic Shift of Energy Levels. Phys. Rev. **72**, 339 (1947)

10. J. S. Schwinger: On Quantum Electrodynamics and the Magnetic Moment of the Electron. Phys. Rev. **73**, 416 (1948)
11. R. P. Feynman: The Theory of Positrons. Phys. Rev. **76**, 749 (1949)
12. R. P. Feynman: Space-Time Approach to Quantum Electrodynamics. Phys. Rev. **76**, 769 (1949)
13. F. J. Dyson: The Radiation Theories of Tomonaga, Schwinger, and Feynman. Phys. Rev. **75**, 486 (1949)
14. F. J. Dyson: The S Matrix in Quantum Electrodynamics. Phys. Rev. **75**, 1736 (1949)
15. B. Odom, D. Hanneke, B. D'Urso, and G. Gabrielse: New Measurement of the Electron Magnetic Moment Using a One-Electron Quantum Cyclotron. Phys. Rev. Lett. **97**, 030801 (2006)
16. G. Gabrielse, D. Hanneke, T. Kinoshita, M. Nio and B. Odom: New Determination of the Fine Structure Constant from the Electron *g* Value and QED. Phys. Rev. Lett. **97**, 030802 (2006)

Secondary Literature

17. S. S. Schweber: *QED and the Men Who Made It, Dyson, Feynman, Schwinger, and Tomonaga* (Princeton University Press, New Jersey 1994)
18. J. Mehra and K. A. Milton: *Climbing the Mountain, The Scientific Biography of Julian Schwinger* (Oxford University Press, Oxford 2000)
19. J. Mehra: *The Beat of a Different Drum, The Life and Science of Richard Feynman* (Oxford University Press, Oxford 1994)

Quantum Entropy

Dominik Janzing

Q

The von Neumann entropy of a quantum system with density operator ρ is given by [2]

$$S(\rho) := -\mathrm{tr}(\rho \log \rho). \tag{1}$$

Let

$$\rho = \sum_{j \in I} p_j |\psi_j\rangle\langle\psi_j| \tag{2}$$

be a decomposition of ρ into mutually orthogonal pure states where I is a countable, but possibly infinite, index set. Then we obtain

$$S(\rho) = \mathcal{H}(p), \tag{3}$$

where

$$\mathcal{H}(p) := -\sum_{j \in I} p_j \log p_j$$

is the Shannon entropy [5] of the probability distribution given by $(p_j)_{j \in I}$. This is exactly the uncertainty of the measurement results when an observable is measured that has $|\psi_j\rangle\langle\psi_j|$ as its spectral projections [6]. For an arbitrary non-degenerate observable with spectral projections $|\phi_j\rangle\langle\phi_j|$ the probabilities of the measurement outcomes are given by

$$q_j := \langle\phi_j|\rho|\phi_j\rangle ,$$

and satisfy $\mathcal{H}(q) \geqslant \mathcal{H}(p)$. Von Neumann entropy is conserved under unitary transformations U since they preserve the eigenvalues. This is in particular true for the dynamical evolution U_t of a closed physical system [7] induced by its Hamiltonian H:

$$\rho_t := U_t \rho U_t^\dagger = e^{-iHt} \rho \, e^{iHt} .$$

However, in many-particle systems with non-trivial interactions, such a formal conservation of entropy is only of limited practical relevance. This is because there may be no feasible non-degenerate measurement for which the uncertainty of measurement results attains $S(\rho_t)$ even though such a measurement may have existed for the initial state ρ. In a many-particle system, observables with spectral projections $U_t|\psi_j\rangle\langle\psi_j|U_t^\dagger$ could correspond to an arbitrarily complex measurement procedure. This can happen, for instance, if the dynamics U_t creates sophisticated quantum correlations between the particles (▶ correlations in quantum mechanics). For every feasible measurement the system then would behave like a system with higher entropy and we observe entropy increase on the *phenomenological* level.

Apart from such a "practical view", it has been argued that complexity aspects are also relevant from the fundamental point of view: Zurek [3] defines the *physical entropy* of a *classical* system as the sum of the Shannon entropy (formalizing the missing knowledge about the state) and the algorithmic information content (algorithmic randomness, i.e. the Kolmogorov complexity) present in the available data about the system. Mora et al. [4] describes a quantum generalization of Kolmogorov complexity and discusses also its thermodynamical relevance.

The interpretation of von Neumann entropy deserves further attention. Equations (2) and (3) may, at first glance, suggest the interpretation that one of the pure states $|\psi_j\rangle$ is present and $S(\rho)$ quantifies the *missing knowledge* about which one is the true one. However, this ignores first that the decomposition of ρ into pure states is in general not unique and, second, that ρ can also be the state of a subsystem of a *pure* entangled state (▶ entropy of entanglement). The scenario below shows that $S(\rho)$ has nevertheless an *information theoretic* meaning in the sense that it quantifies the resources required to transmit a quantum state in the same way as Shannon entropy quantifies the resources required to transmit a classical message.

To sketch this analogy, we assume a sender uses k different symbols $1, \ldots, k$ to transmit a classical message. If the symbol j is chosen with probability p_j and the total message consists of n symbols we expect that the number n_j of occurrences of j satisfy for large n

$$\frac{n_j}{n} \approx p_j \quad \text{i.e.} \quad \sum_j \frac{n_j}{n} \log \frac{n_j}{n} \approx \sum_j p_j \log p_j . \tag{4}$$

Using a precise version of (4), right, coding theory (see [5] for details) defines the set of *typical sequences* and shows that the numbers $N(n)$ of such sequences satisfy

$$\frac{1}{n}\log N(n) \to \mathcal{H}(p)\,. \tag{5}$$

The definition of typical sequences is chosen such that the probability for obtaining an untypical one tends to zero for $n \to \infty$. In this limit, the same message can thus be encoded into $\mathcal{H}(p)$ bits per symbol (provided that all logarithms are defined with respect to the basis 2). One can also show that $\mathcal{H}(p)$ bits per copy are really necessary.

Now we consider a scenario where the sender transmits the quantum state $|\psi_j\rangle$ with probability p_j. A message of length n is then given by a quantum state

$$|\psi\rangle := |\psi_{j_1}\rangle \otimes |\psi_{j_2}\rangle \otimes \cdots \otimes |\psi_{j_n}\rangle\,.$$

If all states $|\psi_j\rangle$ are mutually orthogonal, the above arguments suggest that we can restrict the attention to *typical states* $|\psi\rangle$, i.e. those whose numbers n_j of occurrences of the states $|\psi_j\rangle$ satisfy condition (4), right. They span a ► Hilbert space of dimension $N(n)$ satisfying again the asymptotical condition (5).

However, the more interesting case is when the sender uses *non-orthogonal* signal states. From the point of view of an observer who does not know which one of the states $|\psi_j\rangle$ has been chosen, the sender emits the density operator

$$\rho = \sum_{j=1}^{k} p_j |\psi_j\rangle\langle\psi_j|.$$

The density operator for n signal states is then given by

$$\rho^{\otimes n}.$$

Let $\rho = \sum_{i=1}^{k} q_i |\phi_i\rangle\langle\phi_i|$ be a decomposition of ρ into *mutually orthogonal* states. Even though the pure states $|\phi_j\rangle$ do not have any direct intuitive meaning since this set can be completely disjoint from the set of signal states, it turns out that they are useful for an mathematical analysis of the resource requirements: The density operator $\rho^{\otimes n}$ can be written as a mixture of states of the form

$$|\phi\rangle = |\phi_{i_1}\rangle \otimes |\phi_{i_2}\rangle \otimes \cdots \otimes |\phi_{i_n}\rangle.$$

In analogy to the arguments above, we consider only those states ϕ for which the sequence of indices is typical. They span a subspace whose dimensions $N(n)$ satisfy the asymptotical condition (5) with $\mathcal{H}(q)$ instead of $\mathcal{H}(p)$.

This shows that the number of quantum bits required per copy is asymptotically given by $\mathcal{H}(q) = S(\rho)$ and not by $\mathcal{H}(p)$. In other words, the entropy of the probability measure determining the *choice of the signal states* is not relevant. Instead, the quantum entropy of the corresponding mixture determines the required resources [1] even though the eigenstates of the mixture do not have any direct physical interpretation.

Literature

1. B. Schumacher: Quantum coding. Phys. Rev. A **51**(4), 2738–2747 (1995)
2. M. Nielsen and I. Chuang: *Quantum Information and Computation*. (Cambridge University Press, Cambridge 2000)
3. W. Zurek: Algorithmic randomness and physical entropy. Phys. Rev. A **40**, 4731–4751 (1989)
4. C. Mora, H. Briegel, and B. Kraus: Quantum Kolmogorov complexity and its applications. arXiv.org:quant-ph/0610109.
5. T. Cover and J. Thomas: *Elements of Information Theory*. (Wiley, New York 1991)
6. R. Omnes: *The Interpretation of Quantum Mechanics*. (Princeton University Press, Princeton 1994)
7. J. Jauch: *Foundations of Quantum Mechanics*. (Addison Wesley, New York 1968)

Quantum Eraser

Basil James Hiley

To understand the notion of a quantum eraser we need first to consider the interference produced when light falls on a pair of slits. As long as we think of light as being a wave phenomenon, there is no problem in understanding how the phase difference between the wave arriving at a point on the screen from one slit and the wave arriving from the second slit gives rise to the interference effects.

The problem arises when we learn that the wave consists of photons (► light quantum), a problem that becomes more acute when the incident beam consists only of a few photons arriving per second. If the photon is a localised packet of energy then the question as to which slit the photon passed through becomes inevitable. This question becomes even more pertinent when one realises that particles like ► electrons, neutrons and even atoms produce exactly the same interference patterns using pairs of slits of the appropriate size.

The obvious way to explore this situation further is to see if we can set up some form of experiment to find through which slit each particle actually passes. In this way we might be able to understand how the interference pattern arises. Unfortunately what we find is that for all experiments that give a definite answer for each particle, the interference pattern disappears. This means that once we know which way the particle goes, we lose the interference pattern. Alternatively if we have no means of knowing which way the particles go, then we get a sharp interference pattern. This phenomenon is known as ► 'wave-particle' duality.

One of the earlier ways of explaining the loss of interference was to argue that any attempt to determine which way the particle went would induce a series of random phase changes in the beam. These phase changes arise because in order to 'see' where the particle is, some form of scattering would have to be used. It is this scattering that produces the random phase changes which would clearly destroy the interference pattern [1].

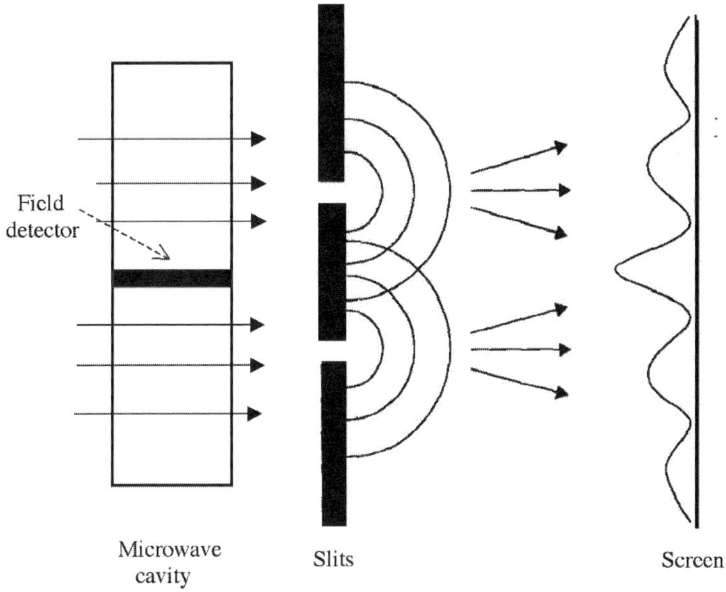

Fig. 1 Two-slits with cavity in place

This explanation underwent a radical reappraisal when it was discovered that it was possible to store which-way information in a microwave cavity without inducing any random phases into a beam of atoms. (► which-way experiments) Thus, rather than subjecting the atoms to a scattering process, they are simply allowed to give up any internal excitation energy to the microwave cavity through which they pass. This process does not produce any phase change to the centre of mass ► wave function. Will we see any interference effects in this case? The answer is 'no'. [2].

Q

This new experimental arrangement allowed for a new possibility. Would it be possible to erase the which-way information? If this is possible, would we then recover interference? Scully and Drühl [3] were the first to show that it should indeed be possible to recover interference effects. The principle is as follows: suppose a pair of microwave cavities are placed in front of the two-slit system as shown in the figure. As an excited atom passes through one of the cavities it will give up its internal energy leaving the cavity in an excited state. If we now repeat this for many atoms we will potentially know through which slit each atom has passed. The result of such an experiment shows that there are no interference fringes.

Suppose now we want to 'erase' the which-way information. We can do this by removing the common wall of the microwave cavities and inserting a radiation detector as shown in the figure. The function of this detector is to become excited when the cavity state is a symmetric combination of the two individual cavity fields, and becomes de-excited when this combination is anti-symmetric. In both cases the which-way information is lost.

Let us repeat the first experiment, recording the arrival position of each atom on the screen. Now at any time after noting the atom's final position, we can remove the common wall of the cavities and insert the radiation field detector. Once the detector responds, we lose the which-way information for that particular atom. We note whether the cavity detector is excited or de-excited in each case. By repeating this procedure we can produce two ensembles of atoms, one corresponding to the positions of the atoms arriving at the screen when the cavity detector is found to be excited and the other corresponding to those positions where the cavity detector is in its de-excited state.

We find each of these ► ensembles exhibit interference fringes, the maxima of one set corresponding to the minima of the other set. In other words by erasing the which-way information we have regained the interference. Furthermore if we superimpose these two patterns, we find the fringes exactly 'cancel' each other, producing a uniform distribution with no evidence of interference. A clear illustration of these effects has been brought out using the Bohm model [4].

This example illustrates a general principle that when which-way data is known, interference disappears, but as the which-way data becomes unavailable, interference appears. It is the process of the destruction of this which-way information that is referred to as the 'quantum erasure'.

In this brief account, we have only discussed the two-slit experiment, but the principle applies to any system that offers binary alternatives such as the Stern–Gerlach magnet (► Stern–Gerlach experiment), polarised light, the Mach-Zehnder interferometer (► Consistent Histories) and so on. A more detailed discussion of these other examples, together with a detailed quantitative account of this type of experiment can be found in Englert and Bergou [5]. In this paper the practical use of the eraser to maximise fringe visibility, is discussed.

We conclude this discussion with a final word of warning about the meaning of the words 'eraser' and 'delayed choice' which have been misunderstood. The situation has not been helped by statements like 'the past is undefined and undefinable without the observation [in the present]' [7]. These words, ambiguous at best, have sometimes been mistakenly taken to mean that somehow the past *dynamical* behaviour of the atoms can be affected by what we decide to do at some later time. This is not the case. Bohr [6] himself makes this very clear. He stresses that when we come to interpret experimental results predicted by the quantum formalism "it is essential that the *whole experimental arrangement* be taken into account".

In the cases we have discussed above, we have two distinct experimental arrangements: (1) the arrival of atoms with two distinct separate cavities in place and (2) the arrival of atoms with one large cavity containing a radiation field detector. The fact that we can remove the common wall cavities and insert a field detector in the first experiment at a later time still means we have two distinct experiments. The word 'eraser' arises simply because we have changed the experimental conditions, the change, of course can be 'delayed' indefinitely provided the cavity modes remain stable. There is no question of the dynamics of the atoms being changed as a result of any delay in changing the experimental conditions.

Literature

1. R. P. Feynman, R. B. Leighton, M. Sands: *The Feynman Lectures on Physics* (Addison-Wesley, Reading, MA 1965, Vol. III, sec. 3)
2. M. O. Scully, B.-G. Englert, H. Walther: Quantum optical tests of complementarity. Nature **351**, 111–116 (1991)
3. M. O. Scully, K. Drühl: Quantum eraser: a proposed photon correlation experiment concerning observation and the "delayed choice" in quantum mechanics. Phys. Rev. **25A**, 2208–2213(1982)
4. B. J. Hiley, R. E. Callaghan: What is erased in the quantum erasure? Found. Phys. **36**, 1869–1883 (2006)
5. B. -G. Englert, J. A. Bergou: Quantitative quantum eraser. Optics Commun. **179**, 337–355 (2000)
6. N. Bohr: *Atomic Physics and Human Knowledge* (Science Editions, New York, 1961, p. 50)
7. J. A. Wheeler: The 'Past' and the 'Delayed Choice' Double-slit Experiment, in *Mathematical Foundations of Quantum Theory,* ed. A. R. Marlow (Academic Press, New York, 1978, pp. 9–48)

Quantum Field Theory

Frank Wilczek

Quantum field theory is the application of quantum mechanics to systems whose degrees of freedom depend continuously on space and time. In the quantum mechanics of a point particle, states are specified by ► wave function $\psi(\boldsymbol{x})$, which gives the probability amplitude to find the particle at the position $\boldsymbol{x}$. In quantum field theory, states are specified by a wave function $\Psi(\phi(x))$ which specifies the probability amplitude for the field ϕ to be in the configuration $\phi(x)$.

Q

Quantum field theory was first developed to enable the application of quantum mechanics to theories that obey the special theory of relativity, specifically Maxwell's electrodynamics and Dirac's electron theory. Relativistic theories of interacting point particles are awkward to construct. The limiting speed of propagation c means that the influence felt by a given particle due to a second particle depends on where that second particle was in the past. Thus to evolve the state of set of particles, it is not sufficient to know their present positions. Fields avoid this difficulty, because the state of the field reflects the propagating influences as they propagate.

Quantum field theory is the framework in which the regnant theories of the electroweak and strong interactions, which together form the Standard Model ► particle physics, are formulated. ► Quantum Electrodynamics (QED), besides providing a complete foundation for atomic physics and chemistry, has supported calculations of physical quantities with unparalleled precision. The experimentally measured value of the magnetic dipole moment of the muon,

$$(g_\mu - 2)_{\text{exp.}} = 233\,184\,600\,(1680) \times 10^{-11}, \tag{1}$$

for example, should be compared with the theoretical prediction

$$(g_\mu - 2)_{\text{theor.}} = 233\,183\,478 \ \ (308) \times 10^{-11}. \tag{2}$$

In ► Quantum Chromodynamics (QCD) we cannot, for the foreseeable future, aspire to comparable accuracy. Yet QCD provides different, and at least equally impressive, evidence for the validity of the basic principles of quantum field theory. Indeed, because in QCD the interactions are stronger, QCD manifests a wider variety of phenomena characteristic of quantum field theory. These include especially running of the effective coupling with distance or energy scale and the phenomenon of confinement. QCD has supported, and rewarded with experimental confirmation, both heroic calculations of multiloop diagrams and massive numerical simulations of (a discretized version of) the complete theory.

The techniques of quantum field theory have also proved fruitful for describing the dynamics of many interacting particles, in the same spirit that hydrodynamics emerges as a fruitful description of systems of many interacting atoms. Impressive applications include ► superconductivity, the low-temperature behavior of the quantum liquids He^3 and He^4, and the theory of second-order phase transitions. Although for reasons of space and focus I will not attempt to do justice to this aspect here, the continuing interchange of ideas between condensed matter and high energy theory, through the medium of quantum field theory, is a remarkable phenomenon in itself. A partial list of historically important examples includes global and local spontaneous symmetry breaking, the ► renormalization group, effective field theory, ► solitons, instantons, and fractional charge and statistics.

Quantum Field Theory and Reality

What are the essential features of quantum field theory?

This question has no sharp answer. Theoretical physicists are very flexible in adapting their tools, and no axiomization can keep up with them. However I think it is fair to say that there are two characteristic, core ideas of quantum field theory. First: The basic dynamical degrees of freedom are operator functions of space and time – quantum fields, that obey appropriate commutation relations. Second: The interactions of these fields are local in space and time. Thus the equations of motion and commutation relations governing the evolution of a given quantum field at a given point in space-time should depend only on the behavior of fields and their derivatives at that point. One might find it convenient to use other variables, whose equations are not local, but in the spirit of quantum field theory there must always be some underlying fundamental, local variables. These ideas, combined with postulates of ► symmetry (e.g., in the context of the standard model, Lorentz and gauge invariance) turn out to be amazingly powerful, as will emerge the discussion below.

The field concept came to dominate physics starting with the work of Faraday in the mid-nineteenth century. Its conceptual advantage over the earlier Newtonian program of physics, to formulate the fundamental laws in terms of forces among atomic particles, emerges when we take into account the circumstance, unknown to Newton (or, for that matter, Faraday) but fundamental in special relativity, that influences travel no faster than a finite limiting speed. For then the force on a given particle at a given time cannot be deduced from the positions of other particles at that time, but must be deduced in a complicated way from their previous positions. Faraday's intuition that the fundamental laws of electromagnetism could be expressed most simply in terms of fields filling space and time was of course brilliantly vindicated by Maxwell's mathematical theory.

The concept of ► locality, in the crude form that one can predict the behavior of nearby objects without reference to distant ones, is basic to scientific practice. Practical experimenters – if not astrologers – confidently expect, on the basis of much successful experience, that after reasonable (generally quite modest) precautions to isolate their experiments from the environment they will obtain reproducible results.

The deep and ancient historic roots of the field and locality concepts provide no guarantee that these concepts remain relevant or valid when extrapolated far beyond their origins in experience, into the subatomic and quantum domain. This extrapolation must be judged by its fruits. That brings us, naturally, to a second question:

What does quantum field theory add to our understanding of the world, that was not already present in quantum mechanics and classical field theory separately?

Undoubtedly the single most profound fact about Nature that quantum field theory uniquely explains is *the existence of different, yet indistinguishable, copies of elementary particles*. Two ► electrons anywhere in the Universe, whatever their origin or history, are observed to have exactly the same properties. We understand this as a consequence of the fact that both are excitations of the same primary reality, the electron field. The same logic, of course, applies to photons (► light quantum) or quarks (see ► Color Charge Degree of Freedom in Particle Physics; Mixing and Oscillations of Particles; Particle Physics; Parton Model; QCD); or even to composite objects such as atomic nuclei, atoms, or molecules. The indistinguishability of particles is so familiar, and so fundamental to all of modern physical science, that we could easily take it for granted. Yet it is by no means obvious. For example, it directly contradicts one of the pillars of Leibniz' metaphysics, his "principle of the identity of indiscernables," according to which two objects cannot differ solely in number. Maxwell thought the similarity of different molecules so remarkable that he devoted the last part of his *Encyclopedia Brittanica* entry on Atoms – well over a thousand words – to discussing it. He concluded that "the formation of a molecule is therefore an event not belonging to that order of nature in which we live ... it must be referred to the epoch, not of the formation of the earth or the solar system ... but of the establishment of the existing order of nature ...".

The existence of classes of indistinguishable particles is the necessary logical prerequisite to a second profound insight from quantum field theory: *the assignment of unique quantum statistics* to each class. Given the ► indistinguishability of a class of elementary particles, and complete invariance of their interactions under interchange, the general principles of quantum mechanics teach us that solutions forming any representation of the permutation symmetry group retain that property in time; but they do not constrain which representations are realized. Quantum field theory not only explains the existence of indistinguishable particles and the invariance of their interactions under interchange, but also constrains the symmetry of the solutions. For bosons only the identity representation is physical (symmetric wave functions), for fermions only the one-dimensional odd representation is physical (antisymmetric wave functions). One also has the ► spin statistics theorem, according to which objects with integer spin are bosons, whereas objects with half odd integer ► spin are fermions. Of course, these general predictions have been verified in many experiments. The fermion character of electrons, in particular, underlies the stability of matter and the structure of the periodic table.

A third profound general insight from quantum field theory is *the existence of antiparticles*. This was first inferred by Dirac on the basis of a brilliant but obsolete interpretation of his equation for the electron field, whose elucidation was a crucial step in the formulation of quantum field theory. In quantum field theory, we reinterpret the Dirac wave function as a position (and time) dependent operator. It can be expanded in terms of the solutions of the ► Dirac equation, with operator coefficients. The coefficients of positive-energy solutions are operators that destroy electrons, and the coefficients of the negative-energy solutions are operators that create positrons (with positive energy). With this interpretation, an improved version of Dirac's hole theory emerges in a straightforward way. (Unlike the original hole theory, it has a sensible generalization to bosons, and to processes where the number of electrons minus positrons changes.) A very general consequence of quantum field theory, valid in the presence of arbitrarily complicated interactions, is the ► CPT theorem. It states that the product of charge conjugation, ► parity, and time reversal is always a symmetry of the world, although each may be – and is! – violated separately. Antiparticles are strictly defined as the CPT conjugates of their corresponding particles.

The three outstanding facts we have discussed so far: the existence of indistinguishable particles, the phenomenon of ► quantum statistics, and the existence of antiparticles, are all essentially consequences of *free* quantum field theory. When one incorporates interactions into quantum field theory, two other profound features of the physical reality get brightly illuminated.

The first of these is *the ubiquity of particle creation and destruction processes*. Local interactions involve products of field operators at a point. When the fields are expanded into ► creation and annihilation operators multiplying modes, we see that such interactions correspond to processes wherein particles can be created, annihilated, or changed into different kinds of particles.

This possibility arose, of course, arose in the primeval quantum field theory, quantum electrodynamics, where the primary interaction arises from a product of the electron field, its Hermitean conjugate, and the photon field. Processes of radiation and absorption of photons by electrons (or positrons), as well as electron–positron pair creation, are encoded in that product. But because the emission and absorption of light is such a common experience, and electrodynamics is such a special and familiar classical field theory, this correspondence between formalism and reality initially did not make a big impression. The first conscious exploitation of quantum field theory's potential to describe processes of transformation was Fermi's theory of beta decay. He turned the procedure around, by inferring from the observed processes of particle transformation the nature of the underlying local interaction of fields. Fermi's theory involved creation and annihilation not of photons, but of atomic nuclei and electrons (as well as neutrinos) – the traditional ingredients of "matter." It began the process whereby classic atomism, involving stable individual objects, was replaced by a more sophisticated and accurate picture. In this picture it is only the fields, and not the individual objects they create and destroy, that are permanent.

The second is *the association of forces and interactions with particle exchange*. When Maxwell completed the equations of electrodynamics, he found that they supported source-free electromagnetic waves. Thus the classical electric and magnetic fields took on a life of their own. Electric and magnetic forces between charged particles are explained as due to one particle acting as a source for electric and magnetic fields, which then influence other charged particles. Given that particles arise as excitations of quantum fields, Maxwell's discovery corresponds to the existence of real photons, while the mediation of forces through fields corresponds to the exchange of virtual photons.

This logic applies generally. Thus the connection between interactions and the exchange of particles is a general feature of quantum field theory. It was used by Yukawa to infer the existence and mass of pions from the range of nuclear forces, in electroweak theory to infer the existence, mass, and properties of W and Z bosons prior to their observation, and in QCD to infer the existence and properties of gluon jets prior to their observation.

The two additional outstanding facts we just discussed: the possibility of particle creation and destruction, and the association of particles with forces, are essentially consequences of classical field theory supplemented by the connection between particles and fields we learn from free field theory. Indeed, classical waves with nonlinear interactions will change form, scatter, and radiate, and these processes exactly mirror the transformation, interaction, and creation of particles. In quantum field theory, they are properties one sees already in *tree graphs*.

The foregoing major consequences of free quantum field theory, and of its formal extension to include nonlinear interactions, were all well appreciated by the late 1930s. The deeper properties of quantum field theory, which will form the subject of the remainder of this paper, arise from the need to introduce *infinitely many degrees of freedom*, and the possibility that all these degrees of freedom are excited as quantum-mechanical fluctuations. From a mathematical point of view, these deeper properties arise when we consider *loop graphs*.

From a physical point of view, the potential pitfalls associated with the existence of an infinite number of degrees of freedom first showed up in connection with the problem which led to the birth of quantum theory, that is the ultraviolet catastrophe of blackbody radiation theory. Somewhat ironically, in view of later history, in that context the crucial contribution of the quantum theory was to remove the disastrous consequences of the infinite number of degrees of freedom possessed by classical electrodynamics. The classical electrodynamic field can be decomposed into independent oscillators with arbitrarily high values of the wavevector. According to the equipartition theorem of classical statistical mechanics, in thermal equilibrium at temperature T each of these oscillators should have average energy kT. Quantum mechanics alters this situation by insisting that the oscillators of frequency ω have energy quantized in units of $\hbar\omega$. Then the high-frequency modes are exponentially suppressed by the Boltzmann factor, and instead of kT receive $[\hbar\omega \exp(-\hbar\omega/kT)]/[1 - \exp(-\hbar\omega/kT)]$. The role of the quantum, then, is to prevent accumulation of energy in the form of very small amplitude excitations of arbitrarily high frequency modes. It is very effective in suppressing the *thermal* excitation of high-frequency modes.

But while removing arbitrarily small amplitude excitations, quantum theory introduces the idea that the modes are always *intrinsically* excited to a small extent, proportional to $\hbar$. This so-called zero point motion is a consequence of the uncertainty principle. For a harmonic oscillator of frequency ω, the ground state energy is not zero, but $\frac{1}{2}\hbar\omega$. In the case of the electromagnetic field this leads, upon summing over its high-frequency modes, to a highly divergent total ground state energy. For most physical purposes the absolute normalization of energy is unimportant, and so this particular divergence does not necessarily render the theory useless.[1] It does, however, illustrate the dangerous character of the high-frequency modes, and its treatment gives a first indication of the leading theme of renormalization theory: we can only require – and generally will only obtain – sensible, finite answers when we ask questions that have direct, operational physical meaning.

The existence of an infinite number of degrees of freedom was first encountered in the theory of the electromagnetic field, but it is a general phenomenon, deeply connected with the requirement of locality in the interactions of fields. For in order to construct the local field $\psi(x)$ at a space-time point x, one must take a superposition

$$\psi(x) = \int \frac{\mathrm{d}^4k}{(2\pi)^4} \mathrm{e}^{\mathrm{i}kx} \tilde{\psi}(k) \tag{3}$$

that includes field components $\tilde{\psi}(k)$ extending to arbitrarily large momenta. Moreover in a generic interaction

$$\int \mathcal{L} = \int \psi(x)^3 = \int \frac{\mathrm{d}^4k_1}{(2\pi)^4} \frac{\mathrm{d}^4k_2}{(2\pi)^4} \frac{\mathrm{d}^4k_3}{(2\pi)^4} \tilde{\psi}(k_1)\tilde{\psi}(k_2)\tilde{\psi}(k_3)(2\pi)^4 \delta^4(k_1 + k_2 + k_3) \tag{4}$$

we see that a low momentum mode $k_1 \approx 0$ will couple without any suppression factor to high-momentum modes k_2 and $k_3 \approx -k_2$. In this sense, local couplings are "hard." Because locality requires the existence of infinitely many degrees of

[1] One would think that gravity should care about the absolute normalization of energy. The zero-point energy of the electromagnetic field, in that context, generates an infinite cosmological constant. This might be canceled by similar negative contributions from fermion fields, as occurs in supersymmetric theories, or it might indicate the need for some other profound modification of physical theory.

freedom at large momenta, with hard interactions, ultraviolet divergences similar to the ones cured by Planck, but driven by quantum rather than thermal fluctuations, are never far off stage. The deeper physical consequences of quantum field theory arise from this circumstance.

First of all, it is much more difficult to construct nontrivial examples of interacting relativistic quantum field theories than purely formal considerations would suggest. One finds that *consistent quantum field theories form a quite limited class, whose extent depends sensitively on the dimension of space-time and the spins of the particles involved.* Their construction is quite delicate, requiring limiting procedures whose logical implementation leads directly to renormalization theory, the running of couplings, and asymptotic freedom. ► Color Charge Degree of Freedom in Particles Physics; QCD.

Secondly, *even those quantum theories that can be constructed display less symmetry than their formal properties would suggest.* Violations of naive scaling relations – that is, ordinary dimensional analysis – in QCD, and of baryon number conservation in the standard electroweak model are examples of this general phenomenon. The original example, unfortunately too complicated to explain fully here, involved the decay process $\pi^o \rightarrow \gamma\gamma$, for which chiral symmetry (treated classically) predicts much too small a rate. When the correction introduced by quantum field theory (the so-called 'anomaly') is retained, excellent agreement with experiment results.

These deeper consequences of quantum field theory, which superficially might appear rather technical, largely dictate the structure and behavior of the so-called standard model – and, therefore, of the physical world.

Formulation

The physical constants $\hbar$ and c are so deeply embedded in the formulation of relativistic quantum field theory that it is standard practice to declare them to be the units of action and velocity, respectively. In these units, of course, $\hbar = c = 1$. With this convention, all physical quantities of interest have units which are powers of mass. Thus the dimension of momentum is (mass)^1 or simply 1, since mass$\times c$ is a momentum, and the dimension of length is (mass)$^{-1}$ or simply -1, since $\hbar c$/mass is a length. The usual way to construct quantum field theories is by applying the rules of ► quantization to a continuum field theory, following the canonical procedure of replacing Poisson brackets by commutators (or, for fermionic fields, anticommutators). The field theories that describe free spin 0 or free spin $\frac{1}{2}$ fields of mass m, μ respectively are based on the Lagrangian densities

$$\mathcal{L}_0(x) = \frac{1}{2}\partial_\alpha\phi(x)\partial^\alpha\phi(x) - \frac{m^2}{2}\phi(x)^2 \tag{5}$$

$$\mathcal{L}_{\frac{1}{2}}(x) = \bar{\psi}(x)(i\gamma^\alpha\partial_\alpha - \mu)\psi(x). \tag{6}$$

Since the action $\int \mathrm{d}^4x\mathcal{L}$ has mass dimension 0, the mass dimension of a scalar field like ϕ is 1 and of a spinor field like ψ is $\frac{3}{2}$. For free spin 1 fields the Lagrangian density is that of Maxwell,

$$\mathcal{L}_1(x) = -\frac{1}{4}(\partial_\alpha A_\beta(x) - \partial_\beta A_\alpha(x))(\partial^\alpha A^\beta(x) - \partial^\beta A^\alpha(x)), \tag{7}$$

so that the mass dimension of the vector field A is 1. The same result is true for nonabelian vector fields (Yang-Mills fields).

Thus far all our Lagrangian densities have been quadratic in the fields. Local interaction terms are obtained from Lagrangian densities involving products of fields and their derivatives at a point. The coefficient of such a term is a coupling constant, and must have the appropriate mass dimension so that the Lagrangian density has mass dimension 4. Thus the mass dimension of a Yukawa coupling y, which multiplies the product of two spinor fields and a scalar field, is zero. Gauge couplings g arising in the minimal coupling procedure $\partial_\alpha \rightarrow \partial_\alpha + \mathrm{i}gA_\alpha$ are also evidently of mass dimension zero.

The possibilities for couplings with nonnegative mass dimension are very restricted. This fact is quite important, for the following reason. Consider the effect of treating a given interaction term as a perturbation. If the coupling κ associated to this interaction has negative mass dimension $-p$, then successive powers of it will occur in the form of powers of $\kappa\Lambda^p$, where Λ is some parameter with dimensions of mass. Because, as we have seen, the interactions in a local field theory are hard, we can anticipate that Λ will characterize the largest mass scale we allow to occur (the cutoff), and will diverge to infinity as the limit on this mass scale is removed. So we expect that it will be difficult to make sense of fundamental interactions having negative mass dimensions, at least in perturbation theory. Such interactions are said to be nonrenormalizable.

The standard model is formulated entirely using renormalizable interactions. If nonrenormalizable interactions occur in an effective description of physical behavior below a certain mass scale, then the theory must change its nature – presumably by displaying new degrees of freedom – at some larger mass scale. The fact that the standard model contains only renormalizable operators signifies that it does not require modification up to arbitrarily high scales (at least on the grounds of divergences in perturbation theory).

Moreover, all the renormalizable interactions consistent with the gauge symmetry and multiplet structure of the standard model do seem to occur – "what is not forbidden, is mandatory". There is a beautiful agreement between the symmetries of the standard model, allowing arbitrary renormalizable interactions, and the symmetries of the world. One understands, for example, why strangeness is violated but baryon number is not. (The only discordant element is the so-called θ term of QCD, which is allowed by the symmetries of the standard model but is measured to be quite accurately zero. A plausible solution to this problem exists. It involves a characteristic very light *axion* field.)

The power counting rules for estimating divergences assume that there are no special symmetries canceling off the contribution of high energy modes. They do not apply, without further consideration, to antisymmetric theories, in which the contributions of boson and fermionic modes cancels, nor to theories derived from supersymmetric theories by soft supersymmetry breaking. In the latter case the scale of supersymmetry breaking plays the role of the cutoff Λ.

The power counting rules, as discussed so far, are too crude to detect divergences of the form $\ln \Lambda^2$. Yet divergences of this form are pervasive and extremely significant, as we shall now discuss.

Running Couplings

The problem of calculating the energy associated with a constant magnetic field, in the more general context of an arbitrary nonabelian gauge theory coupled to spin 0 and spin $\frac{1}{2}$ charged particles, provides an excellent concrete illustration of how the infinities of quantum field theory arise, and of how they are dealt with. It introduces the concept of running couplings in a natural way, and leads directly to qualitative and quantitative results of great significance for physics. The interactions of concern to us appear in the Lagrangian density

$$\mathcal{L} = -\frac{1}{4g^2} G^I_{\alpha\beta} G^{I\alpha\beta} + \bar{\psi}(i\gamma^\nu D_\nu - \mu)\psi + \phi^\dagger(-D_\nu D^\nu - m^2)\phi \tag{8}$$

where $G^I_{\alpha\beta} \equiv \partial_\alpha A^I_\beta - \partial_\beta A^I_\alpha - f^{IJK} A^J_\alpha A^K_\beta$ and $D_\nu \equiv \partial_\nu + \mathrm{i} A^I_\nu T^I$ are the standard field strengths and covariant derivative, respectively. Here the f^{IJK} are the structure constants of the gauge group, and the T^I are the representation matrices appropriate to the field on which the covariant derivative acts. This Lagrangian differs from the usual one by a rescaling $gA \to A$, which serves to emphasize that the gauge coupling g occurs only as a prefactor in the first term. It parameterizes the energetic cost of nontrivial gauge curvature, or in other words the stiffness of the gauge fields. Small g corresponds to gauge fields that are difficult to excite.

From this Lagrangian it would appear that the energy required to set up a magnetic field B^I is just $\frac{1}{2g^2}(B^I)^2$. That is the classical energy, but in the quantum theory it is not the whole story. A more accurate calculation must include the effect of the imposed magnetic field on the ▶ zero-point energy of the charged fields. Earlier, we met and briefly discussed a formally infinite contribution to the energy of the ground state of a quantum field theory (specifically, the electromagnetic field) due to the irreducible quantum fluctuations of its modes, which mapped to an infinite number of independent harmonic oscillators. Insofar as only differences in energy are physically significant, we could ignore that infinity. But the *change* in the zero-point energy in response to a magnetic field is reflected in the work it takes to impose the field, and is a measurable effect.

Postponing momentarily the derivation, let me anticipate the form of the answer, and discuss its interpretation. Without loss of generality, I will suppose that the magnetic field is aligned along a normalized, diagonal generator of the gauge group. This allows us to drop the index, and to use terminology and intuition from electrodynamics freely. If we restrict the sum to modes whose energy is less than a cutoff Λ, we find for the energy

$$\mathcal{E}(B) = \mathcal{E} + \delta\mathcal{E} = \frac{1}{2g^2(\Lambda^2)}B^2 - \frac{1}{2}\eta B^2(\ln(\Lambda^2/B) + \text{finite}) \tag{9}$$

where

$$\eta = \frac{1}{96\pi^2}[-(T(R_o) - 2T(R_{\frac{1}{2}}) + 2T(R_1))] + \frac{1}{96\pi^2}[3(-2T(R_{\frac{1}{2}}) + 8T(R_1))], \tag{10}$$

and the terms not displayed are finite as $\Lambda \to \infty$. The notation $g^2(\Lambda^2)$ has been introduced for later convenience. The factor $T(R_s)$ is the trace of the representation for spin s, and basically represents the sum of the squares of the charges for the particles of that spin. The denominator in the logarithm is fixed by dimensional analysis, assuming $B \gg \mu^2, m^2$.

The most striking, and at first sight disturbing, aspect of this calculation is that a cutoff is necessary in order to obtain a finite result. If we are not to introduce a new fundamental scale and compromise locality, we must remove reference to the arbitrary cutoff Λ from our description of physically meaningful quantities. This is the challenge addressed by the renormalization program. Its guiding idea is the thought that if we are working with experimental probes characterized by energy and momentum scales well below Λ, we should expect that our capacity to affect, or be sensitive to, the modes of much higher energy will be quite restricted. Thus one expects that when attention is restricted to low energy-momentum processes, all explicit reference to the cutoff Λ can be removed.

In our magnetic energy example, for instance, we see immediately that the difference in susceptibilities

$$\mathcal{E}(B_1)/B_1^2 - \mathcal{E}(B_0)/B_0^2 = \text{finite} \tag{11}$$

is independent of Λ as $\Lambda \to \infty$. Thus once we measure the susceptibility, or equivalently the coupling constant, at one reference value of B, the calculation gives sensible, unambiguous predictions for all other values of B.

This simple example illustrates a much more general result, the central result of the classic renormalization program. It goes as follows. A small number of quantities, corresponding to the couplings and masses in the original Lagrangian, that if calculated formally would diverge or depend on the cutoff, are chosen to fit experiment. They define the physical, as opposed to the original, or "bare," couplings. Thus, in our example, we can define the susceptibility to be $\frac{1}{2g^2(B_0)}$ at some reference field B_0. Then we have the physical or renormalized coupling

$$\frac{1}{g^2(B_0)} = \frac{1}{g^2(\Lambda^2)} - \eta \ln(\Lambda^2/B_0). \tag{12}$$

(In this equation I have ignored, for simplicity in exposition, the finite terms. These are relatively negligible for large B_0. Also, there are corrections of higher order in g^2.) This of course determines the "bare" coupling to be

$$\frac{1}{g^2(\Lambda^2)} = \frac{1}{g^2(B_0)} + \eta \ln(\Lambda^2/B_0). \tag{13}$$

In these terms, the central result of perturbative renormalization theory is that after bare couplings and masses are reexpressed in terms of their physical, renormalized counterparts, the coefficients in the perturbation expansion of any physical quantity approach finite limits, independent of the cutoff, as the cutoff is taken to infinity. (To be perfectly accurate, one must also perform wave-function renormalization. This is no different in principle; it amounts to expressing the bare coefficients of the kinetic terms in the Lagrangian in terms of renormalized values.) The question whether this perturbation theory converges, or is some sort of asymptotic expansion of a soundly defined theory, is a separate issue. This loophole is no mere technicality, as we will soon see.

Picking a scale B_0 at which the coupling is defined is analogous to choosing the origin of a coordinate system in geometry. One can describe the same physics using different choices of normalization scale, so long as one adjusts the coupling appropriately. We capture this idea by introducing the concept of a running coupling defined, in accordance with (12), to satisfy

$$\frac{\mathrm{d}}{\mathrm{d}\ln B}\frac{1}{g^2(B)} = \eta. \tag{14}$$

Q

With this definition, the choice of a particular scale at which to define the coupling will not affect the final result.

It is profoundly important, however, that the running coupling does make a real distinction between the behavior at different mass scales, even if the original underlying theory was formally scale invariant (as is QCD with massless quarks), and even at mass scales much larger than the mass of any particle in the theory.

The distinction among scales, in a formally scale-invariant theory, embodies the phenomenon of *dimensional transmutation*. Rather than a range of theories, parametrized by a dimensionless coupling, we have a range of theories differing only in the value of a dimensional parameter, say (for example) the value of B at which $1/g^2(B) = 1$.

Clearly, the qualitative behavior of solutions of (14) depends on the sign of η. If $\eta > 0$, the coupling $g^2(B)$ will get smaller as B grows, or in other words as we treat more and more modes as dynamical, and approach closer to the bare charge. These modes were enhancing, or *antiscreening* the bare charge. This is the case of *asymptotic freedom*.

In asymptotically free theories, we can complete the renormalization program in a convincing fashion. There is no barrier to including the effect of very large energy modes, and removing the cutoff. We can confidently expect, then, that the theory is well-defined, independent of perturbation theory. In particular, suppose the theory has been discretized on a space-time lattice. This amounts to excluding the modes of high energy and momentum. In an asymptotically free theory one can compensate for these modes by adjusting the coupling in a well-defined, controlled way as one shrinks the discretization scale. Very impressive nonperturbative calculations in QCD, involving massive computer simulations, have exploited this strategy. They demonstrate the complete consistency of the theory and its ability to account quantitatively for the masses of hadrons.

In a nonasymptotically free theory the coupling does not become small, there is no simple foolproof way to compensate for the missing modes, and the existence of an underlying limiting theory becomes doubtful.

Now let us discuss how η can be calculated. The two terms in (10) correspond to two distinct physical effects. The first is the convective, diamagnetic (screening) term. The overall constant is a little tricky to calculate, and I do not have space to do it here. Its general form, however, is transparent. The effect is independent of spin, and so it simply counts the number of components (one for scalar particles, two for spin-1/2 or massless spin-1, both with two helicities). It is screening for bosons, while for fermions there is a sign flip, because the zero-point energy is negative for fermionic oscillators.

The second is the paramagnetic spin susceptibility. For a massless particle with spin s and gyromagnetic ratio g_m the energies shift, giving rise to the altered zero-point energy

$$\Delta\mathcal{E} = \int_0^{E=\Lambda} \frac{d^3k}{(2\pi)^3}\frac{1}{2}\left(\sqrt{k^2 + g_m s B} + \sqrt{k^2 - g_m s B} - 2\sqrt{k^2}\right). \tag{15}$$

This is readily calculated as

$$\Delta\mathcal{E} = -B^2 (g_m s)^2 \frac{1}{32\pi^2} \ln\left(\frac{\Lambda^2}{B}\right). \tag{16}$$

With $g_m = 2$, $s = 1$ (and $T = 1$) this is the spin-1 contribution, and with $g_m = 2$, $s = \frac{1}{2}$, after a sign flip, it is the spin-$\frac{1}{2}$ contribution. The preferred moment $g_m = 2$ is a direct consequence of the Yang-Mills and Dirac equations, respectively.

This elementary calculation gives us a nice heuristic understanding of the unusual antiscreening behavior of nonabelian gauge theories. It is due to the large paramagnetic response of charged vector fields. Because we are interested in very high energy modes, the usual intuition that charge will be screened, which is based on the electric response of heavy particles, does not apply. Magnetic interactions, which can be attractive for like charges (paramagnetism) are, for highly relativistic particles, in no way suppressed. Indeed, they dominate numerically.

Though I have presented it in the very specific context of vacuum magnetic susceptibility, the concept of running coupling is much more widely applicable. The basic heuristic idea is that in analyzing processes whose characteristic energy-momentum scale (squared) is Q^2, it is appropriate to use the running coupling at Q^2, i.e., in our earlier notation $g^2(B = Q^2)$. For in this way we capture the dynamical effect of the virtual oscillators which can be appreciably excited, while avoiding the formal divergence encountered if we tried to include all of them (up to infinite mass scale). At a more formal level, use of the appropriate effective coupling allows us to avoid large logarithms in the calculation of Feynman graphs, by normalizing the vertices close to where they need to be evaluated. There is a highly developed, elaborate chapter of quantum field theory which justifies and refines this rough idea into a form where it makes detailed, quantitative predictions for concrete experiments. Calculations of two- and even three-loop graphs with complicated interactions among the virtual particles are needed to do justice to the attainable experimental accuracy.

An interesting feature visible in Fig. 1 is that the theoretical prediction for the coupling focuses at large Q^2, in the sense that a wide range of values at small

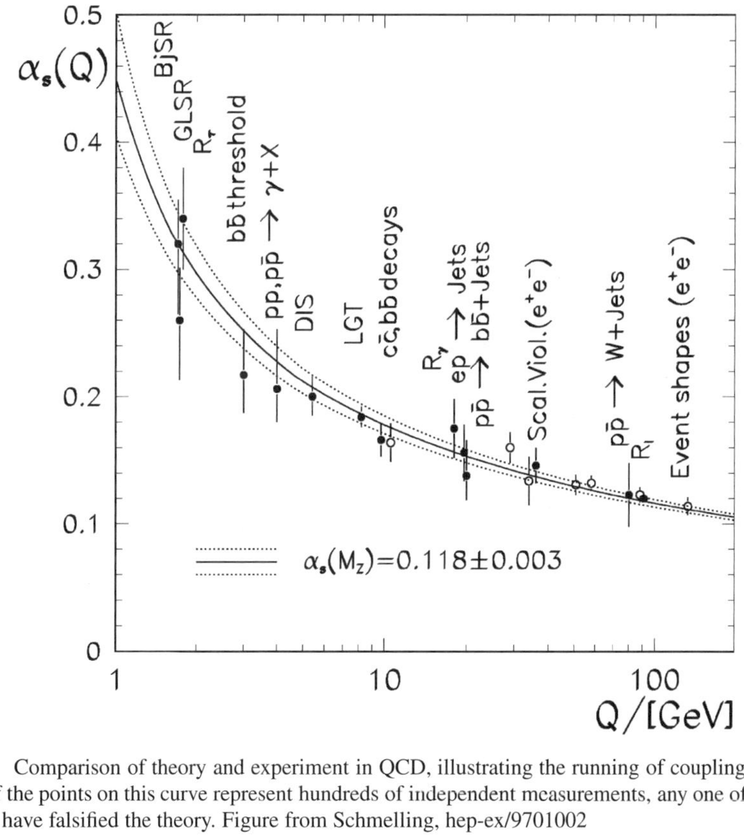

Fig. 1 Comparison of theory and experiment in QCD, illustrating the running of couplings. Several of the points on this curve represent hundreds of independent measurements, any one of which might have falsified the theory. Figure from Schmelling, hep-ex/9701002

Q^2 converge to a much narrower range at larger Q^2. Thus even crude estimates of what are the appropriate scales (e.g., one expects $g^2(Q^2)/4\pi \sim 1$ where the strong interaction is strong, say for $100\,\text{MeV} \lesssim \sqrt{Q^2} \lesssim 1\,\text{GeV}$) allow one to predict the value of $g^2(M_Z^2)$ with $\sim$10% accuracy. The original idea of Pauli and others that calculating the fine structure constant was the next great item on the agenda of theoretical physics now seems misguided. We see this constant as just another running coupling, neither more nor less fundamental than many other parameters, and not likely to be the most accessible theoretically. But our essentially parameter-free approximate determination of the observable strong interaction analogue of the fine structure constant realizes a form of their dream.

The electroweak interactions start with much smaller couplings at low mass scales, so the effects of their running are less dramatic (though they have been observed). Far more spectacular than the modest quantitative effects we can test directly, however, is the conceptual breakthrough that results from application of these ideas to unified models of the strong, electromagnetic, and weak interactions.

The different components.of the standard model have a similar mathematical structure, all being gauge theories. Their common structure encourages the speculation that they are different facets of a more encompassing ► gauge symmetry, in which the different strong and weak color charges, as well as electromagnetic charge, would all appear on the same footing. The multiplet structure of the quarks and leptons in the standard model fits beautifully into small representations of unification groups such as $SU(5)$ or $SO(10)$. There is the apparent difficulty, however, that the coupling strengths of the different standard model interactions are widely different, whereas the symmetry required for unification requires that they share a common value.The running of couplings suggests an escape from this impasse. Since the strong, weak, and electromagnetic couplings run at different rates, their inequality at currently accessible scales need not reflect the ultimate state of affairs. We can imagine that spontaneous symmetry breaking – a soft effect – has hidden the full symmetry of the unified interaction. What is really required is that the fundamental, bare couplings be equal, or in more prosaic terms, that the running couplings of the different interactions should become equal beyond some large scale.

Using simple generalizations of the formulas derived and tested in QCD, we can calculate the running of couplings, to see whether this requirement is satisfied in reality. In doing so one must make some hypothesis about the spectrum of virtual particles. If there are additional massive particles (or, better, fields) that have not yet been observed, they will contribute significantly to the running of couplings once the scale exceeds their mass. Let us first consider the default assumption, that there are no new fields beyond those that occur in the standard model. The results of this calculation are displayed in Fig. 2.

Considering the enormity of the extrapolation this calculation works remarkably well, but the accurate experimental data indicates unequivocally that something is wrong. There is one particularly attractive way to extend the standard model, by including supersymmetry. Supersymmetry cannot be exact, but if it is only mildly

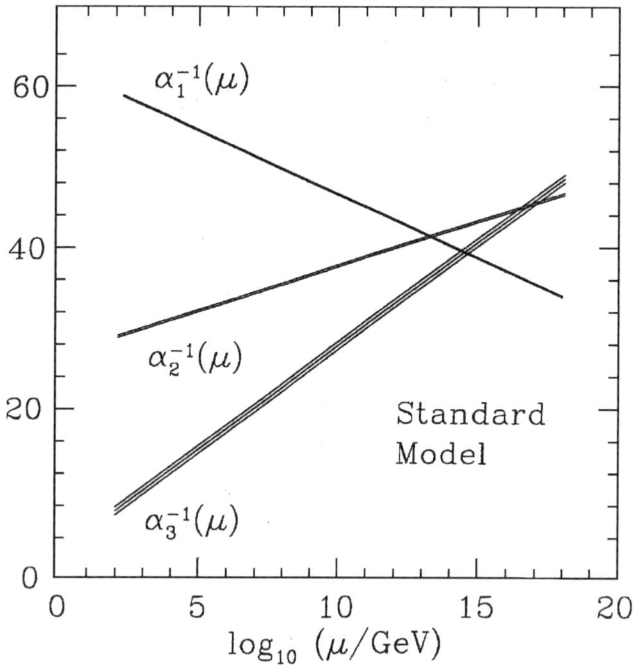

Fig. 2 Running of the couplings extrapolated toward very high scales, using just the fields of the standard model. The couplings do not quite meet. Experimental uncertainties in the extrapolation are indicated by the width of the lines. Figure courtesy of Dienes

broken (so that the superpartners have masses $\lesssim$ 1 TeV) it can help explain why radiative corrections to the Higgs mass parameter, and thus to the scale of weak symmetry breaking, are not enormously large. In the absence of supersymmetry power counting would indicate a hard, quadratic dependence of this parameter on the cutoff. Supersymmetry removes the most divergent contribution, by cancelling boson against fermion loops. If the masses of the superpartners are not too heavy, the residual finite contributions due to supersymmetry breaking will not be too large.

The minimal supersymmetric extension of the standard model, then, makes semi-quantitative predictions for the spectrum of virtual particles starting at 1 TeV or so. Since the running of couplings is logarithmic, it is not extremely sensitive to the unknown details of the supersymmetric mass spectrum, and we can assess the impact of supersymmetry on the unification hypothesis quantitatively. The results, as shown in Fig. 3, are quite encouraging.

A notable result of the unification of couplings calculation, especially in its supersymmetric form, is that the unification occurs at an energy scale which is enormously large by the standards of traditional ▶ particle physics, perhaps approaching 10^{16-17} GeV. From a phenomenological viewpoint, this is fortunate. The most compelling unification schemes merge quarks, antiquarks, leptons, and antileptons into

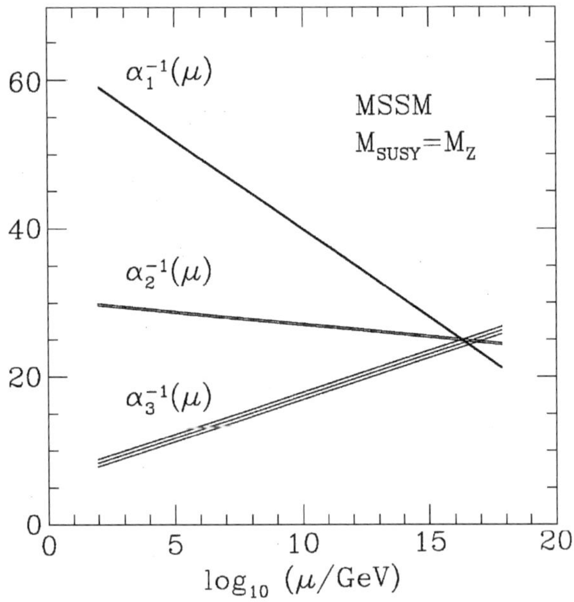

Fig. 3 Running of the couplings extrapolated to high scales, including the effects of supersymmetric particles starting at 1 TeV. Within experimental and theoretical uncertainties, the couplings do meet

common multiplets, and have gauge bosons mediating transitions among all these particle types. Baryon number violating processes almost inevitably result, whose rate is inversely proportional to the fourth power of the gauge boson masses, and thus to the fourth power of the unification scale. Only for such large values of the scale is one safe from experimental limits on nucleon instability. From a theoretical point of view the large scale is fascinating because it brings us from the internal logic of the experimentally grounded domain of particle physics to the threshold of ► quantum gravity.

Literature

For further information about quantum field theory, the reader may wish to consult:

1. T.P. Cheng and L.F. Li, *Gauge Theory of Elementary Particle Physics* (Oxford 1984)
2. M. Peskin and D. Schroeder, *Introduction to Quantum Field Theory* (Addison-Wesley 1995)
3. S. Weinberg, *The Quantum Theory of Fields, I* (Cambridge 1995) and *The Quantum Theory of Fields, II* (Cambridge 1996)

Quantum Gravity (General) and Applications

Claus Kiefer

What is Quantum Gravity?

Quantum theory is a general theoretical framework to describe states and interactions in Nature. It does so successfully for the strong, weak, and electromagnetic interactions. Gravity is, however, still described by a classical theory – Einstein's theory of general relativity, also called geometrodynamics. So far, general relativity seems to accommodate all observations which include gravity; there exist some phenomena which could in principle need a more general theory for their explanation (Dark Matter, Dark Energy, Pioneer Anomaly), but this is an open issue.

Quantum gravity would ultimately be a physical theory, both mathematically consistent and experimentally tested, that accommodates the gravitational interaction into the quantum framework. Such a theory is not yet available. Therefore, one calls quantum gravity all approaches which are candidates for such a theory or suitable approximations thereof. The following sections will first focus on the general motivation for constructing such a theory, and then introduce the approaches which at the moment look most promising.

Why Quantum Gravity?

Q

No experiment or observation is known which definitely needs a quantum theory of gravity for its explanation. There exist, however, various theoretical reasons which indicate that the current theoretical framework of physics is incomplete and that one needs quantum gravity for its completion. Here is a list of such reasons:

- *Singularity theorems*: Under general conditions, it follows from mathematical theorems that spacetime singularities are unavoidable in general relativity. The theory thus predicts its own breakdown. The two most relevant singularities are the initial cosmological singularity ('Big Bang') and the singularity inside black holes. Since the classical theory is then no longer applicable, a more comprehensive theory must be found – the general expectation is that this is a *quantum* theory of gravity.
- *Initial conditions in cosmology*: This is related to the first point. Cosmology as such is incomplete if its beginning cannot be described in physical terms. According to modern cosmological theories, the Universe underwent an era of exponential expansion in its early phase called inflation. While inflation gives a satisfactory explanation for issues such as structure formation, it cannot give, by itself, an account of how the Universe began. Nor is it clear how likely inflation

indeed is. A thorough understanding of initial conditions should shed some light on this as well as on the origin of irreversibility, that is, on the arrow of time.

- *Evolution of black holes*: Black holes radiate with a temperature proportional to $\hbar$, the Hawking temperature, see below. For the final evaporation, a full theory of quantum gravity is needed since the semi-classical approximation leading to the Hawking temperature then breaks down. This final phase could be of astrophysical relevance, provided small relic black holes from the early Universe ('primordial black holes') exist.
- *Unification of all interactions*: All nongravitational interactions have so far been successfully accommodated into the quantum framework. Gravity couples universally to all forms of energy. One would therefore expect that in a unified theory of all interactions, gravity is described in quantum terms, too.
- *Inconsistency of an exact semi-classical theory*: All attempts to construct a fundamental theory where a classical gravitational field is coupled to quantum fields have failed up to now. Such a framework is here called an 'exact semi-classical theory'; it corresponds to the limit where the quantum fields propagate on a classical background spacetime.
- *Avoidance of divergences*: It has long been speculated that quantum gravity may lead to a theory devoid of the ubiquitous divergences arising in quantum field theory. This may happen, for example, through the emergence of a natural cutoff at small distances (large momenta). In fact, modern approaches such as string theory or loop quantum gravity (see below) provide indications for a discrete structure at small scales.

Quantum gravity is supposed to be a fundamental theory which is valid at all scales. There exists, however, a distinguished scale where one would expect that typical quantum-gravity effects can never be neglected. This scale is found if one combines the gravitational constant (G), the speed of light (c), and the quantum of action ($\hbar$) into units of length, time, mass (and energy). In honour of Max Planck, who presented these units first in 1899, they are called *Planck units*. Explicitly, they read

$$l_\mathrm{P} = \sqrt{\frac{\hbar G}{c^3}} \approx 1.62 \times 10^{-33}\,\mathrm{cm}\,, \tag{1}$$

$$t_\mathrm{P} = \sqrt{\frac{\hbar G}{c^5}} \approx 5.40 \times 10^{-44}\,\mathrm{s}\,, \tag{2}$$

$$m_\mathrm{P} = \sqrt{\frac{\hbar c}{G}} \approx 2.17 \times 10^{-5}\,\mathrm{g} \approx 1.22 \times 10^{19}\,\mathrm{GeV}/c^2\,. \tag{3}$$

They are called Planck length, Planck time, and Planck mass, respectively.

Structures in the Universe usually occur at scales which are simple powers of the gravitational 'fine-structure constant'

$$\alpha_\mathrm{g} = \frac{G m_\mathrm{pr}^2}{\hbar c} = \left(\frac{m_\mathrm{pr}}{m_\mathrm{P}}\right)^2 \approx 5.91 \times 10^{-39}\,, \tag{4}$$

where m_{pr} denotes the proton mass. Stellar masses and stellar lifetimes can be derived, in an order-of-magnitude estimate, from this number. Its smallness is responsible for the irrelevance of quantum gravity in usual astrophysical considerations.

Structural Issues of Quantum Gravity

Quantization of gravity means quantization of geometry ▶ quantization. But which structures should be quantized, that is, to which structures should one apply the ▶ superposition principle? Following Chris Isham, one can do this at each order of the following hierarchy of structures:

Point set of events → topological structure → differentiable manifold → causal structure → Lorentzian structure.

Those structures that are not quantized remain as absolute (nondynamical) entities in the formalism. One would expect that in a fundamental theory no absolute structure remains. This is referred to as *background independence* of the theory. Still, however, most of the approaches to quantum gravity contain at least the first three structures as classical entities.

A particular aspect of background independence is the 'problem of time,' which arises in any approach to quantum gravity. On the one hand, time is external in ordinary quantum theory; the parameter t in the ▶ Schrödinger equation is identical to Newton's absolute time – it is *not* turned into an operator and is presumed to be prescribed from the outside. This is true also in special relativity where the absolute time t is replaced by Minkowski spacetime, which is again an absolute structure. On the other hand, time in general relativity is dynamical because it is part of the spacetime described by Einstein's equations. Both concepts cannot be fundamentally true, so a theory of quantum gravity would entail important changes for our understanding of time.

Experimental Status

One of the main problems in searching for a theory of quantum gravity is the lack of a direct experimental hint. For example, in order to probe the Planck scale directly, present-day accelerators would have to be of galactic size. Direct tests are therefore expected to arise from astrophysical or cosmological observations. However, some speculative theories with higher dimensions allow for the possibility of an experimental test at the Large Hadron Collider (LHC), which starts to operate at CERN in 2009.

Experiments are available only for the level of external Newtonian gravity interacting with micro- or mesoscopic systems (▶ Mesoscopic Quantum Phenomena). Examples are neutron and atom interferometry. On the level of quantum field theory on a curved spacetime, a definite, but not yet tested prediction was made: black holes emit thermal radiation. This is the Hawking effect, named after the physicist Stephen Hawking (*1942) who derived it in 1974. For a Schwarzschild black hole of mass M, the temperature is

$$T_{\mathrm{BH}} = \frac{\hbar c^3}{8\pi k_{\mathrm{B}} G M} \approx 6.17 \times 10^{-8} \left(\frac{M_\odot}{M}\right) \mathrm{K}\,, \tag{5}$$

where k_{B} denotes Boltzmann's constant. The black hole shrinks due to Hawking radiation and possesses a finite lifetime. The final phase, where γ-radiation is being emitted, could be observable. The temperature (5) is unobservably small for black holes that result from stellar collapse. One would need primordial black holes produced in the early Universe because they could possess a sufficiently low mass. For example, black holes with an initial mass of 5×10^{14} g would evaporate at the present age of the Universe. In spite of several attempts, no experimental hint for black-hole evaporation has been found.

Since black holes radiate thermally, they also possess an entropy, the 'Bekenstein–Hawking entropy,' which is given by the expression

$$S_{\mathrm{BH}} = \frac{k_{\mathrm{B}} c^3 A}{4G\hbar} = k_{\mathrm{B}} \frac{A}{4 l_{\mathrm{P}}^2}, \tag{6}$$

where A is the surface area of the event horizon. For a Schwarzschild black hole with mass M, this reads

$$S_{\mathrm{BH}} \approx 1.07 \times 10^{77} k_{\mathrm{B}} \left(\frac{M}{M_\odot}\right)^2 . \tag{7}$$

Since the Sun has an entropy of about $10^{57} k_{\mathrm{B}}$, this means that a black hole resulting from the collapse of a star with a few solar masses would experience an increase in entropy by twenty orders of magnitude during its collapse. It is one of the challenges of any theory of quantum gravity to provide a microscopic explanation for this entropy, that is, to derive (6) from a counting of microscopic quantum gravitational states.

Due to the equivalence principle, there exists an effect related to (5) in flat Minkowski space. An observer with uniform acceleration a experiences the standard Minkowski vacuum not as empty, but as filled with thermal radiation with temperature

$$T_{\mathrm{DU}} = \frac{\hbar a}{2\pi k_{\mathrm{B}} c} \approx 4.05 \times 10^{-23}\, a \left[\frac{\mathrm{cm}}{\mathrm{s}^2}\right] \mathrm{K}. \tag{8}$$

This temperature is often called the 'Davies–Unruh temperature,' named after the physicists Paul Davies (*1946) and William Unruh (*1945). It, too, has not yet been experimentally tested, but efforts are being made in this direction.

What are the Main Approaches?

The main present approaches to find a theory of quantum gravity can be classified according to the following scheme.

- *Quantum general relativity*: The most straightforward attempt, both conceptually and historically, is the application of 'quantization rules' to classical general relativity. One further distinguishes the following subapproaches:
  - *Covariant approaches*: These are approaches that employ four-dimensional covariance at some stage of the formalism. Examples include perturbation theory, effective field theories, renormalization-group approaches, and path integral methods (such as Regge calculus or dynamical triangulation). For example, in the path integral one sums over all suitable four-dimensional metrics in order to arrive at a quantum gravitational Green function or wave functional. Pioneers of the covariant approach include Léon Rosenfeld, Matvei Bronstein, and Bryce DeWitt.
  - *Canonical approaches*: Here one makes use of a Hamiltonian formalism and identifies appropriate canonical variables and conjugate momenta. Examples include quantum geometrodynamics (where gravity is described in metric form) and loop quantum gravity (where gravity is described by a connection integrated around a closed loop). They are characterized by a constraint equation of the form
$$H\Psi = 0\,, \tag{9}$$
    where H denotes the full Hamilton operator for the gravitational field as well as all nongravitational fields; Ψ is the full wave functional for these degrees of freedom. In the geometrodynamical approach, this equation is called the *Wheeler–DeWitt equation*, in honour of the physicists John Archibald Wheeler (1911–2008) and Bryce DeWitt (1923–2004), who first discussed this equation in detail. The loop approach goes mainly back to work by Abhay Ashtekar (*1949), Lee Smolin (*1955), and Carlo Rovelli (*1956).

    As can be recognized from the stationary form of equation (9), these theories are explicitly timeless, that is, devoid of any classical time parameter. They thus solve the 'problem of time' by getting rid of time at the fundamental level. This should happen in the other approaches, too, but the situation is there much less clear.

- *String theory*: This is the main approach to construct a unifying quantum framework of all interactions. The quantum aspect of the gravitational field only emerges in a certain limit in which the different interactions can be distinguished from each other. All particles have their origin in excitations of fundamental strings. The fundamental scale is given by the string length; it is supposed to be of the order of the Planck length, although the Planck length is here a derived quantity.

  String theory was originally developed as a theory of hadrons. While its unsuitability in this field became soon clear, it was later devised as a theory for the physics at the Planck scale. Among the pioneers who introduced string theory in the gravitational context are Joël Scherk and John Schwarz.

- Other attempts such as the quantization of topology or the theory of causal sets.

In perturbation theory, the important concept of the *graviton* emerges. In this approximation one decomposes the metric, $g_{\mu\nu}$, into a background part, $\bar{g}_{\mu\nu}$, and a 'small' perturbation, $f_{\mu\nu}$,

$$g_{\mu\nu} = \bar{g}_{\mu\nu} + \sqrt{\frac{32\pi G}{c^4}} f_{\mu\nu}. \tag{10}$$

Only the perturbation is being quantized. The important assumption is the presence of an (approximate) *background* with respect to which standard perturbation theory (formulation of Feynman rules, etc.) can be applied. In this approximate framework the quantum aspects of gravity are encoded in a spin-2 particle propagating on the background – the *graviton*, which arises from $f_{\mu\nu}$. The ensuing perturbation theory is, however, *nonrenormalizable*: at each order in the expansion with respect to G, new types of divergences occur which have to be absorbed into appropriate parameters that in turn have to be fixed by measurement. Nevertheless, one can derive in the low-energy limit concrete effects from perturbation theory. One is the quantum gravitational correction to the Newtonian potential between two masses m_1 and m_2,

$$V(r) = -\frac{Gm_1m_2}{r}\left(1 + 3\frac{G(m_1+m_2)}{2rc^2} + \frac{41}{10\pi}\frac{G\hbar}{r^2c^3}\right). \tag{11}$$

Another is the decay rate of excited states in atomic physics through emission of gravitons; for example, the decay rate in hydrogen from the $3d$ level to the ground state is

$$\Gamma_{\mathrm{g}} = \frac{Gm_{\mathrm{e}}^3 c\alpha^6}{360\hbar^2} \approx 5.7 \times 10^{-40}\ \mathrm{s}^{-1}, \tag{12}$$

where α is the fine-structure constant and m_{e} the electron mass. This corresponds to a life-time of

$$\tau_{\mathrm{g}} \approx 5.6 \times 10^{31}\ \text{years}, \tag{13}$$

which is too large to be measurable. The problem of nonrenormalizability in perturbation theory is avoided by string theory.

Quantum general relativity as well as string theory have found applications for quantum black holes and for quantum cosmology. Both approaches have, for restricted situations, proposed a microscopic explanation for the black-hole entropy (6). The corresponding microscopic states are either those of spin networks (in loop quantum gravity) or D-branes (in string theory). On the other hand, a clear picture of black-hole evaporation is elusive, although there is strong evidence in all approaches that there is no fundamental loss of information during this process. As for quantum cosmology, preliminary results exist for a wide range of topics: singularity avoidance, initial conditions, origin of structure, and the arrow of time. Direct effects may be seen in the anisotropy spectrum of the cosmic background radiation, but the situation is presently unclear. It should also be mentioned that both string theory and loop quantum gravity predict that space is discrete at very small scales (near the string length or the Planck length, respectively), with possible observational relevance.

A central issue is also the recovery of established physical theories as approximations from quantum gravity. Quantum geometrodynamics gives at least on the formal level a picture of how a semi-classical time parameter and the limit of quantum field theory in a background spacetime emerge as approximations (using a type of scheme similar to the Born–Oppenheimer approximation in molecular physics). This includes the classical behavior of spacetime due to decoherence (► decoherence, experimental observation of decoherence; time in quantum mechanics). The situation in loop quantum gravity is not yet fully clear. As for string theory, it has not yet succeeded to achieve one of its major goals – the recovery of the Standard Model of ► particle physics.

Primary Literature

1. L. Rosenfeld, *Über die Gravitationswirkungen des Lichtes*. Zeitschrift für Physik, **65**, 589–599 (1930).
2. M. Bronstein, *Quantentheorie schwacher Gravitationsfelder*. Physikalische Zeitschrift der Sowjetunion, **9**, 140–157 (1936).
3. B. S. DeWitt, *Quantum theory of gravity. I. The canonical theory*. Physical Review, **160**, 1113–1148 (1967).
4. B. S. Dewitt, *Quantum theory of gravity. II. The manifestly covariant theory*. Physical Review, **162**, 1195–1239 (1967).
5. B. S. Dewitt, *Quantum theory of gravity. III. Applications of the covariant theory*. Physical Review, **162**, 1239–1262 (1967).
6. J. Scherk and J. H. Schwarz, *Dual models for non-hadrons*. Nuclear Physics B, **81**, 118–144 (1974).
7. A. Ashtekar, *New variables for classical and quantum gravity*. Physical Review Letters, **57**, 2244–2247 (1986).

Q

General Introductions

8. S. Carlip, *Quantum gravity: a progress report*. Reports on Progress in Physics, **64**, 885–942 (2001). A related online version is http://arxiv.org/abs/gr-qc/0108040.
9. C. Kiefer, *Quantum Gravity. Second Edition* (Oxford University Press, Oxford 2007).
10. C. Kiefer, *Quantum gravity: general introduction and recent developments*. Annalen der Physik, **15**, 129–148 (2006). A related online version is http:arxiv.org/abs/gr-qc/0508120.

Special Topics

11. D. Coule, *Quantum cosmological models*. Classical and Quantum Gravity, **22**, R125–R166 (2005). A related online version is http://arxiv.org/abs/gr-qc/0412026.
12. C. Kiefer and C. Weber, *On the interaction of mesoscopic quantum systems with gravity*. Annalen der Physik, **14**, 253–278 (2005). A related online version is http://arxiv.org/abs/quant-ph/0408010.

13. C. Rovelli, *Quantum Gravity* (Cambridge University Press, Cambridge, 2004).
14. H. D. Zeh, *The Physical Basis of the Direction of Time*. Fifth Edition (Springer, Berlin, 2007). See also http://www.time-direction.de.
15. B. Zwiebach, *A First Course in String Theory* (Cambridge University Press, Cambridge, 2004).
16. Review articles by various authors in the online journal Living Reviews in Relativity, see http://relativity.livingreviews.org/

Collections of Essays on General and Special Topics

17. *Quantum Theory of Gravity*, edited by S. M. Christensen (Adam Hilger, Bristol, 1984).
18. *Quantum Concepts in Space and Time*, edited by R. Penrose and C. J. Isham (Clarendon Press, Oxford, 1986).
19. *Conceptual Problems of Quantum Gravity*, edited by A. Ashtekar and J. Stachel (Birkhäuser, Boston, 1991).
20. *Quantum Gravity: From Theory to Experimental Search*, edited by D. Giulini, C. Kiefer, and C. Lämmerzahl. Lecture Notes in Physics **631** (Springer, Berlin, 2003).
21. *The Structural Foundations of Quantum Gravity*, edited by D. Rickles, S. French, and J. T. Saatsi (Clarendon Press, Oxford, 2006).
22. *Quantum Gravity: Mathematical Models and Experimental Bounds*, edited by B. Fauser, J. Tolksdorf, and E. Zeidler (Birkhäuser, Basel, 2007).

Popular Books

23. J. B. Barbour, *The End of Time* (Oxford University Press, Oxford, 2000).
24. B. Greene, *The Elegant Universe* (Vintage, London, 2005).
25. L. Smolin, *Three Roads to Quantum Gravity* (Perseus Books Group, 2002).

Quantum Hall Effect*

Rolf R. Gerhardts, Jürgen Weis, and Klaus von Klitzing

In 1980, Klaus von Klitzing made the unexpected discovery that, at sufficiently low temperatures and high magnetic fields, the Hall resistance of a two-dimensional electron system assumes quantized values, which turned out to depend only on fundamental constants and integer numbers. For this discovery, which nowadays is used to reproduce the unit of the electrical resistance with an unprecedented accuracy, he was honored in 1985 with the Nobel Prize in Physics. A coherent explanation of the

* This contribution is based on an original German article by Klaus von Klitzing, Rolf Gerhardts, Juergen Weis, '25 Jahre Quanten Hall-Effekt', in Physik Journal (June 2005, pp. 37–44). It was translated by Rolf Gerhardts and is reprinted by permission of the authors and Physik Journal.

fact that, independently of the material and the exact geometry of the Hall sample, these quantized values can be reproduced with such high accuracy, has been found only in recent years.

The Phenomenon and its Discovery

The quantized Hall effect (QHE) was discovered early in February 1980, when Klaus von Klitzing performed a series of experiments at the high-field magnet-laboratories in Grenoble, France, in order to investigate the transport properties of silicon based metal-oxide-semiconductor field-effect-transistors (MOSFET's), which up to now form the basic building blocks of highest-integrated electrical circuits. The aim was to improve on the mobility of charge carriers in these devices. This requires to understand, which kind of scattering processes (caused by surface roughness, interface charges, impurities, etc.) has the strongest effect on the motion of the ► electrons in the thin conducting layer at the interface between silicon and silicon-oxide, which is only a few nanometers thick. To this end, G. Dorda (Siemens AG) and M. Pepper (Plessey Company) had provided specially prepared Si–MOSFET's (Fig. 1), which allowed for *four-point-measurements* on the conducting layer so that, in the presence of a perpendicular magnetic field, its usual (longitudinal) resistance $R_{xx} = U_x/I_x$ and its Hall resistance $R_{xy} = U_y/I_x$ could be determined independently. The electron density in the conducting layer could be changed by a suitable gate voltage. To suppress disturbing scattering processes due to the electron–phonon interaction, the experiment was carried out at low

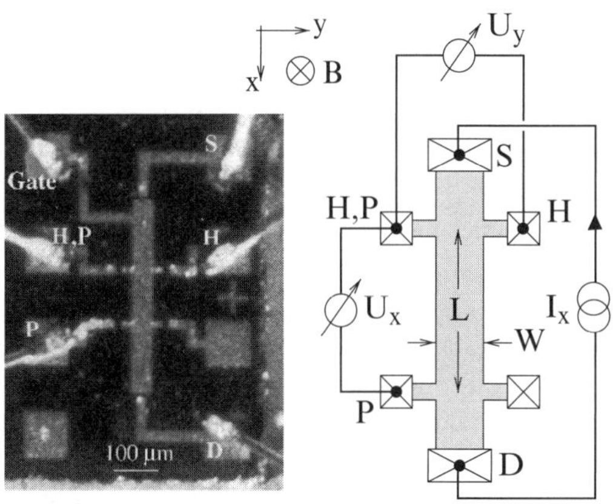

Fig. 1 Typical silicon-MOSFET for the measurement of the xx- and xy-components of the magnetoresistance tensor of the conducting layer underneath the gate. For a fixed current between the source (S) and drain (D) contacts, the potential differences between the contacts P–P and H–H are directly proportional to the resistances R_{xx} and R_{xy}, respectively. A positive gate voltage increases the charge carrier density underneath the gate

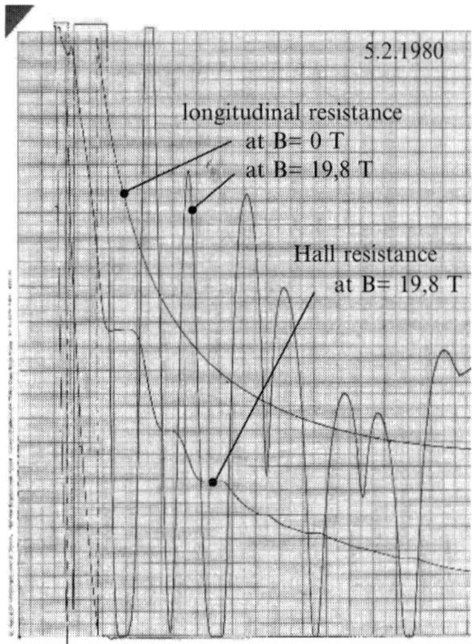

Fig. 2 The first experiment showing the quantized-Hall-effect, performed at liquid-He temperatures. Without magnetic field the electric resistance of the Si–MOSFET (*blue curve*) decreases monotonously with increasing gate voltage, since the electron density increases linearly with the gate voltage. At a magnetic field of 19.8 T, the Hall resistance (*black*) shows pronounced plateaus at values of the gate voltage, for which the longitudinal resistance (*red*) vanishes. The marker points to the quantized Hall plateau around filling factor $\nu = 4$

temperatures (typically 4.2 K) and at high magnetic fields (several Tesla). As function of the electron density (gate voltage) the Hall resistance R_{xy} showed plateaus while simultaneously the longitudinal resistance R_{xx} vanished (see Fig. 2). The important discovery was that the plateau values did not depend on any specific parameters of the experiment, not on source-drain or gate voltage, not on the magnetic field or any geometry factors, and that they can be written as $R_{xy} = h/(ie^2)$, where h is ▶ Planck's constant, e is the elementary charge, and $i = 1, 2, 3, \ldots$ is an integer [1].

There have been many attempts to understand this result, and it is instructive to compare it with the "classical" description of the Hall effect. It has been known since 1966 that the electrons, forced by a positive gate voltage towards the interface between the silicon crystal and an oxide layer, may form a two-dimensional electron system (2DES) [2]. For these electrons the energy of motion perpendicular to the interface is quantized and if, at sufficiently low temperature, only the lowest quantum state is partially occupied and separated from the next quantized state by an energy much larger than the thermal energy, then the motion perpendicular to the interface is frozen out and only the free motion parallel to the interface is possible. This is the situation of a 2DES. The knowledge about transport and optical properties of 2DES's at the time of the discovery of the QHE has been reviewed by Ando, Fowler, and Stern [24].

If one assumes that in the 2DES electric current density and field distribution are homogeneous between the voltage contacts, one obtains from the resistance values the components

$$\varrho_{xx} = R_{xx} W/L, \quad \varrho_{xy} = R_{xy} \tag{1}$$

of the magneto-resistivity tensor $\vec{\vec{\varrho}}$, which relates the local current density $\boldsymbol{j}$ in the 2DES and the local electric field $\boldsymbol{E}$ by $\boldsymbol{E} = \vec{\vec{\varrho}} \cdot \boldsymbol{j}$.

Classically, a high magnetic field $\vec{B} = (0, 0, B)$ perpendicular to an ideal, non-interacting 2DES forces the electrons to move uniformly on circular orbits (cyclotron motion). An additional homogeneous in-plane electric field $\boldsymbol{E} = (E_x, 0, 0)$ leads to a "Hall drift" with velocity $\vec{v}_D = \boldsymbol{E} \times \vec{B}/B^2 = (0, v_D, 0)$, where $v_D = -E_x/B$. Multiplying with the surface density n_s of the 2DES and with the electron charge, we obtain the Hall current density $\boldsymbol{j} = -en_s\vec{v}_D$. Thus, if there is no scattering of the electrons, the classical consideration yields $\varrho_{xx} = 0$ and $\varrho_{xy} = B/(en_s)$. This indicates already that under the condition of the QHE the conduction electrons move without being scattered.

Due to the Landau quantization, the periodic cyclotron motion is restricted to discrete energy values. In the ideal case then the energy spectrum of the 2DES consists of discrete energy levels with gaps, which are given by the cyclotron energy and the ► Zeeman spin-splitting, which both increase with increasing magnetic field. Also the degeneracy of the Landau levels increases with increasing magnetic field: the number of states per Landau level (and per spin direction) and per area is $n_L = B/\Phi_0$, with the magnetic flux density B (which usually is just called magnetic field) and the magnetic flux quantum $\Phi_0 = h/e$. In a homogeneous 2DES with area density n_s the filling factor of the discrete (Landau and spin) levels becomes $\nu = n_s/n_L = (h/e)n_s/B$. For an ideal 2DES without scattering, the calculation of the Hall resistivity is not affected by the Landau quantization, so that one obtains

Q

$$\varrho_{xy} = B/(en_s) = h/(\nu e^2). \tag{2}$$

Thus, the plateau values of the QHE correspond to integer values of the filling factor, $\nu = i$, i.e. to a situation in which a certain number of the discrete but macroscopically degenerate energy levels is completely occupied, while all other levels are empty. In this situation the occupied states are separated from the empty states by a finite energy gap, which at sufficiently low temperatures can not be bridged by thermal excitations, i.e. no (quasi-elastic) scattering, and, as a consequence, no damping or dissipation processes are possible in the 2DES.

Surprisingly these "integer-quantized" values $\varrho_{xy} = h/(ie^2)$ are observed as values of the global Hall resistance $R_\mathrm{H} = R_{xy}$ not only for the discrete values of the ratio n_s/B, which correspond to an integer filling factor of the 2DES, but in wide intervals around these values, provided the temperature is low enough. Figure 2 shows the experimental curves with the characteristic plateaus in the Hall resistance R_{xy} and the corresponding zeroes in the longitudinal resistance R_{xx}, which revealed the quantized Hall effect [1].

The fact, that the values of the resistances R_{xx} and R_{xy} are unchanged over finite intervals of the gate voltage (in the plateau regions), led to the presumption, that there the electrostatically induced electrons occupy "localized states", which do not contribute to the electronic transport. A large amount of work about localization and other "reservoir models", which assume that a part of the induced electrons does not participate in the electronic transport, has been published in the past [25]. But these are interpretations of the QHE, which usually rely on additional assumptions, e.g. that the 2DES is essentially homogeneous and can be described by a position-independent resistivity tensor. If this were correct, one should expect that edge effects lead to a dependence of the measured resistance values on the sample width. This is, however, not the case.

The plateau values of the Hall resistance do not depend on the presence or absence of such localized states, and are with very high accuracy given by the relation $R_{xy} = h/(ie^2)$ $(i = 1, 2, 3, \ldots)$. They are independent of details of the experimental setup, and especially of geometrical details. Since in the plateau regime the longitudinal resistance vanishes, $R_{xx} = 0$, the exactly quantized value is obtained for the Hall resistance, even if the Hall voltage is measured between contacts on both sides of the sample, which are not located exactly opposite to each other (see Fig. 1). The plateau values are even independent of the semiconductor material, which contains the 2DES. For instance, in GaAs/(AlGa)As hetero-structures, where the 2DES occupies states which result from the conduction-band minimum of GaAs near the Γ-point, and each Landau level splits into two levels with opposite ► spin, one observes the same values as in Si-MOSFET's, where each Landau level splits into four states, since in addition to the spin-splitting one has a valley-splitting. [Whereas isotropic, unstressed silicon has six equivalent, degenerate conduction-band minima, only two of them (those with heavy effective masses in the direction perpendicular to the Si/SiO_2 interface) contribute to the bound states occupied by the 2DES, and the degeneracy of their energy levels is lifted, since the interface destroys the inversion symmetry.] This lifted fourfold degeneracy of the Landau levels has been identified in the experiment [1] shown in Fig. 2.

Nowadays the QHE discovered by K. von Klitzing in 1980 is usually called the "Integer Quantum Hall Effect" (IQHE), in order to distinguish it from the "Fractional Quantum Hall Effect" (FQHE), which was discovered in 1982 on high-mobility GaAs/Al_xGa_{1-x}As hetero-structures and shows plateaus of the Hall resistance with values $R_H = h/(fe^2)$, where f is a fraction of simple integer numbers with odd-integer denominator [3]. The most prominent examples are $f = 1/3$ and $f = 2/3$, but many others have been reported, too. [The high mobility was achieved by "modulation doping", a method which separates the donors, needed to provide the electrons for the 2DES, by a spacer from the 2DES, in order to reduce the scattering of the electrons by the ionized-donor potentials.] The FQHE was again an unexpected discovery. Whereas the IQHE was believed to be a single-particle effect, for which the mutual Coulomb interactions between the electrons of the 2DES are unimportant, the FQHE was attributed to such interactions, which may at fractional filling of the Landau levels lead to collective ground-states with strong correlations. For simple fractions such correlated ground-states have been calculated [4] soon after the discovery of the FQHE. In 1998 Dan C. Tsui, Horst L. Störmer and Robert B. Laughlin were awarded the Nobel Prize for the discovery of the FQHE and its explanation.

In the subsequent years the number of publications containing the keyword "quantum Hall effect" in the title or the abstract increased drastically, to about one publication per day at present. In the meantime, the QHE is discussed not only within solid state physics, but also in nearly all other areas of modern physics. The spectrum of published papers extends from "Quantum Computing" in quantum-Hall-systems to "Quantum-Hall-Quarks", and even to a higher-dimensional QHE in string-theory. Up to now more than ten books have been published on the Quantum-Hall-Effect [26].

Quantized Hall Effect and Metrology

The most important equation in connection with the quantized Hall resistance, $U_{\mathrm{H}} = (h/ie^2) \cdot I$, was confirmed in the first experiment with such a high accuracy, that even the finite input impedance (1 MΩ) of the x-y-recorder, used for the voltage measurement, had to be taken into account as a correction. An accurately reproducible electric resistance, independent of the geometry of the device and of microscopic details of its material, was, of course, of great importance for metrological institutes as a new and universal resistance standard. Therefore, this new quantum phenomenon (the occurrence of Planck's constant h makes this obvious) was submitted for publication under the title "Realization of a Resistance Standard based on Fundamental Constants". At that time, however, it seemed more important to improve the value of Sommerfeld's fine-structure constant α, given by $\alpha^{-1} = (h/e^2)(2/\mu_0 c) = 137.036\ldots$, where the magnetic field constant $\mu_0 = 4\pi\, 10^{-7}$ N/A^2 and the velocity of light in vacuum, $c = 299\,792\,458$ m/s, had and have today fixed values. Therefore the publication appeared under the title "New Method for High-Accuracy Determination of the ► Fine-Structure Constant based on Quantized Hall Resistance" [1].

In the meantime the importance of the QHE as the basis of a resistance standard has been fully appreciated [27]. Its applicability relies on the facts, that the plateaus measured (at fixed magnetic field) as a function of the electron density (see Fig. 2), or (at fixed electron density) as a function of the magnetic field (see Fig. 3), are extremely flat, and that the quantized Hall resistance (around filling factor $\nu = 1$) apparently has always the fundamental value $h/e^2 = 25\,812.807\ldots\,\Omega$. After the discovery of this macroscopic quantum effect, the experiment has been repeated in many metrological institutes with much higher accuracy as can be achieved in a research lab. The effect proved to be extremely stable and reproducible. Obviously the remaining inaccuracy of resistance measurements results mainly from the uncertainty in the reproduction of the SI ohm. Due to the internationally accepted definitions of the fundamental SI units second (s), meter (m), kilogram (kg), and ampere (A), all mechanical and electric quantities are well defined.

Q

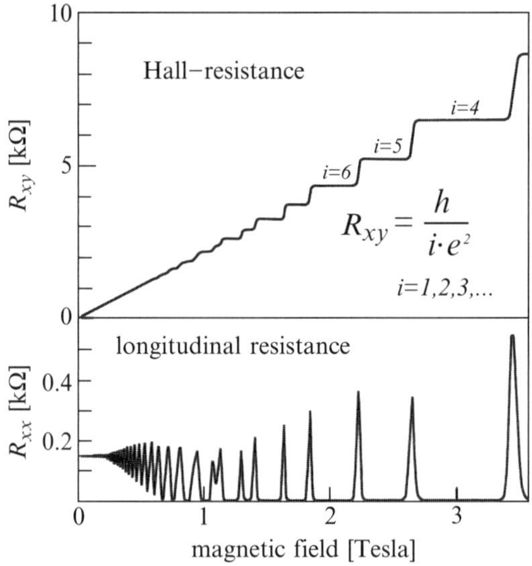

Fig. 3 Typical traces of the Hall resistance R_{xy} and of the longitudinal resistance R_{xx} of a 2DES as measured as a function of the magnetic field B on samples of average quality. The zeros of R_{xx} coincide with the plateaus of R_{xy} at the quantized values $h/(i \cdot e^2)$. At small B-values one observes the classical Drude behavior: $R_{xy} \propto B$, R_{xx} = constant

However, the fundamental unit ampere can be reproduced only with a relatively large error of the order of 10^{-6}, if it is calculated, according to its definition, from the force between two current-carrying wires. As a consequence, the derived unit $1\,\Omega = 1\,\mathrm{s}^{-3}\,\mathrm{m}^2\,\mathrm{kg}\,\mathrm{A}^{-2}$, which depends on all fundamental units, is available only with an error, which is even larger than 10^{-6}.

Nowadays the SI unit ohm is known with a smaller error than the fundamental unit ampere, because a resistance can be realized as the ac-impedance $|Z(\omega)| = 1/(\omega C)$ of a capacitance C. Since the capacitance C of a capacitor depends only on its geometry (with vacuum as dielectric medium), the SI-ohm can be realized by using only the fundamental units of time (to measure the frequency $\omega/2\pi$) and of length (to calculate C for a so called calculable Thompson–Lampard-capacitor [5]). As these units are known with very high accuracy, also the SI-unit ohm can be realized with an error as low as 10^{-7}. Using this and the QHE, one can obtain the fine-structure constant with the same accuracy.

The quantized Hall resistance is, however, more stable and better reproducible than any resistance that has been calibrated in SI-units. Therefore, the Comité Consultatif d'Electricité suggested to take as value of the von-Klitzing constant $R_K = h/e^2$ exactly 25 812.807 Ω, with the notation R_{K-90}. This value $R_{K-90} =$ 25 812.807 Ω has been accepted since 1. 1. 1990 as the reference value for resistance calibrations, and is now denoted as *conventional* von-Klitzing constant [6]. Direct comparisons by different national institutes showed that the reference values deviated [7] by less than $2 \cdot 10^{-9}$, provided the published rules for reliable measurements had been obeyed [8]. Unfortunately this high reproducibility and stability of the quantized Hall resistance can not be used immediately for a correspondingly accurate determination of the fine-structure constant, since the value of the quantized Hall resistance in SI-units is not known so accurately. Only in connection with other experiments, e.g. high-precision measurements (and calculations) of the anomalous magnetic moment of the electron, of the gyro-magnetic ratio of the proton, or of the neutron mass, does one obtain a best fit for the value of the fine-structure constant with an error of only $3.3 \cdot 10^{-9}$. This leads to a value $R_K = (25812.807449 \pm 0.00086)\,\Omega$ for the von-Klitzing constant (CODATA 2002) [27]. Very accurate values of fundamental constants (especially of α) are important in view of speculations about a possible time-dependence of some of the fundamental constants. Experimental indications of a cosmic evolution of the fine-structure constant α are under dispute, but could not be confirmed till now. The rate of change $|\partial\alpha/\partial t|$ is – if non-zero at all – less than 10^{-16} per year.

A combination of quantized Hall effect and Josephson-effect (which allows to express the electric voltage in units of h/e) makes it possible to relate the electric power (which depends on Planck's constant h) with the mechanical power (which depends on the mass m). Measurements with a so called Watt balance yield the best value for Planck's constant [9], provided the mass is accurately known on the basis of the "International Prototype Kilogram" (which is not stable in time). Alternatively, one could fix the value of Planck's constant and thereby obtain a new realization of the unit of mass (just as the fixing of the velocity of light led to a new realization of the unit of length). At present suggestions are under discussion, to fix exactly not only the Planck constant h, but also the elementary charge e (and, thereby, the quantized Hall resistance). This would replace and allow to abandon the definitions of the basic units "kilogram" and "ampere", which have been valid up to now, but are unstable in time (kg) and hard to realize with satisfactory precision (A).

Physics of the Integer-Quantized Hall Effect

Bulk Effects and Edge States

A particle with electric charge q ($q = -e$ for electrons), moving with velocity $\vec{v}$ in a homogeneous magnetic field $\vec{B}$, is subjected to the Lorentz force $\vec{F} = q(\vec{v} \times \vec{B})$, perpendicular to both $\vec{v}$ and $\vec{B}$. In a current-carrying, three-dimensional, laterally confined conducting layer of thickness d in a perpendicular magnetic field B this leads to charge accumulation and depletion at opposite lateral boundaries and, thereby, to a Hall voltage $U_H = R_H(B) \cdot I$ (named after Edwin Hall, who described this effect in 1879 for the first time). Within the Drude model, which describes the charge carriers as a classical gas, the Hall resistance is given by

$$R_H = -B/(qn_q d), \tag{3}$$

and reveals important properties of the conductor: the density n_q of free charge carriers and the sign of their charge ($q = +e$ for holes and $q = -e$ for electrons).

For a two-dimensional electron system the product $n_q d = n_s$ reduces to the area density and the Hall resistance simplifies to $R_{\mathrm{H}} = B/(en_s)$. Indeed the Hall resistance increases at small magnetic fields linearly with increasing B (see Fig. 3), and its slope allows to determine the area density n_s of the 2DES. Only at relatively high magnetic fields do the plateaus with the quantized values $R_{\mathrm{H}} = h/(\nu e^2)$ occur, where ν equals an integer number $i = 1, 2, 3, \ldots$ (or a fraction in the case of the FQHE). Here we see a fundamental difference between the Hall resistance at low and at high magnetic fields: while at low B-values R_{H} depends on material parameters like electron density n_s, the quantized plateau values at high B-values are absolutely independent of material properties.

A corresponding behavior is observed for the longitudinal resistivity. The classical Drude theory yields the B-independent value $\varrho_{xx} = m^\star/(e^2 n_s \tau)$, which depends, in addition to the electron density n_s, on the effective mass $m^\star$ of the electrons and the momentum relaxation time τ, which describes the scattering of electrons, at low temperatures predominantly by randomly distributed impurities. Indeed a B-independent resistance R_{xx} is observed in the experiment at low B-values (see Fig. 3). At somewhat higher B-values Shubnikov-de Haas (SdH) oscillations occur, with an amplitude, which increases with increasing B until the minima of the SdH oscillations reach the value zero. At still higher B-values the QHE sets in, and the plateau values of R_{xy} are accompanied by vanishing R_{xx}, which no longer contains information about the material parameters $m^\star$, n_s, and τ.

The vanishing of R_{xx} in the plateau regimes of the QHE means that the occurrence of the quantized Hall plateaus is accompanied by a dissipationless current flow along the Hall bar. This does, however, not mean that there is no dissipation at all in the system. In fact the two-point resistance, which is measured by the voltage-drop between the current-carrying contacts S (source) and D (drain), equals (in the regime of the QHE) the Hall resistance, i.e. the electric power $R_{\mathrm{H}} I^2$ is dissipated. This Joule heat is produced at opposite corners of the sample near the current-carrying contacts, as could be visualized by means of the fountain effect with liquid helium [10].

The question remains, how can we understand the occurrence of plateaus in the Hall resistance with the quantized values, and the simultaneous disappearance of dissipation in the bulk of the sample? In the following we will concentrate on the case of GaAs-based heterostructures, to avoid complications due to the multivalley conduction-band-structure of silicon. A homogeneous magnetic field B in z-direction, perpendicular to the plane of the 2DES, leads to Landau quantization of the cyclotron motion, so that in the ideal case (neglecting collision broadening effects due to scattering) the electrons occupy Landau levels at discrete energy eigenvalues

$$\varepsilon_{n,\pm} = (n + 1/2)\hbar\omega_c \pm (g^\star/2)\mu_B B, \tag{4}$$

with the cyclotron energy $\hbar\omega_c = \hbar e B/m^\star$, the Landau level index $n = 0, 1, 2, , \ldots,$ the spin ► quantum numbers ± 1, the Bohr magneton $\mu_B = e\hbar/(2m_e)$, and the effective Landé factor $g^\star$. Each of these energy levels is macroscopically degenerate, with $n_L = eB/h$ states per unit area. Since the ► degeneracy of the levels, as well as their distance, increases with increasing B, at constant density the electrons will, with increasing B, be redistributed to lower Landau levels. This leads, as a function of B, to a saw-tooth-like shape of the Fermi energy (i.e., at finite temperature the chemical potential μ_{ch}), which follows in the ideal case the energy of a partly occupied Landau level until this is totally depleted, then it jumps to the next lower level and follows this with increasing B, and so on. If scattering and level broadening effects, and also finite temperature, are taken into account, the increasing parts of the function $\mu_{\text{ch}}(B)$ are no longer strictly linear, but qualitatively the saw-tooth-behavior survives as long as the energy gaps between the collision-broadened Landau levels are much larger than the thermal energy $k_{\text{B}}T$, as could be confirmed experimentally by employing a metallic single-electron transistor as local electrometer [11].

The chemical potential jumps from one Landau level to an adjacent one exactly at those values of the magnetic field B, at which the filling factor $\nu = n_s/n_L$ assumes integer values, $\nu = i$. At these values the occupied states are separated from the empty states by an energy gap, which is much larger than $k_{\text{B}}T$, so that, according to the Pauli principle, no scattering processes are possible. In this situation any reasonable quantum theory of magneto-transport in a 2DES yields [24]

$$R_{xx} = 0, \quad R_{xy} = h/(ie^2), \tag{5}$$

i.e. the values of the free 2DES without any interactions. Does this (trivial) result explain the IQHE? Certainly not! So far we have tacitly assumed a homogeneous 2DES, and then the result (5) applies only to isolated values of the magnetic field. The problem is to understand, why it applies with extreme accuracy to B-intervals of finite width, the plateaus.

Q

Theories, which try to explain the QHE as property of the resistivity tensor of an, on the average, homogenous (and infinite) sample (e.g. localization theories) have already been mentioned. If, at a certain density, such a theory would yield the result (5) in a certain B-interval, application to a sample of finite width W must take into account that this result can not be valid in a depletion region (typically of width $\delta \gtrsim 100$ nm) near the sample edges, where the electron density drops to zero. Then one has to expect to measure deviations from the quantized values, which are of the order δ/W. This is for realistic values of W ($\lesssim$1 mm) many orders of magnitude larger than the accuracy with which the quantized values can be reproduced in experiments. In addition to such theoretical arguments, there are many experimental hints, that the assumption of a homogeneous sample is neither correct nor important for the explanation of the QHE.

There are many experimental indications that, in the plateau regime of the IQHE, the interior of the sample is not important: it can be partly removed or, by suitable gates, tuned to another electron density, without changing the quantized value of the measured Hall-resistance. Also the exact arrangement of the contacts plays a minor role (see Fig. 4). This has been interpreted as indication that the relevant currents

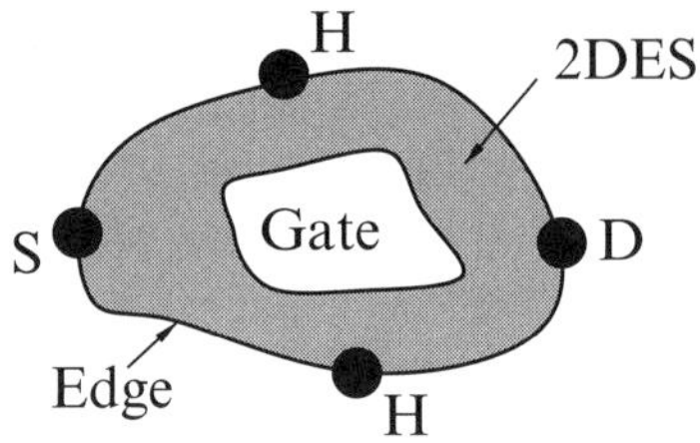

Fig. 4 The measured value of the quantized Hall resistance does not depend on variations of the electron density in the interior of the sample, e.g. depletion or accumulation by a gate (provided the gate does not reach from one edge to the opposite one). Even etching a hole through the sample has no effect. Also the precise position of the voltage contacts (H) is irrelevant, provided the current carrying contacts S and D are located between them

flow near the sample edges. The edge channel picture of the IQHE, elaborated by M. Büttiker [12] since 1988, proved to be very successful for the description of the resistances measured on complicated samples with many gates and contacts, and found its way into textbooks [28]. We will, however, focus on a somewhat different microscopic picture of the IQHE, which evolved from more recent theoretical and experimental investigations of the position dependence of electron and current density in (narrow) Hall bars.

Compressible and Incompressible Regions

As already mentioned, for a long time it was general belief that Coulomb interactions were unimportant for the understanding of the IQHE. However in 1992, D.B. Chklovskii, B.I. Shklovskii, and L.I. Glazman [13] pointed out that, in a real 2DES with lateral confinement, in which the electron density decreases from a finite bulk value to zero near the edges, electronic screening effects become extremely important in high magnetic fields, where the magnetic length $\ell = (\hbar/eB)^{1/2} = (10\,\mathrm{T}/B)^{1/2} \cdot 8.11\,\mathrm{nm}$ is much smaller than the lengths scale, on which electron density and confinement potential vary. About a decade later it turned out that immediate consequences of these screening effects can be measured experimentally, and open a new approach to the understanding of the QHE.

If one neglects screening effects under these conditions, the Landau bands show a spatial dispersion given by the external confinement potential, bending upwards near the edges. If in the bulk of the sample several Landau levels are occupied, the density profile drops like a step-function towards zero at the edges, with wide plateaus (given by the separation of adjacent Landau bands at the Fermi level), which correspond to the integer filling factor of the occupied bands and are separated by steep steps of a width given by the extent of the Landau wavefunctions, which is of the order ℓ. This is unrealistic, since this electron density profile would change strongly with changing magnetic field, which would cost a lot of Coulomb energy.

In the idealized case of small collision broadening and low temperature, the (thermodynamic) density of states $(\partial n_s/\partial \mu_{\mathrm{ch}})$ is extremely high, if the chemical potential falls onto a Landau energy, and is nearly zero, if it falls into a gap between such energies. In the first case screening is nearly perfect, since it costs no energy to change the position of electrons. In the second case no screening is possible since occupied and empty electron states are separated by the large (as compared with $k_B T$) energy gap. In an inhomogeneous sample with sufficiently high bulk density one meets both situations. There are "compressible" regions in which screening is nearly perfect, so that the total, screened potential (i.e. the sum of the external confinement potential and the Hartree potential produced by the spatial distribution of the 2DES) is flat and one of the Landau energy levels is "pinned" (within $k_{\mathrm{B}} T$) to the Fermi level (i.e. the electrochemical potential $\mu^\star_{\mathrm{ch}}$, which is constant, if the inhomogeneous 2DES is in thermal equilibrium). In addition there are "incompressible" regions, in which $\mu^\star_{\mathrm{ch}}$ falls between adjacent Landau bands, so that there no redistribution of electrons is possible and the density is constant, since the filling factor of the Landau levels there has a fixed integer value.

In the case of idealized Hall bars (translation invariance in one direction) these regions become parallel stripes. Compressible stripes, in which adjacent Landau bands are pinned to $\mu^\star_{\mathrm{ch}}$, are usually separated by an incompressible stripe across which the total potential varies by the amount of the energy difference between these two bands. Chklovskii *et al.* [13] have evaluated these ideas for a 2DES in a half-plane geometry for the idealized case of zero level broadening and zero temperature, and under some simplifying assumptions (only in-plane charges, perfect screening where $n_s(y) > 0$), which allowed analytical calculations. For instance, for a 2DES with bulk filling factor ν and metal gate at $y < y_{\mathrm{edge}}$ the distance $y_\nu = |y - y_{\mathrm{edge}}|$ of the (center of the) innermost incompressible stripe with filling factor $\mathrm{int}(\nu)$ and its width a_ν are given by

$$y_\nu = \frac{d_0}{1 - [\mathrm{int}(\nu)/\nu]^2}, \qquad a_\nu = \frac{4 y_\nu}{\nu} \sqrt{\frac{\mathrm{int}(\nu) a^\star_{\mathrm{B}}}{\pi d_0}}, \tag{6}$$

where the length d_0 depends on the average electron density, $a^\star_{\mathrm{B}}$ is the effective Bohr radius, and $\mathrm{int}(\nu)$ is the integer part of ν. Experiments using single-electron transistors as electrometer succeeded to make this stripe-structure in the depletion regime of a 2DES visible [14, 15]. A schematic plot of such a stripe-structure is shown in Fig. 5.

These calculations have soon been applied to a simplified Hall-bar geometry [16] and generalized to a self-consistent thermodynamic equilibrium theory [17, 18], in which the screened potential is calculated from the electron density by solution of Poisson's equation and the electron density is calculated from the total effective potential $V(y)$ in a Hartree-type approximation,

$$n(y) = \sum_n^\infty \sum_{s=\pm} \int \frac{\mathrm{d}Y}{2\pi \ell^2} |\psi^{(s)}_{n,Y}(y)|^2 f\big(\varepsilon_{n,s}(Y) - \mu^\star_{\mathrm{ch}}\big). \tag{7}$$

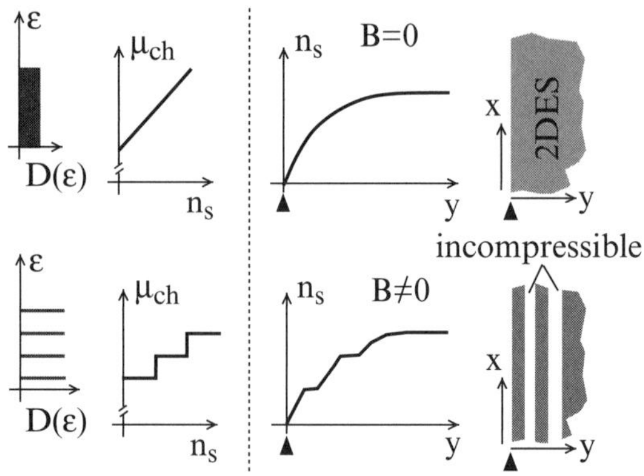

Fig. 5 Sketch of the density of states $D(\varepsilon)$ without and with applied magnetic field B, and of the resulting relation between chemical potential and electron density, $\mu_{\text{ch}}(n_s)$, for a homogeneous 2DES (*left half of the figure*). Also shown are sketches of the density profile near the sample edge and, on the right side, the compressible regions (with states near the Fermi level) are indicated as *dark*, incompressible regions as *white stripes*

Here $\varepsilon_{n,s}(Y)$ are the energy eigenvalues with normalized wavefunctions $\Psi(x, y) = L_x^{-1/2} \exp(\mathrm{i}xk_x)\psi_{n,Y}^{(s)}(y)$ and $Y = \ell^2 k_x$, $f(E) = 1/[1 + \exp(E/k_\mathrm{B}T)]$ is the Fermi-Dirac distribution function, and $\mu_{\text{ch}}^\star$ is the electrochemical potential, which is constant in thermodynamic equilibrium. If the potential $V(y)$ varies slowly on the scale ℓ, one may neglect the spatial extent of the wavefunctions, $|\psi_{n,Y}^{(s)}(y)|^2 \approx \delta(y - Y)$, and replace the energy eigenvalues by $\varepsilon_{n,\pm}(Y) = \varepsilon_{n,\pm} + V_{\text{conf}}(Y)$, where $\varepsilon_{n,\pm}$ are the energy eigenvalues (4) of the homogeneous system without confinement potential. This leads to the often used Thomas-Fermi approximation

$$n(y) = \int \mathrm{d}\varepsilon\, D(\varepsilon) f\big(\varepsilon - \mu_{\text{ch}}(y)\big), \tag{8}$$

where $D(\varepsilon)$ is the density of states (DOS) of the homogeneous 2DES (here the Landau DOS), and $\mu_{\text{ch}}(y) = \mu_{\text{ch}}^\star - V(y)$ is the position-dependent chemical potential. This approach allowed to demonstrate how the incompressible stripes evolve with decreasing temperature [17, 18]. The possible relevance of the incompressible stripes for the QHE was, however, still not clear.

A major breakthrough was achieved when a low-temperature scanning-force-microscope was developed and employed to measure Hall-potential-profiles, i.e. the change of the potential landscape due to a fed-in current I_x, as compared to the thermodynamic equilibrium state ($I_x = 0$) [19, 20]. Figure 6 shows typical Hall-potential-profiles measured on narrow samples of 10 to 15 μm width (because of the restricted scanning area) for different magnetic fields (i.e. average filling factors) in the regime of a quantized Hall plateau (QHP). The profiles show very different position dependences: (a) For B-values well above a QHP the Hall potential drops linearly across the whole sample, i.e. the Hall electric field is constant and

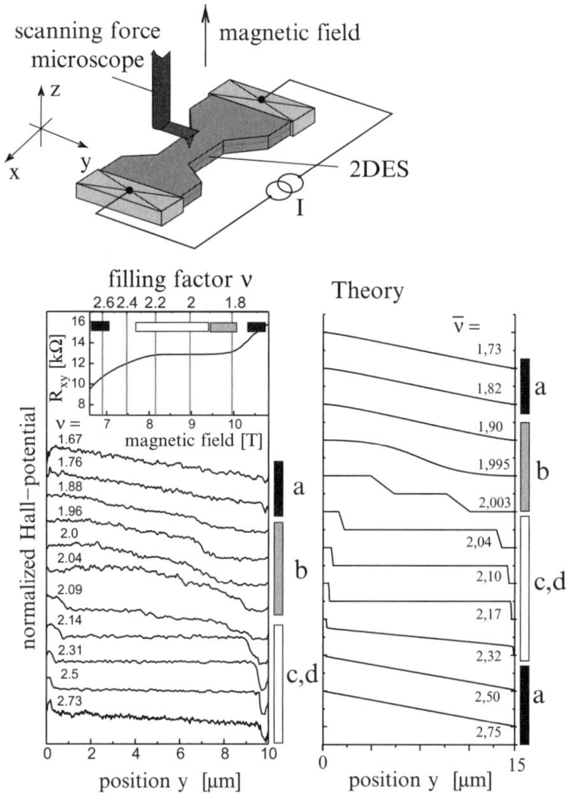

Fig. 6 Hall-potential profiles for several magnetic fields (characterized by filling factors ν), which were measured across the narrow part of the sample sketched in the *upper left part* of the figure. For comparison, the Hall resistance R_{xy} in this magnetic field region is also shown, after [19, 20]. The ν-dependent characteristics of the potential profiles are the following. Type (a): linear potential variation; type (b): non-linear drop in the center, very close to integer filling factor; (c): potential drop only cross incompressible stripes, constant Hall potential in the interior; (d): partial drop near the edges and linear variation in the interior of the sample. *Right panel*: calculated Hall-potential profiles for an idealized 15 μm wide sample (low current, linear response) after [21]

the current is spread over the whole sample (as one would expect from the Drude theory). (b) As the magnetic field is lowered and enters the upper edge of the QHP, the Hall potential drops in a non-linear (and sometimes even non-monotonous) manner in the center of the sample. Although in this region extremely small changes of the magnetic field may lead to considerable changes of the potential profile, one measures the quantized value for the Hall resistance. (c) At lower B-values well inside the QHP the Hall potential is constant in the center of the sample and drops only across two stripes, which move with decreasing B towards the sample edges and become narrower. The current now flows exclusively through these stripes. (d) For B slightly below the lower edge of the QHP still some fraction of the Hall potential drops near the edges, but a linear variation in the center region sets in. With

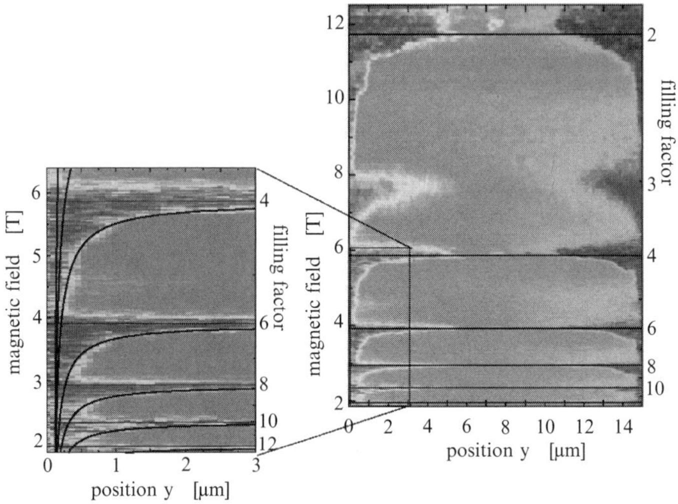

Fig. 7 Color-coded plot of Hall-potential profiles, measured for a large interval of magnetic fields, which covers several quantum-Hall plateaus. As can be seen in the magnification, the Hall voltage drops at the positions of the innermost incompressible stripes, described by (6), after [19,20]

further decreasing B, the fraction dropping near the edges decreases to zero and the linear behavior (a) is recovered, until the upper edge of the QHP with the next higher integer filling factor is reached and behavior (b) sets in. This kind of behavior repeats itself for each QHP, as is shown in Fig. 7. [19,20] But what is the reason for these different types of Hall-potential-profiles?

The position and the B-dependence of the stripes observed in the case (c) seemed to be in good agreement with the position and B-dependence predicted for the incompressible stripes. This motivated model calculations of the current distribution in narrow Hall-bars under high magnetic fields [21, 22]. An external non-equilibrium current $I_x = \int \mathrm{d}y j_x(y)$ was applied to the idealized Hall-bar (with translation invariance in x-direction). The resulting current density $\boldsymbol{j}$ and the "driving electric field" $\boldsymbol{E} = \nabla \mu^{\star}_{\mathrm{ch}}(x, y)$ were assumed to satisfy a local ohmic relation $\boldsymbol{E}(y) = \vec{\vec{\varrho}}(y) \cdot \boldsymbol{j}(y)$. The local resistivity tensor $\vec{\vec{\varrho}}(y) = [\vec{\vec{\sigma}}(y)]^{-1}$ was taken from a calculation of the conductivity tensor for a homogeneous 2DES by replacing its filling factor ν by the local value $\nu(y)$. The feedback of the applied current on the selfconsistent electrostatic potential, which is measured in experiment, was calculated under the assumption of local equilibrium [22]. These calculations, and a critical examination of the validity of the Thomas-Fermi approximation [21], lead to the following picture for an idealized Hall-bar with translation invariant, symmetric external confinement potential $V_{\mathrm{conf}}(y) = V_{\mathrm{conf}}(-y)$.

At high temperatures ($k_{\mathrm{B}}T \gtrsim 0.3\,\hbar\omega_c$) magnetic quantum effects are smeared out and the Drude theory holds: the current is distributed over the whole sample, the Hall electric field is constant, i.e. the Hall potential varies linearly across the sample. At low temperatures ($k_{\mathrm{B}}T \lesssim 0.01\,\hbar\omega_c$) a strong dependence on the magnetic field B is found. If B is so high, that everywhere in the sample the filling factor is less than one, $\nu(y) < 1$, the 2DES is compressible, the current is distributed over the whole sample and the Hall potential varies linearly across the sample (very similar to the result of the Drude theory). As the magnetic field is lowered, the filling factor $\nu(y) = 1$ is reached for a value $B = B_1$ in the center $y = 0$ of the sample. For $B \lesssim B_1$ an incompressible stripe (IS) evolves in the center, which broadens rapidly with decreasing B until it splits into two stripes, since a compressible stripe occurs in the center. With further decreasing B the two IS's move towards the sample edges and become narrower. Their position and width is reasonably well approximated by the analytical expressions (6) as long as they are sufficiently wide. But at a B-value $\hat{B}_1$ the width of the IS's becomes zero, and in the interval $\hat{B}_1 > B > B_2$ no IS's exist, where B_2 is the B-value at which an IS with local filling factor $\nu(y) = 2$ evolves in the middle of the sample. At still lower B-values, this IS broadens, then splits into two IS's, which shrink while moving towards the edges and vanish, before an IS with the next integer value of the filling factor occurs, and so on. At sufficiently high temperature T, the longitudinal resistivity, and as a consequence the current density, is finite everywhere in the 2DES. As T is lowered, at positions with integer values of the filling factor the longitudinal resistivity becomes small while the current density becomes large. This is the situation (d) observed in the experiment: partial drop of the Hall voltage near the edges and linear variation in the center. If, at sufficiently low temperature, IS's with integer filling factor evolve, the total applied current flows in these stripes, so that the longitudinal resistance R_{xx} of the sample vanishes. Since only IS's with the same value of the local filling factor exist, the Hall resistance R_{xy} assumes the quantized value, with an error which becomes exponentially small in the limit of low temperature [21].

This leads to a simple and consistent interpretation of the experimental results [19, 20] on narrow etched Hall-bars, if one takes into account that the donor distribution, and as a consequence the confinement potential, will exhibit fluctuations in both spatial directions [29]. Due to such fluctuations, the IS's will no longer be parallel to the sample edges. They may be bended and their width may fluctuate. If one starts with situation (a) of Fig. 6 and lowers B, the upper edge of the QHP corresponding to filling factor $\nu = i$ will be reached if a percolating IS with this filling factor occurs between source and drain contact, which is not necessarily at $B = B_i$. Whereas for the idealized case at $B \lesssim B_i$ a broad IS is calculated, in reality the corresponding incompressible region may contain compressible islands. These islands will have a large effect on the effective potential in their immediate surroundings, but they will not affect the measured value of the quantized Hall resistance. The effect of such compressible islands is indicated schematically in Fig. 8.

The model calculations also revealed some other features, which are confirmed by the experiments on narrow samples. Since the high-B edge of a QHP is determined by a wide incompressible region in the center of the sample, while the low-B edge is determined by narrow incompressible regions near the sample edges, the latter are much more sensitive to perturbations. For example, with increasing temperature the QHP's melt from the low-B edge, while they are much more stable at

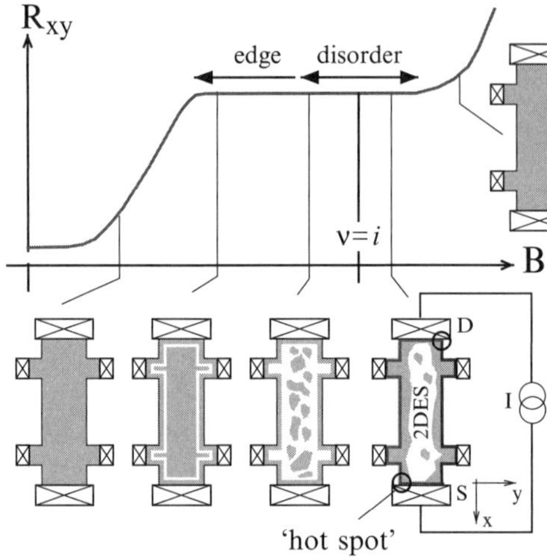

Fig. 8 Schematic sketch of the development of the compressible (*grey*) and of the innermost incompressible (*white*) regions in a real, inhomogeneous 2DES during a sweep of the magnetic field over a quantized Hall plateau. Around the integer value of the (average) filling factor (close to the high-B edge in narrow samples) the plateau is stabilized by disorder and inhomogeneities, near the low-B edge it is stabilized by incompressible stripes near the sample edges

the high-B edge. Increasing the applied current beyond the linear response regime leads to asymmetry of the two incompressible stripes [22], which depends on the current direction. A corresponding asymmetry can be seen in the experimental voltage curves.

Summary

In the nearly three decades since its discovery many models have been developed to explain the quantized Hall effect. The focus was put either on the sample edge or on the bulk, but both regions are of importance for the QHE. As an electrochemical potential difference is applied between source (S) and drain (D) contact (i.e. in x-direction), the potential of S is carried by a compressible region along one edge, the potential of D is transferred by a compressible region along the other edge. The electrochemical potential difference acts thus as Hall voltage across the Hall bar (y-direction). If the interior of the sample between S and D consists of a connected incompressible region with integer filling factor i, maybe interrupted by local islands with another filling factor, the electric field $E_y(y)$ resulting from the Hall voltage drives the Hall current dissipationless (perpendicular to the electric field) through the incompressible region in the interior of the sample. Since the Hall voltage drops only across incompressible regions with the same filling factor i, one measures the quantized value $R_{\mathrm{H}} = h/(ie^2)$ for the Hall resistance. The details of the voltage drop are not relevant. Inhomogeneities of the electron density and, eventually, localization of electrons by potential fluctuations, guarantee that, for moderate changes of the average electron density (at fixed magnetic field) or of the magnetic field (at fixed average electron density), a connected incompressible region with this filling factor i remains present in the interior of the sample: the quantized value of the Hall resistance occurs as a plateau. The current distribution varies strongly during such changes, since the landscape of compressible and incompressible regions changes strongly. The quantized value of the Hall resistance, however, remains unchanged, as long as the compressible regions along opposite sample edges are separated by incompressible regions and, therefore, their electrochemical potentials remain constant. These incompressible regions hinder the exchange of electrons between opposite edges of the sample, i.e. they suppress "backscattering". Between two contacts on the same sample edge no voltage drop can be measured, i.e. $R_{xx} = U_x/I_x = 0$. This does not require that the whole interior of the sample is incompressible. Well developed incompressible stripes near the sample edges are sufficient to suppress this backscattering and to keep the outermost compressible regions on their potentials. Then the total applied current flows through these incompressible stripes, while the compressible regions between these stripes do not contribute to the current transmission. As a consequence, the potential is constant in the interior and drops only across the incompressible stripes. This is the situation near the low-B edges of the quantum Hall plateaus.

These ideas, stimulated by and explaining the scanning force-microscope experiments, [19, 20] make it plausible why even on finite samples with impurities the quantized values of the Hall resistance can be measured with extraordinary precision: they occur when percolating incompressible regions exist. On these incompressible regions the quantized values of the resistivity are realized, and the externally applied current is forced to flow only through these regions (only then the entropy production of the stationary non-equilibrium state is minimized) [29]. An extension of the model calculations to wider samples, and a more rigorous justification of its basic assumptions, seem desirable. Also the mechanisms leading to a breakdown of the quantized Hall effect above a critical current and, related to that, heating effects in the quantized Hall regime [23], require additional work.

The actual research in the field of quantized Hall effect, deals however mainly with correlation effects, which become increasingly important with increasing quality (and mobility) of the two-dimensional electron systems, and lead to the discovery of more and more incompressible many-electron states, visualized by new fractional quantum Hall plateaus. Phenomena like ▶ Bose–Einstein condensation, skyrmion-type excitations, fractional charges, vanishing longitudinal resistance induced by microwave radiation, as well as stripe- and bubble-like phase-textures in higher Landau levels are surprising discoveries of recent years and indicate that also for the future the quantized Hall effect will remain an actual and interesting field of research.

Primary Literature

1. K. von Klitzing, G. Dorda, and M. Pepper: New Method for High-Accuracy Determination of the Fine-Structure Constant Based on Quantized Hall Resistance, Phys. Rev. Lett. **45**, 494 (1980).
2. A. B. Fowler, F. F. Fang, W. E. Howard, and P. J. Stiles: Magneto-Oscillatory Conductance in Silicon Surfaces, Phys. Rev. Lett. **16**, 901 (1966).
3. D. C. Tsui, H. L. Störmer, and A. C. Gossard: Two-Dimensional Magnetotransport in the Extreme Quantum Limit, Phys. Rev. Lett. **48**, 1559 (1982).
4. R. B. Laughlin: Anomalous Quantum Hall Effect: An Incompressible Quantum Fluid with Fractionally Charged Excitations, Phys. Rev. Lett. **50**, 1395 (1983).
5. A. M. Thompson and D. G. Lampard: A New Theorem in Electrostatics and its Application to Calculable Standards of Capacitance, Nature **177**, 888 (1956).
6. T. J. Quinn: News from the BIPM, Metrologia **26**, 69 (1989).
7. F. Delahaye, T. J. Witt, R. E. Elmquist, and R. F. Dziuba: Comparison of Quantum Hall Effect Resistance Standards of the NIST and the BIPM, Metrologia **37**, 173 (2000).
8. F. Delahaye and B. Jeckelmann: Revised Technical Guidelines for Reliable DC Measurements of the Quantized Hall Resistance, Metrologia **40**, 217 (2003).
9. E. R. Williams, R. L. Steiner, D. B. Nevell, and P. T. Olsen: Accurate Measurement of the Planck Constant, Phys. Rev. Lett. **81**, 2404 (1998).
10. U. Klaß, W. Dietsche, K. von Klitzing, and K. Ploog: Imaging of the Dissipation in Quantum-Hall-Effect Experiments, Z. Phys. B - Condensed Matter **82**, 351 (1991).
11. Y. Y. Wei, J. Weis, K. v. Klitzing, and K. Eberl: Single-Electron Transistor as an Electrometer Measuring Chemical Potential Variations, Appl. Phys. Lett. **71**, 2514 (1997).
12. M. Büttiker: Absence of Backscattering in the Quantum Hall Effect in Multiprobe Conductors, Phys. Rev. B **38**, 9375 (1988).
13. D. B. Chklovskii, B. I. Shklovskii, and L. I. Glazman: Electrostatics of Edge States, Phys. Rev. B **46**, 4026 (1992).
14. Y. Y. Wei, J. Weis, K. v. Klitzing, and K. Eberl: Edge Strips in the Quantum Hall Regime Imaged by a Single-Electron Transistor, Phys. Rev. Lett. **81**, 1674 (1998).
15. J. Weis, Y. Y. Wei, and K. v. Klitzing: Single-Electron Transistor Probes Two-Dimensional Electron System in High Magnetic Fields, Physica E **3**, 23 (1998).
16. D. B. Chklovskii, K. A. Matveev, and B. I. Shklovskii: Ballistic Conductance of Interacting Electrons in the Quantum Hall Regime, Phys. Rev. B **47**, 12605 (1993).
17. K. Lier and R. R. Gerhardts: Self-Consistent Calculation of Edge Channels in Laterally Confined Two-Dimensional Electron Systems, Phys. Rev. B **50**, 7757 (1994).
18. J. H. Oh and R. R. Gerhardts: Self-Consistent Thomas-Fermi Calculation of Potential and Current Distributions in a Two-Dimensional Hall Bar Geometry, Phys. Rev. B **56**, 13519 (1997).
19. E. Ahlswede, P. Weitz, J. Weis, K. von Klitzing, and K. Eberl: Hall Potential Profiles in the Quantum Hall Regime Measured by a Scanning Force Microscope, Physica B **298**, 562 (2001).
20. E. Ahlswede, J. Weis, K. von Klitzing, and K. Eberl: Hall Potential Distribution in the Quantum Hall Regime in the Vicinity of a Potential Probe Contact, Physica E **12**, 165 (2002).
21. A. Siddiki and R. R. Gerhardts: Incompressible Strips in Dissipative Hall Bars as Origin of Quantized Hall Plateaus, Phys. Rev. B **70**, 195335 (2004).
22. K. Güven and R. R. Gerhardts: Self-Consistent Local-Equilibrium Model for Density Profile and Distribution of Dissipative Currents in a Hall Bar Under Strong Magnetic Fields, Phys. Rev. B **67**, 115327 (2003).
23. S. Kanamaru, H. Suzuura, and H. Akera: Spatial Distribution of Electron Temperature in Quantum Hall Systems with Compressible and Incompressible Strips, J. Phys. Soc. Jpn. **75**, 064701 (2006).

Secondary Literature

24. T. Ando, A. B. Fowler, and F. Stern: Electronic Properties of Two-Dimensional Systems, Rev. Mod. Phys. **54**, 437 (1982).
25. B. Kramer, S. Kettemann, and T. Ohtsuki: Localization in the Quantum Hall regime, Physica E **20**, 172 (2003).
26. Z. F. Ezawa, *Quantum Hall Effects, Field Theoretical Approaches and Related Topics* (World Scientific, Singapore, 2000); see also www.fkf.mpg.de/klitzing/.
27. H. Bachmair, E. O. Göbel, G. Hein, J. Melcher, B. Schumacher, J. Schurr, L. Schweitzer, and P. Warnecke: The von Klitzing resistance standard, Physica E **20**, 14 (2003); for the most recent CODATA-recommended values of fundamental constants see: http://physics.nist.gov/cuu/constants/index.html.
28. S. Datta, *Electronic Transport in Mesoscopic Systems* (Cambridge University Press, Cambridge, 1995).
29. R. R. Gerhardts: The Effect of Screening on Current Distribution and Conductance Quantisation in Narrow Quantum Hall Systems, phys. stat. sol. (b) **245**, 378 (2008).

Quantum Interrogation

Paul G. Kwiat

The notion of a "negative-result" measurement was first discussed by Renninger [1] and later by Dicke [2]. As a simple example, consider a single photon incident on a beamsplitter, with a 100% efficient detector in the reflected port; if we somehow know that the photon amplitude has already encountered the detector, and yet no detection has taken place, then this non-detection certainly "collapses" the original superposition of the photon ► wave function solely into the transmitted path. Elitzur and Vaidman (EV) [3] suggested a modified system in which a second beamsplitter is used to recombine the two paths (see Fig. 1). In the absence of any object in one arm of the interferometer, complete destructive interference of the two paths leads to a zero probability that a detector at one of the ports fires. On the other hand, the presence of a non-transmitting object necessarily inhibits the destructive interference (as there is then only one path by which the photon can reach the recombining beamsplitter) so that sometimes this "dark" detector will fire. This indicates the presence of the object, even though the detected photon most certainly did not travel the path containing the object, in essence an "interaction-free" measurement.[1] (► Interaction-free Measurement)

This simple scheme was experimentally verified using single photons (► light quantum) (conditionally prepared via parametric down-conversion), achieving a

[1] We prefer the more general term "quantum interrogation", which allows for the possibility that the photon *does* pass through the path with the object, e.g., if the object is semi-transparent or only partially blocks the arm of the interferometer.

Q

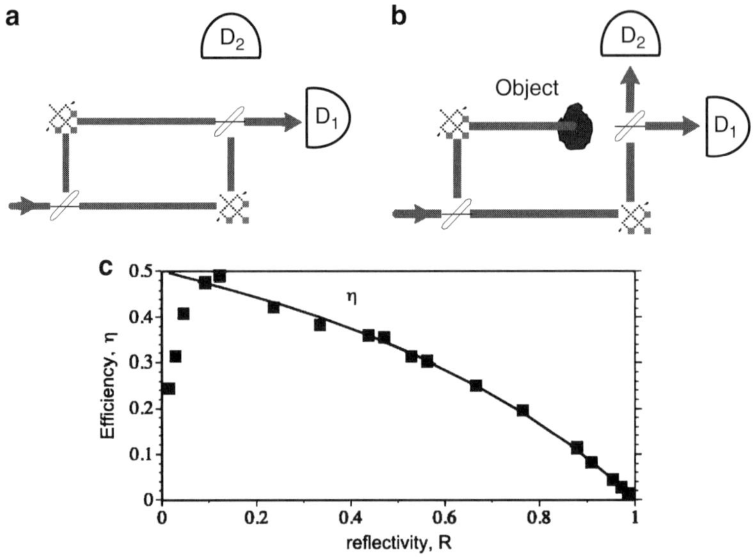

Fig. 1 (**a**). In the absence of an object in the interferometer, each incident photon is detected at D_1, i.e., D_2 never fires. (**b**). A non-transmitting object disrupts the interference, so D_2 now receives photons, unambiguously indicating the presence of the object, even though any detected photon must have taken the bottom path in the interferometer. (**c**). Varying the beamsplitter reflectivities enables one to optimize the efficiency, approaching the 50% limit possible with this technique (data from [5])

$\sim$ 33% efficiency for detecting the presence of an opaque object in an interaction-free way [4]. Another experiment verified that by adjusting the reflectivity of the interferometer beamsplitters, one could achieve an efficiency approaching 50% (Fig. 1c), and by incorporating a focused beam, demonstrated the basic elements of a reduced-absorption imaging method [5]. Similar quantum interrogation experiments have now been performed with neutrons [6] and even proposed as a means to read out superconducting qubits without any energy exchange [7].

Although it was originally thought that 50% efficiency was the best one could achieve, in fact a method based on the ► quantum Zeno effect (QZE) allows much better performance: In principle in a lossless setup, one can detect the presence of a non-transmitting object all the time, with no chance of absorption by the object! The basic idea of the QZE [8] is that repeated strong measurements of a quantum system can continually project it into its initial state, thereby inhibiting the otherwise slow evolution out of this state. As a simple example, consider the arrangement shown in Fig. 2a. A single photon with initial horizontal polarization is cycled N times through a Michelson interferometer (with a polarizing beamsplitter (PBS)). In each cycle, a waveplate is used to rotate the polarization by a small amount $\Delta\theta = \pi/2N$. In the absence of an object in the interferometer, the photon polarization rotates stepwise from horizontal to vertical. On the other hand, the presence of a non-transmitting object in the vertical polarization arm of the interferometer will inhibit

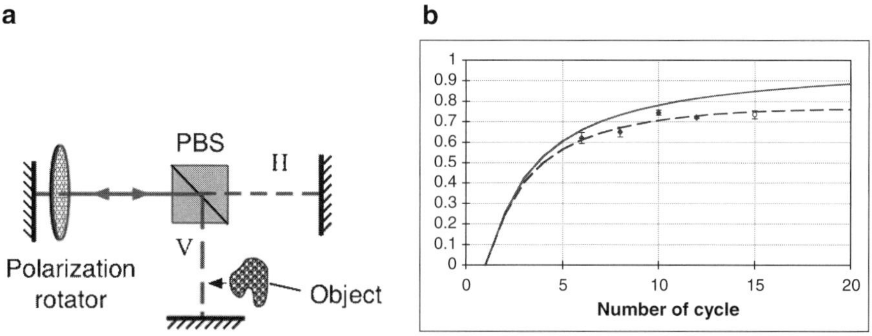

Fig. 2 (**a**). Conceptual scheme incorporating the quantum Zeno effect to realize high-efficiency quantum interrogation. An initially horizontally (H) polarized photon is allowed to circulate N times (experiencing a rotation by $\pi/2N$ each cycle) before being removed and its polarization analyzed. In the absence of the object, the photon will have vertical (V) polarization. In the presence of an object in the V arm, the final polarization of the photon will be H, with a negligibly small probability the photon is absorbed. (**b**). Plot of efficiency vs. number of cycles for ideal lossless system (*solid curve*), and one with ~95% loss (*dashed curve*), corresponding to experimental data (*diamonds*) (from [9])

this evolution. Now after N cycles, the photon has a high probability $\cos^{2N}(\pi/2N)$ of still being horizontally polarized. As N becomes very large (and the corresponding effective coupling between the horizontal and vertical-polarization arms of the interferometer becomes very small) the probability that the photon remains in its initial horizontal polarization state approaches 1, while the probability the photon is ever absorbed by the object approaches zero.[2] Therefore, by simply observing the final state of the polarization of the photon one can determine in an interaction-free way whether or not there was a (non-transmitting) object present. Such a scheme has been implemented [9], achieving efficiencies of ~75%.

A related method involves shining monochromatic light into a highly resonant optical cavity, with mirrors of very high reflectivity $R \approx 1 - \varepsilon$. In the absence of any object in the interferometer, the incident field will, after a transient period, experience full constructive interference for transmission, i.e., essentially all the incident light will be transmitted. On the other hand, if there is an opaque object in the cavity, this will prevent the necessary coherent build up of fields that would otherwise lead to destructive interference for reflection off the entrance mirror; now the incident light simply bounces off the cavity, with probability R. Thus, detection of a reflected (transmitted) photon indicates the presence (absence) of a non-transmitting object in the cavity. Such a scheme has been experimentally realized [10], achieving an interaction-free detection probability up to 88%. By using a scanning system, one can also generalize this technique to 2-D imaging; in particular, Inoue and Bjork were able to "image" the silhouette of a piece of film without exposing the film itself [11].

[2] The presence of loss in the rest of the system actually prevents one from reaching the limit $N \to \infty$, so in any real system the maximum efficiency is strictly <1.

Q

One topic of interest is whether or not the quantum interrogation techniques can be useful when the object is partially transmitting; certainly any sort of imaging would be much more valuable if a "grayscale" for absorption could be obtained. By making enough measurements, it is always possible to distinguish between a transparent object and one with some absorption – multiple passes through the latter object cause it to effectively look more opaque [12]. However, in general two partial transparencies cannot be perfectly distinguished [13].

Finally, one of the more intriguing applications of the methods of quantum interrogation is to the topic of ▶ quantum computation. Mitchison and Jozsa [14] showed that if one can put a quantum computer into a superposition of "running" and "not running", it is possible to gain information about the result even in instances when the algorithm did not run – a "counterfactual quantum computation" (CFQC). The mere fact that the computer *could* have run is enough to disrupt interference (in the same way that the presence of an opaque object disrupts the interference in Fig. 1). This EV-style approach has been experimentally realized [15] using a simple beamsplitter to put an incident photon into a superposition of passing through or not passing through an optical implementation of Grover's search algorithm [16] (▶ quantum computation); an efficiency – likelihood of a CFQC – of 32% was attained. Although the original method only works a fraction of the time and only on certain possible results, a more complicated system based on the QZE approach – many weak measurements (▶ weak value and weak measurements) – was predicted to again recover high efficiencies [15].

Primary Literature

1. M. Renninger: Observations without disturbing the object. Z. Phys. **158**, 417–421 (1960)
2. R. H. Dicke: Interaction-free quantum measurements: A paradox? Am. J. Phys. **49**, 925–930 (1981)
3. A. C. Elitzur, L. Vaidman: Quantum mechanical interaction-free measurements. Found. Phys. **23**, 987–997 (1993)
4. P. G. Kwiat, H. Weinfurter, T. Herzog, A. Zeilinger, M. A. Kasevich: Interaction-free Measurement. Phys. Rev. Lett. **74**, 4763 (1995)
5. A. G. White, J. R. Mitchell, O. Nairz, P. G. Kwiat: 'Interaction-free' Imaging. Phys. Rev. **A 58**, 605 (1998)
6. M. Hafner, J. Summhammer: Experiment on interaction-free measurement in neutron interferometry. Phys. Lett. **A 235**, 563–568 (1997)
7. G. S. Paraoanu: Interaction-free measurements with superconducting Qubits. Phys. Rev. Lett. **97**, 180406 (2006)
8. B. Misra, E. C. G. Sudarshan: The Zeno's paradox in quantum theory. J. Math. Phys. **18**, 756 (1977); A. Peres: Zeno paradox in quantum theory. Am. J. Phys. **48**, 931 (1980)
9. P. G. Kwiat, A. G. White, J. R. Mitchell, O. Nairz, G. Weihs, H. Weinfurter, A. Zeilinger: High-efficiency quantum interrogation measurements via the quantum Zeno effect. Phys. Rev. Lett. **83**, 4725 (1999)
10. T. Tsegaye, E. Goobar, A. Karlsson, G. Björk, M. Y. Loh, K. H. Lim: Efficient interaction-free measurements in a high-finesse interferometer. Phys. Rev. **A 57**, 3987–3990 (1998); N. Namekata, S. Inoue: High-efficiency interaction-free measurements using a stabilized Fabry-Perot cavity. J. Phys. **B 39**, 3177–3183 (2006)

11. S. Inoue and G. Bjork: Experimental demonstration of exposure-free imaging and contrast amplification. J. Opt. B: Quantum Semiclass. Opt **2**, 338–345 (2000)
12. P. G. Kwiat: "Experimental and theoretical progress in interaction-free measurements". Phys. Scripta **T76**, 115 (1998)
13. G. Mitchison, S. Massar: Absorption-free discrimination between semitransparent objects. Phys. Rev. **A 63**, 032105 (2001)
14. G. Mitchison, R. Jozsa: Counterfactual computation. Proc. R. Soc. Lond. **A457**, 1175–1193 (2001)
15. O. Hosten, M. T. Rakher, J. T. Barreiro, N. A. Peters, P. G. Kwiat: Counterfactual quantum computation via quantum interrogation. Nature **439**, 939 (2006)
16. L. K. Grover: Quantum mechanics helps in searching for a needle in a haystack. Phys. Rev. Lett. **79**, 325–328 (1997)

Secondary Literature

17. P. G. Kwiat, H. Weinfurter, A. Zeilinger: Quantum seeing in the dark. Sci. Am. **275** 52–58 (1996)

Quantum Jump Experiments

Howard J. Carmichael

The notion of ► quantum jumps entered quantum physics in 1913, in the year Niels Bohr (1885–1962) proposed a quantized model of the ► Rutherford atom and a prescription for obtaining the Rydberg formula for the emission spectrum of atomic hydrogen ► Bohr's atom model. It is inherent in the simple relation [1]

Q

$$W_{\tau_2} - W_{\tau_1} = h\nu,$$

which equates a difference in electron binding energies in initial and final atomic stationary states to the energy of an emitted quantum of radiation of frequency ν. Atomic stationary states are labeled by an integer τ, and Bohr speaks of "the passing of the system from a state corresponding to $\tau = \tau_1$ to one corresponding to $\tau = \tau_2$"; this passing is the *quantum jump*, Planck's constant and it proceeds with the emission of a quantum of radiation of energy $h\nu$, where h is ► Planck's constant; the reverse jump accompanies absorption. In 1916 Einstein (1879–1955) raised the quantum jump to the level of a genuine principle of quantum dynamics. By proposing probabilistic rules for the absorption and emission (spontaneous and stimulated) of radiation quanta, Einstein managed to arrive at a dynamical explanation of the Planck formula for the spectrum of ► black-body radiation [2]. This so-called A and B theory [13, 14] continues in wide use today, providing the basis for rate-equation models of the interaction of light and quantized matter; although, its founding upon the quantum jump, adopted as a fundamental event, is superceded by the quantum mechanics of Schrödinger (1887–1961) and Heisenberg (1901–1976).

In quantum mechanics the Bohr–Einstein quantum jump is generalized as the *quantum transition*, the probabilistic change from an initial (prepared) state to a final (observed) state – from ket vector $|i\rangle$ to ket vector $|f\rangle$. In 1985 Cook and Kimble [3] suggested an experiment to demonstrate the original ▶ quantum jump between atomic stationary states, building upon the *electron shelving* idea of Hans Dehmelt [4] and recently developed methods for trapping and cooling single atomic ions. Dehmelt received the Nobel prize for developing the ion trap in 1989. His electron shelving idea was proposed in 1975, as an amplifying mechanism for the detection of a weak transition in single-atom ▶ spectroscopy. It is illustrated by the energy-level diagram of Fig. 1. Two radiative transitions in a mercury ion are represented. The 194 nm-transition is *strong* and dipole-allowed, while the transition at 281.5 nm is a metastable dipole-non-allowed transition and *weak*. If both are excited by near-resonant radiation, the dominant effect will be the scattering of a steady stream of photons (▶ light quantum) – some 10^8 per second – on the strong transition. An equally important feature is to be noted, though; occasional transitions (quantum jumps) occur on the weak transition, and these "shelve" the electron in the $5d^9 6s^2\,{}^2D_{5/2}$ stationary state, temporarily turning off the strong-transition fluorescence. The fluorescence is therefore predicted to be intermittent, and its abrupt turning off and on records quantum jumps on the weak 281.5 nm-transition. With a metastable lifetime on the order of 0.1 s, interruptions in the strong-transition fluorescence are readily observed, even if only 0.1% of the photon stream can be collected and counted. A series of observations were made in 1986 with single trapped barium [5, 6] and mercury [7] ions, and in 1995 quantum jumps of a single terrylene ($C_{30}H_{16}$) molecule were observed through intermittent fluorescence [8].

Related but slightly different methods were used to observe quantum jumps in other systems. In 1999 Peil and Gabrielse observed quantum jumps between Landau levels [15] of an electron bound in a cyclotron orbit [9]. In their experiment there

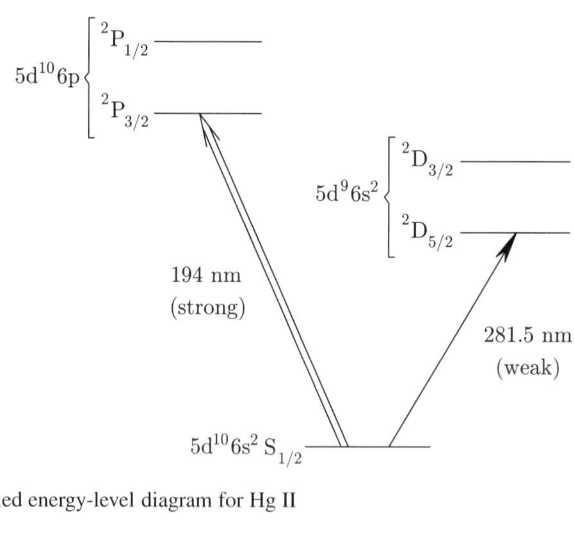

Fig. 1 Simplified energy-level diagram for Hg II

is no fluorescence signal monitoring the initial and final stationary states. To realize an equivalent monitoring, a coupling of the cyclotron motion – which takes place in a plane perpendicular to an applied magnetic field – to a harmonic oscillation along the axis of the magnetic field is used. The resonance frequency of the axial motion depends upon the cyclotron energy; it therefore shifts abruptly when the electron makes a quantum jump. The scheme realizes a so-called QND (quantum nondemolition) measurement [16] of the cyclotron energy which is observed continuously over time. A similar QND method was used to observe quantum jumps of a radiation oscillator (a mode of the electromagnetic field) in a superconducting microwave cavity [10]. In this experiment the quantum jumps record the "birth" (energy increase) or "death" (energy decrease) of a photon in the cavity. Compared with the observation of quantum jumps through intermittent fluorescence, here the roles of atom and photon are reversed, with the number of photons monitored by Ramsey interferometry [17] carried out on a stream of Rydberg atoms passing through the cavity. A frequency shift that depends on photon number is recorded through a phase shift of the Ramsey fringe.

In quantum mechanics, evolution according to the ▶ Schrödinger equation is continuous and nothing jumps [17]. The interpretation of quantum jump experiments must therefore face the question: in what sense is the discontinuous jump of Bohr and Einstein observed? Figure 2 illustrates a segment of intermittent fluorescence from a simulation of an electron shelving experiment. Gaps in the record of photons scattered on the strong transition (marked by vertical lines) indicate periods where the electron is shelved in the $5d^96s^2\,{}^2D_{5/2}$ stationary state (Fig. 1). In a naive interpretation, the electron jumped into this state at the beginning of each gap. An analysis like that of Cook and Kimble [3] which holds for excitation by incoherent

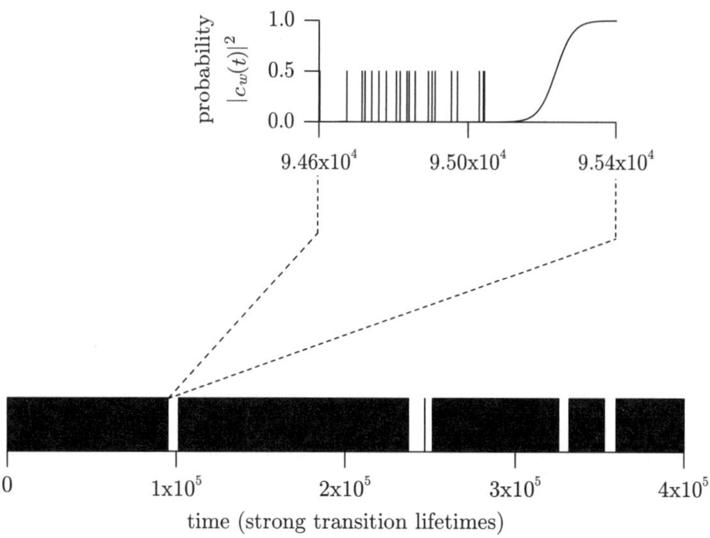

Fig. 2 Quantum trajectory simulation of intermittent fluorescence

radiation incorporates the jump explicitly, as it uses Einstein's phenomenological model to describe the emission and absorption of radiation quanta. Such an analysis is inappropriate for coherent (laser) excitation, however, since coherence in excitation implies the creation, according to the ▶ Schrödinger equation, of a coherent ▶ superposition of atomic stationary states:

$$c_g(t)|g\rangle + c_s(t)|s\rangle + c_w(t)|w\rangle,$$

where g, s, w denote ground, strong, weak, and $c_g(t)$, $c_s(t)$, $c_w(t)$ are some time-dependent complex numbers. A resolution of the appearance of quantum jumps with the superposition of atomic stationary states is reached by incorporating the measurement process, i.e., the recording of the strong-transition fluorescence, into the Schrödinger evolution. The central element is the notion of a *null measurement* – here the non-appearance of an anticipated photon scattered on the strong transition. Porrati and Putterman [11] pointed out the importance of this idea, and it is the central ingredient of the quantum trajectory treatment of photon scattering [18–20] used to generate Fig. 2. In quantum trajectory theory one simulates a record of scattered photon times while simultaneously evolving the state of the ion as a superposition of stationary states. The exploded time-scale in Fig. 2 shows what is revealed about the start of a gap in the monitored strong-transition fluorescence. After a last photon is recorded (of course known to be "last" in retrospect only), the probability $|c_w(t)|^2$ that the ion occupies the shelved state eventually begins to grow and evolves continuously to $|c_w(t)|^2 = 1$. The interpretation is that $|c_w(t)|^2$ represents an *expectation* that the ion is in the shelved state, an expectation conditioned upon the information available in the monitored fluorescence. As scattered photons continue *not* to appear, the expectation eventually grows to a certainty; no actual "jump" into the shelved state is confirmed. Typically, the period of uncertainty corresponds to the time required for the scattering of a few tens of photons on the strong transition. Similar commentary applies to all quantum jump experiments: though a quantum jump is inferred, the abruptness of the observed change of state is set by the finite time resolution of the measurement and no violation of the continuous quantum mechanics of Schrödinger and Heisenberg is confirmed.

Primary Literature

1. N. Bohr: On the Constitution of Atoms and Molecules, Philosophical Magazine **26**, 1–25 (1913)
2. A. Einstein: Strahlungsemission und -absorption nach der Quantentheorie, Verhandlungen der Deutschen Physikalishen Gesellschaft **18**, 318–323 (1916); *Quantentheorie der Strahlung*, Physikalische Zeitschrift **18**, 121–128 (1917)
3. R. J. Cook, H. J. Kimble: Possibility of Direct Observation of Quantum Jumps. Physical Review Letters **10**, 1023–1026 (1985)
4. H. G. Dehmelt: Bulletin of the American Physical Society **20**, 60 (1975)
5. W. Nagourney, J. Sandberg, H. Dehmelt: Shelved Optical Electron Amplifier: Observation of Quantum Jumps. Physical Review Letters **56**, 2797–2799 (1986)

6. Th. Sauter, W. Neuhauser, R. Blatt, P. E. Toschek: Observation of Quantum Jumps. Physical Review Letters **57**, 1696–1698 (1986)
7. J. C. Berquist, R. G. Hulet, W. M. Itano, D. J. Wineland: Observation of Quantum Jumps in a Single Atom. Physical Review Letters **57**, 1699–1702 (1986)
8. Th. Basché, S. Kummer, C. Bräuchle: Direct Spectroscopic Observation of Quantum Jumps in a Single Molecule. Nature **373**, 132–124 (1995)
9. S. Peil, G. Gabrielse: Observing the Quantum Limit of an Electron Cyclotron: QND Measurements of Quantum Jumps between Fock States. Physical Review Letters **83**, 1287–1290 (1999)
10. S. Gleyzes, S. Kuhr, C. Guerlin, J. Bernu, S. Deléglise, U. B. Hoff, M. Brune, J.-M. Raimond, S. Haroche: Quantum Jumps of Light Recording the Birth and Death of a Photon in a Cavity. Nature **446**, 297–300 (2007)
11. M. Porrati, S. Putterman: Wave-Function Collapse due to Null Measurements: The Origin of Intermittent Atomic Fluorescence. Physical Review A **36**, 929–932 (1987); *ibid* **39**, 3010–3030 (1989)

Secondary Literature

12. H. C. Ohanian: *Physics* (Norton, New York 1985, 928–936)
13. H. Haken, H. C. Wolf: *The Physics of Atoms and Quanta* (Springer, Berlin 2007, Ch. 8, 54–57)
14. R. Loudon: *The Quantum Theory of Light* (Oxford University Press, Oxford 2000, 16–19, 360–365)
15. C. Cohen-Tannoudji, B. Diu, F. Laloe: *Quantum Mechanics, Volume One* (Wiley, New York, 1977, 742–764)
16. V. B. Braginsky, F. Ya. Khalili: Quantum Nondemolition Measurements: From Toys to Tools. Reviews of Modern Physics **68**, 1–11 (1996)
17. N. F. Ramsey: Experiments with Separated Oscillatory Fields and Hydrogen Masers. Science **248**, 1612–1619 (1990)
18. H. D. Zeh: There are No Quantum Jumps, nor are There Particles!. Physics Letters A **172**, 189–192 (1989)
19. M. B. Plenio, P. L. Knight: The Quantum-Jump Approach to Dissipative Dynamics in Quantum Optics. Reviews of Modern Physics **70**, 101 (1998)
20. H. J. Carmichael: *An Open System Approach to Quantum Optics* (Springer, Berlin 1993, Chaps. 7–10); *Statistical Methods in Quantum Optics 2: Non-Classical Fields* (Springer, Berlin 2007, Chaps. 17–19)
21. M. O. Scully, M. S. Zubairy: *Quantum Optics* (Cambridge University Press, Cambridge 1997, 260–267)

Q

Quantum Jumps

Klaus Hentschel

Niels Bohr's (1885–1962) ▶ atomic model initially provoked much opposition. "Bohr's work on the quantum theory of the Balmer formula (in the *Phil. Mag.*) has driven me to despair", the Leyden theoretician Paul Ehrenfest (1880–1933) wrote his colleague Hendrik Antoon Lorentz (1853–1928) on 25 August 1913. "If this is the way to reach the goal, I must give up doing physics." The assumption of 'quantum jumps', i.e., an ▶ electron's sudden and unpredictable transition between two stable orbits around the nucleus, was an integral part of Bohr's model. Bohr's mentor Ernest Rutherford (1871–1937) in Manchester raised doubts about it in a letter to Bohr, dated 20 March 1913: "How does an electron decide what frequency it is going to vibrate at when it passes from one stationary state to the other? It seems to me that you would have to assume that the electron knows beforehand where it is going to stop." Knowing that was indeed imperative in Bohr's semi-classical model of emission and absorption, because the electromagnetic wave of frequency v (linked to the energy difference ΔE between two stationary states by $E = hv$) must 'start radiating' as soon as the electron 'jumps' (cf. also Fig. 1). According to Rutherford, Bohr's effort to combine a discontinuous quantum process of emission and absorption with a classical continuum model of radiation as electromagnetic waves thus raised deep problems concerning causality ▶ indeterminism. These problems stayed with semi-classical ▶ quantum theory to its bitter end and were even aggravated in the quantum mechanics of 1925/26.

Bohr's solution was simply to declare classical electrodynamics out of order. The problem that any charged particles (such as ▶ electrons on their 'orbits' around the positively charged nucleus) must continually loose energy (Larmor's theorem) was thus done away with.[1] He was so bold as to stipulate that the atom only radiates during 'jumps' between energy levels and refused to go into further detail about the physical processes involved. Instead he sought a suitable phenomenological description, concentrating on ▶ observables before and after a given measurement. Because particularly for the ▶ Stark effect and ▶ Zeeman effect the number of combinatorically possible transitions between energy levels exceeds the number of observed spectral lines, ▶ selection rules had to be imposed to reduce the number of admissible 'jumps'. As long as the interaction between different electrons of one atom is

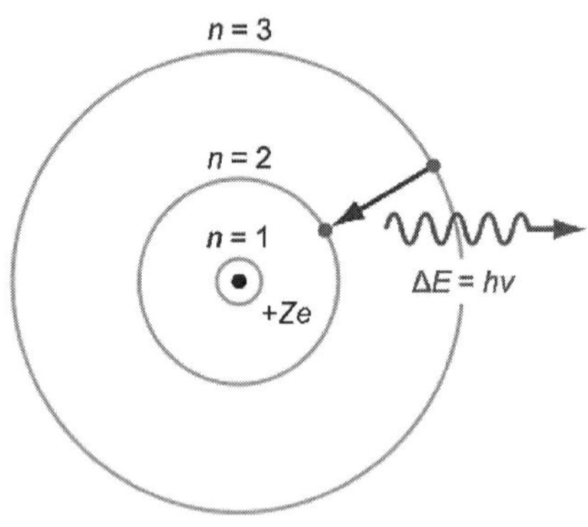

Fig. 1 Bohr atom model with quantum jump of the electron from the $n = 3$ to the $n = 2$ orbit. The energy difference ΔE between the two orbits is emitted as photonic energy of hv. Source: Wikimedia Commons

[1] After hearing a talk on Bohr's atomic model in the Zurich physics colloquium, Max von Laue (1879–1960) rose and said: "That's all nonsense; Maxwell's equations are correct under all circumstances, and an electron orbiting around a positive nucleus is bound to radiate." (Quoted in [2], 86)

not too large, only those transitions take place where just one of the electrons makes a “jump”, i.e., only one alters its orbital quantum number l by ±1 (see, e.g., [1, 2nd ed.], p. 85). After initial protest, the scientific community learned to live with problems of interpretation by simply ignoring them as best as possible and developing a rather instrumentalistic attitude (► quantum theory, crisis period). A deeper understanding of selection rules and other features of the semi-classical atomic models only became possible after the discovery of ► spin and the advent of quantum mechanics in 1925/26. Formerly useful mental models like electron ‘orbits’, ‘jumps’, etc. no longer made sense because of ► Heisenberg’s uncertainty relation. See also ► quantum jump experiments.

Literature

1. G. Herzberg: *Atomic Spectra and Atomic Structure* (New York: Prentice-Hall 1937, 2nd edn. New York: Dover Publications 1944)
2. M. Jammer: *The Conceptual Development of Quantum Mechanics* (New York: McGraw-Hill 1966)

Quantum Logic

Peter Mittelstaedt

The Early History of Quantum Logic

Q

Already in his pioneering work “Mathematische Grundlagen der Quantenmechanik” of 1932, J. von Neumann mentioned that projection operators (► projection) in ► Hilbert space correspond to elementary propositions in quantum mechanics, and that also the logical connectives ∧ (and), ∨ (or), and ¬ (not) can adequately be expressed in terms of projection operators. Compared to classical logic, the calculus of propositions, that is based on projection operators, is essentially restricted by the mutual commensurability or incommensurability of the propositions in question.

This calculus was investigated more in detail in the work of G. Birkhoff and J. von Neumann in 1936 [1]. In terms of lattice theory, the authors could show, that the lattice of quantum mechanical propositions is given by an orthocomplemented lattice that is not distributive and in general also not modular. The title of their paper “The logic of quantum mechanics” indicates the similarity and distinctness between this “logic” and the well known classical proposition logic, which is given by a Boolean lattice L_C.

The lattice of quantum mechanical propositions that correspond to projection operators in Hilbert space, was further elaborated by Piron [2] and Jauch [3] and found to be an orthocomplemented, orthomodular lattice L_Q with a zero-element 0 and a unit element I. In addition, the lattice of projection operators is atomic and it fulfils the covering property. Together with these additional properties, the lattice will be denoted here by $L_Q{}^*$.

Is Quantum Logic a Genuine Logic?

The propositional logic L_Q which corresponds to the lattice L_Q is often called "quantum logic". At first, this terminology is merely based on the analogy with the classical propositional logic L_C, which corresponds to a Boolean lattice, i.e. to an orthocomplemented, distributive lattice L_C. However, at this stage of the discussion, it is by no means clear, whether the structure L_Q is at all a logic in the genuine sense of this concept. This problem can be treated by recourse to the operational justification of intuitionistic and classical logic by means of a semantic that refers to calculi [4] or to dialogs [5]. Making use of the mutual incommensurability of elementary quantum mechanical propositions one finds, that these elementary propositions are only "restrictedly available" in a calculus [6] or in a dialog [7]. On the basis of a "quantum dialog game" with a "restricted availability" of elementary propositions, a calculus L_{Qi} for an "intuitionistic quantum logic" can be established. It can be shown that the calculus L_{Qi} is consistent and complete with respect to the semantic of quantum dialogs [8]. Under the additional assumption that elementary propositions are value definite, i.e. fulfil generally the law $A \vee \neg A = I$ of the excluded middle (*tertium non datur*) – we arrive at the calculus L_Q of full quantum logic.

The Lindenbaum-Tarski algebra of the calculus L_Q is an orthocomplemented, orthomodular lattice with universal bounds 0 and I, which we denoted here by L_Q. The Lindenbaum-Tarski algebra of the calculus L_{Qi} of intuitionistic quantum logic is also a lattice L_{Qi}, but this structure [1] is of less interest, since there are no physical reasons to dispense with the value definiteness of elementary propositions. The calculus L_Q can be further elaborated. If we assume, that the elementary propositions refer to one single quantum system, then we arrive at a calculus L_Q^*, the Lindenbaum-Tarski algebra of which is the lattice $L_Q{}^*$ mentioned above with the additional properties of atomicity and the covering property [9].

Irrespective of the successful logical reconstruction of the lattice $L_Q{}^*$ and the completeness and consistency of the logical calculi L_{Qi} and L_Q, Jauch and Piron [10] had argued that the lattice L_Q must not be considered as a logic, since the operation of material implication "$A \rightarrow B$" cannot be expressed by $\neg A \vee B$ as in a Boolean lattice. The material implication is indispensable for the application of the *modus ponens law*in logical inference. However, it could be shown [11] that the slight generalisation $\neg A \vee (A \wedge B)$ of the formula mentioned fulfils in L_Q almost all requirements that are fulfilled by $\neg A \vee B$ in L_C. Moreover, it could be shown

[1] The lattice L_{Qi} is described in detail in P. Mittelstaedt (1978), chapter V.

that in the lattice L_{Qi} of intuitionistic quantum logic for any two elements A and B there exists a uniquely defined generalised material implication A→B, which can, however, not be expressed by the other connectives ∧, ∨, and ¬ [12]. If value definiteness of elementary propositions is presupposed, the proposition A→B agrees with the proposition ¬A ∨ (A ∧ B) mentioned above.

Hence, the calculi L_Q and L_{Qi} fulfil the most important requirements of a logical calculus. The difference between these calculi and the calculi L_C and L_i of classical and intuitionistic logic is, that in the traditional calculi for any two propositions A and B the compound propositions A → (B → A) and B → (A → B) are formally true, whereas in the quantum logical systems L_Q and L_{Qi} these propositions are formally true if and only if the propositions A and B are commensurable. In L_Q the difference to L_C can also be expressed by the fact that for two propositions A and B the distributive law A = (A ∧ B) ∨ (A ∧ ¬B) is formally true in L_C but not in L_Q [2], [18].

The Bottom-Top Reconstruction of Quantum Mechanics

On the basis of the logical reconstruction of the lattice L_Q^* described above, a bottom-top reconstruction of quantum mechanics in Hilbert space was envisaged by several authors. Starting from a formal language of quantum physics it seemed to be possible to proceed in a few steps to quantum logic, to the lattice L_Q^* and finally to the lattice L_H of closed subspaces of Hilbert space. The last step was strongly motivated by the Piron-McLaren theorem[3] [2, 13, 14] which states that a lattice L_Q^* (of length at least 4) is isomorphic to the lattice $L_H(D)$ of closed subspaces of a Hilbert space $H(D)$ over a division ring D, where D is given by the real, the complex, or the quaternion numbers. If the real and the quaternion numbers could be excluded by experimental evidence, we would arrive at the Hilbert space H over the complex numbers and thus at quantum mechanics in Hilbert space.

However, the lattice L_Q^* does not restrict the choice of the division ring per se to the real, the complex and the quaternion numbers. Quite surprisingly, Keller [15] proved a negative result in 1980. There are lattices L_Q^* that fulfil all the conditions of the Piron-McLaren theorem but nevertheless allow for non-classical Hilbert spaces over non-Archimedean division rings. This unexpected result was considered by some authors as demonstrating the fundamental impossibility of the quantum logic approach to quantum mechanics in a Hilbert space over the complex numbers. Hence, the bottom-top reconstruction of quantum mechanics mentioned, was supposed to be impossible[4] [19, 20]. However, this discouraging conclusion has been contradicted by an important result by Solèr that allows for a purely lattice-theoretical characterisation of classical Hilbert spaces. In fact, every lattice

2 Cf. P. Mittelstaedt, (1978) and (2005), Chapter 13.

3 Cf. Piron (1964), McLaren (1965), and Varadarajan (1968).

4 For more details cf. Dalla Chiara et al. 2001, pp. 48–50 and Dalla Chiara et al. 2004, pp. 72–74.

which satisfies in addition to the conditions of the Piron-McLaren theorem also the so-called "angle bisecting condition" [16], is isomorphic to a classical Hilbert lattice [17].

Although this mathematical result provides some hope to achieve one day the main goal of quantum logic, the bottom-top reconstruction of classical Hilbert lattices, this goal is still far away. The missing link is an operational condition for quantum mechanical propositions that finally leads – within the lattice-theoretical formulation of quantum logic – to the "angle-bisecting condition" mentioned above. Only if this "operational Solèr condition" can be formulated and justified by plausible physical reasoning, the quantum logical reconstruction of quantum mechanics could be considered as finally established.

Primary Literature

1. G. Birkhoff, J. von Neumann: *The logic of quantum mechanics*. Ann. Math. 37, 832–43 (1936).
2. C. Piron: *Axiomatique quantique*. Helv. Phys. Acta 37, 439–68 (1964).
3. J.M. Jauch: *Foundations of Quantum Mechanics*. Addison-Wesley, Reading. MA (1968).
4. P. Lorenzen: *Einführung in die operative Logik und Mathematik*. Springer, Heidelberg (1955).
5. P. Lorenzen: *Metamathematik*. BI-Wissenschaftsverlag, Mannheim (1962).
6. P. Mittelstaedt: *Untersuchungen zur Quantenlogik*. Verlag der Bayerischen Akademie der Wissenschaften, München (1959).
7. P. Mittelstaedt: *Philosophische Probleme der Modernen Physik*. BI-Wissenschaftsverlag, Mannheim (1963).
8. E.-W. Stachow: *Completeness of Quantum Logic*. J. Philos. Logic 5, 237–80 (1976).
9. E.-W. Stachow: *Structures of a quantum language for individual systems*, In: Recent Developments in Quantum Logic, eds. P. Mittelstaedt, E.-W. Stachow: BI-Wissenschaftsverlag, Mannheim, 129–45 (1984).
10. J. M. Jauch, C. Piron: *What is Quantum Logic?* In: Quanta, Essays in Theoretical Physics dedicated to Gregor Wentzel, eds. P.G.O. Freund, C. G. Goebel, Y. Nambu: University of Chicago Press, 166–81 (1970).
11. P. Mittelstaedt: *On the interpretation of the lattice of subspaces of the Hilbert space as a propositional calculus,* Zeitschrift für Naturforschung, 27a, 1358–362 (1972).
12. P. Mittelstaedt: *Quantum Logic*. D. Reidel Publishing Company, Dordrecht (1978).
13. M. D. McLaren: *Nearly modular orthocomplemented lattices,* Trans. Am. Math. Soc. 114, 401–16 (1965).
14. V. S. Varadarajan: *Geometry of Quantum Theory*. vol.1 Van Nostrand, Princeton.
15. H. A. Keller: *Ein nichtklassischer Hilbertraum,* Mathematische Zeitschrift 172, 41–9 (1980).
16. R. P. Morash: *Angle bisection and orthoautomorphism in Hilbert lattices*. Can. J. Math. 25, 261–72 (1973).
17. M. P. Solèr: *Characterisation of Hilbert spaces by orthomodular lattices*. Commun. Algebra 23 (1), 219–43 (1995).

Secondary Literature

18. P. Mittelstaedt with P. Weingartner: *Laws of Nature*. Springer, Heidelberg (2005).
19. M. L. Dalla Chiara, R. Giuntini: *Quantum Logics*. arXiv:quant-ph/0101028 v2 (2001).
20. M. L. Dalla Chiara, R. Giuntini, R. Greechie: *Reasoning in Quantum Theory*. Kluwer, Dordrecht (2004).

Quantum Mechanics

See: Born rule; Heisenberg picture; Schrödinger picture; Schrödinger equation; Uncertainty relation; Orthodox interpretation; relativistic quantum mechanics; wave mechanics.

Quantum Numbers

Klaus Hentschel

Within the context of the ► atomic model by Niels Bohr (1885–1962), observable spectrum lines of frequency ν are described as ► quantum jumps of bound ► electrons between quantized energy levels E_n according to the rule: $\nu_{nm} = (E_n - E_m)/h$, with h = a quantum of action, ► Planck's constant. With the small correction for the so-called reduced mass, the energy E_n of each electron orbit around the atomic core is given as:

$$E_n = \frac{2\pi^2 e^4 m M}{h^2(m + M)} \cdot \frac{1}{n^2}$$

m = the mass of the electron; M = the mass of the atomic core; n is the *first* (*or "main" quantum number*) mainly determining the energy level of each electron, aside from small corrections mostly relevant to precision ► spectroscopy and described by other subsequently introduced quantum numbers.

As the analogy between the planetary orbits around a massive sun and electron orbits around the positively charged nucleus already implies, these electron orbits would generally not be circles but ellipses. However, within the framework of classical mechanics, all ellipses generated from the circle by adiabatic transformations are energetically equivalent to the circle, so Bohr initially thought that other orbit forms would be reducible to simple circular orbits. But Arnold Sommerfeld (1868–1951), a theoretical physicist trained as a mathematician and familiar with Einstein's theory of relativity, noted that electrons on highly eccentric orbits increase speed when approaching the nucleus. Relativistically, this leads to a slight increase in their mass and thus to a slight drop in energy of the respective orbit against a circular orbit.

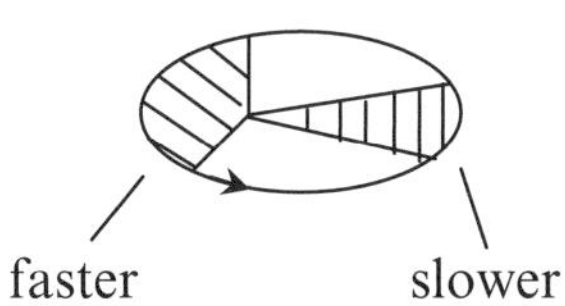

In order to describe this, Sommerfeld introduced another *azimuthal quantum number l* (*sometimes also called k or* n_φ), describing the degree of eccentricity of the electron orbit, with a the largest and b the smallest diameter (see, for instance, [5]). Classically, all eccentricities $\varepsilon = b/a$ are permissible, but within the early ► quantum theory another ► quantization condition is imposed and only certain orbits are allowed for which

$$J_\varphi = \int_0^{2\pi} p_\varphi d\rho = n_\varphi \cdot h \equiv l \cdot h$$

and the eccentricity $\varepsilon = b/a$

$$\frac{b}{a} = \frac{n_\varphi}{n} = \frac{l}{n} \quad \text{with } l < n$$

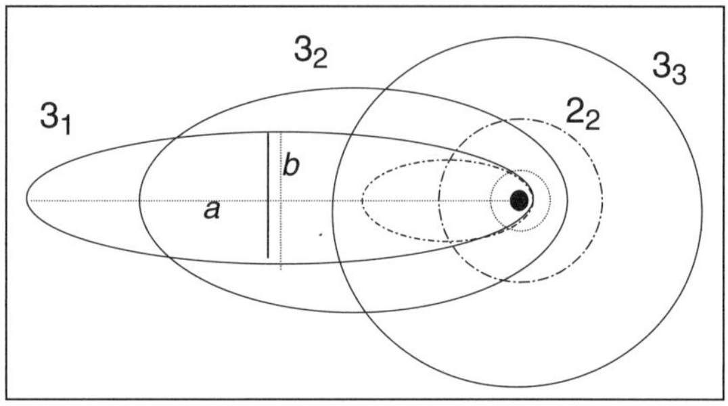

When an external field is imposed on the atom, these ellipses can orient themselves in various ways with respect to the field (for instance, a magnetic field causing the ► Zeeman effect or an electric field leading to the ► Stark effect). Again, classically, all angles between orbit and external field would be permissible, but in quantum theory only certain angular orientations α are allowed (see also ► Stern–Gerlach experiment and ► vector model). Systematic analysis of data from the ► spectroscopy of Zeeman multiplets showed that all permissible orientations could be labelled with one additional *magnetic quantum number m*, with $|m| \leqslant l$, thus $m = -l, -l+1, -l+2 \ldots, 0, 1, 2 \ldots l-2, l-1, l$; and for the angle α: $\cos\alpha = m/1$ and $|m| \leqslant |l| \leqslant |n|$.

As is explained in more detail in the article on ► spin, in January 1925 Wolfgang E. Pauli (1900–1958) first expressed this mechanically indescribable ambiguity as a new *quantum number* μ, later redubbed $s = \pm 1/2$ (for doublets). Each electron was described by a set of four ► quantum numbers:

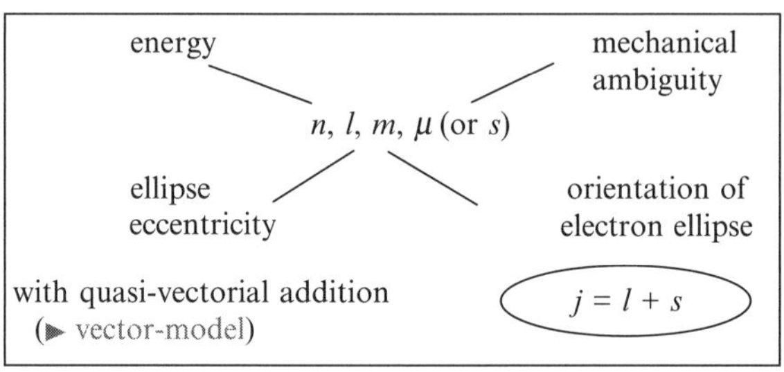

With this set of *four different quantum numbers n, l, m,* and s (sometimes alternatively n, l, j, and s), it was possible to classify all electrons in bound states around an atom's positively charged core. In order to achieve a perfect fit with the number of atoms in each row of the periodic table, Pauli had to introduce another constraint on the shell structure: no two electrons of an atom may have all the four quantum numbers in common, the *Pauli principle* (► exclusion principle):

> There can never be two or more equivalent electrons in the atom in which the values of all [four] quantum numbers... concur within a strong field... If in the atom there is an electron for which these quantum numbers... have specific values, then this state is occupied. [2, p. 776]

The electron configuration of each atom was constructed of shells, starting from the lowest possible energy level, i.e., the lowest main quantum numbers $n = 1, 2, \ldots$, and so on. For each given n, there will be $n - 1$ different eccentricities l, and for each l, there will be $2l + 1$ different space orientations, and finally two different spin orientations.

For $n = 1, l = 0$, therefore, only two electrons are in the lowest shell; for $n = 2$, l will either be 1, with $m = -1, 0$, or -1, or l will be 0. Altogether, because spin orientation yields another factor 2, we have $2 \times (3+1) = 8$ electrons in the next shell, for $n = 3$, the resulting total will be $2 \times (5+3+1) = 18$ and so on. We thus see that the resulting series of so-called *golden numbers* 2, 8, 18, 32, ..., perfectly fits the structure of the periodic table of the elements, with only two chemical elements in the first row (hydrogen and helium), eight in the second row (starting with lithium and ending with neon), etc. Bohr and Pauli had succeeded in deriving the usual period lengths of the periodic table. The arrangement of the periodic system of the elements thus seemed to make a little more sense again, at least as far as the main groups were concerned. But it came at the cost of a "classically indescribable kind of ambiguity"; and Pauli's prohibition of any duplication among the quantum numbers occupying a given state was no better justifiable according to classical theory and only understood within the context of the ► Fermi–Dirac statistics of later quantum mechanics.

Primary Literature

1. W. Pauli: Über den Einfluß der Geschwindigkeitsabhängigkeit der Elektronenmasse auf den Zeemaneffekt. Zeitschrift für Physik **31**, 373–385 (received Dec. 2, 1924, publ. Feb. 19, 1925)
2. W. Pauli: Über den Zusammenhang des Abschlusses der Elektronengruppen im Atom mit der Komplexstruktur der Spektren. Zeitschrift für Physik **31**, 765–783 (received Jan. 16, publ. Mar. 21, 1925)
3. W. Pauli: *Wissenschaftlicher Briefwechsel mit Bohr, Einstein, Heisenberg u.a.*, ed. by Karl von Meyenn, vol. 1 (1919–1929) (Springer: Berlin 1979)
4. S. Goudsmit & G. E. Uhlenbeck: Ersetzung der Hypothese vom unmechanischen Zwang durch eine Forderung bezüglich des innern Verhaltens jedes einzelnen Elektrons. Die Naturwissenschaften **13**, 953–954 (1925)
5. A. Sommerfeld: *Atombau und Spektrallinien* (Vieweg: Braunschweig 1st ed. 1919 trans. *Atomic Structure and Spectral Lines*, Methuen: London 1923)

Secondary Literature

6. F. Hund: *Geschichte der Quantentheorie*, BI: Mannheim 1984, 4th ed. 1996
7. M. Jammer. *The Conceptual Development of Quantum Mechanics*, (McGraw Hill: New York 1966, Chaps. 2–3)
8. N. Robotti: Quantum numbers and electron spin. Archives Internationales d'Histoire des Sciences **40**, 305–331 (1990)

Q

Quantum State Diffusion Theory (QSD)

Mauricio Suárez

Quantum state diffusion (QSD) is possibly the most sophisticated collapse interpretation on offer today. It is closely related to the Ghirardi–Rimini–Weber (GRW) style-theories (▶ GRW), but it assumes that free particles are idealisations. According to QSD all physically real particles are subject to a degree of interaction with their environment. The fundamental equation of QSD is the linear master equation, which looks just like the ▶ Schrödinger equation, but with additional terms besides the usual Hamiltonian [1, pp. 44–45]:

$\mathrm{d}\rho/\mathrm{d}t = -\mathrm{i}/h[\mathbf{H}, \rho] + \sum_j (\mathbf{L}_j\, \rho\, \mathbf{L}_j{}^* - {}^1\!/_2\, \mathbf{L}_j{}^*\mathbf{L}_j\, \rho - {}^1\!/_2\, \rho \mathbf{L}_j{}^*\, \mathbf{L}_j)$, where the Lindblad operators $\mathbf{L}_j$ may or not be Hermitian.

The two limiting cases are:

1. LINDBLAD: The environmental interaction dominates and the Hamiltonian internal dynamics is negligible (these are "wide open systems" ▶ decoherence):

$$\mathrm{d}\rho/\mathrm{d}t = \sum_j (\mathbf{L}_j\, \rho\, \mathbf{L}_j{}^* - {}^1\!/_2\, \mathbf{L}_j{}^*\, \mathbf{L}_j\, \rho - {}^1\!/_2\, \rho\, \mathbf{L}_j{}^*\, \mathbf{L}_j).$$

2. SCHRÖDINGER: The environmental interaction is negligible and the Hamiltonian dynamics dominates ("completely isolated systems"):

$$\mathrm{d}\rho/\mathrm{d}t = -\mathrm{i}/h\, [\mathbf{H},\, \rho]\,.$$

So QSD recovers the Schrödinger equation for the idealisation of a completely isolated system. In general, however, the full linear master equation applies, and the resulting diffusion process for the quantum state on the Bloch sphere is similar to ▶ Brownian motion in 3-d physical space. A measurement is typically modelled within QSD as a wide open system interaction with a macroscopic measuring device [2]. Thus QSD predicts a transition from a pure state (▶ states, pure and mixed) to a ▶ mixed state for the pointer position, which it claims solves the measurement problem. See also ▶ Bohmian mechanics; Measurement theory; Metaphysics in Quantum Mechanics; Modal Interpretation; Objectification; Projection Postulate.

Literature

1. Ian Percival: *Quantum State Diffusion* (Cambridge University Press, Cambridge 1999)
2. N. Gisin, I. C. Percival: The quantum-state diffusion model applied to open systems. J. Phys. A: Math Gen. **25**, 5677–5691 (1992)

Quantum State Reconstruction

Stefan Weigert

Quantum state reconstruction, or state reconstruction for short, aims at identifying an unknown quantum state (► states in quantum mechanics) on the basis of experimentally accessible data. The Quantum Optics community usually refers to this inverse problem as *quantum (state) tomography* while the expression *quantum state estimation* is often used in the field of ► Quantum computation. Reconstruction procedures depend on the physical context defined by the system carrying the unknown state, the experimentally accessible ► observables, the size of the ► ensemble of systems prepared in the unknown state, and the precision of the measured data.

A two-level system (such as a spin-1/2, a qubit, or the two polarizations of a photon) prepared in a state with density matrix $\hat{\rho}$ is sufficient to illustrate the idea of state reconstruction. With two non-negative eigenvalues summing to one, the density matrix is a *positive* operator, and it depends on *three* real parameters. In the Bloch representation, the parameters combine to a real vector $\boldsymbol{n}$ with length $|\boldsymbol{n}| \leqslant 1$,

$$\hat{\rho} = \frac{1}{2}\left(\mathbb{I} + \boldsymbol{n} \cdot \hat{\boldsymbol{\sigma}}\right) ,$$

where $\mathbb{I}$ denotes the identity operator, and the components of the spin operator $\hat{\boldsymbol{\sigma}}$ are given by the ► Pauli matrices $\hat{\sigma}_x$, $\hat{\sigma}_y$, and $\hat{\sigma}_z$. This parametrization of the density matrix $\hat{\rho}$ is immediately useful for state reconstruction since the components of the vector $\boldsymbol{n}$ coincide with the expectation values of the ► Pauli matrices in the state $\hat{\rho}$,

$$n_j = \mathrm{Tr}[\hat{\sigma}_j \hat{\rho}] \equiv \langle \hat{\sigma}_j \rangle_\rho , \quad j = x, y, z .$$

Q

The three observables $\hat{\sigma}_x$, $\hat{\sigma}_y$, and $\hat{\sigma}_z$ are *informationally complete*: any state $\hat{\rho}$ of the two-level system is determined uniquely by the values of the measured expectations $\langle \hat{\sigma}_x \rangle_\rho$, $\langle \hat{\sigma}_y \rangle_\rho$, and $\langle \hat{\sigma}_z \rangle_\rho$. No *pair* of observables allows one to reconstruct the state of a two-level system but many other *triples* (and larger sets) of observables exist which are also informationally complete. This flexibility is highly desirable from an experimental point of view. Specific reconstruction procedures will take into account any additional information: if a system is known to reside in a pure state (► states, pure and mixed), for example, it will be sufficient to measure a smaller number of expectation values.

The reconstruction of a quantum state in a laboratory is necessarily based on expectation values which are known only *approximately*: any ensemble used to measure an expectation value such as $\langle \hat{\sigma}_x \rangle_\rho$ is *finite*, and any measuring apparatus invariably introduces uncertainties. Consequently, the collected data will be compatible with a continuous family of quantum states. The reconstruction is complicated by the fact that unacceptable density matrices with *negative* eigenvalues may arise upon inverting the information contained in experimentally observed mean values. To determine the 'best' candidate among the acceptable states requires additional selection criteria such as the maximum-likelihood method, for example.

In 1933, W. Pauli raised he question [1] whether the probability distributions $|\langle q|\psi\rangle|^2 dq$ (to find a particle located near position q) and $|\langle p|\psi\rangle|^2 dp$ (to find the particle with a momentum close to p) determine a single pure state $|\psi\rangle$. This is an early instance of quantum state reconstruction, with a negative answer: in general, there is a family of pure states, called Pauli partners, which give rise to the *same* Pauli data.

E. Schrödinger suggested in 1935 to think of the ▶ wave function as a *catalogue of expectations*, that is, a tool which succinctly holds the information about the expectation value of any observable [2]. *In nuce*, this remark contains the concept of quantum state reconstruction. Knowing the expectation values of all observables effectively means to know the quantum state, and only a technical problem remains to be solved, namely to identify an informationally complete set of observables, or *quorum*. Given such a quorum it becomes possible to express Schrödinger's equation in terms of expectation values only – thereby eliminating any reference to the wave function or density matrix of the system [3].

The tomography of classical objects has inspired a successful method of quantum state reconstruction. Quantum tomography is based on the Wigner function (▶ *Wigner distribution*), an intuitively appealing way to represent the state $\hat{\rho}$ of a quantum particle. This real function resembles a classical probability distribution for two real variables q and p although it may take negative values and, therefore, cannot be observed experimentally. It is not difficult, however, to derive *marginals* from the Wigner function which are legitimate probability distributions. As shown in 1989, suitable families of marginals provide sufficient information to recover the Wigner function and, *a fortiori*, the unknown state $\hat{\rho}$ [4]. The marginals can be measured through optical homodyning, a well-established technique of quantum optics, as has been demonstrated experimentally in 1993 [5].

Regarding the efficiency of different reconstruction schemes, some quantitative results are known for states residing in a d-dimensional space. Given a finite ensemble of quantum systems in one and the same state, the statistical error is *minimal* if measurements are performed with respect to $d + 1$ sets of *mutually unbiased* bases, each containing d observables [6]. So far, the required set of observables has been found to exist only if the dimension d equals the power of a prime number.

To extract maximal information about an unknown state of which N copies of are provided, it is often advantageous to go beyond the traditional framework of ▶ projective measurements, using ▶ positive operator-valued measurements instead. Within the field of quantum cloning (▶ no-cloning theorem), the quality of a given reconstruction procedure is measured by the *fidelity* which compares the estimated state to the original one.

Primary Literature

1. W. Pauli: Die allgemeinen Prinzipien der Wellenmechanik. In *Handbuch der Physik*, ed. by H. Geiger and K. Scheel (Springer, Berlin 1933, Vol. 24, 98)
2. E. Schrödinger: Die gegenwärtige Situation in der Quantenmechanik. Naturwiss. 23, 823–828 (1935)
3. S. Weigert: Quantum Time Evolution in Terms of Non-Redundant Probabilities. Phys. Rev. Lett. 84, 802–805 (2000)
4. K. Vogel and H. Risken: Determination of quasiprobability distributions in terms of probability distributions for the rotated quadrature phase. Phys. Rev. 40, 2847–2849 (1989)
5. D. T. Smithey, M. Beck, M. G. Raymer, and A. Faridani: Measurement of the Wigner distribution and the density matrix of a light mode using optical homodyne tomography: application to squeezed states and the vacuum. Phys. Rev. Lett. 70, 1244–1247 (1993)
6. W. K. Wootters and B. D. Fields: Optimal state-determination by mutually unbiased bases. Ann. Phys. 191, 363–381 (1989)

Secondary Literature

7. U. Leonhardt: *Measuring the Quantum State of Light* (Cambridge University Press, Cambridge 1997)
8. M. Paris and J. Řeháček (eds.): *Quantum State Estimation* (Springer 2004)
9. L. M. Artiles, R. D. Gill, and M. I. Guţă: An invitation to quantum tomography. J. R. Statist. Soc. B 67, 109–134 (2005)

Quantum Statistics

Arianna Borrelli

Q

In *quantum statistics*, the behaviour of quantum systems with a large number of degrees of freedom (e.g. an assembly of many particles) is investigated with the help of statistical considerations [1]. Although in principle analogous to classical statistical mechanics, the statistics of quantum systems requires more caution than the classical one.

There are two main differences between the classical and the quantum case, and they are linked to the ► Heisenberg uncertainty principle and to the ► indistinguishability of quantum particles of the same kind. According to the uncertainty principle, even the most complete description of the state of a quantum system will not allow unique predictions for the values of all observable quantities. This intrinsically quantistic uncertainty has to be carefully combined with the classical uncertainty due simply to our ignorance of the state of the system. This task is accomplished by employing the formalism of ► state operator and ► density matrix.

Moreover, when two or more quantum particles of the same kind (e.g. photons ► light quantum; ► electrons) are present in a system, the number of the system's possible states must be determined by a counting procedure different from the one employed in the classical case. This requirement is variously described as"indistinguishabiliy","▶ identity" or "permutivity" of quantum particles.

The earliest forms of quantum statistics to emerge were two counting procedures for indistinguishable particles which established themselves as physically significant around the middle of the 1920s: the statistics of ▶ Bose–Einstein (1924) and that of ▶ Fermi-Dirac (1925–1926). From the late 1920s onward, with the development of the formalism of the state operator, a more general formulation of quantum statistics became possible [2, 3].

In quantum mechanics, having maximum information about a system means knowing that it is in a pure state (▶ states, pure and mixed) described by a specific state vector $|\psi\rangle$ in ▶ Hilbert space. In this case, only quantum uncertainty enters the picture. Otherwise, the system is said to be in a ▶ mixed state characterized by a probability distribution over all possible state vectors $|\psi_\alpha\rangle$, and it is described by a state operator ρ. Given an ▶ orthonormal basis $|i\rangle$, any vector $|\psi_\alpha\rangle$ can be written as

$$|\psi_\alpha\rangle = \sum_i a_i^\alpha |i\rangle.$$

A mixed state can thus be defined by a probability distribution $P(\alpha)$ over the sets $\{a^\alpha\}$. The relevant state operator ρ is then represented in the basis $|i\rangle$ by the density matrix:

$$\rho_{ij} = \sum_\alpha P(\alpha)\, a_i^\alpha a_j^{\alpha\,*} = \langle a_i a_j^* \rangle,$$

where $\langle\rangle$ represents the average according to the distribution $P(\alpha)$. The diagonal elements ρ_{ii} of the density matrix give the probability of finding the system in the state $|i\rangle$. Using state operator and density matrix, the average value of any observable can be computed keeping into account both quantum and statistical uncertainty at the same time [4].

To perform quantum statistical computations, it is necessary to make some initial assumptions on $P(\alpha)$. In analogy to the classical case, all possible pure states of a system are considered equally probable if no other information is available (postulate of equiprobability). For quantum systems in thermal equilibrium, the density matrix ρ_{ij} is assumed to be diagonal when the chosen basis vectors $|i\rangle$ are energy eigenstates. If the energy of the system is conserved, this means that ρ will not change with time. A sufficient condition for having $\langle a_i a_j^* \rangle = 0$ for $i \neq j$ is for the relative phases of the coefficients a_i to be distributed randomly (postulate of random phases). See also ▶ Generalization of Quantum Statistics.

Primary Literature

1. K. Huang: *Statistical Mechanics* (Wiley, New York 1963, 183–205)
2. R. C. Tolman: *The Principles of Statistical Mechanics* (Oxford University Press, Oxford 1938)
3. D. ter Haar: *Elements of Statistical Mechanics* (Rinehart, New York 1956)
4. K. Blum: *Density Matrix Theory and Applications* (Plenum, New York 1981)

Quantum Theory, 1914–1922

▶ Bohr's Atomic Model
▶ Specific Heats
▶ Spectroscopy

Quantum Theory, Crisis Period 1923–Early 1925

Klaus Hentschel

Niels Bohr's (1885–1962) atomic model as one of the cornerstones of pre-1925 ▶ quantum theory was incredibly successful for a whole decade, from 1913 to roughly 1922. ▶ The Bohr–Sommerfeld atomic model allowed a qualitative understanding of the basic spectrum series of hydrogen ▶ Bohr's atom model and hydrogen-like atoms. It was also possible to extend this basic model to incorporate additional, subtle effects such as the correction of the Rydberg constant due to the effective mass calculation of atomic nucleus plus ▶ electron, or relativistic corrections due to the very high orbital velocity of strongly bound electrons close to the nucleus. The ▶ semi-classical models also explained the observed splitting of spectrum lines in electric and magnetic fields (▶ Stark effect, ▶ Zeeman effect). X-ray spectra also fell into place with the work by Henry Moseley (1887–1915) and others on ▶ quantum jumps of electrons from inner orbits (see [1, 5, 7, 9–12]). Around 1920, Bohr and his collaborators in Copenhagen were busy explaining how to build up the periodic system using the idea of successively filling available places in an electron orbit or shell ([1] vol. 4, [13]). Closed shells were linked to the 'golden' or 'magic' numbers 2, 8, 18, 32. An even more intricate form of this 'number mysticism', as some of the actors jestingly called this playing with fitting formulae devoid of physical interpretation, seemed to allow at least a partial mapping of the complicated spectrum line splittings observed in the anomalous Zeeman effect, for instance (▶ Landé g-factors and further refs. given there).

By the early 1920s various problems emerged, however, that turned out not to be treatable within the framework of Bohr's and Sommerfeld's quantum theory, despite the relentless efforts of the ▶ Sommerfeld school in Munich and competing groups in Göttingen, Copenhagen, and Leiden. The spectrum line intensities of the Zeeman and Stark ▶ multiplets, for instance, could not be calculated satisfactorily, nor did many of the heavier atoms seem to follow the patterns of hydrogen-type atoms. The model could thus not be extended further and an impasse seemed to have been reached ([14–16]). Worse still, persistent anomalies surfaced pointing to aggravating discrepancies between theory and experimental data, which had already reached a relative margin of error of 10^{-8} and better in precision ▶ spectroscopy.

In 1922 Werner Heisenberg (1901–76), then still studying under Arnold Sommerfeld (1868–1951) in Munich, started trying to account for even-numbered multiplets and other subtleties in this "Zeeman botany in quantum sauce" (another ironic term of the time) by introducing semi-integral ► quantum numbers n. The usual formula for multiplicity $m = 2n+1$ of uneven multiplets was thus formally extended to even ones like the infamous doublets in alkali spectra ([6, 8]). But what did these half-integral quantum numbers correspond to? In late 1924, Wolfgang Pauli (1900–58) started to toy with the idea of a "mechanically unaccountable duplicity" or ambiguity (*klassisch nicht beschreibbare Zweideutigkeit*), a strange precursor to the idea of ► spin (which only emerged at the end of 1925—too late to rescue the old ► quantum theory from its internal problems, but of crucial importance for later quantum mechanics).

It was at this time that Heisenberg wrote to his teacher Sommerfeld: "This state of physics really doesn't appeal to me" (4 Jan. 1923, in [2], p. 134; cf. also [6]). Two years later, the situation had deteriorated even further. Wolfgang Pauli was utterly disgusted. He wrote to Ralph de Kronig (21 May 1925, in [4], p. 216):

> "Physics is very much stuck in a rut again at the moment; it is far too hard for me, at least, and I wish I were a film comedian or the like and had never heard of physics".

Always a bit ahead of others about the state of affairs, Pauli wrote to Sommerfeld on Dec. 6, 1924 ([4], p. 182, [2] p. 177):

> "The conceptual models are in serious crisis now, you know, of a principal nature, which I believe will end in another radical sharpening of the contrast between classical and quantum theory. [...] the concept of definite, clear electron orbits within the atom are [probably] hardly maintainable. One gets the impression from all models now that we're speaking an inadequate language for the simplicity and beauty of the quantum world."

Bohr wrote even more pointedly in late 1924: "I have the feeling that we stand at a turning point, since now the extent of the entire swindle has been characterized so exhaustively" (22 Dec. 1924, German transl. of the orig. Danish in [4], p. 195; all English translations by Ann M. Hentschel).

The increasing frustration and mounting uncertainty about the further trajectory emboldened physicists to venture down unconventional paths. Even unheard-of, radically new ways out of the dilemma were tried. It became permissible to break with everything, even with former sacred cows like integral values for quantum numbers (Heisenberg in 1924) and the law of conservation of energy (see the entry on the short-lived ► Bohr–Kramers–Slater theory of 1924).

But what could one cling to in this search for a new framework? What could be the stable foundations of an otherwise radically new quantum theory? The answer that Heisenberg and Pauli gave was crystal clear, naive though it was: empirical facts, i.e. in their understanding, experimentally verifiable, multiply confirmed statements about observable quantities such as energy intervals, frequencies or line intensities (all directly based data from ► spectroscopy), ionisation levels and lowest binding energies (data from ► scattering experiments, gas-ionisation and spark spectra, for instance).

This new, somewhat positivistic insistence on ► observables, as they were soon called, was not surprising. Pauli, in particular, had grown up in the 'anti-metaphysical' context of fin-de-siècle Vienna. Actually, he was the god-son of Vienna's foremost apostle of phenomenalist thinking, the physicist-philosopher Ernst Mach (1836–1916). Pauli was thus the first to stop referring to electron orbits, perhaps reminding himself of what Mach had always asked when someone in his presence talked about atoms as something immediately given: "Hab's aans g'sehn?" Have you ever seen one? Like atoms, electron orbits around the atomic core also were only indirectly inferred from a complicated chain of hypothetico-inductive reasoning and were thus by no means directly perceptible. Who could guarantee that electron orbits actually existed? So Pauli decided to scrap this 'metaphysical' concept and to concentrate on observables:

> "The relativistic doublet formula seems to me to show beyond doubt now that not just the dynamical concept of force [Hertz] but also the kinetic concept of motion in classical theory will have to undergo profound modifications. (That is why I also avoided the term 'orbit' in my paper throughout.) As this concept of motion is based on the correspondence principle, above all theoreticians must work on clarifying it. I think that energy and momentum values of stationary states are something much more real than 'orbits'. [...]
>
> We must not bind the atoms in the chains of our prejudices – to which, in my opinion, also belongs the assumption that electron orbits exist in the sense of ordinary mechanics – but we must, on the contrary, adapt our concepts to experience" (Pauli to Bohr, 12 Dec. 1924, [4], 188f.)

For a while Heisenberg remained skeptical about this radical suggestion and tried other avenues (including the half-integer quantum numbers), but he failed to reach closer agreement with the observed intensities of spectrum lines. In June 1925 he gave up and decided to implement Pauli's demand for "a profound modification of the classical concept of motion". In describing the state of a mechanical system, he consistently only used observable oscillation frequencies and amplitudes and represented them by an integral of quantities in quantum theory. As Max Born (1882–1970) was quick to point out, Heisenberg was applying a type of mathematics totally new to him: matrix algebra.

Q

In his pathbreaking paper about 'a quantum-theoretical reinterpretation of kinematical and mechanical relations', Heisenberg wrote in July 1925 [3]:

> "In this situation it seems more advisable to completely abandon all hope of observing the hitherto unobservable quantities (like location, revolving time of the electron), [...] and to try to develop a quantum-theoretical mechanics analogous to the classical mechanics, in which only relations between observables occur."

Heisenberg is more explicit in a letter to Pauli from 9 July 1925, in which he enclosed his manuscript for critique before submitting it for publication ([4], p. 231):

> "It really is my conviction that an interpretation of the Rydberg formulas in the sense of circular or elliptic orbits in classical geometry do not make the slightest physical sense and my whole pathetic efforts go toward completely stamping out the concept of orbits, which cannot be observed anyway, and to replace them suitably."

Thus ► matrix mechanics was born and with it the first step toward a new generation of theories all somewhat equivalent to each other, also including Schrödinger's

► wave mechanics, Born's and Wiener's operator mechanics and Dirac's q-algebra. They are subsumed under the label quantum mechanics, which this dictionary uses throughout to label the bundle of new theories that emerged since the summer of 1925.

Why is this relatively short episode in the much longer history of quantized theories so important to deserve its own entry here? First of all, it contains some of the most exciting moments the history of twentieth century physics has to offer. Secondly, this episode is significant also from a philosophical point of view. To understand the course of events leading from the old, stable and semi-classical quantum theory of 1913 to 1922 to the new, equally successful and even more stable paradigm of quantum mechanics of post 1925, Thomas Kuhn's (1922–1996) model of scientific revolutions comes to mind. It describes such transitions between stable, but mutually incompatible paradigms. According to Kuhn, this transition should be preceded by a crisis of the old paradigm, with ever growing numbers of anomalies and mounting frustration among practitioners of the old craft. This is precisely what happened here, so this episode actually provides one of the best fits in the history of science for the general pattern described by Kuhn's model of scientific revolutions. In particular, the final stage of the old quantum theory between late 1922 and early 1925 encompasses various characteristics of a deep crisis of a reigning but threatened paradigm in Kuhn's sense:

- A hectic proliferation of various different ad-hoc models and schemes,
- Futile efforts to find correspondence rules between these various ad-hoc schemes,
- An inability to supplant the traditional phenomenological approach with causal reasoning,
- A sort of 'anything goes' mentality as a result of these problems,
- Deep disappointment with the current state of the discipline.

The fit within Kuhn's scheme is incomplete, though. Rather than being fully incommensurable, the old quantum theory and the new quantum mechanics were more intricately related to each other (see ► correspondence principle, ► quantum statistics). But many years were needed before this was fully understood.

Primary Literature

1. N. Bohr: *Collected Works* (Amsterdam: North Holland, ed. by Leon Rosenfeld, Eric Rüdinger, Finn Aaserud et al.), Vol. 2: *Works on Atomic Physics 1912–1917* (publ. 1981), vol. 3: *The Correspondence Principle 1918–23* (publ. 1976), vol. 4: *The Periodic System 1920–23* (publ. 1977), and vol. 5: *The Emergence of Quantum Mechanics, mainly* 1924–26 (publ. 1984)
2. M. Eckert & K. Märker (eds.): *Arnold Sommerfeld – Wissenschaftlicher Briefwechsel* (GNT, Stuttgart 2004, Vol. 2: 1919–1951)
3. W. Heisenberg: Über die quantentheoretische Umdeutung kinematischer und mechanischer Beziehungen, *Zeitschrift für Physik* **33** (1925), 879–893; Engl. transl. in B. L. van der Waerden (ed.): *Sources of Quantum Mechanics* (Dover, New York 1968, pp. 261–276)
4. K. von Meyenn (ed.): *Wolfgang Pauli – Wissenschaftliche Korrespondenz* (Springer, Berlin 1979, Vol. I: 1919–1929)

5. A. Sommerfeld: *Atombau und Spektrallinien* (Vieweg, Braunschweig 1st ed. 1919 contrasted against the 5th edn. 1931), also translated into English as *Atomic Structure and Spectral Lines* (Methuen, London: 1923, from the 3rd German edition in 1923 and from the 5th ed. in 1934)

Secondary Literature

6. D. Cassidy: *Werner Heisenberg and the crisis in quantum theory*, 1920–1925 (Ph.D. thesis, Purdue University 1976)
7. O. Darrigol: *From c-Numbers to q-Numbers* (University of California Press, California 1992, Chaps. 8–10)
8. K. Hentschel: Heinrich Hertz's *Mechanics*: A model for Werner Heisenberg's April 1925 paper on the anomalous Zeeman effect', in *Heinrich Hertz: Classical Physicist, Modern Philosopher*, ed. by David Baird et al. (Reidel, Dordrecht 1998, pp. 183–223)
9. H. Hettema: Bohr's theory of the atom 1913–23: A case study in the progress of scientific research programmes, *Studies in the History and Philosophy of Modern Physics* **26**, 307–323 (1995)
10. H. Hettema & T. Kuipers: Sommerfeld's Atombau: A case study in potential truth approximation, in: *Cognitive Patterns in Science and Common Sense* (Rodopi, Amsterdam 1995, 273–297)
11. M. Jammer: *The Conceptual Development of Quantum Mechanics* (Wiley, New York 1966, Chaps. 3–5)
12. H. Kragh: The fine-structure of hydrogen and the gross structure of the physics community, *Historical Studies in the Physical Sciences* **15**, 67–125 (1984/85)
13. H. Kragh: Bohr's atomic theory and the chemists, 1913–1925, *Rivista di storia della scienza* **2**, 463–486 (1985)
14. I. Lakatos: Falsification and the Methodology of Scientific Research Programmes, in Imre Lakatos & Alan Musgrave (eds.): *Criticism and the Growth of Knowledge* (Cambridge University Press, Cambridge 1970, esp. pp. 138–154)
15. H. Radder: An imminent criticism of Lakatos' account of the degenerating phase of Bohr's atomic theory, *Journal for General Philosophy of Science* **13**, 99–109 (1982)
16. G. Zoubek & B. Lauth: Zur Rekonstruktion des Bohrschen Forschungsprogrammes, *Erkenntnis* **37**, 223–273 (1992)

Quantum Theory, Early Period (1900–1913)

Clayton Gearhart

The quantum made its first tentative appearance in physics in 1900, in Max Planck's (1858–1947) work on black body radiation. But only in 1913 did Niels Bohr (1885–1962) apply it to the spectrum of hydrogen. How did quantum theory develop in the intervening years? One may conveniently distinguish two themes: a quantum theory of matter, often in equilibrium with a Maxwellian electromagnetic field; and the more radical theory of ► light quanta, introduced by Albert Einstein (1879–1955) in 1905 and pursued almost exclusively by him for many years. A more general theme was the mysterious nature of the new quantum world that emerged in the early years of the twentieth century.

Planck had set the stage for the ► quantization of matter late in 1900, when he accurately described ► black-body radiation by assuming that "energy elements" of size $h\nu$ were partitioned among a collection of "resonators," oscillating electric dipoles in equilibrium with an electromagnetic field. He had adopted these finite energy elements, which he borrowed from an 1877 paper by Ludwig Boltzmann (1844–1906), in order to explain the latest measurements from the nearby *Physikalisch-Technische Reichsanstalt* (► Black Body). But in his three short papers, Planck said nothing about their physical interpretation. Did he intend them to have merely a formal significance? Did he believe they were consistent with earlier physical theory? Or had already he begun to grasp their implications, however dimly and tentatively? His contemporaries found it hard to understand him, as have later historians. Nevertheless, over the next decade it became clear that his energy elements, or quanta as they came to be called, represented a sharp and irretrievable break with earlier theory. The "quantum revolution" that over the last century has fundamentally altered our understanding of nature was underway.

In 1907, Einstein found a new arena for Planck's resonators: Using the statistical mechanics that he had developed starting in 1902, he calculated the ► specific heat of a solid at low temperatures, picturing the solid as a collection of quantized resonators. He found that the molar specific heat fell off from its equipartition value of $3R$ at high temperatures, where R is the gas constant, and approached zero as the temperature approached absolute zero. In 1907, Einstein could appeal only to limited data for the specific heat of diamond. But over the next several years his theory was brilliantly confirmed by the experiments on the specific heats of solids conducted by Walther Nernst (1864–1941) and his students in Berlin.

Nernst had begun these measurements seeking confirmation for his 1906 Heat Theorem, which concerned the equilibrium point of chemical reactions. But as he learned that his measurements also supported Einstein's predictions, he became an enthusiastic promoter of quantum theory. He played a leading role in organizing the first Solvay Conference, which met in Brussels in November, 1911 and brought together about twenty of Europe's leading physicists to ponder the implications of the new quantum ideas. This conference in turn helped persuade the physics community of their importance.

Thus by the end of 1911, Planck's resonators – quantized simple harmonic oscillators – were widely seen as essential to an understanding of both black-body radiation and the specific heats of solids. About the same time, a second material system emerged: the rotator, a rotating "dumbbell" consisting of two point masses that could be either rigidly connected, or joined by a spring. Physicists and physical chemists applied this model to both molecular spectra and the specific heats of diatomic gases.

Once again, Nernst and his assistants led the way. In a February 1911 paper, published well before the first Solvay Conference, Nernst argued that the quantum theory might shed light on long-standing puzzles in the specific heats of gases. Why, for example, do the specific heats of monatomic gases show no rotational degrees of freedom? Why, for most diatomic gases, did additional degrees of freedom gradually appear, well above room temperature.

Nernst speculated that the rotational energy of diatomic gases might show quantum effects by falling off at low temperatures, and singled out hydrogen as a particularly promising candidate for investigation. Early in 1912, Arnold Eucken (1884–1950), one of Nernst's assistants who had been closely involved in the experiments on the specific heats of solids, published measurements of the specific heat of hydrogen gas down to 35 K. In what must have been a thoroughly gratifying result, Eucken found that the specific heat fell sharply from 5/2 *RT* per mole to 3/2 *RT*, just what one would expect if the rotational degrees of freedom were freezing out.

In the same 1911 paper, Nernst developed a theoretical framework for rotating diatomic molecules. Surprisingly from a modern point of view, he did *not* quantize the rotator. Instead, he argued that rotating molecules would exchange harmonic oscillator quanta with quantized Planck resonators with which they were in equilibrium. Nernst's theory was flawed, but Einstein adopted a corrected version and outlined it briefly in his 1911 Solvay report.

In 1912, Niels Bjerrum (1879–1958), a Danish chemist working in Nernst's laboratory, applied quantum concepts to molecular spectra. Building on earlier work by Lord Rayleigh (1842–1919) and Paul Drude (1863–1906), he argued that vibrational absorption peaks appearing in the infrared should be broadened due to the effects of rotation. In contrast to Nernst, Bjerrum quantized the energies and frequencies of the rotators, perhaps following a tentative suggestion by Hendrik Antoon Lorentz (1853–1928) at the first Solvay Conference. Bjerrum's conjecture was confirmed in 1913, when Eva von Bahr (1874–1962), a Swedish physicist working in Heinrich Rubens' (1865–1922) laboratory in Berlin, found sharp peaks in the absorption spectrum of hydrogen chloride (HCl). These peaks not only confirmed the quantization of rotational motion, but provided yet another strong piece of evidence for quantum theory generally. Bjerrum and others thought that these peaks corresponded directly to quantized molecular rotation frequencies. This point of view persisted for many years, even after Niels Bohr interpreted the frequencies of *atomic* spectral lines as the *differences* between the energies of atomic energy states.

A third problem emerged from efforts to apply both Nernst's Heat Theorem and quantum theory to ideal gases, in order to find the equilibrium point of chemical reactions. Some scientists tried to quantize translational motion directly. Others assumed only that gases were in equilibrium with quantized solids. These efforts resulted in multiple derivations of the Sackur–Tetrode equation and calculations of the "entropy constant" by Otto Sackur, Hugo Tetrode, Otto Stern, Planck, and others. Some of the earliest work involving indistinguishable particles in quantum theory grew out of these efforts, which continued for many years beyond 1913. This paragraph no more than touches on a long and complex history.

All of these problems involved a quantum theory of matter, in which Maxwell's theory of electricity and magnetism still held sway. Einstein, however, in a 1905 paper that he called "very revolutionary" in a letter to his friend Conrad Habicht, put forward the radical suggestion that light consists of "a finite number of energy quanta that are localized in points of space, move without dividing, and can be absorbed or created only as a whole." He justified this point of view through an extended analogy between the entropies of an ideal gas and of black-body radiation, and pointed to several experimental effects, among them the ► photoelectric effect, that, he argued, could best be explained by these independent particle-like ► light quanta.

But light is a wave in Maxwell's theory, and experiments on diffraction and interference could be explained only by wavelike behavior. Most physicists were therefore reluctant to challenge Maxwell's highly successful theory of electromagnetism. Einstein was virtually alone in his advocacy of light quanta for nearly twenty years, until Arthur Compton's experiments made them inescapable in the early 1920s (► Compton effect).

Nevertheless, light quanta and their connection to black-body radiation remained at the center of Einstein's thoughts. An essential tool, as he probed the nature of this new quantum world, was the analysis of fluctuations that had first appeared in his pre-1905 papers on statistical mechanics. In 1909, he considered fluctuations in the energy of electromagnetic radiation described by the Planck radiation law, as well as fluctuations in the momentum of a mirror in equilibrium with such radiation. The resulting equations had two terms: One was consistent with fluctuations due to wave interference, the other with Einstein's particle-like light quanta. Einstein spoke of "a kind of fusing of the wave and emission theories of light."

In 1910 Einstein and Ludwig Hopf (1884–1939) extended this analysis to momentum fluctuations in a gas of resonators in equilibrium with a Maxwellian electromagnetic field. But this time, in a complex calculation that reduced the role of equipartition to a bare minimum, they took the radiation energy density as an unknown and applied equipartition *only* to the translational motion of the gas – a seemingly incontestable assumption. They found that the resulting energy density obeyed the impossible Rayleigh–Jeans law (► Black-body radiation). The challenge posed by Planck's new radiation law seemed more inescapable than ever. Fluctuations also figured in the famous 1916 paper in which Einstein introduced his famous A and B coefficients in a new and influential derivation of Planck's radiation law.

Fluctuations played a more ambiguous role in 1913, when Einstein and Otto Stern (1888–1969) proposed a theory to describe the specific heat of hydrogen, developing Einstein's brief sketch at the first Solvay Conference (see above). They were also investigating the implications of Planck's new zero-point energy, introduced in 1911 as part of his "second quantum theory" (► Black-body radiation, ► Zero-point energy). Following Nernst, Einstein and Stern did *not* quantize the rigid rotator. Instead, they assumed that all rotators at a given temperature had the same rotational frequency, and equated the kinetic energy, $\frac{1}{2}J\,(2\pi\nu)^2$, where J is the moment of inertia and ν the rotational frequency of the rotator, to the average energy of a Planck resonator with the same frequency, $h\nu/\left(e^{h\nu/kT}-1\right)+h\nu/2$, where the second term is the zero-point energy, and h and k are, respectively, Planck's and Boltzmann's constants. The rotational frequency is thus a perfectly continuous function of temperature. A calculation with no zero-point energy yielded an impossible curve for the specific heat. But a second calculation with a zero-point energy of $h\nu/2$ resulted in excellent agreement with Eucken's data – ironically, far better than anyone else would achieve for well over a decade.

Einstein and Stern said almost nothing about a physical interpretation. But a second and almost unrelated section of their paper makes clear that they did *not* adopt Nernst's picture of a rotator exchanging harmonic oscillator quanta with Planck's resonators. There they repeated Einstein and Hopf's calculation, again featuring a gas of resonators in equilibrium with a Maxwellian electromagnetic field. But now they added a zero-point energy $h\nu$ (*not* $h\nu/2$) to the average resonator energy. And this time, instead of the impossible Raleigh–Jeans law, they found Planck's radiation law, from which the average energy of the Planck resonators could be extracted *without* first quantizing those resonators!

Einstein and Stern touched only lightly on the implication that zero-point energies might lie behind quantum phenomena, "without recourse to any kind of discontinuities," as they put it. They hoped that further work might remove the discrepancy between the different zero-point energies in the two calculations, but nevertheless said it was "doubtful that other difficulties could be overcome without the assumption of quanta."

Within a few months, Einstein had abandoned this approach. And in spite of the good agreement with Eucken's measurements, no one else took it up. Indeed, only a few months later, Paul Ehrenfest (1880–1933) followed Lorentz's lead and published an account of the specific heat of hydrogen in which the rotators were quantized, much as Bjerrum had done for molecular spectra a year earlier.

Einstein himself could easily have taken this route. Lorentz, however tentatively, had shown the way at the first Solvay Conference, and the calculation itself was virtually identical to Einstein's 1907 calculation of the specific heats of solids. That he did not do so, and instead followed the route outlined above, shows just how fluid and uncertain the state of quantum theory remained, more than a decade after Planck's first tentative introduction of the quantum into physics.

Q

Primary Literature

1. W. Nernst: *The New Heat Theorem* (2nd ed. transl. Guy Barr, Dover, New York 1969; first edition, 1917; second edition, 1924; the 1969 Dover edition is usually the easiest to find)
2. J. Stachel et al., eds.: *Collected Papers of Albert Einstein* (Princeton: Princeton University Press 1987 and later, especially Vols. 1–4)

Secondary Literature

3. A. Assmus: The molecular tradition in early quantum theory. Historical Studies in the Physical and Biological Sciences **22**, 209–231 (1992)
4. A. Assmus: The Americanization of molecular physics. Historical Studies in the Physical and Biological Sciences **23**, 1–34 (1992)
5. D. Kormos Barkan: Walther Nernst and the Transition to Modern Physical Science (Cambridge University Press, Cambridge 1999)

6. O. Darrigol: Statistics and combinatorics in early quantum theory, II: Early symptoma of indistinguishability and holism. Historical Studies in the Physical and Biological Sciences **21**, 237–298 (1991)
7. A. Desalvo: From the Chemical Constant to Quantum Statistics: A Thermodynamic Route to Quantum Mechanics. Physis **29**, 465–537 (1992)
8. C. Fujisaki: From Delandres to Kratzer – Development of the understanding of the origin of infrared band spectra, I (1880–1913). Historia scientarum **24**, 53–75 (1983) and From Delandres to Kratzer – Development of the understanding of the origin of infrared band spectra, II (1913–1920). Historia scientarum **25**, 57–86 (1983)
9. M. Jammer: The Conceptual Development of Quantum Mechanics (McGraw-Hill, New York 1966)
10. M.J. Klein: Einstein and the Wave-Particle Duality. Natural Philosopher **3**, 3–49 (1964)
11. J. Mehra and H. Rechenberg: The Historical Development of Quantum Mechanics, Vol. 1, Springer, Berlin 1982
12. P.W. Milonni: The Quantum Vacuum: An Introduction to Quantum Electrodynamics (Academic Press, New York 1994, esp. Chap. 1); and P. W. Milonni and M. L. Shih, Zero-point energy in early quantum theory. American Journal of Physics **59**, 684–698 (1991)
13. A. Pais: Subtle is the Lord... The Science and the Life of Albert Einstein (Oxford University Press, Oxford 1982)
14. J. Uffink: Insuperable difficulties: Einstein's statistical road to molecular physics. Studies in History and Philosophy of Modern Physics **37**, 36–70 (2006)

Please see also the References for ▶ Black-body radiation and ▶ Specific heats.

Quantum Zeno Effect

Erich Joos

The *Quantum Zeno Effect* describes the slowing down of the evolution of a quantum system under repeated measurements. In the limit of arbitrarily dense measurements motion would be completely inhibited.

The now popular name "quantum Zeno effect" (or "Zeno paradox") was introduced by Misra and Sudarshan in 1977 [1]. The effect has been described independently by many authors. (It can even be traced back to von Neumann's 1932 treatise "Mathematical foundations of quantum theory".) Other names used are "Turing's paradox", "watched pot behavior", or "watchdog effect".

The quantum Zeno effect only appears "paradoxical" or surprising, if the influence of measurements on a quantum system is not properly taken into account. Many systems (in particular, exponentially decaying systems) are not influenced at all by repeated measurements. This can be understood by a closer analysis of the dynamics of repeated measurements [2].

Let a system be prepared in its "undecayed" state $|u\rangle$ at some initial instant $t = 0$. Unitary evolution leads to a ▶ superposition of this undecayed state with some orthogonal ("decayed") states $|d_k\rangle$, with amplitudes a_u and a_{d_k}, respectively,

$$|\Psi(t)\rangle = \exp(-Ht)\,|u\rangle$$
$$= a_u(t)\,|u\rangle + \sum_{d_k \neq u} a_{d_k}(t)\,|d_k\rangle\,.$$

The ("survival") probability of finding the system still "undecayed" (i.e. in the state $|u\rangle$) at a later time $t > 0$ is

$$P(t) = |a_u(t)|^2$$
$$= |\,\langle u|\exp(-\mathrm{i}Ht)\,|u\rangle\,|^2.$$

Expanding the exponential in powers of t gives

$$P(t) = 1 - (\Delta H)^2 t^2 + O(t^4)$$

with

$$(\Delta H)^2 = \langle u|\,H^2\,|u\rangle - \langle u|\,H\,|u\rangle^2\,.$$

If the measurement performed on the same unstable system is carried out not just once, but is repeated N times in the interval $[0, t]$, the probability that it will be found undecayed in all N measurements is then given by

$$P_N(t) \approx \left[1 - (\Delta H)^2 \left(\frac{t}{N}\right)^2\right]^N > 1 - (\Delta H)^2 t^2 = P(t).$$

The non-decay probability is always increased, that is, the decay is suppressed; in the limit of arbitrarily dense measurements it comes to a complete halt,

$$P_N(t) = 1 - (\Delta H)^2 \frac{t^2}{N} + \ldots \stackrel{N\to\infty}{\longrightarrow} 1.$$

Q

Thus under continuous measurement the system would not move at all.

A demonstration of the quantum Zeno effect was performed by Itano et al. in an experiment [3] with Be^+ ions confined in a Penning trap (see Fig. 1).

In this setup the population of two levels is measured by coupling them to a third atomic level which decays rapidly by emitting fluorescence light. The first two levels represent the "measured object", the third level together with the emitted photons (▶ light quantum) play the role of the measurement device.

As is evident from the above derivation of the quantum Zeno effect, the quadratic time dependence of transition probabilities in the short-time limit is important [4, 5]. This approximation is valid for a sinusoidally oscillating system (as in the Itano experiment), but may often be only a poor approximation. The most important counterexample is represented by exponentially decaying systems, where it has long been known that the quadratic limit holds only for a very short timescale (now sometimes called *Zeno time*). Indeed, if the decay probability were *exactly* exponential,

$$P(t) = \exp(-\Gamma t),$$

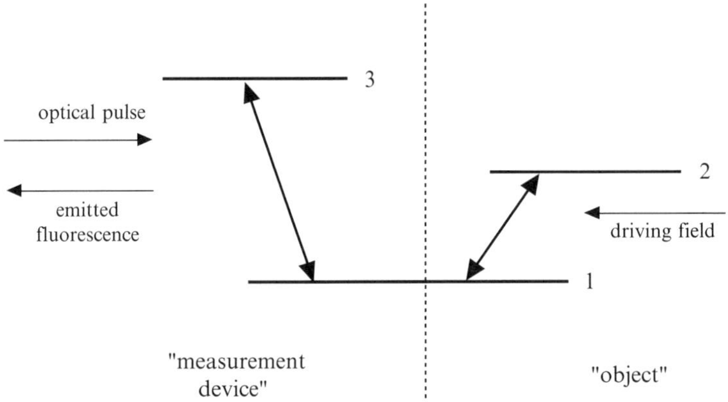

Fig. 1 Level structure for an experiment demonstrating the quantum Zeno effect. The Rabi oscillations of the transition between levels 1 and 2 (driven by a resonant radiofrequency field) are monitored by exciting the optical transition $1 \rightarrow 3$ resulting in light emission from level 3 through spontaneous emission. In this way the $1 \leftrightarrow 3$ transition together with the emitted light acts like a (nearly ideal) measurement device discriminating between levels 1 and 2

there would be no Zeno effect at all, since trivially

$$P_N(t) = \left(\exp\left(-\Gamma \frac{t}{N}\right)\right)^N = \exp(-\Gamma t).$$

Clearly, the Zeno effect is a consequence of measurement dynamics, again emphazing the well-known fact that a quantum measurement cannot be simply understood as information increase. A related discussion refers to the so-called ▶ *interaction-free measurements* [6], which in fact represent *strong* measurement-like interaction and can be understood as a special case of the quantum Zeno effect [7, 8]. One should also note, that "negative-result measurements" (where a measurement device does not "fire") also contribute to the Zeno effect [9].

A more precise description of the dynamics behind the Zeno effect can be achieved by replacing the phenomenological collapse rule by a dynamical model for the measurement process [2,7,10,11,12]. From this perspective, the Zeno effect can be viewed as the limiting case of very strong ▶ *decoherence*, that is, very strong measurement-like interaction of a quantum system with other degrees of freedom [13]. (▶ Experimental observation of decoherence). Since decoherence destroys phase relations at the system of interest, its motion (which in unitary quantum theory completely relies on coherence) would come to a standstill, if coherence were completely absent. If the density matrix ρ is exactly diagonal for all times, $\rho_{\alpha\beta} = \rho_{\alpha\alpha}\delta_{\alpha\beta}$, the von Neumann equation immediately yields Zeno freezing:

$$\mathrm{i}\frac{\mathrm{d}}{\mathrm{d}t}\rho_{\alpha\alpha} = \sum_\beta \left(H_{\alpha\beta}\rho_{\beta\alpha} - \rho_{\alpha\beta}H_{\beta\alpha}\right) \equiv 0.$$

Measurement models allow not only the discussion of the apparent contradiction between Zeno effect and exponential decay (described by rate equations) [2], but also a more realistic treatment of the small-time behavior, where system-dependent features may lead to interesting effects (such as the so-called "anti-Zeno effect" [7, 13].

The Zeno effect may find application in the field of quantum computing, where it could possibly be used to constrain the motion of a system to certain subspaces of its ► Hilbert space [14]. It may also be of relevance for the stability of molecules, where (already small) transition rates between spatial configurations may be further reduced by the influence of the natural environment.

Primary Literature

1. B. Misra, E.C.G. Sudarshan: The Zeno's paradox in quantum theory. J. Math. Phys. **18**, 756–763 (1977)
2. E. Joos: Continuous measurement: Watchdog effect versus Golden rule. Phys. Rev. **D29**, 1626–1633 (1984)
3. W.M. Itano, D.J. Heinzen, J.J. Bollinger, D.J. Wineland: Quantum Zeno effect. Phys. Rev. **A41**, 2295–2300 (1990)

Secondary Literature

4. A. Peres: Zeno paradox in quantum theory. Am. J. Phys. **48**, 931–932 (1980)
5. W. M Itano: Perspectives on the quantum Zeno paradox. quant-ph/0612187 (2006)
6. P.G. Kwiat, H. Weinfurter, A. Zeilinger: Quantum seeing in the dark. Sci. Am. **275**, 52–58 (1996)
7. E. Joos, H.D. Zeh, C. Kiefer, D. Giulini, J. Kupsch, I.-O. Stamatescu : *Decoherence and the Appearance of a Classical World in Quantum Theory*, Sect. 3.3.1, 2nd edn. (Springer, Berlin 2003)
8. P.G. Kwiat, A.G. White, J.R. Mitchell, O. Nairz, G. Weihs, H. Weinfurter, A. Zeilinger: High-efficiency quantum interrogation measurements via the quantum Zeno effect. Phys. Rev. Lett. **83**, 4725–4728 (1999)
9. D. Home, M.A.B. Whitaker: Negative-result experiments, and the requirement of wavefunction collapse. J. Phys. **A25**, 2387–2394 (1992)
10. L.E. Ballentine: Comment on 'Quantum Zeno effect'. Phys. Rev. **A43**, 5165–5167 (1991)
11. K. Kraus: Measuring processes in quantum mechanics I. Continuous observation and the Watchdog effect. Found. Phys. **11**, 547–576 (1981)
12. L.S. Schulman: Continuous and pulsed observations in the quantum Zeno effect. Phys. Rev. **A57**, 1509–1515 (1998)
13. E. Joos: Dynamical consequences of strong entanglement. Séminaire Poincaré **1**, 133–144 (2005), quant-ph/0512079
14. P. Facchi, S. Pascazio: Quantum Zeno subspaces. Phys. Rev. Lett. **89**, 080401 (2002)

Quarks

See ► Color Charge Degree of Freedom in Particle Physics; Mixing and Oscillations of Particles; Particle Physics; Parton Model; QCD; QFT.

Quasi-Classical Limit

N.P. Landsman

The *quasi-classical limit* of quantum mechanics refers, roughly speaking, to the limit $\hbar \to 0$. Of course, $\hbar$ is a dimensionful *constant*, but in practice one studies the semi-classical regime of a given quantum theory by forming a dimensionless combination of $\hbar$ and other parameters; this combination then re-enters the theory as if it were a dimensionless version of $\hbar$ that can indeed be varied.

The oldest example of this procedure is Planck's radiation formula ► black body radiation; Planck's constant. Indeed, the observation of Einstein [5] and Planck [8] that in the limit $\hbar\nu/kT \to 0$ this formula converges to the classical equipartition law may well be the first use of the $\hbar \to 0$ limit of quantum theory; note that Einstein put $\hbar\nu/kT \to 0$ by letting $\nu \to 0$ at fixed T and $\hbar$, whereas Planck took $T \to \infty$ at fixed ν and $\hbar$.

Another example is the one-particle ► Schrödinger equation, where one may pass to dimensionless parameters by introducing a typical energy scale ϵ and a typical length scale λ. In terms of the dimensionless variable $\tilde{x} = x/\lambda$, the rescaled Hamiltonian H/ϵ is then dimensionless and contains $\hbar$ through the dimensionless variable $\tilde{\hbar} = \hbar/\lambda\sqrt{2m\epsilon}$. In particular, large mass means effectively small $\hbar$.

Finally, as perhaps first remarked by Bogoliubov [1], averages of N single-particle operators satisfy commutation relations in which $\hbar$ has been replaced by $\hbar/N$, so that the limit $\hbar \to 0$ is effectively equivalent to the limit $N \to \infty$. This remark lies at the basis of the quantum theory of macroscopic observables (see [19] and references therein).

The quasi-classical limit has two separate aims, which should be sharply distinguished conceptually (although there is considerable overlap in the mathematical techniques that are used):

1. The approximation of solutions to the quantum-mechanical equations of motion (e.g. the Schrödinger equation) by solutions of the corresponding classical equations.
2. The derivation of classical mechanics, and more generally the explanation of the appearance of the classical world, from quantum theory.

The first application is mathematically sophisticated but is conceptually quite straightforward. The best-known technique is the *WKB approximation*, which goes back to Wentzel [11], Kramers [7] and Brillouin [3] in 1926. In the case of the time-independent Schrödinger equation, one postulates that the wave function has the form

$$\Psi(x) = a_{\hbar}(x)\mathrm{e}^{\frac{\mathrm{i}}{\hbar}S(x)}, \tag{1}$$

where S is independent of $\hbar$, substitutes this Ansatz into the Schrödinger equation, and expands in powers of $\hbar$. At lowest order this yields the (time-independent) Hamilton-Jacobi equation $H(\partial S/\partial x, x) = E$, where H is the classical Hamiltonian. This equation is supplemented by the so-called (homogeneous) transport equation

$$\left(\frac{1}{2}\Delta S + \sum_k \frac{\partial S}{\partial x^k}\frac{\partial}{\partial x^k}\right) a_0 = 0. \tag{2}$$

Higher-order terms in $\hbar$ yield further, inhomogeneous transport equations for the expansion coefficients $a_j(x)$ in $a_{\hbar} = \sum_j a_j \hbar^j$. These can be solved in a recursive way, starting with (2). There are various problems with this method, the main ones being convergence and the fact that in most cases of interest the Ansatz (1) is only valid locally (in x), leading to problems with caustics. These problems have been addressed in a sophisticated field of mathematics called *microlocal analysis* [15, 18, 21]. The WKB method is of little use for chaotic systems and has to be replaced by techniques surrounding the so-called *Gutzwiller trace formula*; see [16, 14].

Another insight dating back to the early days of (mature) quantum theory is ► *Ehrenfest's Theorem* from 1927 [4], which states that for any wave function Ψ (in the domain of the position operator and of $\partial V(x)/\partial x^j$, where V is the potential) one has

$$m\frac{\mathrm{d}^2}{\mathrm{d}t^2}\langle x^j\rangle(t) = -\left\langle \frac{\partial V(x)}{\partial x^j}\right\rangle(t), \tag{3}$$

where the brackets $\langle\cdots\rangle(t)$ denote expectation values in the time-dependent state $\Psi(t)$. This looks like Newton's second law, with the tiny but crucial difference that this law should have $(\partial V/\partial x^j)(\langle x\rangle(t))$ on the right-hand side. For further developments in this direction see [17], as well as the literature on microlocal analysis just cited. In particular, Egorov's Theorem in microlocal analysis is closely related to Ehrenfest's: it states that for a large class of Hamiltonians and classical observables f one has $Q(f)(t) = Q(f_t) + O(\hbar)$. Here $Q(f)$ is the Weyl quantization of f (► Quantization) and the left-hand side evolves according to the quantum equation of motion, whereas the right-hand side follows the classical one.

The last early idea we mention is the Wigner function (► *Wigner distribution*), introduced in 1932 [12]. Namely, each wave function Ψ (or, more generally, each density matrix) defines a function W_{Ψ} on classical phase space, defined by

$$W_{\Psi}(p,q) = \int_{\mathbb{R}^n} \mathrm{d}^n v\, \mathrm{e}^{ipv}\,\overline{\Psi(q+\tfrac{1}{2}\hbar v)}\Psi(q-\tfrac{1}{2}\hbar v). \tag{4}$$

Q

This function has the property

$$(\Psi, Q(f)\Psi) = \int_{\mathbb{R}^{2n}} \frac{\mathrm{d}^n p \mathrm{d}^n q}{(2\pi)^n} W_\Psi(p,q) f(p,q), \tag{5}$$

where $(\, ,\,)$ is the inner product in the Hilbert space $L^2(\mathbb{R}^n)$ and $Q(f)$ is the Weyl quantization of f as before. Thus the Wigner function transforms quantum-mechanical expectation values into classical ones, with the proviso that W_Ψ may fail to be positive and therefore cannot strictly be interpreted as a classical phase space distribution. Nonetheless, it is an extremely effective tool for studying the $\hbar \to 0$ limit [13].

The second application of the quasi-classical limit, i.e. to the explanation of the classical world, is a very deep and largely unsolved problem (cf. [19]) for a survey). To their credit, also here many of the key ideas date back to the founders of quantum mechanics.

Bohr's ▶ *correspondence principle* [2, 10] was, in its original form, not concerned with the classical limit of electronic orbits (but rather with the emitted radiation, which for wide orbits behaves approximately classically). However, at a later stage it was transformed into the general idea that large quantum numbers should give rise to classical behaviour. Applied to atoms, this idea works if it is combined with Schrödinger's suggestion that particle behaviour emerges from ▶ wave mechanics by looking at ▶ wave packets [9] (see [20] for a modern account). In particular, semi-classical motion emerges if a localized wave packet is formed as a superposition of tens of thousands of energy eigenfunctions with similarly large ▶ quantum numbers. Such a wave packet initially follows a time-evolution with almost classical periodicity, but subsequently spreads out after a number of orbits. During this second stage the (Born) probability distribution approximately fills the classical orbit. On a much longer time scale one sees *wave packet revival*, in that the wave packet recovers its initial localization. Then the whole cycle starts once again. See [22] for a popular account and [23] for a technical review. Another successful application of the correspondence principle is to the classical limit of quantum partition functions [24].

Heisenberg's famous 1927 paper [6] not only contained his uncertainty relations, but also suggested that the classical world emerged from quantum mechanics through observation: *'Die Bahn entsteht erst dadurch, daß wir sie beobachten.'* (*'The trajectory only comes into existence because we observe it.'*) This idea has to be combined with the quasi-classical limit in order to have the beginning of an explanation of classical physics from quantum theory. Here modern methods of ▶ decoherence and ▶ consistent histories play an important role.

Primary Literature

1. N.N. Bogoliubov: On a new method in the theory of superconductivity. Nuov. Cim. **7**, 794–805 (1958).
2. N. Bohr: Über die Anwendung der Quantentheorie auf den Atombau. I. Die Grundpostulaten der Quantentheorie. Z. Phys. **13**, 117–165 (1923).

3. L. Brillouin: La mécanique ondulatoire de Schrödinger; une méthode générale de resolution par approximations successives. C.R. (Paris) **183**, 24–26 (1926).
4. P. Ehrenfest: Bemerkung über die angenäherte Gültigkeit der klassischen Mechanik innerhalb der Quantenmechanik. Z. Phys. **45**, 455–457 (1927).
5. A. Einstein: Über einen die Erzeugung und Verwandlung des Lichtes betreffenden heuristischen Gesichtpunkt. Ann. Phys. **17**, 132–178 (1905).
6. W. Heisenberg: Über den anschaulichen Inhalt der quantentheoretischen Kinematik und Mechanik. Z. Phys. **43**, 172–198 (1927).
7. H.A. Kramers: Wellenmechanik und halbzahlige Quantisierung. Z. Phys. **39**, 828–840 (1926).
8. M. Planck: *Vorlesungen Über die Theorie der Wärmestrahlung* (J.A. Barth, Leipzig, 1906).
9. E. Schrödinger: Der stetige Übergang von der Mikro- zur Makromekanik. Die Naturwissenschaften **14**, 664–668 (1926).
10. J.H. van Vleck: The absorption of radiation by multiply periodic orbits, and its relation to the correspondence principle and the Rayleigh-Jeans law. Part I, II. Phys. Rev. **24**, 330–346, 347–365 (1924).
11. G. Wentzel: Eine Verallgemeinerung der Quantenbedingungen für die Zwecke der Wellenmechanik. Z. Physik. **38**, 518–529 (1926).
12. E.P. Wigner: On the quantum correction for thermodynamic equilibrium. Phys. Rev. **40**, 749–759 (1932).

Secondary Literature

13. M.V. Berry: Semi-classical mechanics in phase space: a study of Wigner's function. Phil. Trans. Roy. Soc. (London) **287**, 237–271 (1977).
14. M. Brack & R.K. Bhaduri: *Semiclassical Physics* (Westview, Boulder, 2003).
15. V. Guillemin & S. Sternberg: *Geometric Asymptotics* (American Mathematical Society, Providence, 1977).
16. M.C. Gutzwiller: *Chaos in Classical and Quantum Mechanics* (Springer, New York, 1990).
17. K. Hepp: The classical limit of quantum mechanical correlation functions. Commun. Math. Phys. **35**, 265–277 (1974).
18. V. Ivrii: *Microlocal Analysis and Precise Spectral Asymptotics* (Springer, New York, 1998).
19. N.P. Landsman: Between classical and quantum, in *Handbook of the Philosophy of Science Vol. 2: Philosophy of Physics*, ed. by J. Butterfield and J. Earman, pp. 417–554 (North-Holland, Elsevier, Amsterdam, 2007).
20. R.G. Littlejohn: The semiclassical evolution of wave packets. Phys. Rep. **138**, 193–291 (1986).
21. A. Martinez: *An Introduction to Semiclassical and Microlocal Analysis* (Springer, New York, 2002).
22. M. Nauenberg, C. Stroud, & J. Yeazell: The classical limit of an atom. Sci. Am. **June**, 24–29 (1994).
23. R.W. Robinett: Quantum wave packet revival. Phys. Rep. **392**, 1–119 (2004).
24. B. Simon: The classical limit of quantum partition functions. Commun. Math. Phys. **71**, 247–276 (1980).

R

Radioactive Decay Law (Rutherford–Soddy)

Friedel Weinert

The formulation of the radioactive decay law, in 1902, by Ernest Rutherford (1871–1937) and Frederick Soddy (1877–1956) was part of a number of discoveries around the turn of the century, which paved the way to the establishment of quantum mechanics, as the physics of the atom. In November 1895, W. Röntgen (1845–1923) discovered ► X-rays; in 1896 A. H. Becquerel (1852–1908) discovered radioactivity during an investigation of phosphorescence in uranium salts; finally in 1897 J.J. Thomson (1856–1940) discovered the ► electron. Rutherford and Soddy based their formulation of the radioactive law on the 'emanation theory' of radioactive decay. According to this theory, radioactivity is an 'atomic' phenomenon, which is accompanied by 'chemical' changes. Note that in 1902, Rutherford had not yet inferred from large-angle ► scattering experiments that the atom had a nucleus (► Rutherford atom). One chemical element, Rutherford and Soddy explained, was transformed into another by emitting charged particles: α-particles or β-particles. Around that time Rutherford already knew that radioactivity manifested itself in the form of 'alpha rays' or 'beta rays', which proved to consist of particles. Prior to his discovery of the nucleus model of the atom (1911), Rutherford regarded alpha particles as ionized helium atoms. α-particles are helium nuclei with an exit velocity of approximately 10^7m s^{-1} (with energies ranging between 4–9 MeV) and positive charge so that they experience deflections in electric and magnetic fields. β-particles are ► electrons with emission velocities, which range between 10^8m s^{-1} and $0.999c$, and negative charge so that they, too, experience deflections in electric and magnetic fields. (Beta decay reveals a continuous energy spectrum up to a maximum E_0, depending on the type of nucleus involved; the kinetic energy Q can range from a few keV into the region of MeV.) Rutherford and Soddy emphasized that the 'chemical' changes had their seat within the atom and not on the molecular level. Today radioactivity denotes the ability of certain nuclei to undergo transformations through the emanation or emission of radiation. (Rutherford and Soddy were aware that this process can include γ-radiation – light of very short wavelength –, which is not deflected in electric or magnetic fields.) Rutherford and Soddy could not say what caused the emission of the subatomic particles from the atomic nuclei. The radioactive elements, their theory stipulated, 'must be undergoing spontaneous transformation' [1, 493]. In terms of the classical notion of determinism, the emanation theory did not permit the precise prediction of the time and trajectories of emitted particles. The theory was based on the formulation of statistical laws, which give rise to ► indeterminism.

D. Greenberger et al. (eds.), *Compendium of Quantum Physics: Concepts, Experiments, History and Philosophy*, © Springer-Verlag Berlin Heidelberg 2009

Fig. 1 Rutherford's curve

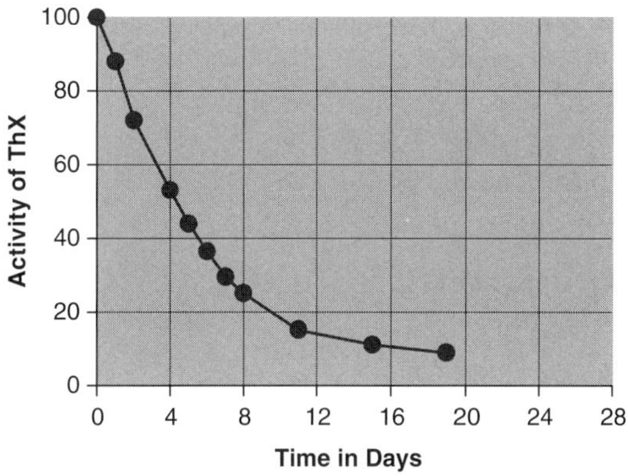

The decay law states the probability of decay in a given ensemble (▶ ensembles in quantum mechanics), N_0, of radioactive material for a given period of time, t. Note that α-decay occurs in nuclei with high atomic weight, A ($A = N + Z$, where N is the number of neutrons and Z the number of protons); β-decay only occurs in nuclei, in which the number of neutrons, N, is greater than the number of protons, Z.) In the original words of Rutherford and Soddy, 'if I_0 represents the initial activity and I_t the activity after time t, (then)

$$\frac{I_t}{I_0} = \mathrm{e}^{-\lambda t}$$

where λ is a constant and e the base of natural logarithm' [1, 482]. The decay constant λ can be rewritten as $\lambda = \ln / T_{1/2}$, where $T_{1/2}$ is the half-life, i.e. the period in which half of the given N_0 of radioactive material will decay. (Fig. 1)

As we know today, the half-life of radioactive elements ranges from seconds to millions of years. The decay law is not statistical in the nineteenth century sense of reflecting our degree of ignorance of the specific boundary conditions, under which individual atoms in an ensemble of radioactive elements will decay, but in the twentieth century sense of reflecting a genuinely indeterministic process in nature, which gives rise to statements about the average decay rate of a given ensemble of atoms. This means that the decay rate of individual atoms equals the decay rate of the ensemble. The statistical nature of this law is illustrated, using Rutherford's original data, as in Fig. 1.

The discovery of the radioactive decay law was an important step on the road to a questioning of the classical notions of causality and determinism, as they were often presupposed in classical physics. ▶ Indeterminism.

R

Literature

1. E. Rutherford, F. Soddy: The Cause and Nature of Radioactivity I, II. Philosophical Magazine **IV**, 370–96, 569–85 (1902)

2. J. L. Heilbron: The Scattering of α and ß Particles and Rutherford's Atom. Archive for History of Exact Sciences **4**, 247–307 (1968)
3. St. Weinberg: *The Discovery of Subatomic Particles* (Harmondsworth, Penguin Books 1993)

Relativistic Quantum Mechanics

Helge Kragh

Attempts to establish a relativistic quantum mechanics – an integration of ► quantum theory and the theory of relativity – predate the emergence of quantum mechanics in the 1920s. Shortly after Niels Bohr had proposed his ► atom model in 1913, he and a few others realized that ► quantum theory might be improved by using relativity rather than classical mechanics. These efforts culminated in 1916–17 when Arnold Sommerfeld in Munich devised a modified version of Bohr's model by incorporating the relativistic variation of the mass of an electron moving around an atomic nucleus. That is, rather than assuming the mass to be constant, Sommerfeld adopted the expression $m(v) = m_0(1-v^2/c^2)^{-1/2}$, where v is the electron's velocity and c the velocity of light. The result was an expression of the energy levels in hydrogen-like atom ► Bohr's atom model that predicted a fine structure with a separation in frequency proportional to $\alpha^2 Z^2$, where α is the fine-structure constant and Z the nuclear charge (or atomic number). Sommerfeld's theory received experimental support from measurements in both the optical and the X-ray region, and the confirmation was widely seen as a triumph of the Bohr-Sommerfeld atomic model as well as the special theory of relativity.

A few physicists believed that gravitation theory, in the form of Einstein's general theory of relativity, had to be incorporated in atomic theory. The Kepler motion of ► electrons around an atomic nucleus was analyzed by means of general relativity by Georg Jaffé, Mandoval Vallarta and others in 1922–25; however, their works were ignored by most mainstream physicists who believed that general relativity was of no importance in atomic physics. In a paper of 1922, Erwin Schrödinger applied Hermann Weyl's extension of Einsteinian general relativity to atomic theory. Although Schrödinger's paper would later come to appear as prescient, at the time his work attracted no more attention than other theories in the same tradition.

Louis de Broglie's innovative theory of 1922–23, which postulated the existence of ► matter waves, was solidly founded on the (special) theory of relativity. According to de Broglie, quantum theory and special relativity theory were unified by the relativistic formula $mc^2 = h\nu = hc/\lambda$, or $\lambda = h/p$ (where λ is the wavelength associated with the momentum p of some particle, whether a ► light quantum or an ► electron). In late November 1925 Schrödinger reached the decision that to transform de Broglie's hypothesis into a wave theory of atomic structure he would need a wave equation governing the behaviour of de Broglie's somewhat mysterious matter waves. Since de Broglie's hypothesis was thoroughly relativistic, naturally Schrödinger sought for a wave equation that satisfied the requirements of the special theory of relativity: its form had to be Lorentz invariant. At new year's time he had found such an equation for the amplitude connected with the electron, and after hard mathematical work he succeeded in solving it in the case of the hydrogen atom.

Schrödinger calculated (what came to be known as) the energy eigenvalues and from these he derived the energy spectrum of hydrogen. Although his calculations gave a fine structure for the red H_α line, it did not fit with the experimentally confirmed Sommerfeld theory: Schrödinger's ► wave mechanics yielded a fine-structure separation of the H_α doublet nearly thrice the observed value. Consequently he was forced to use the non-relativistic approximation, and it was this form – since then known as the ► Schrödinger equation – that he reported in his famous series of papers in the spring of 1926. The relativistic eigenvalue equation for an electron in the electrostatic field of potential φ reads

$$\hbar^2 c^2 \Delta\psi + [(E - e\varphi)^2 - m_0 c^4]\psi = 0$$

where $\hbar = h/2\pi$. Shortly after the appearance of Schrödinger's ► wave mechanics, the equation was derived by several physicists, including Oskar Klein, Wolfgang Pauli, Vladimir Fock, Walter Gordon, de Broglie, and Schrödinger himself. Klein, ignorant about Schrödinger's unpublished derivation, may have been the first to derive the equation, which he framed in the context of a five-dimensional unification of wave mechanics, electromagnetism and general relativity. Whatever the parentage, Schrödinger's relativistic equation came to be known as the Klein-Gordon equation. The corresponding time-dependent equation for a free electron reads

$$\hbar^2 c^2 \Delta\psi + \hbar^2 \partial^2\psi/\partial t^2 = m_0 c^4 \psi$$

The equation is Lorentz invariant and reduces to the ordinary Schrödinger equation in the limit $v/c \rightarrow 0$. But is it the right equation for an electron? R

There were two problems that indicated that this was not the case. First, the equation did not result in the right doublet splitting of the lines in the hydrogen spectrum. Second, it did not incorporate the electron's ► spin, which by the fall of 1926 had become accepted by most physicists and somehow had to be understood in terms of quantum mechanics. The problems seemed to have no solutions within the Klein-Gordon framework, but in Germany an alternative approach was followed, namely by including relativistic effects as corrections to the non-relativistic theory. This method led to a partial success in the spring of 1926, when Pascual Jordan and Werner Heisenberg, developing ideas due to Wolfgang Pauli, succeeded to derive the fine-structure formula in a first-order approximation. They added to the usual Hamiltonian not only a perturbation term describing the relativistic correction to the kinetic energy but also a term referring to the spin of the electron. However, in spite of its empirical success the phenomenological Jordan–Heisenberg–Pauli theory was not entirely satisfactory. Since relativity was added as a first-order correction, the theory was not genuinely relativistic; moreover, the spin effect was introduced in an ad hoc manner, being grafted to the theory rather than explained by it. An entirely satisfactory theory would not only be able to account for the doublet phenomena but also explain them in the sense of deducing them from the basic principles of relativity and quantum mechanics.

The quantum-mechanical understanding of spin improved with the theories independently proposed by Pauli and Charles Darwin in the spring of 1927. However, these theories did not go substantially beyond the phenomenological level of the Jordan-Heisenberg-Pauli theory and they failed to combine spin and relativity. In spite of their importance, they did not offer a solution to the still more delicate problem of integrating quantum mechanics with the theory of relativity. Such a solution, based on an entirely novel approach, came in early 1928.

Paul Dirac reasoned that according to the general principles of quantum mechanics the formal structure of the Schrödinger equation – meaning the expression $H\psi = i\hbar\partial\psi/\partial t$ – must be retained in any future unification of relativity and the quantum theory of electrons. This ruled out the Klein-Gordon equation and implied that the relativistic wave equation had to be of the first order in the space derivatives. Dirac's reasoning suggested the starting procedure

$$i\hbar\partial\psi/\partial t = c\sqrt{m_0^2c^2 + p_1^2 + p_2^2 + p_3^2}\,\psi$$

where $p_1 = -i\hbar\partial/\partial x$, etc. By "playing around with mathematics" he found a way to linearize the square root, i.e. to write it in the form $\alpha_1 p_1 + \alpha_2 p_2 + \alpha_3 p_3 + \alpha_4 m_0 c$. The α-coefficients were matrices of the same kind as those Pauli had introduced in his spin theory, but they had four rows and columns (whereas Pauli's were 2×2 matrices).

Dirac's paper, entitled "The Quantum Theory of the Electron," appeared in the *Proceedings of the Royal Society* in January 1928. It is noteworthy that originally he did not think of the electron's spin. It was only after having found the wave equation that he discovered that it, in an extended form where the electromagnetic field was taken into account, included a term representing the magnetic moment of the electron. Since this quantity is given by the spin vector, the electron's spin appeared as a consequence of the theory. Dirac proved that his equation satisfied Lorentz invariance, and he also showed that its first approximation led to the approximate fine-structure formula. He did not attempt to find the exact solution but supposed that it would result in the same energy spectrum that Sommerfeld had found more than a decade earlier. This was indeed the case, such as shown by Gordon and Darwin in the spring of 1928.

Dirac's theory of the electron was received with enthusiasm by his colleagues and had a revolutionary effect on quantum physics. It was primarily for this work that he was awarded the Nobel Prize in 1933. Although the theory was very much the result of Dirac's genius, it (or something like it) would most likely have been found by other physicists even had he not presented the theory in January 1928. Several physicists tried at the time to construct a relativistic spin quantum theory, and some of them, such as Jordan and Hendrik Kramers, came close to the goal. Kramers obtained an approximate quantum description of a relativistic spinning electron in terms of a second-order wave equation and later proved that his equations were equivalent to Dirac's linear equation.

The new theory of relativistic quantum mechanics was quickly explored by physicists and mathematicians. For example, the Dirac matrices and the properties of the Dirac wave function were studied by Hermann Weyl, Bartel L. van der Waerden, John von Neumann and others. Several theoretical physicists – including Weyl, Fock and Georges Lemaître – transformed the wave equation in forms that could be incorporated into the framework of general relativity. Gregory Breit showed in 1928 that the Dirac matrices can be understood as velocities in the sense that $dx_\mu/dt = c\alpha_\mu$ ($\mu = 1, 2, 3$). Because of the property $\alpha_\mu^2 = 1$ the result seemed to lead to the paradoxical conclusion that a free electron will always move with the velocity of light ($v = \pm c$), a paradox that was taken up by Schrödinger in 1930 in his theory of the so-called *Zitterbewegung* of the electron (a microscopic, rapidly oscillatory motion superposed on the electron's "macroscopic" velocity).

Dirac's theory of the electron also inspired cosmological thinking, if only indirectly. Arthur Eddington was greatly impressed by the ▶ Dirac equation which he elevated to a status of universal significance and used to derive relationships between cosmic and atomic constants. Based on his own interpretation of the Dirac equation, he calculated the value of the fine structure constant and related it to the number of protons in the universe. The general idea of integrating quantum mechanics, cosmology and general relativity was pursued also by the Russian physicist Matvei Bronstein who in 1933–36 discussed unified "*cGh* physics" and examined the quantum limits of general relativity at what later would be called the Planck length, $l_P = (hG/c^3)^{1/2}$. However, Bronstein's works attracted little attention at the time.

From an empirical point of view, Dirac's theory faced successes as well as problems. On the one hand, it proved successful in the study of relativistic scattering processes, first investigated by Nevill Mott in Cambridge and Klein and Yoshio Nishina in Copenhagen. On the other hand, some of the predictions that followed from Dirac's theory disagreed with experiment. For example, the theory, believed to apply also to protons, predicted a value of the proton's magnetic moment that was nearly three times smaller than the measured value. It also led Mott to predict that free electrons should be polarized, yet experiments failed to detect the effect. (After more than a decade's confusion, it turned out that the early experiments were wrong. Free electrons are polarized, in agreement with the Mott-Dirac prediction.)

The most serious problem of the Dirac equation was the " ± difficulty" referring to the fact that the equation formally included solutions with negative energies. Of the four components of the wave function, two referred to positive-energy states and two to negative-energy states. In late 1929 Dirac believed he had found a solution to the problem. He assumed a world of negative-energy states occupied by an infinite number of electrons and argued that the few unoccupied states – the "holes" – would appear as observable physical entities, particles with positive energy and positive charge. He originally suggested that the holes were protons, but was unable to account for their large mass and also the stability of ordinary matter (where protons and electrons would presumably annihilate to gamma rays). This first theory of antiparticles was universally met with skepticism. It caused Schrödinger to propose an alternative relativistic theory of the electron which avoided the $\pm$ difficulty and retained the empirically confirmed results of Dirac's theory. However, Schrödinger's theory was shortlived. Not only did it face experimental difficulties, it also failed to obey strict Lorentz invariance, and for these reasons it was not considered a valid alternative to Dirac's theory.

Dirac's shortlived idea of representing protons as antielectrons was philosophically appealing because of its unitary character. In 1930 all matter was believed to consist of protons and electrons; thus, if the proton was a vacant negative-energy state – an electron in disguise – Dirac would in effect have reduced the known elementary particles to just one fundamental entity, the electron. However, what he referred to as "the dream of philosophers" remained a dream. In a remarkable paper of 1931, mainly dealing with the possible existence of magnetic monopoles, he admitted that the proton could not be the antiparticle of the electron. As an alternative he suggested the existence of a new elementary particle with the same mass and spin as the electron, but of opposite charge. He thought that such hypothetical particles existed somewhere in nature and that they might be produced in collision processes involving two gamma photons (► light quantum). Moreover, because the proton was now a separate species of particle, it would probably have its own antiparticle, a negatively charged proton.

The hypothesis of antielectrons was considered speculative, but the situation changed dramatically in 1932–33 when Dirac's particle was detected in cosmic ray experiments. Although Carl Anderson found cloud chamber tracks from positive electrons in 1932, at first he failed to identify them correctly and it was only in 1933 that he realized that he had discovered the positive electron or "positron," as he called it. However, Anderson did not identify his positron with Dirac's antielectron, which he probably was unaware of. The correct identification positive electron = positron = antielectron came later in 1933 when Patrick Blackett and Guiseppe Occhialini analyzed cosmic ray data. Naturally, the discovery of the positron greatly enhanced the status of Dirac's theory of antiparticles, and that in spite of widespread opposition to his interpretation in terms of holes. In 1934 Robert Oppenheimer and Wendell Furry, and independently Enrico Fermi, showed that antiparticles could be accounted for by quantum field theory without introducing the Dirac "sea" of unobservable negative-energy particles.

The great success of the Dirac equation caused interest in the older Klein-Gordon equation to fade away. That the Klein–Gordon equation is really as good as any quantum-mechanical equation, was made clear only in 1934 when Pauli and Victor Weisskopf revived the Klein–Gordon theory. If interpreted correctly, namely as a field theory for Bose-Einstein particles, there is nothing wrong with the Klein-Gordon theory, Pauli and Weisskopf argued. They proved that concepts such as pair creation, annihilation and antiparticles could be established without accepting the idea of a vacuum filled with negative-energy particles. Ever since, the Klein-Gordon equation has proved an indispensable tool in quantum field theory. See also ► algebraic quantum mechanics; operational quantum mechanics.

Primary Literature

1. P. Dirac: The Quantum Theory of the Electron. Proceedings of the Royal Society A **117**, 610–624 (1928)
2. P. Dirac: Quantised Singularities in the Electromagnetic Field. Proceedings of the Royal Society A **133**, 60–72 (1931)
3. W. Heisenberg, P. Jordan: Anwendung der Quantenmechanik auf das Problem der anomalen Zeemaneffekte. Zeitschrift für Physik **37**, 263–277 (1926)
4. O. Klein, Oskar (1926). Quantentheorie und fünfdimensionale Relativitätstheorie. Zeitschrift für Physik **37**, 895–906 (1926)
5. W. Pauli, V. Weisskopf: Über die Quantisierung der skalaren relativistischen Wellengleichung. Helvetica Physica Acta **7**, 709–731 (1934)

Secondary Literature

6. N. R. Hanson: *The Concept of the Positron* (Cambridge University Press, Cambridge 1963)
7. H. Kragh: Equation with the Many Fathers: The Klein-Gordon Equation in 1926. American Journal of Physics **52**, 1024–1033 (1984)
8. H. Kragh: Unifying Quanta and Relativity? Schrödinger's Attitude to Relativistic Quantum Mechanics, in M. Bitbol, O. Darrigol, eds., *Erwin Schrödinger. Philosophy and the Birth of Quantum Mechanics* (Editions Frontieres, Paris 1992, 315–338)
9. J. Mehra: *The Golden Age of Theoretical Physics* (World Scientific, Singapore 2001, 1030–1091)
10. M. Stöckler: Philosophische Probleme der relativistischen Quantenmechanik (Duncker & Humblot, Berlin 1984)

Renormalization

Arianna Borrelli

Procedures of *renormalization* are used in ► quantum field theory to deal with divergent integrals appearing in perturbative calculations of order higher than the lowest one. These ill-defined expressions would seem to render perturbative computations meaningless, thus depriving quantum field theory of an essential tool for obtaining phenomenological predictions. However, in some theories it is possible to circumvent this problem and formally compensate for the divergencies, obtaining for observable quantitites finite predictions which closely match experimental data. This was shown to be possible for ► QED in the late 1940s, when the development of renormalization procedures resulted in agreement between theoretical estimates and experimental measurements of the fine and hyperfine structure of the hydrogen spectrum (► spectroscopy; Bohr's atom model).

The central idea of renormalization is to systematically isolate and remove the divergencies by means of a redefinition (renormalization) of the nonperturbed field equation and of its parameters, usually referred to as "bare" masses and charges. Bare parameters are not observable and can therefore be assumed to have exactly those values – finite or infinite – which are needed to compensate for the divergencies. If all infinities can be eliminated by imposing only a finite number of renormalization conditions based on experimental data, the theory is said to be renormalizable. Renormalizability is a nontrivial feature of a theory, because it implies that the potentially infinite number of divergencies occurring in its perturbative expansion can be eliminated at all orders by iterating the same subtraction scheme. Beside QED, other renormalizable theories are ► QCD and the Standard Model ► quantum field theory; particle physics for electro-weak interactions.

Renormalization is a successful technique for deriving phenomenological predictions, but some foundational questions regarding it remain open [6, 8, 10, 12]. The divergent expressions are integrals over the four-momenum p_μ of functions which, for $p_\mu \rightarrow \infty$, do not converge to zero rapidly enough to be integrable (ultraviolet divergencies). Therefore, their presence could be taken to mean that QED and other theories work for low energies, but fail at high energies, where they should be replaced by models in which no divergencies occur, e.g. string theories ► quantum gravity. On the other hand, the fact that the divergencies turn out to be renormalizable might be physically significant. In this case, renormalizability would be a feature which quantum field theories should be expected to possess. Historically, the principle of renormalizability has played a central role in determining the development of quantum field theories. Finally, there is the problem that proofs of renormalizability are based on pertubative arguments, but evidence that the relevant perturbative expansions actually converge is lacking – in fact, there are indications that this might not always be the case.

Renormalization procedures can be carried out in a number of different ways [3]. The first step is always what is called "regularization". Regularizing a theory means modifying it in such a way, that divergent expressions become finite. For example, integrals may be modified so, as to extend only up to some high-energy cutoff Λ, or the number of space–time dimensions of the theory may be changed from 4 to $d = 4 - \epsilon$, thus rendering logarithmically divergent integrals finite in the ultraviolet region. Once the regularized, but potentially divergent, expressions have been isolated and eliminated according to some predetermined scheme, the regularization parameter (e.g. Λ, ϵ) can be removed, formally recovering the original theory minus the divergencies. In regularizing a theory, special care must be taken to preserve all its ► symmetries. However, this is not always possible, so that in the end renormalization may result in anomalous terms (anomalies) violating some symmetry of the nonrenormalized model. Anomalies are not just formal artefacts of the theory: for example, the axial anomaly has been shown to contribute to the decay rate for $\pi^0 \rightarrow \gamma + \gamma$.

The final, finite results of regularization and renormalization procedures depend in part on arbitrary choices, from which however observable prediction are expected to be independent. Formally, this means that renormalized expressions have to satisfy specific renormalization group equations, a condition which in turn provides physically relevant information, for example that, in QCD, interactions between quarks decrease in intensity in the limit of very short ranges (asymptotic freedom). ► Color Charge Degree of Freedom in Particles Physics; Mixing and Oscillations of Particles; Particle Physics; Parton Model QCD; QFT.

The occurrence of divergencies in quantum field theory had been noted already in the 1930s, but it was only in 1947–1948 that a number of scientists came to the idea that, by subtracting the infinities, one might obtain physically meaningful results. The development of renormalization theory was an essential part of the construction of QED, whose main actors were Sin-itiro Tomonaga (1906–1979), Julian Schwinger (1918–1994), Richard Feynman (1919–1988), and Freeman Dyson (1923–) [9]. Important stimuli for the development of renormalization theory came from a conference held on Shelter Island in 1947, where Hendrik Kramers (1894–1952) showed how mass renormalization could be used to circumvent divergencies, and where new experimental results on the hydrogen spectrum were presented. In 1949, Dyson outlined a proof of renormalizability of QED [1], which was complemented by other authors in the 1950s and 1960s. After the success of QED, attempts were made to formulate renormalizable quantum-field-theoretical models for weak interactions. In 1971, Gerard 't Hooft (1946–), working within the research program of his tutor Martin Veltmann (1931–), proved that this could be done using nonabelian ► gauge theories [2]. In 1999, the two scientists shared the Physics Nobel Prize for this result. In the early 1970s, renormalization group techniques were employed to show that QCD possesses the property of asymptotic freedom, helping establish it as a model for strong interactions. For this achievement, David J. Gross (1941–), H. David Politzer (1949–) and Frank Wilczek (1951–) were awarded the 2004 Nobel Prize in Physics.

Primary Literature

1. F. J. Dyson: The S-Matrix in quantum electrodynamics. Phys. Rev. **75**, 1736–1755 (1949)
2. G. 't Hooft: Renormalizable Lagrangiangs for massive Yang-Mills fields. Nucl. Phys. **B35**, 167–188 (1971)

Secondary Literature

3. T. -P. Cheng, L.-F. Li: *Gauge Theory of Elementary Particle Physics* (Clarendon Press, Oxford, 1984, 31–85, 279–291)
4. S. Aramaki: Formation of the renormalization theory in quantum electrodynamics. Historia Scientiarum **32**, 1–42 (1987)
5. S. Aramaki: Development of the renormalization theory in quantum electrodynamics (I) and (II). Historia Scientiarum **36**, 97–116, **37**, 91–113 (1989)
6. P. Teller: Infinite renormalization. Phil. Sci. **56**, 238–257 (1989)
7. L. M. Brown (ed.): *Renormalization. From Lorentz to Landau (and Beyond)* (Springer, New York, 1993)
8. T. Y. Cao, S. S. Schweber: The conceptual foundations and the philosophical aspects of renormalization theory. Synthese **97**, 33–108 (1993)
9. S. S. Schweber: *QED and the men who made it: Dyson, Feynman, Schwinger, and Tomonaga* (Princeton University Press, Princeton, 1994)

10. N. Huggett, R. Weingard: The renormalization group and effective field theories. Synthese **102**, 171–194 (1995)
11. G 't Hooft: Renormalization of gauge theories, in: *The Rise of the Standard Model. Particle Physics in the 1960s and 1970s*, ed. by L. Hoddeson, L. Brown, M. Riordan and M. Dresden (Cambridge University Press, Cambridge, 1997, 179–198)
12. D. Anselmi: A new perspective on the philosophical implications of quantum field theory. Synthese **135**, 299–328 (2003)
13. D. J. Gross: The discovery of asymptotic freedom and the emergence of QCD. Proceedings of the National Academy of the USA **102**, 9099–9108 (2005)

Rigged Hilbert Spaces in Quantum Physics

J-P. Antoine, R. Bishop, A. Bohm, and S. Wickramasekara

Introduction

A rigged Hilbert space (RHS) is the mathematical space underlying ▶ Dirac notation of quantum mechanics. There are two versions of RHS's used in quantum theory, the Schwartz space version and the Hardy space version. The Schwartz space version gives mathematical meaning to bras, kets and the Dirac basis vector expansion, as well as describes the quantum mechanical ▶ observables by an algebra of everywhere defined (continuous) operators. The Hardy space version provides the mathematics that unifies quantum scattering, resonance and decay phenomena in an exact theory. It gives meaning to Lippmann–Schwinger kets and Gamow vectors, and results in an exact lifetime-width relation $\tau = \hbar/\Gamma$, which in the Hilbert space theory was only justified as a Weisskopf–Wigner approximation. This theory of resonances leads to a semigroup time evolution, thus overcoming the problems with causality and exponential catastrophe. The relativistic version of Hardy space theory leads to semigroup representations of Poincaré transformations into the forward lightcone. These representations allow, for the first time, the mass and width a relativistic resonance, such as the Z°-boson, to be unambiguously defined from fundamental principles.

Prehistory: From Matrices and Differential Operators to Algebras of Observables and Dirac Kets

In their early work, Born and his school (Heisenberg [1], Jordan [2], Wiener [3] and others), developed an approach to quantum mechanics using matrices for physical observables, commonly called ▶ matrix mechanics. Alternatively, Schrödinger [4] developed a wave equation for quantum mechanics using differential operators. Dirac [5, 6] realized that the algebraic relations for the "dynamical variables" were the important features that determined the properties of the operators. This observation suggested starting with an algebra of observables represented by abstract linear operators and then looking for a linear space in which they could act.

For certain algebras of observables, this linear space would be a finite dimensional scalar product space, e.g., the $(2j + 1)$-dimensional space $\mathcal{R}^j$ for angular momentum states. Linear operators corresponding to observables were represented by Hermitian matrices on this scalar product space, e.g., the $(2j+1) \times (2j+1)$ angular momentum matrices on $\mathcal{R}^j$. Hilbert had generalized finite dimensional scalar product spaces to infinite dimensions so that the vectors ϕ would be represented as linear combinations of basis vectors $|n)$

$$\phi = \sum_{n=1}^{\infty} |n)(n|\phi), \tag{1a}$$

with coordinates $(n|\phi)$ that are square summable sequences:

$$(\phi, \phi) = \sum_{n=1}^{\infty} |(n|\phi)|^2 < \infty. \tag{1b}$$

In this way, an infinite dimensional complex vector space Ψ with a scalar product (ψ, ϕ) was introduced. Therewith, the convergence of infinite sums and continuity of linear operators in Ψ became as important questions as the algebraic relations between observables. The observables such as energy, momentum, position and angular momentum, which were defined by their algebraic (commutation) relations, have as their mathematical image linear operators H (energy), $\boldsymbol{P}$ (momentum), $\boldsymbol{Q}$ (position) and $\boldsymbol{J}$ (angular momentum) on this vector space Ψ. The elements ϕ of Ψ are interpreted as representing physical states, and the matrix elements squared $|(n|\phi)|^2 = (\phi|n)(n|\phi)$ as quantum mechanical probabilities. For instance, if $|n) = |E_n)$ is an eigenvector of the observable H with eigenvalue E_n, i.e., $H|E_n) = E_n|E_n)$, then $|(E_n|\phi)|^2$ is the probability of obtaining the value E_n in a measurement of energy H in the state ϕ. Once infinite dimensional vector spaces were introduced, it became evident that great care must be exercised when dealing with linear operators. For instance, whether for a given $\phi \in \Psi$, the vectors such as $\boldsymbol{P}\phi$ and $\boldsymbol{Q}\phi$ also fulfill the defining condition (1b) is a subtle question that required serious analysis.

Quantum mechanics has not only discrete eigenvalues, like the E_n in the components $(E_n|\phi)$ of the vectors ϕ, but also continuous values $E, 0 \leqslant E < \infty$, leading to a continuum of components $\phi(E) = (E|\phi)$, the energy wave functions. As another example, the solution of the Schrödinger differential equation $\psi(\boldsymbol{x})$ is a function of continuous position $\boldsymbol{x} \in \mathbb{R}^3$ and its Fourier transform $\tilde{\psi}(\boldsymbol{p})$ is a function of momentum $\boldsymbol{p} \in \mathbb{R}^3$.

To include continuous energies and other continuous observable values, it is necessary to generalize (1a) and (1b) to continuous superpositions:

R

$$\phi = \int \mathrm{d}E|E)(E|\phi), \qquad (\psi, \phi) = \int \mathrm{d}E\,(\psi|E)(E|\phi), \tag{2a}$$

In analogy to (1b), one is tempted to require that the $(E|\phi)$ are square integrable functions

$$(\phi, \phi) = \int_0^\infty \mathrm{d}E(\phi|E)(E|\phi) = \int_0^\infty \mathrm{d}E|\phi(E)|^2 < \infty, \tag{2b}$$

and similarly for the position and momentum wavefunctions

$$(\psi, \psi) = \int \mathrm{d}\boldsymbol{x}\,\psi^*(\boldsymbol{x})\psi(\boldsymbol{x}) < \infty, \quad (\tilde{\psi}, \tilde{\psi}) = \int \mathrm{d}\boldsymbol{p}\,\tilde{\psi}^*(\boldsymbol{p})\tilde{\psi}(\boldsymbol{p}) < \infty. \tag{2c}$$

The interpretation of $|(E_n|\phi)|^2$ as probability motivates the interpretation of the quantity $|(E|\phi)|^2$ as probability density, for which, as for other densities in physics, one expects to use a smooth function. This is the theory that Hilbert, von Neumann and Nordheim [7] were working on in the 1926–1927 period.

If the integrands in (2) representing probability densities are smooth (or even piecewise continunous), then the integrals (2) are the usual Riemann integrals. However, the space of Riemann square integrable functions is not topologically complete (with respect to the norm topology defined by (2b)) [8], a property that leads to serious mathematical difficulties. In order to obtain a complete space (i.e., every Cauchy sequence of vectors has a limit element in the space), von Neumann chose for integrals of (2) Lebesgue integrals. The resulting topologically complete, infinite dimensional vector space is called a (realization of the) Hilbert space $\mathcal{H}$, which contains the algebraic inner product space as a (dense) subspace, $\Psi \subset \mathcal{H}$. This Hilbert space theory was an enormous mathematical accomplishment. It led to a demonstration of the equivalence between the mathematical frameworks of ► matrix mechanics and ► wave mechanics (in the sense that each is a concrete realization of an abstract Hilbert space) and to the first mathematical theory of quantum physics [9].

However, there are some conceptual and computational difficulties with the Hilbert space theory, of which the following two are particularly significant. First, with Lebesgue integrable functions, the concept of a well defined value of the function $\phi(E) = (E|\phi)$ at a given E does not have a meaning as it does for continuous functions. This in turn means that the symbol $|E)$ cannot be given a meaning at each value of E for $0 \leqslant E < \infty$. Thus, in the position representation, although Schrödinger had assumed that ► wave function must be continuous on both physical and metaphysical grounds, the Hilbert space theory implicitly rejected these assumptions and associated wave mechanics with the much larger space of functions, which includes such pathological functions as those that are discontinuous everywhere. Second, not all quantum mechanical observables (e.g., not both $\boldsymbol{P}$ and $\boldsymbol{Q}$) could be represented by continuous operators defined everywhere in $\mathcal{H}$.

Undisturbed by von Neumann's arguments, Dirac proposed a formalism for quantum physics with great computational capacity and broad predictive power. The essential features of Dirac's formalism, often referred to as the bra-ket formalism, are the following:

1. Physical observables are represented by linear operators in a scalar product space Ψ and these operators form an algebra. Therefore, it makes sense to arbitrarily add and multiply operators to form new operators.
2. For a given quantum physical system, there exist complete systems of commuting observables (CSCO) in the algebra of observables. The system of eigenvectors for a chosen CSCO furnishes a basis for the space Ψ, i.e., every vector $\phi \in \Psi$ can be expanded with respect to the eigenvectors of the CSCO.
   For instance, let H, $\boldsymbol{J}^2$ and J_3 be such a CSCO for a spherically symmetric Hamiltonian H (where the J_i are the angular momentum operators). This CSCO has common eigenvectors $|Ejj_3\rangle$:

$$H|Ejj_3\rangle = E|Ejj_3\rangle, \tag{3a}$$

$$\boldsymbol{J}^2|Ejj_3\rangle = j(j+1)|Ejj_3\rangle, \quad J_3|Ejj_3\rangle = j_3|Ejj_3\rangle. \tag{3b}$$

   The energy eigenvalues may be discrete E_n so that every $\phi \in \Psi$ can be expanded as

$$\phi = \sum_{E_n jj_3} |E_n jj_3)(E_n jj_3|\phi), \tag{4a}$$

   or continuous $0 \leqslant E \leqslant \infty$ so that

$$\phi = \sum_{jj_3} \int_0^\infty \mathrm{d}E|Ejj_3\rangle\langle Ejj_3|\phi\rangle, \tag{4b}$$

   or both so that

$$\phi = \sum_{E_n jj_3} |E_n jj_3)(E_n jj_3|\phi) + \sum_{jj_3} \int_0^\infty \mathrm{d}E|Ejj_3\rangle\langle Ejj_3|\phi\rangle. \tag{4c}$$

For discrete E_n, the $|E_n jj_3)$ are the usual eigenvectors fulfilling the orthogonality conditions

$$(E_{n'} j' j_3'|E_n jj_3) \equiv \big(|E_{n'} j' j_3'), |E_n jj_3)\big) = \delta_{n'n}\delta_{j'j}\delta_{j_3' j_3}, \tag{5}$$

where $\delta_{n'n}$, $\delta_{j'j}$ and $\delta_{j_3' j_3}$ are the Kronecker deltas. For continuous E, the $|Ejj_3\rangle$ are the Dirac kets. They are not in the space Ψ or the Hilbert space $\mathcal{H} \supset \Psi$. They are new eigenvectors which, instead of (5), fulfill the "Dirac orthogonality condition"

$$\langle E' j' j_3'|Ejj_3\rangle = \delta(E' - E)\delta_{j'j}\delta_{j_3' j_3}, \tag{6}$$

where $\delta(E' - E)$ is defined as the mathematical object that fulfills the identity

$$\int \mathrm{d}E'\delta(E' - E)\langle E' jj_3|\phi\rangle = \langle Ejj_3|\phi\rangle \tag{7}$$

for all “well-behaved energy wave functions” $\langle Ejj_3|\phi\rangle = \phi_{jj_3}(E) \equiv \phi(E)$. The comparison of (7) with the equation

$$\sum_{n'} \delta_{n'n}(E_{n'}jj_3|\phi) = (E_n jj_3|\phi) \tag{8}$$

fulfilled by $\delta_{n'n}$ shows that $\delta(E' - E)$, often called the Dirac delta function, is the analogue of Kronnecker’s $\delta_{n'n}$ for continuous variables.

The property (7) for $\delta(E' - E)$ cannot be fulfilled by any proper function of E'. Instead, it was mathematically defined by (7) for a class of functions $\{\phi(E)\}$ and called a distribution by Schwartz [10]. Subsequently, this led to a new area of mathematics called distribution theory and ultimately to RHS’s.

From Dirac Kets to Gamow Vectors: Schwartz Space vs. Hardy Space Triplets

Dirac’s quasi-mathematical formalism used many postulated or tacitly assumed properties that are not definable for elements of the Hilbert space. For instance, the eigenkets (3a) with continuous eigenvalues, introduced by Dirac in [5,6] and further developed in his books [11] (the first and third editions in 1930 and 1947, respectively), were not mathematically well defined. However, textbooks have continued to use both Dirac delta functions and kets ever since Dirac’s bra-ket formalism. Though it lacked a rigorous mathematical foundation, this formalism has been used by physicists because of its many postulated features and its calculational convenience: the observables are treated like an algebra of linear operators on the entire space of physical states Ψ and, hence, could be handled like continuous operators; every Hermitian observable has a complete set of eigenkets (4); the wave functions are well-behaved smooth functions; each state vector ϕ corresponds to *one* wave function $\phi(E) = \langle E|\phi\rangle$ rather than to a whole equivalence class of functions which may differ from one another on a set of Lebesgue measure zero (for instance, on all rational numbers). These features constitute an enormous simplification over von Neumann’s Hilbert space theory.

There is a wide range of choice for the set of wavefunctions $\{\phi(E)\}$ admissible within the Dirac formalism. This leaves the Dirac formalism largely undefined but also flexible. The standard choice, if one is at all concerned with these mathematical subtleties, is the space of infinitely differentiable functions that, along with all their derivatives, vanish at infinity faster than any inverse polynomial. This function space, now called the Schwartz space $\mathcal{S}$, also plays an important role in the distribution theory of Schwartz [10]. With the development of distribution theory, the delta symbol in (6), which was completely outside of any rigorous mathematical framework for almost two decades after its introduction, could be given a mathematical meaning as a continuous antililear functional on the Schwartz function space $\mathcal{S}$.

The theory of distributions of Schwartz was an important inspiration to Gel'fand and his collaborators for developing a new mathematical structure during 1955–1959 [12], which they called a rigged Hilbert space (RHS). Later, along with Maurin [13], they proved the Dirac basis vector expansion (4) as the nuclear spectral theorem.

A rigged Hilbert space is a triplet of spaces

$$\Phi \subset \mathcal{H} \subset \Phi^{\times}, \tag{9}$$

where $\mathcal{H}$ is a Hilbert space, Φ is a dense subspace of $\mathcal{H}$, endowed with a locally convex topology τ_{Φ} that is stronger than the norm topology inherited from $\mathcal{H}$ (i.e., a stronger notion of convergence), and $\Phi^{\times}$ is the space of continuous antilinear functionals on Φ. Each space in (9) is dense in the next one, and all embeddings are linear and continuous.

The original motivation for introducing RHS's in quantum mechanics was to provide a rigorous formulation of Dirac formalism. This was done in the 1960s, independently by Antoine [14–16], Bohm [17, 18], Roberts [19, 20], and jointly by Kristensen, Meljbo and Poulsen [21], with many later contributions, e.g., [22–28]. The essential result of these papers was to show that, with a suitably constructed rigged Hilbert space, physical states can be represented by elements of the space Φ and observables by an algebra of continuous linear operators in Φ. The construction then allows basis vectors $|E\rangle$ of (2) and (6), which are undefined in the Hilbert space theory for continuous E, to be well defined as elements of the dual space $\Phi^{\times}$. A detailed mathematical analysis of these developments may be found in the next entry [29].

As mentioned above, the standard choice for allowed wavefunctions $\langle E|\phi\rangle$ are Schwartz functions, i.e., an RHS where the space Φ is realized by the Schwartz function space $\mathcal{S}$. The Schwartz RHS provides the mathematical foundation of the quantum theory that describes the structure and spectra of stationary states, and the time symmetric evolution of states which is given by a one parameter group $U(t)$. With a suitable generalization of this construction, it is possible to obtain differentiable representations of all finite dimensional compact and non-compact Lie groups [14–20, 30–33]. Particularly relevant among these are the symmetry groups of spacetime, both non-relativistic and relativistic.

However, the Schwartz RHS is not sufficient for a quantum theory of scattering and decay where one analytically continues the S-matrix into the complex energy plane [34–36]. In the empirical description of resonance phenomena, one uses the energy (or, in the relativistic case, the invariant mass) values of the complex plane and works with Gamow vectors [37] which are associated with the complex eigenvalues of the Hamiltonian. One also uses Lippmann–Schwinger kets with $\pm i\epsilon$ energy in the denominator [38–40]. The Schwartz RHS accommodates neither Lippmann–Schwinger kets nor exponentially decaying Gamow kets and thus cannot provide a relation between the lifetime of decay τ and the width Γ (or, the complex pole position) of a resonance.

To obtain a mathematical theory that unifies quantum resonance and decay phenomena, one needs to take a step beyond the confines of Dirac's formalism or the Schwartz RHS theory. What is remarkable is that this step beyond the Schwartz space theory can be taken within the general mathematical framework of RHS's. Specifically, this theory requires a careful mathematical distinction between the set of *prepared in-states* and the set of *observed out-states* (more precisely, out-observables). In the discussions on the foundations of quantum theory, a distinction is made between the notions of states ϕ, which are prepared by a preparation apparatus, and observables $\Lambda = |\psi\rangle\langle\psi|$, which are registered by a detector. In terms of these states and observables, the theory predicts the Born probabilities $|(\phi, \Lambda(t)\phi)|^2$ for an observable Λ in a state ϕ. These probabilities are to be compared with the normalized detector counts of events $\frac{N(t)}{N}$. In scattering theory, one makes a distinction between in-states ϕ^+ and out-states ψ^- for which one uses separate basis vector expansions:

$$\phi^+ = \int_0^\infty \mathrm{d}E\,|E^+\rangle\langle^+E|\phi^+\rangle \quad \text{and} \quad \psi^- = \int_0^\infty \mathrm{d}E\,|E^-\rangle\langle^-E|\psi^-\rangle, \tag{10}$$

where $|E^\pm\rangle = |E \pm \mathrm{i}\epsilon\rangle$ are considered to be two different Lippmann–Schwinger kets fulfilling the two different Lippmann–Schwinger equations.

However, in the mathematical foundations of quantum mechanics, the set of state vectors $\{\phi\}$ is identified with the set of observable vectors $\{\psi\}$, usually by associating both with the same Hilbert space $\mathcal{H}$. Similarly in scattering theory, the kets $|E^\pm\rangle$ of expansions (10) are thought of as two sets of basis vectors for the same vector space. In contrast, in the RHS's theory of scattering and decay phenomena, one generalizes the Schwartz RHS theory of Dirac's formalism to a theory with two RHS's, one for the set of prepared in-states $\{\phi^+\}$,

$$\{\phi^+\} = \Phi_- \subset \mathcal{H} \subset \Phi_-^\times \ni |E^+\rangle \tag{11+}$$

and the other for the set of detected out-observable vectors $\{\psi^-\}$,

$$\{\psi^-\} = \Phi_+ \subset \mathcal{H} \subset \Phi_+^\times \ni |E^-\rangle \tag{11−}$$

where $\mathcal{H}$ is the same Hilbert space. One now distinguishes mathematically between states $\{\phi^+\} = \Phi_-$ and observables $\{\psi^-\} = \Phi_+$ and relates them to Lippmann–Schwinger kets $|E^+\rangle \in \Phi_-^\times$ and $|E^-\rangle \in \Phi_+^\times$, respectively. Thus, the RHS theory elevates the physical content of the notions of state and observable vectors into a mathematical principle.

From this pair of RHS's for state and observable vectors, a mathematically consistent theory of resonance scattering and decay phenomena can be obtained by letting the spaces Φ_- and Φ_+ to be defined in their energy representation by Hardy spaces on the lower and upper complex semiplanes, respectively [41–44]. In particular, the energy wavefunctions $\langle^+E|\phi^+\rangle = \phi^+(E)$ and $\langle^-E|\psi^-\rangle = \psi^-(E)$ in (10) are smooth, rapidly decreasing Hardy functions on the lower and upper complex semiplanes. The basis kets $|E^{\pm}\rangle$ can now be well-defined as elements of the dual spaces $\Phi_{\mp}^{\times}$, and therewith Dirac-type basis vector expansions (10) of ϕ^{+} and ψ^{-} can be rigorously obtained in terms of $|E^{\pm}\rangle$ by way of the nuclear spectral theorem. The theory based on RHS's (11) also contains exponentially decaying Gamow vectors and Breit–Wigner resonance amplitudes as well-defined mathematical concepts [45]. This Hardy space theory has been subsequently extended to *relativistic* resonances and decaying states [46]. One of the important outcomes of the relativistic extension is the unique and unambiguous definition it provides for the mass and width of a relativistic resonance, a much debated problem since the early 1990s.

One particularly important aspect in which the Hardy-type RHS's differ from the Schwartz-type RHS's entails the class of allowed representations of symmetry groups, including non-compact spacetime symmetry groups. In the Schwartz-type construction, the unitary representations of Lie groups in the Hilbert space $\mathcal{H}$ can be restricted to Φ and extended to $\Phi^{\times}$ to obtain differentiable representations in these spaces [33]. Thus, quantum mechanical symmetry transformations represented by groups can be well accommodated in the Schwartz-type RHS's, and many of the elements of the algebra of observables arise as the derivatives of these representations in Φ and $\Phi^{\times}$. In contrast, Hardy-type RHS's do not furnish representations of the spacetime symmetry *groups*. In particular, in the non-relativistic version, the time evolution in $\Phi_{\pm}$ is given by one parameter semigroups $U_{\pm}(t)$ with $t \geqslant 0$. In the relativistic version, the spacetime evolution in $\Phi_{\pm}$ is given by semigroups $U_{\pm}(I, a)$, where a are spacetime four vectors with $a_0 \geqslant 0$ and $a^2 \geqslant 0$, i.e., by representations of the Poincaré semigroup into the forward lightcone [46, 47]. These semigroup representations encode the fundamental causal structure of physics. The search for a consistent mathematical theory that unifies resonance and decay phenomena unwittingly leads to quantum mechanical causality.

Summary and Conclusion

Originally, the RHS was an offspring of the Dirac formalism of quantum mechanics. After the pioneers of quantum physics had arrived at an algebra of observables [1–5], von Neumann was the first to give a rigorous mathematical meaning to quantum theoretical notions, such as states and observables [9], using the Hilbert space of Lebesgue square integrable functions and self-adjoint operators in it [7, 9]. This was a monumental achievement of the human intellect, but it resulted in a rather complicated mathematical structure mainly because it involved physically unintuitive Lebesgue integration and unbounded operators. The vast majority of practicing physicists remained unaware of these mathematical subtleties and complications. In their practical calculations, physicists treated the Hilbert space theory of quantum physics like a theory of continuous (bounded) operators in a linear scalar product space and carried out all integrals as Riemann integrals. Although most physicists were not using the full mathematical formalism of the

Hilbert space, some properties that could not be derived without the precise mathematics of the Hilbert space did enter the standard body of knowledge. One such example is the unitary (hence reversible) time evolution that could be derived as the solution to the dynamical Schrödinger equation only under the precise Hilbert space structure. Nevertheless, physicists took this to be universally true and incorporated it into their practical calculations.

Irrespective of von Neumann's Hilbert space theory, Dirac [5, 11] proposed and developed (in two stages, in the first edition of [11] in 1930 and the third edition in 1947) his bra-ket formalism. In this formalism, every physical observable is represented by an everywhere defined "Hermitian" operator that has a complete set of eigenvectors with discrete or continuous eigenvalues, and every state vector is a (discrete and/or continuous) superposition of these eigenvectors (4). For continuous eigenvalues, in analogy to the Kronecker-δ, Dirac introduced the δ symbol that bears his name today. Ever since its introduction, most physicists have used the Dirac formalism as their theory of quantum mechanics.

Schwartz (1950) gave a proper mathematical content to the Dirac-δ and other similiar "generalized functions" with his theory of distributions [10]. Later, Grothendieck (1966) introduced a specific topological vector space called nuclear vector space [48]. On this basis, Gel'fand and his school [12] and Maurin [13] developed the Rigged Hilbert Space. The main mathematical purpose of these Schwartz-type RHS's (9) was to provide a theory of unitary representations of non-compact Lie groups. The generator of each non-compact subgroup of such a representation has continuous eigenvalues of the type envisioned by Dirac, and RHS's provide the tools to handle the eigenvalue problem for these generators. In particular, with RHS's, Dirac kets could be defined as elements of $\Phi^\times$, i.e., continuous antilinear functionals on Φ, and Dirac's basis vector expansion (2) and (4) could be proved as the nuclear spectral theorem. Within the Schwartz-type RHS's (9), the Schrödinger and Heisenberg dynamical equations can be solved as vector valued differential equations in Φ (or in $\Phi^\times$). The resulting time evolution of states and observables is given by a continuous one parameter group of operators, just as in the Hilbert space.

Going from the one parameter time evolution group to more general non-compact Lie groups, the topology (the meaning of convergence) of the space Φ is defined by a countable family of scalar products $(\phi, \psi)_n = (\phi, \Delta^n \psi)$, where Δ is the Laplacian of the group, also known as the Nelson operator [49], and $(\phi, \psi)_{n=0} = (\phi, \psi)$ is the Hilbert space inner product [44]. This topology is stronger than the Hilbert space topology, and with respect to it, the generators of the group, and therefore the enveloping algebra, are represented by continuous operators in Φ. By duality, there is also a representation of the enveloping algebra as well as the group by continuous operators in the space $\Phi^\times$, where the topology is the weak-* topology. Eigenkets of the generators of non-compact subgroups of these representations exist as elements of the space $\Phi^\times$, e.g., the eigenkets $|\boldsymbol{x}\rangle$ of position operators $\boldsymbol{Q}$, or $|\boldsymbol{p}\rangle$ of the momentum operators $\boldsymbol{P}$ with eigenvalues $\boldsymbol{x} \in \mathbb{R}^3$ and $\boldsymbol{p} \in \mathbb{R}^3$, respectively. In contrast, it is not possible to obtain a representation of the enveloping algebra of a non-compact group by continuous operators in a Hilbert space.

Structure and spectra of microphysical systems is one aspect of quantum theory for which the Schwartz-type RHS's provide a complete solution. The other aspect of quantum theory is scattering, resonance and decay phenomena along with the dynamics governing their evolution. In heuristic treatments of scattering, the mathematical subtleties of the Hilbert space theory were ignored. Instead, solutions of the Schrödinger equation with purely outgoing boundary conditions were advocated [50,51]. Mathematically undefined kets $|E^{\pm}\rangle$ with infinitesimal imaginary part $\pm i\epsilon$ of energy were used to obtain, respectively, the incoming and outgoing solutions of Lippmann–Schwinger equations [38–40]. Resonance and decaying states were intuitively associated with an asymmetric "irreversible" time evolution [52–54].

While these heuristic methods were adequate for some physical applications, when they were compared with the precise mathematical consequences of the Hilbert space, one was necessarily led to contradictions. For instance, heuristic Gamow vectors [37] and rigorous unitary time evolution are mutually contradictory, as exemplified by the exponential catastrophe [55]. Furthermore, the deviations from the exponential decay law [56,57], another mathematical consequence of the structure of the Hilbert space, leads to inconsistencies with Einstein causality [58].

Thus it was clear that for a description of resonance and decay phenomena, it was necessary to go beyond the time symmetric mathematical theory based on the Hilbert space, or on the Schwartz-type RHS theory. But many of the empirical notions, like Gamow states and Lippmann–Schwinger kets, have been very successful for the description of scattering and decay. Therefore, what was needed was a mathematical structure that incorporated and legitimized these useful heuristic notions of resonance scattering and decay. Hardy-type RHS's precisely provide this mathematical framework in the same way as the Schwartz-type RHS's had provided the framework for Dirac's formalism.

With the Hardy RHS's (11±), it is possible to define mathematical entities having the same useful properties as the heuristic Gamow vectors and Lippmann–Schwinger kets. Because of shared characteristics, the new entities were called by the same names. In the Hardy RHS's, these new mathematically well defined entities provide a rigorous mathematical theory that unifies resonance scattering and decay phenomena and predicts the lifetime-width relation $\tau = \frac{\hbar}{\Gamma}$ as an exact identity, not just as an approximation based on the Weisskopf–Wigner methods. In the relativistic version, the theory provides a unique, unambiguous, gauge invariant definition of mass and width of a resonance [46].

R

The new theory of Hardy-type RHS's retains the useful heuristic features of previous descriptions of resonance scattering and decay phenomena and eliminates the contradictory mathematical consequences based on the Hilbert space theory. Salient among the latter is unitary evolution, which is now replaced by an asymmetric, semigroup evolution. Though it emerges in the mathematical theory as a consequence of the axioms suggested by the experimental and phenomenological properties of resonances and decaying states, the semigroup evolution can be looked at primarily as a manifestation of the fundamental causal structure of the physical world [59–61].

See also ► Time in quantum mechanics.

Literature

1. W. Heisenberg: Z. Phys. 33, 879 (1925)
2. M. Born, W. Heisenberg, P. Jordan: Z. Phys. 35, 557 (1926)
3. M. Born, N. Wiener: Z. Phys. 36, 174 (1926)
4. E. Schrödinger: Ann. Phys. 81, 109 (1926) and references
5. P. A. M. Dirac: Proc. R. Soc. 109, 642 (1926)
6. P. A. M. Dirac: Proc. R. Soc. 110, 561 (1926)
7. D. Hilbert, J. von Neumann, L. Nordheim: Math. Ann. 98, 1 (1927)
8. The word topology has two different meanings (Webster's dictionary): (1.) The study of those properties of geometric forms that remain invariant under certain transformations. (2.) The study of limits in sets considered as collections of points, and related to it, the collection of open sets making a topological space, i.e., a space in which a meaning of convergence of infinite sequences is defined. Physicists usually associate the first meaning with the word. In this paper, we use the second meaning.
9. J. von Neumann: *Mathematical Foundations of Quantum Mechanics* (Springer, Berlin, 1932), Princeton University Press (1955)
10. L. Schwartz: *Théorie des Distributions, I–II*, (Hermann, Paris, 1957, 1959). New edition, revised and augmented [First published in 1950–1951]
11. P. A. M. Dirac: *The Principles of Quantum Mechanics* (Clarendon Press, Oxford, 1930), 3rd edn. (Clarendon Press, Oxford, 1947)
12. I. M. Gel'fand, N. Y. Vilenkin: *Generalized Functions IV: Applications of Harmonic Analysis* (Academic Press, New York, 1964). [Translation of the 1961 Russian edition]
13. K. Maurin: *General Eigenfunction Expansions and Unitary Representations of Topological Groups*. (Polish Scientific Publishers, Warsaw, 1968)
14. J-P. Antoine: *Formalisme de Dirac et problèmes de symétrie en Mécanique Quantique*, Thèse de doctorat, Université Catholique de Louvain (1966)
15. J-P. Antoine: J. Math. Phys. 10, 53 (1969)
16. J-P. Antoine: J. Math. Phys. 10, 2276 (1969)
17. A. Böhm: *Rigged Hilbert Spaces*, Int. Centre for Phys. Bult. 9 (1964)
18. A. Bohm: in *Boulder Lectures in Theoretical Physics IX A: Mathematical Methods of Theoretical Physics* (Wiley, New York, 1967)
19. J. E. Roberts: J. Math. Phys. 7, 1097 (1966)
20. J. E. Roberts: Commun. Math. Phys. 3, 98 (1966)
21. P. Kristensen, L. Meljbo, E. Thue Poulsen: Commun. Math. Phys. 1, 175 (1965)
22. D. Babbitt: Rep. Math. Phys. 3, 37 (1972)
23. D. Fredricks: Rep. Math. Phys. 8, 277 (1975)
24. K. Napiorkowski: Bull. Acad. Pol. Sci. 22, 1215 (1974)
25. K. Napiorkowski: Bull. Acad. Pol. Sci. 23, 251 (1975)
26. S. J. L. van Eijndhoven, J. de Graaf: *A Mathematical Introduction to Dirac's Formalism* (North Holland, Amsterdam, 1986)
27. O. Melsheimer: J. Math. Phys. 15, 902 (1974)
28. O. Melsheimer: J. Math. Phys. 15, 917 (1974)
29. J-P. Antoine, A. Bohm, S. Wickramasekara: *Rigged Hilbert Spaces for the Dirac formalism of quantum mechanics*, this volume.
30. A. Bohm: J. Math. Phys. 8, 1551 (1967), Appendix B.
31. B. Nagel: in *Studies in Mathematical Physics (Proc. Istanbul 1970)*, ed. by A. O. Barut. (Reidel, Dordrecht and Boston, 1970), pp. 135–54
32. G. Lindblad, B. Nagel: Ann. Inst. H. Poincaré 13, 27 (1970)
33. S. Wickramasekara, A. Bohm: J. Phys. A: Math. Gen. 35, 807 (2002)
34. See, for example, M. L. Goldberger, K. M. Watson: *Collision Theory* (Wiley, New York, 1964)
35. R. J. Eden, P. V. Landshoff, P. J. Olive, J. C. Polkinghorne: *The Analytic S-Matrix*, Cambridge U.P., Cambridge, 1966
36. R. G. Newton: *Scattering Theory of Waves Particles*, 2nd edn. (Springer, Berlin, 1982)

37. G. Gamow: Z. Phys. 51, 204 (1928)
38. B. A. Lippmann, J. Schwinger: Phys. Rev. 79, 469 (1950)
39. M. Gell-Mann, H. L. Goldberger: Phys. Rev. 91, 398 (1953)
40. W. Brenig, R. Haag: Fortschr. Phys. 7, 183 (1959)
41. A. Bohm: Lett. Math. Phys. 3, 455 (1978)
42. A. Bohm: J. Math. Phys. 33, 2813 (1981)
43. M. Gadella: J. Math. Phys. 24, 1462 (1983)
44. A. Bohm, M. Gadella: *Dirac Kets, Gamow Vectors and Gel'fand Triplets*, Springer Lecture Notes in Physics, vol. 348 (Springer, Berlin, 1989)
45. A. Bohm, N. L. Harshman: *Rigged Hilbert spaces and time asymmetric quantum theory*, this volume.
46. A. Bohm, H. Kaldass, S. Wickramasekara: Fortschr. Phys. 51, 569 (2003); 51, 604 (2003)
47. L. S. Schulman: Ann. Phys. 59, 201 (1970)
48. A. Grothendieck: Memoirs Am. Math. Soc., Nr. 16, Providence, RI, 1966
49. E. Nelson: Ann. Math. 70, 572 (1959)
50. R. Peierls: Proc. R. Soc. Lond., Series A, 166, 277 (1938), In Proc. 1954 Glasgow Conf. on Nuclear and Hadron Physics (Pergamon Press, New York, 1959)
51. E. Hernandez, A. Mondragon: Phys. Rev. C 29, 722 (1984)
52. E. Merzbacher: *Quantum Mechanics* (Wiley, New York, 1970) Chap. 18
53. C. Cohen-Tannoudji, B. Diu, F. Laloë: *Quantum Mechanics* (Wiley, New York, 1977) Vol. II, p. 1345, pp. 1353–354
54. T. D. Lee: *Particle Physics and Introduction to Field Theory* (Harwood Academic, New York, 1981), Chap. 13. [The irreversible nature of quantum mechanical decay has been mentioned in these and other textbooks and volumes of lecture notes.]
55. A. Bohm, M. Gadella, G. B. Mainland: Am. J. Phys. 57, 1103 (1989)
56. L. A. Khalfin: Sov. Phys. JETP 6 1063 (1958)
57. L. Fonda, G. C. Ghirardi, A. Rimini: Rep. Prog. Phys. 41, 589 (1978)
58. E. Fermi: Rev. Mod. Phys. 4, 87 (1932); M. I. Shirokov: Yad. Fiz. 4, 1077 (1966); G. C. Hegerfeldt: Phys. Rev. Lett. 72, 596 (1994); A. Bohm, N. L. Harshman, H. Walther: Phys. Rev. A. 66, 01207 (2002)
59. J. J. Halliwell, J. Pérez-Mercader, W. H. Zurek (eds.): *Physical Origins of Time Asymmetry* (Cambridge University, London, 1994)
60. In particular, see M. Gell-Mann, J. B. Hartle: *Time Symmetry and Asymmetry in Quantum Mechanics and Quantum Cosmology*, [58], pp. 311–45
61. A. Bohm, H.-D. Doebner, P. Kielanowski (eds.): *Irreversibility and Causality: Semigroups and Rigged Hilbert Spaces*, LNP 504 (Springer, Berlin, 1998)

Rigged Hilbert Spaces for the Dirac Formalism of Quantum Mechanics

J-P. Antoine, A. Bohm, and S. Wickramasekara

Introduction

As explained in the preceding entry [1], the original motivation for introducing Rigged Hilbert Spaces (RHS) in quantum mechanics was to provide a rigorous formulation of the ► Dirac notation. This was done in the 1960s, independently by

Antoine [2–4] Bohm [5, 6], Roberts [7, 8], and jointly by Kristensen, Meljbo and Poulsen [9], with many later contributions, e.g. [10–14] (Actually, the idea to use an RHS in the formulation of quantum mechanics was suggested to JPA in 1962 by Bargmann, then on sabbatical in Zürich).

We recall that a rigged Hilbert space is a triplet of spaces

$$\Phi \subset \mathcal{H} \subset \Phi^{\times}, \tag{1}$$

where $\mathcal{H}$ is a Hilbert space, Φ is a dense subspace of $\mathcal{H}$, endowed with a locally convex topology τ_{Φ} that is finer than the norm topology inherited from $\mathcal{H}$ (i.e., a stronger notion of convergence), and $\Phi^{\times}$ is the space of continuous antilinear functionals $F(\phi)$ on Φ. By duality, each space in (1) is dense in the next one and all embeddings are linear and continuous. Standard examples of rigged Hilbert spaces are the Schwartz distribution spaces over $\mathbb{R}$ or $\mathbb{R}^N$, namely $\mathcal{S} \subset L^2 \subset \mathcal{S}^{\times}$ or $\mathcal{D} \subset L^2 \subset \mathcal{D}^{\times}$ [15–17].

As discussed in [1], Dirac's formalism undergoes some rather subtle modifications to achieve rigor; nevertheless, its formal features that are used in quantum theory are largely reproduced by a RHS $\Phi \subset \mathcal{H} \subset \Phi^{\times}$ with Φ given, in the simplest case, by the abstract Schwartz space. To show how and why this RHS structure provides a rigorous meaning to Dirac's formalism, we have to describe the mathematics involved in more detail. In particular, we must describe how to choose the space Φ for a given physical system, then make the link with the measurement process and finally discuss the realization of symmetries in this new framework.

Mathematical Properties of the RHS

Given the Hilbert space $\mathcal{H}$ of (1), the choice of the space Φ is not yet fixed, but it depends on the system at hand. In general, Φ is required to fulfill the following conditions:

(1) Φ should be *complete* with respect to τ_{Φ}; that is, every Cauchy sequence converges to an element of Φ.
(2) Φ should be *reflexive*; that is, the dual of the dual of Φ can be identified with Φ, $(\Phi^{\times})^{\times} \simeq \Phi$. In most cases, Φ can be obtained as the intersection of a *countable* family of Hilbert spaces, $\Phi = \bigcap_{n \in \mathbb{N}} \mathcal{H}_n$. It is then a Fréchet space.
(3) Φ should be *nuclear*. In the case where $\Phi = \bigcap_{n \in \mathbb{N}} \mathcal{H}_n$, this means that, for each n, there is an $m > n$ such that the embedding $\mathcal{H}_m \to \mathcal{H}_n$ is a Hilbert–Schmidt operator.

Next, we must fix our notation. For $F \in \Phi^{\times}$, $F(\phi)$ will denote the value of F at the vector $\phi \in \Phi$. If $F \in \mathcal{H}$, we normalize the duality form by requiring $F(\phi) = (\phi|F)$, where $(\cdot|\cdot)$ denotes the scalar product of $\mathcal{H}$ (recall that F is antilinear). This motivates the notation $F(\phi) \equiv \langle \phi|F \rangle$ for *any* $\phi \in \Phi$ and $F \in \Phi^{\times}$, with the obvious convention $\langle F|\phi \rangle = \langle \phi|F \rangle^*$, such that $\langle \phi|F \rangle = (\phi|F)$ for $F \in \mathcal{H} \subset \Phi^{\times}$. That is, the functional $\langle \phi|F \rangle$ is an extension of the Hilbert space scalar product.

The motivation for the nuclearity property (3) above is that it allows one to exploit the *nuclear spectral theorem* of Gel'fand and Maurin [16, 17], which says the following: Let A be a closed linear operator in $\mathcal{H}$, which maps Φ into itself continuously with respect to τ_Φ. Then A may be transported by duality to a linear operator $A^\times : \Phi^\times \to \Phi^\times$, which is an extension of the usual adjoint operator $A^\dagger$ in the Hilbert space, namely:

$$A^\times F(\phi) = F(A\phi), \text{ for all } \phi \in \Phi \text{ and for all } F \in \Phi^\times, \tag{2}$$

which we also write

$$\langle \phi | A^\times F \rangle = \langle A\phi | F \rangle, \ \forall \phi \in \Phi, \ F \in \Phi^\times. \tag{3}$$

For such an operator, the vector $\xi_\lambda \in \Phi^\times$ is called a *generalized eigenvector* of A, with eigenvalue $\lambda \in \mathbb{C}$, if it satisfies

$$\langle \phi | A^\times \xi_\lambda \rangle \equiv A^\times \xi_\lambda(\phi) = \lambda^* \xi_\lambda(\phi) \equiv \lambda^* \langle \phi | \xi_\lambda \rangle, \text{ for all } \phi \in \Phi. \tag{4}$$

This equality can also be written in the Dirac notation as

$$A^\times |\xi_\lambda\rangle = \lambda^* |\xi_\lambda\rangle, \quad |\xi_\lambda\rangle \in \Phi^\times. \tag{5}$$

Now assume that A has a self-adjoint extension A_0 in $\mathcal{H}$ with a non-degenerate spectrum, and that Φ is nuclear and complete. In this case, $A^\times$ is an extension of both A and A_0 (collectively, $\widehat{A}$). Then the nuclear spectral theorem asserts that A (or $\widehat{A}$) possesses a complete orthonormal set of generalized eigenvectors $\xi_\lambda \in \Phi^\times$, $\lambda \in \mathbb{R}$. This means that, for any two $\phi, \psi \in \Phi$, one has

$$\begin{aligned} \langle \phi | \psi \rangle &= \int_{\mathbb{R}} \xi_\lambda(\phi)\, \xi_\lambda(\psi)^* \, \mathrm{d}\mu(\lambda) \\ &\equiv \int_{\mathbb{R}} \langle \phi | \xi_\lambda \rangle \, \langle \xi_\lambda | \psi \rangle \, \mathrm{d}\mu(\lambda) \end{aligned} \tag{6}$$

for some measure μ on $\mathbb{R}$. For quantum mechanical operators A, the measure μ may be split into a discrete and an absolutely continuous part such that (6) can be written as

$$\langle \phi | \psi \rangle = \sum_i \langle \phi | \lambda_i \rangle \langle \lambda_i | \psi \rangle + \int \langle \phi | \lambda_\rho \rangle \langle \lambda_\rho | \psi \rangle \rho(\lambda) \mathrm{d}\lambda, \tag{7}$$

where the $\{\lambda_i\}$ are the discrete eigenvalues of A in $\mathcal{H}$, $|\xi_\lambda\rangle\langle\xi_\lambda|\mathrm{d}\mu(\lambda) = |\lambda_\rho\rangle\langle\lambda_\rho|\rho(\lambda)$ $\mathrm{d}\lambda$, where $\rho(\lambda)$ is a non-negative integrable function and the integral extends over the absolutely continuous Hilbert space spectrum of A. Then, the Dirac kets are $|\lambda\rangle = |\lambda_\rho\rangle\sqrt{\rho(\lambda)}$.

R

The net result of this theorem is to put the eigenvalues and the points of the continuous spectrum of A on the same footing — exactly what is usually assumed in the Dirac formulation of quantum mechanics. Indeed, using Dirac's notation, (6) and (7) are written as a decomposition of the identity:

$$I = \int_{\mathbb{R}} |\xi_\lambda\rangle\langle\xi_\lambda| \, \mathrm{d}\mu(\lambda) = \sum_i |\lambda_i\rangle\langle\lambda_i| + \int_{\mathbb{R}} \mathrm{d}\lambda \, |\lambda\rangle\langle\lambda|, \quad \langle\lambda|\lambda'\rangle = \delta(\lambda - \lambda'), \tag{8}$$

with the proviso that this quantity makes sense only between two vectors of Φ. In other words, I must be understood as the (linear) embedding of Φ into $\Phi^\times$, or equivalently as a sesquilinear form on $\Phi \times \Phi$.

Actually the symbol $|\xi_\lambda\rangle\langle\xi_\lambda|$ in (8) may be interpreted as a genuine projection operator from Φ onto the λ-component in the decomposition, combining von Neumann's direct integral approach with the nuclear spectral theorem. According to von Neumann, the self-adjoint operator A determines a decomposition of $\mathcal{H}$ into a direct integral of one-dimensional spaces $\mathcal{H}(\lambda)$:

$$\mathcal{H} \simeq \int_{\mathbb{R}}^{\oplus} \mathcal{H}(\lambda) \, \mathrm{d}\mu(\lambda), \tag{9}$$

which "diagonalizes" A:

$$f \sim \{f(\lambda)\}, \; f(\lambda) \in \mathcal{H}(\lambda), \text{ with } \|f\|^2 = \int_{\mathbb{R}} |f(\lambda)|^2 \, \mathrm{d}\mu(\lambda), \tag{10}$$

$$Af \sim \{\lambda f(\lambda)\}. \tag{11}$$

As already mentioned, the difficulty with this formulation is that $\mathcal{H}(\lambda)$ is *not* a subspace of $\mathcal{H}$ if λ is a point of μ-measure zero. This is why there are no true eigenvectors associated to the points of the continuous spectrum.

However, if the space Φ in (1) is nuclear, then the map $\tau_\lambda : \phi \mapsto \phi(\lambda)$, $\phi \in \Phi$, $\phi(\lambda) \in \mathcal{H}(\lambda)$, is continuous and nuclear for μ-almost all λ. Therefore, one may write

$$\tau_\lambda \phi = \phi(\lambda) = \langle\phi|\xi_\lambda\rangle h(\lambda), \text{ where } \xi_\lambda \in \Phi^\times, \; h(\lambda) \in \mathcal{H}(\lambda). \tag{12}$$

Then the dual mapping $\tau'_\lambda : \mathcal{H}(\lambda) \to \Phi^\times$ is continuous as well and it allows us to identify each vector $\xi \in \mathcal{H}(\lambda)$ with a functional $\widetilde{\xi} = \tau'_\lambda \xi \in \Phi^\times$. Finally, the combined map $\chi_\lambda = \tau'_\lambda \tau_\lambda$, which is a nuclear operator mapping Φ into $\Phi^\times$, acts as a projection operator onto the eigensubspace $\Phi^\times_\lambda$ corresponding to the eigenvalue λ.

If the spectrum of the self-adjoint operator A_0 has non-trivial multiplicity, i.e., $\dim \mathcal{H}(\lambda) > 1$, as in the case of a spherically symmetric Hamiltonian described in (4.c) of [1], then the whole machinery still goes through. The map τ_λ of (12) reads as

$$\tau_\lambda \phi = \phi(\lambda) = \sum_n \langle\phi|\xi_{\lambda,n}\rangle h_n(\lambda), \tag{13}$$

where $\xi_{\lambda,n} \in \Phi^{\times}$ and $\{h_n(\lambda),\ n = 1, 2, \dots \dim \mathcal{H}(\lambda)\}$ is a basis of $\mathcal{H}(\lambda)$. Thus the expansion (8) becomes

$$I = \int_{\mathbb{R}} \sum_n |\xi_{\lambda,n}\rangle\langle\xi_{\lambda,n}| \, \mathrm{d}\mu(\lambda)$$
$$= \sum_{i,n} |\lambda_{i,n}\rangle\langle\lambda_{i,n}| + \int_{\mathbb{R}} \mathrm{d}\lambda \sum_n |\lambda, n\rangle\langle\lambda, n|, \quad \langle\lambda, n|\lambda', n'\rangle = \delta(\lambda - \lambda')\delta_{n,n'}. \tag{14}$$

Yet a word of caution is necessary here. If one is interested only in the spectral properties of A, one may require that the spectrum of A in $\Phi^{\times}$ consists exactly of the points of its spectrum in $\mathcal{H}$. If this is the case, one says that (1) is a *tight rigging* for A. Tight riggings are by no means guaranteed for a given operator A, as can be seen from the sufficient conditions given in [18–21]. On the other hand, there are important cases where one actually needs generalized eigenvalues that do *not* belong to the Hilbert space spectrum of A. As we can see in [1] and [22], scattering theory is a major example, where resonances are associated with complex eigenvalues of the Hamiltonian, with Gamow vectors as generalized eigenvectors. As operators in the ▶ Hilbert space, these Hamiltonians are self-adjoint and as such their Hilbert space spectra are real. Therefore, a tight rigging would not permit a description of resonance states by complex eigenvectors. However, in the more general case, it is possible to construct rigged Hilbert spaces such that self-adjoint Hamiltonians have complex generalized eigenvalues [23–25].

As mentioned in [1], the von Neumann approach to quantum mechanics has conceptual and mathematical difficulties. For instance, many of the ▶ operators representing physical observables, such as position and momentum fulfilling the commutation relations $\left[Q_i, P_j\right] = \mathrm{i}\delta_{ij} I$, are necessarily non-continuous operators. (In general, the generators of unitary representations of non-compact subgroups of a Lie group are non-continuous, i.e., unbounded.) Unbounded operators cannot be defined on the whole Hilbert space and as such there are subtle issues associated with choosing an appropriate dense domain in which they can be well-defined. This is the reason why Dirac's notion of an algebra of observables is difficult to realize in the Hilbert space theory. Furthermore, not all ▶ self-adjoint operators can be interpreted as physical ▶ observables and not all elements of the Hilbert space can be interpreted as states. In the Hilbert space, there are *physical* vectors that represent preparable states and many other vectors that do not. Furthermore, there are *generalized* vectors associated with quantum measurements, which are not elements of the Hilbert space.

As the point of departure for the rigged Hilbert space theory, one can consider the construction of the space Φ in (1) in such a way that physical observables are defined as continuous, bounded operators in Φ. One starts with the Hilbert space theory and identifies a common, dense, invariant domain $\mathcal{D}$ on which the "relevant" observables of the theory are defined. In particular, one chooses a distinguished ("labeled" [7]) family $\mathcal{O}$ of observables, which have both a meaningful physical interpretation (in terms of measurements, say) and a mathematical definition (as self-adjoint operators with a dense invariant domain $\mathcal{D}$ in $\mathcal{H}$). Hence, $\mathcal{O}$ is an algebra of operators on $\mathcal{D}$. Then, one equips this domain with a suitable ("projective") topology that makes all the elements of $\mathcal{O}$ continuous operators and calls the resulting topological vector space Φ. Taking duals, one obtains a RHS $\Phi \subset \mathcal{H} \subset \Phi^{\times}$, *defined* by the system. A simple example is the algebra generated by P, Q and I for the harmonic oscillator [26]. For unitary representations of all finite dimensional non-compact Lie groups, an RHS can be constructed in a similar way [27]. The space Φ constructed this way is nuclear for a large class of representations [27]. Therefore, the nuclear spectral theorem applies and yields a rigorous formulation of Dirac's bra-and-ket formalism for which the Dirac kets appear as generalized eigenvectors of operators (generators) for non-compact subalgebras. These results are routinely used by physicists, but they cannot be justified solely in Hilbert space.

The simplest class of examples in non-relativistic quantum mechanics is that of a particle, either free or in a nice potential V. The labeled observables are position $\mathbf{Q}$, momentum $\mathbf{P}$ and energy $H = \mathbf{P}^2/2m + V(Q)$. The corresponding RHS is $\mathcal{S}(\mathbb{R}^3) \subset L^2(\mathbb{R}^3) \subset \mathcal{S}^{\times}(\mathbb{R}^3)$. The most well-recognized representative of this class of examples is the harmonic oscillator potential mentioned above [26].

As a byproduct of the RHS formulation, a new interpretation of quantum measurements suggests itself. Given the RHS just constructed, it seems natural to interpret Φ as the space of *physical* states, i.e., states that can be prepared in actual experiments (notice that, since the Hamiltonian H is certainly an element of $\mathcal{O}$, all the states in Φ automatically have a finite energy, since they belong to the domain of H). Now an element of $\Phi^{\times}$ is an antilinear functional on Φ, i.e., a procedure that associates a number to each state, while preserving the linear structure. This is clearly related to a measurement apparatus or a reference frame.

Group Representations

Ever since the work of Wigner and others [28–34], unitary representations of groups have been used to describe ▶ symmetry transformations in quantum physics. In particular, the famous symmetry representation theorem of Wigner [33] and Bargmann [34] asserts that the symmetry transformations of a physical system are represented in the state vector space, taken to be a Hilbert space $\mathcal{H}$, by unitary (or antiunitary) operators. Therefore, if G is the (Lie) group of symmetry transformations of a physical system, then what is of interest in quantum theory is a unitary representation U of G in the Hilbert space $\mathcal{H}$ of the system. For instance, symmetry transformations of non-relativistic and relativistic spacetime are described in quantum physics by unitary representations of the Galilei group and Poincaré group, respectively.

If U is a unitary representation of a symmetry group G, then $U(g)$, $g \in G$, should transform physical states into physical states, continuously, and similarly for observables $|F_E\rangle\langle F_E|$ or $|\psi\rangle\langle\psi|$ representing measurement apparatuses. Thus one should have two other realizations of U, in addition to U itself, namely:

• One in Φ, denoted U_Φ. This representation is the restriction of U in $\mathcal{H}$ to Φ: $U_\Phi(g)\phi = U(g)\phi$ for $g \in G, \phi \in \Phi$.

• And one in $\Phi^\times$, denoted $U_\Phi^\times$. This is the extension of $U^\dagger$ from $\mathcal{H}$ to $\Phi^\times$: $U_\Phi^\times(g)F = U^\dagger(g)F$ for $g \in G,\ F \in \mathcal{H}$.

The two representations U_Φ and $U_\Phi^\times$ are contragredient of each other. That is,

$$\begin{aligned}\langle U_\Phi^\times(g^{-1})F|U_\Phi(g)\phi\rangle &= \langle F|\left(U_\Phi^\times(g^{-1})\right)^\times U_\Phi(g)\phi\rangle \\ &= \langle F|U_\Phi(g^{-1})U_\Phi(g)\phi\rangle = \langle F|U(g^{-1})U(g)\phi\rangle \\ &= \langle F|\phi\rangle,\ \forall\, g \in G, \phi \in \Phi,\ F \in \Phi^\times, \end{aligned} \tag{15}$$

which corresponds to the unitarity of U acting in $\mathcal{H}$:

$$\left(U^\dagger(g^{-1})f|U(g)h\right) = (U(g)f|U(g)h) = (f|h)\,,\ \forall\, g \in G \text{ and } f, h \in \mathcal{H}. \tag{16}$$

As is easily verified, this definition implies that $U_\Phi^\times$ is an extension of both $U_\Phi^\dagger$ and $U^\dagger$, as it should in view of (1).

As we have said, in general the space Φ is supposed to be a reflexive Fréchet space. For consistency, we must assume that the representation U_Φ is continuous in Φ, that is, the map $g \mapsto U_\Phi\phi$ is continuous from G to Φ, for every $\phi \in \Phi$. Then the contragredient representation $U_\Phi^\times$ is automatically continuous in $\Phi^\times$ [35]. Notice that rigged Hilbert spaces of this type have also been used in pure group theory, namely in the decomposition of unitary representations of non-compact groups like SU(1,1) or $SO_o(2,1)$ [36, 37].

From Group to Semigroup Representations

R

According to the results of Wigner and Bargmann for Hilbert space and the subsequent extensions of these results to rigged Hilbert spaces as outlined above, spacetime transformations of a quantum system are given by representations of the Galilei or Poincaré group. As a special case, we now consider the one-parameter time evolution group $U^\dagger(t)$. Applied to an in-state vector ϕ^+ (see [1, Sec.III]), we get

$$\phi^+(t) = \mathrm{e}^{-\mathrm{i}Ht}\phi^+ = U^\dagger(t)\phi^+, \quad \text{with } -\infty < t < \infty. \tag{17}$$

This follows as the solution of the Schrödinger equation (21) when it is solved under the Hilbert space boundary condition. Similarly, the time evolution of an operator Λ^- representing an observable is given by

$$\begin{aligned}&\Lambda^-(t) = \mathrm{e}^{\mathrm{i}Ht}\Lambda^-\mathrm{e}^{-\mathrm{i}Ht} = U(t)\Lambda^-U^\dagger(t) \\ &\left(\text{or, } \psi^-(t) = \mathrm{e}^{\mathrm{i}Ht}\psi^- \quad \text{for } \Lambda^- = |\psi^-\rangle\langle\psi^-|\right), \quad \text{with } -\infty < t < \infty, \end{aligned} \tag{18}$$

as follows from the solution of the Heisenberg equation (22) under the Hilbert space boundary condition. The $U(t)$ as well as $U^\dagger(t) = U^{-1}(t)$ with $-\infty < t < \infty$ form a group of unitary operators in Hilbert space. With this unitary time evolution, the Born probabilities for an observable Λ^- in the state $\phi^+(t)$ (or, equivalently, of $\Lambda^-(t)$ in ϕ^+) can be calculated as

$$\mathcal{P}_{\phi^+(t)}(\Lambda^-) = \underset{(\blacktriangleright\ \text{Schrödinger picture})}{\mathrm{Tr}\Big(U^\dagger(t)|\phi^+\rangle\langle\phi^+|U(t)\Lambda^-\Big)} = \underset{(\blacktriangleright\ \text{Heisenberg picture})}{\mathrm{Tr}\Big(|\phi^+\rangle\langle\phi^+|U(t)\Lambda^- U^\dagger(t)\Big)}$$
$$\textit{for all } -\infty < t < +\infty. \tag{19}$$

The Bargmann–Wigner theorem is based on the assumption that for every transformation of the observable relative to the state, there exists an inverse transformation *also of the observable relative to the state*. This assumed symmetry of time translations is encoded in the unitary group $U(t)$ of (18). However, such inverse time translations are physically impossible since an observable can be measured on a state ϕ^+ *only* after the state has been prepared, say at a finite time t_0 (which can be set to $t_0 = 0$). That means that the probability for an observable $\Lambda^-(t)$ in the state ϕ^+,

$$\mathcal{P}_{\phi^+}(\Lambda^-(t)) = \mathrm{Tr}\Big(|\phi^+\rangle\langle\phi^+|U(t)\Lambda^- U^\dagger(t)\Big) \tag{20}$$

makes sense only for $t \geqslant t_0 = 0$ at which time the state ϕ^+ has been prepared.

This is a manifestation of causality and it implies that the time evolution of an observable relative to the state is physically defined only for $t \geqslant 0$. Therefore, the time translations of the state relative to the observable, or equivalently, also of the observable relative to the state, should be represented in the mathematical theory by a *semigroup*, rather than a unitary group. From this observation we infer that the time evolution must be given by semigroups $U(t),\ t \geqslant 0$ in (17) and $U^\dagger(t),\ t \geqslant 0$ in (18) [38].

Although there is no direct experimental reason in favor of unitary group representations for time evolution, unitary evolutions are intrinsic to the mathematical structure of Hilbert space. In particular, a theorem due to Stone and von Neumann states that every self-adjoint operator A generates a unitary group $U(\alpha)$, $-\infty < \alpha < +\infty$, such that $\mathrm{i}\frac{\mathrm{d}U(\alpha)}{\mathrm{d}\alpha}\Big|_{\alpha=0} = A$. For the time evolution of the state $\phi^+(t)$, this means (17) is the solution to the Schrödinger dynamical equation

$$\mathrm{i}\hbar\frac{\mathrm{d}\phi^+(t)}{\mathrm{d}t} = H\phi^+(t) \quad \text{under the boundary condition } \phi^+(t) \in \mathcal{H}, \tag{21}$$

while for the time evolution of the observable $\Lambda^-(t) = |\psi^-(t)\rangle\langle\psi^-(t)|$, (18) is the solution to the Heisenberg dynamical equation

$$\mathrm{i}\hbar\frac{\mathrm{d}\psi^-(t)}{\mathrm{d}t} = -H\psi^-(t) \quad \text{under the boundary condition } \psi^-(t) \in \mathcal{H}. \tag{22}$$

Put differently, if one solves these dynamical differential equations under the boundary conditions of the Hilbert space, then one always gets unitary group solutions. Similarly, if we solve these equations under the boundary conditions of the Schwartz RHS. That is, the solutions to the differential equations (21) and (22) under the conditions $\phi^+ \in \Phi$ and $\psi^- \in \Phi$ are still of the form (17) and (18) with $-\infty < t < +\infty$ (though, due to the topological structure of Φ and $\Phi^\times$, the continuity and differentiability of these representations are different from those of unitary representations in the Hilbert space). Thus, although it has the many advantages outlined above, the RHS (1) of Dirac's formalism given by Schwartz RHS does not overcome the causality problem of Hilbert space quantum mechanics. The solution to this problem, along with a consistent theory of quantum scattering and decay phenomena is given by an extension of the RHS framework using smooth, rapidly decaying Hardy functions in place of the Schwartz functions of (1). This Hardy space theory is described in the accompanying article [22].

Literature

1. J-P. Antoine, R. Bishop, A. Böhm, S. Wickramasekara: *Rigged Hilbert Spaces in Quantum Physics*, this volume.
2. J-P. Antoine: Formalisme de Dirac et problèmes de symétrie en Mécanique Quantique, Thèse de doctorat, Université Catholique de Louvain (1966)
3. J-P. Antoine: J. Math. Phys. 10, 53–69 (1969)
4. J-P. Antoine: J. Math. Phys. 10 2276–290 (1969)
5. A. Böhm: *Rigged Hilbert Spaces*, Int. Centre for Theor. Phys., Bult. 9, (1964)
6. A. Bohm: in *Rigged Hilbert Space and Mathematical Description of Physical Systems*, in Boulder Lectures in Theoretical Physics IX A: Mathematical Methods of Theoretical Physics (Wiley, New York, 1967)
7. J. E. Roberts: J. Math. Phys. 7, 1097–104 (1966)
8. J. E. Roberts: Commun. Math. Phys. 3, 98–119 (1966)
9. P. Kristensen, L. Meljbo, E. Thue Poulsen: Commun. Math. Phys., 1, 175 (1965)
10. S. J. L. van Eijndhoven, J. de Graaf: *A Mathematical Introduction to Dirac's Formalism* (North Holland, Amsterdam, 1986)
11. O. Melsheimer: J. Math. Phys. 15, 902–16 (1974)
12. O. Melsheimer: J. Math. Phys. 15, 917–25 (1974)
13. C. Foiaş: Rev. Math. Pures Appl. 7, 241–82 (1962)
14. C. Foiaş: Rev. Math. Pures Appl. 7, 571–602 (1962)
15. L. Schwartz: *Théorie des Distributions, I–II.* (Hermann, Paris, 1957, 1959). New edition, revised and augmented [First published in 1950–951]
16. I. M. Gel'fand, N. Y. Vilenkin: *Generalized Functions IV: Applications of Harmonic Analysis.* (Academic Press, New York, 1964). [Translation of the 1961 Russian edition]
17. K. Maurin: *General Eigenfunction Expansions and Unitary Representations of Topological Groups.* (PWN Polish Scientific Publishers, Warsaw, 1968)
18. D. Babbitt: Rep. Math. Phys. 3, 37–42 (1972)
19. D. Fredricks: Rep. Math. Phys. 8, 277–93 (1975)
20. K. Napiorkowski: Bull. Acad. Pol. Sci. 22, 1215–218 (1974)
21. K. Napiorkowski: Bull. Acad. Pol. Sci. 23, 251–52 (1975)
22. A. Bohm, N. L. Harshman: *Rigged Hilbert spaces and time asymmetric quantum theory*, this volume
23. A. Bohm: Lett. Math. Phys. 3, 455–61 (1979)

24. A. Bohm: J. Math. Phys. 22, 2813–823 (1981)
25. M. Gadella: J. Math. Phys. 24, 1462–469 (1983)
26. A. Bohm, M. Gadella: *Dirac Kets, Gamow Vectors and Gelfand Triplets*, Springer Lecture Notes in Physics, vol. 348. Springer, New York (1989)
27. A. Bohm: J. Math. Phys., 8, 1551–558 (1967), Appendix B
28. E. P. Wigner: Z. Phys. 40, 492 (1927)
29. E. P. Wigner: Z. Phys. 43, 642 (1927)
30. E. P. Wigner: Z. Phys. 45, 602 (1927)
31. F. Hund: Z. Phys. 43, 788 (1927)
32. H. Weyl: *The Theory of Groups and Quantum Mechanics, 2nd edn.* (Dover, 1950), (First edition, 1931)
33. E. P. Wigner: *Group Theory and Its Applications to Quantum Mechanics and Atomic Spectra, 1st edn.*, 1931. (Academic Press, New York, 1952)
34. V. Bargmann: J. Math. Phys. 5, 862–68 (1964)
35. F. Bruhat: Bull. Soc. Math. France 84, 97–205 (1956)
36. B. Nagel: in *Studies in Mathematical Physics (Proc. Istanbul 1970)*, pp. 135–54, ed. by A.O. Barut (ed.) (Reidel, Dordrecht and Boston 1970)
37. G. Lindblad, B. Nagel: Ann. Inst. H. Poincaré 13, 27–56 (1970)
38. In the relativistic case, the time evolution semigroup generalizes to a semigroup of spacetime translations into the forward light cone, which is invariant under the group of orthochronous Lorentz transformations. In this sense, semigroup time evolution can be viewed as an expression of causality [39, 40]
39. A. Bohm, H. Kaldass, S. Wickramasekara: Fortschr. Phys. 51, 569–603 (2003)
40. A. Bohm, H. Kaldass, S. Wickramasekara: Fortschr. Phys. 51, 604–34 (2003)

Rigged Hilbert Spaces and Time Asymmetric Quantum Theory

A. Bohm and N.L. Harshman

Rigged Hilbert Spaces and Dirac's Bra-Ket Formalism

The rigged Hilbert space (RHS) is a triplet of linear topological spaces

$$\Phi \subset \mathcal{H} \subset \Phi^{\times}, \tag{1}$$

which is obtained from a linear space with scalar product Ψ by completing it with respect to three topologies. A topology τ specifies the definition of convergence, and when a space is completed with respect to a topology τ, the τ-limit elements of Cauchy sequences are adjoined to the space. For example, in (1) the space $\mathcal{H}$ is an abstract Hilbert space, i.e. it is the completion of Ψ with respect to the topology $\tau_{\mathcal{H}}$ given by the norm $||\phi|| = \sqrt{(\phi, \phi)}$. The space Φ is the completion with respect to a stronger topology τ_{Φ} and the space $\Phi^{\times}$ is the space of continuous functionals on Φ.

The linear space with scalar product Ψ is the space that most physics texts and papers call "the Hilbert space". The space $\Phi^\times$ is the space one needs in order to give a mathematical meaning to the formalism that Dirac introduced in the first edition (1930) of his book [1], and which he simplified in the third edition (1947) [2] ► Dirac notation. The space $\mathcal{H}$ is the space that von Neumann introduced in his Hilbert space formulation of quantum mechanics in 1932 [3], where he remarked that Dirac's formalism [1] is "scarcely surpassed in brevity and elegance" but "in no way satisfies the requirements of rigor." An example of a Hilbert space, also called a realization of the abstract Hilbert space, is the space of Lebesgue square integrable functions L^2. Unfortunately, in this space one cannot define the scalar product of functions with the commonly used Riemann integrals; instead one must use Lebesgue integrals to obtain the complete Hilbert space L^2 (and not just a realization of the linear space Ψ).

Von Neumann's Hilbert space provides a mathematically rigorous formulation of quantum mechanics, but it has some physically unintuitive features. For instance, a state is represented not by a single wave function, but by a class of Lebesgue square integrable functions that differ from each other on a set of measure zero, which could even be the set of rational numbers. In contrast, physicists measure probability distributions at only a finite number of points and then interpolate the data with smooth functions. This practice suggests that states are better represented by functions $\phi(E)$ that have the following properties: they are continuous, infinitely differentiable, and they and their derivatives decrease for $E \to \infty$ faster than any inverse power of E. These properties *define* the Schwartz function space.

The standard example, used in quantum mechanics [4–6], group representations [7–11] and axiomatic quantum field theory [12], is the following RHS (1): the space Φ is realized by the space of Schwartz functions on the positive real line $\mathcal{S}(\mathbb{R}_+)$. The space $\mathcal{H}$ is realized by Lebesgue square-integrable functions $L^2(\mathbb{R}_+)$ and the space $\Phi^\times$ is realized by the space of tempered distributions $\mathcal{S}^\times(\mathbb{R}_+)$, which includes generalized functions like the Dirac delta defined below.

In quantum mechanical applications of RHSs, the space Φ is identified as the space of *physical* states, i.e. those states that can be prepared and measured by experiments. The ► observables that act as linear operators on the physical states should be represented by *continuous* linear operators in Φ and the set of these observables are represented by an algebra of ► operators. That means observables can be multiplied and added without worrying about domain questions since they are all defined everywhere in Φ. This feature is of enormous importance for practical calculations, but cannot be implemented in the Hilbert space even for the basic algebra of observables generated by position Q, momentum P, and energy H operators that fulfill the Heisenberg commutation relations

$$(QP - PQ) = [Q, P] = \mathrm{i}\hbar \tag{2a}$$

and

$$H = \frac{P^2}{2m} + V(Q). \tag{2b}$$

R

Within the Dirac formalism it is tacitly assumed that observables can be added and multiplied. This means that the space Φ must be constructed such that the observables form an algebra of linear operators defined on the linear space of states Φ. (3)

Observables are measured by numerical values; therefore the operators that represent them should have eigenvalues and eigenvectors. One can prove, however, that for certain operators, such as P and Q in (2a), there are *no* eigenvectors in the Hilbert space. Nonetheless, Dirac postulated that the observables (like P, Q, and H in(2a), but also more generally) have a *complete* set of eigenvectors, the Dirac kets. These kets were postulated to have the following two properties:

(i) On them, the observables have a set of eigenvalues that are discrete, continuous, or a combination of continuous and discrete:

$$H|E\rangle = E|E\rangle, \text{ with } 0 \leqslant E < \infty \text{ and/or } E \in \{E_1, E_2, ..., E_n, ...\} \quad (4a)$$

$$P|p\rangle = p|p\rangle, \text{ with } -\infty < p < \infty \quad (4b)$$

$$Q|x\rangle = x|x\rangle, \text{ with } -\infty < x < \infty. \quad (4c)$$

This means the kets are labeled by the eigenvalues such as x, p, and E.

(ii) These vectors provide a basis system and every vector $\psi \in \Phi$ can be uniquely represented by a linear combination of these basis vectors.

As an example of the second point, consider the case that H has only a discrete set of eigenvectors $|E_n\rangle$. Then every ψ is expanded as

$$|\psi\rangle = \sum_n |E_n\rangle c_n = \sum_n |E_n\rangle\langle E_n|\psi\rangle, \quad (5a)$$

where the coordinates or components $c_n = \langle E_n|\psi\rangle$ are complex numbers. The basis vectors $|E_n\rangle$ are orthonormal (orthogonal and normalized) if H is self-adjoint ($H = H^\dagger$), i.e.

$$(|E_i\rangle, |E_j\rangle) = \langle E_i|E_j\rangle = \delta_{ij} = \begin{cases} 1 & i = j \\ 0 & i \neq j \end{cases} . \quad (5b)$$

The norm of every state vector ψ is finite and is calculated as

$$(\psi, \psi) = \sum_n \langle\psi|E_n\rangle\langle E_n|\psi\rangle = \sum_n |\langle E_n|\psi\rangle|^2 = \sum_n |c_n|^2 < \infty. \quad (5c)$$

This holds, for instance, if H in (2b) has the particular form of a quantum oscillator with mass m and spring constant k:

$$H = \frac{P^2}{2m} + \frac{1}{2}kQ^2 \quad (6)$$

with $E_n = \hbar\omega(n + 1/2)$ $(n = 0, 1, \ldots)$ and where $\omega = \sqrt{k/m}$. The equations (5) are the infinite dimensional generalizations of the basis vector expansion of a three dimensional vector $\boldsymbol{x} = \sum_{i=1}^{3} \boldsymbol{e}_i x^i$.

In general one cannot find for every self-adjoint operator such as H or P a *complete* set of eigenvectors such that (5) holds. However, in the RHS (1) realized by the Schwartz space

$$\mathcal{S}(\mathbb{R}) \subset L^2(\mathbb{R}) \subset \mathcal{S}^\times(\mathbb{R}), \tag{7}$$

for every vector $\psi \in \Phi$ and every self-adjoint operator, the continuous analogues of (5a) hold:

$$|\psi\rangle = \int_{-\infty}^{\infty} \mathrm{d}p\, |p\rangle\langle p|\psi\rangle = \int_{-\infty}^{\infty} \mathrm{d}p\, |p\rangle\psi(p), \tag{8a}$$

$$|\psi\rangle = \int_{-\infty}^{\infty} \mathrm{d}x\, |x\rangle\langle x|\psi\rangle = \int_{-\infty}^{\infty} \mathrm{d}x\, |x\rangle\psi(x), \tag{8b}$$

$$|\psi\rangle = \sum_n |E_n\rangle\langle E_n|\psi\rangle + \int_0^{\infty} \mathrm{d}E\, |E\rangle\langle E|\psi\rangle = \sum_n |E_n\rangle c_n + \int_0^{\infty} \mathrm{d}E\, |E\rangle\psi(E) \tag{8c}$$

The kets $|x\rangle$, $|p\rangle$, and $|E\rangle$ exist as generalized eigenvectors of the operators Q, P, and H (or any other self-adjoint operator representing an observable). They are elements of $\Phi^\times$ and the eigenvalue equations (4) are defined to mean

$$\langle\psi|H^\times|E\rangle \equiv \langle H\psi|E\rangle = E\langle\psi|E\rangle \text{ for all } \psi \in \Phi, \tag{9}$$

and similarly for $|p\rangle$, $|x\rangle$, etc. The coordinates or wave functions $\langle E|\psi\rangle = \langle\psi|E\rangle^*$ are elements of the Schwartz space $\mathcal{S}(\mathbb{R})$, not of $L^2(\mathbb{R})$, and the generalized basis vector expansions (8) do not hold for every $\psi \in \mathcal{H}$ but *only* for every $\psi \in \Phi$. The operator $H^\times$, called the conjugate operator of H, is defined generally for all linear continuous operators A on ϕ by

$$\langle A\psi|F\rangle = \langle\psi|A^\times|F\rangle \tag{10}$$

for all $\psi \in \Phi$ and all $F \in \Phi^\times$. Since the space Φ is constructed such that the physical observables are represented by continuous operators on the space Φ, the conjugate $A^\times$ is a continuous operator in $\Phi^\times$. The conjugate $A^\times$ is an extension of the Hilbert space adjoint $A^\dagger$: $A^\dagger \subset A^\times$. The observables form an algebra of operators in Φ as well as in $\Phi^\times$. In contrast, in the Hilbert space $\mathcal{H}$, one cannot have a continuous algebra of observables even for the canonical commutation relations (2a). For an example of how to construct the Schwartz space for the operator algebra of (2a) and (6) see [13], Sect. 1.

The continuous analogue of (5c) is now

$$(\psi, \psi) = \int\int \mathrm{d}E\, \mathrm{d}E' \langle\psi|E\rangle(|E\rangle, |E'\rangle)\langle E'|\psi\rangle, \tag{11}$$

but for this to make sense one needs the continuous analogue $(|E\rangle, |E'\rangle)$ of the Kronecker delta δ_{nm} of (5b). This is obtained if one takes the generalized scalar product $(|E'\rangle, \psi)$ (precisely, the functional $\langle E'|$ at any element $\psi \in \Phi$):

$$\langle E'|\psi\rangle = \int_0^\infty \mathrm{d}E\, \langle E'|E\rangle\langle E|\psi\rangle. \tag{12}$$

By treating the new symbol $\langle E'|E\rangle$ introduced in (12) as though it were a scalar product like (5b) but extended to continuous values, the object $\langle E'|E\rangle$ has the property that it maps the function $\psi(E) \in \mathcal{S}(\mathbb{R}_+)$ by integration over the positive real axis to its specific value $\psi(E')$ at a particular energy E'. A mathematical object with such a property did not exist in the mathematics of the 1920s and 1930s, but only achieved rigorous definition when Schwartz created the theory of distributions or generalized functions 20 years later [14]. Dirac's formalism was unhindered by all these mathematical complications. He postulated the properties (4) and (8), and since (5b) held for the discrete case, he introduced the Dirac delta "function"

$$\langle E'|E\rangle = \delta(E - E') \tag{13}$$

and stipulated that it fulfill (12). It is not truly a function, but it is a distribution and an element of $\mathcal{S}^\times(R_+)$.

The requirements expressed in (3), (12), and (13) form the basis of Dirac's formalism for quantum mechanics. Inspired by this, first Schwartz created the theory of distributions [14]. Then, extending this work, the Gel'fand school [15, 16] introduced into mathematics the RHS for the spectral analysis of ► self-adjoint and unitary operators. Their nuclear spectral theorem is the mathematical version of Dirac's continuous basis vector expansion (8). The RHS is the mathematical structure in which various assumption of Dirac's formalism, e.g. (2a), (4), (8c), (12), and (13), can be realized [4–11].

Thus, the Dirac formalism has been given a mathematical meaning by the Schwartz-RHS. The RHS's of quantum physical systems are constructed such that the fundamental observables, like momentum, energy, and position (and many more, such as angular momentum and intrinsic observables like charges and isospin usually connected with groups of transformations of space time and of charge spaces) are represented by an algebra of continuous operators. Then one chooses a complete commuting system of observables. For the oscillator this is just one operator, for example H, P or Q, and the Dirac basis vector expansion for the operator is like (5a), (8a), or (8b) for the oscillator. For other quantum systems, for example a particle in a spherically symmetric potential of the three dimensional space, the complete system of commuting observables consists of three operators, either the momentum operators P_1, P_2, and P_3 or the Hamiltonian H and angular momentum operators $\mathbf{J}^2$ and J_3, and possibly some other set of observables that measure, for example, the internal properties and whose eigenvalues are collectively labeled as η. Then the Dirac basis vector expansion (nuclear spectral theorem) is

$$\psi = \sum_{jj_3\eta} \int \mathrm{d}E \, |E, j, j_3, \eta\rangle\langle E, j, j_3, \eta|\psi\rangle. \tag{14}$$

The energy wave functions $\langle E, j, j_3, \eta|\psi\rangle = \psi_{jj_3\eta}(E)$ are Schwartz space functions if we use for the RHS the abstract Schwartz space.

Hardy Space Triplets for Resonance Scattering and Decay

The Schwartz-RHS gives a mathematical justification for the Dirac formalism. It defines the Dirac kets, justifies the algebraic manipulation of the observables, and proves the continuous basis vector expansion (8). However, it does not provide a mathematical theory of scattering, resonances, and decaying states, and neither does the Hilbert space formulation. The description of resonances and decay phenomena in standard quantum mechanics is provided by the Weisskopf–Wigner approximations [17, 18] and it is well-known to experts that "there does not exist...a rigorous theory to which these various methods can be considered and approximation" [19]. This is connected with the Stone–von Neumann theorem [20, 21] which states that the solutions of the ▶ Schrödinger equation in $\mathcal{H}$ are given by the time-symmetric unitary group $U^\dagger(t) = \exp(-\mathrm{i}Ht)$ (or by the unitary group $U(t) = \exp(\mathrm{i}Ht)$) for *all* times $-\infty < t < \infty$. In the RHS formulation using the Schwartz space, the space Φ (and not $\mathcal{H}$) is the set of physical states. That means one has to solve the Schrödinger equation

$$\mathrm{i}\frac{\mathrm{d}\phi(t)}{\mathrm{d}t} = H\phi(t) \tag{15}$$

under the boundary condition that $\phi \in \Phi$. Note that Φ and $\mathcal{H}$ have different definitions of convergence, therefore the limits involved in taking the derivative of ϕ in the space Φ is different from taking the limit in $\mathcal{H}$. The $\tau_{\mathcal{H}}$-limit is defined by one norm, whereas the τ_Φ-limit is stronger and given by countably infinite number of norms. Thus the solutions of (15) in Φ do not have to be the same as the solutions in $\mathcal{H}$ [22], but for the Schwartz-RHS, the solutions to (15) also have the group time evolution property:

$$\phi(t) = \mathrm{e}^{-\mathrm{i}Ht}\phi, \text{ for all } -\infty < t < \infty \text{ and for all } \phi \in \Phi. \tag{16}$$

Therefore, the Hilbert space axiom of standard (von Neumann) quantum theory,

$$\text{set of physical states} = \{\phi\} = \mathcal{H} = \text{Hilbert space}, \tag{17}$$

as well as the Schwartz space axiom of the mathematical theory for the Dirac formalism,

$$\text{set of physical states} = \{\phi\} = \Phi = \text{abstract Schwartz space}, \tag{18}$$

lead to the same reversible time evolution. The time evolution of the prepared states fulfills (16) and there will exist a state $\phi(t)$ for every $t > 0$ and also for every $t < 0$.

The physical quantities measured in experiments with quantum systems are the Born probabilities. For instance, the probability to measure an observable $\Lambda = |\psi\rangle\langle\psi|$ in the state ϕ is given by the Born probability $|\langle\psi|\phi(t)\rangle|^2$, and according to (16), this is predicted for every time $-\infty < t < \infty$. However, this contradicts causality; a quantum mechanical state must be prepared first at some time t_0 before the observable can be measured in this state at times $t > t_0$. That means Born probabilities $|\langle\psi|\phi(t)\rangle|^2$ can be measured only for $t > t_0$. Consequentially, the evolution (16) makes physical sense only for $t \geqslant t_0$. In other words, instead of the unitary group solution, one should find solutions that obey semigroup evolution

$$\phi(t) = \mathrm{e}^{-\mathrm{i}H(t-t_0)}\phi, \text{ for only } t > t_0. \tag{19}$$

Such solutions do not exist in the Schwartz space Φ or in the Hilbert space $\mathcal{H}$.

The time t_0 before which "the state is defined completely by the preparation" has already been mentioned by Feynman [23]. Gell-Mann and Hartle [24] applied this idea to the probabilities of histories for the expanding universe considered as a closed quantum system. They did not derive (19); they restricted the time evolution in (16) to $t > t_0$ (where t_0 is the time of the big bang) by fiat, violating the Hilbert space and Schwartz space axioms (17) and (18). Other examples of systems with a physically well-defined t_0 are quasi-stable particles produced by the strong interactions that decay on a much slower time scale via the weak interaction [25]. That the decay of excited atoms and of elementary particles is a time asymmetric (sometimes also called irreversible) process has also been remarked in textbooks [26–28].

In the Hilbert space formulation of quantum mechanics [3], one cannot distinguish between vectors ϕ describing states and vectors ψ describing observables like $\Lambda = |\psi\rangle\langle\psi|$ (or more general observables like $A = \sum_n a_n|\psi_n\rangle\langle\psi_n|$). One assumes that

$$\text{set of states} = \{\phi\} = \mathcal{H} = \text{set of observable vectors} = \{\psi\} \tag{20}$$

and the time evolution for both is given by a unitary group for all times t. In the Dirac formalism based on (18), one also identifies the set of state vectors and observables vectors:

$$\{\phi\} = \Phi = \{\psi\} \tag{21}$$

and one has a single basis vector expansion such as (14) (or (8)) and one space of continuous antilinear functionals (kets) $|E\rangle = |E, j, j_3, \eta\rangle \in \Phi^\times$. In contrast, two sets of basis vectors are used in the heuristic conventional treatment of scattering theory. These are the plane wave in-states $|E^+\rangle$ and out-"states" $|E^-\rangle$ that are solutions of the Lippmann–Schwinger (LS) equation and are given by

$$|E^\pm\rangle = |E \pm \mathrm{i}\epsilon\rangle = |E\rangle + \frac{1}{E - H \pm \mathrm{i}\epsilon} V|E\rangle, \tag{22}$$

where $\epsilon \to +0$, $(H - V)|E\rangle = E|E\rangle$, and V represents the scattering potential [29–31]. The $\pm\,\mathrm{i}\epsilon$ in the LS equation (22) implies that the energy wave functions $(\psi^-(E))^* = \langle^- E|\psi^-\rangle^* = \langle\psi^-|E^-\rangle$ and $\phi^+(E) = \langle^+ E|\phi^+\rangle$ are the boundary values of analytic functions in the lower complex energy semi-plane (for complex energy $z = (E + \mathrm{i}\epsilon)^* = E - \mathrm{i}\epsilon$, immediately below the real axis on the second sheet of the S-matrix). In analogy to the Dirac expansion (8c), the $|E^{\pm}\rangle$ are taken as basis systems for the Dirac basis vector expansions

$$|\phi^+\rangle = \int_0^\infty \mathrm{d}E\,|E^+\rangle\langle^+E|\phi^+\rangle, \tag{23a}$$

and

$$|\psi^-\rangle = \int_0^\infty \mathrm{d}E\,|E^-\rangle\langle^-E|\psi^-\rangle. \tag{23b}$$

The $\pm\,\mathrm{i}\epsilon$ in the phenomenological LS equation (22) suggests that the energy wave functions $\langle^-E|\psi^-\rangle$ are Schwartz functions that can be analytically continued into the upper half complex energy plane (second sheet of the S-matrix) and the $\langle^+E|\phi^+\rangle$ are Schwartz functions analytic in the lower complex plane. Since the sets of vectors $\{\phi^+\}$ and $\{\psi^-\}$ are defined by the sets of wave functions $\{\langle^+E|\phi^+\rangle\}$ and $\{\langle^-E|\psi^-\rangle\}$, it suggests that there are two RHS's involved. One RHS

$$\{\phi^+\} = \Phi_- \subset \mathcal{H} \subset \Phi_-^\times \tag{24a}$$

is used for the set of state vectors $\{\phi^+\}$ (in-states), which are defined by the preparation apparatus, such as an accelerator. Another RHS

$$\{\psi^-\} = \Phi_+ \subset \mathcal{H} \subset \Phi_+^\times \tag{24b}$$

is used for the set of observable vectors $\{\psi^-\}$ (out-states, or better, out-observables), which are defined by the registration apparatus, such as a detector. The vectors ϕ^+ and ψ^- are very similar to the in- and out-states in the S-matrix element of traditional scattering theory [32–34]:

$$\langle^-\psi|\phi^+\rangle = (\psi^-, \phi^+) = (\psi^{\mathrm{out}}, S\phi^{\mathrm{in}}) = (\psi^{\mathrm{out}}, \phi^{\mathrm{out}}). \tag{25}$$

To specify the properties of the wave functions $\langle^-E|\psi^-\rangle$ and $\langle^+E|\phi^+\rangle$ and therewith the spaces Φ_+ and Φ_- of vectors ψ^- and ϕ^+, one checks under which mathematical conditions on the spaces $\{\langle^-E|\psi^-\rangle\}$ and $\{\langle^+E|\phi^+\rangle\}$ one can derive reasonable physical consequences from the hypothesis (24). A reasonable physical consequence would be a unification of resonance scattering and decay phenomena. One starts with the definition of a resonance by the S-matrix pole at the complex energy $z_\mathrm{R} = E_\mathrm{R} - \mathrm{i}\Gamma/2$. From the pole, one seeks the requirements that will allow the derivation of two important signatures of time asymmetry: the Breit–Wigner amplitude for resonance scattering

$$a_j^{\mathrm{BW}}(E) = \frac{R_i}{E - (E_\mathrm{R} - \mathrm{i}\Gamma/2)} \tag{26a}$$

and the exponentially decaying Gamow vector ϕ^{G} for the unstable states. The Gamow vector must be a ket $\phi^{\mathrm{G}} = |(E_{\mathrm{R}} - \mathrm{i}\Gamma/2)^-\rangle = |(E_{\mathrm{R}} - \mathrm{i}\Gamma/2), j, j_3, \eta^-\rangle \in \Phi_+^\times$ (it is not in $\mathcal{H}$, where exponentially decaying states are precluded [35]) with the eigenvalue property

$$\langle H\psi^-|(E_{\mathrm{R}}-\mathrm{i}\Gamma/2)^-\rangle \equiv \langle\psi^-|H^\times|(E_{\mathrm{R}}-\mathrm{i}\Gamma/2)^-\rangle = (E_{\mathrm{R}}-\mathrm{i}\Gamma/2)\langle\psi^-|(E_{\mathrm{R}}-\mathrm{i}\Gamma/2)^-\rangle \tag{26b}$$

for all $\psi^- \in \Phi_+$. Here $H^\times$ is the unique extension of $H^\dagger = H$ to the space $\Phi_+^\times$. Further, ϕ^{G} must have the exponential semigroup evolution

$$\begin{aligned}\langle \mathrm{e}^{-\mathrm{i}Ht}\psi^-|(E_{\mathrm{R}} - \mathrm{i}\Gamma/2)^-\rangle &\equiv \langle\psi^-|\mathrm{e}^{-\mathrm{i}H^\times t}|(E_{\mathrm{R}} - \mathrm{i}\Gamma/2)^-\rangle \\ &= \mathrm{e}^{\mathrm{i}E_{\mathrm{R}}t}\mathrm{e}^{-\Gamma t/2}\langle\psi^-|(E_{\mathrm{R}} - \mathrm{i}\Gamma/2)^-\rangle \end{aligned} \tag{26c}$$

for all $\psi^- \in \Phi_+$ but only for $t \geqslant 0$ since the decaying state must first be prepared at a time $t = t_0 = 0$ before it can decay.

The results (26a)–(26c) can be obtained if one assumes that in addition to being Schwartz functions, the energy wave functions can be analytically continued into either the upper- or lower-half complex energy plane (second sheet) [36]. Precisely, the analytically continued wave functions $\psi^-(z) = \langle^- z|\psi^-\rangle$ and $\phi^+(z) = \langle^+ z|\phi^+\rangle$ are smooth Hardy functions [1] on the complex semiplanes $\mathbb{C}_+$ and $\mathbb{C}_-$, respectively:

$$\phi^+(E) = \langle^+ E|\phi^+\rangle \in (\mathcal{H}^2_- \cap \mathcal{S})|_{\mathbb{R}_+} \tag{27a}$$

$$\psi^-(E) = \langle^- E|\psi^-\rangle \in (\mathcal{H}^2_+ \cap \mathcal{S})|_{\mathbb{R}_+}. \tag{27b}$$

The mismatch in signs between the wave function and the smooth Hardy spaces is for historical reasons: the '$\pm$' of the wave functions is the convention in scattering theory and has an independent origin from the '$\pm$' of the Hardy space analyticity requirements.

[1] A precise definition of the smooth Hardy space is that a function $\psi^-(E)$ is in $\mathcal{H}^2_+ \cap \mathcal{S}$ if and only if: (i) $\psi^-(E)$ belongs to the Schwartz space $\mathcal{S}$, (ii) $\psi^-(E)$ admits analytic continuation, $\psi^-(z) = \psi^-(E + \mathrm{i}y)$, to the upper half plane ($y > 0$), and (iii) For any straight line in the upper half plane parallel to the real line, there exists a positive number $K > 0$ such that for all positive $y > 0$ the integral $\int_{-\infty}^{\infty} |\psi^-(E + \mathrm{i}y)|^2 \, \mathrm{d}x < K$ is uniformly bounded by K, which means that the bound is valid for a particular K and any $y > 0$. This integral is the usual Riemann integral and the constant K depends on the specific function $\psi^-(E)$. The definition for $\mathcal{H}^2_- \cap \mathcal{S}$ is identical, just replacing the upper half plane by the lower half plane. Since any function in $\mathcal{H}^2_\pm \cap \mathcal{S}$ is an analytic continuation of a function on the real line, it is automatically determined by its values on any interval in the real line and viceversa. In particular, any function in $\mathcal{H}^2_\pm \cap \mathcal{S}$ is totally determined by its values on the positive half line and conversely. The spaces $\mathcal{H}^2_\pm \cap \mathcal{S}|_{\mathbb{R}^+}$ are the spaces of functions in $\mathcal{H}^2_\pm \cap \mathcal{S}$, restricted to the positive semiaxis, i.e., in the functions in $\mathcal{H}^2_\pm \cap \mathcal{S}|_{\mathbb{R}^+}$, we have ignored their values on the negative part of the real line. This shows a one to one onto correspondence between the spaces $\mathcal{H}^2_\pm \cap \mathcal{S}$ and $\mathcal{H}^2_\pm \cap \mathcal{S}|_{\mathbb{R}^+}$.

With the pair of Hardy function spaces (27) one can construct a pair of Gel'fand triplets of function spaces

$$(\mathcal{H}^2_\mp \cap \mathcal{S})|_{\mathbb{R}^+} \subset L^2(\mathbb{R}_+) \subset \left((\mathcal{H}^2_\mp \cap \mathcal{S})|_{\mathbb{R}^+}\right)^\times \tag{28}$$

and show that these Hardy function spaces are locally convex nuclear spaces [37]. Therefore, the Dirac basis vector expansions (23) are fulfilled as the nuclear spectral theorem for the Hardy space triplets (24). The time asymmetry (19) is the mathematical consequence of the Paley–Wiener theorem [38] in the same way the unitary group evolution is the consequence of the Stone–von Neumann theorem [20, 21].

Therewith (27) is an axiom for the mathematical theory of quantum physics that distinguishes mathematically between prepared (in-)states described by the RHS (24a) and registered observables described by the RHS (24b) in the same way as the experimentalists distinguish between the preparation apparatus of a state and the detector of an observable. It provides a unified description of resonance and decay phenomena and it leads to asymmetric, semigroup time evolution. Without the mathematical notion of the RHS this time asymmetric quantum theory could not have been conceived. See also ► Time in quantum mechanics.

The authors would like to acknowledge fruitful discussion with Manuel Gadella.

Literature

1. P. A. M. Dirac: *The Principles of Quantum Mechanics* (Clarendon Press, Oxford, 1930)
2. P. A. M. Dirac: *The Principles of Quantum Mechanics*, 3rd edn. (Clarendon Press, Oxford, 1947)
3. J. von Neumann: *Mathematischen Grundlagen der Quantenmechanik* (Springer, Berlin, 1932) [English translation: Princeton University Press, 1995]
4. P. Kristensen, L. Mejlbo, E. Thue Poulsen: *Tempered distributions in infinitely many dimensions: I.* Canonical field operators. Commun. Math. Phys. 1, 175–214 (1965)
5. A. Bohm: Rigged Hilbert space and mathematical description of physical systems. In *Mathematical Methods of Theoretical Physics*, Lecture Series in Theoretical Physics, vol. 9A, ed. by Wesley E. Brittin, Asim O. Barut and Marcel Guenin (Gordon and Breach, New York, 1967)
6. J. E. Roberts: *The Dirac Bra and Ket formalism*. J. Math. Phys. 7, 1097–104 (1966)
7. A. Bohm: *Associative algebra in the problem of mass formulas*. J. Math. Phys. 8, 1551–558 (1966)
8. J-P. Antoine: *Dirac formalism and symmetry problems in quantum mechanics. I*: general Dirac formalism. J. Math. Phys. 10, 53–69 (1969); *Dirac formalism and symmetry problems in quantum mechanics: II.* symmetry problems. J. Math. Phys. 10, 2276–290 (1969)
9. B. Nagel: *Lecture Notes on Group Representations*. Preprint, College de France (1970)
10. G. Lindblad, B. Nagel: *Continuous bases for unitary irreducible representations of SU(1, 1)*. Annales de l'institut Henri Poincaré (A) Physique théorique 13, 27–56 (1970)
11. O. Melsheimer: *Rigged Hilbert space formalism as an extended mathematical formalism for quantum systems. I.* General theory. J. Math. Phys. 15, 902–16 (1974); *Rigged Hilbert space formalism as an extended mathematical formalism for quantum systems. II.* Transformation theory in nonrelativistic quantum mechanics. J. Math. Phys. 15, 917–25 (1974)
12. R. F. Streater, A. S. Wightman: *PCT, Spin Statistics, and All That*, Chap. 2 (Princeton, New Jersey, 1964)

13. A. Bohm, M. Gadella: *Dirac Kets, Gamow Vectors, and Gel'fand Triplets*, Lecture Notes in Physics, vol. 348 (Springer, Berlin, 1989)
14. L. Schwartz: *Théorie des Distributions* (Hermann, Paris, 1950)
15. I. M. Gel'fand, N. Ya. Vilenkin: *Generalized Functions*, vol. 4 (Moscow, 1961) (English translation: Academic Press, New York, 1964)
16. K. Maurin: *Generalized Eigenfunction Expansions and Unitary Representations of Topological Groups* (Polish Scientific Publishers, Warsaw, 1968)
17. V. Weisskopf, E. P. Wigner: *Berechnung der natürlichen Linienbreite auf Grund der Diracsichen Licht theorie*. Z. f. Physik 63, 54–73 (1930); Uber die natürlich Linienbreite in der Strahlung des harmonischen Oszillators. 65, 18–27 (1930)
18. M. L. Goldberger, K. M. Watson: *Collision Theory* (Wiley, New York, 1964) Chap. 8
19. M. Levy: *On the description of unstable particles in quantum field theory*. Nuovo Cimento 13, 115–38 (1959)
20. M. H. Stone: *On one-parameter unitary groups in Hilbert space*. Ann. Math. 33, 634–48 (1932)
21. J. von Neumann: *Uber Einen Satz Von Herrn M. H. Stone*. Ann. Math. 33, 567–73 (1932)
22. S. Wickramasekara, A. Bohm: *Symmetry representations in the rigged Hilbert space formulation of quantum mechanics*. J. Phys. A 35, 807–29 (2002), example 2.2
23. R. P. Feynman: *Space–time approach to nonrelativistic quantum electrodynamics*. Rev. Mod. Phys. 20, 367–87 (1948)
24. M. Gell-Mann, J. B. Hartle: *Time symmetry and asymmetry in quantum mechanics and quantum cosmology*. In Physical Origins of Time Asymmetry, ed. by J. J. Halliwell, et al. (Cambridge University Press, London, 1994) pp. 311–45; Strong Decoherence. LANL archive, `gr-qc/9509054` (1995)
25. A. Bohm: *Time asymmetric quantum physics*. Phys. Rev A 60, 861–76 (1999)
26. E. Merzbacher: *Quantum Mechanics*, Chap. 18 (Wiley, New York, 1970)
27. C. Cohen-Tannoudji, B. Diu, F. Laloë: *Quantum Mechanics*, Vol. II (Wiley, New York, 1977), p. 1345, 1353–354
28. T. D. Lee: *Particle Physics and Introduction to Field Theory*, Chap. 13 (Harwood Academic, New York, 1981)
29. B. A. Lippmann, J. Schwinger: *Variational principles for scattering processes: I*. Phys. Rev. 79, 469–80 (1950)
30. M. Gell-Mann, M. L. Goldberger: *The formal theory of scattering*. Phys. Rev. 91, 398–408 (1953)
31. W. Brenig, R. Haag. *Allgemeine Quantentheorie der Stossprozesse*. Forthschr. Phys. 7, 183–242 (1959)
32. A. Bohm: *Quantum Mechanics: Foundations and Applications*, 3rd edn., 3rd printing (Springer, New York, 2001)
33. R. G. Newton: *Scattering Theory of Waves and Particles*, 2nd edn, Chap. 7. (Springer, New York, 1982)
34. A. Bohm, H. Kaldass, S. Wickramasekara: *Time asymmetric quantum theory*: II. Relativistic resonances from S-matrix poles. Fortschr. Phys. 51, 569–603 (2003)
35. L.A. Khalfin: *Contribution to the decay theory of a quasi-stationary state*. Sov. Phys. Dokl. 2, 340 (1957) [English translation Sov. Phys. JETP **6**, 1053–63 (1958)]
36. A. Bohm: *Gamow state vectors as functionals over subspaces of the nuclear space*. Lett. Math. Phys. 3, 455–61 (1978); *Resonance poles and Gamow vectors in the rigged Hilbert space formulation of quantum mechanics*. J. Math. Phys. 22, 2813–823 (1981)
37. M. Gadella: *A rigged Hilbert space of Hardy-class functions: Applications to resonances*. J. Math. Phys. 24, 1462–469 (1983)
38. R. Paley, N. Wiener: *Fourier Transform in the Complex Domain* (American Mathematical Society, New York, 1934)

Russell–Saunders Coupling

Klaus Hentschel

The ► vector model provides various ways of calculating the vectorial sum of all the contributing angular momenta l_i and ► spins $s_i = 1/2$ for atoms with more than one ► electron. (► Spin; Stern–Gerlach experiment; Vector model). Either all the l_i are first summed up to one L, and then combined with $S = \sum_i s_i$, or all the l_i and s_i are first summed up separately to j_i with $J = \sum_i j_i$. The noncommutativity of ► operators makes these two procedures in general non-equivalent, yielding different combinatorics, and thus different energy levels and transitions. The first possibility is called Russell–Saunders coupling (also referred to as L-S coupling or strong coupling because it assumes that the interaction of L and S to form a joint J for each electron is much stronger than between different ► electrons). For magnetic dipole radiation, the ► selection rules are: $\Delta J = \pm 1$ or 0, and similar for ΔL and ΔM with the additional constraint that a transition from $M = 0$ to $M = 0$ is forbidden for $\Delta J = 0$. The selection rule $\Delta S = 0$ leads to a prohibition of intercombinations. Russell-Saunders coupling is valid for the lighter, hydrogen-like atom ► Bohr's atom model, for which the multiplet splitting is small compared to the energy difference of the levels with the same electron configuration but different L. For heavier atoms and for the energetically higher terms, ► jj-coupling yields the better approximation. Transition cases between the two couplings also occur (see, e.g. [2], 175f.).

Literature

1. G. Herzberg: *Atomic Spectra and Atomic Structure* (New York: Prentice-Hall 1937, 2nd ed. New York: Dover Publications 1944) (http://wwwchem.uwimona.edu.jm:1104/courses/RScoupling.html; accessed July 20, 2006)
2. C. Candler: *Atomic Spectra and the Vector Model* (Princeton: Van Nostrand 1937, esp. chaps. 11, 16–18)

R

Rutherford Atom

J.L. Heilbron

The identification of the "corpuscle" (later renamed ► "electron") by J.J. Thomson (1856–1940) in 1897 inspired the design of atomic models by the British school of mechanistic physics. The obvious initial assumption, based on relative weights, was that atoms consisted of thousands of elementary particles whose measured ratio of charge to mass (e/m) was about 2000 times that of the hydrogen atom ► Bohr's atom model $(e/m)_H$ as known from electrolysis. Further information came from the spontaneous emission of rays from radioactive substances. These were the alpha and beta rays distinguished by Rutherford in 1898 and identified as material particles through bending in a magnetic field: by 1900 it was known that $(e/m)_\beta = (e/m)$, and therefore that the beta ray probably was identical with the corpuscle, and by 1904 that $(e/m)_\alpha = (1/2)(e/m)_H$ and therefore (if e_α was not smaller than e) that the alpha particle was heavier than a hydrogen atom.

Around 1900 physicists began to direct alpha and beta particles from naturally radioactive substances onto various targets to see what would happen. Thomson evaluated the results for beta particles on the assumption that the observed deviations in their paths arose from a large number of very small pushes exerted on them by individual corpuscles constituting the target atoms. However, observation did not agree with theory on the original assumption that the number of ► electrons n in an atom of relative atomic weight A ($A_H = 1$) was around $1000A$. By 1906 Thomson had discovered that to bring his theory of multiple scattering into approximate agreement with the facts, he had to assume that $n \approx 3A$.

This result was important, for two reasons. For one, it gave the positive charge in or on the atom a more substantial role than it had in ► Thomson's atom model, where it was merely a property of the assembled electrons. Then he had ascribed the entire weight of the atom to its electrons; by reducing their number by three orders of magnitude, he had necessarily to ascribe most of the weight of the atom to its positive charge. Still, there should be many electrons even in very light atoms. Rutherford had proved by 1908 that $e_\alpha = -2e$ and that the alpha particle is a helium atom minus two electrons. It followed from these results and the previous findings $e \approx 3A$ and $(e/m)_\alpha = (1/2)(e/m)_H$ that $A_{He} \approx 4$ and $n_\alpha \approx 10$, that is, that the alpha particle retained some ten electrons and, presumably, was a structure of atomic dimensions.

Against this background, Thomson's former student Ernest Rutherford (1871–1937), by then (1910) professor of physics at the University of Manchester, investigated the scattering of alpha particles ► large-angle scattering. He took up the subject not from a desire to devise a better atomic model, but in order to improve his method of deducing the charge carried by an alpha particle. The experiments were not entirely reproducible owing, in Rutherford's opinion, to the scattering of alpha particles from the walls of the tubes that confined them. He assigned two of his assistants, Hans Geiger (1882–1945) and Ernest Marsden (1889–1970), to determine the extent of the scattering in order to be able to correct for it. ► Large-angle scattering; scattering experiments.

Geiger and Marsden showed that one of every 8000 alpha particles was turned through more than 90 degrees by collisions on a platinum or gold target. Rutherford's intuition refused to accept that Thomson's atoms could reflect alpha particles, supposed to be of atomic size, by a series of collisions with atomic electrons. There were not nearly enough of them ($n \approx 3A$!). In his first efforts to calculate the probability of a reflection, Rutherford drew the alpha particle as if it were an atom; but as

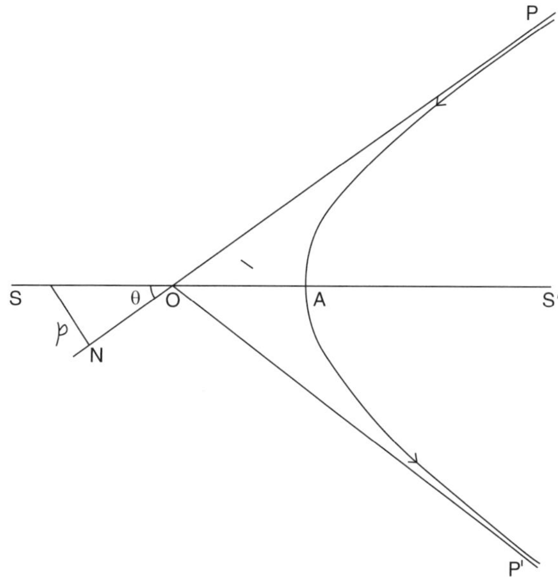

Fig. 1 Rutherford's diagram for large-angle single scattering (Source: Wikimedia Commons). The force centre, considered at rest, is at S; the hyperbola PAP′, of which S is the external focus, is the path of the alpha particle; p, the "impact parameter," is the perpendicular dropped from the focus to the original direction of the incoming particle. If the force between the particle and the nucleus is attractive, the same trajectory can be produced with the internal focus S′ as force centre. For a time Rutherford thought that the nucleus might be negative, and the force on the alpha particle an attraction. (Rutherford's draftsman erred in making the distance OA, where O is the crossing of the asymptotes, less than OS; S′ lies to the right of A at a distance OS′ = OS.)" Source: Wikipedia Commons

he progressed, he seems to have assimilated it to a beta particle, that is, to a charged mass point. This tacit assumption in effect introduced the ► nuclear model for the helium atom; for, if an alpha particle was a point with a double positive charge, the helium atom, evidently of atomic dimensions, must have two electrons in orbits very large in comparison with the central charge.

Assuming that a platinum or gold target had the same structure as a helium atom, but with a central point charge of $100e$, Rutherford could derive the Geiger-Marsden result on the further supposition that the widely scattered alpha particles received their entire deviation in a single stroke from a large central charge occupying a very small volume, and not from many slight deflections in encounters with the atomic electrons. That recovered at the high end of the periodic table the relation $n \approx A/2$ required for helium in order that the alpha particle be a point charge in the scattering calculation ($A_{\mathrm{Pt}} \approx 200$). Thus the primary evidence Rutherford offered for his nuclear model rested not only on the Geiger-Marsden experiments but also on a novel "single-scattering" approach opposed to the "multiple scattering" theory developed by Thomson.

The difference in atomic weight between elements in a row in the periodic table of the elements averages about two units; hence, according to Rutherford's approximation $n \approx A/2$, $\Delta n \approx \Delta A/2 \approx 1$. That deduction inspired and anchored the concept of atomic number. Assigning then to each element an atomic number Z equal to its place in the table, and assuming that chemists had not missed any elements (or had left the right number of spaces for ones unknown), $\Delta Z = 1 \approx \Delta n$. In Rutherford's theory, Ze represents the central atomic charge or, as it soon was called, nucleus; the charge on the hydrogen nucleus $Z_{\mathrm{H}}e$ should be e if no fractional electronic charge exists. Apparently chemists had succeeded in arranging the elements by their weight only by luck, only because, in general, the sequence of A is usually that of Z. Anomalies occur at K/A, Ni/Co, and I/Te, where arrangement by A inverts the chemical order. In the nuclear model, Z, which numbers the electrons in a neutral atom, indicates chemical properties. Organizing the table by Z rather than A removed the three anomalies and brought the discovery that atomic weight does not control chemistry. A given chemical behaviour might be compatible with a range of atomic weights. Hence the coeval discovery of the principle of isotopy in the existence of radioactive elements with the same chemistry and different atomic weights found a perfect representation in the Rutherford atom. The electronic structure and Z determined chemical behaviour, the weight of the nucleus its radioactivity.

Rutherford's group at Manchester included several people who worked out the consequences of isotopy and atomic number, particularly Henry Moseley (1887–1915), George de Hevesy (1885–1960), and Niels Bohr (1885–1962). Bohr also made good use of a consequence of the nuclear atom that most physicists thought its chief demerit. The hydrogen atom with a single orbiting electron is radically unstable: if obliged to follow the ordinary laws of mechanics and electrodynamics, the electron would either fall into the nucleus by radiating away its energy or be driven out of the atom by any passing electromagnetic disturbance. Bohr insisted on this plain fact, which the plenary atoms of Thomson disguised, to ground his argument that microphysics required a principle foreign to ordinary physics in order to account for the stability of atoms. His dictum that atomic electrons whose motions satisfied a certain condition incorporating Planck's quantum would be stable against radiation loss and mechanical perturbations made it possible to bring the precise measurements of ► spectroscopy to bear on the nuclear model and thereby to open up the subatomic quantum world. See also ► Atomic models, Bohr's atom model.

Primary Literature

1. E. Rutherford: The scattering of alpha and beta particles by matter and the structure of the atom. Philosophical magazine **21,** 669–88 (1911), in J. Chadwick, ed., *The collected papers of Lord Rutherford of Nelson*. Vol. 2 (George Allen and Unwin, London 1963, 238–54). This volume contains all of Rutherford's published papers concerning the invention of the nuclear atom.

Secondary Literature

2. J.L. Heilbron: The scattering of alpha and beta particles and Rutherford's atom. Archive for History of Exact Science **4**, 247–307 (1968)
3. J.L. Heilbron: H.G.J. Moseley: *The life and letters of an English physicist* (University of California Press, Berkeley 1974)
4. J.L. Heilbron: *Ernest Rutherford and the explosion of atoms* (Oxford University Press, New York 2003)
5. F. Weinert: Theories, Models and Constraints. Studies in History and Philosophy of Science **30/2,** 303–333 (1999)

R

S

Scattering Experiments

Brigitte Falkenburg

The scattering experiments of subatomic physics probe subatomic structure. The scattering of well-known "probe particles" at some unknown structures gives rise to two kinds of results: (i) ► *large-angle scattering* and the discovery of pointlike structures inside the atom (► *Rutherford atom*); (ii) the production of stable and unstable particles, which are identified from characteristic particle tracks, scattering events and resonances. In a scattering experiment, a particle beam of well-known mass m and charge q is extracted with well-known momentum $\boldsymbol{p}$ and energy E from a particle accelerator. In a fixed-target m experiment, the particles are scattered at some bulk of matter. In a collider experiment, two particle beams are crossed in a small interaction zone and scattered off each other. The scattering results are obtained by measuring the kinematic and dynamic properties of scattered particles and counting their relative frequencies. The effective cross section obtained from these relative frequencies corresponds to the transition probability of a quantum mechanical scattering process.

History

In the earliest scattering experiments, radiation from a radioactive source was collimated and sent to some target [1]. The measurement results were obtained by counting the relative frequency of particles scattered into direction θ. Around 1908, Ernest Rutherford (1871–1937) and his assistants scattered α-particles at thin aluminium or gold foil and detected them using a simple scintillation method. As discovered in 1903, a screen laminated with zinc sulfide starts to phosphoresce in total darkness when it is exposed to α-rays. Observed with a magnifying glass, this glow could be resolved into a variety of single light flashes. In 1909, they found unexpected *large-angle scattering*. In 1911, Rutherford postulated the atomic nucleus as a pointlike scattering center inside the atom described by a Coulomb potential ► Rutherford Atom. Niels Bohr (1885–1962) developed this into his ► *Bohr's atomic model*.

Scattering experiments at particle accelerators started in the 1950s [2]. In a particle accelerator, charged particles (► electrons, protons, or heavy ions) are accelerated by means of electric and/or magnetic fields to high energies. The first cyclotron, designed by Ernest Lawrence (1901–1958) in 1929 and running in 1930, had magnetic poles of diameter 10 cm. A later 9-in. model accelerated protons

D. Greenberger et al. (eds.), *Compendium of Quantum Physics: Concepts, Experiments, History and Philosophy*, © Springer-Verlag Berlin Heidelberg 2009

beyond 1 MeV. After the second world war, the era of the big machines began. The size of the machines was rapidly increased in order to increase the beam energy. In the 1950s, the first proton synchrotrons were built. They generated beams of 1–10 GeV. In the 1970s, 500 GeV were reached. Current machines (2007) generate beams of the order of 1–14 TeV, the LHC (large hadron collider) at CERN will reach 14 TeV.

The bubble chamber, developed in 1952, was a particle detector and a hydrogen target, too. In order to increase the efficiency, later experiments used heavy targets equipped with electronic particle detectors (photomultipliers, spark chambers, drift chambers, Čerenkov counters, etc.) These detectors made the collider experiments possible.

The scattering experiments in the 1950s–1970s showed that the atomic nucleus is not pointlike [2, 3]. In the 1950s, the form factors of protons and neutrons were measured. In addition, the resonances of many unstable particles were detected, giving rise to an increasing "particle zoo" which was tamed in terms of group theory. In 1967, *large-angle scattering* recurred in a collider experiment at the Stanford Linear Accelerator (SLAC), indicating the quark constituents of the nucleon. In 1974, the J/Ψ-resonance confirmed the current standard model of particle physics by establishing the prediction of a "charmed" particle. (► Quantum field theory). The high energy scattering experiments of the 1980s and 1990s measured further particles of the standard model, namely the b and t quark and the vector bosons $W^{\pm}$, Z^0 of the electro-weak interaction. The current scattering experiments at the LHC are designed to finding the last "missing link" of the standard model, the Higgs boson, and particles beyond the standard model [4].

Scattering Models

1. Classical model: Charged massive particles, described as mass points, are scattered at some potential without or with energy transfer, giving rise to elastic or inelastic scattering. For elastic scattering, the trajectory of a particle is described by the *impact parameter* b which depends on the scattering angle θ and the kinetic energy E of the particle before and after the scattering. θ and E can be measured. b is characteristic of classical scattering, it is the minimum distance of the scattered particle to the scattering center or potential source (see ► Rutherford atom).

2. *Quantum mechanical model:* A particle beam of well-defined energy is prepared as a ► wave function in a momentum state that corresponds to a plane wave. The scattering process is described by the quantum mechanics of scattering, in Born approximation plus eventually some higher order(s) of perturbation theory. The scattering is described in the *wave picture* ► Davisson–Germer experiment; Stern–Gerlach experiment; Schrödinger equation, whereas the measurement results are described in the *particle picture* ► *Franck–Hertz experiment*. Here, the usual ► *probabilistic interpretation* of quantum mechanics applies. The quantum mechanical expectation value for a certain scattering outcome gives the probability of this

kind of scattering process, which empirically corresponds to the relative frequency of particle detections of this kind, for a large number of scattering events.

3. Relation between both models: In the quantum mechanical model, there is neither a trajectory nor an impact parameter of the individual scattered particles. For the Coulomb potential, however, the quantum mechanical and the classical model give exactly the same probabilistic prediction, namely Rutherford's formula.

Beam Energy and Spatial Resolution

Why were the machines made bigger and bigger in order to generate beams of higher and higher energies? Due to the formal analogy between the wave equations of quantum mechanics and optics, the spatial resolution of a particle beam is analogous to that of the microscope: The smaller the wavelength of the rays of an observation instrument, the smaller the structures that can be observed. Indeed the beam momentum p corresponds to a ► *de Broglie wavelength* $\lambda = 2\pi\hbar/p$. With increasing beam energy E or beam momentum p, the de Broglie wavelength λ of the scattered "probe" particles decreases. Therefore, the higher the beam energy is, the better the spatial resolution of a scattering experiment will be, and hence the smaller the spatial structures which may be measured.

The idea behind the design of collider experiments is also to increase the scattering energy. A fixed target is at rest in the laboratory, whereas a collider experiment brings two beams of high energy and opposite momentum into collision. In the

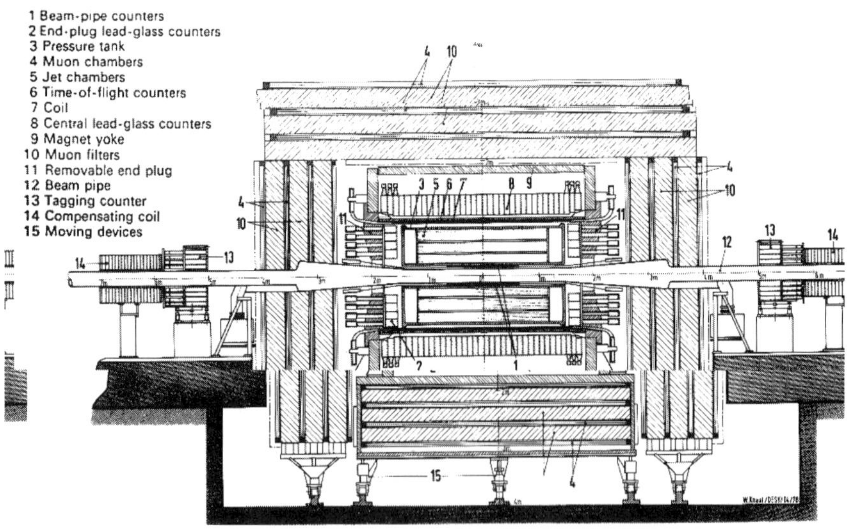

Fig. 1 Collider experiment: JADE detector for the measurement e^+e^- collisions, DESY [4, 3rd edn., p. 63; by permission of the author.]

center-of-gravity frame of the scattering process, in a collider experiment the beam energy is much higher.

The Effective Cross Section

The characteristic quantity of the scattering is the *effective cross section* or *(scattering) cross section*. In the classical model, it is calculated from the angle dependence of the impact parameter b. It has the dimension of an area and is expressed in units of *Barn* (1 Barn $= 10^{-24}$ cm^2). In particle physics, the *differential* and *total* cross section are distinguished.

The *differential cross section* $\mathrm{d}\sigma/\mathrm{d}\Omega$ is proportional to the probability of the scattered particles per scattering angle θ respectively a corresponding infinitesimal solid angle $\mathrm{d}\Omega$. As an empirical quantity, $\mathrm{d}\sigma/\mathrm{d}\Omega$ is measured from the relative frequency of particles which are scattered into a finite solid angle $\Delta\Omega$. In the theoretical model, $\mathrm{d}\sigma/\mathrm{d}\Omega$ is defined from the number of particles N^{sc} scattered into the differential solid angle $\mathrm{d}\Omega$ that belongs to the scattering angle θ, per differential surface $\mathrm{d}F$ and per scattering center and taken in the formal limit of infinitely many incoming particles (N_{C} = number of scattering centers):

$$\frac{\mathrm{d}\sigma}{\mathrm{d}\Omega} = \lim_{N_{\mathrm{in}}\to\infty} \frac{N^{\mathrm{sc}}}{N_{\mathrm{in}} N_{\mathrm{C}}} \cdot \frac{\mathrm{d}F}{\mathrm{d}\Omega}.$$

The formal limit expresses the difference between probability and relative frequency, i.e., the unavoidable gap between a probabilistic quantity and its empirical basis. Here, probability is understood as the limit of relative frequency for very big event numbers.

The number N^{in} of incoming particles per differential surface $\mathrm{d}F$ is usually unknown, just as the number of scattering centers N^{ST}. Without these numbers, $\mathrm{d}\sigma/\mathrm{d}\Omega$ is only known up to some normalization factor. In the classical model, $\mathrm{d}\sigma/\mathrm{d}\Omega$ depends on the impact parameter b as follows:

S

$$\frac{\mathrm{d}\sigma}{\mathrm{d}\Omega} = \frac{b}{\sin\theta} \cdot \left|\frac{\mathrm{d}b}{\mathrm{d}\theta}\right|$$

For the Coulomb potential $V(\boldsymbol{r}) = C/r$, Rutherford's formula is obtained (*large angle scattering*).

The *total cross section* σ is obtained by integrating $\mathrm{d}\sigma/\mathrm{d}\Omega$ over all solid angles. It expresses the probability of a certain kind of scattering process. It is a measure for the "hit ratio" of some kind of particle reaction. In a simple mechanical model, the total cross section may be illustrated as the *effective surface of the reaction*, that is, as the area of an extended and impenetrable scattering center, at which negligibly small probe particles bounce off just like balls at the slats of a garden fence. The expression "effective cross section" or "scattering cross section" stems from this simple mechanical model. In general, σ or $\mathrm{d}\sigma/\mathrm{d}\Omega$ depends not only on geometric quantities, but in addition on the kinetic energy of the probe particles and an eventual energy transfer, in accordance with the relation between the beam energy and the spatial resolution explained above.

In quantum mechanics, the effective cross section is defined as an abstract probabilistic quantity. In a scattering experiment, the effective cross section is measured from the relative frequency of scattering events of a given type. In the effective cross section of a kind of particle reaction, quantum field theory meets experiment. For a given kind of subatomic scattering process, σ and $\mathrm{d}\sigma/\mathrm{d}\Omega$ are proportional to the quantum mechanical transition probability respectively to the corresponding element of the S-matrix. The cross section of a particle reaction is calculated from the S-matrix of the interaction term of a quantum field theory. The S-matrix gives the transition probabilities of initial quantum states to final quantum states. The initial quantum states correspond to the incoming particles of a scattering experiment, i.e., the beam particles and the target nuclei. The final quantum states correspond to the outgoing diffracted wave or the scattered particles which are detected.

Data Analysis

The data analysis of a scattering experiment proceeds in the following steps [5]. Position measurements are made. They give rise to ► *particle tracks*. The particle tracks are interpreted in terms of scattering events. The particle tracks and scattering events are analyzed in terms of mass, charge, momentum, and various other kinematic and dynamic quantities. The numbers of scattering events of a certain dynamic type are counted. In a high energy scattering experiment, relativistic kinematics is used. From the relative numbers of scattering events for a given momentum transfer finally the differential or total cross section of a certain particle reaction is determined. Standard statistical methods are used to correct systematic errors of the measurement, as far as possible, and to determine the measurement errors.

Primary Literature

1. E. Rutherford, J. Chadwick, C. D. Ellis: *Radiations from Radioactive Substances* (Cambridge University Press, London, 1930) Repr.: Cambridge University Press, 1951)

Secondary Literature

2. A. Pais: *Inward Bound*, (Clarendon Press, Oxford, 1986, pp. 190–192)
3. M. Riordan: *The Hunting of the Quark*, (Simon & Schuster, New York, 1987, Chaps. 6–8)

4. D. H. Perkins: *Introduction to High Energy Physics* (Addison-Wesley, Menlo Park, CA, 3rd edn. 1987, 4th edn. 2000)
5. B. Falkenburg: *Particle Metaphysics. A Critical Account of Subatomic Reality* (Springer, Heidelberg, 2007, Chaps. 3, 5)

Schrödinger Equation

Marianne Breinig

In non-relativistic quantum mechanics, the state of a physical system at a fixed time t_0 is defined by specifying a ket $|\psi(t_0)\rangle$ belonging to the space ε. ε is a complex, separable ► Hilbert space, a complex linear vector space in which an inner product is defined and which possesses a countable, ► orthonormal basis. In the ► Schrödinger picture the time evolution of the state vector is governed by a partial differential equation called the Schrödinger equation,

$$(\mathrm{i}\hbar\partial/\partial t)|\psi(t)\rangle = \mathrm{H}(t)|\psi(t)\rangle,$$

which is a recipe for calculating $|\psi(t)\rangle$ when $|\psi(t_0)\rangle$ is known. Here H is the ► Hamiltonian operator for the system.

The Schrödinger equation was developed by Erwin Schrödinger (1887–1961) in 1926 in coordinate representation, where the state vector is represented by a wave function $\psi(\boldsymbol{r},t)$. Schrödinger's original aim was to find a consistent mathematization of De Broglie's intuitive vision of the ► electrons as standing ► matter waves around the nucleus. Like Louis de Broglie (1892–1987), Schrödinger hoped that the ► quantization of electron orbits would thus be reinterpretable as the result of the condition that the electron waves around the nucleus are mutually reinforcing themselves, i.e. as a periodicity constraint between the orbit $2\pi r$ equal to integral multiples of their hypothetical wavelength λ. Because he knew from spectroscopic fine structure effects (► spectroscopy and Bohr-Sommerfeld's model ► Bohr's model) that the velocity of the electrons around the nucleus was actually quite high, he first attempted a relativistically invariant description, taking account of the velocity-dependence of electron mass. A preserved Schrödinger-manuscript written during a ski holiday in Arosa during Dec. 1925/ Jan. 1926 shows that he thus first ended up with an equation surprisingly close to the Klein-Gordon-equation, found one year later by Oskar Klein (1894–1977) and Walter Gordon (1893–1939) and then instrumental in Dirac's ► relativistic quantum mechanics of 1930. But in early 1926, Schrödinger gave up his effort to formulate a theory of electrons as standing waves in relativistically invariant terms and instead tried a simpler non-relativistic variant (on the detailed reconstruction of Schrödinger's pathway to the equation named after him, differing strongly from the much more formal and

S

downright obtuse derivation presented his first papers [1], see his selected correspondence with various physicists [5] and various historical studies [6–10], particularly [7] and [8] for a close analysis of the surviving Schrödinger manuscripts and his detour via ► relativistic quantum mechanics).

Assuming that the electrons, interpreted as de Broglie matter waves, would satisfy a classical wave equation, $\Delta\psi + k^2\psi = 0$ for the amplitude ψ of their wave motion around the nucleus, Schrödinger then inserted

$$k = \frac{2\pi}{\lambda}, \Rightarrow \Delta\psi + \frac{4\pi^2}{\lambda^2}\psi = 0,$$

with λ as hypothetical wavelength of the electron matter waves.

After inserting the de Broglie relation between electron wavelength λ and momentum mv, $\lambda = h/mv$, he must have obtained

$$\Delta\psi + \frac{4\pi^2 m^2 v^2}{h^2}\psi = 0.$$

Inserting of the classical relation between total energy E, potential energy U and kinetic energy $T = (1/2)mv^2$ then yielded a simple non-relativistic wave-equation

$$E - U = \frac{1}{2}mv^2, \quad \Delta\psi + \frac{8\pi^2 m}{h^2}(E - U)\psi = 0.$$

Until June 1926, Schrödinger still believed the ψ-function to be a real-valued function until he realized that he definitely needed complex-valued solutions of the type $\psi(t) \sim \psi_0 \cdot \exp(2\pi\, iE\, t/h)$.

Because Schrödinger's wave-mechanics, as it was soon called, promised a much more intuitive understanding, allowed a much simpler and faster calculation of solutions for various standard potentials $V(r)$ and also yielded solution for problems uncalculable for the competing formalisms of Heisenberg's ► matrix mechanics and Born & Wiener's operator mechanics with which it was then also proven to be physically equivalent in 1926/27, the vast majority of physicists soon only used Schrödinger's approach which dominated the further development of quantum mechanics until 1930.

In more general terms, for a single particle the Schrödinger equation has the form

$$(-\hbar^2/(2m))\nabla^2\psi(\boldsymbol{r}, t) + U(\boldsymbol{r}, t)\psi(\boldsymbol{r}, t) = \mathrm{i}\hbar\partial\psi(\boldsymbol{r}, t)/\partial t.$$

The state vector $|\psi(t)\rangle$ encodes all the information the rest of the world, called the observer, can have about the system at time t. A ► measurement changes the information the observer has about the system and therefore changes the state vector. Between measurements, the state vector changes deterministically.

In free space ($U(\boldsymbol{r}, t) = 0$), plane waves of the form $\psi(\boldsymbol{r}, t) = A\exp(\mathrm{i}(\boldsymbol{k}\cdot\boldsymbol{r} - \omega t))$ are possible solutions of the Schrödinger equation as long as $\hbar\omega = \hbar^2 k^2/(2\,m)$. But a plane wave is not square-integrable, it is not a proper wave function. Since the Schrödinger equation is a linear equation, the ▶ superposition principle applies, and a linear combination of plane wave solutions is also a solution.

$$\psi(\boldsymbol{r}, t) = \sum_{k} a_k \exp(\mathrm{i}(\boldsymbol{k} \cdot \boldsymbol{r}) - \omega t),$$

as long as for each $\boldsymbol{k}$ we have $\hbar\omega_k = \hbar^2 k^2/(2m)$.

Since $\boldsymbol{k}$ is a continuous variable, the most general solution is not a sum, but an integral;

$$\psi(\boldsymbol{r}, t) = \int g(\boldsymbol{k}) \exp(\mathrm{i}(\boldsymbol{k} \cdot \boldsymbol{r}) - \omega t)\, \mathrm{d}^3 k,$$

where the function g($\boldsymbol{k}$) can be complex, $g(\boldsymbol{k}) = |g(\boldsymbol{k})| \exp(\mathrm{i}\alpha(\boldsymbol{k}))$, and where $\alpha(\boldsymbol{k})$ determines the phase of the plane wave. Such a ▶ wave function is called a three-dimensional ▶ wave packet and can represent any non-pathological square-integrable wave function. Proper wave functions of free particles are wave packets.

The Schrödinger equation can be solved analytically and exactly only for a few simple systems. Approximation methods and numerical techniques are usually combined to find approximate solutions. If the Hamiltonian does not contain time explicitly, then separation of time and space coordinates is possible. Any state vector $|\psi(t)\rangle$ can be expanded in terms of the ▶ eigenstates of the Hamiltonian $\{|\psi_n\rangle\}$, where $\mathrm{H}|\psi_n\rangle = E_n|\psi_n\rangle$.

$$|\psi(t)\rangle = \sum_n c_n(t)|\psi_n\rangle.$$

The eigenvalue equation for the ▶ Hamiltonian operator $\mathrm{H}|\psi_n\rangle = \mathrm{E}_n|\psi_n\rangle$ is called the *time-independent Schrödinger equation*. In ▶ wave mechanics it is solved in coordinate representation.

Since the Schrödinger equation is a linear equation, there exists an operator, called the evolution operator $U(t, t_0)$, that transforms $|\psi(t_0)\rangle$ into $|\psi(t)\rangle$.

The properties of this evolution operator follow from its definition and the Schrödinger equation.

S

- $U(t_0, t_0) = I$.
- $(\mathrm{i}\hbar\partial/\partial t)U(t, t_0)|\psi(t_0)\rangle = \mathrm{H}\,U(t, t_0)|\psi(t_0)\rangle$ for any $|\psi(t_0)\rangle$.
  Therefore $(\mathrm{i}\hbar\partial/\partial t)U(t, t_0) = \mathrm{H}\,U(t, t_0)$.
  Properties (a) and (b) completely define the evolution operator.
- $|\psi(t)\rangle = U(t, t')|\psi(t')\rangle$, $|\psi(')\rangle = U(t', t'')|\psi(t'')\rangle$. Therefore $|\psi(t)\rangle = U(t, t')U(t', t'')|\psi(t'')\rangle = U(t', t'')|\psi(t'')\rangle$.
  We can generalize to $U(t_n, t_1) = U(t_n, t_{n-1})U(t_{n-1}, t_{n-2}) \ldots U(t_3, t_2)\,U(t_2, t_1)$.
  Let $t'' = t$. Then $U(t, t')U(t', t) = \mathrm{I}$, and interchanging the role of t' and t, $U(t', t)U(t, t') = \mathrm{I}$. Therefore $U(t, t')^{-1} = U(t', t)$.
- The Schrödinger equation yields
  $\mathrm{d}|\psi(t)\rangle = |\psi(t + \mathrm{d}t)\rangle - |\psi(t)\rangle = -(\mathrm{i}/\hbar)\mathrm{H}(t)|\psi(t)\rangle\mathrm{d}t$. Therefore
  $|\psi(t + \mathrm{d}t)\rangle = [\mathrm{I} - (\mathrm{i}/\hbar)\mathrm{H}(t)]|\psi(t)\rangle\mathrm{d}t$, and $U(t + \mathrm{d}t, t) = \mathrm{I} - (\mathrm{i}/\hbar)\mathrm{H}(t)\mathrm{d}t$
  is the infinitesimal evolution operator.
  $U^{\dagger}(t + \mathrm{d}t, t) = \mathrm{I} + (\mathrm{i}/\hbar)\mathrm{H}(t)\,\mathrm{d}t$ since I and H are Hermitian.

$U^{\dagger}(t+\mathrm{dt},\ t)\ U(t+\mathrm{dt},\ t)\ =\ U(t+\mathrm{dt},\ t)\ U^{\dagger}(t+\mathrm{dt},\ t)\ =\ \mathrm{I}$, since for an infinitesimal operator we neglect terms higher than first order in dt. The infinitesimal evolution operator is a ▶ unitary operator. Therefore $U(t,\ t_0)$, which is a product of infinitesimal evolution operators, is unitary.

$$U^{\dagger}(t,\ t_0) = U^{-1}(t,\ t_0) = U(t_0,\ t).$$

- If H does not explicitly depend on time, then we can solve $(\mathrm{i}\hbar\partial/\partial t)U(t,\ t_0) = \mathrm{H}\,U(t,\ t_0)$ for $U(t,\ t_0)$. We find $U(t,\ t_0) = \exp(-\mathrm{i}\mathrm{H}(t\text{–}t_0)/\hbar)$.

If H does not explicitly depend on time, the evolution operator simplifies finding the time dependence of the state vector after the eigenstates of H have been found. The state vector at time t_0 is expanded in terms of these eigenstates and the wave vector at time t is calculated from

$$|\psi(t_0)\rangle\ =\ \sum_n \mathrm{c}_n(t_0)|\psi_n\rangle,\quad |\psi(t)\rangle\ =\ \sum_n \mathrm{c}_n(t_0)\exp(-\mathrm{i}\mathrm{E}_n(t-t_0)/\hbar|\psi_n\rangle.$$

In the Schrödinger picture the time development of the state vector is entirely deterministic provided that the system is left undisturbed

$$|\psi(t_2)\rangle = U(t_2,\ t_1)|\psi(t_1)\rangle.$$

In coordinate representation this yields for a single particle

$$\langle \boldsymbol{r}_2|\psi(t_2)\rangle = \langle \boldsymbol{r}_2|U(t_2,\ t_1)|\psi(t_1)\rangle = \int \mathrm{d}^3r_1\langle \boldsymbol{r}_2|U(t_2,\ t_1)|\boldsymbol{r}_1\rangle\langle \boldsymbol{r}_1|\psi(t_1)\rangle\rangle,$$

or

$$\psi(\boldsymbol{r}_2, t_2) = \mathrm{d}^3r_1\langle \boldsymbol{r}_2|U(t_2,\ t_1)|\boldsymbol{r}_1\rangle\psi(\boldsymbol{r}_1, t_1).$$

$\langle \boldsymbol{r}_2|U(t_2,\ t_1)|\boldsymbol{r}_1\rangle\ =\ K(\boldsymbol{r}_2, t_2; \boldsymbol{r}_1, t_1)$ is called the *propagator* for the Schrödinger equation and can be interpreted as the probability amplitude that a particle that at t_1 is located precisely at $\boldsymbol{r}_1$ will be found at $\boldsymbol{r}_2$ at time t_2. The propagator is the Green's function for the time-dependent Schrödinger equation

$$[(-\hbar^2/(2m))\nabla_2^2+U(\boldsymbol{r}_2)-\mathrm{i}\hbar\partial\psi(\boldsymbol{r}, t)/\partial t_2]\,K(\boldsymbol{r}_2, t_2; \boldsymbol{r}_1, t_1) = \mathrm{i}\hbar\delta(t_2-t_1)\delta(\boldsymbol{r}_2-\boldsymbol{r}_1).$$

We may write $U(t_2,\ t_1)\ =\ U(t_2,\ t_{an})\ U(t_{an},\ t_{an-1})\ \ldots\ U(t_{a2},\ t_{a1})\ U(t_{a1},\ t_1)$, i.e. we may divide the time interval $t_2 - t_1$ into subintervals. Then, by inserting the closure relation for each subinterval we obtain

$$K(2, 1)\ =\int \mathrm{d}^3\mathrm{r}_{an}\int \mathrm{d}^3\mathrm{r}_{an-1}\ldots\int \mathrm{d}^3\mathrm{r}_{a2}\int \mathrm{d}^3\mathrm{r}_{a1}\,K(2, a_n)\,K(a_n, a_{n-1})\ldots$$
$$K(a_2, a_1)\,K(a_1, 1).$$

$K(2,1)$ can be interpreted to be the coherent superposition of the probability amplitudes associated with all possible space-time paths starting at 1 and ending at 2.

This concept of the propagator as the coherent ► superposition of the probability amplitudes has led to Feynman's postulates, a new formulation of the postulate concerning the evolution of a physical system.

Primary Literature

1. E. Schrödinger: *Quantisierung als Eigenwert-Problem.* Annalen der Physik. (4) 79 (1926), 361–76, 489–527, 80, S. 437–90, 81 (1926), S. 109–40 (all also reprinted in Schrödinger's Gesammelte Werke. Braunschweig: Vieweg 1984, vol. 3); Engl. transl.: An undulatory theory of the mechanics of atoms and molecules. Phys. Rev. 28 (1926), 1049–070.

Secondary Literature

2. C. Cohen-Tannoudji, B. Diu, F. Laloë: *Quantum Mechanics.* Volume One (Wiley, 1977)
3. J.J. Sakurai: *Modern Quantum Mechanics.* Revised Edition (Addison Wesley, 1994)
4. B.H. Brandsden, C.J. Joachain: *Quantum Mechanics.* 2nd edition (Prentice Hall, 2000)
5. K. Przibram (Hrsg.) Schrödinger, Planck, Einstein, Lorentz: *Briefe zur Wellenmechanik* (Springer, Wien 1963)
6. V.V. Raman, P. Forman: *Why was it Schrödinger who developed de Broglie's ideas*? Historical Studies in the Physical Sciences 1 (1969), 291–303
7. L. Wessels: *Schrödinger's route to wave mechanics. Studies in the History and Philosophy of Science.* 10 (1977), 311–40
8. H. Kragh: *Erwin Schrödinger and the wave equation: the crucial phase.* Centaurus. 26 (1982), 154–97
9. M. Jammer: *The Conceptual Development of Quantum Mechanics.* Baltimore: Johns Hopkins University Press 1966, chap. 5
10. J. Mehra and H. Rechenberg: *The Historical Development of Quantum Theory.* New York: Springer, vol. 4

Schrödinger's Cat

Henry Stapp

Erwin Schrödinger and Werner Heisenberg were the originators of two approaches, known respectively as "► wave mechanics" and "► matrix mechanics", to what is now called "quantum mechanics" or "quantum theory". The two approaches appear to be extremely different, both in their technical forms, and in their philosophical underpinnings. Heisenberg arrived at his theory by effectively renouncing the idea of trying to represent a physical system, such as a hydrogen ► Bohr's atom model for example, as a structure in space–time, but instead, following the lead of Einstein's 1905 theory of relativity, representing only empirically observable properties, such as the transition amplitudes between the stationary states of the atom. These amplitudes can be arranged in square arrays of numbers. In Heisenberg's scheme these arrays, and other like them, are combined according to certain rules that were later recognized by Max Born to be the rules of matrix multiplication. The whole scheme is abstract and mathematical, and avoids using any space–time picture of what is going on at the atomic level. Schrödinger, on the other hand, represented the electron in an atom by a cloudlike wave surrounding the nucleus. This is a space–time structure that, superficially at least, is more in line with the classical physical theories of the eighteenth and nineteenth centuries.

Niels Bohr invited Schrödinger to come to Copenhagen to present his ideas, and to discuss this subject with himself, Heisenberg, and others. Schrödinger arrived in Copenhagen on October 1st, 1926, and was immediately intensively engaged by Bohr and the others in a "debate" that lasted for days. Eventually, Schrödinger become ill, and was confined in Bohr's home to a bed, upon which Bohr sat, continuing the discussion. Schrödinger finally exclaimed "If all the quantum jumping is here to stay, then I am sorry that I ever became involved in quantum mechanics". Bohr replied, "But we are glad that you did!"

This "quantum jumping" (► quantum jumps) was the key issue. Schrödinger [1] represented the electron in an atom by a wave that obeyed an equation similar to the one obeyed by the waves occurring in classical electromagnetic theory, or by the waves on the ocean. (► Schrödinger picture) He believed that his waves would have a "realistic" interpretation similar to what had come before in physics. But the Copenhagen group argued that his wave must be viewed as an abstraction that could be used to compute results of measurements, but that could not be "real" in the same sense that the waves in classical physics could be imagined to be real. In particular, the wave had to undergo sudden jumps when a measurement was performed that revealed new knowledge or information. (► Wave function collapse; see also ► ensembles in quantum mechanics). (Copenhagen interpretation. See ► Born rule; Consistent Histories; Metaphysics in Quantum Mechanics; Nonlocality; Orthodox Interpretation; Transactional Interpretation).

The problem was how to understand these "jumps". They are required to occur because if one accepts that our measuring devices, along with our own bodies and brains, and the entire surrounding physical universe, are made of atoms, then this whole lot, taken as a whole, should be subject to the laws of atomic physics. But these laws entail that the states, first of our measuring devices, and then of our bodies and brains, will generally evolve into continuous smears that represent a mixtures of "all possibilities" for what *might* happen, in stark contrast to the particular possibilities that we experience as actually happening. For example, in the case of a radio-active nucleus surrounded by an instrument that detects, and signals, the detection of the decay of the nucleus, the evolving quantum state of the world will eventually contain contributions associated with the *continuum of times* at which the decay might possibly be detected by the instrument. And the state will contain also contributions associated with the *continuum of times* at which the brains of the observers of the instrument *might possibly* register the signal associated with the detection, rather than just the part corresponding to the time that observer actually experiences the signal.

The straight-forward way out of this difficulty would be to introduce into the physical theory some *new physical process* that would, at some level between "atomic size" and "macroscopic size", manage to bring the properties at the macroscopic quantum scale into line with what we normally see. That would mean that the present orthodox theory, which lacks the specification of any such process, would be fundamentally incomplete, and that correct predictions would depend in the end on the details of this currently unspecified process.

Bohr, Heisenberg, and Pauli, thinking along the lines initiated by Einstein, recognized that a neater solution, much more in line with Ockham's razor, could be constructed by stipulating, economically, that *no such new physical process exists*, and by then using, instead of such a process, the fact the space–time structures that are needed for the description of relationships between our observations are the space–time structures occurring in our observations themselves. The theory is then formulated as a set of rules connecting our observation to the symbols in the quantum mathematical formalism.

Bringing the knowledge of observers into the theory in this essential and explicit way is a huge departure from the ideas of classical physics, where the external physical world is imagined to have, independently of all observers, the space–time properties that observers can "see" if they happen to look. Their observations play no essential role. Of course, Einstein had broken the ice with his focusing on the readings on clocks and rulers that idealized observers could "see". But behind the quantum shift was also the emphasis on the (long-standing) philosophical view that the proper mission of science is to provide us with useful tools, rather than with the philosophical satisfaction of believing that we know the truth about nature. Classical mechanics deceived scientists and philosophers for more than two centuries into believing that it provided them with an essentially true picture of reality. The founders of quantum theory sought to avoid making the same mistake.

The quantum shift in perspective was proclaimed in the opening words of Bohr's 1934 book:

> The task of science is both to extend the range of our experience and reduce it to order.

Later on he elaborates:

> In our description of nature the purpose is not to disclose the real essence of phenomena, but only to track down as far as possible relations between the multifold aspects of our experience. ([2] p. 18)

> ...the formalism does not allow pictorial representation along accustomed lines, but aims directly at establishing relations between observations obtained under well defined conditions. ([3] p. 71)

> ...we must recognize above all, that even when phenomena transcend the scope of classical physical theories, the account of the experimental arrangement and the recording of observations must be given in plain language, suitably supplemented by technical physical terminology. This is a clear logical demand, since the very word "experiment" refers to a situation where we can tell others what we have done and what we have learned. ([3] p. 72)

These quotes emphasize the fact that, according to the Bohr/Copenhagen view, a space–time description comes into the quantum mechanical theory through us; through our own descriptions of our probing actions and the feedbacks we receive. There is in orthodox (Copenhagen) quantum mechanics no specification of any observer-independent process that endows even large measuring devices with the essentially-classical space–time structure that we all intuitively believe each macroscopic device possesses, even when we are not seeing or otherwise sensing it. Thus if a system is confined to a black box that blocks our being able to have any knowledge of its contents, beyond what follows from our knowledge of the preparation, then the quantum theoretic representation of the contents of the box will be just the quantum state that evolves from the prepared state via the Schrödinger equation governed evolution.

It is within this context that Schrödinger proposed his diabolical experiment. (Fig. 1) He places his cat in a black box containing a radio-active source that triggers a device that has a 50% chance to release the contents of a pellet of cyanide that, if released, will kill the otherwise health cat. Under these conditions, the evolution in accordance with the Schrödinger equation of what's in the box will eventually generate a state representing a 50–50 mixture of one part corresponding to a dead cat and another part corresponding to an alive cat. Since no one can observe what is happening inside the box, and since the theory does not allow any endowing of any space–time properties except via observation and Schrödinger evolution, the theory is left in the posture of having to retain, until someone looks inside the box, both the dead-cat part and the alive-cat part. (Interaction with the environment renders certain interference experiments unfeasible, but does not eliminate either part.)

This situation seems highly counter-intuitive. But it poses no problem for the Copenhagen view, which specifies that the entire theory is naught but a set of rules designed to allow predictions about relationships between observations to be calculated (See ▶ Matrix Mechanics for an example.) The cat situation is in accord with what Bohr and company had said all along: Schrödinger's wave is an abstraction

Fig. 1 *Source*: B.S. De Witt and N. Graham (eds.): *The Many Worlds Interpretation of Quantum Mechanics* (Princeton 1973, 156). Reproduced by permission of Princeton University Press

that can be used to compute expectations about human experiences, but it cannot rationally be imagined to be real in sense that the waves in classical physical theories could be imagined to be real.

Heisenberg suggested, later on, that the quantum state could be interpreted as an "objective tendency" for an observational event to occur. There is no problem with the idea that there is in the box a "state" that represents both a tendency for the cat to be found completely dead when some person looks, and also an equally weighted tendency for the cat to be found to be completely healthy, with no tendency for any other possibility to be found.

Because science is regarded as a cooperative *human* endeavour, cats are not included among the "we" who "can tell others what we have done and what we have learned".

The rational coherence of Bohr's position rest squarely on his premise that the purpose of science is to provide us with useful practical tools, not to explain essences. Schrödinger's cat highlights this fact.

Literature

1. E. Schrödinger: Die Gegenwärtige Situation in der Quantummechanik. Naturwissenschaften 23, 807–812, 823–849 (November 1935)
2. N. Bohr: *Atomic theory and the description of nature* (Cambridge University Press, Cambridge 1934)
3. N. Bohr: *Atomic physics and human knowledge* (Wiley, New York 1959)

Schrödinger Picture

Marianne Breinig

S

In non-relativistic quantum mechanics, the state of a physical system at a fixed time t_0 is defined by specifying a ket $|\psi(t_0)\rangle$ belonging to the space ε. ε is a complex, separable ► Hilbert space, a complex linear vector space in which an inner product is defined and which possesses a countable, ► orthonormal basis. The vectors in such a space have the properties mathematical objects must have in order to be capable of describing a quantum system.

In the Schrödinger picture the time evolution of the state vector is governed by the ► Schrödinger equation,

$$(\mathrm{i}\hbar\partial/\partial t)|\psi(t)\rangle = \mathrm{H}(t)|\psi(t)\rangle,$$

which is a recipe for calculating $|\psi(t)\rangle$ when $|\psi(t_0)\rangle$ is known. The Schrödinger equation is linear. The correspondence between $|\psi(t)\rangle$ and $|\psi(t_0)\rangle$ is therefore linear.

There exists a linear operator that transforms $|\psi(t_0)\rangle$ into $|\psi(t)\rangle$.

$$|\psi(t)\rangle = \mathrm{U}(t,\ t_0)|\psi(t_0)\rangle.$$

The operator U(t, t_0) is called the *evolution operator*. The evolution operator is a ► unitary operator. If H does not explicitly depend on time, then $\mathrm{U}(t,\ t_0) = \exp(-\mathrm{i}\mathrm{H}(t\text{–}t_0)/\hbar)$. If $|\psi(t_0)\rangle$ is expanded in terms of eigenstates of H, i.e. if

$$|\psi(t_0)\rangle = \Sigma_n\, a_n(t_0)|\mathrm{E}_n\rangle,$$

where $\mathrm{H}|E_n\rangle = E_n|E_n\rangle$, then

$$|\psi(t)\rangle = \sum_n a_n(t_0)\ \exp(-\mathrm{i}E_n(t-t_0)/\hbar)|E_n\rangle = \sum_n a_n(t)\,|E_n\rangle.$$

Time evolution is a unitary transformation. All unitary transformations are changes of representation. We distinguish between active and passive unitary transformations. Active transformation change all state vectors while leaving the basis vectors unchanged. ► Operators are defined through their action on the basis vectors and therefore do not change under an active transformation. Passive transformations change the basis vectors and therefore change the operators, but leave the state vectors
unchanged.

In the Schrödinger picture the time evolution of a physical system is a continuous, active unitary transformation. The state vector is transformed, it evolves in time. The basis vectors are not changing. All operators are constant in time, unless they contain time explicitly. The Schrödinger equation describes the evolution of a physical system in a particular representation.

Primary Literature

1. C. Cohen-Tannoudji, B. Diu, F. Laloë: *Quantum Mechanics*, Volume One (Wiley, New York, 1977, 312–314)

Selection Rules

Klaus Hentschel

According to Niels Bohr's (1885–1962) ► Bohr's atomic model, spectral lines occur when ► electrons perform ► quantum jumps between stable orbits around the positively charged nucleus. For simple atoms like hydrogen and helium, Bohr's model achieved pretty good agreement between theoretical predictions or retrodictions and experimental data. The situation was more complicated for some of the heavier atoms, however, or when external electric or magnetic fields were present (see ► Stark effect and ► Zeeman effect). In such cases, by no means all combinatorically possible transitions between the various energy levels are actually observable. Many theoretically possible spectral lines were missing, thus leading to quite complicated spectral patterns. In order to explain such observed patterns and the absence of other spectrum lines, Bohr and his co-workers as well as some members of the ► Sommerfeld school in Munich introduced what are called selection rules (*Auswahlregeln*).

In terms of the Bohr/Sommerfeld ► quantum numbers n, m, l and j, one of these rules stipulated that the magnetic quantum number m, linked to the number of components into which a spectral line split in the ► Zeeman effect, has to change by units of ± 1 or remain unchanged, i.e., $\Delta m = \pm 1$ or 0. In addition, the transition $m = 0 \rightarrow m = 0$ is also forbidden. Similar constraints were also established for l and j, i.e., $\Delta l = \pm 1$ and $\Delta j = \pm 1$. During the semi-classical phase of ► quantum theory, such phenomenological rules were physically uninterpretable. Physicists simply stipulated them *ad hoc*, in a consciously instrumentalistic attitude (► quantum theory, crisis period). A deeper physical understanding of these selection rules in terms of the conservation of total angular momentum – an exact ► symmetry strictly obeyed by all quantum systems – only became possible after the introduction of the concept of ► spin in late 1925. ► See also Stern–Gerlach experiment; Vector model. Because electrons are spin 1/2 particles and ► light quanta (photons) have spin 1, the emission of a photon against the electron's axis of rotation is compensated by the spin-flip of the electron in order to preserve the overall angular momentum of the system (hence $\Delta m = \pm 1$). A transition with $\Delta m = 0$ is possible only if the emission is tilted with respect to the electron's axis of rotation, thus explaining the different state of polarization of the emitted photons and the requirement that in this case m has to differ from 0 (i.e., the electron has to precess around the axis; cf., e.g., [1], pp. 84ff., 153ff.).

Similar *ad hoc* rules to explain "restrictions on the nature and scope of possible measurements" were also introduced into elementary particle theory by Wick, Wightman and Wigner [2], Heisenberg [3] et al., there called ► super-selection rules.[1] In some versions of Everett's ► many world interpretation, probabilistically defined selection rules also exist for quantum histories.

S

[1] According to Wick *et al.* [2, p. 103], "a superselection rule operates between subspaces [of the total Hilbert space], if there are neither spontaneous transitions between their state vectors (i.e., if a selection rule operates between them), and if, in addition to this, there are no measurable quantities with finite matrix elements between their state vectors."

Literature

1. G. Herzberg: *Atomic Spectra and Atomic Structure* (New York: Prentice-Hall 1937, 2nd edn. New York: Dover Publications 1944)
2. G.C. Wick, A.S. Wightman, E.P. Wigner: The intrinsic parity of elementary particles. Physical Review (2nd ser.) **88,** 101–105 (1952)
3. W. Heisenberg: Quantum theory of fields and elementary particles. Reviews of Modern Physics **29,** 269–278 (1957)
4. A. Sommerfeld: *Atombau und Spektrallinien* (Braunschweig: Vieweg 1919; Engl. transl. *Atomic Structure and Spectral Lines*, London: Methuen 1923)
5. M. Weis: Selection rules, online as http://math.ucr.edu/home/baez/spin/node10.html

Self-Adjoint Operator

Werner Stulpe

Self-adjoint operator, a sharpening of the concept of a symmetric operator. A linear ► operator A acting in a complex ► Hilbert space $\mathcal{H}$ and defined on a dense linear submanifold D_A is called *symmetric* or *Hermitian* if $\langle\phi|A\psi\rangle = \langle A\phi|\psi\rangle$ for all $\phi, \psi \in D_A$. A densely defined operator in $\mathcal{H}$ is symmetric if and only if the scalar products $\langle\phi|A\phi\rangle$, $\phi \in D_A$, are real.

The *adjoint* A^* of a densely defined linear operator A is defined as follows. The domain D_{A^*} of A^* consists of all vectors $\phi \in \mathcal{H}$ for which there exists a vector $\chi_\phi \in \mathcal{H}$ such that $\langle\phi|A\psi\rangle = \langle\chi_\phi|\psi\rangle$ for all $\psi \in D_A$; since D_A is dense in $\mathcal{H}$, χ_ϕ is uniquely determined, and $A^*\phi = \chi_\phi$ concludes the definition of A^*. In particular, $\langle\phi|A\psi\rangle = \langle A^*\phi|\psi\rangle$ for all $\psi \in D_A$ and all $\phi \in D_{A^*}$. The adjoint is a closed (► operator) linear operator, but the submanifold D_{A^*} need not be dense in $\mathcal{H}$; D_{A^*} is dense in $\mathcal{H}$ if and only if A is closable in which case $\overline{A} = A^{**}$ (by definition, $A^{**} = (A^*)^*$). A densely defined linear operator A is called self-adjoint if $A = A^*$, i.e., $\langle\phi|A\psi\rangle = \langle A\phi|\psi\rangle$ for all $\phi, \psi \in D_A = D_{A^*}$.

A densely defined linear operator is symmetric if and only if A^* is an extension of A (briefly written as $A \subseteq A^*$), that is, A^* coincides with A on D_A, but possibly has a larger domain. It can be shown that a symmetric operator satisfies $A \subseteq A^{**} \subseteq A^*$ where A^{**} is the closure (► operator) of A. Thus, for a closed symmetric operator, $A = A^{**} \subseteq A^*$ holds true, and for a self-adjoint operator, $A = A^{**} = A^*$. A symmetric operator is called *essentially self-adjoint* if its closure $\overline{A} = A^{**}$ is self-adjoint; an essentially self-adjoint operator satisfies $A \subseteq A^{**} = A^*$. A necessary and sufficient criterion for the self-adjointness of a symmetric operator A is that $R_{A+\mathrm{i}I} = R_{A-\mathrm{i}I} = \mathcal{H}$, where I is the unit operator and $R_{A+\mathrm{i}I}$, for instance, the range of the operator $A + \mathrm{i}I$ which is defined on D_A; a criterion for the essential self-adjointness of A is that $R_{A+\mathrm{i}I}$ and $R_{A-\mathrm{i}I}$ are dense in $\mathcal{H}$.—For a linear operator with domain $D_A = \mathcal{H}$ the concepts of symmetry and self-adjointness are equivalent; a symmetric or self-adjoint operator defined on $\mathcal{H}$ is necessarily bounded (► operator).

An (unbounded) symmetric operator need not have a self-adjoint extension; if a self-adjoint extension exists, it is in general not unique. A symmetric operator A has exactly one self-adjoint extension if and only if A is essentially self-adjoint. Self-adjointness is a crucial property of an operator since only self-adjoint operators always have a spectral decomposition as pointed out below. The Hamiltonian operators of quantum mechanics (▶ Hamiltonian operator) are often given as essentially self-adjoint differential expressions.

(Spectral decomposition, see ▶ Density operator; Ignorance interpretation; Measurement theory; Objectification; Operator; Probabilistic Interpretation; Propensities in Quantum Mechanics; Wave Mechanics).

As an example, let the simple differential operator $P_0 = -\mathrm{i}\frac{\mathrm{d}}{\mathrm{d}x}$ be defined on $D_{P_0} = \{\psi \in L^2([a,b],\mathrm{d}x) \mid \psi$ absolutely continuous, $\psi' \in L^2([a,b],\mathrm{d}x)$, $\psi(a) = \psi(b) = 0\}$ (absolutely continuous functions are in particular differentiable almost everywhere). The domain D_{P_0} is a dense submanifold of $L^2([a,b],\mathrm{d}x)$ and can alternatively be characterized according to $D_{P_0} = \{\psi \in L^2([a,b],\mathrm{d}x) \mid \partial\psi \in L^2([a,b],\mathrm{d}x),\ \psi(a) = \psi(b) = 0\}$ where $\partial\psi$ is the derivative of ψ in the sense of distributions; the linear operator P_0 in $L^2([a,b],\mathrm{d}x)$ is unbounded and closed. By integration by parts, P_0 is symmetric. The adjoint P_0^* is again given by $P_0^* = -\mathrm{i}\frac{\mathrm{d}}{\mathrm{d}x}$, but on the domain $D_{P_0^*} = \{\psi \in L^2([a,b],\mathrm{d}x) \mid \partial\psi \in L^2([a,b],\mathrm{d}x)\}$ which is larger than D_{P_0}; P_0^* is also closed, but not symmetric. So P_0 is not self-adjoint; nevertheless, P_0 has infinitely many self-adjoint extensions, namely, $P_\alpha = -\mathrm{i}\frac{\mathrm{d}}{\mathrm{d}x}$ on $D_{P_\alpha} = \{\psi \in L^2([a,b],\mathrm{d}x) \mid \partial\psi \in L^2([a,b],\mathrm{d}x),\ \psi(a) = \mathrm{e}^{\mathrm{i}\alpha}\psi(b),\ \alpha \in \mathbb{R}\}$.

The multiplication operator Q_0 on $D_{Q_0} = L^2([a,b],\mathrm{d}x)$ defined by $(Q_0\psi)(x) = x\psi(x)$ is bounded and self-adjoint where $\|Q_0\| = \max\{|a|,|b|\}$. The multiplication operator Q in $L^2(\mathbb{R},\mathrm{d}x)$, defined on $D_Q = \{\psi \in L^2(\mathbb{R},\mathrm{d}x) \mid \mathrm{id}_{\mathbb{R}}\psi \in L^2(\mathbb{R},\mathrm{d}x)\}$ by $Q\psi = \mathrm{id}_{\mathbb{R}}\psi$, i.e., by $(Q\psi)(x) = x\psi(x)$, is unbounded and self-adjoint. The differential operator $P = -\mathrm{i}\frac{\mathrm{d}}{\mathrm{d}x}$ in $L^2(\mathbb{R},\mathrm{d}x)$, defined on $D_P = \{\psi \in L^2(\mathbb{R},\mathrm{d}x) \mid \partial\psi \in L^2(\mathbb{R},\mathrm{d}x)\}$, is also unbounded and self-adjoint. The same holds for the Laplace operator $H_0 = -\Delta = -\left(\frac{\partial^2}{\partial x_1^2} + \frac{\partial^2}{\partial x_2^2} + \frac{\partial^2}{\partial x_3^2}\right)$ in $L^2(\mathbb{R}^3,\mathrm{d}x)$, defined on $D_{H_0} = \{\psi \in L^2(\mathbb{R}^3,\mathrm{d}x) \mid \partial_i\psi,\ \partial_i\partial_j\psi \in L^2(\mathbb{R}^3,\mathrm{d}x),\ i,j = 1,2,3\}$. As a final example, the *Schrödinger operator* $\mathring{H} = -\Delta + V(x) = -\Delta + V(x_1,x_2,x_3)$ in $L^2(\mathbb{R}^3,\mathrm{d}x)$ where V is a suitable real-valued function, can be defined on the dense linear submanifold $C_0^\infty(\mathbb{R}^3)$ (consisting of the infinitely differentiable complex-valued functions on $\mathbb{R}^3$ with compact support); under relatively general conditions on the function V, $\mathring{H}$ is essentially self-adjoint, its self-adjoint extension H has the domain $D_H = D_{H_0}$, and the spectrum (see the fourth of the following paragraphs) of H is bounded from below.

In the sequel, let $\mathcal{H}$ be a separable ▶ Hilbert space; for a nonseparable Hilbert space some of the following statements must slightly be modified. The eigenvalues (▶ operator) of a symmetric or self-adjoint operator, if there are any, are real, there are at most countably many ones, and the corresponding eigenspaces are orthogonal (▶ Hilbert space) to each other; in general, such an operator does not have a complete orthonormal system of eigenvectors. A compact (► operator) self-adjoint operator A does have countably many eigenvalues with 0 as only possible accumulation point. Each nonzero eigenvalue is of finite *multiplicity*, i.e., the corresponding eigenspace is finite-dimensional. The eigenspaces are mutually orthogonal, moreover, A has a complete orthonormal system of eigenvectors which is obtained by choosing an ► orthonormal basis in each eigenspace and joining these bases. Correspondingly, a compact self-adjoint operator has the *spectral decomposition* $A = \sum_{i=1}^{\infty} \lambda_i P_{\phi_i}$ where $\lambda_1, \lambda_2, \ldots$ are the nonzero eigenvalues of A, counted according to their multiplicity and arranged according to $|\lambda_1| \geqslant |\lambda_1| \geqslant \ldots > 0$, $\phi_1, \phi_2, \ldots$ is an orthonormal system of corresponding eigenvectors, $P_{\phi_i} = |\phi_i\rangle\langle\phi_i|$ are the corresponding one-dimensional orthogonal projections (► projection), and the sum converges in the operator norm (► operator) or is finite (in the latter case, 0 must be an eigenvalue of infinite multiplicity, provided that $\mathcal{H}$ is infinite-dimensional).

A bounded self-adjoint operator need not have any eigenvalue. Instead, every bounded or unbounded self-adjoint operator has a spectral decomposition. A *spectral measure* E is a mapping that assigns an orthogonal ► projection $E(B)$ to each Borel set B of the real line $\mathbb{R}$ such that (i) $E(\emptyset) = 0$, $E(\mathbb{R}) = I$ and (ii) $E\left(\bigcup_{i=1}^{\infty} B_i\right)\phi = \sum_{i=1}^{\infty} E(B_i)\phi$ for every sequence of mutually disjoint Borel sets $B_1, B_2, \ldots$ and all $\phi \in \mathcal{H}$; as a consequence, the projections $E(B_i)$ are orthogonal to each other. Furthermore, the mapping associating each Borel set B with the number $\langle\psi|E(B)\psi\rangle$ is a probability measure if $\psi \in \mathcal{H}$ is a unit vector. The *spectral theorem for self-adjoint operators* now states that there is a one-one correspondence between the self-adjoint operators A in $\mathcal{H}$ and the spectral measures such that (i) $D_A = \left\{\psi \in \mathcal{H} \,\middle|\, \int_{\mathbb{R}} \lambda^2 \langle\psi|E(\mathrm{d}\lambda)\psi\rangle < \infty\right\}$ and (ii) $\langle\psi|A\psi\rangle = \int_{\mathbb{R}} \lambda \langle\psi|E(\mathrm{d}\lambda)\psi\rangle$ for all $\psi \in D_A$; the self-adjoint operator is uniquely determined by the scalar products $\langle\psi|A\psi\rangle$, $\psi \in D_A$. The representation (ii) of A is called its *spectral decomposition.*

The concept of spectral measure is closely related to the concept of spectral family. A *spectral family* F is a function assigning an orthogonal projection $F(\lambda)$ to each real number λ such that (i) $F(\lambda) \leqslant F(\mu)$ for $\lambda \leqslant \mu$, (ii) $\lim_{\lambda\to\infty} F(\lambda)\phi = \phi$ and $\lim_{\lambda\to-\infty} F(\lambda)\phi = 0$ for all $\phi \in \mathcal{H}$, and (iii) $\lim_{\epsilon\to 0} F(\lambda+\epsilon)\phi = F(\lambda)\phi$ for all $\lambda \in \mathbb{R}$ and all $\phi \in \mathcal{H}$; the function associating each real number λ with the number $\langle\psi|F(\lambda)\psi\rangle$ is a cumulative distribution function if ψ is a unit vector. A spectral measure E defines a spectral family according to $F(\lambda) = E((-\infty, \lambda])$, conversely, there exists exactly one spectral measure such that $E((\lambda, \mu]) = F(\mu) - F(\lambda)$. Using the spectral family corresponding to the spectral measure of a self-adjoint operator, the integrals in the spectral theorem can be considered as Riemann–Stieltjes integrals, e.g., $\langle\psi|A\psi\rangle = \int_{-\infty}^{\infty} \lambda \,\mathrm{d}\langle\psi|F(\lambda)\psi\rangle$.

The *spectrum* σ_A of a self-adjoint operator A is a subset of $\mathbb{R}$ that can be characterized by the spectral measure E of A or by the corresponding spectral family F. A real number λ belongs to the spectrum of A if and only if, for every $\epsilon > 0$, $E((\lambda-\epsilon, \lambda+\epsilon)) \neq 0$ (equivalently, λ is a point of increase of F). A real number is an eigenvalue of A if and only if $E(\{\lambda\}) \neq 0$ (equivalently, F is discontinuous at λ in the sense that $\lim_{\epsilon\to 0} F(\lambda-\epsilon)\phi \neq F(\lambda)\phi$ for some $\phi \in \mathcal{H}$); λ is a point of the spectrum that is not an eigenvalue if and only if $E((\lambda-\epsilon,\lambda+\epsilon))\neq 0$ for every $\epsilon>0$ and $E(\{\lambda\})=0$ (equivalently, λ is a point of increase of F and F is continuous at λ). Finally, λ is not a point of the spectrum if and only if there exists an $\epsilon>0$ such that $E((\lambda-\epsilon,\lambda+\epsilon))=0$ (equivalently, λ is a point of constancy of F). The spectrum of a self-adjoint operator is a closed subset of the real line and for bounded self-adjoint operators a compact set.

For a self-adjoint operator A with spectral measure E and for a complex-valued Borel-measurable function f on the real line, a closed operator $f(A)$ can be defined by (i) $D_{f(A)}=\left\{\psi\in\mathcal{H}\,\middle|\,\int_{\mathbb{R}}|f(\lambda)|^2\langle\psi|E(\mathrm{d}\lambda)\psi\rangle<\infty\right\}$ and (ii) $\langle\psi|f(A)\psi\rangle=\int_{\mathbb{R}}f(\lambda)\,\langle\psi|E(\mathrm{d}\lambda)\psi\rangle$ where $D_{f(A)}$ is dense in $\mathcal{H}$. The association of the functions f and the operators A with the operators $f(A)$ is called the *functional calculus of the self-adjoint operators*. If f is real-valued, then $f(A)$ is self-adjoint, and the spectral measure of $f(A)$ is given by $E^{f(A)}(B)=E(f^{-1}(B))$ where $f^{-1}(B)=\{\lambda\in\mathbb{R}\,|\,f(\lambda)\in B\}$. If is f bounded, $f(A)$ is bounded. If A is bounded, the set of all operators $f(A)$ where f is a continuous complex-valued function, is the C^*-algebra generated by A, i.e., the smallest C^*-algebra containing A; this C^*-algebra is a commutative C^*-algebra of (bounded) operators. The continuous functions f need not be bounded since, in the definition of $f(A)$, it is sufficient to integrate over the spectrum of A which is compact if A is bounded. If A is a bounded self-adjoint operator, the set of all $f(A)$ where f is a bounded measurable function, is the von Neumann algebra generated by A (► algebraic quantum mechanics).

Another version of the spectral theorem states that every self-adjoint operator is unitarily equivalent to a *multiplication operator* acting in some Hilbert space of square-integrable functions or in a direct sum of such Hilbert spaces. A vector $\chi\in\mathcal{H}$ is called a *cyclic vector for* A, A being a self-adjoint operator, if the submanifold generated by the vectors $E(B)\chi$ where B is a Borel set of $\mathbb{R}$ and E the spectral measure of A, is dense in $\mathcal{H}$. Let A be a self-adjoint operator with a cyclic vector χ (which need not exist in general), and let μ_χ be the measure defined on the Borel sets of $\mathbb{R}$ by $\mu_\chi(B)=\langle\chi|E(B)\chi\rangle$. Then there exists a ► unitary operator U from $\mathcal{H}$ onto the Hilbert space $L^2(\mathbb{R},\mu_\chi)$ of the μ_χ-quadratically integrable functions such that (i) $D_A=\left\{\psi\in\mathcal{H}\,\middle|\,\int_{\mathbb{R}}\lambda^2|(U\psi)(\lambda)|^2\,\mu_\chi(\mathrm{d}\lambda)<\infty\right\}$ and (ii) $(UAU^{-1}\phi)(\lambda)=\lambda\phi(\lambda)$ where $\phi=U\psi$ for some $\psi\in D_A$. The realization of A as a multiplication operator in $L^2(\mathbb{R},\mu_\chi)$ is not unique (since the cyclic vector χ is not unique) and is called a *spectral representation of* A. If the finite measure μ_χ is equivalent to the Lebesgue measure $\mathrm{d}\lambda$, i.e., if μ_χ and $\mathrm{d}\lambda$ have the same sets of measure zero, then A can be represented as a multiplication operator in the Hilbert space $L^2(\mathbb{R},\mathrm{d}\lambda)$ of the Lebesgue-quadratically integrable functions. For a self-adjoint operator A with no cyclic vector, there is also a spectral representation. In this case there exist countably many vectors $\chi_1,\chi_2,\ldots\in\mathcal{H}$ and a unitary operator U from $\mathcal{H}$ onto the direct sum $L^2(\mathbb{R},\mu_{\chi_1})\oplus L^2(\mathbb{R},\mu_{\chi_2})\oplus\ldots$ such that UAU^{-1} is again a multiplication operator, i.e., $(UAU^{-1}\phi)(\lambda)=\lambda\phi(\lambda)$ where $\phi=U\psi$ is an element of the direct sum and $\psi\in D_A$.

As an example, let A be a (bounded or unbounded) self-adjoint operator in $\mathcal{H}$ with a spectrum consisting entirely of eigenvalues; equivalently, A has a complete orthonormal system of eigenvectors. Let λ_i be the eigenvalues and P_i the orthogonal projections onto the corresponding eigenspaces. Then the spectral measure E^A of A is given by $E^A(B)\phi = \sum_{\{i|\lambda_i \in B\}} P_i\phi$, B being a Borel set of $\mathbb{R}$ and $\phi \in \mathcal{H}$, and the spectral decomposition reads $A\psi = \sum_i \lambda_i P_i\psi$, $\psi \in D_A$. The self-adjoint differential operators P_α mentioned above are of the type of the operator A. As another example, the spectral measure E^Q of the self-adjoint operator Q introduced above is given by $E^Q(B)\phi = \chi_B\phi$ where $\chi_B(x) = 1$ for $x \in B$, $\chi_B(x) = 0$ for $x \notin B$, and $\phi \in L^2(\mathbb{R}, \mathrm{d}x)$; the spectrum σ_Q of Q is $\mathbb{R}$. The spectral measure of the operator Q_0 reads $E^{Q_0}(B)\phi = \chi_{B\cap[a,b]}\phi$ where $\phi \in L^2([a,b], \mathrm{d}x)$; $\sigma_{Q_0} = [a,b]$. Finally, the differential operator P is unitarily equivalent to the multiplication operator Q, more precisely, $P = F^{-1}QF$ and $D_P = \{\psi \in L^2(\mathbb{R}, \mathrm{d}x) \mid \psi = F^{-1}\phi,\ \phi \in D_Q\}$ where F is the ► unitary operator of the Fourier transform, so $E^P(B) = F^{-1}E^Q(B)F$ and $\sigma_P = \mathbb{R}$. Furthermore, $FPF^{-1} = Q$ is an instance of the general statement of the preceding paragraph.

The spectral representation of a self-adjoint operator A can be related to a generalized eigenvector problem of A encompassing the so-called *improper eigenvalues and eigenvectors*. For a rigorous treatment of quantum mechanics, the concept of improper eigenvalues and eigenvectors is not necessary; however, for calculational purposes it is sometimes useful to work with this concept which is mostly done in a formal, heuristic manner. For instance, if A has no proper eigenvalues, the points λ of the spectrum of A are improper eigenvalues where the improper eigenvectors ϕ_λ, $A\phi_\lambda = \lambda\phi_\lambda$, are not elements of the Hilbert space $\mathcal{H}$. Moreover, there exists a complete orthonormal system of improper eigenvectors. If, in addition, A has a cyclic vector $\chi \in \mathcal{H}$ and the measure μ_χ defined above is absolutely continuous w.r.t. the Lebesgue measure, then the improper eigenvalues are of multiplicity 1, a complete orthonormal system of improper eigenvectors ϕ_λ satisfies $\langle\phi_\lambda|\phi_\mu\rangle = \delta(\lambda-\mu)$, δ being the δ-distribution, and every vector $\psi \in \mathcal{H}$ can be written as $\psi = \int_{\sigma_A} \alpha(\lambda)\phi_\lambda\,\mathrm{d}\lambda$ where σ_A is the spectrum of A and $\alpha(\lambda) = \langle\phi_\lambda|\psi\rangle$, $\alpha \in L^2(\sigma_A, \mathrm{d}\lambda)$.—A sound mathematical basis for the concept of improper eigenvalues and eigenvectors is presented in [8]; beyond that, the improper eigenvectors can, under some conditions on A, be interpreted as *eigenfunctionals* in the context of the so-called *Gelfand triples* [9].

If A is a self-adjoint or only symmetric operator and if $\phi_1, \phi_2, \ldots$ is a complete orthonormal system in $\mathcal{H}$ that belongs to the domain D_A, then the statement $\psi = A\chi$, $\chi \in D_A$, can be expressed by the matrix representation (► operator) $\beta_i = \sum_j a_{ij}\alpha_j$, $i = 1, 2, \ldots$, where $\alpha_j = \langle\phi_j|\chi\rangle$, $\beta_i = \langle\phi_i|\psi\rangle$, and $a_{ij} = \langle\phi_i|A\phi_j\rangle$. The matrix elements satisfy $a_{ij} = \overline{a}_{ji}$, i.e., they form a *Hermitian matrix*.

Literature

1. F. Riesz, B. Sz.-Nagy: *Functional Analysis* (Frederick Ungar Publishing, New York 1955; reprint Dover Publications, Mineola 1991)
2. M. Reed, B. Simon: *Methods of Modern Mathematical Physics* (Academic Press, New York), *I: Functional Analysis* (1972, second edition 1980), *II: Fourier Analysis, Self-Adjointness* (1975)
3. N. Dunford, J. T. Schwartz: *Linear Operators* (Interscience Publishers, New York), *Part I: General Theory* (1958), *Part II: Spectral Theory* (1963)

4. W. Młak: *Hilbert Spaces and Operator Theory* (Springer, Dordrecht 1991)
5. D. Werner: *Funktionalanalysis* (Springer, Berlin 1995, third edition 2000)
6. G. Ludwig: *Foundations of Quantum Mechanics I* (Springer, New York 1983), Appendix IV
7. E. Prugovečki: *Quantum Mechanics in Hilbert Space* (Academic Press, New York 1971, second edition 1981; reprint Dover Publications, Mineola 2006)
8. O. M. Nikodym: *The Mathematical Apparatus for Quantum Theories* (Springer, Berlin 1966)
9. A. Böhm: *The Rigged Hilbert Space and Quantum Mechanics*, Springer Lecture Notes in Physics **78** (Springer, Berlin 1978)

Semi-classical Models

Markus Arndt

The Notion of Semi-classicality

Within the literature on quantum physics the word "semi-classical" is used both very often and with different meanings. But three situations are most commonly encountered: Firstly, quantum systems that are approximated by classical models at high ► quantum numbers. Secondly, the mathematical description of composite systems which can be simplified by dividing the problem into a classical and a quantum sector. And finally, open quantum systems which reveal classical properties in their interaction with a complex environment.

The various definitions of semi-classicality apply to a vast range of physical systems, covering quantum optics [1], atomic physics [2], molecular physics [3], mesoscopic and solid state physics [4] or even ► quantum gravity [5]. A recent and comprehensive resource letter by Gutzwiller [6] provides nearly four hundred commented references to important papers on that subject. And a number of these papers have been collected and reprinted in [7].

S

Systems at High Quantum Numbers

It has been proven in countless experiments, that quantum physics is the correct theory for describing the world of elementary particles, atoms and molecules. It is also widely believed, that quantum theory is equally correct in the macroscopic world. However, in many cases the use of classical models is simpler and already fully sufficient for the description of observed phenomena. This is why Niels Bohr suggested the ► *correspondence principle*, which should connect the two worlds in the limit of sufficiently high quantum numbers [8]. In this sense, quantum theory should become "semi-classical".

A good example for this is the hydrogen atom: When Bohr built his first atom model ► Bohr's atom model [8], aiming at the quantitative understanding of atomic spectra, he started from the assumption that ▸ electrons were circulating around the nucleus on trajectories similar to those of planets around the sun. However, he had to complement this classical model by the quantum hypothesis that the electron can only travel with discrete angular momenta. Such a trajectory picture is incompatible with energy conservation, as circulating charges inevitably emit electromagnetic radiation, but Bohr's analysis allowed to explain the observed atomic spectra surprisingly well.

In 1926, Schrödinger solved the inconsistencies of this "semi-classical" view by assigning a *stationary complex* ▸ *wave function*, of amplitude A and phase ϕ, to the atomic electrons. The square modulus $|A|^2$ of this function then describes the probability to find the electron in a particular state. The hydrogen ground state is then correctly represented by a spherical wavefunction rather than a circular race track for electrons. The quantum picture thus differs markedly from Bohr's first "semi-classical" view.

However, it turns out that the quantum and the classical description approach each other again in Rydberg atoms, i.e. in atoms excited to high electron energies [9]. When the atom's valence electron is excited to a high electronic quantum number n, a high orbital angular momentum l and a high magnetic quantum number m, with $l = |m| = n - 1$, the electron's wavefunction is again rather well localized on a tight torus which resembles a lot the original idea of a classical electron trajectory.

Such "circular" Rydberg states are the most classical atomic states that can actually be prepared in the lab. They couple only weakly to the nuclear core but very efficiently to external fields. They are therefore very interesting in laboratory demonstrations of fundamental quantum information phenomena [10].

A second example for classical physics as a limiting case of quantum theory can also be identified for continuous variable systems. Similarly to the case of optics, where wave optics is approximated by geometrical ray optics for sufficiently short wavelengths, one may also find a classical approximation for the motion of a quantum object at high momentum and correspondingly short ▸ de Broglie wavelength.

This idea is implemented in the Wenzel–Kramers–Brillouin (WKB) method, which is a "semi-classical" technique for solving the ▸ Schrödinger equation:

$$i\hbar\frac{\partial}{\partial t}\psi(\boldsymbol{x},t) = \left[-\frac{\hbar^2}{2m}\Delta + U(\boldsymbol{x})\right]\psi(\boldsymbol{x},t). \tag{1}$$

If we rewrite the wavefunction of a propagating particle in the exponential form

$$\psi(\boldsymbol{x},t) \equiv A(\boldsymbol{x}) \cdot e^{iS(\boldsymbol{x},t)/\hbar}. \tag{2}$$

and insert this into (1), we find two expressions for the real and imaginary part, and in particular:

$$\frac{\partial S}{\partial t} + \frac{(\nabla S)^2}{2m} + U = \frac{\hbar^2}{2m}\frac{\Delta A}{A}. \tag{3}$$

Equation (3) is usually identified as a classical limit of the quantum description, since for $\hbar \to 0$ it corresponds to the Hamilton–Jacobi equation. In classical physics it describes both the flow of interaction free particles in an external potential $U(\boldsymbol{x})$, and the physics of ray optics as a limiting case for the propagation of electromagnetic waves. Of course it is not physically possible to reduce a fundamental constant to zero, but classical mechanics represents a good approximation to quantum physics as long as the phase and amplitude of the wave vary sufficiently slowly.

Division into Classical and Quantum Sectors

Frequently, semi-classical models represent also a mathematical simplification of a problem which can be achieved by dividing a complex system into at least two parts. One of these parts is sufficiently simple and sufficiently important to be treated by quantum theory, while the other subsystem may still be described using classical physics.

For instance, "semi-classical gravity" usually describes an approach to ▶ quantum gravity in which matter fields are taken to be quantum while the gravitational field is treated classically.

A typical example from quantum optics is the atom-photon interaction, which can be treated at different levels of classicality. In general, the Hamiltonian of the atom–light system reads:

$$\hat{H}_{\text{tot}} = \hat{H}_{\text{atom}} + \hat{H}_{\text{field}} + \hat{H}_{\text{int}}.$$

But depending on the experimental situation, it may be sufficient to choose the mathematical treatment to be fully classical, semi-classical or fully quantum mechanical.

a. Classical matter and classical light: In most situations of our everyday life, we can rely on a purely classical treatment of both the atoms and the light. This is for instance the case when we irradiate a solid lump with light from a lamp. As soon as we know the intensity and color of the light, as well as the absorption coefficient and the heat capacity of the solid, we can for instance determine the temperature increase in the irradiated solid. This does not require any detailed knowledge of the underlying quantum properties.

S

b. Classical light field coupled to quantized internal atomic states: Of course, matter is actually composed of discrete atoms, and each of them has an infinite set of quantized energy levels. But in the interaction with monochromatic light it is often justified to approximate atoms as two-level quantum systems, when the photon energy E_{L} is resonant with the energy difference between the excited state $|e\rangle$ and the ground state $|g\rangle$: $E_{\text{L}} = \hbar\omega_{\text{L}} \simeq E_{\text{e}} - E_{\text{g}} = \hbar\omega_{\text{A}}$. In this simplified situation the atom is described by the Hamiltonian

$$H_{\text{atom}} = \frac{1}{2}\hbar\omega(|e\rangle\langle e| - |g\rangle\langle g|). \tag{4}$$

The presence of the monochromatic light field of amplitude $\mathcal{E} = \mathcal{E}_0 \cdot \cos(\omega t)$ is included in the time-dependent external potential

$$\hat{H}_{\text{int}} = -\hat{d} \cdot \mathcal{E},$$

where the dipole operator $\hat{d} = d|g\rangle\langle e| + d^*|e\rangle\langle g|$ describes the quantum transition between the two atomic levels. It is proportional to the dipole matrix element $d = -e\langle e|\boldsymbol{r}|g\rangle$, and thus a real quantum entity. But the electric field amplitude $\mathcal{E} = \sqrt{2I/c\varepsilon_0}$ can still be related to the light intensity using classical electrodynamics. This procedure is very often justified, as an intensity as little as 1 mW already corresponds to a photon flux of about $\sim 10^{16}$ photons per second. The quantum granularity of the photon field, i.e. the addition or removal of a few photons from the beam, can then be safely neglected. The semi-classical Hamiltonian then reads

$$\hat{H}_{\text{tot}} = \hat{H}_{\text{atom}} + \hat{H}_{\text{int}}.$$

This atom–light interaction model is for instance relevant in most practical situations related to the description of atomic spectra or optical atom traps [11].

c. Quantum atom and quantum light: A full quantum treatment becomes necessary, when only a few photons (▶ light quantum) are strongly coupled to a few atomic levels. A typical example is that of two-level atoms inside a cavity, i.e. in experiments testing cavity ▶ quantum electrodynamics [10, 12]. The presence of the cavity dramatically enhances the interaction between photon and atom. A single photon inside the cavity may then suffice to cause internal or external state changes of the atom. The photonic Hamiltonian is then described by

$$H_{\text{field}} = \hbar\omega\hat{a}^{\dagger}\hat{a}, \tag{5}$$

with the photon ▶ creation and annihilation operators $\hat{a}^{\dagger}$ and $\hat{a}$. The interaction Hamiltonian now includes the electric field of a single photon of frequency ω within the volume V:

$$\hat{\mathcal{E}} = \sqrt{(\hbar\omega)/(\varepsilon_0 V)}(\hat{a} + \hat{a}^{\dagger}).$$

Open Quantum Systems, Coupled to a Complex Environment

Open quantum systems, i.e. systems in interaction with a complex environment, are also often denoted as semi-classical systems. Here the name refers to the fact that most of the unique quantum features – such as ▶ superposition and ▶ entanglement – seem to vanish after contact with the experimentally uncontrollable many-body system.

Decoherence theory [13, 14] elucidates how the coupling between a quantum system S and its environment E reduce the coherence within the quantum system and leads to the appearance of classical properties. Recent experimental examples from quantum optics, are coherent photon states in a lossy cavity [15] or molecular ▸ de Broglie wavelength interacting with their environment through thermal photons or collisions with residual gas atoms [16].

Interestingly, ▸ decoherence only leads to "semi-classical" phenomena: no quantum phase relation is actually lost. The quantum correlations (▸ correlations in quantum mechanics) only extend to and get entangled with an enormously larger system of many particles in a complex environment. And this is the reason why we cannot trace and retrieve them any more. In this sense, the apparent classicality turns out to be a result of our finite information handling capacities but one might still think of the underlying world as being ruled by quantum theory.

Literature

1. P. Meystre, Murray III Sargent: *Elements of Quantum Optics* (Springer, Heidelberg, 2006)
2. B. G. Englert: *Semiclassical Theory of Atoms*, Lecture Notes in Physics (Springer, Heidelberg, 1988)
3. M. S. Child: *Semiclassical Mechanics with Molecular Applications*, Int. Ser. of Monographs on Chemistry (Clarendon Press, Oxford, 1991)
4. M. Brack, R. Bhaduri: *Semiclassical Physics*, Frontiers in Physics (Addison-Wesley, Reading, MA, 1997)
5. P. Fré, V. Gorini, G. Magli, U. Moschella (eds.): *Classical and Quantum Black Holes, Studies in High Energy Physics, Cosmology and Gravitation* (Institute of Physics Publishing, Bristol, 1999)
6. M. C. Gutzwiller, Resource letter ICQM-1: the interplay between classical and quantum mechanics. Am. J. Phys. **66**, 304–324 (1998)
7. M. C. Gutzwiller: *The Interplay Between Classical and Quantum Mechanics* (American Association of Physics Teachers, College Park, MD, 2001)
8. C. Cohen-Tannoudji, B. Diu, F. Laloe: *Quantum Mechanics* (Wiley, New York, 2nd edn., 1991)
9. T. F. Gallagher: Review Article: Rydberg Atoms, Reports on Prog. Phys. **51**, 143–188 (1988)
10. J. M. Raimond, M. Brune, S. Haroche: Manipulating quantum entanglement with atoms and photons in a cavity. Rev. Mod. Phys. **73**, 565–582 (2001)
11. S. Chu: Cooling and trapping of atoms and biomolecules. Opt. Soc. Am. **559**, 199 (1998)
12. H. Walther: Cavity quantum electrodynamics. Rep. Prog. Phys. **69**, 1325–1382 (2006)
13. W. H. Zurek: Decoherence, Einselection, and the quantum origins of the classical. Rev. Mod. Phys. **75**, 715–775 (2003)
14. D. Giulini, E. Joos, C. Kiefer, J. Kupsch, O. Stamatescu, H. D. Zeh: Decoherence and the Appearance of the Classical World in Quantum Theory (Springer, Berlin, 2003)
15. M. Brune, E. Hagley, J. Dreyer, X. Matre, A. Maali, C. Wunderlich, J. M. Raimond, S. Haroche: Observing the Progressive Decoherence of the "Meter" in a Quantum Measurement. Phys. Rev. Lett. **77**, 4887–4890 (1996)
16. M. Arndt, K. Hornberger, A. Zeilinger: Probing the limits of the quantum world. Phys. World **18**, 35–40 (2005)

Shor's Algorithm

See ► quantum computation.

Solitons

A. Seeger

Historical Background

The *soliton concept* has in common with other mathematical concepts such as *vectors, tensors, matrices* that it arises not only in mathematics but also in numerous other fields, including physics, chemistry, biology as well as various branches of engineering science (see, e.g., [1]). It is therefore not surprising that, depending on the field, the same name denotes different objects or properties and that a simple definition comprising the entire current usage cannot be given.

The name *soliton* has its origin in hydraulics. In 1834, the Scottish engineer–scientist John Scott Russell (1808–1882), while studying the movement of ships on the Union Canal between Edinburgh and Glasgow, discovered what he described as a 'large, *solitary*, progressive wave' [2]. This 'heap of water' originated when a fast-moving ship was suddenly stopped. The swell, however, travelled along the channel with essentially constant shape and with a velocity

$$V = [g(H + h_0)]^{1/2} \tag{1}$$

that depended only on the height H of the water level and the amplitude h_0 of the swell ($g = 9.81\,\mathrm{m\,s^{-2}}$). Russell realised the difference between the '*solitary wave of translation*', as he also called the phenomenon, and the more common *oscillatory* waves which do not involve transport of matter over long distances. He was convinced of the fundamental nature of his discovery but could not give a convincing theoretical explanation based on, say, Stokes' equations of fluid dynamics. After many attempts by British and French scientists – either fruitless or only partially successful – the French mathematician Joseph Valentin Boussinesq (1842–1929) solved the problem in 1872 by demonstrating that a fourth-order partial differential equation for the height $h(x, t)$ of the water level in a *shallow* canal has the solution

$$h = h_0\,\mathrm{Sech}^2[(3h_0/4H^3)^{1/2}(x - V\,t)], \tag{2}$$

the speed V being given by (1) [3]. Equation (2) accounted very well indeed for Russell's numerous observations as well as later experimental work.

In 1895, apparently without being aware of the work of Boussinesq and other French scientists, the Dutch mathematician Dieterik Johannes Korteweg (1848–1941) and his Ph.D. student, Gustav de Vries, showed that the solitary wave described by (1, 2) is a solution of a simpler evolution equation with *first-order* t- and *second-order* x-derivatives, now known as KdV equation [4]. It could have easily been derived from the *forth-order* Boussinesq equation by *assuming* that the shape of the wave was time-independent and by considering only waves travelling in one direction in space (say, in the $+x$-direction). In numerical studies of the KdV equation, Zabusky and Kruskal [5] noted that the 'solitary wave' (2) possesses certain *persistence properties*. When two such waves with different speeds meet, they get temporarily modified but eventually emerge unchanged from the collision. Clearly inspired by Russell's nomenclature, Zabusky and Kruskal introduced the expression 'soliton' in the title of their paper, which dealt not with fluids but with so-called collisionless plasmas. Subsequent analytical investigations showed that the persistence is a consequence of a *non-linear superposition theorem* obeyed by the solutions of the KdV equation. From the mathematical point of view, the validity of such a theorem is an indispensable feature of a solitonic system.

It is a widespread but unjustified claim that the developments just described mark the *discovery* of solitonic behaviour. (As one of many examples in the literature, see Fokas and Zakharov [25]: "The fascinating new world of solitons and of integrable behaviour was discovered by Kruskal and Zabusky".) For a thorough and objective discussion of the subject the reader is referred to the thesis of M. Heyerhoff [6], which also covers the nineteenth century work on Russell's 'solitary wave' and relevant work on differential geometry referred to below. The essential aspects of solitonic behaviour were discovered in the period 1951–53 in a study not of the KdV equation but of the *Bour–Enneper equation* [7]. (For the name and its alternative *Sine-Gordon equation* see the next-but-one section.) These analytical investigations preceded the corresponding work on the KdV equation by more than a decade. In contrast to the Galilei-invariant KdV, the Bour–Enneper equation is *Lorentz invariant* and therefore of particular interest for quantum field theories. Hence, the present essay concentrates on it.

Non-linear Wave Equations with Particle Solutions

In the summer of 1924, Prince Louis de Broglie (1892–1987, Nobel Prize for physics 1929), working in Paris in the private laboratory of his elder brother Maurice de Broglie, proposed that the motion of relativistic particles is guided by 'phase waves' [8]. In their search for a wave equation that fits de Broglie's ideas, Klein [26], Schrödinger [27], Fock [28], Gordon [29], and Kudar [30] proposed as a Lorentz-invariant wave equation for particles of mass m the *linear* partial differential equation

$$\Delta\psi - \partial^2\psi/\partial^2(ct)^2 = (2\pi/h)^2 m^2 c^2 \psi, \qquad (3)$$

where $h = 6.626 \times 10^{-34}$ Js denotes Planck's constant and c the 'limiting speed' (in the present case the speed of light in vacuum). Depending on the context, (3) is known as one-dimensional Helmholtz equation, Schrödinger–Gordon equation [9], or Klein–Gordon equation ► relativistic quantum mechanics. For various reasons – one of them being the failure to account for the electron ► spin and the effects going with it – (3) and its extension to charged particles in an electromagnetic field [9] were soon found to be unsuitable for the description of ► electrons. For the one-electron problem it was successfully replaced by the ► Dirac equation. Nevertheless, the Schrödinger–Gordon equation remained of interest for the description of spin-zero bosons, in particular of the electrically charged pi-mesons, π^+ and π^-.

A natural question to ask is whether non-linear generalisations of (3) can be found that yield 'particle-like' solutions. This is indeed the case. If on the right-hand side of (3) the dependent variable ψ is replaced by $(2\pi mc/h) f(\psi)$, where the non-linear function $f(\psi)$ satisfies certain conditions to be specified presently, restriction of the spatial variation of ψ to the x-dimension followed by the substitution

$$z = (1 - V^2/c^2)^{-1/2} (x - ct) \tag{4}$$

leads to

$$\mathrm{d}\psi^2/\mathrm{d}z^2 = f(\psi) \tag{5}$$

with the solution

$$\int [F(\psi)]^{-1/2} \mathrm{d}\psi = \pm 2^{1/2} z, \quad F(\psi) = \int f(\psi) \mathrm{d}\psi. \tag{6}$$

Suppose now that $f(\psi)$ is a differentiable function with simple zeros, with $\mathrm{d}f(\psi)/\mathrm{d}\psi > 0$ at more than one zero, e.g. at $\ldots < \psi_{-1} < \psi_0 < \psi_1 < \ldots$. The constant of integration in (6_2) may be chosen in such a way that $F(\psi)$ has double zeros at ψ_0 and ψ_1. With this choice, a solution $\psi = \psi(z)$ obtained by inverting (6_1) represents a *kink* or an *antikink*, depending on the choice of the sign in (6_1), both with the *kink height*

$$a_\mathrm{K} := \psi_1 - \psi_0. \tag{7}$$

The name *kink* for this type of configuration was introduced by Shockley [31] in the context of dislocations in crystals but has since found more widespread usage. As shown in Fig. 1, positive kinks are transitions from an (almost) constant solution $\psi = \psi_0$ at large negative z to an (almost) constant solution $\psi = \psi_1$ at large positive z. More generally, in the language to be introduced below, kinks connect adjacent ground states of systems with degenerate ground states.

Kinks resulting from non-linear generalisations of (3) may travel with constant speed $|V|$ either in the $+x(V > 0)$ or in the $-x(V < 0)$ direction as if they were relativistic particles subject to the Lorentz contraction. From the field-theory point of view, they are *excitations* of the ground states of the system whose energy,

$$E = (1 - V^2/c^2)^{-1/2} E_\mathrm{K}, \tag{8}$$

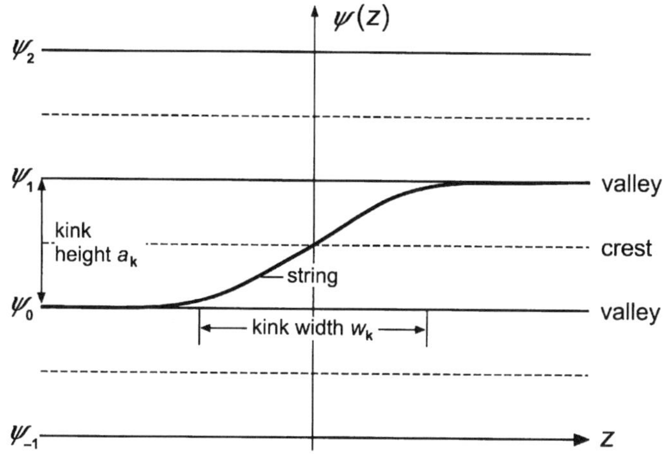

Fig. 1 Shape of a +-kink connecting the dislocation segments lying in the valleys $\psi = \psi_0$ and $\psi = \psi_1$

is concentrated in a narrow spatial region, called *kink width* $w_{\rm k}$. If between ψ_0 and ψ_1 the function $F(\psi)$ has just one maximum, $F_{\max}$, a natural measure of the kink width is

$$w_{\rm k} := [(1 - V^2/c^2)/2F_{\max}]^{1/2}\, a_{\rm k}. \tag{9}$$

A general expression for the *rest energy* $E_{\rm k}$ of a kink will be given in (30).

It is trivial that the sum of two or more solutions of the equations obtained by replacing the right-hand side of (3) with a *non-linear* function $f(\psi)$ cannot be *exact* solutions of these equations. It may be an *approximate* solution as long as the individual solutions do not overlap, i.e. if neighbouring kinks are many kink widths apart. Until about 1950 it was undisputed consensus among physicists that analogous statements hold for *all* finite-amplitude excitations of non-linear systems. It was believed that such excitations could not permanently co-exist and that their coupling through the non-linearity necessarily results in gradual dissipation of their kinetic energy (in the case of kinks to phonon-type excitations).

S

A Non-linear Wave Equation with Soliton Solutions

Towards the middle of 1950, the present writer noted that in the context of the differential geometry of surfaces with constant negative Gaussian curvature, known as *pseudospherical surfaces* [10, 11], the non-linear partial differential equation

$$\partial^2\omega/\partial\xi\,\partial\eta = \sin\omega \tag{10}$$

had been extensively studied in the second half of the nineteenth century. (The name comes from the fact that while the Gaussian curvature of spheres is $K \equiv +1$, that of pseudo-spherical surfaces is $K \equiv -1$. In the application to pseudo-spherical surfaces, the co-ordinate lines $\xi = const.$ and $\eta = const.$ are the asymptotic lines on these surfaces, $\omega = \omega(\xi, \eta)$ denoting the angle between these lines.) He realised that quite a few of the nineteenth-century results might have far-reaching consequences in physics, in particular in the theory of dislocations [12]. The transformation

$$x = \xi +, \eta, t = \xi - \eta \tag{11}$$

gives us

$$\partial^2\omega/\partial x^2 - \partial^2\omega/\partial t^2 = \sin\omega. \tag{12}$$

With $\omega = 2\pi\psi/a_{\mathrm{K}}$, the dimensionless generalisation (12) of the Klein–Gordon equation satisfies the conditions for the existence of kink solutions. These are easily found to read

$$\omega = \pm 4\,\mathrm{arctg}\{(1 - V^2)^{-1/2}(x - Vt)\}. \tag{13}$$

The kink velocity $V$is measured in units of the limiting speed c, which need not necessarily be identical with the speed of light.

The significance of (10), (12) for pseudospherical surfaces was noted by the German mathematician Alfred Enneper (1830–1885) in 1868/70 [32]. Already in 1862, (10) had been encountered by the French mathematician Edmond Bour (1832–1866) in another branch of the differential geometry of surfaces [33]. Hence, the name *Bour–Enneper equation* for (10), (12) appears more appropriate than the wide-spread denomination Sine-Gordon equation, particularly since the equation has no relationship to W. Gordon and the name Sine-Gordon was originally intended to be a private joke (see Heyerhoff [6], Chap. 4). In hindsight, from the point of view of physics the key discovery was made by the Swedish mathematician Albert Victor Bäcklund (1845–1922). In 1882 he demonstrated [13] that from a known solution $\omega = \omega_0(\xi, \eta)$ of the *second-order* differential equation (10) further solutions $\omega = \omega_1(\xi, \eta)$ may be obtained by integrating the following system of *first-order* differential equations:

$$\begin{aligned} \frac{1}{2}\frac{\partial(\omega_1 - \omega_0)}{\partial\xi} &= \frac{1 + \sin\sigma}{\cos\sigma}\sin\left[\frac{\omega_1 + \omega_2}{2}\right], \\ \frac{1}{2}\frac{\partial(\omega_1 + \omega_0)}{\partial\eta} &= \frac{1 - \sin\sigma}{\cos\sigma}\sin\left[\frac{\omega_1 - \omega_2}{2}\right]. \end{aligned} \tag{14}$$

The integrability condition of this system is

$$\partial^2\omega_0/\partial\xi\partial\eta = \sin\omega_0, \tag{15}$$

i.e. the equation which, by assumption, is satisfied by $\omega_0(\xi,\eta)$. By means of the substitution

$$y = \mathrm{tg}(\omega_1/4) \tag{16}$$

the system (14) may be transformed into the following pair of Riccati equations:

$$\partial y/\partial \xi = (a\,y^2 + b\,y + c),\ \partial y/\partial \eta = (a'y^2 + b'y + c'). \tag{17}$$

The explicit form of the coefficients a, a', etc. are given, e.g., in [14], together with the corresponding expressions in the (x, t) co-ordinate system. Since these expressions are substantially more complicated than (14), it is indeed advisable to perform intermediate calculations in the *light-cone co-ordinates* (ξ, η) rather than in the 'physical' co-ordinates (x, t). The rationale of this is that the lines $\xi =$ const., $\eta =$ const. are the *characteristics* of (14).

The system (14) constitutes a so-called *total differential equation*. Its integrability condition (15) is necessary and sufficient for the general solution of the system (14) to be of the form

$$\Phi_\sigma(\omega_1; \xi, \eta) = C_1, \tag{18}$$

where C_1 is a constant of integration. Thus, the solutions $\omega = \omega_1$ of (15) that may be derived from a given 'starting solution' ω_0, called *Bäcklund transforms* of ω_0, constitute a two-parameter family with parameters σ and C_1. As is well known, by means of the substitution $y = Y'/Y$, where Y' denotes the partial derivatives of Y, the Riccati-type system (17) may be transformed into a set of two *linear* equations for $Y(\xi, \eta)$. This indicates that the Bäcklund transforms of a given solution are *superposable*, although not linearly but according to a law that is related to the addition theorem of the tangent function. For a non-linear partial differential equation this is a highly exceptional property.

The (non-linear) superposability of Bäcklund transforms is made explicit by the relationship

$$\mathrm{tg}[(\omega_3 - \omega_0)/4] = \frac{\cos[(\sigma_1 + \sigma_2)/2]}{\sin[(\sigma_1 - \sigma_2)/2]}\mathrm{tg}[(\omega_1 - \omega_2)/4] \tag{19}$$

Here ω_0 denotes the *starting solution*, ω_1 and ω_2 are its *Bäcklund transforms* with parameters σ_1 or σ_2, respectively, and ω_4 is the solution of (10), (12) resulting from the '*superposition*' of ω_1 and ω_2. We illustrate the power of the preceding approach by two simple examples [7]. Further examples can be found in the literature [7, 14].

(1) Take $\omega_0 \equiv 0 \pmod{2\pi}$ as starting solution. Its Bäcklund transforms are

$$\omega_j = 4\mathrm{arctg}\,\{\gamma_j \exp[(x - t\ \sin\sigma_j)/\cos\sigma_j]\}\ (j = 1, 2). \tag{20}$$

(19) gives us

$$\omega_3 = \mathrm{arctg}\left\{\frac{\cos\left[\frac{\sigma_1+\sigma_2}{2}\right]}{\sin\left[\frac{\sigma_1-\sigma_2}{2}\right]}\ \frac{\gamma_1 \exp\varepsilon_1 - \gamma_2 \exp\varepsilon_2}{1 + \gamma_1\gamma_2 \exp(\varepsilon_1 + \varepsilon_2)}\right\} \tag{21}$$

with

$$\varepsilon_j \equiv (x - t\sin\sigma_j)/\cos\sigma_j = (1 - V_j^2)^{-1/2}\,(x - V_j t). \tag{22}$$

In (20), γ_j denotes constants of integration that replace the constants C_j $(j = 1, 2)$ on the right-hand side of (18).

The way in which the solution ω_3 was constructed suggests that it represents two kinks that move with speeds V_1 and V_2. In analogy to the Einstein–Minkowski description in relativity theory, we may ascribe to each of them a *world line* in the $x - t$ plane, i.e. a relationship between their locations $x_j (j = 1, 2)$ and time t. As long as the two kinks are sufficiently far apart, their world lines are straight with slopes V_1 and V_2. If $V_1 \neq V_2$, at some time t the two kinks collide and interact with each other. Figure 2 illustrates this for the collision at $t \approx 0$ of a kink that was originally at rest at the position $x = -\Delta x_1/2$ with a second kink that approaches the region of collision at $x \approx 0$ with the speed $V_2 > 0$. The 'world region of interaction', in which it is difficult or even impossible to discern the individual kinks, is indicated as a circle in the $x - t$ plane. The amazing feature of Fig. 2 is that, in striking contrast to the pre-1950 expectations referred to in Sect. 2, there is *no* transfer of kinetic energy to other excitation modes of the system, e.g. to non-harmonic oscillations. Furthermore, the collision does *not* alter the distribution of the total energy and of the particle momentum among the kinks. After the collision there are still two kinks with velocities $V = 0$ and $V = V_2$. Since kinks on a given dislocation line are indistinguishable (a feature they have in common with elementary particles), we cannot distinguish between the view-point that "there has been no exchange of kinetic energy between the kinks" or the classical-mechanics description "moving kink has come to rest after having transferred its entire kinetic energy to its collision partner". In any case, the statement "the kinks do not interact at all" would be wrong. The world lines at large positive t are *not* the prolongation of the

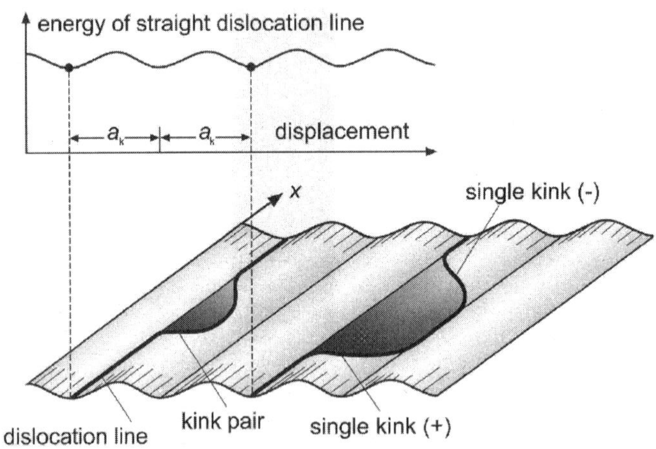

Fig. 2 Two stages of the movement of a dislocation line in a periodic potential of period a_k. On the left, a dislocation begins to overcome the energy barrier locally by forming an incipient kink pair. Under the action of an applied stress a positive and a negative kink move in opposite x-directions, thereby shifting the dislocation line gradually from one valley to the next (right). The shading indicates the areas swept out by the dislocation in these steps

lines at large negative t but have been shifted with respect to them by Δx_1 and Δx_2. The original German name of Δx_j was '*Treffstrecke*' [7], translated into English as *recoil distance* [13]. The concept is closely related to the concept of '*time delay*' in the theory of quantum scattering processes [34]. According to Donth (see [13] for further details and illustrations) the recoil distances in kink–kink collisions are

$$|\Delta x_j| = {}^1\!/\!_2 \cos\sigma_j \ln|\cos\{{}^1\!/\!_2(\sigma_1+\sigma_2)\}/\sin\{{}^1\!/\!_2(\sigma_1-\sigma_2\}|\ (j=1,2). \tag{23}$$

(2) Starting again from $\omega_0 \equiv 0$ (mod2π), we choose σ_1, σ_2 as complex conjugates and write

$$\sigma_1 = \sigma' + \mathrm{i}\sigma'', \sigma_2 = \sigma' - \mathrm{i}\sigma'', \tag{24}$$

where $\sigma'$and σ'' are real numbers. Without loss of generality, the choice $\gamma_1 = \gamma_2 = 1$ in (20) leads to

$$\omega_3 = 4\mathrm{arctg}[H\ \mathrm{Sech}(B_1x + B_2t)\ \sin(D_1x + D_2t)]. \tag{25}$$

The (real) parameters H, B_1, B_2, D_1, and D_2 are given in terms of σ' and σ'' in [7, 13]. For a general choice of the parameters, (25) represents ► *wave packet* with phase velocity

$$V_{\mathrm{ph}} = -\mathrm{Cosh}\,\sigma''/\sin\sigma' \tag{26}$$

and group velocity

$$V_{\mathrm{gr}} = -\sin\sigma'/\mathrm{Cosh}\,\sigma'' = 1/V_{\mathrm{ph}}. \tag{27}$$

If we choose $\sigma' = 0$, the group velocity vanishes, and we get the so-called *breather mode*

$$\begin{aligned}\omega_3 &= 4\mathrm{arctg}[\mathrm{Sech}(x/\mathrm{Cosh}\,\sigma'')\sin(t\ \mathrm{Tgh}\sigma'')/\mathrm{Sinh}\,\sigma] \\ &= 4\mathrm{arctg}\left\{\frac{\left(1-\Omega^2\right)^{1/2}}{\Omega}\frac{\sin(\Omega t)}{\cosh\left[\left(1-\Omega^2\right)^{1/2}x\right]}\right\}.\end{aligned} \tag{28}$$

The breather mode of the Bour–Enneper equation is a *localised oscillation* with circular frequency $\Omega = \mathrm{Tgh}\sigma''$ and amplitude 4arccos Ω. In the limiting case $\Omega \ll 1$, it describes a kink–antikink pair with a total energy slightly less than that of two separate kinks at rest (in the dimensionless units of the Bour–Enneper equation equal to 16) and zero total momentum. Starting from rest, the two kinks attract each other, move towards each other, and annihilate. At this stage the total energy has been transformed into kinetic energy. From thereon the process is reversed. The kinks are recreated and move away from each other until they reach the position of maximal separation. This configuration may be obtained from the starting configuration by interchanging the two members of the kink–antikink pair. Energetically, the situation is the same as at the start, hence the cycle just described is repeated with opposite sign of ω. The period of the entire oscillation cycle is thus $2\pi/\Omega$.

In the limit $\sigma'' \to \infty$ the above solutions reduce to the *travelling-wave* solutions of the Klein–Gordon equation, from which *Eigenschwingungen* ('normal-mode vibrations') can be formed. (Non-linear generalisations of *standing* small-amplitude vibrations with soliton properties can also be obtained from the Bour–Enneper equation [35–37]. The breather solution is one example.) The familiar normal-mode vibrations and the solitonic modes have in common that they can be superposed indefinitely without destroying their identity. In the original publication [7] this property led to the denomination "*Eigenbewegungen*" ("normal motions") for the solitonic solutions of the Bour–Enneper equation, with a subdivision into "translatorische Eigenbewegungen" (σ_j real) and "oszillatorische Eigenbewegungen" (σ_j pairwise complex conjugate). The corresponding English names *translational solitons* and *oscillatory solitons* have not yet found general usage.

Appearance and Significance of the Bour–Enneper Equation in Physics

The first branch of physics in which (12) appeared was *crystal plasticity* [38]. Up to the present, the application to *kinks in dislocations* is of particular importance [15, 39]. It may be illustrated by a model whose '*ground state*' is a flexible string lying in one of the 'valleys' of a horizontal corrugated iron sheet that is imagined to be large enough for border effects to be negligible. The string represents a dislocation with line tension γ_d and effective mass m_d per unit length, the corrugated iron the so-called *Peierls relief* [39, 40], a periodic variation of the energy of a dislocation as a function of its location in the crystal lattice. An external shear stress that tends to push pre-existing dislocations through the Peierls relief may be modelled by slightly tilting the sheet. The plastic deformation of a crystal caused by the applied stress proceeds by moving segments of the dislocations into an adjacent valley, thus creating *kink–antikink pairs* as shown in Fig. 2.

Owing to the periodicity of the Peierls relief, the model just described is a system with *degenerate* ground states since at zero stress its energy is independent of the valley in which the dislocation/string happens to be located. Kinked dislocations may be considered as *excited states* that connect two distinct ground states. The excitation energy (the kink formation energy E_k) consists of two contributions. (1) A kink increases the potential energy of a dislocation because the segment connecting the two valleys is lifted up the hill separating them. (2) The total dislocation length is increased, hence work has to be done against the line tension.

Once a kink pair has been formed, the plastic deformation will proceed further, since the applied stress will drive the kinks apart and cause them to slip along the dislocation line. In this way, the shifting of a dislocation line from one Peierls valley to an adjacent one is effected by overcoming an energy barrier that is much lower than that required for shifting the entire dislocation line as a whole. This is analogous to the overcoming of the shear strength of perfect crystals by the formation of dislocations.

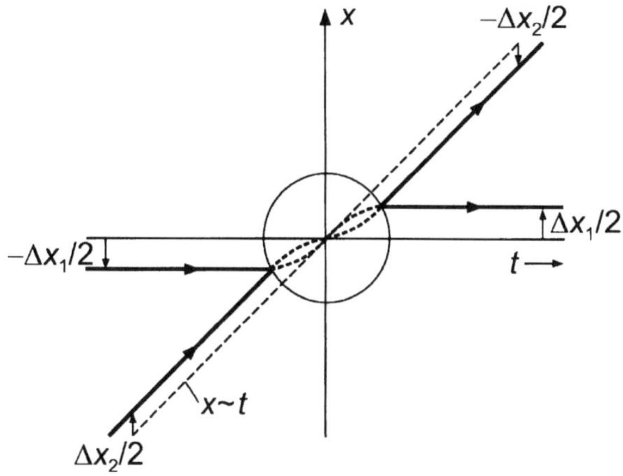

Fig. 3 World-line diagram representing the collision at time $t \approx 0$ of a kink at rest at the position $x = -\Delta x_1/2$ with a kink of the same sign moving with high speed in the $+x$-direction. The circle represents the 'world region of interaction', in which the identity of the kinks is partially lost. After the collision both kinks re-appear unchanged, one of them having been displaced by Δx_1 to the position $x = \Delta x_1/2$. The other kink resumes its former speed but along a world line that has been shifted by Δx_2

For a quantitative treatment of the present model, the simplest assumption is that the energy of the string varies as $U_\mathrm{P} = U_0 \sin^2(\pi u/a_\mathrm{k})$, where u is the displacement of the string and a_k the period of the '*Peierls potential*' U_P (cf. Fig. 2). With the assumption $|\mathrm{d}u/\mathrm{d}x| \ll 1$ (realistic in metals, Fig. 2 being foreshortened), this leads to

$$\gamma_\mathrm{d}\partial^2 u/\partial x^2 - m_\mathrm{d}\partial^2 u/\partial t^2 = \mathrm{d}U_\mathrm{P}/\mathrm{d}u = (\pi U_0/a_\mathrm{k})\sin(2\pi u/a_\mathrm{k}), \tag{29}$$

hence, with appropriate normalisation, to the Bour–Enneper equation (10), (12). Among the soliton solutions of (29) are single kinks of height a_k, sequences of equidistant kinks, and standing or running waves of finite amplitude. Kink–antikink pairs in unstable equilibrium, appearing in overcoming of the Peierls barriers (Fig. 2), may be obtained by adding on the right-hand side of (29) a constant term accounting for the applied stress [14].

S

In summary, the importance of (12), (15), (29) in physics is due to the following features.

1. The equations admit kink and antikink solutions, a *topological* property that they share with other non-linear equations as discussed in Sect. 2.
2. They possess *solitonic* solutions. Their characteristic is that they may be superimposed notwithstanding the strong non-linearity on the right-hand side of the equations and, thus, possess particle properties. The *interactions* of these 'particle solutions' mediated by the non-linearity are *minimal* in the sense that from collisions the solutions emerge without having altered their 'shape' or their

momentum. Since 'particles' of the same 'charge' are *indistinguishable* (see Sect. 5), the only permanent effect of a collision is a parallel displacement of their world lines.

3. The coincidence of properties (1) and (2) justifies calling the solutions (13) *topological solitons*. The difference from the non-topological KdV solitons (2) has far-reaching consequences when the mathematical results are to be applied to 'real' situations, since there will always be small violations of the assumptions on which the Bour–Enneper or the Korteweg–deVries equations are based. Whereas topological solitons remain kinks even if their kinetic energy is gradually transferred to oscillatory modes, Russell's "heap of water" will gradually be dispersed even if the conditions leading to the Boussinesq solution (2) are only mildly violated.
4. Equations (12), (15), (29) are *Lorentz-invariant* and possess particle–antiparticle solutions, in contrast to the Galilei-invariant KdV equation. Hence they may serve as models of relativistic field theories.
5. Owing to their particle-like properties, topological solitons are suitable objects of *statistical thermodynamics* and *quantum theory* [16–19]. Selected examples will be given in the next section.

Energetics, Statistical Thermodynamics, and Quantum Theory

Within the framework of the model outlined above, the *rest energy* of a single kink, E_k, may be calculated without having to evaluate the solution $\psi(z)$. Since the contributions (1) and (2) referred to in the preceding section turn out to be equal for any choice of $F(\psi)$ satisfying the requirements of Sect. 2, E_k can be expressed explicitly in terms of $F(\psi)$ as [39]

$$E_\mathrm{k} = (2\gamma_\mathrm{d}\, F_\mathrm{max})^{1/2} a_\mathrm{k} \int\limits_{\upsilon=0}^{\upsilon=1} [F(a_\mathrm{k}\upsilon)/F_\mathrm{max}]^{1/2} \mathrm{d}\upsilon. \tag{30}$$

In the special case of (12), the dimensionless integral in (30) equals π^{-1}.

Kink generation not only increases the *energy* of the system but also affects its phonon frequencies and hence its *entropy*. These changes may be calculated by considering small deviations $\varphi(x,t)$ from an exact solution of the underlying partial differential equation, e.g. the kink solution $\omega_\mathrm{k}(x,t)$ of (12). First-order perturbation theory leads to the linear equation

$$\partial^2\varphi/\partial x^2 - \partial^2\varphi/\partial t^2 - \cos\omega_\mathrm{k}\ \varphi(x,t) = 0. \tag{31}$$

Transforming $\omega_\mathrm{k}(x,t)$ to the time-independent z-frame (4) permits the ansatz

$$\varphi(z,t) = \varphi(z)\exp(\pm\, \mathrm{i}\Omega t) \tag{32}$$

which gives us a time-independent Schrödinger equation for $\varphi(z)$ and a dispersion relation for the wave-like solutions of (31). Comparison with the dispersion relation of the one-dimensional Helmholtz equation,

$$\Omega = (k^2 + 1)^{1/2}, \tag{33}$$

where k is a dimensionless wave number, allows us to calculate the vibration frequencies of a kinked string. From these the 'entropy of formation' of kinks, S_k, is obtained by evaluating the partition function of a set of harmonic oscillators [14, 20].

The dispersion relation (33) is identical with that of Yukawa's meson theory [21]. Its quantisation leads to particles with *finite rest mass*, Yukawa's U-particles, which in nuclear physics are now identified with π-mesons. In the context of kinks in dislocations the corresponding quanta are called *'heavy phonons'* [20]. In a quantum picture, the short-range interaction between kinks in the same dislocation may thus be described as due to the exchange of the heavy phonons between colliding kinks. As will be discussed in the next paragraph, their *'light'* counterparts, acoustic phonons with zero rest mass, are responsible for the long-range interaction between kinks but have virtually no effect on S_k.

Since dislocation lines are embedded in 3-dimensional elastic media and surrounded by long-range strain fields, modelling them as elastic strings may be inadequate in some circumstances. The *long-range interaction* between kinks is an important example. It arises from the deviation of a dislocation from a straight line caused by a kink. The resulting modification of the dislocation strain fields leads to a pseudo-Coulomb interaction between kinks in the same dislocation line, resulting in an interaction energy $\pm\gamma_0 a_k{}^2/2q$ between two kinks in the same, otherwise straight, dislocation separated by the distance q [41]. (γ_0 is closely related to the line tension γ_d introduced in (29), the ratio γ_d/γ_0 being of the order of magnitude unity but never less than one.) Thus, we may carry further the analogy between elementary particles and kinks by considering the quantity $(\gamma_0/2)^{1/2}a_k$ as a *pseudo-charge* of the kinks. The massless acoustic phonons of the elastic medium then play the same role as the photons (▶ light quantum) in the electrostatic interaction in elementary ▶ particle physics. The change-over from the pseudo-Coulomb interaction at large kink separations to the Yukawa-type interaction at small separations has been experimentally confirmed in detail in experiments involving the formation of kink–antikink pairs during the plastic deformation of metals [42].

In working out the *equilibrium density* of the kinks from the change of the free energy of the system, we have to take into account that, owing to the translational invariance of (4), (5), equation (31) has always a zero-frequency mode. It corresponds to the motion of kinks along the Peierls valley direction. This *Goldstone mode* contributes to the free energy as a one-dimensional gas of non-interacting particles of mass $m_k = E_k/c^2$, called *soliton gas*. Here the *'limiting speed of the energy'* c is the speed with which the heavy phonons of large wavelengths propagate along straight dislocation lines lying in Peierls valleys.

In contrast to the phonon modes just discussed, periodic soliton solutions with *finite* amplitudes such as the *breather* solution (28) cannot be quantised as harmonic oscillators. The appropriate procedure is the quantisation in terms of *action variables* [22]. We illustrate this by considering the breather solutions *at rest.* They form a one-parameter family,

$$\omega = 4\,\mathrm{arctg}\{\mathrm{tg}(I/16)\,\mathrm{Sech}[x\,\sin(I/16)]\,\sin[t\cos(I/16)]\}, \tag{34}$$

with the relationship

$$\Omega = \cos(I/16) \tag{35}$$

between the breather frequency Ω and the parameter I. The breather energy is given by [7, 14]

$$E_{\mathrm{breather}} = 2\,E_{\mathrm{k}}(1-\Omega^2)^{1/2} = 2E_{\mathrm{k}}\sin(I/2E_{\mathrm{k}}), \tag{36}$$

hence the breather frequency by

$$\Omega_{\mathrm{breather}} = \partial E_{\mathrm{breather}}/\partial I. \tag{37}$$

Equation (37) is the classical relationship between frequency, energy, and action variable [23] if we identify $2\pi I$ with the action variable of a closed orbit in classical mechanics. We may thus map the breather motion, which originated from a field-theory description, on the one-dimensional motion of a mass point [24]. This allows us (1) to quantise breather modes by the Bohr–Sommerfeld–Einstein quantisation rule, (2) to make use of the adiabatic invariance of the action variable of a periodic system subject to a perturbation that varies at most slowly during the period $2\pi/\Omega$, and (3) to treat thermally activated rate processes involving breather-type motions by means of Kramers' rate theory [17, 43]. On the other hand, the field-theory description permits the coupling between breather and phonon modes and thus the radiation damping of driven breathers to be treated quantitatively [24].

The current difficulties in formulating a theory of elementary particles intended to comprise not only the strong and the electroweak force but also gravitation are widely attributed to the fact that the established theories treat the elementary particles as point-like. Among the motivations for the development of string theories of elementary-particle physics ▶ quantum gravity is the desire to replace the conventional point particles by extended entities, the 'strings' or 'brans'. Attempts to avoid the concept of *point-like* elementary particles by considering non-linear field equations have a long history, going back at least as far as to the work of 1912 of the German physicist Gustav Mie (1868–1957) and connected, in particular, with the name of Albert Einstein (1879–1955). In the present context, the most interesting work is that of the British physicist Tony Hilton Royle Skyrme (1922–1987), summarised competently in a biography by Dalitz [19]. Skyrme came across (12) in 1958 [44] when studying the Strong Interaction between nucleons. In computer experiments with Perring [45, 46] published in 1962, i.e. well before the analogous work of Zabusky and Kruskal [5] on the KdV equation, he rediscovered the breather solution of the Bour–Enneper equation and the collision properties found analytically already in the early 1950s [7]. It took a further decade until the significance of Skyrme's ideas for elementary-particle physics was fully recognised.

In the analogy between kinks and elementary particles, *topological solitons* do play the role of '*extended particles*'. Their 'world lines' should therefore be replaced by '*world tubes*' with a diameter of the order of magnitude of the kink width w_{K} [cf. (9)]. Divergences that may appear in approximate expressions can be avoided by recourse to more fundamental descriptions based on, say, *atomic* models of crystals. Since the future of the string theories is still open, it is too early to speculate to what extent the soliton properties of the Bour–Enneper equation might help in visualising the outcome.

Primary Literature

1. S. F. Trullinger, V.E. Zakharov, V. I. Pokrovsky (eds): *Solitons* (North-Holland, Amsterdam 1986)
2. J. S. Russell: *Report on Waves*. Fourteenth Meeting of the British Association for the Advancement of Science (John Murray, London 1844, pp. 311–390)
3. J. V. Boussinesq: *Théorie des ondes et des remus qui se propagent le long d'un canal rectangulaire horizontal, en communiquant au liquide contenu dans le canal des vitesses sensiblement pareilles de la surface au fond*. J. Math. Pures Appl. (2) **17**, 55 (1872)
4. D. J. Korteweg, G. de Vries, Philos. Mag. **39**, 422 (1895)
5. N. J. Zabusky, M. D. Kruskal, Phys. Rev. Letters **15**, 240 (1965)
6. M. Heyerhoff: *Die frühe Geschichte der Solitonentheorie*. Dr. rer. nat. thesis (Universität Greifswald, Greifswald 1997)
7. A. Seeger, H. Donth, A. Kochendörfer, Z. Physik **134**, 173 (1953)
8. L. de Broglie: *Sur la définition générale de la correspondance entre onde et mouvement*. C. R. Acad. Sci. Paris **197**, 39 (1924) – For a more detailed account see de Broglie's thesis in Ann. Physique **3**, 22 (1925). E. Schrödinger (Ann. d. Physik **79**, 734 (1926)) acknowledges his indebtedness to this paper
9. G. Wentzel: *Einführung in die Theorie der Wellenfelder* (Franz Deutike, Vienna 1943)
10. L. Bianchi: *Vorlesungen über Differentialgeometrie* (2nd ed., B. G. Teubner, Leipzig 1910)
11. L. P. Eisenhart: *An Introduction to Differential Geometry* (Princeton University Press, Princeton 1947)
12. A. Seeger: *Theorie der Gitterfehlstellen, Part C, in: Encyclopedia of Physics*, ed. by S. Flügge, Vol. VII/1 (Springer, Berlin etc. 1955, pp. 476–636)
13. A. Seeger: *Solitons in Crystals*, in: Continuum Models of Discrete Systems (CMDS3, SM Study No.15), ed. by E. Kröner and K.-H. Anthony (University of Waterloo Press 1980, p. 253–327)
14. A. Seeger, P. Schiller: *Kinks in Dislocation Lines and Their Effects on the Internal Friction in Crystals*, in: Physical Acoustics, Vol. IIIA, ed. by W. P. Mason, (Academic Press, New York 1966, Chap. 8)
15. A. V. Bäcklund: *Om ytor med konstant negativ krökning* (Lunds Universitets Års-skrift XIX, IV, 1882–83, pp. 1–48)
16. M. A. Collins: *Solitons in Chemical Physics*, in: Advances in Chemical Physics, Vol. LIII, ed. by L Prigogine and St. A. Rice, J. Wiley, New York etc.1983, pp. 225–339
17. A. Seeger, *Solitons and Statistical Thermodynamics*, in: Trends in Applications of Pure Mathematics to Mechanics (Lecture Notes in Physics Vol. 249), ed. by E. Kröner and K. Kirchgässner (Springer, Berlin etc. 1986, pp. 114–155)
18. O. M. Braun, Y. S. Kivshar: *The Frenkel–Kontorova Model* (Springer, Berlin etc. 2004)
19. R. N. Dalitz: *An Outline of Life and Work of Tony Hilton Royle Skyrme*, Intern. J. Mod. Physics A3, 2719 (1988)
20. A. Seeger: *Structure and Diffusion of Kinks in Monatomic Crystals*, in: Dislocation 1984, ed. by P. Veyssière, L. Kubin, J. Castaing, Ed. CNRS (Paris 1984, pp. 141–178)

21. H. Yukawa, Proc. Phys.-Math. Soc. Japan 17, 488 (1935)
22. A. Einstein, Verh. d. Deutsch. Physikal. Gesellsch. 19, 82 (1917)
23. V. I. Arnol'd: *Mathematical Methods in Classical Mechanics* (Springer, New York etc. 1978)
24. R. Döttling, J. Eßlinger, W. Lay, A. Seeger: *Nonlinear Coherent Structures* (Lecture Notes in Physics Vol 353), ed. by M. Barthes and J. Léon, Springer, Berlin etc. 1990, p. 193

Secondary Literature

25. A. S. Fokas, V. E. Zakharov, in: *Important Developments in Soliton Theory*, ed. by A. S. Fokas and V. E. Zakharov (Springer, Berlin 1993, p. 1)
26. O. Klein, Z. Physik **37**, 895 (1926)
27. E. Schrödinger, Ann. d. Physik **79**, 489 (1926)
28. V. A. Fock, Z. Physik **38**, 242 (1926); **39**, 226 (1926)
29. W. Gordon, Z. Physik **40**, 117 (1926)
30. J. Kudar, Ann. d. Physik **81** 632 (1926). – János Kudar (b. 1902 in Ráckeve, Hungary) is less known than the authors of [26–29]. A short biography may be found in R. Rürup: *Schicksale und Karrieren* (Wallstein Verlag, Göttingen 2008, p. 246)
31. W. Shockley, Trans. AIME **194**, 829 (1952)
32. A. Enneper, Nachr. d. königl. Gesellschaft der Wissenschaften u. d. Georg August Universität Göttingen 1868, 252; 1870, 493
33. E. Bour, Journ. de l' École Imperial Polytechnique **19**, Cahier 39, 1 (1862)
34. Ph. A. Martin, Acta Phys. Austriaca, Suppl. XXIII, 157 (1981)
35. R. Steuerwald: *Über Ennepersche Flächen und Bäcklundsche Transformation*, Abhandl.d. Bayerischen Akad. d. Wissenschaften, math.-naturwissenschaftl. Abtlg, Neue Folge, Heft 40, München 1936
36. A. Seeger, Z. Naturforschg 8a, 47 (1953)
37. A. Seeger, Z. Wesolowski, Int. J. Eng. Sci. 19, 1535 (1981)
38. A. Seeger, Proc. R. Soc. London A 371, 173 (1980)
39. A. Seeger, Z. Metallkde 93, 760 (2002)
40. R. Peierls, Proc. Phys. Soc. (London) 52, 34 (1940)
41. H. O. K. Kirchner, Acta Phys. Austriaca 48, 115 (1978)
42. L. Hollang, M. Hommel, A. Seeger, phys. stat. sol. (a) 160, 329 (1997)
43. H. A. Kramers, Physica 7, 284 (1940)
44. T. H. R. Skyrme, Proc. Roy Soc. A 247, 260 (1958)
45. J. K. Perring, T. H. R. Skyrme, Nucl. Phys. 31, 550 (1962)
46. T. H. R. Skyrme, Nucl. Phys. 31, 556 (1962)

Sommerfeld School

Michael Eckert

The development of scientific specialties is often related to scientific schools-from a historical perspective as much as from an epistemological vantage point. Quantum mechanics is not exceptional in this regard; its emergence was to a large extent a product of the scientific schools of Niels Bohr in Copenhagen, Max Born in

Göttingen and Arnold Sommerfeld in Munich. A school is primarily a locally defined group under the influence of a charismatic teacher. Often this influence results in a common way of thinking, so that the school becomes also a thought collective in an epistemological sense. Not so, however, for Sommerfeld's school. From an epistemological perspective, Sommerfeld pupils like Peter Debye and Werner Heisenberg, for example, hardly belong to a common thought collective. Nevertheless, both are prominent representatives of Sommerfeld's school and contributed decisively to quantum theory.

Arnold Sommerfeld (1868–1951) began his career as a mathematician in Königsberg, Göttingen, Clausthal and Aachen, before he was appointed in 1906 professor of theoretical physics at Munich. Before the First World War, there were few ordinary chairs and institutes for this specialty. Sommerfeld was keen to demonstrate what mathematics is able to accomplish in physics. It was his declared intent to turn his Munich institute into a "nursery" of theoretical physics. His mathematical approach towards physics was more versatile than that of other theorists. Sommerfeld's "physical mathematics," as his own teacher, Felix Klein, called it, allowed him to encompass a broad range of problems. Many theory-minded physicists were attracted to the new center. Albert Einstein and Paul Ehrenfest, for example, expressed a desire to study under Sommerfeld, although these hopes never materialized. The first generation of Sommerfeld pupils to acquire their doctoral degrees in the Munich "nursery" included Peter Debye (1908), Ludwig Hopf (1909), Wilhelm Lenz (1911), Peter Paul Ewald (1912) and Alfred Landé (1914). Their theses dealt with the theory of diffraction, turbulence, wireless telegraphy, crystal optics and quantum theory.

In 1915, Sommerfeld made ► Bohr's atomic model a major focus of research at his institute. His advanced students were absent on war duty, but the school spirit was kept alive by the exchange of letters. Sommerfeld collaborated with Paul Epstein, who because of his Russian nationality was under police surveillance in Munich but was allowed to work at Sommerfeld's institute. The Tübingen spectroscopist Friedrich Paschen also provided him with precise spectroscopic data. With their help Sommerfeld extended Bohr's model to a theory able to explain the ► fine-structure of atomic spectra and the ► Stark effect. This early success lent credit to Bohr's model at a time when it was still being regarded with skepticism. In 1916, Adalbert Rubinowicz arrived from Poland in order to become Sommerfeld's assistant. Rubinowicz solved the problem of how to select from among the multitude of electronic transitions between atomic orbits those which are actually observed. His "► selection rules" were based on the conservation laws of energy and angular momentum. Sommerfeld considered Rubinowicz's approach superior to Bohr's, who based the same results on the ► correspondence principle. In 1918, Sommerfeld began an extensive correspondence with Manne Siegbahn about x-ray spectra. Walther Kossel from the Technical University in Munich became Sommerfeld's closest collaborator on x-ray ► spectroscopy, by which Sommerfeld was able to extend the range of his theory to comprehend "atomic structure and spectral lines." Thus he entitled a book on this subject in 1919, which was soon regarded as the "bible of atomic physics."

The heyday of Sommerfeld's school was during the early 1920s with the arrival of his two prodigies, Wolfgang Pauli and Werner Heisenberg. Pauli, then still a student, was entrusted with a review article on the theory of relativity for the *Enzyklopädie der Mathematischen Wissenschaften*. In 1921, Pauli finished his study with a semi-classical theory of the ionized hydrogen molecule (one electron orbiting two centers). Heisenberg, too, was in his beginning semesters when Sommerfeld gave him a chance to prove his mettle. In 1921, he chose him as his close collaborator for analyzing recent spectroscopic data on the ► Zeeman effect. Heisenberg interpreted these data in terms of a "core model," an attempt to explain the (anomalous) Zeeman effect with half-integer ► quantum numbers. A few years later, some features of Heisenberg's model could be transferred to a new quantum feature, the half-integral ► spin. Despite these early efforts, which made Heisenberg a well-known name within the then still small community of quantum theorists, Sommerfeld posed him another challenge as the topic for Heisenberg's doctoral thesis: the theory of turbulence. Both Heisenberg and Pauli continued their promising careers under the tutelage of Max Born in Göttingen and Niels Bohr in Copenhagen. Throughout the 1920s and early 1930s, advanced students frequently traveled from one center to another, so it is not possible to trace their achievements to any specific school.

The advent of quantum mechanics by the mid 1920s stirred debates among theorists about the basic principles of physics. Unlike his master pupil Heisenberg, Sommerfeld contributed little to these debates. For him, quantum mechanics served primarily as an opportunity for solving heretofore inaccessible problems rather than for reflecting on the foundations of physics. In 1927, Sommerfeld paved the way for a quantum mechanical solid-state theory by applying the new ► Fermi-Dirac statistics to the classical free electron gas model of metals. Subsequently, he coauthored with his former student, Hans Bethe, a comprehensive article on the electron theory of metals for the *Handbuch der Physik*. Throughout the decade between Heisenberg's and Schrödinger's pioneering publications in 1926 and Sommerfeld's retirement in 1935, Sommerfeld's institute was an attractive center for applications of quantum mechanics. Some of his doctoral students and foreign research fellows, who learned from Sommerfeld's lectures and seminars how to apply Schrödinger's ► wave mechanics, became famous for their contributions in quite different areas, ranging from molecular and solid-state physics to astrophysics (such as Herbert Fröhlich, Walter Heitler, Linus Pauling, Isidore I. Rabi, Albrecht Unsöld, Heinrich Welker, to mention only a few representative names). The legacy of Sommerfeld's school becomes apparent from his two volumes on *Atomic Structure and Spectral Lines*, as far as quantum theory is concerned, and otherwise from the six volumes of his lectures on theoretical physics. It is by versatility rather than by any focus on a particular theme that Sommerfeld and his school has exerted a lasting influence.

Specific Heats

Clayton Gearhart

The equipartition theorem states that the average energy associated with each separable, quadratic term in the Hamiltonian results in a thermal energy of $1/2\ RT$ per mole, where R is the gas constant and T the absolute temperature. This theorem, which emerged early in the history of kinetic theory in the nineteenth century, was quickly found to be in sharp disagreement with experiment, particularly for gases.

Thus, for a monatomic ideal gas with three translational and three rotational degrees of freedom, the equipartition theorem predicts that the thermal energy per mole is 3 RT, and the specific heat at constant volume C_V is $3R$. This result is often expressed in terms of γ, the ratio of the specific heats at constant pressure and volume. For this case, $\gamma = C_P/C_V = 4/3$, since for one mole of an ideal gas, $C_P = C_V + R$. The same result obtains for a diatomic gas if the two gas atoms are rigidly connected. If they are instead connected by a massless spring (with quadratic terms in both kinetic and potential energy), one finds $C_V = 4R$, and $\gamma = 5/4$.

Experiments told a different story. Experiments on monatomic gases over a wide range of temperatures consistently found $C_V = 3/2R$, or $\gamma = 5/3$, corresponding to three translational degrees of freedom. Apparently, monatomic gases did not rotate. Experiments at room temperature on common diatomic gases such as oxygen and nitrogen yielded $\gamma = 7/5$, corresponding to three translational and two rotational degrees of freedom. One rotational degree of freedom was missing; and apparently the molecules did not vibrate. At higher temperatures, however, the specific heat steadily increased, suggesting an inexplicable *gradual* onset of additional degrees of freedom. To make matters worse, atomic and molecular spectra hinted at additional internal degrees of freedom that did not contribute to specific heats.

Nineteenth-century physicists were perplexed and alarmed by these discrepancies. James Clerk Maxwell (1831–1879) in 1875 said that they constituted "the greatest difficulty yet encountered by the molecular theory." Lord Kelvin (1824–1907) in 1901 considered them one of the "two clouds" hanging over nineteenth-century physics. Ludwig Boltzmann (1844–1906) in his *Lectures on Gas Theory* argued that the energy of rotation about an axis of symmetry would not change in collisions, or would at best change very slowly. And Max Planck, in the preface to his 1897 thermodynamics text, spoke of "Obstacles, at present insurmountable" standing in the way of kinetic theory.

The situation with solids was more promising. As early as 1818, the French scientists Pierre Louis Dulong (1785–1838) and Alexis Thérèse Petit (1791–1820) showed that the specific heats of most solids were about 6 cal mole^{-1}K^{-1}, or $3R$, a value that, as Boltzmann pointed out, agreed nicely with the equipartition law. The few exceptions occasioned little concern: The specific heat of diamond, for example, was about 1.5 cal mole^{-1}K^{-1} at room temperatures, but fell to 0.76 at 220 K, and approached the equipartition value only at temperatures well above 1000 K.

This state of affairs changed as a result of two developments. First, the ability to liquefy gases such as oxygen and nitrogen in the late 1870s, and hydrogen in the late 1890s, permitted scientists to measure the specific heats of matter at low temperatures. This advance was in large part due to Sir James Dewar (1842–1923) in England, and later, to Heike Kamerlingh Onnes (1853–1926) in Leiden and Walther Nernst (1864–1941) in Germany. Second, the development of quantum theory in the early years of the twentieth century showed a way out of the dilemmas posed by the equipartition theorem (► Quantum theory, early period, ► Black-body radiation).

Thus in 1907, Albert Einstein used Max Planck's quantized resonators to predict that the specific heats of solids should fall off from the value $3R$ at room temperature to zero at low temperatures – the equipartition theorem, which assumes continuous energies, no longer holds in quantum theory. For confirmation, Einstein could point only to the specific heat of diamond. But over the next several years his theory was brilliantly confirmed by the experiments on the specific heats of solids conducted by Walther Nernst and his students in Berlin. They developed new and innovative experimental techniques, including platinum thermometers and vacuum calorimetry, as they learned to measure specific heats accurately over a wide range of temperatures down to the temperature of liquid hydrogen. By 1910, Nernst and his students had measured the specific heats of numerous solids, and shown that they did indeed approach zero at low temperatures, much as predicted by Einstein's theory. More quantitatively accurate theories were soon developed by Max Born (1882–1970) and Theodore von Kármán (1881–1963), and by Peter Debye (1884–1966).

Nernst also took the lead in measuring the specific heats of gases. In 1911, he noted that quantum theory might well be the key to understanding the discrepancies between the equipartition theorem and the measured specific heats. He proposed hydrogen as a particularly promising candidate for investigation, and the following year his assistant, Arnold Eucken (1884–1950), used a vacuum calorimeter to show that the specific heat of hydrogen gas at constant volume fell from just under the equipartition value of $5/2\ R$ at room temperature to $3/2\ R$ at about 40 K. The rotational degrees of freedom had frozen out due to quantum effects, much as Nernst had predicted.

Over the next 15 years, numerous theorists attempted to find quantitatively accurate theories for the specific heat of hydrogen. These attempts were notably unsuccessful until the development of modern quantum mechanics beginning in 1925. Finally, in 1927, the American physicist David Dennison (1900–1976) showed how to use the quantum mechanical theory of indistinguishable particles to find an accurate description of the specific heat of hydrogen.

It is remarkable that so commonplace a quantity as the specific heat should have played such a central role in early quantum theory. Moreover, the experimental and theoretical study of specific heats played an important part in the physics, chemistry, and technology of the nineteenth and twentieth centuries in ways that extend far beyond quantum theory, although a full treatment is beyond the scope of this essay.

Primary Literature

1. L. Boltzmann: *Vorlesungen über Gastheorie*, 2 vols, 1896/1898 (Barth, Leipzig 1896; reprinted in Roman U. Sexl, ed., Gesamtausgabe, Band 1, Vieweg, Braunschweig 1981); translated in Stephen G. Brush: *Lectures on Gas Theory* (University of California Press, Berkeley 1964; reprinted Dover, New York 1995)
2. W. Nernst: *Theoretical Chemistry* (MacMillan, London 1911). There are many editions of this text, both in German and in English translation, from the first edition (1893) to the 16th (1926)
3. J. R. Partington, W. G. Shilling: *The specific heats of gases* (E. Benn, London 1924)

Secondary Literature

4. S. G. Brush: *The Kind of Motion We Call Heat*, 2 vols. (North Holland, New York 1976)
5. D. Cahan: *An Institute for an Empire*: The Physkalisch-Technische Reichsanstalt, *1871–1918* (Cambridge University Press, Cambridge 1989)
6. M. J. Klein: Einstein, Specific Heats, and the Early Quantum Theory. Science **14**, 173–180 (1965)
7. K. Mendelssohn: *The Quest for Absolute Zero* (Taylor & Francis, London 1977)

Spectral Decomposition

See ► Density operator; Ignorance interpretation; Measurement theory; Objectification; Operator; Probabilistic Interpretation; Propensities in Quantum Mechanics; Self-adjoint operator; Wave Mechanics.

Spectroscopy

S

Klaus Hentschel

Spectroscopic data, together with ► scattering experiments, were probably the most important experimental input to the development of ► quantum theory and early quantum mechanics. Not a discipline in its own right (see [6]), spectroscopy was practiced within chemistry, optics and astrophysics and has a history extending far back. Discontinuous features in the spectra of sunlight and from the flames of various substances were the subject of intense study throughout the nineteenth century. As early as 1815, the Munich optician Joseph Fraunhofer (1787–1826) published a detailed map of the solar spectrum exhibiting about 350 dark lines. He realized these dark lines could serve as useful markers for specific colors in the otherwise continuous solar spectrum. His specific application was for gauging precision measurements of the refractive indices of various types of glass being manufactured at the glass-works under his purview (see [7]). Other maps of wider range and greater detail followed (see [8]). In 1859, Gustav Robert Kirchhoff (1824–1887) and Robert Wilhelm Bunsen (1811–1899) discovered the exact coincidence of dark absorption lines in the solar spectrum with bright emission lines in the spectra of various chemical elements heated to incandescence. Only then could the two very different types of spectra be correctly interpreted as due to absorption and emission. The Bunsen-burner flame was colorless, so a correlation between the presence of certain spectrum lines with specific chemical elements in samples of unknown constitution became feasible: Spectrum analysis was born and quickly matured into one of the most active research fields of the latter half of the nineteenth century. Detailed tables were compiled: Heinrich Kayser (1853–1940) and Henry Augustus Rowland (1848–1901), for instance, catalogued tens of thousands of spectrum lines and related them to the known chemical elements. Roughly a dozen new elements (e.g., caesium, rubidium and indium) were discovered by investigating the tell-tale spectrum lines not yet correlated with any element. Detailed examinations of the spectra of various gases in a discharge tube revealed groups of lines of similar appearance nonrandomly distributed over the spectrum, which came to be known as series and bands (Fig. 1).

Sharp, principal and diffuse series were distinguished (whence the later S, P and D designations) in the hydrogen spectrum, in particular. Many of these series spectra were also detected in the spectrum of the sun and other stars. Employing geometric analogy, the Basel mathematics teacher Johann Jakob Balmer (1825–1898) hit upon a formula for the wavelength λ of these series lines as a function of an integer variable n: $\lambda_n = h \cdot n^2/(n^2 - 4)$. He realized that other series were possible if $4 = 2^2$ is replaced with other squares in the denominator. Such series were later identified by Friedrich Paschen (1908), Frederick S. Brackett (1922) and August Herman Pfund (1924) in the infrared, and by Theodore Lyman (1914) in the ultraviolet. An inquiry into the relation between these various series lines led Walter Ritz (1878–1909) to suggest the so-called combination law, according to which the difference between any two series-line frequencies yields another series line. But deeper understanding of these various pieces of the puzzle had to await the rise of Niels Bohr's ► atomic model.

"As soon as I saw Balmer's formula," Niels Bohr (1885–1962) later said, "the whole thing was clear to me." The Danish spectroscopist Hans Marius Hansen

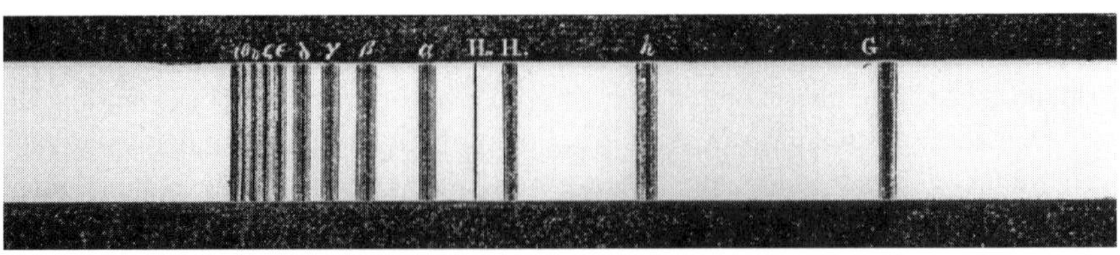

Fig. 1 The first known series of hydrogen as depicted by William Huggins (1880)

(1886–1956) had just told him about the Balmer series of hydrogen in February 1913, freshly returned from a postdoc stay at Göttingen, where he had been conducting experiments with the ▶ Zeeman effect on lithium together with Woldemar Voigt (1850–1919). Bohr made a last-minute revision to his paper for the *Philosophical Magazine* to start with a discussion of emission and absorption lines in the hydrogen spectrum and a derivation of the Rydberg constant R according to his new atomic model (see, e.g., [11]). From ▶ Bohr's atom model, Bohr had already derived energy E as a function of nucleus mass m and charge Ze(with n a natural number):

$$E_n = R \cdot mZ^2e^4/n^2$$

In order to obtain the formula for the Balmer spectrum-line frequencies v:

$$v = v_0\frac{m^2 - 4}{m^2} = v_0 - \frac{4v_0}{m^2},$$

Bohr just had to apply Einstein's assumption that $E = hv$ and:

$$v \sim E_1 - E_2 = \text{const}\left(\frac{1}{n_1^2} - \frac{1}{n_2^2}\right)$$

The frequencies of series lines were thus *not* directly correlated with the oscillatory motion of ▶ electrons around the nucleus, as had always been assumed; they were rather related to *differences* between the initial and final energy level E. The emission or absorption of a spectral line was equivalent to a ▶ quantum jump by an electron between stable orbits at different energy levels. This reinterpretation of spectra so comprehensible to us today was a veritable *Gestalt switch* as defined by Thomas Kuhn (1922–1996). "When [Einstein] heard this he was extremely astonished and told me: 'Then the frequency of the light does not depend at all on the frequency of the electron... this is an enormous achievement. The theory of Bohr must then be right.' " (From G. Hevesy's letter to Bohr, 23 Sep. 1913 [1, vol. 2, p. 533]).

Within a matter of years, Bohr's considerations totally transformed spectroscopy. Instead of plotting spectrum maps, spectroscopists reinterpreted all spectrum lines as transitions between different energy levels and constructed term diagrams (like Fig. 2). Each spectrum line provided a clue to the existing stable energy levels of electron orbits around the nucleus of a given element and the allowed transitions between them. In 1913 Henry Moseley (1887–1915) managed to explain series regularities in X-ray spectra. He showed that their frequencies ν were also dependent on nuclear charge Z, but not as $\nu \sim Z^2$ as in the Balmer series, but $\sim (Z-1)^2$. This strict regularity led to the discovery of several new chemical elements: technetium, promethium and rhenium. In the following year Bohr realized that his formula for the hydrogen series lines could also be adapted to the helium spectrum if $Z = 1$ is replaced by $Z = 2$. Thus the long-known Pickering spectrum series in certain stellar spectra was explained and soon also observed in a discharge tube filled with

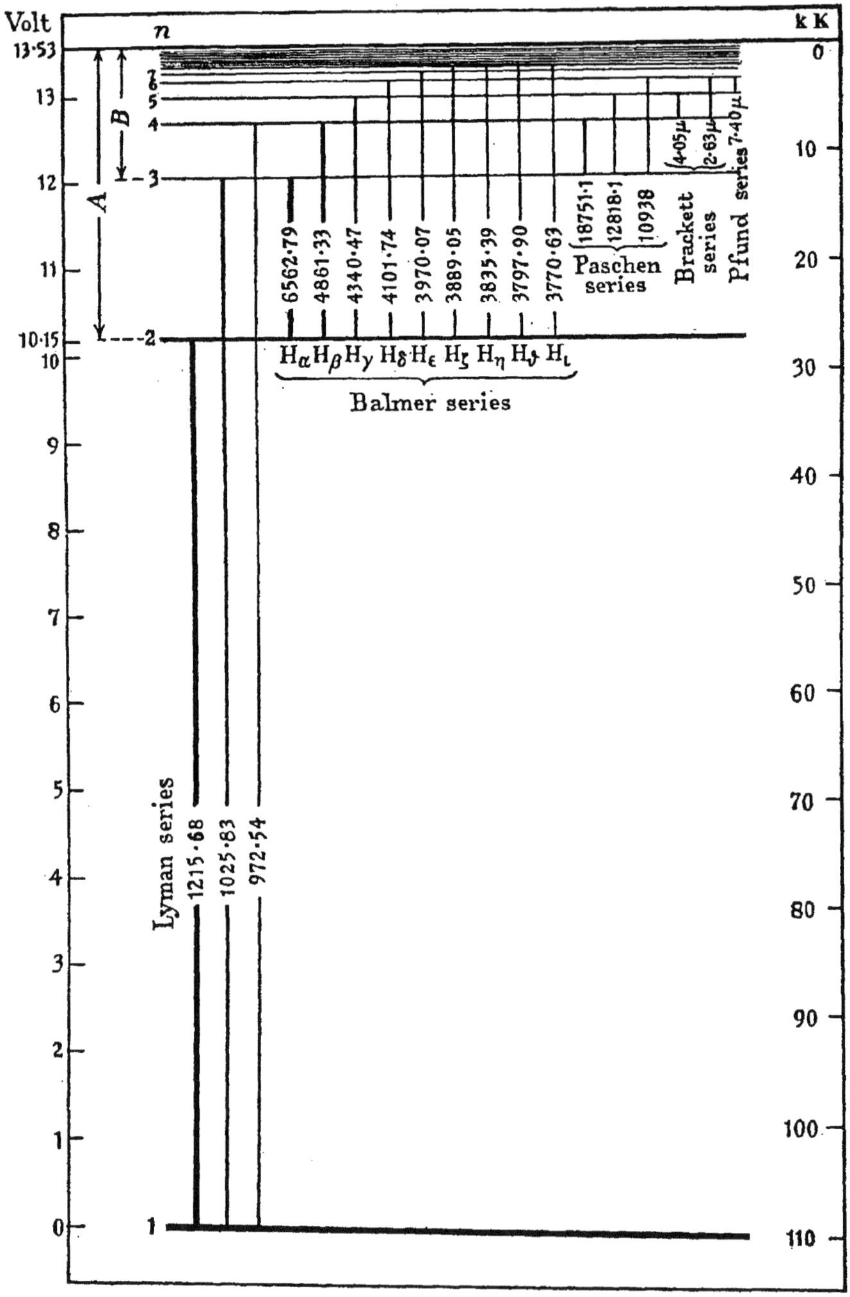

Fig. 2 Energy levels of the hydrogen atom with the series lines (Candler 1937, p. 7)

pure helium gas. By 1915, Bohr himself and others also succeeded in explaining the observed line splitting of atoms radiating in magnetic and electric fields, the ► Zeeman and ► Stark effects. Since, in general, there were fewer spectrum lines observed than were combinatorically possible, special ► selection rules were set for transitions. Only with the advent of the concept of ► spin in late 1925 were these phenomenological rules better understood as arising from angular momentum conservation, with electrons being spin 1/2 particles and the ► light quantum (or photon) carrying spin 1. A merely descriptive spectroscopy was thus replaced by explanatory hypotheses based on ► Bohr's atomic model. Quantum mechanics as formulated in 1925/26 yielded formulas for the spectral series and other regularities fully equivalent to the semi-classical Bohr–Sommerfeld atomic model in first order, and only slightly differing in higher orders of perturbation theory (see, e.g., [13]). Spectroscopic data were again crucial in its development. See also ► Spin echo.

Primary Literature

1. N. Bohr: *Collected Works* (Amsterdam, Vol. **2** 1981 and Vol. **4** 1977)
2. J.J. Balmer: Notiz über die Spectrallinien des Wasserstoffs. Verhandlungen der Naturforschenden Gesellschaft Basel (1885), 548–560, 750–752 and Annalen der Physik (3rd ser.) **25**, 80–87
3. J.J. Balmer: A new formula for the wave-lengths of spectral lines. Astrophysical Journal **5**, 199–209 (1897), (with the geometric analogy that had originally guided him in his search for the formula)
4. W. Huggins: On the photographic spectra of stars. Philosophical Transactions of the Royal Society **171**, 669–690 and plate 33 (1880); summary in Nature **21**, 269–270 (1880)
5. A. Sommerfeld: *Atombau und Spektrallinien* (Braunschweig: Vieweg 1919); Engl. transl. *Atomic Structure and Spectral Lines* (London: Methuen 1923)

Secondary Literature

6. K. Hentschel: Spectroscopy or spectroscopies? Nuncius **17/2**, 589–614 (2003)
7. M. Jackson: *Spectrum of Belief – Joseph von Fraunhofer and the Craft of Precision Optics* (Cambridge, MA: MIT 2000)
8. K. Hentschel: *Mapping the Spectrum. Techniques of Visual Representation in Research and Teaching* (Oxford University Press 2002)
9. J.C.D. Brand: *Lines of Light. The Sources of Dispersive Spectroscopy*, 1800–1930 (Luxembourg: Gordon and Breach Publishers 1995)
10. C. Candler: *Atomic Spectra and the Vector Model* (Princeton: Van Nostrand 1937)
11. J. Heilbron, T.S. Kuhn: The genesis of the Bohr atom. Historical Studies in the Physical Sciences **1**, 211–290 (1969)
12. G. Herzberg: *Atomic Spectra and Atomic Structure* (New York: Prentice-Hall 1937, 2nd ed. New York: Dover Publications 1944)
13. M. Jammer: *The Conceptual Development of Quantum Mechanics* (New York: McGraw-Hill 1966)

Spin

Klaus Hentschel

According to quantum mechanics, spin—the intrinsic angular momentum of an electron, nucleus, or elementary particle at rest—is a decidedly nonclassical concept. The ► spin statistics theorem of ► quantum statistics distinguishes bosons and fermions obeying ► Bose–Einstein statistics or ► Fermi–Dirac statistics, respectively, depending on whether the particle's spin is an even or odd multiple of $\hbar/2$, with $\hbar = h/2\pi$ (h being ► Planck's constant). The convoluted history of the concept of spin nevertheless reaches back into the final ► crisis period of the old ► quantum theory, linked to the semi-classical ► atomic model by Niels Bohr (1885–1962), Arnold Sommerfeld (1868–1951) and their collaborators (the ► Sommerfeld school).

In the early 1920s, precise experimental data from ► spectroscopy, particularly regarding the anomalous ► Zeeman effect, forced researchers to deviate from the rule imposed by the Bohr–Sommerfeld atomic model that all ► quantum numbers must be integers. Experiments by Miguel A. Catalán (1894–1957) made evident that many spectrum lines were finely split by magnetic fields into so-called multiplets with $2l + 1$ equidistant components, l being the azimuthal quantum number. These multiplets were thus described by a new magnetic quantum number m, and the rule $|m| \leqslant l$ stating that permissible states have to be between $+m, m-1, m-2\ldots 0, -1, -2\ldots$ and $-m$. This yields $2m + 1$ different states, a perfect fit with the observed $(2m + 1)$-multiplet. Semi-classically, m could be interpreted as the component of l in the direction of the exterior magnetic field (both in units of $h/2\pi$), so the orientation of the electron orbits relative to the magnetic field was space quantized—only a few discrete orientations were permitted. Likewise, transitions between states had to be restricted to $\Delta m = \pm 1, 0$ by a superimposed ► selection rule. What about doublet lines with only two visible components? Applying the standard multiplet rule would lead directly to $l = 1/2$, implying $m = \pm 1/2$, hence half-integer quantum numbers. Alfred Landé (1888–1976) was the first to dare to operate with half-integer ► quantum numbers in search of an explanation for doublets in alkali spectra and other anomalies in the ► Zeeman effect [see 14, 15].

But how to interpret these strange half-integral quantum numbers? In 1922, the young Werner Heisenberg (1901–1976), then still in the clutches of the ► Sommerfeld school in Munich, speculated that this half-integer value would result from a time-average over an integer multiple of a quantized angular momentum, contributing 50% to the outer shell and 50% to the atomic core [1]. Heisenberg & Sommerfeld [2] also tried to explain the anomalous Zeeman effect in terms of a magnetic interaction of the outermost bound electron (the so-called *Leuchtelektron*) with the magnetic momentum of the stronger-bound ► electrons closer to the atomic core (the *Rumpfelektronen*). However, this model would lead one to expect a strong correlation between ► Landé's g factors and the atomic charge number Z of the respective element, which was at odds with observation.

Another young student of Sommerfeld, Wolfgang Pauli (1900–1958), devised a different, equally bold approach to explain such doublet structures. Pauli [3] concluded that the "Rumpf"-electrons of the closed shell should have no effective angular momentum at all. Instead, he imposed a mysterious "mechanically indescribable ambiguity" on the outermost electron ("a characteristic ambiguity of the "Leuchtelektron" not describable by classical theory") as a hypothetical alternative explanation to the doublet structure. This ambiguity also led to two possible orientations for the outermost electron relative to the external magnetic field. This in turn yielded the doublet splitting of alkali spectra and similar atoms.

In January 1925 Pauli first expressed this mechanically indescribable ambiguity as a new quantum number $\mu = \pm 1/2$ (for doublets), and $\Delta\mu = 0$ or ± 1 as a new ► selection rule. Each electron was thus described by a set of four ► quantum numbers n, l, m and μ (sometimes alternatively called n, l, j and s). The electron configuration of each atom was constructed of shells, starting from the lowest possible energy level. Pauli's new constraint imposed on the shell structure that no two electrons of an atom have all the four quantum numbers in common: the Pauli principle (or ► exclusion principle):

> "There can never be two or more equivalent electrons in the atom in which the values of all [four] quantum numbers... concur within a strong field... If in the atom there is an electron for which these quantum numbers... have specific values, then this state is occupied." [4, p. 776; cf. 17]

In this way Pauli succeeded in deriving the usual period lengths of 2, 8, 18, 32,... from the periodic table. The arrangement of the periodic system of the elements thus seemed to make a little more sense again, at least as far as the main groups were concerned. But it came at the cost of a "classically indescribable kind of ambiguity"; and Pauli's prohibition of any duplication among the quantum numbers occupying a given state, was no better justifiable according to classical theory and only understood within the context of the ► Fermi–Dirac statistics of later ► quantum mechanics.

So we are already very close to the discovery of electron spin, and yet still so far away. Pauli refused to address the problem of *how* this ambiguity would be comprehended within the classical model (e.g., as an intrinsic angular momentum): He argued that this feature was "classically indescribable" because the electron's rotational velocity around its own axis was too large (according to Pauli it was greater than c). Instead Pauli, godchild of the positivist Ernst Mach, held a very instrumentalistic conception—he just introduced one more model into the discussion that he himself did not quite believe in:

> "It scarcely needs emphasis that further development of the theory must show to what extent such a conception hits the mark and whether it can be elaborated further. This interpretation faces major obstacles, particularly with regard to its natural connection with the correspondence principle. Furthermore, there is surely much correct about the conventional view, which reflects certain features of the phenomena better than the one tentatively suggested

here. In a following note it will be shown, on the other hand, however, that the latter interpretation proves to be more physically useful in describing other aspects of the phenomena. Perhaps the final solution to the problems set forth here will lie in the direction of a middle road between these two interpretations." [5, correspondence, early 1925]

The constantly growing set of quantum numbers and phenomenologically determined criteria like Sommerfeld's ► selection rules and ► Landé's g-factors led to acceptable agreement between theory and experiment. Nevertheless it left an unpleasant aftertaste of mere *ad hoc* description without any deeper understanding of the reasons behind all these rules. Physicists described their predicament humorously as "term zoology" and "Zeeman botany". Sommerfeld spoke of "number mysteries"; Runge ironically referred to "witches times-tables of quantum physics".

But not everyone thought like Pauli. In early 1925, Ralph de L. Kronig (1904–1995) concluded from a letter by Pauli that the electron must have an intrinsic angular momentum in order to explain the peculiar ambiguity not describable according to classical conceptions. Pauli repudiated this idea off-hand on the following arguments:

1. A factor 2 was missing between the calculated doublet splitting and observational data.
2. The magnetic moment of an atomic nucleus was too small.
3. The rotation velocity of such a spinning electron was incredibly high. Calculated on the basis of classical assumptions, it yielded superluminal velocities along the electron's periphery ► superluminal communication.

Completely unaware of this exchange which prevented Kronig from pursuing this idea further, two young postdocs in Leyden, George Eugene Uhlenbeck (1900–1988) and Samuel Abraham Goudsmit (1902–1978), took as an explanation of the anomalous Zeeman effect the assumption that

(1) Each individual electron bears a magnetic moment $\boldsymbol{M}$ that can be generated from an intrinsic rotation with angular momentum (spin $\boldsymbol{S}$)

$$\boldsymbol{M} = 2 \cdot \frac{e}{2mc} \boldsymbol{S}$$

(2) Quantitatively, this magnetic moment is twice the amount expected in a naive semi-classical model

Thus the magnetic and mechanical moment should differ by a factor 2 from the value valid for an atomic system with a point charge of $e/2mc$, that is, the quotient of the Bohr magneton. The resulting fact that the total angular momentum of $\boldsymbol{J}$ and μ (total magnetic moment) were not parallel explained why the distances between various magnetic levels in the anomalous Zeeman effect differed in size depending on the term (► vector model) (Fig. 1).

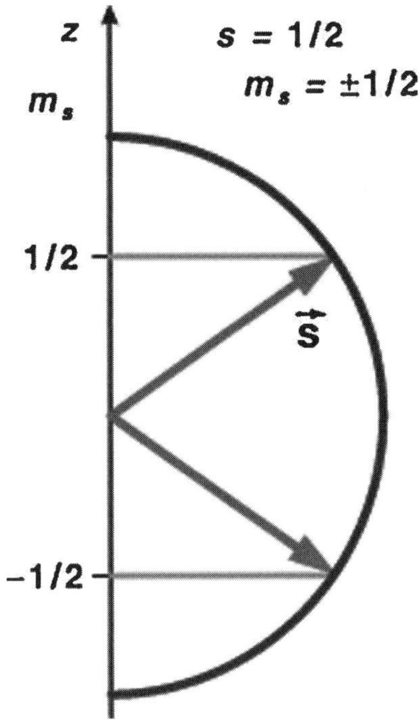

Fig. 1 Vector model of electron spin. Source: Stöckler Taschenbuch der Physik 2000, 769. Reprinted by permission of the publisher

The consideration by Uhlenbeck and Goudsmit in the summer of 1925 was basically very simple. Pauli had already noticed in 1924/25 that there are four quantum numbers; but within a semi-classical framework, for a single electron, this could only mean:

4 degrees of freedom = 3 translational degrees + 1 internal degree of freedom

S

For a point-like or extremely small particle this in turn pointed to an intrinsic rotation!

The first reaction to the paper by Uhlenbeck and Goudsmit on record was by Hendrik A. Lorentz (1853–1928). In a letter from Oct. 19, 1925 he noticed (as Pauli had with respect to Kronig's earlier proposal) that there were problems with the rotational velocity v of such a spinning electron, because

$$\mu \sim \frac{e}{m} \cdot \left(\frac{v}{r}\right)$$

which led to $v \sim 10 \cdot c$, or approximately ten times the velocity of light, which is physically impossible. But the brief note by the two Dutch physicists had already been irretrievably submitted to *Die Naturwissenschaften*. In reply to their worried request for advice, their mentor Paul Ehrenfest (1880–1933) consoled them with the words: "You are both young enough to afford a stupidity like that." [9, 10, 11]

So the bold hypothesis of an electron spin found its way into print even though no one dared to believe it at that point. The remaining quantitative problem with the missing factor 2 for the doublet separation (which Pauli had already pointed out to Kronig) was only clarified in early 1926. Lewellyn Hilleth Thomas (1903–1992) explained it as arising from a missed Lorentz transformation from the spinning electron's frame of reference against the laboratory system. By that time, the 'old' semi-classical quantum theory by Bohr, Sommerfeld and their pupils had already been replaced by the modern quantum mechanics of Heisenberg and Schrödinger, which led to a much deeper—nonclassical—understanding of spin from the symmetries and statistics of the quantum systems.

Although spin was thus first 'discovered' at the end of 1925 and only acknowledged by the scientific community in 1926, that is, *after* the development of quantum mechanics, it was nevertheless a product of the old semi-classical style of modeling that still took angular momentum, orbits and mechanical models seriously. The Pauli principle and spin remain integral parts of the *new* quantum mechanics but their historical *roots* lay in the old Bohr-Sommerfeld form of quantum theory. Like the ► electron, the concept of spin thus also had a pretty complicated early 'biography' [20], but it is still very much alive today. See also ► partity; quantum field theory; spin echo.

Primary Literature

1. W. Heisenberg: *Zur Quantentheorie der Linienstruktur und der anomalen Zeemaneffekte*. Zeitschrift für Physik **8**, 273–297 (1922)
2. W. Heisenberg, A. Sommerfeld: *Die Intensität der Mehrfachlinien und ihrer Zeemankomponenten*. Zeitschrift für Physik **11**, 131–154 (1922)
3. W. Pauli: *Über den Einfluß der Geschwindigkeitsabhängigkeit der Elektronenmasse auf den Zeemaneffekt*. Zeitschrift für Physik **31**, 373–385 (received Dec. 2, 1924, publ. Feb. 19, 1925)
4. W. Pauli: *Über den Zusammenhang des Abschlusses der Elektronengruppen im Atom mit der Komplexstruktur der Spektren*. Zeitschrift für Physik **31**, 765–783 (received Jan. 16, publ. Mar. 21, 1925)
5. W. Pauli: *Wissenschaftlicher Briefwechsel mit Bohr, Einstein, Heisenberg u.a.*, ed. by Karl von Meyenn, vol. **1** (1919–1929) (Springer, Berlin 1979)
6. S. Goudsmit, G.E. Uhlenbeck: *Ersetzung der Hypothese vom unmechanischen Zwang durch eine Forderung bezüglich des innern Verhaltens jedes einzelnen Elektrons*. Die Naturwissenschaften **13**, 953–954 (1925)
7. N. Bohr, R. de L. Kronig, J.C. Slater: *Spinning electrons and the structure of spectra*. Nature **117**, 264–5, 550, 587 (1925)
8. L. H. Thomas: *The motion of the spinning electron*. Nature **117**, 514 (dated Feb. 20, publ. April 10, 1926)
9. S. Goudsmit: *Die Entdeckung des Elektronenspins*. Physikalische Blätter **21**, 445–453 (1965)
10. G. Uhlenbeck, S. Goudsmit: in Physics Today (June 1976), 40–48.
11. S.A. Goudsmit: *The discovery of the electron spin* (talk in 1971): http://www.lorentz.leidenuniv.nl/history/spin/goudsmit.html (last accessed May 28, 2007).

Secondary Literature

12. F. Hund: *Geschichte der Quantentheorie* (BI, Mannheim 1984, 4th ed. 1996, Chaps. 9–10)
13. M. Jammer: *Conceptual History of Quantum Mechanics* (McGraw Hill, New York 1966, sect. 3.3–3.4)
14. P. Forman: *The doublet riddle and atomic physics circa 1924*. Isis **59**, 156–174 (1968)
15. D. Serwer: Unmechanischer Zwang Pauli: *Heisenberg and the rejection of the mechanical atom*, 1923–1925. Historical Studies in the Physical Sciences **8**, 189–256 (1977)
16. O. Darrigol: *C-numbers and Q-numbers* (Univ. of California Press, Berkeley 1992, Chaps. 8–9)
17. B.L. van der Waerden: *Exclusion principle and spin*, in M. Fierz, V.F. Weisskopf (eds.) *Theoretical Physics in the Twentieth Century* (Interscience, New York 1960, pp. 199–244)
18. N. Robotti: *Quantum numbers and electron spin*. Archives Internationales d'Histoire des Sciences **40**, 305–331 (1990)
19. M. Morrison: *Spin—all is not what it seems*. Studies in the History and Philosophy of Modern Physics **38**, 527–557 (2007)
20. T. Arabatzis: *Representing Electrons, A Biographical Approach to Theoretical Entities* (University of Chicago Press, Chicago 2006), esp. Chap. 8.

Spin Echo

Antoine Weis

Spin echo is a technique, introduced in 1950 by Erwin Hahn, for suppressing inhomogeneous line broadening effects in ▶ magnetic resonance spectroscopy. The width of a magnetic resonance line (in the low rf power limit $\omega_1^2 \ll \gamma_1\gamma_2$) is determined by the transverse relaxation time $T_2 = 1/\gamma_2$ (cf (3) of ▶ magnetic resonance). An inhomogeneous magnetic field $\boldsymbol{B}_0$ produces an inhomogeneously broadened line which can be understood as the superposition of many lines with narrow widths γ_2. The spin echo technique overcomes the loss of spectral resolution due to the inhomogeneous broadening.

S

Consider a system of N spins, initially aligned along z. At time $t = 0$ the spins are tipped by a $\pi/2$-pulse to the y direction, and the (inhomogeneous) magnetic field $\boldsymbol{B}_0 = B_0(x, y)\,\widehat{z}$ drives their precession in the x–y plane. Because of the field inhomogeneity ΔB, the different spins precess at different angular frequencies (Fig. 1), and the macroscopic transverse polarization components $\boldsymbol{P}_{x,y} = \sum_{i=1}^{N} \left\langle \boldsymbol{S}_{x,y}^{(i)} \right\rangle$ decay because of the collective dephasing (Fig. 2, left).

Although the ensemble averaged polarization vanishes for times larger than the inhomogeneous dephasing time $T_2^* \propto 1/\Delta B$, the phase memory of the individual spins will survive for a longer time $T_2 \gg T_2^*$ and the spins can be made to rephase following the application of a π-pulse at time $t = T$ after the initial $\pi/2$-pulse that started the dephasing. Such a pulse rotates all the spin vectors by 180° around the x-axis, which, for spins in the x–y plane, is equivalent to a reversal of their

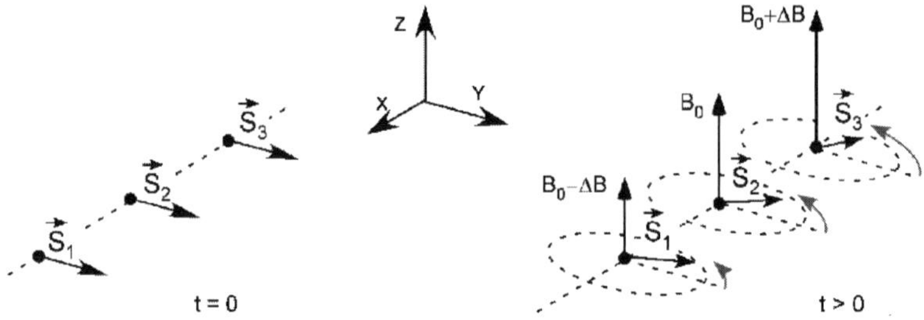

Fig. 1 Precession of $N = 3$ initially aligned spins in an inhomogeneous magnetic field

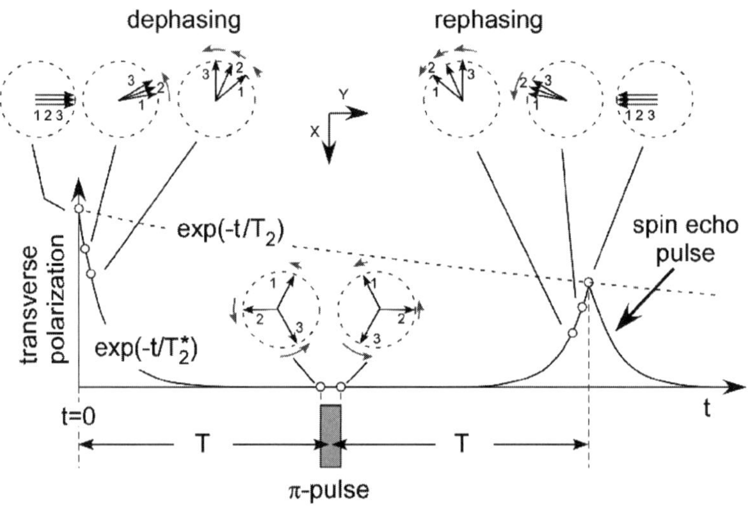

Fig. 2 Decay of the transverse spin polarization in the x–y plane due to inhomogeneous dephasing (*left*). The π-pulse after time T reverses the y-components of the individual spins and the spins rephase to a maximum transverse polarization at time $t = 2T$

y-components (Fig. 2, center). As a consequence the faster precessing spins (here S_3 and S_2) will catch up again with the slower spins, so that after the time $t = 2T$ all spins are again completely in phase, yielding a maximal transverse polarization, which for later times of course will decay again because of the inhomogeneity. The reappearance of a finite polarization from an apparently depolarized sample is called a spin echo.

The echo pulse amplitude is smaller than the starting amplitude, i.e., the initial transverse polarization due to the (homogeneous) T_2 relaxation. From the variation of the echo amplitude as a function of the time interval T one can thus infer T_2.

An interesting variant of spin echo spectroscopy was developed for neutrons and has become known under the name of neutron spin echo spectroscopy. The investigation of inelastic neutron scattering via phase shifts requires highly monochromatic neutrons. This requirement is rendered obsolete by using the echo technique which rephases neutrons of different velocities, so that all velocities contribute to the signal, yielding a large gain in statistics and sensitivity.

Similar echo phenomena can be observed in any multilevel quantum system subject to inhomogeneous relaxation, such as, e.g., in two-level atom and ions, for which echoes occur in the optical spectral range, where they are then called photon echoes. See also ► magnetic resonance; spectroscopy; spin.

Primary Literature

1. E. L. Hahn: Spin Echoes. Phys. Rev. **80**, 580–594 (1950)
2. N. A. Kurnit, I. D. Abella, S. R. Hartmann: Observation of a Photon Echo. Phys. Rev. Lett. **13**, 567–568 (1964)
3. D. Abella, N. A. Kurnit, S. R. Hartmann: Photon Echoes. Phys. Rev. **141**, 391–406 (1966)
4. F. Mezei: A new concept in polarized thermal neutron techniques. Z. Phys. **A 255**, 146 (1972)

Secondary Literature

5. F. Mezei, F. Pappas, T. Gutberlet (eds.): *Neutron Spin Echo Spectroscopy: Basics, Trends and Applications*. (Springer Lecture Notes in Physics, Springer, Berlin, Heidelberg 2003)

Spin Statistics Theorem

Arianna Borrelli

S

The term *spin-statistics theorem* is used to indicate theoretical explanations of the connection exhibited by non-relativistic quantum systems of identical particles between the particles' ► spin and their quantum-statistical behaviour. In such systems, particles of integer spin follow ► Bose–Einstein statistics, while particles of half-integer spin obey ► Fermi–Dirac statistics. High-precision experiments have not revealed any violations of this rule [8]. In the framework of relativistic ► quantum field theory, it is possible to show that, under the assumption that all particles are either bosons or fermions (symmetrization postulate), the spin-statistics connection is a consequence of basic physical postulates such as relativistic ► invariance, positive energy or time-reversal invariance.

From 1936 until today, a number of proofs of the connection between spin and statistics have been offered, with varying degrees of rigour and generality and imposing on the theory different physical requirements and limitations [10–12]. The proof which eventually entered textbook-tradition was given by Wolfgang Pauli (1900–1958) in 1940, and relied on results obtained previously by his assistant Markus Fierz (1912–2006) (1939) [1]. In the 1960s, the term "spin-statistics theorem" established itself to indicate these demonstrations, even though they are usually not equivalent to each other. The term was introduced by Raymond F. Streater (1936–) an Arthur S. Wightman (1922–) in their summary of axiomatic quantum field theory (1964) [7].

In its quantum-relativistic formulation, the theorem states that, when quantizing a field $\psi(x)$ (i.e. when formally transforming it into an ▶ operator), one is not free to choose at will between commutation and anticommutation relations, but has to impose the one or the other according to the way in which the field $\psi(x)$ transforms under a change of the relativistic reference frame (Lorentz transformation). If the "wrong" choice is made, the quantized theory will not fulfil physically significant requirements such as positive energy, positive probability, invariance under time-reversal, or the condition that the influence of interactions should not propagate faster than light (▶ locality).

In the non-relativistic limit, the Lorentz transformation properties of $\psi(x)$ determine its behaviour with respect to space rotations, and therefore the spin of the corresponding particles: scalar fields have spin 0, vectors have spin 1, Dirac-spinors have ▶ spin $\frac{1}{2}$, and so on. The choice between commutation and anticommutation relations translates into the ▶ symmetry or antisymmetry of the non-relativistic many-particle ▶ wave function, and determines whether the particles will obey Bose–Einstein or Fermi–Dirac statistics. The connection between spin and statistics observed in non-relativistic quantum systems is thus shown to be a consequence of imposing physical requirements in the relativistic framework.

All versions of the spin-statistics theorem have to make some initial assumptions on the mathematical form of the theory. For example, some authors deal only with the lowest spin values (0, $\frac{1}{2}$, 1), some only with noninteracting particles, others weaken the requirement of relativistic invariance. The proofs of the spin-statistics connection reflect both the history of quantum field theory and the different approaches to it, variously giving priority to rigorous axiomatic structure, maximum generality, minimal requirements, or the simplicity of the arguments.

Early proofs, including Pauli's 1940 paper, relied on mathematical procedures whose legitimacy was only proved years later, and sometimes also on manipulations which are today regarded as illegitimate. From the late 1940s onward, with the development of the mathematical apparatus of quantum field theory, more rigorous and elaborated proofs were formulated. Interest in the subject has remained lively and, in 2000, a conference was devoted to "The spin-statistics connection and commutation relations", summarizing the many theoretical and experimental developments in the field, with particular attention to possible violations of the symmetrization postulate.

In his 1940 paper, Pauli proved the spin-statistics connection for noninteracting fields corresponding to any spin value by requiring positive energy and locality [1, 10, 12]. He assumed the generic field $\psi(x)$ to obey linear differential equations whose solutions could be expressed as a superposition of plane waves $e^{i(k^{\mu}x_{\mu})}$. Using the classification of the representations of the Lorentz group introduced by Bartel van der Waerden (1903–1996), Pauli was able to classify the behaviour of all possible candidates to the role of energy-momentum operator and show that, if $\psi(x)$ corresponded to half-integer spin values, the energy function would not be positive definite. From this he concluded, as Fierz had done before him, that a field $\psi(x)$ with half-integer spin had to be quantised with anticommutation relations so that, by using the ensuing ► exclusion principle, an infinite number of negative-energy states could be regarded as being already occupied. In this way, one would in the end recover a physical system with positive energy.

To prove the second part of the theorem, Pauli implemented locality by requiring that ► operators derived from $\psi(x)$ and associated to physical quantities should commute for spacelike separations, i.e. for events which, in some reference frame, occur at the same time in two different places. He showed that, when a field with integer spin was quantized according to anticommutation rules, this condition would lead to a relation implying that the field is identically zero. This result was based on a mathematical argument whose legitimacy was proved only years later.

In 1949, Richard Feynman (1919–1988) used his newly developed computational techniques for ► QED to show that the spin-statistics-connection follows from the requirement that probability values must be $\leqslant 1$ [2]. In 1964, Steven Weinberg proved the spin-statistics theorem both for fermions and for bosons by requiring that quantized fields should either commute or anticommute for spacelike separations [6, 13].

In the context of axiomatic quantum field theory, much attention has been devoted to the spin-statistics theorem and to its relationship with the invariance of theories with respect to the combination of time-reversal, charge-conjugation and parity transformation (► CPT-theorem). Julian Schwinger (1918–1994) endeavoured to determine the conditions under which both the spin-statistics theorem and the CPT-theorem would obtain (1958) [3]. Gerhard Lüders (1920–1995) and Bruno Zumino (1923–) (1958) and, contemporarily but independently, Nicholas Burgoyne (1932–1958) instead proved the spin-statistics theorem on the basis of postulates such as Lorentz invariance, positive energy and positive metric of the ► Hilbert space, and then used it as a starting point to prove the CPT-theorem [4, 5].

Works on the spin-statistics theorem have relied on increasingly complex mathematical arguments, and some authors have attempted to find what they felt would be a "simple" demonstration. Ian Duck (1933–) and George Sudarshan (1931–) have historically reviewed the subject from this point of view, and Sudarshan has offered a proof based on rotational invariance (1997) [11]. In the last decades, various authors have investigated the spin-statistics connection outside the boundaries of standard relativistic quantum field theory, for example in non-relativistic quantum mechanics, supersymmetry or superstrings, often relying on topological arguments. Most recently, a formulation of the spin-statistics theorem for classical mechanics has been proposed (J. A. Morgan 2004) [9].

Primary Literature

1. W. Pauli: The connection between spin and statistics. Phys. Rev. **58**, 716–722 (1940)
2. R. P. Feynman: The theory of positrons. Phys. Rev. **76**, 749–759 (1949)

3. J. Schwinger: Spin, statistics and the TCP theorem. Proc. Natl. Acad. Sci. U S A **44**, 223–228 (1958)
4. G. Lüders, B. Zumino: Connection between spin and statistics. Phys. Rev. **110**, 1450–1453 (1958)
5. N. Burgoyne: On the connection between spin and statistics. Nuovo Cimento **8**, 607–609 (1958)
6. S. Weinberg: Feynman rules for any spin. Phys. Rev. **133**, B1318–B1332 (1964)
7. R. F. Streater, A. S. Wightman: *PCT, Spin and Statistics and All That*, 2nd edn. (Princeton University Press, Princeton, 2000, 146–161)
8. R. C. Hilborn, G. M. Tino (eds.): *Spin-Statistics Connection and Commutation Relations. Experimental Tests and Theoretical Implications* (American Institute of Physics, Melville, 2000)
9. J. A. Morgan: Spin and statistics in classical mechanics. Am. J. Phys. **72**, 1408–1417 (2004)

Secondary Literature

10. S. Tomonaga: *The Story of Spin* (The University of Chicago Press, Chicago 1997, 131–149)
11. I. Duck, E. C. G. Sudarshan: *Pauli and the Spin-Statistics Theorem* (World Scientific, Singapore, 1997)
12. C. P. Enz: *No Time to be Brief. A Scientific Biography of Wolfgang Pauli* (Oxford University Press, Oxford, 2002, 309–336)
13. M. Massimi, M. Redhead: Weinberg's proof of the spin-statistics theorem. Stud. Hist. Phil. Mod. Phys. **34**, 621–650 (2003)

Squeezed States

Martin Bodo Plenio

In this section we will discuss some basic properties of so-called squeezed quantum states. These states are characterized by the property that they will exhibit fluctuations for some physical observable quantities that are smaller than the fluctuations when the same quantity is measured on the vacuum state. Such states, often for optical fields, have applications in various areas of physics ranging from enhanced measurement precisions to quantum information processing.

A pure squeezed state [1] may be represented as a ► wave function in position space where it takes the form

$$\langle x|\psi_{\mathrm{sq}}\rangle = \psi_{\mathrm{sq}}(x) = [2\pi(\Delta x)^2]^{-1/4}\exp\left[-\left(\frac{x-\langle x\rangle}{2\Delta x}\right)^2 + \mathrm{i}\frac{\langle p\rangle x}{\hbar}\right], \quad (1)$$

where

$$(\Delta x)^2 = \langle x^2\rangle - \langle x\rangle^2 \quad \text{where} \quad \langle f(x)\rangle = \int_{-\infty}^{\infty} |\psi(x)|^2 f(x)\mathrm{d}x. \quad (2)$$

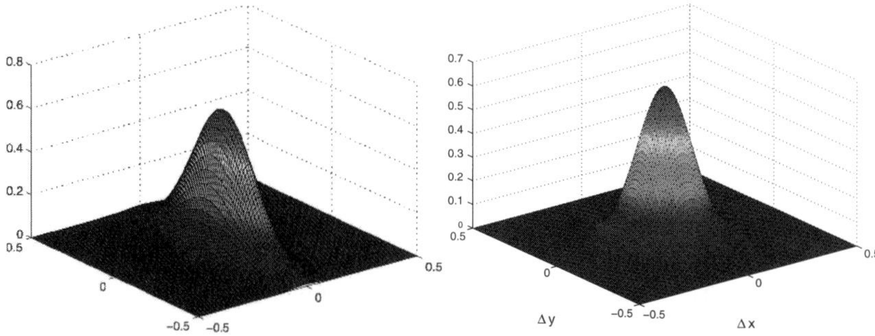

Fig. 1 The Wigner functions of the vacuum state which is a specific example of a coherent state (*left hand side*) and of a squeezed state (*right hand side*) whose variance in one quadrature component is suppressed below the vacuum level by a factor of 3 at the expense of increasing the variance in the other quadrature component by a factor of 3

$(\Delta x)^2$ and $(\Delta p)^2$ represent the uncertainties in the measurement of the observables x and p. For particles these may be position and momentum while for light fields these are the in-phase and out-of-phase components, also known as position and momentum quadrature components. The above formulae share great similarity with ► coherent states. In fact, coherent states represent the special case when $\Delta x = \Delta p = \sqrt{\hbar/2}$. Thus, for squeezed states, the uncertainty of one quadrature component, e.g. position x may be reduced at the expense of the other, e.g. momentum p. Coherent and squeezed states may be visualized neatly employing the ► Wigner distribution, and two examples are shown in Fig. 1. The fact that the variance, for example in position, is reduced below the vacuum level has applications in precision measurements as the reduced uncertainty allows for a more precise determination of position. Squeezed states gained considerably more attention when it was suggested that squeezed light might be used to achieve better sensitivity in the interferometric detection of gravitational waves [2]. This stimulated the development of experimental methods for the generation of squeezed states of light [3]. The generation of squeezed states of light requires non-linear optical effects such as parametric oscillation and second harmonic generation. As these non-linearities are often weak this makes the generation of substantial levels of squeezing difficult to achieve.

S

Another area in which squeezed states are of increasing relevance is that of optical quantum communication in the continuous variable regime [4, 5]. Here, a fundamental aim is the generation of ► entanglement in the form of two-mode squeezed states such that each light mode is accessible to a different possible distant party. In the Fock state representation these states take the form

$$|re^{i\phi}\rangle = \sqrt{\frac{1}{\cosh r}} \sum_{n=0}^{\infty} \frac{\sqrt{(2n)!}}{n!} \left(-\frac{1}{2}e^{i\phi} \tanh r\right)^n |n\rangle|n\rangle. \tag{3}$$

This state exhibits strong correlations as for example a photon number measurement in one mode determines the outcome of a photon number measurement in the other mode. Two-mode squeezed states form the basic resource for ► quantum communication protocols such as quantum state teleportation. Various methods for the generation of such states exist. A simple method consists of sending two single mode squeezed states of the type described above onto the two inputs of a beam-splitter making sure that one squeezed state exhibits squeezing along the x quadrature while the other exhibits exactly the same degree of squeezing but along the p quadrature. The output of the beam-splitter will then be a two-mode squeezed state as in (3).

The distribution of two-mode squeezed states, and therefore entanglement, generally suffers from noise and the development of methods to combat the effects of noise and to improve the squeezing and entanglement in such states is an active area of research today [6, 7].

Literature

1. R. Loudon, P. L. Knight: Squeezed light. J. Mod. Opt. **34**, 709 (1987)
2. C. M. Caves: Quantum-mechanical radiation-pressure fluctuations in an interferometer. Phys. Rev. Lett. **45**, 75 (1980)
3. D. F. Walls, G. Milburn: *Quantum Optics* (Springer, Berlin, 1994)
4. J. Eisert, M. B. Plenio: Introduction to the basics of entanglement theory in continuous-variable systems. Int. J. Quant. Inf. **1**, 479 (2003)
5. M. B. Plenio, S. Virmani: An introduction to entanglement measures. Quant. Inf. Comp. **7**, 1 (2007)
6. D. E. Browne, J. Eisert, S. Scheel, M. B. Plenio: Driving non-Gaussian to Gaussian states with linear optics. Phys. Rev. A **67**, 062320 (2003)
7. J. Eisert, D. E. Browne, S. Scheel, M.B. Plenio: Distillation of continuous-variable entanglement with optical means. Ann. Phys. **311**, 431 (2004)

Standard Model

See ► Quantum field theory; Particles Physics.

Stark Effect

Klaus Hentschel

In late 1913 Johannes Stark (1874–1957), the professor of experimental physics at the technical university of Aachen who would later champion the Aryan physics movement, discovered the effect of electric fields on spectral lines. This phenomenon is usually referred to as the Stark effect, though some Italian authors prefer to call it 'Stark-Lo Surdo effect', because Antonio Lo Surdo (1880–1949) independently also found this long-sought electric analogue to the magnetic ▶ Zeeman effect. Both discoverers worked with specially constructed discharge tubes. Stark's tube allowed stable electric fields of up to 100,000 V cm^{-1}. In numerous experiments during the course of the next decade, Stark demonstrated the following:

- The spectrum lines in the Balmer series of hydrogen split up into several components
- The number of these components increases with the series number
- Splitting and polarization of Balmer lines is symmetric to the original line
- The splitting seemed to be asymmetric for some other elements
- The distances between the hydrogen spectral-line components (in units of frequency or wave-number) are all integer multiples of a smallest line distance
- The splitting interval Δ increases proportionally with the electric field F (i.e., $\Delta \sim F$ for not too small or too large F)
- For very small electric fields and atoms not subject to a permanent dipole moment, Δ actually increases by the second power of F ('quadratic Stark effect')
- For very strong electric fields $F \sim$ 1,000,000 V cm^{-1}, the splitting is asymmetric, as was found experimentally by two Japanese physicists in 1918 and derived theoretically by Arnold Sommerfeld in 1921 ('Stark effect of second order')

Mathematical techniques from perturbation theory to make corrections for Kepler ellipses induced by remote third masses were already well developed at the time. Applying these techniques, Paul Sophus Epstein (1871–1939) in Munich (a member of the ▶ Sommerfeld School) and the astrophysicist Karl Schwarzschild (1873–1916) in Potsdam succeeded independently of each other in incorporating this effect in the ▶ atomic model of Niels Bohr (1885–1962) and Arnold Sommerfeld (1868–1951).

In analogy to the ▶ Zeeman effect, they interpreted the Stark effect as a splitting of energy levels of initial and final states, in this case induced by the external electric field, i.e., as a vanishing of the degeneracy in normal hydrogen. Put intuitively, eccentric orbits of the ▶ electrons start to differ in energy from less eccentric orbits due to the external electric field. The problem is described mathematically in parabolic coordinates (ξ, η, ψ), with ψ as the angle off the z-axis which is parallel to the external electric field F.

$$\frac{y^2}{\xi} + 2x = \xi \qquad \frac{y^2}{\eta} - 2x = \eta$$

The main ▶ quantum number n is then the sum of three quantum numbers n_ξ, n_η, n_ψ linked to the three degrees of freedom of the system. Because ψ is a cyclic coordinate, $n_\psi \geqslant 1$, i.e., $n_\psi = 0$ is forbidden (analogous to the discussion of ▶ fine structure). Intuitively put, this means that the electron *has* to revolve around the z-axis).

Thus the main quantum number $n = n_\xi + n_\eta + n_\psi = n_\xi + n_\eta + m + 1$, with $n_\psi = 1, 2, 3, \ldots \Leftrightarrow m = 0, 1, 2, \ldots$ and m the so-called azimuthal quantum number $m = n_\psi - 1$. After elaborate calculations (cf. [1, Chap. 6, Sect. 2]), one obtains for the energy of the orbit as a function of the quantum numbers and the field:

$$-E(n, n_\eta, n_\xi, F) = \frac{2\pi^2 \mu Z^2 e^4}{h^2 n^2} + \frac{3hF}{8\pi^2 \mu Z e} n(n_\eta - n_\xi)$$

The first expression on the right-hand side of the equation recovers the normal Balmer term; the second term describes the energetic splitting $\sim F$ and $\sim n(n_\eta - n_\xi)$. After insertion of initial state (1) and final state (2), the splitting $\Delta\nu$ of spectral lines in terms of frequency results as

$$\Delta\nu = \frac{3h \cdot F}{8\pi^2 \mu Z e}[n_2(n_\eta - n_\xi)_2 - n_1(n_\eta - n_\xi)_1]$$

These formulas thus correctly describe Δ as proportional to the field F, and symmetric to $\Delta\nu = 0$, because for each allowed transition $(n_\eta, n_\xi, m)_1$, $(n_\eta, n_\xi, m)_2$, there also exists an inverse transition. Additional ► selection rules had to be set so as not to get too many components: $\Delta m = 0$ or 1, with the additional constraint of excluding $m = 0 \rightarrow m = 0$ sufficing to explain the observed number of components and splitting patterns. The observed polarization of the $\Delta m = \pm 1$ components also agreed well with what was expected classically for light emitted from moving charges: circular polarization for observations vertical to the field. The outcome was a perfect ► semi-classical model to explain the normal Stark effect for hydrogen and similar simple atoms. In 1920, Bohr's assistant Henrik Anthony Kramers (1894–1952) showed that Epstein's and Schwarzschild's approximation was only good as long as the exterior electric fields were large compared with the relativistic fine structure of the unperturbed energy levels. For small F and atoms without permanent dipole moment, Δ was proportional to the 2nd power of the electric field. This 'quadratic Stark effect' and a smooth transition from the quadratic to the normal Stark effect for an increasing F were confirmed experimentally by Rudolf Ladenburg (1882–1952) in Breslau in 1924.

The Stark effect of second order was found by the two guest researchers at the laboratory of the Mt. Wilson observatory, Toshio Takamine (1885–1959) and Noboru Kokubu. Experimenting with an unusually high electric field of 147,000 V cm^{-1}, they found an asymmetric shift of 0.8 Å of the middle component towards the red instead of the normal symmetric splitting. Upon hearing about their result from Bohr, Arnold Sommerfeld used second-order perturbation calculations to derive this asymmetric shift. This Stark effect of second order was also responsible for the so-called pole effect, an asymmetric line broadening well-known to spectroscopists (cf. [8,9], pp. 357–366). After the advent of quantum mechanics, Erwin Schrödinger (1887–1961) was the first to show that very similar results could be derived for the Stark effect within this new framework. The resulting formulas were virtually equivalent for the normal Stark effect, whereas small deviations between the old semi-classical Sommerfeld formulas and the new

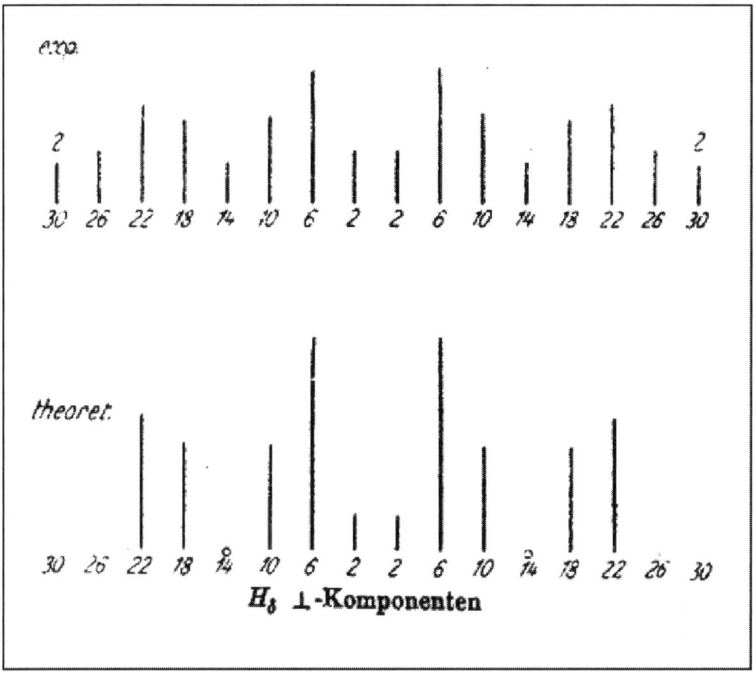

Fig. 1 Comparison between experimental results (above) and theoretical calculations (below) for the splitting of the hydrogen H_δ line in an electric field (observed vertically). From [3, p. 473]

quantum mechanical formulas existed for the second and third-order Stark effect. By 1929 it had become clear that quantum mechanics yielded better agreement with experimental precision measurements (see, e.g., [10]), even though it took much longer for a perfect match between theory and experiment to be reached.

S

Primary Literature

1. A. Sommerfeld: *Atombau und Spektrallinien*, Braunschweig: Vieweg 1919 (Engl. transl. *Atomic Structure and Spectral Lines*, London: Methuen 1923)
2. A. Lo Surdo: Sui fenomeno analogo a quello di Zeeman nel campo elettrico. Rendicondi dell Accademia Nazionale dei Lincei (5th ser.) **22**, 664–666 (1913), **23**, 252–253, 326 (1914)
3. E. Schrödinger: Quantisierung als Eigenwertproblem. 3. Mitteilung: Störungstheorie mit Anwendung auf den Stark-Effekt der Balmerlinien. Annalen der Physik (4th ser.) **80**, 437–490 (1926)
4. J. Stark: Beobachtungen über den Effekt des elektrischen Feldes auf Spektrallinien. Die Naturwissenschaften **1**, 1182 (1913), **2**, 145–148, 542f (1914)
5. J. Stark: Observations of a separation of spectral lines by an electric field. Nature **92**, 401 (1913)

6. J. Stark: *Elektrische Spektralanalyse chemischer Atome* (Leipzig: Hirzel 1914)
7. J. Stark, P. Epstein: *Der Starkeffekt* (Stuttgart: Battenberg 1965, reprint of papers by both with commentary by Armin Hermann)

Secondary Literature

8. K. Hentschel: An unwelcome discovery: The pole effect in the electric arc as a threat to precision spectrometry in the early 20th century. Archive for History of Exact Sciences 51/3, 199–271 (1997)
9. K. Hentschel: *Zum Zusammenspiel von Instrument, Experiment und Theorie*, Hamburg: Kovač 1998, 334–358
10. R. Ladenburg: Einfluß elektrischer Felder auf Spektrallinien (Starkeffekt), in Müller-Pouillet: *Lehrbuch der Physik*, Vol. **2.2**, Chap. 40, 2231–2290 (1929)
11. M. Leone, A. Paoletti, N. Robotti: A simultaneous discovery: The case of Johannes Stark and Antonino Lo Surdo. Physics in Perspective **6**, 271–294 (2004)

States in Quantum Mechanics

Leslie E. Ballentine

The most general meaning of the term *state* is *a manner of existing, a combination of attributes belonging to a thing* (paraphrased from the Oxford English Dictionary). In physics the term *state* has various, more specific meanings in thermodynamics, in classical mechanics, and in quantum mechanics, but all include the notion that a knowledge of the state is sufficient to make predictions about the future behavior of the system.

A pure state (▶ states, pure and mixed) is one that is specified or controlled as precisely as possible. In classical mechanics a pure state is specified by a point in phase space, i.e. by the values of all position and momentum variables. In quantum mechanics a pure state is specified by a ▶ wave function or state vector in ▶ Hilbert space. In both classical and quantum mechanics the motion of the state is deterministic, in the sense that the specification of the initial state determines a mathematically unique trajectory of future states.

However, the dissimilarities between the classical and quantum pure states are even more significant. The specification of the classical state uniquely determines all observable properties of the system, as functions of the position and momentum variables. But the connection between the quantum state and observation is only probabilistic; the state vector does not determine the values of the ▶ observables, but only the probabilities of the various possible values. The same classical state leads necessarily to the same observable events, but a new preparation of the same quantum state may lead to quite different observable outcomes. Thus, even though the time evolution of the state vector is deterministic, the appearance of events is not. It is the connection between the quantum state and the observable events that is indeterministic, notwithstanding the deterministic nature of the ▶ Schrödinger equation. That the link between the state and the observable events is only statistical, is the most significant difference of quantum mechanics from classical mechanics. But if we recognize that the prediction of future events must be probabilistic, then the quantum state fulfills the basic notion of *state* as being sufficient for predictions about the future behavior of the system. Indeed, the *probabilities* for all observable properties are uniquely determined by the quantum state. ▶ Probability in quantum mechanics.

When we consider general states, comprising both pure and ▶ mixed states, then the analogy between classical and quantum states becomes closer. A general classical statistical state is described by a probability distribution on phase space, $\rho_c(q, p)$, where q and p are the coordinates and momenta. A pure state is recovered in the extreme limit in which all probability is concentrated on a single point; all other probability distributions are mixed states. A general quantum state is described by a ▶ state operator (also called a ▶ density matrix), ρ, which is a positive Hermitian operator with unit trace. A pure state with state vector $|\psi\rangle$ is obtained if $\rho = |\psi\rangle\langle\psi|$. The use of general states makes the comparison between classical and quantum mechanics easier because the results of both theories are expressed in terms of probabilities and averages. The average value of an observable, represented by the quantum operator A or the classical function $A(q, p)$, is given by $\mathrm{Tr}(\rho A)$ in quantum mechanics, and by $\int \rho_c(q, p)A(q, p)\mathrm{d}q\mathrm{d}p$ in classical mechanics. The equations of motion for the quantum and classical state functions are very similar. The former involves the commutator with the Hamiltonian, $\partial\rho/\partial t = -(\mathrm{i}/\hbar)[H, \rho]$, and the latter involves the Poisson bracket, $\partial\rho_c/\partial t = \{H_c, \rho_c\}_{\mathrm{PB}}$. By contrast, Newton's equation for a single classical orbit bears no similarity to Schrödinger's equation for a state vector.

But more important than these formal similarities and differences is a very substantial difference. Two classical orbits that begin close together in phase space can, in time, become widely separated from each other, but two state vectors that are initially close in Hilbert space will not separate at all because the unitarity of the time-development operator implies that $\langle\psi_1(t)|\psi_2(t)\rangle$ remains constant. This fact was once considered to be a serious obstacle to the emergence of classical mechanics as a limiting case of quantum mechanics. But the obstacle disappears if the proper analog of a quantum state is a classical statistical state, since the overlap of two nearby classical probability distributions, $\int \rho_{c1}(q, p, t)\rho_{c2}(q, p, t)\mathrm{d}q\,\mathrm{d}p$, is independent of t, as is the overlap of two quantum state operators, $\mathrm{Tr}(\rho_1(t)\rho_2(t))$. Thus the classical limit of a quantum state should be regarded as an ▶ ensemble of classical trajectories, rather than a single trajectory [1].

S

The concept of a quantum state in the modern theory is very different from that in the early quantum theory (▶ Quantum theory, early period). N. Bohr postulated that, from the continuum of classical atomic orbits, a ▶ quantization condition selected a discrete subset of *permitted* states, between which discontinuous ▶ quantum jumps took place. The old Bohr-orbit theory proved to be inadequate, and was replaced by Schrödinger's wave equation (▶ wave mechanics; Schrödinger equation). But the notion of permitted states lingered on, with the stationary solutions (energy eigenstates) of Schrödinger's equation taking on the role of the "permitted" orbits of the old theory. That notion is quite obsolete. There are now plenty of experiments [2–4] that demonstrate the physical significance of the nonstationary solutions to Schrödinger's equation, which are therefore every bit as "permitted" as are the stationary solutions.

Literature

1. L. E. Ballentine: *Quantum Mechanics – A Modern Development* (World Scientific, Singapore 1998) – a general reference on quantum states
2. J. A. Yeazell, C. R. Stroud: Observation of spatially localized atomic electron wave packets. Phys. Rev. Lett. **60**, 1494 (1998)
3. L. Marmet, H. Held, G. Raithel, J. A. Yeazell, H. Walter: Observation of Quasi–Landau wave packets. Phys. Rev. Lett. **72**, 3779 (1994)
4. J. Wals, H. H. Fielding, J. F. Christian, L. C. Snoek, W. J. van den Zande, H. B. van Linden van den Heuvell: Observation of Rydberg wave packet dynamics in a Coulombic and magnetic field. Phys. Rev. Lett. **72**, 3783 (1994)

States, Pure and Mixed, and Their Representations

Leslie E. Ballentine

The concept of *state* in quantum mechanics, considered abstractly, is a means of calculating probabilities and averages for all ▶ observables. States can be given many different mathematical representations. The most familiar are the ▶ wave function $\Psi(x)$ and the state vector $|\Psi\rangle$ in ▶ Hilbert space, but these describe only *pure states*. A general quantum state is represented by a ▶ *state operator*, ρ, (also called a ▶ *statistical* operator, or ▶ a *density matrix*), which is a positive Hermitian operator with unit trace. That is to say, it must satisfy the three conditions,

$$\mathrm{Tr}\rho = 1\,,\ \ \rho = \rho^{\dagger}\,,\ \ \langle u|\rho|u\rangle \geqslant 0 \ \text{ for all vectors } |u\rangle. \tag{1}$$

The average value of an observable, represented by the operator A, is given by $\langle A\rangle = \mathrm{Tr}(\rho A)$. ▶ Gleason's theorem shows that, under broad but non-trivial conditions, this state operator provides the most general means of introducing a probability measure in Hilbert space. The distinction between "pure case" (reiner Fall) and "mixed case" (Gemenge) was introduced by Hermann Weyl (1885–1955) [1].

A *pure state* with state vector $|\psi\rangle$ is obtained if ρ is a one-dimensional projection operator, $\rho = |\psi\rangle\langle\psi|$, in which case the expression for the average reduces to

$\langle A \rangle = \langle \Psi | A | \Psi \rangle$. A pure state operator can be identified by a variety of mathematical conditions [2]. The most useful is that, in addition to (1), it also satisfies $\mathrm{Tr}(\rho^2) = 1$. The set of all operators satisfying (1) is a convex set, with the pure states being the extremal members.

Non-pure states are commonly called ▶ *mixed states* because they can be represented as convex combinations of pure states (which need not be orthogonal),

$$\rho = \sum_i w_i |\Psi_i\rangle\langle\Psi_i| \,, \quad 0 \leqslant w_i \leqslant 1. \tag{2}$$

However, this representation of ρ as a mixture of pure states cannot be taken literally, since every non-pure state has infinitely many different representations as a mixture of the form (2). These have been fully classified [3].

The distinction between *pure* and *mixed* states should not be confused with the distinction between *eigenstates* and *superpositions*. Let A be a Hermitian operator that represents some physical observable. Corresponding to A there is a set of solutions to the equation

$$A|a_i\rangle = a_i|a_i\rangle \tag{3}$$

The vectors $\{|a_i\rangle\}$ are called the *eigenvectors* of the operator A, and the real numbers $\{a_i\}$ are called the *eigenvalues*. Eigenvectors are mathematically special because the action of the operator A leaves their direction unchanged, and only multiplies them by the eigenvalue. According to the fundamental postulates of quantum mechanics, the eigenvalues are the possible values of the observable, and for the state ρ the probability of obtaining the particular value a_i in a measurement of the observable will be $\langle a_i|\rho|a_i\rangle$. (We assume, for simplicity, that the set of eigenvalues is discrete, and that the eigenvectors are normalized so that $\langle a_i|a_i\rangle = 1$.)

Now suppose that the state operator ρ is chosen to be the projection operator $\rho = |a_i\rangle\langle a_i\rangle$, or equivalently, that the state vector is $|\psi\rangle = |a_i\rangle$. Evidently, the measurement will yield the value a_i with probability one. Such a state is called an *eigenstate* of the observable A. Conversely [4], if the measurement yields the value a_i with probability one, and if the set of eigenvalues is *nondegenerate* ($a_i \neq a_j$ for $i \neq j$), then the state must be the eigenstate represented by the vector $|a_i\rangle$.

S

A *superposition* state vector can be formed as a linear combination of eigenvectors,

$$|\psi\rangle = \sum_i c_i|a_i\rangle \tag{4}$$

This is sometimes, misleadingly, refered to as a 'mixture' of the eigenstates. Such terminology is to be deplored, since $|\psi\rangle$ is a pure state. The term 'mixture' should be reserved for state operators of the form (2).

The state operator can be given a matrix representation (the *density matrix*) by choosing a particular set of basis vectors, the "position" and "momentum" representations being common examples. By writing the "position" density matrix as $\langle q - \frac{1}{2}x|\rho|q + \frac{1}{2}x\rangle$, and Fourier transforming with respect to the difference variable x while keeping the centroidal variable q constant, we obtain the *Wigner function*

(► Wigner distribution), which is a representation of the state operator that is intermediate between the position and momentum representations, and bears a partial similarity to a classical phase space distribution.

Literature

1. H. Weyl: Quantenmechanik und Gruppentheorie. Zeitschrift für Physik **46**, 1–46 (1927)
2. L. E. Ballentine: *Quantum Mechanics – A Modern Development* (World Scientific, Singapore 1998, Sect. 2.3)
3. L. P. Hughston, R. Jozsa, W. K. Wooters: A Complete Classification of Quantum Ensembles having a given Density Matrix. Phys. Lett. A **183**, 14 (1993)
4. Ref. [2], Sect. 2.4.

State Operator

The most general representation of a quantum state. See the articles *States, Pure and Mixed, and their Representations* and *States in Quantum Mechanics*. The terms ► *statistical operator* and ► *density matrix* are also used.

Statistical Operator

An alternative term for the ► *state operator*, used mainly in quantum statistical mechanics.

Stern–Gerlach Experiment

Friedel Weinert

The *Stern–Gerlach experiments* (SG experiments) were prepared and carried out by Otto Stern (1888–1969) and his junior collaborator Walther Gerlach (1889–1979) between 1921 and 1925. [1–6] According to modern textbook interpretations the experiments established experimentally the so-called ► quantization of angular momentum and therefore the discreteness of the magnetic moment of atomic particles ► Spin; Vector model. This phenomenon is known as 'space quantization' (*Richtungsquantelung*) of angular momentum. As indicated below, the actual historical context, in which the experiments were carried out, is more complex. Quantization of angular momentum means that particles like ► electrons orbit the nucleus only in certain permitted planes. The experiments demonstrated, for the first time, the idea, proposed by Arnold Sommerfeld (1868–1951) in 1916, of the quantization of the orbital planes of the electron in the atom. The orbital planes of electrons do not only possess discrete sizes and shapes. These orbital planes must also be inclined in certain ways. They must have discrete spatial orientations in relation to a co-ordinate system like an external magnetic field. The size, shape and orientation of the orbital planes are indicated by ► quantum numbers (n, l, m_l). In addition it became clear in 1925 that a quantum number for intrinsic angular momentum, s, was needed. These quantum numbers specify the state of the atoms in an atom beam. When a beam of atoms is sent through a non-uniform magnetic field, this discrete spatial orientation will be revealed on a screen mounted behind the magnet. Stern and Gerlach therefore ran these experiments on beams of silver atoms in inhomogeneous magnetic fields. The purpose of the SG experiments is to maximize the effect of magnetic field gradients, $\partial B_z/\partial z$, on the silver atoms. It is necessary for the magnetic field to be inhomogeneous so that the magnetic moments of the particles feel a net force acting on them. In fact, in a non-uniform magnetic field, with gradient $\partial B_z/\partial z$, the magnetic dipole moments, μ, experience both a torque, which makes them align with the magnetic field, B_z, but also a net force, which leads to their displacement. In a typical Stern–Gerlach experiment, the magnetic field will split the beam into two parts and send the silver atoms either into the upper or the lower beam. Two scenarios can be distinguished:

1. The beam of silver atoms – silver atoms have 47 electrons – is sent through the magnet but the magnet is switched off. A screen mounted behind the magnet will record the impact of the atoms. When the magnet is switched off, one central trace will be recorded after the passage of the atom beam ($l = 0, m_l = 0$) because no deflection is experienced by the atoms in this state (Fig. 1).
2. The magnet is now switched on when the beam of atoms is sent through. Depending on the precise state of the atom beam, specified by its quantum numbers,

S

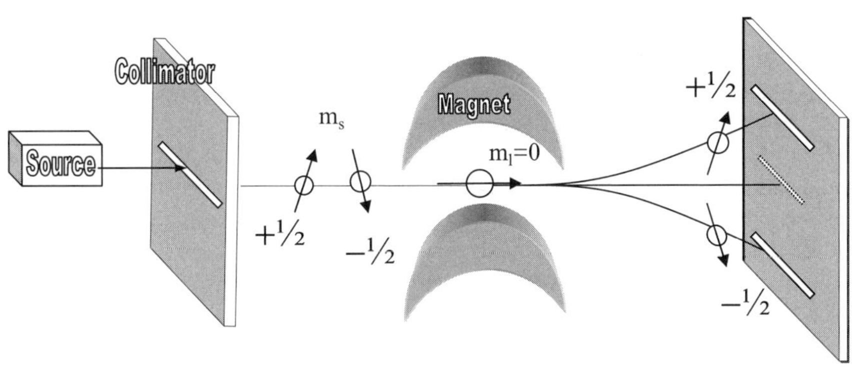

Fig. 1 The Stern–Gerlach experiments 1921–25

and assuming the simplest case, two traces will appear on the screen. The *effect* of the magnet will be an intensity shift. When the magnet was switched off the intensity maximum was in the centre of the screen. But with the magnet switched on, this central intensity maximum will become a minimum. The central trace will disappear and two clearly separated traces will appear, deflected upwards and downwards respectively (Fig. 1). With the magnet switched on, the magnet will *cause* the atom beam to split exactly into two halves (under appropriate conditions). This shift will happen only if the magnetic gradient is large enough to cause the displacement of the magnetic moments.

On the modern theory, an electron has orbital angular momentum, L, and spin angular momentum, S. The total angular momentum, J, is the sum of L and S:

$$J = L + S. \quad (1)$$

Generally, the magnetic moment, μ, is related to J through the expression

$$\mu = \frac{-e}{2m} J. \quad (2)$$

The SG experiment detected two traces, in violation of equation (2). The silver atoms were in their ground state (orbital angular momentum $l = 0$, $m_l = 0$ and hence no deflection is expected; spectroscopic notion $^2S_{1/2}$) but the splitting was due to the magnetic moment of the spin angular momentum of the electron ($m_s = \pm 1/2\hbar$) in the z-direction (direction of the magnetic field). When $l = 0$, it follows from expression (1) that we are left with the value for $S = 1/2\hbar$ for intrinsic spin, so that the beam splits into two and leaves two traces.

The historical situation was more complicated than this textbook account. [13] Strictly speaking, Stern and Gerlach believed that they had found Sommerfeld's quantization of angular momentum, L. They did not realize that the observed space quantization was due to the magnetic moment of the spinning electron (hence S). The two experimenters believed that their experiments had decidedly disproved the classical Lamor theory, which was based on continuous values for magnetic moments. They thought their experiments confirmed Sommerfeld's old quantum theory (1916), which postulated, in addition to the usual quantum numbers for the size and shape of orbits, a spatial orientation of the 'Keplerian' orbits of the electrons around the nucleus. The discovery of spin angular momentum of the electron came in 1925, when George Eugene Uhlenbeck (1900–1988) and Samuel Abraham Goudsmit (1902–1978) proposed the concept of ► spin. Contrary to frequently made claims in modern physics textbooks, Stern and Gerlach were not surprised by their results (splitting of beam into two traces) because this is just what Sommerfeld's theory told them to expect. Today many features of the Stern–Gerlach and the ► double-slit experiments reappear in so-called ► *which-way* experiments.

The Stern–Gerlach experiments are also interesting from a *philosophical* point of view. First, they demonstrate the relative robustness of experimental results and their relative independence from the theoretical conceptions, on which they are based. Secondly, they tell us that the often-quoted *acausality* of quantum mechanical processes is not supported by the SG experiments. It is not difficult to apply Mill's 'method of difference', a form of eliminative induction, to this situation to establish its causal nature. The only difference between otherwise two identical situations, including the preparation of the atoms in identical atomic states, specified by the quantum numbers, lies in the behaviour of the magnet. If it is not switched on and there is no magnetic field, one central trace appears; if it is switched on and a magnetic field is applied to the passing atoms, two traces appear in the simplest case ($l = 0$). The set of causal conditions is closed. There are no other interfering factors to be considered. We are therefore justified in concluding that the creation of the non-uniform magnetic field is the cause, given the initial state of the atoms, of the splitting of the atomic beam into two parts. As is customary in quantum mechanics, no claim is made about the behaviour of the individual atoms making up the beam. Since the initial orientation of their magnetic moment is random it is not possible to predict, which way they will turn under the influence of the magnet. But statistical predictions can be made about the behaviour of the whole beam. The rules of quantum mechanics specify how atom beams in different states behave. For instance, if $l \neq 0$ an odd number of traces will appear on the screen. The SG experiments show that causal relations obtain in the quantum domain but they are not deterministic. Hence causality and the pair ► indeterminism-determinism must be distinguished.

Primary Literature

1. O. Stern: Ein Weg zur experimentellen Prüfung der Richtungsquantelung im Magnetfeld. Zeitschrift für Physik **7**, 249–53 (1921)
2. W. Gerlach, O. Stern: Der experimentelle Nachweis des magnetischen Moments des Silberatoms. Zeitschrift für Physik **8**, 110–1 (1921)
3. W. Gerlach, O. Stern: Der experimentelle Nachweis der Richtungsquantelung im Magnetfeld. Zeitschrift für Physik **9**, 349–52 (1922)
4. W. Gerlach, O. Stern: Das magnetische Moment des Silberatoms. Zeitschrift für Physik **9**, 353–5 (1922)
5. W. Gerlach, O. Stern: Über die Richtungsquantelung im Magnetfeld. Annalen der Physik **74** (1924)
6. W. Gerlach, O. Stern: Über die Richtungsquantelung im Magnetfeld II. Annalen der Physik **76** (1925)
7. A. Einstein, P. Ehrenfest: Quantentheoretische Bemerkungen zum Experiment von Stern und Gerlach. Zeitschrift für Physik **11**, 31–4 (1922); reprinted in P. Ehrenfest: *Collected Scientific Papers*. Ed. M. Klein (North Holland, Amsterdam 1959, 452–55)

Secondary Literature

8. R. G. Griffiths: *Consistent Quantum Theory* (Cambridge University Press, Cambridge 2003, Chaps. 17–17)
9. R. Harré: *Great Scientific Experiments* (Oxford University Press, Oxford 1983, Part III, B)
10. R. I. G. Hughes: *The Structure and Interpretation of Quantum Mechanics*. (Harvard University Press, Cambridge, MA 1989, 1–8)

11. M. Jammer: *The Conceptual Development of Quantum Mechanics* (McGraw Hill, New York 1966)
12. M. Morrison: Spin: All is not what it seems. Studies in History and Philosophy of Modern Physics **38**, 529–57 (2007)
13. F. Weinert: Wrong Theory – Right Experiment: The Significance of the Stern–Gerlach Experiments. Studies in History and Philosophy of Modern Physics **26**, 75–86 (1995)
14. W. Walter: Otto Stern: Leistung und Schicksal. Gesellschaft Deutscher Chemiker, Fachgruppe Geschichte der Chemie: Mitteilungen **3**, 69–82 (1989)
15. R. Heinrich, H.-R. Bachmann: *Walther Gerlach* (Deutsches Museum, Munich1989, esp. 48–54)

Superconductivity

Kostas Gavroglu

Electrical Resistance in the Very Cold

The first systematic studies of the dependence of electrical resistance on temperature had been undertaken by L.P. Cailletet (1832–1913), E. Bouty (1846–1922) and Z.F. Wroblewski (1845–1888) in 1885. Their researches led them to the assertion that it would not be unreasonable to expect a zero value for the resistance for a temperature higher than −273°C. The next set of exhaustive measurements of the electrical resistance of various metals were performed by James Dewar (1842–1923) and John Ambrose Fleming (1849–1945). In 1896 they completed a study of the resistance of mercury at liquid air temperature, and their results indicated that the resistance of mercury could vanish at zero degrees Kelvin.

After having liquefied helium in 1908, Heike Kamerlingh Onnes (1853–1926), in 1911, at Leiden, measured the resistance of platinum and that of pure mercury at helium temperatures. He found that at 3K the value of the resistance of pure mercury became 0.0001 times the value of the resistance of solid mercury at 0°C, extrapolated from the melting point. Later that year the phenomenon was reaffirmed at 4.19K. By 1913 it was realized that impurities did not play any role in hindering the disappearance of the ordinary resistance, and the phenomenon was for the first time called the "superconductivity" of mercury [22]. In 1914 Kamerlingh Onnes discovered that an external magnetic field could disturb superconductivity by "generating resistance" in lead and tin. It was, also, found that superconductivity was destroyed when current above a certain threshold value passed through the superconductor.

Eduard Riecke (1845–1915) and Paul Drude (1863–1906) [12] treated the electric current in a metal as a drift of an electron gas under the influence of an electric field. H.A. Lorentz's (1853–1928) theory of electrical conduction had as a starting point the statistical theory of Maxwell and Boltzmann, and he investigated the dynamics of the collision processes. Nevertheless, his theory could not account for the rapid fall of resistance at extremely low temperatures.

In 1924 Lorentz drew attention to a remark originally made by Maxwell concerning perfect electrical conductors: If a conductor has no resistance there will be no electric field inside it even when there is a current flowing. The physical meaning of this result was that any change of the external magnetic field induced currents on the surface of the metal, and the magnetic field of these currents inside the metal compensated the change of external field, thus keeping the field "frozen-in" the metal. This physical assumption was regarded as being so self-evident that there was no systematic experimental study of the phenomenon.

It was Felix Bloch (1905–1983) who in 1928 proposed a satisfactory electron theory of conduction on the basis of ► wave mechanics. The ► electrons in a metal were considered to be uncoupled, though the field in which any one electron moved was found by an averaging process over the other electrons. If the metal was at absolute zero, its lattice determined a periodic potential field for the electronic motions, and the electrical resistance by the immobile lattice was zero. The resistance consisted of the "impurity resistance" and the resistance due to the thermal motion of the atoms. According to Bloch's analysis of the motion of an electron in a perfect lattice, all the electrons in a metal could be considered to be "free", but it did not necessarily follow that they were all conduction electrons. This theory accounted for metals, semi-conductors and insulators but not for superconductors. Bloch tried unsuccessfully to solve the problem in 1928–1929. He showed that the most stable state of a conductor, in the absence of an external magnetic field, was a state with no currents. But, superconductivity *was* a stable state displaying persistent currents without external fields: "This brought me to the facetious statement that all theories of superconductivity can be disproved, later quoted in the more radical form of "Bloch's theorem"; Superconductivity is impossible." [4]

In 1932, W.H. Keesom (1876–1976) with J.N. van den Ende found a jump of the ► specific heat at the critical temperature of tin. This prompted Paul Ehrenfest (1880–1933), to introduce the notion of phase transition of second order. A.J. Rutgers suggested its application to superconductivity. C.J. Gorter proceeded to calculate the difference in the Gibbs function of a superconductive sample in zero magnetic field and of the same sample in the normal state. At about the same period Lev Landau (1908–1968) attempted to show that the resulting superconductive state can have lower free energy than the state of random motion. Assuming uniform saturation current density, Landau showed that it is possible to find a balanced system of local currents which will be electrodynamically stable.

The End of Old Certainties

At the beginning of November 1933 there appeared a short letter in *Naturwissenschaften* by Walther Meissner (1882–1974) and R. Ochsenfeld (1901–1993) which presented strong evidence that, contrary to every expectation and belief of the past twenty years, a superconductor expelled the magnetic field. Superconductors were found to be diamagnetic. The letter noted several experimental arrangements, involving a pair of solid tin or lead cylinders or a cylindrical lead tube. In each case the sample was cooled below its transition point in a constant magnetic field. When the transition point was reached a sharp increase of flux was registered. Meissner and Ochsenfeld concluded that the magnetic flux in the specimen did not remain constant, but the lines of force were driven out of the superconductor, thereby increasing the flux in its neighbourhood. It appeared that the magnetic field was pushed out after the transition to the superconducting state and the magnetic flux became zero. The phenomenon of transition to the superconducting state turned out to be a reversible phenomenon: It did not matter whether the transition to the superconducting state had been realized in the presence of an external magnetic field or in the absence of such a field.

Gorter immediately sent a note to *Nature*, suggesting $B = 0$ to be a general characteristic of superconductivity. This meant that the condition $B = 0$ assumed in the thermodynamical treatment was not a restrictive hypothesis. In other words, after the Meissner-Ochsenfeld result, a superconductor could be regarded as a perfect conductor as well as a perfect diamagnet.

The Theory of Fritz and Heinz London

The first successful theory of superconductivity was formulated by Fritz London (1900–1954) and Heinz London (1907–1970). The Londons assumed that the diamagnetism must be taken to be an intrinsic property of an ideal superconductor, and not merely a consequence of perfect conductivity. They proposed that superconductivity demanded an entirely new relation in which the current was connected not with the electric, but with the magnetic field. The breakthrough came when they realised that the original acceleration equation proposed by Heinz in his doctorate and which involved a relation between time derivatives of the current and the magnetic field, could be integrated without having to add a constant of integration. Such an assumption would lead to the electrodynamics of a superconductor which were consistent both with the zero resistance and the Meissner effect. By the end of September 1934, Fritz and Heinz London had formulated the phenomenological theory of the electrodynamics of a superconductor which was published in the *Proceedings of the Royal Society* on November 13, 1934.

The "obvious" thing to do with the Meissner-Ochsenfeld result was to try to fit it into Maxwell's electrodynamics, but with the permeability changing to zero, the equations became indeterminate. The first such attempt to supplement Maxwell's equation was made by F. Becker, F. Sauter and C. Heller. They argued that in a superconductor, or rather in a body without any resistance, one cannot have any change of magnetic field, and they pointed out that, because of the inertia of the electrons, an applied electric field would accelerate them steadily. But the Londons objected to such an approach, feeling that the equations proposed by Becker, Heller and Sauter implied more than "is verified by experiment". What they proposed can be summarised as follows.

Maxwell's second equation of the electrodynamic field took the form

$$\mathrm{d}H/\mathrm{d}t = -c\,\mathrm{curl}\boldsymbol{E} = 0 \qquad (1)$$

and after integration

$$\boldsymbol{H} = \boldsymbol{H_0}$$

where H was the field in the specimen when the latter lost its resistance. If there are n electrons per cm^3 of mass m, charge e and velocity v, the current density $j = nev$, and

$$E = (4\pi\lambda^2/c^2)\delta\boldsymbol{J}/\delta t \qquad (2)$$

where λ is a constant. Taking curls on both sides of (1) and using Faraday's law

$$(4\pi\lambda^2/c^2)\mathrm{curl}\boldsymbol{J} = -\delta\boldsymbol{H}/\delta t \qquad (3)$$

Substituting in Maxwell's equation curl $\boldsymbol{H} = (4\pi/c)\boldsymbol{J}$,

$$\lambda^2\nabla^2\delta\boldsymbol{H}/\delta t = \delta\boldsymbol{H}/\delta t \qquad (4)$$

Integrating with respect to time, (4) became

$$\lambda^2\nabla^2(\boldsymbol{H} - \boldsymbol{H}_0) = (\boldsymbol{H} - \boldsymbol{H}_0) \qquad (5)$$

where $\boldsymbol{H}_0$ is an arbitrary field–the field which happened to be inside the body when it last lost its resistance. The general solution of (4), therefore, meant that, practically, the original field persisted in the superconductor for ever. Fritz and Heinz, however, noted that equation (1) implied more.

From the magnetic properties of a perfect conductor the simpler result $\delta\boldsymbol{H}/\delta t = 0$ (1) was obtained instead of (4). The novelty of (4) was in showing that the value $\delta\boldsymbol{H}/\delta \mathrm{t} = 0$ (or $\boldsymbol{H} = \boldsymbol{H}_0$) was also to be found only at a depth inside the metal greater than λ. Indeed, the solutions of this equation decreased exponentially as one receded from the surface, where they were fitted into the values of the external field. There was no point in developing this form of the theory any further, for equation (3) merely led to equation $\boldsymbol{H} = \boldsymbol{H}_0$ with the modification that the magnetic field penetrated the body to a small but finite depth. The Londons proposed that the connection between magnetic field **H** and current density $\boldsymbol{J}_s$ for the pure superconductive case may be given by the equation

$$(4\pi\lambda^2/c^2)\mathrm{curl}\mathbf{J} = -\boldsymbol{H} \qquad (6)$$

Equation (6) can be obtained by time integration from (3) if it is assumed that the constant of integration is zero ($\boldsymbol{H}_0 = 0$) and it was considered as a completion of Becker, Heller and Sauter's formalism by fixing the integration constant of the magnetic field according to the Meissner effect.

Equation (6) led to

$$\lambda^2\nabla^2\boldsymbol{H} = \boldsymbol{H} \qquad (7)$$

For large specimens, the characteristic feature of the solutions of this equation is that they decay exponentially into the interior of the specimen. At a distance λ from the surface the field is practically zero. Meissner's experimental result is represented by (6) with one restriction, namely that the magnetic flux decreases, not abruptly on the surface, but continuously in a very small interval below the surface. Equations (2) and (6) described the zero resistance and the Meissner effect respectively.

Equation (6) says more than (3), so far as it includes the Meissner effect. Proceeding from (6) to (3) by differentiating with respect to time, it is not possible to deduce (2). Nevertheless, the following weaker statement is obtained from (3).

$$\mathrm{curl}((4\pi\lambda^2/c^2)\mathbf{J} - \mathbf{E}) = 0$$

which shows that

$$(4\pi\lambda^2/c^2)\mathbf{J} - \mathbf{E} = \mathrm{grad}\mu$$

where μ is a scalar. On the other hand (2) leads not to (6) but only to its time derivative (3). Thus, the propositions (2) and (6) "posses, so to speak, the same degree of generality". [24]

It is not, then, unreasonable to take (6) to be "more fundamental" than (2), and this was an indication that a supercurrent could be regarded as a kind of diamagnetic current. In examining the relation between the behaviour of a ring and the Meissner effect, Fritz London showed that (6) can be expressed in such a way as to provide some clues for what was required of a fundamental theory of superconductivity. He suggested that the entire superconductor behaves as a "single big diamagnetic atom". He then went on to argue that if the ground state eigenfunction is "rigid" and, thus, not modified very much by an applied magnetic field, the current density will be proportional to the vector potential and, thus, give the equation which describes the Meissner effect.

Fritz and Heinz London supposed "the electrons to be coupled by some form of interaction. Then the lowest state of the electron may be separated by a finite distance from the excited ones" [24]. This may the earliest suggestion of an energy gap. In 1935 Fritz London showed that the average momentum of the electrons did not change in a superconductor when the field was applied, and he suggested that the reason may be a long range order which maintained the local average value of the momentum constant over large distances in space. This order would be maintained even in the presence of the magnetic field. The ordered ground state was regarded as a single quantum state extending throughout the metal. It was these considerations which led London to present for the first time his views about superconductivity as a macroscopic quantum phenomenon.

When London talked of a "macroscopic" interpretation he meant a phenomenological theory whose interpretation depended on a "microscopic" mechanism which set it apart from that used to explain ordinary conduction. The differentiating characteristic of this new microscopic mechanism was the *macroscopic* dimensions of the stationary waves.

Some Further Developments in the Theory of Superconductivity

The need to clarify the character of the electron-electron interaction was becoming more and more urgent. This was so, especially, since it was, still, very difficult to understand why the independent electron model of metals worked so well.

One of the first definite proposals for such an interaction was due to W. Heinsenberg (1901–1976). In 1947 Heisenberg suggested that the singular part of the Coulomb interaction could lead to superconductivity. Heisenberg assumed that in an electrically neutral metal, the first-order perturbation caused by this interaction vanished and that only the second-order perturbation was significant. For the lowest temperatures, Heisenberg suggested that there might be a very large number of "current threads" which are randomly distributed and did not give rise to a macroscopic current. However, if these current threads form a monocrystal by freezing, then the macrocurrent will persist in such a system. From such considerations, Heisenberg was able to derive the basic equations of the Londons.

In 1950 V.L. Ginzburg and Landau proposed a model where the energy needed to produce a change in the superconducting state over any distance was explicitly included in the theory. They worked out the thermodynamics of their model by defining a parameter ω, which was a measure of order in the superconducting phase and which was zero above the transition temperature. They, then, identified ω with the square of an effective ► wave function Ψ, which they set equal to the concentration of the superconducting electrons. Ψ did not describe a single particle, but the motion of the superconducting condensate as a whole. Their theory predicted correctly the dependence of critical field upon the temperature. When the effective wave function was considered constant, the Ginzburg–Landau theory gave the London equations.

Since the discovery of superconductivity there had been a widely and firmly held belief that the ion masses, being so much larger than the electron masses, could not play an important role in the establishment of the superconductive state. H. Frohlich in 1950 conceived the idea that just the "opposite of the 'dictum' contains the truth." [10] The ► quantum field theoretical treatment showed that the kinetic energy of the ions attached to a moving electron may be much smaller than the kinetic energy of the electron. Frohlich applied the field theoretic methods to the interaction of the electrons in a metal with the lattice vibrations, and he found that the interaction would lead to an attraction between the electrons. At the same time, and independently of what was predicted by these theoretical developments, experiments were undertaken to determine whether or not there was, in fact, a dependence of critical temperature on isotopic mass. These experiments showed, surprisingly at the time, that the critical temperature varied inversely with the square root of the isotopic mass.

Frohlich's 1950 paper was followed by John Bardeen (1908–1991) attempt to formulate an interaction between electrons and phonons as the basis of a theory of superconductivity. His model had many similarities with that of Frohlich. Nevertheless, the predicted condensation energy was too large. Also in such a case, where the properties of the state were dramatically altered, the use of perturbation theory was unjustified and M.R. Schafroth had shown that the theory could not lead to the Meissner effect. Nevertheless, most members of the community believed that in the assumption that electron-phonon interaction should be somehow responsible for superconductivity.

In 1952 Frohlich used a canonical transformation in order to circumvent the deadlocks of perturbation theory. He proposed an effective electron-phonon interaction without taking into account the Coulomb repulsion whose effects, it was possible, to shadow the Frohlich interaction. In 1955 Bardeen and D. Pines showed that this was not the case: for pairs of electrons whose energies were within a characteristic phonon energy of the Fermi surface, this attractive interaction would dominate the repulsive screened Coulomb interaction.

At about the same time Ginsburg and Schafroth, noted that such electron pairs would obey ▶ Bose-Einstein statistics. Schanfroth together with Blatt and Butler, in 1957 suggested that by choosing the form of the interaction between the particles, a Fermi gas would behave like a gas made up of charged bosons.

An important result was derived by L.N. Cooper in 1956. He showed that if there is an effective attractive interaction, a pair of quasi-particles above the Fermi sea will form a bound state no matter how weak the interaction. Thus, in the presence of attractive interactions, the Fermi sea which describes the ground state of the normal metal is unstable against the formation of such bound pairs.

Bardeen, Cooper and J.R. Schrieffer made the decisive step for the formulation of a microscopic theory of superconductivity. In 1957 they showed how to generalize the Cooper pair states to the many-body problem at absolute zero and derived an expression for the ground state energy difference between normal and superconducting states and for the energy gap at $t = 0$ K. They, then, extended their theory to obtain the excitation spectrum and made detailed calculations for various thermal and transport properties at temperatures above absolute zero showing that it was the second-order phase transition between electrons which caused them to couple to the phonons. In 1972 all three received the Nobel Prize for physics. John Bardeen became the first person to have received the Nobel Prize twice in the same field.

Literature

1. J. Bardeen: On superconductivity, in S. Flugge ed., *Encyclopedia of Physics*, Volume XV (1956)
2. J. Bardeen, L. N. Cooper, J. R. Schrieffer: *Theory of superconductivity*. Physical Review **108**, 1175–12000 (1957)
3. J. M. Blatt: *Theory of Superconductivity* (Academic, New York 1964)
4. F. Bloch: Some remarks on theory of superconductivity. Physics Today, **19**, 27–36 (1966)
5. H. Casimir: Superconductivity and Superfluidity, in J. Mehra ed. *The Physicist's Conception of Nature*, (Reidel, Dordrecht 1973)
6. P. F. Dahl: *Superconductivity, its Historical Roots and Development from Mercury to the Ceramic Oxides* (American Institute of Physics, New York 1973)
7. B. S. Deaver, W. M. Fairbank: Experimental evidence for quantized flux in superconducting cylinders. Physical Review Letters **7**, 43–46 (1961)

8. J. Dewar, J. A. Fleming: The electrical resistance of metals and alloys at temperatures approaching the absolute zero. Philosophical Magazine **XXXVI**, 271–299 (1896)
9. H. Frohlich: Isotope effect in superconductivity. Proceedings of the Physical Society (London) **A63**, 778 (1950)
10. H. Frohlich: The theory of superconductive state. Reports on Progress in Physics **24**, 1–23 (1961)
11. K. Gavroglu: *Fritz London (1900–1954), A Scientific Biography*. Afterword by John Bardeen (Cambridge University Press, Cambridge 1995)
12. K. Gavroglu, Y. Goudaroulis: *Methodological Aspects in the Development of Low Temperature Physics 1881–1957: Concepts out of Context(s)* (Kluwer, Dordrecht 1989)
13. K. Gavroglu, Y. Goudaroulis: The Remarkable Work of 'Le Gentleman du zero absolu', in *Through Measurement to Knowledge*: *The Selected Papers of Heike Kamerlingh Onnes 1853–1926* (Kluwer, Dordrecht 1991, 1–94)
14. K. Gavroglu, Y. Goudaroulis: Some Methodological and Historical Considerations in Low Temperature Physics. The Case of Superconductivity 1911–1957. Annals of Science **41**, 135–149 (1984)
15. C. J. Gorter: Some remarks on the thermodynamics of superconductivity. Archives du Musée Teyler, ser. III, Vol.VII, 378–387 (1933a)
16. C. J. Gorter, H. B. G. Casimir: On supraconductivity I. Physica **1**, 306–320 (1934)
17. W. Heisenberg: The electron theory of superconductivity. *Two Lectures* (Cambridge University Press, Cambridge 1949)
18. E. Jurkowitz: The Old School and the History of Science. Ph.D. Dissertation University of Toronto (1994)
19. H. Kamerlingh Onnes: The importance of accurate measurements at very low temperatures. CPL (Supplement) 9 (1904)
20. H. Kamerlingh Onnes: The liquefaction of helium. CPL, 108 (1908)
21. H. Kamerlingh Onnes: Further experiments with liquid helium, D. on the change of the electrical resistance of pure metals at very low temperatures. V. The disappearance of the resistance of mercury. CPL 122b and CPL 124c (1911)
22. H. Kamerlingh Onnes: Report on the researches made in the Leiden cryogenics laboratory between the second and third international congress of refrigeration. CPL (Supplement) 34b, 37–70 (1913)
23. M. Laue von: Theorie der Supraleitung. Naturwissenschaften **34**, 441–442 (1947)
24. F. London, H. London: The electromagnetic equations of the supraconductor. Proceedings of the Royal Society **149**, 71–88 (1935)
25. F. London: Macroscopical interpretation of supraconductivity. Proceedings of the Royal Society **152**, 24–34 (1935)
26. F. London: *Superfluids*. Vol.1 (Wiley, New York 1950)
27. J. C. McLennan: Opening address in a discussion on superconductivity and other low temperature phenomena. Proceedings of the Royal Society of London **A152**, 1–8 (1935)
28. W. Meissner, R. Ochsenfeld: Ein neuer Effect bei Eintritt der Supraleitfahigkeit. Die Naturwissenschaften **21**, 787–788 (1933)
29. W. Meissner: The magnetic effects occurring on transition to the supraconductive state. Proceedings of the Royal Society (London) **A152**, 13–15 (1935)

Superfluidity

Kostas Gavroglu

The Peculiar Properties of Helium

Ever since 1911 – three years after the liquefaction of helium – when Heike Kamerlingh Onnes (1853–1926) discovered that helium had a maximum density at about 2 K, there were various indications that at that temperature "something happens to helium." By the end of the 1930s the phenomena associated with liquid helium below 2.19 K would defy all the attempts to describe, let alone understand, the behaviour of liquid helium by classical hydrodynamics.

In 1930 Keesom (1876–1976) and van der Ende [1], quite accidentally, observed that liquid helium-II (liquid helium below 2.19 K) passed with remarkable ease through extremely small leaks – something which was not possible for higher temperatures, even for gaseous helium. This observation indicated an enormous drop of the viscosity when helium was below 2.19 K. During 1932, Keesom and Clusius reported that the ► specific heat curve had "an extremely sharp maximum" although there was no latent heat for the transition from helium-I (liquid helium above 2.19 K) to helium-II, but they could not figure out the "inner causes" for such a transition. Keesom decided to repeat the same measurements more accurately and in the paper he wrote with his daughter Anna Petronella, they proposed, after Paul Ehrenfest's (1880–1933) suggestion, for the first time the term "lambda point" to indicate the transition from helium-I to helium-II. They, then, attempted to measure the heat conduction in helium-II. They found that below the lambda-point "the heat conductivity is about 200 times that of copper at ordinary temperatures, or about 14 times that of very pure copper at liquid hydrogen temperatures. Hence liquid helium-I was by far the best heat conducting substance we know."[1]

When some years later, in 1935, the viscosity of helium was measured by Wilhelm, Misener and Clark in Toronto and in 1938 by Keesom and MacWood [2] in Leiden using the method of rotating disks, it was found that the change in viscosity was continuous, and even though it became less with the fall of temperature, it did not differ appreciably from that of helium-I. But the difference when compared to the results derived by the capillary method was about one million. Such an enormous difference in viscosity by the two different, yet equivalent methods could not be understood in the framework of classical hydrodynamics. More accurate viscosity measurements by Pyotr Kapitza (1894–1984) confirmed the earlier results and he used the term *superfluidity* to characterize this strange behaviour of helium.

"Perhaps the strangest of all the properties" was reported by Allen (1908–2001) and Jones in February 1938. Allen and Jones [3] wanted to extend the heat

[1] Keesom and Keesom (1936), 360.

conductivity experiments to lower and lower temperature differences and for that matter used an apparatus consisting of a reservoir capillary. When they supplied heat to the inner vessel, they saw that the inner helium level, far from being depressed, seemed to rise above that of the reservoir. The rise increased with heat input and, for constant input, with falling temperature. This was the "thermomechanical effect", a mass flow of helium opposing the heat current. In one of their experiments they used a powder-filled bulb, open at the bottom and with a narrow orifice at the top. When they heated the powder by shining a light on it, they observed a jet of liquid helium rising from the upper end to a height of several centimeters. The phenomenon was named fountain effect. Extremely small temperature differences between the reservoir and inner vessel were sufficient to produce a very large convection. It seemed, thus, impossible to treat the hydrodynamical and calorific properties of liquid helium-I independently.

In 1939 Daunt and Mendelssohn in Oxford and Kikoin (1908–1984) and Lasarew in Kharkov found that liquid helium flowed from one container to another inside it (or outside it depending on the relative height of the liquid helium surface) by means of a film of thickness of the order of 100 atoms formed on the walls. Such films, of course, are formed by any liquid which wets a solid surface, but the viscosity of an ordinary liquid is such that the film forms slowly and there is practically no movement. Helium-II is the only fluid which, owing to its superfluidity, forms a swiftly moving film.

A Strange Phenomenon Explained by an Even Stranger Mechanism

In November of 1937 the Centenary Conference for Van der Waals took place in Amsterdam. Among the speakers of the Conference was Mayer who had attempted to solve the general problem for any law of central force between the molecules. Kahn and Uhlenbeck showed that Mayer's treatment could be shown to be formally analogous to Einstein's equations for the ideal Bose gas – for which Einstein had predicted a condensation phenomenon. It was this work by Mayer which directed Fritz London's attention to the Einstein condensation paper.

S

In Fritz London's (1900–1954) proposed model each helium atom moved nearly free in the self-consistent periodic field formed by the other atoms similar to the way ► electrons move in a metal according to Bloch's theory – but with a crucial difference. The helium atoms obeyed ► Bose–Einstein statistics, whereas the electrons in metals obeyed ► Fermi–Dirac statistics. As a first step London disregarded the self-consistent field altogether and considered the ideal Bose–Einstein gas. Einstein had already discussed in 1924 a peculiar condensation phenomenon of this gas. But, because of Uhlenbeck's observation in his doctoral thesis "in the course of time the degeneracy of the Bose–Einstein gas has rather got the reputation of having only an imaginary existence."[2]

[2] London (1938a), 644.

Liquid helium-II, despite its high degree of "order," instead of being close to a "liquid" or solid crystal, is, owing to its extremely large volume, much closer to a gas than to an ordinary liquid. This gas-like nature combined with the high degree of order of helium-II prompted London to look closely into the possibilities provided by the phenomenon of Bose–Einstein condensation. But, since all real gases had been condensed in temperatures higher than the temperature where the ideal Bose–Einstein gas started this condensation phenomenon, the mechanism appeared to be "devoid of any practical significance."[3]

Fritz London's short paper in *Nature* was published on April 9, 1938 [4]. He started with a critique of Frohlich's scheme to account for the lambda transition as an order–disorder transition and directed his attention to an entirely different interpretation of this strange phenomenon. For an ideal Bose–Einstein gas the condensation phenomenon represented a discontinuity of the derivative of the specific heat. Such a discontinuity was experimentally observed for liquid helium.

In his paper published in the *Physical Review* in December 1938 [5] London attempted to provide an explanation for the transport properties. Below a certain temperature that depends on the mass and density of the particles, a finite fraction of them begins to *collect* in the lowest energy state, that is they assume zero momentum. The remaining particles have a velocity distribution similar to a classical gas, flying about as *individuals*. Since both components – the "condensed" and the "excited"– occupy the total volume of the container as if one was dissolved into the other, there is no condensation in the ordinary sense. "But if one likes analogies, one may say that there is actually a condensation, but *only in momentum space and not in ordinary space*"[4]. There was, then, an equilibrium of two phases. One contained the molecules of momentum zero and occupying in the space of momenta, a zero volume. The second phase contained molecules with a distribution over all the momenta as it was found in temperatures higher than the transition temperature. No separation of phases was to be found in ordinary space.

The Two-Fluid Model

Laszlo Tisza, a Hungarian born physicist, proceeded in 1938 to formulate the two-fluid model for superfluidity. Tisza's first step was to examine the concept of viscosity in liquids and gases in view of the discrepancy between the methods of measurement of viscosity and he concluded that this was not a kinetic coefficient of an unusual value, but the breakdown of the viscosity concept: there was no Navier–Stokes equation with a viscosity parameter! Tisza's paper in *Nature* on May 21, regarded helium-II [6] as a mixture of two (completely interpenetrating) components, the normal and the superfluid. These components or "fluids" are distinguished by different hydrodynamical behaviour, in addition to the difference in their heat

[3] London (1938b), 947.

[4] London (1938b), 951.

contents. A very narrow capillary (acting as an 'entropy filter') was permeable only to the superfluid flow, but not to the normal fluid. While the uncondensed normal fluid is supposed to retain the properties of an ordinary liquid (it is identical with helium-I), the condensed superfluid fraction of helium-II is meant to be incapable of taking part in dissipation processes. At absolute zero, the entire liquid is supposed to be a superfluid consisting of condensed atoms, while at the transition temperature this component vanishes. An oscillating disk in helium-II experienced friction by the normal fluid while a fine capillary allowed the superfluid to pass without experiencing friction. Similarly, an interpretation was provided for the thermo-mechanical effect. Since in this model the temperature of a volume of helium-II simply meant a relative concentration of the two fluids, a change in this concentration could be registered as either a cooling or a heating. Absorption of heat had the effect of increasing the concentration of the viscous component and also the osmotic pressure at the expense of the superfluid which was sucked into the cell.

This explanation led to the prediction of the inverse effect, namely that helium forced through a fine capillary should be richer in superfluid and, therefore, exhibit a drop in temperature. This effect known as "mechano-caloric effect" was observed in 1939 by Daunt and Mendelssohn. The anomalously high heat transport in helium-II was also consistent with the assumptions of the two-fluid model. The important thing here was that the superfluid and viscous components may have different flow velocities, giving rise to an "internal convection" which was connected with an energy transfer without any mass transfer. This internal convection accounted for the super heat-conductivity. Tisza predicted that the thermomechanical effect ought to have an inverse: a superfluid transfer from vessel A to B should lead to heating A and cooling B. This was readily verified.

A few months later in another short note presented to the *Academie des Sciences* in Paris, Tisza went much further; he recognized that this model implied a very strange feature, namely that in liquid helium-I the temperature would obey a wave equation. Tisza called these waves "temperature waves" – they would later be known as "second sound" and the temperature dependence of their velocity would be a decisive test of the validity of the two fluid model.

The Soviet Union, Kapitza and Landau

The phenomenal development of low temperature physics in the Soviet Union is justifiably tied to the career of Pyotr Kapitza. In fact, excluding some areas of applied physics, low temperature physics became the trademark of Soviet physics – especially during the war years. Kapitza and Lev Landau (1908–1968) were the towering figures. In 1941 Kapitza [7] published the results of his extensive measurements on the behaviour of the two kinds of helium. He put forward the hypothesis that the abnormal heat conductivity was due to heat transferred by convection currents. It was calculated that to explain the values of thermal conductivity observed by Keesom and Keesom in 1936 [8], the convection velocity must be assumed to be about $50\,\mathrm{m\,s^{-1}}$. Kapitza decided to measure this velocity, but his experiments yielded a heat transfer at least 20 times greater than that measured by Keesom. Consequently the convection velocity had to be of the order of $1{,}000\,\mathrm{m\,s^{-1}}$! It became quite obvious that the then accepted mechanisms for heat transfer could not be of much help in explaining such large convection velocities.

But it was "an accidental observation" [7][5] which gave their work an impetus in a totally new direction. Kapitza found that the pressure pulsations transmitted from the helium pipeline of the laboratory into the helium in the capillary caused substantial changes in the thermal conductivity. Kapitza suggested the possibility of two spatially separated mass currents, flowing into the bulb of the surface layer of the inner perimeter of the tube and outflowing through the center of the tube. In order to explain the great thermal conductivity of helium-II on the basis of this pattern of movement, Kapitza suggested that there is a difference between the heat function of helium in this film and in the free state, and thus the difference in heat content between the two mass currents was accounted for by the Van der Waals forces of the capillary wall on the surface of the layer of the liquid. This hypothesis led to the prediction that the thermal conductivity of helium would be strictly normal in the absence of surface phenomena. Subsequent experiments showed that the entropy of liquid helium flowing through the narrow tubes was, indeed, zero. This had been already predicted by both Tisza and London, but Kapitza thought that these schemata could not provide a "rigid theoretical basis"[6] for his observations and pointed to the theory of liquid helium proposed by Landau and published in the same year as his experiments.

Landau attempted to construct a ► quantum theory of liquids by direct ► quantization of the hydrodynamical variables such as the density, the current and the velocity without explicit reference to the interatomic forces. He considered the quantized states of the whole liquid instead of the single atoms, and started with the state of the fluid at absolute zero, which he considered as its ground state. Excitation of vorticity would represent departure from the zero temperature states. Departure from the ground state could also arise from the excitation of one or more units of sound-wave energy or "phonons." In this way, Landau constructed the energy spectrum of a liquid from two types of excitations; to the phonons of the solid body he added a spectrum of "rotons" which defined the elementary excitations of the vortex spectrum. Thus in Landau's theory, helium became a background liquid in which excitations moved, and there existed only one fluid: liquid helium. In a way, the ground states and the excitations played the role of the superfluid and the normal state respectively. The excitations were the normal state because they may be scattered and reflected, and hence, showed viscosity. The fluid associated with the ground state was superfluid because it could not absorb a phonon from the walls of the tube or a roton unless it was flowing with a velocity greater than the velocity of sound or a "critical velocity" respectively. Below the lesser of these two velocities the flowing helium would not interact with the walls and, hence,

---

[5] Kapitza (1980), 24.

[6] Ibid., 638.

would be superfluid – unless, as Landau pointed out, some other mechanism, as yet undetermined, limited the flow.

Landau's formalism led to two different equations for the propagation of sound, and, hence, to two velocities for sound. One of them was related to the usual velocity due to compressibility while the other depended strongly on the temperature. This was the same phenomenon as Tisza's thermal or temperature waves. Landau named them "second sound." The first and unsuccessful attempt to generate and detect second sound waves was made with acoustic apparatus by Shalnikov and Sokolov. They failed and that was interpreted by London to mean that Landau's theory was "born refuted." The failure to observe second sound acoustically was explained in 1944 by Lifshitz (1915–1985) who made a more detailed theoretical analysis of second sound waves and showed that if one used the usual mechanical methods for generating sound, then "second sound" is masked by the ordinary sound. But a plate with a periodically varying temperature would radiate only the "second sound." Using such a "radiator," Peshkov in 1944 was able to demonstrate the existence of standing thermal waves for the first time.

These results were communicated to the International Conference which took place at Cambridge in the spring of 1946 [9] devoted to low temperature physics and elementary ► particle physics, even though, the scientists from the Soviet Union did not attend the Conference and where London gave the opening paper titled "The present state of the theory of liquid helium." London insisted that both superconductivity and superfluidity were manifestations of quantum mechanisms on a macroscopic scale and that the decisive test between his and Tisza's approach and Landau's schema would be the study of the low temperature properties of helium-3 where the absence of superfluidity would be ascribed to the role of statistics.

"Second Sound" at Very Low Temperatures

Peshkov's new measurements for the second sound velocities between 1.36 and 2.19 K, were not in agreement with Landau's prediction for this temperature range. Landau proceeded to modify the energy spectrum of the phonons after the results. The measurements appeared to agree with Tisza's predictions, but the predictions of Tisza and Landau were approximately similar down to 1K, but sharply diverged below 1K. The velocity first went through a maximum for which both theories gave identical results, and then went through a minimum rising sharply as the temperature approached absolute zero.

Pellam's measurements in 1949 below 1.4K, showed an increase in velocity and differed considerably from Peshkov's. It was the experiments of Maurer and Herlin in 1949 [10] that settled the issue of the temperature dependence of the second sound velocity below 1K. Using the pulse method initiated by Peshkov, they were able to lower the temperature to 0.85K and observe an increase of velocity starting at about 1.1K. The results were quite unambiguous and could have been used to corroborate Landau's approach – if it weren't for the new experiments, completed at the same time, which tried to detect superfluidity in a pure liquid sample of He^3, and found negative indications down to 1.05K. Maurer and Herlin appeared to believe that the results were not necessarily contradicting the predictions of the Bose–Einstein hypothesis, but they felt that further refinements should be introduced in the model to account for the second sound velocity results. Few months later Pellam and Scott [11] would also observe the increase of the second sound velocity in the very low temperature range, and would be of the same opinion as to the relevance of these measurements to distinguish between the two competing theories.

The measurements that corroborated Landau's theory came from the Mond Laboratory at Cambridge. In 1950 Atkins and Osborne [12], using two different demagnetizations, were able to measure velocities down to 0.17K. They found that there was a marked increase that could be extrapolated to 0K and found it to equal Landau's prediction.

The Importance of Liquid He^3

In 1949 measurements on the viscosity of He^3 were reported from Argonne National Laboratory. The viscosity was measured by letting He^3 pass through a fine slit, and it did not show any discontinuity down to 1.05K. London felt that these measurements had confirmed the dependence of superfluidity on statistics and decided to send a review article to *Nature*. He no longer insisted on the difference of the second-sound velocity at temperatures around 1K, but rather on the implications of the statistics. He believed that what the reported absence of superfluidity of He^3 settled, was the issue concerning the necessity of the assumption of Bose–Einstein statistics for any theory professing to provide an explanation for the properties of helium-II. William Fairbank (1917–1986) examined the extent to which He^3 behaved as an ideal Fermi–Dirac gas, by measuring the strengths of the He^3 nuclear magnetic resonance signals as the temperature of the liquid He^3 was reduced. When measurements were resumed below 1.2K there was a definite departure from the predictions of the Curie law and the liquid appeared to behave as an ideal Fermi–Dirac gas having a degeneracy temperature of 0.45K. Furthermore, one of the best known results derived by Fairbank was the discovery of the flux quantization, predicted by London, by detecting macroscopic quantization of the magnetic field outside a superconductor.

By 1956 Richard Feynman (1918–1988) was able present a theory synthesizing the views of London and Landau. Considering all previous theories as phenomenological, his microscopic theory did not "supplant the phenomenological theories. It turns out to support them."[7] He showed that despite the strong forces of interaction between helium atoms which could have undermined the ideal gas approximation by London, they did indeed allow the Bose–Einstein condensation. He also showed that some of Landau's assumptions which were rather empirical could be justified quantum mechanically and that the rotons were a kind of quantum mechanical analog of a microscopic vortex ring.

[7] Feynman (1953a), 1291.

Literature

1. W. Keesom, J. N. van der Ende: The specific heat of solids obtainable with liquid helium. IV. Measurements of the atomic heats of tin and zinc. CPL **219b** (1932)
2. W. Keesom, G. E. MacWood: The viscocity of liquid helium. Physica **5**, 737–744 (1938)
3. J. F. Allen, H. Jones: New phenomena connected with heat flow in helium II. Nature **141**, 243–244 (1938)
4. F. London: The λ-phenomenon of liquid helium and the Bose-Einstein degeneracy. Nature **141,** 643–644 (1938)
5. F. London: On the Bose-Einstein condensation. Physical Review **54**, 947–954 (1938)
6. L. Tisza: Transport phenomena in helium-II. Nature **141**, 913 (May 21, 1938)
7. P. Kapitza: Problems of liquid helium. A Report at the General Assembly of the USSR Academy of Sciences, 28 December 1940. Reprinted in P. Kapitza: *Experiment, Theory, Practice, Articles and Addresses* (Reidel Publishers, Boston 1980)
8. W. Keesom, A. P. Keesom: On the heat conductivity of liquid helium. Physica **3**, 359–360 (1936)
9. V. Peshkov: Determination of the velocity of propagation of the second sound in Helium II. Journal of Physics (USSR) **10**, 389–398 (1946)
10. R. D. Maurer, M. A. Herlin: Second sound velocity in helium II. Physical Review **76**, 948–950 (1949)
11. J. R. Pellam, R. B. Scott: Second sound velocity in paramagnetically cooled liquid helium II. Physical Review **76,** 869–870; 948–950 (1949)
12. K. R. Atkins, D. V. Osborne: The velocity of second sound below 1 K, Philosophical Magazine **41**, 1078–1081 (1950)
13. E. L. Andronikashvilii: *Reflections on Liquid Helium,* translated by R. Berman (American Institute of Physics 1989)
14. B. S. Deaver, W. M. Fairbank: Experimental evidence for quantized flux in superconducting cylinders. Physical Review Letters **7,** 43–46 (1961)
15. R. Feynman: Atomic theory of the lambda transition in helium. Physical Review **91**, 1291–1301 (1953)
16. R. Feynman: Atomic theory of liquid helium near absolute zero. Physical Review **91**, 1301–1310 (1953)
17. K. Gavroglu: *Fritz London (1900–1954), A Scientific Biography.* Afterward by John Bardeen (Cambridge University Press, Cambridge 1995)
18. K. Gavroglu, Y. Goudaroulis: Some Methodological and Historical Considerations in Low Temperature Physics II: The case of Superfluidity. Annals of Science **43**, 137–146 (1986)
19. K. Gavroglu, Y. Goudaroulis: *Methodological Aspects in the Development of Low Temperature Physics 1881–1957: Concepts out of Context(s)* (Kluwer, Dordrecht 1989)
20. K. Gavroglu, Y. Goudaroulis: The Remarkable Work of 'Le Gentleman du zero absolu', *in Through Measurement to Knowledge: The Selected Papers of Heike Kamerlingh Onnes 1853–1926* (Kluwer, Dordrecht 1991, 1–94)
21. A. Griffin: *Excitations in a Bose-Condensed Liquid* (Cambridge University Press 1994)
22. H. Kamerlingh Onnes: The importance of accurate measurements at very low temperatures. **CPL** (Supplement) 9 (1904)
23. H. Kamerlingh Onnes: The liquefaction of helium. **CPL**, 108 (1908)
24. H. Kamerlingh Onnes: Report on the researches made in the Leiden cryogenics laboratory between the second and third international congress of refrigeration. **CPL** (Supplement) 34b, 37–70 (1913)
25. P. L. Kapitza: *The Collected Papers of P.L. Kapitza*, edited by D. ter Haar (Pergamon 1965)
26. W. H. Keesom: Quelques remarques en rapport avec l'anomalie de la chaleur specifique de l'helium au point lambda, CPL (Supplement) **71e**, 47–52 (1932)
27. W. H. Keesom: *Helium* (Elsevier 1942)
28. W. Keesom, A. P. Keesom: On the anomaly in the specific heat of liquid helium. CPL **221d**, 19–26 (1932)

29. W. Keesom, M. Wolfke: Two different liquid states of helium. CPL **190b**, 17–22 (1927)
30. A. Kozhevnikov: Piotr Kapitza and Stalin's government: a study in moral choice. Historical Studies in the Physical and Biological Sciences **22**, 131–163 (1992)
31. L. D. Landau: The theory of superfluidity of helium II. Journal of Physics (USSR), **5**, 71–90 (1941); On the hydrodynamics of helium II, Ibid. **8**, 1–3 (1944)
32. L. D. Landau: On the theory of superfluidity of Helium II. Journal of Physics (USSR), **11**, 9–92 (1947)
33. L. D. Landau: On the theory of superfluidity. Physical Review **75**, 884–885 (1949)
34. E. M. Lifshitz: Radiation of sound in helium II. Journal of Physics **8**, 110–114 (1944)
35. E. Maxwell: Isotope effect in the superconductivity of mercury. Physical Review **78**, 477 (1950)
36. L. Tisza: The theory of liquid helium. Physical Review **72**, 838–854 (1947)

Superluminal Communication in Quantum Mechanics

Daniel J. Gauthier

One consequence of the special theory of relativity is that no information can be transmitted between two parties in a time shorter than it would take light, propagating through vacuum, to travel between the parties. That is, the speed of information transfer is less than or equal to the speed of light in vacuum c. Hypothetical faster-than-light (superluminal) communication is very intriguing because causality would be violated [8]. Causality is a principle where an event is linked to a previous cause; superluminal communication would allow us to change the outcome of an event after it has happened. I'm sure all of us at one point in our lives would like a cell-phone with superluminal capabilities!

Soon after Einstein published the theory of relativity, scientists began the search for examples where objects or entities travel faster than c. There are many known examples of superluminal motion [8], yet explaining, in simple terms, why such motions do not violate the special theory or allow for superluminal communication can be exceedingly difficult. Also, approximations used to solve models of the physical world can lead to subtle errors, sometimes resulting in predictions of superluminal signaling. For these reasons, studying superluminal signaling can be an interesting exercise because it often reveals unexpected aspects of our universe or the theories we use to describe its behavior.

The possibility of superluminal motions in classical physics have been known for over a century. For example, the group velocity of a pulse of light propagating through a dispersive dielectric can exceed c, where the group velocity gives (approximately) the speed of the peak of the pulse [10]. There exists a simple mathematical proof demonstrating that such behavior cannot be used for superluminal communication, but this proof sheds little insight on recent experiments that report clear evidence fast group velocities. One current explanation is that points of non-analyticity are created on the optical waveform at each moment when new information is encoded on the optical carrier, and that these points travel precisely at *c* [6]. Other points on the waveform (such as the pulse peak) convey no new information that cannot already be determined from the non-analyticity point and hence fast motion of the waveform in between points of non-analyticity do not violate the special theory. Another example of apparent superluminal motion occurs in certain expanding galaxies, known as superluminal stellar objects [12]. This motion can be explained by considering motions of particles whose speed is just below *c* (i.e., highly relativistic) and moving nearly along the axis connecting the object and the observer. Hence, these are not superluminal motions after all.

Quantum mechanics also appears to provide a mechanism for superluminal communication because of its nonlocal characteristic. A measurement performed on a system ▶ wave function collapse at all locations simultaneously [11], an effect that does not occur in classical physics and hence deserves further consideration with regards to superluminal communication.

One *gedanken* experiment that has received recent attention involves correlated particles generated by an Einstein–Podolsky–Rosen (▶ EPR problem) source. For concreteness, let's consider a system that generates two correlated photons (▶ light quantum) that travel in opposite directions and have zero total angular momentum. Furthermore, two observers, Alice and Bob, are located on opposite sides and at large distances from the source. They are equipped with optical components that can analyze the state of polarization of the arriving photons. Bob is slightly further away from the source than Alice, and we want to establish a one-way superluminal communication link from Alice to Bob.

In one scenario, Alice places a special type of polarizing beam splitter that spatially separates one state of linear polarization (say vertical, V) from the other state of polarization (horizontal, H). The output ports of the polarizing beam splitter are directed to single-photon detectors. Bob has an identical apparatus, which is at a great distance from Alice, and he aligns the axis of his polarizing beam splitter the same as Alice's. Because of the fact that their total angular momentum of the photons is zero, whenever Alice measures V, the wavefunction collapses and Bob is assured of measuring H essentially instantaneously after Alice performs her measurement. Similarly, Bob will measure V whenever Alice measures H. In this configuration, the polarization beam splitters and single-photon detectors perform measurements in the "linear" basis.

Alice and Bob can also perform measurements in the "circular" basis, where the analysis apparatus will determine whether the photons are left circular (LC) polarized or right circular (RC) polarized. This measurement can be performed by placing a birefringent plate – known as a quarter-wave plate – in front of the polarizing beam splitters, where the optical axis of the plate is orientated at 45 degrees to the axis of the linear polarizing beam splitter. The birefringent plate converts incident circularly polarized light into either H or V linearly polarized light, which is subsequently analyzed by the polarizing beam splitter and detectors. With the waveplate in the system, Bob is assured to measure LC (RC) whenever Alice measures RC (LC).

The communication scheme is based on a change of measurement basis. By inserting the waveplate in the setup or not, Alice can force Bob's photon to be either linear or circular polarized. Thus, Alice can transmit binary information to Bob by inserting – or not – the waveplate in her apparatus. All he has to do is to determine with certainty whether Alice was using the linear or circular basis. The first hitch with this scheme is a well known classical result – the only way to measure whether a optical beam is linear or circular polarized is to analyze it both with linear and circular polarizers. In other words, Bob would have to send the photon through the linear-basis apparatus *and* the circular-basis apparatus. Unfortunately, one apparatus destroys the incident photon as a result of the measurement and hence it is unavailable to send on the other.

A clever way to get around this problem is for Bob to "clone" the incident photon so there are two copies, where one copy will be sent to a linear-basis apparatus and the other is sent to a circular-basis apparatus. The process of stimulated emission of radiation is thought to clone an incident photon, so scientists first considered placing an optical amplifier in the path of the photon (an optical amplifier increases the number of photons via the stimulated emission process) [5, 9]. Unfortunately, an optical amplifier adds additional photons – via the process of spontaneous emission – to the beam path and these additional photons have an arbitrary state of polarization [4]. These "junk" photons destroy the benefits of the amplifier and hence prevent Alice from communicating with Bob via the nonlocal characteristics of quantum mechanics.

The problem with the superluminal communication scheme is much deeper that it appears from the discussion above. The very linearity of quantum mechanics prevents the cloning of an arbitrary quantum state, a result of the ► *no-cloning theorem*. Thus, any device – not just an optical amplifier – fails to clone the incident photon and hence the communication scheme fails [2, 4, 7].

Other researchers have wondered whether an imperfect copy of the incident photon would be sufficient for superluminal communication. The best or optimal quantum copying machine has been identified [1]; even with the best possible copying apparatus, the quantum communication scheme just barely fails. This failure is nicely summarized by Gisin [3] in his 1998 paper: "Once again, quantum mechanics is right at the border line of contradicting relativity, but does not cross it. The peaceful coexistence between quantum mechanics and relativity is thus re-enforced." See also ► Einstein locality; locality; nonlocality.

Primary Literature

1. V. Bužek, M. Hillery: Quantum copying: Beyond the no-cloning theorem. Phys. Rev. A **54**, 1844 (1996)
2. D. Dieks: Communication by EPR devices. Phys. Lett. **92A**, 271 (1982)
3. N. Gisin: Quantum cloning without signaling. Phys. Lett. A **242**, 1 (1998)

4. R.J. Glauber: Amplifiers, attenuators, and Schrödinger's Cat,' in *New Techniques and Ideas in Quantum Measurement Theory*, ed. by D. M. Greenberg, Ann. N.Y. Acad. Sci. **480**, 336 (1986)
5. N. Herbert: FLASH – A superluminal communicator based upon a new type of quantum measurement. Found. Phys. **12**, 1171 (1982)
6. M. D. Stenner, D. J. Gauthier, M. A. Neifeld: The speed of information in a 'fast light' optical medium. Nature **425**, 695 (2003)
7. W. K. Wootters, W. H. Zurek: A single quantum cannot be cloned. Nature **299**, 802 (1982)

Secondary Literature

8. M. Fayngold: *Special Relativity and Motions Faster Than Light* (Wiley-VCH Verlag GmbH, Weinheim 2002, Sect. 6.15)
9. N. Herbert: *Faster Than Light* (New American Library Books, Penguin Inc., New York, 1998)
10. P. W. Milonni: *Fast Light, Slow Light, and Left-Handed Light* (Institute of Physics Publishing, Bristol 2005, Chap. 3)
11. H. Paul: *Introduction to Quantum Optics, From Light Quanta to Quantum Teleportation* (Cambridge University Press, Cambridge, 2004, Sect. 11.5)
12. J. A. Zensus, T. J. Pearson (eds.): *Superluminal Radio Sources* (Cambridge University Press, Cambridge, 1987)

Superposition Principle (Coherent and Incoherent Superposition)

Marianne Breinig

S

In non-relativistic quantum mechanics, the state of a physical system at a fixed time t is defined by specifying a ket $|\psi(t)\rangle$ belonging to the space ε. ε is a complex, separable ► Hilbert space, a complex linear vector space in which an inner product is defined and which possesses a countable ► orthonormal basis. Every measurable physical quantity is called an observable and is described by a Hermitian operator acting in ε. The only possible results of a measurement are the eigenvalues of the Hermitian operator associated with the measurement, and immediately after the measurement the state ket is a corresponding eigenstate. Every Hermitian operator has at least one basis of orthonormal eigenvectors. Every state vector $|\psi(t)\rangle$ can therefore be written as a linear superposition of eigenvectors of any observable. If two Hermitian operators commute a common eigenbasis can be found. If they do not commute, then no common eigenbasis exists.

Let $\{|a_n\rangle\}$ be an orthonormal basis of eigenvectors of the operator A,

$$\mathrm{A}\,|a_n\rangle = a_n|\,a_n\rangle. \tag{1}$$

For simplicity assume that the eigenvalues are not degenerate. Let $|\psi_1\rangle$ and $|\psi_2\rangle$ be two normalized eigenvectors of the operator B with eigenvalues b_1 and b_2, respectively.

$$\mathrm{B}|\psi_1\rangle = b_1|\psi_1\rangle, \mathrm{B}|\psi_2\rangle = b_2|\psi_2\rangle. \tag{2}$$

If B is the Hamiltonian H, then $b_1 = E_1$ and $b_2 = E_2$. If A and B do not commute, i.e. $[\mathrm{A}, \mathrm{B}] \neq 0$, then $|\psi_1\rangle$ and $|\psi_2\rangle$ are linear superpositions of eigenvectors of A. Assume that $[\mathrm{A}, \mathrm{B}] \neq 0$ and that a measurement at $t = 0$ determines $|\psi(0)\rangle = |\psi_1\rangle$. If B is the Hamiltonian, then the measurement determines that the system is in a stationary state. The probability that a subsequent measurement of A will yield the eigenvalue a_n is $\mathrm{P}_1(a_n) = |\langle a_n|\psi_1\rangle|^2$. Similarly, if $|\psi(0)\rangle = |\psi_2\rangle$ then $\mathrm{P}_2(a_n) = |\langle a_n|\psi_2\rangle|^2$. Now consider a system in a *normalized pure state* (► states, pure and mixed)

$$|\psi\rangle = \lambda_1|\psi_1\rangle + \lambda_1|\psi_2\rangle, \langle\psi|\psi\rangle = 1, |\lambda_1|^2 + |\lambda_2|^2 = 1. \tag{3}$$

If B is the Hamiltonian, then the system is not in a stationary state, it is in a coherent superposition of stationary states.

The probability that a measurement of B will yield b_1 is $|\langle\psi_1|\psi\rangle|^2 = |\lambda_1|^2$. The probability that a measurement of B will yield b_2 is $|\lambda_2|^2$. The probability that a measurement of A will yield a_n is

$$\begin{aligned} \mathrm{P}(a_n) &= |\langle a_n|\psi\rangle|^2 \\ &= \langle a_n|\psi\rangle\langle\psi|a_n\rangle = |\lambda_1|^2\mathrm{P}_1(a_n) + |\lambda_2|^2\mathrm{P}_2(a_n) \\ &\qquad + 2\mathrm{Re}(\lambda_1\lambda_2^* \times \langle a_n|\psi_1\rangle\langle\psi_2|a_n\rangle) \\ &\neq |\lambda_1|^2\mathrm{P}_1(a_n) + |\lambda_2|^2\mathrm{P}_2(a_n). \end{aligned} \tag{4}$$

The last term in the expression for $\mathrm{P}(a_n)$ describes interference effects. If a system is in a pure state which is a coherent superposition of eigenstates of an observable B and we measure an observable A which does not commute with B, then we must take interference effects into account when predicting the result of a measurement. We may consider $\mathrm{P}(a_n) = |\langle a_n|\psi\rangle|^2$ as the square of the probability amplitude $\langle a_n|\psi\rangle = \langle a_n|\lambda_1\psi_1\rangle + \langle a_n|\lambda_2\psi_2\rangle$. The probability amplitude is the weighted sum of the probability amplitudes $\langle a_n|\psi_1\rangle$ and $\langle a_n|\psi_2\rangle$. To obtain the probability $\mathrm{P}(a_n)$ for a linear superposition of states, we take the square of the weighted sum of the probability amplitudes, not the sum of the squares.

A pure state is not a *statistical mixture of states*. The concept of a statistical mixture of states (► mixed state) is used when dealing with incomplete information about the initial state of a system. Assume it is only known that the system is in one of the eigenstates $\{|\psi_k\rangle\}$ of the operator B and that it has the probability p_k ($\sum_k p_k = 1$) of being in the pure state $|\psi_k\rangle$. If B is the Hamiltonian, the system then is in an incoherent superposition of stationary states. If the system is in a statistical mixture of the states $|\psi_1\rangle$ and $|\psi_2\rangle$ with weights $p_1 = |\lambda_1|^2$ and $p_2 = |\lambda_2|^2$ respectively, then the probability of measuring a_n is $P(a_n) = |\lambda_1|^2\,\mathrm{P}_1(a_n) + |\lambda_2|^2$

$P_2(a_n)$. Interference effects are absent for an incoherent superposition or a statistical mixture of states. We cannot describe a statistical mixture using an "average state vector". In general, when dealing with a statistical mixture, probabilities enter at two levels. The initial information about the system is given in terms of probabilities, and the predictions of Quantum Mechanics are probabilistic.

A simple example:

Let the operator B be the Hamiltonian of the system, $B = H$, $b_1 = E_1$, $b_2 = E_2$, and let $|\psi(0)\rangle = \lambda_1|\psi_1\rangle + \lambda_1|\psi_2\rangle$. Then $|\psi(t)\rangle = \lambda_1 \exp(-\mathrm{i}E_1 t/\hbar)|\psi_1\rangle + \lambda_2 \exp(-\mathrm{i}E_2 t/\hbar)|\psi_2\rangle$, and

$$\begin{aligned} P(a_n) = {} & |\lambda_1|^2 P_1(a_n) + |\lambda_2|^2 P_2(a_n) \\ & + 2\mathrm{Re}(\lambda_1\lambda_2^* \times \exp(-\mathrm{i}(E_1 - E_2)t/\hbar)\langle a_n|\psi_1\rangle\langle\psi_2|a_n\rangle) \end{aligned}$$

$P(a_n)$ now is time dependent and oscillates with a frequency $\nu_{12} = (E_1 - E_2)/h$. We observe quantum beats.

Primary Literature

1. C. Cohen-Tannoudji, B. Diu, F. Laloë: *Quantum Mechanics*, Volume One (Wiley, New York 1977)

Secondary Literature

2. P.A.M. Dirac: *The Principles of Quantum Mechanics*, 4th edn. (Oxford University Press, Oxford 1958)

Superselection Rules

S

Domenico Giulini

General Notion

The notion of *superselection rule* (henceforth abbreviated SSR) was introduced in 1952 by Wick (1909–1992), Wightman, and Wigner (1902–1995) [9] in connection with the problem of consistently assigning intrinsic parity to elementary particles. They understood an SSR as generally expressing "restrictions on the nature and scope of possible measurements".

The concept of SSR should be contrasted with that of an ordinary ▸ selection rule (SR). The latter refers to a dynamical inhibition of a certain transition, usually due to the existence of a conserved quantity. Well known SRs in Quantum Mechanics concern radiative transitions of atoms. For example, in case of electric dipole radiation, they take the form $\Delta J = 0, \pm 1$ (except $J = 0 \to J = 0$) and $\Delta M_J = 0, \pm 1$. They say that the ▶ quantum numbers J, M_J associated with the atom's total angular momentum may at most change by one unit. But this is only true for electric dipole transitions, which, if allowed, represent the leading-order contribution in an approximation for wavelengths much larger than the size of the atom. The next-to-leading-order contributions are given by magnetic dipole and electric quadrupole transitions, and for the latter $\Delta J = \pm 2$ *is* possible. This is a typical situation as regards SRs: They are valid for the leading-order modes of transition, but not necessarily for higher order ones. In contrast, a SSR is usually thought of as making a more rigorous statement. It not only forbids certain transitions through particular modes, but altogether as a matter of some deeper lying principle; hence the "Super". In other words, transitions are not only inhibited for the particular dynamical evolution at hand, generated by the given ▶ Hamiltonian operator, but for all conceivable dynamical evolutions.

More precisely, two states ψ_1 and ψ_2 are separated by a SR if $\langle \psi_1 \mid H \mid \psi \rangle = 0$ for the given Hamiltonian H. In case of the SR mentioned above, H only contains the leading-order interaction between the radiation field and the atom, which is the electric dipole interaction. In contrast, the states are said to be separated by a SSR if $\langle \psi_1 \mid A \mid \psi_2 \rangle = 0$ for *all* (physically realisable) ▶ observables A. This means that the relative phase between ψ_1 and ψ_2 is not measurable and that coherent superpositions of ψ_1 and ψ_2 cannot be verified or prepared. It should be noted that such a statement implies that the set of (physically realisable) observables is strictly smaller than the set of all ▶ self-adjoint operators on ▶ Hilbert space. For example, $A = \mid \psi_1 \rangle \langle \psi_2 \mid + \mid \psi_2 \rangle \langle \psi_1 \mid$ is clearly self-adjoint and satisfies $\langle \psi_1 \mid A \mid \psi_2 \rangle \neq 0$. Hence the statement of a SSR always implies a restriction of the set of observables as compared to the set of all (bounded) self-adjoint operators on Hilbert space. In some sense, the existence of SSRs can be formulated in terms of observables alone (see below).

Since all theories work with idealisations, the issue may be raised as to whether the distinction between SR and SSR is really well founded, or whether it could, after all, be understood as a matter of degree only. For example, dynamical ▶ decoherence is known to provide a very efficient mechanism for generating apparent SSRs, without assuming their existence on a fundamental level [11] [14].

Elementary Theory

In the most simple case of only two *superselection sectors*, a SSR can be characterised by saying that the ▶ Hilbert space $\mathcal{H}$ decomposes as a direct sum of two orthogonal subspaces, $\mathcal{H} = \mathcal{H}_1 \oplus \mathcal{H}_2$, such that under the action of each observable vectors in $\mathcal{H}_{1,2}$ are transformed into vectors in $\mathcal{H}_{1,2}$ respectively. In other words, the action of observables in Hilbert space is reducible, which implies that $\langle \psi_1 \mid A \mid \psi_2 \rangle = 0$ for each $\psi_{1,2} \in \mathcal{H}_{1,2}$ and all observables A. This constitutes an inhibition to the ► superposition principle in the following sense: Let $\psi_{1,2}$ be normed vectors and $\psi_+ = (\psi_1 + \psi_2)/\sqrt{2}$, then

$$\langle \psi_+ \mid A \mid \psi_+ \rangle = \tfrac{1}{2}\big(\langle \psi_1 \mid A \mid \psi_1 \rangle + \langle \psi_2 \mid A \mid \psi_2 \rangle\big) = \mathrm{Tr}(\rho A), \tag{1}$$

where

$$\rho = \tfrac{1}{2}\big(\mid \psi_1 \rangle\langle \psi_1 \mid + \mid \psi_2 \rangle\langle \psi_2 \mid\big). \tag{2}$$

Hence, considered as state (expectation-value functional) on the given set of observables, the ► density matrix ρ corresponding to ψ_+ can be written as non-trivial convex combination of the (pure) density matrices for ψ_1 and ψ_2 and therefore defines a ► mixed state rather than a pure state (► states, pure and mixed). Relative to the given observables, coherent ► superpositions of states in $\mathcal{H}_1$ with states in $\mathcal{H}_2$ do not exist.

In direct generalisation, a characterisation of *discrete SSRs* can be given as follows: There exists a finite or countably infinite family $\{P_i \mid i \in I\}$ of mutually orthogonal ($P_i P_j = 0$ for $i \neq j$) and exhaustive ($\sum_{i \in I} P_i = 1$) ► projection operators ($P_i^\dagger = P_i$, $P_i^2 = P_i$) on Hilbert space $\mathcal{H}$, such that each observable commutes with all P_i. Equivalently, one may also say that states on the given set of observables (here represented by density matrices) commute with all P_i, which is equivalent to the identity

$$\rho = \sum_i P_i \rho P_i. \tag{3}$$

We define $\lambda_i := \mathrm{Tr}(\rho P_i)$ and let $I' \subset I$ be the subset of indices i for which $\lambda_i \neq 0$. If we further set $\rho_i := P_i \rho P_i / \lambda_i$ for $i \in I'$, then (3) is equivalent to

$$\rho = \sum_{i \in I'} \lambda_i \rho_i, \tag{4}$$

showing that ρ is a non-trivial convex combination if I' contains more than one element. The only pure states are the projectors onto rays within a single $\mathcal{H}_i$. In other words, only vectors (or rays) in the union (not the linear span) $\bigcup_{i \in I} \mathcal{H}_i$ can correspond to pure states. If, conversely, *any* non-zero vector in this union defines a pure state, with different rays corresponding to different states, one speaks of an *abelian superselection rule*. The $\mathcal{H}_i$ are then called *superselection sectors* or *coherent subspaces* on which the observables act irreducibly. The subset Z of observables commuting with all observables is then given by $Z := \big\{\sum_i a_i P_i \mid a_i \in \mathbb{R}\big\}$. They are called *superselection-* or *classical observables*.

In the general case of *continuous SSRs* $\mathcal{H}$ splits as direct integral of an uncountable set of Hilbert spaces $\mathcal{H}(\lambda)$, where λ is an element of some measure space Λ, so that

$$\mathcal{H} = \int_\Lambda \mathrm{d}\mu(\lambda)\mathcal{H}(\lambda) \tag{5}$$

S

with some measure $d\mu$ on Λ. Observables are functions $\lambda \mapsto O(\lambda)$, with $O(\lambda)$ acting on $\mathcal{H}(\lambda)$. Closed subspaces of $\mathcal{H}$ left invariant by the observables are precisely given by

$$\mathcal{H}(\Delta) = \int_{\Delta} d\mu(\lambda)\mathcal{H}(\lambda)\,, \tag{6}$$

where $\Delta \subset \Lambda$ is any measurable subset of non-zero measure. In general, a single $\mathcal{H}(\lambda)$ will not be a subspace (unless the measure has discrete support at λ).

In the literature, SSRs are discussed in connection with a variety of superselection-observables, most notably univalence, overall mass (in non-relativistic QM), electric charge, baryonic and leptonic charge, and also time.

Algebraic Theory of SSRs

In ► Algebraic Quantum Mechanics, a system is characterised by a C^*–algebra $\mathcal{C}$. Depending on contextual physical conditions, one chooses a faithful representation $\pi : \mathcal{C} \to \mathcal{B}(\mathcal{H})$ in the (von Neumann) algebra of bounded operators on Hilbert space $\mathcal{H}$. After completing the image of π in the weak operator-topology on $\mathcal{B}(\mathcal{H})$ (a procedure sometimes called *dressing* of $\mathcal{C}$ [12]) one obtains a vonNeumann sub-algebra $\mathcal{A} \subset \mathcal{B}(\mathcal{H})$, called the *algebra of (bounded) observables*. The physical observables proper correspond to the self-adjoint elements of $\mathcal{A}$.

The *commutant* $\mathcal{S}'$ of any subset $\mathcal{S} \subseteq \mathcal{B}(\mathcal{H})$ is defined by

$$\mathcal{S}' := \{A \in \mathcal{B}(\mathcal{H}) \mid AB = BA\,, \forall B \in \mathcal{S}\}\,, \tag{7}$$

which is automatically a von Neumann algebra. One calls $\mathcal{S}'' := (\mathcal{S}')'$ the von Neumann algebra generated by $\mathcal{S}$. It is the smallest von Neumann sub-algebra of $\mathcal{B}(\mathcal{H})$ containing $\mathcal{S}$, so that if $\mathcal{S}$ was already a von Neumann algebra one has $\mathcal{S}'' = \mathcal{S}$; in particular, $\big(\pi(\mathcal{C})\big)'' = \mathcal{A}$.

SSRs are now said to exists if and only if the commutant $\mathcal{A}'$ is not trivial, i.e. different from multiples of the unit operator. Projectors in $\mathcal{A}'$ then define the sectors. Abelian SSRs are characterised by $\mathcal{A}'$ being abelian. $\mathcal{A}'$ is often referred to as *gauge algebra*. Sometimes the algebra of physical observables is *defined* as the commutant of a given gauge algebra. That the gauge algebra is abelian is equivalent to $\mathcal{A}' \subseteq \mathcal{A}'' = \mathcal{A}$ so that $\mathcal{A}' = \mathcal{A} \cap \mathcal{A}' =: \mathcal{A}^c$, the centre of $\mathcal{A}$. An abelian $\mathcal{A}'$ is equivalent to *Dirac's requirement*, that there should exist a complete set of commuting observables [7] (cf. Chap. 6 of [14]).

In ► Quantum Logic a quantum system is characterised by the lattice of propositions (corresponding to the closed subspaces, or the associated projectors, in Hilbert-space language). The subset of all propositions which are compatible with all other propositions is called the *centre of the lattice*. It forms a Boolean sub-lattice. A lattice is called *irreducible* if and only if its centre is trivial (i.e. just consists of 0, the smallest lattice element). The presence of SSRs is now characterised by a non-trivial centre. Propositions in the centre are sometimes called *classical*.

SSRs and Conserved Additive Quantities

Let Q be the operator of some charge-like quantity that behaves additively under composition of systems and also shares the property that the charge of one subsystem is independent of the state of the complementary subsystem (here we restrict attention to two subsystems). This implies that if $\mathcal{H} = \mathcal{H}_1 \otimes \mathcal{H}$ is the Hilbert space of the total system and $\mathcal{H}_{1,2}$ those of the subsystems, Q must be of the form $Q = Q_1 \otimes 1 + 1 \otimes Q_2$, where $Q_{1,2}$ are the charge operators of the subsystems. We also assume Q to be conserved, i.e. to commute with the total Hamiltonian that generates time evolution on $\mathcal{H}$. It is then easy to show that a SSR for Q persists under the operations of composition, decomposition, and time evolution: If the density matrices $\rho_{1,2}$ commute with $Q_{1,2}$ respectively, then, trivially, $\rho = \rho_1 \otimes \rho_2$ commutes with Q. Likewise, if ρ (not necessarily of the form $\rho_1 \otimes \rho_2$) commutes with Q, then the reduced density matrices $\rho_{1,2} := \mathrm{Tr}_{2,1}(\rho)$ (where Tr_i stands for tracing over $\mathcal{H}_i$) commute with $Q_{1,2}$ respectively. This shows that if states violating the SSR cannot be prepared initially (for whatever reason, not yet explained), they cannot be created though subsystem interactions [10]. This has a direct relevance for ► measurement theory, since it is well known that an exact von Neumann measurement of an observable P_1 in system 1 by system 2 is possible only if P_1 commutes with Q_1, and that an approximate measurement is possible only insofar as system 2 can be prepared in a superposition of Q_2 eigenstates [2].

As already indicated, the foregoing reasoning does not explain the actual existence of SSRs, for it does not imply anything about the *initial* nonexistence of SSR violating states. In fact, there are many additive conserved quantities, like momentum and angular momentum, for which certainly no SSRs is at work. The crucial observation here is that the latter quantities are physically always understood as *relative* to a system of reference that, by its very definition, must have certain localisation properties which exclude the total system to be in eigenstate of *relative* (linear and angular) momenta. Similarly it was argued that one may have superpositions of relatively charged states [1]. A more completeaccount of this conceptually important point, including a comprehensive list of references, is given in Chap. 6 of [14].

SSRs and Symmetries

Symmetries in quantum mechanics are often implemented via *unitary ray- representations* rather than proper unitary representations (here we discard anti-unitary ray-representations for simplicity). A unitary ray-representation is a map U from the symmetry group G into the group of ► unitary operators on Hilbert space $\mathcal{H}$ such that the usual condition of homomorphy, $U(g_1)U(g_2) = U(g_1 g_2)$, is generalised to

$$U(g_1)U(g_2) = \omega(g_1, g_2)\, U(g_1 g_2)\,, \tag{8}$$

where $\omega : G \times G \to \mathrm{U}(1) := \{z \in \mathbb{C} \mid |z| = 1\}$ is the so-called *multiplier* that satisfies

$$\omega(g_1, g_2)\omega(g_1 g_2, g_3) = \omega(g_1, g_2 g_3)\omega(g_2, g_3)\,, \tag{9}$$

for all g_1, g_2, g_3 in G, so as to ensures associativity: $U(g_1)\big(U(g_2)U(g_3)\big) = \big(U(g_1)U(g_2)\big)U(g_3)$. Any function $\alpha : G \to \mathrm{U}(1)$ allows to redefine $U \mapsto U'$ via $U'(g) := \alpha(g)U(g)$, which amounts to a redefinition $\omega \mapsto \omega'$ of multipliers given by

$$\omega'(g_1, g_2) = \tfrac{\alpha(g_1)\alpha(g_2)}{\alpha(g_1 g_2)}\,\omega(g_1, g_2)\,. \tag{10}$$

Two multipliers ω and ω' are called *similar* if and only if (10) holds for some function α. A multiplier is called *trivial* if and only if it is similar to $\omega \equiv 1$, in which case the ray-representation is, in fact, a proper representation in disguise.

The following result is now easy to show: Given unitary ray-representations $U_{1,2}$ of G on $\mathcal{H}_{1,2}$, respectively, with non-similar multipliers $\omega_{1,2}$, then no ray-representation of G on $\mathcal{H} = \mathcal{H}_1 \oplus \mathcal{H}_2$ exists which restricts to $U_{1,2}$ on $\mathcal{H}_{1,2}$ respectively. From this a SSR follows from the requirement that the Hilbert space of pure states should carry a ray-representation of G, since such a space cannot contain invariant linear subspaces that carry ray-representations with non-similar multipliers.

An example is given by the SSR of univalence, that is, between states of integer and half-integer ▶ spin. Here G is the group $SO(3)$ of proper spatial rotations. For integer spin it is represented by proper unitary representations, for half integer spin with non-trivial multipliers. Another often quoted example is the Galilei group, which is implemented in non-relativistic quantum mechanics by non-trivial unitary ray-representations whose multipliers depend on the total mass of the system and are not similar for different masses.

Such derivations have sometimes been criticised (e.g. in [15]) for depending crucially on one's prejudice of what the symmetry group G should be. The relevant observation here is the following: Any ray-representation of G can be made into a proper representation of a larger group $\bar{G}$, which is a *central extension* of G. But no superselection rules follow if $\bar{G}$ rather than G were required to be the acting symmetry group on the set of pure states. For example, in case of the rotation group, $G = SO(3)$, it is sufficient to take $\bar{G} = SU(2)$, its double (and universal) cover. For G the 10-parameter inhomogeneous Galilei group it is sufficient to take for $\bar{G}$ an extension by the additive group $\mathbb{R}$, which may even be motivated on classical grounds [6].

SSRs in Local Quantum Field Theories

In ▶ quantum field theory SSRs can arise from the restriction to (quasi) local observables. Charges which can be measured by fluxes through closed surfaces at arbitrarily large spatial distances must then commute with all observables. A typical example is given by the total electric charge, which is given by the integral over space of the local charge density ρ. According to Maxwell's equations, the latter equals the divergence of the electric field $\boldsymbol{E}$, so that Gauß' theorem allows to write

$$Q = \lim_{R \mapsto \infty} \int_{\|\boldsymbol{x}\|=R} (\boldsymbol{n} \cdot \boldsymbol{E}) \mathrm{d}\sigma \,, \tag{11}$$

where $\boldsymbol{n}$ is the normal to the sphere $\|\boldsymbol{x}\| = R$ and $\mathrm{d}\sigma$ its surface measure. If A is a local observable its support is in the *causal complement* of the spheres $\|\boldsymbol{x}\| = R$ for sufficiently large R. Hence, in the quantum theory, A commutes with Q. It is possible, though technically far from trivial, that this formal reasoning can indeed be justified in Local ► Quantum Field Theory [8]. For example, one difficulty is that Gauß' law does not hold as an operator identity.

In modern local quantum-field theory [13], representations of the quasi-local algebra of observables are constructed through the choice of a preferred state on that algebra (GNS-construction), like the Poincaré invariant vacuum state, giving rise to the *vacuum sector*. The superselection structure is restricted by putting certain selection conditions on such states, like e.g. the Doplicher–Haag–Roberts (DHR) selection criterion for theories with mass gap (there are various generalisations [13]), according to which any representation should be unitarily equivalent to the vacuum representation when restricted to observables whose support lies in the causal complement of a sufficiently large (causally complete) bounded region in spacetime. Interestingly this can be closely related to the existence of ► gauge groups whose equivalence classes of irreducible unitary representations faithfully label the superselection sectors. Recently, a systematic study of SSRs in "locally covariant quantum field theory" was started in [5]. Finally we mention that SSRs may also arise as a consequence of non-trivial spacetime topology [3].

Environmentally Induced SSRs

The ubiquitous mechanism of ► decoherence effectively restricts the *local* verification of coherences [14]. For example, scattering of light on a particle undergoing a ► double-slit experiment *delocalises* the relative-phase information for the two beams along with the escaping light. Hence effective SSRs emerge locally in a practically irreversible manner, albeit the correlations are actually never destroyed but merely delocalised. The emergence of effective SSRs through the dynamical process of decoherence has also been called *einselection* [11]. For example, this idea has been applied to the problem of why certain molecules naturally occur in eigenstates of chirality rather than energy and ► parity, i.e. why sectors of different chirality seem to be superselected so that chirality becomes a classical observable. This is just a special case of the general question of how classical behaviour can emerge in Quantum Theory. It may be asked whether *all* SSRs are eventually of this dynamically emergent nature, or whether strictly fundamental SSRs persist on a kinematical level [14]. The complementary situation in theoretic modelling may be characterised as follows: Derivations of SSRs from axiomatic formalisms lead to exact results on models of only approximate validity, whereas the dynamical approach leads to approximate results on more realistic models.

SSRs in Quantum Information

In the theory of ► quantum communication a somewhat softer variant of SSRs is defined to be a restriction on the allowed local operations (completely positive and trace-preserving maps on density matrices) on a system [4]. In general, it therefore leads to constraints on (bipartite) ► entanglement. Here the restrictions considered are usually not thought of as being of any fundamental nature, but rather for mere practical reasons. For example, without an external reference system for the definition of an overall spatial orientation, only "rotationally covariant" operations $\mathcal{O} : \rho \mapsto \mathcal{O}(\rho)$ are allowed, which means that $\mathcal{O}$ must satisfy

$$\mathcal{O}\big[U(g)\rho U^{\dagger}(g)\big] = U(g)\mathcal{O}(\rho)U^{\dagger}(g) \quad \forall g \in SO(3)\,, \tag{12}$$

where U is the unitary representation of the group $SO(3)$ of spatial rotations in Hilbert space. Insofar as the local situation is concerned, this may be rephrased in terms of the original setting of SSRs, e.g. by regarding $SO(3)$ as gauge group, restricting local observables and states to those commuting with $SO(3)$. On the other hand, one also wishes to consider situations in which, for example, a local bipartite system (Alice and Bob) is given a state that has been prepared by a third party that is *not* subject to the SSR.

Primary Literature

1. Y. Aharonov, L. Susskind: *Charge superselection rule*. Phys. Rev., **155**(5), 1428–1431 (1967)
2. H. Araki, Y. Mutsuo M. *Measurement of quantum mechanical operators*. Phys. Rev., **120**(2), 622–626 (1960).
3. A. Ashtekar, A. Sen. *On the role of space–time topology in quantum phenomena: Superselection of charge and emergence of nontrivial vacua*. J. Math. Phys., **21**(3), 526–533 (1980).
4. S. D. Bartlett, H. M. Wiseman. *Entanglement constrained by superselection rules*. Phys. Rev. Lett., **91**(9), 097903 (2003)
5. R. Brunetti, G. Ruzzi. *Superselection sectors and general covariance. I.* Commun. Math. Phys., **270**(1), 69–108 (2007)
6. D. Giulini. *On Galilei invariance in quantum mechanics and the Bargmann superselection rule*. Ann. Phys. (New York), **249**(1), 222–235 (1996)
7. Josef Maria Jauch. *Systems of observables in quantum mechanics*. Helvetica Phys. Acta **33**, 711–726 (1960)
8. Franco Strocchi and Arthur Strong Wightman. *Proof of the charge superselection rule in local relativistic quantum field theory*. J. Math. Phys., **15**(12), 2198–2224 (1974) Erratum: J. Math. Phys., **17**(10), 1930–1931 (1976)
9. G. C. Wick, A. S. Wightman, E. P. Wigner. *The intrinsic parity of elementary particles*. Phys. Rev., **88**(1), 101–105 (1952)

10. G. C. Wick, A. S. Wightman, E. P. Wigner. *Superselection rule for charge.* Phys. Rev., **1**(12), 3267–3269 (1970)
11. W. H. Zurek. *Decoherence, einselection, and the quantum origins of the classical.* Rev. Mod. Phys., **75**, 715–775 (2003)

Secondary Literature

12. N.N. Bogolubov et al: *General Principles of Quantum Field Theory*, volume 10 of 'Mathematical Physics and Applied Mathematics'. Kluwer Academic Publishers, Dordrecht, 1990
13. R. Haag: *Local Quantum Physics: Fields, Particles, Algebras. 'Texts and Monographs in Physics'*. Springer Verlag, Berlin, second revised and enlarged edition, 1996.
14. E. Joos et al: *Decoherence and the Appearence of a Classical World in Quantum Theory.* Springer Verlag, Berlin, second revised and enlarged edition, 2003.
15. S. Weinberg: *The Quantum Theory of Fields. Volume I Foundations.* Cambridge University Press, Cambridge, 1995

Symmetry

K. Mainzer

Symmetry concepts play a central role in physics [13]. The ► invariance (► covariance) properties of a system under specific symmetry transformations can either be related to the conservation laws of physics or be able to establish the structure of the fundamental interactions. This is the most essential aspect of symmetry as it concerns the basic principles of physics and the interactions themselves and not only the properties of a particular system [14].

In geometry, figures or bodies are called symmetrical when they possess common measures or proportions. Thus the Platonic bodies can be rotated and turned at will without changing their regularity. Similarity transformations, for example, leave the geometric form of a figure unchanged, i.e. the proportional relationships of a circle, equilateral triangle, rectangle, etc. are retained, although the absolute dimensions of these figures can be enlarged or decreased. Therefore one can say that the form of a figure is determined by the similarity transformations that leave it unchanged (invariant). In mathematics, a similarity transformation is an example of an automorphism [12]. In general an automorphism is the mapping of a set (e.g. points, numbers, functions) onto itself that leaves unchanged the structure of this set (e.g. proportional relations in Euclidean space). Automorphisms can also be characterized algebraically in this way: (1) Identity I that maps every element of a set onto itself, is an automorphism. (2) For every automorphism T an inverse automorphism T' can be given, with $T \cdot T' = T' \cdot T = I$. (3) If S and T are automorphisms, then so is the successive application $S \cdot T$. A set of elements with a composition that fulfils these three axioms is called a group. The symmetry of a mathematical structure is determined by the group of those automorphisms that let it unchanged (invariant).

S

Symmetry transformations can be classified in two classes: continuous and discrete transformations. Continuous transformations are in turn divided into global and local transformations. By definition, a symmetry transformation is said to be continuous if the set of parameters, which are necessary to describe the transformation, range over a continuous set of values. Examples of continuous transformations are the translation in space, the rotation around a given axis, and the translation in time. These symmetry transformations are global because once the transformation of a given point in space has been fixed, then the transformation at all other points in space is also fixed. Basic principles of physics like linear momentum conservation, angular momentum conservation and energy result from the symmetry properties of the interactions under global space and time continuous transformations [15]. According to Emmy Noether's theorem [1], a Lagrangian theory possesses N conserved quantities, if the theory (i.e. the Lagrangian function) is invariant under a N-parameter continuous transformation. Noether's theorem is not only a cornerstone of classical physics, but, by the ► correspondence principle, of quantum physics as well.

The state space of a quantum system is a ► Hilbert space of finite or countably-infinite dimension. A quantum state is a one-dimensional subspace of the state space H. Any normalized vector in the one-dimensional subspace of a state can be used to represent this state, and is called a state vector. The original formulation of quantum mechanics assumed a one-to-one correspondence of one-dimensional subspaces of the state space with physical states, implying the unrestricted validity of the ► superposition principle for state vectors. This requirement is equivalent to the exclusion of ► superselection rules. A statement that selects some vectors, adding that they are physically unrealizable as state vectors is called a superselection rule. If there are superselection rules, then there exist subspaces of the state space that cannot be connected to each other by any observable. Not all ► self-adjoint operators on the state space are therefore ► observables [16].

Ignoring superselection rules, the states of a quantum system span a projective Hilbert space. Every vector ψ in the Hilbert space H determines a one-dimensional subspace, called the ray $\bar{\psi}$. The inner product of two rays $\bar{\psi}$ and $\bar{\varphi}$ is defined by

$$\langle\bar{\psi} \mid \bar{\varphi}\rangle = \frac{|\langle\psi \mid \varphi\rangle|}{\|\psi\| \cdot \|\varphi\|}$$

The set of all rays in H is called the projective Hilbert space $\bar{H}$ associated with the Hilbert space H. A symmetry transformation of quantum mechanics is an automorphism of the projective Hilbert space $\bar{H}$ associated with the state space H. Thus the symmetry of quantum mechanics is given by the automorphism group Aut($\bar{H}$). A theorem of Eugene P. Wigner [2] asserts that the automorphism group Aut($\bar{H}$) can be represented by the group of ► unitary operators acting on the state space H. Let H_1 and H_2 be Hilbert spaces and F be a mapping from H_1 into H_2. Then F is called linear if $F\,(a\psi + b\varphi) = aF\psi + bF\varphi$

F is called antilinear if $F(a\psi + b\phi) = a^* F\psi + b^* F\phi$
F is called isometric if $\|F\psi\| = \|\psi\|$

for all ψ and φ from H_1 and all complex numbers a and b. If the range of a linear isometric operator $F: H_1 \rightarrow H_2$ is the whole space H_2, then F is called unitary. An antiunitary operator $F: H_1 \rightarrow H_2$ is an antilinear isometric operator having the range H_2. Wigner's theorem implies that two realizations of quantum mechanics whose state spaces are connected by a unitary or antiunitary transformation are from a logical point of view equivalent. Historically, the fact that symmetries in quantum mechanics are described by projective unitary representations has been known since Hermann Weyl. In his book on *Gruppentheorie und Quantenmechanik* (1928) he stated: 'The pure case or state is (...) more properly represented by the ray than by the vector, and we must therefore operate in the ray field in system space rather than in the vector field.' [2] Wigner published his theorem in his textbook (1931) without full proof. A complete proof was given by V. Bargmann (1954) [4].

In quantum physics, all the properties of a system can be derived from the state or ► wave function associated with that system. The absolute phase of a wave function cannot be measured, and has no practical meaning, as it cancels out the calculations of the probability distribution. Only relative phases are measurable in an interference experiment. Therefore it is possible to change the phase of a wave function without leading to any observable effect. Formally a phase transformation of the wave function $\psi(x, t)$ can be written as

$$\psi(x, t) \rightarrow \psi'(x, t) = e^{i\alpha}\psi(x, t)$$

with the parameter (phase) α of the transformation. If α is constant, i.e. the same for all points in space-time, the equation expresses the fact that once a phase convention has been made at a given point in space-time, the same convention must be adopted at all other points. This is an example of a global transformation applied to the field $\psi(x, t)$. If $\alpha = \alpha(x, t)$ is a function of space and time, then such a transformation will not leave invariant any equation of $\psi(x, t)$ with space or time derivatives. This is in particular true for the ► Schrödinger equation or any relativistic wave equation for a free particle. In order to satisfy the invariance under a local phase transformation it is necessary to modify the equations in some way, which describe the form of interaction. Such modifications will introduce additional terms, which describe the interaction of the particle with external fields. The question if and which force of interaction is realized can only be decided empirically. This is the gauge principle or principle of local symmetry. Historically, the principle of gauge invariance (► gauge symmetry) dates back to a (false) idea of Weyl who assumed a deeper dependence between the laws of matter and electromagnetism [5].

A discrete symmetry transformation is described by parameters ranging over a discrete set of values. Examples are symmetry operations that leave unaffected a crystal by reflections through planes, inversions with respect to a centre point and rotations around a given axis with angles $2\pi/n (n = 2, 3, 4$ or $6)$ corresponding to the periodicity of the crystal lattice. In elementary ► particle physics, there are three discrete transformations for interactions between leptons and quarks: the charge conjugation C, the parity transformation P, and the time reversal T. In a charge conjugation operation

$$C : q_\alpha \to -q_\alpha$$

All the particles of a system are replaced by their antiparticles and therefore all charges q_α change sign. The parity transformation

$$P : \boldsymbol{r} \to -\boldsymbol{r}$$

corresponds to a space inversion relative to a point. In a system of Cartesian coordinates, a point with coordinates (x, y, z) transforms into $(-x, -y, -z)$ under the parity operation. The position vector **r** changes sign under a space inversion. The time reversal operation

$$T : t \to -t$$

corresponds to the inversion of the time variable t. The laws of physics are invariant with respect to T. Symmetry of time means that it is physically impossible to distinguish between forward and backward moving in time. Quantum theory of fields requires the invariance of the fields and interactions under the combined transformations of the three operations *CPT*. The *CPT*-theorem was proved by Wolfgang Pauli in 1957 [6]. If one of the three symmetries is violated, then, according to the ► *CPT*-theorem, one of the other two symmetries has also to be violated. For example, the violation of parity P requires that C or T be violated. If the invariance under the combination of two transformations holds, then the invariance under the third transformation must also hold. For example, the invariance under *CP* implies the invariance under T and vice-versa. The decay of Kaons is the only known example of time violation T which is enforced by a *CP*-violation. Further on, the *CPT* invariance implies that the masses and the lifetime of a particle is identical to those of antiparticles. *CPT* invariance has been empirically confirmed to very high precision [17].

Before 1956, it was assumed that ► parity was a fundamental symmetry of physical processes. In 1956, Tsung Dao Lee and Chin Ning Yang examined the question of whether processes driven by the weak interaction would distinguish left or right [7]. Their famous experiments performed in the beta decay of ^{60}Co, and in the weak decays of pions and muons, $\pi^+ \to \mu^+ + \nu_\mu$ and $\mu^+ \to e^+ + \nu_e + \bar{\nu}_\mu$ not only provided the empirical support to the suggestions of Lee and Yang but also showed that parity violation was an universal property of the weak interaction.

The observation of parity violation was soon incorporated in the theory of weak interaction and is now a part of modern unified theory of electro-weak interactions, the Standard Model ► quantum field theory; particle physics. Actually, the fundamental physical forces of interaction can be characterized by local gauge symmetries. The unitary group $U(n)$ and the special unitary group $SU(n)$ refer to the unitary transformation of a n-dimensional complex coordinate space [12]. In the standard model, gravitation, electromagnetic, weak and strong interaction are represented by local Poincaré-, $U(1)$-, $SU(2)$-, and $SU(3)$- gauge groups. The research program of unified theories tries to unify the fundamental forces step by step in states of higher energy characterized by unified local symmetries. In 1954, the Yang-Mills theory tried at first to unify proton and neutron by a gauge theory of isospin-symmetry [8]. But the Yang-Mills theory only predicted massless gauge particles of interaction in contradiction to empirical observations. Later on, J. Goldstone [9] and P. Higgs [10] introduced the mechanism of spontaneous symmetry breaking (Higgs' mechanism) in order to give appropriate gauge particles the desired mass. The intuitive idea is that a symmetric theory can have asymmetric consequences. For example, the equations of a ball and the wheel of a roulette are symmetric with respect to the rotation axis, but the ball always keeps lying in an asymmetric position. In a first step, electromagnetic and weak forces could already be unified at very high energies in an accelerator ring. For energies of more than 100 Gigaelectron–Volts and distances less than 10^{-16} cm, there would be a perfect $U(1) \times SU(2)$ symmetry, in which the $W^{\pm}$ and Z^{o} field quanta would be exchanged as rapidly as the photon. Their transformations are described by the same symmetry group $U(1) \times SU(2)$. At a critical value of lower energy the symmetry spontaneously breaks apart into two partial symmetries $U(1)$ of electromagnetic force and $SU(2)$ of weak interaction. The gauge particles of weak interaction get their mass by the Higgs mechanism, the photon of electromagnetic interaction remains massless.

After the successful unification of electromagnetic and weak interaction physicists try to realize the "big" unification of electromagnetic, weak and strong forces, and in a last step the "superunification" of all four forces. There are several research strategies of superunifications such as supergravity and superstring theories. Mathematically they are described by extensions of richer structures of local symmetries and their corresponding gauge groups. On the other hand the variety of elementary particles is actualized by spontaneous symmetry breaking. The concept of local symmetry and symmetry breaking play an immense role in cosmology. During cosmic expansion and cooling temperature, the initial unified supersymmetry of all forces broke apart into the subsymmetries of physical interactions, and the corresponding elementary particles were crystallized in phase stages leading to more variety and complexity.

The phases of cosmic expansion are determined by properties of symmetry breaking. For example, in the case of weak interaction, neutrinos occur only as a left-handed helix, but not as a right-handed one which means parity violation. This kind of antisymmetry or dissymmetry seems also to be typical for molecular structures of life. Protein analysis shows that amino acids have an antisymmetrical carbon atom and occur only in the left-handed configuration. Weak interaction takes part in the chemical bonds. Thus, cosmic parity violation of weak interaction is assumed to cause the selection of chiral molecules. The reason is that the left-handed (L) and right-handed (D) examples of chiral molecules can be distinguished by a tiny parity violating energy difference ΔE_{pv}. The energetically stable examples (e.g., L-form of amino acids) are preserved. But, this assumption is only based on theoretical calculations (e.g., Hartree–Fock procedures in physical chemistry). We still miss exact measurements of experiments because of the tiny small parity violation energy difference ΔE_{pv} (e.g., $4 \cdot 10^{-14}$ $(hc)\text{cm}^{-1}$ (H_2O_2), $1 \cdot 10^{-12}$ $(hc)\text{cm}^{-1}$ (H_2S_2)), although there are proposed experiments with spectroscopic methods [11].

From a philosophical point of view, the epistemic question arises whether symmetry only concerns syntactic and semantic properties of scientific theories and their models, or whether they are real structures of the world. Empirical structuralism defends a strict empiristic view [18]: symmetry only refer to syntactical and semantic properties of mathematical structures which are inventions of the human mind. But if they are only syntactical and semantic constructions, why do observations, measurements and predictions display these regularities? It seems to be a wonder or miracle. Hilary Putnam put it in the "no miracle-argument" of scientific realism: "The positive argument for realism is that it is the only philosophy that doesn't make the success of science a miracle" [19]. Structural realism assumes that mathematical structures refer to real structures of the world, independent of syntactical and semantic representations in the human mind. The question is which mathematical terms and models refer to ontological structures [20]. In general, the gauge principle only determines the form of the coupling term of physical interaction. But the existence of a physical force is an empirical question which, of course, cannot be derived from an a priori demand of local symmetry. A gauge group characterizes a physical interaction mathematically in terms of local symmetry. It is epistemically remarkable that only gauge-invariant quantities have observable effects. Local phase transformations do not change any measurable observable. Therefore, the gauge principle or demand for local symmetry can epistemically be considered as a filter of observables in a theory of physical interactions ([21] cf. [22]).

Primary Literature

1. E. Noether: Invariante Variationsprobleme. Nachr. Ges. Wiss. Göttingen, Math. Phys. Kl. (1918) 235–257
2. H. Weyl: *Gruppentheorie und Quantenmechanik* (Hirzel, Leipzig 1928; English trans. *The Theory of Groups and Quantum Mechanics*, Dover, New York 1950, 75)
3. E.P. Wigner: *Gruppentheorie und ihre Anwendung auf die Quantenmechanik der Atomspektren* (Vieweg, Braunschweig 1931; English trans. Group Theory, Academic, New York 1959)
4. V. Bargmann: On Unitary Ray Representation of Continuous Groups. Ann. Math. **59**, 1–46 (1954)
5. H. Weyl: Gravitation und Elektrizität. Sitzungsberichte der Königlich Preußischen Akademie der Wissenschaften zu Berlin (1918) 465–480; H. Weyl: Gravitation and the Electron. Proc. Nat. Acad. Sci. U.S.A. 15, 323–334 (1929)
6. W. Pauli: *Niels Bohr and the Development of Physics* (Pergamon Press, London 1957)
7. T.D. Lee, C.N. Yang: Questions of Parity Conservation in Weak Interactions. Phys. Rev. **104**, 254 (1956)
8. C.N Yang, R.L. Mills: Conservation of Isotopic Spin and Isotopic Gauge Invariance. Phys. Rev. **96**, 191–195 (1954)
9. J. Goldstone: Field theories with 'Superconductor' solutions. N. Cimento **19**, 154–164 (1961)
10. P.W. Higgs: Broken symmetries, massless particles, and gauge fields. Phys. Lett. **12**, 132–133 (1964)
11. M. Quack: Molecular spectra, reaction dynamics, symmetries and life. Chimia **57**, 147–160 (2003)

Secondary Literature

12. K. Mainzer: *Symmetrien der Natur* (De Gruyter, Berlin 1988, English trans. *Symmetries of Nature*, De Gruyter, New York 1996); K. Brading, E. Castellani (eds.): *Symmetries in Physics* (Cambridge University Press, Cambridge 2003)
13. K. Mainzer: *Symmetry and Complexity. The Spirit and Beauty of Nonlinear Science* (World Scientific, Singapore 2005)
14. M.G. Doncel, A. Hermann, L. Michel, A. Pais (Eds.): *Symmetries in Physics (1600-1980)* (Servei de Publicacions, UAB, Barcelona 1987)
15. E. Schmutzer: *Symmetrien und Erhaltungssätze der Physik* (Akademie, Berlin 1972)
16. H. Primas: *Chemistry, Quantum Mechanics and Reductionism* (Springer, Berlin 1983)
17. H. Genz, R. Decker: *Symmetrie und Symmetriebrechung in der Physik* (Vieweg, Braunschweig 1991, p. 160)
18. B.C. van Fraassen: *Laws and Symmetry* (Clarendon Press, Oxford 1989)
19. H. Putnam: *Mathematics, Matter and Method.* Philosophical Papers Vol. 1 (Cambridge University Press, Cambridge 1975, p. 73)
20. M. Kuhlmann, H. Lyre, A. Wayne (eds.): *Ontological Aspects of Quantum Field Theory* (World Scientific, Singapore 2002)
21. K. Mainzer: Symmetry and Complexity in Dynamical Systems. Eur. Rev. Acad. Eur. **13** (suppl. 2), 29–48 (2005)
22. K. Brading, E. Castellani (Eds.): *Symmetries in Physics* (Cambridge University Press, Cambridge 2003)

T

Time in Quantum Theory

H.D. Zeh

In quantum mechanics, time is understood as an external ('classical') concept. So it is assumed, as in classical physics, to exist as a controller of all motion – either as absolute time or in the form of proper times defined by a classical spacetime metric. In the latter case it is applicable to local quantum systems along their world lines. According to this assumption, time can be read from appropriate classical or quasi-classical 'clocks'.

This conception has to be revised only when general relativity, where the spatial metric becomes a dynamical object, is itself quantized [1] – as required for consistency (see IV). The thereby achieved 'quantization of time' does not necessarily lead to a discretization of time – just as the ► quantization of free motion does not require a discretization of space. On the other hand, the introduction of a fundamental gravitational constant in addition to ► Planck's constant and the speed of light leads to a natural *Planck time* unit, corresponding to $5.40\ 10^{-44}$ sec. This may signal the need for an entirely novel conceptual framework – to be based on as yet missing empirical evidence. A formal (canonical) quantization of time would also be required in non-relativistic *Machian* ('relational') dynamical theories [4], which consistently replace the concept of time by some reference motion. If quantum theory is universally valid, *all* dynamical processes (including those that may serve as clocks or definers of time) must in principle be affected by quantum theory. What does this mean for the notion of time?

Historically, the dynamics of quantum systems seemed to consist of individually undetermined stochastic ► 'quantum jumps' between otherwise 'stationary' states (energy eigenstates) – see [2] for an early review of the formalism and the attempt of an interpretation. Such stochastic events are observed in quantum measurements, in particular. For this reason, von Neumann [3] referred to the time-dependent ► Schrödinger equation as a 'second intervention', since Schrödinger had originally invented it to describe consequences of time-dependent external 'perturbations' of a quantum system. Note, however, that atomic clocks are *not* based on any stochastic quantum events, even though they have to be designed as open systems in order to allow their permanent reading (representing 'measurements' of the clock – see IV).

In a consistent ► Schrödinger picture, *all* dynamics is described as a time dependence of the quantum *states*, while the ► observables are fixed formal kinematical concepts (see also Sect. 2.2 of [5]). The time dependence according to the ► Schrödinger equation can be completely understood as an interference phenomenon between different stationary states $|m\rangle$, which possess individually meaningless phase factors $\exp(\mathrm{i}\omega_m t)$. Their superpositions are able to describe time-dependent quantum states $|\alpha(t)\rangle$ in the form

$$|\alpha(t)\rangle := \int dq\,\psi_\alpha(q,t)|q\rangle = \sum_m c_m \exp(i\omega_m t)|m\rangle.$$

The ► wave function $\psi_\alpha(q,t)$ is here used to define the time-dependent state $|\alpha(t)\rangle$ in abstract ► Hilbert space (cf. ► rigged Hilbert spaces). The Hilbert space basis $|q\rangle$ diagonalizes an appropriate observable Q. The time dependence of a quantum state is in fact meaningful only *relative* to such a fixed basis, as demonstrated by means of the wave function in the above definition.

In non-relativistic quantum mechanics, the time parameter t that appears in the Schrödinger wave function $\psi(q,t)$ is identified with Newton's absolute time. So it is presumed to exist regardless of how or whether it is measured. The letter q represents all variables $q_i (i = 1 \ldots I)$ that form the required configuration space. The special case of a point mass, where $q \equiv x, y, z$ corresponds to a single space point, has often led to confusion of the wave function with a time-dependent spatial *field* (relativistically a field on spacetime). It is essentially this misconception that has led to the meaningless search for a *time operator* T in analogy to the position operator of a particle. However, time t is here *not* a dynamical variable. In N-particle mechanics, for example, the configuration space variables q are equivalent to N space points (that is, $I = 3N$ variables). In quantum field theory, the amplitudes of all fields $\Phi(x,y,z)$ at all space points even form a continuum. These field variables are thus distinguished from one another by their spatial arguments, which thereby assume the role of 'indices' to Φ, just as i for the variables q_i [6]. Therefore both, space and time, are assumed to be absolutely defined classical preconditions for kinematics and dynamics – even though they appear in the formalism in different ways.

If the variables q are field amplitudes, the canonical quantization of n fields leads to a time-dependent wave functional $\Psi[\Phi_1(x,y,z),\ldots,\Phi_n(x,y,z),t]$, rather than to n field operators on spacetime. This conclusion holds relativistically, too (see III). The corresponding Hilbert space readily includes superpositions of different 'particle' numbers ('occupation numbers'). For bosons, the latter are simply oscillator quantum numbers for the eigenmodes (first postulated by Planck, and later explained by Schrödinger by the numbers of nodes of their wave functions). The ultimate universal local Hilbert space basis is hoped to be found in unified field theory.

Schrödinger's general wave function $\psi(q,t)$ may be Fourier transformed with respect to *all* its arguments – in spite of their different interpretations. This transformation defines wave numbers k in the *configuration* space and frequencies ω. They may be rescaled into canonical momenta (in general different from conventional, that is, spatial momenta) and energies by means of Planck's constant. The Fourier transformation gives rise to a *formal* 'time operator', $T := i\partial/\partial\omega$, that allows one to define a continuous shift operation for frequencies: $U(\Delta\omega) := \exp(i\Delta\omega T)$. It does *not* in general transform a solution of the Schrödinger equation into another solution, since this would require a continuous and unbounded energy spectrum. Pairs of Fourier variables are subject to the Fourier theorems,

$$\Delta q \Delta k \geqslant 1 \text{ and } \Delta t \Delta \omega \geqslant 1,$$

which apply to all functions $\psi(q,t)$ – regardless of the existence of any dynamical law or a Hamiltonian H. These ▶ Heisenberg uncertainty relations between corresponding variables must have *physical* consequences when applied to solutions of the Schrödinger equation. Those based on the Fourier theorem relating time and frequency are usually interpreted as representing a 'time-energy uncertainty relation' (see [7]). Well known, for example, is the spectral line width required for metastable states. A 'time uncertainty' can also be defined by the finite duration of a preparation or measurement process.

II. The situation is somewhat obscured in the ▶ Heisenberg picture. In the algebraic Born-Heisenberg-Jordan quantization procedure, 'observables' were introduced in formal analogy to the classical *dynamical variables*, such as $q(t)$ and $p(t)$, while quantum states were *not* regarded as dynamical objects. Observables would assume definite *values* only in appropriate measurements or discrete 'quantum events' (von Neumann's first intervention – historically related to Bohr's quantum jumps between his discrete classical orbits). Time *durations* are then often defined operationally by means of pairs of such events – not according to the Schrödinger dynamics. The latter is here merely regarded as a tool for calculating probabilities for the occurrence of events, which are then assumed to represent the only *real* quantum phenomena.

Note that in the Heisenberg picture certain properties of quantum states *seem* to represent some hidden time dependence. For example, the kinetic energy operator in the Schrödinger picture (the Lapacean) measures the curvature of the wave function $\psi(q,t)$ at given time t – not any quantitiy related to motion, such as *classical* kinetic energy. Its non-vanishing minimum (achieved for a wave function that does not change sign) is in the Heisenberg picture interpreted as representing 'zero point fluctuations' of the corresponding variables q.

This picture has led to much confusion – including the search for a 'time observable' T that would depend on the specific system Hamiltonians H by obeying commutation relations

$$[T, H] = \mathrm{i}\hbar,$$

in analogy to position and momentum observables (see the Introduction of [8] for a review). However, since realistic Hamiltonians possess a ground state, their spectra are bounded from below, and a time operator obeying this commutation relation cannot possess a spectrum represented by all real numbers (as pointed out by Wolfgang Pauli [2]). It may nonetheless be related to time intervals between certain pairs of events that can be measured at a system characterized by the Hamiltonian H.

A formal equivalence between the Schrödinger and a Heisenberg picture for the purpose of calculating expectation values of measurement results is known to hold for isolated, unitarily evolving systems (which are exceptions in reality). For *asymptotically* isolated objects participating in a scattering process one may use the interaction picture, where part of the Hamiltonian dynamics is absorbed into the observables characterizing asymptotic states. This includes the 'dressing' of quantum fields. However, *macroscopic* systems always form open systems; they never become isolated, even when dressed. Such systems may approximately obey effective non-unitary dynamics (master equations). In principle, this dynamics has to be derived from the unitary (Schrödinger) evolution of an entangled *global* quantum state, that would have to include all 'external interventions'. Under realistic assumptions this leads to permanently growing ▶ entanglement with the environment – locally observed as ▶ decoherence [5].

This extremely fast and in practice irreversible process describes a *dislocalization* of quantum superpositions. It thereby mimics ▶ quantum jumps (events): components which represent different macroscopic properties (such as different pointer positions or different registration times of a detector) are almost immediately dynamically decoupled from one another. None of them is selected by decoherence as the *only* existing one. Pauli, when arguing in terms of the Heisenberg picture, regarded such events as occurring 'outside the laws of nature', since they withstood all attempts of a local dynamical description. In the global Schrödinger picture, the time-asymmetry of this dynamical decoupling of components ('branching') can be explained in terms of the time-symmetric dynamics by means of an appropriate initial condition for the wave function of the universe – the same condition that may also explain thermodynamical and related time asymmetries ('arrows of time') [9]. In essence, this initial condition requires that non-local entanglement did not yet exist just after the big bang, and therefore has to *form* dynamically ('causally'). The resulting asymmetry in time may give rise to the impression of a *direction of time*.

III. In ▶ quantum field theory, a Schrödinger equation that controls the dynamics of the field functionals may well be relativistic – containing only local interactions with respect to the space-dependent field variables (in this way facilitating the concept of a *Hamiltonian density* in space). A wave function(al) obeying a relativistic Schrödinger equation never propagates faster than light with respect to the underlying presumed absolute spacetime. Recent reports of apparently observed superluminal phenomena (▶ superluminal communication) were either based on inappropriate clocks, or on questionable interpretations of the wave function. For example, the exact energy eigenstate of a particle, bound to an attractive potential in a state of negative energy $E = -|E|$, would extend to spatial infinity according to $\exp(-\sqrt{|E|}r)$ outside the range of the potential. It has therefore been claimed to be able in principle to cause effects at any distance within any finite time [10]. However, if the wave function of the bound system forms dynamically (according to the Schrödinger equation rather than by quantum jumps), it can only subluminally *approach* the exact eigenstate with its infinite exponential tail. This time-dependence requires a minimum energy spread that is in accord with the time-frequency Fourier theorem. Similar arguments hold relativistically also for particle number eigenstates, which cannot have sharp spatial boundaries because of Casimir-type effects (▶ Casimir effect) (in principle observable for moving mirrors); all bounded systems must relativistically be in superpositions of different particle numbers.

In the theory of relativity, proper times assume the role of Newton's absolute time for all *local* systems, that is, for those approximately following world lines in spacetime. However, quantum states are generically nonlocal (entangled), and they

T

do not consist of or define local subsystem states. One may then introduce auxiliary time coordinates (arbitrary spacetime foliations) in order to define the dynamics of *global* states on these artificial 'simultaneities'. A Hamiltonian (albeit of very complex form – in general including a whole field of Coriolis-type forces with effective 'particle' creation and annihilation terms) would nonetheless *exist* in this case. As these artificial simultaneities may be assumed to propagate just locally, one speaks of 'many-fingered time'. Dynamical evolution in quantum theory is in general *locally non-unitary* (to be described by a master equation) because of the generic nonlocal entanglement contained in the unitarily evolving global quantum state. Unitary evolution may therefore be confirmed only in exceptional, quasi-isolated (microscopic) systems.

IV. According to Mach's ideas, no concept of absolute time should be required or meaningful. Any time concept could then be replaced by simultaneity *relations* between trajectories of different variables (including appropriate clocks) – see [4] and Chap. 1 of [9]. *Classically,* timeless trajectories $q_i(\lambda)$, where λ is an arbitrary parameter, are still defined. Mach's principle requires only that the fundamental dynamical laws are invariant under reparametrizations of λ. *In quantum theory,* the wave function cannot even depend on such a time-ordering parameter, since there are no trajectories any more that could be parametrized. This fact excludes even dynamical successions of spatial geometries (the dynamical states of general relativity), which would form a foliation of spacetime. On the other hand, any appropriate variable q_0 that is among the arguments of a time-less wave function $\psi(q)$ may be regarded as a more or less appropriate global physical clock. According to the ► superposition principle, *superpositions* of different values q_0 – that is, of different 'physical times' – would then have to exist as real physical states (just as the superpositions of different values of any other physical variable).

In conventional quantum mechanics, superpositions of different times of an event are well known. For example, a coherently decaying metastable state (that can be experimentally confirmed to exist by means of interference experiments in the case of decay fragments only weakly interacting with their environment) is a superposition of different decay times. Similarly, the quantum state for a single variable x and a clock variable u, say, would have to be described by a wave function $\psi(x, u)$. This means that the *classical* dependence of x on clock time u, defined by their time-less trajectory $x(u)$, is replaced by the less stringent ► *entanglement* between x and u that is defined by such a wave function [11]. The clock variable u becomes quasi-classical only when it is pertinently decohered, such that superpositions of different times u always remain dislocalized (locally inaccessible). The same conclusion holds for the mentioned superposition of different decay times if its corresponding partial waves (► wave packet forming thin spherical shells in space unless reflected somewhere) are decohered from one another.

Atomic clocks, in particular, are based on the time-dependent superposition of two close atomic energy eigenstates (defining 'beats'). These oscillating states would immediately be destroyed by decoherence whenever they were measured (read). Therefore, they have to be dynamically correlated with the ► coherent state of a maser field that is in resonance with them. This time-dependent coherent state is known to be ‘robust’ against decoherence – including genuine measurements [12]. So it permits the construction of a quasi-classical atomic clock that can be read. Exactly classical clocks would be in conflict with the uncertainty relations between position and momentum of their ‘hands’.

The above-described consequences of Mach’s principle with respect to time do indeed apply *in general relativity* to a closed universe. Spatial geometries on a time-like foliation of spacetime, which would classically determine all proper times [13], are now among the *dynamical variables* q (arguments of the wave function) – similar to the mentioned clock variable u. Moreover, *material* clocks intended to ‘measure’ these proper times within a given precision would have to possess a minimum mass in order to comply with the uncertainty relations [14], while this mass must then in turn disturb the spacetime metric.

A time *coordinate* t in general relativity is a physically meaningless parameter (such as λ – not u – in the above examples). Invariance of the theory under reparametrization, $t \to f(t)$, requires a ‘Hamiltonian constraint’: $H = 0$ [1, 15]. In its quantum mechanical form, $H\Psi = 0$, this leads to the trivial Schrödinger dynamics $\partial\Psi/\partial t = 0$, where Ψ is now a wave functional on a configuration space consisting of spatial geometries and matter fields. As this consequence seems to remain valid for all unified theories that contain ► quantum gravity, one has to conclude that *there is no time* on a fundamental level; all dynamics is encoded in the static entanglement described by Ψ. Surprisingly, though, the time-less *Wheeler-DeWitt equation* [1],

$$H\psi = 0,$$

(also called an Einstein-Schrödinger equation) becomes hyperbolic for Friedmann type universes – similar to a relativistic wave equation on spacetime (see Sect. 2.1 of [9]). This allows one to formulate a complete boundary condition for Ψ in the form of an ‘intrinsic initial condition’ [16]. It requires Ψ and its first derivative to be given on a ‘time-like’ hypersurface, defined according to the hyperbolic form of the kinetic energy operator contained in H (now a d’Alembertian), in this universal configuration space (DeWitt’s ‘superspace’). For example, such initial data can be freely chosen at a small value of the expansion parameter a of the universe. A low-entropy condition at $a \to 0$ then leads to an ‘intrinsic arrow of time’: total entropy on time-like hypersurfaces must grow (for statistical reasons) as a function of the size of the universe – regardless of any external concept of time.

Quasi-classical time can here only be recovered within the validity of a Born-Oppenheimer approximation with respect to the square root of the inverse Planck mass [15], while spatial geometry, which defines all fundamental physical clocks, is strongly entangled with, and thus decohered by, matter [17]. In analogy to the coherent set of apparent light rays that approximately describe the propagation of *one* extended light wave in space in the limit of short wave lengths (geometric optics), quasi-classical times are defined *separately* for all quasi-trajectories in superspace. Each of them then defines a dynamically autonomous quasi-classical world (an ‘Everett branch’ of the global wave function in unitary description) – including a specific quasi-classical spacetime. As ► ‘Schrödinger cat’ states evolve abundantly out of microscopic superpositions in measurement-type interactions, there cannot be just one quasi-classical world (analogous to just one light ray in geometric optics) according to the Schrödinger dynamics. Material clocks, such as atomic clocks, require further (usually not quite as strong) decoherence to become quasi-classical.

Primary Literature

1. B. S. DeWitt: Quantum theory of gravity, Phys. Rev. **160**, 1113 (1967)
2. W. Pauli: In Handbuch der Physik, ed. by H. Geiger and K. Scheel, 2nd edn., Vol. 24 (Springer, Berlin 1933)
3. J. von Neumann: *Mathematische Grundlagen der Quantenmechanik* (Springer, Berlin 1932)

Secondary Literature

4. J.B. Barbour: *The End of Time* (Weidenfeld & Nicolson, London 1999)
5. E. Joos, H. D. Zeh, C. Kiefer, D. Giulini, J. Kupsch, and I.-O. Stamatescu: *Decoherence and the Appearance of a Classical World in Quantum Theory,* 2nd edn. (Springer, Berlin, Heidelberg 2003)
6. J. Hilgevoord: The uncertainty principle for energy and time, Am. J. Phys. **64**, 1451 (1996)
7. P. Busch: The time energy uncertainty relation, In [8]
8. J. G. Muga, R. Sals Mayato, and I. L. Egusquiza: *Time in Quantum Mechanics* (Springer, Berlin 2002)
9. H. D. Zeh: *The Physical Basis of the Direction of Time*, 5th edn. (Springer, Berlin, Heidelberg 2007)
10. G. Hegerfeldt: Causality problems for Fermi's two-atom system, Phys. Rev. Lett. **72**, 596 (1994)
11. D. N. Page and W. K. Wootters: Evolution without evolution: Dynamics described by stationary observables, Phys. Rev. **D27**, 2885 (1983)
12. O. Kübler and H. D. Zeh: Dynamics of quantum correlations, Ann. Phys. N.Y. **76**, 405 (1973)
13. R. F. Baierlein, D. H. Sharp, and J.A. Wheeler: Three-dimensional geometry as carrier of information about time, Phys. Rev. **126**, 1864 (1962)
14. H. Salecker and E.P. Wigner: Quantum limitations of the measurement of space-time distances, Phys. Rev. **109**, 571 (1958)
15. C. Kiefer: *Quantum Gravity*, 2nd edn. (Clarendon Press, Oxford 2007)
16. H. D. Zeh: Emergence of classical time from a universal wave function, Phys. Lett. **A116**, 9 (1986)
17. E. Joos: Why do we observe a classical spacetime? Phys. Lett. **A116**, 6 (1986)

Trace

Roderich Tumulka

Trace of an operator: The sum of the diagonal elements of the operator's matrix representation. The "trace" is a number that can be associated with an operator T on ► Hilbert space, and is usually denoted $\mathrm{tr}(T)$, $\mathrm{tr}\,T$, $\mathrm{Tr}(T)$, or $\mathrm{Tr}\,T$. It can be a complex number, or $+\infty$, or can be undefined (because it is of the type $\infty - \infty$). The set of operators whose trace is a finite complex number is called the *trace class*.

Definition (1) The trace of an $n \times n$ matrix $A = (a_{ij})_{i,j \leqslant n}$ is defined as the sum of the entries on the main diagonal:

$$\mathrm{tr}(A) = \sum_{i=1}^{n} a_{ii}. \tag{1}$$

(In the sum convention of general relativity, this is written $a_i{}^i$.) For an $n \times m$ matrix with unequal number of rows and columns there is no concept of trace.

(2) For a (linear) operator T on a finite-dimensional vector space, $\mathrm{tr}(T)$ is defined as the trace of its matrix representation relative to an arbitrary basis. It can be shown that the value of $\mathrm{tr}(T)$ does not depend on the choice of the basis.

(3) For an infinite matrix $A = (a_{ij})_{i,j \in \mathbb{N}}$, the trace is defined as the series (infinite sum)

$$\mathrm{tr}(A) = \sum_{i=1}^{\infty} a_{ii}, \tag{2}$$

provided it converges.

(4) For an operator T on a (separable) Hilbert space $\mathscr{H}$, one would like to define its trace as the trace of its matrix representation relative to an arbitrary orthonormal basis $\{\phi_1, \phi_2, \ldots\}$, that is

$$\mathrm{tr}(T) = \sum_{n=1}^{\infty} \langle \phi_n | T \phi_n \rangle. \tag{3}$$

However, the series may not converge, or may converge for one ► orthonormal basis and not for another. That is why one splits the definition in two steps [1]. If T is a positive operator (i.e., $\langle \psi | T \psi \rangle \geqslant 0$ for every $\psi \in \mathscr{H}$) then its trace is defined by (3), which is either a nonnegative real number or $+\infty$; it can be shown that this value does not depend on the choice of the orthonormal basis. This definition is extended to non-positive operators as follows. An operator T belongs to the *trace class* if the positive operator $|T| = \sqrt{T^*T}$ has finite trace (where T^* denotes the adjoint operator of T); for such T we can define the trace by (3), as it can be shown that the series converges (to a finite complex number) and its value is independent of the orthonormal basis. Every trace class operator is bounded.

Properties (1) The trace is linear:

$$\mathrm{tr}(S+T)=\mathrm{tr}(S)+\mathrm{tr}(T),\qquad \mathrm{tr}(\lambda T)=\lambda\,\mathrm{tr}(T) \tag{4}$$

for all operators S, T from the trace class and all $\lambda \in \mathbb{C}$. ($S+T$ and λT belong to the trace class, too.)

(2) The trace is invariant under cyclic permutation of factors:

$$\mathrm{tr}(AB\cdots YZ)=\mathrm{tr}(ZAB\cdots Y). \tag{5}$$

(We assume here that at least one of the factors $A, B, \ldots, Z$ belongs to the trace class and the others are bounded; in that case, also $AB\cdots YZ$ belongs to the trace class.) In particular $\mathrm{tr}(AB)=\mathrm{tr}(BA)$ and $\mathrm{tr}(ABC)=\mathrm{tr}(CAB)$, which is, however, not always the same as $\mathrm{tr}(CBA)$.

(3) If an operator T can be diagonalized, i.e., if there exists an orthonormal basis of eigenvectors, then $\mathrm{tr}(T)$ is the sum of the eigenvalues, counted with multiplicity (= degree of degeneracy).

(4) The trace of the adjoint operator T^* is the complex-conjugate of the trace of T: $\mathrm{tr}(T^*)=\mathrm{tr}(T)^*$.

(5) The trace of a self-adjoint operator T (in the trace class) is real: $\mathrm{tr}(T) \in \mathbb{R}$. A self-adjoint operator lies in the trace class if and only if it is bounded, its spectrum is discrete, all nonzero eigenvalues have finite multiplicity, and the sum of the eigenvalues (with multiplicity) is finite (i.e., converges absolutely).

(6) The trace of a positive operator $T \geqslant 0$ is nonnegative: $\mathrm{tr}(T) \geqslant 0$.

Trace Formula in Quantum Theory When an observable, given by the self-adjoint operator T, is measured on a system with density matrix ρ then the probability that the outcome Z lies in the set $\Delta \subseteq \mathbb{R}$ is

$$\mathbb{P}(Z \in \Delta)=\mathrm{tr}(\rho\, P_\Delta) \tag{6}$$

with P_Δ the spectral projection of T corresponding to the spectrum in Δ.

Literature

1. Reed, M., Simon, B.: *Methods of Modern Mathematical Physics. Vol. I: Functional Analysis* (Academic Press, New York and London 1972)

Transactional Interpretation of Quantum Mechanics

John G. Cramer

Interpretations of quantum mechanics provide an account of the meaning of the quantum formalism and guidance on how to use the formalism to connect with nature and to make predictions on the outcome of experiments. The first interpretation was the Copenhagen interpretation, developed by Heisenberg and Bohr the late 1920s. It has become the orthodox view of the meaning of the quantum formalism, but it has lead to an uncomfortably large number of interpretational paradoxes (► Errors and paradoxes in quantum mechanics) associated with relativity conflicts, ► wave-particle duality, wave function collapse, and quantum ► nonlocality.

The *transactional interpretation of quantum mechanics* [1, 2] is a leading alternative to the Copenhagen interpretation. The transactional interpretation (TI) is explicitly nonlocal and is able to explain *all* of the interpretational paradoxes. It is relativistically invariant, so that it can be used with the relativistic wave equations as well as the ► Schrödinger equation. It uses the retarded (Ψ) and advanced (Ψ^*) wave function solutions of these equations in a "handshake" that provides a rationale for understanding the formal structure of quantum ► wave mechanics and for treating quantum ► wave functions as physically present in space. In fact, the advanced-retarded transactions are "visible" in the quantum wave-mechanics formalism.

The logical development of the transactional interpretation starts with the time-symmetric classical electromagnetism of Dirac [3], and Wheeler and Feynman [4,5], which describes electromagnetic processes as exchanges between retarded (normal) and advanced (time-reversed) electromagnetic waves. The transactional interpretation applies the time-symmetric Wheeler–Feynman view to the quantum mechanical wave function solutions of the electromagnetic wave equation. The lessons learned about electromagnetic quantum waves are then extended to wave functions describing the behavior of massive particles (e.g., ► electrons, protons, etc.) by applying the same interpretation to their relativistic wave equations. Finally, the Schrödinger equation is included as a nonrelativistic reduction of the relativistic wave equations in the limit of small velocities.

The transactional interpretation views each quantum event as a "handshake" or "transaction" process extending across space–time that involves the exchange of advanced and retarded waves to enforce the conservation of certain quantities (energy, momentum, angular momentum, ...). It asserts that each quantum transition forms in four stages: (1) *emission*, (2) *response*, (3) *stochastic choice*, and (4) *repetition to completion*.

The first stage of a quantum event, illustrated in Fig. 1, is the *emission* of an "offer wave" by the "source," which is the object supplying the quantities transferred. The offer wave is the time-dependent retarded quantum wave function Ψ, as used in

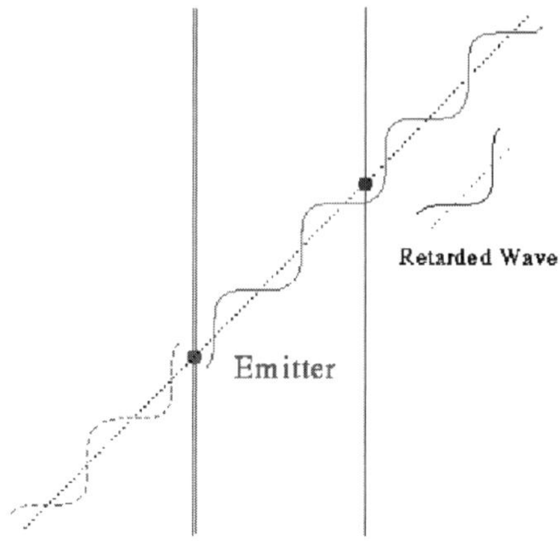

Fig. 1 Schematic view of emission stage

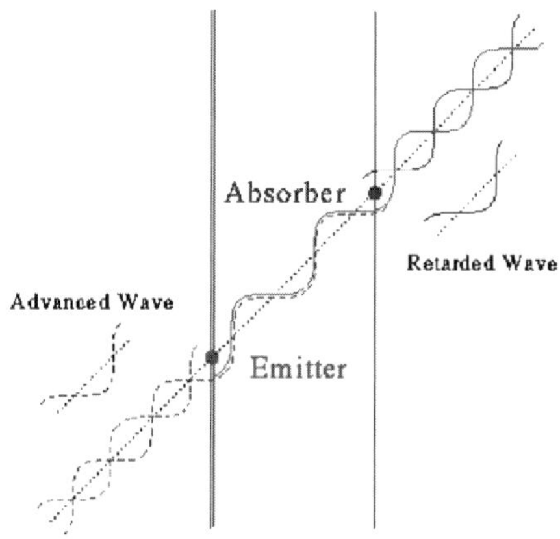

Fig. 2 Schematic view of response stage

standard quantum mechanics. It spreads through space–time until it encounters the "absorber," the object receiving the conserved quantities.

The second stage of a quantum event is the *response* to the offer wave by any potential absorber (there may be many in a given event). Such an absorber produces an advanced "confirmation wave" Ψ^*, the complex conjugate of the quantum offer wave function Ψ. The confirmation wave travels in the reverse time direction and arrives back to the source at precisely the instant of emission with an amplitude of $\Psi\Psi^*$. In transactions involving "entangled" waves, i.e., emission of two or more waves linked by a conservation law (e.g., conservation of momentum or angular momentum), the corresponding confirmation waves must match so that the conservation law is implemented (Fig. 2).

The third stage of a quantum event is the *stochastic choice* exercised by the source in selecting one from among the possible transactions. It does this in a linear probabilistic way based on the strengths $\Psi\Psi^*$of the advanced-wave "echoes" it receives from the potential absorbers. However, in order to avoid transactional inconsistencies pointed out by Maudlin [6], the probabilistic decision must be hierachical, with the decision to select or not select transactions from small space–time intervals occurring "before" any transactions from larger space–time intervals are allowed to form.

The final stage of a quantum event is the *repetition to completion* of this process by the source and absorber, reinforcing the selected transaction repeatedly until the conserved quantities are transferred and the potential quantum event becomes a real event.

The application of the transactional interpretation in resolving interpretational quantum paradoxes is discussed in detail in references [1] and [7]. Briefly, conflicts with relativity are eliminated because the TI is relativistically invariant. Paradoxes associated with wave-particle duality and the ► Heisenberg uncertainty relations are resolved and clarified because the offer wave is wavelike and can be quite general, but the completed transaction is particle-like and must localize and project out specific components of the offer wave function. Collapse paradoxes are resolved because formation of the transaction provides an account of the process called "wave function collapse" in the Copenhagen interpretation (Fig. 3). And perhaps most important, the TI accounts of the quantum nonlocality of entangled states as resulting from dual transactions for the entangled states that are required to be consistent at the emission location, enforcing conservation laws and explaining the nonclassical "EPR" link between widely separated measurements on entangled particles.

Because all of the consistent interpretations of quantum mechanics describe the same quantum formalism, and that formalism makes all of the testable predictions, there is no way of using experimental tests to choose between interpretations. It is possible that an interpretation can be falsified by finding it to be inconsistent with the quantum formalism [8]. In the absence of such falsification, however, the choice between interpretations must be made on the basis of other criteria: parsimony, absence of paradoxes, ease of use, and facility for using the interpretation to speculate and extrapolate.

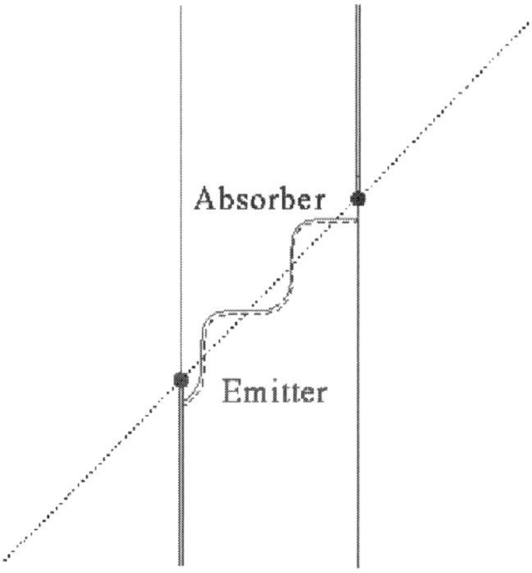

Fig. 3 Schematic view of completed transaction

T

If rated on the basis of these criteria, the transactional interpretation gets a very high score. It does well with parsimony because "extra" assumptions of the Copenhagen interpretation, in particular, the ► Born rule and wave function collapse, are implicit in the transactional interpretation and do not require extra assumptions [1]. As mentioned above, the transactional interpretation resolves essentially all of the interpretational paradoxes raised by the Copenhagen interpretation. It is easy to use because waves and transactions, assumed to be physically present in space, can be diagrammed (see [1] and [7] for examples). Its use for speculation and extrapolation is more subjective, but many practicing physicists have reported finding it useful in areas like quantum optics and ► quantum computation.

Therefore, the transactional interpretation should be seriously considered as a useful and powerful alternative to the orthodox Copenhagen interpretation. See ► Born rule; Consistent Histories; Metaphysics in Quantum Mechanics; Nonlocality; Orthodox Interpretation; Schrödinger's Cat.

Primary Literature

1. J. G. Cramer: *The transactional interpretation of quantum mechanics*. Reviews of Modern Physics 58, 647–688 (1986)
2. J. G. Cramer: *An overview of the transactional interpretation*. International Journal of Theoretical Physics 27, 227 (1988)
3. P. A. M. Dirac: *Classical theories of radiating atoms*. Proceedings of the Royal Society of London A167, 148–169 (1938)
4. J. A. Wheeler, R. P. Feynman: *Interaction with the absorber as the mechanism of radiation*. Reviews of Modern Physics 17, 157–181 (1945)
5. J. A. Wheeler, R. P. Feynman: *Classical electrodynamics in terms of direct interparticle action*. Reviews of Modern Physics 21, 425–433 (1949)
6. T. Maudlin: *Quantum Non-Locality and Relativity* (Blackwell Publishers, Oxford 2002)
7. J. G. Cramer: *A transactional analysis of interaction free measurements*. Foundations of Physics Letters 19, 63–73 (2006), e-Print Archive: quant-ph/0508102
8. J. G. Cramer: *A farewell to Copenhagen?*. Analog Magazine (December 2004). Online at http://www.analogsf.com/0412/altview.shtml.

Secondary Literature

9. J. G. Cramer: *The Plane of the Present and the New Transactional Paradigm of Time*, in: Time and the Instant, edited by Robin Drurie (Clinamen Press, UK, 2001, Chapter 9)
10. J. Gribbin: *Schrödinger's Kittens and the Search for Reality* (Back Bay Books, USA, 1996)

Tunneling

Günter Nimtz and Brian Clegg

Tunneling represents the most fundamental process in physics. According to our present understanding tunneling started the universe about 13 billions years ago. Nowadays we know that tunneling is involved in radioactivity and in nuclear fusion – the latter effect is heating the sun. Tunneling is the process of molecular inversion motion in chemistry and is important in modern microelectronic devices. Physicists introduced the name tunneling for a classical forbidden process, which the theory of quantum mechanics explained around 1927: A ball, for instance, cannot overcome a hill if its kinetic energy is less than the hill's gravitational potential energy. In this case the ball rolls back. However, quantum mechanics explained that the ball has a tiny probability of getting to the other side of the hill. Similarly, an α-particle leaves the attractive nuclear potential well despite having a small energy, thereby producing radioactivity. In figures 1 and 2 an α-particle is illustrated as a wave packet embedded in a valley between two hills, which represent the attractive nuclear forces. The energy of the particle is assumed to be too small to overcome the tops of the hills. However, radioactivity, which was observed a 100 years ago, i.e., the decay of an atomic nucleus, is explained by quantum mechanics as a probability that a low energy particle is observed at the other side of the hill.

The explanation of alpha-decay as quantum mechanic tunneling followed around 1928 by George Gamow and simultaneously, but independently, by Edward U. Condon and Ronald W. Gurney. Incidentally, in 1927, Friedrich Hund was the first to notice the possibility of the phenomenon of tunneling, which he called barrier penetration, in a calculation of the ground state in a double-well potential. The phenomenon arises, for example, in the inversion transition of the ammonia molecule.

Radioactivity is accompanied by the release of energy, which is the source of nuclear power stations. The opposite process takes place in the sun and enabled nuclear fusion by tunneling of protons, penetrating the repelling Coulomb forces. This process ends up producing Helium and setting heat free. It provide the heat source of the sun and produces the terrific power of the atomic hybrid hydrogen bomb.

In quantum mechanics, see for instance Merzbacher [9] and Gasiorowicz [10], the one-dimensional stationary ► Schrödinger equation describes the tunneling mechanism of ► wave packet by the relations

$$\frac{d^2\Psi}{d^2x^2} + 2m/\hbar^2(W - U_0) = 0, \tag{1}$$

$$k^2 = k_0^2 - (2mU_0/\hbar^2), \tag{2}$$

$$k_0^2 = (2mW/\hbar^2), \tag{3}$$

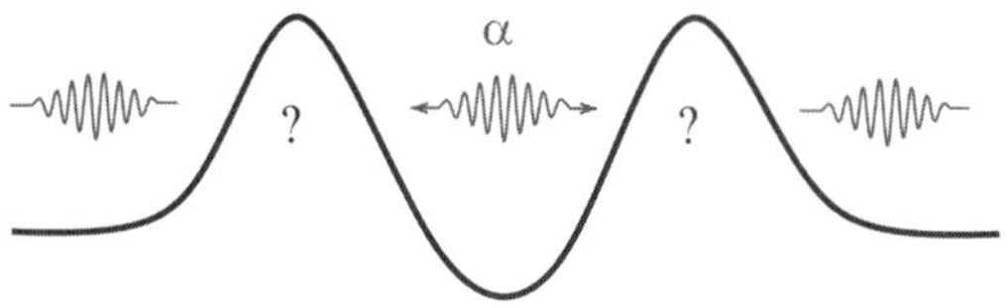

Fig. 1 Illustrating the α-particle decay of a nucleus. The α-particle is embedded between the 'hills' of the attractive nuclear forces. However, there is a small probability to leave the well by tunneling. What happens inside the hill?

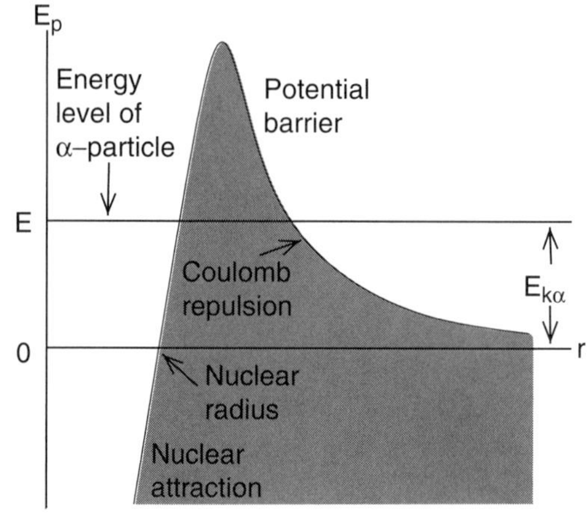

Fig. 2 Details of the right hand side of figure 1. The force components of the nuclear valley in which an α-particle is embedded are given. There is a minuscule probability of tunneling through the potential barrier

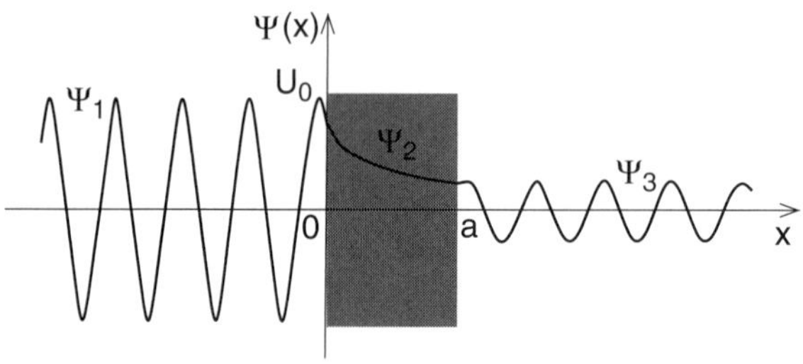

Fig. 3 Illustrating (1) of the wave function $\Psi(r)$. Between $x = 0$ and a is located the potential barrier and the tunneling region

where Ψ is the wave function of the wave particle in question, W the particle energy, U_0 the barrier height, m the particle mass, $\hbar$ the ▶ Planck's constant, k and k_0 are the wave numbers (i.e., 2π times the reciprocal wave lengths) in the potential barrier and in free space, respectively. Figure 3 displays the solution of the wave function $\Psi(r)$. In the case of $W < U$ the wave number k is imaginary. This special solution of the Schrödinger equation is called tunneling. With k being imaginary, the time becomes zero or equivalently the wave packet velocity becomes infinite inside a barrier. The tunneling solution of the Schrödinger equation represents an action at a distance: an incoming signal leaves the barrier at the same instant.

Zero-time tunneling was calculated for ▶ electrons by Hartman, by Low and Mende, and by Leavens and McKinnon, for instance [1–3]. A critical analysis of the many tunneling time expressions since 1930 is presented in Ref. [4]. The conclusion of this theoretical investigation is that the phase time result originally obtained by Wigner and Hartman are the best expressions to calculate a tunneling time. This statement was confirmed in photon and phonon experiments and recently by Eckle et al. in the electron ionization tunneling process in helium [5, 6].

The zero-time behavior in barriers was observed first in photonic tunneling experiments by Enders and Nimtz [7]. Such experiments represent the optical analogy to quantum mechanical tunneling as was discussed by Sommerfeld [11]. The tunneling process is not completely described by the Maxwell theory for electromagnetic waves, where the tunneling solutions are called evanescent modes. The more sophisticated ▶ quantum electrodynamics describes photonic tunneling by virtual photons (▶ light quantum) in agreement with experiments as reported recently [8].

Thus a particle with an energy smaller than that of the surrounding barrier can penetrate it, i.e., can tunnel through it with a minuscule but finite probability. Amazingly, the particle does not spend time inside the barrier, the barrier represents a zero-time space. The particle enters and leaves the barrier space at the same instant. The zero-time tunneling is a near field effect, which is observable over distances comparable with the extension of the particle. Tunneling violates the *relativistic (Einstein) causality*, which does not allow a signal to travel faster than the velocity c of light in vacuum and it violates the Einstein relation $W^2 = c^2 p^2$, where W is the energy and p is the photon momentum. However, tunneling does not allow the construction of time machines. So-called *primitive causality* is not violated: effect always follows cause, an ironic result considering the noncausal nature of quantum mechanics as was proved in Ref. [7].

Primary Literature

1. T. Hartman: Tunneling of a wave packet. J. Appl. Phys. **33**, 3427–33 (1962)
2. F. E. Low, P. F. Mende: A note on the tunneling time. Ann. Phys. NY **210**, 380–87 (1991)
3. C. R. Leavens, W. R. McKinnon: An exact determination of the mean dwell time based on the qunatum clock of Salecker and Wigner. Phys. Lett. A **194**, 12–20 (1994)
4. S. Collins, D. Lowe and J. R. Barker: The quantum mechanical tunnelling time problem. J. Phys. C **20**, 6213–32 (1987)
5. G. Nimtz and A. A. Stahlhofen: Universal tunneling time for all fields, Ann. Phys. (Berlin) **17**, 374–79 (2008)
6. P. Eckle et al.: Attosecond Ionization and Tunneling Delay Time Measurements in Helium. SCIENCE **322**, 1525–29 (2008)
7. G. Nimtz: Do evanescent modes violate relativistic causality? Lect. Notes Phys. **702**, 506–31 (2006)
8. A. A. Stahlhofen, G. Nimtz: Evanescent modes are virtual photons. Europhys. Lett. **76**, 189–95 (2006)

Secondary Literature

9. E. Merzbacher: *Quantum Mechanics*, 2nd ed. (Wiley, New York 1970)
10. S. Gasiorowicz: *Quantum Physics* (Wiley, New York 1996)
11. A. Sommerfeld: *Lectures on Theoretical Physics, Optics* (Academic, New York 1954, 27–28)

Two-State Vector Formalism

L. Vaidman

The two-state vector formalism (TSVF) [1] is a time-symmetric description of the standard quantum mechanics originated in Aharonov, Bergmann and Lebowitz [2]. The TSVF describes a quantum system at a particular time by two quantum states: the usual one, evolving forward in time, defined by the results of a complete measurement at the earlier time, and by the quantum state evolving backward in time, defined by the results of a complete measurement at a later time.

According to the standard quantum formalism, an ideal (von Neumann) measurement at time t of a non-degenerate variable A tests for existence at this time of the forward evolving state $|A = a\rangle$ (it yields the outcome $A = a$ with certainty if this was the state) and creates the state evolving towards the future:

$$|\Psi(t')\rangle = e^{-\frac{i}{\hbar}\int_t^{t'} H\mathrm{d}t}|A = a\rangle, \quad t' > t. \tag{1}$$

(In general, the Hamiltonians $H(t)$ at different times do not commute and a time ordering has to be performed.)

In the TSVF this ideal measurement also tests for backward evolving state arriving from the future $\langle A = a|$ and creates the state evolving towards the past:

$$\langle\Phi(t'')| = \langle A = a|e^{\frac{i}{\hbar}\int_t^{t''} H\mathrm{d}t}, \quad t'' < t. \tag{2}$$

Apart from some differences (discussed below) following from the asymmetry of the memory arrow of time, one can perform similar manipulations of the forward and backward evolving states. In particular, neither can be cloned and both can be teleported.

Given complete measurements, $|A = a\rangle$ at t_1 and $|B = b\rangle$ at t_2, the complete description of a quantum system at time t, $t_1 < t < t_2$, is the *two-state vector* [3]:

$$\langle\Phi| \ |\Psi\rangle, \tag{3}$$

where the states $\langle\Phi|$ and $|\Psi\rangle$ are obtained using $(1, 2)$.

The two-state vector provides the maximal information regarding the way the quantum system can affect at time t any other system. In particular, the two-state vector describes the influence on a measuring device coupled with the system at time t. An ideal measurement of a variable O yields an eigenvalue o_n with probability given by the Aharonov, Bergman, Lebowitz (ABL) rule:

$$\mathrm{Prob}(o_n) = \frac{|\langle\Phi|\mathbf{P}_{O=o_n}|\Psi\rangle|^2}{\sum_j |\langle\Phi|\mathbf{P}_{O=o_j}|\Psi\rangle|^2}. \tag{4}$$

This is, essentially, a conditional probability. We consider an ensemble (▶ ensembles in quantum mechanics) of pre- and post-selected quantum systems with the desired outcomes of the measurements at t_1 and t_2. Only those systems (and all of them) are taken into account. Intermediate measurement (or the absence of it) might change the probabilities of the outcomes of the post-selection measurement at time t_2, but this is irrelevant: it only changes the size of the pre- and post-selected ensemble given the size of the pre-elected ensemble at t_1.

Note that the ABL rule simplifies the calculation of probabilities of the outcome of intermediate measurements. In the standard approach we need to calculate the time evolutions between time t and t_2 of all states corresponding to all possible outcomes of the intermediate measurement, while in the TSVF we have to calculate evolution of only one (backward evolving) state.

The pre- and post-selected quantum system (described by the two-state vector) has very different features relative to the system described by a single, forward evolving quantum state. The Heisenberg Uncertainty Principle does not hold: non-commuting ▶ observables might be simultaneously well defined, i.e. each observable might have a dispersion-free value provided that it was the only one measured at time t. As an example, consider a ▶ spin-$\frac{1}{2}$ particle in a field free region. Assume that σ_z was measured at t_1, σ_x at t_2 and both were found to be 1. When at time t, an outcome of a measurement of a variable (if measured) is known with certainty, it is named *an element of reality* [8]. Thus, in the above example, both $\sigma_z = 1$ and $\sigma_x = 1$ are such elements of reality.

For pre- and post-selected systems there might be apparently contradicting elements of reality. Consider now a spin-$\frac{1}{2}$ particle which can be located in two boxes, A and B, which is described by the two-state vector:

T

$$\langle\Phi|\ |\Psi\rangle = \frac{1}{3}\left(\langle A,\uparrow_z| + \langle A,\downarrow_z| - \langle B,\uparrow_z|\right) \quad \left(|A,\uparrow_z\rangle + |A,\downarrow_z\rangle + |B,\uparrow_z\rangle\right), \tag{5}$$

(where $|A,\uparrow_z\rangle$ represents the particle in box A with spin $\uparrow_z$). Then, there are two elements of reality: "the particle in box A with spin up" and "the particle in box A with spin down". Indeed, the measurement of the projection $\mathbf{P}_{A\uparrow}$ has the outcome $\mathbf{P}_{A\uparrow} = 1$ with certainty, and the outcome of the other projection (if measured instead) is also certain: $\mathbf{P}_{A\downarrow} = 1$. This can be readily verified using the ABL rule or the standard formalism.

Obviously, the measurement of the product of the projections is certain too: $\mathbf{P}_{A\uparrow}\,\mathbf{P}_{A\downarrow} = 0$, so this example shows also *the failure of the product rule*: at time t we know with certainty that if A is measured, the outcome is a, and if B is measured instead, the outcome is b, but nevertheless, the measurement of AB is not ab. (The product rule does hold for the standard, pre-selected quantum systems.)

This example is mathematically equivalent to the three-box paradox [4] in which a single pre- and post-selected particle can be found with certainty both in box A if searched there and in box B if searched there instead. These bizarre properties of elements of reality generated much controversy about the *counterfactual* usage of the ABL rule (▶ Counterfactuals in Quantum Mechanics). It should be stressed that "elements of reality" should not be understood in the ontological sense, but only in the operational sense, given by their definition.

The most important outcome of the TSVF is the discovery of *weak values* of physical variables [5]. When at time t, another system couples weakly to a variable O of a pre- and post-selected system $\langle\Phi|\ \ |\Psi\rangle$, the effective coupling is not to one of the eigenvalues, but to the weak value:

$$O_{\mathrm{w}} \equiv \frac{\langle\Phi|O|\Psi\rangle}{\langle\Phi|\Psi\rangle}. \tag{6}$$

The weak value might be far away from the range of the eigenvalues, and this can lead to numerous surprising effects, described in the entry ▶ Weak Value and Weak Measurement.

There is an important connection between weak and strong measurements. If the outcome of a strong measurement $O = o_i$ is known with certainty, the weak measurement has to yield the same value, $O_{\mathrm{w}} = o_i$. The inverse is true for dichotomic variables: if the weak value is equal to one of the two eigenvalues, a strong measurement should give this outcome with certainty.

In both strong and weak measurements, the outcome manifests via the shift of the pointer variable. For strong measurements it might be random, but for weak measurements it is always certain (and equals to the weak value). Sometimes it is called "weak-measurement elements of reality" [9].

A generalization of the concept of the two-state vector (with natural generalizations of the ABL rule and weak value) is a "superposition" of two-state vectors which is called a *generalized two-state vector* [4]:

$$\sum_i \alpha_i \langle\Phi_i|\ |\Psi_i\rangle. \tag{7}$$

A quantum system described by a generalized two-state vector requires pre- and post-selection of the system together with an ancilla which is not measured between the pre- and post-selection.

Systems described by generalized two-states vectors might have more unusual properties. The ▶ Heisenberg uncertainty relation breaks down in even more dramatic way: we can have a set of many non-commuting observables having dispersion-free values and not just the trivial case of two, one observable defined by pre-selection and another by post-selection. An extensively analyzed example of this kind is "the mean king problem" [6, 7] in which we have to know all observables of the set of the non-commuting observables for all possible outcomes of the post-selection measurement.

Another natural multiple-time non-local generalization is to consider $2N$-state vector (or generalized $2N$-state vector) which provides a complete description of how a (composite) system can affect other systems coupled to it in N space-time points. Preparing and testing such $2N$-state vectors require multiple-time and non-local measurements. (Note that causality puts some constrains on such measurements [10].) An incomplete description in which we associate only one (forward or backward) evolving state with some space-type points is also of interest. For example, two spin-$\frac{1}{2}$ particles in an entangled "state" which evolves forward in time for one particle and backward for the other particle, can be completely correlated:

$$\frac{1}{\sqrt{2}}\left(|\uparrow\rangle_A \langle\uparrow|_B + |\downarrow\rangle_A \langle\downarrow|_B\right). \tag{8}$$

Here, the measurements of the spin in components in any direction yield the same result for both particles. There is no pre-selected quantum system with such property.

The TSVF is a time symmetric approach. However, there are some differences between forward and backward evolving quantum states: we can always create a particular forward evolving quantum state, say $|A = a\rangle$. We measure A, and if the outcome is a different eigenvalue than a, we perform an appropriate transformation to the desired state. We cannot, however, create with certainty a particular backward evolving quantum state, since the correction has to be performed before we know the outcome of the measurement. The difference follows from the time asymmetry of the memory arrow of time. This asymmetry is not manifest in the ABL rule and the weak value, because the outcome of measurement is the *shift* of the pointer during the measurement interaction and this is invariant under changing the direction of time evolution. The shift is between zero and the outcome of the measurement and this is where the memory arrow of time introduces the asymmetry. The state "zero" is always in the earlier time: we do not "remember" the future and thus we cannot fix the final state of the measuring device to be zero.

T

The TSVF is equivalent to the standard quantum mechanics, but it is more convenient for analyzing the pre- and post-selected systems. It helped to discover numerous surprising quantum effects. The TSVF is compatible with almost all interpretations of quantum mechanics but it fits particularly well the ▶ many-worlds interpretation. The concepts of "elements of reality" and "weak-measurement elements of reality" obtain a clear meaning in worlds with particular post-selection, while they have no ontological meaning in the scope of physical universe which incorporates all the worlds. Finally, the TSVF provides a framework for a modification of quantum mechanics [11] in which the backward evolving state is actually exists now, and it is not just a useful tool for describing pre- and post-selected systems. In this radical proposal there is no collapse and there are no multiple worlds.

Literature

1. Y. Aharonov and L. Vaidman, The Two-State Vector Formalism, an Updated Review, in Time in Quantum Mechanics, Vol 1., Lect. Notes Phys. **734**, 399 (2008).
2. Y. Aharonov, P. G. Bergmann and J. L. Lebowitz, Time Symmetry in the Quantum Process of Measurement, Phys. Rev. B **134**, 1410 (1964).
3. Y. Aharonov and L. Vaidman, Properties of a Quantum System During the Time Interval Between Two Measurements, Phys. Rev. A **41**, 11 (1990).
4. Y. Aharonov and L. Vaidman, Complete Description of a Quantum System at a Given Time, J. Phys. A **24**, 2315 (1991).
5. Y. Aharonov, D. Albert, and L. Vaidman, How the Result of Measurement of a Component of the Spin of a Spin-$\frac{1}{2}$ Particle Can Turn Out to Be 100? Phys. Rev. Lett. **60**, 1351 (1988).
6. L. Vaidman, Y. Aharonov and D. Albert, How to ascertain the values of σ_x, σ_y, and σ_z of a spin-$1/2$ particle, Phys. Rev. Lett. **58**, 1385 (1987).
7. B.G. Englert and Y. Aharonov, The Mean King's Problem: Prime Degrees of Freedom, Phys. Lett. A, 284 (2001).
8. L. Vaidman, Lorentz-Invariant "Elements of Reality" and the Joint Measurability of Commuting Observables, Phys. Rev. Lett. **70**, 3369 (1993).
9. L. Vaidman, Weak-Measurement Elements of Reality, Found. Phys. **26**, 895 (1996).
10. L. Vaidman and I. Nevo, Nonlocal Measurements in the Time-Symmetric Quantum Mechanics, Int. J. Mod. Phys. B **20**, 1528 (2006).
11. Y. Aharonov and E. Gruss, Two-time interpretation of quantum mechanics, quant-ph/0507269 (2005).

U

Uncertainty Principle, Indetermincay Relations

See ▶ Heisenberg uncertainty relations.

Unitary Operator

Werner Stulpe

Unitary operator, a sharpening of the concept of an isometric operator. A linear ▶ operator J defined on a complex (real) Banach space $\mathcal{X}$ (▶ Hilbert space) with values in some complex (real) Banach space $\mathcal{Y}$ is called *isometric* or an *isometry* if it preserves the norm, i.e., $\|J\phi\| = \|\phi\|$ for all $\phi \in \mathcal{X}$. An isometric operator is bounded (▶ operator) with norm $\|J\| = 1$, invertible, and the range R_J is a closed (▶ Hilbert space) submanifold of $\mathcal{Y}$ which is, even in the case $\mathcal{Y} = \mathcal{X}$, in general smaller than $\mathcal{Y}$ (if $\mathcal{X}$ and $\mathcal{Y}$ have the same finite dimension, then $R_J = \mathcal{Y}$). The inverse operator J^{-1} is an isometry with domain $D_{J^{-1}} = R_J$ and the range $R_{J^{-1}} = \mathcal{X}$. Two Banach spaces $\mathcal{X}$ and $\mathcal{Y}$ are called *(norm-) isomorphic* if there exists an isometry from $\mathcal{X}$ to $\mathcal{Y}$ such that $R_J = \mathcal{Y}$.

An isometric operator J defined on a complex (real) Hilbert space $\mathcal{H}$ with values in some complex (real) Hilbert space $\mathcal{K}$ automatically preserves the scalar products also, i.e., $\langle J\phi|J\psi\rangle = \langle\phi|\psi\rangle$ for $\phi, \psi \in \mathcal{H}$. Such an operator is called *unitary* [1–6] if $\mathcal{H}$ and $\mathcal{K}$ are complex Hilbert spaces and if its range is $\mathcal{K}$. That is, a linear operator U from some complex Hilbert space $\mathcal{H}$ to some other complex Hilbert space $\mathcal{K}$ is *unitary* if (i) $D_U = \mathcal{H}$, (ii) $\langle U\phi|U\psi\rangle = \langle\phi|\psi\rangle$ for $\phi, \psi \in \mathcal{H}$, and (iii) $R_U = \mathcal{K}$. The inverse U^{-1} is also unitary where, in the case of $\mathcal{H} = \mathcal{K}$, $U^{-1} = U^*$ holds (the assumption $\mathcal{H} = \mathcal{K}$ is not necessary, but corresponds to the definition of the adjoint operator given in the section ▶ operator).

The following example shows that an isometric operator acting in a complex Hilbert space is in general not unitary. Let $\phi_1, \phi_2, \ldots$ be a complete orthonormal system of an infinite-dimensional separable ▶ Hilbert space $\mathcal{H}$. For every vector $\psi \in \mathcal{H}$, $\psi = \sum_{i=1}^{\infty} \alpha_i\phi_i$, $\sum_{i=1}^{\infty} |\alpha_i|^2 < \infty$, define $J\psi = \sum_{i=1}^{\infty} \alpha_i\phi_{2i}$; J is isometric since $\|J\psi\|^2 = \sum_{i=1}^{\infty} |\alpha_i|^2 = \|\psi\|^2$, but J is not unitary since $R_J \neq \mathcal{H}$. In particular, the Hilbert space $\mathcal{H}$ is isomorphic to the subspace (closed submanifold) spanned by $\phi_2, \phi_4, \ldots$. An important example of a unitary operator is the Fourier transform in the Hilbert space $L^2(\mathbb{R}, \mathrm{d}x)$ of the square-integrable functions on $\mathbb{R}$. For functions $\phi \in L^2(\mathbb{R}, \mathrm{d}x)$ that are also integrable (i.e., for $\phi \in L^2(\mathbb{R}, \mathrm{d}x) \cap L^1(\mathbb{R}, \mathrm{d}x)$), one can define the *Fourier transform* $\tilde{\phi}$ *of* ϕ by $\tilde{\phi}(k) = \frac{1}{\sqrt{2\pi}} \int_{\mathbb{R}} \phi(x) e^{-ikx} \mathrm{d}x$ and the

Fourier transform F by $F\phi = \tilde{\phi}$. Since $\tilde{\phi} \in L^2(\mathbb{R}, \mathrm{d}x)$ and F is a densely defined, norm-preserving linear operator, F can uniquely be extended to an isometry defined on $L^2(\mathbb{R}, \mathrm{d}x)$ with values in $L^2(\mathbb{R}, \mathrm{d}x)$; moreover, since the range of this isometry is $L^2(\mathbb{R}, \mathrm{d}x)$, F becomes a unitary operator (*Fourier–Plancherel theorem*). The preservation of the scalar product reads explicitly $\int_{\mathbb{R}} \overline{\phi(x)}\psi(x)\,\mathrm{d}x = \int_{\mathbb{R}} \overline{\tilde{\phi}(k)}\tilde{\psi}(k)\,\mathrm{d}k$ where $\phi, \psi \in L^2(\mathbb{R}, \mathrm{d}x)$.

The (pure) states and ► observables of a sort of quantum systems are traditionally described by the unit vectors of a Hilbert space $\mathcal{H}$ and by the self-adjoint operators A acting in $\mathcal{H}$, respectively. Given a unitary operator U from $\mathcal{H}$ to some other Hilbert space $\mathcal{K}$, the state vectors $\psi \in \mathcal{H}$, $\|\psi\| = 1$, can be transformed according to $\psi' = U\psi$ and the observables according to $A' = UAU^{-1}$. Under this transformation, the physically meaningful expectation values remain invariant: $\langle\psi'|A'\psi'\rangle = \langle\psi|A\psi\rangle$. The representation of the states and observables by unit vectors and self-adjoint operators in $\mathcal{H}$ is *unitarily equivalent* to the representation by vectors and operators in $\mathcal{K}$. This is applied in the context of *representations of quantum mechanics* (e.g., configuration-space or momentum-space representation, matrix representations) as well as in the context of *pictures of quantum dynamics* (Schrödinger, Heisenberg, and interaction picture).

Given a ► self-adjoint operator A in $\mathcal{H}$ with spectral measure E, for each $t \in \mathbb{R}$ a unitary operator $\mathrm{e}^{\mathrm{i}tA}$ is defined by $\langle\psi|\mathrm{e}^{\mathrm{i}tA}\psi\rangle = \int_{\mathbb{R}} \mathrm{e}^{\mathrm{i}t\lambda}\langle\psi|E(\mathrm{d}\lambda)\psi\rangle$, $\psi \in \mathcal{H}$. The family of the unitary operators $U_t = \mathrm{e}^{\mathrm{i}tA}$, $t \in \mathbb{R}$, satisfies (i) $U_0 = I$, (ii) $U_{s+t} = U_tU_s = U_sU_t$ for all $s, t \in \mathbb{R}$, and (iii) $\|U_t\phi - \phi\| \to 0$ for all $\phi \in \mathcal{H}$ as $t \to 0$. A family of unitary operators U_t with $t \in \mathbb{R}$ and the properties (i)–(iii) is called a *strongly continuous one-parameter group of unitary operators*. To each such one-parameter group there exists a uniquely determined self-adjoint operator A such that $U_t = \mathrm{e}^{\mathrm{i}tA}$ (*Stone's theorem*). Thus, there is a one-one correspondence between the self-adjoint operators A in $\mathcal{H}$ and the strongly continuous one-parameter groups of unitary operators U_t; A is called the *infinitesimal generator of* U_t, $t \in \mathbb{R}$. The derivative $\frac{\mathrm{d}}{\mathrm{d}t}U_t\phi\Big|_{t=0} = \lim_{h\to 0}\frac{U_h\phi-\phi}{h}$, the limit being taken in the norm of $\mathcal{H}$, exists if and only if $\phi \in D_A$ where $\frac{\mathrm{d}}{\mathrm{d}t}U_t\phi\Big|_{t=0} = \mathrm{i}A\phi$. Moreover, for all $\phi \in D_A$ and all $t \in \mathbb{R}$, $U_t\phi \in D_A$ and $\frac{\mathrm{d}}{\mathrm{d}t}U_t\phi = \mathrm{i}AU_t\phi$. If the self-adjoint operator A is bounded, then in addition $U_t = \mathrm{e}^{\mathrm{i}tA} = \sum_{n=0}^{\infty}\frac{(\mathrm{i}t)^n}{n!}A^n$ holds, the infinite sum converging w.r.t. the operator norm. Furthermore, the one-parameter group U_t, $t \in \mathbb{R}$, is norm-continuous and $\frac{\mathrm{d}}{\mathrm{d}t}U_t = \mathrm{i}AU_t$, the derivative also being taken in the operator norm.

The energy observable of a sort of quantum systems is described by its ► Hamiltonian operator H. The self-adjoint operator H also determines the time development of the states; in fact, $-\frac{1}{\hbar}H$ is the generator of the time translations, i.e., every state $\psi_0 \in \mathcal{H}$, $\|\psi_0\| = 1$, at time $t = 0$ determines the state at any time $t \in \mathbb{R}$ according to $\psi_t = \mathrm{e}^{-\frac{\mathrm{i}}{\hbar}Ht}\psi_0$. If $\psi_0 \in D_H$, then $\psi_t \in D_H$ for all $t \in \mathbb{R}$, and ψ_t satisfies $\mathrm{i}\hbar\dot{\psi}_t = H\psi_t$; the latter ordinary differential equation in Hilbert space is the abstract version of ► *Schrödinger's equation*.

In quantum mechanics, symmetry transformations (► symmetry) are also represented by unitary operators. For instance, in the Hilbert space $L^2(\mathbb{R}, \mathrm{d}x)$ (► Hilbert space) of ▶ wave function of particles moving in one spatial direction, a unitary operator U_a is defined by $(U_a\psi)(x) = \psi(x-a)$ where $\psi \in L^2(\mathbb{R}, \mathrm{d}x)$ and $a \in \mathbb{R}$; U_a describes the translation of the states ψ, $\|\psi\| = 1$, by a. The strongly continuous one-parameter group $\{U_a\}_{a\in\mathbb{R}}$ has the infinitesimal generator $-P = \mathrm{i}\frac{\mathrm{d}}{\mathrm{d}x}$, the differential operator P (▶ self-adjoint operator) is, up to the factor $\hbar$, the momentum operator in the one-dimensional configuration-space representation. In this representation the multiplication operator Q (▶ self-adjoint operator) is the position operator, and $\frac{1}{\hbar}Q$ is the infinitesimal generator of a one-parameter group $\{U_b\}_{b\in\mathbb{R}}$ of unitary operators; U_b describes the boost of the momentum of the states by b. In the Hilbert space $L^2(\mathbb{R}^3, \mathrm{d}x)$ of ▶ wave function on three-dimensional configuration space, a spatial rotation of the states is described by the unitary operator defined by $(U_R\psi)(x) = \psi(R^{-1}x)$ where R is a rotation of $\mathbb{R}^3$ and $\psi \in L^2(\mathbb{R}^3, \mathrm{d}x)$. The family $\{U_R\}_{R\in SO(3)}$ is a unitary representation of the rotation group $SO(3)$. Euclidean transformations which associate every $x \in \mathbb{R}^3$ with $Rx + a$, $a \in \mathbb{R}^3$, give rise to the unitary operators $U_{R,a}$ defined by $(U_{R,a}\psi)(x) = \psi(R^{-1}(x-a))$.

The action of a unitary operator U can, since U is bounded, represented in matrix form (▶ operator). As a consequence of $U^{-1} = U^*$, the matrix elements $u_{ij} = \langle\phi_i|U\phi_j\rangle$, $\phi_1, \phi_2, \ldots$ being a complete orthonormal system in $\mathcal{H}$, satisfy $\sum_j \bar{u}_{ij}u_{kj} = \delta_{ik}$ as well as $\sum_j \bar{u}_{ji}u_{jk} = \delta_{ik}$; i.e., the matrix elements constitute a *unitary matrix*.

In the context of Hilbert spaces, partial isometries are sometimes of interest. Given two Hilbert spaces $\mathcal{H}$ and $\mathcal{K}$, a *partial isometry from* $\mathcal{H}$ *to* $\mathcal{K}$ is a linear operator J from $\mathcal{H}$ to $\mathcal{K}$ such that (i) $D_J = \mathcal{H}$, (ii) $\|J\phi\| = \|\phi\|$ for all ϕ belonging to some subspace $\mathcal{X}$ of $\mathcal{H}$, and (iii) $J\phi = 0$ for all $\phi \in \mathcal{X}^\perp$. So $\mathcal{H} = \mathcal{X} \oplus \mathcal{X}^\perp$, $\mathcal{K} = R_J \oplus R_J^\perp$, $\psi = \phi + \chi$ where $\psi \in \mathcal{H}$, $\phi \in \mathcal{X}$, and $\chi \in \mathcal{X}^\perp$; $J\psi = J\phi$, J acts as an isometry on $\mathcal{X}$ and as a unitary operator between $\mathcal{X}$ and R_J (note that, as a closed submanifold, R_J itself is a Hilbert space).

Literature

1. F. Riesz and B. Sz.-Nagy: *Functional Analysis* (Frederick Ungar Publishing, New York 1955; reprint Dover Publications, Mineola 1991)
2. M. Reed and B. Simon: *Methods of Modern Mathematical Physics I: Functional Analysis* (Academic, New York 1972, second edition 1980)
3. W. Młak: *Hilbert Spaces and Operator Theory* (Springer, Dordrecht 1991)
4. D. Werner: *Funktionalanalysis* (Springer, Berlin, Heidelberg, 1995, third edition 2000)
5. G. Ludwig: *Foundations of Quantum Mechanics I* (Springer, New York 1983), Appendix IV
6. E. Prugovečki: *Quantum Mechanics in Hilbert Space* (Academic, New York 1971, second edition 1981; reprint Dover Publications, Mineola 2006)

V

Vector Model

Klaus Hentschel

The vector model was developed around 1920 to describe the intricate coupling of angular momentum L (▶ Spin; Stern–Gerlach experiment) and ▶ spin S to electric and magnetic fields (either inside the atom or to external fields imposed by experimenters in ▶ spectroscopy). Both L and S are modeled as vectors in three-dimensional space; their vectorial sum, the total angular momentum, is $J = L + S$.

According to space quantization ▶ Stern–Gerlach experiment as first postulated by Arnold Sommerfeld (1868–1951) in 1916, not all possible orientations of these vectors relative to the electric or magnetic field (defining the direction of the z-axis) are allowed. The projection of the angular momentum L onto the z-axis ought to be multiples of $\hbar$. This restriction also leads to similar restrictions of the orientation of J and explains the symmetric splitting of spectral lines into multiplets in the normal ▶ Zeeman effect and ▶ Stark effect in the most natural way. For atoms with more than one electron, various ways of calculating the vectorial sum $J$of all the contributing angular momenta l_i and spins $s_i = 1/2$ are possible. Either all the l_i are summed up first to one L, and then combined with $S = \sum_i s_i$, or all the l_i and s_i are first summed up separately to j_i with $J = \sum_i j_i$ (as shown in Fig. 1). Because

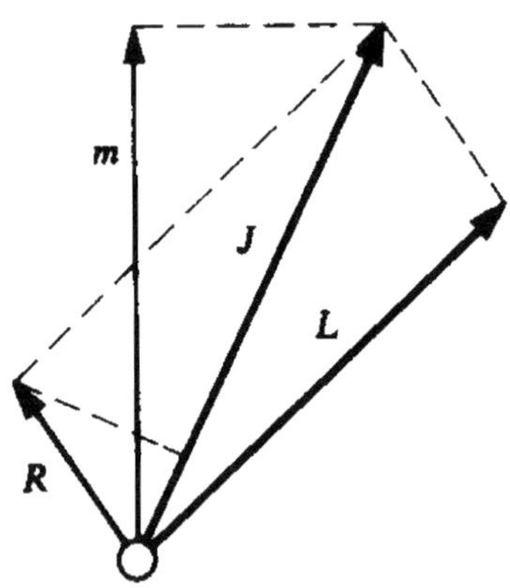

Fig. 1 Landé's vector model: The orbit angular momentum vector $\boldsymbol{L}$ and the atomic core momentum vector $\boldsymbol{R}$ (later redubbed spin S) add up vectorially to the total momentum vector $\boldsymbol{J}$. $\boldsymbol{R}$, $\boldsymbol{L}$ and $\boldsymbol{J}$ have to be imagined precessing around the external magnetic field (whose axis is by convention always *drawn vertically upwards*). The component of $\boldsymbol{J}$ parallel to the magnetic field determines the magnetic moment $\boldsymbol{m}$ of the atom which can only take quantized values because of ▶ space quantization. *Source*: Friedrich Hund, *Geschichte der Quantentheorie* (Mannheim: BI Wissenschaftsverlag, 1984, 118; by permission of the publisher)

D. Greenberger et al. (eds.), *Compendium of Quantum Physics: Concepts, Experiments, History and Philosophy*, © Springer-Verlag Berlin Heidelberg 2009

of the noncommutativity of ▶ operators, these two procedures are in general not equivalent with each other. The first is called ▶ Russell-Saunders-coupling, valid for the lighter atoms, the latter ▶ jj-coupling yielding the better approximation for heavier atoms and for the energetically higher terms.

It turned out that in order to get satisfactory agreement with observable line splittings, the length of the vector L actually had to be proportional to the square root of $L(L + 1)$, with similar expressions for other vectors such as S and J. For Alfred Landé (1888–1976), who first suggested this in 1919 within the framework of Bohr's and Sommerfeld's semi-classical ▶ Bohr atom model, this procedure was admittedly fully *ad hoc*. Problems with this model even triggered a crisis of ▶ quantum theory between ca. 1923 and early 1925. Strange half ▶ quantum numbers were postulated by Werner Heisenberg (1901–76) and Wolfgang Pauli (1900–58) in early 1925, foreshadowing the concept of spin only to emerge in late 1925. A deeper understanding of this strange "numerology" in the "Zeeman salad" (both expressions by representatives of the ▶ Sommerfeld school) had to await the development of formal quantum mechanics in 1925/26, in which the square of any ▶ observable A is defined as the two-fold action of an operator $\hat{A}$ on a state vector, yielding its eigenvalue $a$in the first step, and $a + 1$ in the second, thus $\hat{A}^2$ yields $a(a + 1)$ and not a^2.

Primary Literature

1. C. Candler: *Atomic Spectra and the Vector Model* (Princeton 1937, 2nd ed. London 1964), esp. Chaps. 11, 16–18
2. G. Herzberg: *Atomic Spectra and Atomic Structure* (New York 1937; Reprint Dover Publications 1944)

Secondary Literature

3. P. Forman: Alfred Landé and the anomalous Zeeman Effect, 1919–1921. Historical Studies in the Physical Sciences 2, 153–261 (1970)

W

Wave Function

Helge Kragh

The wave function of a quantum-mechanical system is the quantity that allows calculation of the various outcomes of an experiment or observation involving the system. It characterizes the system's physical state. The wave function ψ was introduced as a central element in Erwin Schrödinger's ► wave mechanics in the spring of 1926, whereas a similar quantity did not exist in the earlier versions of quantum mechanics due to Werner Heisenberg, Max Born, Pascual Jordan and Paul A.M. Dirac. But it was soon demonstrated that the various versions are mathematically equivalent and that the wave function can be translated into ► matrix mechanics as a *state vector*.

Schrödinger introduced in a formal way the wave function in the very beginning of the first communication of "Quantisierung als Eigenwertproblem," where he just called it "a new unknown ψ." It appeared in his fundamental wave equation and had to satisfy certain mathematical criteria, but its physical meaning was unclear. What is waving? What is it waving in? It was tempting to ask such questions, but it was soon realized that they carried no meaning. Schrödinger initially required ψ to be real, but in his fourth communication he admitted that the "mechanical field scalar ψ" was in general a complex quantity. This alone indicated that the wave functions could not be given a physical existence in the same sense as, say, water waves. In addition, the wavelike processes defined by ψ took place in the system's configuration space, not in the ordinary space.

Schrödinger initially thought of particles as represented by ► wave packets, and then, when the idea did not work, attempted to describe the electrical charge in terms of ψ. This interpretation, too, had to be abandoned, and later in 1926 Max Born proposed the ► probability interpretation that since then has been generally accepted. According to Born, ψ has not itself any direct physical meaning, although the absolute square $|\psi|^2 = \psi^*\psi$ has. The quantity represents neither a particle nor a charge density, but a probability density: $|\psi|^2 \mathrm{d}V$ is the probability that the system is in the state ψ and localized in the volume element $\mathrm{d}V$.

Ever since the birth of wave mechanics it has been discussed which kinds of physical systems can be assigned a wave- or ψ-function. Niels Bohr always emphasized that measuring apparatus and like macroscopic objects are "classical" and cannot be described by a wave function, whereas Schrödinger famously assigned a wave function to a cat locked up in a sealed box (► Schrödinger's cat). Arthur Stanley Eddington was willing even to describe the universe in terms of ψ, an idea which later was taken up in so-called quantum cosmology by Bryce DeWitt, James Hartle, Stephen Hawking and others.

Primary Literature

1. E. Schrödinger: Quantisierung als Eigenwertproblem. Annalen der Physik **79**, 361–376 (1926)
2. M. Born: Quantenmechanik der Stossvorgänge. Zeitschrift für Physik **38**, 803–827 (1926)
3. J. B. Hartle, St. W. Hawking: The Wave Function of the Universe. Physical Review **D 28**, 2960–2975 (1983)

Secondary Literature

4. M. Beller: *Quantum Dialogue. The Making of a Revolution* (University of Chicago Press, Chicago 1999)

Wave Function Collapse

I.-O. Stamatescu

Under "collapse of the wave function" (or "state vector reduction") one understands the 'sudden' change of the system's state in a measurement. This change is not reducible to classical "information gain", but is a genuine quantum mechanical concept, directly related to the concept of quantum state. It is especially relevant if we consider that quantum mechanics describes the behaviour of individual systems. In the following we shall first describe the role of the collapse as a formal concept in this context, then we shall discuss some variants of physical approaches to collapse. We shall comment on the notion of "individual systems" in quantum mechanics at the end of this article.

Collapse in the formalism of quantum theory. (Figure 1).

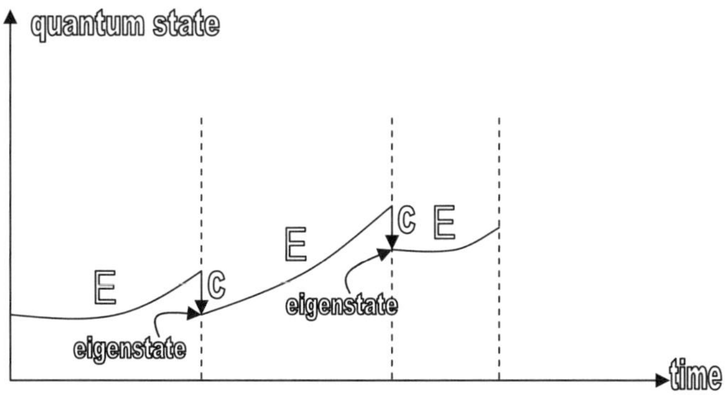

Fig. 1 Time evolution, E, of ψ and collapse, C, adapted from R. Penrose, *The Road to Reality* (2005, 823)

The notion of state of a system is a fundamental concept in physics. In classical physics all quantities which can be measured upon the system (▶ "observables": e.g., positions and momenta of a point particle) can, in principle, be simultaneously assigned precise values and this uniquely defines the state. There is therefore a one to one relation between states and observations. In quantum theory, however, only a subset of observables can be fixed at any given moment. A maximally determined state obtains by fixing a maximal set of simultaneously measurable ("compatible") observables, e.g., the position components. But there will be other observables, here the momenta, which do not posses definite values in this state. Relating states to observations is therefore a more special and not trivial procedure.

This also implies that the concept of ▶ measurement becomes essential. Here we shall only refer to an ideal measurement, which is understood as any physical arrangement by which a particular observable concerning the system of interest is fixed to some well defined value. But if the initial state of the system was such that it did not determine this particular observable beforehand, this indeterminacy will show up as irreproducibility of the result when repeating the experiment under the same conditions (same apparatus and identically "prepared" systems). Only the relative frequency of these results can be associated to a probability distribution determined by the initial state (quantum effects show up here as interference terms and non-trivial correlations when performing correlated measurements, which cannot be understood classically ▶ correlations in quantum mechanics). After the measurement, however, the state of the system must be such that the measured observable is no longer undetermined but has now been fixed to the measured value, hence the state has changed abruptly and randomly with the given probability distribution. We speak of *collapse of the state* anterior to the measurement onto the state in which the measurement leaves the system.

The formalism of quantum theory allows to write any given state as a ▶ superposition of other states, in particular of such states where the observable of interest has well defined values. Collapse, or state reduction means then the survival after measurement of only that state out of the superposition for which the value of the observable matches the result of the measurement.

In as much, therefore, that we can speak of individual systems and measurements, collapse is a logically necessary ingredient in the formalism. The representation of states as vectors in a ▶ Hilbert space makes the above considerations transparent and well defined: linear combinations of vectors realize the superposition of states, with the coefficients giving the weights and their square modulus the corresponding probabilities. Here collapse appears as a sudden and generically random change in the state vector, as opposed to the continuous, deterministic transformations of the latter due to the various physical interactions the system may be subjected to. Accordingly, in this setting the axioms of quantum mechanics include a measurement and collapse postulate (von Neumann's "first intervention"), besides the definition of states as vectors in a Hilbert space (which incorporates the superposition principle), the definition of observables and expectation values and the dynamical evolution equations (von Neumann's "second intervention").

In the following we shall be slightly more formal. The reader who does not want to be bothered with technical detail may go directly to the *Physical approaches*.

The quantum mechanical Hilbert space is a generically infinitely-dimensional linear space over the complex field, with an inner scalar product and the associated norm and distance and which is complete under this norm. The states of a physical system are represented as vectors in this space and physical interventions upon the system as ▶ operators acting on these vectors. In particular observables are represented as hermitean operators, in accordance with the reality of measurements. We can use ortho-normalized bases and any vector can be decomposed in such a basis as

$$|\psi\rangle = \sum_n c_n \, |\varphi_n\rangle, \quad \langle\varphi_m|\varphi_n\rangle = \delta_{mn}, \tag{1}$$

where we used in the Dirac bracket notation (▶ Dirac notation) for the vectors and scalar products (for all these concepts see the corresponding articles). In the following we shall only consider so-called pure states (▶ states, pure and mixed) and use normalized vectors $\|\psi\| = 1$ with $\|\cdot\|$: the Hilbert space norm. The expectation of any operator A in the state $|\psi\rangle$ is then $\langle\psi|A|\psi\rangle$ and all information about possible observations onto the system in this state is contained in the "density operator" ("▶ density matrix")

$$\rho = |\psi\rangle\,\langle\psi| = \sum_{n,m} c_n \, c_m^* \, |\varphi_n\rangle\,\langle\varphi_m|. \tag{2}$$

with the help of which we can obtain expectation values for any observable.

If we choose the basis vectors $|\varphi_n\rangle$ above to be *eigenstates* of some observable A

$$A\,|\varphi_n\rangle \; = a_n \, |\varphi_n\rangle, \tag{3}$$

then a measurement of A upon the system in state $|\psi\rangle$ will produce some value, say a_{n_0}, with probability $\langle\varphi_{n_0}|\,\rho\,|\varphi_{n_0}\rangle = |c_{n_0}|^2$ and leave the system in the state φ_{n_0}. This means an abrupt change of the state vector which can be seen as a sudden "rotation" of the latter aligning it with one of its components, chosen *randomly* with the mentioned probability:

$$|\psi\rangle = \sum_n c_n \, |\varphi_n\rangle \;\; \longrightarrow \;\; |\psi'\rangle = |\varphi_{n_0}\rangle. \tag{4}$$

W

This "reduction of the state vector" (collapse, or von Neumann's "first intervention") is to be contrasted with the *deterministic* dynamical evolution of the state vector due to physical interactions (von Neumann's "second intervention"), realized by a ▶ unitary operator acting continuously in time, (written in differential form this is the ▶ Schrödinger equation):

$$|\psi(t)\rangle = U(t, t_0)\,|\psi(t_0)\rangle. \tag{5}$$

Physical approaches to collapse

The conceptual differences between von Neumann's first and second interventions have led to many interpretational problems. In standard quantum theory the collapse of the wave function is associated with the measurement but the moment of its occurrence (the "Heisenberg cut") can be anywhere between the actual interaction of the system with the apparatus and the conscious registration of the result. If the observer is considered external this appears to introduce a subjective element in the theory, with corresponding ambiguities (► "Wigner's friend"). These problems have prompted many attempts to give the collapse a more physical ground. These attempts can be divided in three classes: "no collapse" (in deterministic extensions which reproduce quantitatively quantum theory), "apparent collapse" (in quantum theory itself within a certain interpretation) and "dynamical collapse" (in the frame of theories which *approximate* quantum theory).

The first class essentially corresponds to the ► hidden variables theories. In this case there is no collapse at all, the state precisely determines every observable and the spread of results in a repeated experiment is due to the different values taken by the "hidden variables" which make that we in fact deal with different initial states each time, the difference escaping however our control (is hidden). An elaborated theory hereto has been set up by D. Bohm 1952 and has been further developed thereafter. It is a celebrated theorem established by J. S. Bell 1964 that demanding agreement with quantum theory requires *non-local* hidden variables. This is brought to a quantitative test in the so called Bell inequalities ► Bell's theorem for correlated measurements which should be fulfilled for *local* hidden variable theories. Experiments up to date appear to violate these inequalities and show agreement with the quantum mechanical predictions. Non-local hidden variables, though allowed by this test, contradict a basic principle of physics – ► locality. This, and difficulties in pursuing this program for realistic physical theories diminishes the attractiveness of hidden variable theories.

In the second case the accent is on illuminating the physics of the measurement process. We shall here discuss the so called environmental decoherence argument as raised by H. D. Zeh 1970 and W. H. Zurek 1981. The measurement is realized by some physical interaction with an "apparatus" understood as a quantum system. The discussion uses the observation that quantum systems which in some way form a compound have to be considered as "entangled", which means that in a generic state of the compound system the component systems do not possess a separate state. This is a generic feature of quantum theory and means among others that, in principle, the notion of isolated system is only an approximation whose goodness depends on the physical situation. Now, a measurement implies an ► entanglement between the system and the apparatus. Moreover, since the latter essentially is a macroscopic system, it unavoidably will be entangled with an environment which is not accessible to our observations (e.g., light scattered from the pointers and leaving the experimental arrangement). Observations upon the system imply therefore an averaging over the states of the environment which are associated with different "pointer" states of the apparatus and are macroscopically different. This leads to the loss of observable interference between the different states of the apparatus. This simulates therefore a classical statistics.

To be more specific (again, these technical aspects can be skipped), if $\varphi_n^{\{1\}}$, $\varphi_n^{\{2\}}$ are bases for the two component systems in a binary compound (say, two atoms in a molecule) a generic state of the latter is

$$|\Psi\rangle = \sum_{m,n} c_{mn}\, |\varphi_m^{\{1\}}\rangle\, |\varphi_n^{\{2\}}\rangle$$
$$= \sum_n c_n\, |\psi_n^{\{1\}}\rangle\, |\varphi_n^{\{2\}}\rangle, \tag{6}$$

where for the second equation we used a certain redefinition of the states. This total wave function generally does not factorize, hence it does not allow any of the two systems to be in a definite state. With '1' designating an apparatus and '2' a system to be measured (6) is also a model for the physical interactions during a measurement process:

$$|\Psi\rangle = \sum_{m,n} c_{mn}\, |\varphi_m^{\{\text{app}\}}\rangle\, |\varphi_n^{\{\text{sys}\}}\rangle$$
$$= \sum_n c_n\, |\psi_n^{\{\text{app}\}}\rangle\, |\varphi_n^{\{\text{sys}\}}\rangle. \tag{7}$$

The apparatus is entangled both with our system and with the environment. Let us consider the apparatus as being such that the total wave function can be written as

$$|\Psi\rangle = \sum_n c_n\, |\phi_n^{\{\text{env}\}}\rangle\, |\psi_n^{\{\text{app}\}}\rangle\, |\varphi_n^{\{\text{sys}\}}\rangle, \tag{8}$$

where the environmental states $|\phi_n^{\{\text{env}\}}\rangle$ differ macroscopically and are therefore orthogonal. Since we have no access to the situation of the environment (we cannot make correlated experiments involving the states of the environment), according to the quantum mechanical formalism any information we can obtain about the system is contained in the "reduced density matrix" where the environmental situation has been "traced out":

$$\rho_{\text{red}} = \sum_k \langle\phi_k^{\{\text{env}\}}|\, |\Psi\rangle\, \langle\Psi|\, |\phi_k^{\{\text{env}\}}\rangle$$
$$= \sum_n |c_n|^2\, |\psi_n^{\{\text{app}\}}\rangle\, |\varphi_n^{\{\text{sys}\}}\rangle\, \langle\psi_n^{\{\text{app}\}}|\, \langle\varphi_n^{\{\text{sys}\}}|. \tag{9}$$

At variance to the general case (2), ρ_{red} is diagonal, which implies that we cannot observe the typical quantum mechanical interference between the different possible issues of the measurement.

This consequence – the *simulation of a classical statistics* – of the "unavoidable entanglement" with an uncontrollable environment stays at the basis of the effect called ► decoherence which is a *specific quantum mechanical effect implying no further hypothesis*. It is always present, independently of interpretations, of measurement models, etc. and is well defined in each physical situation. Its relevance for the measurement is to "de-correlate" the various possible results, as shown above, which therefore *appear* as distributed according to a classical ensemble. This does not replace collapse (which requires the choice of just one of these possible results, accompanied by the corresponding acquirement by the system of the corresponding wave function, after the interaction with the apparatus has ceased). However, it makes possible an alternative point of view, that of an "apparent collapse". The basis for this point of view is the so called "relative state interpretation" of quantum mechanics proposed by H. Everett III 1957, according to which all possible outcomes of each measurement coexist but that due to the local nature of the observations their histories form different branches of the evolution of the total system (in end effect, the world). (► Many worlds interpretation). The role of decoherence effects at measurement is now to ensure that no local observations can put into evidence correlations between the different branches, which are thus completely "unaware" of each other. From the point of view of one given branch the other components of the wave function appear therefore as irretrievably lost. Although the system is still entangled with the rest of the universe and therefore does not possess in principle a wave function for itself, any observations upon the system within one branch give the same results *as if* formal collapse had occurred (the observer is viewed as part of the quantum world and thus his consciousness follows the same branching pattern). This perspective calls for cosmological arguments. A picture of these steadily branching histories is however difficult to realize and, for instance in the so called "many-worlds" representation, somewhat unintuitive. Related interpretations are provided, e.g., in the ► consistent histories approach of R.B. Griffith 1984 and M. Gell-Mann and J. B. Hartle 1990.

Finally, the class 3 models define collapse as a genuine physical effect. This obtains as a supplementary postulate, which, in the formulation of G. C. Ghirardi, A. Rimini and T. Weber 1975, (► GRW Theory) states that the wave function of any spatial degree of freedom collapses spontaneously in a random manner, thereby fixing this degree of freedom to a value randomly chosen with the distribution given by the wave function before collapse ("spontaneous collapse" or "spontaneous localization" hypothesis). There are also other possibilities to achieve a dynamical collapse, for instance turning the Schrödinger equation into a stochastic differential equation through the addition of a non-linear noise term as proposed by P. Pearle 1976. In this case the collapse is only approximate, the collapsed wave function retaining an exponentially falling tail. The main features are, however, similar, namely:

- Even if for each degree of freedom the collapse occurs extremely rarely, the apparatus being a *macroscopic* object will be steadily subject to collapses. Since the (*microscopic*) system to be measured becomes entangled with the apparatus, *see (7)*, the collapse acting in the latter and retaining some term, say $|\psi_{n_0}^{\{\mathrm{app}\}}\rangle$

of the superposition automatically selects the corresponding component vector of the system, $|\varphi_{n_0}^{\{sys\}}\rangle$, fixing in this way the corresponding observable and leaving the system in a pure state. Therefore this model *explains* measurement.

- Collapse as a physical random process is not compatible with quantum mechanics in the sense that it leads to measurable deviations from the predictions of the latter. The details (parameters) of this process can be, however, so tuned, that these effects are detectable only for macroscopic systems, where they are welcome, but not for microscopic systems, where to a good precision the standard quantum mechanical predictions should hold.
  To be more specific, in the discrete random collapse model, for instance, with a frequency of spontaneous collapses of, e.g., 10^{-17}s^{-1} the wave function of a microscopic system will collapse about once in 10^{10} years, the age of the universe, while a macroscopic body with typically 10^{23} degrees of freedom would undergo a collapse as often as 10^6 times per second. This is compatible both with the behaviour of atoms, with the action of an apparatus and with the localized appearance of macroscopic objects, for which the successive spontaneous localizations of internal degrees of freedom soon pins down the center of mass of the body. Similar effects are obtained in the noisy dynamics models.
- The collapse is assumed to act on spatial degrees of freedom ("spontaneous localization") which is reasonable since usual interactions are local. It seems difficult, however, to obtain relativistic generalizations of the model, in particular for local quantum field theories.

Replacing the formal postulate of "collapse in the measurement" by the postulate of "general stochastic evolution" of the wave function appears somewhat arbitrary and one would like to have corroboration from further observations. This, however, appears very difficult, since the predicted new physics has similar signature with environmental decoherence and would be masked by the latter even if present. As long as we have no independent evidence for such a universal stochastic dynamics its postulate remains however *ad hoc*.

Note that none of these proposals really solves the problem, namely to provide a non-formal explanation for the collapse and the measurement process of *standard quantum mechanics*: either we modify the theory in an *in principle* measurable way (even if we may tune the parameters to ensure that the difference does not show up *in practice*), or we only provide an "as if" effect (even if the difference to true collapse might be of only cosmological relevance). This has prompted Bell to speak of "good for all practical purposes" in connection with some of these (and others) "solutions". Finally, non-local hidden variables might not be seen as a real alternative. But even if not solving the problem the various theoretical studies contributed very much to illuminate it.

W

As already mentioned, the problem of collapse is relevant in an interpretation of quantum theory pertaining to individual events. Many of the conceptual problems can be discarded in a statistical interpretation which states that wave function, collapse, etc. are only mathematical instruments which allow us to make statistical predictions, and the latter are the only place where theory meets the real world. It may appear, however, that this "economical" point of view unnecessarily impoverishes the theory. In fact statistics is not a real "thing" or event in itself, but is a conclusion drawn from the observation of many single events. The theory does refer to the latter individually and in some special cases does this in an unambiguous way, for instance when it predicts probability 0 or 1 for a certain event. These are incentives to assume that it does account for individual events generally, even if we cannot make an intuitive picture of this reference. It would seem, in some sense, quite a miracle and in fact unintuitive to have the extraordinary explanatory power of quantum theory based on a lucky choice of theoretical "instruments" completely detached from reality. This does not mean that wave functions, etc. should exist as such in reality, but that there are things and a structure in reality which support such abstractions. On the other hand it seems rather difficult to grasp this structure. Its features, as they might be suggested by the theory, do not appear unambiguous and easily understandable. The foregoing discussion of the collapse illustrates these problems.

Bell's inequalities. **(See also ► Bell's theorem)**

The non-classical character of the correlation in the expectations concerning correlated measurements on two entangled subsystems which do not possess states of their own, i.e., if it is not possible to rewrite (6) as a product of two factors, can be quantitatively exhibited in corresponding experiments. Assume we measure the properties A, A' on system '1' *and* B, B' on '2', that is, we use the observables (hermitean operators) $\{O\} = \{A \otimes B,\ A' \otimes B,\ \cdots\}$ and construct the quantity:

$$\Delta(A, A'; B, B') \equiv |\mathcal{E}(AB) - \mathcal{E}(AB')| + |\mathcal{E}(A'B) - \mathcal{E}(A'B')|, \tag{10}$$

where $\mathcal{E}$ denote the corresponding expectations in the given state of the total system:

$$\mathcal{E}(O) = \langle\Psi|O|\Psi\rangle. \tag{11}$$

Then we have (we choose $\|O\| \leqslant 1$, i.e., $\|O\psi\| \leqslant \|\psi\|, \forall\, \psi$):

$$\Delta(A, A'; B, B') = |\langle\Psi|A(B - B')|\Psi\rangle| + |\langle\Psi|A'(B + B')|\Psi\rangle| \tag{12}$$

$$= |\langle A\Psi|(B - B')\Psi\rangle| + |\langle A'\Psi|(B + B')\Psi\rangle|$$

$$\leqslant \|A\Psi\|.\|(B - B')|\Psi\| + \|A'\Psi\|.\|(B + B')\Psi\|$$

$$\leqslant \|(B - B')\Psi\| + \|(B + B')\Psi\|$$

$$\leqslant \sqrt{2[\|(B - B')\Psi\|^2 + \|(B + B')\Psi\|^2]} \tag{13}$$

$$= \sqrt{4[\|B\Psi\|^2 + \|B'\Psi\|^2]} \leqslant 2\sqrt{2}. \tag{14}$$

If we were dealing with a classical problem, that is the expectations were taken with respect to a classical ensemble:

$$\mathcal{E}_c(O) = \int O\mathrm{d}\mu, \tag{15}$$

with $d\mu$ a (positive semidefinite) probability measure and $\{O\}$ *real* valued functions (assumed to be less than 1 in absolute value) we would have instead:

$$\Delta_c(A, A'; B, B') = |\mathcal{E}_c(A(B - B'))| + |\mathcal{E}_c(A'(B + B'))| \tag{16}$$

$$\leqslant \mathcal{E}_c(|A|.|B - B'|) + \mathcal{E}_c(|A'|.|B + B'|) \leqslant \mathcal{E}_c(|B - B'|)| + \mathcal{E}_c(|B + B'|)$$

$$= \mathcal{E}_c(|B - B'| + |B + B'|) \leqslant 2, \tag{17}$$

since the general inequality:

$$\|a\| + \|b\| \leqslant \sqrt{2(\|a\|^2 + \|b\|^2)} \tag{18}$$

which was used in (13) could be replaced by the equality:

$$|a| + |b| = |a + b.\mathrm{sgn}(ab)| \tag{19}$$

if a, b are *real numbers*. The inequality (12,14) can be saturated if B, B' (A, A') do not commute and the subsystems are non-trivially correlated, i.e., $|\Psi\rangle$ does not factorize and the subsystems are not in pure states. Notice that (16,17) would also hold if our quantum mechanical problem were reducible to a classical one (local hidden variables). These are the well known *Bell's inequalities*, 1980, and the experimental evidence to date seems to violate the bound (16,17) and to support (12,14).

Primary Literature

1. D. Bohm, Phys. Rev. **85** (1952) 166
2. H. D. Zeh, Found. Phys. **1** (1970) 69
3. W. H. Zurek, Phys. Rev. **D24** (1981) 1516
4. H. Everett III, Rev. Mod. Phys. **29** (1957) 454
5. R.B. Griffith, J. Stat. Phys. **36** (1984) 219
6. M. Gell-Mann and J. B. Hartle, in *Complexity, Entropy and the Physics of Information* W. H. Zurek Ed. (Addison Wesley 1990)
7. G. C. Ghirardi, A. Rimini and T. Weber, Il Nuovo Cim. **30B** (1975) 133
8. P. Pearle, Phys. Rev. **D13** (1976) 857

Secondary Literature

Discussion of these problems beyond that in the standard textbooks can be found, e.g., in

9. B. d'Espagnat: *Conceptual Foundations of Quantum Mechanics* (Benjamin 1976); *Veiled Reality* (Addison Wesley 1995)
10. J. S. Bell: *Speakable and Unspeakable in Quantum Mechanics* (Cambridge University Press 1987) which contains a collection of J. S. Bell's papers on the subject

11. D. Bohm: *Quantum Theory* (Dover 1989)
12. E. Joos, H. D. Zeh, C. Kiefer, D. Giulini, J. Kupsch and I.-O. Stamatescu: *Decoherence and the Apperence of a Classical World in Quantum Theory* (Springer, 3rd edition 2002)

which also includes an extensive bibliography. A collection of early papers is

13. J. A. Wheeler and W. H. Zurek: *Quantum Theory and Measurement* (Princeton University Press 1983)

Wave Mechanics

Marianne Breinig

In 1926 Erwin Schrödinger published a consistent mathematical theory of quantum mechanics, which became known as *wave mechanics*. He developed a partial differential equation, the ► Schrödinger equation, which now is considered the basic equation of non-relativistic quantum mechanics. Although wave mechanics was soon shown to be equivalent to ► matrix mechanics, the competing theory of quantum mechanics developed by Werner Heisenberg in 1925, many physicists favored wave mechanics, because they considered it more intuitive and because the ► Schrödinger equation was often easier to solve than the Heisenberg equation.

The *Schrödinger equation*,

$$(-\hbar^2/(2m))\nabla^2\psi(\boldsymbol{r},t) + U(\boldsymbol{r},t)\psi(\boldsymbol{r},t) = \mathrm{i}\hbar\partial\psi(\boldsymbol{r},t)/\partial t,$$

describes the time evolution of the wave function $\psi(\boldsymbol{r},t)$ which characterizes a non-relativistic particle of mass m, without internal structure, whose potential energy is given by $U(\boldsymbol{r},t)$. It can be generalized to a many-body equation

$$\sum_i [(-\hbar^2/(2m_i))\nabla_i^2\psi(\boldsymbol{r}_1,\boldsymbol{r}_2,\ldots,t)] \\ +U(\boldsymbol{r}_1,\boldsymbol{r}_2,\ldots,t)\psi(\boldsymbol{r}_1,\boldsymbol{r}_2,\ldots,t) = \mathrm{i}\hbar\partial\psi(\boldsymbol{r}_1,\boldsymbol{r}_2,\ldots,t)/\partial t.$$

Consider a single particle. The ► wave function $\psi(\boldsymbol{r},t)$ contains all the information the rest of the world, called the observer, can have about the particle at time t, without interacting with the particle. An interaction is called a ► measurement. It changes the information the observer has about the particle and therefore changes the wave function. Between measurements the wave function evolves deterministically.

The wave function is interpreted as the probability amplitude of the particle's presence. $|\psi(\boldsymbol{r},t)|^2$ is the *probability density*. (► Born rule) The probability that a particle at time t will be found in a volume element d^3r located about $\boldsymbol{r}$ is

$dP(\boldsymbol{r},t) = |\psi(\boldsymbol{r},t)|^2 d^3r$. For a single particle the total probability of finding it anywhere in space at time t is equal to 1. (In non-relativistic Quantum Mechanics material particles, unlike photons (▶ light quantum), are neither created nor destroyed.) Therefore

$$\int_{\text{all space}} |\psi(\boldsymbol{r},t))|^2 d^3r = 1.$$

A proper wave function must be square-integrable and therefore normalizable.

The Schrödinger equation implies *local conservation of probability*. The probability current density is given by

$$\boldsymbol{j}(\boldsymbol{r},t) = \frac{1}{m}\mathrm{Re}\left[\psi^*(\boldsymbol{r},t)\frac{\hbar}{\mathrm{i}}\nabla\psi(\boldsymbol{r},t)\right],$$

and the equation

$$-\frac{\partial}{\partial t}|\psi(\boldsymbol{r},t)|^2 = \nabla\cdot\boldsymbol{j}(\boldsymbol{r},t),$$

which expresses local conservation of probability, can be obtained multiplying the Schrödinger equation by $\psi^*(\boldsymbol{r},t)$ and its complex conjugate by $-\psi(\boldsymbol{r},t)$ and adding the two equations.

To make predictions about the outcome of a measurement, we must operate on the wave function with an ▶ operator. Every measurable quantity or *observable* is associated with a Hermitian operator. For example, the operator for the x-component of the momentum p_x is the differential operator $(\hbar/\mathrm{i})\partial/\partial x$. We have to take the partial derivative of the wave function with respect to x and then multiply by $(\hbar/\mathrm{i})$. The operator for the energy E is $\mathrm{i}\hbar\partial/\partial t$. It is also a differential operator. The operator for the position x is x. We have to multiply the wave function by x. If the operator for a particular observable A operates on a wave function $\psi(\boldsymbol{r},t)$ and the result of this operation is the wave function $\psi(\boldsymbol{r},t)$ multiplied by a real constant, then the wave function is said to be an eigenfunction of the operator and the constant is one of its eigenvalues. A measurement of the observable at time t will for certain yield the eigenvalue. There will be no uncertainty about the outcome of the measurement. If the operator for a particular observable A operates on a wave function $\psi(\boldsymbol{r},t)$ and the result of this operation is NOT the wave function $\psi(\boldsymbol{r},t)$ multiplied by a real constant, then the wave function is NOT an eigenfunction of the operator and there is uncertainty about the outcome of a measurement. The result of every measurement of an observable will be one of its eigenvalues. But if the wave function $\psi(\boldsymbol{r},t)$ is NOT an eigenfunction of the operator, then all we can predict is the probability of measuring any of the possible eigenvalues. We then can predict the average value of repeated measurements on identically prepared systems, but we cannot predict the outcome of an individual measurement.

Given the normalized wave function $\psi(\boldsymbol{r},t)$, the expression for the mean value of an observable A is $\langle \mathrm{A} \rangle = \int d^3r\,\psi^*(\boldsymbol{r},t)\,\mathrm{A}\psi(\boldsymbol{r},t)$.

The root mean square deviation $\Delta A = \sqrt{<A^2> - <A>^2}$ characterizes the dispersion of the measurement around $<A>$. It is a measure of the spread that one should expect in the result of a measurement of the observable A.

The *principle of spectral decomposition* states that any wave function $\psi(\boldsymbol{r}, t)$ can be expanded in terms of the eigenfunctions of any observable. Let $\{\psi_a^i(\boldsymbol{r})\}$ denote the set of orthonormal eigenfunctions of the observable A, and let A $\psi_a^i(\boldsymbol{r}) = a\psi_a^i(\boldsymbol{r})$. If the eigenvalue a is degenerate, then the superscript i denotes different eigenfunctions with the same eigenvalue a. Any wave function $\psi(\boldsymbol{r}, t)$ can be written as

$$\psi(\boldsymbol{r}, t) = \sum_{a,i} c_a^i(t)\psi_a^i(\boldsymbol{r}), \text{ with } \sum_{a,i} |c_a^i(t)|^2 = 1.$$

The $c_a^i(t)$ are the expansion coefficients. If the observable A is measured, the result of the measurement will belong to the set of eigenvalues $\{a\}$. Spectral decomposition, see ▶ Density operator; Ignorance interpretation; Measurement theory; Objectification; Operator; Probabilistic Interpretation; Propensities in Quantum Mechanics; Self-adjoint operator.

The probability that a measurement of A at time t will yield the eigenvalue a' is

$$P_{a'} = \sum_i |c_{a'}^i(t)|^2,$$

If a measurement of A yields a', then the wave function immediately after the measurement is $\psi_{a'}(\boldsymbol{r}, t) = \sum_i c_{a'}^i(t)\psi_{a'}^i(\boldsymbol{r})$.

The Schrödinger equation describes how the wave function evolves between measurements. To determine the wave function $\psi(\boldsymbol{r}, t_0)$ at some initial time t_0, we have to measure a complete set of commuting observables, i.e., a set of observables that have a unique set of common eigenfunctions. The results of the measurements at t_0 then specify the wavefunction $\psi(\boldsymbol{r}, t_0)$ completely.

The Schrödinger equation for a particle moving in one dimension through a region where its potential energy is a function of position only has the form

$$(-\hbar^2/(2m))\nabla^2\psi(\boldsymbol{r}, t) + U(\boldsymbol{r})\psi(\boldsymbol{r}, t) = \mathrm{i}\hbar\partial\psi(\boldsymbol{r}, t)/\partial t.$$

We are often interested in finding the eigenfunctions of the energy operator $\mathrm{i}\hbar\partial/\partial t$, i.e., we are interested in finding the wave functions of a particle whose energy can be predicted with certainty. For an eigenfunction of the energy operator we have

$$\mathrm{i}\hbar\partial\psi(\boldsymbol{r}, t)/\partial t = E\psi(\boldsymbol{r}, t).$$

Therefore

$$\psi(\boldsymbol{r}, t) = \psi(\boldsymbol{r})\exp(-\mathrm{i}Et/\hbar).$$

For eigenfunctions of the energy operator the Schrödinger equation becomes time independent.

$$(-\hbar^2/(2m))\nabla^2\psi(\boldsymbol{r}) + U(\boldsymbol{r})\psi(\boldsymbol{r}) = E\psi(\boldsymbol{r}, t).$$

The operator $(-\hbar^2/(2m))\nabla^2 + U(\boldsymbol{r})$ is called the Hamiltonian operator H, and the *time-independent Schrödinger equation* is often abbreviated as

$$\mathrm{H}\psi(\boldsymbol{r}) = E\psi(\boldsymbol{r}).$$

The possible solutions $\psi(\boldsymbol{r})$ of the time-independent Schrödinger equation are the eigenfunctions of the ► Hamiltonian operator. The corresponding wave functions $\psi(\boldsymbol{r}, t)$ are obtained by just multiplying $\psi(\boldsymbol{r})$ by $\exp(-\mathrm{i}Et/\hbar)$, where E is the appropriate eigenvalue for each eigenfunction of H. The wave function of a particle whose energy E can be predicted with certainty is of the form $\psi(\boldsymbol{r}, t) = \psi(\boldsymbol{r})$ $\exp(\text{-}\mathrm{i}Et/\hbar)$.

The probability density then is given by

$$|\psi(\boldsymbol{r}, t)|^2 = \psi(\boldsymbol{r})\exp(-\mathrm{i}Et/\hbar)\psi^*(\boldsymbol{r})\exp(\mathrm{i}Et/\hbar) = |\psi(\boldsymbol{r})|^2.$$

The probability of finding the particle with well defined energy at a particular position $\boldsymbol{r}$ is therefore independent of time. The probability current density is zero. The particle is said to be in a *stationary state*.

The Schrödinger equation is a linear equation. There exists a linear operator that transforms $\psi(\boldsymbol{r}, t_0)$ into $\psi(\boldsymbol{r}, t)$.

$$\psi(\boldsymbol{r}, t) = U(t, t_0)\psi(\boldsymbol{r}, t_0).$$

The operator $U(t, t_0)$ is called the *evolution operator*. The evolution operator is a unitary operator. If H does not explicitly depend on time, then the Schrödinger equation yields $U(t, t_0) = \exp(-\mathrm{i}\mathrm{H}(t - \mathrm{t}_0)/\hbar)$. If an arbitrary wavefunction $\psi(\boldsymbol{r}, t_0)$ is expanded in terms of eigenfunctions of H, i.e., if

$$\psi(\boldsymbol{r}, t_0) = \sum_n c_n\psi_n(\boldsymbol{r}),$$

with $\mathrm{H}\psi_n(\boldsymbol{r}) = E_n\psi_n(\boldsymbol{r})$, then

$$\psi(\boldsymbol{r}, t) = \sum_n c_n exp(-iE_n(t - t_0)/\hbar)\psi_n(\boldsymbol{r}) = \sum_n c_n(t)\psi_n(\boldsymbol{r}).$$

This yields the wave function at any time t.

A simple example:

Assume we want to solve the Schrödinger equation in one dimension,

$$(-\hbar^2/(2m))\partial^2\psi(x)/\partial x^2 + U(x)\psi(x) = E\psi(x).$$

Defining $k_1^2 = 2mE/\hbar^2$, $k_0(x)^2 = 2mU(x)/\hbar^2$, and $k(x)^2 = k_1^2 - k_0(x)^2$ we can simplify the notation.

$$\partial^2\psi(x)/\partial x^2 + k(x)^2\psi(x) = 0,$$

Let us solve this equation for the "infinite square well." We assume $U(x) = 0$ for $x = 0$ to L, and $U(x) =$ infinite everywhere else. A particle cannot penetrate a region with infinite potential energy, there is no chance that we can find it there, and its wave function in that region is zero. We put the particle in a one-dimensional box, out of which it has no chance of escaping. In the region from $x = 0$ to $x = L$ the potential energy $U(x) = 0$. The particle can freely move inside the box. Therefore $k_0(x) = 0$ and $k(x)^2 = k_1^2$. Possible wave functions for the particle must satisfy the equation

$$\partial^2\psi(x)/\partial x^2 + k^2\psi(x) = 0,$$

and they must be zero at $x = 0$ and $x = L$, because the eigenfunctions of H must be continuous and the wave function is zero outside the region from $x = 0$ to $x = L$. Real solutions of the Schrödinger equation which are zero at $x = 0$ and $x = L$ are $\psi(x) = A\sin(kx)$, with $kL = n\pi$, with $n = 1, 2, 3, \ldots$. The possible values of k are $k_n = n\pi/L$, the possible values of the energy are $E_n = \hbar^2 k_n^2/(2m) = n^2\pi^2\hbar^2/(2mL^2)$. The potential and the first five possible energies a particle can have are shown in Fig. 1. units are used (Fig. 1).

The energy of a particle in an infinite square well is quantized. If we measure the energy we can only measure one of the eigenvalues, $E_n = n^2\pi^2\hbar^2/(2mL^2)$, $n = 1, 2, 3, \ldots$. The confinement of the particle leads to energy ▶ quantization. If we measure E_n, then right after the measurement the wave function of the particle is

$$\psi_n(x,t) = A_n \sin(n\pi x/L)\exp(-iE_n t/\hbar).$$

The square of the normalized wave function $|\psi_n(x,t)|^2 = |\psi_n(x)|^2 = A_n^2 \sin^2(n\pi x/L)$ is equal to the probability per unit length of finding the particle with energy E_n at position x. To normalize the wave function we have to choose $A_n^2 = 2/L$. Then $\int_{-\infty}^{+\infty} |\psi(x,t)|^2 dx = 1$, and the total probability of finding the particle inside the well is 1.

A particle in an infinite square well does not have to be in an eigenstate of the energy operator. If we measure the position of a particle in the well and find it at

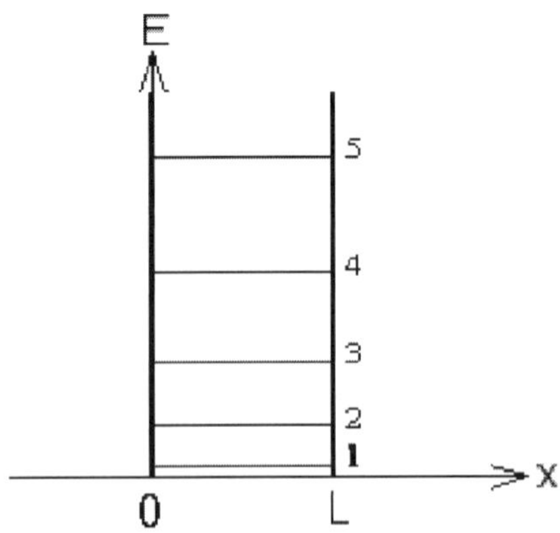

Fig. 1 Energy levels of a particle in a 1D "infinite square well"

some position x, then right after the measurement the particle is in an eigenstate of the position operator. Its energy is unknown, we can at most determine its average energy and the probability of measuring one of its eigenenergies in a subsequent measurement. Right after our measurement, the particle is in a ► superposition of energy eigenstates. Let us investigate one of those superpositions.

Assume a particle of mass m moves in one dimension in a square well with walls of infinite height a distance L apart and that the particle is known to be in a state consisting of an equal admixture of the two lowest energy eigenstates of the system.

$P(x,t) = |\psi(x,t)|^2$ is the probability per unit length of finding the particle at position x as a function of time.

$\psi(x,t) = 2^{-1/2}[\psi_1(x,t) + \psi_2(x,t)]$, with

$$\psi_1(x,t) = (2/L)^{1/2} \sin(\pi x/L) \exp(-(i/\hbar)E_1 t),$$
$$\psi_2(x,t) = (2/L)^{1/2} \sin(2\pi x/L) \exp(-(i/\hbar)E_2 t),$$
$$\text{and } E_1 = \pi^2\hbar^2/(2mL^2),\ E_2 = 4\pi^2\hbar^2/(2mL^2).$$

Therefore

$$\begin{aligned}|\psi(x,t)|^2 &= (1/2)|\psi_1(x,t) + \psi_2(x,t)|^2 \\ &= (1/L)|\sin(\pi x/L)\exp(-(i/\hbar)E_1 t) + \sin(2\pi x/L)\exp(-(i/\hbar)E_2 t)|^2 \\ &= (1/L)[\sin^2(\pi x/L) + \sin^2(2\pi x/L) \\ &\quad +2\sin(\pi x/L)\sin(2\pi x/L)\cos((E_2 - E_1)t/\hbar)].\end{aligned}$$

$P(x,t)$ is no longer independent of time, the probability per unit length of finding the particle at x is changing with time. The probability current density at position x is

$$\begin{aligned}j(x,t) &= (\hbar/m)\mathrm{Re}((-i)\psi^*(x,t)\nabla(\psi(x,t)) \\ &= (\pi\hbar/(mL^2))\sin(\pi x/L)(1 - \cos(2\pi x/L))\sin((E_2 - E_1)t/\hbar)\end{aligned}$$

and we can verify that $-\partial|\psi(x,t)|^2/\partial t = \partial j(x,t)|/\partial x$.

Primary Literature

1. E. Schrödinger: An undulatory theory of the mechanics of atoms and molecules. Phys. Rev. **28** (1926), 1049–1070
2. J.J. Sakurai: *Modern Quantum Mechanics*, Revised Edition (Addison Wesley 1994, 98–109)

W

Secondary Literature

3. B.H. Brandsden, C.J. Joachain: *Quantum Mechanics*, 2nd Edition (Prentice Hall, 2000)

Wave Packet

Helge Kragh

A wave packet is a concentrated train of (quantum) waves of various wavelengths or momenta with the property that the packet is confined within a small region of space. Such a packet can be constructed by adding a very large number of waves so chosen that their sum interferes destructively everywhere except in a small region. If harmonic waves of different momenta are superposed, the packet can be expressed in the form $\psi(x) = \int A(k)e^{ikx}dk$ where $k = p/\hbar$ and $A(k)$ is the amplitude corresponding to the wave number k.

Although speculative attempts to identify atoms with systems of standing waves can be found back in the nineteenth century, in a quantum context it was Schrödinger who invented wave packets and related them to atomic particles. In his second communication on ► wave mechanics Schrödinger discussed the possibility of constructing a wave group or packet equivalent to a pointlike particle, such as an electron, and in a subsequent paper of 1926 he provided a more elaborate discussion in which he introduced the ► superposition principle. Analyzing the case of a one-dimensional harmonic oscillator, Schrödinger constructed for the first time a wave packet as an exact solution of the ► Schrödinger equation. Making use of the superposition principle, he constructed a wave packet of the form $\psi = \sum a^n \psi_n / n!$, where a is a large number, ψ_n are the eigenstates, and $0 \leqslant n \leqslant \infty$. The resulting wave packet, he showed, remains compact as time goes on and it has an energy which is exactly the same as the one of the classical oscillator. Schrödinger's wave packet was a "minimum uncertainty wave packet," the first example of what later became known as "► coherent states." He believed that this result would be valid also for electrons moving in atomic orbits and, if so, that it indicated that perhaps electrons and other particles *are* wave packets. At the end of his paper he foresaw that it was only a matter of time until "the representation by wave mechanics of the hydrogen atom" ► Bohr's atom model would be achieved.

However, in letters to Schrödinger from June 1926, Lorentz demonstrated that a permanent wave packet cannot be constructed for an atomic electron and that Schrödinger's success with the harmonic oscillator was accidental. "In the present form of your theory you will be unable to construct wave packets that can represent electrons moving in very high Bohr orbits," Lorentz wrote. It is unknown how Schrödinger reacted, but most likely Lorentz' critique contributed to a change in his ontology: by the fall of 1926 Schrödinger concluded that his original belief in the primacy of waves was not an integral part of wave mechanics.

Some of Lorentz's objections were independently made by Heisenberg in his famous paper of 1927 in which he introduced the ► Heisenberg uncertainty principle, which he derived by means of arguments based on wave packets. According to Heisenberg, "Schrödinger's reasoning is only viable for the case of the harmonic oscillator treated by him; in all other cases a wave packet spreads out in the course of time over the whole immediate neighborhood of the atom." He observed that the peculiar properties of the wave packet Schrödinger had found was a consequence of the fact that the energy levels of the harmonic oscillator are equally spaced (namely, given by $E_n = (n + 1/2)\hbar\omega$). Moreover, Heisenberg found that the size of the probability wave packet – $\psi\psi^*$ rather than ψ – representing a freely moving particle would increase indefinitely with the time.

Wave packets were not only important in the chain of arguments that led Heisenberg to his uncertainty relations, they also played a crucial role in Bohr's physical interpretation of quantum theory and his formulation of the ▶ complementarity principle in the fall of 1927 where he used wave packets to represent both ▶ light quanta and ▶ electrons. The problem with the wave packet picture illustrated to Bohr that "the contrast between the wave theory superposition principle and the assumption of the individuality of particles" was irremediable. At that time, Schrödinger had abandoned his wave ontology and no longer thought of electrons as constituted by wave packets.

The papers by Schrödinger and Heisenberg were discussed by several physicists in 1927–1928, including George Darwin, Earle Kennard and Arthur Ruark, who all recognized that electrons cannot be represented just as wave packets. Or, as Kennard expressed it, "the electron must always be assigned a greater degree of reality than that of a wave packet."

As indicated by the title of Schrödinger's paper of 1926, "The Continuous Transition from Micro- to Macromechanics," his aim was to understand the behaviour of macroscopic bodies from quantum principles. Although wave packets would not do as representations of subatomic particles, in 1927 Paul Ehrenfest showed that there were no corresponding problems with spreading wave packets (Fig. 1) in the case of macroscopic bodies. As an example he calculated the time it would take for a particle of mass m and represented by a probability wave packet of width Δ to spread

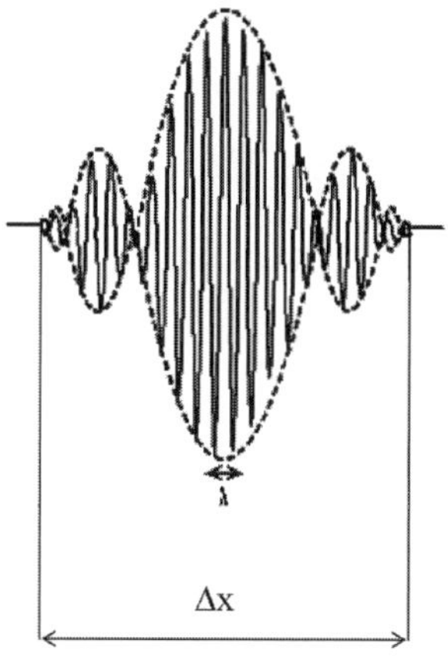

Fig. 1 Example of a wave packet. Source: Wikimedia Commons

out to double its initial size. His result was $T \cong \Delta\sqrt{m/\hbar}$. Because of the smallness of ► Planck's constant ($\hbar = 1.05 \times 10^{-34}$ Js) this means that the doubling time is nearly infinite for a macroscopic particle. For a particle of linear size $\Delta = 0.001$ cm and mass $m = 1$ g, the doubling time is about 10,000 times the age of the universe.

Another important work, relating to Schrödinger's and Ehrenfest's, was due to Peter Debye, who showed that ► wave packet, simulating mass and charge points, can be constructed also without using the special expansion coefficient that Schrödinger had used in his treatment of the harmonic oscillator. Debye discussed in 1927 the behaviour of wave packets of one degree of freedom for any kind of force, and found that their maxima move in accordance with the classical laws. His work was one of many that aimed at showing the correspondence-like connection between quantum mechanics and classical physics.

Primary Literature

1. E. Schrödinger: Der stetige Übergang von der Mikro- zur Makromechanik. Die Naturwissenschaften **14**, 664–666 (1926)
2. W. Heisenberg: Über den anschaulichen Inhalt der quantentheoretischen Kinematik und Mechanik. Zeitschrift für Physik **43**, 172–198 (1927)
3. K. Przibram ed.: *Briefe zur Wellenmechanik* (Springer, Vienna 1963)

Secondary Literature

4. M. Jammer: *The Philosophy of Quantum Mechanics* (Wiley, New York 1974)
5. F. Steiner: Schrödinger's Discovery of Coherent States. Physica **B 151**, 323–326 (1988)
6. H. Kragh, B. Carazza: Classical Behavior of Macroscopic Bodies from Quantum Principles: Early Discussions. Archive for History of Exact Sciences **55**, 43–56 (2000)

Wave-Particle Duality: Some History

Bruce R. Wheaton

Our modern understanding of light is the result of dispute since the scientific revolution of the seventeenth century. The roots of that contention, however, precede the contributions of Aristotle, and I daresay the final story has yet to be written.

Following Plato and his student Aristotle, what we see in our lives are "secondary" qualities that originate from an unseen world of "primary" events. In their view whatever the primary causes of sound should seem similar to the water, and of matter to the rocks we encounter in life. The earlier philosophers tended to find guidance from a single entity: Thales from water with its waves lapping the shores; Anaximines sowed the seeds of all within all, a proto-atomic hypothesis later developed by Demokratus and Leucretius. For Aristotle, light was special. It can coexist in the celestial and earthly world, thus it could not be compounded of Empedokles' four elements. He refers to light as a process, an "actualization" of a latent property. Light thus occurs instantaneously, since it is *not* propagated. That light delineates straight lines underlay surveying and observational astronomy, and made both accessible to geometry, like mechanics, in the ancient world.

Aristotle's worldview dominated natural speculation throughout the middle ages. But the distinction between the discrete and the continuous is an important philosophical issue that has driven epistemological discussion in the west since the pre-Socratics. Its modern locus in quantum physics is only the most recent manifestation. It informed discussion of the contrast amongst Empedokles' elements; figures in Aristotle's Platonic distinction between what we observe and the underlying primary qualities of things; of theological issues in the middle ages; of renaissance mathematics upon the introduction of numerical *al-jebra* in conflict with Greek continuity; of nascent optics; of electron/field physics after Maxwell; and its modern quantum guise will be diverted and changed in the future. These conflicting views, a Hegelian dichotomy, had competed for hegemony in western natural philosophy since before Aristotle.

Even with the remarkable advances in medieval study of optical properties of lenses for eyeglasses, the telescope, the microscope; discovery of Snell's law of refraction (1621); even later successful attempts to measure the speed of light (Roemer 1676); one finds little inquiry into the nature, rather than the properties, of light even in writings of masters like Averroes, Witelo, and Kepler. Descartes, for example, pictured the cosmos a plenum in which light *is* the pressure exerted by motion of its parts at a distance from the eye. Before refined devices existed to measure the quantitative properties of light, the issue remained one of smoke and mirrors.

But with the revolution in science of the seventeenth century, all changed. Materialism rose ascendant, so observed secondary properties, even of light, tended to be ascribed to unperceived atoms. Thus natural philosophers of the eighteenth century set themselves the goal of verifying what many took to be Isaac Newton's corpuscular theory of light (henceforth CT) in its finest manifestations.

Newton's *Opticks* (1704 and later editions) capped his efforts beginning in 1672 to extend mathematical analysis to include refraction, diffraction, and color. Newton ascribed the observed periodicity ("Newton's rings") to "fits of transmission" by what otherwise must be something like particles of light, particles that differ in their three spatial dimensions; and he assigned different particle-like characteristics to each color of light as its "connate property." By this he explained the peculiarity of the beam splitting in two on transmission through calcite, long-known as a useful navigating tool called "Iceland spar."

Leonhard Euler's *Nova theoria lucis et colorum* (1746) represents the crest of the opposing undulatory theory (UT) in the eighteenth century. He proposed a truly periodic wave where light frequencies parallel the harmonies of sound. But in this period when wave interference was barely recognized, the ability of any wave to yield observed rectilinear propagation raised grave difficulties, and Euler's ideas were not widely embraced even on the continent. The battleground would be the fine points of light in its interaction with matter. Rectilinear propagation and reflection favored the CT, in accord with senses of taste and touch. Refraction and diffraction constituted as seemingly fatal a difficulty for any CT as rectilinear propagation posed for the UT, based on senses of sight and hearing.

Christaan Huygens (1629–95), struck by the incompatibility of geometric continuity with algebraic discreteness, had offered an elegant explanation of both properties in 1678 that light is best portrayed as an irregular sequence of discontinuous impulses propagating in a medium (*not* UT.) Newton's authority had bullied most philosophers of the eighteenth century to overlook Huygens' penetrating objections. The devil clearly lay in the details and the battle soon focused on *polarization* which seemed explainable on both accounts to the kinetic ontology of the time, now to be described.

The "Laplacian school" in early nineteenth century France accorded conceptually coherent explanation of reflection, refraction, diffraction, and polarization in terms of gravity-like forces acting *within* atomic 'atmospheres' of subtle caloric fluid. Using crystals as analyzers, Etienne Malus found that sunlight can be polarized just by reflecting off materials like glass and water. This eliminated the atomic atmosphere necessary to Pierre-Simon Laplace's position and dulled Ockham's razor to an extent that began to offend, even in France.

Educated in Scotland, English dissenter Thomas Young studied medicine in Göttingen and took interest in hearing and the acoustical waves of sound as discussed by, among others, Euler. His detailed studies of the physiology of the eye soon turned up so many parallels between observable properties of sound and light that he was led to Newtonian heresy before 1800. Young proposed an UT he sought to authenticate as the "true" Newtonian view, an ambiguity, like the sense of smell, somewhere between particles and waves.

Young developed many practical demonstrations for public lectures in London of his belief that, like sound, light is a longitudinal wave. He demonstrated that, like acoustic sound, hydrodynamic water waves passed through a double aperture show marked interference effects, producing no disturbance where the crests of one align with the troughs of the other, as in Fig. 1 ► double-slit experiment. And he extended this analogy to light with little idea of the medium in which it propagated, but thereby calculating the approximate wavelength of light by 1803.

Throughout these public claims, Young apologized that the water waves, being transverse, were only an approximation to the longitudinal waves of light and sound. His qualitative results made their way despite the Napoleonic wars to the director of the Bureau des Longitudes, François Arago (1786–1853) who passed the issue to his *cadet*. Augustin Fresnel (1788–1827), son of a mason in Normandy, given the most rigorous scientific education available anywhere in the world at the Ecole Polytéchnique, found the CT untenable in principle. Mathematically adept, Fresnel saw through the mathematical haze to the physical failure of the Laplacian program. Because Fresnel had come to his revelation in ignorance of Young's but armed with differential equations, a crucial difference emerged. From the first Fresnel admitted

Fig. 1 Young's ripple tank results of interference of equal frequencies from A & B: low at C & E, high at D & F. From Young, Course XX, 267 (1807)

a transverse component in polarization-induced color changes in thin crystals, he came by 1821 to realize that the transverse tail must be wagging the longitudinal dog. Polarization is most realistically treated mathematically as interference of *two* waves moving along the same line but separated by a 90° ($\lambda/4$) phase shift; their interference in an analyzing crystal produces the observed result.

This implied that wave direction could as readily be thought to lie orthogonal to the physical motion of the aetherial medium. Indeed, were the transverse component to rotate rapidly enough about that direction of wave motion, the polarizing asymmetry would vanish and appear as unpolarized light. Lacking the acoustic base from whence Euler and Young proceeded, Fresnel's mathematical analysis of interference could now stand on purely transverse waves. Figure 2 is his version of Young's experiment, except here the two sources A and B are the diffracted waves at the edges of obstacle AB. In Young's hydrodynamic image the water goes up and down while the wave proceeds along the surface; he had been apologizing for a decade about the inaccuracy of his ripple tank, so Fresnel's transverse waves came as a lightning bolt.

Fresnel's 1816 "Mémoire sur la diffraction de la lumière" is the foundation of the classical UT of light; it led to remarkable tools like ► spectroscopy to analyze the chemical nature of the stars. That paragon combination of theory and experiment, Heinrich Hertz declaimed in 1889 that "for all practical purposes, the wave theory of light is a certainty." Despite the immense advances that acceptance of the UT's enlightened legacy brought, it too would shift out of favor in the twentieth century.

Cracks in the UT began to appear, almost unnoticed, in the 1880s with the famous aether-drift experiments of Americans Albert Michelson and Edward Morley that seemed to find no aether in which light could propagate. But the most challenging troubles followed concurrent improvements in vacuum technology that led to cathode discharge tubes and to discovery of ► X-rays in 1895. The most relevant explanation of this "new form of radiation" was a resuscitation of Huygens'

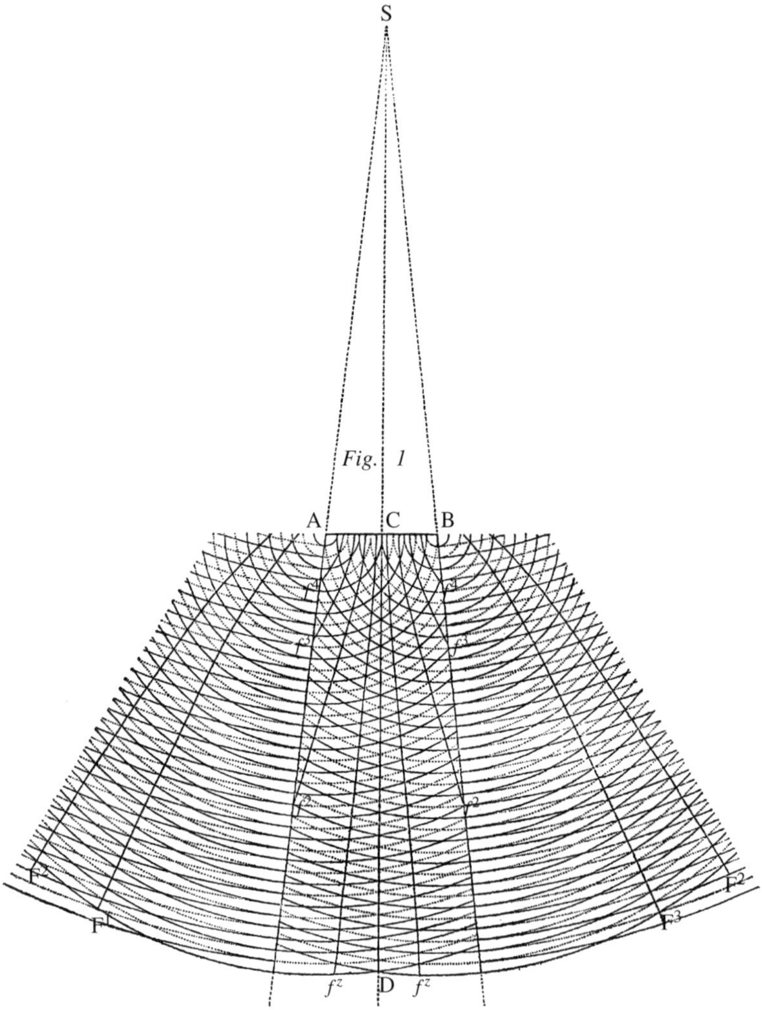

Fig. 2 Fresnel's mathematical reconstruction of Young's double-slit experiment, where the sources A & B are the waves from a single source diffracted at the edges of obstacle C. Fresnel was able to show that the lines of equal interference (like F^1 & F^2 are hyperbolic). From Verdet, ed. Oeuvres d'Augustin Fresnel, vol. 1, p. 95, Paris: Imp. imp., 1866

disconnected impulse model of light, now from the pen of George Stokes. Today we have an acoustic analogy to this early view of X-rays (and Huygens' of light): the sonic boom. Constructed by superposition of wake vibrations in the continuum of the atmosphere, it has nonetheless a localized effect on the ground. You hear it as if it were a pistol shot. It combines the UT and the CT in a trice and its possibilities, other than the pregnant Cherenkov radiation (► Bremsstrahlung), have been largely ignored by physicists and left to SST designers.

Primary Literature

1. Aristotle: *De anima*, 418–419
2. I. Newton: *Opticks*. A brief preliminary version (Moxon, London 1670) caused so much rancorous contention with Robert Hooke that Newton withheld publication to (Smith and Walford, London 1704) after Hooke's death. With important later additions, the prime source is the 4th edn. of 1730 readily available (Dover, New York 1979)
3. Th. Young: Bakerian lecture: Experiments and calculations relative to physical optics. Royal Society of London Philosophical Transactions **94**, 1–16 (1804)
4. Th. Young: *Course of lectures on natural philosophy* (Johnson, London 1807, Vol. 1, on hydrodynamics p. 290; on sound pp. 389–96; on light pp. 464ff; the double-slit is plate XX, figure 267). The remarkable Baconian encyclopaedic approach of the man can be divined merely by reading his short introduction and analytical table of contents
5. A. Fresnel: *Oeuvres completes* (ed. Verdet Paris: Imp. imp., 1866 vol. 1, 9–33; 89–122*; 129–170). *This appeared in Annales de chemie et physique **1**, 239–81 (1816); all were published in the *Comptes rendus* of the Academie, but not until after Fresnel was elected a member in 1823, four years before his death from tuberculosis

Secondary Literature

6. S. Boffi, M. D'Anna: *Le radici del dualismo onda-corpuscolo* (Bibliopolis, Napoli 1999)
7. R. McCormmach: J. J. Thomson and the structure of light. British Journal of the History of Science **3**, 362–387 (1967)
8. J. C. Poggendorff: Über den relativen Werth der beiden Lichttheorien. Ann Pk, **28**, 381–383 (1833). Footnote to article by Brewster
9. P. Rowland: *Waves versus corpuscles – The revolution that never was* (PD, Liverpool 1992)
10. B. R. Wheaton: *The tiger and the shark – Empirical roots of wave-particle dualism* (Cambridge University Press, Cambridge 1983) An extensive literature exists on history of optics and light by *inter alia:* Buchwald, Cantor, Lindberg, Ronchi, Sabra, Shapiro

Wave-Particle Duality: A Modern View

Bruce R. Wheaton

W

Our understanding of light is the result of dispute since the scientific revolution of the 17th century. Students of physics today are taught "wave-particle duality": belief based on otherwise conflicting experiments that electromagnetic radiation is a periodic wave that, at high frequencies, exhibits increasingly localized concentration of energy. It is a wave with particle characteristics: *something* akin to energy that under some circumstances exhibits interference like periodic waves, and under others acts like a stream of bullets.

In optics, Newton's corpuscular theory (CT) of light was later challenged by a purely periodic undulatory theory (UT) espoused by Young and Fresnel. With the discovery in 1895 of ► x-rays, the then accepted UT came under new attack, particularly in their now-measurable electrical effect on gases. J. J. Thomson, was alarmed that light, like x-rays, seemed to ionize precious few of the atoms it encountered. Were either a UT product propagating spherically, more atoms should be ionized than he could find. He suggested light itself might be "directed radiation" sometimes called "needle rays," and began to wonder around 1909 whether very weak light would still show classical ► double-slit experiment. The experiment, performed by Geoffrey Taylor with yellow light of such low intensity the photographic exposure took a week, nonetheless produced the classic pattern of fringes.

It seemed that only evidence for interference of x-rays would clear up the matter and decide in favor of the UT, but it was not to be. Walther Friedrich & Paul Knipping's claim to find x-ray crystal interference in 1912 coincided with the abandonment of the last classical attempt to explain the optical ► photoelectric effect. On the one hand 1912 brought the UT into greater coherence with x-rays. On the other it forced acceptance of the new quantum transformation relation (QTR) on the *absorption* of light by metals; that is, of Einstein's widely-rejected ► light-quantum from 1905.

It was one thing to claim that light is emitted in quantum units, but an entirely different matter to understand how it could possibly be absorbed only in quanta. How does an atom 'know' that it has absorbed enough UT light? It seemed impossible, but Einstein might be right that light is in some way corpuscular. What tipped the balance in the early 1920s also came from ► x-rays. When they ionize a gas, ► electrons are released. But two paradoxes had been found in this process. (► Errors and paradoxes in quantum mechanics). If x-rays are spherically propagating electromagnetic effects, they spread their effect over increasingly larger spherical shells centered on their point of production. If there is enough energy at any point in a shell to ionize an atom, all atoms at that distance should be ionized, yet too few electrons were being found: a paradox of "quantity." The ones released should only receive $1/4\pi d^3$ the total energy in the shell at distance d, yet those few electrons had far too much kinetic energy: a paradox of "quality."

Clearly the ► light-quantum could no longer be ignored. The most influential experiments were done on generalized x-ray scattering results in the U.S. by Arthur Compton (► Compton effect), on the x-ray ► photoeffect in France by Maurice de Broglie, and on similar γ-ray phenomena in Britain by Charles Drummond Ellis (1895–1980). In all cases the corpuscular behavior of electromagnetic radiation prevailed: see ► matter waves.

In 1928 Werner Heisenberg reconciled and codified the incommensurability inherent in the new quantum mechanics in the form of his "indeterminacy principle" ► Heisenberg uncertainty relations. Although he formulated it to rationalize the non-commuting properties necessitated by his ► matrix mechanics, in its most fundamental form regarding position and momentum it speaks directly to wave-particle duality. To be monochromatic, a wave must extend to infinity. When interpreted as a probability, such a ► wave function spreads the likelihood of finding the associated particle also over infinite dimension. Correspondingly, if the position of, say, an electron is precisely fixed by experiment, its ► wave packet is constrained to a small spatial dimension, and the Fourier expansion of such a small 'wavelet' leaves its frequency indeterminate. L. De Broglie's ► matter-wave mechanics relationship $\lambda = h/p$ fixes the momentum of the electron to that frequency, so $\Delta x \Delta p \sim h$ as Heisenberg's principle requires.

These considerations were most troubling to atom-architect Niels Bohr, whose adherence to classical principles was as rock beneath his physics. He rationalized the wave-particle divide in a tribute to Volta in 1927 as characteristic of different, co-existing physical systems that "complement" one another at their intersection. Others, like Einstein, would not go even that far and rejected the anti-deterministic consequences required by the new quantum mechanics. A series of objections followed over the years: ► Einstein-Podolsky-Rosen paradox; Bohm's qualitative ► hidden variables; L. de Broglie's theory of the double-solution; all intended (without success) to show that determinism persists, perhaps only hidden to human perception, and that wave-particle duality is a chimera. In 1964 John Bell quantified Bohm's hidden variable hypothesis, showing that were measurement of the state of one particle formerly entangled with that of another to fix that of the other before its measurement, certain Bell inequalities, must hold. (► Bell's theorem). Careful experiments on correlated ► spin determinations of parts of former molecules, and on locations of formerly associated photons (► light quantum) failed to exhibit those inequalities, hence corroborating the orthodox quantum mechanical view.

The latter prescient example is double-slit interference, like Young showed to be true for light. (See part 1 *supra.*) If it were possible, without disturbing the interference pattern, simultaneously to determine through which slit the "particle" traveled, the thrust of Heisenberg's principle could be parried. Einstein proposed a double-slit thought experiment in which the recoil of the slits themselves might signal which was penetrated, and it was promptly challenged by Bohr, acting to defend what came to be called the "Copenhagen interpretation." ► Born rule; Consistent Histories; Metaphysics in Quantum Mechanics; Nonlocality; Orthodox Interpretation; Schrödinger's Cat; Transactional Interpretation.

From the 1927–8 electron crystal scattering results by Davisson (USA) and G. Thomson (UK) right up to the 1960s, classical double-slit interference of electrons remained in the "Gedankenexperiment" realm. Then Jönsson in Tübingen found a clever means to produce slit masks of unprecedented minuteness (ca 1 μ). Figure 1 shows the result for double-slit interference of an electron beam, the first direct corroboration that the Young result still obtains.

These considerations have led more recently to attempts to determine which aperture an electron has passed through *without disturbing* the wave interference pattern that results. Bohr had argued persuasively that, according to his ► correspondence principle, this was not possible, even after Einstein posited his recoiling slit thought experiment to do so. However with recent development of micromasers, a proposal (Scully et al. 1991) to detect "which way" (which slit) an excited rubidium atom (^{85}Rb) passes through a system of micromaser cavities might answer: one of the two masers will detect an emitted microwave photon and leave which-way information,

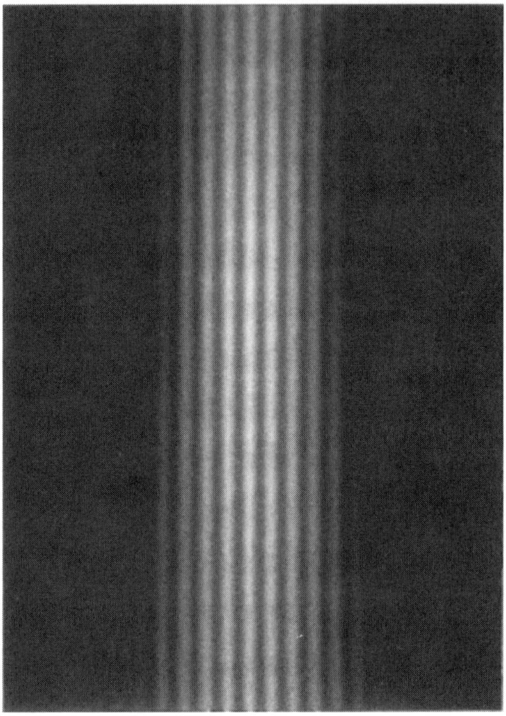

Fig. 1 Electron-optical two-slit interference. Source: Zeitschrift für Physik 155 (1959), 427–74

see Fig. 2. (► Which-way experiments). On this view, the non-interference pattern expected by Bohr is the ► superposition of *two* identical interference patterns 180° ($\lambda/2$) apart in phase: one due to photons whose "which way" slit is determined, the other caused by those whose "which way" information is not determined. According to the experimenters it is due to 'the correlation of the centre-of-mass wave function to the photon degrees of freedom in the cavities that is responsible for the loss of interference.' [10, p. 114]

More recently, refined experiments resulted in a curious inversion of the Braggs' classic 1913 research program to determine material crystal structure using incident x-rays. In 1998 excited rubidium atoms were projected onto a "lattice" of standing-beam light-waves. [4] When a second quantum system was added to the microwave interferometer it was able to store pathway information in the atom beam with the result that the interference pattern disappeared. While the effect appeared to be below the Heisenberg threshold, the conclusion was that it was due to correlations (an environmental form of ► "entanglement") between the microwave detector and quantum-kinetic motion within the rubidium beam itself. These possibilities have naturally led to controversy, raising the interesting question of whether ► complementarity trumps indeterminacy (► Heisenberg's uncertainty relation), and final conclusions remain, if at all, in the future.

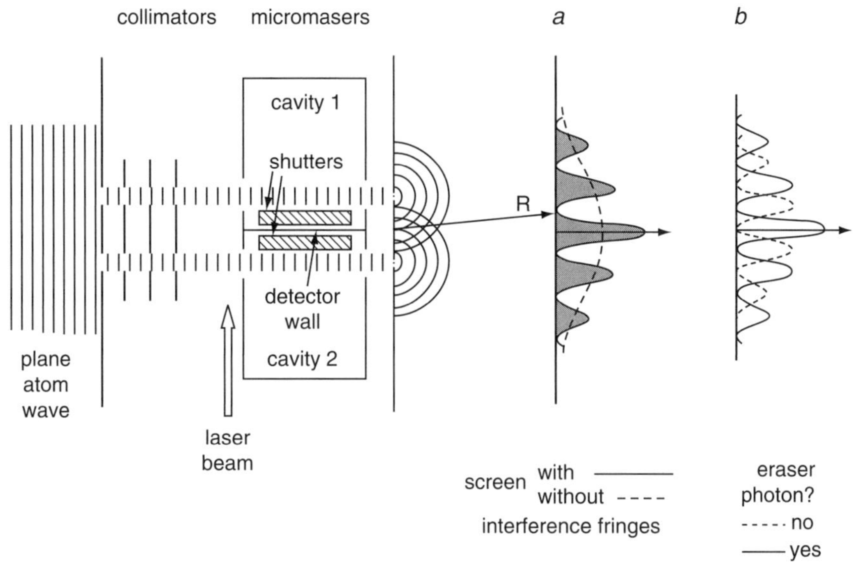

Fig. 2 Quantum erasure configuration. Source: Nature 351(1991), 115. Reprinted by permission of Nature Magazine

Primary Literature

1. N. Bohr: Das Quantenpostulat und die neuere Entwicklung der Atomistik. Die Naturwissenschaften **16**, 245–57 (1928)
2. M. de Broglie: La relation $h\nu = \varepsilon$ dans les phénomènes photoélectriques. Atomes et électrons (Gauthier, Paris 1923, 80–100)
3. L. de Broglie: Thèses. Recherches sur la théorie des quanta (Masson, Paris 1924)
4. S. Dürr, T. Nonn, G. Rempe: Origin of quantum-mechanical complementarity probed by 'which-way' experiment in an atom interferometer. Nature **395**, 33–7 (1998)
5. C. Jönsson: Elektroneninterferenzen an mehreren künstlich herstellten Feinspalten. Zeitschrift für Physik **161**, 454–74 (1961); English translation in part in Am. Jl physics, **42**, 4–11 (1974)

Secondary Literature

6. S. Boffi, M. D'Anna: Le radici del dualismo onda-corpuscolo (Bibliopolis, Napoli 1999)
7. H. Carmichael: Quantum fluctuations of light: A modern perspective on wave/particle duality. J. Evans, A. Thorndike, eds.: Quantum mechanics at the crossroads (Springer, Heidelberg 2006, 183–212)
8. W. Heisenberg: Die Entwicklung der Quantentheorie, 1918–1928. Die Naturwissenschaften 17, 490–6 (1929)
9. M. Klein: Einstein and the wave-particle duality. Natural philosopher **3**, 3–49 (1964)
10. M. Scully, B. Englert, H. Walther: Quantum optical tests of complementarity. Nature **351**, 111–6 (1991)
11. B. R. Wheaton: The Tiger and the Shark – Empirical roots of wave-particle dualism (Cambridge University Press, Cambridge 1983)

12. B. R. Wheaton: Atomic waves in private practice. J. Evans, A. Thorndike, eds.: Quantum mechanics at the crossroads (Springer, Heidelberg 2006, 39–71)
13. B. Falkenburg: Wave-particle duality, in B. Falkenburg: Particle metaphysics: A critical account of subatomic reality (Springer, Berlin 2007, 265–316)

Weak Value and Weak Measurements

Lev Vaidman

The weak value of a variable O is a description of an effective interaction with that variable in the limit of weak coupling. For a pre- and post-selected system described at time t by the two-state vector $\langle\Phi| \ |\Psi\rangle$ [1], the weak value is [2]:

$$O_{\mathrm{w}} \equiv \frac{\langle\Phi|O|\Psi\rangle}{\langle\Phi|\Psi\rangle}. \tag{1}$$

Contrary to classical physics, variables in quantum mechanics might not have definite values at a given time. In the complete description of a usual (pre-selected) quantum system, the state $|\Psi\rangle$ yields probabilities p_i for various outcomes o_i of (an ideal) measurement of the variable O. Numerous measurements on an ▶ ensemble of identical systems yield an average – expectation value of O: $\sum p_i o_i$. Since $p_i = |\langle O = o_i \ |\Psi\rangle|^2$, the expectation value can be expressed as $\langle\Psi|O|\Psi\rangle$. If the coupling to the measuring device is very small, this expression is related directly to the response of the measuring device, and the measurement does not reveal the eigenvalues o_i and their probabilities p_i. Specifically, $\langle\Psi|O|\Psi\rangle$ is the shift of the quantum state of the pointer variable of the measuring device, which, otherwise, is not distorted significantly due to the measurement interaction.

For pre- and post-selected quantum system, the response of the measuring device or any other system coupled weakly to the variable O, is the shift of the quantum state by the weak value (1). The coupling can be modeled by the von Neumann measurement interaction

$$H = g(t)PO, \tag{2}$$

where $g(t)$ defines the time of the interaction, $\int g(t) = 1$, and P is conjugate to the pointer variable Q. The weakness of the interaction is achieved by choosing the ▶ wave function of the measuring device so that P is small. Small value of P requires also a small uncertainty in P, and thus a large uncertainty of the pointer variable Q in the initial state and consequently, a large uncertainty in the measurement. Therefore, usually, we need a large ensemble of identical pre- and post-selected quantum systems in order to measure the weak value.

For rare post-selection, when $|\langle\Phi|\Psi\rangle| \ll 1$, the weak value (1) might be far away from the range of the eigenvalues of O, so it clearly has no statistical meaning as an "average" of o_i. If we model the initial state of the pointer by a Gaussian $\Psi_{\mathrm{in}}^{\mathrm{MD}}(Q) = (\Delta^2\pi)^{-1/4} e^{-Q^2/2\Delta^2}$ with large Δ ensuring small P, the final state, to a good approximation, is the shifted Gaussian $\Psi_{\mathrm{fin}}^{\mathrm{MD}}(Q) = (\Delta^2\pi)^{-1/4} e^{-(Q-O_{\mathrm{w}})^2/2\Delta^2}$. The standard measurement procedure with weak coupling reveals only the real part of the weak value, which is, in general, a complex number. Its imaginary part can be measured by observing the shift in P, the conjugate to the pointer variable [3,4].

The real part of the weak value is the outcome of the standard measurement procedure at the limit of weak coupling. Unusually large outcomes, such as ▶ spin 100 for a spin$-\frac{1}{2}$ particle [2], appear from peculiar interference effect (sometimes called Aharonov–Albert–Vaidman (AAV) effect) according to which, the superposition of the pointer wave functions shifted by small amounts yields similar wave function shifted by a large amount. The coefficients of the superposition are universal for a large class of functions for which the Fourier transforms is well localized around zero.

In the usual cases, the shift is much smaller than the spread Δ of the initial state of the measurement pointer. But for some variables, e.g., averages of variables of a large ensemble, for very rare event in which all members of the ensemble happened to be in the appropriate post-selected states, the shift is of the order, and might be even larger than the spread of the quantum state of the pointer [5]. In such cases the weak value is obtained in a single measurement which is not really "weak".

One can get an intuitive understanding of the AAV effect, noting that the coupling of the weak measurement procedure does not change significantly the forward and the backward evolving quantum states. Thus, during the interaction, the measuring device "feels" both forward and backward evolving quantum states. The tolerance of the weak measurement procedure to the distortion due to the measurement depends on the value of the scalar product $\langle\Phi|\Psi\rangle$.

Since the quantum states remain effectively unchanged during the measurement, several weak measurements can be performed one after another and even simultaneously. "Weak-measurement elements of reality" [6], i.e., the weak values, provide self consistent but sometimes very unusual picture for pre- and post-selected quantum systems. Consider a three-box paradox in which a single particle in three boxes is described by the two-state vector

$$\frac{1}{3}\left(\langle A| + \langle B| - \langle C|\right) \quad \left(|A\rangle + |B\rangle + |C\rangle\right), \tag{3}$$

where $|A\rangle$ is a quantum state of the particle located in box A, etc. Then, there are the following weak-measurements elements of reality regarding projections on various boxes: $(\mathbf{P}_A)_{\mathrm{w}} = 1$, $(\mathbf{P}_B)_{\mathrm{w}} = 1$, $(\mathbf{P}_C)_{\mathrm{w}} = -1$. Any weak coupling to the particle in box A behaves as if there is a particle there and the same is true for box B. Finally, a weak measuring device coupled to the particle in box C is shifted by the same value, but in the opposite direction. The coupling to the projection onto all three boxes, $\mathbf{P}_{A,B,C} = \mathbf{P}_A + \mathbf{P}_B + \mathbf{P}_C$ "feels" one particle: $(\mathbf{P}_A + \mathbf{P}_B + \mathbf{P}_C)_{\mathrm{w}} = (\mathbf{P}_A)_{\mathrm{w}} + (\mathbf{P}_B)_{\mathrm{w}} + (\mathbf{P}_C)_{\mathrm{w}} = 1$.

There have been numerous experiments showing weak values [7–11], mostly of photon polarization and the AAV effect has been well confirmed. Unusual weak values were used for explanation peculiar quantum phenomena, e.g., superluminal velocity of tunneling particles [12,13]. (► Superluminal communication; tunneling).

When the AAV effect was discovered, it was suggested that the type of an amplification effect which takes place for unusually large weak values might lead to practical applications. Twenty years later, the first useful application has been made: Hosten and Kwiat [14] applied weak measurement procedure for measuring spin Hall effect in light. This effect is so tiny that it cannot be observed without the amplification.

Literature

1. The Two-State Vector Formalism, this volume, pp. 802–6.
2. Y. Aharonov, D. Albert, L. Vaidman: How the result of measurement of a component of the spin of a spin-1/2 particle can turn out to be 100. Phys. Rev. Lett. **60**, 1351 (1988)
3. Y. Aharonov, L. Vaidman: Properties of a quantum system during the time interval between two measurements. Phys. Rev. **A 41**, 11 (1990)
4. R. Jozsa: Complex weak values in quantum measurement. Phys. Rev. **A 76**, 044103 (2007)
5. Y. Aharonov, D. Albert, A. Casher, L. Vaidman: Surprising quantum effects. Phys. Lett. **A 124**, 199 (1987)
6. L. Vaidman: Weak-measurement elements of reality. Found. Phys. **26**, 895 (1996)
7. N. W. M. Ritchie, J. G. Story, R. G. Hulet: Realization of a measurement of a weak value. Phys. Rev. Lett. **66**, 1107 (1991)
8. A. D. Parks, D. W. Cullin, D. C. Stoudt: Observation and measurement of an optical Aharonov–Albert–Vaidman effect. Proc. R. Soc. Lond. **A 454**, 2997 (1998)
9. N. Brunner, A. Acin, D. Collins, N. Gisin, V. Scarani: Optical Telecom networks as weak quantum measurements with postselection. Phys. Rev. Lett. **93**, 180402 (2003)
10. K. J. Resch, J. S. Lundeen, A. M. Steinberg: Experimental realization of the quantum box problem. Phys. Lett. **A 324**, 125 (2004)
11. G. L. Pryde, J. L. O'Brien, A. G. White, T. C. Ralph, H. M. Wiseman: Measurement of quantum weak values of photon polarization. Phys. Rev. Lett. **94**, 220405 (2005)
12. N. Brunner, V. Scarani, M. Wegmuller, M. Legre, N. Gisin: Direct measurement of superluminal group velocity and signal velocity in an optical fiber. Phys. Rev. Lett. **93**, 203902 (2004)
13. D. R. Solli, C. F. McCormick, R. Y. Chiao, A. Popescu, J. M. Hickmann: Fast light, slow light, and phase singularities: a connection to generalized weak values. Phys. Rev. Lett. **92**, 043601 (2004)
14. O. Hosten, P. Kwiat: Observation of the spin Hall effect of light via weak measurements. Science **319**, 787 (2008)

Werner States

Antonio Acín

In our macroscopic world, correlations are established by means of a set of classical instructions, that could be agreed in advance or come from a source. Using these pre-established instructions, distant parties that are unable to communicate can behave in a correlated manner. Assume for instance a scenario where two distant parties are asked different questions from a set of m possible questions with n possible answers. We denote by x and y the question asked to Alice and Bob, while a and b label their responses. The correlations between the parties will be described by a joint probability distribution $p(a, b|x, y)$. If the parties received in advance correlated instructions, denoted by λ, but are not able to communicate, the probability distributions can generically be written as

$$p_c(a, b|x, y) = \sum_{\lambda} p(\lambda) p(a|x, \lambda) q(b|y, \lambda). \tag{1}$$

In what follows, correlations of this form are called local, since they can be reproduced by means of a (local) model that uses only classical correlations, given by λ, and local responses, namely $p(a|x, \lambda)$ and $q(b|y, \lambda)$.

Are these correlations modified if the parties share a quantum state of two particles, ρ_{AB}, instead of classical instructions? Here, after receiving the question, the parties apply a local measurement, which depends on the question, on each particle and decide the response depending on the obtained result. Any probability distribution that can be obtained in this way can be written, using the standard ► Born rule for probabilities, as

$$p_q(a, b|x, y) = \mathrm{Tr}(\rho_{AB} M_a^x \otimes M_b^y), \tag{2}$$

where M_a^x and M_b^y are the operators describing the measurements by Alice and Bob. Interestingly, not all the probability distributions having this quantum origin can be written as (1), which means that ► correlations in quantum mechanics are more powerful than their classical counterparts.

All this discussion is nothing but a reformulation of the well-known fact that quantum states violate Bell's inequalities [1]. Indeed, beyond their clear fundamental importance, Bell's inequalities can also be understood as constraints satisfied by all probability distributions achievable by means of shared classical correlations (1). ► Bell's Theorem, then, represents a seminal result for the understanding of quantum mechanics, but also shows that quantum states can be used to establish correlations between distant parties that are not achievable by classical means. A quantum state is said to display non-local correlations when it leads to the violation of a Bell's inequality.

A natural question then emerges: Do all quantum states contain non-local correlations? It is relatively easy to see that (i) all entangled pure states (▶ states, pure and mixed) that are not of product form, $|\psi\rangle \neq |\alpha\rangle\,|\beta\rangle$ violate a Bell's inequality [2], while (ii) measurements on separable states, i.e. states that can be written as a mixture of product states $\rho_{\mathrm{AB}} = \sum_i p_i\,|\alpha_i\rangle\,|\beta_i\rangle\,\langle\alpha_i|\,\langle\beta_i|$, always allow a local description. Remarkably, there exist entangled mixed states, i.e. states that are not separable, whose measurement correlations can also be described by a local model. Thus, these states, despite being entangled, do not violate any Bell's inequality. The first examples of such states were derived in 1989 by Werner [3]. These states are now known as Werner states and play a fundamental role in foundations of quantum mechanics and quantum information theory.

Werner states, ρ_{W}, are those states belonging to a composite space $\mathcal{C}^d \otimes \mathcal{C}^d$ that remain unchanged when the two parties apply the same unitary operation, $(U \otimes U)\rho_{\mathrm{W}}(U \otimes U)^\dagger = \rho_{\mathrm{W}}$. For the sake of simplicity, we restrict here the considerations to the simplest case of two-dimensional systems, $d = 2$. In this case, Werner states are given by the mixture of a singlet state, $|\psi^-\rangle = (|01\rangle - |10\rangle)/\sqrt{2}$, and completely depolarized noise,

$$\rho_{\mathrm{W}} = p\,|\psi^-\rangle\langle\psi^-| + (1-p)\frac{\mathbb{1}}{4}. \tag{3}$$

Werner proved that these states are entangled whenever $p > 1/3$. If Alice and Bob perform local ▶ spin measurements on directions $\hat{n}_{\mathrm{A}}$ and $\hat{n}_{\mathrm{B}}$, the obtained correlations read

$$p\,(a,b|\hat{n}_{\mathrm{A}},\hat{n}_{\mathrm{B}}) = \frac{1 - p \times ab \times \hat{n}_{\mathrm{A}} \cdot \hat{n}_{\mathrm{B}}}{4}. \tag{4}$$

Here, $\hat{n}_{\mathrm{A}}$ and $\hat{n}_{\mathrm{B}}$ represent the labels for the local measurements by Alice and Bob, while the measurement outcomes are $a, b = +1, -1$. The goal is to be able to reproduce this probability distribution by means of classical correlations. Werner built a local model achieving this. It works as follows: the classical correlations are given by normalized real vectors, $\hat{n}_\lambda \in \mathcal{R}^3$. Alice's response is governed by the overlap between the received vector and the vector defining her measurement, $p_{\mathrm{W}}(+1|\hat{n}_{\mathrm{A}},\lambda) = (1 + \hat{n}_{\mathrm{A}} \cdot \hat{n}_\lambda)/2$, as in the quantum case. Bob's response is equal to $+1$ if $\hat{n}_{\mathrm{A}} \cdot \hat{n}_\lambda < 0$, otherwise is -1. Putting all these things together, one can see that the obtained correlations are the same as in the quantum case (4) with $p = 1/2$. Therefore, Werner states with $1/3 < p \leqslant 1/2$ have a local description despite being entangled.

It is clear that Werner's result represents a seminal and surprising achievement: the fact that a state is entangled is not sufficient to display non-local correlations. Since Werner's original derivation, a few results have been able to generalize his findings to other situations. Among them, there is the extension of Werner's model to completely general measurements [4] or to tripartite states [5]. At this point, it is worth mentioning that even if the correlations between measurement outcomes on a quantum state admit a local description, this state may have some hidden forms of ▶ non-locality: for instance, it may display non-local correlations after sequences of local measurements [6,7] or be useful when performing quantum teleporation [8] ▶ quantum communication. To conclude, the relation between ▶ entanglement and non-locality is fascinating and full of open questions!

Literature

1. J. S. Bell: On the Einstein Podolsky Rosen paradox. Physics **1**, 195 (1964)
2. N. Gisin: Bell's inequality holds for all non-product states. Phys. Lett. A **154**, 201 (1991)
3. R. F. Werner: Quantum states with Einstein–Podolsky–Rosen correlations admitting a hidden-variable model. Phys. Rev. A **40**, 4277 (1989)
4. J. Barrett: Nonsequential positive-operator-valued measurements on entangled mixed states do not always violate a Bell inequality. Phys. Rev. A **65**, 042302 (2002)
5. G. Toth, A. Acín: Genuine tripartite entangled states with a local hidden-variable model. Phys. Rev. A **74**, 030306 (2006)
6. S. Popescu: Bell's inequalities and density matrices: revealing "hidden" nonlocality. Phys. Rev. Lett. **74**, 2619 (1995)
7. N. Gisin: Hidden quantum nonlocality revealed by local Filters. Phys. Lett. A **210**, 151 (1996)
8. S. Popescu: Bells inequalities versus teleportation: What is nonlocality? Phys. Rev. Lett. **72**, 797 (1994)

Which-Way or Welcher-Weg-Experiments

Paul Busch and Gregg Jaeger

The issue of the ▶ wave-particle duality of light and matter is commonly illustrated by the ▶ double-slit experiment, in which a quantum object of relatively well defined momentum (such as a photon, electron, neutron, atom, or molecule) is sent through a diaphragm containing two slits, after which it is detected at a capture screen. It is found that an interference pattern characteristic of wave behaviour emerges as a large number of similarly prepared quantum objects is detected on the screen. This is taken as evidence that it is impossible to ascertain through which slit an individual quantum object has passed; if that were known in every individual case and if the quantum objects behaved as free classical particles otherwise, an interference pattern would not arise.

The notion that a description of atomic objects in terms of definite classical particle trajectories is not in general admissible is prominent in Werner Heisenberg's seminal paper [1] of 1927 on the ▶ Heisenberg uncertainty principle; there he notes: "I believe that one can fruitfully formulate the origin of the classical 'orbit' in this way: the 'orbit' comes into being only when we observe it." In the same year, in his famous Como lecture, Niels Bohr introduced the ▶ complementarity principle, which entails that definite particle trajectories cannot be defined or observed for atomic objects because according to it their spatiotemporal and causal descriptions are mutually exclusive [2]. Bohr cited the uncertainty relation as a symbolic expression of complementarity but recognized that this relation also offered room for *approximately defined* simultaneous values of position and momentum. Still in the same year, at the 1927 Solvay conference, Albert Einstein questioned the impossibility of determining the path taken by an individual particle in a double-slit interference experiment [21]; he proposed an experimental scheme wherein he considered it possible to infer through which slit the particle passed, without thereby destroying the interference pattern by measuring the recoil of the double-slitted diaphragm. This was the first instance of a *welcher-weg* or *which-way experiment*. As Bohr reported in his 1949 tribute to Einstein [3], he was able to demonstrate that Einstein's proposal was in conflict with the principles of quantum mechanics.

In subsequent years, different variants of such a *welcher-weg* experiment were considered as thought experiments illustrating the mutual exclusive options of either determining the path of a quantum object or observing its interference behaviour. Although Einstein's proposal of measuring the recoil of the double-slit system to infer the path was shown by Bohr to lead to an uncertainty of the slit location sufficient to blur the interference pattern, Feynman [22] later argued that any attempt to observe the path of an electron by shining light on it will lead to random momentum kicks on it in line with the uncertainty principle, thus washing out the interference.

A more rigorous quantum mechanical model and analysis of Einstein's which-way thought experiment was undertaken by Wootters and Żurek in 1979 [4]. The initial slit through which the photons are sent is suspended with a spring, and its centre-of-mass motion is described quantum mechanically as that of particle subjected to a harmonic potential. This allows for a choice of measurements that can be performed on the slit once the photon (► light quantum) has passed it and proceeds through the double-slit system towards the final screen. If an (approximate) measurement of the position of the slit is made, it is found that the photons impinging on the final screen build up an interference pattern; on the other hand, if the momentum of the initial slit is determined sufficiently precisely so as to allow the determination of the photon's path, the interference pattern does not develop. The fact that both choices are possible after the photon has passed the screen is due to quantum correlations (► entanglement) developing between states of the photon and the initial screen; the experiment can thus be considered an instance of Wheeler's ► *delayed-choice experiment* [5]. (For a recent experimental realization, see [6].)

Wootters and Żurek also gave an information-theoretic characterization of the trade-off between the quality of the path determination and the concurrent degradation of the interference contrast. They noted that even at 99% path certainty, there is still an interference pattern with a crest to valley ratio of 3/2. In this way, they demonstrated that Bohr's initially strict notion of complementarity is compatible with the notion of *graded* or *quantitative* complementarity (to which Bohr had already hinted in 1927 [2]), under which the exclusivity of the experimental options for path determination and interference observation are characterized more precisely and reconciled in a certain sense. This conclusion was subsequently corroborated by demonstrations of the joint *approximate* measurability of noncommuting observables, such as complementary path and interference observables measured in the context of Mach-Zehnder interferometry. (Examples and references can be found in the review [23].) In the 1980s, the discovery of novel information-theoretic uncertainty relations (e.g., [7–9]) and a related Mach-Zehnder interferometric which-way experiment performed with laser light [8] boosted interest in the investigation of quantitative wave-particle duality.

In the Wootters-Żurek model, path information is obtained by effecting a momentum exchange between the photon and the initial slit screen. In 1991, Scully, Englert and Walther proposed a radically new variant [10]. In their experiment, each laser-excited atom of a beam passes through an initial double-slit diaphragm and its possible paths are then directed through two auxiliary microwave cavities that can be configured so as to allow the path information to be obtained before it exits another double-slit diaphragm (see Figure 1). This allows entanglement to arise between atomic-path and cavity occupation states. The interaction involved is too weak to lead to any significant momentum transfer, which therefore cannot account for the destruction of the final interference pattern. As also shown in [10], the interference pattern can be restored if a suitable observable of the auxiliary system not commuting with the path indication operator is precisely measurable in an alternative configuration. Because the path information that would be present is then no longer available, this phenomenon is called *quantum erasure*; it was first described by Scully and Drühl in 1982 [11]; an experimental realization incorporating the delayed-choice feature was reported in [12].

The Scully–Englert–Walther apparatus allowing one to switch between two such configurations is a modification of the ► double-slit experiment. By appropriately switching between configurations, information associated with one or the other non-commuting observable is erased. In the standard double-slit experiment, in the configuration with both slits open, strong quantum interference is observed for the input pure state (► states, pure and mixed) $|\psi\rangle = \frac{1}{\sqrt{2}}(|\psi_1\rangle + |\psi_2\rangle)$, where $|\psi_i\rangle$ is the state corresponding to entry with certainty into slit $i = 1, 2$, even when elementary particles enter one by one; there are two paths that the initially prepared members of the ► ensemble could take from preparation to the measurement. In another configuration where only one of the two slits is available at a time, so that complete path information is obtained, then no interference pattern appears on the detection screen; there is only one path history possible from preparation to point of detection for each particle. In these two configurations, non-commuting ► observables are measured, one in each case.

The Scully–Englert–Walther experiment adds an auxiliary system capable of becoming entangled with the primary quantum system. The enlarged apparatus allows alternation between the above two cases, with the option to make the choice of configuration at any time before the final screen is contacted. The auxiliary system can definitively indicate, although indirectly, which slit was entered by the primary system by exploiting state entanglement [10, 14]. The primary and auxiliary systems are arranged so as to interact in such a way that phenomena which *would have* occurred in one configuration are not exhibited in the other. The incoming quantum ensemble is that of a beam of Rydberg atoms rather than of elementary particles, a laser is introduced as the first apparatus element and is oriented perpendicularly to the atom beam so as to allow its excitation, an auxiliary system consisting of

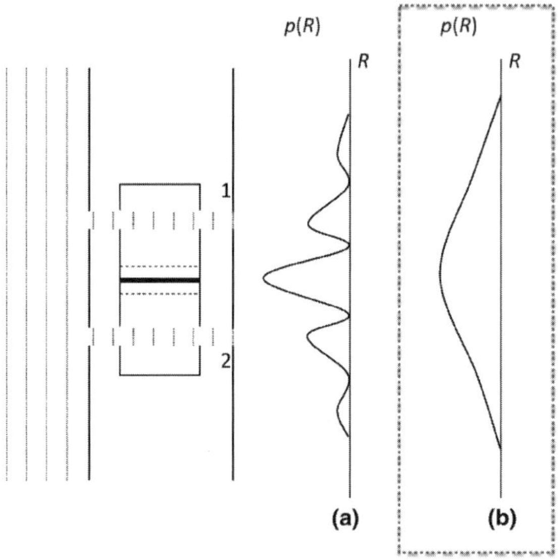

Fig. 1 Apparatus for quantum erasure: A modified version of the standard ► double-slit apparatus, where two intermediate microcavities with internal shutters (dark dashed lines) and a radiation absorber (thick solid line) have been introduced and excited atoms are input that de-excite with certainty within one of the cavities. (**a**) Atom detections when shutters are opened; path information is unavailable because radiation is indiscriminately absorbed. (**b**) Atom detections when the radiation absorber is unreachable, so that radiation is selectively contained in one cavity or the other; path information, which is incompatible with interference, is available. Opening the shutters, even after each atom has passed the double-slitted diaphragm, effectively erases path information, which is irretrievable from the common radiation absorber, taking case (**b**) to case (**a**)

a pair of micro-cavities is placed after it, and an additional double-slit diaphragm placed after the cavities, as shown Fig. 1. The two micro-cavities are each of a length such that the atoms will de-excite with extremely high probability between the their entrances and exits. Each cavity will therefore capture any radiation emitted from atoms entering it, allowing the atoms of the beam to become entangled with the cavity pair before entering the remainder of the system. The two cavities constituting the auxiliary system are adjacent but separated by a wall covered on each side by shutters which, when opened, allow captured radiation to be absorbed from either cavity without the discriminating from where it came. Rapid switching of the shutters between open and closed positions allows the choice of configuration to be delayed until very near the time each atom strikes the screen.

In order to allow path information to be stored, the laser of this new apparatus is sufficiently powerful that, when turned on, it will excite every one of the beam atoms from its ground state to its excited state. The state of the atomic system is thus prepared as $|\psi(r)\rangle|j\rangle = \frac{1}{\sqrt{2}}(|\psi_1(r)\rangle + |\psi_2(r)\rangle)|j\rangle$, where the position coordinate of the elementary particles of the standard experiment is replaced by that of the atomic center-of-mass position coordinate r and the atomic internal states are written $|j\rangle$, $j = 0, 1$, the ground and excited states, respectively. Without the laser on, all atoms are in the ground state $|0\rangle$. The atom beam is then described by the pure product state $|\psi(r)\rangle|0\rangle$, so that its squared magnitude, the probability density of detected atoms at the final screen position $r = R$ is

$$p(R) = \frac{1}{2}\Big[\big(||\psi_1(R)\rangle|^2 + ||\psi_2(R)\rangle|^2\big) + \big(\langle\psi_2(R)|\psi_1(R)\rangle + \langle\psi_1(R)|\psi_2(R)\rangle\big)\Big]\langle 0|0\rangle,$$

with $\langle 0|0\rangle = 1$, that is, one finds the sort of interference pattern observed in the standard double-slit experiment when both slits are available. With the laser is turned on and the shutters kept closed, with the atoms prepared in $|\psi(R)\rangle|1\rangle$, atomic radiation is deposited into one of the cavities and the state of the enlarged system must be considered, namely,

$$\begin{aligned}|\Psi\rangle &= \frac{1}{\sqrt{2}}\big(|\psi_1\rangle|0\rangle|1_{C1}0_{C2}\rangle + |\psi_2\rangle|0\rangle|0_{C1}1_{C2}\rangle\big)\\ &= \frac{1}{\sqrt{2}}\big(|\psi_1\rangle|1_{C1}0_{C2}\rangle + |\psi_2\rangle|0_{C1}1_{C2}\rangle\big)|0\rangle\end{aligned}$$

where the subscripts $\{\mathrm{C_i}\}$ indicate the cavity pair with eigenstates $|k_{\mathrm{C_1}}l_{\mathrm{C_2}}\rangle$, with $k = 0, 1$ indexing the occupation eigenvalue of cavity 1 feeding slit 1 and $l = 0, 1$ indexing that of cavity feeding slit 2.

Thus, with the laser turned on and cavity shutters kept closed, the external atomic state and the occupation state of the two-cavity system become entangled, whereas the internal atomic state factors out. The probability density for arrival of atoms at point R on the screen is that shown in case (b) of Figure 1:

$$\begin{aligned}p = \frac{1}{2}\big[&(||\psi_1\rangle|^2 + ||\psi_2\rangle|^2) + \langle\psi_1|\psi_2\rangle\langle 1_{C1}0_{C2}|0_{C1}1_{C2}\rangle\\ &+\langle\psi_1|\psi_2\rangle)\langle 0_{C1}1_{C2}|1_{C1}0_{C2}\rangle\ \big]\langle 0|0\rangle\end{aligned}$$

where here, as in the previous equation, the position argument R in $p(R)$, $|\Psi(R)\rangle$, and $|\psi_i(R)\rangle$ has been omitted but is implied. Then, $\langle 1_{C1}0_{C2}|0_{C1}1_{C2}\rangle = 0$ and $\langle 0_{C1}1_{C2}|1_{C1}0_{C2}\rangle = 0$ imply that the terms including them are zero. The observed interference pattern of atoms striking the final screen is thus $p(R) = \frac{1}{2}||\psi_1(R)\rangle|^2 + \frac{1}{2}||\psi_2(R)\rangle|^2$, a simple probability sum corresponding to state mixture; the introduction of the cavities which selectively interacting with passing atoms depending on their proximity to each slit allows for distinguishability in principle of the paths of the atoms as long as their interior shutters are kept closed. The atomic detection pattern can be understood to occur because the enlarged system contains entangled subsystems. However, the path information encoded in this *de facto* two-cavity memory can readily be erased by switching instead to the configuration in which the internal shutters of the two cavities are opened, which allows the stored radiation to reach the photon absorber. In that case, because the radiation in the cavities from which path information might be retrievable is instead lost from them to the absorber, taking both cavity states to their ground states $|0_{\mathrm{C_1}}0_{\mathrm{C_2}}\rangle$, which then factor out:

$$|\Psi\rangle = \frac{1}{\sqrt{2}}\big(|\psi_1\rangle|0\rangle|0_{C_1}0_{C_2}\rangle + |\psi_2\rangle|0\rangle|0_{C_1}0_{C_2}\rangle\big) = \frac{1}{\sqrt{2}}\big(|\psi_1\rangle + |\psi_2\rangle\big)|0\rangle|0_{C_1}0_{C_2}\rangle.$$

The path information is therefore no longer encoded in them. Interference reappears, as in case (a) of Fig. 1:

$$p = \frac{1}{2}\Big[\big(||\psi_1\rangle|^2 + ||\psi_2\rangle|^2\big) + \big(\langle\psi_2|\psi_1\rangle + \langle\psi_1|\psi_2\rangle\big)\Big]\langle 0|0\rangle\langle 0_{C_1}0_{C_2}|0_{C_1}0_{C_2}\rangle.$$

The first realization of a *welcher-weg* experiment with individual atoms similar to the proposal of Scully, Englert and Walther was obtained by Dürr, Nonn and Rempe in 1998 [15]. It is shown there that neither mechanical momentum transfers nor the position-momentum uncertainty relation are relevant for the explanation of the destruction of interference. Nevertheless duality relations have been found that describe a quantitative trade-off between the quality of path determination and interference visibility [16–18] which have been shown to be instances of appropriate uncertainty relations [23].

A neutron-interferometric double resonance experiment involving neutrons and photons allowing simultaneous observation of interference and individual energy losses have also been used to test Einstein's related 'Einweg' assumption, in discussions with Bohr, that particles take single definite paths despite these paths being unknown to experimenters [19, 24]. For a penetrating philosophical discussion of the issues and debates arising from the seminal paper of Scully, Englert and Walther [10] the reader is referred to [25].

Primary Literature

1. Heisenberg, W.: Über den anschaulichen Inhalt der quantentheoretischen Kinematik und Mechanik. Zeitschrift für Physik **43**, 172-198 (1927). The quotation is taken from the English translation in [20].
2. Bohr, N.: The quantum postulate and the recent development of atomic theory. Nature **121**, 580-590 (1928).
3. Bohr, N.: Discussion with Einstein on Epistemological Problems in Atomic Physics, in Albert Einstein: Philosopher-Scientist, ed. P. Schilpp (Library of Living Philosophers, Evanston, 1949).
4. Wootters, W.K./Zurek, W.H.: Complementarity in the double-slit experiment: Quantum nonseparability and a quantitative statement of Bohr's principle. Physical Review D **19**, 473-484 (1979).
5. Wheeler, J.A.: The past and the delayed-choice double-slit experiment, in Marlow, A.R. (ed.), Mathematical Foundations of Quantum Theory (Academic Press, New York, 1978).
6. Jacques, V./Wu, E./Grosshans, F./Treussart, F./Grangier, P./Aspect, A./Roch, J.-F.: Experimental realization of Wheelers delayed-choice GedankenExperiment, arxiv:quant-ph/0610241 (2006).
7. Deutsch, D.: Uncertainty in Quantum Measurements. Physical Review Letters **50**, 631-633 (1983).
8. Mittelstaedt, P./Prieur, A./Schieder, R.: Unsharp particle-wave duality in a photon split beam experiment. Foundations of Physics **17**, 891-903 (1987).
9. Greenberger, D.M./Yasin, A.: Simultaneous wave and particle knowledge in a neutron interferometer. Physics Letters A **128**, 391-394 (1988).

10. Scully, M.O./Englert, B.-G./Walther, H.: Quantum optical tests of complementarity. Nature **351**, 111-116 (1991).
11. Scully, M.O./Drühl, K.: Quantum eraser - A proposed photon correlation experiment concerning observation and delayed choice in quantum mechanics. Physical Review A **25**, 2208-2213 (1982).
12. Kim, Y.-H./Yu, R./Kulik, S.P./Shi, Y./Scully, M.O.: Delayed 'Choice' Quantum Eraser. Physical Review Letters **84**, 1-5 (2000).
13. Wheeler, J. A.: Law without law, in Wheeler, J.A./Zurek, W.H. (eds.), Quantum Theory and Measurement (Princeton, New Jersey, 1983, pp. 182-213).
14. Scully, M.: Quantum Optics (Cambridge University Press, Cambridge, 1997, Section 20.1).
15. Dürr, S./Nonn, T./Rempe, G.: Origin of quantum-mechanical complementarity probed by a 'which-way' experiment in an atom interferometer. Nature **395**, 33-37 (1998).
16. Jaeger, G./Shimony, A./Vaidman, L.: Two interferometric complementarities. Physical Review A **51**, 54-67 (1995).
17. Englert, B.-G.: Fringe Visibility and which-way information: An inequality. Physical Review Letters **77**, 2154-2157 (1996).
18. Englert, B.-G./Bergou, J.A.: Quantitative quantum erasure. Optics Communications **179**, 337-355 (2000).
19. Vigier, J.-P. /Rauch, H.: Proposed neutron interferometry test of Einstein's 'Einweg' assumption in the Bohr-Einstein controversy. Physics Letters A **151**, 269-275 (1990).

Secondary Literature

20. Wheeler, J.A./Zurek, W.H. (eds.), Quantum theory and measurement (Princeton, New Jersey, 1983).
21. Bacciagaluppi, G./Valentini, A.: Quantum mechanics at the crossroads: Reconsidering the 1927 Solvay Conference (Cambridge University Press, UK, 2009).
22. Feynman, R./Leighton, R./Sands, M.: The Feynman Lectures on Physics Vol. III (Addison Wesley, 1965).
23. Busch, P./Shilladay, C.R.: Complementarity and uncertainty in Mach-Zehnder interferometry and beyond. Physics Reports **435**, 1-31 (2006).
24. Rauch, H./Werner, S.: Neutron interferometry (Oxford, 2000).
25. Falkenburg, B.: Particle Metaphysics – A Critical Account of Subatomic Reality (Springer 2007, Sec. 7.4 and 7.5.).

Wigner Distribution

R.F. O'Connell

In contrast to classical physics, the language of quantum mechanics involves ► operators and ► wave functions (or, more generally, ► density operators). However, in1932, Wigner formulated quantum mechanics in terms of a distribution function $W(q, p)$, the marginals of which yield the correct quantum probabilities for q and p separately [1]. Its usefulness stems from the fact that it provides a re-expression of quantum mechanics in terms of classical concepts so that quantum mechanical expectation values are now expressed as averages over phase-space distribution functions. In other words, statistical information is transferred from the density operator to a quasi-classical (distribution) function. Wigner [1] presented a specific form for $W(p, q)$, while recognizing that other possibilities exist, depending on the conditions which are imposed on W. Wigner's choice has the virtue of mathematical simplicity but it has the feature that it may take negative values, with the result that several authors have investigated non-negative distribution functions. However, we regard negative values of W as a manifestation of its quantum nature and the fact that it "... cannot be really interpreted as the simultaneous probability for coordinates and momenta..." [1] Wigner's original paper was concerned with using W for the specific purpose of calculating the quantum correction for thermodynamic equilibrium. The recognition of its more general applicability stems mainly from the work of Groenewold [2] and Moyal [3], who investigated the correspondence between physical quantities and quantum operators and showed, in particular, that the correspondence is not unique and moreover, that the distribution functions obtained by the Weyl correspondence [4] are the Wigner functions. Moyal also showed how the time dependence of W and other such functions (– which arise from alternative association rules other than Wigner-Weyl but which lead to the same physical results) may be determined without using the ► Schrödinger equation. In fact, Moyal's paper was a landmark contribution as, in essence, "...it establishes an independent formulation of quantum mechanics in phase space" [5]. As for all quantum formulations, Ballentine [6] has shown that the development of the classical limit of the Wigner distribution is a subtle process, especially in view of the fact that, in general, $W(q, p)$ has negative parts. Turning to specifics, we present some basic results developed in the original pioneering papers [1–4, 28] but conveniently presented in a comprehensive review by Hillery et al. [7]. Thus, in one-dimensional space (generalization to n dimensions being straightforward), for a ► mixed state represented by a density matrix $\hat{\rho}$,

$$W(q, p) = \frac{1}{\pi\hbar} \int_{-\infty}^{\infty} \mathrm{d}y \langle q - y|\hat{\rho}|q + y\rangle \mathrm{e}^{2\mathrm{i}py/\hbar}, \tag{1}$$

whereas, for a pure state (► states, pure and mixed) represented by a wave function $\psi(q)$,

$$W(q, p) = \frac{1}{\pi\hbar} \int_{-\infty}^{\infty} \mathrm{d}y \psi^*(q + y)\psi(q - y)\mathrm{e}^{2\mathrm{i}py/\hbar}. \tag{2}$$

However, in order to calculate correct expectation values and ensemble averages (► ensembles in quantum mechanics), it is also necessary to specify the classical function $A(q, p)$ corresponding to a quantum operator $\hat{A}$ as

$$A(q, p) = \int \mathrm{d}z\, \mathrm{e}^{\mathrm{i}pz/\hbar} \langle q - \frac{1}{2}z|\hat{A}|q + \frac{1}{2}z\rangle, \tag{3}$$

so that $\int\int \mathrm{d}q\,\mathrm{d}p\,A(q,p) = 2\pi\hbar\ \mathrm{Tr}(\hat{A})$. This ensures that

$$\int \mathrm{d}q \int \mathrm{d}p\, A(q,p)B(q,p) = (2\pi\hbar)\ \mathrm{Tr}(\hat{A}\hat{B}), \tag{4}$$

and

$$\int \mathrm{d}q \int \mathrm{d}p\, A(q,p)W(q,p) = \mathrm{Tr}(\hat{\rho}\hat{A}(\hat{q},\hat{p})), \tag{5}$$

so that, in particular, we see that $W(q,p)$ derived from the density matrix, is $(2\pi\hbar)^{-1}$ times the phase space operator which corresponds to the same matrix. Following these original papers, [1–4,28] there were many papers devoted to extending the framework and overall understanding of distribution functions. In addition, distributions other than those of Wigner were introduced, notable those of Kirkwood, Cahill and Glauber, Glauber, Sudarshan and Husimi (all of which are reviewed in [8], where it is noted that some of these are everywhere non-negative) and Cohen [9] and all require classical functions different from that given in (3) in order to ensure consistency. It is clear that all distribution functions are not measurable, despite some claims to the contrary in the literature, where in fact what is observed are the marginal q probabilities from which values of $W(q,p)$ are inferred but one could equally have inferred values for other distribution functions. The earliest applications of the Wigner function were in the arena of statistical mechanics but, more recently, among the diverse areas in which the W function was found to be useful we mention hydrodynamics [10], plasmas [11], quantum corrections for transport coefficients [12], collision theory [13] and signal analysis [14]. However, we feel that the overwhelming majority of applications are to be found in quantum systems where fluctuations and dissipation are playing an important role. In this context, the 1984 review of the W function by Hillery et al. [7] made extensive reference to its relevance in quantum optics, which is underlined by the more recent books of Scully and Zubairy [15] and Schleich [16]. Complementary to this work is the application of the W function to a variety of problems in quantum statistical mechanics, where effects associated with the analysis of quantum systems in a heat bath (including the radiation field heat bath) are of the essence. As examples of the usefulness of the W function in this context we note its role in obtaining the simplest approach to solving the initial value quantum Langevin equation and, concomitantly, the solution to an exact master equation [17] and also its role in the investigation of ► Schrödinger cat superpositions [18]. However there are limitations to the usefulness of the W function (some of which were discussed by Moyal [3]), notably for particles with ► spin and for relativistic particles. Finally, we mention the excellent and comprehensive overview of selected papers on quantum mechanics in phase space, with emphasis on the Wigner function [5].

Primary Literature

1. E. Wigner: On the quantum correction for thermodynamic equilibrium. Phys. Rev. **40**, 749–759 (1932)
2. H. Groenewold: On the principles of elementary quantum mechanics. Physica **12**, 405–460 (1946)
3. J. Moyal: Quantum mechanics as a statistical theory. Proc. Camb. Phil. Soc. **45**, 99–124 (1949)
4. H. Weyl: Quantenmechanik und Gruppentheorie. Z. Phys. **46**, 1–46 (1927)

Secondary Literature

5. C. K. Zachos, D. B. Fairlie, T. L. Curtright: *Quantum Mechanics in Phase Space* (World Scientific, 2005)
6. L. E. Ballentine: Quantum Mechanics: a Modern Development, Chap. 15 (World Scientific, 2005)
7. M. Hillery, R. O'Connell, M. Scully, E. Wigner: Distribution functions in physics fundamentals. Phys. Rep. **106**, 121–167 (1984)
8. R. F. O'Connell: Quantum distribution functions in non-equilibrium statistical mechanics, in *Frontiers of Nonequilibrium Statistical Physics* (Plenum Publishing Corporation, 1986), pp. 83–95
9. L. Cohen: Generalized phase-space distribution functions. J. Math.Phys. **7**, 781–786 (1966)
10. J. H. Irving, R. W. Zwanzig: The statistical mechanical theory of transport processess. V. Quantum Hydrodynamics. J. Chem. Phys. **19**, 1173–1180 (1951)
11. W. E. Brittin, W. R. Chappell: The Wigner distribution function and second quantization in phase space. Rev. Mod. Phys. **34**, 620–627 (1962)
12. S. Choi, J. Ross: Quantum corrections for transport coefficients. J. Chem. Phys. **33**, 1324 (1960)
13. R. Carruthers, F. Zachariasen: Quantum collision theory with phase-space distributions. Rev. Mod. Phys. **55**, 245–285 (1983)
14. L. Cohen: *Time–Frequency Analysis* (Prentice-Hall, Englewood Cliffs, NJ, 1995)
15. M. O. Scully, M. S. Zubairy: *Quantum Optics* (Cambridge University Press, Cambridge, England, 1997)
16. W. P. Schleich: *Quantum Optics in Phase Space* (Wiley, England, 2001)
17. G. W. Ford, R. F. O'Connell: Exact solution of the Hu–Paz–Zhang master equation. Phys. Rev. D **64**, 105020, 1–13 (2001)
18. G. W. Ford, R. F. O'Connell: Wigner distribution analysis of a Schrödinger cat superposition of displaced equilibrium coherent states. Acta Phys. Hungarica Quant. Electron. B **20**, 91–94 (2004)

Wigner's Friend

Henry Stapp

Eugene Wigner published, in 1961, a widely reprinted article [1] entitled "Remarks on the Mind–Body Problem" in which he stresses the basic role played by consciousness in quantum theory. But if consciousness is basic then the question immediately arises: Whose consciousness? To explore this issue Wigner considers a situation in which his "friend", rather than he himself, is observing the effects of an atomic process, the radiation of a visible photon.

In order to formulate the problem Wigner first explains the entry of consciousness into physical theory:

When the province of physical theory was extended to encompass microscopic phenomena, through the creation of quantum mechanics, the concept of consciousness came to the fore again: it was not possible to formulate the laws of quantum mechanics without reference to the consciousness. [2] All that quantum mechanics purports to describe are probability connections between subsequent impressions (also called 'apperceptions') of consciousness, and even though the dividing line between the observer, whose consciousness is being affected, and the observed physical object can be shifted towards one or the other to a considerable degree [3], it cannot be eliminated.

His reference [2] is to von Neumann's work (► orthodox interpretation) on the shifting of the boundary between those aspects of nature that are described in the mathematical language of quantum theory, and those that are described in the psychological language by means of which we describe our actual and possible conscious experiences. The job of quantum theory is to make predictions about connections between such experiences. His reference [3] was to Heisenberg's famous pronouncement:

> The conception of objective reality ... evaporated into the ... mathematics that represents no longer the behavior of elementary particles but rather our knowledge of this behavior.
>
> The concept of "our knowledge" is reasonably clear insofar as "we are able to communicate to others what we have done and what we have learnt" [4].

But in practice different people often know different things.

The thought experiment considered by Wigner involves, essentially, an atomic state that emits a visible photon into an optical system that directs the rays emitted from the atom in certain directions into the retina of the eye of Wigner's friend, and directs the rays emitted in other direction to some other place. The ► wave function of the atom plus the photon will be a ► superposition of components corresponding to different directions of the photon emission. If the interaction of the photon with the retina, and of the retina with the brain of the friend – who is presumed to be attending to what she is seeing – is now included in the physical description, then the state of his friend's brain generated by the purely physical laws of motion would include a part that corresponds to her observing the flash and another part corresponding to her not observing the flash. When Wigner asks his friend whether she saw the flash, then, upon his registering of her response, the wave function (quantum state) that represents his knowledge of her brain and body will suddenly jumps to one state or the other. Yet before he learned about her reaction his representation of her state was in a combination of the "I observed a flash" and "I observed no flash" alternatives.

Wigner is willing to admit that, if the purely physically described laws entail it, then an unobserved inanimate measuring device could exist in a state that represents a combination of two macroscopically different states. However, although solipsism may be a *logical* possibility, "everyone believes that the phenomena of sensation are widely shared by organisms that we consider to be living". And, accordingly, his friend will surely report that she did [or did not] experience the flash [as the case may be] *before* she reported that fact to him. Wigner concludes from these considerations that his friend was "not in a state of suspended animation" before *he* learned about her state: he concludes that her quantum state became one or the other of these two alternatives when *she* became conscious of the flash, not when *he* came to know what she reported.

Wigner asserts that "The preceding argument for the difference in the roles of inanimate tools of observation and observers with consciousness – hence for a violation of physical laws where consciousness plays a role – is entirely cogent so long as one accepts the tenets of orthodox quantum theory and all their consequences."

Wigner proposes, then, that "the being with a consciousness must have a different role in quantum mechanics than the inanimate measuring device." He proposes, in essence, that the occurrence of a conscious experience is an objective reality that is correlated to a change in an objective wave function. "Our knowledge" can then be interpreted to be the aggregate of the conscious knowledge of *all* systems that possess consciousness (Fig. 1). This allows quantum theory to be regarded as an objective theory that describes the interaction between an objective physical aspect that is described in terms of the mathematical language of quantum theory, and an objective mental aspect that is described in terms of the concepts of thoughts, ideas, and feelings – i.e., in terms of the concepts of psychology. This move allows what had originally been a fundamentally anthropocentric, pragmatic, subjective theory to be elevated into a nonanthropocentric objective theory of an objective reality having physically described aspects and psychologically described aspects related in the specific way specified by the ► orthodox interpretation quantum theory spelled out by John von Neumann [2].

Primary Literature

1. E. Wigner: *Remarks on the Mind-Body Problem*, in I. J. Good, ed.: *The Scientist Speculates* (Heinemann, London 1961, 284–302); Basic Books, New York (1962). Reprinted in E. Wigner: *Symmetries and Reflections* (Indiana University Press, Bloomington, Indiana 196, 171–184) and in J.A. Wheeler and W. H. Zurek eds.: *Quantum Theory and Measurement* (Princeton University Press 1983, 168–81)
2. J. Von Neumann: *Mathematische Grundlagen der Quantenmechanik* (Springer, Berlin 1932, Ch. VI) (*Translated as Mathematical Foundations of Quantum Mechanics*, Princeton University Press, 1955)
3. Heisenberg: *The representation of nature in contemporary physics*. Daedalus, 95–108 (Summer 1958)
4. N. Bohr: *Essays 1958/1962 on Atomic Physics and Human Knowledge* (New York, Wiley 1962, 3)

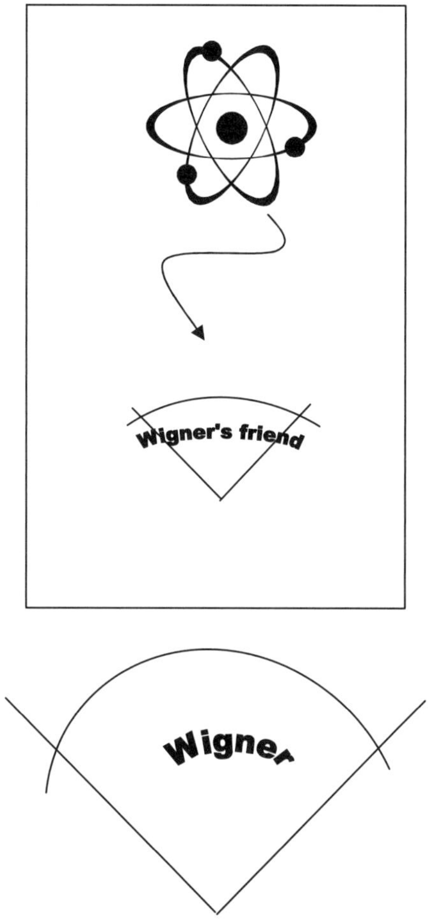

Fig. 1 (**a**) An illustration of Wigner's argument that the role of 'a conscious being' is different from that of an inanimate measuring device. The first step is to assume that the state of the atom plus the photon is the superposition: $\alpha\Psi_1 + \beta\Psi_2$. (**b**) The second step is to treat Wigner's friend as an unobserved inanimate measuring device that has two states: either it registers the photon, χ_1 or it does not χ_2. According to the orthodox interpretation of quantum mechanics the state of the combined system after interaction is a linear superposition of states: $\alpha(\Psi_3 \times \chi_1) + \beta(\Psi_4 + \chi_2)$; or if the interaction with the environment is taken into account, the mixture of $(\Psi_3 \times \chi_1)$, with probability $|\alpha|^2$ plus $(\Psi_4 \times \chi_2)$, with probability $|\beta|^2$. [Ψ_3 is the atomic part of Ψ_1 and $\Psi_4$is the atomic part of Ψ_2.] Thus the device prior to any observation of it has part corresponding to the photon's being registered, and a part corresponding to the photon's not being registered. (**c**) But now suppose that the initially unobserved (by Wigner) observational device is a conscious human being, e.g., Wigner's friend. Wigner asks the question, and his friend answers that she saw the flash [or did not see the flash] before she let Wigner know whether or not she saw it. Wigner concludes his friend was not in a state of suspended animation prior to when he learned which state she was in. He concludes that the state of the combined system of atom plus his conscious friend, after she had experienced the outcome, was either definitely or $(\Psi_3 \times \chi_1)$ or definitely $(\Psi_4 \times \chi_2)$, not a combination of the two. Wigner's proposal is a move away from the Copenhagen idea that the quantum state represents knowledge available to a community of communicating observers, who have a common knowledge that is useful for making predictions about their combined future experiences. Wigner suggests that each conscious being is able to collapse one single objective quantum state, regardless of whether the information is actually physically shared. It is a move away from an essentially subjective pragmatic interpretation toward a more objective absolute one

Secondary Literature

5. D. Z. Albert: *Quantum Mechanics and Experience* (Harvard University Press, Cambridge, MA, London, England 1992, 76–79)
6. L. Sklar: *Philosophy of Physics* (Oxford University Press, Oxford 1992, 185–187)

X

X-Rays

Bruce R. Wheaton

Modern physics began with the discovery of X-rays in 1896 by Wilhelm Conrad Röntgen (1845–1923), an event well described. Less known is the important role X-rays played in the earliest introduction of quantum concepts. Their early impulse interpretation forced consideration in 1896 of *quantity* of impulses, unlike during the prior century of thought about radiation. This set the stage for a sea change in concepts of radiation.

Improvements in vacuum technology from the 1850s had led to cathode discharge tubes and X-rays. These were an "entirely new form of radiation" that could pass right though opaque matter. Many hypotheses emerged in explanation, the most profound a resuscitation of Christiaan Huygens (1629–95) disconnected impulse model of light, now from the pen of George Gabriel Stokes (1819–1903). Each collision of a cathode-beam electron at the anode gives rise to a single such impulse propagating away, only the vast number of impacts gives rise to the seeming continuous flow of the X-rays. They lack periodicity just as would be expected of white light comprised of a continuum of frequencies.

Within 4 years Dutch physicists demonstrated diffraction of X-rays from a slit, implying a wavelength of 1 Å (10^{-4} that of light), which seemed to argue against the accepted impulse model. This challenged the young Arnold Sommerfeld (1868–1951) in Göttingen, who in 1900 showed impulses could diffract but would show no fringes. He concluded that a continuum of electromagnetic disturbances exists, from periodic waves of light to aperiodic impulses of X-rays and the γ-rays discovered that year by Paul Villard (1860–1934). By 1905 it was clear that X-rays propagate with the speed of light.

X-rays passing through a gas release electrons in numbers and velocities easily measured. But there seemed to be too few (quantity) and those had more energy (quality) than was expected. Both paradoxes led many to the view that, unlike light, X-rays do not spread their energy isotropically into the aether, but concentrate it in specific directions. And the case for γ-rays was even stronger, so that several of their investigators began to argue forcefully that γs are actual material particles.

In response, Charles Barkla's (1877–1944) experiments on secondary X-rays stimulated from elements by X-rays showed them to be polarized and have periodic properties "characteristic" of the scattering material as one of two components; the other an inhomogeneous X-ray component soon to be called the *Bremsenanteil*. This led to lively controversy in the English literature between William Henry Bragg (1862–1942) and Barkla, in the German between Johannes Stark (1874–1957) and Sommerfeld about the physical nature of X-rays.

D. Greenberger et al. (eds.), *Compendium of Quantum Physics: Concepts, Experiments, History and Philosophy*,

On the surface, all seemed resolved in favor of periodic waves when, in 1912, X-rays directed through a crystal showed unmistakable interference effects. But the new crystal metric simultaneously provided the most accurate yet indication that X-rays transfer energy only in quantum units.

The next decade, largely in response to the successes of the ▶ Bohr atom, saw little consideration of the "nature" of X-rays except amongst experimentalists. Millikan was astonished in 1916 to corroborate Albert Einstein's (1879–1955) equation for the ▶ photoelectric effect. Precise new techniques developed to measure β-particle ▶ electrons from ▶ radioactive decay law were applied to secondary electrons released by X-rays and by γ-rays. The newly invented Coolidge X-ray tube provided rays of unprecedented stability for precise tests. And in William Duane's (1872–1935) Harvard laboratory in 1918 his student came very close to corroborating Einstein's photoeffect law for X-rays.

But in the periphery of physics in post-war Europe, these issues carried weight. In particular, the interns in the private laboratory of Maurice de Broglie (1875–1960) in Paris took "atoms of light" very seriously indeed. The X-ray photoeffect, now amenable to precise quantitative study with the β-ray spectrometer, became subject of intense research by Alexandre Dauvillier (1892–1979). His results convinced de Broglie that X-rays "must be corpuscular" or "energy must be concentrated in points on the surface of the wave." The elder de Broglie presented his findings at the third Solvay Congress in Paris 1921, where (with corroborating γ-ray findings from Charles Ellis (1895–1980)) they dominated discussion at the entire meeting.

It is well-known that Maurice's younger brother Louis de Broglie (1892–1987) turned this seeming paradox into his hypothesis of ▶ matter waves in 1923. His reconciliation of ▶ wave-particle duality led directly to Erwin Schrödinger's (1887–1961) ▶ wave mechanics, one of the two statements of the new ▶ quantum mechanics of 1926. Schrödinger's arose from radiation theory, Werner Heisenberg's (1901–1976) ▶ matrix mechanics from concerns with atomic theory; another corollary of wave-particle duality.

Primary Literature

1. M. de Broglie: *Les rayons x* (Blanchard, Paris 1922)
2. W. Friedrich, P. Knipping, M. Laue: Interferenz-Erscheinungen bei Röntgenstrahlen. Sitzungsberichte der bayerischen Akademie der Wissenschaften, München **42**, 303–22 (1912); reprinted with additions in Annalen der Physik (4th ser.) **41**, 971–88 (1913)
3. E. Marx: Die Geschwindigkeit der Röntgenstrahlen. Physikalische Zeitschrift 6, 768–77 (1905)
4. W. Röntgen: Űber eine neue Art von Strahlen. Sitzungsberichte der Physikalisch-Medizinischen Gesellschaft zu Würzburg (1895), 132–41; reprinted in Annalen der Physik (3rd ser.) **64**, 1–11 (1898)
5. A. Sommerfeld: Theoretisches über die Beugung der Röntgenstrahlen. Physikalische Zeitschrift **1**, 105–11 (1899); **2**, 55–60 (1900)
6. J. Thomson: On a theory of the structure of the electric field and its application to Röntgen radiation and to light. Philosophical Magazine (6th ser.) **19**, 301–13 (1910)

Secondary Literature

7. O. Glasser: *Wilhelm Conrad Röntgen und die Geschichte der Röntgenstrahlen* (Springer, Berlin 1931; English translation Thomas, Springfield 1934)
8. J. Jenkin, R. Leckey, J. Liesegang: The development of X-ray photoelectron spectroscopy: 1900–1960. Journal of electron spectroscopy and related phenomena **12**, 1–35 (1977)
9. H. Klickstein: *Wilhelm Conrad Röntgen* – On a new kind of rays (Mallinkrodt, St. Louis 1966)
10. B. R. Wheaton: Impulse X-rays and radiant intensity: The double-edge of analogy. Historical Studies in the Physical Sciences **11**, 367–90 (1981)
11. B. R. Wheaton: *The tiger and the shark* – Empirical roots of wave-particle dualism (Cambridge University Press, Cambridge 1983)

Z

Zeeman Effect

Klaus Hentschel

Pieter Zeeman (1865–1943) had been searching for the influence of magnetic fields on spectral lines since 1892. Michael Faraday's (1791–1867) demonstration of the rotation of the plane polarization of light in electric fields had led Faraday himself and several other experimenters to expect such an influence. But Zeeman only succeeded in late 1896, after having installed a strong Rühmkorff electromagnet and a large concave grating, which latter he had obtained personally from its inventor Henry Augustus Rowland (1848–1901). For discovering the effect bearing his name, Zeeman obtained the Nobel Prize for physics of 1902, together with the theoretical physicist Hendrik Antoon Lorentz (1853–1928), who provided its classical theoretical interpretation.

Initially, in late October 1896, Zeeman could only observe a diffuse line broadening that had actually been predicted by Joseph Larmor's (1857–1942) ▶ electron theory. But in November, Zeeman was able to confirm a prediction by his Leiden colleague, Lorentz, concerning the polarization of the two fringes. In the spring of 1897, Zeeman first recorded distinct splittings of spectral lines into doublets and triplets. These features became understandable by interpreting the splitting as due to a precession of ▶ electrons under the influence of the external magnetic field. As negatively charged particles, electrons have to precess around the axis of a magnetic field H at the so-called Larmor frequency $\nu_L = 1/2\ e/m\ H/c$. There were three possibilities: the external magnetic field was either (i) parallel or (ii) antiparallel or (iii) orthogonal to the electron's axis of precession. All other cases could be explained as linear ▶ superpositions of these three basic cases. In case (i), the energy of the electron is increased, in (ii) decreased, and in (iii) unchanged. Hence a splitting into three components ought to result, and the splitting should be proportional to the strength of the magnetic field. Even the size of the observed triplet splitting was of the right order of magnitude, given a specific charge e/m of the electron of roughly $1.6 \cdot 10^7$ e.m.u. J.J. Thomson had just determined this through electric and magnetic deviation of ▶ cathode rays and inferred the existence of "corpuscles" in them.

So, this normal Zeeman effect was explained fairly well by classical Larmor–Lorentz electron theory. In Niels Bohr's (1885–1962) atomic model, this normal Zeeman triplet could also be derived. Because of the external magnetic field, not all elliptic orbits of similar eccentricity were energetically equivalent any more. Depending on the inclination of the electron's orbit with respect to the magnetic field, the energy is slightly increased, decreased or unchanged (for orthogonal orientation). Space quantization (▶ Stern–Gerlach experiment) restricts this orbit inclination to only a few permitted angles, labeled by a new 'magnetic' ▶ quantum number $m = -1$, 0 or $+1$, thus leading to a splitting into three energy levels. As Arnold Sommerfeld (1861–1949) showed in 1916, other symmetric splittings into an odd number of components could be handled similarly, with $M = 2J + 1$ as the so-called multiplicity of the normal Zeeman splitting (cf. Fig. 1). Using the ► correspondence principle, Bohr's assistant Hendrik Anthony Kramers (1894–1952) also tried to derive the relative intensities of the various multiplet components, but agreement with observations was insufficient.

In the winter of 1897/98, Thomas Preston (1860–1900) in Dublin, Alfred Cornu (1841–1902) in Paris and Albert Michelson (1852–1931) in Chicago, independently found "anomalous" splittings of spectral lines into quartets, sextets, octets, and even more complicated patterns. Such splittings, which soon became known as the anomalous Zeeman effect, remained absolutely mysterious in the classical electron theory and deeply problematic for ► Bohr's atomic model as well.

It was also unclear why the anomalous Zeeman effect changed over to the normal effect under very large magnetic field strengths, as Friedrich Paschen (1865–1947) and Ernst Back (1881–1959) found in 1912. Around 1920, Carl Runge (1856–1927) in Göttingen and Alfred Landé (1888–1976) in Tübingen did manage to describe the complicated anomalous Zeeman patterns phenomenologically. Carl Runge showed that the splittings $\Delta\nu$ followed a numerological rule with q_1 and q_2 integer numbers smaller than the "Runge denominators" r_1 and r_2: (see Fig. 2 for an example)

$$\Delta\nu = \frac{q}{r} \cdot \Delta\nu_\mathrm{L} = \frac{q_1}{r_1}\Delta\nu_\mathrm{L} - \frac{q_2}{r_2}\Delta\nu_\mathrm{L} \Rightarrow \frac{q}{r} = \frac{q_1 r_2 - q_2 r_1}{r_1 r_2}$$

Landé introduced the ► Landé g-factors with strange coefficients $\sim m(m+1)$, etc., but both of these approaches remained *ad hoc*. Persistent problems with the anomalous Zeeman effect substantially contributed to the crisis of ► quantum theory

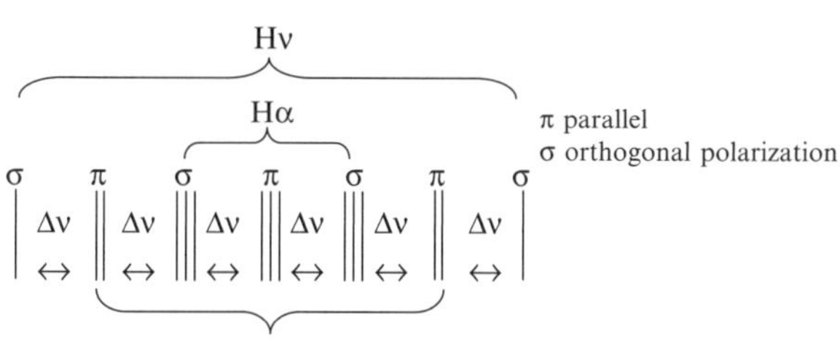

Fig. 1 Sommerfeld's 1916 description of the normal Zeeman effect for the splittings of the hydrogen Balmer series lines H_α, H_β and H_γ (► spectroscopy) including their state of polarization relative to the direction of the magnetic field)

Fig. 2 Example of a complicated anomalous Zeeman splitting (for Runge denominators $r_i = 3$ and 5 in Runge's rule, leading to $q = 0, \pm1, \pm2, \pm3, \pm5, \pm6, \pm8, \pm9, \pm10, \pm12, \pm13, \pm15$), i.e., 23 components!

c.1923 – early 1925. Only after the introduction of the concept of ► spin in late 1925 and the development of quantum mechanics could the observed splittings and relative intensities for the anomalous Zeeman effect be properly derived and physically understood as the result of gyroscopic forces of the electron's magnetic moment $\mu = -eh/2mc$, i.e. one full Bohr magneton and not half a Bohr magneton, as would be expected from classical electron theory (see [9, pp. 97ff., 108] [6]).

Primary Literature

Discussion of these problems beyond that in the standard textbooks can be found, e.g., in

1. N. Bohr: On the effect of electric and magnetic fields on spectral lines. Philosophical Magazine (6th ser.), **27**, 506–524 (1914)
2. H.A. Kramers: Intensities of spectral lines. Det kgl. Danske Videnskaps Selskap, mathematisk-fysike Meddeleser (8) **3/3**, 1–103 (1919)
3. A. Sommerfeld: *Atombau und Spektrallinien* (Braunschweig 1919; Engl. transl. *Atomic Structure and Spectral Lines*, London 1923)
4. P. Zeeman: *Researches in Magneto-Optics* (Macmillan, London 1913)

Secondary Literature

5. T. Arabatzis: The discovery of the Zeeman effect. Studies in History and Philosophy of Science **23**, 365–388 (1992)
6. C. Candler: *Atomic Spectra and the Vector Model* (Princeton University Press, Princeton 1937, esp. chap. 6)
7. P. Forman: Alfred Landé and the anomalous Zeeman effect, 1919–1921. Historical Studies in the Physical Sciences **2**, 153–261 (1970)
8. K. Hentschel: Die Entdeckung des Zeemaneffekts als Beispiel für das komplexe Wechselspiel von wissenschaftlichen Instrumenten, Experiment & Theorie. Physikalische Blätter **52**, 1232–1235 (1996)
9. G. Herzberg: *Atomic Spectra and Atomic Structure* (Prentice Hall, New York 1937; Reprint Dover Publications 1944)

Zero-Point Energy

Peter W. Milonni

The concept of *zero-point energy* first appeared in 1912, when Max Planck (1858–1947) published his "second theory" of ► black-body radiation [1]. In this theory the energy of a harmonic oscillator of frequency ν in thermal equilibrium at temperature T is equal to

$$E(T) = \frac{h\nu}{e^{h\nu/kT} - 1} + \frac{1}{2}h\nu \,, \tag{1}$$

where h and k are, respectively, the Planck and Boltzmann constants. The second term on the right is the zero-point energy, i.e., the energy at zero temperature, where all motion should cease and the energy should be zero according to classical physics. The assumptions about the emission and absorption of radiation that led to Planck's expression were not justified by the fully developed quantum theory that came later, but (1) turned out to be correct. Zero-point energy was invoked shortly after Planck's work by Einstein and Stern [2], who used it to explain the observed temperature dependence of the specific heat of molecular hydrogen, and by Debye [3], who noted that zero-point energy of the atoms of a crystal lattice would cause a reduction in the intensity of the diffracted radiation in X-ray diffraction even as the temperature approached absolute zero. In 1924 Mulliken [4] provided direct evidence for the zero-point energy of molecular vibrations by comparing the band spectra of $B^{10}O$ and $B^{11}O$: the isotopic difference in the transition frequencies between the ground vibrational states of two different electronic levels would vanish if there were no zero-point energy, in contrast to the observed spectra. A year later the zero-point energy of a harmonic oscillator was deduced from Heisenberg's ► matrix mechanics [5] and shortly thereafter from ► Schrödinger's equation. The energy levels of a harmonic oscillator of frequency ν are given according to quantum theory by

$$E_n = (n + \frac{1}{2})h\nu \,, \quad n = 0, 1, 2, 3, \dots \,. \tag{2}$$

For an oscillator with spring constant k and mass m, $\nu = \sqrt{k/4\pi^2 m}$ and the zero-point energy $E_0 = \sqrt{h^2 k/16\pi^2 m}$ is seen to be largest for small masses. Thus, because of their small masses, He^3 and He^4 do not solidify at small pressures as $T \to 0$ because their zero-point motion prevents crystallization.

Zero-point energy is important in the quantum theory of radiation, according to which each field mode of frequency ν has zero-point energy $\frac{1}{2}h\nu$. This allows the interpretation of the van der Waals interaction between two atoms, for instance, in terms of a change in the zero-point energy of the electromagnetic field. More generally the presence of matter modifies the zero-point field energy in a way that depends on the nature and distribution of the matter, and this can result in small but measurable forces between macroscopic bodies. The best known example of this consequence of zero-point field energy is the Casimir force between uncharged, perfectly conducting plates. ► Casimir effect.

Although zero-point energy is an integral part of basic quantum theory [6–8], it leads to a profound difficulty when considered in the context of general relativity. Any energy density of the vacuum contributes to a cosmological constant of the type introduced by Einstein in order to obtain static solutions to his field equations. The zero-point energy density of the vacuum, due to all quantum fields, is extremely large, even when we cut off the largest allowable frequencies based on plausible physical arguments. It implies a cosmological constant larger than the limits imposed by observation by about 120 orders of magnitude. This "cosmological constant problem" remains unresolved.

Primary Literature

1. M. Planck: Über die Begründung des Gesetzes der schwarzen Strahlung. Ann. d. Phys. **37**, 642 (1912)
2. A. Einstein, O. Stern: Einige Argumente für die Annahme einer molekularen Agitation beim absoluten Nullpunkt. Ann. d. Phys. **40**, 551 (1913)
3. P. Debye: Interferenz von Röntgenstrahlen und Wärmebewegung. Ann. d. Phys. **43**, 49 (1914)
4. R. S. Mulliken: The band spectrum of boron monoxide. Nature **114**, 349 (1924)
5. W. Heisenberg: Über quantentheoretische Umdeutung kinematischer und mechanischer Beziehungen. Z. Phys. **33**, 879 (1925)

Secondary Literature

6. L. Abbott: The mystery of the cosmological constant. Sci. Am. **258**, 82 (May, 1988)
7. P. W. Milonni, M.-L. Shih: Zero-point energy in early quantum theory. Am. J. Phys. **59**, 684 (1991)
8. K. A. Milton: *The Casimir Effect – Physical Manifestations of Zero-Point Energy* (World Scientific, Singapore 2001)

English/German/French Lexicon of Terms[1]

English	German	French
Angular momentum	Drehimpuls	moment angulaire
Annihilation operator	Vernichtungsoperator	opérateur d'annihilation
Bell inequalities	Bellsche Ungleichung	Inégalités de Bell
Blackbody radiation	Hohlraumstrahlung, Schwarzkörperstrahlung	rayonnement du corps noir
Brownian motion	Brownsche Bewegung, Brownsche Molekularbewegung	Mouvement brownien
Collapse of wavefunction	Kollaps *or* Reduktion der Wellenfunktion	réduction de la fonction d'onde
Creation operator	Erzeugungsoperator	opérateur de création
Decaying states	zerfallende Zustände	états se désintégrant
Delayed choice experiment	Experiment mit verzögerter Wahl	Experience à choix retardé
Detached observer	aussenstehender Beobachter	observateur détaché
Double-slit or two-slit experiment	Doppelspalt – Experiment	Expérience des fentes d'Young or Expérience à doubles fentes
Entanglement	Verschränkung	Intrication
Excitation states	Anregungszustände	états d'excitation
Excited states	angeregte Zustände	états excités
Gauge theories	Eichtheorien	théorie de jauge
Hidden parameters	verborgene Variable	variables cachés
Improper mixture	Gemisch	mélange impropre
Large-angle scattering	Rückwärtsstreuung	diffusion à grand angle

[1] Many thanks to Michel Le Bellac for his help with French terms.

English	German	French
Many-worlds interpretation	Viele-Welten-Interpretation	interprétation multimondes
Measurement problem	Messproblem	problème de la mesure
Mixture of states	Gemenge	vrai mélange
Observable, non-commuting	Observable, nichtvertauschbare	Observable non-commutantes
Observable, physical quantity, measurable quantity	Observable, physikalische Grösse, Messgrösse, beobachtbare Grösse	Observable propriété physique, propriété mesurable
Occam's razor	Occams Rasiermesser	rasoir d'Occam
Operator, self-adjoint	Operator, selbstadjungierter	opérateur autoadjoint
Pauli exclusion principle	Pauli-Prinzip, Paulisches Ausschliessungsprinzip	Principe de Pauli
Pilot wave	Führungswelle	onde pilote
Plum pudding model	Rosinenkuchenmodell	modèle du gâteau aux raisins
Pure state	reiner Fall	état pur
Quantum eraser	Quantenlöscher/Quantenradierer	gomme quantique
Relative states interpretation	relative Zustände Interpretation	théorie de la relativité des états
Schrödinger equation	Schrödinger-Gleichung	èquation de Schrödinger
Smeared-out states	verschmierte Zustände	états étalés ou non-localisés
Space quantization	Richtungsquantelung	quantification de l'espace
Spin	Spin	Spin
State	Zustand	État
State reduction	Zustandsreduktion	réduction d'état
Superposition	Superposition or kohärente Überlagerung	superposition
Superselection Rules	Superauswahlregeln	règle de supersélection
Trace	Spur	Trace
Tunnel effect	Tunneleffekt	effet tunnel
Wave function	Wellenfunktion	fonction d'onde
Wave packet	Wellenpaket	paquet d'ondes
Wave-particle duality	Welle-Teilchen Dualismus	dualité onde-particule
Which way experiments	welcher-weg Experimente	Mesure de chemin (suivi)

Selected Resources for Historical Studies

The following resources are recommended as starting points for those actively researching the history of quantum physics and quantum mechanics:

1. Paul Forman, John Heilbron, Thomas S. Kuhn and Lily Allen (Eds.): *Sources for History of Quantum Physics*, American Philosophical Society, Memoirs vol. 68 (1967), also available online as http://www.amphilsoc.org/library/guides/ahqp/
2. Bruce Wheaton (Ed.): *Inventory of Sources for History of 20th Century Physics: Report and Microfiche Index to 700.000 Letters*, Stuttgart: GNT, 1993 (the most complete finding aid for unpublished letters to and from twentieth century physicists).
3. Bartel van der Waerden (Ed.) *Sources of Quantum Mechanics, Edited with a Historical Introduction*, New York: Dover, 1968 (contains English translation of many key papers in the history of quantum theory and quantum mechanics).
4. Max Jammer, Friedrich Hund, Helmut Rechenberg and Jagdish Mehra, among others, have published books of various length, detail and quality about the history of quantum mechanics which are all still available in print. Jammer's *Conceptual Development of Quantum Mechanics* (New York: AIP 1989 [1st ed. 1966]) or Friedrich Hund's *History of Quantum Theory,* London: Harrap 1975 (German orig. 1972) are a good start for beginners even though they are not up to date in all historical details. More specific themes are covered in greater depth in studies, for instance, by Bruce Wheaton: *The Tiger and the Shark: Empirical Roots of Wave-Particle-Dualism* (Cambridge: Cambridge University Press 1992), Olivier Darrigol: *From C-numbers to Q-numbers. The Classical Analogy in the History of Quantum Mechanics* (Berkeley: University of California Press, 1992), and James Cushing: *Quantum Mechanics and the Copenhagen Hegemony* (Chicago: University of Chicago Press 1994). For the experimental basis of early quantum theory, the best study remains Hans Kangro*: Early History of Planck's Radiation Law* (London: Taylor & Francis 1976).
5. www.nobel.org for the cv's, the laudatios and talks by all Nobel prize laureates.

6. Guide to the archival collections in the Niels Bohr Library of the American Institute of Physics, College Park, MD: American Institute of Physics, 1994 and supplement 1996 as well as their online finding aids to be found at http://www.aip.org/history/
7. http://www.aip.org/history/ with various excellent online exhibitions, for instance on Marie Curie, Albert Einstein, Werner Heisenberg, and Andrej Sakharov as well as on the discovery of the electron, cyclotrons and superconductivity, to name just a few; furthermore, there are links to the International Archival Catalog (ICOS), an excellent visual archive of photographs and films, oral history interviews, and links to other Archival Finding Aids (with name and subject search).
8. http://www.alberteinstein.info/ with digitized Einstein manuscripts and a searchable archival database. The multivolume *Collected Papers of Albert Einstein*, appearing at Princeton University Press have already reached the early 1920s and include all of his papers and nearly all of his correspondence in annotated form.
9. Other collected works are available on Niels Bohr (Amsterdam: North Holland, 1972–2006), Erwin Schrödinger (Vienna: Austrian Academy of Sciences), Werner Heisenberg (Berlin: Springer, 1984–1993) and Eugene Paul Wigner (New York: Springer, 1992–1998), to name just a few prominent examples.
10. The online version of the Sommerfeld correspondence edition: http://www.lrz-muenchen.de/~Sommerfeld/ with summaries of all known letters to and from Arnold Sommerfeld. The full text of c.600 selected letters can be found in the two-volume edition by Michael Eckert and Karl Märker (Eds.) *Arnold Sommerfeld: Wissenschaftlicher Briefwechsel*, Stuttgart: GNT, 2 Vols.: 2000 and 2004, see http://www.gnt-verlag.de/de/?id=53
11. Wolfgang Pauli's *Scientific Correspondence* has been edited by Karl von Meyenn in a multivolume edition (Berlin: Springer, 1979–2005), now covering the time span 1919–1958. A select edition of letters on wave mechanics has been edited by Karl Przibram under the title: *Briefe zur Wellenmechanik: Schrödinger, Planck, Einstein, Lorentz* (Vienna: Springer 1963). For many other quantum physicists, such painstaking editorial work has yet to be done, however.
12. http://www.malvine.org/ This online search engine for unpublished correspondence and manuscripts by all kinds of authors, not just physicists, allows name searches in archives and repositories all over Europe.

The Contributors

Antonio Acín is a Telecommunication Engineer from the Universitat Politècnica de Catalunya and has a degree in Physics from the Universitat de Barcelona (UB). He got his PhD in Theoretical Physics in 2001 from the UB. After a post-doctoral stay in Geneva, in the group of Prof. Gisin (GAP-Optique), he joined the ICFO-The Institute of Photonic Sciences of Barcelona as a post-doc in 2003. He first became Assistant Professor there in 2005 and later Associate Professor (April 2008). More recently, he has become an ICREA (a Catalan Research Institution) Research Professor, September 2008, and has been awarded a Starting Grant by the European Research Council (ERC). E-mail: antonio.acin@icfo.es

Diederik Aerts is professor at the 'Brussels Free University (Vrije Universiteit Brussel - VUB)' and director of the 'Leo Apostel Centre (CLEA)', an interdisciplinary and interuniversity (VUB, UGent, KULeuven) research centre, where researchers of different disciplines work on interdisciplinary projects. He is also head of the research group 'Foundations of the Exact Sciences (FUND)' at the VUB and editor of the international journal 'Foundations of Science (FOS)'. His work centers on the foundations of quantum mechanics and its interpretation, and recently he has investigated the applications of quantum structures to new domains such as cognitive science and psychology. E-mail: diraerts@vub.ac.be

Hanne Andersen holds a MSc in physics and comparative literature from the University of Copenhagen and a PhD in philosophy of science from the University of Roskilde, Denmark, and she is currently associate professor at the Department for Science Studies at the University of Aarhus, Denmark. Among her publications are On Kuhn (Wadsworth 2001) and The Cognitive Structure of Scientific Revolutions (co-authored with Peter Barker and Xiang Chen, Cambridge University Press 2006). E-mail: hanne.andersen@ivs.au.dk

Jean-Pierre Antoine (1940) received his MS and his PhD in Physics from Université catholique de Louvain, in 1962 and 1966, respectively. After postdoctoral positions at Princeton University (NJ), University of Pittsburgh (PA), and Université de Genève (Switzerland), he spent his academic career at the Université catholique de Louvain. He was also invited professor at many foreign universities. At the Université catholique de Louvain, he was Head of the Institute of Theoretical and Mathematical Physics (FYMA), Chairman of the Department of Physics, and member of various university committees. He is a member of the European Physical Society, the American Physical Society, the International Association of Mathematical Physics, the American Mathematical Society. E-mail: Jean-Pierre.Antoine@uclouvain.be

Theodore Arabatzis is Associate Professor of History and Philosophy of Physics at the University of Athens. He has written on late nineteenth and early twentieth century physics and chemistry. He has also published on philosophical and historiographical issues concerning conceptual change, scientific realism, and experimentation. He is the author of *Representing Electrons: A Biographical Approach to Theoretical Entities* (2006). E-mail: tarabatzis@phs.uoa.gr

Markus Arndt is presently full professor of Quantum Nanophysics at the University of Vienna.

His research is focused on quantum optics and interferometry with clusters and large molecules, aiming at understanding the quantum/classical transition and applications of molecule interferometry for metrology and lithography. His started his research as a Diploma student with H. Walther in LMU Munich, continued as a PhD student with T. W. Hänsch at the Max-Planck Institute for Quantum Optics in Garching, was a Postdoc with J. Dalibard at the *Ecole Normale Supérieure* in Paris and with A. Zeilinger at the University of Innsbruck. His research on molecule interferometry was awarded the START prize (BM:BWK, FWF), Kohlrausch prize (ÖPG) and Erich Schmid prize (ÖAW). E-mail: markus.arndt@univie.ac.at

Leslie Ballentine is a Professor Emeritus at Simon Fraser University, in Burnaby, B.C., Canada. Born in 1940 at Wainwright, Alberta, he completed a B.Sc. and M.Sc. at the University of Alberta before going to Cambridge, England for his Ph.D. His earlier research activities were in condensed matter physics and statistical mechanics, but those have gradually given way to his current interests, which include the foundations of quantum mechanics, quantum chaos, and the emergence of classical behavior from quantum mechanics. E-mail: Leslie_Ballentine@sfu.ca

Jeffrey A. Barrett is Professor of Logic and Philosophy of Science at UC Irvine. He is author of *The Quantum Mechanics on Minds and Worlds* Oxford University Press (2001), the online *Stanford Encyclopedia of Philosophy* article on Everett's relative-state formulation of quantum mechanics, and many journal articles on the quantum measurement problem more generally. E-mail: j.barrett@uci.edu

Charlotte Bigg is an historian of physics, chemistry and astrophysics in the nineteenth and twentieth centuries. After completing her PhD on the history of spectroscopy at the University of Cambridge in 2002, she worked at the Max Planck Institute for the History of Science in Berlin and at ETH Zurich in Switzerland. She is currently a research scientist at the Centre Alexandre Koyré/Centre National de la Recherche Scientifique in Paris. E-mail: bigg@mpiwg-berlin.mpg.de

Dr. Robert C. Bishop is Associate Professor of Physics and Philosophy and the John and Madeleine McIntyre Professor of History and Philosophy of Science at Wheaton College. He held research postdocs in Germany and taught at The London School of Economics and Political Science, University of Oxford and Rice University before coming to Wheaton. His research interests are in the foundations of physics and interrelations between physics, consciousness and free will. E-mail: Bishop@wheaton.edu

Arno Bohm is Professor of Physics at the University of Texas at Austin. His research covers group theoretical methods in particle physics, the idea of the spectrum generating group, relativistic spectrum generating group and relativistic collective models. He is also engaged in research on the quantum geometric phase and foundations and applications in molecule physics (1989–1993). He worked on the rigged Hilbert space formulation of quantum mechanics (1964–1966), with applications to decay and Golden Rule, time evolution semigroup (1975–1981), and irreversible quantum theory (1993–2001). E-mail: bohm@physics.utexas.edu

Arianna Borrelli has worked in research both as a theoretical physicist and as a historian of pre-modern natural philosophy and modern science. She is at present collaborating in a joint project on the "History and Foundations of Quantum Mechanics" of the Max Planck Institute for the History of Science and the Fritz Haber Institute of the Max Planck Society. E-mail: aborrelli@mpiwg-berlin.mpg.de

Marianne Breinig was certified as a teacher of Mathematics and Physics by the J. W. Goethe Universität, Frankfurt, Germany in 1972. In 1975 she received an MA and in 1979 a PhD in Physics from the University of Oregon. After 2 years as a Postdoctoral Research Associate she joined the faculty at the University of Tennessee in 1981. Since 1988 she has also held a position as Adjunct R and D Participant in the Physics Division of the Oak Ridge National Laboratory. She focuses her efforts in the areas of atomic collision physics and physics teaching and instructional technology. E-mail: mbreinig@utk.edu

Hans J. Briegel obtained his PhD in Theoretical Physics from the Ludwig-Maximilians-Universität (LMU) Munich in 1994. He worked as a postdoctoral researcher at Texas A&M University, Harvard University, and Innsbruck University. From 1998 to 2003 he was at the LMU Munich, where he received his Habilitation. In 2003, he joined the faculty of physics at the University of Innsbruck as a Full Professor. He is also a Scientific Director at the Institute for Quantum Optics and Quantum Information of the Austrian Academy of Sciences in Innsbruck. His current research interests include the theory of quantum information and computation, molecular biophysics, and the foundations of quantum mechanics. E-mail: hans.briegel@uibk.ac.at

Dagmar Bruss completed her doctorate in theoretical particle physics in 1994 at the University of Heidelberg, Germany. She joined the University of Oxford, England, as a European Research Fellow from 1996 to 1997, where she started to work in the newly emerging field of quantum information theory. She spent another European Fellowship in 1998 at ISI Torino, and then joined the University of Hannover, Germany, where she completed her Habilitation in 2002. Since 2004 Dagmar Bruss has been Full Professor at the Institute for Theoretical Physics at Heinrich-Heine-University of Düsseldorf, Germany. Her main research interests are quantum information theory, in particular foundations, entanglement theory, quantum cloning and state estimation, quantum cryptography, and quantum optical implementations. E-mail: bruss@thphy.uni-duesseldorf.de

Jeffrey Bub (PhD, London University, 1966, in mathematical physics) is distinguished university professor in the Department of Philosophy and the Institute for Physical Science and Technology at the University of Maryland. He has numerous publications on the conceptual foundations of quantum mechanics and is the author of two books: *The Interpretation of Quantum Mechanics* (Reidel, 1974) and *Interpreting the Quantum World* (Cambridge University Press, 1997/1999), which won the prestigious Lakatos Award in 1998. His current research is focused on foundational questions arising in the field of quantum information and computation. He was awarded the Kirwan Faculty Research and Scholarship Prize in 2005 for his work in this area. E-mail: jbub@umd.edu

Paul Busch is Professor of Mathematical Physics at the University of York (since 2005). His research interests lie in the foundations of quantum physics and quantum information, with a focus on the quantum theory of measurement, quantum theory and relativity, and the notion of unsharp quantum reality. He studied physics, mathematics and philosophy at the University of Cologne and received his doctorate in theoretical physics there in 1982. He has held postdoctoral research positions at the University of Cologne and the Max Planck Institute for biophysical Chemistry, Göttingen, and long-term visiting positions in Florida, Heidelberg, Harvard, and at the Perimeter Institute. He has published over 90 papers and is coauthor of two monographs, *The Quantum Theory of Measurement* (Springer, 1991/1996) and *Operational Quantum Physics* (Springer, 1995/1997). Since 2005 he has served as associate editor of the journal *Foundations of Physics*. In 2008 he was elected Member of L'Académie de Philosophie des Sciences. E-mail: pb516@york.ac.uk

Craig Callender is professor of philosophy at the University of California, San Diego. He received his PhD in Philosophy from Rutgers University in 1997, writing a dissertation on the problem of the direction of time. He then worked at the London School of Economics in the Department of Logic, Philosophy and Scientific Method as lecturer and senior lecturer from 1996–2000. In 2001 he moved to UCSD, where he continues to work on the philosophical foundations of statistical mechanics, quantum mechanics and relativity. He is particularly interested in the intersection between philosophy of time and physics. E-mail: ccallender@ucsd.edu

Howard Carmichael earned a PhD from the University of Waikato in 1977. He joined the faculty of the University of Arkansas in 1983 and moved to the University of Oregon in 1989 where he was Professor of Physics from 1991 to 2001. He is currently a member of the Physics Department at the University of Auckland where he holds the Dan Walls Chair in theoretical physics. Professor Carmichael is a Fellow of the American Physical, the Optical Society of America, and the Royal Society of New Zealand. E-mail: h.carmichael@auckland.ac.nz

Brian Clegg was born in Rochdale, Lancashire, UK (now at Swindon, UK). After graduating from Cambridge University (specializing in experimental physics), he gained an MA in Operational Research at Lancaster University, and went on to join British Airways, where he formed a new department developing hi-tech solutions for the airline. In 1994 he left BA and has since written a range of science books for a general readership including *The God Effect* and *A Brief History of Infinity*. Brian is editor of the www.popularscience.co.uk book review site and a Fellow of the Royal Society of Arts. E-mail: brian@brianclegg.net

John G. Cramer was born in Houston, Texas, USA. He attended Rice University in Houston, receiving BA, MA, and PhD degrees, the latter in experimental nuclear physics. He is currently Professor of Physics at Washington University. He became director of the University of Washington Nuclear Physics Laboratory in 1984 and oversaw the construction of the DOE-funded $10 million Superconducting Booster Project, which was completed on time and under budget. In 1990 he stepped down as NPL Director and turned his research in a new direction focused primarily on ultra relativistic heavy ion physics. Since then, he has been an active participant in the NA35 and NA49 Collaborations at CERN and the STAR Collaboration at the RHIC Facility at Brookhaven National Laboratory. Since the 1980s he has also worked in the foundations and interpretation of quantum mechanics. He introduced the transactional interpretation of quantum mechanics, a leading rival of the orthodox Copenhagen interpretation. E-mail: cramer@phys.washington.edu

Cécile DeWitt-Morette Jane and Roland Blumberg, Centennial Professor in Physics, Emerita, University of Texas at Austin, Doctorat d'Etat (Université de Paris, 1947). Founder of the *Ecole de Physique* (Les Houches). Professional appointments in France and in the U.S.. Visiting appointments in several countries (Ireland, Denmark, Brazil, England, India, Portugal, Germany, Iran). Author of "Analysis, Manifolds and Physics" (2 volumes) with Yvonne Choquet-Bruhat, "Functional Integration; Action and Symmetries" with Pierre Cartier; over 100 research articles, many of them related to functional integration since 1951. E-mail: cdewitt@physics.utexas.edu

Detlev Dürr was born in 1951, Germany. He obtained his PhD in Physics from the University of Münster in 1978 and his habilitation in mathematics at Bochum in 1984. He was a Postdoc in the Department of Mathematics (Rutgers 1979–1985 with interruptions) and a Heisenberg fellow (DFG) from 1985 to 1989. He has been Professor of Mathematics at the 'Mathematisches Institut LMU München' since 1989. His research area is Physics/Mathematical Physics/Mathematics. Books: *Bohmian Mechanics* (Springer 2009); *Chance in Physics*, Springer Lecture Notes in Physics, Editor (2001); *The Quantum Universe*, IOP Special Edition, Guest-Editor (2007). E-mail: duerr@mathematik.uni-muenchen.de

Michael Eckert *Diplom* degree (1976) in physics at the Technical University, Münich; doctorate there and at the University of Bayreuth in theoretical physics (1979 Dr. rer. nat.); staff researcher at the Deutsches Museum (1981–88); reader of mathematics, physics and computer science for the Bavarian textbook publishers in Münich (1989–95); staff researcher at the University of Münich (1995–2000) and at the Deutsches Museum (2001 to date). Project topics: history of solid-state physics, microelectronics and nuclear energy in the FRG, editor of the Sommerfeld works, history of fluid-aerodynamics, biography of A. Sommerfeld. His recent books include *Arnold Sommerfeld: Wissenchaftlicher Briefwechsel* (2 volumes: 2000, 2004); *The Dawn of Fluid Dynamics* (2006). E-mail: M.Eckert@lrz.uni-muenchen.de

Brigitte Falkenburg studied physics in Erlangen, Berlin and Heidelberg and philosophy in Bielefeld. She gained a D. Phil. in 1985 and, a year later, a doctorate in experimental particle physics. Since 1997 she has been full professor for Theoretical Philosophy and the Philosophy of Science and Technology at the Technical University of Dortmund. She is a member of the *Acadèmie Internationale de la Philosophie des Sciences*, and the *European Academy of Science* and speaker of the Philosophy of Physics Working Group of the German Physical Society. She has won prizes for her work on philosophy of science and has published numerous books and papers on various aspects of physics and metaphysics. E-mail: falkenburg@fk14.tu-dortmund.de

Roman Frigg is a Senior Lecturer in philosophy at the London School of Economics. He holds a PhD from the University of London. His main research interests lie in general philosophy of science and philosophy of physics. He has published papers on quantum mechanics, the foundations of statistical mechanics, randomness, chaos, complexity theory, probability, scientific models, computer simulations, and he is currently working on book on models and theories. Further information can be found on his website at www.romanfrigg.org. E-mail: r.p.frigg@lse.ac.uk

Daniel J. Gauthier received the B.S. (1982), M.S. (1983), and Ph.D. (1989) degrees from the Institute of Optics, University of Rochester. He is a Fellow of the Optical Society of America and the American Physical Society and is Professor and Chair of Physics at Duke University. His research interests include applications of slow light in classical and quantum information processing. E-mail: gauthier@phy.duke.edu

Kostas Gavroglu is professor of history of science at the Department of History and Philosophy of Athens at the University of Athens, Greece. He is also the Director of the Historical Archive of the University of Athens. His research interests are the history of artificial cold (mainly superconductivity and superfluidity), the history of quantum chemistry and the reception of ideas in the Balkans during the 19th century. Among his books is the biography of Fritz London (Cambridge University Press, 1995). For further details see http://www.phs.uoa.gr/dlab/gavroglou. E-mail: kgavro@phs.uoa.gr

Clayton Gearhart is Professor of Physics at St. John's University in Collegeville, Minnesota. His PhD thesis comprised an experimental study of liquid helium. His current main interests include the history of thermodynamics, statistical mechanics, and early quantum theory. E-mail: cgearhart@csbsju.edu

Rolf Gerhardts received his academic education and degrees (Dr. rer. nat. in 1969 and, after a post-doctoral year at Stanford, Habilitation in 1975) at Cologne. After about 11 years at the FU Berlin, since 1982 as Professor, he moved to the MPI for Solid State Research in Stuttgart in 1986. Since 1997 he has been 'apl. Professor' at the University of Würzburg. E-mail: R.Gerhardts@fkf.mpg.de

Domenico Giulini was born in Heidelberg (Germany) where he also did his undergraduate studies in Physics and Mathematics before going to Cambridge (England), where he finished his Ph.D. in 1990 on a topic in mathematical general relativity under the supervision of Gary Gibbons. Between 1997 and 2001 he worked at the University of Zürich with Norbert Straumann and between 2007 and 2009 at the Max-Planck-Institute for Gravitational Physics (Albert-Einstein Institute) at Potsdam. He is currently Professor at the Institute of Theoretical Physics at the University of Hannover. E-mail: domenico.giulini@itp.uni.hannover.de

Sheldon Goldstein is a Professor of Mathematics, Physics and Philosophy at Rutgers University in New Brunswick, New Jersey. He worked for many years on probability theory and the rigorous foundations of statistical mechanics. In particular, he has investigated the arrow of time and the notion of entropy, why it tends to increase, and why physical systems tend to approach equilibrium. In more recent years his main concern has been with the foundations of quantum mechanics, particularly Bohmian mechanics and nonlocality. The goal of this research is to make sense, good clean sense, of quantum theory. E-mail: oldstein@math.rutgers.edu

Oscar Wallace Greenberg is Professor of Physics at the University of Maryland, College Park. He received his PhD at Princeton University in 1957. He is a Captain, US Air Force, Retired. He is a Fellow of the American Physical Society and the Washington Academy of Sciences. Greenberg introduced the color quantum number for quarks, a fundamental part of the standard model of elementary particles. He also introduced the accepted model of baryons, the symmetric quark model and has made several other essential contributions to high-energy and quantum physics. E-mail: owgreen@physics.umd.edu

Daniel Greenberger studied at MIT, writing his thesis with Laszlo Tisza, and did his graduate work at the University of Illinois. He then spent 2 years in the army and a year at Ohio State University, and won an NSF postdoctoral fellowship which he spent at Berkeley in the group of Geoffrey Chew. He then went on to CCNY, where he has been since, with breaks for Oxford, MIT, Vienna, and Munich. He has been a Fulbright fellow in Vienna, and spent a year at Garching, with Herbert Walther, with a senior Humboldt award. He is also an editor of several journals, and a fellow of the American Physical Society, and was elected a foreign member of the Austrian Academy of Sciences. He has also organized several conferences, and he had a two-volume festschrift dedicated to him by *Foundations of Physics*, as well as a conference this summer in Vienna. E-mail:greenbgr@sci.ccny.cuny.edu

Robert B. Griffiths is the Otto Stern University Professor of Physics at Carnegie-Mellon University, where he has been a faculty member since 1964. His degrees are: A.B. (1957) from Princeton University; M. S. (1958) and Ph.D. (1962) from Stanford University, all in Physics. He is a Fellow of the American Physical Society and a Member of the National Academy of Sciences of the U.S.A. He received the Dannie Heineman Prize for Mathematical Physics from the American Physical Society in 1984. He is the author or coauthor of 130 papers on various topics in theoretical physics, mainly statistical mechanics and quantum mechanics. Recently he has published a book on the fundamental principles of quantum theory. E-mail: rgrif@cmu.edu

Nathan Harshman received his BS in physics and English from Duke University in 1995 and his PhD in Theoretical Physics from University of Texas at Austin in 2001. After 2 years as Wiess Instructor of Physics at Rice University, he was appointed Assistant Professor of Physics at American University in Washington, DC. His general interests are composite and unstable quantum systems and the intersection of quantum information theory and particle physics. Currently he is working on entanglement in continuous variable systems, and in particular on entanglement generation in particle scattering. E-mail: harshman@american.edu

Richard Healey has been a Professor of Philosophy at the University of Arizona since 1991. He grew up in England but has taught in the USA since 1980. His research interests center on the conceptual foundations and implications of physical theories for the nature of space, time, causation, realism and holism. This research has been supported by several NSF grants. His perennial attempts to understand quantum mechanics resulted in the publication (in 1989) of *The Philosophy of Quantum Mechanics*. In 2007 he published a second book, *Gauging What's Real*, which explores the conceptual foundations of the gauge theories physicists currently use to describe the fundamental forces of nature. E-mail: rhealey@email.arizona.edu

J.L. Heilbron formerly professor of history and the vice chancellor at the University of California, Berkeley is an honorary fellow of Worcester College, Oxford. His contributions to the historiography of twentieth century physics, which center on the quantum theory of the atom and early nuclear physics, include biographies of Moseley, Planck, and Rutherford, and studies of Lawrence and his laboratory. He now works primarily on the early modern period. E-mail: johnheilbron@berkeley.edu

Christian Heinicke studied physics and philosophy in Cologne and Bonn. He received his Ph.D. in theoretical physics from Cologne University in 2005 (Thesis on *Exact solutions in Einstein's theory of gravity and beyond*). He than was research assistant at the institute of theoretical physics in Cologne. His research interests focus on the structure of non-Riemannian spacetimes and its philosophical implications. Since 2008 he works as teacher at the Gymnasium Bayruther Str. in Wuppertal. E-mail: chh@thp.uni-koeln.de

Carsten Held is Professor of Philosophy of Science at the Universität Erfurt. His research concentrates on the foundations of quantum mechanics and on problems at the interface of philosophy of science and philosophy of language. He is the author of *Die Bohr-Einstein-Debatte* (Schöningh 1998) and *Frege und das Grundproblem der Semantik* (mentis 2005). E-mail: carsten.held@uni-erfurt.de

Meir Hemmo is a professor of philosophy in the University of Haifa. He received his PhD degree from Cambridge University in 1996. His main research areas are philosophy of physics, in particular the foundations of quantum mechanics and statistical mechanics, probability, and some issues in metaphysics. He published papers on the measurement problem in quantum mechanics, the problem of the direction of time in statistical mechanics and the interpretation of probability. E-mail: meir@research.haifa.ac.il

Klaus Hentschel studied physics (with a diploma in theoretical high-energy physics), philosophy (MA) and history of science (Ph.D. and *habilitation*) at the university of Hamburg. He is currently full professor for history of science and technology in Stuttgart. He has published widely on the history of physics and philosophy of science, including *Physics and National Socialism*, several books on the history of spectroscopy, the Einstein Tower, the interplay of scientific instrument-making with experimentation and theory formation as well as on 'invisible hands' in scientific research. He has received numerous awards and is a member of Leopoldina, the German Academy of Scientists. See www.uni-stuttgart.de/hi/gnt/hentschel. E-mail: Klaus.hentschel@po.hi.uni-stuttgart.de

Basil J. Hiley is Emeritus Professor of Physics at Birkbeck, University of London. His main interests are in the foundations of quantum theory and relativity. He had a very long and fruitful collaboration with David Bohm developing alternative approaches to quantum theory. At present he is investigating a process-based theory of pre-space with a specific aim of developing a quantized form of general relativity. With the late David Bohm he has co-authored *The Undivided Universe*, Routledge, London, 1993. He is also co-editor with David Peat *of Quantum Implications, Essays in Honour of David Bohm*, Routledge, London 1987. See www.bbk.ac.uk/tpru. E-mail: b.hiley@bbk.ac.uk

Dieter Hoffmann is Research Scholar at the Max Planck Institute for the History of Science in Berlin. His research interests cover the History of Physics in the nineteenth and twentieth Centuries, especially Institutional History of Quantum Theory and Modern Metrology; History of Science in the GDR. E-mail: dh@mpiwg-berlin.mpg.de

Nick Huggett received his Ph.D. in philosophy from Rutgers University in 1995. After working at Brown University for a year, he joined the faculty at UIC in 1996. He has published several books and numerous essays on a wide range of topics within philosophy and physics. He teaches a course on the philosophy of time. E-mail: huggett@uic.edu

Tom Imbo received his Ph.D. in theoretical high energy physics from the University of Texas at Austin in 1988. After spending the 1988–1989 academic year at Texas as a postdoctoral fellow, Tom was elected a Junior Fellow in the Harvard University Society of Fellows where he remained in residence through the summer of 1992. Since Fall 1992 he has been a member of the physics faculty at the University of Illinois at Chicago. E-mail: imbo@uic.edu

Dominik Janzing was born in 1966. He studied physics in Tübingen (Germany) and Cork (Ireland) and received a Ph.D. in mathematics from the University of Tübingen in 1998. From 1998 to 2006 he was a postdoc and senior scientist at the Computer Science department of the University of Karlsruhe (TH) where worked on quantum thermodynamics, quantum control, as well as quantum complexity theory and its physical foundations. E-mail: janzing@ira.uka.de

Gregg Jaeger is an assistant professor at Boston University, from which he obtained his Ph.D. in Physics 1995 and has taught in both the College of General Studies (Natural Sciences) and the College of Arts and Sciences (Mathematics). He carried out postdoctoral research in experimental physics at the U.S. National Institute of Science and Technology and in the history and philosophy of science at the Dibner Institute at the Massachusetts Institute of Technology. His work now focuses primarily on quantum information science. E-mail: jaeger@bu.edu

Erich Joos earned his Ph.D. in Theoretical Physics at the University of Heidelberg in 1983. After conducting research at the Universities of Heidelberg, Austin (Texas) and Hamburg, he is now living as an independent scholar with his family in Schenefeld near Hamburg, Germany. His current research interests have focused on decoherence theory, interpretations and foundations of quantum theory, in particular the interface between classical mechanics and quantum theory. In 1985 he published the first paper (with H.D. Zeh) quantifying the classical-quantum transition via decoherence. E-mail: ej@erichjoos.de

David Kaiser is an associate professor in the Program in Science, Technology, and Society at the Massachusetts Institute of Technology, and a lecturer in MIT's Department of Physics. He completed Ph.D.s in physics and in history of science at Harvard University. Kaiser is author of the award-winning book, "Drawing Theories Apart: The Dispersion of Feynman Diagrams in Postwar Physics" (University of Chicago Press 2005), as well as editor of several books on the history of modern physical sciences. His work has been featured in such magazines as "Harper's", "Science", and "Scientific American", on NOVA television programs and National Public Radio broadcasts, and in specialist journals in physics and history. E-mail: dikaiser@mit.edu

Claus Kiefer studied physics and astronomy at the Universities of Heidelberg and Vienna. He obtained his Ph.D. in 1988 at the University of Heidelberg (with H. D. Zeh). From 1989 to 1993 he was Postdoc at the University of Zürich; from 1993–2001 Lecturer at the University of Freiburg (Habilitation 1995). Since 2001 he has been Professor at the University of Cologne.

Interests: Quantum Gravity, Cosmology, Black Holes, Foundations of Quantum Theory. For more information, see http://www.thp.uni-koeln.de/gravitation/. E-mail: kiefer@thp.uni-koeln.de

Helge Kragh is professor of history of science and technology at the University of Aarhus, Denmark. He has worked in the history of the physical sciences, including chemistry and astronomy, and also in the history of technology. For the last several years his research focus has been on the development of modern physical cosmology. E-mail: ivhhk@ivh.au.dk

Paul G. Kwiat is the Bardeen Chair in Physics, at the University of Illinois in Urbana-Champaign. A Fellow of the American Physical Society and the Optical Society of America, he has given invited talks at numerous national and international conferences, and has authored over 120 articles. His research includes the phenomena of quantum interrogation, quantum erasure, and optical implementations of quantum information protocols, particularly using entangled – and hyper-entangled – photons from parametric down-conversion. E-mail: kwiat@illinois.edu

Pekka Lahti was born 1948 (Turku, Finland). He obtained his doctoral degree in theoretical physics (1979) at the University of Turku. His current positions are docent for theoretical physics and senior research fellow.

He has authored over 100 scientific papers and co-authored of four monographs. E-mail: peklahti@utu.fi

Nicolaas P. (Klaas) Landsman studied physics, mathematics, and astronomy at the University of Amsterdam (1981–1985). His initial research was in quantum field theory at finite temperature and the quark-gluon plasma. His M.Sc. thesis, published as *Physics Reports 145, 141–249* (1987), has been cited almost 700 times. He earned his Ph.D. in Theoretical Physics in 1989. Most of his mathematical ideas from the period 1989–1997 have appeared in his book *Mathematical Topics Between Classical and Quantum Mechanics*

(Springer 1998). Landsman became a Full Professor of Mathematical Physics in 2001 at Amsterdam and moved to the Radboud University Nijmegen in 2004. He is an official advisor to the Dutch Ministry of Education, Culture and Science. E-mail: landsman@math.ru.nl

Michel Le Bellac is Professor Emeritus at the University of Nice, France. He graduated from the Ecole Normale Supérieure in Paris. His field of research is theoretical elementary particle physics. He is author of several textbooks, among them "Quantum and Statistical Field Theory" (1992, Oxford), "Thermal Field Theory" (1996) and "A Short Introduction to Quantum Information and Computation" (2006), both published by Cambridge University Press. E-mail: Michel.Le_Bellac@inln.cnrs.fr

Anthony J. Leggett was born in London, England in March 1938. He attended Balliol College, Oxford where he majored in Literae Humaniores (classical languages and literature, philosophy and Greco-Roman history), and thereafter Merton College, Oxford where he took a second undergraduate degree in Physics. He completed a D.Phil. (Ph.D.) degree in theoretical physics under the supervision of D.ter Haar. In 1983 he became John D. and Catherine T. Macarthur Professor at the University of Illinois at Urbana-Champaign, a position he currently holds. In 2006 he was also appointed to the Mike and Ophelia Lazaridis Distinguished Research Chair at the University of Waterloo. He was awarded the Nobel Prize in Physics (with A.A. Abrikosov and V.L. Ginzburg) in 2003. His principal research interests lie in the areas of condensed matter physics, particularly high-temperature superconductivity, and the foundations of quantum mechanics. E-mail: aleggett@uiuc.edu

Jon Magne Leinaas received his education at the University of Oslo, where he received his doctorate in 1980. Postdoctoral fellowships at Nordita and CERN. Associate professor at the University of Stavanger, and since 1989 physics professor at the University of Oslo. Research interests in quantum physics and quantum field theory, with a focus on quantum effects in low-dimensional systems. A current interest in the geometry of quantum entanglement. E-mail: j.m.leinaas@fys.uio.no

Holger Lyre long-term Visiting Professor at the Philosophy Department, University of Bielefeld, Germany. 1993 diploma degree in physics, University of Dortmund, 1996 PhD in philosophy, University of Bochum, and 2003 Habilitation, University of Bonn. Main research fields: philosophy of science (in particular: structural realism), philosophy of physics (in particular: foundations of gauge theories), and philosophy of mind (in particular: mental externalism).

Book publications as author: *Lokale Symmetrien und Wirklichkeit.* Mentis, Paderborn, 2004. *Informationstheorie. Eine philosophisch-naturwissenschaftliche Einführung.* Fink, Munich, 2002. *Quantentheorie der Information.* Springer, Vienna, 1998. E-mail: lyre@uni-bonn.de

Prof. Dr. Klaus Mainzer (born 1947) studied mathematics, physics, and philosophy (1968–1972), Ph.D. (1973) and habilitation (1979) in philosophy, assistant and lecturer (1973–1979) at the university of Münster; Heisenberg-scholarship (1980); 1981–1988 professor for philosophy (foundations and history of science) at the University of Constance, vice-president of the university of Constance (1985–1988); 1988–2008 chair for philosophy of science, dean (1999–2000), director of the Institute of Philosophy (1989–2008) and of the Institute of Interdisciplinary Informatics (1998–2008) at the University of Augsburg; since 2008 chair for philosophy of science and technology, director of the Carl von Linde-Academy at the Technical University of Munich; member of several academies, president of the German Society of Complex Systems and Nonlinear Dynamics (since 1996), chairman of the Japanese-German Society of Integrative Science (Leibniz-Association e.V.); author of numerous books and articles. E-mail: Mainzer@cvl-a.tum.de

Michela Massimi teaches at University College, London. Her undergraduate degree was in philosophy (University of Rome "La Sapienza") and in 2002 she received a PhD at the London School of Economics. Before joining UCL she was a research fellow at Cambridge, UK, where she lectured on the philosophy of science and the philosophy of physics. E-mail: m.massimi@ucl.ac.uk

Storrs McCall received a B.A. from McGill University, and a B.Phil. and D.Phil. from Oxford. He taught for 6 years at the University of Pittsburgh, and for 5 years at Makerere University in Kampala, Uganda, as the first person to teach philosophy in East Africa. His research interests lie in logic, in the philosophy of space and time, in decision theory, in the interface between mind and brain, and in the philosophy of mathematics. He is the author of *Aristotle's Modal Syllogisms* and *A Model of the Universe*, and is the editor of *Polish Logic, 1920–1939.* He is presently Professor of Philosophy at McGill. E-mail: storrs.mccall@mcgill.ca

Peter Milonni is a Laboratory Fellow (ret.) and Laboratory Associate in the Theoretical Division (Complex Systems Group) at the Los Alamos National Laboratory. His research has been primarily in the areas of quantum optics and electrodynamics. He is the author of several books, including "Lasers" (with J.H. Eberly) and "The Quantum Vacuum: An Introduction to Quantum Electrodynamics", and was the 2008 recipient of the Max Born Award from the Optical Society of America. E-mail: pwm@lanl.gov

Kimball A. Milton received his Ph.D. under the direction of Julian Schwinger at Harvard in 1971, and served under Schwinger's direction as a postdoc at UCLA until 1979. He then spent 2 years at Ohio State, and 4 years at Oklahoma State University, before becoming Professor at the University of Oklahoma in 1986. He is now George Lynn Cross Research Professor, OU's highest research honor. E-mail: milton@nhn.ou.edu

Peter Mittelstaedt was born in 1929. He obtained his Diploma in Theoretical Physics (nuclear physics) with W. Heisenberg in Göttingen in 1954 and his PhD in Theoretical Physics (nuclear physics) under W. Heisenberg in 1956. Peter Mittelstaedt was a Research fellow at CERN in Geneva in (1958), a research assistant at the Max-Planck-Institute for Physics und Astrophysics in Munich (1959–62, 1963–65), and a research fellow at MIT in Cambridge (MA) (1962–63). He obtained his *Habilitation* at the University of Munich with a paper on *Quantum Logic* in 1961. Peter Mittelstaedt is professor emeritus of Theoretical Physics, University of Cologne and author of a number of books. E-mail: mitt@thp.uni-koeln.de

Umar Mohideen is professor of physics at the University of California, Riverside, California. He obtained his PhD in 1992 from Columbia University, New York. Before joining the University of California, he did his post doctoral fellowship at AT&T Bell Laboratories in Murray Hill, New Jersey. His scientific interests are: quantum optics (particularly the Casimir effect) and single molecular interactions. He is the author of more than 58 papers in refereed journals. He is Fellow of the American Physical Society and Fellow of the American Association for the Advancement of Science. E-mail: umar.mohideen@ucr.edu

Michael Martin Nieto was born in Los Angeles, California in 1940 and received a B. A. with highest honors from the University of California, Riverside, in 1961. He obtained a Ph.D. from Cornell University in 1966. He took a permanent position at Los Alamos National Laboratory in 1972, becoming a Lab Fellow in 2003. His main interest are in the theoretical physics of elementary particles, quantum mechanics, and (astrophysical) tests of gravity. Among his awards are Alexander von Humboldt Senior Research Award, Germany, and the Distinguished Alumnus Award of the University of California, Riverside. E-mail: mmn@lanl.gov

Günter Nimtz studied electrical engineering. He graduated from the University of Vienna and became a professor of physics at the University of Cologne in 1983. Much of his research focussed on quantum tunnelling. He achieved emeritus status in 2001. At present, he is teaching and doing fundamental research at the University of Koblenz-Landau. E-mail: G.Nimtz@ Uni-Koeln.DE

R. F. O'Connell Boyd Professor, Dept. of Physics, LSU, R. F. O'Connell received a B.Sc. degree (1953) from the National University of Ireland (after which he worked for 4 years as a Telecommunications Engineer), a Ph.D. from the University of Notre Dame (1962), and a D.Sc. (1975) from the National University of Ireland. Before joining Louisiana State University in 1964, he worked at the Dublin Institute for Advanced Studies and at IBM. E-mail: oconnell@phys.lsu.edu

Martin Plenio received his PhD from the University of Göttingen in 1994 and is now a full professor of quantum physics in the physics department and director of the quantum information programme at the Institute for Mathematical Sciences of Imperial College. He has made important contributions to entanglement theory, precision metrology, and the implementation of quantum simulation and quantum information processing. He is author of 130 articles in refereed journals and recipient of the Maxwell medal and prize of the UK Institute of Physics; he presented the Royal Society Clifford Paterson Lecture for 2008. E-mail: m.plenio@imperial.ac.uk

Helmut Rauch is professor emeritus at the Vienna University of Technology. His scientific interests are: neutron physics, quantum optics, foundations of quantum mechanics and reactor physics. He invented (together with U. Bonse and W. Treimer) the perfect crystal neutron interferometer. With S.A. Werner he published the book *Neutron Interferometry* (Clarendon Press, Oxford 2000). Honours include the Erwin Schrödinger Award of the Austrian Academy of Sciences and the Ludwig Wittgenstein Award of the Austrian Research Association. E-mail: rauch@ati.ac.at

Dr. Daniel Rohrlich received his Ph.D. in physics in 1986 from the State University of New York at Stony Brook. He works as a researcher and lecturer at Ben Gurion University in Beersheba, Israel. He is the author, with Yakir Aharonov, of *Quantum Paradoxes: Quantum Theory for the Perplexed* (Wiley-VCH), 2005. Other work includes a path integral for spin [*Nucl. Phys.* **B299**, 471 (1988), with H. B. Nielsen], generalizations and measures of entanglement [*Found. Phys.* **24**, 379 (1994), *Phys. Rev.* **A56**, R3319 (1997), with S. Popescu] and experiments on quantum dots. E-mail: rohrlich@bgu.ac.il

Simon Saunders is Professor of Philosophy of Physics and Fellow of Linacre College at the University of Oxford. He has worked in the foundations of quantum field theory, quantum mechanics, symmetries, and thermodynamics and statistical mechanics. He was an early proponent of the view of branching in the Everett interpretation as an 'effective' process based on decoherence. His most recent works include 'On the explanation of quantum statistics', (with F. Muller) 'Discerning Fermions', and (with D. Wallace) 'Branching and uncertainty'. E-mail: simon.saunders@philosophy.ox.ac.uk

Arne Schirrmacher studied physics, biology and philosophy at the universities of Hamburg, Oxford and Munich and earned his doctorate in mathematical physics. Turning fully to history of science at a research stay at the University of California at Berkeley, he then joined the Berlin Max Planck Institute for the History of Science. He has worked on a number of projects on the history of modern physics at the research institute of the Deutsches Museum in Munich since 1996, also spending one year at the Institute for the History of Science at Göttingen University. E-mail: A.Schirrmacher@deutsches-museum.de

Maximilian Schlosshauer's work is focussed on quantum decoherence and the quantum-to-classical transition. After completing his undergraduate education at Freiburg University, Germany, he graduated from Lund University, Sweden, with a Master of Science degree in 2001. He did his doctoral research with Arthur Fine at the University of Washington in Seattle and received his Ph.D. in Physics in 2005. He is author of the Springer monograph "Decoherence and the Quantum-to-Classical Transition" and has also worked on problems in theoretical biophysics. E-mail: schlosshauer@nbi.dk

Erhard Scholz has been Professor for History of Mathematics, University of Wuppertal since 1989. Guest professor Institute for History of Science, University Göttingen 1993. Member of Interdisciplinary Centre for Science and Technology Studies since 2005. Interests include: Mathematics in the nineteenth and twentieth centuries; Interrelation between theoretical mathematics and applications in natural sciences; Philosophical aspects of mathematical concept formation. Publications related to topic of this volume: "Introducing groups into quantum mechanics." *Hist. Math.* (2006). "Weyl entering the 'new' quantum mechanics discourse." In C. Joas, C. Lehner, J. Renn (eds.). *HQ-1: Conf. on the Hist. of Quant. Phys.* (2007). E-mail: Erhard.Scholz@math.uni-wuppertal.de

Alfred Seeger was born in Stuttgart in 1927. He obtained his PhD at the age of 23 in 1951. Subsequently he worked at Bristol university with pioneers of solid state physics, amongst them Sir Nevill Francis Mott, who was awarded the Nobel Prize in Physics in 1977. In 1955 he taught physics at the University of Cambridge. In 1959 he was appointed to the newly established chair for solid state physics at the Technical University of Stuttgart and became a scientific member of the Max Planck Institute for Metal Research. From 1965 to his retirement he was director of the Physics Institute at the MPI for Metal Research.

Abner Shimony was born in Columbus, Ohio on March 10, 1928. He worked at the Massachusetts Institute of Technology, Philosophy Department, 1959–1968 and Boston University, Philosophy and Physics Departments, Associate Professor and Professor, 1968–1994. His present position is Professor Emeritus of Philosophy and Physics, Boston University. He is engaged in many peace organizations. His work was honored in the Festschrift "Quantum Mechanical Studies For Abner Shimony", 2 volumes, Kluwer Academic Publishers, 1997. His publications include many articles in Philosophy and Physics journals and "Search for a Naturalistic World View", 2 volumes, CUP, 1993 (collected essays). E-mail: abner.shimony@gmail.com

Ana Simões teaches history of science at the Unit for the History and Philosophy of Science, Faculty of Sciences, University of Lisbon. She is also the coordinator of the Research Center for the History of Science of the University of Lisbon. Her main interests lie on the history of the physical sciences, including quantum chemistry, and in the history of science in Portugal, eighteenth to the twentieth century. She is presently completing a book on the history of quantum chemistry, together with Kostas Gavroglu. E-mail: aisimoes@fc.ul.pt

John C.H. Spence is Regents Professor of Physics at Arizona State University, holding a joint appointment with Lawrence Berkeley National Laboratory. He obtained his PhD in physics from Melbourne University in 1973, followed by postdoctoral work at Oxford UK. He was awarded the Distinguished Scientist Medal of the Microscopy Society of America in 2006 and a Humboldt award in 1991. He is a member of the Department of Energys BESAC committee. John has worked in many areas of condensed matter and diffraction physics, and is the author of texts on High Resolution Electron Microscopy and (with J.M. Zuo) on Electron Microdiffraction. His current main interest is in femtosecond X-ray diffraction in biophysics. E-mail: john.spence@asu.edu

Ion-Olimpiu Stamatescu born in Ploiesti, Romania 1941, Diploma in Physics University of Bucharest 1962, Ph.D. University of Heidelberg 1972. Since 1977 professor at the University Heidelberg, since 1988 senior research fellow at FESt (Protestant Institute for Interdisciplinary Research), Heidelberg, retired 2006. Scientific interests: theory of elementary particles, cognition and artificial intelligence and foundation of physics, co-editor or co-author of many books. Member of the Deutsche Physikalische Gesellschaft, of the Acad. Internationale de Philosophie des Sciences (corr.) and of the European Society for the Study of Science and Theology. E-mail: stamates@ThPhys.Uni.Heidelberg.DE

Henry Stapp has spent his entire career working in frontier areas of theoretical physics. In the late sixties he worked on foundational problems with Werner Heisenberg in Munich, and in the early seventies with John Wheeler in Austin. His studies of the quantum measurement problem have led to a strong interest in the apparent effects of our conscious thoughts upon physical processes occurring in our brains. His conclusions in this area are described in many technical articles, and, in less technical terms, in his two books, "Mind, Matter, and Quantum Mechanics" and "Mindful Universe: Quantum Mechanics and the Participating Observer". E-mail: HPStapp@lbl.gov

Werner Stulpe Earning a doctorate in (theoretical) physics at the Technische Universität, Berlin under K.-E. Hellwig; 1987–1988 Postdoc research stay at the University of Denver, Colorado, under Stanley Gudder; 1997 Postdoctoral lecture qualification ("Habilitation") at the Technische Universität Berlin for the field of theoretical physics; 2003 Appointment to a Mathematics Professorship at Aachen University of Applied Sciences. E-mail: stulpe@fh-aachen.de

Mauricio Suárez is associate professor in Logic and Philosophy of Science at Madrid's Complutense University. His research interests are in the philosophical foundations of physics and general scientific epistemology, and he has published widely in both areas. He is the editor of *Probabilities, Causes and Propensities in Physics*, forthcoming with *Springer*. Within the philosophy of quantum mechanics he is perhaps best known for his attempt to revive Henry Margenau's *latency*, or propensity, interpretation of quantum observables. E-mail: msuarez@filos.ucm.es

Roderich Tumulka is an Assistant Professor in the Department of Mathematics at Rutgers, the State University of New Jersey, in New Brunswick, New Jersey (USA). He was born in Frankfurt am Main (Germany) and studied mathematics and physics in Frankfurt and Munich. As a post-doctoral researcher, he worked in Genoa (Italy), Bures-sur-Yvette (France) and Tuebingen (Germany). E-mail: tumulka@math.rutgers.edu

Lev Vaidman born in 1955, moved from Leningrad to Israel and received the B. Sc. degree from Hebrew University in Jerusalem and then, after the army service, the M. Sc. degree from Weizmann Institute in Rehovot. He started his work in foundations of quantum mechanics at Tel-Aviv University under supervision of Yakir Aharonov and obtained his Ph.D. in 1987. After spending 3 years at the University of South Carolina he came back to the Physics department of Tel-Aviv University where he became full professor in 2005. E-mail: vaidman@post.tau.ac.il

Klaus von Klitzing studied and graduated (Dr. rer. nat) in Physics at the TU Braunschweig. He obtained his Habilitation in Würzburg (1978). After visits as research associate in England, France, and the USA he became Professor at the TU Munich, where he stayed from 1980 to 1984. In 1985 he was appointed as Director at the MPI for Solid State Research. In 1985 Klaus von Klitzing was awarded the Nobel Prize in Physics for the discovery of the quantum-Hall-effect, at the age of 42. E-mail: K.Klitzing@fkf.mpg.de

Stefan Weigert researches and teaches as a Reader in Mathematical Physics at the University of York, UK, which he joined in 2005. He studied physics at the RWTH Aachen, DE, and at the University of Utrecht, NL. He received his venia docendi in 1996 from the Universität Basel, CH, where he earlier obtained his PhD degree. After a year at the Lawrence Berkeley Laboratory, USA, he held positions at the Université de Neuchâtel, CH, and at the University of Hull, UK. Most of his research relates to phenomena, concepts, and applications of non-relativistic quantum theory. In recent years, he has published in quantum information and PT symmetry. E-mail: slow500@york.ac.uk

Gregor Weihs is Full Professor of Photonics at the Institute for Experimental Physics at the University of Innsbruck in Austria. He received his MSc degree from the same university in 1994. After positions at the University of Vienna, Stanford University, and Tokyo University he joined the Institute for Quantum Computing and the Department of Physics and Astronomy at the University of Waterloo where he became the Canada Research Chair in Quantum Photonics. His research interests are in quantum optics, photonics, and the foundations of quantum physics. E-mail: weihs@iqc.ac

Friedel Weinert is Professor of Philosophy at Bradford University in the UK. His main interest lies in the investigation of the history and philosophy of science and in particular in the interactions between science and philosophy. The exploration of the philosophical consequences of science – both the natural and the social sciences – has led to the edited volumes *Laws of Nature* (1995) and the *Compendium of Quantum Physics* (2009) and to the authored books *The Scientist as Philosopher* (2004) and *Copernicus, Darwin & Freud* (2009). E-mail: f.weinert@brad.ac.uk

Antoine Weis obtained his Ph.D. in physics in1984: ETH-Zurich (Switzerland). From 1989 to 1990 he was project Leader, at the ETH-Zurich (Switzerland), and from 1990 to 1996 scientific collaborator at the Max-Planck-Institut fr Quantenoptik, Garching (Germany). In 1996 he obtained his Habilitation in Physics, LMU München (Germany). From 1996 to 1999 he was Associate professor at the University of Bonn (Germany) and from 1999 to today he has been Full professor, University of Fribourg (Switzerland). E-mail: Antoine.Weis@unifr.ch

Jürgen Weis After studying Physics in Ulm, Jürgen Weis moved to Stuttgart, where he got his Dr. rer. nat. in 1994. After a post-doc time at AT&T Bell Labs he joined the MPI for Solid State Research in 1996. In 2002 he received his Habilitation, in 2006 he was appointed to 'apl. Professor' at the University of Stuttgart. For his work on semiconductor quantum dots he received the Rudolf-Kaiser-Award in 2004. E-mail: j.weis@fkf.mpg.de

Bruce R. Wheaton taught physics and history of science at the University of California, Berkeley over a period of 22 years. He is author of, *inter alia, The tiger and the shark: Empirical roots of wave-particle dualism; Inventory of sources for history of twentieth-century physics: report and microfiche index to 70,000 letters*. Otherwise he enjoys playing Bach on the harpsichord he built, and once taught physics to the Westminster Choir. As Principal of Technology and Physical Science History Associates (google "Atapsha") he lectures internationally and coordinates publication of research on the cultural significance of science to a public that knows little of its desperate need for it. E-mail: wheatonbr@berkeley.edu

Sujeev Wickramasekara is currently assistant professor of physics at Grinnell College, Iowa. He received his doctoral degree in physics from the University of Texas at Austin in 2000 and bachelor's degree in physics (summa cum laude) from the University of Southern California. Prior to Grinnell College, he has worked as a postdoctoral fellow as the University of Texas at Austin and as Wiess Instructor of Physics and Astronomy at Rice University. E-mail: WICKRAMA@Grinnel.EDU

Frank Wilczek grew up in Queens, NY and attended the University of Chicago. After getting his Ph.D. from Princeton, he spent time on the faculty there and at the Institute for Advanced Study, as well as at UCSB's Institute for Theoretical Physics, now the KITP. Frank is currently the Herman Feshbach professor of physics at MIT. Frank has received many prizes for his work in physics, including the Nobel Prize of 2004 for work he did as a graduate student at Princeton University, when he was only 21-years-old. He is author of *Longing for Harmonies* (1987, with Betsy Devine) and *Lightness of Being* (2008). E-mail: wilczek@mit.edu

Christopher Witte studied physics and mathematics at the TU Berlin (Germany) and the KU Leuven (Belgium). Since then he has been working as a research assistant (at TU) on fundamental aspects of quantum theory and as editor of scientific books. E-mail: cwitte@physik.tu-berlin.de

Andrzej Kajetan Wróblewski is professor emeritus at the University of Warsaw. He completed his Ph.D. in 1961 and became associate professor (1965) and full professor (1971) at that University. His scientific interests are experimental elementary particle physics, especially the mechanism of hadron production, and also history of physics. An elected member of several learned societies, he has also been honoured with the Smoluchowski Medal of the Polish Physical Society and honorary doctorates from the universities of Siegen, Chapman (Orange, USA), and Glasgow (UK). E-mail: akw@fuw.edu.pl

Daniela Wünsch In 2000, after studying physics Daniela Wuensch graduated in the history of science at the University of Stuttgart. Her books "The Rise of the 5th Dimension. Life and Work of Theodor Kaluza (1885–1954)" (2007) and "The Path of Science in the Labyrinth of Culture" (2008) have been published by Termessos in Göttingen. She works and teaches in Göttingen, Germany. E-mail: dwuensch@theodor-kaluza.de

Nino Zanghí is Professor of Theoretical Physics in the Department of Physics at the Università di Genova (Italy). His research concerns mostly the conceptual and mathematical foundations of statistical mechanics and quantum physics. Noteworthy is his work, in collaboration with D. Dürr and S. Goldstein, on the emergence of quantum randomness in Bohmian mechanics. E-mail: zanghi@ge.infn.it

H. Dieter Zeh studied physics in Braunschweig (Brunswick) and Heidelberg, where he began work on theoretical nuclear physics. After continuing his research in California, he returned to the University of Heidelberg, where he later became professor of theoretical physics. His studies of collective motion in nuclei led him to address the quantum-to-classical transition in general, and in particular the quantum measurement problem, which is in turn related to many aspects of irreversibility (*arrows of time*). During this work, Zeh recognized and formulated the universal and unavoidable role of uncontrollable quantum entanglement, thus becoming a founder of the area now known as decoherence. E-mail: zeh@uni-heidelberg.de